THE NATURE OF NATURE

THE NATURE OF NATURE

Examining the Role of
Naturalism in Science

Edited by
Bruce L. Gordon
and William A. Dembski

Wilmington, Delaware

Copyright © 2011 Bruce L. Gordon, William A. Dembski

All rights reserved. No part of this publication may be reproduced or transmitted in any form or by any means, electronic or mechanical, including photocopy, or any information storage and retrieval system now known or to be invented, without permission in writing from the publisher, except by a reviewer who wishes to quote brief passages in connection with a review written for inclusion in a magazine, newspaper, broadcast, or online publication.

Library of Congress Cataloguing-in-Publication data available upon request.

ISI Books
Intercollegiate Studies Institute
3901 Centerville Road
Wilmington, DE 19807-1938
www.isibooks.org

Manufactured in the United States of America

Contents

Foreword: Beyond Naturalism to Science — xi
Steve Fuller

Introduction: The Nature of Nature Confronted — xix
Bruce L. Gordon and William A. Dembski

PART I:
NATURALIZING SCIENCE:
SOME HISTORICAL AND PHILOSOPHICAL CONSIDERATIONS

Introduction — 1

1. The Rise of Naturalism and Its Problematic Role in Science and Culture — 3
   Bruce L. Gordon

2. Science without God: Natural Laws and Christian Beliefs — 62
   Ronald L. Numbers

3. Varieties of Methodological Naturalism — 82
   Ernan McMullin

4. Sauce for the Goose: Intelligent Design, Scientific Methodology, and the Demarcation Problem — 95
   Stephen C. Meyer

PART II:
THE EPISTEMOLOGICAL AND ONTOLOGICAL FOUNDATIONS OF NATURALISM

Introduction — 135

5. Evolution versus Naturalism — 137
   Alvin C. Plantinga

6.	More on the Illusion of Defeat	152
	William J. Talbott	
7.	Evolutionary Naturalism: Epistemically Unseated or Illusorily Defeated?	166
	A. It's No Illusion! *Alvin C. Plantinga*	166
	B. The End of an Illusion? *William J. Talbott*	172
8.	A Quantum-Theoretic Argument against Naturalism	179
	Bruce L. Gordon	
9.	The Incompatibility of Naturalism and Scientific Realism	215
	Robert C. Koons	
10.	Truth and Realism	228
	Alvin I. Goldman	
11.	Must Naturalists Be Realists?	247
	Michael Williams	
12.	The Role of Concepts in Our Access to Reality	256
	Nicholas Wolterstorff	

PART III:
THE ORIGIN OF BIOLOGICAL INFORMATION AND THE EMERGENCE OF BIOLOGICAL COMPLEXITY

Introduction		273
13.	On the Origins of Life	276
	David Berlinski	
14.	DNA: The Signature in the Cell	293
	Stephen C. Meyer	
15.	Mysteries of Life: Is There "Something Else"?	346
	Christian de Duve	
16.	Life's Conservation Law: Why Darwinian Evolution Cannot Create Biological Information	360
	William A. Dembski and Robert J. Marks II	

Contents

17. Regulated Recruitment and Cooperativity in the Design of Biological Regulatory Systems — 400

    Mark Ptashne

18. The Nature of Protein Folds: Quantifying the Difficulty of an Unguided Search through Protein Sequence Space — 412

    Douglas D. Axe

19. The Limits of Non-Intelligent Explanations in Molecular Biology — 429

    Michael J. Behe

20. The Chain of Accidents and the Rule of Law: The Role of Contingency and Necessity in Evolution — 443

    Michael Shermer

21. Molecular Convergence: Repeated Evolution or Repeated Designs? — 460

    Fazale R. Rana

PART IV: COSMOLOGICAL ORIGINS AND FINE-TUNING

Introduction — 485

22. Eternal Inflation and Its Implications — 487

    Alan Guth

23. Naturalism and the Origin of the Universe — 506

    William Lane Craig

24. Cosmic Evolution, Naturalism and Divine Creativity, *or* Who Owns the Robust Formational Economy Principle? — 535

    Howard J. Van Till

25. Living in the Multiverse — 547

    Steven Weinberg

26. Balloons on a String: A Critique of Multiverse Cosmology — 558

    Bruce L. Gordon

27. Habitable Zones and Fine-Tuning — 602

    Guillermo Gonzalez

PART V: MATHEMATICS

Introduction   641

28. Mathematical Naturalism   643

    Philip Kitcher

29. Mathematics—Application and Applicability   667

    Mark Steiner

PART VI: EVOLUTIONARY PSYCHOLOGY, NEUROSCIENCE, AND CONSCIOUSNESS

Introduction   691

30. Toward Mapping the Evolved Functional Organization of Mind and Brain   695

    John Tooby and Leda Cosmides

31. On the Origins of the Mind   713

    David Berlinski

32. Consciousness   726

    John R. Searle

33. Consciousness and Neuroscience   746

    Francis Crick and Christof Koch

34. Supervenience and the Downward Efficacy of the Mental: Nonreductive Physicalism and the Christian Tradition   770

    Nancey Murphy

35. Conscious Events as Orchestrated Space-Time Selections (with a new addendum)   790

    Stuart Hameroff and Roger Penrose

36. Quantum Interactive Dualism: The Libet and Einstein-Podolsky-Rosen Causal Anomalies   815

    Henry P. Stapp

37. The Physical Sciences, Neuroscience, and Dualism   835

    James P. Moreland

Contents

PART VII:
SCIENCE, ETHICS, AND RELIGION

Introduction — 853

38. Evolution and Ethics — 855
    Michael Ruse

39. Naturalism's Incapacity to Capture the Good Will — 865
    Dallas Willard

40. Naturalism, Science, and Religion — 880
    Michael Tooley

41. Theism Defended — 901
    William Lane Craig

Acknowledgments — 920
Contributors — 921
Sources and Permissions — 928
Index of Names — 931
Index of Subjects — 941

Foreword

Beyond Naturalism to Science

Steve Fuller

I can appreciate a book called *The Nature of Nature* because I have called myself a "reflexive naturalist."[1] If naturalism is the default metaphysical position of the natural sciences, then that fact must itself be explained by the theories and methods of the natural sciences. But how does one do that without begging crucial questions? For example, it is tempting to argue that the natural sciences are simply a somewhat more disciplined version of commonsense, which itself is an extension of the problem-solving skills that all animals deploy in adapting to their environments. Such a view was explicitly promoted in the twentieth century by the American pragmatists John Dewey and Willard Quine, both of whom invoked "naturalism" to name their own positions. It had the virtue of bringing science down to earth by making it seem to be part of our evolutionary heritage. But in the process, this view presumed that the natural sciences' ideological self-understanding was largely correct. In other words, science was explained as an instance of something of which it was already trying to provide a theoretical account. Yet the epistemic distinctiveness of science, whereby it breaks with animal modes of intelligence and even everyday forms of human reasoning, remained unaddressed, if not obscured, in the process.

If science were indeed little more than incrementally enhanced ordinary perception—say, the indefinite aggregation of beings with indefinitely powerful eyes—then the history of science would not have been as it is. Long before Thomas Kuhn, it was recognized that science periodically reconstitutes society's taken-for-granted ideas, regardless of whether that serves to stabilize or destabilize our place in nature. For example, the relativity and quantum revolutions in twentieth-century physics shifted science's epistemic center of gravity very far from commonsense and in ways that have placed humanity's survival increasingly at risk. Notwithstanding a persistent undercurrent of protest, sometimes in the name of humanity but increasingly in the name of nature, we have acted as if the benefits of science outweigh the costs. But what is the source of this overriding faith? After all, the practical benefits continue to be controversial—as in, say, nuclear energy, stem-cell research, and nanotechnology.

Nevertheless, science's theoretical benefits remain clear. In particular, the revolutions in physics appear unequivocally good because they significantly advance the distinctly human quest for "universal" knowledge in at least three senses of that term, all of which cut against the spirit of naturalism:

(1) Science aspires to knowledge of all things, under conditions potential and actual, regardless of their relevance to our everyday lives, let alone personal or species survival. Indeed, most of what science purports to be about will never be experienced directly by anyone, and only a very small fraction of the reality charted by science will be encountered even by expert scientists. Another, more empowering way to make the same point is to say that what we may come to know is not restricted to what we have known. In that sense, science transcends history. A scientific inquiry into "the nature of life" or "the nature of mind" may begin by examining living organisms and conscious humans, but it then poses the question in broader, more abstract terms, countenancing the range of material properties that living and conscious beings might possess, wherever they might be.

(2) Science aspires to articulate all things in a common language, no matter how different these things may appear to our senses. Newton first achieved this feat by mathematically unifying our understanding of the motions of the planets and objects on Earth. John Locke, one of the original Newtonians, canonized the difference between how things are and how they seem to be in terms of "primary" and "secondary" qualities. In the wake of Newton, mathematics' quest to articulate the primary qualities of things has led the discipline to develop in directions far removed from ordinary tasks of measurement and calculation, from non-Euclidean geometries to the paradoxes of set theory. Corresponding to that has been the introduction of specialized uses for ordinary terms as well as some entirely new terms and conceptualizations that have periodically challenged our basic understanding of space, time, and cause. In this context, for better or worse, symbolic logic has been promoted as the language of universal thought.

(3) Science aspires to be knowledge for everyone, a universal human legacy. This is the sense in which science most controversially defies naturalism. The palpable diversity of human beings—be it by origins or appearance—is not sufficient to discount their experience and ideas from the constitution of scientific knowledge. On the contrary, a mark of the "unscientific" standing of a form of knowledge is that it cannot command universal human assent by virtue of its intimate connection to personal experience. Of course, this is not to deny that different people understand things differently. Rather, it is to say that for such an understanding to count as *science,* it must be made relevant to those who have not and probably will not experience it for themselves. The principle applies equally to someone possessing arcane technical expertise and rare indigenous knowledge. Here the positivist philosophical distinction between the "context of discovery" and the "context of justification" is apt: The more widely justifiable a form of knowledge is, the greater its entitlement to scientific status.

All three senses of science's "universality" defy any easy naturalistic explanation because they force us to conceptualize knowledge from a standpoint rather removed from our ordinary embodied experience. Thomas Nagel famously identified this standpoint as "the view from nowhere," which certainly captures the spirit of points (1) and (2) above.[2] But it captures point (3) as well, since the range of experience—both real and imagined—of any given individual human is in fact quite limited vis-à-vis that potentially available to all of humanity. This fact helps to explain the hazardous utility of our reliance on what psychologists have variously called "heuristics," "stereotypes," "biases" and "prejudices." These corner-cutting features of cognition make the most out of what immediately strikes us as salient in our experience. And were science simply a species of what Larry Laudan provocatively dubbed "plebeian induction," then science could be understood as an elaborate exercise in corner-cutting, which would mean routinely discounting the testimony of those whose experience we cannot readily assimilate.[3] But in fact science requires what Laudan called "aristocratic induction," that is, the ability to see beyond our default expectations to something deeper and even counterintuitive. This point applies to how we regard not only the natural world but also those who regard the natural world.

The Abrahamic religions—Judaism, Christianity, and Islam—provide the only clear historic basis for conceptualizing science as "universal" in this robust sense of aspiring to knowledge of all things by and for all people. That basis is the biblical doctrine that humans are unique as beings created in the image and likeness of God. To be sure, the doctrine has been variously interpreted, but it has been generally read as a call to epistemic empowerment—at the very least, that humans can know much more than what is necessary to maintain their animal existence. Islam started to take this call seriously as a vehicle of proselytism, which resulted in the compilation and harmonization of sacred (i.e. Abrahamic) and secular (i.e. Greco-Roman) writings into a complete body of knowledge closed under Qur'anic principles.

To be sure, this practice created enormous tensions within Islam, which reached critical proportions in the work of the twelfth-century Cordoban jurist Ibn-Rushd, known to Christians as Averroes, the great commentator on Aristotle in contrast to whom Thomas Aquinas defined many of his major positions. Averroes questioned the ultimate unity of knowledge, defending instead a "double truth" doctrine inspired by Plato, whereby science and religion would be distinguished as, respectively, truth for the elites and truth for the masses. While this view has kept Averroes a heretic within Islam to this day, it did not prevent his works from exercising enormous influence within Christendom, first as the means by which the Aristotelian corpus was rediscovered in Europe, and second—and more potently—as the source of the idea that science could be used to reform religion in aid of a purer sense of epistemic unity. To be sure, Christendom also treated this as a heresy—indeed, one called "Averroism." As it turns out, its most famous exponent was Galileo, who studied in a city, Padua, that for over two centuries had been the hotbed of Averroism.

I have encapsulated this strand in Western intellectual history because, regardless of the various obstacles that organized religion has placed in the way of science over the centuries, the best explanation for the shape and persistence of science's fundamental questions has been theological. The point is repeatedly driven home by the authors in this volume in their investigations into the various questions of "origins," "ends" and "design" that have framed scientific inquiry into the nature of life, mind, and the universe as a whole. From the empiricist standpoint privileged by naturalism, it is difficult to motivate, let alone resolve, these questions, as they all invite us to consider reality from a perspective not normally our own. Unsurprisingly, naturalism's most trenchant adherents—from the ancient Epicureans to moderns like Hume and Darwin—studiously avoided reaching any firm conclusions on these matters and denied that science could ever contribute substantially to them. Indeed, every argument they raised against natural theology could always be turned against "natural science," understood as the name of systematic inquiry into fundamental explanations. (Hume's own reservations about Newton certainly had this character.) After all, such inquiry invariably forces us to countenance entities, processes, and events that transcend the normal run of experience. Naturalists routinely underestimate the decisive role that experiments play in science: Experiments are less about enhancing our natural modes of understanding than providing access to new modes of understanding that we might otherwise lack, were it not for the intelligent design of the laboratory.

Once we recognize the inadequacy of an unreflexive naturalism to explain the aspirations—and success—of science, the meanings of many of the key metaphysical, if not ideological, positions associated with the scientific worldview start to look less secure. In particular, materialism starts to look problematic. Marx's co-conspirator, Friedrich Engels, got it exactly right when he characterised the nineteenth-century revolution in thermodynamics as heralding a "dematerialized materialism" in both science and society.[4] Engels was expressing approval that the English word "work" had come to mean human labor (*Arbeit* in German) and mechanical power (*Kraft*), both of which could be understood in the same mathematical terms as the release of energy over a certain distance for a certain duration. For Engels, a true believer in conservation principles, thermodynamics proved that any increase in production would require an increase in productivity—that is, an efficiency saving in energy expenditure. The only way that could happen would be through a more intelligently organized workforce. Though hardly a spiritualist, Engels nevertheless recognized that humanity's species-distinctive sense of intellectual progress lay in its increasing ability to do more with less over time. This intuition continues to be respected in the pivotal role played by new technology in models of economic growth.

But exactly how does the requisite intelligence come about—through some sort of self-generated reorganization of the current system or through the introduction of a wholly new factor from outside the system? The economists' way of posing the question has acquired more general currency: Is technological innovation "endogenous" or "exogenous" in origin? Throughout its history as a political practice, Marxism has struggled with a version of this problem—namely, how to manage the transition from capitalism to socialism: Should one wait patiently until the material factors and social relations of production rearrange themselves in some appropriate manner (e.g., during a war or a depression), or should one strike decisively with a top-down strategy that activates elements of a system that might otherwise remain dormant indefinitely? In sum, we have the difference between social democracy and revolutionary communism: Kautsky the politician in office versus Lenin the leader in exile as rival heirs to the legacy of Marx and Engels.

For those who like their conceptual distinctions to bear less political freight, analogous cases are readily found. Consider cybernetics, a field that aimed to understand system maintenance in terms of information theory, an application of thermodynamics to communication under suboptimal conditions, as in the noise generated by walkie-talkies in wartime. Cybernetics captured the imaginations of a broad range of interdisciplinary polymaths in the third quarter of the twentieth century, including the mathematician Norbert Wiener and the anthropologist Gregory Bateson, who debated whether the difference between, say, good and evil, right and wrong, ordered and disordered could be exhaustively explained by the self-regulated rearrangement of the components ("servomechanisms") within a given system.[5] Wiener was more inclined to this position than Bateson, who leveraged some ideas derived from the paradoxes of set theory to argue that a higher-order system may be needed that operates in ways that are at odds with the system on which it acts.

In the contemporary intellectual landscape, we might contrast Stuart Kauffman's "order for free" defense of a self-organizational approach to evolution that downplays the role of natural selection (or, for that matter, any *deus ex machina*, including a deity) with William Dembski's "no free lunch" argument against the self-sufficiency of evolutionary algorithms, which implicitly allows for a theologically inspired idea of intelligent design as the source of genuinely new information in nature. This debate is well represented by the Dembski and Marks piece in this volume. However one ultimately stands on the various issues at stake, it is clear that we have moved a long way from the idea that nature can be understood as if it were the product of no intelligence at all. The long march from the blind chance and mindless necessity of classical materialism to the intelligently organized work that Engels adumbrated for a dematerialized materialism began with the explicit introduction of the engineering mentality into physics through thermodynamics, which quickly spread to political economy and physiology, the dominant biomedical science well into the twentieth century. Later in the century, this trajectory acquired forward momentum from the mathematical formulation of information theory and the invention of the solid-state transistor, culminating in the discovery of DNA as the biological precedent for this feat of miniaturized storage and transmission of information, a point that has been long stressed by Stephen Meyer, who is represented by two pieces in this volume.

From the standpoint of intelligent design theory, the true scientific revolution in biology that allowed the discipline finally to break away from natural history (as both physics and economics had already done) and catch up with developments in the physical sciences would be recounted as follows. First, respect would be paid to the mid-nineteenth-century Bohemian monk Gregor Mendel, who first discerned interesting mathematical patterns in inheritance that, after some neglect, became the basis for population genetics in the early twentieth century. He is reasonably counted as the Copernicus figure in this revolution. However, the story starts to acquire serious momentum with the molecularization of genetics in the 1930s, when Warren Weaver, a pioneer in information theory and director of the natural sciences division of the Rockefeller Foundation, funded the retooling of Cambridge's Cavendish Laboratories to enable the discovery of the structure of DNA, the downstream effects of which continue to this day in the mechanical sequencing of genomes from various species.

In terms of this alternative historiography, the Darwinian "tree of life" image of phylogeny may no longer be treated as a self-sufficient account of the actual descent of the species, as suggested by the *au courant* hypothesis of the "molecular clock," whereby differences in the genomes of different species are taken as indicators of when they divided from a common ancestor in evolutionary history. Rather, phylogeny may come to be understood as a prototype for, so to speak, a periodic table of biological elements, whereby the macromolecules of the genes are correlated with the expression of traits that may recur at several moments and in different creatures in natural history. In retrospect, Darwin's evolutionary approach to taxonomy may come to be seen as a temporary diversion from the original strategy put forward by Carolus Linnaeus in the eighteenth century to classify life forms according to a design-based logic, the legacy of which continues in the Linnaean coinage, *Homo sapiens* ("man the wise"), a shorthand attempt at a functional definition of our species.

But does all this mean that evolution itself is a chimera? Not necessarily. But it does leave wide open the means and ends (if any) of evolution. Simon Conway Morris's magnum opus *Life's Solution: Inevitable Humans in a Lonely Universe* (2003) makes this point in a particularly compelling way.[6] But before considering how Morris's suggestive account of evolution effectively upends the Darwinian account, almost in spite of itself, it is worth taking a brief look at one current account of evolution that takes the thermodynamic turn in the physical sciences very seriously yet stays firmly within the confines of naturalism. This theory, put forward a quarter of a century ago by two systematic taxonomists, equates the trajectory of evolution with an overall increase in biological entropy.[7] Whatever direction evolution seems to be taking is simply a reflection of the options for life becoming more limited, as nature gradually expends itself. Thus, any semblance of progress is an illusion created by the path-dependency of a process that leaves in its wake a lengthening trail of extinct precursors. This rather gloomy view of things envisages humans as possibly the last surviving species in an always already dying world. Although neither Darwin nor the biologist of recent times whose work was closest in spirit to

Darwin's, Stephen Jay Gould, adhered to entropy theory, they had their own rather pessimistic sense of evolution's naturalistic endgame.

In striking and rather more hopeful contrast, we have Conway Morris, an Anglican Christian who nevertheless joins the atheist Richard Dawkins in rejecting the exaggerated sense of contingency that Gould associated with evolutionary history. But where Dawkins seems satisfied that natural selection is sufficient to explain the adaptation of species to their environments, Conway Morris argues that larger forces are also in play that, in effect, overdetermine evolutionary outcomes. Contrary to the spirit of Gould's famous thought experiment, replaying the tape of natural history would very likely produce more or less the same run of species. In support of this claim, Conway Morris reads the palaeobiological record as implying the long-term convergence of life forms. And while the array of evidence he mobilizes is impressively diverse and up to date, the claim itself is far from novel. That life forms tend to converge over time, rather than diverge, is the oldest and most persistent scientific objection to Darwin's construal of evolution. Indeed, over the past century, it has been associated with a variety of heretical Catholic biologists from St. George Mivart to Pierre Teilhard de Chardin.[8]

Conway Morris's self-positioning in this debate is rather curious. While taking great pains to distance himself from both creationism and intelligent design theory (which he tends to conflate), he nevertheless subscribes to the view that over the eons "life" has "navigated" the space of biological possibilities to reach similar solutions to similar problems of environmental adaptation. The image of navigation is not new to this general discussion. Not only is it embedded in the very word "cybernetics" (from the Greek for "helmsman"), but, more to the point, it recurs in Neo-Lamarckian accounts of evolution that stress both the uniqueness and the ultimacy of humanity. Taken to its extreme, the image suggests that everything worth preserving about evolutionary history eventually finds its way into our own biological makeup—perhaps in the future with the help of biotechnology? Conway Morris, of course, does not draw such a provocative conclusion, but it can be found in the work of Teilhard and that of his biggest mainstream biological defender, the geneticist Theodosius Dobzhansky, who was perhaps most responsible for converting the Neo-Darwinian synthesis into the dominant paradigm in biology. As an Orthodox Christian, Dobzhansky was attuned to the idea of *theosis*, namely, the eventual re-identification of humanity with the deity at the end of evolution. Perhaps this volume will help remind us of this fact the next time we hear Dobzhanksy's most famous sound bite, "Nothing in biology makes sense except in the light of evolution."[9]

Foreword

Notes

1. S. Fuller, "Epistemology Radically Naturalized: Recovering the Normative, the Experimental, and the Social." In R. Giere, ed. *Cognitive Models of Science*. Minneapolis: University of Minnesota Press, 1992, 429–59.
2. T. Nagel, *The View from Nowhere*. Oxford: Oxford University Press, 1986.
3. L. Laudan, *Science and Hypothesis*. Dordrecht: Reidel, 1981.
4. A. Rabinbach, *The Human Motor: Energy, Fatigue, and the Origins of Modernity*. New York: Harper Collins, 1990.
5. S. Heims, *Constructing a Social Science for America: The Cybernetics Group 1946–1953*. Cambridge MA: MIT Press, 1991.
6. Simon Conway Morris. *Life's Solution: Inevitable Humans in a Lonely Universe*. Cambridge: Cambridge University Press, 2003 (see especially Chapter 10, "Evolution Bound: The Ubiquity of Convergence").
7. D. R. Brooks and E. O. Wiley, *Evolution as Entropy: Toward a Unified Theory of Biology*. Chicago: University of Chicago Press, 1986.
8. S. Fuller. *Dissent over Descent: Intelligent Design's Challenge to Darwinism*. Cambridge: Icon Books, 2008.
9. T. Dobzhansky, *The Biology of Ultimate Concern*. New York: New American Library, 1967.

Introduction

The Nature of Nature Confronted

Bruce L. Gordon and William A. Dembski

We all take for granted that scientific theorizing and investigation epitomize rational activity. But what gives science its foundation as a rational, truth-conducive enterprise? Why should we even suppose that nature is intelligible to the human mind? In short, what are the metaphysical and epistemological *presuppositions* that *justify* scientific activity? And what bearing does the answer to such questions have on the tenability of philosophical or methodological naturalism, and vice versa?

Once such questions about the ultimate justification of science are raised, it becomes clear that any conclusions regarding the nature of nature derived from scientific practice and rational reflection will depend on the answers we give to these questions. A central issue in this interplay between presuppositions and conclusions, one made all the more pressing by recent scientific advances, is whether the universe is self-existent, self-sufficient, and self-organizing, or whether instead it is grounded in a reality that transcends space, time, matter, and energy. More pointedly, does our universe find its ultimate explanatory principle in *matter* or *mind*? Any perspective on the implications of scientific research for resolving this issue must address whether there are limits to what science can tell us, and whether there are essential constraints that circumscribe scientific methodology and the entities to which it may appeal in its explanations.

So shall we say that science is only concerned with giving natural explanations of the natural world and that any *other* sorts of explanations, whatever their merits, are *not* scientific? Or shall we say that *any* rigorous reasoning based on empirical evidence and theory construction is scientific? Setting aside the suggestion that some natural phenomena may not *have* a rational explanation,[1] it is clear that saying either of these things leaves open the logical possibility that certain features of the world have rational explanations that are *not* natural. It then becomes a matter of semantics whether any such rational but *non*-natural explanations should be regarded as "scientific." Were it not for the cultural cachet and authority our era grants to all things scientific, and the regrettable tendency to assimilate rationality to science *simpliciter*, this matter of semantics would be insignificant. As things stand, however, this game of words has at times become a war of words in a pitched battle symptomatic of a deeper intellectual and cultural divide.

While the editors of this volume stand on one side of this divide, the essays in this volume advance a rational and balanced discussion of these questions. To further this goal, we have refrained from extensive commentary on and analysis of each essay, instead providing a short introduction to each of the seven parts of the book and then letting the authors speak for themselves, so that the readers may judge for themselves. In short, we thought it best to let the readers enter the interplay of ideas and the titanic clash of worldviews in what follows, using their own best judgment as a guide. Before taking leave, however, a brief account of the academic conference that served as the genesis of this compendium is in order.

"The Nature of Nature" Revisited

"The Nature of Nature" conference was the first and last fruit of the former Michael Polanyi Center for Complexity, Information and Design (MPC) at Baylor University. The Polanyi Center was also the first, and to this day only, intelligent design think tank at a major research university.[2] It was so named with the permission and blessing of Michael Polanyi's son, John Polanyi, a Nobel Prize–winning chemist at the University of Toronto. The MPC hosted the "Nature of Nature" conference during April 12–15, 2000, on the Baylor campus. The origin of the conference was as follows: Late in the fall of 1999, the director of the MPC, William Dembski, received a substantial grant from the Templeton Foundation. He and the MPC's associate director, Bruce Gordon, had close ties to Seattle's Discovery Institute (in particular, to its Center for Science and Culture) as well as to *Touchstone Magazine* (which had just recently published a special issue on intelligent design coedited by Dembski).[3] Together, they approached these organizations about funding a conference devoted to a key theme in the MPC planning document: the nature of nature.[4] All three organizations enthusiastically supported the proposed conference. Along with support from Baylor's Institute for Faith and Learning, the conference was ready to proceed. Its official description read as follows:

> Is the universe self-contained or does it require something beyond itself to explain its existence and internal function? Philosophical naturalism takes the universe to be self-contained, and it is widely presupposed throughout science. Even so, the idea that nature points beyond itself has recently been reformulated with respect to a number of issues. Consciousness, the origin of life, the unreasonable effectiveness of mathematics at modeling the physical world, and the fine-tuning of universal constants are just a few of the problems that critics have claimed are incapable of purely naturalistic explanation. Do such assertions constitute arguments from incredulity—an unwarranted appeal to ignorance? If not, is the explanation of such phenomena beyond the pale of science? Is it, perhaps, possible to offer cogent philosophical and even scientific arguments that nature does point beyond itself? The aim of this conference is to examine such questions.

When John Wilson, the editor of *Books & Culture*, saw the conference description and the lineup of speakers, he remarked that it promised to be the most important academic conference of the past twenty years. Many of the participants ended up feeling the same way. By any standard, the conference was an unqualified success. Paid attendance was huge. The roster of conference speakers was exceptional, not only for bringing together the very top thinkers about naturalism along with various luminaries in the scientific community, but also for providing just the right balance on all sides of this controversial topic.[5]

Introduction

"The Nature of Nature" conference was the Polanyi Center's glorious beginning. As already intimated, it was also the center's swan song. The conference ended Saturday, April 15. That following Tuesday, April 18, the Baylor Faculty Senate voted 27 to 2 to shut the center down. The *Chronicle of Higher Education,* reporting the following day, quoted Baylor neuroscientist Charles Weaver: "I have never seen faculty as upset over any issue, and I've been here 12 years now. It's just sheer outrage." All activities of the center were immediately suspended. Robert Baird, chair of the faculty senate, told the *Baylor Lariat* (the school newspaper), that the creation of the MPC was "one of the most divisive issues to have arisen on the Baylor campus during my 32 years on faculty."

Rather than close the center right then, the Baylor administration instituted a "peer-review committee" to examine the center and make recommendations for its future. In October 2000, while recognizing the legitimacy of the academic work associated with the center and the appropriateness of Baylor as a context for it, the review committee recommended (with, in our view, an inadequate grasp of Polanyi's work and legacy) dropping the Polanyi name and absorbing the center into the Institute for Faith and Learning: "the Committee believes that the linking of the name of Michael Polanyi to programs relating to intelligent design is, on the whole, inappropriate. Further, the Polanyi name has come by now in the Baylor context to take on associations that lead to unnecessary confusion."[6] The Baylor administration complied with all the committee's recommendations. The Michael Polanyi Center thus ceased to exist.

With all the turmoil surrounding the Polanyi Center (its demise was widely reported across the United States), any hope of procuring a conference proceedings for "The Nature of Nature" event seemed dashed. Both Dembski and Gordon remained on at Baylor as contract faculty for several years following the Polanyi Center's dismantlement. And for a time, the Baylor administration hoped to salvage something from the wreckage, even holding out the hope that Dembski and Gordon might be rehabilitated in the eyes of the recalcitrant Baylor faculty who had forced the center's closure. But as the years passed, such hopes vanished. The Baylor administration became further embattled in its strategy to reverse the university's drift toward secularization and by financial concerns related to the capital projects (facilities expansion) it had undertaken. At this juncture, continued support of Dembski and Gordon was, in the words of the administration, "politically unviable." Ironically, around the time that Dembski and Gordon left Baylor (May and August 2005, respectively), Robert Sloan, who had hired them, was also forced to step down from the Baylor presidency. It was a clean sweep for the opposition and a clear setback for academic freedom.[7] Notwithstanding, a decade after "The Nature of Nature" conference, harbingers of a brighter future for Baylor are evident. Many faculty are committed to sound scholarship and open dialogue with novel research programs like intelligent design—ideas that have an obvious place at an institution like Baylor—and the new president, Kenneth Starr, no stranger to controversy, seems willing to give such faculty his support.

In 2004, toward the end of Dembski's and Gordon's tenure at Baylor, contact was renewed with Richard Spencer, who at the time was Child Family Professor of Engineering at the University of California, Davis, and who remains on the faculty there as a professor of electrical and computer engineering. Spencer had attended "The Nature of Nature" conference as part of a contingent from his local church (Grace Valley Christian Center) led by the Senior Pastor, Rev. P.G. Mathew. At the conference, Spencer was an active and incisive participant in the discussions following the plenary sessions, and as a consequence he had come to Dembski's and Gordon's attention. After contact was renewed, subsequent discussions with Spencer and the church administration about the fate of ID research at Baylor led to the birth of a vision for resurrecting the conference proceedings. Doing so, however, proved easier said than done. Gordon's work as the primary contact and liaison with the original conference speakers had involved a lot of negotiation and some last-minute substitutions. In addition, over four years had passed since the event had taken place, and most of the original talks, many of which were delivered without preparation of a formal paper, were not available. Those talks which were available in written form were not necessarily representative of the current views of the participants or the current state of research. All of these difficulties were further compounded by the continual controversy swirling around intelligent design as a research program.

Nonetheless, this project was not to be deterred. Grace Valley arranged a generous grant for Bruce Gordon to serve as the primary editor, with William Dembski as coeditor, for a suitably updated compendium based on the conference.[8] Such a book would incorporate the latest thinking by key conference participants along with some important voices who were not able to attend the original event. The Center for Science and Culture at Discovery Institute subsequently also gave its considerable support to the effort. The result, a major landmark in the ongoing debate over Nature's nature, is the volume you hold in your hands.

Introduction

Notes

1. For a discussion of this possibility, see the first section of Bruce Gordon's essay "The Rise of Naturalism and Its Problematic Role in Science and Culture" in Part I, and the discussion of Humean supervenience in his Part II essay titled "A Quantum-Theoretic Argument against Naturalism."
2. The opening paragraph from the five-year planning document that led to the founding of the Michael Polanyi Center (a document submitted in March 1999 to Robert Sloan, who was then president of Baylor University) read as follows:

> Naturalism currently dominates science, both in the secular and in the Christian academy. According to naturalism, science is best practiced without reference to anything "non-natural." Granted, science's proper object of study is nature. Naturalism, however, assumes an impoverished view of nature that artificially limits nature to brute material processes subject to no intelligent guidance or control. The problem with naturalism is not that it limits science to the study of nature, but with its hidden assumptions about *the nature of nature*. Is nature a seamless causal web controlled solely by undirected natural processes—what Jacques Monod called "chance and necessity"? Or do intelligent causes also play a fundamental and ineliminable role within nature?

The MPC began quietly as one among several academic centers at Baylor, building bridges to the Baylor community through such activities as a faculty book discussion group on the nature of science and its relation to the historic religious foundations of Baylor as an institution. William Dembski was the center's founder and director, Bruce Gordon its associate director.

The MPC was named after Michael Polanyi, a world-class physical chemist who turned to philosophy later in his career. Among the things that prompted Polanyi's move to philosophy from science was his desire to foster free and open scientific inquiry. Polanyi had visited the Soviet Union in the 1930s at the height of the Stalinist repression and saw how scientific inquiry could be subverted through authoritarianism and ideology. In response, Polanyi stressed the need for science to be free of artificial strictures and to regularly reexamine its presuppositions. It was in this spirit that Baylor's Michael Polanyi Center was founded.

When the website for the MPC debuted in January 2000, a controversy immediately ensued that fell within the parameters of an ongoing battle between the Baylor administration and a certain segment of the Baylor faculty regarding the vision for Baylor as a Christian, and specifically Baptist, institution of higher learning. The administration was interested in restoring the historic Christian foundations of Baylor and encouraging scholarship and research that would advance such concerns; the opposed segment of the faculty, represented by the faculty senate, was more interested in compartmentalizing Baylor's identity as a Baptist institution and sealing it off from any contact with academic work. The ultimate battle, of course, related to whether Baylor would remain a Christian institution of higher learning or drift into secularization as had so many denominationally founded schools before it. For those interested in the history of the Michael Polanyi Center and a chronological account of the controversy surrounding it, the full planning document, as well as other documents and news reports related to events associated with the center, may be found at http://www.designinference.com/documents/2007.12.MPC_Rise_and_Fall.htm (last accessed August 15, 2009).
3. The July/August 1999 *Touchstone* was a special double-issue focused on intelligent design. The essays in this issue, with an additional contribution by Bruce Gordon, were subsequently published as a book edited by William Dembski and James Kushiner titled *Signs of Intelligence: Understanding Intelligent Design* (Grand Rapids, MI: Brazos Press, 2001).
4. The key line in the planning document as it relates to the conference theme ran: "The problem with natu-

ralism is not that it limits science to the study of nature, but with its hidden assumptions about *the nature of nature*." See footnote 2.

5. The complete conference schedule and roster of speakers can be found as part of the general history of the Michael Polanyi Center at http://www.designinference.com/documents/2007.12.MPC_Rise_and_Fall.htm (last accessed August 15, 2009).

6. Again, see http://www.designinference.com/documents/2007.12.MPC_Rise_and_Fall.htm (last accessed August 15, 2009).

7. This affront to academic freedom, unfortunately, is a common experience among academic representatives and defenders of intelligent design as a research program. Though the debacle surrounding the Polanyi Center was not covered in the film, similar cases involving academic persecution of ID researchers at a variety of institutions formed the subject of a Ben Stein documentary titled *Expelled: No Intelligence Allowed*, which was released in spring 2008. Those interested in this issue and the defense of academic freedom in origins research can find more information at the website: http://www.academicfreedompetition.com/.

8. See also the grateful remarks of the editors in the acknowledgments.

Part I

Naturalizing Science: Some Historical and Philosophical Considerations

Introduction

This section provides the historical context for the debate about naturalism that runs through the pages of this volume—a debate that begins with an examination of naturalism's philosophical credentials, and ranges across the spectrum of the sciences from physics and cosmology to the origin and development of life, the nature of mathematical knowledge, the nature of the human mind and consciousness, and, finally, questions of ethics and religion. The theme of the historical essays, which serve as an introduction to this debate, is the origin and rise of philosophical and methodological naturalism in Western thought, and the role these ideas have come to play—for good or ill—in contemporary science and culture.

In the lead essay, "The Rise of Naturalism and Its Problematic Role in Science and Culture," philosopher of science **Bruce Gordon** begins by examining the implications of philosophical naturalism for the very possibility of rational explanation. Philosophical naturalism, he argues, undermines knowledge and rationality altogether, ultimately leading to the instrumentalization of belief and the fragmentation of culture. With the stage thus set, Gordon chronicles the rise of naturalism in conjunction with the creeping secularization of Western civilization, and discusses the confluence of factors that have led to the embrace of an exclusive humanism by the cultural elite—all the while noting that the foundations for modern science were laid in a Judeo-Christian context where order was sought in nature because it was expected, and it was expected because nature itself was regarded as the product of the mind of God. Gordon then corrects various misconceptions about the "warfare" between science and religion that allegedly began with the Enlightenment, and elucidates the role played by Darwinism in the process of secularization and the establishment of philosophical naturalism's hegemony in the academy, providing an analysis of the deleterious effects Darwinian ideas have had in the broader culture. The essay concludes with an argument for the reconstitution of science on a transcendent foundation, examining the effectiveness of two ways in which this might be done.

The remaining essays focus more explicitly on the rise of methodological naturalism in science and whether it has, on the whole, been a positive development. In "Science without God: Natural Laws and Christian Beliefs," historian of science **Ronald Numbers** chronicles the history of methodological naturalism, and the role that theists played in promoting it, as a constraint on scientific theorizing and investigation. He discusses the motives that various devout Christians in the scientific community had for supporting it as a path to understanding nature. For them at least, Numbers argues, methodological naturalism did not lead to secularization, and it has not invariably led to secularization in society at large. In "Varieties of Methodological Naturalism," philosopher of science **Ernan McMullin** discusses the historical relationship between ontological and methodological naturalism, arguing that there are deep problems with the former when taken as a substantive philosophical position. He then carefully analyzes the content of methodological naturalism, revealing at least three forms of it that need to be disentangled in contemporary discussions, and contending that two of them are worthy of support, even by theists.

In the final essay, "Sauce for the Goose: Intelligent Design, Scientific Methodology, and the Demarcation Problem," philosopher of science **Stephen Meyer** examines the question of whether science has a definitional essence, and defends the reasonability of intelligent design as a scientific research program by multiple standards of significance. He then dissects methodological naturalism as a meta-criterion for scientificity, arguing that its justification is circular at best, and that the constraint it represents could have a deleterious effect on scientific progress. The historical sciences in particular, he concludes, are better off employing metaphysically neutral criteria, and following the evidence where it leads.

1

The Rise of Naturalism and Its Problematic Role in Science and Culture

Bruce L. Gordon

> There is only the fight to recover what has been lost
> And found and lost again and again: and now under conditions
> That seem unpropitious. But perhaps neither gain nor loss.
> For us, there is only the trying. The rest is not our business.
> —T. S. Eliot, *The Four Quartets (East Coker)*

It is worthwhile reflecting on how philosophical naturalism rose to its contemporary place of hegemony not just in the sciences, but in the academy in general. It was not always so. The institution of the university was an invention of medieval Christianity and modern science itself was birthed out of a Judeo-Christian worldview, a truth that has been lost in the current landscape of whiggish tales about the backwardness of the Middle Ages and the "warfare" between science and religion that supposedly began with the Enlightenment.[1] A corrective is in order. I propose, therefore, to begin with a concise reflection on the very possibility of rational explanation in the context of naturalism, arguing that it is a woefully deficient context for the scientific enterprise both metaphysically and epistemologically. I will then develop a historico-philosophical etiology of the rise of naturalism and correct a variety of egregious historical misconceptions, all by way of a general argument that the current ontological and methodological foundations for the pursuit of scientific truth are misconceived, counterproductive, and in dire need of reconstitution on transcendent grounds.

1. Rational Explanations Within and Without Naturalism

Among those holding to the universality of rational explanations, some would maintain that while it may be a logical possibility that rational but nonnatural explanations exist, as a matter of fact, there are no examples of such. Natural explanation holds sway not only over the sciences, but over everything else as well. This is the viewpoint affirmed by philosophical naturalism. In its reductive form—which insists that everything must ultimately be explained at the level of physics—it devolves into scientism, the belief that the heuristic methodology of the physical sciences is applicable in all fields of inquiry and all real knowledge is the result of such investigations. Aside from its self-referential incoherence, scientism establishes a hermetic boundary between facts and values that strips all values of their factuality and all facts of any objective noninstrumental valuation. The end result is moral nihilism and the instrumentalization of rationality to subjective ends incapable of objective evaluation in terms of their intrinsic merit.

Disenchanted with reductionism, nonreductive naturalists adhere to the universal scope and rationality of natural explanations, while asserting that consciousness and rationality are *mental* properties that supervene on and emerge from *physical* circumstances to which they are nonetheless irreducible. This attempt to combine a materialist monism about entities with a pluralism of supervenient or emergent properties—accounting for consciousness, intentionality, rationality, normativity, and a variety of other things that pose prima facie difficulties for naturalism—is indicative of the highly malleable character of naturalist doctrine. This native elusivity, combined with strategies of retrenchment, is designed to insulate the fundamental thesis of naturalism—the causal closure of the material realm—from disconfirming evidence. The literature on supervenience and emergence offering variations of this sleight of hand is voluminous and we cannot survey it here, but such a survey is unnecessary since we can more easily provide two *principled* arguments for the falsity of both reductive and nonreductive naturalism (anomalous monism).[2]

First of all, quite apart from the research difficulties engendered by multiple neurophysiological realizability of function, there is no way that conscious apprehension of meaning (semantics) could be generated from neurochemical syntax. Consider that the semantic content of a graphical representation in natural language is *transcendently imposed* upon it by an *intelligent assignment* of meaning to the symbols within the structure of attendant grammatical rules. Brain electrochemistry can be no different. It consists in a molecular arrangement of neurons and synaptic traffic that bears no meaning in itself, but rather requires a *transcendent meaning correlate* that is not intrinsic to the brain as a biological system but rather the property of a consciousness that is *distinct* from it. John Searle's "Chinese Room Argument" thus serves not just as an illustration of the falsity of functionalism and an explanation of why computers will never have *conscious* experiences;[3] with all due respect to Searle's claims on behalf of biological naturalism,[4] it *also* serves as an illustration of the falsity of nonreductive materialist monism and an explanation of why personal consciousness, while *correlated* with proper brain function, is *ontologically and operationally distinct* from it. So by all means, let us establish the neurophysiological correlates of thought insofar as we can, but not be so naïve as to think such an achievement would establish materialist monism; on the contrary, an immaterial consciousness *must* exist if our beliefs have semantic content, which they undeniably do. The rejection of an ontological distinction between matter and mind requires the self-refuting *belief* that there are no such things as beliefs. It leaves us with an eliminative materialism that renders consciousness an illusion and therefore precludes the very possibility of rationality.

A parallel argument can be given that *belief* in naturalism, whether reductive or nonreductive, is epistemically self-defeating. Various forms of this argument have been offered by C. S. Lewis, Richard Taylor, and Alvin Plantinga. Since the most recent version of Plantinga's sophisticated evolutionary argument against naturalism is available to the reader in this collection, I distill here the essence of the Lewis-Taylor-Plantinga insight.[5]

The prospect of human knowledge depends upon the veridicality of our perceptions and the validity of our reasoning processes. If the certainty resulting from cognitive perception and valid inference provides a genuine grasp of how reality must be independent of our minds, then knowledge is possible, but if the certainty so obtained is a mere feeling and *not* a genuinely reliable insight into reality, then we do not have knowledge. Now, if naturalism is true, human beings came about as the result of undirected processes of evolution that had *no goal* in mind. In such case, our cognitive faculties are the end result of mindless causes and historical accidents that take *no account* of truth or logic, just the exigencies of survival. Under such conditions, *any* complex of beliefs and desires that conduces to survival would suffice. What we believe to be true under such conditions is therefore an accidental historical byproduct of purely natural events that *bear no intrinsic relation* to the actual truth of the beliefs we hold; it is an expression of how our brains just happen to work. That our beliefs should actually be true under such conditions seems quite unlikely; at the very least, whether our beliefs are true or false cannot be ascertained. If naturalism is true, therefore, our reasoning processes are so discredited that they cannot support the truth of *any* of the beliefs we happen to hold, especially those rather distant from immediate experience, such as the belief in naturalism itself. Belief in naturalism is therefore *epistemically self-defeating*, and since there is for the naturalist no remedy to this situation, it is irrational to be a philosophical naturalist because it destroys the possibility of rationality altogether.

This leaves us to philosophical naturalists who are most appropriately categorized as "pragmatists." While eschewing anything beyond the natural realm, they deny the universality of rational explanation. In their view, all explanations are radically contextual, and nothing would license the assumption that all contexts—even scientific ones—are mutually reconcilable. This stance, which represents a kind of *irrationalism*, is ironically the most consistent realization of the metaphysical and epistemological implications of philosophical naturalism. Adoption of the pragmatic stance ultimately leads to the fragmentation and instrumentalization of *all* rationality, scientific rationality included. Whatever science *is*, under the rubric of pragmatic naturalism, it is *not* the rational search for a unified truth about the natural world. It is merely one instrumentality among many in a relativistic world of personal and societal agendas that have no objective standard in respect of their merit. Without such a universal standard, "truth" has no purchase point that would grant it objective ascendancy, "knowledge" becomes mere power in the service of appetite, and *everything* gets politicized. While many academic pragmatists give no personal evidence of an oppressive and egregious agenda—indeed, many are so concerned about prejudice that, as Richard Weaver once observed, they are at war with simple predication[6]—nonetheless, there is nothing intrinsic to this metaphysic, or rather *lack* of metaphysic, which would restrain the pragmatist from atrocities or exhort him to defend against them.[7] Again, nihilism reigns and the voice of relativized "reason" dissolves into static, full of sibilant sound, signifying nothing. Rationalistic or irrationalistic, reductive or nonreductive, it is therefore clear that through its de-transcendentalization of both nature and reason, philosophical naturalism destroys any rational basis for science, and ultimately contributes to the destabilization and fragmentation of civil society.

So much for naturalistic accounts of rationality; but there are also those, among whom I include myself, who would maintain that while rational explanation has a legitimate claim to universality, natural explanation does *not*. Indeed, they would assert that natural explanations are only possible because they are grounded in an intelligent order that is *transcendently imposed* upon the natural realm—without belief in the existence of such an order, scientific practice would seem little better than reading patterns into tea leaves or chicken entrails. The fact that scientific activity is less arbitrary and more useful than such fabrications may be taken as evidence for the existence of just such a transcendently imposed order; indeed, this presupposition is the *transcendental ground* of the very possibility of science as a rational, truth-conducive enterprise. Within this rubric, natural explanations may be either adequate or inadequate to their intended explanandum, and it becomes possible that rigorous analysis may show certain features of the universe and of biological systems to be best explained as the result of intelligent causes that *transcend* nature. Furthermore, in respect of human rationality itself, like also begets like, and the reason that our cognitive faculties are capable of perceiving truth and reasoning correctly is that they too—when functioning properly in their intended environment—operate as intelligently designed systems that have the formation of true beliefs as their purpose.[8] Only under such conditions does our unspoken faith in human reason make sense. It could not be otherwise, for as we have just seen, the embrace of naturalism utterly destroys the possibility of *any* knowledge or rationality.[9]

2. Naturalism Rising: Historical Causes and Societal Consequences

Given the pathological character of naturalism, it is worth examining how it gained hegemony in the academy at large and the sciences in particular, and it is important to correct a few myths about the origins of modern science and the alleged "warfare" between science and religion.

In his monumental tome *A Secular Age*, Charles Taylor attempts to trace the multiple genealogies and etiologies of modern secularism in an effort to understand how a society in which it was virtually impossible *not* to believe in God became one in which such belief is relativized and privatized among the masses, and discouraged as "anti-intellectual" by the cultural elite.[10] In doing so, he distinguishes three senses of "secular" (but is particularly concerned with the last). These are: (1) the removal of God from public spaces through the sociocultural and political privatization of religion; (2) the decline of private belief in God and regular participation in associated religious practices; and (3) a change in the societal conditions of belief, whereby belief in God is relativized as merely one option among many and rendered more difficult to embrace.[11] The story that Taylor tells is rich, subtle, and complex, not to mention profound and unsettling in its implications. Broadly speaking, there are two factors of special relevance to this massive restructuring of the conditions of belief. The first relates to what Taylor calls the "cosmic imaginary," and the second to what he calls the "social imaginary." Profound changes have taken place in the way that Western cultural elites unreflectively envisage the universe and our place within it. We have moved from an ordered and personal cosmos in which humanity has a special purpose and place to an impersonal universe with no special place for humanity and no purposes beyond those we imagine for ourselves. There has also been a profound change in the way that the modern and postmodern West conceives of its social life and perceives its identity in relationship to other cultures and civilizations. I set aside this second factor so that I might focus on the first, which is more central to our concerns.

As we shall see, in much the same manner that Judeo-Christian belief once gave Western civilization its conscience and moral structure, its sense of duty and responsibility, its love of freedom within a respect for the rule of law, its model of self-sacrifice in response to need or in confrontation with evil, its basic principles of decency and its self-confident backbone in world affairs, so too Judeo-Christian theistic belief laid the historical foundations for Western science and technological success.[12] That this should be so is hardly surprising, for as we have seen, belief in an intelligently ordered cosmos is the transcendental ground for the very possibility of science as a rational and truth-conducive enterprise. The central irony is that having used this ladder to scale the heights, many contemporary scientists are now bent on discarding it in a naïve, Wittgensteinian fashion. What they fail to recognize is that apart from such a ladder, the whole edifice of modern science and human accomplishment becomes an inaccessible city in the clouds. This does not mean these scientists will cease to walk its streets, of course—merely that they will do so illegitimately, acknowledging neither the ladder they have used nor the borrowed theistic capital that paid for their climb and built the city itself. While it is quite reasonable to pursue scientific research without *explicit* reference to the transcendent framework in which such an endeavor makes sense, nevertheless, under the aegis of unbelief and the denial of an intelligently ordered reality, scientific practice becomes somewhat of a performative contradiction, and its utility a deep mystery.

2.1 Framing the "Cosmic Imaginary"

So how did these desultory conditions of belief gain cultural currency? In Taylor's terminology, how did the "cosmic imaginary" transform into the "buffered selves" and "closed world structures" of the "immanent frame"? (I borrow selectively from the richness of Taylor's vocabulary and analysis, but will alter the focus and trajectory of the discussion.) As Taylor points out, the modern cosmic imaginary is not able to be captured within any one categorization of views: It runs the gamut from the most hard-nosed of atheistic materialisms to the most fundamentalistic of Christian orthodoxies; it encompasses New Age syncretism and the revival of ancient paganisms; it apparently includes every shade of irreligion or religiosity presently or historically imaginable. This does not mean, of course, that all these views are rationally defensible, just that they are all present. Within this broader background, we are specifically concerned with the rise of materialist philosophy and the hegemony of philosophical naturalism among the cultural elites and in the scientific subculture in particular. A narrative that Taylor recognizes in this regard, but on which he does not dwell, focuses on the rise of nominalism over against realism in the medieval dispute concerning universals, and the impetus this gave to empiricism and an explanatory preoccupation with material mechanism. We turn, then, to the chain of development in Western thought that begins at this seemingly arcane juncture.

Nominalism

Nominalism is the doctrine that abstract concepts, general terms, or universals do not exist as independent realities but only as names within human language. Historically, the doctrine first arose in conjunction with a strong form of theological voluntarism, advocated by William of Ockham (c.1285–1347), which made God's intellect subordinate to his will so as to liberate divine action from any necessity. This led to a denial that God has an essential nature, and opened the door to divine arbitrariness and universal possibilism, the view that there are no necessary truths, not even logical ones, constraining divine action. While this extreme nominalism is plainly a self-defeating position, the theological concerns that gave rise to it are intuitive enough. Ockham dissented from the Thomistic view that God's moral commands were rationally and metaphysically necessitated by his nature, maintaining that such a view inappropriately restricts God's freedom and detracts from the sovereignty of his will; rather, the basis of moral law resides solely in the divine will, which could have differently constituted the difference between right and wrong.

These Ockhamite concerns focused on the divine command theory of ethics can be generalized to what Plantinga has called the "sovereignty-aseity" intuition, in which (1) God has created everything distinct from himself, (2) God is not dependent on anything distinct from himself, (3) everything distinct from God is dependent on him, and (4) everything is within God's control.[13] Given the unacceptable theological consequences of nominalism and the broader logical difficulties it readily generates, it seems clear that God's will must be subject to his nature and not free in an absolute sense; the Thomistic approach to the sovereignty-aseity problem is clearly the superior one. We need not be concerned that God's will is constrained by anything outside himself, however, if we conceive of abstract objects as dependent upon him in an appropriate way. The problem may be solved by embracing a kind of Neoplatonism that understands universals and other abstracta as having primary *conceptual existence* in the mind of God, not as formal entities existing in some mind-independent platonic realm. In such case, their ontological dependence on God is clear. On this view, the necessity inherent in truths involving abstract objects is expressive of necessities inherent in the divine nature itself, and there is no problem in recognizing that the divine will is subject to the divine nature: It could not be otherwise, for God cannot deny himself—that is, he cannot negate who he is or act in a manner inconsistent with his character.

A full articulation of this alternative to both nominalism and classical Platonism is beyond the scope of present concern.[14] What *is* important for us is the influence of nominalism on subsequent thought. As the story has commonly been told, nominalism is embraced in order to preserve God's freedom, but unconstrained divine freedom implies the absolute contingency of the natural order, from which follows the requirement of a strict empiricism in natural investigations.[15] As Foster originally articulated the thesis, the "*voluntary* activity of the Creator (i.e., that in his activity which exceeds determination by reason) terminates on the *contingent* being of the creature... But the contingent is knowable only by sensuous experience. If, therefore, the contingent is essential to nature, experience must be indispensable to the science of nature."[16] As the story usually goes, the divine voluntarism that gained ascendancy with the medieval nominalists was then carried forward in the Protestant Reformation, particularly in its Calvinist guise, by the transformation of the medieval distinction between God's absolute versus ordained power (*potentia dei absoluta et ordinata*) into a distinction between God's ordinary versus his extraordinary providence.[17] In other words, this distinction "operationalized" divine power by maintaining that God preserved the possibility of altering the ordained order of events by the exercise of his absolute power. While this seems unexceptional and even common-sensical in and of itself, it is asserted that this operationalized version of the power distinction entered the English theological tradition through the influence of Calvin and was "elaborated in England by both Puritan and Anglican divines and by members of the Royal Society as well."[18] In particular, it is suggested that it had a strong influence on Newton and his contemporaries in respect of their understanding of God's relationship to the world, and derivatively, the observational development of classical mechanics. For rather than nature's patterns proceeding from the internal dictates of a hierarchical order in which everything had a necessary and "natural" place discernible by reason alone, the orderliness of nature was instead imposed from without in accordance with the contingent freedom of the divine will, and hence in a manner requiring empirical investigation. "The prime mover" in the "metaphysical adjustments necessary for the inception of the classical or Newtonian science," was therefore, in the words of Francis Oakley, "the renewed and disturbing pressure upon Greek modes of thought of the Semitic idea of an omnipotent Creator-God."[19]

Now this is all well and good, and there is undoubtedly some core of truth to it, but it is not the whole story behind the Judeo-Christian genesis of modern science. Most especially, while nominalism had deleterious effects in later workings out (a point we will return to in a moment), neither divine voluntarism nor the Christian impetus to the empirical study of nature had logical or historical need of it. A weaker form of divine voluntarism would *not* require that God's intellect be subordinated to his will, thereby making his acts *arbitrary* expressions of absolute power. Rather, God's will can be *contingent*—in the sense that not all of his acts are rationally or morally *necessitated* by his nature—but nonetheless constrained by his nature and rational purposes so that his acts are not arbitrary. In short, one can assert that all of created reality is ontologically and actively *dependent* on the will of God—arbitrated by God—without maintaining that God's will is not an expression of his nature, or, as nominalism would have it, that God lacks an objective nature altogether. The neoplatonic conception outlined earlier is perfectly compatible with this understanding of nature's dependence on the divine will and the practical necessity of empirical investigation to determine its manifestation. The central point to be grasped in this discussion, however, is that it is *not*, as whiggish secular conceit would have it, that modern science came into being only as our approach to nature broke free of Christian thought forms, but rather, that it came into being precisely *because* of the Judeo-Christian conception of nature that provided its justification.

Some further myths regarding the nominalist-voluntarist thesis should be dispelled. Contrary to the assertion that John Calvin (1509–64) was a conduit for the nominalistic expression of divine voluntarism in early modern thought, Calvin himself repudiated this position as an excess within medieval scholasticism. In the *Institutes,* he rejects any arbitrariness in divine power, stating "we do not advocate the fiction of 'absolute might'; because this is profane, it ought rightly to be hateful to us. We fancy no lawless god who is a law unto himself."[20] What is more, we discern throughout this historical period nominalists who were *not* empiricists (Descartes is a prime example, and Leibniz [1646–1716] manifested nominalist tendencies), as well as advocates of empirical science who were *not* nominalists (such as Johannes Kepler [1571–1630], the Cambridge platonists Henry More [1614–87] and Ralph Cudworth [1617–88], and perhaps most notably, Isaac Newton [1642–1727] and his defender Samuel Clarke [1675–1729]).[21]

Nevertheless, while *nominalistic* divine voluntarism was one factor among many contributing to the rise of empirical science (and a logically dispensable one at that), the instrumental success of the empirical approach *did* contribute to the ascendancy of nominalism over logical realism by emphasizing the reality perceived by the senses over the reality perceived by the intellect. This in turn led to a focus on *material causation* and *material mechanism* as explanatory principles, though Newtonian gravitational theory, modern quantum theory, and intelligent design theory are notable exceptions to this preoccupation. As a matter of history rather than intellectual necessity, therefore, universals and other abstracta came to be perceived in the modern consciousness as mere names adaptable to our convenience. This had profound consequences, for the *anti-essentialism* that nominalism represents—in conjunction with an unbridled empiricism and preoccupation with material mechanism—gave birth to philosophical naturalism, and contributed to its intellectual rise; and the *constructivism* that nominalism represents, recasting all categories and categorization as mere instrumental invention, gave birth to the relativisms that have come to dominate the intellectual landscape. Thus was nature de-transcendentalized, denatured, disenchanted, and instrumentalized; thus was humanity stripped of its essence, dehumanized, and turned into an accident of time and chance; and thus was morality demoralized and made into arbitrary social convention, transforming us into beings without a moral compass, blown this way and that by animal passions and cultural vicissitudes we no longer have intrinsic reason or will to resist. Insofar as we have not completely succumbed to nihilistic consequences, it is the residual felt grace of transcendent reality that preserves us. We would do well to strengthen its influence both intellectually and viscerally if we wish to escape the leveling of the relativisms *du jour* and stave off intellectual and sociocultural disintegration.

The Advent of the "Buffered Self"

In the progression from the Middle Ages to the present in Western society, this move from the intellect to the senses has been paralleled by a change in the conceptualization of the self and material reality from spiritual imagination to sensual preoccupation. In Taylor's terminology, we have moved from a porous conception of the self that is vulnerable to the spirits, demons, and powers of an enchanted world to a buffered view that sees the self as separate from an external world structure that is wholly contained in an immanent frame. Taylor casts the Reformation period in Christian history—from the Lollards and Hussites through the Protestant reformers to the Jesuit Counter-Reformation—as the intermediary step that led from an unsophisticated theism encompassing superstitious animism and sacred priestly magic to the completely disenchanted world of modern humanism and unbelief.[22]

The mediating reforms of this period are characterized by Taylor as a radical simplification of religious practice, in which the power of God was no longer seen as acting through sacred relics or sacred locations or mystical priestly rites or anything else amenable to our control, but rather directly through faith in God's mercy provided in Christ and an inner transformation wrought by the Spirit of God. He casts this as a form of "disenchantment," since all of the sacramentals of the old religion, its magical elements, have been purged because they arrogate power to us rather than to God.[23] The "good magic" of church relics and priestly powers was rejected as a form of idolatry and all rites used as magical manipulations of the supernatural realm condemned as evil practices subject to satanic influence. This led to an intensification of old beliefs in witches and exacerbated the conditions leading to the witch-hunts of the late medieval and Reformation periods.

In Taylor's account, the inevitable reaction and recoil from these excesses had the ultimate effect of undermining the worldview framework under which the witchcraft persecutions made sense.[24] He further observes that this effect was compounded by the positive impact of a new conception of Christian liberty in which salvation came by faith in Christ, service was freely rendered to God with one's whole heart, and people were no longer terrorized by superstitious fears and therefore seeking protection in magical objects or rituals.[25] This freed society from its former taboos, magic, and sacramentals, and allowed the world to be reconceptualized in accordance with a rational reordering concentrated in the will of God. In the social sphere, this rational reordering of the world was threefold in conception: a disciplined personal life, a well-ordered society, and an inner motive for both located in the glory of God, not personal gain or human convenience.[26] Taylor then argues that the marked success this rational reordering ultimately achieved allowed for the move to an exclusive humanism by lopping off the reference to God at two points: first, by conceiving of societal order purely as a matter of human flourishing, and not as a matter of obeying or glorifying God; and second, by regarding the power to pursue order in our personal and social lives as a purely human capacity, achievable without divine assistance or religious motivations.[27]

This transition, evident in the thinking of early humanists like Erasmus (1466–1536), was advanced by a form of neostoicism associated with Justus Lipsius (1547–1606) that had a strong influence on educational institutions on both sides of the Catholic-Protestant divide, affecting

Catholic, Lutheran, and Calvinist thought; it was further extended through the influence of Descartes (1596–1650). Alarmed by the societal disorder they had catalyzed, the reformers found much that was attractive in Lipsius's emphasis on the ability of the human will, through reason, to exercise control over the passions and assume responsibility in obedience to principles rooted in God's Providence. Translated into the realm of politics, Lipsius's approach involved a rationalization of the state and the means of governance through principles of autocratic rule, societal discipline, and military defense. The profound influence of such ideas is evident in that the early modern state was indeed structured in accordance with these principles.[28] Descartes was certainly influenced by neostoicism through Lipsius's French followers Pierre Charron (1541–1603) and Guillaume du Vair (1556–1621), as well as through his Jesuit education, but he extended Lipsius's emphasis on the human will in an entirely new direction. While neostoicism saw the hegemony of reason as objectively present in forms derivative of divine Providence and at work in reality so that rational self-possession was achieved through an insight into the order of things and an acceptance of one's place in it, Descartes rejected the objectivity of form in nature, advocating a *nominalistic* divine voluntarism and a rationalist mechanization of matter. He therefore conceived the hegemony of reason simply in terms of its control over the desires, that is, its instrumentalization of them to ends we judge to be worthy.[29] Cartesian ethics therefore requires a rigorous distinction between the mind and the body, because virtue consists in the mind's exercise, through the will, of sovereign control over the body and the passions that arise from the relation between soul and body. In the Cartesian metaphysic, all thought and meaning are *internal* to individual minds and everything *outside* the mind is bare mechanism. So "[b]oth science and virtue require that we disenchant the world, that we make the rigorous distinction between mind and body. . . . We have to set up a firm boundary, the one . . . which defines the buffered self. For Descartes, seeing reality as pure mechanism is the way of establishing that boundary, it is indispensable to it."[30]

Of course, the thoroughgoing philosophical naturalist will not see Cartesianism as genuine disenchantment, because it banishes neither supernaturalism nor mind-body dualism. Enchantment remains in the form of what the behaviorist Gilbert Ryle, in his disparagement of the Cartesian picture, called the "ghost in the machine."[31] In the course of nature, broadly conceived, God is the ghost in its mechanisms, his will functioning in an unconstrained (formless) capacity as the basis of its operation; in the case of human beings, it is the mind or soul that is the ghost in the machine, animating the body, in now infamous Cartesian conception, through the medium of the pineal gland. It must be observed, however, that under the aegis of theism, neither the model for natural regularity nor the model for the relationship between mind and body have to be articulated by way of the Cartesian picture; superior accounts are available.[32] Nevertheless, Descartes' picture of mind-body interactions was explicitly constructed to provide a kind of human *autonomy* that frees our minds and our wills from *metaphysical* dependence on God in a way that other theistic options do not. In his own words, written to Queen Christina of Sweden in 1647:

> Now free will is in itself the noblest thing we can have because it makes us in a certain manner equal to God and exempts us from being his subjects; and so its rightful use is the greatest of all the goods we possess, and further, there is nothing that is more our own or that matters more to us. From all this it follows that nothing but free will can produce our greatest contentments.[33]

It is here that we see not just a self buffered from occult powers in nature, but a self whose autonomy is protected and insulated even from God. So while the philosophical naturalist will still denounce enchantment, a significant break has taken place: Nature has been reduced to mere mechanism, and the human will has been freed from metaphysical dependence on God. What Descartes leaves us with is the disengaged rational stance characteristic of modernity and ostensibly epitomized by its most prestigious product: natural science. The pretense of modernity is that this disengagement is motivated merely by the quest for epistemic objectivity; the reality is that "a big part of the motivation resides in the prestige and admiration surrounding the stance itself, with the sense of freedom, power, control, invulnerability, dignity, which it radiates."[34] From this standpoint of alleged objectivity and invulnerability, of disengaged autonomy and rational control, only one more "lopping off" is required on the path to an exclusive humanism and metaphysical naturalism: purging the ghost from the machine altogether by banishing God from the world and the soul from the body, thus completing the causal closure of the immanent frame.[35]

2.2 A Matter of History: Setting the Record Straight

A crucial aspect of this cultural path toward acceptance of the causal closure of the immanent frame and the denial of transcendence is a narrative which began to spread in the sixteenth century, proclaiming the benighted character of the medieval period and the vast desert it is alleged to constitute in the history of human accomplishment. This narrative easily transmogrified into a general theme asserting the antipathy between religious belief—particularly Christian belief—and intellectual or scientific progress. We therefore turn to a tale of the supposedly implacable animosity between science and religion and the attendant claim that Christianity has held back the progress of Western civilization. I believe Taylor underestimates the unchallenged authority this story has acquired in the public mind, and the power it exerts both in the educational establishment and on popular opinion in support of a destructive philosophical naturalism—it needs to be discredited.

So let us, as concisely as possible, take down this myth and establish a framework for a proper history and healthy conception of the relationship between science and religion. We have already seen how the metaphysical conceptions undergirding the modern investigation of nature emerged from medieval theological discussion. The contingency of creation in combination with the rationality of the Creator led to the recognition that the rational structure of the created order could only be discerned through empirical investigation. Despite this continuity between medieval scholastic metaphysics and the epistemic justification of early modern scientific rationality, the narrative of the irremediable opposition between modern science and medieval Christianity took root around the same time that Descartes was radicalizing human autonomy and framing modern epistemology in terms of the disengaged rational stance. This was due in part to the fact that certain key figures of the time portrayed themselves as making a heroic break with the medieval past rather than recognizing their indebtedness to it, while others were animated by anticlericalism and a hatred of Christianity.

The Historical Origins of the "Dark Ages" and "Warfare" Myths

We find evidence of this disparagement of medieval scholarship in the writings Francis Bacon (1561–1626), who contended in the *Novum Organum* that

The Rise of Naturalism

[O]nly three revolutions and periods of learning can properly be reckoned: one among the Greeks, the second among the Romans, and the last among us, that is to say, the nations of Western Europe, and to each of these hardly two centuries can justly be assigned. The intervening ages of the world, in respect of any rich or flourishing growth of the sciences, were unprosperous. For neither the Arabians nor the Schoolmen need be mentioned, who in the intermediate times rather crushed the sciences with a multitude of treatises than increased their weight.[36]

This disregard for medieval scholasticism was sharpened into palpable animosity a century later by the antitheistic French encyclopedists, particularly Voltaire (1694–1778), Rousseau (1712–78), and Condorcet (1743–94). Voltaire wrote of the "general decay and degeneracy" characteristic of the Middle Ages, and the "cunning and simplicity . . . brutality and artifice" of the medieval mind.[37]

Speaking of the medieval period, Rousseau opined that

Europe had relapsed into the barbarism of the first ages. A few centuries ago the peoples of that part of the world, who today live such enlightened lives, lived in a state worse than ignorance. Some nondescript scientific jargon, even more contemptible than ignorance, had usurped the name of knowledge, and posed a nearly invincible obstacle to its return.[38]

Finally, Condorcet was even more virulent in his condemnation:

Disdain for the humane sciences was one of the first characteristics of Christianity. It had to avenge itself against the outrages of philosophy, and it feared that spirit of doubt and inquiry, that confidence in one's own reason which is the bane of all religious belief. The natural sciences were odious and suspect, for they are very dangerous to the success of miracles, and there is no religion that does not force its devotees to swallow a few physical absurdities. So the triumph of Christianity was the signal for the complete decadence of philosophy and the sciences.[39]

Though perhaps willful on the part of the *philosophes*, the depth of ignorance on display here should be a source of amazement to the historically informed. Clearly, other motivations are in play.

The viewpoint advanced by such narratives was widely disseminated in the late nineteenth century by Jacob Burckhardt (1818–97), a well-respected Swiss historian of the Renaissance, who insisted that "the Middle Ages . . . spared themselves the trouble of induction and free inquiry."[40] While J. W. Draper's *History of the Conflict between Religion and Science* (1874) also deserves mention for advancing this tide of ignorance, the supremely influential and crowning manifestation of this nescient deluge was Andrew Dickson White's two-volume opus, *A History of the Warfare of Science with Theology in Christendom* (1896). I will correct a variety of its misrepresentations momentarily, but the general tenor of the account is admirably summarized in this passage:

The establishment of Christianity . . . arrested the normal development of the physical sciences for over fifteen hundred years. The cause of this arrest was twofold: First, there was created an atmosphere in which the germs of the physical sciences could hardly grow—an atmosphere in which all seeking in nature for truth as truth was regarded as futile. . . . Then, too, there was established a standard to which all science that did struggle up through this atmosphere must be made to conform—a standard which favoured magic rather than science, for it was a standard of dogmatism obtained from literal readings in the Jewish and Christian Scriptures. . . . For twelve centuries, then, the physical sciences were thus discouraged or perverted by the dominant orthodoxy. Whoever studied nature studied it either openly to find illustrations of the sacred text, useful in the "saving of souls," or secretly to gain the aid of occult powers, useful in securing personal advantage.[41]

What is all the more dismaying with respect to the influence had by this narrative is that the historical dissimulation it represents was to a large degree intentional. White, the cofounder and first president of Cornell University, openly admitted he wrote the book to exact revenge on Christian critics of his plan to establish Cornell as a coeducational secular institution.[42]

White's narrative persists today in the work of a variety of professional historians, in school textbooks, and in popular presentations on the history of science. Hence Daniel Boorstin (1983: 100; quoted in Stark 2007: 129) remarks that "the leaders of orthodox Christendom built a grand barrier against the progress of knowledge . . . After Christianity conquered the Roman Empire and most of Europe . . . we observe a Europe-wide phenomenon of scholarly amnesia, which afflicted the continent from A.D. 300 to at least 1300."[43] Or more recently, Charles Freeman (2003: xviii) has offered matter-of-factly that, by the fifth century of the Christian era, "not only [had] rational thought been suppressed, but there [had] been a substitution for it of 'mystery, magic, and authority.'"[44] These false pronouncements are reinforced by the misinformation that many of us were taught in school, much of which is derived from White, even when he was not the first to assert it. For instance, it is common to emerge from a grade school education with the belief that Columbus had to fight against churchmen who thought the Earth was flat in order to get funding for his voyage, that Giordano Bruno was burned at the stake for his astronomical work, that the rise of Christianity was instrumental to the fall of Rome and plunged Europe into darkness, that Copernicus was an isolated genius whose advocacy of heliocentrism required him to hide from Church authorities, that Vesalius risked the wrath of the Church by performing human dissections, and that Galileo was condemned by the pope just because of his view that the Earth moves around the Sun. None of these statements are accurate, but they are widely believed, and the list of inculcated falsehoods can be multiplied. When these misrepresentations are combined with popular books and PBS specials like Carl Sagan's *Cosmos*, which portrays the "Dark Ages" as a time of overwhelming superstition and ignorance due to the deleterious effects of Christianity, "it is little wonder," as David Lindberg observes, "that the ignorance and degradation of the Middle Ages has become an article of faith among the general public, achieving the status of invulnerability merely by virtue of endless repetition."[45] It is also little wonder that irreligion is on the rise among the young and impressionable, and that our culture is drifting ever deeper into a presumptive and profoundly ignorant secularity.

Correcting Some Common Historical Misrepresentations

Before I give a condensed account of the achievements of the Middle Ages and the continuity between medieval and modern science, we should dispatch with the litany of all too prevalent grade-school falsehoods just mentioned.[46]

The Rise of Naturalism

(1) Every educated person living at the time of Christopher Columbus (1451–1506), especially the hierarchy of the Roman Catholic Church, knew that the world was round.[47] This had been common knowledge among the educated since ancient times and it was maintained through the Middle Ages. White's "wise men of Spain" did not oppose Columbus because they believed the Earth to be flat; they objected because they knew its circumference to be larger than Columbus thought and that his voyage would take far longer than he anticipated.[48]

(2) Giordano Bruno's execution in 1600, while unfortunate, was not, as White portrays it, the result of his reviving unwelcome currents in Greek thought and anticipating Laplace's nebular hypothesis, but rather for pursuing hermetic sorcery and espousing a heretical speculative theology involving the existence of an infinite number of worlds.[49]

(3) Edward Gibbon (1737–94) popularized the idea that, after Constantine's conversion and proclamation of Christianity as the official religion of the Roman Empire, the influence of Christianity precipitated the fall of Rome by its concern for the afterlife rather than the present life, and its inherent pacifism, thereby plunging the world into darkness and ignorance. This is, of course, nonsense. Christianity played a unifying and strengthening role with regard to Roman identity, particularly in the Western Empire, and the Roman Christians of that time were not pacifists—they were every bit as pugnacious as the pagans. While the ultimate causes of the fall of Rome are manifold—though most are related to several centuries of subsiding Roman military strength and growing military capacity among various Germanic groups like the Goths, Huns, Vandals, Burgundians, and Franks—Christian influence is not to be numbered among them.[50] Furthermore, as I shall recount below, Christian influence led to the preservation and expansion of knowledge during the Middle Ages.

(4) Nicolaus Copernicus (1473–1543) was not an isolated genius advocating an essentially new hypothesis, nor was he living in fear because of his advocacy of heliocentrism, though we again have White to blame for this myth.[51] Copernicus had received an excellent education at the universities in Krakow, Bologna, Padua, and Ferrara. Further, he was *taught* the tools and ideas that led to a heliocentric model, since these had been developed by scholastic scientists over the preceding two centuries.

William of Ockham, of nominalist fame, had challenged the view that angels pushed the heavenly bodies to overcome friction by arguing that the space outside the atmosphere may be a vacuum in which friction was absent. Ockham's views were extended by colleagues at Oxford, notably Walter Burley (1275–1357) and Walter Heytesbury (1330–71), but had the most impact at the University of Paris, where Jean Buridan (c.1295–1358) articulated a theory of impetus that was a clear anticipation of Newton's first law of motion (inertia). Buridan also defended the hypothesis that the Earth turned on its axis, creating the appearance that the heavenly bodies "rise" and "set," recognizing that this was a far more parsimonious view given the velocities at which distant objects would otherwise have to travel. Buridan's work was defended against objections and mathematically extended by his successors at the University of Paris, Nicole d'Oresme (1325–82) and Albert of Saxony (c.1316–90), and then by Nicolas of Cusa (1401–64), who realized (as had his intellectual predecessors) that regardless of whether your position was on the Earth or some distant heavenly object, it would always appear from that vantage point that you were at a motionless center, with the heavens in motion around you. It followed that the perception the Earth was still may not, in fact, be the case.

All Copernicus did to advance the discussion was suppose it was not the case, place the Sun at the center instead, and construct a mathematical model in which the planetary orbits were circular. Because the planetary orbits are in fact elliptical—a discovery that fell to Johannes

Kepler—Copernicus's theory actually required more epicycles than Ptolemy's. Nonetheless, it constituted the first explicit mathematical model that was not geocentric. As to the significance of this construction, as I. Bernard Cohen states, "The idea that a Copernican revolution in science occurred goes counter to the evidence . . . and is an invention of later historians."[52]

(5) In respect of the place of Andreas Vesalius (1514–64) in the history of medicine, we must again confront one of White's deliberate misrepresentations. As White would have us believe,

> In the search for real knowledge [Vesalius] risked the most terrible dangers, and especially the charge of sacrilege, founded upon the teachings of the Church for ages . . . forbidding all dissection, and . . . threatening excommunication against those practicing it. Through this sacred conventionalism Vesalius broke without fear; despite ecclesiastical censure, great opposition in his own profession, and popular fury, he studied his science by the only method that could give useful results. No peril daunted him. To secure material for his investigations, he haunted gibbets and charnel-houses, braving the fires of the Inquisition and the virus of the plague. First of all men he began to place the science of human anatomy on its solid modern foundations . . .[53]

And on and on, lie upon half-truth upon lie. The truth of the matter is this: Dissections had already been taking place for two centuries at the very universities where Vesalius learned and practiced his medical procedures. The framing assumption at these Christian institutions of higher learning was that what makes us human is our soul, not our physiology, and so dissection of cadavers had no more theological implications than such studies of animal bodies. By late in the thirteenth century, Mondino de'Luzzi (c.1270–1326) had already written a textbook on dissection. He performed a public dissection at the University of Bologna in 1315, and subsequently dissections were common at Italian universities. Spanish universities conducted public dissections starting in 1391, and the first dissection in Vienna was done in 1404. Even Copernicus took part in a human dissection at the University of Padua in 1504 during a brief period there as a medical student. In short, the "introduction [of human dissection] into the Latin West, made without serious objection from the Church, was a momentous occurrence;" and truthfully, "although Vesalius profoundly altered the attitude toward biological phenomena, he yet prosecuted his researches undisturbed by ecclesiastical authorities."[54] What is more, rather than being a humanist as he is often portrayed, Vesalius was also a thoroughgoing Christian, who died while returning from a pilgrimage to the Holy Land.[55]

(6) Last, we come to Galileo Galilei (1564–1642) and perhaps the most famous controversy of all. What really happened? If we are looking for the truth, we will have to look beyond Bertolt Brecht's stage play about the conflict. It *is* true that Galileo did find himself at cross-purposes with the Catholic Church in his later years, was forced to recant his belief that the Earth orbits the Sun, and was sentenced to live in seclusion for the last nine years of his life. The further truth of the matter, however, is that Galileo's troubles had rather more to do with his difficult personality, lack of sound judgment, and various internecine disputes among Catholic theologians over biblical interpretation catalyzed by the hermeneutical narrowing of the Counter-Reformation, than simply with his scientific views. It is well known that Pope Urban VIII (1623–44) presided over the Galileo affair; what is less well known is that he and Galileo were good friends earlier in life when the future pope was still Cardinal Matteo Barberini, and that Barberini greatly enjoyed the sarcastic humor Galileo directed against the Jesuits. He was less amused years later, after advising Galileo to sidestep the current round of theological disputation over the heliocentric hypothesis by making it clear that he spoke hypothetically as a mathematician, not a theologian, to find his words in the *Dialogue Concerning Two New Sciences* placed in the mouth of the half-wit Simplicio, who voiced all the errors the book was written to correct. Moreover, Galileo deceived the pope about the time of the book's release, which resulted in a firestorm of controversy that necessitated a response. Still, when summoned to Rome for his views, as a devout Catholic Galileo went, even though he had been offered asylum in Venice; and when sentenced to prison by the Inquisition, despite Barberini's wounded pride, Galileo's old friend commuted his sentence to house arrest.

While the Galileo affair did catalyze an unfortunate tightening of restrictions on intellectual freedom, four things should be noted. First, within the broader Catholic environment, Galileo's views on biblical hermeneutics had a wide range of advocates, and many were untroubled by the suggestion that the Earth revolved around Sun. This latitude was expressed in various ways. Friar Marin Mersenne (1588–1648), for example, pressed those in his circle of influence to recognize that God could place the Earth anywhere he wanted, and it was the job of scientists to figure out where he had put it. With more sensitivity to the question of biblical hermeneutics, the Carmelite friar Paolo Antonio Foscarini (1565–1616) had argued that the heliocentric view was perfectly compatible with Scripture, since God often accommodates himself to our mode of understanding in relation to how things appear to us. By speaking in a common manner regarding appearances, nothing false is communicated, for this is indeed how things appear; yet by investigation of nature we may deepen our understanding. Given Galileo's increasing realization that the Copernican theory was superior, he took a view of Scripture similar to that of Foscarini. Thus, even though the Church, in a period of Counter-Reformational exegetical literalism and rigidity, officially ruled against this hermeneutical strategy, it may be more broadly recognized that there was no necessary conflict here and many scholastics appreciated this fact.

Second, Galileo acted rashly and with little thought for the consequences either to himself or to the scientific enterprise as a whole. He had been warned that the times were dangerous and he needed to proceed circumspectly, yet he needlessly provoked a clamp-down by the Inquisition.

Third, despite the activities of the Inquisition, scientific books written by Catholics tended to circulate freely. For instance, the famous Quiroga Index (1583) had little impact on the accessibility of scientific works. Galileo's works, written much later, never found their way onto a list of forbidden books. The Inquisition's primary concern was with works judged to promote superstition—this was how Giordano Bruno got himself into an irretrievable mess—so their attack focused mostly on works dealing with astrology and alchemy.

And fourth, the Galileo affair was not essentially a conflict either between science and Christianity or enlightenment humanism and Christianity. Galileo was not an enlightenment humanist; he was a devout Catholic who further understood that there was no necessary conflict between his scientific investigations and Scripture. In short, Galileo's *own* religious views did not inhibit his scientific work at all, they rather advanced it, for he recognized God's design in the natural world and argued that "the book of nature is a book written by the hand of God in the language of mathematics."[56]

On Medieval Progress and the Continuity of Scientific Development

The point of these corrections is not merely to deconstruct the myth that science and Christianity have been at each other's throats since the "Renaissance," the "Enlightenment," and the "Scientific Revolution" left the "Dark Ages" behind, but to make the case that these historical categories are artificial and prejudicial constructs which ignore the continuity between medieval and modern thought, and ignore the fact that the flowering of scientific research and humane studies that began to blossom exponentially around the sixteenth century was an outgrowth and consequence of the Judeo-Christian worldview in general, and scholastic theology and research in particular. The standard story that the catalyst for this blossoming was the fall of Constantinople in 1453 and a consequent influx of Byzantine scholars and ancient texts into the West that led to a recovery of ancient learning is another myth that does not stand up to scrutiny. The Western scholastics had already been reading, translating, and arguing about all of the important authors of antiquity for centuries. Much more relevant was the invention of the printing press by Johannes Gutenberg around 1455, and the printing of the first Bible. There soon followed an avalanche of printed books, almost all of them religious, with the Bible foremost among them. This was the catalyst for an extraordinarily rapid expansion of literacy since people now had something to read—and in their own native tongue, rather than Latin. This development, more than any other, fed the Protestant Reformation and the general flowering of knowledge and learning that ensued throughout Europe.[57]

Of course, the invention of the printing press was nearing the chronological end of a long list of technological accomplishments that steadily increased throughout the Middle Ages.[58] Military technology advanced with iron chain-mail armor, knights carrying lances mounted on horses with Norman saddles and stirrups, iron horseshoes, appropriation of the Chinese invention of gunpowder for novel use in cannon and gunnery, and so on. Agricultural technology was moved forward by a new and effective way of harnessing horses and by the invention of waterwheels, camshafts, and mills. Time-keeping and sea travel advanced with the invention of mechanical clocks, innovations in ship building, and the compass; no sooner was the latter invented than it was refined, conjoined with astronomical knowledge, and effectively applied to marine navigation. The list is endless. As Rodney Stark describes it, "Christianity did not plunge Europe into an era of ignorance and backwardness. Rather, so much technical progress took place during this era that, by no later than the thirteenth century, European technology surpassed anything to be found elsewhere in the world."[59]

All of this was aided and accelerated by the profound political and socioeconomic innovations of the Middle Ages. Between the sixth and the tenth centuries, chattel slavery gave way to serfdom. This difference was not merely terminological. Serfs had rights and a substantial degree of autonomy: They married freely and their families were not subject to sale or dispersal; they paid rent and controlled their own time and work hours; in some places, their work obligations were limited and their actions were more akin to hired labor than any semblance of slavery; and even though serfs were bound to their lords by certain obligations, so too their lords had obligations to them, and to a higher authority. Such mutual obligation was integral to the structure and nature of the feudal system.[60] Some scholars maintain that Christianity had nothing to do with this transition, but this is clearly false. While the New Testament accepted the regrettable de facto existence of the institution of slavery, it established the equal humanity and moral worth of both master and slave (Ephesians 6:5–9; Philemon; etc). This tension between the existence of slavery and the Christian recognition of the equality of all humanity in the eyes of God grew ever more intense.[61] The Church extended its sacraments to all slaves and then, eventually, imposed a ban on the enslavement of Christians and Jews. Given the cultural influence and authority of the Church in medieval Europe, this ban had the effect of universal abolition.[62]

Another happy innovation of the medieval period flowing from Christian influence was the establishment of private property rights under the rule of law. The Bible takes private property rights as a given, morally condemning theft and fraud. From Augustine (354–430) through Albertus Magnus (c.1193–1280), Thomas Aquinas (1225–74), and William Ockham, and right up to the present, the majority of Christian theologians have understood the institution of private property as a divinely mandated human right, and its protection by those in authority as a divine injunction. This conception was codified in the Magna Carta, several articles of which give a long and detailed list of property rights that are protected against royal usurpation. Article 61 created the House of Lords, which was charged with, among other things, preserving the rights and liberties guaranteed by this charter of freedoms under a rule of law to which even the king and the nobles were subject. Stephen Langton, the archbishop of Canterbury from 1207 until his death in 1228, was the indispensable figure in the struggle against the abuses of power by King John of England, and with the aid of the English barons forced the king to sign the Magna Carta on June 15, 1215. Among other things, this charter accelerated the economic freedoms that gave birth to full-blown capitalist commerce in Christian Europe and a growth in the *general* wealth and prosperity of its whole population that was unrivaled in the history of the world—until America, as a constitutional republic founded on an explicitly Judeo-Christian ethico-political philosophy committed to religious and economic freedom under the rule of law, did Europe one better.

While the virtues of honest work and frugality, and the recognition that future returns follow from present efforts, were recognized and affirmed from the start in Christianity, it was not until the thirteenth century that Christian theologians developed a mature understanding of the moral aspects of a capitalist economy—issues like lending, credit, investment, interest, banking, free markets, salaried labor, profits, etc. This whole process got started because, as technological innovation in moving heavy loads over land had progressed and transport over longer distances had become practical, formerly self-sufficient monastic communities and estates began to focus their labor and specialize, which led in turn to a fairly rapid shift in the late ninth century from a barter to a cash economy.[63]

As capitalism progressed—where "capitalism" may roughly be understood in terms of risking the investment of wealth or resources in productive activity for the purpose of gain—devout Christians found it necessary to reflect on the relationship between their faith and their increasing prosperity. More depth was required than Augustine's fourth-century view that commerce was not inherently evil and it was the responsibility of the individual to live honestly.[64] For instance, in a cash economy, how are fair prices decided? Albertus Magnus, in his *Commentary on the Sentences of Peter Lombard*, had reached the recognition that "just price" is no more and no less than what "goods are worth according to the estimation of the market at the time of sale."[65] Thomas Aquinas, who was a student of Magnus, elaborated in the *Summa Theologica* that "the just price of things is not absolutely definite" but a function of supply and demand settled at a given time by free negotiation between sellers and buyers.[66] The question of interest on loans was more complicated since Christianity had initially inherited from Judaism a prohibition on usury (see Deuteronomy 23:19–20), though the prohibition in Judaism was restricted to fellow Jews, so interest could be charged to foreigners. In modern terms, the scholastics came to recognize that opportunity costs and risks were associated with the lending of capital that merited just compensation, and by careful reasoning this principle was extended to a wide variety of increasingly complicated situations.[67]

All of this economic activity, which had begun with the monastic estates, led to the creation of banks, accounting procedures, and the invention of insurance. By the fourteenth century, there were no less than 173 major banks in northern Italy, and commercial banking was spreading quickly throughout Europe.⁶⁸ Banks had begun as secure depositories for money and its safe and accurate exchange over long distances, but soon began to use their cash reserves to issue credit that could be used to start new enterprises. Surplus money was no longer uselessly hoarded but put into circulation for the purpose of creating more wealth, and interest was paid to depositors who assumed some risk by placing their money with banks that issued loans. For similar reasons, insurance was invented in the 1300s as a means of spreading risk to avert catastrophic individual loss, and insurance companies offset their own risk by charging fees. This buying and selling of risk by banks and insurance companies catalyzed the creation of many new business ventures and a vast increase in prosperity throughout Europe. A rising tide lifts all boats. So it was that the benefits of modern capitalism emerged out of medieval monastic economic activity and the scholastic analyses that freed it from its earlier theological shackles.⁶⁹

These socioeconomic and political developments, which were brought to life and nurtured in the cradle of the Judeo-Christian worldview that dominated medieval European society, provided fertile metaphysical, epistemic, sociocultural, and economic ground for scientific theorizing and experimentation. We have already touched on some of the developments in medieval science that were brought to fruition by scholastics in that quintessential Christian invention for the advancement of learning and the dissemination of research and scholarship: the university. We have also seen how the medieval invention of the printing press galvanized literacy and the spread of learning, contributing an indispensable condition for the exponential expansion of the sciences and humanities in the sixteenth century. But is there anything *beyond* this that catalyzed the growth of scientific knowledge? Is there any point of *discontinuity* between the "scientific revolution," so called, and the steady progress in scholastic science that gave rise to it?

Three hypotheses have been put forward in the attempt to capture a key difference between medieval and modern science: the role of *mathematics* in modern science, the role of the *experimental method* in modern science, and the rise of the *mechanical philosophy* in modern science.⁷⁰ As we shall see, all three of these hypotheses are deficient, though the last perhaps less so than the others. The *first* proposal is that the key change that came about in the modern period was the unification of the power of mathematics with physical theorizing that arose through overcoming an Aristotelian prohibition on the transgression of disciplinary boundaries. Not so. While it is true that Aristotle distinguished between mathematics and physics and articulated hierarchical disciplinary boundaries, he consistently rejected prohibitions against crossing the boundary between physics and mathematics.⁷¹ Moreover, the historical record of the Middle Ages tells a different tale, revealing astronomy, geometrical optics, dynamics, and the theory of weights as successful examples of medieval science resulting from the unification of mathematics and physics.⁷² It was these achievements that were appropriated and further developed by Copernicus, Galileo, Kepler, and a vast array of other early modern scientists.

The *second* suggestion is that the experimental method played a novel role in the scientific achievements of modernity that distinguished them from the medieval practice of science. Again, the actual record of scientific methodological *practice* in the Middle Ages shows this to be false. Ptolemy (c.90–168) was extensively involved in astronomical observation and optical experimentation.⁷³ The Alexandrian Christian platonist philosopher John Philoponus (c.490–570) performed imprecise experiments to ascertain the truth of the Aristotelian contention that the speed of descent was proportional to the weight of a dropped body, discovering—contrary to Aristotle—that there was very little difference.⁷⁴ During the historical period when medieval Islam was scientifically productive, Ibn al-Haytham (c.965–1039) performed experiments to ascertain the truth of optical theories and Kamāl al-Dīn al-Fārisī (c.1267–1320) devised the first mathematically adequate explanation of the rainbow on the basis of experiments using a camera obscura to analyze light rays passed through transparent glass globes filled with water, an experiment that was duplicated around the same time in medieval Christendom by Theodoric of Freiberg, a Dominican friar.[75] Astronomic observations were made by Levi ben Gerson (1288–1344) and Johannes de Muris (1290–1351), using a variety of instruments to refute certain aspects of Ptolemaic planetary models and to observe solar eclipses for the purpose of showing certain predictions of Alfonsine astronomical tables to be incorrect.[76] Roger Bacon (c.1220–92) argued extensively for empirical methodologies that would subject scientific arguments to experimental tests and pursued a variety of scientific experiments himself.[77] The thirteenth-century physician Petrus Perigrinus de Maricourt, a contemporary of Bacon, did extensive experimentation with magnets to determine their properties and anticipated many of the discoveries that would be credited to William Gilbert in the seventeenth century.[78] Another contemporary of Bacon, the Franciscan friar Paul of Taranto, though an alchemist, conducted a variety of laboratory experiments attempting to achieve the transmutation of substances. It is fair to say that such researches paved the way for modern chemistry.[79] The bottom line is that scientific experimentation was widely recognized as useful from late antiquity throughout the Middle Ages, and experiments were performed when it was recognized that doing so could help to confirm or disconfirm a scientific claim; the experimental method, therefore, was *not* a distinctive of modern science.[80]

Finally, we are left to consider whether the rise of the mechanical philosophy provides a basis for the flowering of science in the modern period and separates it from medieval thought.[81] This is the most plausible of the three hypotheses regarding the difference between medieval and modern science, but we can still discern a fundamental continuity with medieval thought, even while recognizing that mechanical explanation is *not*, any more than mathematical description or controlled experimentation, a universal constraint on modern scientific theorizing.[82] The mechanical philosophy is often said to have reduced all natural causality to efficient and material causes, setting aside the Aristotelian categories of formal and final cause. There is a methodological sense in which this is true, and another, metaphysical sense, in which it is not. When mechanism is invoked, the *formal* cause remains in the *design* of the mechanism and the *final* cause remains in the *purpose* the mechanism was created to serve; the difference arises from the non-Aristotelian way in which the mechanical philosophy is *applied*, since formal and final causes are in the background and the focus is on the search for an *explanation* in terms of the material implementation of an efficient mechanism. And this returns us, after a fashion, to the Cartesian conception of nature as *mere* mechanism that we encountered earlier. David Lindberg seems to suggest that the mechanical philosophy was somehow a direct revival of the Greek atomism of Leucippus, Democritus, and Epicurus mediated by the Latin writings of Lucretius, Diogenes Laertius, and Cicero.[83] But while random collisions of atoms in the void certainly constitute a series of events conforming to efficient material causality, the metaphysical picture lacks any order, and thus fails to have the requisite metaphysical and epistemological connection to the developmental flow of Christian thought from which the mechanical philosophy is derived. The connection between Greek atomism and the mechanical philosophy of the modern period, evinced in the thought of Pierre Gassendi (1592–1655) and others, therefore seems one of opportunistic appropriation more than formative conceptualization. That this is so is evident from the fundamental structural changes and transcendent teleological directedness by which Gassendi modified Epicurean metaphysics.

The move beyond Aristotelian scholasticism was already underway in the philosophical and theological discussions of the time and did not need Greek atomism to catalyze it. We must instead return to our earlier account.

We described the path to the partial "disenchantment" of the world that the mechanical philosophy represents as arising through changes in the cosmic imaginary that sprang from the nominalistic de-transcendentalization of nature associated with radical divine voluntarism—which is particularly relevant in the case of Descartes—and the parallel cultural buffering of the self through mediating Reformational changes that rejected the old sacramentals and expanded the conception of the sacred to all of reality, thus leading in the long term to the unintended effect of a broad desensitization to the sacred by rendering it commonplace (see footnote 23). We also noted that the role of nominalism in this development, even though embraced by Descartes, was not an essential part of the path to the mechanical philosophy, and it was not embraced by many others who pursued mechanical explanations.

So what is the missing piece in this puzzle? There is *another* historical aspect to the appearance of mechanical explanations that has not, to my knowledge, been much emphasized at all. It should be. This is the immense flowering of technological progress and mechanical invention throughout the medieval period that we mentioned above. The realm of technological invention is an aspect of human ingenuity to which intellectual historians are often blind, but which radically influences our experience of the world. The conceptual theological connection that needs to be made here is simple: human beings invent machines to aid in the accomplishment of human purposes, and human beings are created in the image of God. Since God is the Creator and artificer of reality as a whole, and since he designed the world for the accomplishment of his purposes through the transcendent imposition of order upon it, we should expect to find divine analogues of human technology in nature itself. This is clearly part of the perspective of Robert Boyle (1627–91), and it is arguably an element in the thought of many advocates of the mechanical philosophy. In fact, this current of thought is found at least as far back as Nicole d'Oresme (1325–82), who remarked that God's creation "is much like that of a man making a clock and letting it run and continue its own motion by itself."[84] The move to mechanical explanation was thus the natural historical confluence of medieval reflection on divine providence with the progress made throughout the Middle Ages in machine technology. And when this idea of mechanical explanation was applied to natural investigation, which seems inevitable in that cultural context, it was found to be fruitful beyond anything that could have been imagined. So the exponential seventeenth-century flowering of physics did not really constitute a revolutionary break with medieval science at all. It was rather a *continuous* logical outworking—derived from developments in scholastic philosophical theology and medieval technological invention—that reached its *consummation* in this historical period.

An irony is that the scientist regarded as the poster child for the fruitfulness of the mechanical philosophy, Isaac Newton (1642–1727), did not himself adhere exclusively to the metaphysics of efficient material causation associated with it. Newton redirected the mechanical philosophy from a preoccupation with corpuscularianism and contact mechanisms to a consideration of forces, the mathematical treatment of which does not require any account of their provenance—a circumstance on which he famously remarked in the General Scholium to the *Principia*, "Hypotheses non fingo." Indeed, one of his supreme scientific achievements, the theory of universal gravitation, was *not* a mechanical explanation at all, but rather a fruitful mathematical *description* of phenomena on the basis of a postulated gravitational force that acted at a distance and permitted very precise predictions of astronomical events. As a theist,

Newton was not bothered by this state of affairs in the least, since he regarded God as being integrally involved in the quotidian course of nature. What should therefore be observed—especially in respect of Newton's theory of universal gravitation—is that it was precisely his *rejection* of methodological naturalism in conjunction with his mathematical genius that enabled him to follow the course that he did, and to revolutionize physics and astronomy. As a consequence, gravitational theory proceeded quite happily without a "mechanism" for well over two centuries, until Einstein provided one in the general theory of relativity: Mass-energy affects the structure of spacetime through gravitational waves propagating at the speed of light. Of course, quantum theory *undoes* all this. Feynman diagrams as aids to computation and visualizability are *mere* expedients. What is basic to quantum field theory and essential to its empirical adequacy are (1) nonlocal action-at-a-distance that defies any conception of efficient material causality, and (2) statistical descriptions and principles of superposition that cast aside anything even vaguely reminiscent of individuatable material substances with intrinsic identities.[85] And if, as most physicists believe, the reconciliation of general relativity with quantum theory will come through the quantization of gravity, all of these basic quantum-theoretic consequences will reassert themselves in gravitational phenomena.[86]

What is clear from this lengthy discussion is that, in all significant respects, modern science is continuous with the development of medieval science, and faith in the very *possibility* of science was derivative of medieval theology. Recognition of this fact is not new, but it seems perpetually to be getting lost in the mists of unwarranted assumptions about medieval benightedness. So far removed is this insight from modern secular consciousness that, as Rodney Stark observes,[87] Alfred North Whitehead (1861–1947) shocked his Harvard audience during his 1925 Lowell lectures (and subsequently the Western academic world as a whole when they were published) by remarking:

> I do not think, however, that I have even yet brought out the greatest contribution of medievalism to the formation of the scientific movement. I mean the inexpugnable belief that every detailed occurrence can be correlated with its antecedents in a perfectly definite manner, exemplifying general principles. Without this belief the incredible labours of scientists would be without hope. It is this instinctive conviction, vividly poised before the imagination, which is the motive power of research: that there is a secret, a secret which can be unveiled. How has this conviction been so vividly implanted in the European mind?
>
> When we compare this tone of thought in Europe with the attitude of other civilisations when left to themselves, there seems but one source of its origin. It must come from the medieval insistence on the rationality of God, conceived as with the personal energy of Jehovah and with the rationality of a Greek philosopher. Every detail was supervised and ordered: the search into nature could only result in the vindication of the faith in rationality. Remember that I am not talking about the explicit beliefs of a few individuals. What I mean is the impress on the European mind arising from the unquestioned faith of centuries. By this I mean the instinctive tone of thought and not a mere creed of words.[88]

As shocked as his audience was, Whitehead was completely right, and modern secularist conceit stands as much in need today of being reminded of its intellectual indebtedness to Christianity as it was then. What is more, as we have seen, the philosophical naturalist needs to be reminded that by forsaking a worldview grounded in a transcendent order he has completely lost his entitlement to a rationally structured reality accessible to the human intellect.

2.3 The Impact of Darwinism

Through the end of the nineteenth century, the vast majority of scientists were at least deists, and more likely Jewish or Christian theists.[89] In fact, in 1872, Charles Darwin's cousin, Francis Galton (1822–1911), one of the founders of quantitative psychology and an outspoken atheist and enthusiastic advocate of eugenics, conducted a survey of "English men of science" to determine whether youthful religion had a deterrent impact on the freedom of their scientific research.[90] To his dismay, not only did over 90 percent of the respondents—including Charles Darwin himself—respond in the negative, almost every respondent indicated a church affiliation. In 1914, in another effort to establish the irreligiosity of scientists, the American psychologist James Leuba conducted a more rigorous survey.[91] He found that 41.8 percent of his sample group believed in a God who answered prayer, another 41.5 percent had a more deistic view, and 16.7 percent had no belief in God whatsoever. If Leuba's results are restricted to "leading scientists," however, the number holding some substantial form of religious belief dropped to 30 percent. When Leuba's exact survey was repeated in 1996 the results remained much the same, except that the number of leading scientists—represented by members of the National Academy of Sciences—having strong religious beliefs had dropped below 10 percent.[92] A broader survey of religious belief by scholarly field, which included 60,028 academics, was conducted by the Carnegie Commission in 1969. It indicated levels of religious belief among natural scientists in the 55 to 60 percent range, with about two-thirds of these being orthodox. The percentage of religious believers in the social sciences was much lower, however, averaging around 45 percent, with the worst field being anthropology, where only 29 percent had any manner of religious belief at all, and only 11 percent were orthodox.[93] The moral of these surveys is that while religious belief persists among scientists and academics in general, it has declined precipitously since the nineteenth century, especially in its orthodox form, and especially among those regarded as being at the "top" of their field. So what happened to bring this about?

Undoubtedly, we have already grasped a major part of the story in our account of the tremendous changes in the social and cosmic imaginaries that have overtaken Western civilization in the last century, changes that have given rise to the widespread and unreflective acceptance among the intellectual elite of the causal closure and solitary existence of the immanent frame. This belief has obviously been reinforced by the social impact of the false narratives that emphasize a "warfare" between science and religion, a misperception given considerable impetus by Andrew Dickson White's vengeful writings at the end of the nineteenth century. But we should also not underestimate the impact of Charles Darwin's (1809–82) theory of universal common descent, which purports to explain speciation in the history of life solely by means of natural selection acting on random variation in populations. "Random," of course, means exactly that: objectively undirected and therefore without discernible purpose. That this was Darwin's intent seems clear, for he remarked in 1876 that "there seems to be no more design in the variability of organic beings, and in the action of natural selection, than in the course in which the wind blows."[94] Indeed, until Darwin's theory gained wide acceptance, an eventuality given a boost by the discovery of the genetic basis of inheritance and a neo-Darwinian synthesis in which natural selection supervenes on random genetic mutations, design had been regarded as the formative principle in biology. What Darwin is thought by many to have provided (as Richard Dawkins goes to great lengths to emphasize[95]), is a way of explaining the *appearance* of design in organisms without recourse to *actual* design. In light of this, Dawkins has no compunction about claiming that "although atheism might have been logically tenable before Darwin, Darwin made it possible to be an intellectually fulfilled atheist."[96] It is for this reason and others that Daniel Dennett describes Darwinism as a "universal acid" that "eats through just about every traditional concept, and leaves in its wake a revolutionized world-view."[97] By Dennett's account, of course, that revolutionized worldview is a radical philosophical naturalism in which traditional religion is preserved only as a curiosity in a "cultural zoo."[98]

This 3-D (Darwin, Dawkins, and Dennett) disavowal of design and purpose is, to be sure, fundamentally a philosophical stance. Even if the neo-Darwinian picture were accepted, the assertion that genetic mutation as the basis of variation is ontologically random and that environmental selection is absolutely blind is an unverifiable and gratuitous postulation.[99] The gratuitousness of this assertion has not stopped its appearance in a wide variety of textbooks used in secondary and tertiary education, however, which, as an expression of the mindset of the authors of these educational materials, speaks to the hegemony of philosophical naturalism in public institutions of higher learning and the broader cultural influence of Darwinism.[100] It is also revelatory of Darwinism's impact as regards the fundamental changes in the cosmic imaginary that have taken place in the last century. If we are looking for the historical locus at which formal and final causes were purged from the "scientific" view of reality, it clearly was *not* with the mechanical philosophy, which retained formal causes in the design plan of the mechanism and final causes in the purposes they were intended to serve; rather, as we have seen, formal intent realized through purposeful implementation was banished by the advent of Darwinism.

What is more, while *methodological naturalism* had a long history[101] and a growing foothold in science—even though it was emphatically rejected by Newton, whose presence towers over physics even today—the effect of Darwin's insertion of it into biology in terms of sowing the seeds of philosophical naturalism cannot be underestimated, in part because the presence of discrete intentional design in the biological realm had been one of the mainstays of natural theology.[102] Darwin did more than introduce methodological naturalism into biology, however; he contended that it was an indispensable criterion for any theory to be regarded as *scientific*.[103] As William North Rice, professor of geology at Wesleyan University and a Methodist, stated the matter: "The great strength of Darwinian theory lies in its coincidence with the general spirit and tendency of science. It is the aim of science to narrow the domain of the supernatural, by bringing all phenomena within the scope of natural laws and secondary causes."[104]

Rice was not alone among Christians in advancing this conception of science. There were a good many Christian thinkers who regarded methodological naturalism as a principle for *discovering* the laws by which God governed creation,[105] and this understanding has provided—and still provides—a context for justifying methodological naturalism within the broad framework of a Christian metaphysics. It is an intellectual stance that is as unnecessary by way of its heuristic restrictions, however, as it has been unfortunate in terms of its broader effects. In conjunction with Darwinism, methodological naturalism has had the unintended consequence of screening the theological basis of natural science from the practice of science in a way that led to a definitive intellectual "lopping off" of theistic metaphysics. This completed the causal closure of the immanent frame in the realm of natural science—which, by the end of the nineteenth century, was regarded as the paradigm intellectual activity and the model for epistemic rigor that represented the standard to which all other academic disciplines should aspire—and it contributed mightily to the hegemonic embrace of philosophical naturalism and exclusive humanism in the academy. This was particularly manifest in the fledgling human sciences of anthropology, psychology, and sociology, which, in their quest to be truly "scientific," felt bound to emulate the natural sciences and endorse the principle of methodological naturalism. In their desire for scientific "respectability," anthropologists, psychologists, and sociologists so thoroughly naturalized the study of humanity that we became nothing but material products of our environment—we were "denatured"—and thus lost our essential humanity.

The conceptual space opened by the de-transcendentalization accompanying the ascendancy of nominalism, which led to a focus on the material and the sensual, and ultimately, a repudiation of the immaterial and intellectual as the basis of reality, once the anti-essentialistic implications of Darwinism were fully understood, left us as empty and accidental selves with no natures to be grasped and no fundamental humanity to which we could appeal. David Hull articulates this insight in a slightly different vocabulary:

> The implications of moving species from the metaphysical category that can appropriately be characterized in terms of "natures" to a category for which such characterizations are inappropriate are extensive and fundamental. If species evolve in anything like the way that Darwin thought they did, then they cannot possibly have the sort of natures that traditional philosophers claimed they did. If species in general lack natures, then so does *Homo sapiens* as a biological species. If *Homo sapiens* lacks a nature, then no reference to biology can be made to support one's claims about "human nature" . . . Because so many moral, ethical, and political theories depend on some notion or other of human nature, Darwin's theory brought into question all these theories. The implications are not entailments. One can always dissociate "*Homo sapiens*" from "human being," but the result is a much less plausible position.[106]

Such a position is especially implausible without an immaterial human soul and a rational intellectual basis on which metaphysical—and hence physical—reality is ordered.

Having jettisoned this transcendent reality in favor of one based solely on matter, sensory experience, and efficient material causation, the social sciences, by their embrace of Darwinism, also embraced a thoroughgoing metaphysical naturalism. Is it any wonder, then, that by 1969 only 11 percent of American anthropologists held to *any* form of orthodox religious belief?

Darwinism has therefore played an essential role in the broader culture in moving us from the conception of an ordered and personal cosmos in which humanity has a special purpose and place, to an impersonal universe with no special place for humanity and no purposes beyond those we imagine for ourselves. And because we have no essential human nature, and because reality itself has no essential order, we can imagine *anything* for ourselves that gives us pleasure, for there is no higher good. By means of continual educational inculcation, therefore, Darwinism—as an expression of metaphysical purposelessness—has been an indispensable contributor to the spread of secularism in Western society, an undeniable force in our sense of cultural and existential arbitrariness, a logical antecedent to our inevitable embrace of moral relativism, hedonism, and utilitarianism, and a prodigious catalyst for the broader cultural experience of meaninglessness, especially among the younger generations.

3. Reconstituting the Foundations of Science

Setting the disastrous cultural consequences of Darwinian *naturalism* aside, there can be two responses to neo-Darwinian *biology* on the part of those who affirm transcendent purpose in the universe and transcendent constraints on morality.

The first is an accommodation that seeks to embed it in a larger context, or modify it in some way, so as to preserve what is necessary by way of a transcendent metaphysics and morality. This response is predicated on satisfaction with the basic picture that neo-Darwinism offers of evolutionary mechanisms and universal common descent, and a desire not to be in tension with the current received wisdom in the field of evolutionary biology. It is also often accompanied by a cautionary tale offering some form of the "subtraction story" we encountered earlier—the idea that the epistemic credentials of religion have been on the retreat since the Enlightenment, and we certainly don't want a repeat of the "Galileo affair" in respect of neo-Darwinism, so it's always best to adapt theological conceptions to the "scientific consensus." While we have seen that there is little to recommend such subtraction stories, we may further observe that, if one feels compelled on other grounds to be a theistic evolutionist in a way that is distinct from an intelligent design theorist who accepts informationally front-loaded or actively directed common descent, then at least two suitable means are at hand for embedding neo-Darwinism in a metaphysics of transcendence.

The first is the path of Simon Conway Morris, professor of evolutionary paleobiology at Cambridge University and an orthodox Christian.[107] Contrary to the absolute contingency of evolutionary development asserted by the late Harvard zoologist Stephen Jay Gould,[108] Conway Morris argues that the paleontological evidence points distinctly to a phenomenon of evolutionary convergence of structure and morphology across a wide range of separate evolutionary lineages.[109] The phenomenon of evolutionary convergence, taken by itself, suggests the following picture: while the action of neo-Darwinian mechanisms observed over a short period of time gives the appearance of randomness, over geological periods of time they ineluctably give rise to a convergent order. This convergent order may be taken as evidence for law-like constraints on the operation of natural selection; it enables us to speak, as Conway Morris does, of "inevitable humans in a lonely universe."[110] While it is certainly not detectable evidence of a transcendent intelligence operative in the biological realm, if we take this picture at face value, it is congruent with a transcendent order that is divine in origin and bespeaks the possibility of evolutionary creation. From a Christian perspective, all that remains then is to bring this scenario into relationship with orthodox theology. Reasonable accounts of how this can be done may be found in Denis Lamoureux's book *Evolutionary Creation: A Christian Approach to Evolution* or—although written from an intelligent design perspective—William Dembski's *The End of Christianity: Finding a Good God in an Evil World*, which is able to accommodate a continuous evolutionary scenario within a more traditional and concordist theological framework. Appropriately contextualized, therefore, Christian theology has no necessary conflict with evolutionary biology.

Paleontological evidence for evolutionary convergence aside—evidence that is equally intelligible on the basis of common modular design—the difficulty associated with this view of things is its indispensable premise that the origin of life and the transition among organismal body plans is a highly probable occurrence, if not an inevitable one, against the background of known physical, chemical, and biochemical laws, in conjunction with everything known about genetics and developmental biology. But nothing could be further from the truth. Not only is a plausible account of how neo-Darwinian evolution could take place getting farther out of reach as our biological knowledge expands, certain essential features of it are mathematical impossibilities.[111] Granted that this inconvenient truth poses an even greater problem for neo-Darwinian *naturalism*, is there nonetheless some way that a staggeringly improbable neo-Darwinian scenario could be incorporated in a theistic metaphysic? Of course—after a fashion—and this brings us to the second way of embedding neo-Darwinism within a metaphysic of transcendence: divine middle knowledge, or Molinism as it is sometimes called, a theological discovery of the sixteenth-century Jesuit theologian of the Counter-Reformation, Luis de Molina (1535–1600), as he reflected on the biblical text, engaged in dialogue with the Jansenists, and responded to the Protestant reformers.[112]

The basic Molinist conception is as follows: logically prior to creation, God has complete knowledge of all of the universes that he can bring into being, and what would happen—inclusive of all the genuinely contingent events that he does *not* himself directly cause—if he brought any one of these universes into existence. Suppose now that there is not just a *logically* possible, but a *metaphysically* possible universe which, while extraordinarily improbable, nonetheless instantiates undirected chemical evolution in the origin of life and undirected neo-Darwinian processes in the increase of biological complexity, and it does so in a manner that unfolds in cosmological and geological time precisely the way that our universe has developed. By hypothesis, this is a metaphysically possible universe, so we know that God *can* create it. Given his middle knowledge of what would happen if he *did* create it, and his determination that by doing so he will realize his purposes, we may suppose that God *could have chosen* to create in this way, in which case our existence would be explained by the fact that God created our universe with these extraordinarily improbable properties. So it appears that there is a way to reconcile a transcendent metaphysic with a neo-Darwinian universe even when its development is highly *improbable*.

But there are two problems with this scenario for those who wish to embrace it with the intent of avoiding both the detectability of intelligent design and direct divine involvement in the historical course of nature. The first problem is that the extraordinary improbability of such a universe leads to a design inference, though this time the locus of design resides in God's choice of which universe to actualize among all of the metaphysically possible universes he could have brought into being. The second problem is that our universe is one in which the mathematical descriptions of quantum theory hold, and this means that it is irremediably causally *incomplete* at the physical level. It therefore requires the existence of an immaterial metaphysical cause to bring about the requisite causal closure.[113] What follows from this is that Molinism, as a theory of divine knowledge, is an insufficient *metaphysical* basis for the historical development of a universe that is causally incomplete in the way that ours is. Any conception of providence based on middle knowledge must therefore be supplemented with an account of divine *action*. But this means that God is actively involved in the history of our universe in precisely the nitty-gritty way that avid neo-Darwinians in the community of theists wish to avoid,[114] so why bother being a neo-Darwinian? *C'est la vie*.

This leads us, then, to the second response to neo-Darwinian *biology* from defenders of transcendence. It is to argue that, as a consequence of the eight observations below, we are warranted in concluding that neo-Darwinian gradualism, which relies on undirected chemical evolution to explain the origin of the first biological replication and translation system, and maintains that natural selection operating on random genetic mutations was subsequently sufficient for the evolution of increasingly complex life, is a demonstrably insufficient and untenable explanation for this purpose. Furthermore, from our knowledge of the *existence* of intelligence as a presently operative cause, and our constant and repeated experience that it is the *only cause known to be sufficient* to the production of complex specified information, we are conspicuously warranted in concluding on grounds of causal existence and uniqueness that intelligent design is the *best explanation* of the origin and development of life; this conclusion respects a form of abductive inference that is ubiquitous in the historical sciences.[115]

(1) Various indisputable facts about biochemistry, molecular biology, and genetic information demonstrate the physico-mathematical impossibility of producing the first life by undirected chemical evolution or self-organizing physical chemistry.[116] It is furthermore mathematically demonstrable that neo-Darwinian mechanisms are unable to generate the complex specified genetic and epigenetic information necessary to life's function.[117]

(2) The existence of irreducibly complex molecular machines and biochemical systems—systems that consist of multiple integrated parts of interacting functionality, absence of any one of which would render the system nonfunctional—cannot be explained by Darwinian mechanisms. The reason is simple: These biological structures cannot come into being through a sequence of steps that maintain the functionality that is necessary for natural selection to operate; they must therefore have come into being all at once, or they would not exist at all.[118] Darwin himself recognized that if such systems existed, his theory would be untenable: "If it could be demonstrated that any complex organ existed which could not possibly have been formed by numerous, successive, slight modifications, my theory would absolutely break down."[119]

(3) There are intractable obstacles to the gradual evolutionary transition between body plans evident from genetics, embryology, developmental biology, and comparative morphology. In particular, the existence of epigenetic or ontogenetic information is now widely recognized in embryology. DNA alone does not determine the way that larger systems of proteins are formed, and it plays even less of a role in determining how cell types, tissue types, and organs arrange themselves into body plans. What really do influence embryological development are the three-dimensional structure of the cell membrane and cytoskeleton, as well as the spatial architecture of the zygote during embryogenesis. Developmental biologists do not yet know where all of the extra information that is needed resides. What is clear, however, is that the unfolding of a body plan in ontogenesis involves multiple layers of coordinated hierarchical complex specified information.[120]

(4) There are ubiquitous incompatibilities among molecular and morphological homologies that lead to wildly inconsistent phylogenies that make nonsense of any suggestion of a universal tree of life and common descent.[121]

(5) There is overwhelming evidence that long periods of stasis punctuated by saltations—most dramatically in the Cambrian explosion and the mammalian radiation—form the structure of the paleontological record.[122]

(6) The "ancient repetitive elements" and "pseudogenes" thought to be vestigial DNA remnants are, as predicted by design theorists, being discovered to have function. In some of these cases, elements formerly thought to be pseudogenes have been found, in fact, to be operational genes.[123] More generally, the over 98 percent of the genome *not* dedicated to coding for proteins, and formerly dismissed as mostly evolutionary "junk," is being found to serve a wide variety of purposes.[124] These purposes include, but are not restricted to:

(a) The regulation of DNA transcription, replication and repair, as well as the rate of its translational expression;[125]

(b) Assisting the proper folding and maintenance of chromosomes and controlling their interactions with the nuclear membrane;[126]

(c) Functioning as indicators of loci for programmed rearrangements of genetic material;[127]

(d) Controlling RNA processing, editing, and splicing;[128]

(e) Regulating embryological development;[129] and

(f) Assisting in immunodefense functions.[130]

The general sense that is emerging is that *non*–protein-coding regions of the genome serve much the same purpose as an operating system in a computer that is capable of super-intending over multiple simultaneous operations.[131] If we now revisit the much-touted claim that human and chimpanzee DNA are at least 98 percent the same, what we need to realize is that this figure refers to regions that can be aligned after numerous macro- and micro-rearrangements, as well as insertions or deletions, have been "corrected" by software programs or by hand. An array of scrambled sequences distinguish human and chimp DNA, not to mention species-specific "junk" elements. When one takes into account that almost *all* of the mammalian genome—not just the approximately 1.5 percent that encodes proteins—is transcribed into various functional non–protein-coding RNAs, then such species-specific rearrangements take on a new significance. If you've ever wondered how humans and chimps can supposedly share 98 percent of their DNA and be such radically different species, a good part of the answer appears to be in the RNA-based operating system—most of which was formerly dismissed as junk.[132]

(7) The fact that constant and repeated experience confirms to us that the only presently acting cause *sufficient* to the production of complex specified information is *intelligence* means that intelligent design is the best explanation available for such information when it is discovered in the universe and in living systems.[133]

(8) The independent presence of complex specified information in cosmological and intra-cosmological fine-tuning warrants a *universal* design inference that makes short work of any "principle of mediocrity."[134]

In short, if you dispense with the epistemologically and metaphysically destructive presumption of philosophical naturalism, plus the needlessly restrictive presupposition of methodological naturalism, and then follow the negative and positive evidence to the *best* available explanation, you will be led to embrace intelligent design.[135] This is a reassuring convergence, because, as we have seen, presupposition of the metaphysical framework of transcendent intelligent design is essential to the very possibility of science as a rational and truth-conducive enterprise, and it is also the historical foundation on which modern science was built. It is safe to say that developments in cosmology and biology over the last twenty-five years have led us to the point of restoring transcendent intelligent design to its rightful place as a *heuristic* in scientific research, as well as being the metaphysical and epistemological foundation that justifies scientific practice.

APPENDIX:
The Origin of Life Requires an Intelligent Cause

How does one distinguish the product of intelligence from the product of chance? One way is to give a rigorous mathematical characterization of design in terms of conformity of an event to a pattern of very small probability that is constructible on the basis of knowledge that is independent of the occurrence of the event itself. When the event in question is the origin of replication and translation, this specificity is evident in a variety of correspondences necessary to biological functionality. In assessing the probability of such an event one has to take into account what William Dembski has called "specificational resources" and "replicational resources."[136] Specificational resources are essentially the number of possibilities for biologically functional patterns, and replicational resources are the maximum number of attempts the universe, with its structure of laws and constants, has to generate one of these patterns.

We can get a handle on specificational resources in the biological context by remembering that the protein-coding segments of DNA must produce proteins that fold properly in three dimensions, and then focusing on the prevalence of properly folding proteins in the sequence-space of amino acids. In doing this, it is important to recognize that the sequence of triplet-codons along the sugar-phosphate backbone of DNA is *not* determined by any biochemical laws or self-organizational properties in physical chemistry any more than the chemistry of ink bonding to paper explains the daily content of the *Wall Street Journal*.[137] Furthermore, the chemical independence exhibited by the nucleotide sequences relative to the backbone of the molecule is *essential* to the information-carrying capacity of DNA. In other words, there is an *in principle* reason that the biological information which forms the basis of life has no causal explanation in terms of physico-chemical or biochemical laws or self-organizational scenarios.

With this in mind, we turn to the prevalence of functional protein folds in amino acid sequence space and rigorous studies like those conducted by Douglas Axe.[138] Axe's work leads to the conclusion that we may reasonably expect a *single* protein folding domain to have a probability in amino acid sequence space of about 1 in 10^{74}. Keep in mind that this is just *one* folding domain (superfamily). If we take one of the simplest known self-replicating systems, *mycoplasma genitalium*,[139] we are looking at about 268 superfamilies (folding domains) that are required to get it off the ground, each of which has a probability on the order of 1 in 10^{74}. The probability of the nonintelligent production of this minimal self-reproducing unit is therefore on the order of 1 in $10^{(74 \times 268)}$ = 1 in $10^{18,632}$. A more reasonable estimate for a suite of biologically functional proteins that might get life as we know it off the ground requires about 1,000 superfamilies (folding domains), in consequence of which we'd be looking at probabilities on the order of 1 in $10^{74,000}$.

We may now ask for an upper bound on the computational capacity of our universe. Seth Lloyd at MIT has produced a nice result showing that 10^{120} is the maximum number of bit operations that the observable universe could have performed in its 13.7 billion year history.[140] If we now let $\varphi_s(T)$ represent the specificational resources for a minimal suite of functional superfamilies, let $\mathbf{P}(T|\mathbf{H})$ represent the probability of the chance occurrence (defined by chance hypothesis $\mathbf{H}$) of an event exhibiting pattern T, and note that 10^{120} places an upper bound on the replicational resources of our universe, then if $10^{120} \times \varphi_s(T) \times \mathbf{P}(T|\mathbf{H}) < \frac{1}{2}$, it is less likely than not on the scale of the whole universe, with all possible probabilistic resources factored into the analysis, that the pattern can be explained by chance. Expressing this result in information-theoretic form yields what Dembski[141] calls the "context-independent specified complexity" of the event represented by the pattern T, namely:

$$\kappa = -\log_2[10^{120} \times \varphi_S(T) \times \mathbf{P}(T|\mathbf{H})].$$

In other words, if $10^{120} \times \varphi_S(T) \times \mathbf{P}(T|\mathbf{H}) < \frac{1}{2}$, then $\kappa > 1$. So context-independent specified information is present when $\kappa > 1$, and under these conditions, a design inference is warranted.

Now, if we take 268 to be the lower-bound on the number of folding domains needed for a self-replicating system that could serve as the aboriginal biological unit needed for natural selection to begin, we find that

$$10^{120} \times \varphi_S(T) \times \mathbf{P}(T|\mathbf{H}) = 10^{120} \times 10^{2.43} \times 10^{-18,632} \approx 10^{-18,509.57} \ll \frac{1}{2}, \text{ or}$$
$$\kappa = -\log_2[10^{-18,509.57}] \approx 61{,}493.59 \gg 1.$$

If we retain the more realistic estimate for a suite of biologically functional proteins that might suffice to get life as we know it off the ground, we have instead that

$$10^{120} \times \varphi_S(T) \times \mathbf{P}(T|\mathbf{H}) = 10^{120} \times 10^{3} \times 10^{-74,000} = 10^{-73,877} \ll \frac{1}{2}, \text{ yielding}$$
$$\kappa = -\log_2[10^{-73,877}] \approx 245{,}414 \gg 1.$$

Either way, there is no realistic possibility that these things happened by nonintelligent means.

The Rise of Naturalism

Notes

1. By a "whiggish" tale, of course, I mean a self-congratulatory just-so story explaining how we got to be so smart when our intellectual forebears were so benighted.
2. See also the general critique of supervenience and emergence models as explanatorily vacuous in the context of fundamental physical theory in my essay on quantum theory in Part II of this volume.
3. Searle 2004: 89–92, 100–2.
4. See John Searle's essay in Part VI of this volume.
5. See also C. S. Lewis's *Miracles* (New York: Macmillan, 1960 [1947], 12–25), Richard Taylor's *Metaphysics*, 2nd ed. (Englewood Cliffs: Prentice-Hall, 1974 [1963], 115–19), Alvin Plantinga's *Warrant and Proper Function* (New York: Oxford University Press, 1993, pp.194–237) and *Warranted Christian Belief* (New York: Oxford University Press, 2000, 227–40, 281–84, 350–51), and lastly, James Beilby's edited collection *Naturalism Defeated? Essays on Plantinga's Evolutionary Argument against Naturalism* (Ithaca: Cornell University Press, 2002).
6. Of course, this problematization of simple predication is itself symptomatic of the pragmatist's denial of the existence and/or epistemic accessibility of absolutes. Once the link between truth and ontology is thus severed and a "deflationary" approach adopted (Davidson 1984, 1990; Field 1986; Horwich 1990; Quine 1969; Williams 1986, 1999), the outcomes of scientific investigations are reduced to matters of localized epistemic practices, of sociology and politics, and philosophy of science itself becomes a branch of cultural studies (Rouse 1987, 1996). In short, the leveling functions of pragmatism and modern liberalism (progressivism) are one and the same: since "truth" on this view is not metaphysically monolithic but rather a function of changing perceptions induced by the interplay of societal power structures, identity politics and the ideologies *du jour* hold sway, cultural fragmentation ensues, and order is kept—if it is kept—not by a common recognition of metaphysical reality and a transcendent moral structure that can be instantiated in the rule of law, but rather by an increased investment in unconstrained and arbitrary powers of the state as a panacea for the human condition.
7. Nicholas Wolterstorff illustrates this truth clearly in his discussion of Richard Rorty's pragmatic liberalism in the epilogue (pp.385–393; cf. also 316–322) to his theory of justice as grounded in rights that he develops in *Justice: Rights and Wrongs* (Princeton: Princeton University Press, 2008).
8. For an extended defense of this thesis, see Alvin Plantinga's books *Warrant and Proper Function* (New York: Oxford University Press, 1993) and *Warranted Christian Belief* (New York: Oxford University Press, 2000), as well as the *festschrift* edited by Jonathan Kvanvig, *Warrant in Contemporary Epistemology: Essays in Honor of Plantinga's Theory of Knowledge* (Lanham: Rowman & Littlefield Publishers, Inc., 1996). A broader discussion of the metaphysical and epistemological inadequacy of naturalism along with its deleterious consequences can be found in William L. Craig and J. P. Moreland, eds. *Naturalism: A Critical Analysis* (New York: Routledge, 2000) and Michael C. Rea, *World Without Design: The Ontological Consequences of Naturalism* (Oxford: Clarendon Press, 2002).
9. It is therefore deeply ironic that theism, contrary to the ill-conceived arguments of "new atheists" like Richard Dawkins, Daniel Dennett, Sam Harris, Christopher Hitchens, Natalie Angier and Susan Jacoby, rather than being the antithesis of reason, proves instead to be its ontological foundation and operative salvation. For an amusing reflection on the historical misrepresentations put forward by the new atheist contingent, and their misguided attempts to appropriate the authority of science for their cause, see David Berlinski's book *The Devil's Delusion: Atheism and Its Scientific Pretensions* (New York: Basic Books, 2009).
10. A parallel story, focused on our institutions of higher learning, is told by George Marsden in his helpful book *The Soul of the American University: From Protestant Establishment to Established Non-Belief*

(Oxford: Oxford University Press, 1994). See also James Tunstead Burtchaell's *The Dying of the Light: The Disengagement of Colleges and Universities from Their Christian Churches* (Grand Rapids: Eerdmans, 1998).

11. We will not dwell on the first sense of secularization—the sociocultural and political privatization of religion—but it too has been quite harmful and has arisen in the American context through a gross distortion of the doctrine that there must be separation of the church and the state. In the classical sense, the separation of church and state simply meant that the government would not mandate membership in a state church nor collect taxes in support of an official state religion. Disestablishment was a wise constitutional measure devised to overcome the difficulties posed by denominational pluralism in an overwhelmingly Judeo-Christian society. It was never intended as a legislative fulcrum for the privatization of religion and its banishment from the public square; this latter interpretation is an invention (and goal) of modern secularism, which is not itself "neutral," but rather intent on advancing its own sectarian interests. I cannot undertake a discussion of these matters here, but I highly recommend Richard John Neuhaus's (mildly dated) classic *The Naked Public Square: Religion and Democracy in America*, Second Edition (Grand Rapids: Wm. B. Eerdmans Publishing Company, 1986); more current and also quite helpful is Hunter Baker's *The End of Secularism* (Wheaton: Crossway Books, 2009).

12. The essential contributions of Christianity to the rise of Western civilization and the great benefits it has brought to the world are stories not often told in these times. The general conceit is that such stories are both impolitic and false, and we have outgrown them. Nothing could be further from the truth; we ignore them at our peril. One of the outstanding merits of Christianity is its discernment of the metaphysical basis for and moral structure of the individual conscience, and its exhortation of the individual to follow it. As Paul Johnson (1976: 516) observes, it is this conscience that " . . . is the father of all other freedoms. For conscience is the enemy of tyranny and the compulsory society; and it is the Christian conscience which has destroyed the institutional tyrannies Christianity itself has created—the self-correcting mechanism at work. The notions of political and economic freedom both spring from the workings of the Christian conscience as a historical force; and it is thus no accident that all the implantations of freedom throughout the world have ultimately a Christian origin." Those further interested in the salutary effects of Christianity on the development of Western civilization would do well to consider Rodney Stark's books *For the Glory of God: How Monotheism Led to Reformations, Science, Witch-Hunts, and the End of Slavery* (Princeton: Princeton University Press, 2003) and *The Victory of Reason: How Christianity Led to Freedom, Capitalism, and Western Success* (New York: Random House, 2005). Also helpful are Edward Grant's *The Foundations of Modern Science in the Middle Ages: Their Religious, Institutional, and Intellectual Contexts* (Cambridge: Cambridge University Press, 1996) and Jonathan Hill's *What Has Christianity Ever Done for Us? How It Shaped the Modern World* (Downers Grove: IVP, 2005).

Given the imperfections in human nature and our capacity for evil, the history of mankind *with* Christianity has been sobering enough; but *without* it, how much more horrific might our history be? Christianity has been a civilizing agent; a de-Christianized world is not a happy prospect. What may be expected from a secular Darwinian ethic, for instance, is admirably illustrated by Charles Darwin's *Descent of Man* (1871) and chronicled in Richard Weikart's careful and scholarly work *From Darwin to Hitler: Evolutionary Ethics, Eugenics, and Racism in Germany* (New York: Palgrave Macmillan, 2004), John West's invaluable study *Darwin Day in America: How Our Politics and Culture Have Been Dehumanized in the Name of Science* (Wilmington: ISI Books, 2007), Wesley Smith's books *The Culture of Death: The Assault on Medical Ethics in America* (San Francisco: Encounter Books, 2000) and *A Rat is a Pig is a Dog is a Boy: The Human Cost of the Animal Rights Movement* (New York: Encounter Books, 2010), and Benjamin Wiker's *Moral Darwinism: How We Became Hedonists* (Downers Grove: IVP, 2002).

13. Plantinga 1980.

14. The interested reader will find these themes ably explored by Alvin Plantinga (*Does God Have a Nature?* Milwaukee: Marquette University Press, 1980), Thomas Morris and Christopher Menzel ("Absolute

Creation," *American Philosophical Quarterly*, vol. 23, 1986: 353–62), and by Menzel in a separately authored paper ("Theism, Platonism, and the Metaphysics of Mathematics," *Faith and Philosophy*, vol. 4, 1987: 365–82). Also of interest is Mark Steiner's paper in Part V of this volume.

15. Foster 1934; Oakley 1961; Heimann 1978; Osler 1994.

16. Foster, 1934: 311.

17. Klaaren 1977: 39–52; Moonan 1994.

18. Dobbs 1991: 111.

19. Oakley 1961: 452.

20. Calvin 1559 [1960]: III.xxiii.3.

21. Other examples confounding the nominalist voluntarist thesis abound. The reader is directed to the much more nuanced discussion of voluntarist versus intellectualist influences in respect of the Judeo-Christian origins of modern science offered by Peter Harrison (2002), as well as the broad view of the impetus that medieval scholasticism gave to modern science presented by Edward Grant (1996) and Rodney Stark (2003, 2005).

22. Taylor 2007: 72–88, 130–36, *et passim*.

23. But one can also understand the Reformation period as achieving an *expansion* of the sacred, as Taylor recognizes (2007: 79), since creation as a whole remains God's handiwork and God is now recognized as working *everywhere* in the circumstances of the ordinary lives of the faithful—work, marriage, leisure, and so on—to sanctify them. Taylor explores this theme of the sacredness of ordinary life and vocation more extensively in an earlier work (1989: 215–27). I suppose, however, in the present context of providing an etiology of secularization, one could anticipate that such a sacralization of reality and life as a whole might for some have the effect of trivializing the sacred through a universalization of its presence. To that which is everywhere present we may eventually become blind, insensitive and altogether forgetful, so the ubiquity of the sacred is culturally consummated by its nullification—by filling our field of vision it eventually becomes a background that we ignore. Familiarity breeds, if not contempt, insensitivity. There is perhaps some truth to this, though desacralization has many other causes interwoven in the etiological tapestry of secularization.

24. Taylor 2007: 80. Rodney Stark (2003: 201–88) presents a different view. While Taylor may be correct that these excesses pushed a few in the direction of unbelief, the hysteria driving the witch-hunts was brought to a halt by the Christian community itself—further evidence of the self-correcting mechanism of the Christian conscience at work (see Paul Johnson's insight, quoted in footnote 12)—not by early "Enlightenment voices" who spurned Christianity altogether in an unwarranted rejection of all belief in the supernatural as pernicious and false. As Stark demonstrates in regard to the European witch-hunts, "[w]itchcraft beliefs and persecutions did not succumb to the arrival of an age of science and reason. No! Just as their predecessors had deduced Satanism as the mechanism behind non-Church magic, it was deeply committed and well-trained scholastics, responding to the evidence of their senses, who stripped it of its evidential basis" (2005: 287).

25. Taylor 2007: 81. While this might be true in some cases, as a general indictment of Catholic sacramentals it is overly harsh.

26. Taylor 1989: 227–30; 2007: 82.

27. Taylor 2007: 84. That something along the historical lines of this conceptual narrative has happened and continues to be worked out is undeniable. In the long run it is human nature, once success is achieved, to arrogate credit to ourselves and downplay reliance on the factors that made it possible. Individual exceptions occur, of course, but when society as a whole is involved, the reorientation described here takes on an air of inevitability. Those who, like the present author, would resist the historical flow that has led to the intellectual hegemony of naturalism in the academy and the balkanization of our culture, must not only stand athwart history yelling "Stop!," as William Buckley observed, but provide cogent intellectual reminders of our origins, how it is that we got where we are, the deleterious social consequences of continuing on this path,

and the intellectual grounds for the opposing metaphysic that must be retrieved. More pointedly, both intellectually and socioculturally, we have *not* and will *never* arrive at a point where we may "lop off the reference to God" as a ladder useful to our climb but no longer needed; rather, in doing so we will always be sawing off the branch on which we're sitting in the vain expectation that we'll continue to be suspended in mid-air. Apart from a deliberate re-attachment of this branch, therefore, a precipitous fall seems inevitable.

28. Oestreich 1982.
29. Taylor 1989: 147–48; Taylor 2007: 130–34; Gillespie 2008: 40–41.
30. Taylor 2007: 131.
31. Ryle 1949 (1980).
32. For further exploration of this subject and a sophisticated defense of mind-body dualism, see J.P. Moreland's contribution to Part VI of this volume, or the more extensive discussion in his book *Consciousness and the Existence of God: A Theistic Argument* (New York: Routledge, 2008). See also various essays in Kevin Corcoran, ed., *Soul, Body and Survival: Essays on the Metaphysics of Human Persons* (Ithaca: Cornell University Press, 2001), and Peter van Inwagen and Dean Zimmerman, eds., *Persons: Human and Divine* (Oxford: Clarendon Press, 2007).
33. "Outre que le libre arbitre est de soy la chose la plus noble qui puisse estre en nous, d'autant qu'il nous rend en quelque façon pareils à Dieu & semble nous exempter de luy estre sujets, & que, par consequent, son bon usage est le plus grand de tous nos biens, il est aussi celuy qui est le plus nostre & qui nous importe le plus, d'où il suit que ce n'est que de luy que nos plus grands contentemens peuvent proceder." (*Oeuvres de Descartes*, ed. Charles Adam et Paul Tannery. Paris: Vrin, 1973, vol. 5, 85; English translation in *Descartes: Philosophical Letters*, translated by Anthony Kenny. Oxford: Oxford University Press, 1970: 228; quoted in Taylor 2007: 133).
34. Taylor 2007: 286. This too is the way the *secular* mind portrays this transition, namely, as a move to intellectual maturity and dignity. From the *Judeo-Christian* standpoint, of course, it rather constitutes a rebellion—conscious or unconcious—against transcendent moral authority, followed by a *post facto* effort at self-justification.
35. As Taylor summarizes the situation, "This stance of general disengagement was in great tension with orthodox Christianity with its sense of God as personal agency, in relation to us. This sense gives rise to the notion that the grasp of things which arises from our engagement in relation with God, in communion with Him, and others under Him, will be different and superior to what we can grasp outside of, or withdrawn from this relation. If the spill-over effect from disengaged science goes far enough, it threatens this crucial feature of the understanding of faith" (2007: 286–87).
36. Bacon, *Novum Organum*, Book One, §78.
37. Voltaire 1761: vol. 1, 82; quoted in Lindberg 2007: 358.
38. Rousseau 1750 [1987]: 3.
39. Condorcet 1795 [1955]: 72.
40. Burckhardt 1860 [1954]: 371; quoted in Lindberg 2007: 358.
41. White 1896: vol. 1, 375–76.
42. Brooke and Cantor 1998: 18; Stark 2003: 123; *see also* Lindberg and Numbers 1986; Russell 1991.
43. Boorstin, 1983: 100; quoted in Stark 2007: 129.
44. Freeman 2003: xviii.
45. Lindberg 2007: 358.
46. An entertaining discussion of a variety of such historical misconceptions is contained in Numbers, ed. (2009).
47. Grant 1971, 1994; Hamilton 1996; Russell 1991; Stark 2003.
48. White 1896: vol. 2, 108–9.

49. White 1896: vol. 1, 57; Yates 1964, 1979; Stark 2003.
50. Bloch 1940 [1961]; Bury 1889; Drake 1996; Ferrill 1986; Grant 1978; Luttwak 1976; Millar 1967; Pirenne 1922 [1955], 1936 [1958]; Potter 2004; Stark 2003; and Wolfram 1997.
51. White 1896: vol. 1, 121ff.
52. Cohen 1985; cf. also Armitage 1951; Clagett 1961; Crosby 1997; Danielson 2000; Gingerich 1975; Grant 1994, 1996; Rosen 1971; and Stark 2003.
53. White 1896: vol. 2, 50.
54. Grant 1996: 205; Singer 1925 [1970]: 129.
55. See also Armitage 1951; Mason 1962; O'Malley 1964; Porter 1998; and especially Stark 2003.
56. See Basile 1983; Biagoli 1993; Blackwell 1991; Brooke and Cantor 1998; Caroti 1987; Homza 2006; Kamen 1997; Langford 1971; McGrath 1999; Pagano 1984; Shea 1986; and Stark 2003.
57. Eisenstein 1979; Hirsch 1967; Johnson 1976; Ozment 1980.
58. Gies and Gies 1994; Gimpel 1976; White 1940, 1962.
59. Stark, 2003: 134.
60. Bloch 1940 [1961], 1975; Davis 1966; Stark 2003, 2005.
61. Bonnassie, 1991.
62. Stark 2005. When slavery reared its ugly head anew in European society in the modern era, Christianity again was the powerful moral agent of its abolition—in England through the dogged Christian persistence of William Wilberforce (1759–1833), and in America through a bloody civil war, the moral justification for which, despite the unfortunate damage done to states' rights under the tenth amendment, could not be more clearly expressed than in the words of Lincoln's *Second Inaugural Address*:

> ... The Almighty has His own purposes. "Woe unto the world because of offenses; for it must needs be that offenses come, but woe to that man by whom the offense cometh." If we shall suppose that American slavery is one of those offenses which, in the providence of God, must needs come, but which, having continued through His appointed time, He now wills to remove, and that He gives to both North and South this terrible war as the woe due to those by whom the offense came, shall we discern therein any departure from those divine attributes which the believers in a living God always ascribe to Him? Fondly do we hope, fervently do we pray, that this mighty scourge of war may speedily pass away. Yet, if God wills that it continue until all the wealth piled by the bondsman's two hundred and fifty years of unrequited toil shall be sunk, and until every drop of blood drawn with the lash shall be paid by another drawn with the sword, as was said three thousand years ago, so still it must be said "the judgments of the Lord are true and righteous altogether."

The battle for civil rights in the twentieth century, as represented by Martin Luther King (1929–68), was also deeply Christian in its original motivations. Unfortunately, this original movement was co-opted, subverted and corrupted by identity politics and the ironically—but predictably—regressive effects of the socioeconomic policies of the "Great Society" initiative under the Johnson administration (see Banfield 1974 [1990]; Murray 1984; Goldberg 2007; and Sowell 1984, 1995b).

63. Stark 2005: 60.
64. Baldwin 1959.
65. Quoted in Stark 2005: 65, from de Roover 1958: 422.
66. Ibid.
67. See John Noonan's *The Scholastic Analysis of Usury* (Cambridge: Harvard University Press, 1957) for an exhaustive account of this topic.
68. Stark 2005: 112–13.

69. Three observations need to be made before we move on to discuss the continuity between medieval and modern science. First of all, as should be clear from this account, the thesis of Max Weber's *The Protestant Ethic and the Spirit of Capitalism* (New York: Charles Scribner's Sons, 1904 [1958]) that Protestantism gave birth to capitalism and the Industrial Revolution is quite false. The idea that large scale capitalist society was a sociological impossibility prior to the Reformation is refuted simply by the fact of its existence (Trevor-Roper 1969 [2001]: 20–21; Stark 2003: 118–19). This beneficial development was neither a Catholic nor Protestant sectarian achievement, but the product of Christian ingenuity *simpliciter*.

Second, for those laboring under the false impression that socialism and the welfare state are morally superior to democratic capitalism and private charities, I encourage reflection on the law of unintended consequences in the context of economic policy. I strongly recommend George Gilder's *Wealth and Poverty* (New York: Basic Books, 1980) as a classic modern statement of the moral case for capitalism. With respect to private charitable work and charitable giving and its effects, Arthur C. Brooks's revealing study *Who Really Cares? The Surprising Truth about Compassionate Conservatism* (New York: Basic Books, 2006) will help set the record straight, as will Marvin Olasky's books *The Tragedy of American Compassion* (Washington DC: Regnery Gateway, 1992), *Renewing American Compassion* (Washington DC: Regnery Publishing, Inc., 1997) and *Compassionate Conservatism: What it is, What it Does, and How it Can Transform America* (New York: Free Press, 2000). For those with specifically Christian concerns about capitalism—though everyone can benefit—I also highly recommend Jay Richards's book *Money, Greed, and God: Why Capitalism is the Solution and Not the Problem* (San Francisco: HarperOne, 2009). Richards undertakes to dispel eight myths about capitalism that arise, for example, from contrasting capitalism with an unrealizable ideal rather than its actual alternatives; from focusing on good intentions rather than the unintended consequences of our actions; from assuming that the essence of capitalism is greed, that trade requires a winner and a loser, and that wealth isn't created but just transferred; from the belief that charging interest on money is invariably exploitive; from confusing aesthetic judgments with economic arguments; and from Malthusian fear-mongering that population trends will continue unabated unless actively checked and that current natural resources will run out because they will always be the ones that we use. He concludes with an appendix that gives lie to Darwinian interpretations of capitalist economics and places free market activity in the context of the providential order. In his recent book, *The Israel Test* (Minneapolis: Richard Vigilante Books, 2009), George Gilder adds another interesting twist to the moral analysis of capitalism, arguing that widespread hostility toward Israel, like hostility toward the United States, arises primarily from jealous anger toward capitalist creativity and the success and freedom it breeds. In short, both Israel and the United States are hated for their virtues, their freedoms, and their successfulness. He therefore suggests that historical anti-Semitism, even when masked by Marxist or fascist "moral" critiques of capitalism, is rooted in envy, greed, resentment, and a fundamental misunderstanding that sees economics as a zero-sum game. If one factors out a long history of *purely* ethnic and religious hatred toward Judaism and Jews that is rooted in Islam, it seems clear that Gilder is on to something important, for the huge pool of remaining animosity consists mostly of the hatred and envy of socialists, Marxists and fascists, i.e., the political left, for the successes that have come from Jewish entrepreneurial genius, along with a refusal to recognize the broader benefits that these triumphs have brought to the world. (In this regard, Dan Senor's and Saul Singer's new book *Start-up Nation: The Story of Israel's Economic Miracle* (New York: Twelve, 2009) also provides inspiring insight into Jewish ingenuity and its universal benefits under free market conditions.) Ultimately, leftist anger and jealousy rests on a deeply defective conception of "fairness" that entails a leveling function for government that can only be achieved by an unjustifiable, damaging, and all too frequently tyrannical intervention in human affairs, that is, an unconscionable restricting of human freedoms. Once these basic ideas are appreciated, going on to an explicit understanding of the very real relationship between economic and political freedom is absolutely essential. In this regard, I recommend Friedrich Hayek's *The Road to Serfdom* (Chicago: The University of Chicago Press, 1944) and *The Fatal Conceit: The Errors of Socialism* (Chicago: University of Chicago Press, 1988), as well as Milton Friedman's classic *Capitalism and Freedom* (Chicago: The University of Chicago Press, 1962) and Milton & Rose Friedman's *Free to Choose: The Classic Inquiry into the Relationship between Freedom and Economics* (San Diego: Harcourt, Inc., 1980 [1990]). To conclude this lengthy second point, for those still blaming capitalism in the form of "greedy investment bankers and mortgage brokers" for the 2008 collapse of the American housing market and the deep recession that has ensued, a helpful study of the false economic incentives and signals created by social-policy-driven government regulation of the mortgage industry and lack of essential accountability with respect to government regulation itself can be found in Thomas Sowell's book *The Housing Boom and Bust* (New York: Basic Books, 2009; see also Sowell 2007, 2008). This is not to deny that elements of the banking and mortgage industries were *complicit* in what happened, but impetus for their irresponsible behavior came from disastrous government mandates pushing mortgages to unqualified buyers and the failure to separate commercial banking from investment operations (see Fleckenstein 2008). In short, lacking a proper understanding of human nature and therefore misjudging the likely consequences of its actions, government regulated banking and finance in ways it should *not* have done, and failed to regulate it in ways it *should* have done, multiplying moral hazard in every direction (in this regard, see also Sowell 1987, 1995b).

As a third and last observation, the reader will note that I have not discussed the remarkable medieval progress in music, art, literature and education, though in respect of education the Christian invention of the university has been mentioned. For a brief survey of these subjects, see Rodney Stark's *The Victory of Reason: How Christianity Led to Freedom, Capitalism, and Western Success*. New York: Random House, 2005: 51–53

70. Lindberg 2007: 360–67.

71. See, for example, Aristotle's *Physics*, II.2, 19b23–31 and the discussion in Heath 1949: 11–12, 98–100, 211–14.

72. Lindberg 2007: 254–313, 357–67.

73. Pedersen 1974; A. Mark Smith 1982, 1996: 229–39.

74. Cohen and Drabkin 1958: 217–20.

75. Sabra 2003; Boyer 1959: 125–30; Wallace 1959: 174–224.

76. Samsó, "Levi ben Gerson"; Poulle, "John of Murs".

77. Hackett 1997: 277–315; Lindberg 1996.

78. Grant 1974: 368–76.

79. W. Newman 2006.

80. Lindberg 2007: 364.

81. Ashworth 2003; Burtt 1924 [1954]; Deason 1986; Lindberg 2007: 364–67; Osler 2002; Westfall 1958

82. For a more extensive discussion of this theme in relation to the demarcation problem, see Stephen Meyer's essay "Sauce for the Goose," which concludes Part I of this compendium.

83. Lindberg, 2007: 364–65.

84. Quoted in Crosby 1997: 83 and Stark 2003: 147.

85. I explore the metaphysical implications of these facts at length in my essay "A Quantum-Theoretic Argument against Naturalism" in Part II of this volume. For reflections on the historical debate over action-at-a-distance, see McMullin (1989).

86. Roger Penrose (1996, 2000, 2005: 816–68) puts a different spin on this question, arguing that gravity will be found to play an indispensable role in quantum state reduction. Joy Christian (2001) also holds this view.

87. Stark 2003: 147–48.

88. Whitehead 1925 [1967]: 13.

89. A slender but charming volume giving brief life stories of forty-eight scientists who were Christians, from John Philoponus in the sixth century to Arthur Eddington in the early twentieth, is Dan Graves's *Scientists of Faith: Forty-Eight Biographies of Historic Scientists and Their Christian Faith*. Grand Rapids: Kregel Resources, 1996.

90. Galton 1875.
91. Leuba 1916 (1921).
92. Larson and Witham 1997.
93. See Stark 2003: 194.
94. Darwin 1876 [1958]: 87.
95. Dawkins 1986.
96. Ibid., 6.
97. Dennett 1995: 63.
98. Ibid., 519–20.
99. The recognition that purposelessness is unverifiable is not equivalent to the assertion that it cannot be falsified. We are not playing the same game as those critics of intelligent design theory who claim that it is not testable and then go on to assert that, furthermore, it has been shown to be false! Indeed, while a positive demonstration of a lack of purpose is unattainable, all that is needed to falsify it is *one* example of a biological system that not only cannot be explained on a neo-Darwinian basis, but which bears the requisite *positive* characteristics of intelligent design (ID). If ID theorists are correct, there are many such systems in biology, and cosmology bears evidence of intelligent design as well. A number of papers in this volume address the issue of intelligent design in specific contexts; we will make some general observations about it in the next and final section of this essay.
100. Some documentation is in order since skepticism is the common reaction to being told that the purposelessness of existence is the message being actively communicated in standard textbooks on biological evolution aimed at the young. I thank Casey Luskin for the legwork he has done on this. Here are some prime examples of the fostering of nihilism through public education:

"By coupling undirected, purposeless variation to the blind, uncaring process of natural selection, Darwin made theological or spiritual explanations of the life processes superfluous" (Douglas Futuyma, *Evolutionary Biology*, 3rd edition. Sunderland: Sinauer Associates, Inc., 1998: 5).

"Darwin knew that accepting his theory required believing in *philosophical materialism*, the conviction that matter is the stuff of all existence and that all mental and spiritual phenomena are its by-products. Darwinian evolution was not only purposeless but also heartless—a process in which the rigors of nature ruthlessly eliminate the unfit. Suddenly, humanity was reduced to just one more species in a world that cared nothing for us. The great human mind was no more than a mass of evolving neurons. Worst of all, there was no divine plan to guide us" (Joseph S. Levine and Kenneth R. Miller, *Biology: Discovering Life*, 1st edition. Lexington: D.C. Heath and Company, 1992: 152; 2nd edition, 1994: 161; emphasis in original).

"Evolution works without either plan or purpose . . . *Evolution is random and undirected*" (Kenneth R. Miller and Joseph S. Levine, *Biology*, 1st edition. Upper Saddle River: Prentice-Hall, 1991: 658; 3rd edition, 1995: 658; 4th edition, 1998: 658; emphasis in original).

"It is difficult to avoid the speculation that Darwin, as has been the case with others, found the implications of his theory difficult to confront. . . . The real difficulty in accepting Darwin's theory has always been that it seems to diminish our significance. Earlier, astronomy had made it clear that the Earth is not the center of the solar universe, or even of our own solar system. Now the new biology asked us to accept the proposition that, like all other organisms, we too are the products of a random process that, as far as science can show, we are not created for any special purpose or as part of any universal design" (Helena Curtis and N. Sue Barnes, *Invitation to Biology*, 3rd edition. New York: Worth Publishers, 1981: 474–75).

"The advent of Darwinism posed even greater threats to religion by suggesting that biological relationships, including the origin of humans and of all species, could be explained by natural selection without the intervention of a god. Many felt that evolutionary randomness and uncertainty had replaced a deity having conscious, purposeful, human characteristics. The Darwinian view that evolution is a historical process and present-type organisms were not created spontaneously but formed in a succession of selective events that occurred in the past, contradicted the common religious view that there could be no design, biological or otherwise, without an intelligent designer. . . . The variability by which selection depends may be random, but adaptations are not; they arise because selection chooses and perfects only what is adaptive. In this scheme a god of design and purpose is not necessary. . . . Religion has been bolstered by paternalistic social systems in which individuals depend on the beneficences of those more powerful than they are, as well as the comforting idea that humanity was created in the image of a god to rule over the world and its creatures. Religion provided emotional solace . . . Nevertheless, faith in religious dogma has been eroded by natural explanations of its mysteries, by a deep understanding of the sources of human emotional needs, and by the recognition that ethics and morality can change among different societies and that acceptance of such values need not depend on religion" (Monroe W. Strickberger, *Evolution*, 3rd edition. Sudbury: Jones and Bartlett Publishers, 2000: 70–71).

Such examples could be multiplied indefinitely, but you get the idea. Philosophical naturalism rules the academy and it is being promulgated in the name of science through educational materials directed at young people who are not in a position to be able to separate philosophical from scientific claims, let alone to evaluate the scientific claims themselves. (This is one reason it is important to have a textbook available that sensitizes students to philosophical issues and presents the arguments for and *against* neo-Darwinism. At the moment, the only available textbook serving this purpose is S. C. Meyer, S. Minnich, J. Moneymaker, P.A. Nelson, and R. Seelke, *Explore Evolution: The Arguments for and against Neo-Darwinism*. London: Hill House Publishers, 2007.) Even so, when the Kansas State Board of Education sought in 2005 to mitigate such language and have some of the evidence against neo-Darwinism introduced into the teaching of evolution, no less than 38 Nobel laureates under the auspices of the Elie Wiesel Foundation for Humanity, of all things, signed a joint statement to the KSBE informing them that "evolution is understood to be the result of an unguided, unplanned process of random variation and natural selection" (http://media.ljworld.com/pdf/2005/09/15/nobel_letter.pdf). Following the National Association of Biology Teacher's 1997 removal from their description of the evolution of life an assertion that it was an "unsupervised, impersonal, unpredictable and natural process," ninety-nine academics, including over seventy evolutionary biologists, sent a letter of protest to the NABT asserting that evolution indeed is "an impersonal and unsupervised process . . . The NABT leaves open the possibility that evolution is in fact supervised in a personal manner. This is a prospect that every evolutionary biologist should vigorously and positively deny" (http://www.metanexus.net/Magazine/ArticleDetail/tabid/68/id/2790/Default.aspx). So it is clear that evolutionary biologists insist, despite the in principle unverifiability of the assertion, that the process of evolution be understood metaphysically as undirected and purposeless. In fact, they are inclined to regard this perspective as an achievement. Francisco J. Ayala, an ex-Catholic priest and a professor in the UC Irvine evolutionary biology department, titled a recent article he published in *The Proceedings of the National Academy of Sciences* "Darwin's greatest discovery: Design without designer" (*PNAS* 104 (suppl. 1, May 15, 2007): 8567–73). This is the view of the scientific establishment, and it is the view they want the world to embrace.

101. Numbers 1977; Dilley 2007.
102. Roberts and Turner 2000: 28–29, 47, 91–92, *et passim*.
103. Darwin 1859 [1964]: 488; see also the epigraphs opposite the title page.

104. Rice 1867: 608.

105. See the other essays in Part I of this volume, especially the essay "Science without God: Natural Laws and Christian Beliefs" by Ronald Numbers. As a pertinent philosophical point to consider, when Charles Lyell introduced the principle of uniformitarianism into geology (Lyell 1830–33), seeking to explain the former changes of the Earth's surface by causes now in operation, while he took himself to be espousing a form of methodological naturalism, the principle he articulated need *not* be understood in this way. One of the causes now in operation is intelligence, and given that its operation has well-defined and distinguishing marks, it too may be admitted into the range of presently operative and identifiable causes useful to the historical sciences. On this latter point, see Stephen Meyer's essay "Sauce for the Goose: Intelligent Design, Scientific Methodology, and the Demarcation Problem" in this volume.

106. Hull 1989: 74–75.

107. Morris 1998, 2003.

108. Gould, 1990, 2003.

109. For a list of such convergences, see Conway Morris 2003: 457–61.

110. Morris, 2003.

111. Arthur 1997; Axe 2000, 2004; Axe, Dixon, and Lu 2008; Behe 1996, 2003, 2007; Berlinski 1998; Blanco *et al* 1999; Dembski 1998, 2002, 2004a, 2005; Dembski and Ruse 2004; Dembski and Wells 2008; Denton 1985, 1998; Durston *et al* 2007; Eden 1967; Gilbert *et al* 1996; Koonin 2000, 2007; Meyer 2003, 2004, 2009; Meyer *et al* 2007; Miklos 1993; Minnich and Meyer 2004; Moorhead and Kaplan 1967; Müller and Newman 2003; Nelson and Wells 2003; S. Newman 2006; Polanyi 1967, 1968; Schützenberger 1967; Thaxton *et al* 1984; Thomson 1992; Valentine 1995, 2004; Wagner 2001; Wagner and Altenberg 1996; Wagner and Stadler 2003; Yockey 1978, 1992; etc. See also the work being done at Biologic Institute: http://www.biologicinstitute.org/.

112. Molina 1588 [1953], 1588 [1988]; Flint 1998; Hasker *et al* 2000; see also Fischer, ed. 1989; Morris, ed. 1988.

113. This may be hard to swallow at first sight, but it encapsulates the ultimate significance of the rigorous extended argument developed in my essay "A Quantum-Theoretic Argument against Naturalism" in Part II of this volume. I also recommend consideration of Alexander Pruss's defense of the principle of sufficient reason (Pruss 2006).

114. I make this argument in a much more detailed fashion in my forthcoming essay "Quantum Theory and Middle Knowledge: An Occasionalist Rapprochement" (under submission).

115. Meyer 1990, 2000, 2006, 2009.

116. Axe 2000, 2004, and also his contribution to this volume; Bradley 2004; Dembski 1998, 2002, 2004b, 2005; Dembski and Wells 2008: 165–266; Denton 1985: 233–325; Eden 1967; Koonin 2000, 2007; Meyer 2003, 2004b, 2009; Polanyi 1967, 1968; Thaxton, Bradley and Olsen 1984; Trevors and Abel 2004; see also the appendix to this essay.

117. Axe, Dixon, and Lu 2008; Behe and Snoke 2004; Berlinski 1998; Dembski and Marks 2009a, 2009b, see also their essay in this volume; Durrett and Schmidt 2008; Durston, Abel and Trevors 2007; Schützenberger 1967; Voie 2006; von Sternberg 2008. Some of this literature deals with various failed attempts to model neo-Darwinian evolution computationally. For a visual illustration of the failings of Thomas Schneider's *ev* program, which purports to be a model for the generation of biological information, see http://www.evoinfo.org/Resources/EvWare/index.html (last accessed September 2, 2009). A dissection of the inadequacies of the much-lauded AVIDA program can be found in the essay by Dembski and Marks in this volume, along with a general discussion of how a variety of computational programs allegedly modeling undirected evolution smuggle exogenous information into their search algorithms. The paper by Axe, Dixon and Lu (2008) provides a model with tight analogies to protein evolution that allows for much more realistic computational

The Rise of Naturalism

studies to be done. For more research along these lines, see http://www.biologicinstitute.org and http://www.evoinfo.org (last accessed September 2, 2009).

118. Alberts 1998; Behe 1996, 1998, 2003, 2004, 2007; Dembski and Wells 2008: 145–161; DeRosier 1998; Dutcher 1995; Endow 2003; Halkier 1992; Lönnig 2004; Meyer *et al* 2007: 115–123; Minnich and Meyer 2000; Voie 2006.

119. Darwin 1859 [1964]: 189.

120. Arthur 1997; Durrett and Schmidt 2008; Erwin, Valentine, and Jablonski 1997; Goodwin 1985; Harold 1995; Lönnig 2004; Meyer 2009: 473–477; Meyer *et al* 2007: 65–71; Moss 2004; Müller and Newman 2003; Nijhout 1990; Sapp 1987; Thomson 1992; Valentine 2004; von Sternberg 2008; Wagner 2001; Wagner and Stadler 2003.

121. Adoutte *et al* 2000; Aguinaldo *et al* 1997; Baguñà and Garcia-Fernàndez 2003; Bapteste *et al* 2005; Benton 2001; Brocchieri 2001; Cao *et al* 1998; De Jong 1998; Dembski and Wells 2008: 113–144; de Pouplana 2004; Doolittle 1999a, 1999b, 2000, 2002, 2005; Duvall and Ervin 2004; Farré *et al* 2009; Fenske *et al* 2003; Fischer and Eisenberg 1999; Forterre and Philippe 1999a, 1999b; Gordon 1999; Graur and Martin 2004; Hall 1996; Hedges and Sibley 1994; Jones and Blaxter 2005; Lake, Jain and Rivera 1999; Lopez and Bapteste 2009; Lynch 1999; Margulis 2006; Martin and Embley 2004; Meyer *et al* 2007: 39–63; Mindell and Meyer 2001; Nelson and Wells 2003; Nielsen and Martinez 2003; Panganiban and Rubenstein 2002; Patterson *et al* 1993; Philippe and Forterre 1999; Rivera and Lake 2004; Rokas *et al* 2003; Rokas *et al* 2005; Rokas and Carroll 2006; Sander and Schmidt-Ott 2004; Santos and Tuite 2004; Siew and Fischer 2003a, 2003b; Telford and Budd 2003; Valentine, Jablonski and Erwin 1999; Wills 2002; Woese 1998, 2002

122. Ager 1976; Aris-Brosou and Yang 2003; Carroll 1997/98; Dembski and Wells 2008: 57–92; Eldredge 1985; Eldredge and Gould 1972; Gordon and Olson 1995; Gould and Eldredge 1993; Kemp 2005; Lönnig 2004; Meyer *et al* 2003a, 2003b; Meyer 2004a, 2004b; Meyer *et al* 2007: 15–38, 128–140; Olson 1981; Raup 1979; Simmons 2005; Valentine 2004; Valentine and Jablonski 2003; Valentine, Jablonski and Erwin 1999

123. Goh *et al* 2008; Kandouz *et al* 2004; Tam et al 2008; Watanabe *et al* 2008; Piehler *et al* 2008

124. Von Sternberg 2002; Meyer 2009: 407ff.

125. Han, Szak, and Boeke 2004; Janowski *et al* 2005; Goodrich and Kugel 2006; Li *et al* 2006; Pagano *et al* 2007; Van de Lagemaat *et al* 2003; Donnelly, Hawkins, and Moss 1999; Dunn, Medstrand, and Mager 2003; Burgess-Beusse *et al* 2002; Medstrand, Landry, and Mager 2001; Mariño-Ramírez *et al* 2005; von Sternberg and Shapiro 2005; Morrish *et al* 2002; Tremblay, Jasin, and Chartrand 2000; Grawunder, *et al* 1997; Wilson, Grawunder, and Liebe 1997; McKenzie and Brennan 1996; Arnaud *et al* 2000; Rubin, Kimura, and Schmid 2002; Bartel 2004; Mattick and Makunin 2005.

126. Henikoff, Ahmad, and Malik 2001; Bell, West, and Felsenfeld 2001; Pardue and DeBaryshe 1999; Henikoff 2000; Figueiredo *et al* 2002; Schueler *et al* 2001; Jordan *et al* 2003.

127. Green 2007; Figueiredo *et al* 2002.

128. Chen *et al* 2008; Jurka 2004; Lev-Maor *et al* 2003; Kondo-Iida *et al* 1999; Mattick and Makunin 2006

129. Dunlap *et al* 2002; Hyslop *et al* 2005; Peaston et al 2004.

130. Mura *et al* 2004; Kandouz *et al* 2004.

131. Mattick and Gagen 2001; von Sternberg and Shapiro 2005.

132. Cohen 2007; Farré et al 2009; Pollard 2009; Taylor 2009; von Sternberg 2009. I thank Rick Sternberg for a very helpful discussion of these matters.

133. Dembski 1998, 2002, 2003, 2004b, 2005; Dembski and Marks 2009a, 2009b; Dembski and Wells 2008: 165–205; Meyer 1990, 2000, 2006, 2009; see also the contributions of Dembski, Marks, and Meyer to this volume.

134. Barrow and Tipler 1986; Brownlee and Ward 2000; Collins 2003, 2009, forthcoming; Copan and Craig 2004; Craig 1998; Craig and Sinclair 2009; Craig and Smith 1993; Davies 1982; Gonzalez 2005,

2009; Gonzalez, Brownlee, and Ward 2001a, 2001b; Gonzalez and Richards 2004; Sinclair 2009; see also the contributions by Craig, Gonzalez, and Gordon in Part IV of this volume.

135. In furtherance of this conclusion, the reader is referred to the essays by Douglas D. Axe, David Berlinski, William Lane Craig, William A. Dembski, Guillermo Gonzalez, Bruce L. Gordon, Robert Koons, Robert J. Marks II, Stephen C. Meyer, James P. Moreland, Alvin C. Plantinga, Fazale Rana, Mark Steiner, Dallas Willard, and Nicholas Wolterstorff in this volume.

136. Dembski 2005.

137. Polanyi 1967, 1968; Thaxton *et al* 1984; Meyer 2009.

138. Axe, 2000, 2004.

139. See http://supfam.mrc-lmb.cam.ac.uk/SUPERFAMILY/cgi-bin/gen_list.cgi?genome=mg (last accessed August 27, 2009). I thank Doug Axe for a discussion of minimally complex self-reproducing systems found in nature. While *mycoplasma genitalium* provides an example of such, it is not self-sustaining and is highly dependent on its surrounding biological environment for a variety of products it cannot produce itself.

140. Lloyd 2002.

141. Dembski 2005.

References

Adoutte, A., G. Balavoine, N. Lartillot, O. Lespinet, B. Prud'homme, and R. De Rosa (2000) "The New Animal Phylogeny: Reliability and Implications."*Proceedings of the National Academy of Sciences USA* 97: 4453–56.

Ager, D. V. (1976) "The nature of the fossil record." *Proceedings of the Geological Association* 87: 131–59.

Aguinaldo, A.M.A., J.M. Turbeville, L.S. Linford, M.C. Rivera, J.R. Garey, R.A. Raff, and J.L. Lake. (1997) "Evidence for a Clade of Nematodes, Arthropods and Other Molting Animals." *Nature* 387: 489–93.

Alberts, Bruce. (1998) "The cell as a collection of protein machines: preparing the next generation of molecular biologists." *Cell* 92: 291–94.

Aris-Brosou, S., and Z. Yang. (2003) "Bayesian models of episodic evolution support a late Precambrian explosive diversification of the Metazoa." *Molecular Biology and Evolution* 20: 1947–54.

Armitage, Angus. (1951) *The World of Copernicus*. New York: Mentor Books.

Arnaud, Phillipe, Chantal Goubely, Thierry Pe'Lissier, and Jean-Marc Deragon. (2000) "SINE Retroposons Can Be Used In Vivo as Nucleation Centers for De Novo Methylation." *Molecular and Cellular Biology* 20: 3434–41.

Arthur, W. (1997) *The origin of animal body plans*. Cambridge: Cambridge University Press.

Ashworth, William B., Jr. (2003) "Christianity and the Mechanistic Universe," in David C. Lindberg and Ronald L. Numbers, eds. *When Science and Christianity Meet*. Chicago: University of Chicago Press, 61–84.

Axe, Douglas D. (2000) "Extreme functional sensitivity to conservative amino acid changes on enzyme exteriors." *Journal of Molecular Biology* 301: 585–96.

———. (2004) "Estimating the prevalence of protein sequences adopting functional enzyme folds." *Journal of Molecular Biology* 341: 1295–1315.

Axe, Douglas D., Brendan W Dixon, and Phillip Lu. (2008) "*Stylus:* A system for evolutionary experimentation based on a protein/proteome model with non- arbitrary functional constraints." *PLoS ONE* 3: e2246 (*doi:10.1371/journal.pone.0002246*).

Bacon, Francis. (1620 [1999]) *Selected Philosophical Works* (edited by Rose-Mary Sargent). Indianapolis: Hackett Publishing Company, Inc.

Baguñà, Jaume, and Jordi Garcia-Fernàndez. (2003) "Evo-devo: the long and winding road." *International Journal of Developmental Biology* 47: 705–13.

Baker, Hunter. (2009) *The End of Secularism*. Wheaton: Crossway Books.

Baldwin, John W. (1959) *The Medieval Theories of Just Price*. Philadelphia: The American Philosophical Society.

Banfield, Edward. (1958) *The Moral Basis of a Backward Society*. New York: The Free Press.

———. (1974 [1990]) *The Unheavenly City Revisited*. Prospect Heights: Waveland Press, Inc.

Bapteste, E., E. Susko, J. Leigh, D. MacLeod, R. L. Charlebois, and W. F. Doolittle. (2005) "Do Orthologous Gene Phylogenies Really Support Tree-Thinking?" *BioMed Central Evolutionary Biology* 5: 33.

Barrow, J., and F. Tipler. (1986) *The Anthropic Cosmological Principle*. New York: Oxford University Press.

Bartel, David. (2004) "MicroRNAs: Genomics, Biogenesis, Mechanism, and Function." *Cell* 116: 281–97.

Barzun, Jacques. (2000) *From Dawn to Decadence: 500 Years of Western Cultural Life, 1500 to the Present*. New York: HarperCollins Publishers, Inc.

Basile, B. (1983) "Galileo e il teologo 'Copernico' Paulo Antonio Foscarini," in *Rivista di letteratura italiana* 1: 63–96.

Behe, Michael J. (1996) *Darwin's Black Box: The Biochemical Challenge to Evolution*. New York: The Free Press.

———. (1998) "Intelligent Design Theory as a Tool for Analyzing Biochemical Systems," in William A. Dembski, ed. *Mere Creation: Science, Faith, and Intelligent Design*. Downers Grove: InterVarsity Press, 177–94.

———. (2003) "Design in the Details: The Origin of Biomolecular Machines," in John Angus Campbell and Stephen C. Meyer, eds. *Darwinism, Design, and Public Education*. East Lansing: Michigan State University Press, 287–302.

———. (2004) "Irreducible Complexity: Obstacle to Darwinian Evolution," in W. Dembski and M. Ruse, eds. *Debating Design: From Darwin to DNA*. Cambridge: Cambridge University Press, 352–70.

———. (2007) *The Edge of Evolution: The Search for the Limits of Darwinism*. New York: The Free Press.

Behe, Michael J. and David W. Snoke. (2004) "Simulating Evolution by Gene Duplication of Protein Features that Require Multiple Amino Acid Residues." *Protein Science* 13: 2651–64.

Beilby, James, ed. (2002) *Naturalism Defeated? Essays on Plantinga's Evolutionary Argument against Naturalism*. Ithaca: Cornell University Press.

Bell, C., A. G. West, and G. Felsenfeld. (2001) "Insulators and Boundaries: Versatile Regulatory Elements in the Eukaryotic Genome." *Science* 291: 447–50.

Benton, Michael J. (2001) "Finding the tree of life: matching phylogenetic trees to the fossil record through the 20th century." *Proceedings of the Royal Society of London B* 268: 2123–30.

Berlinski, David. (1998) "Gödel's Question," in William Dembski, ed. *Mere Creation: Science, Faith & Intelligent Design*. Downers Grove: InterVarsity Press, 402–26.

———. (2009) *The Devil's Delusion: Atheism and Its Scientific Pretensions*. New York: Basic Books.

Biagoli, M. (1993) *Galileo, Courtier: The Practice of Science in the Culture of Absolutism*. Chicago: University of Chicago Press.

Blackwell, R. J. (1991) *Galileo, Bellarmine and the Bible*. Notre Dame: University of Notre Dame Press.

Blanco, F., I. Angrand, and L. Serrano. (1999) "Exploring the conformational properties of the sequence space between two proteins with different folds: an experimental study." *Journal of Molecular Biology* 285: 741–53.

Bloch, Marc. (1940 [1961]) *Feudal Society* (2 volumes). Chicago: University of Chicago Press.

———. (1975) *Slavery and Serfdom in the Middle Ages*. Berkeley: University of California Press.

Bloom, Allan. (1987) *The Closing of the American Mind*. New York: Simon and Schuster.

Bonnassie, Pierre. (1991) *From Slavery to Feudalism in South-Western Europe*. Cambridge: Cambridge University Press.

Boorstin, Daniel (1983) *The Discoverers*. New York: Random House.

Boyer, Carl B. (1959) The Rainbow: From Myth to Mathematics. New York: Yoseloff.

Bradley, Walter. (2004) "Information, Entropy, and the Origin of Life," in W. Dembski and M. Ruse, eds. *Debating Design: From Darwin to DNA*. Cambridge: Cambridge University Press, 331–51.

Brocchieri, Luciano. (2001) "Phylogenetic inferences from molecular sequences: review and critique." *Theoretical Population Biology* 59: 27–40.

Brooke, John, and Geoffrey Cantor. (1998) Reconstructing Nature: The Engagement of Science and Religion. Oxford: Oxford University Press.

Brooks, Arthur C. (2006) Who Really Cares? The Surprising Truth about Compassionate Conservatism. New York: Basic Books.

Brownlee D., and P. Ward. (2000) Rare Earth. New York: Copernicus Books.

Burckhardt, Jacob (1860 [1954]) The Civilization of the Renaissance in Italy (translated by S.G.C. Middlemore). New York: Modern Library.

Burgess-Beusse, B., C. Farrell, M. Gaszner, M. Litt, V. Mutskov, F. Recillas-Targa, M. Simpson, A. West, and G. Felsenfeld. (2002) "The Insulation of Genes from External Enhancers and Silencing Chromatin." *Proceedings of the National Academy of Sciences USA* 99: 16433–437.

Burtchaell, James T. (1998) *The Dying of the Light: The Disengagement of Colleges and Universities from Their Christian Churches.* Grand Rapids: Eerdmans.

Burtt, E. A. (1924, rev. ed. 1932 [1954]) *The Metaphysical Foundations of Modern Physical Science.* Garden City: Doubleday.

Bury, J. B. (1889) *A History of the Later Roman Empire, from Arcadius to Irene (395 A.D. to 800 A.D.).* New York: Macmillan and Co.

Calvin, John. (1559 [1960]) *Institutes of the Christian Religion* (in 2 volumes). Translated by Ford Lewis Battles, edited by John T. McNeill. Philadelphia: Westminster John Knox Press.

Cao, Y., P.J. Waddell, N. Okada, and M. Hasegawa. (1998) "The complete mitochondrial DNA sequence of the shark Mustelus manazo: Evaluating rooting contradictions to living bony vertebrates." *Molecular Biology and Evolution* 15: 1637–46.

Caroti, S. (1987) "Un sostenitore napoletano della mobilità della terra: Il padre Paolo Antonio Foscarini," in *Galileo e Napoli,* edited by F. Lomonaco and M. Torrini. Naples: Guida, 81–121.

Carroll, R. L. (1997/98) "Limits to knowledge of the fossil record." *Zoology* 100: 221–31.

Chen, Ling-Ling, Joshua N. DeCerbo, and Gordon G. Carmichael. (2008) "Alu Element-Mediated Gene Silencing." *EMBO Journal:* 1–12.

Christian, Joy (2001) "Why the quantum must yield to gravity," in Craig Callender and Nick Huggett, eds. *Physics Meets Philosophy at the Planck Scale: Contemporary Theories in Quantum Gravity.* Cambridge: Cambridge University Press, 305–38.

Clagett, Marshall. (1961) *The Science of Mechanics in the Middle Ages.* Madison: University of Wisconsin Press.

Cohen, I. Bernard. (1985) *Revolution in Science.* Cambridge: The Belknap Press of Harvard University Press.

Cohen, Jonathan. (2007) "Relative Differences: The Myth of 1%." *Science* 316: 1836.

Cohen, Morris R. and I. E. Drabkin, eds. (1958) *A Source Book in Greek Science.* Cambridge: Harvard University Press.

Collins, Robin. (2003) "Evidence for Fine-Tuning," in N. Manson, ed. *God and Design: The Teleological Argument and Modern Science.* New York: Routledge, 178–99.

———. (2009) "The teleological argument: an exploration of the fine-tuning of the universe," in William Lane Craig and J. P. Moreland, eds. *The Blackwell Companion to Natural Theology.* Oxford: Blackwell Publishers, 202–81.

———. (forthcoming) *The Well-Tempered Universe.*

Condorcet, A.-N. (1795 [1955]) *Sketch for a Historical Picture of the Progress of the Human Mind* (translated by June Barraclough). New York: Noonday Press.

Conway Morris, Simon. (1998) *The Crucible of Creation: The Burgess Shale and the Rise of Animals.* Oxford: Oxford University Press.

———. (2003) *Life's Solution: Inevitable Humans in a Lonely Universe.* Cambridge: Cambridge University Press.

Copan, Paul, and William Lane Craig. (2004) "Scientific Evidence for Creation *Ex Nihilo,*" in P. Copan and W. L. Craig, *Creation out of Nothing: A Biblical, Philosophical. And Scientific Exploration.* Grand Rapids: Baker Academic, 219–48.

Copan, Paul and William Lane Craig, eds. (2009) *Contending with Christianity's Critics: Answering New Atheists & Other Objectors.* Nashville: B&H Publishing Group.

Craig, William Lane. (1998) "Design & the Cosmological Argument," in William A. Dembski, ed. *Mere Creation: Science, Faith, and Intelligent Design.* Downers Grove: InterVarsity Press, 332–59.

Craig, William Lane, and J. P. Moreland, eds. (2000) *Naturalism: A Critical Analysis.* New York: Routledge.

Craig, William Lane, and James D. Sinclair. (2009) "The *kalam* cosmological argument," in William Lane Craig and J. P. Moreland, eds. *The Blackwell Companion to Natural Theology.* Oxford: Blackwell Publishers, 101–201.

Craig, William Lane, and Quentin Smith. (1993) *Theism, Atheism, and Big Bang Cosmology.* Oxford: Clarendon Press.

Crosby, Alfred W. (1997) *The Measure of Reality: Quantification and Western Society, 1250–1600.* Cambridge: Cambridge University Press.

Daniel, Stephen H. (2001) "Berkeley's Christian Neoplatonism, Archetypes, and Divine Ideas." *Journal of the History of Philosophy* 39 (2): 239–58.

Danielson, Dennis R. (2000) *The Book of the Cosmos: Imagining the Universe from Heraclitus to Hawking.* Cambridge: Perseus Publishing.

Darwin, Charles. (1859 [1964]) *On the Origin of Species by Means of Natural Selection, or the Preservation of Favoured Races in the Struggle for Life* (facsimile edition, edited by Ernst Mayr). Cambridge: Harvard University Press.

———. (1871 [1981]) *The Descent of Man,* and *Selection in Relation to Sex.* Princeton: Princeton University Press.

———. (1876 [1958]) *The Autobiography of Charles Darwin 1809–1882.* (Edited and with an appendix and notes by his grand-daughter Nora Barlow). London: Collins.

Davidson, Donald. (1984) *Inquiries into Truth and Interpretation.* Oxford: Clarendon Press.

———. (1990) "The Structure and Content of Truth." *The Journal of Philosophy* 87(6): 279–328.

Davies, Paul. (1982) *The Accidental Universe.* Cambridge: Cambridge University Press.

Davis, David B. (1966) *The Problem of Slavery in Western Culture.* Ithaca: Cornell University Press.

Dawkins, Richard. (1986) *The Blind Watchmaker: Why the evidence of evolution reveals a universe without design.* New York: W.W. Norton & Company.

Deason, Gary B. (1986) "Reformation Theology and the Mechanistic Conception of Nature," in David C. Lindberg and Ronald L. Numbers, eds. *God and Nature: Historical Essays on the Encounter between Christianity and Science.* Berkeley: University of California Press, 167–91.

De Jong, W. W. (1998) "Molecules remodel the mammalian tree." *Trends in Ecology and Evolution,* 13(7): 270–74.

Dembski, William A. (1998) *The Design Inference: Eliminating Chance through Small Probabilities.* Cambridge: Cambridge University Press.

———. (2002) *No Free Lunch: Why Specified Complexity Cannot Be Purchased without Intelligence.* Lanham: Rowman & Littlefield.

———. (2003) "Reinstating Design within Science," in John Angus Campbell and Stephen C. Meyer, eds. *Darwinism, Design, and Public Education.* East Lansing: Michigan State University Press, 403–17.

———. (2004a) *The Design Revolution: Answering the Toughest Questions About Intelligent Design.* Downers Grove: InterVarsity Press.

———. (2004b) "The Logical Underpinnings of Intelligent Design," in W. Dembski and M. Ruse, eds. *Debating Design: From Darwin to DNA.* Cambridge: Cambridge University Press, 311–30.

———. (2005) "Specification: The Pattern that Signifies Intelligence," *http://www.designinference.com/documents/2005.06.Specification.pdf.*

———. (2009) *The End of Christianity: Finding a Good God in an Evil World.* Nashville: Broadman & Holman Academic.

Dembski, William A., and Robert J. Marks II. (2009a) "Conservation of Information in Search: Measuring the Cost of Success." *IEEE Transactions on Systems, Man and Cybernetics A, Systems & Humans* 39(5): 1051–61.

———. (2009b) "The Search for a Search: Measuring the Information Cost of Higher Level Search." *The International Journal of Information Technology and Intelligent Computing* (forthcoming).

Dembski, William A. and Michael Ruse. (2004) *Debating Design: From Darwin to DNA.* Cambridge: Cambridge University Press.

Dembski, William A., and Jonathan Wells. (2008) *The Design of Life: Discovering Signs of Intelligence in Biological Systems.* Dallas: Foundation for Thought and Ethics.

Denton, Michael. (1985) *Evolution: A Theory in Crisis.* Bethesda: Adler & Adler, Publishers, Inc.

———. (1998) *Nature's Destiny: How the Laws of Biology Reveal Purpose in the Universe.* New York: The Free Press.

Dennett, Daniel C. (1995) *Darwin's Dangerous Idea: Evolution and the Meanings of Life.* New York: Simon & Schuster.

de Pouplana, Llius Ribas, ed. (2004) *The Genetic Code and the Origin of Life.* New York: Kluwer Academic/Plenum Publishers.

de Roover, Raymond. (1958) "The Concept of the Just Price: Theory and Economic Policy." *The Journal of Economic History* 18: 418–34.

DeRosier, D. J. (1998) "The turn of the screw: The bacterial flagellar motor." *Cell* 93: 17–20.

Dictionary of Scientific Biography, 16 volumes. (1970–1980) New York: Scribner's.

Dilley, Stephen C. (2007) *Methodological Naturalism, History, and Science.* Tempe: Arizona State University, Ph.D. dissertation.

Dobbs, Betty Jo Teeter. (1991) *The Janus faces of genius: The role of alchemy in Newton's thought.* Cambridge: Cambridge University Press.

Doolittle, W. Ford. (1999a) "Lateral Genomics." *Trend in Biochemical Sciences* 24: M5-M8.

———. (1999b) "Phylogenetic Classification and the Universal Tree." *Science* 284: 2124–28.

———. (2000a) "The nature of the universal ancestor and the evolution of the proteome." *Current Opinion in Structural Biology* 10: 357–58.

———. (2000b) "Uprooting the Tree of Life." *Scientific American* 282: 90–95.

———. (2002) "Microbial genomes multiply." *Nature* 416: 698.

———. (2005) "If the Tree of Life Fell, Would We Recognize the Sound?" in Jan Sapp, ed. *Microbial Phylogeny and Evolution: Concepts and Controversies.* New York: Oxford University Press, 119–33.

Donnelly, S. R., T. E. Hawkins, and S. E. Moss. (1999) "A Conserved Nuclear Element with a Role in Mammalian Gene Regulation." *Human Molecular Genetics* 8: 1723–28.

Drake, H. A. (1996) "Lambs into Lions: Explaining Early Christian Intolerance," in *Past and Present* 153: 3–36.

Dunlap, K. A., M. Palmarini, M. Varela, R. C. Burghardt, K. Hayashi, J. L. Farmer, and T. E. Spencer. (2006) "Endogenous Retroviruses Regulate Periimplantation Placental Growth and Differentiation." *Proceedings of the National Academy of Sciences USA* 103: 14390–395.

Dunn, C. A., P. Medstrand, and D. L. Mager. (2003) "An Endogenous Retroviral Long Terminal Repeat Is the Dominant Promoter for Human B1,3-galactosyltransferase 5 in the Colon." *Proceedings of the National Academy of Sciences USA* 100: 12841–846.

Durrett, R., and D. Schmidt (2008) "Waiting for two mutations: with applications to regulatory sequence evolution and the limits of Darwinian evolution." *Genetics* 180(2): 1501–9.

Durston, K., D. Chiu, D. Abel, and J. Trevors. (2007) "Measuring the functional sequence complexity of proteins." *Theoretical Biology and Medical Modelling* 4: 47 (doi:10.1186/1742–4682–4–47).

Dutcher, S. K. (1995) "Flagellar assembly in two hundred and fifty easy-to-follow steps." *Trends in Genetics* 11: 398–404.

Duvall, Melvin R., and Autumn Bricker Ervin (2004) "18S gene trees are positively misleading for monocot/dicot phylogenetics." *Molecular Phylogenetics and Evolution* 30: 97–106.

Eden, Murray. (1967) "Inadequacies of Neo-Darwinian Evolution as a Scientific Theory," in P. S. Morehead and M. M. Kaplan, eds., *Mathematical challenges to the Darwinian interpretation of evolution* (Wistar Institute Symposium Monograph). New York: Allen R. Liss, 5–12, 109–11.

Eisenstein, Elizabeth L. (1979) *The Printing Press as an Agent of Change.* Cambridge: Cambridge University Press.

Eldredge, Niles. (1985) *Time Frames: The Rethinking of Darwinian Evolution and the Theory of Punctuated Equilibria.* New York: Simon & Schuster.

Eldredge, Niles and Stephen Jay Gould. (1972) "Punctuated Equilibria: An Alternative to Phyletic Gradualism," in T. J. M. Schopf, ed. *Models in Paleobiology.* San Francisco: Freeman, Cooper and Company, 82–115.

Endow, Sharon A. (2003) "Kinesin motors as molecular machines." *BioEssays* 25: 1212–19.

Erwin, D. H., J. Valentine, and D. Jablonski. (1997) "The Origin of Animal Body Plans." *American Scientist* 85: 126–37.

Farré, M., M. Ponsà, and M. Bosch. (2009) "Interstitial telomeric sequences (ITSs) are not located at the exact evolutionary breakpoints in primates." *Cytogenetic and Genome Research* 124(2): 128–31.

Fenske, Christine, Gottfried J. Palm, and Winfried Hinrichs. "How unique is the genetic code?" *Angewandte Chemie International Edition* 42: 606–10.

Ferrill, Arthur. (1986) *The Fall of the Roman Empire: The Military Explanation.* London: Thames and Hudson.

Field, Hartry. (1986) "The Deflationary Conception of Truth," in G. MacDonald and C. Wright, eds. *Fact, Science, and Morality.* Oxford: Blackwell.

Figueiredo, L. M., L. H. Freitas-Junior, E. Bottius, J.-C. Olivo-Marin, and A. Scherf. (2002) "A Central Role for *Plasmodium Falciparum* Subtelomeric Regions in Spatial Positioning and Telomere Length Regulation." *EMBO Journal* 21: L815-L824.

Fischer, Daniel, and David Eisenberg. (1999) "Finding Families for Genomic ORFans." *Bioinformatics* 15: 759–62.

Fischer, John Martin, ed. (1989) *God, Foreknowledge, and Freedom.* Stanford: Stanford University Press.

Fleckenstein, William A. (2008) *Greenspan Bubbles: The Age of Ignorance at the Federal Reserve.* New York: McGraw-Hill.

Flint, Thomas P. (1998) *Divine Providence: The Molinist Account.* Ithaca: Cornell University Press.

Forterre, P., and H. Philippe. (1999a) "The Last Universal Common Ancestor (LUCA), Simple or Complex?" *Biological Bulletin* 196: 373–77.

———. (1999b) "Where Is the Root of the Universal Tree of Life?" *BioEssays* 21: 871–79.

Foster, M. B. (1934) "The Christian doctrine of creation and the rise of modern natural science." *Mind* 43: 446–68.

Freeman, Charles. (2003) *The Closing of the Western Mind: The Rise of Faith and the Fall of Reason.* New York: Knopf.

Friedman, Milton. (1962) *Capitalism and Freedom.* Chicago: The University of Chicago Press.

Friedman, Milton, and Rose Friedman. (1980 [1990]) *Free to Choose: The Classic Inquiry into the Relationship between Freedom and Economics.* San Diego: Harcourt, Inc.

Galton, Francis. (1875) *English Men of Science: Their Nature and Nurture.* New York: D. Appleton and Company.

Geivett, R. Douglas, and Gary R. Habermas, eds. (1997) *In Defense of Miracles: A Comprehensive Case for God's Action in History.* Downers Grove: IVP.

George, Robert P. (2001) *The Clash of Orthodoxies: Law, Religion, and Morality in Crisis.* Wilmington: ISI Books.

Gies, Francis and Joseph Gies. (1994) *Cathedral, Forge, and Waterwheel: Technology and Invention in the Middle Ages.* New York: HarperCollins.

Gilbert, S. F., J. M. Opitz, and R. A. Raff. (1996) "Resynthesizing evolutionary and developmental biology." *Developmental Biology* 173: 357–72.

Gilder, George. (1980) *Wealth and Poverty*. New York: Basic Books.

———. (2009) *The Israel Test*. Minneapolis: Richard Vigilante Books.

Gimpel, Jean. (1976) *The Medieval Machine: The Industrial Revolution of the Middle Ages*. New York: Penguin Books.

Gillespie, Michael Allen (2008) *The Theological Origins of Modernity*. Chicago: University of Chicago Press.

Gingerich, Owen. (1975) "'Crisis' versus Aesthetic in the Copernican Revolution," in *Vistas in Astronomy* 17: 85–93.

Goh, S.-H., Y. T. Lee, N. V. Bhanu, M. C. Cam, R. Desper, B. M. Martin, R. Moharram, R. B. Gherman, and J. L. Miller. (2005) "A Newly Discovered Human Alpha Globin Gene." *Blood* DOI 10.1182/blood-2005-03-0948.

Goldberg, Jonah. (2007) *Liberal Fascism: The Secret History of the American Left from Mussolini to the Politics of Meaning*. New York: Doubleday.

Gonzalez, G. (2005) "Habitable Zones in the Universe." *Origins of Life and Evolution of Biospheres* 35: 555–606.

Gonzalez, G. (2009) "Planet Formation and the Galactic Habitable Zone." *Icarus*, in press.

Gonzalez, G., Brownlee, D. and Ward, P. (2001a) "The Galactic Habitable Zone: Galactic Chemical Evolution." *Icarus* 152: 185–200.

Gonzalez, G., Brownlee, D. and Ward, P. (2001b) "The Galactic Habitable Zone." *Scientific American* 285 (October): 60–67.

Gonzalez, Guillermo, and Richards, Jay. (2004) *The Privileged Planet: How Our Place in the Cosmos is Designed for Discovery*. Washington, DC: Regnery Publishing, Inc.

Goodrich, J. A., and J. F. Kugel. (2006) "Non-coding-RNA Regulators of RNA Polymerase II Transcription." *Nature Reviews Molecular and Cell Biology* 7: 612–16.

Goodwin, Brian. (1985) "What are the Causes of Morphogenesis?" *BioEssays* 3: 32–36.

Gordon, Malcolm S. (1999) "The concept of monophyly: a speculative essay." *Biology and Philosophy* 14: 331–48.

Gordon, Malcolm S., and Everett C. Olson. (1995) *Invasions of the Land: The Transition of Organisms from Aquatic to Terrestrial*. New York: Columbia University Press.

Gould, Stephen Jay. (1990) *Wonderful Life: The Burgess Shale and the Nature of History*. New York: W. W. Norton & Company, Inc.

———. (2002) *The Structure of Evolutionary Theory*. Cambridge: Harvard University Press.

Gould, Stephen Jay, and Niles Eldredge. (1993) "Punctuated equilibrium comes of age." *Nature* 366: 223.

Graur, D., and W. Martin. (2004) "Reading the entrails of chickens: molecular timescales of evolution and the illusion of precision." *Trends in Genetics* 20: 80–86.

Grant, Edward. (1971) *Physical Science in the Middle Ages*. New York: Wiley.

———. (1994) *Planets, Stars, and Orbs: The Medieval Cosmos, 1200–1687*. Cambridge: Cambridge University Press.

———. (1996) *The Foundations of Modern Science in the Middle Ages: Their Religious, Institutional, and Intellectual Contexts*. Cambridge: Cambridge University Press.

Grant, Edward, ed. (1974) *A Source Book in Medieval Science*. Cambridge: Harvard University Press.

Grant, Michael. (1978) *A History of Rome*. London: Faber and Faber.

Graves, Dan. (1996) *Scientists of Faith: Forty-Eight Biographies of Historic Scientists and Their Christian Faith*. Grand Rapids: Kregel Resources.

Grawunder, U., M. Wilm, X. Wu, P. Kulesza, T. E. Wilson, M. Mann, and M. R. Lieber. (1997) "Activity of DNA Ligase IV Stimulated by Complex Formation with XRCC4 Protein in Mammalian Cells." *Nature* 388: 492–95.

Green, David G. (2007) "The Role of Translocation and Selection in the Emergence of Genetic Clusters and Modules." *Artificial Life* 13: 249–58.

Hackett, Jeremiah, ed. (1997) *Roger Bacon and the Sciences: Commemorative Essays.* Leiden: Brill.

Halkier, T. (1992) *Mechanisms in Blood Coagulation Fibrinolysis and the Complement System.* Cambridge: Cambridge University Press.

Hall, Brian K. (1996) "Baupläne, phylotypic stages, and constraint: Why there are so few types of animals." *Evolutionary Biology* 29: 215–53.

Hamilton, Richard F. (1996) *The Social Misconstruction of Reality: Validity and Verification in the Scholarly Community.* New Haven: Yale University Press.

Han, Jeffrey S., Suzanne T. Szak, and Jef D. Boeke. (2004) "Transcriptional Disruption by the L1 Retrotransposon and Implications for Mammalian Transcriptomes." *Nature* 429: 268–74.

Harold, Franklin M. (1995) "From Morphogenes to Morphogenesis." *Microbiology* 141: 2765–78.

Harrison, Peter. (1998) *The Bible, Protestantism, and the Rise of Natural Science.* Cambridge: Cambridge University Press.

———. (2002) "Voluntarism and Early Modern Science." *History of Science* 40: 63–89.

Hasker, William, David Basinger, and Eef Dekker, eds. (2000) *Middle Knowledge: Theory and Applications.* Frankfurt: Peter Lang, Europäischer Verlag der Wissenschaften.

Hayek, Friedrich. (1944) *The Road to Serfdom.* Chicago: The University of Chicago Press.

———. (1988) *The Fatal Conceit: The Errors of Socialism.* Chicago: University of Chicago Press.

Heath, Thomas L. (1949) *Mathematics in Aristotle.* Oxford: Clarendon Press.

Hedges, S. Blair, and Charles G. Sibley. (1994) "Molecules vs. morphology in avian evolution: The case of the 'pelicaniform' birds." *Proceedings of the National Academy of Sciences USA* 91: 9861–65.

Hedley, Douglas and Sarah Hutton, eds. (2008) *Platonism at the Origins of Modernity: Studies on Platonism and Early Modern Philosophy.* New York: Springer.

Heimann, Peter. (1978) "Voluntarism and immanence: Conceptions of nature in eighteenth century thought." *Journal of the History of Ideas* 39: 271–83.

Henikoff, Steven. (2000) "Heterochromatin Function in Complex Genomes." *Biochimica et Biophysica Acta* 1470: 01–08.

Henikoff, Steven, Kami Ahmad, and Harmit S. Malik. (2001) "The Centromere Paradox: Stable Inheritance with Rapidly Evolving DNA." *Science* 293: 1098–1102.

Hill, Jonathan. (2005) *What Has Christianity Ever Done for Us? How it Shaped the Modern World.* Downers Grove: IVP.

Hirsch, Rudolf (1967) *Printing, Selling and Reading, 1450–1550.* Weisbaden: Harrassowitz.

Homza, Lu Ann (2006) *The Spanish Inquisition, 1478–1614: An Anthology of Sources.* Indianapolis: Hackett Publishing Company, Inc.

Horwich, Paul. (1990) *Truth.* Oxford: Basil Blackwell, Ltd.

Hull, David L. (1988) *Science as a Process: An Evolutionary Account of the Social and Conceptual Development of Science.* Chicago: University of Chicago Press.

———. (1989) *The Metaphysics of Evolution.* Albany: SUNY Press.

Hyslop, L., M. Stojkovic, L. Armstrong, T. Walter, P. Stojkovic, S. Przyborski, M. Herbert, A. Murdoch, T. Strachan, and M. Lakoa. (2005) "Downregulation of NANOG Induces Differentiation of Human Embryonic Stem Cells to Extraembryonic Lineages." *Stem Cells* 23: 1035–43.

Janowski, Bethany, A., K. E. Huffman, J. C. Schwartz, R. Ram, D. Hardy, D. S. Shames, J. D. Minna, and D. R. Corey. (2005) "Inhibiting Gene Expression at Transcription Start Sites in Chromosomal DNA with Antigene RNAs." *Nature Chemical Biology* 1: 216–22.

Johnson, Paul. (1976) *A History of Christianity.* New York: Atheneum.

Jones, Martin, and Mark Blaxter. (2005) "Animal Roots and Shoots." *Nature* 434: 1076–77.

Jordan, I. K., I. B. Rogozin, G. V. Glazko, and E. V. Koonin. (2003) "Origin of a Substantial Fraction of Human Regulatory Sequences from Transposable Elements." *Trends in Genetics* 19: 68–72.

Jurka, Jerzy. (2004) "Evolutionary Impact of Human Alu Repetitive Elements." *Current Opinion in Genetics and Development* 14: 603–8.

Kamen, Henry. (1997) *The Spanish Inquisition: An Historical Revision.* London: Weidenfeld & Nicolson.

Kandouz, M., A. Bier, G. D. Carystinos, M. A. Alaoui-Jamali, and G. Batist. (2004) "Connexin43 Pseudogene Is Expressed in Tumor Cells and Inhibits Growth." *Oncogene* 23: 4763–70.

Kemp, T. S. (2005) *The Origin & Evolution of Mammals.* New York: Oxford University Press.

Kimball, Roger. (1990) *Tenured Radicals: How Politics Has Corrupted Our Higher Education.* New York: HarperCollins Publishers.

Klaaren, Eugene. (1977) *Religious Origins of Modern Science.* Grand Rapids: Eerdmans.

Kondo-Iida, E., K. Kobayashi, M. Watanabe, J. Sasaki, T. Kumagai, H. Koide, K. Saito, M. Osawa, Y. Nakamura, and T. Toda. (1999) "Novel Mutations and Genotype–Phenotype Relationships in 107 Families with Fukuyama-Type Congenital Muscular Dystrophy (FCMD)." *Human Molecular Genetics* 8: 2303–9.

Koonin, Eugene. (2000) "How many genes can make a cell?: the minimal genome concept." *Annual Review of Genomics and Human Genetics* 1: 99–116.

———. (2007) "The cosmological model of eternal inflation and the transition from chance to biological evolution in the history of life," *Biology Direct* (http://www.biology-direct.com/content/2/1/15).

Kvanvig, Jonathan L., ed. (1996) *Warrant in Contemporary Epistemology: Essays in Honor of Plantinga's Theory of Knowledge.* Lanham: Rowman & Littlefield Publishers, Inc.

Lake, J.A., R. Jain, and M.C. Rivera. (1999) "Mix and Match in the Tree of Life." *Science* 283: 2027–28.

Lamoureux, Denis O. (2008) *Evolutionary Creation: A Christian Approach to Evolution.* Eugene: Wipf & Stock.

Langford, Jerome L. (1971) *Galileo, Science and the Church* (revised edition). Ann Arbor: University of Michigan Press.

Larson, Edward J., and Larry Witham. (1997) "Scientists Are Still Keeping the Faith." *Nature* 386 (April 3): 435.

Lasch, Christopher. (1979) *The Culture of Narcissism: American Life in an Age of Diminishing Expectations.* New York: W. W. Norton & Company, Inc.

Leuba, James. (1916 [1921]) *The Belief in God and Immortality.* Chicago: Open Court Publishing Co.

Lev-Maor, G., *et al.* (2003) "The Birth of an Alternatively Spliced Exon: 3' Splice-Site Selection in Alu Exons." *Science* 300: 1288–91.

Lewis, C. S. (1947 [1960]) *Miracles.* New York: Macmillan.

Li, Long-Cheng, S. T. Okino, H. Zhao, H., D. Pookot, R. F. Place, S. Urakami, H. Enokida, and R. Dahiya. (2006) "Small dsRNAs Induce Transcriptional Activation in Human Cells." *Proceedings of the National Academy of Sciences USA* 103: 17337–342.

Lindberg, David C. (1996) *Roger Bacon and the Origins of the "Perspectiva" in the Middle Ages: A Critical Edition and English Translation of Bacon's "Perspectiva," with Introduction and Notes.* Oxford: Clarendon Press.

———. (2007) *The Beginnings of Western Science: The European Scientific Tradition in Philosophical, Religious, and Institutional Context, Prehistory to A.D. 1450* (2nd Edition). Chicago: University of Chicago Press.

Lindberg, David C., and Ronald L. Numbers, eds. (1986) *God and Nature: Historical Essays on the Encounter between Christianity and Science.* Berkeley: University of California Press.

———. (2003) *When Science and Christianity Meet*. Chicago: University of Chicago Press.

Lloyd, Seth. (2002) "Computational Capacity of the Universe." *Physical Review Letters* 88(23): 237901 (http://arxiv.org/PS_cache/quant-ph/pdf/0110/0110141v1.pdf).

London, Herbert. (2008) *America's Secular Challenge: The Rise of a New National Religion*. New York: Encounter Books.

Lönnig, Wolf-Ekkehard. (2004) "Dynamical genomes, morphological stasis, and the origin of irreducible complexity," in V. Parisi, V. de Fonzo, and F. Aluffi-Pentini, eds. *Dynamical Genetics*. Trivandrum: Research Signpost, 101–19.

Lopez, Philippe, and Eric Bapteste. (2009) "Molecular phylogeny: reconstructing the forest." *Comptes Rendu Biologies* 332(2–3): 171–82.

Luttwak, Edward N. (1976) *The Grand Strategy of the Roman Empire*. Baltimore: Johns Hopkins University Press.

Lyell, Charles. (1830–1833) *Principles of Geology: Being an Attempt to Explain the Former Changes of the Earth's Surface, by Reference to Causes Now in Operation* (3 volumes). London: Murray.

Lynch, Michael. (1999) "The Age and Relationships of the Major Animal Phyla." *Evolution* 53: 319–25.

Margulis, Lynn. (2006) "The Phylogenetic Tree Topples," *American Scientist* 94(3): 194.

Mariño-Ramírez, L., K. C. Lewis, D. Landsman, and I. K. Jordan. (2005) "Transposable Elements Donate Lineage-Specific Regulatory Sequences to Host Genomes." *Cytogenetic and Genome Research* 110: 333–41.

Marsden, George M. (1994) *The Soul of the American University: From Protestant Establishment to Established Non-Belief*. Oxford: Oxford University Press.

Martin, William, and T. Martin Embley. (2004) "Early Evolution Comes Full Circle." *Nature* 431: 134–37.

Mason, Stephen F. (1962) *A History of the Sciences* (revised edition). New York: Macmillan.

Mattick, John S., and Michael J. Gagen. (2001) "The Evolution of Controlled Multitasked Gene Networks: The Role of Introns and Other Noncoding RNAs in the Development of Complex Organisms." *Molecular Biology and Evolution* 18: 1611–30.

Mattick, John S., and I. V. Makunin. (2005) "Small Regulatory RNAs in Mammals." *Human Molecular Genetics* 14: R121–R132.

———. (2006) "Non-coding RNA." *Human Molecular Genetics* 15: R17–R29.

McGrath, Alister E. (1998) *The Foundations of Dialogue in Science & Religion*. Oxford: Blackwell Publishers, Inc.

McKenzie, Richard W., and Mark D. Brennan. (1996) "The Two Small Introns of the Drosophila Affinidisjuncta Adh Gene Are Required for Normal Transcription." *Nucleic Acids Research* 24: 3635–3642.

McMullin, Ernan. (1989) "The Explanation of Distant Action: Historical Notes," in James T. Cushing and Ernan McMullin, eds. *Philosophical Consequence of Quantum Theory: Reflections on Bell's Theorem*. Notre Dame: University of Notre Dame Press, 272–302.

McMullin, Ernan, ed. (1988) *Construction and Constraint: The Shaping of Scientific Rationality*. Notre Dame: University of Notre Dame Press.

Medstrand, P., J.-R. Landry, and D. L. Mager. (2001) "Long Terminal Repeats Are Used as Alternative Promoters for the Endothelin B Receptor and Apolipoprotein C-I Genes in Humans." *Journal of Biological Chemistry* 276: 1896–1903.

Menzel, Christopher. (1987) "Theism, Platonism, and the Metaphysics of Mathematics." *Faith and Philosophy* 4: 365–82.

Meyer, Stephen C. (1990) *Of Clues and Causes: A Methodological Interpretation of Origin of Life Studies*. Cambridge University: Ph.D. dissertation.

———. (2000) "The Scientific Status of Intelligent Design: The Methodological Equivalence of Naturalistic and Non-Naturalistic Origins Theories," in M.J. Behe, W.A. Dembski, and S.C. Meyer, *Science and Evidence for Design in the Universe.* San Francisco: Ignatius Press, 151–211.

———. (2003) "DNA and the Origin of Life: Information, Specification, and Explanation," in John Angus Campbell and Stephen C. Meyer, eds. *Darwinism, Design, and Public Education.* East Lansing: Michigan State University Press, 223–85.

———. (2004a) "The Cambrian Information Explosion: Evidence for Intelligent Design," in W. Dembski and M. Ruse, eds. *Debating Design: From Darwin to DNA.* Cambridge: Cambridge University Press, 371–91.

———. (2004b) "The origin of biological information and the higher taxonomic categories." *Proceedings of the Biological Society of Washington* 117(2): 213–39.

———. (2006) "A Scientific History—and Philosophical Defense—of the Theory of Intelligent Design." *Religion-Staat-Gesellschaft* 7(2): 203–47.

———. (2009) *Signature in the Cell: DNA and the Evidence for Intelligent Design.* San Francisco: HarperOne.

Meyer, Stephen C., Scott Minnich, Jonathan Moneymaker, Paul A. Nelson, and Ralph Seelke. (2007) *Explore Evolution: The Arguments for and against Neo-Darwinism.* London and Melbourne: Hill House Publishers.

Meyer, Stephen C., Marcus Ross, Paul Nelson, and Paul Chien (2003a) "The Cambrian Explosion: Biology's Big Bang," in John Angus Campbell and Stephen C. Meyer, eds. *Darwinism, Design, and Public Education.* East Lansing: Michigan State University Press, 323–402.

———. (2003b) "Stratigraphic First Appearance of Phyla Body Plans, Phyla-Subphyla Body Plans, and the Probability of Other Body Plans Originating in the Cambrian Explosion (Appendices C-E)" in John Angus Campbell and Stephen C. Meyer, eds. *Darwinism, Design, and Public Education.* East Lansing: Michigan State University Press, 593–611.

Miklos, G. L. G. (1993) "Emergence of organizational complexities during metazoan evolution: perspectives from molecular biology, palaeontology and neo-Darwinism." *Memoirs of the Association of Australasian Palaeontologists* 15: 7–41.

Millar, Fergus, ed. (1967) *The Roman Empire and Its Neighbours.* London: Weidenfeld & Nicolson.

Mindell, David P., and Axel Meyer. (2001) "Homology evolving." *Trends in Ecology and Evolution* 16: 343–440.

Minnich, Scott A., and Stephen C. Meyer. (2004) "Genetic Analysis of Coordinate Flagellar and Type III Regulatory Circuits in Pathogenic Bacteria," in M. W. Collins and C. A. Brebbia, eds. *Design and Nature II: Comparing Design in Nature with Science and Engineering.* Southampton: Wessex Institute of Technology, 295–304.

Molina, Luis de. (1588 [1953]) *Liberi Arbitrii cum Gratiae Donis, Divina Praescientia, Providentia, Praedestinatione et Reprobatione Concordia,* edited by J. Rabeneck. Madrid: Collegium Maximum, Societatis Iesu selecti scriptores.

———. (1588 [1988]) *On Divine Foreknowledge (Part IV of the* Concordia), translated with an introduction and notes by Alfred J. Freddoso. Ithaca: Cornell University Press.

Moonan, Lawrence (1994) *Divine power: The medieval power distinction and its adoption by Albert, Bonaventure, and Aquinas.* Oxford: Oxford University Press.

Moorhead, Paul S. and Martin M. Kaplan. (1967) *Mathematical Challenges to the Neo-Darwinian Interpretation of Evolution.* New York: Alan R. Liss, Inc.

Moreland, J. P. (2008) *Consciousness and the Existence of God.* New York: Routledge.

———. (2009) *The God Question: An Invitation to a Life of Meaning.* Eugene: Harvest House Publishers.

Morris, Thomas, ed. (1988) *Divine and Human Action: Essays in the Metaphysics of Theism.* Ithaca: Cornell University Press.

Morris, Thomas, and Menzel, Christopher. (1986) "Absolute Creation." *American Philosophical Quarterly* 23: 353–62.

Morrish, Tammy A., Nicolas Gilbert, Jeremy S. Myers, Bethaney J. Vincent, Thomas D. Stamato, Guillermo E. Taccioli, Mark A. Batzer, and John V. Moran. (2002) "DNA Repair Mediated by Endonuclease-Independent LINE-1 Retrotransposition." *Nature Genetics* 31: 159–65.

Moss, Lenny. (2004) *What Genes Can't Do.* Cambridge: MIT Press.

Müller, G. B. and S. A. Newman (2003) "Origination of organismal form: the forgotten cause in evolutionary theory," in G. B. Müller and S. A. Newman, eds. *Origination of organismal form: beyond the gene in developmental and evolutionary biology.* Cambridge: MIT Press, 3–12.

Mura, M., P. Murcia, M. Caporale, T. E. Spencer, K. Nagashima, A. Rein, and M. Palmarini. (2004) "Late Viral Interference Induced by Transdominant Gag of an Endogenous Retrovirus." *Proceedings of the National Academy of Sciences USA* 101: 11117–122.

Murray, Charles. (1984) *Losing Ground: American Social Policy 1950–1980.* New York: Basic Books.

———. (2003) *Human Accomplishment: The Pursuit of Excellence in the Arts and Sciences, 880 B.C. to 1950.* New York: HarperCollins Publishers, Inc.

Nelson, Paul A., and Jonathan Wells. (2003) "Homology in Biology: Problem for Naturalistic Science and Prospect for Intelligent Design," in John Angus Campbell and Stephen C. Meyer, eds. *Darwinism, Design, and Public Education.* East Lansing: Michigan State University Press, 303–22.

Neuhaus, Richard John. (1986) *The Naked Public Square: Religion and Democracy in America* (Second Edition). Grand Rapids: Wm. B. Eerdmans Publishing Company.

Newman, Stuart A. (2006) "The Developmental Genetic Toolkit and the Molecular Homology-Analogy Paradox." *Biological Theory* 1(1): 12–16.

Newman, William R. (2006) *Atoms and Alchemy: Chymistry and the Experimental Origins of the Scientific Revolution.* Chicago: University of Chicago Press.

Nielsen, Claus, and Pedro Martinez. (2003) "Patterns of gene expression: homology or homocracy?" *Development, Genes and Evolution* 213: 149–54.

Nijhout, H. Frederik (1990) "Metaphors and the Role of Genes in Development." *BioEssays* 12: 441–46.

Noonan, John T. (1957) *The Scholastic Analysis of Usury.* Cambridge: Harvard University Press.

Numbers, Ronald L. *(1977) Creation by Natural Law: Laplace's Nebular Hypothesis in American Thought.* Seattle: University of Washington Press.

———. (1998) *Darwinism Comes to America.* Cambridge: Harvard University Press.

Numbers, Ronald L., ed. (2009) *Galileo Goes to Jail and Other Myths about Science and Religion.* Cambridge: Harvard University Press.

Oakley, Francis. (1961) "Christian Theology and the Newtonian Science: The Rise of the Concept of the Laws of Nature." *Church History* 30 (4): 433–57.

Oestreich, Gerhard. (1982) *Neostoicism and the early modern state.* Cambridge: Cambridge University Press.

Olasky, Marvin. (1992) *The Tragedy of American Compassion.* Washington DC: Regnery Gateway.

———. (1997) *Renewing American Compassion: How Compassion for the Needy Can Turn Ordinary Citizens Into Heroes.* Washington DC: Regnery Publishing.

———. (2000) *Compassionate Conservatism: What it is, What it Does, and How it Can Transform America.* New York: Free Press.

Olson, Everett C. (1981) "The problem of missing links: today and yesterday." *The Quarterly Review of Biology* 56: 405–42.

O'Malley, Charles D. (1964) *Andreas Vesalius of Brussels, 1514–1564.* Berkeley: University of Califoria Press.

Osler, Margaret. (1994) *Divine will and the mechanical philosophy: Gassendi and Descartes on contingency and necessity in the created world.* Cambridge: Cambridge University Press.

———. (2002) "Mechanical Philosophy," in Gary B. Ferngren, ed. *Science and Religion: A Historical Introduction.* Baltimore: Johns Hopkins University Press, 143–52.

Ozment, Steven. (1980) *The Age of Reform 1250–1550: An Intellectual and Religious History of Late Medieval and Reformation Europe.* New Haven: Yale University Press.

Pagano, A. M. Castelnuovo, F. Tortelli, R. Ferrari, G. Dieci, and R. Cancedda. (2007) "New Small Nuclear RNA Gene-like Transcriptional Units as Sources of Regulatory Transcripts." *PLoS Genetics* 3: e1.

Pagano, S. M., ed. (1984) *I documenti del processo di Galileo Galilei.* Vatican City: Pontifical Academy of Sciences.

Panganiban, Grace, and John L.R. Rubenstein. (2002) "Developmental functions of the Distal-less/Dlx homeobox genes." *Development* 129: 4371–86.

Pardue, M.-L., and P. G. DeBaryshe. (1999) "Drosophila Telomeres: Two Transposable Elements with Important Roles in Chromosomes." *Genetica* 107: 189–96.

Patterson, C., D.M. Williams, and C.J. Humphries. (1993) "Congruence between Molecular and Morphological Phylogenies," *Annual Review of Ecology and Systematics* 24, 153–88.

Peaston, E., A. V. Evsikov, J. H. Graber, W. N. de Vries, A. E. Holbrook, D. Solter, and B. B. Knowles. (2004) "Retrotransposons Regulate Host Genes in Mouse Oocytes and Preimplantation Embryos." *Developmental Cell* 7: 597–606.

Pedersen, Olaf. (1974) *A Survey of the Almagest.* Acta Historica Scientiarum Naturalium et Medicinalium, v01.30. Odense: Ondense University Press.

Penrose, Roger. (1996) "On gravity's role in quantum state reduction." *General Relativity and Gravitation* 28: 581–600.

———. (2000) "Wavefunction collapse as a real gravitational effect," in A. Fokas, T.W.B. Kibble, A. Grigouriou, and B. Zegarlinski, eds. *Mathematical Physics 2000.* London: Imperial College Press, 266–82.

———. (2005) *The Road to Reality: A Complete Guide to the Laws of the Universe.* New York: Alfred A. Knopf.

Pestritto, Ronald J. (2009) *Woodrow Wilson and the Roots of Modern Liberalism.* Lanham: Rowman & Littlefield Publishers, Inc.

Pestritto, Ronald J., and Atto, William J., eds. (2008) *American Progressivism: A Reader.* Lanham: Lexington Books.

Philippe, H., and P. Forterre. (1999) "The Rooting of the Tree of Life is Not Reliable." *Journal of Molecular Evolution* 49: 509–23.

Piehler, A. P., M. Hellum, J. J. Wenzel, E. Kaminski, K. B. Haug, P. Kierulf, and W. E. Kaminski. (2008) "The Human ABC Transporter Pseudogene Family: Evidence for Transcription and Gene-Pseudogene Interference." *BMC Genomics* 9: 165.

Pirenne, Henri. (1936 [1958]) *A History of Europe from the End of the Roman World in the West to the Beginnings of the Western States.* New York: Doubleday Anchor.

———. (1922 [1955]) *Mohammed and Charlemagne.* New York: Barnes and Noble.

Plantinga, Alvin. (1980) *Does God Have a Nature?* Milwaukee: Marquette University Press.

———. (1993) *Warrant and Proper Function.* New York: Oxford University Press.

———. (2000) *Warranted Christian Belief.* New York: Oxford University Press.

Plantinga, Alvin, and Tooley, Michael. (2008) *Knowledge of God.* Oxford: Blackwell Publishing.

Polanyi, Michael. (1967) "Life transcending physics and chemistry." *Chemical and Engineering News* 45(35): 54–66.

———. (1968) "Life's irreducible structure." *Science* 160: 1308–12.

Pollard, Katherine S. (2009) "What makes us human? Comparisons of the genomes of humans and chimpanzees revealing those rare stretches of DNA that are ours alone." *Scientific American* May 300 (5): 44–49.

Porter, Roy. (1998) *The Greatest Benefit to Mankind: A Medical History of Humanity.* New York: W.W. Norton & Company.

Potter, David S. (2004) *The Roman Empire at Bay* A.D. *180–395.* London: Routledge.

Poulle, Emmanuel. "John of Murs." *Dictionary of Scientific Biography,* v. 7, 128–33.

Pruss, Alexander R. (2006) *The Principle of Sufficient Reason: A Reassessment.* Cambridge; Cambridge University Press.

Quine, Willard van Orman (1969) *Ontological Relativity and Other Essays.* New York: Columbia University Press.

Rae, Scott B. (2000) *Moral Choices: An Introduction to Ethics* (2nd edition). Grand Rapids: Zondervan.

Raup, David M. (1979) "Conflicts between Darwin and Paleontology." *Field Museum of Natural History Bulletin* 50: 22–29.

Rawls, John. (1971) *A Theory of Justice.* Cambridge: The Belknap Press of Harvard University Press.

Rea, Michael. (2002) *World without Design: The Ontological Consequences of Naturalism.* Oxford: Clarendon Press.

Rice, William North (1867) "The Darwinian Theory of the Origin of the Species." *New Englander* 26: 603–35.

Richards, Jay. (2009) *Money, Greed, and God: Why Capitalism is the Solution and Not the Problem.* San Francisco: HarperOne.

Rivera, Maria C., and James A. Lake. (2004) "The Ring of Life Provides Evidence for a Genome Fusion Origin of Eukaryotes." *Nature* 431: 152–55.

Roberts, Jon H., and Turner, James. (2000) *The Sacred and the Secular University.* Princeton: Princeton University Press.

Rokas, Antonis and Sean B. Carroll. (2006) "Bushes in the Tree of Life," *PLOS Biology* 4(11): 1899–1904.

Rokas, A., N. King, J. Finnerty, and S.B. Carroll. (2003) "Conflicting phylogenetic signals at the base of the metazoan tree." *Evolution and Development* 5: 346–59.

Rokas, A., D. Krüger, and S. B. Carroll. (2005) "Animal Evolution and the Molecular Signature of Radiations Compressed in Time." *Science* 310: 1933–38.

Rorty, Richard. (1989) *Contingency, Irony, and Solidarity.* Cambridge: Cambridge University Press.

Rosen, Edward. (1971) *Three Copernican Treatises,* 3rd edition. New York: Octagon Books.

Rouse, Joseph. (1987) *Knowledge and Power: Toward a Political Philosophy of Science.* Ithaca: Cornell University Press.

―――. (1996) *Engaging Science: How to Understand Its Practices Philosophically.* Ithaca: Cornell University Press.

Rousseau, Jean-Jacques. (1750 [1987]). "Discourse on the Sciences and the Arts" in *The Basic Political Writings* (translated by Donald A. Cress). Indianapolis: Hackett Publishing Company, Inc.

Rubin, C. M., R. H. Kimura, and C. W. Schmid. (2002) "Selective Stimulation of Translational Expression by Alu RNA." *Nucleic Acids Research* 30: 3253–61.

Russell, Jeffrey Burton. (1991) *Inventing the Flat Earth: Columbus and Modern Historians.* New York: Praeger.

Ryle, Gilbert. (1980 [1949]) *The Concept of Mind.* New York: Penguin Books.

Sabra, Abdelhamid I. (2003) "Ibn al-Haytham's Revolutionary Project in Optics: The Achievement and the Obstacle," in Hogendijk, Jan P., and Abdelhamid I Sabra, eds., *The Enterprise of Science in Islam: New Perspectives.* Cambridge: MIT Press, 85–118.

Samsó, Julio. "Levi ben Gerson." *Dictionary of Scientific Biography,* v01.8, 279–82.

Sander, Klaus, and Urs Schmidt-Ott. (2004) "Evo-Devo aspects of classical and molecular data in a historical perspective." *Journal of Experimental Zoology B* (Molecular and Developmental Evolution) 302: 69–91.

Santos, Manuel A.S., and Mick F. Tuite. (2004) "Extant Variations in the Genetic Code," in Lluis Ribas de Pouplana, ed. *The Genetic Code and the Origin of Life.* New York: Kluwer Academic/Plenum Publishers, 183–200.

Sapp, J. (1987) *Beyond the Gene.* New York: Oxford University Press.

Schueler, Mary G., Anne W. Higgins, M. Katharine Rudd, Karen Gustashaw, Huntington F. Willard. (2001) "Genomic and Genetic Definition of a Functional Human Centromere." *Science* 294: 109–15.

Schützenberger, Marcel. (1967) "Algorithms and the Neo-Darwinian Theory of Evolution," in P. S. Morehead and M. M. Kaplan, eds., *Mathematical challenges to the Darwinian interpretation of evolution* (Wistar Institute Symposium Monograph). New York: Allen R. Liss, 73–75, 121.

Searle, John. (2004) *Mind: A Brief Introduction.* Oxford: Oxford University Press.

Senor, ,Dan and Saul Singer (2009) *Start-up Nation: The Story of Israel's Economic Miracle.* New York: Twelve, 2009.

Shea, William R. (1986) "Galileo and the Church," in Lindberg, David C., and Ronald L. Numbers, eds. *God and Nature: Historical Essays on the Encounter between Christianity and Science.* Berkeley: University of California Press, 114–35.

Siew, Naomi, and Daniel Fischer. (2003a) "Analysis of Singleton ORFans in fully sequenced microbial genomes." *Proteins: Structure, Function, and Genetics* 53: 241–51.

_____. (2003b) "Twenty thousand ORFan microbial protein families for the biologist?" *Structure* 11: 7–9.

Simmons, Nancy B. (2005) "An Eocene Big Bang for Bats." *Science* 307: 527–28.

Sinclair, James D. (2009) "At Home in the Multiverse? Critiquing the Atheist Many-Worlds Scenario," in Paul Copan and William Lane Craig, eds. *Contending with Christianity's Critics: Answering New Atheists & Other Objectors.* Nashville: B&H Publishing Group, 6–25.

Singer, Charles. (1925 [1970]) "Historical Relations of Religion and Science," in *Science, Religion and Reality.* Joseph Needham, editor. Port Washington: Kennikat Press, 87–148.

Smith, A. Mark. (1982) "Ptolemy's Search for a Law of Refraction: A Case Study in the Classical Methodology of "Saving the Appearances' and Its Limitations." *Archive for History of the Exact Sciences* 26: 221–240.

_____, translator. (1996) *Ptolemy's Theory of Visual Perception: An English Translation of the 'Optics' with Introduction and Commentary.* Transactions of the American Philosophical Society, v01.86, pt.2. Philadelphia: American Philosophical Society.

Smith, Wesley J. (2000) *The Culture of Death: The Assault on Medical Ethics in America.* San Francisco: Encounter Books.

_____. (2010) *A Rat is a Pig is a Dog is a Boy: The Human Cost of the Animal Rights Movement.* New York: Encounter Books.

Sowell, Thomas. (1984) *Civil Rights: Rhetoric or Reality?* New York: William Morrow.

_____. (1987) *A Conflict of Visions: Ideological Origins of Political Struggles.* New York: Basic Books.

_____. (1995a) *Race and Culture: A World View.* New York: Basic Books.

_____. (1995b) *The Vision of the Anointed: Self-Congratulation as a Basis for Social Policy.* New York: Basic Books.

_____. (1997) *Migrations and Cultures: A World View.* New York: Basic Books.

_____. (1999a) *Conquests and Cultures: An International History.* New York: Basic Books.

_____. (1999b) *The Quest for Cosmic Justice.* New York: Simon and Schuster.

_____. (2007) *Basic Economics: A Common Sense Guide to the Economy* (3rd edition). New York: Basic Books.

_____. (2008) *Applied Economics: Thinking Beyond Stage One* (2nd edition). New York: Basic Books.

_____. (2009) *The Housing Boom and Bust.* New York: Basic Books.

Stark, Rodney. (2003) *For the Glory of God: How Monotheism Led to Reformations, Science, Witch-Hunts, and the End of Slavery.* Princeton: Princeton University Press.

_____. (2005) *The Victory of Reason: How Christianity Led to Freedom, Capitalism, and Western Success.* New York: Random House.

Tam, O. H., A. A. Aravin, P. Stein, A. Girard, E. P. Murchison, S. Cheloufi, E. Hodges, M. Anger, R. Sachidanandam, R. M. Schultz, and G. J. Hannon. (2008) "Pseudogene-Derived Small Interfering RNAs Regulate Gene Expression in Mouse Oocytes." *Nature* 453: 534–38.

Taylor, Charles. (1989) *Sources of the Self: The Making of the Modern Identity.* Cambridge: Harvard University Press.

———. (2007) *A Secular Age.* Cambridge: The Belknap Press of Harvard University Press.

Taylor, Jeremy. (2009) *Not a Chimp: The hunt to find the genes that make us human.* Oxford: Oxford University Press.

Taylor, Richard. (1974 [1963]) *Metaphysics* (Second Edition). Englewood Cliffs: Prentice-Hall.

Telford, Maximilian J., and Graham E. Budd. (2003) "The place of phylogeny and cladistics in Evo-Devo research." *International Journal of Developmental Biology* 47: 479–90.

Thaxton, C. B., W. L. Bradley, and R. L. Olsen. (1984) *The mystery of life's origin: reassessing current theories.* New York: Philosophical Library.

Thomson, K. S. (1992) "Macroevolution: The morphological problem." *American Zoologist* 32: 106–12.

Tremblay, A., M. Jasin, and P. Chartrand. (2000) "A Double-Strand Break in a Chromosomal LINE Element Can Be Repaired by Gene Conversion with Various Endogenous LINE Elements in Mouse Cells." *Molecular and Cellular Biology* 20: 54–60.

Trevor-Roper, H. R. (1969 [2001]) *The Crisis of the Seventeenth Century: Religion, the Reformation, and Social Change.* Indianapolis: Liberty Fund.

Trevors, J. T., and D. L. Abel. (2004) "Chance and necessity do not explain the origin of life." *Cell Biology International* 28: 729–39.

Valentine, J. W. (1995) "Late Precambrian bilaterians: grades and clades," in W. M. Fitch and F. J. Ayala, eds., *Temporal and mode in evolution: genetics and paleontology 50 years after Simpson.* Washington, D.C.: National Academy Press, 87–107.

———. (2004) *On the origin of phyla.* Chicago: University of Chicago Press.

Valentine, J. and D. Jablonski. (2003) "Morphological and developmental macroevolution: a paleontological perspective." *International Journal of Developmental Biology* 47: 517–22.

Valentine, J., D. Jablonski and D. Erwin. (1999) "Fossils, Molecules, and Embyros: New Perspectives on the Cambrian Explosion."*Development* 126: 851–59.

Van de Lagemaat, L. N., J. R. Landry, D. L. Mager, and P. Medstrand. (2003) "Transposable Elements in Mammals Promote Regulatory Variation and Diversification of Genes with Specialized Functions." *Trends in Genetics* 19: 530–36.

Voie, Øyvind Albert. (2006) "Biological function and the genetic code are interdependent." *Chaos, Solitons and Fractals* 28: 1000–4.

Voltaire, François Marie Arouet de. (1761) *Works* (translated by T. Smollett, T. Francklin, *et al.*), Volume 1. London: J. Newberry *et al.*

Von Sternberg, Richard M. (2002) "On the Roles of Repetitive DNA Elements in the Context of a Unified Genomic-Epigenetic System." *Annals of the New York Academy of Sciences* 981: 154–88.

———. (2008) "DNA codes and information: Formal structures and relational causes." *Acta Biotheoretica* 56(3): 205–32 (doi:10.1007/s10441–008–9049–6).

———. (2009) "Guy Walks Into a Bar and Thinks He's a Chimpanzee: The Unbearable Lightness of Chimp-Human Genome Similarity." *Evolution News & Views,* May 14 (http://www.evolutionnews.org/2009/05/guy_ walks_into_a_bar_and_think.html; last accessed September 3, 2009).

Von Sternberg, Richard M., and James A. Shapiro. (2005) "How Repeated Retroelements Format Genome Function." *Cytogenetic and Genome Research* 110: 108–16.

Wagner, G. P. (2001) "What is the promise of developmental evolution? Part II: A causal explanation of evolutionary innovations may be impossible." *Journal of Experimental Zoology* (Mol. Dev. Evol.) 291: 305–9.

Wagner, G. P. and L. Altenberg. (1996) "Complex Adaptations and the Evolution of Evolvability." *Evolution* 50(3): 967–76.

Wagner, G. P. and P. F. Stadler. (2003) "Quasi-independence, homology and the Unity-C of type: a topological theory of characters." *Journal of Theoretical Biology* 220: 505–27.

Wallace, William A. (1959) *The Scientific Methodology of Theodoric of Freiberg.* Fribourg: Fribourg University Press.

Watanabe T., Y. Totoki, A. Toyoda, M. Kaneda, S. Kuramochi-Miyagawa, Y. Obata, H. Chiba, Y. Kohara, T. Kono, T. Nakano, M.A. Surani, Y. Sakaki, and H. Sasaki. (2008) "Endogenous siRNAs from Naturally Formed dsRNAs Regulate Transcripts in Mouse Oocytes." *Nature* 453: 539–43.

Weaver, Richard M. (1948) *Ideas Have Consequences.* Chicago: The University of Chicago Press.

Weikart, Richard. (2004) *From Darwin to Hitler: Evolutionary Ethics, Eugenics, and Racism in Germany.* New York: Palgrave Macmillan.

———. (2009) *Hitler's Ethic: The Nazi Pursuit of Evolutionary Progress.* New York: Palgrave Macmillan.

West, John. (2006) *Darwin's Conservatives: The Misguided Quest.* Seattle: Discovery Institute Press.

———. (2007) *Darwin Day in America: How Our Politics and Culture Have Been Dehumanized in the Name of Science.* Wilmington: ISI Books.

Westfall, Richard S. (1958) *Science and Religion in Seventeenth Century England.* New Haven: Yale University Press.

White, Andrew Dickson. (1896) *The History of the Warfare of Science with Theology in Christendom* (2 volumes). New York: Appleton.

White, Lynn. (1940) "Technology and Invention in the Middle Ages," in *Speculum* 15: 141–56.

———. (1962) *Medieval Technology and Social Change.* Oxford: Oxford University Press.

Whitehead, Alfred North. (1925 [1967]) *Science and the Modern World.* New York: The Free Press.

Wiker, Benjamin. (2002) *Moral Darwinism: How We Became Hedonists.* Downers Grove: IVP.

Wiker, Benjamin, and Witt, Jonathan. (2006) *A Meaningful World: How the Arts and Sciences Reveal the Genius of Nature.* Downers Grove: IVP Academic.

Williams, Michael. (1986) "Do We (Epistemologists) Need a Theory of Truth?" *Philosophical Topics* 14: 223–42.

———. (1999) "Meaning and Deflationary truth." *Journal of Philosophy* 96: 545–64.

Wills, Matthew A. (2002) "The tree of life and the rock of ages: Are we getting better at estimating phylogeny?" *BioEssays* 24: 203–7.

Wilson, T. E., U. Grawunder, and M. R. Liebe. (1997) "Yeast DNA Ligase IV Mediates Non-Homologous DNA End Joining." *Nature* 388: 495–98.

Wilson, Woodrow. (1912 [1961]) *The New Freedom*, with an introduction and notes by William Leuchtenberg. Englewood Cliffs: Prentice-Hall, Inc.

Woese, Carl. (1998) "The Universal Ancestor." *Proceedings of the National Academy of Sciences USA* 95: 6854–59.

———. (2002) "On the Evolution of Cells." *Proceedings of the National Academy of Sciences USA* 99: 8742–47.

Wolfram, Herwig. (1997) *The Roman Empire and Its Germanic Peoples.* Berkeley: University of California Press.

Wolterstorff, Nicholas. (2008) *Justice: Rights and Wrongs.* Princeton: Princeton University Press.

Yates, Francis A. (1964) *Giordano Bruno and the Hermetic Tradition.* Chicago: University of Chicago Press.

———. (1979) *The Occult Philosophy in the Elizabethan Age.* London: Routledge & Kegan Paul.

Yockey, H. P. (1978) "A calculation of the probability of spontaneous biogenesis by information theory." *Journal of Theoretical Biology* 67: 377–98.

———. (1992) *Information theory and molecular biology.* Cambridge: Cambridge University Press.

2

SCIENCE WITHOUT GOD:
NATURAL LAWS AND CHRISTIAN BELIEFS

RONALD L. NUMBERS

Nothing has come to characterize modern science more than its rejection of appeals to God in explaining the workings of nature.[1] Numerous scientists, philosophers of science, and science educators have made this claim. In 1982 a United States federal judge, eager to distinguish science from other forms of knowledge, especially religion, spelled out "the essential characteristics of science." At the top of his list appeared the notion that science must be "guided by natural law." No statement, declared the judge, could count as science if it depended on "a supernatural intervention." Five years later the U.S. Supreme Court affirmed the judge's reasoning.[2]

Students of nature have not always shunned the supernatural. It took centuries, indeed millennia, for naturalism to dominate the study of nature, and even at the beginning of the twenty-first century, as we shall see, a tiny but vocal group of "theistic scientists" is challenging what they regard as the arbitrary exclusion of the supernatural from science. In exploring how naturalism came to control the practice of science, I hope to answer some basic questions about the identity and motives of those who advocated it. In particular I want to illuminate the reasons why naturalism, described by some scholars as the great engine driving the secularization of Western society, attracted so much support from devout Christians, who often eagerly embraced it as the method of choice for understanding nature. Naturalization, as we shall see, did not lead inevitably to secularization.

First, however, we need some clarification about terms. Historians have employed the word "naturalism" to designate a broad range of views: from a purely methodological commitment to explaining the workings of nature without recourse to the supernatural, largely devoid of metaphysical implications about God, to a philosophical embrace of materialism tantamount to atheism. When Thomas H. Huxley (1825–95) coined the term "scientific naturalism" in 1892, he used it to describe a philosophical outlook that shunned the supernatural and adopted empirical science as the only reliable basis of knowledge about the physical, social, and moral worlds. Although such metaphysical naturalism, rooted in the findings of science, has played an important role in the history of philosophy and religion, its significance in the history of scientific practice has remained small compared to what has recently come to be called methodological naturalism, the focus of this chapter.[3]

Science without God

1. Naturalism and Natural Philosophy

Recorded efforts to explain naturally what had previously been attributed to the whimsy of gods date back to the Milesian philosophers of the ancient Greek world, who, six centuries before the birth of Christianity, declared such phenomena as earthquakes, lightning, and thunder to be the result of natural causes. A little later, Hippocratic physicians expanded the realm of the natural to include most diseases, including epilepsy, "the sacred disease." As one Hippocratic writer insisted, "Each disease has a natural cause and nothing happens without a natural cause." The first-century Roman philosopher Lucius Annaeus Seneca, ever suspicious of supernatural causation, calmed the fears of fellow citizens by assuring them that "angry deities" had nothing to do with most meteorological or astronomical events: "Those phenomena have causes of their own."[4]

As these scattered examples show, belief in natural causes and the regularity of nature antedated the appearance of Christianity, with its Judaic notion of God as creator and sustainer of the universe. Although inspired by a man regarded as divine and developed in a milieu of miracles, Christianity could, and sometimes did, encourage the quest for natural explanations. Long before the birth of modern science and the appearance of "scientists" in the nineteenth century, the study of nature in the West was carried out primarily by Christian scholars known as natural philosophers, who typically expressed a preference for natural explanations over divine mysteries. During the philosophical awakening of the twelfth century, for instance, Adelard of Bath (ca. 1080–ca. 1150), a much-traveled Englishman familiar with the views of Seneca, instructed his nephew on the virtues of natural explanations:

> I will take nothing away from God: for whatever exists is from Him and because of Him. But the natural order does not exist confusedly and without rational arrangement, and human reason should be listened to concerning those things it treats of. But when it completely fails, then the matter should be referred to God.

A number of other medieval churchmen expressed similar views, on occasion extending the search for natural explanations to such biblical events as Noah's Flood, previously regarded as a miracle.[5]

By the late Middle Ages the search for natural causes had come to typify the work of Christian natural philosophers. Although characteristically leaving the door open for the possibility of direct divine intervention, they frequently expressed contempt for soft-minded contemporaries who invoked miracles rather than searching for natural explanations. The University of Paris cleric Jean Buridan (ca. 1295–ca. 1358), described as "perhaps the most brilliant arts master of the Middle Ages," contrasted the philosopher's search for "appropriate natural causes" with the common folk's erroneous habit of attributing unusual astronomical phenomena to the supernatural. In the fourteenth century the natural philosopher Nicole Oresme (ca. 1320–82), who went on to become a Roman Catholic bishop, admonished that, in discussing various marvels of nature, "there is no reason to take recourse to the heavens, the last refuge of the weak, or demons, or to our glorious God as if He would produce these effects directly, more so than those effects whose causes we believe are well known to us."[6]

Enthusiasm for the naturalistic study of nature picked up in the sixteenth and seventeenth centuries as more and more Christians turned their attention to discovering the so-called secondary causes that God employed in operating the world. The Italian Catholic Galileo Galilei

(1564–1642), one of the foremost promoters of the new philosophy, insisted that nature "never violates the terms of the laws imposed upon her." In a widely circulated letter to the Grand Duchess Christina, written in 1615, Galileo, as a good Christian, acknowledged the divine inspiration of both Holy Scripture and the Book of Nature—but he insisted that interpreters of the former should have no say in determining the meaning of the latter. Declaring the independence of natural philosophy from theology, he asserted "that in disputes about natural phenomena one must begin not with the authority of scriptural passages but with sensory experience and necessary demonstrations."[7]

Far to the west, in England, the Anglican philosopher and statesman Francis Bacon (1561–1626) was preaching a similar message of independence, warning of "the extreme prejudice which both religion and philosophy hath received and may receive by being commixed together; as that which undoubtedly will make an heretical religion, and an imaginary and fabulous philosophy." Christians, he advised, should welcome rather than fear the truth that God operates the world largely, though not exclusively, through natural laws discoverable by human effort. Although conceding that too great an emphasis on natural law might undermine belief in God, he remained confident that further reflection would "bring the mind back again to religion."[8]

The danger Bacon perceived did not take long to materialize. As natural philosophers came to view nature as "a law-bound system of matter in motion," a vast machine running with little or no divine intervention, they increasingly focused on the regularities of nature and the laws of motion rather than on God's intrusions. When the French Catholic natural philosopher René Descartes (1596–1650) boldly constructed a universe of whirling ethereal fluids and speculated how the solar system could have been formed by the action of these vortices operating according to the God-ordained laws of nature, he acquired considerable notoriety for nearly pushing God out of the cosmos altogether. His pious fellow countryman Blaise Pascal (1623–62) accused Descartes, somewhat unfairly, of trying to dispense with God altogether, according Him only "a flip of the finger in order to set the world in motion." Fearing clerical retribution in the years after Galileo's trial, Descartes disingenuously declared his own cosmogony to be "absolutely false."[9]

The English chemist Robert Boyle (1627–91)—as ardent an advocate of the mechanical philosophy as Descartes yet as pious as Pascal—viewed the discovery of the divinely established laws of nature as a religious act. A devout Protestant with great reverence for the Bible, Boyle regarded revelation as "a foreign principle" to the study of the "laws or rules" of nature. He sought to explain natural phenomena in terms of matter in motion as a means of combating pagan notions that granted nature quasi-divine powers, not as a way to eliminate divine purpose from the world. According to the historians Edward B. Davis and Michael Hunter, viewing the cosmos as a "compounded machine" run according to natural laws struck Boyle as being "more consistent with biblical statements of divine sovereignty than older, non-mechanistic views" of an intelligent nature. "By denying 'Nature' any wisdom of its own, the mechanical conception of nature located purpose where Boyle believed it belonged: over and behind nature, in the mind of a personal God, rather than in an impersonal semi-deity immanent within the world." God's customary reliance on natural laws (or secondary causes) did not, in Boyle's opinion, rule out the possibility of occasional supernatural interventions, when God (the primary cause) might act "in special ways to achieve particular ends." This view became common among Christian men of science, as well as among clerics.[10]

No one contributed more to the popular image of the solar system as a giant mechanical device than the University of Cambridge professor of mathematics Isaac Newton (1642–1727), a man of deep, if unorthodox, religious conviction who unblushingly attributed the perfections of the solar system to "the counsel and dominion of an intelligent and powerful Being." Widely recognized as the greatest natural philosopher of all time for his discovery of the role of gravity in the operation of the universe, he insisted that natural knowledge should be based on observations and experiments, not hypotheses. Although he chided Descartes for his attempt to explain the solar system by "mere Laws of Nature," he himself believed that God typically used them "as instruments in his works." In private correspondence he even speculated in Cartesian fashion about how God might have used natural laws to construct the solar system from a "common Chaos."[11]

Endorsed by such publicly religious natural philosophers as Boyle and Newton, the search for natural laws and mechanical explanations became a veritable Christian vocation, especially in Protestant countries, where miraculous signs and wonders were often associated with Catholic superstition. As one Anglican divine complained in 1635, some Protestants were even beginning to question the efficacy of prayer, believing that God, working through second causes, "hath set a constant course in nature."[12]

For ordinary folk the most compelling instances of supernaturalism giving way to naturalism occurred not in physics or chemistry but in such areas as meteorology and medicine, and in explanations of epidemics, eclipses, and earthquakes. Already by the sixteenth century, supernatural explanations of disease had largely disappeared from medical literature except in discussions of epidemics and insanity, which remained etiological mysteries, and venereal diseases, the wages of sin. In writing about the common afflictions of humanity—fractures, tumors, endemic diseases, and such—physicians seldom mentioned God or the devil. Even when discussing the plague, the most dreaded disease of all, they tended merely to acknowledge its supernatural origin before passing quickly to its more mundane aspects. The great French surgeon Ambroise Paré (1510–90), for example, explicitly confined himself to "the natural causes of the plague," saying that he would let divines deal with its ultimate causes. Priests and theologians may have placed greater emphasis on supernatural causes and cures, but in general they too easily accommodated new medical knowledge by maintaining that God usually effected his will through natural agencies rather than by direct intervention. Theological interests thus seldom precluded searching for natural causes or using natural therapies.[13]

The most dramatic, and in some ways revealing, episode in the naturalization of disease occurred in the British colonies of North America in the early 1720s. Christians had long regarded smallpox, a frighteningly deadly and disfiguring disease, as God's ultimate scourge to punish sinners and bring them to their knees in contrition. Thus, when an epidemic threatened to strike New England in 1721, the governor of Massachusetts called for a day of fasting and repenting of the sins that had "stirred up the Anger of Heaven against us." However, the Puritan Cotton Mather (1663–1728), one of the town of Boston's leading ministerial lights, offered an alternative to repentance—inoculation with an attenuated but live form of smallpox—in hopes of preventing the disease by natural means. Having heard rumors of successful inoculations against smallpox in Africa and the Middle East, Mather, a fellow of the Royal Society of London and a natural philosopher in his own right, proposed that the untested, potentially lethal, procedure be tried in Boston. The best-trained physician in town, William Douglass (1691–1752), fearing that inoculation would spread rather than prevent the disease and resenting the meddling of ministers in medical matters, urged Mather to rely instead on "the all-wise Providence of God Almighty" and quit trying to thwart God's will. Mather and five other clerics countered that such reasoning would rule out all medical intervention. Cannot pious persons, they asked, give into the method or practice without having their devotion and subjection to the All-wise Providence of God Almighty call'd in question? . . . Do we not in the use of all means depend on God's blessing? . . . For, what hand or art of Man is there in this Operation more than in bleeding, blistering and a Score more things in Medical Use? which are all consistent with a humble trust in our Great preserver, and a due Subjection to His All-wise Providence.

Besides, added Mather, Dr. Douglass risked violating the biblical commandment against killing by refusing to use inoculation to save lives. After post-epidemic calculations demonstrated the efficacy of inoculation, smallpox, previously a divine judgment, became a preventable disease. Few Christians lamented the metamorphosis. And in generations to come their descendents would give thanks to God as medical science brought cholera, diphtheria, yellow fever, and even venereal diseases under natural control.[14]

The same process occurred in meteorology. Benjamin Franklin (1706–90), who as a teenager in Boston had backed Douglass in his quarrel with Mather over smallpox inoculation, found himself on the opposite side of a similar debate a few decades later, after announcing the invention of a device to prevent another of God's judgments on erring humanity: lightning. When a French cleric denounced lightning rods as an inappropriate means of thwarting God's will, the American printer turned scientific celebrity scornfully replied: "he speaks as if he thought it presumption in man to propose guarding himself against the *Thunders of Heaven!* Surely the Thunder of Heaven is no more supernatural than the Rain, hail or Sunshine of heaven, against the Inconvenience of which we guard by Roofs & Shades without Scruple." Reflective Christians quickly accepted Franklin's logic, and before long lightning rods were adorning the steeples of churches throughout Europe and North America, protecting them not from God's wrath but from a dangerous and capricious natural occurrence.[15]

Reactions to the great earthquakes of 1727 and 1755 further illustrate the inroads of scientific naturalism on popular culture. On the night of October 29, 1727, a violent earthquake shook the northern colonies of America, producing widespread damage to property. Terrified residents, humbled by this apparent display of divine anger, set aside fast days and begged God to forgive them for such sins as Sabbath-breaking, pride, and drunkenness. To promote repentance among his parishioners, Thomas Prince (1687–1758), the Puritan pastor of Boston's Old South Church, preached a sermon titled *Earthquakes the Works of God and Tokens of His Just Displeasure*. In it he conceded that the ignition of gases in the Earth's interior might have touched off the tremors—but then argued that such secondary explanations only demonstrated "how the mighty GOD works invisibly by sensible Causes, and even by those that are extremely little and weak, produces the greatest and most terrible Effects in the World." Cotton Mather similarly insisted on God's active role in producing such catastrophes. "Let the Natural Causes of Earthquakes be what the Wise Men of Enquiry please," he wrote. "They and their Causes are still under the government of Him that is the GOD of Nature." Twenty-eight years later, on November 18, 1755, a second earthquake jolted New England. This time nearly two months passed before community leaders called for a public fast. When the aging Reverend Prince reissued his earlier sermon on earthquakes as tokens of God's displeasure, Professor John Winthrop IV (1714–79) of Harvard College calmed the timorous with the assurance that, although God bore ultimate responsibility for the shaking, natural causes had produced the tremors. "I think Mr. Winthrop has laid Mr. Prince flat on [his] back, and seems to take some pleasure in his mortification," gloated one of the professor's admirers.[16]

Science without God

Years ago the historian Keith Thomas claimed that as the mechanical philosophy pushed God further and further into the distance, it "killed the concept of miracles, weakened the belief in the physical efficacy of prayer, and diminished faith in the possibility of direct divine inspiration." Undoubtedly, some people experienced this effect. The revelations of natural philosophy helped to convince the liberal Boston minister Charles Chauncy (1705–87), for example, that

> God does not communicate either being or happiness to his creatures, at least on this Earth, by an immediate act of power, but by concurring with an established course of nature. What I mean is, he brings creatures into existence, and makes them happy, by the intervention of second causes, operating under his direction and influence, in a stated, regular uniform manner.

But for every liberal such as Chauncy there were scores of Christians who continued to believe in miracles, prayer, and divine inspiration—while at the same time welcoming the evidence that epidemics, earthquakes, and lightning bolts derived from natural causes. And for every natural philosopher who lost his faith to science, many more found their beliefs untouched by the search for natural causes of physical events. For them, the search for natural laws led to a fuller understanding of God, not disbelief.[17]

2. The Decline of Natural Philosophy and the Beginnings of Modern Science

No single event marks the transition from godly natural philosophy to naturalistic modern science, but sometime between roughly the mid-eighteenth and mid-nineteenth centuries students of nature in one discipline after another reached the conclusion that, regardless of one's personal beliefs, supernatural explanations had no place in the practice of science. As we have seen, natural philosophers had often expressed a preference for natural causes, but few, if any, had ruled out appeals to God. In contrast, virtually all scientists (a term coined in the 1830s but not widely used until the late nineteenth century), whether Christians or non-Christians, came by the latter nineteenth century to agree that God-talk lay beyond the boundaries of science.[18]

The roots of secular science can be traced most clearly to Enlightenment France, where the spirit of Descartes lingered. Although not a materialist—he believed in God and the existence of immaterial souls—Descartes had pushed naturalism to the point of regarding animals as mere machines. This extreme form of naturalism scarcely influenced the course of scientific investigation, especially outside of France, but it did spur some French Cartesians to go even further than Descartes. The French physician Julien Offray de La Mettrie (1709–51), for example, suggested that humans are nothing but "perpendicularly crawling machines," a claim that even the French found sensational. While acknowledging God to be the author of the Book of Nature, La Mettrie insisted that "experience and observation," not revelation, should be "our only guides." Such methods might not lead to absolute truth about human nature, but they provided "the greatest degree of probability possible on this subject." Like many men of science to follow, he was willing to trade theological certainty for such scientific probability.[19]

Much more influential on scientific practice was La Mettrie's countryman Georges-Louis Leclerc de Buffon (1707–88), an ardent admirer of Newton and one of the most prominent natural historians in the eighteenth century. Buffon called for an emphasis on the regularities of nature and a renunciation of all appeals to the supernatural. Those studying physical subjects, he argued, "ought as much as possible, to avoid having recourse to supernatural causes." Philosophers "ought not to be affected by causes which seldom act, and whose action is always sudden and violent. These have no place in the ordinary course of nature. But operations uniformly repeated, motions which succeed one another without interruption, are the causes which alone ought to be the foundation of our reasoning." Buffon professed not to care whether such explanations were true—so long as they appeared probable. A theist, though not a practicing Christian, Buffon acknowledged that the Creator had originally set the planets in motion, but considered the fact of no value to the natural philosopher. Buffon's methodological convictions inspired him to propose a natural history of the solar system, based on the notion that a passing comet, "falling obliquely on the sun," had detached the matter that later formed the planets and their satellites. Although his speculations never caught on, some Christian critics justifiably criticized him for trying "to exclude the agency of a divine Architect, and to represent a world begun and perfected merely by the operation of natural, undesigning causes."[20]

A far more successful account of the origin of the solar system came from the Frenchman Pierre-Simon de Laplace (1749–1827), who had abandoned training as a cleric for a career in mathematics and astronomy and eventually became one of the leading men of science in Europe. Finding Buffon's methodology attractive but his theory physically implausible, Laplace in 1796 proposed that the planets had been formed from the revolving atmosphere of the primitive Sun, which, as it cooled and contracted, had abandoned a succession of Saturn-like rings, which had coalesced to form the planets. On the occasion of a visit in 1802 to the country estate of Napoleon Bonaparte, Laplace entertained his host with an account of his so-called nebular hypothesis. When the French leader asked why he had heard no mention of God, Laplace supposedly uttered the much-quoted words "Sire, I have no need of that hypothesis." The only firsthand account of the exchange simply reports Napoleon's disappointment with Laplace's explanation that "a chain of natural causes would account for the construction and preservation of the wonderful system." Either way, there was no mistaking the impious message.[21]

Laplace's thoroughly naturalistic hypothesis, authored by a notorious unbeliever, represented the secularization of natural philosophy at its baldest. Not surprisingly, some Christians denounced Laplace for his transparently atheistic science. This was especially true in the English-speaking world, where the tradition of natural theology remained strong and where French science was widely viewed as tending toward godless materialism. It seemed clear to the Scottish divine Thomas Chalmers (1780–1847), for example, that if "all the beauties and benefits of the astronomical system [could] be referred to the single law of gravitation, it would greatly reduce the strength of the argument of a designing cause." One of the classic arguments for the existence of God rested on the observation that an object appearing to have been made for a particular purpose had been produced by an intelligent and purposeful designer. If the solar system looked as if it were not the result of necessity or of accident, if it appeared to have been made with a special end in mind, then it must have had a designer, namely God. But what happened to the argument if the arrangement had resulted simply from the laws of nature operating on inert matter?[22]

John Pringle Nichol (1804–59), a minister-turned-astronomer at the University of Glasgow and an avid popularizer of the nebular hypothesis, offered one plausible answer. He dismissed Chalmers's argument that "we can demonstrate the existence of a Deity more emphatically from that portion of creation of which we remain ignorant, than from any portion whose processes we have explored," as downright dangerous to Christianity. Such fears, Nichol believed, stemmed from a misunderstanding of the term "law." To him, laws simply designated divine order: "LAW of itself is no substantive or independent power; no causal influence sprung of blind necessity, which carries on events of its own will and energies without command."[23]

As more and more of the artifacts of nature, such as the solar system, came to be seen as products of natural law rather than divine miracle, defenders of design increasingly shifted their attention to the origin of the laws that had proved capable of such wondrous things. Many Christians concluded that these laws had been instituted by God and were evidence of his existence and wisdom. In this way, as John Le Conte (1818–91) of the University of California pointed out in the early 1870s, the cosmogony of Laplace helped to bring about a transformation in the application of the principle of design "from the region of facts to that of laws." The nebular hypothesis thus strengthened, rather than weakened, the argument from design, opening "before the mind a stupendous and glorious field for meditation upon the works and character of the Great Architect of the Universe."[24]

Christian apologists proved equally adept at modifying the doctrine of divine providence to accommodate the nebular hypothesis. Instead of pointing to the miraculous creation of the world by divine fiat, a "special" providential act, they emphasized God's "general" providence in creating the world by means of natural laws and secondary causes. While the Creator's role in the formation of the solar system thus changed, it neither declined nor disappeared, as some timid believers had feared. Daniel Kirkwood (1814–95), a Presbyterian astronomer who contributed more to the acceptance of the nebular hypothesis in America than anyone else, argued that if God's power is demonstrated in sustaining and governing the world through the agency of secondary causes, then it should not "be regarded as derogating from his perfections, to suppose the same power to have been exerted in a similar way in the process of its formation."[25]

God's reliance on secondary causes in the daily operation of the world made it seem only reasonable to suppose that he had at least sometimes used the same means in creating it. "God generally effects his purposes . . . by intermediate agencies, and this is especially the case in dead, unorganized matter," wrote one author:

> If, then, the rains of heaven, and the gentle dew, and the floods, and storms, and volcanoes, and earthquakes, are all products of material forces, exhibiting no evidence of miraculous intervention, there is nothing profane or impious in supposing that the planets and satellites of our system are not the immediate workmanship of the great First Cause. . . . God is still present; but it is in the operation of unchangeable laws; in the sustaining of efficient energies which he has imposed on the material world that he has created; in the preservation of powers, properties, and affinities, which he has transferred out of himself and given to the matter he has made.[26]

To at least one observer, Laplace's cosmogony offered an even more convincing demonstration of divine providence than did the traditional view:

> How much more sublime and exalted a view does it give us of the work of creation and of the Great Architect, to contemplate him evolving a system of worlds from a diffused mass of matter, by the establishment of certain laws and properties, than to consider him as taking a portion of that matter in his hand and moulding it as it were into a sphere, and then imparting to it an impulse of motion.[27]

So eager were many Christians to baptize the nebular hypothesis that they even read it back into the first chapter of Genesis. The Swiss-American Arnold Guyot (1807–84), a highly respected evangelical geographer at Princeton College, took the lead in educating fellow Christians on the close correspondence between the nebular hypothesis and the Mosaic narrative. Assuming the "days" of Genesis 1 to represent great epochs of creative activity, he argued that if the formless "waters" created by God at the beginning actually symbolized gaseous matter, then the light of the first day undoubtedly had been produced by the chemical action resulting from the concentration of this matter into nebulae. The dividing of the waters on the second day symbolized the breaking up of the nebulae into various planetary systems, of which ours was one. During the third epoch the Earth had condensed to form a solid globe, and during the fourth, the nebulous vapors surrounding our planet had dispersed to allow the light of the Sun to shine on the Earth. During the fifth and sixth epochs God had populated the Earth with living creatures. "Such is the grand cosmogony week described by Moses," declared the Christian professor. "To a sincere and unprejudiced mind it must be evident that these great outlines are the same as those which modern science enables us to trace, however imperfect and unsettled the data afforded by scientific researchers may appear on many points."[28]

During the second half of the nineteenth century Guyot's harmonization of the nebular hypothesis and the Bible became a favorite of Christian apologists, especially in America, even winning the endorsement of such staunchly orthodox theologians as Charles Hodge (1797–1878) of Princeton Theological Seminary. "The best views we have met with on the harmony between science and the Bible," Hodge wrote in his immensely influential *Systematic Theology*, "are those of Professor Arnold Guyot, a philosopher of enlarged comprehension of nature and a truly Christian spirit."[29] In Christian classrooms across the United States and Canada, at least, students learned that Laplace's nebular hypothesis, despite its author's own intention, testified to the wisdom and power of God and spoke to the truth of Scripture.

A Preliminary Discourse on the Study of Natural Philosophy (1830) by English astronomer John Herschel (1792–1871) was described by one scholar as "the first attempt by an eminent man of science to make the methods of science explicit." Frequently extrapolating from astronomy, the paradigmatic science of the time, Herschel asserted that sound scientific knowledge derived exclusively from experience—"the great, and indeed only ultimate source of our knowledge of nature and its laws"—which was gained by *observation* and *experiment*, "the fountains of all natural science." Natural philosophy and science (he used the terms interchangeably) recognized only those causes "having a real existence in nature, and not being mere hypotheses or figments of the mind." Although this stricture ruled out supernatural causes, Herschel adamantly denied that the pursuit of science fostered unbelief. To the contrary, he insisted that science "places the existence and principal attributes of a Deity on such grounds as to render doubt absurd and atheism ridiculous." Natural laws testified to God's existence; they did not make him superfluous.[30]

Efforts to naturalize the history of the Earth followed closely on the naturalization of the skies—and produced similar results. When students of Earth history, many of them Protestant ministers, created the new discipline of geology in the early nineteenth century, they consciously sought to reconstruct Earth history using natural means alone. By the 1820s virtually all geologists, even those who invoked catastrophic events, were eschewing appeals to the supernatural. When the British geologist Charles Lyell (1797–1875) set about in the early 1830s to "free the science from Moses," the emancipation had already largely occurred. Nevertheless, his landmark *Principles of Geology* (1830–33) conveniently summed up the accepted methods of doing geology, with the subtitle, *Being an Attempt to Explain the Former Changes of the Earth's Surface, by Reference to Causes Now in Operation,* conveying the main point. As Lyell described his project to a friend, it would in good Herschelian fashion "endeavour to establish the principles of reasoning in the science," most notably the idea "that no causes whatever have from the earliest time to which we can look back to the present ever acted but those now acting & that they never acted with different degrees of energy from that which they now exert."[31]

Lyell applauded his geological colleagues for following the lead of astronomers in substituting "fixed and invariable laws" for "immaterial and supernatural agents":

> Many appearances, which for a long time were regarded as indicating mysterious and extraordinary agency, are finally recognized as the necessary result of the laws now governing the material world; and the discovery of this unlooked for conformity has induced some geologists to infer that there has never been any interruption to the same uniform order of physical events. The same assemblage of general causes, they conceive, may have been sufficient to produce, by their various combinations, the endless diversity of effects, of which the shell of the Earth has preserved the memorials, and, consistently with these principles, the recurrence of analogous changes is expected by them in time to come.

The community of geologists, comprising mostly Christian men of science, thus embraced "the undeviating uniformity of secondary causes"—with one troubling exception.[32]

Like so many of his contemporaries, Lyell, a communicant of the Church of England, for years stopped short of extending the domain of natural law to the origin of species, especially of humans. At times he leaned toward attributing new species to "the intervention of intermediate causes"; on other occasions he appealed to "the direct intervention of the First Cause," thus transferring the issue from the jurisdiction of science to that of religion. He used his *Principles of Geology* as a platform to oppose organic evolution, particularly the theories of the late French zoologist Jean-Baptiste Lamarck (1744–1829), and he professed not to be "acquainted with any physical evidence by which a geologist could shake the opinion generally entertained of the creation of man within the period generally assigned."[33]

The person most responsible for naturalizing the origin of species—and thereby making the problem a scientific matter—was Lyell's younger friend Charles Darwin (1809–82). As early as 1838 Darwin had concluded that attributing the structure of animals to "the will of the Deity" was "no explanation—it has not the character of a physical law & is therefore utterly useless." Within a couple of decades many other students of natural history (or naturalists, as they were commonly called) had reached the same conclusion. The British zoologist Thomas H. Huxley, one of the most outspoken critics of the supernatural origin of species, came to see references to special creation as representing little more than a "specious mask for our ignorance." If the advocates of special creation hoped to win a hearing for their views as science, he argued, then they had an obligation to provide "some particle of evidence that the existing species of animals and plants did originate in that way." Of course, they could not. "We have not the slightest scientific evidence of such unconditional creative acts; nor, indeed, could we have such evidence," Huxley noted in an 1856 lecture; "for, if a species were to originate under one's very eyes, I know of no amount of evidence which would justify one in admitting it to be a special creative act independent of the whole vast chain of causes and events in the universe." To highlight the scientific vacuity of special creation, Darwin, Huxley, and other naturalists took to asking provocatively whether "elemental atoms flash[ed] into living tissues? Was there vacant space one moment and an elephant apparent the next? Or did a laborious God mould out of gathered Earth a body to then endue with life?" Creationists did their best to ignore such taunts.[34]

In his revolutionary *Origin of Species* (1859) Darwin aimed primarily "to overthrow the dogma of separate creations" and to extend the domain of natural law throughout the organic world. He succeeded spectacularly—not because of his clever theory of natural selection (which few biologists thought sufficient to account for evolution), nor because of the voluminous evidence of organic development that he presented, but because, as one Christian reader bluntly put it, there was "literally nothing deserving the name of Science to put in its place." The American geologist William North Rice (1845–1928), an active Methodist, made much the same point. "The great strength of the Darwinian theory," he wrote in 1867, "lies in its coincidence with the general spirit and tendency of science. It is the aim of science to narrow the domain of the supernatural, by bringing all phenomena within the scope of natural laws and secondary causes."[35]

In reviewing the *Origin of Species* for the *Atlantic Monthly,* the Harvard botanist Asa Gray (1810–88) forthrightly addressed the question of how he and his colleagues had come to feel so uncomfortable with a "supernatural" account of speciation. "Sufficient answer," he explained, "may be found in the activity of the human intellect, 'the delirious yet divine desire to know,' stimulated as it has been by its own success in unveiling the laws and processes of inorganic Nature." Minds that had witnessed the dramatic progress of the physical sciences in recent years simply could not "be expected to let the old belief about species pass unquestioned." Besides, he later explained, "the business of science is with the course of Nature, not with interruptions of it, which must rest on their own special evidence." Organic evolution, echoed his friend George Frederick Wright (1838–1921), a geologist and ordained Congregational minister, accorded with the fundamental principle of science, which states that

> we are to press known secondary causes as far as they will go in explanation of facts. We are not to resort to an unknown (i.e., supernatural) cause for explanation of phenomena till the power of known causes has been exhausted. If we cease to observe this rule there is an end to all science and all sound sense.[36]

All of the above statements welcoming Darwinism as a legitimate extension of natural law into the biological world came from Christian scientists of impeccable religious standing: Rice, a Methodist; Gray, a Presbyterian; Wright, a Congregationalist. Naturalism appealed to them, and to a host of other Christians, in part because it served as a reliable means of discovering God's laws. As the Duke of Argyll, George Douglas Campbell (1823–1910), so passionately argued in his widely read book *The Reign of Law* (1867), the natural laws of science represented nothing less than manifestations of God's will. Christians could thus celebrate the rule of natural law as "the delight, the reward, the goal of Science." Even the evangelical theologian Benjamin B. Warfield (1851–1921), a leading defender of biblical inerrancy in turn-of-the-century America, argued that teleology (that is, belief in a divinely designed world) "is in no way inconsistent with . . . a complete system of natural causation."[37]

The adoption of naturalistic methods did not drive most nineteenth-century scientists into the arms of agnosticism or atheism. Long after God-talk had disappeared from the heartland of science, the vast majority of scientists, at least in the United States, remained Christians or theists. Their acceptance of naturalistic science sometimes prompted them, as Jon H. Roberts has pointed out, "to reassess the relationship between nature and the supernatural." For example, when the American naturalist Joseph Le Conte (1821–1901), John Le Conte's brother, moved from seeing species as being "introduced by the miraculous interference of a personal intelligence" to viewing them as the products of divinely ordained natural laws, he rejected all "anthropomorphic notions of Deity" for a God "ever-present, all-pervading, ever-acting." Nevertheless, Le Conte remained an active churchgoing Christian.[38]

The relatively smooth passage of naturalism turned nasty during the last third of the nineteenth century, when a noisy group of British scientists and philosophers, led by Huxley and the Irish physicist John Tyndall (1820–93), began insisting that empirical, naturalistic science provided the only reliable knowledge of nature, humans, and society. Their anticlerical project, aimed at undermining the authority of the established Anglican Church and dubbed "scientific naturalism" by Huxley, had little to do with naturalizing the practice of science but a lot to do with creating positions and influence for men such as themselves. They sought, as the historian Frank M. Turner has phrased it, "to expand the influence of scientific ideas for the purpose of secularizing society rather than for the goal of advancing science internally. Secularization was their goal; science, their weapon."[39]

For centuries, men of science had typically gone out of their way to assure the religious of their peaceful intentions. In 1874, however, during his presidential address to the British Association for the Advancement of Science, Tyndall declared war on theology in the name of science. Men of science, he threatened,

> would wrest from theology, the entire domain of cosmological theory. All schemes and systems which thus infringe upon the domain of science must, in so far as they do this, submit to its control and relinquish all thought of controlling it. Acting otherwise proved always disastrous in the past, and it is simply fatuous today.

In contrast to most earlier naturalists, who had aspired simply to eliminate the supernatural from science while leaving religion alone, Tyndall and his crowd sought to root out supernaturalism from all phases of life and to replace traditional religion with a rational "religion of science." As described by one devotee, this secular substitute rested on "an implicit faith that by the methods of physical science, and by these methods alone, could be solved all the problems arising out of the relation of man to man and of man towards the universe." Despite the protests of Christians that the scientific naturalists were illegitimately trying to "associate naturalism and science in a kind of joint supremacy over the thoughts and consciences of mankind," the linkage of science and secularization colored the popular image of science for decades to come.[40]

The rise of the social sciences in the late nineteenth century in many ways reflected these imperialistic aims of the scientific naturalists. As moral philosophy fragmented into such new disciplines as psychology and sociology, many social scientists, insecure about their scientific standing, loudly pledged their allegiance not only to the naturalistic methods of science but to the philosophy of scientific naturalism as well. Most damaging of all, they turned religion itself into an object of scientific scrutiny. Having "conquered one field after another," noted an American psychologist at the time, science now entered "the most complex, the most inaccessible, and, of all, the most sacred domain—that of religion." Under the naturalistic gaze of social scientists the soul dissolved into nothingness, God faded into an illusion, and spirituality became, in the words of a British psychologist, "'epiphenomenal,' a merely incidental phosphorescence, so to say, that regularly accompanies physical processes of a certain type and complexity." Here, at last, Christians felt compelled to draw the line.[41]

3. Reclaiming Science in the Name of God

By the closing years of the twentieth century, naturalistic methods reigned supreme within the scientific community, and even devout Christian scientists scarcely dreamed of appealing to the supernatural when actually doing science. "Naturalism rules the secular academic world absolutely, which is bad enough," lamented one concerned layman. "What is far worse is that it rules much of the Christian world as well." Even the founders of scientific creationism, who brazenly rejected so much of the content of modern science, commonly acknowledged naturalism as the legitimate method of science. Because they narrowed the scope of science to exclude questions of origins, they typically limited it to the study of "present and reproducible phenomena" and left God and miracles to religion. Given the consensus on naturalism, it came as something of a surprise in the late 1980s and 1990s when a small group of so-called theistic scientists and camp followers unveiled plans "to reclaim science in the name of God." They launched their offensive by attacking methodological naturalism as atheistic or, as one partisan put it, "absolute rubbish"—and by asserting the presence of "intelligent design" (ID) in the universe.[42]

The roots of the intelligent design argument run deep in the soil of natural theology, but its recent flowering dates from the mid-1980s. The guru of ID, Berkeley law professor Phillip E. Johnson (b. 1940), initially sought to discredit evolution by demonstrating that it rested on the unwarranted assumption that naturalism was the only legitimate way of doing science. This bias, argued the Presbyterian layman, unfairly limited the range of possible explanations and ruled out any consideration of theistic factors. Johnson's writings inspired a Catholic biochemist at Lehigh University, Michael J. Behe (b. 1952), to speak out on the inadequacy of naturalistic evolution for explaining molecular life. In *Darwin's Black Box* (1996), Behe maintained that biochemistry had "pushed Darwin's theory to the limit . . . by opening the ultimate black box, the cell, thereby making possible our understanding of how life works." The "astonishing complexity of subcellular organic structure" led him to conclude—on the basis of scientific data, he asserted, "not from sacred books or sectarian beliefs"—that intelligent design had been at work. "The result is so unambiguous and so significant that it must be ranked as one of the greatest achievements in the history of science," he gushed. "The discovery [of intelligent design] rivals those of Newton and Einstein, Lavoisier and Schroedinger, Pasteur and Darwin" and by implication elevated its discoverer to the pantheon of modern science.[43]

The partisans of ID hoped to spark "an intellectual revolution" that would rewrite the ground rules of science to allow the inclusion of supernatural explanations of phenomena. If Carl Sagan (1934–96) and other reputable researchers could undertake a Search for Extra-Terrestrial Intelligence (SETI) in the name of science, they reasoned, why should they be dismissed as unscientific for searching for evidence of intelligence in the biomolecular world? Should logical analogies fail to impress, ID advocates hoped that concerns for cultural diversity might win them a hearing. "In so pluralistic a society as ours," pleaded one spokesman, "why don't alternative views about life's origin and development have a legitimate place in academic discourse?"[44]

This quixotic attempt to foment a methodological revolution in science created little stir within the mainstream scientific community. Most scientists either ignored it or dismissed it as "the same old creationist bullshit dressed up in new clothes." The British evolutionary biologist Richard Dawkins (b. 1941) wrote it off as "a pathetic cop-out of [one's] responsibilities as a scientist." Significantly, the most spirited debate over intelligent design and scientific naturalism took place among conservative Christian scholars. Having long since come to terms with doing science naturalistically, reported the editor of the evangelical journal *Perspectives on Science and Christian Faith,* "most evangelical observers—especially working scientists—[remained] deeply skeptical." Though supportive of a theistic worldview, they balked at being "asked to add 'divine agency' to their list of scientific working tools."[45]

As the editor's response illustrates, scientific naturalism of the methodological kind could—and did—coexist with orthodox Christianity. Despite the occasional efforts of unbelievers to use scientific naturalism to construct a world without God, it has retained strong Christian support down to the present. And well it might, for, as we have seen, scientific naturalism was largely made in Christendom by pious Christians. Although it possessed the potential to corrode religious beliefs—and sometimes did so—it flourished among Christian scientists who believed that God customarily achieved his ends through natural means.[46]

I began this chapter by asserting that nothing characterizes modern science more than its rejection of God in explaining the workings of nature. That statement is, I believe, true. It would be wrong, however, to conclude that the naturalization of science has secularized society generally. As late as the 1990s nearly 40 percent of American scientists continued to believe in a personal God, and, despite the immense cultural authority of naturalistic science, the overwhelming majority of Americans maintained an active belief in the supernatural. During the waning years of the twentieth century, 47 percent of Americans affirmed the statement that "God created man pretty much in his present form at one time within the last 10,000 years," and an additional 35 percent thought that God had guided the process of evolution. Only 11 percent subscribed to purely naturalistic evolution. A whopping 82 percent of Americans trusted "in the healing power of personal prayer," with 77 percent asserting that "God sometimes intervenes to cure people who have a serious illness." Science may have become godless, but the masses—and many scientists—privately clung tenaciously to the supernatural.[47]

Notes

1. I am indebted to Louise Robbins and Richard Davidson for their research assistance in the preparation of this chapter and to Edward B. Davis, Bernard Lightman, David C. Lindberg, David N. Livingstone, Robert Bruce Mullin, Margaret J. Osler, Jon H. Roberts, Michael H. Shank, and John Stenhouse for their criticisms and suggestions.

2. Opinion of Judge William R. Overton in *McLean v. Arkansas Board of Education,* in Marcel C. La Follette, ed., *Creationism, Science, and the Law: The Arkansas Case* (Cambridge: MIT Press, 1983), 60. See also Michael Ruse, "Creation-Science Is Not Science," ibid., 151–54; Stephen Jay Gould, "Impeaching a Self-Appointed Judge," *Scientific American,* July 1992, 118–20; *Science and Creationism: A View from the National Academy of Sciences* (Washington, DC: National Academy Press, 1984), 8; *Teaching about Evolution and the Nature of Science* (Washington, DC: National Academy Press, 1998, chap. 3.

3. Bernard Lightman, "'Fighting Even with Death': Balfour, Scientific Naturalism, and Thomas Henry Huxley's Final Battle," in Alan P Barr, ed., *Thomas Henry Huxley's Place in Science and Letters: Centenary Essays* (Athens, GA: University of Georgia Press, 1997), 338. Historians have also used naturalism to identify a philosophical point of view, common centuries ago but rare today, that attributes intelligence to matter. The phrase "methodological naturalism" seems to have been coined by the philosopher Paul de Vries, then at Wheaton College, who introduced it at a conference in 1983 in a paper subsequently published as "Naturalism in the Natural Sciences," *Christian Scholar's Review* 15 (1986), 388–96. De Vries distinguished between what he called "methodological naturalism," a disciplinary method that says nothing about God's existence, and "metaphysical naturalism," which "denies the existence of a transcendent God."

4. G. S. Kirk and J. E. Raven, *The Presocratic Philosophers* (Cambridge: Cambridge University Press, 1957), 73; G. E. R. Lloyd, *Magic, Reason, and Experience. Studies in the Origin and Development of Greek Science* (Cambridge: Cambridge University Press, 1979), 11, 15–29; John Clarke, *Physical Science in the Time of Nero* (London: Macmillan, 1910), 228; Edward B. Davis and Robin Collins, "Scientific Naturalism," in Gary B. Ferngren, ed., *The History of Science and Religion in the Western Tradition: An Encyclopedia* (New York: Garland, 2000), 201–7.

5. David C. Lindberg, *The Beginnings of Western Science: The European Scientific Tradition in Philosophical, Religious, and Institutional Context, 600 B.C. to A.D. 1450* (Chicago: University of Chicago Press, 1992), 198–202, 228–33. In recent years the history of medieval natural philosophy has generated heated debate. See Andrew Cunningham, "How the Principia Got Its Name; or, Taking Natural Philosophy Seriously," *History of Science* 29 (1991), 377–92; Roger French and Andrew Cunningham, *Before Science: The Invention of the Friars' Natural Philosophy* (Aldershot: Scholar Press, 1996); Edward Grant, "God, Science, and Natural Philosophy in the Late Middle Ages," in Lodi Nauta and Arjo Vanderjagt, eds., *Between Demonstration and Imagination: Essays in the History of Science and Philosophy Presented to John D. North* (Leiden: Brill, 1999), 243–67.

6. Bert Hansen, *Nicole Oresme and the Marvels of Nature* (Toronto: Pontifical Institute of Mediaeval Studies, 1985), 59–60, 137. On Buridan's reputation, see Edward Grant, *The Foundations of Modern Science in the Middle Ages: Their Religious, Institutional, and Intellectual Contexts* (Cambridge: Cambridge University Press, 1996), 144.

7. Maurice A. Finocchiaro, ed., *The Galileo Affair: A Documentary History* (Berkeley and Los Angeles: University of California Press, 1989), 93.

8. Perez Zagorin, *Francis Bacon* (Princeton, NJ: Princeton University Press, 1998), 48–51; Paul H. Kocher, *Science and Religion in Elizabethan England* (San Marino, CA: Huntington Library, 1953), 27–28, 75.

9. Ronald L. Numbers, *Creation by Natural Law: Laplace's Nebular Hypothesis in American Thought* (Seattle:

University of Washington Press, 1977), 3–4; Aram Vartanian, *Diderot and Descartes: A Study of Scientific Naturalism in the Enlightenment* (Princeton, NJ: Princeton University Press, 1953), 52 (quoting Pascal). See also Stephen Gaukroger, *Descartes: An Intellectual Biography* (Oxford: Clarendon Press, 1995), 146–292; Margaret J. Osler, *Divine Will and the Mechanical Philosophy: Gassendi and Descartes on Contingency and Necessity in the Created World* (Cambridge: Cambridge University Press, 1994); and Margaret J. Osler, "Mixing Metaphors: Science and Religion or Natural Philosophy and Theology in Early Modern Europe," *History of Science* 36 (1998), 91–113. On nature as "a law-bound system of matter in motion," see John C. Greene, *The Death of Adam: Evolution and Its Impact on Western Thought* (Ames, IA: Iowa State University Press, 1959).

10. Robert Boyle, *A Free Enquiry into the Vulgarly Received Notion of Nature,* Edward B. Davis and Michael Hunter, eds. (Cambridge: Cambridge University Press, 1996), ix–x, xv, 38–39. See also Rose-Mary Sargent, *The Diffident Naturalist: Robert Boyle and the Philosophy of Experiment* (Chicago: University of Chicago Press, 1995), 93–103. On Christianity and the mechanical philosophy, see Osler, "Mixing Metaphors"; and John Hedley Brooke, *Science and Religion: Some Historical Perspectives* (Cambridge: Cambridge University Press, 1991), chap. 4, "Divine Activity in a Mechanical Universe."

11. Isaac Newton, *Mathematical Principles of Natural Philosophy and His System of the World,* trans. Andrew Motte, revised by Florian Cajori (Berkeley and Los Angeles: University of California Press, 1960), 543–44; H. W. Turnbull, ed., *The Correspondence of Isaac Newton,* 7 vols. (Cambridge: Cambridge University Press, 1959–77), 2: 331–34. For a statement that the solar system could not have been produced by natural causes alone, see Isaac Newton, *Papers and Letters on Natural Philosophy and Related Documents,* I. Bernard Cohen, ed. (Cambridge: Cambridge University Press, 1958), 282. See also David Kubrin, "Newton and the Cyclical Cosmos: Providence and the Mechanical Philosophy," *Journal of the History of Ideas* 28 (1967): 325–46; Richard S. Westfall, *Never at Rest: A Biography of Isaac Newton* (Cambridge: Cambridge University Press, 1980); and Cunningham, "How the *Principia* Got Its Name."

12. Kocher, *Science and Religion,* 97; Keith Thomas, *Religion and the Decline of Magic: Studies in Popular Beliefs in Sixteenth and Seventeenth Century England* (London: Weidenfeld and Nicolson, 1971).

13. This paragraph is based largely on Ronald L. Numbers and Ronald C. Sawyer, "Medicine and Christianity in the Modern World," in Martin E. Marty and Kenneth L. Vaux, eds., *Health/Medicine and the Faith Traditions: An Inquiry into Religion and Medicine* (Philadelphia: Fortress Press, 1982), 133–60, esp. 138–39. On the naturalization of mental illness, see Michael MacDonald, "Insanity and the Realities of History in Early Modern England," *Psychological Medicine* 11 (1981): 11–15; Michael MacDonald, "Religion, Social Change, and Psychological Healing in England, 1600–1800," in W. J. Sheils, ed., *The Church and Healing: Studies in Church History* 19 (Oxford: Basil Blackwell, 1982), 101–25; Ronald L. Numbers and Janet S. Numbers, "Millerism and Madness: A Study of 'Religious Insanity' in Nineteenth Century America," in Ronald L. Numbers and Jonathan M. Butler, eds., *The Disappointed: Millerism and Millenarianism in the Nineteenth Century* (Bloomington, IN: Indiana University Press, 1987), 92–117.

14. John Duffy, *Epidemics in Colonial America* (Baton Rouge, LA: Louisiana State University Press, 1953), 30–32; Otho T. Beall Jr. and Richard H. Shryock, *Cotton Mather: First Significant Figure in American Medicine* (Baltimore: Johns Hopkins Press, 1954), 104–5; Perry Miller, *The New England Mind: From Colony to Province* (Cambridge, MA: Harvard University Press, 1953), 345–66; Maxine van de Wetering, "A Reconsideration of the Inoculation Controversy," *New England Quarterly* 58 (1985): 46–67. On the transformation of cholera from divine punishment to public-health problem, see Charles E. Rosenberg, *The Cholera Years: The United States in 1832, 1849, and 1866* (Chicago: University of Chicago Press, 1962).

15. I. Bernard Cohen, *Benjamin Franklin's Science* (Cambridge, MA: Harvard University Press, 1990), chap. 8, "Prejudice against the Introduction of Lightning Rods."

16. Theodore Hornberger, "The Science of Thomas Prince," *New England Quarterly* 9 (1936): 31; William D. Andrews, "The Literature of the 1727 New England Earthquake," *Early American Literature* 7 (1973): 283.

17. Thomas, *Religion and the Decline of Magic*, 643; Theodore Hornberger, *Scientific Thought in the American Colleges, 1638–1800* (Austin, TX: University of Texas Press, 1945), 82.

18. Andrew Cunningham, "Getting the Game Right: Some Plain Words on the Identity and Invention of Science," *Studies in History and Philosophy of Science* 19 (1988): 365–89. See also Sydney Ross, "Scientist: The Story of a Word," *Annals of Science* 18 (1962): 65–85, which dates the earliest use of the term "scientist" to 1834.

19. Vartanian, *Diderot and Descartes*, 206 (quoting La Mettrie); Kathleen Wellman, *La Mettrie: Medicine, Philosophy, and Enlightenment* (Durham, NC: Duke University Press, 1992), 175, 186–88. See also Aram Vartanian, ed., *La Mettrie's L'Homme Machine: A Study in the Origins of an Idea* (Princeton, NJ: Princeton University Press, 1960); and Jacques Roger, *The Life Sciences in Eighteenth Century French Thought*, ed. Keith R. Benson, trans. Robert Ellrich (Stanford, CA: Stanford University Press, 1997).

20. Georges-Louis Leclerc de Buffon, *Natural History: General and Particular*, trans. William Smellie, 7 vols. (London: W. Strahan & T. Cadell, 1781), vol. 1: 34, 63–82; Numbers, *Creation by Natural Law*, 6–8. On Buffon's religious beliefs, see Jacques Roger, *Buffon: A Life in Natural History*, trans. Sarah Lucille Bonnefoi, ed. L. Pearce Williams (Ithaca, NY: Cornell University Press, 1997), 431. On the naturalism of Buffon and his contemporaries, see Kenneth L. Taylor, "Volcanoes as Accidents: How 'Natural' Were Volcanoes to 18th Century Naturalists?" in Nicoletta Morello, ed., *Volcanoes and History* (Gerlova: Brigati, 1998), 595–618.

21. Roger Hahn, "Laplace and the Vanishing Role of God in the Physical Universe," in Harry Woolf, ed., *The Analytic Spirit. Essays in the History of Science in Honor of Henry Guerlac* (Ithaca, NY: Cornell University Press, 1981), 85–95; Roger Hahn, "Laplace and the Mechanistic Universe," in David C. Lindberg and Ronald L. Numbers, eds., *God and Nature: Historical Essays on the Encounter between Christianity and Science* (Berkeley and Los Angeles: University of California Press, 1986), 256–76.

22. Thomas Chalmers, *On the Power, Wisdom, and Goodness of God as Manifested in the Adaptation of External Nature to the Moral and Intellectual Constitution of Man*, 2 vols. (London: William Pickering, 1835), vol. 1: 30–32. On English attitudes toward French science, see Adrian Desmond, *The Politics of Evolution: Morphology, Medicine, and Reform in Radical London* (Chicago: University of Chicago Press, 1989), chap. 2. This and the following five paragraphs are extracted in large part from Numbers, *Creation by Natural Law*, esp. 79–83, 93.

23. J. P N[ichol], "State of Discovery and Speculation Concerning the Nebulae," *Westminster Review* 25 (1836): 406–8; J. P Nichol, *Views of the Architecture of the Heavens*, 2nd ed. (New York: Dayton & Newman, 1842), 103–5. On Nichol, see also Simon Schaffer, "The Nebular Hypothesis and the Science of Progress," in James R. Moore, ed., *History, Humanity, and Evolution: Essays for John C Greene* (Cambridge: Cambridge University Press, 1989), 131–64.

24. John Le Conte, "The Nebular Hypothesis," *Popular Science Monthly* 2 (1873): 655; "The Nebular Hypothesis," *Southern Quarterly Review* 10 (1846): 228 (Great Architect). It is unclear from the transcript of his lecture whether Le Conte was expressing his own view or quoting the words of someone else.

25. [Daniel Kirkwood], "The Nebular Hypothesis," *Presbyterian Quarterly Review* 2 (1854): 544.

26. "The Nebular Hypothesis," *Southern Quarterly Review*, n.s. 1 (1856): 110–11.

27. "The Nebular Hypothesis," *Southern Quarterly Review* 10 (1846): 240.

28. Arnold Guyot, *Creation; or, The Biblical Cosmogony in the Light of Modern Science* (Edinburgh: T and T Clark, [1883]), 135. For an early version of Guyot's harmonizing scheme, see Arnold Guyot, "The Mosaic Cosmogony and Modern Science Reconciled," (New York) *Evening Post*, 6, 12, 15, and 23 March 1852 (no pagination). See also James D. Dana, "Memoir of Arnold Guyot, 1807–1884," National Academy of Sciences, *Biographical Memoirs*, vol. 2 (1886), 309–47.

29. Charles Hodge, *Systematic Theology*, 3 vols. (New York: Scribner, 1871–73), vol. 1: 573. See also Ronald L. Numbers, "Charles Hodge and the Beauties and Deformities of Science," in John W. Stewart and James

H. Moorhead, eds., *Charles Hodge Revisited: A Critical Appraisal of His Life and Work* (Grand Rapids, MI: Eerdmans, 2002), 77–101.

30. John Frederick William Herschel, *A Preliminary Discourse on the Study of Natural Philosophy*, new ed. (Philadelphia: Lea & Blanchard, 1839), 6, 27–28, 59, 108; Walter E Cannon, "John Herschel and the Idea of Science," *Journal of the History of Ideas* 22 (1961): 215–39; William Minto, *Logic: Inductive and Deductive* (New York: C. Scribner's Sons, 1904), 157, quoted in Laurens Laudan, "Theories of Scientific Method from Plato to Mach: A Bibliographical Review," *History of Science* 7 (1968): 30. On the meaning of science, see Richard Yeo, *Defining Science: William Whewell, Natural Knowledge, and Public Debate in Early Victorian Britain* (Cambridge: Cambridge University Press, 1993).

31. Martin Rudwick, "Uniformity and Progression: Reflections on the Structure of Geological Theory in the Age of Lyell," in Duane H. D. Roller, ed., *Perspectives in the History of Science and Technology* (Norman, OK: University of Oklahoma Press, 1971), 209–27; Roy Porter, *The Making of Geology: Earth Science in Britain, 1660–1815* (Cambridge: Cambridge University Press, 1977), 2; James R. Moore, "Charles Lyell and the Noachian Deluge," *Journal of the American Scientific Affiliation* 22 (1970): 107–15; Leonard G. Wilson, *Charles Lyell: The Years to 1841: The Revolution in Geology* (New Haven, CT: Yale University Press, 1972), 256. On Lyell's indebtedness to Herschel, see Rachel Laudan, *From Mineralogy to Geology: The Foundations of a Science, 1650–1830* (Chicago: University of Chicago Press, 1987), 203–4. See also Martin J. S. Rudwick, "The Shape and Meaning of Earth History," in Lindberg and Numbers, eds., *God and Nature*, 296–321; and James R. Moore, "Geologists and Interpreters of Genesis in the Nineteenth Century," ibid., 322–50.

32. Charles Lyell, *Principles of Geology, Being an Attempt to Explain the Former Changes of the Earth's Surface, by Reference to Causes Now in Operation*, 3 vols. (London: John Murray, 1830–1833), vol. 1: 75–76. On the role of Christian ministers in naturalizing Earth history, see, e.g., Nicolaas A. Rupke, *The Great Chain of History: William Buckland and the English School of Geology (1814–1849)* (Oxford: Clarendon Press, 1983); and Rodney L. Stiling, "The Diminishing Deluge: Noah's Flood in Nineteenth Century American Thought," Ph.D. diss., University of Wisconsin-Madison, 1991.

33. James A. Secord, introduction to *Principles of Geology*, by Charles Lyell (London: Penguin, 1997), xxxii–xxxiii. On the naturalization of physiology, see Alison Winter, "The Construction of Orthodoxies and Heterodoxies in the Early Victorian Life Sciences," in L Bernard Lightman, ed., *Victorian Science in Context* (Chicago: University of Chicago Press, 1997), 24–50.

34. Howard E. Gruber, *Darwin on Man: A Psychological Study of Scientific Creativity, together with Darwin's Early and Unpublished Notebooks*, transcribed and annotated by Paul H. Barrett (New York: Dutton, 1974), 417–18; Leonard Huxley, ed., *The Life and Letters of Thomas Henry Huxley*, 2 vols. (New York: Appleton, 1900), vol. 2: 320; Mario A. di Gregorio, *T. H. Huxley's Place in Natural Science* (New Haven, CT: Yale University Press, 1984), 51; Charles Darwin, *On the Origin of Species*, with an introduction by Ernst Mayr (Cambridge, MA: Harvard University Press, 1966), 483; Adrian Desmond, *Huxley: From Devil's Disciple to Evolution's High Priest* (Reading, MA: Addison-Wesley, 1997), 256. According to Desmond, Huxley first described atoms flashing into elephants in his "Lectures," *Medical Times and Gazette* 12 (1856): 482–83, which includes the statement about lack of evidence for creation. For American references to atomic elephants, see Ronald L. Numbers, *Darwinism Comes to America* (Cambridge, MA: Harvard University Press, 1998), 52. As Neal C. Gillespie has shown, natural theology "had virtually ceased to be a significant part of the day-to-day practical explanatory structure of natural history" well before 1859. Gillespie, "Preparing for Darwin: Conchology and Natural Theology in Anglo-American Natural History," *Studies in History of Biology* 7 (1980): 95.

35. Darwin, *On the Origin of Species*, 466; S. R. Calthrop, "Religion and Evolution," *Religious Magazine and Monthly Review* 50 (1873): 205; [W. N. Rice], "The Darwinian Theory of the Origin of Species," *New Englander* 26 (1867): 608.

36. Asa Gray, *Darwiniana: Essays and Reviews Pertaining to Darwinism* (New York: D. Appleton, 1876), 78–79, from a review first published in 1860 (sufficient answer); Asa Gray, *Natural Science and Religion: Two Lectures Delivered to the Theology School of Yale College* (New York: Charles Scribner's Sons, 1880), 77 (business of science); George F. Wright, "Recent Works Bearing on the Relation of Science to Religion, *No. II*," *Bibliotheca Sacra* 33 (1876): 480.

37. Duke of Argyll, *The Reign of Law,* 5th ed. (New York: John W. Lovell, 1868), 34; B. B. Warfield, review of *Darwinism To-Day,* by Vernon L. Kellogg, in *Princeton Theological Review* 6 (1908): 640–50. I am indebted to David N. Livingstone for bringing the Warfield statement to my attention. For more on Warfield, see David N. Livingstone and Mark A. Noll, "B. B. Warfield (1851–1921): A Biblical Inerrantist as Evolutionist," *Isis* 91 (2000): 283–304. See also Desmond, *Huxley,* 555–56. On the scientific context of Argyll's views, see Nicolaas A. Rupke, *Richard Owen: Victorian Naturalist* (New Haven, CT: Yale University Press, 1994).

38. Jon H. Roberts, *Darwinism and the Divine in America: Protestant Intellectuals and Organic Evolution, 1859–1900* (Madison, WI: University of Wisconsin Press, 1988), 136; Numbers, *Darwinism Comes to America,* 40–41; Joseph Le Conte, "Evolution in Relation to Materialism," *Princeton Review,* n.s. 7 (1881): 166, 174. Jon H. Roberts and James Turner, *The Sacred and the Secular University* (Princeton, NJ: Princeton University Press, 2000), 28–30, discusses "the triumph of methodological naturalism." Of the eighty anthropologists, botanists, geologists, and zoologists elected to the National Academy of Sciences between its creation in 1863 and 1900, only thirteen were known agnostics or atheists; ibid., 41. James H. Leuba's pioneering survey of the beliefs of American scientists, in 1916, turned up only 41.8 percent who affirmed belief "in a God in intellectual and affective communication with humankind, i.e. a God to whom one may pray in expectation of receiving an answer." Fifty years later the figure had declined to 39.3 percent of respondents. See Edward J. Larson and Larry Witham, "Scientists Are Still Keeping the Faith," *Nature* 386 (1997): 435–36.

39. Frank M. Turner, *Between Science and Religion: The Reaction to Scientific Naturalism in Late Victorian England* (New Haven, CT: Yale University Press, 1974), 16. See also Frank M. Turner, *Contesting Cultural Authority: Essays in Victorian Intellectual Life* (Cambridge: Cambridge University Press, 1993); and Bernard Lightman, *The Origins of Agnosticism: Victorian Unbelief and the Limits of Knowledge* (Baltimore: Johns Hopkins University Press, 1987).

40. Frank M. Turner, "John Tyndall and Victorian Scientific Naturalism," in W. H. Brock, N. D. McMillan, and R. C. Mollan, eds., *John Tyndall: Essays on a Natural Philosopher* (Dublin: Royal Dublin Society, 1981), 172; Turner, *Between Science and Religion,* 11–12; Lightman, "'Fighting Even with Death,'" 323–50. See also Ruth Barton, "John Tyndall, Pantheist: A Rereading of the Belfast Address," *Osiris,* 2nd ser., 3 (1987): 111–34.

41. John C. Burnham, "The Encounter of Christian Theology with Deterministic Psychology and Psychoanalysis," *Bulletin of the Menninger Clinic* 49 (1985): 321–52, quotations on 333, 337; Turner, *Between Science and Religion,* 14–16. For a comprehensive discussion of naturalism and the rise of the social sciences, see Roberts and Turner, *The Sacred and the Secular University,* 43–60. On earlier efforts to naturalize psychology, see Roger Cooter, *The Cultural Meaning of Popular Science: Phrenology and the Organization of Consent in Nineteenth Century Britain* (Cambridge: Cambridge University Press, 1984).

42. Phillip E. Johnson, foreword to *The Creation Hypothesis: Scientific Evidence for an Intelligent Designer,* ed. J. P Moreland (Downers Grove, IL: InterVarsity Press, 1994), 7–8; Larry Vardiman, "Scientific Naturalism as Science," *Impact* #293, inserted in *Acts & Facts* 26 (November 1997); Paul Nelson at a conference on "Design and Its Critics," Concordia University, Wisconsin, 24 June 2000. On scientific creationists and naturalism, see Ronald L. Numbers, *The Creationists* (New York: Knopf, 1982), 90, 96, 207–8.

43. Michael J. Behe, *Darwin's Black Box: The Biochemical Challenge to Evolution* (New York: Free Press, 1996), 15, 33, 193, 232–233; "The Evolution of a Skeptic: An Interview with Dr. Michael Behe, Biochemist and Author of Recent Best-Seller, *Darwin's Black Box*," *Real Issue,* November/December 1996, 1, 6–8. For

a brief historical overview of the intelligent design movement, see Numbers, *Darwinism Comes to America,* 15–21.

44. William A. Dembski, "What Every Theologian Should Know about Creation, Evolution, and Design," Center for Interdisciplinary Studies, *Transactions,* May/June 1995, 1–8. See also William A. Dembski, *Intelligent Design: The Bridge between Science and Theology* (Downers Grove, IL: InterVarsity Press, 1999). For a philosophical critique of methodological naturalism from a Christian perspective, see Alvin Plantinga, "Methodological Naturalism?" *Perspectives on Science and Christian Faith,* September 1997, 143–54. For a philosophical critique of intelligent design, see Robert T Pennock, *Tower of Babel: The Evidence against the New Creationism* (Cambridge, MA: MIT Press, 1999).

45. David K. Webb, letter to the editor, *Origins & Design,* spring 1996, 5; J. W. Haas Jr., "On Intelligent Design, Irreducible Complexity, and Theistic Science," *Perspectives on Science and Christian Faith,* March 1997, 1, who quotes Dawkins.

46. John Hedley Brooke has similarly challenged the common notion "that the sciences have eroded religious belief by explaining physical phenomena in terms of natural law"; see his "Natural Law in the Natural Sciences: The Origins of Modern Atheism?" *Science and Christian Belief* 4 (1992), 83.

47. Larson and Witham, "Scientists Are Still Keeping the Faith," 435–36; Eugenie C. Scott, "Gallup Reports High Level of Belief in Creationism" *NCSE Reports,* Fall 1993, 9; John Cole, "Gallup Poll Again Shows Confusion," ibid., Spring 1996, 9; Claudia Wallis, "Faith and Healing," *Time,* 24 June 1996, 63.

3

Varieties of Methodological Naturalism

Ernan McMullin

The terms "naturalism" and "naturalist" acquired new meanings in the late eighteenth century somewhere in the neighborhood of two older sets of terms, "atheism"/"atheist" and "materialism"/"materialist." Perhaps it was because the older pairs had by then acquired for many people a strongly pejorative connotation that the more neutral pair gradually came into use. Not that the equivalence was exact: a naturalist, for example, under one plausible construal could also be a pantheist. In any event, naturalism in the strongest sense could be summed up in the portentous opening lines of Carl Sagan's made-for-TV work, *Cosmos:* The Cosmos is all there is or ever was or ever will be.[1] Naturalism in this sense of that fluid term is thus the claim that Nature (capital "N") is the sum total of existents. There is nothing beyond the physical universe, beyond the vast complex of beings linked together causally in space and time, beings that for reasons that go all the way back to Aristotle we call natures (small "n"). This would rule out the existence of anything *super*natural, anything that lies outside this pattern of nature. In particular, it would rule out the independent existence of a Supreme Being on whom the physical universe is causally dependent. Naturalism is not itself a scientific claim; it would require philosophic argument in its support.

Over recent centuries, a complex approach to nature has gradually developed that has had enormous success. It relies on observation, experiment, and idealization, a language often mathematical in its grammar, hypothesis testing, and much else. A more indirect way of defining naturalism would thus be as the view that all which exists is accessible to this set of methods. Anything not thus accessible, then, could not rightfully be held to exist—again a philosophic, not a scientific, position. The tie between naturalism and the modern sciences of nature is thus quite close. The extraordinary achievements of the latter have become one of the principal arguments in favor of the former.

But the tie between the two has also led numerous philosophers to argue that it can make the naturalist position itself incoherent. If the sciences are made the *sole* source of acceptable knowledge of the real, how is the naturalist claim, quite evidently not itself a scientific one, to be secured? The union of the two sides of naturalism, ontological and methodological, as a single position appears to be self-refuting. In order to evade this challenge it would seem that the methodological constraint would either have to be dropped from the definition or else, perhaps, redefined to allow philosophy to count as "science." I will not pursue this issue further here: it has been exhaustively covered elsewhere in the literature.[2]

Our concern here will be with the methodological version of naturalism only. This version has attained some prominence in the busy science-and-theology literature in recent years and it is in this context primarily that I propose to discuss it. Instead of making a direct ontological claim as to what does or does not exist, methodological naturalism bears instead on the proper approach to be followed in the pursuit of knowledge, taking the methods of the natural sciences in one way or another as the model. For many of those writing about the relationship of science and theology, the term "methodological naturalism" stands for the ways in which science exercises undue epistemic authority vis-à-vis theology, and hence it has become the prompt for heavy attack.[3] What I shall propose is that several different issues have gradually become mixed together in the resultant debate. From the perspective of the religious believer, it would greatly help to disentangle three different positions that could be described as "methodological naturalism," two of which merit support, I shall argue.

1. Strong Methodological Naturalism

The strongest version of methodological naturalism would be the prescription that the *only* legitimate way to gain valid knowledge of the real is to follow the methodology of the natural sciences. This does not make an explicit ontological assertion that whatever is inaccessible to the sciences is unreal, thus leaving agnosticism as an open option. Still, it does come close to ontological naturalism. A version somewhat narrower in scope is of more interest to us here.

This would be the prescription that the only valid source of knowledge *of the natural world* is the natural sciences. This is how we will define strong methodological naturalism (SMN). How would it affect theology and its relationship with the sciences? In his book *Rocks of Ages,* Stephen Jay Gould proposes a particularly vivid form of SMN while at the same time emphasizing that it need not be accompanied by ontological naturalism. He argues for what he somewhat unexpectedly calls a "loving concordat" between science and religion, made possible by the "blessedly simple solution" he is proposing: a demarcation of their proper territories that, if accepted, would rule out conflict between them. He borrows from Catholic theology the notion of a *magisterium* or teaching authority, and then goes on to define NOMA, his principle of non-overlapping magisteria:

> The magisterium of science covers the empirical realm: what is the universe made of (fact) and why does it work this way (theory). The magisterium of religion extends over questions of ultimate meaning and moral value. These two magisteria do not overlap, nor do they encompass all inquiry. . . . To cite the old cliché: science gets the age of rocks, and religion the rock of ages.[4]

Gould admits right away that this neat way of dividing up responsibilities is not likely to meet entire approval on either side. It would exclude, he notes, the notion of miracle as a "special" action on the part of God in the natural order "knowable only through revelation and not accessible to science."[5] It would also exclude any attempt on the part of those more inclined to science than to religion to derive moral truths from scientific findings. But he is quite convinced that the only epistemically responsible mode of approach to the natural world is that of science. And he is, apparently, equally convinced that the spread between fact and value cannot be bridged, so that science cannot provide ultimate meaning to human life nor serve as a ground for moral judgment. There are obvious echoes here of positivism, both the classical Comtean sort and the logical positivism of the century just past.

Gould's proposal, and with it SMN, faces a number of challenges from the perspective of a religious believer. First is that the restriction of religion to the realm of meaning and moral value leaves the status of God as an existent being in doubt. But unless God *is*, unless God is somehow involved in human affairs, it is not at all clear how religion can effectively serve the roles of conveyor of ultimate meaning and arbiter of moral value that Gould assigns to it. Perhaps a naturalized religion? But the restriction set by SMN would not leave even deism open.

A further issue raised by SMN is that of scope. The magisterium of science is apparently to extend to *all* matters that bear on the natural world, completing a process of exclusion that began many centuries ago. There was a time when the Book of Genesis served as a major source of natural knowledge. It is primarily to commentaries on Genesis, indeed, that one turns in order to discover what the views on nature were in the Latin West in the early Middle Ages. In the relative absence of other sources, it is not surprising that the Bible was searched for clues, and debates arose over the "waters above the firmament" and other biblical references of the sort. But the recovery of Aristotelian natural knowledge in the thirteenth century gradually shifted the balance. And a critical moment came when Kepler and Galileo each made a persuasive case for the view that the Scriptures were not intended to be a source of astronomical knowledge.[6] Their arguments would apply by extension to physics, chemistry, geology, and to most parts of the natural sciences that have flourished since their day. In this context, then, SMN would have considerable purchase.

The matter is less clear, however, in other areas. An obvious one is cosmology, recently very much in the limelight. The discovery that the cosmos appears to be "fine-tuned" relative to the possibility of its harboring complex life has set off a lively discussion in which philosophers and theologians have taken part, though of course the impetus prompting the discussion comes originally from the physical sciences. Were the relative strengths of the four fundamental forces to be only slightly different than what they are, for example, the universe would not have been life-bearing, so it is argued. A full discussion of this intriguing topic would lead far afield.[7] In summary, however, there seem to be four possible ways of handling the obvious question that arises.

First is to attribute the coincidence to chance. Luckily for us, so this "solution" would run, there just happens to be a universe of this sort, highly improbable though it would be among the vast number of physically permissible universes. Why not? the supporters of this alternative ask. Second, one could simply withhold judgment in the expectation that later developments in theoretical physics may eliminate the apparent fine-tuning. Third, it may be a selection effect. If a sufficiently large number of universes possessing the requisite range of the relevant parameters were in fact to exist, it would not be surprising that among them there would be one that would satisfy the conditions required for life to develop in it. Finally, if the universe were to be the work of a Creator for whom the development of life, particularly perhaps human life, was a priority, it would be no surprise that it would be of the requisite kind.

If the first two alternatives were to be set aside, the choice would be between two that are ontologically challenging in quite different ways: an enormous set of actually existing universes other than our own, or a Creator concerned that his creation should have the capacity to develop life. The first alternative is at the boundary edge of science as matters stand; the second is an immediate corollary of the Christian doctrine of creation. The choice between them is easy for someone who rejects that doctrine. But how should a Christian choose? It is important to note that an appeal to the creation alternative does not involve any "special" action on the Creator's part, any supplementation of nature at a crucial moment in the evolutionary process, for example. From the Christian perspective it would seem arbitrary, at the very least, to rule out this alternative on methodological grounds, however in the end the choice would be decided. Here is a clear case where SMN proves too sweeping.

Another exception to SMN is knowledge of one particular nature, human nature. Religious believers either take for granted, or else argue vehemently in support of, a whole variety of views regarding human nature and human origins. Believers in the religions of the Book could not permit such knowledge claims to be disallowed. Astronomy, physics, chemistry—fine. But not when it comes to issues of human freedom, moral responsibility, and the rest. There the magisteria of science and of theology clearly overlap and the potential for conflict is evident. Second-order issues of the sort dealt with in anthropology and psychology, like the origins of religious belief itself, would also be part of this overlap.[8]

Philosophers also could raise an objection to SMN in this same context. One of the liveliest areas of debate in contemporary philosophy is the relationship of human mind and human body. Philosophers of a reductionist persuasion would be likely to regard SMN as quite acceptable. But others—a considerable majority at the present moment—would insist on a role for philosophy in determining the ontological state of affairs, whether dualist, non-reductive physicalist, or other. Here again SMN goes too far by effectively ruling out such a role. It is interesting that in the science-and-theology discussions, the relevance of that third factor, philosophy, is rarely acknowledged.

The current debate about methodological naturalism took its origin in the intense controversies in the U.S. over the teaching of evolution in the schools. The campaign to couple with it an alternative account of origins has gone through several phases, first the "creationism" that appealed directly to Genesis as warrant, next the "creation-science" that succeeded it, while laying claim this time to a scientific warrant, then the demand for "equal time," partly on the grounds that Darwinian theory was being taught as though it were the only possible account despite its inadequacies, thus making it no longer a religiously neutral claim. Finally came the "intelligent design"(ID) alternative that now challenges evolutionary orthodoxy. Their defenders objected to the way in which their views were ruled out from the beginning by what many of them came to describe as the "methodological naturalism" of their opponents, sometimes dubbed by them "provisional atheism." What they usually had in mind by this was the *exclusive* competence claimed by scientists regarding the knowledge of nature, what we have called SMN. Proponents of ID in particular, seeking to avoid reference to explicitly religious themes, focus on what they perceive as gaps in the evolutionary account that are insurmountable (they claim) by any agency other than the action of a designing intelligence. They protest strongly that it is, implicitly, SMN that is at fault here. They argue that it is simply being assumed by their opponents that the *only* legitimate approach to these issues is one that conforms to the standard methodology of the natural sciences.

To the extent that this is in fact the case, this criticism is clearly warranted from the perspective of someone who believes that the universe is the work of a Creator. If someone investigating the origin of the first living cells asserts that the action of a designing intelligence can be ruled out from the beginning as an explanation so that there absolutely *must* have been a path to the first cell that is accessible in principle to the standard natural sciences, this does smack of ontological naturalism. It is not a religiously neutral claim. It at least *conceivable*, the believer could insist, that the Creator could have brought about the appearance of the first cell in a way that bypassed the normal processes of nature. To say this is not necessarily to attach any degree of probability to such an occurrence. It is just to say that it is not impossible, because to say that it *is* impossible is, in effect, to deny the possible role of a Creator.

Those who object to methodological naturalism understood as SMN are therefore justified in regarding it as too sweeping from a variety of perspectives. Might there be somewhat more restricted versions that would accomplish more or less the same ends from the point of view of those who find some aspects of methodological naturalism worth retaining, versions that evade the objections to the strong version we have just surveyed? There seem to be at least two. They both take the form of methodological directives.

2. Qualified Methodological Naturalism: Version One (QMN_1)

Instead of conferring an exclusive prerogative on the natural sciences in anything to do with the knowledge of nature, what if we make it a strong *presupposition* of their sufficiency for the task instead? That would be enough to justify the priority routinely given the sciences in investigations of nature and the working assumption animating research into even the most intractable topics that an explanation will eventually be found. It does not actually *exclude* the possibility of, say, the agency of a designing intelligence at "difficult" spots in the evolutionary sequence, but it would allow that possibility to be disregarded (though not denied) by evolutionary scientists as they push onwards. Nor does it exclude the possibility that there could be special topic-areas, like cosmology and the mind-body problem, where philosophy or theology might also have a say.

Qualified Methodological Naturalism (QMN_1) has thus the status of a presupposition, defeasible according to the religious believer, but still sufficient to sustain the actual research behavior of the working scientist. As a methodological prescription, it would appear on the face of it to be something on which nonbelievers and (most) believers could agree. For the former, it might seem weaker than the situation warrants but perhaps acceptable in the light of the fact that the stronger SMN would take its defenders outside the methodology of the sciences whose supposed sufficiency in regard to "natural" topics was their own starting point. What QMN_1 sacrifices relative to the wider coverage of SMN is the exclusionary assumption that set the pot boiling in the first place in the science-and-theology discussions.

In support of QMN_1 one can call on the long history of the natural sciences. This is a pragmatic argument, the sort that is appropriate to a methodological presupposition. As one looks back in time, one can see how methods of investigation of nature sharpened and steadily became more complex as time passed. In the past four centuries, the human reach gradually extended far beyond the immediate deliverances of the human senses, to the distant in space, the distant in scale, the distant in past time.[9] What begins as speculative hypothesis is tested against observation with the aid of powerful instruments, and the increasing reliability of the knowledge gained is attested to in all sorts of indirect ways. Anomalies are the guidepost to further theoretical development. What seems hopelessly discordant eventually falls into place; puzzles that long defy solution eventually yield to patient probing and imaginative stretch. In short, presupposing the sufficiency of the natural sciences as we know them for the task of investigating nature would seem to have the strongest possible historical warrant.

Can one find support for QMN_1 on the side of theology? Here one finds a significant difference between two great traditions in Christian theology, a difference that may help, to some small extent at least, to explain why ID finds so much more support on the side of evangelical Christians than on that of Catholics. In the Thomist tradition, which has done much to shape Catholic theology, a modified version of the Aristotelian notion of nature is fundamental to an understanding of the world around us. Natures are centers of regular activity. Departures from regularity may result from interaction with other natures but in such cases a causal account will still be constructible involving those other natures. Underlying all this is the conserving action of the Creator who has endowed these natures with the causal powers that distinguish one from the other.[10] One must, therefore, distinguish between two sorts of causality here: a primary sort pertaining to the action of the Creator holding each thing in being, and a secondary sort governing the actions of the created natures themselves as they interact with one another.

Built into the Augustinian doctrine of creation in particular is the belief that the potentialities for all that would come later in the development of the natural world were present from the first moment of the creation, the appearance of the human soul excepted.[11] There would be no need for a "special" action on the Creator's part to supplement these potentialities at a later time to make possible some transition that was not within the compass of the original potentialities.[12] This would, of course, suggest the accessibility of the natural order to the sciences that build on the regularities that those potentialities ensure. Natural science is, by definition, an investigation of *natures,* of things that act in the reliably regular ways that underpin the methods of the scientist: the charting of such regularities, the testing of predictions, the postulation of underlying causal structures that would explain these regularities, and all the rest. In the order of nature, then, the natural sciences are sufficient for their task; the presupposition that defines QMN_1 would thus be justified.

Still, departures from that order are possible where the sciences will fail: the Creator is free, after all, to act "beyond nature" (*praeter naturam*). To overcome the negative consequences of the gift of human freedom, the Creator assumes a different relation to the natural order in the context of human need, bringing about departures from that order: miracles that serve as signs not only because of their exceptional character but because of the context in which they occur. They belong to the "order of grace," a salvation history that still goes on. Because they depart from the order of nature—because, that is, of their singularity—they do not yield to the methods of the scientist. But their context will mark them out. Such events leave QMN_1 in place but mark the exceptional contexts where that methodological presupposition may not hold.

In the tradition of Reformed theology, the Thomist conception of nature, with its Aristotelian antecedents, is suspect. Echoing the nominalist criticisms of the fourteenth century, that conception is held to compromise the freedom of the Creator. It imposes too strong a constraint on that freedom: it is as though every created thing *has* to act according to its nature, which would mean, in effect, that God has no freedom in the matter. On the contrary, though a thing may act regularly in a certain way, each of those acts has to be, as it were, separately willed by God; there can be no hint of the necessary about the recurrences they display.[13] Strictly speaking, then, there is no ontology of natures. Thus, the presupposition enshrined in QMN_1 also is suspect. Even a presupposition in this context carries the wrong message.

The nominalists recognized that our knowledge of the physical world depended on those recurrences; these could not be regarded as entirely contingent, or induction understood as generalization over experience would be impossible. So they postulated a distinction between the "absolute power" of God, free of any constraint, and God's "ordained power" (*potentia Dei ordinata*), a power that would uphold the universe as, on the whole, an ordered place. And then they further postulated that God has evidently chosen to exercise the latter power, for the most part at least, in the shaping of the physical world and its operations. Thus, they believed, would the absolute character of the Creator's agency be safeguarded.

In all this discussion, one cannot help but think that though the theological underpinnings of the two accounts are quite different, the universes they describe would not greatly differ. But there is a difference, nevertheless, and it manifests itself particularly in regard to QMN_1. Those who share the Reformed perspective might be reluctant to endorse a presupposition of that sort. In practice, for example, they would be more open to the possibility that the Creator departed from the normal regularities in the shaping of crucial transitions in evolutionary history. They might even believe that the Creator would be *likely* to signal God's constant active presence in that way.[14] Whereas from the Thomist perspective, the presupposition would run the other way. Indeed, for those who share that perspective such an anticipation would risk confusing the order of nature and the order of grace. QMN_1 might therefore serve as a touchstone here to distinguish the two perspectives. And the divergence between them could help to explain in part the vigor of the attack on a generalized "methodological naturalism" by those in the Reformed tradition who express a degree of sympathy for one or another appeal to "special" design on the Creator's part.

3. Qualified Methodological Naturalism: Version Two (QMN_2)

Two contexts in the current science-and-theology debates suggest another related version of methodological naturalism, a qualified version like the foregoing. In both of these contexts, efforts to broaden the usage of the terms "science" and "scientific" have met with strong resistance from those who insist that the terms should be restricted to their normal contemporary usage. Protagonists of Intelligent Design labor to have their arguments for the role of design in the fashioning of the natural kinds labeled as properly "scientific." Likewise, proponents of the epistemology that has come to be associated with Reformed theology argue for the legitimacy of such expressions as "theistic science," where the "science" in question rests in part on premises of distinctively Christian inspiration.[15] Both groups criticize those who oppose such a broadening of extension for what they term a "naturalistic" bias—that is, for relying implicitly on an objectionable sort of methodological naturalism.

How might such a version of naturalism be defined? Perhaps as maintaining the fundamental character of the distinction between the knowledge of nature afforded by the natural sciences (in the usual sense of the term "science") and knowledge that is certified as such in some other way. The possibility of the latter is not denied (SMN has been set aside), so that this is a qualified version of methodological naturalism, QMN_2. Furthermore, the relevance of this distinction is best maintained by retaining the normal sense of the disputed term.

One immediate objection to QMN_2 is that experience has shown that it is difficult, if not impossible, to specify a set of necessary and sufficient conditions that would serve to define a sharp boundary between science and non-science. The borderland is inevitably broad, as is the case with many terms of this sort in popular usage. Nevertheless, it is possible to draw a clear distinction between paradigm examples on both sides of the divide, say, physics and Christian theology. For such as these, QMN_2 can unambiguously apply, while leaving open the possibility of borderline instances where the application of the principle might have to be regarded as tentative.

Two features of "science" in this paradigm sense might be singled out as pragmatic warrant for QMN_2. First is the evidential universality of the claims that science makes, not in the sense that agreement is quickly and easily achieved but that agreed means are at hand for the resolution of controversy.[16] The diverse procedures of the natural sciences have gradually developed and sharpened over the centuries and are aimed mainly at the establishment and explanation of natural regularities.[17] The procedures are universally accessible and are broadly agreed on as rational in view of the sort of knowledge sought.

It is this character in particular that those who maintain the importance of QMN_2 tend to emphasize. In their view, the "science" of "theistic science" lacks this sort of universality, a crucial difference. When a Hindu and a Christian dispute some claim about the nature of time or about human nature, say, and where their disagreement is prompted in part by a theological difference, there is an indubitable lack of universality in the claim that either can make. Religious believers themselves emphasize the importance in this context of faith and of the operation of grace. Though faith and reason are not antithetic, the credibility of a theological assertion is not usually held to be dependent on its universality, that is, on its ability to impose itself on others outside the faith. It seems entirely justifiable to draw attention to this fact, to note the difference between this and the situation in physics or biology, where even though disagreement can be deep and long-lasting, there are ways to move it towards ultimate resolution. It also seems justifiable to mark the difference by means of a distinguishing term—and one is already at hand.

An objection from the opposite perspective is that the term "science" in its original derivation from the Latin term *scientia* was broader in its normal coverage than it is today. It referred to all forms of knowledge held at that time to count legitimately as knowledge, and thus was applicable in such areas as metaphysics and theology. Refusing to allow this broader meaning could, by implication, suggest a denial of epistemic legitimacy to these other areas, hence implicitly a version of SMN.[18]

It must be granted that there is some merit to this objection. There is, however, a strong countervailing argument. The use of the term "science" in the broader sense inevitably raises expectations that cannot be fulfilled. Thus, someone who promotes ID as a "science" will be asked how to test its claims by means of prediction capable of being falsified or to specify how the postulated agency actually works, and so on. The expectation would be that the ID claim should, if properly "scientific," lead to a more and more detailed knowledge of how exactly the alleged design was brought about. And this, of course, could not be done.

This objection brings out the importance of other features besides universality that are commonly associated with the sciences of nature. There is testability: the capacity to try out ways in which the claim might be evaluated. There is the progressive character of the sciences. Through systematic observation that is usually instrument-aided, natural regularities are discovered that lie further and further beyond the range of the human senses. This, in turn, leads to the postulating and testing of underlying structures capable of explaining why those regularities hold. In this way, a process of ontological discovery keeps advancing.

In the currently active debate in the U.S. about ID, two quite different issues are inextricably mingled: the merits of ID as an explanation of specific features of the development of living complexity over Earth's history, and second, the political propriety of teaching those merits in biology classes in public high schools alongside orthodox neo-Darwinian theory as an alternative explanation of those features. It was originally in large part because of this latter issue that proponents of ID insisted on the appellation "science" for their explanation, since only under that label could their arguments be aired in biology classes in the public-school classroom. But this immediately led opponents of their proposal to insist on enforcing the conventional criteria of testability and the rest, criteria that the ID hypothesis could not meet. The proponents of ID were, it would seem, led in significant part for political reasons to project a category for their explanation that in the end it simply did not fit.

Their retort in turn was that the recognition of something as the product of a designing intelligence is a common feature of many sciences—of paleontology and of archaeology, for example. But in response, what is customary in those sciences is a series of follow-up questions that have to be faced before the design hypothesis can be deemed credible, such questions as: how likely would it be that an intelligence of the requisite sort would have been available in that context? When an incised piece of mammoth ivory was unearthed some years ago in the far north of Siberia, a debate arose about a set of scratches on its surface. The fact that it could be dated to a period well before any other signs of human presence in that area became a factor to be reckoned with in the final assessment as to whether the marks were the work of human intelligence. It was not a matter of employing a "filter": excluding the possibility of natural causes (including chance) first, leaving intelligence as the only remaining alternative. All the alternatives had to be considered together. If the ivory fragment could have been dated back, with assurance, to a period before the advent of human beings on Earth, that would presumably have made a major difference!

That is why some discussion of the likelihood of there being an intelligent agent of the required sort, as well as of that agent's being involved, would have to be undertaken in the ID discussion. This could be done, of course, drawing on philosophy alone or on philosophy plus theology. But then the methods of the sciences will not apply to the assessment of the arguments used. This, of itself, does not invalidate the arguments, of course: SMN has already been deemed unacceptable. But the effort to have the ensuing arguments classified as "science" has inevitably led to confusion and a negative verdict on ID in general. Only if the agency of natural causes could be ruled out with certainty in the chosen contexts might the verdict of "design by unknown agents" stand without further exploration. And this outcome, the scientists involved (mindful of QMN_1) are never likely to concede, as they pursue their efforts to illuminate the inescapably shrouded history of the evolutionary process.

The discussion above of QMN_2 has focused in the main on the distinction between the natural sciences and theology; philosophy got only passing mention. The distinction between the sciences and metaphysics raises quite different issues. Since the sciences quite commonly have metaphysical presuppositions and these presuppositions are tested by, among other things, the success or otherwise of the sciences dependent on them (think of the long-held prohibition of action-at-a-distance), the distinction between the two is not so easily drawn. And the criterion of universality that was so crucial in marking off theology would not serve the same role nearly so well for metaphysics or for philosophy generally. But it is the science-theology relationship that is our main concern here.[19]

One further point: The intent of the effort to enlarge the coverage of the term "science" in this context was in part to ensure that other ways of knowing would not be slighted and that the importance of each individual's bringing these different ways together for him- or herself in a single assessment would be acknowledged. Both of these aims are meritorious: QMN_2 in no way challenges them. Unfortunately, we do not have an adequate term to cover the product of this sort of wide-ranging assessment. "Science" won't do, for the reasons we have seen. "Worldview" seems to be the alternative most often preferred but seems a lame choice in some contexts.[20] An agreed term is badly needed here, precisely to bring out the fact that on occasion the individual has to make an epistemic judgment (as, for example, in the fine-tuning example in cosmology) where more determinants than the sciences may have to be taken into account.

In summary, QMN$_2$ seems on pragmatic grounds to be an eminently defensible methodological principle in the context of science-and-theology issues. The priority given to the sciences relative to their universality and their agreed methods of assessment ought not, however, be allowed to undermine the epistemic merits of other ways of knowing that cannot claim these advantages. These merits are best safeguarded in other ways than by appropriating a label that has demonstrably led to so much discord.

4. Reflections

The debates about methodological naturalism that have been so prominent a feature of recent science-and-theology literature could be greatly simplified by drawing the distinctions outlined above and discussing their merits separately, as I have done, instead of attacking or defending something called "methodological naturalism" indiscriminately as though it were a single well-defined position. When defenders of the epistemic rights of theology attack what they call "methodological naturalism," it is first and foremost SMN they have in mind. When scientists defend what *they* call "methodological naturalism," it is most often QMN$_1$ and QMN$_2$ that they are seeking to secure, though they may well have the more extreme SMN in mind, too. Or their use of language may suggest a profession of SMN, even though this may have been unintended. All in all, more precision is called for all round.

The organized campaign in support of ID in the schools has recently (2005) signaled a change in strategy once more. This would be to drop, at least for the moment, the efforts to have the ID arguments recognized as "science," setting aside (in our terms) opposition to QMN$_2$, in effect, and focusing exclusively on highlighting the shortcomings of the orthodox evolutionary theory, as the campaigners see them. The assumption would presumably be that this would point implicitly to the need of an alternative explanation in the relevant contexts, namely a designing Intelligence. On the face of it, there would seem to be a good chance that this could be held to be legal by the courts. And it would, of course, put those who still objected in the invidious position of excluding by fiat the sort of critical discussion that is supposed to be absolutely central to science proper. Indeed, much of the recent sentiment in favor of the ID proposals seems to stem from a suspicion of this sort.

Opponents are still not likely to concede. They could point to the extreme difficulty in organizing and monitoring a fair presentation of the arguments pro and con, especially at the high school level. They would argue that the ID alternative would never be far out of sight in the class discussion. Above all, there would at this point be a strong emotional reaction to any proposal whose lineage (so far as these critics were concerned, at least) could be traced back ultimately to the original Genesis-based program of Henry Morris and his collaborators of half a century ago. In the years intervening, initiatives to modify the public-school biology curriculum have been portrayed to the public by scientific organizations, in terms often overblown, as an out-and-out attack on science generally. While on the other side, opposition to the teaching of evolution has gradually taken on a symbolic role as lightning rod for a much wider set of social discontents. Not an arena, then, where the scientific merits and current shortcomings of broadly neo-Darwinian evolutionary theory are likely to receive an unbiased hearing. . . . And all of this skewed further by the undoubted artificiality of applying a supposedly sharp legal distinction between the interests of church and state to something as closely interwoven as a high school curriculum!

I am making no claim that the distinctions I have attempted to draw here could of themselves lead to a resolution of a confrontation that becomes ever more socially destructive. By now, the emotional divide runs too deep for any easy bridging. I have argued against SMN as too sweeping and have urged the merits of both QMN_1 and QMN_2. Focusing on this tripartite proposal might perhaps serve to let the various participants in this too-often rancorous discussion understand one another's positions a little better. Even this could perhaps serve to moderate the tone of a debate that does no good to the causes of either science or religion.

Notes

1. For a more prosaic definition along the same lines, see, for example, David Armstrong (1980), "Naturalism, Materialism, and First Philosophy," in *Contemporary Materialism,* ed. P. K. Moser and J. D. Trout (London: Routledge, 1995), 35–50 (at 35).
2. Michael Rea, for example, makes it the central theme of his *World without Design: The Ontological Consequences of Naturalism* (Oxford: Clarendon Press, 2002). He argues that in the end there is no coherent way to define naturalism as a "substantive philosophical position." At best, it can only be construed as a "research program." See also Paul K. Moser and David Yandell, "Farewell to Philosophical Naturalism," in *Naturalism: A Critical Analysis,* W. L. Craig and J. P. Moreland, eds. (London: Routledge, 2000), 3–23. There is, of course, a substantial literature of long standing on the naturalist side of the issue.
3. The *locus classicus* is the much-reprinted paper "Methodological Naturalism?" by Alvin Plantinga, *Facets of Faith and Science,* Jitse van der Meer, ed. (Lanham, MD: University Press of America, 1996), vol. 1, 177–221.
4. Stephen Jay Gould, *Rocks of Ages* (New York: Ballantine, 1999), 6.
5. *Rocks of Ages,* 85.
6. See McMullin, "Galileo on Science and Scripture," in *The Cambridge Companion to Galileo,* Peter Machamer, ed. (Cambridge: Cambridge University Press, 1998), 271–347.
7. See, for example, McMullin, "Tuning Fine Tuning" in *Fitness of the Cosmos for Life,* J.D. Barrow et al., eds. (Cambridge: Cambridge University Press, 2008).
8. For a fuller discussion, see McMullin, "Biology and the Theology of Human Nature," in *Controlling Our Destinies: Perspectives on the Human Genome Project,* Phillip Sloan, ed. (Notre Dame, IN: University of Notre Dame Press, 2000), 367–93.
9. Ernan McMullin, "Enlarging the Known World," *Physics and Our View of the World,* J. Hilgevoord, ed. (Cambridge: Cambridge University Press, 1994), 79–113.
10. This, of course, is a substantial modification of the original Aristotelian account. The Thomist account itself has to be revised in the light of its dependence on the long-superseded model of natural science deriving from Aristotle's *Posterior Analytics.*
11. "Evolution and Creation," Introduction to *Evolution and Creation,* E. McMullin, ed. (Notre Dame, IN: University of Notre Dame Press, 1985), 1–56 (see 11–16).
12. See "The Integrity of God's Creation," section 6 of McMullin, "Plantinga's Defense of Special Creation," *Christian Scholar's Review* 21 (1991): 55–79; reprinted in *Intelligent Design Creationism and Its Critics,* R. Pennock, ed. (Cambridge, MA: MIT Press, 2001), 165–96.
13. Plantinga expresses this succinctly: "Natural laws . . . are perhaps best thought of as regularities in the way [God] treats the stuff he has made, or perhaps as counterfactuals of divine freedom," "Methodological naturalism?" 203.
14. Plantinga writes: "According to Scripture, God has often treated what he has made in a way different from the way in which he ordinarily treats it; there is therefore no initial edge to the idea that he would be more likely to have created life in all its variety in the broadly deistic way. In fact, it looks to me as if there is an initial probability on the other side; it is a bit more probable, before we look at the scientific evidence, that the Lord created life and some of its forms—in particular , human life—specially." When Faith and Reason Clash: Evolution and the Bible," *Christian Scholar's Review* 21 (1991), 8–31; reprinted in Pennock, *Intelligent Design Creationism,* 113–45 (see 130). What Plantinga means by "deistic" here is, equivalently, the supposition that the created world was endowed from the beginning with resources sufficient to produce the variety of the living forms that would come later. This would make a "deist" of St. Augustine, his favorite theologian!
15. There is a long history here that goes back to John Calvin, is mediated by Dutch philosopher-theologians Abraham Kuyper and Herman Dooyeweerd of a century ago, and is forcefully represented by Alvin Plantinga

at the present time: "In all the areas of academic endeavor, we Christians must think about the matter at hand from a Christian perspective; we need Theistic Science" ("When Faith and Reason Clash," 141). See the essays making up Part 1 of *Facets of Faith and Science,* Jitse van der Meer, ed., volume 2, for an extended historical and analytic discussion of this tradition.

16. See Ernan McMullin, "Scientific Controversy and Its Termination," in *Scientific Controversies: Case Studies in the Resolution and Closure of Debates,* H. T. Engelhardt and A. Caplan, eds. (Cambridge: Cambridge University Press, 1987), 49–91.

17. In the case of the historical natural sciences, however, sciences like geology, paleontology, evolutionary biology, the intent can be rather different: it may be to establish singulars such as the dates of specific events past or the explanations of such events, making use of laws and theories drawn from the nonhistorical sciences.

18. John Calvin argued for an even stronger version of this objection. From the Bible he derived the claim that having the right God is necessary for obtaining truth or knowledge in the first place, and thus in effect for having access to science proper. The existence of God is something that believers *know,* in the fullest sense of the term "know." Dependence on the Creator is part of the essence of every created thing: thus, if someone proposes to state the essence of a natural kind without acknowledging this dependence, the definition is necessarily incomplete. One can discern the influence of this view on the modern Reformed tradition deriving from Kuyper. See Roy A. Clouser, "On the General Relation of Religion, Metaphysics and Science," *Facets of Faith and Science* 2: 57–79 (see 60–70).

19. After an extended review of the possibilities, Del Ratzsch, one of the most astute defenders of a "Christian natural science," reaches an interesting conclusion: "It seems to me then that it is perfectly possible for there to be a uniquely Christian science, that such a science could be rationally defensible and indeed that there could in principle even be a rationally or spiritually obligatory uniquely Christian science. Whether such a possibility in principle is a real possibility, much less a current actuality, or even that Christians should yearn for such, does not seem to me obvious and at the least deserves a case." In "Tightening Some Loose Screws: Prospects for a Christian Natural Science," *Facets of Faith and Science* 2: 175–90 (see 189).

20. Stephen Wykstra prefers the term "worldview" when he is defending what he calls the "integrationist conception of science," the methodological claim that "scientists should *integrate* religious outlooks into scientific theorizing." In two careful essays, he examines the important role that religious beliefs have played in the earlier history of science. He is cautious in regard to the extent that this would show "whether worldviews [of this kind] have relevance to options in guiding commitments that are open issues for theorizing in our time—not merely in Newton's." "Have Worldviews Shaped Science? A Reply to Brooke," in *Facets of Faith and Science* 1: 91–111 (see 92); "Should Worldviews Shape Science?" *op. cit.,* 2: 162.

4

Sauce for the Goose:
Intelligent Design, Scientific Methodology, and the Demarcation Problem

Stephen C. Meyer

In December 2005, Judge John E. Jones III ruled that a Dover, Pennsylvania, school district could not tell its biology students about a book in the school library that explained the theory of intelligent design. The judge based his decision on the testimony of expert witnesses—two philosophers, Robert Pennock and Barbara Forrest—who argued that the theory of intelligent design is not scientific *by definition*.[1] Since it is not scientific, the judge reasoned, it must be religious. As he put it in his ruling, "Since ID is not science the conclusion is inescapable that the only real effect of the ID Policy is the advancement of religion."[2] Therefore, he ruled, telling students about the theory of intelligent design would violate the establishment clause of the U.S. Constitution.

Many people have heard about the theory of intelligent design only from news reports about the Dover trial in 2005.[3] Naturally, such reports about the trial and the judge's decision have strongly influenced public perceptions of the theory. For many people, if they know anything at all about the theory, they know—or think they know—that intelligent design is "religion masquerading as science."

I encounter this perception nearly every time I speak about the evidence for intelligent design, whether on university campuses or in the media. When I present the evidence for intelligent design, critics do not typically try to dispute my specific empirical claims. They do not dispute that DNA contains specified information, or that this type of information always comes from a mind, or that competing materialistic theories have failed to account for the DNA enigma. Nor do they even dispute my characterization of the historical scientific method or that I followed it in formulating my case for intelligent design as the best explanation for the evidence. Instead, critics simply insist that intelligent design "is just not science," sometimes even citing Judge Jones as their authority.

Since Jones is a lower-level district judge who entered the trial with no apparent background in either science or the history and philosophy of science, and since he made several clear factual errors[4] in his ruling, it would be easy to dismiss his opinion. Jones rendered this response all the more tempting by telling one reporter, apparently in all seriousness, that during the trial he planned to watch the old Hollywood film *Inherit the Wind* for historical background.[5] *Inherit the Wind* is a thinly veiled fictional retelling of the 1925 Scopes "Monkey

Trial." But as historian of science Edward Larson has shown in his Pulitzer Prize-winning *Summer for the Gods,* the drama is grossly misleading and historically inaccurate. Clearly, Jones had little, if any, relevant expertise from which to make a judgment about the merits or scientific status of intelligent design.

His opinion, however, reflected a much broader consensus among scientific and academic critics of intelligent design. Indeed, it was later discovered that Jones lifted more than 90 percent of his discussion of "Whether ID Is Science" in his lengthy opinion virtually verbatim from an American Civil Liberties Union brief submitted to him before his ruling. The ACLU brief, in turn, recapitulated the most common reasons for challenging the scientific status of intelligent design based upon the testimony of the ACLU's own expert witnesses.[6] Thus, the Jones opinion and the witnesses who influenced it effectively expressed an entrenched view common not only among members of the media, but within the scientific establishment at large.

But why isn't the theory of intelligent design scientific? On what basis do critics of the theory make that claim? And is it justified?

1. A Matter of Definitions?

As a philosopher of science, I've always thought there was something odd and even disingenuous about the objection that intelligent design is not scientific. The argument shifts the focus from an interesting question of truth to a trivial question of definition. To say that an idea or theory does or does not qualify as science implies an accepted definition of the term by which to make that judgment. But to say that a claim about reality "is not science" according to some definition says nothing about whether the claim is true—unless it can be assumed that only scientific theories are true. A definition of science does not, by itself, tell us anything about the truth of competing statements, but only how to classify them (whether as scientific or something else, such as philosophical, historical, or religious statements).

So, at one level, I regarded the debate about whether intelligent design qualifies as science as essentially a semantic dispute, one that distracts attention from significant questions about what actually happened in the past to cause life to arise. Does life exhibit evidence of intelligent design or just apparent design? Did life arise by undirected processes, or did a designing intelligence play a role? Surely such questions are not settled by defining one of the competing hypotheses as "unscientific" and then refusing to consider it.

At another level, the debate is tacitly about the basis of the theory itself. Since the term "science" connotes a rigorous experimental or empirical method for studying nature, denying that an idea is scientific implies that rigorous empirical methods played no role in its formulation. To emphasize this impression, many critics of intelligent design insist that the theory is not testable and, for this reason, is neither rigorous nor scientific.[7] Because many people assume that only "the" scientific method produces justified conclusions, the charge that the theory isn't science seems to justify dismissing it as merely a subjectively based opinion or belief. The objection "ID isn't science" is code for "It isn't true," "It's disreputable," and "There is no evidence for it."

That is why the claim that intelligent design is not science—repeated often and with great presumed authority—has led many to reject it before considering the evidence and arguments for it. I realized that in order to make my case—and open minds to the evidence in favor of it—I needed to defend the theory of intelligent design against this charge. To do so, indeed to defend any theory against this charge and to do so with intellectual integrity, requires one to navigate some treacherous philosophical waters. To claim that intelligent design is science implicitly invokes a definition of science—some understanding of what science is. But which definition?

Because of my background, I knew that historians and philosophers of science—the scholars who study such questions—do not agree about how to define science.[8] Many doubt there is even a single definition that can characterize all the different kinds of science. In the philosophy of science this is known as the "demarcation problem," the problem of defining science and distinguishing (or "demarcating") it from "pseudoscience," metaphysics, history, religion, or other forms of thought or inquiry.

Typically philosophers of science have tried to define science and distinguish it from other types of inquiry (or systems of belief) by studying the methods that scientists use to study nature. But that's where the trouble started. As historians and philosophers of science studied the methods that scientists use, they realized that scientists in different fields use different methods.

This, incidentally, is why historians and philosophers of science are generally better qualified to adjudicate the demarcation question than scientific specialists—such as inorganic chemists, for example. As they say of the catcher in baseball, the philosopher and historian of science has a view of the whole field of play, meaning he or she is less likely to fall into the error of defining all of science by the practices used in one corner of the scientific world. I already had some inkling of this from my work as a geophysicist. I was aware that historical and structural geology use distinct (if partially overlapping) methods. But as I delved into the demarcation question, I discovered that different sciences use a wide variety of methods.

Some sciences perform laboratory experiments. Some do not. Some sciences name, classify, and organize natural objects; some sciences seek to discover natural laws; others seek to reconstruct past events. Some sciences seek to formulate causal explanations of natural phenomena. Some provide mathematical descriptions of natural phenomena. Some sciences construct models. Some explain general or repeatable phenomena by reference to natural laws or general theories. Some study unique or particular events and seek to explain them by reference to past (causal) events.

Some sciences test their theories by making predictions; some test their theories by assessing their explanatory power; some test their theories by assessing both explanatory power and predictive success. Some methods of scientific investigation involve direct verification; some employ more indirect methods of testing. Some test theories in isolation from competing hypotheses. Some test theories by comparing the predictive or explanatory success of competing hypotheses. Some branches of science formulate conjectures that cannot yet be tested at all. Some sciences study only what can be observed. Some sciences make inferences about entities that cannot be observed. Some sciences reason deductively; some inductively; some abductively. Some use all three modes of inference. Some sciences use the hypothetico-deductive method of testing. Some use the method of multiple competing hypotheses.

This diversity of methods has doomed attempts to find a single definition (or set of criteria) that accurately characterizes all types of science by reference to their methodological practices. Thus, philosophers of science now talk openly about the "demise" of attempts to demarcate or define science by reference to a single set of methods.[9]

To say that an idea, theory, concept, inference, or explanation is or isn't scientific requires a particular definition of science. Yet if different scientists and philosophers of science could not agree about what the scientific method is, how could they decide what did and did not qualify as science? And how could I argue that the theory of intelligent design is scientific, if I could not say what I meant by "science"? Conversely, how could critics of intelligent design assert that intelligent design is not science without articulating the standard by which they made this judgment? How could any headway in this debate be made without an agreed-upon definition?

I discovered that though it was difficult to define science by reference to a single definition or set of methodological criteria, it was not difficult to define science in such a way that either acknowledged the diversity of methodological practices or refused to specify which method made a discipline scientific. Such an approach allows science to be defined more broadly as, for instance, "a systematic way of studying nature involving observation, experimentation, and/or reasoning about physical phenomena." So far, so good. The difficulty has come when scholars tried to equate science with *a particular* systematic method of studying nature to the exclusion of other such methods.

The situation was not hopeless, however. I discovered that although it was impossible to describe the rich variety of scientific methods with a single definition, it was possible to characterize the methodological practices of specific disciplines or types of science. This made sense. It was precisely the diversity of scientific methods that made defining science as a whole difficult in the first place. Focusing on a single established scientific method as the relevant standard of judgment eliminated the practical problem of deciding how to assess the scientific status of a theory without an established definition of science. Furthermore, from my own studies, I knew the methodological practices of the sciences directly relevant to the questions I was pursuing—the sciences that investigate the causes of particular events in the remote past. Stephen Jay Gould called these sciences the *historical sciences*.[10] I knew that the inference to design followed from a rigorous application of the logical and methodological guidelines of these disciplines. If one carefully follows these guidelines in constructing a case for design, one is entitled to conclude there is a good (if definition-dependent) reason to regard intelligent design as a scientific—and, specifically, historically scientific—theory. In fact, there are several such reasons.

Reason 1: The Case for ID Is Based on Empirical Evidence

The case for intelligent design, like other scientific theories, is based upon empirical evidence, not religious dogma. Contrary to the claims of Robert Pennock,[11] one of the expert witnesses in the Dover trial, design theorists have developed specific empirical arguments to support their theory. To name just one example, I have developed an argument for intelligent design based on the discovery of digital information in the cell.[12] In addition, other scientists now see evidence of intelligent design in the "irreducible complexity" of molecular machines and circuits in the cell,[13] the pattern of appearance of the major groups of organisms in the fossil record,[14] the origin of the universe and the fine-tuning of the laws and constants of physics,[15] the fine-tuning of our terrestrial environment,[16] the information processing system of the cell, and even in the phenomenon known as "homology" (evidence previously thought to provide unequivocal support for neo-Darwinism).[17] Critics may disagree with the conclusions of these design arguments, but they cannot reasonably deny that they are based upon commonly accepted observations of the natural world. Since the term "science" commonly denotes an activity in which theories are developed to explain observations of the natural world, the empirical, observational basis of the theory of intelligent design provides a good reason for regarding intelligent design as a scientific theory.

Reason 2: Advocates of ID Use Established Scientific Methods

The case for intelligent design follows from the application of not one, but two separate systematic methods of scientific reasoning—methods that establish criteria for determining when observed evidence supports a hypothesis. The primary method, the method of multiple competing hypotheses, may be used to justify an inference to intelligent design as the best explanation for the origin of biological information. This method is a standard method of scientific reasoning in several well-established scientific disciplines. Advocates of intelligent design have also developed another method that complements the method of multiple competing hypotheses.

In *The Design Inference* (and in subsequent works), William Dembski established criteria by which intelligently designed systems can be identified by the kinds of patterns and probabilistic signatures they exhibit. On the basis of these criteria, Dembski developed a comparative evaluation procedure—his explanatory filter[18]—to guide our analysis and reasoning about natural objects and artifacts and to help investigators decide among three different types of explanations: chance, necessity, and design.[19] As such, it constitutes a rigorous, systematic, evidence-based method for detecting the effects of intelligence, again suggesting a good reason to regard intelligent design as scientific in accord with common definitions of the term. In more recent work, Dembski has collaborated with Robert Marks to show that Darwinian evolution cannot generate new biological information, but rather, insofar as it is operative in biology, is itself inherently teleological and derivative of an intelligent source characterizable in precise information-theoretic terms.[20]

Reason 3: ID Is a Testable Theory

Most scientists and philosophers of science think that the ability to subject theories to empirical tests constitutes an important aspect of any scientific method of study. But for a theory to be testable, there must be some evidential grounds by which it could be shown to be incorrect or inadequate. And, contrary to the repeated claims of its detractors, the theory of intelligent design *is* testable. In fact, it is testable in several interrelated ways.

First, like other scientific theories concerned with explaining events in the remote past, intelligent design is testable by comparing its explanatory power to that of competing theories. Darwin used this method of testing in *On the Origin of Species*. In the presentation of the case for intelligent design in my book *Signature in the Cell,* I tested the theory in exactly this way by comparing the explanatory power of intelligent design against that of several other classes of explanation. That the theory of intelligent design can explain the origin of biological information (and the origin of the cell's interdependent information processing system) better than its materialistic competitors shows that it has passed an important scientific test.

This comparative process is not a hall of mirrors, a competition without an external standard of judgment. The theory of intelligent design, like the other historical scientific theories it competes against, is tested against our knowledge of the evidence in need of explanation and our knowledge of the cause-and-effect structure of the world. Evaluations of "causal adequacy" guide historical scientific reasoning and help to determine which hypothesis among a competing group of hypotheses has the best explanatory power. Considerations of causal adequacy provide an experience-based criterion by which to test—accept, reject, or prefer—competing historical scientific theories. When such theories cite causes that are known to produce the effect in question, they meet the test of causal adequacy; when they fail to cite such causes, they fail to meet this test.[21]

Since empirical considerations provide grounds for rejecting historical scientific theories or preferring one theory over another, such theories are clearly testable. Like other historical scientific theories, intelligent design makes claims about the causes of past events, thus making it testable against our knowledge of cause and effect. Moreover, because experience shows that an intelligent agent is not only a known, but also the *only* known cause of specified, digitally encoded information, the theory of intelligent design passes two critical tests: the tests of causal adequacy and causal existence as the explanation for any observed occurrence of such information. Precisely because intelligent design uniquely passes these tests, I argue that it stands as the best explanation for the origin of DNA.

Finally, though *historical* scientific theories typically do not make predictions that can be tested under controlled laboratory conditions, they do sometimes generate discriminating predictions about what we should find in the natural world—predictions that enable scientists to compare them to other historical scientific theories. The theory of intelligent design has generated a number of such discriminating empirical predictions. These predictions not only distinguish the theory of intelligent design from competing evolutionary theories; they also serve to confirm the design hypothesis rather than its competitors.[22]

Reason 4: The Case for ID Exemplifies Historical Scientific Reasoning

There is another good, if convention-dependent, reason for classifying intelligent design as a scientific theory. Not only do scientists use systematic methods to infer intelligent design; the specific methods they use conform closely to established patterns of inquiry in the historical scientific disciplines—disciplines that try to reconstruct the past and explain present evidence by reference to past causes rather than trying to classify or explain unchanging laws and properties of nature. Indeed, the theory of intelligent design and the patterns of reasoning used to infer and defend it exemplify each of the four key features of a historical science.

A Distinctive Historical Objective

Historical sciences focus on questions of the form, "What happened?" or "What caused this event or that natural feature to arise?" rather than questions of the form, "How does nature normally operate or function?" or "What causes this general phenomenon to occur?"[23] Those who postulate the past activity of an intelligent designer do so as an answer, or as a partial answer, to distinctively historical questions. The theory of intelligent design attempts to answer a question about what caused certain features in the natural world to come into existence—such as the digitally encoded, specified information present in the cell. It attempts to answer questions of the form "How did this natural feature arise?" as opposed to questions of the form "How does nature normally operate or function?"

A Distinctive Form of Inference

The historical sciences use inferences with a distinctive logical form. Unlike many non-historical disciplines, which typically infer generalizations or laws from particular facts (induction), historical sciences employ *abductive* logic to infer a past event from a present fact or clue. Such inferences are also called "retrodictive." As Gould put it, the historical scientist infers "history from its results."[24] Inferences to intelligent design exemplify this abductive and retrodictive logical structure. They infer a past unobservable cause (in this case, an instance of creative mental action or agency) from present facts or clues in the natural world, such as the specified information in DNA, the irreducible complexity of certain biological systems, and the fine-tuning of the laws and constants of physics.[25]

A Distinctive Type of Explanation

Historical sciences usually offer causal explanations of particular events, not law-like descriptions or theories describing how certain kinds of phenomena—such as condensation or nuclear fission—generally occur. In historical explanations, past causal events, not laws or general physical properties, do the main explanatory work.[26] To explain a dramatic erosional feature in eastern Washington called the Channeled Scablands, a historical geologist posited an event: the collapse of an ice dam and subsequent massive flooding. This and other historical scientific explanations emphasize past events as causes for subsequent events and/or present features of the world.

The theory of intelligent design offers such a distinctively historical form of explanation. Theories of design invoke the act or acts of an agent and conceptualize those acts as causal events, albeit ones involving mental rather than purely physical entities. Advocates of design postulate past causal events (or a sequence of events) to explain the origin of present evidence or clues, just as proponents of chemical evolutionary theories do.

Use of the Method of Multiple Competing Hypotheses

Historical scientists do not mainly test hypotheses by assessing the accuracy of the predictions they make under controlled laboratory conditions. Using the method of multiple competing hypotheses, historical scientists test hypotheses by comparing their explanatory power against that of their competitors. And advocates of the theory of intelligent design also use this method.

In sum, the theory of intelligent design seeks to answer characteristically historical questions, it relies upon abductive/retrodictive inferences, it postulates past causal events as explanations of present evidence, and it is tested indirectly by comparing its explanatory power against that of competing theories. Thus, the theory of intelligent design exhibits each of the main features of a historical science, suggesting another reason to regard it as scientific.

REASON 5: ID ADDRESSES A SPECIFIC QUESTION IN EVOLUTIONARY BIOLOGY

There is another closely related reason to regard intelligent design as a scientific theory. It addresses a key question that has long been part of historical and evolutionary biology: How did the appearance of design in living systems arise? Both Darwin and contemporary evolutionary biologists such as Francisco Ayala, Richard Dawkins, and Richard Lewontin acknowledge that biological organisms *appear* to have been designed.[27] Nevertheless, for most evolutionary theorists, the appearance of design is considered illusory, because they are convinced that the mechanism of natural selection acting on random variations (and/or other similarly unguided mechanisms) can fully account for the appearance of design in living organisms.[28]

In *On the Origin of Species,* Darwin sought to show that natural selection has creative powers comparable to those of intelligent human breeders. In doing so, he sought to refute the design hypothesis by providing a materialistic explanation for the origin of the appearance of design in living organisms. Following Aleksandr Oparin, chemical evolutionary theorists have sought to provide similarly naturalistic accounts for the appearance of design in the simplest living cells.

Is the appearance of design in biology real or illusory? Clearly, there are two possible answers to this question. Neo-Darwinism and chemical evolutionary theory provide one answer, and the competing theory of intelligent design provides the opposite answer. By almost all accounts the classical Darwinian answer to this question—"The appearance of design in biology does not result from actual design"—has long been considered a scientific proposition. But what is the status of the opposite answer? If the proposition "Jupiter is made primarily of methane gas" is a scientific proposition, then the proposition "Jupiter is not made primarily of methane gas" would seem to be a scientific proposition as well. The negation of a proposition does not make it a different type of claim. Similarly, the claim "The appearance of design in biology does not result from actual design" and the claim "The appearance of design in biology does result from actual design" are not two different kinds of propositions; they are two different answers to the same question, a question that has long been part of evolutionary biology and historical science. If one of these propositions is scientific, then it would seem that the other is scientific as well.[29]

Reason 6: ID Is Supported by Peer-Reviewed Scientific Literature

Critics of the theory of intelligent design often claim that its advocates have failed to publish their work in peer-reviewed scientific publications. For this reason, they say the theory of intelligent design does not qualify as a scientific theory.[30] According to these critics, science is what scientists do. Since ID scientists don't do what other scientists do—namely, publish in peer-reviewed journals—they are not real scientists and their theory isn't scientific either.

Critics of the theory of intelligent design made this argument before and during the Dover trial in support of the ACLU's case against the Dover school board policy. For example, Barbara Forrest, a philosophy professor from Southeastern Louisiana State University and one of the expert witnesses for the ACLU, asserted in a *USA Today* article before the trial that design theorists "aren't published because they don't have any scientific data."[31] In her expert witness report in support of the ACLU, Forrest also claimed that "there are no peer-reviewed ID articles in which ID is used as a biological theory in mainstream scientific databases such as MEDLine."[32] Judge Jones apparently accepted such assertions at face value. In his decision, he stated not once, but five separate times, that there were no peer-reviewed scientific publications supporting intelligent design.[33]

But Dr. Forrest's carefully qualified statement gave an entirely misleading impression. In 2004, a year in advance of the trial, I published a peer-reviewed scientific article advancing the theory of intelligent design in a mainstream scientific journal. The publication of this article evoked a huge backlash at the Smithsonian Institution, where the journal, *The Proceedings of the Biological Society of Washington,* was published. Moreover, controversy about the editor's decision and his subsequent treatment spilled over into both the scientific and the mainstream press, with articles about it appearing in *Science, Nature,* the *Wall Street Journal,* and the *Washington Post,* among other places.[34] Both Dr. Forrest and Judge Jones had every opportunity to inform themselves about the existence of at least one peer-reviewed scientific article in support of intelligent design.

In any case, as my institute informed the court in an *amicus curiae* brief, my article was by no means the only peer-reviewed or peer-edited scientific publication in support of the theory of intelligent design.[35] By 2005, scientists and philosophers advocating the theory of intelligent design had already developed their theory and the empirical case for it in peer-reviewed scientific books published both by trade presses[36] and by university presses.[37] Michael Behe's groundbreaking *Darwin's Black Box* was published by the Free Press in New York. William Dembski's *The Design Inference* was published by Cambridge University Press. Both were peer-reviewed. In addition, design proponents have also published scientific articles advancing the case for intelligent design in peer-reviewed scientific books and anthologies published by university presses[38] and in scientific conference proceedings published by university presses and trade presses.[39] Advocates of intelligent design have also published work advancing their theory in peer-reviewed philosophy of science journals and other relevant interdisciplinary journals.[40] Moreover, since the publication of my article in 2004, several other scientific articles supporting intelligent design (or describing research guided by an ID perspective) have been published in mainstream peer-reviewed scientific journals.[41]

Of course, critics of intelligent design may still judge that the number of published books and articles supporting the theory does not yet make it sufficiently mainstream to warrant teaching students about it. Perhaps.[42] But that is a judgment about educational policy distinct from deciding the scientific status, or still less, the merits of the theory of intelligent design itself. Clearly, there is no magic number of supporting peer-reviewed publications that suddenly confers the adjective "scientific" on a theory; nor is there a tribunal vested with the authority to make this determination. If there were a hard-and-fast numerical standard as low as even one, no new theory could ever achieve scientific status. Each new theory would face an impossible catch-22: for a new theory to be considered "scientific" it must have appeared in the peer-reviewed scientific literature, but any time a scientist submitted an article to a peer-reviewed science journal advocating a new theory, it would have to be rejected as "unscientific" on the grounds that no other peer-reviewed scientific publications existed supporting the new theory.

Critics of intelligent design have actually used a similarly circular kind of argument to claim that ID is not science. Before 2004, critics argued that the theory of intelligent design was unscientific, because there were no published articles supporting it in peer-reviewed scientific journals (ignoring the various peer-reviewed books that existed in support of ID). Then once a peer-reviewed scientific journal article was published supporting intelligent design, critics claimed that the article should not have been published, because the theory of intelligent design is inherently unscientific.[43] Indeed, critics accused the editor who published my article of editorial malfeasance, because they thought he should never have considered sending the article out for peer review in the first place.[44] Why? Because, according to these critics, the perspective of the article should have immediately disqualified it from consideration. In short, critics argued that "intelligent design is not scientific because peer-reviewed articles supporting the theory have not been published" and that "peer-reviewed articles supporting intelligent design should not be published because the theory is not scientific," apparently never recognizing the patent circularity of this self-serving, exclusionary logic.

Logically, the issue of peer review is a red herring—a distracting procedural side issue. The truth of a theory is not determined or guaranteed by the place of, or procedures followed, in its publication.[45] Many great scientific theories were first advanced and published without undergoing formal peer review. Though modern peer-review procedures often do a good job of catching and correcting factual mistakes, they also can enforce ideological conformity, stifle innovation, and resist novel theoretical insights. Scientific experts can make mistakes in judgment and, being human, they sometimes reject good new ideas because of prejudicial attachments to older, more familiar ones. The history of science is replete with examples of established scientists summarily dismissing new theories that later proved able to explain the evidence better than previously established theories. In such situations, proponents of new theories have often found traditional organs of publication closed to them. Thus, it is neither surprising nor damning to intelligent design that currently many scientific journals are implacably opposed to publishing articles supporting the theory.

Yet if science is what scientists do, and if publishing peer-reviewed scientific books and articles is part of what scientists do that makes their theories scientific (as critics of ID assert), then there is another good, convention-dependent reason to regard intelligent design as scientific. The scientists who have developed the case for intelligent design have begun to overcome the prejudice against their ideas and have published their work in peer-reviewed scientific journals, books, conference volumes, and anthologies.[46]

So we see that the issue of whether intelligent design qualifies as a scientific theory depends upon the definition of science chosen to decide the question. But consideration of both common definitions of science and the specialized methodological practice of the historical sciences has shown that there are many good—if definition-dependent—reasons for considering intelligent design as a scientific theory.

Maybe there is some other, better definition of science that should be considered; perhaps some specific feature of a scientific theory that intelligent design does not possess, or some specific criterion of scientific practice that its advocates do not follow. Intelligent design meets the criterion of testability, despite what many critics of the theory asserted, but perhaps there are other criteria that it cannot meet. If so, then perhaps these definitional criteria establish a good reason for disqualifying intelligent design from consideration as science after all. Certainly, many critics of intelligent design have argued that the theory lacks many key features of a bona fide scientific theory—that it fails to meet criteria by which science could be defined and distinguished from non-science, metaphysics, or religion. In light of this, we need to examine why critics of the theory—including the judge in the Dover case—have insisted that, despite the arguments just outlined, intelligent design does *not* qualify as a scientific theory.

2. Intelligent Design and Explanation by Natural Law

Michael Ruse has argued that "there are no powers, seen or unseen, that interfere with or otherwise make inexplicable the normal working of material objects."[47] Since the scientific enterprise is characterized by a commitment to "unbroken regularity" or "unbroken law,"[48] scientific theories must explain events or phenomena *by reference to natural laws*.[49] And since intelligent design invokes an event—the conscious activity of a designing agent—rather than a law of nature to explain the origin of biological form and information, Ruse argued that it was scientifically "inappropriate." Ruse also seemed to think that if an intelligent designer had acted during the history of life, then its actions would have necessarily violated the laws of nature, since intelligent agents typically interfere with the otherwise "normal workings of material objects." Since, for Ruse, the activity of an intelligent designer violates the laws of nature, positing such activity—rather than a law—would violate the rules of science.

In response,[50] I have pointed out that the activity of a designing intelligence does not necessarily break or violate the laws of nature. Human agents design information-rich structures and otherwise interfere with the "normal workings of material objects" all the time. When they do, they do not violate the laws of nature; they alter the conditions upon which the laws act. When I arrange magnetic letters on a metallic display board to spell a message, I alter the way in which matter is configured, but I do not alter or violate the laws of electromagnetism. When agents act, they initiate new events within an existing matrix of natural law without violating those laws.[51]

I have also pointed out that Ruse's key demarcation criterion, if applied strictly, *cut just as much against Darwinian and chemical evolutionary theories* (as well as many other scientific theories) as it did against intelligent design. For example, natural laws often *describe* but do not *explain* natural phenomena. Newton's law of universal gravitation described, but did not explain, what caused gravitational attraction. A strict application of Ruse's second criterion would therefore imply that Newton's law of gravity is "unscientific," since it does not offer an *explanation* by natural law.

Many historical scientific theories do not offer an explanation *by natural law*. Instead, they postulate past events (or patterns of events) to explain other past events as well as presently observable evidence. Historical theories explain mainly by reference to events or causes, not laws. For example, if a historical geologist seeks to explain what caused the unusual height of the Himalayas, he or she will cite particular events or factors that were present in the case of the Himalayan mountain-building episode that were not present in other such episodes. Knowing the laws of physics that describe the forces at work in all mountain-building events will not aid the geologist in accounting for the contrast between the Himalayas and other mountain ranges. To explain what caused the Himalayas to rise to such heights, the geologist does not need to cite a general law, but instead to give evidence of a distinctive set of past events or conditions.[52] Evolutionary theories, in particular, often emphasize the importance of past events in their explanations.[53] Aleksandr Oparin's chemical evolutionary theory, for example, postulated a series of events (a scenario), not a general law, in order to explain how the first living cells arose.

Of course, past events and historical scenarios are assumed to take place in a way that obeys the laws of nature. Moreover, our knowledge of cause-and-effect relationships (which we can sometimes formulate as laws) will often guide the inferences that scientists make about what happened in the past and will influence their assessment of the plausibility of competing historical scenarios and explanations. Even so, many historical scientific theories make *no* mention of laws at all. Laws at best play only a secondary role in historical scientific theories. Instead, *events* play the primary explanatory role.

The theory of intelligent design exemplifies the same style of scientific explanation as other historical scientific theories. Intelligent design invoked a past event—albeit a mental event—rather than a law to explain the origin of life and the complexity of the cell. As in other historical scientific theories, our knowledge of cause and effect ("information habitually arises from conscious activity") supports the inference to design. A law (conservation of information)[54] also helps to justify the inference of an intelligent cause as the best explanation. Advocates of intelligent design use a law ("since there is no informational 'free lunch,' the origin of complex specified information always requires intelligent input") to infer a past causal event, the act of a designing mind. But that act or *event* explains the evidence in question. Though laws play a subsidiary role in the theory, a past event (or events) explains the ultimate origin of biological information.

But if explaining events primarily by reference to prior events, rather than laws, does not disqualify other historical scientific theories, including evolutionary theories, from consideration as science, then by the same logic it should not disqualify the theory of intelligent design either. Oddly, in a discussion of population genetics—part of the explanatory framework of contemporary Darwinian theory—Ruse himself noted that "it is probably a mistake to think of modern evolutionists as seeking universal laws at work in every situation."[55] But if laws can play no role or only a subsidiary role in other historical theories, then why was it "inappropriate" for a law to play only a supportive role in the theory of intelligent design?

Conversely, if invoking a past event, rather than a law, made intelligent design unscientific, then by the same token it should make materialistic evolutionary theories unscientific as well. Either way, Ruse's key criterion for scientific status does not provide a basis for discriminating the scientific status of the two types of theories. Both are equivalent in their capacity to meet Ruse's definitional standard.

3. Defeaters Defeated

I have been thinking about the "intelligent design isn't science" objection for a number of years now. As I have done so, I have come to a radical conclusion: not only were there many good—if convention-dependent—reasons for classifying intelligent design as a historical scientific theory, but there were *no* good—that is, non-question-begging—reasons to define intelligent design as *un*scientific. Typically those who argued that "intelligent design isn't science" invoked various demarcation criteria. I have encountered numerous such arguments. Critics claim that intelligent design does not qualify as a scientific theory because: (1) it invokes an unobservable entity,[56] (2) it is not testable,[57] (3) it does not explain by reference to natural law,[58] (4) it makes no predictions,[59] (5) it is not falsifiable,[60] (6) it cites no mechanisms,[61] and (7) it is not tentative.[62]

As I studied these arguments I discovered a curious pattern. Invariably, if the critics applied their definitional criteria—such as observability, testability, or "must explain by natural law"—in a strict way, these criteria not only disqualified the design hypothesis from consideration as science; they also disqualified its chief rivals—other historical scientific theories—each of which invoked undirected evolutionary processes.

Conversely, I discovered that if these definitional criteria were applied in a less restrictive way—perhaps one that took into account the distinctive historical aspects of inquiry into the origin of life—then these criteria not only established the scientific bona fides of various rivals of intelligent design; they confirmed the scientific status of the design hypothesis as well. In no case, however, did these demarcation criteria successfully differentiate the scientific status of intelligent design and its competitors. Either science was defined so narrowly that it disqualified both types of theory, or it was defined so broadly that the initial reasons for excluding intelligent design (or its competitors) evaporated. If one theory met a specific criterion, then so did the other; if one theory failed to do so, then its rival also failed—provided the criteria were applied in an evenhanded and non-question-begging way. Intelligent design and its materialistic rivals were equivalent in their ability to meet various demarcation criteria or methodological norms. Given this methodological equivalence, and given that materialistic evolutionary theories were already widely regarded as scientific, I couldn't see any reason to classify intelligent design as unscientific. The defeaters didn't work.

Because these "defeaters" are used against intelligent design all the time, it's important to see why they fail. So, in what follows, I examine some additional demarcation arguments that are commonly used against intelligent design. (I will not provide an exhaustive demonstration of this equivalence since that would require a book-length argument the details of which only philosophers of science could endure. For those with the requisite endurance, however, various materials are available.)[63]

3.1 OBSERVABILITY

According to critics of intelligent design, the unobservable character of a designing intelligence renders it inaccessible to empirical investigation and, therefore, makes it unscientific. For example, in 1993 biophysicist Dean Kenyon was removed from teaching his introductory biology class at San Francisco State University after he discussed his reasons for supporting intelligent design with his students. His department colleagues believed their actions against him were justified because they believed that he had been discussing an unscientific theory with his class. Some of Kenyon's colleagues argued that the theory of intelligent design did not qualify as a scientific theory because it invoked an *unobservable* entity, in particular, an unseen designing intelligence. In making this argument, Kenyon's colleagues assumed that scientific theories must invoke only *observable* entities. Since Kenyon discussed a theory that violated this convention, they insisted that neither the theory he discussed, nor he himself, belonged in the biology classroom.[64]

Others who defended the action of the biology department, such as Eugenie Scott of the National Center for Science Education, used a similar rationale. She insisted that the theory of intelligent design violated the rules of science because "you can't put an omnipotent deity in a test tube (or keep it out of one)."[65] Molecular biologist Fred Grinnell has similarly argued that intelligent design can't be a scientific concept, because if something "can't be measured, or counted, or photographed, it can't be science."[66]

But was that really the case? Does a reference to an unobservable entity provide a good reason for defining a theory as unscientific? Does my postulation of an unobservable intelligence make my case for intelligent design unscientific?

The answer to that question depends, again, upon how science is defined. If scientists (and all other relevant parties) decide to define science as an enterprise in which scientists can posit only observable entities in their theories, then clearly the theory of intelligent design would not qualify as a scientific theory. Advocates of intelligent design infer, rather than directly observe, the designing intelligence responsible for the digital information in DNA.

But this definition of science would render many other scientific theories, including many evolutionary theories of biological origins, unscientific by definition as well. Many scientific theories infer or postulate unobservable entities, causes, and events. Theories of chemical evolution invoke past events as part of the scenarios they use to explain how the modern cell arose. Insofar as these events occurred millions of years ago, they are clearly not observable today. Darwinian biologists, for their part, have long defended the putatively unfalsifiable nature of their claims by reminding critics that many of the creative processes to which they refer occur at rates too slow to observe in the present and too fast to have been recorded in the fossil record. Furthermore, the existence of many transitional intermediate forms of life, the forms represented by the nodes on Darwin's famous branching tree diagram, are also unobservable.[67] Instead, unobservable transitional forms of life are *postulated* to explain observable biological evidence—as Darwin himself explained. But how is this different from postulating the past activity of an unobservable designing intelligence to explain observable features of the living cell? Neither Darwinian transitional forms, neo-Darwinian mutational events, the "rapid branching" events of Stephen Jay Gould's theory of punctuated equilibria, the events comprising chemical evolutionary scenarios, nor the past action of a designing intelligence are directly observable. With respect to direct observability, each of these theories is equivalent.

Thus, if the standard of observability is applied in a strict way, neither intelligent design nor any other theory of biological origins qualifies as a scientific theory. But let's consider the flip side. What if the standard of observability is applied in a more flexible and, perhaps, realistic way? What if science is defined as an enterprise that examines the observable natural world, but does not necessarily explain empirical observations by reference to observable entities?

Does it make sense to define science in this more flexible way? It does. Many entities and events posited in scientific theories cannot be observed directly either in practice, or sometimes even in principle. Instead, scientists often infer the existence of unobservable entities in order to explain observable events, evidence, or phenomena. Physical forces, electromagnetic or gravitational fields, atoms, quarks, past events, subsurface geological features, biomolecular structures—*all* are unobservable entities inferred from observable evidence. In 2008, under the border between France and Switzerland, European scientists unveiled the Large Hadron Collider. This supercollider will enable physicists to "look" for various elementary particles, including the elusive Higgs boson. None of the particles they hope to find are observable in any direct sense. Instead, physicists try to detect them by the energetic signatures, traces, or decay products they leave behind.

Scientists in many fields detect unobservable entities and events by their effects. They often infer the unseen from the seen. Nevertheless, such entities and events are routinely considered to be part of scientific theories. Those who argue otherwise confuse the event or evidence in need of explanation (which in scientific investigations is nearly always observable in some way) with the event or entity doing the explaining (which often is not).

The presence of unobservable entities in scientific theories creates a problem for those who want to use observability as a demarcation criterion by which to disqualify intelligent design from consideration as scientific. Many theories that are widely acknowledged to be scientific invoke unobservable entities. But if these theories, which include materialistic theories of biological origins, can invoke unobservable entities or events to explain observable evidence and still qualify as scientific, then why can't the theory of intelligent design do so as well?

3.2 Testability Revisited

I earlier mentioned that the theory of intelligent design is testable by empirical means. But for years, I have talked to people—scientists, theologians, philosophers, lawyers, journalists, callers on talk shows—who purport to know that the theory of intelligent design cannot be tested.[68] Sometimes these critics say that intelligent design is untestable because the designing intelligence is *unobservable,* thus combining two demarcation criteria, observability and testability. Other times these critics assert that intelligent design cannot be tested because they assume that ID advocates *must posit an omnipotent deity.* And yet other critics say that intelligent design is untestable because the actions of intelligent agents (of any kind) are inherently *unpredictable,* and testability depends upon the ability to make predictions.

These common objections to the testability and thus the scientific status of intelligent design have dissuaded many people from even considering evidence for intelligent design. So what should we make of these demarcation arguments against intelligent design? Do they provide a good reason for denying that intelligent design is a scientific theory? Do they show, despite the considerations we have offered to the contrary, that intelligent design cannot be tested? Let's take a closer look.

3.3 Unobservables and Testability

Robert Pennock, one of the witnesses in the Dover trial, argued that the unobservable character of a designing intelligence precludes the possibility of testing intelligent design scientifically because, as he explained, "science operates by empirical principles of *observational* testing; hypotheses must be confirmed or disconfirmed by reference to . . . accessible empirical data."[69] Eugenie Scott also seemed to argue that intelligent design cannot be tested *because* it invokes an unobservable entity. In the article I cited above, in which she defended the actions of Kenyon's detractors at San Francisco State, Scott also linked the criterion of observability to testability. After saying, "You can't put an omnipotent deity in a test tube," she went on to say: "As soon as creationists invent a 'theo-meter,' maybe then we can test for miraculous intervention. You can't (scientifically) study variables you can't test, directly or indirectly."[70]

In this version of the argument, critics insist that the unobservable character of a designing intelligence renders the theory inaccessible to empirical investigation, making it both untestable and unscientific. So both "observability" and "testability" are asserted as necessary to scientific status and the converse of one (unobservability) is asserted to preclude the possibility of the other (testability). Superficially this version of the argument seems a bit more persuasive than demarcation arguments that simply invoke observability by itself to disqualify design. Yet it does not stand up to close inspection either.

In the first place, there are many testable scientific theories that refer to unobservable entities. For example, during the race to elucidate the structure of the genetic molecule, both double helix and triple helix models were considered, since both could explain the X-ray images of DNA crystals.[71] Although neither structure could be observed directly, the double helix of Watson and Crick eventually won out because it could explain other observations that the triple helix model could not. The inference to one unobservable structure (the double helix) was accepted because it was judged to possess greater explanatory power than its competitor.

Claims about unobservables are routinely *indirectly* tested against observable evidence in scientific experiments. In many fields of science the existence of an unobservable entity or event is inferred from the positive outcome of testing for the consequences that would result if that hypothesized entity or event (i.e., an unobservable) were accepted as actual, or by the explanatory power the postulation of such an entity or event provides when contrasted with competing hypotheses. Thus, many sciences infer to the best explanation—where the explanation presupposes the reality of an unobservable entity or event—including theoretical physics, geology, molecular biology, genetics, cosmology, psychology, physical and organic chemistry, and evolutionary biology.

Secondly, the historical sciences, in particular, commonly use indirect methods of testing, methods that involve assessing the causal powers of competing unobservable events to determine which would, if true, possess the greatest explanatory power. Notably, Darwin defended the scientific status of his theory by arguing that an assessment of the relative explanatory power of his theory of common descent—a theory about the unobservable past—was a perfectly legitimate and acceptable method of scientific testing.[72]

Finally, intelligent design is testable in precisely this fashion—by examining its explanatory power and comparing it to that of competing hypotheses. The unobservable intelligence postulated by the theory of intelligent design does *not* preclude tests of its existence as long as *indirect* methods of testing hypotheses—such as evaluating comparative explanatory power—are recognized as scientific. If, however, science is defined more narrowly so that only the direct observation of a causal factor counts as a confirmatory test of a causal hypothesis, then neither intelligent design nor a host of other theories qualify as scientific.

Either way, the theory of intelligent design and various evolutionary theories of origins are *equivalent* in their ability to meet the joint criteria of observability and testability. If critics of intelligent design construe these criteria as forbidding both the indirect testing of hypotheses postulating unobservables and inferences to the existence of such unobservables if such tests are favorable, then both intelligent design and its competitors fail to be scientific. On the other hand, if critics construe these criteria to allow inferences to, and indirect testing of, unobservable entities and events, then both intelligent design and many competing evolutionary theories qualify as scientific theories. Either way, these criteria fail to discriminate between intelligent design and many other theories that are already accepted as scientific, and so they fail to provide a good reason for disqualifying intelligent design from consideration as a scientific theory. What is sauce for the goose is sauce for the gander.

3.4 Testability, Omnipotence, and the Supernatural

Robert Pennock argues that there is something else about the unobservable designing intelligence posited by intelligent design that makes it untestable. Specifically, Pennock claims that intelligent design is untestable because it invokes an unobservable *supernatural* being with unlimited powers. He argues that since such a being has powers that could be invoked to "explain any result in any situation," all events are *consistent* with the actions of such a being. Therefore, no conceivable event could disprove the hypothesis of intelligent design. As Ken Miller asserts, "The hypothesis of design is compatible with any conceivable data, [and] makes no testable predictions."[73]

This argument fails for two reasons. First, it misrepresents the theory of intelligent design. The theory of intelligent design does *not* claim to detect a supernatural intelligence possessing unlimited powers. Though the designing agent responsible for life may well have been an omnipotent deity, the theory of intelligent design does not claim to be able to determine that. Because the inference to design depends upon our uniform experience of cause and effect in this world, the theory cannot determine whether or not the designing intelligence putatively responsible for life has powers beyond those on display in our experience. Nor can the theory of intelligent design determine whether the intelligent agent responsible for information in life acted from the natural or the "supernatural" realm. Instead, the theory of intelligent design merely claims to detect the action of some *intelligent* cause (with power, at least, equivalent to those we know from experience) and affirms this because we know from experience that only conscious, intelligent agents produce large amounts of specified information. The theory of intelligent design does not claim to be able to determine the identity or any other attributes of that intelligence, even if philosophical deliberation or additional evidence from other disciplines may provide reasons to consider, for example, a specifically theistic design hypothesis.[74]

Pennock's argument also fails because the theory of intelligent design *is* subject to empirical testing and refutation. Indeed, intelligent design actually makes a much stronger claim than the caricature of it he critiqued during the trial. Pennock critiques the hypothesis that "an omnipotent deity *could* explain the origin of life." But the theory of intelligent design actually advocated differs from that hypothesis since it does not merely affirm that intelligence constitutes a *possible* explanation of certain features of life. Instead, it asserts that intelligent design constitutes the *best* explanation of a particular feature of life because of *what we know about the cause-and-effect structure of the world*—specifically, because of what we know about what it takes to produce large amounts of specified information. For this reason, the design hypothesis is *not* "compatible with any conceivable data" or observations whatsoever.

If it were shown, for example, that the cause-and-effect structure of the world were different than what advocates of intelligent design claim—if, for example, someone successfully demonstrated that "large amounts of functionally specified information *do* arise from purely chemical and physical antecedents," then my design hypothesis, with its strong claim to be the best (clearly superior) explanation of such phenomena, would fail. Intelligent design would remain as a merely possible explanation (much as chance does now). But the claim that intelligent design provides the *best* (most causally adequate) explanation for the origin of biological information would be refuted. Similarly, if it could be shown that key indicators of intelligence—such as specified information—were *not* present in living systems, the basis of the design hypothesis in its present strong form would evaporate. Thus, Pennock and Miller incorrectly portray the theory of intelligent design as being consistent with any empirical situation. The theory of intelligent design is, in fact, testable—just as we have argued.

3.5 Testability and Predictability

When critics of intelligent design are confronted with refutations of a particular demarcation argument, they typically shift their ground and formulate other arguments either by invoking a different demarcation criterion or by applying the original criterion in a more demanding way. For example, after explaining how intelligent design can be tested and how it does make certain kinds of predictions, I commonly hear the objection that the theory of intelligent design is not scientific because it cannot make *other* kinds of predictions. Critics correctly point out, for example, that we cannot predict with complete accuracy what intelligent agents will do since, presumably, intelligent agents possess the capacity to act of their own free volition. Since ID invokes the action of an unpredictable intelligent agent, and since scientific theories must make predictions, theories invoking the activity of intelligent agents are not scientific—or so the argument goes.

Yet standard materialistic theories of evolution (whether chemical or biological) do not make predictions of this kind either. Specifically, evolutionary theory does not make predictions about the future course of evolution. It makes no prediction about the kind of traits or species that random mutations and natural selection will produce in the future. As Ken Miller notes, "The outcome of evolution is not predictable."[75] Even so, most evolutionary biologists think that these theories are scientific—and for good (if convention-dependent) reasons. Evolutionary theories provide explanations of past events and present evidence, and they make predictions about the patterns of evidence that scientists should find *in their future investigations,* for example, of the genome or the fossil record.

In the same way, the theory of intelligent design does not make predictions about when (or whether) the designing intelligence responsible for life will act in the future. Yet it does explain past events and present evidence, and it also makes discriminating predictions about the kind of evidence scientists should find in their future investigations.[76] Thus, neither type of origins theory qualifies as scientific if the "ability to generate predictions" is treated as a condition of scientific status and interpreted in a strict way, though both types of theories qualify as scientific if this criterion is equated with scientific status and interpreted in a more flexible way.

As I studied the various demarcation arguments against intelligent design, I repeatedly found this same pattern. Invariably, the criteria that supposedly showed that intelligent design is inherently unscientific either disqualified both intelligent design *and* its materialistic rivals, or if the criteria were applied more flexibly, legitimated both types of theories—provided, that is, that the criteria were not applied in a question-begging way. Again, what was sauce for the goose was sauce for the gander.

As this pattern became more pronounced with each of the definitional criteria examined, I became more convinced that there was *no good reason* to exclude intelligent design from consideration as a scientific explanation for the origin of biological information. Since—by convention—materialistic theories of biological origin were considered scientific, and since the theory of intelligent design met various criteria of scientific status just as well as these rival theories, it seemed clear that the theory of intelligent design, by the same conventions, must be considered scientific as well.

Consider the case of so-called junk DNA—the DNA that does not code for proteins found in the genomes of both one-celled organisms and multicellular plants and animals. The theory of intelligent design and materialistic evolutionary theories (both chemical and biological) differ in their interpretation of so-called junk DNA. Since neo-Darwinism holds that new biological information arises as the result of a process of mutational trial and error, it predicts that nonfunctional DNA would tend to accumulate in the genomes of eukaryotic organisms (organisms whose cells contain nuclei). Since most chemical evolutionary theories also envision some role for chance interactions in the origin of biological information, they imply that non-functional DNA similarly would have accumulated in the first simple (prokaryotic) organisms—as a kind of remnant of whatever undirected process first produced functional information in the cell. For this reason, most evolutionary biologists concluded upon the discovery of non-protein-coding DNA that such DNA was "junk." In their view, discovery of the non-protein-coding regions confirmed the prediction or expectation of naturalistic evolutionary theories and disconfirmed an implicit prediction of intelligent design.

As Michael Shermer argues, "Rather than being intelligently designed, the human genome looks more and more like a mosaic of mutations, fragmented copies, borrowed sequences, and discarded strings of DNA that were jerry-built over millions of years of evolution."[77] Or as Ken Miller argues: "The critics of evolution like to say that the complexity of the genome makes it clear that it was designed. . . . But there's a problem with that analysis, and it's a serious one. The problem is the genome itself: it's not perfect. In fact, it's riddled with useless information, mistakes, and broken genes. . . . Molecular biologists actually call some of these regions 'gene deserts,' reflecting their barren nature."[78] Or as philosopher of science Philip Kitcher puts it, "If you were designing the genomes of organisms, you would not fill them up with junk."[79]

ID advocates advance a different view of non-protein-coding DNA.[80] The theory of intelligent design predicts that most of the non-protein-coding sequences in the genome should perform *some* biological function, even if they do not direct protein synthesis. ID theorists do not deny that mutational processes might have degraded or "broken" some previously functional DNA, but we predict that the functional DNA (the signal) should dwarf the non-functional DNA (the noise), and not the reverse. As William Dembski explained and *predicted* in 1998: "On an evolutionary view we expect a lot of useless DNA. If, on the other hand, organisms are designed, we expect DNA, as much as possible, to exhibit function."[81] *The dis-*

covery in recent years that non-protein-coding DNA performs a diversity of important biological functions has confirmed this prediction. It also decisively refutes prominent critics of intelligent design—including Shermer, Miller, and Kitcher—who have continued to argue (each as recently as 2008) that the genome is composed of mostly useless DNA.[82]

Contrary to their claims, recent scientific discoveries have shown that the non-protein-coding regions of the genome direct the production of RNA molecules that regulate the use of the protein-coding regions of DNA. Cell and genome biologists have also discovered that these supposedly "useless" non-protein-coding regions of the genome: (1) regulate DNA replication,[83] (2) regulate transcription,[84] (3) mark sites for programmed rearrangements of genetic material,[85] (4) influence the proper folding and maintenance of chromosomes,[86] (5) control the interactions of chromosomes with the nuclear membrane (and matrix),[87] (6) control RNA processing, editing, and splicing,[88] (7) modulate translation,[89] (8) regulate embryological development,[90] (9) repair DNA,[91] and (10) aid in immunodefense or fighting disease[92] among other functions. In some cases, "junk" DNA has even been found to code functional genes.[93] Overall, the non-protein-coding regions of the genome function much like an operating system in a computer that can direct multiple operations simultaneously.[94] Indeed, far from being "junk," as materialistic theories of evolution assumed, the non-protein-coding DNA directs the use of other information in the genome, just as an operating system directs the use of the information contained in various application programs stored in a computer. In any case, contrary to the often-heard criticism that the theory makes no predictions, intelligent design not only makes a discriminating prediction about the nature of "junk DNA," recent discoveries about non-protein-coding DNA also *confirm* the prediction that it makes.[95] In the appendix to my book, *Signature in the Cell*, I describe several *other* discriminating predictions that the theory of intelligent design makes.[96] Despite this, many continue to assert the falsehood that ID is not testable.

4. Methodological Naturalism, or How Dover Was Decided

In light of the preceding discussion about the specific failure of demarcation arguments, how is it that a federal court in 2005 was able to determine that ID is not science? By what definition did the judge in the now famous *Kitzmiller vs. Dover* case use to make this determination?

As it turned out, Judge Jones did a clever thing. He didn't reject intelligent design as science because it failed to meet a *neutral* definition of science or some methodological norm. At the urging of the ACLU, he circumvented the whole demarcation problem *by defining science as the exclusion of intelligent design*—only he didn't call it that. Instead, following the ACLU's expert witnesses and brief, he called the exclusionary principle "methodological naturalism."[97] He then equated science with adherence to that principle and rejected intelligent design because it violated it.

But what is the principle of methodological naturalism? Methodological naturalism asserts that to qualify as scientific, a theory must explain all phenomena by reference to purely material—that is, nonintelligent—causes. As philosopher Nancey Murphy explains, methodological naturalism forbids reference "to creative intelligence" in scientific theories.[98]

So, did the judge find a demarcation criterion or methodological norm that could discriminate between intelligent design and materialistic theories of evolution? Clearly, he did. If science is defined as Judge Jones defined it, intelligent design does not qualify as a scientific theory. But *should* science be defined that way? Did the judge offer a *good* reason for excluding intelligent design from consideration as science?

He did not. Instead, he provided an entirely arbitrary, circular, and question-begging justification for the exclusion of design. I knew, as did many other philosophers of science, that demarcation arguments based upon neutral methodological norms such as testability could not justify a prohibition against intelligent causes in science. The judge in the Dover case supposedly offered a reason for this prohibition, but his reason turned out to be just a restatement of the prohibition by another name. According to Judge Jones, the theory of intelligent design cannot be part of science because it violates the principle of methodological naturalism. But that principle turns out to be nothing more than the claim that intelligent causes—and thus the theory of intelligent design—must be excluded from science.[99] According to this reasoning, intelligent design isn't science because it violates the principle of methodological naturalism. And what is methodological naturalism? It is a rule prohibiting consideration of intelligent design within science.

Thus, despite appearances to the contrary, Judge Jones did not offer a good reason—a theoretically neutral norm or definition of science—by which to justify the exclusion of intelligent design from science. Instead, he simply asserted a prohibition against the consideration of intelligent design, invoked the same prohibition by another name, and then treated it as if it were a reason—a methodological principle—justifying the prohibition itself.

Fortunately we don't look to federal judges to settle great questions of scientific and philosophical import. Did life arise as the result of purely undirected material causes or did intelligence play a role?[100] Surely a court-promulgated definition of science, especially one so logically problematic as methodological naturalism, does not answer that question.[101]

No doubt Judge Jones felt justified in offering such a thin and circular justification for his definition of science because he knew many scientists agreed with him. And, indeed, the majority of scientists may well accept the principle of methodological naturalism. So, if science is what scientists do, and if many or most scientists do not think that hypotheses invoking intelligent causes have a place in their theories, then perhaps intelligent design doesn't qualify as a scientific theory after all. According to this line of thinking, Judge Jones did not impose an arbitrary definition of science. Instead, his ruling merely expressed a preexisting consensus about proper scientific practice from within the scientific community. As the judge himself wrote in the ruling, methodological naturalism is simply a "centuries-old ground rule" of science.[102]

So why shouldn't scientists continue to accept methodological naturalism as a strict rule governing scientific practice? Maybe we should just accept this convention and move on. Of course, some scientists may decide to do exactly that. But if they do, it's important to recognize what that decision would and would not signify about the design hypothesis. Scientists who decide to define explanations involving creative intelligence as unscientific cannot then treat the failure of such hypotheses to meet their definition of science as a tacit refutation of, or reason to reject, such hypotheses. Why? It remains logically possible that an "unscientific" hypothesis (by the criterion of methodological naturalism) might constitute a *better* explanation of the evidence than the currently best "scientific" hypothesis. In light of the evidence discussed in this essay and presented in my book *Signature in the Cell,* I would contend that, whatever its classification, the design hypothesis provides a *better* explanation than any of its materialistic rivals for the origin of the specified information necessary to produce the first life. *Reclassifying an argument does not refute it.*

In any case, there is no compelling reason for the currently dominant convention among scientists to continue. Conventions are just that. Without a good reason for holding them, they may do nothing more than express an unexamined prejudice and block the path of inquiry. When good reasons for rejecting conventions come along, reasonable people will set them aside—and there are now good reasons to set this convention aside.

First, scientists have not always restricted themselves to naturalistic hypotheses, contrary to the claims of one of the expert witnesses in the Dover trial. Newton, for example, made design arguments within his scientific works, most notably in the *Principia* and in the *Opticks*. Louis Agassiz, a distinguished paleontologist and contemporary of Darwin, also made design arguments within his scientific works, insisting that the pattern of appearance in the fossil record strongly suggested "an act of mind." Defenders of methodological naturalism can claim, at best, that it has had normative force during *some* periods of scientific history. But this concedes that canons of scientific method change over time—as, indeed, they do. From Newton until Darwin, design arguments were a common feature of scientific research. After Darwin, more materialistic canons of method came to predominate. Recently, however, this has begun to change as more scientists are becoming interested in the evidence for intelligent design.

Second, many scientific fields currently posit intelligent causes as scientific explanations. Design detection is already part of science. Archaeologists, anthropologists, forensic scientists, cryptographers, and others now routinely infer intelligent causes from the presence of information-rich patterns or structures or artifacts. Furthermore, astrobiologists looking for extraterrestrial intelligence (SETI) do not have a rule against inferring an intelligent cause. Instead, they are open to detecting intelligence, but have not had evidence to justify making such an inference. Thus, the claim that all scientific fields categorically exclude reference to creative intelligence is actually false.

Even some biologists now contest methodological naturalism. Granted, many evolutionary biologists accept methodological naturalism as normative within their discipline. Nevertheless, biologists intrigued by the design hypothesis reject methodological naturalism because it prevents them from considering a possibly true hypothesis. Indeed, a central aspect of the current debate over design is precisely about whether methodological naturalism should be regarded as normative for biology today. Most evolutionary biologists say it should remain normative; scientists advocating intelligent design disagree. But critics of intelligent design cannot invoke methodological naturalism to settle this debate about the scientific status of intelligent design, because methodological naturalism is itself part of what the debate is about.

Third, defining science as a strictly materialistic enterprise commits scientists to an unjustified—and possibly false—view of biological origins. It is at least logically possible that a personal agent—a conscious goal-directed intelligence—existed before the appearance of the first life on Earth. Moreover, as evidenced by the work of William Dembski and Robert Marks, or by our explanation of how intelligent design functions as a historical science using the method of multiple competing hypotheses, there are now rigorous scientific methods by which the activity of intelligent agents can be inferred or detected from certain kinds of effects. Thus, if a personal agent existed before the advent of life on Earth, then it is also at least possible that the activity of such an agent could be detected using one of these methods. In light of this, prohibitions against the design hypothesis in investigations of the origin of life amount to an assumption that no intelligence of any kind existed or could have acted prior to that event. But this assumption is *entirely unjustified,* especially given the *absence of evidence* for a completely materialistic account of abiogenesis.

Finally, allowing methodological naturalism to function as an absolute "ground rule" of method for all of science would have a deleterious effect on the practice of certain scientific disciplines, especially the historical sciences.[103] In origin of life research, for example, methodological naturalism artificially restricts inquiry and prevents scientists from exploring and examining some hypotheses that might provide the most likely, best, or causally adequate explanations. To be a truth-seeking endeavor, the question that origin of life research must address is not, "Which materialistic scenario seems most adequate?" but rather, "What actually caused life to arise on Earth?" Clearly, one possible answer to that latter question is this: "Life was designed by an intelligent agent that existed before the advent of humans." If one accepts methodological naturalism as normative, however, scientists may never consider this possibly true hypothesis. Such an exclusionary logic diminishes the significance of claiming theoretical superiority for any remaining hypothesis and raises the possibility that the best "scientific" explanation (by the lights of methodological naturalism) may not, all things considered, be the best explanation at all.

Scientific theory evaluation is an inherently comparative enterprise. Theories that gain acceptance in artificially constrained competitions can claim to be neither "best" nor "most probably true." At most such theories can be considered "the best, or most probably true, among an artificially limited set of options." Openness to the design hypothesis would seem necessary, therefore, to any fully rational historical biology—that is, to one that seeks the truth, "no holds barred."[104] A historical biology committed to following the evidence wherever it leads will not exclude hypotheses *a priori* because of their possible metaphysical implications. Instead, it will employ only metaphysically neutral criteria—such as causal adequacy—to evaluate competing hypotheses. Yet this more open (and arguably rational) approach would now seem to affirm the theory of intelligent design as the best, most causally adequate, scientific explanation for the origin of the information necessary to build the first living organism.[105]

Notes

1. Jones, *Kitzmiller et al. v. Dover Area School District,* 400 F.Supp.2d 707, 720–21, 735 (M. D. Pa. 2005).
2. Jones, *Kitzmiller et al. v. Dover Area School District,* 400 F.Supp.2d 707, 764 (M. D. Pa. 2005).
3. Prominent news reports include Claudia Wallis, "The Evolution Wars," *Time,* August 7, 2005; Evan Ratliff, "The Crusade Against Evolution," *Wired,* October 2004; and Jodi Wilgoren, "Politicized Scholars Put Evolution on the Defensive," *New York Times,* August 21, 2005.
4. Judge Jones falsely claimed that: (1) the theory of intelligent design affirms a "supernatural creation"—a position that ID proponents who testified in court denied during the trial; (2) proponents of intelligent design make their case solely by arguing against Darwinian evolution; (3) no peer-reviewed scientific publications in support of the theory of intelligent design have been published in the scientific literature; and (4) the theory of intelligent design has been refuted. See DeWolf, et al., *Traipsing into Evolution;* DeWolf, West, and Luskin, "Intelligent Design Will Survive *Kitzmiller v. Dover*"; Luskin, "Will Americans United Retract Their Demonstrably False Claims?"
5. See Worden, "Bad Frog Beer to 'Intelligent Design.'"
6. See West and DeWolf, "A Comparison of Judge Jones's Opinion in *Kitzmiller v. Dover* with Plaintiffs' Proposed 'Findings of Fact and Conclusions of Law.'"
7. There are numerous examples of this claim. Some include: "The claim that life is the result of a design created by an intelligent cause cannot be tested and is not within the realm of science" (Skoog, "A View from the Past," 1–2); "ID has never produced an empirically testable hypothesis" (Forrest and Gross, *Creationism's Trojan Horse,* 235); "The hypothesis of design is compatible with any conceivable data, makes no testable predictions, and suggests no new avenues for research" (Miller, *Only a Theory,* 87). For examples of some testable predictions made by intelligent design theory, see the appendix to my book *Signature in the Cell* (2009).
8. "There is no demarcation line between science and non-science, or between science and pseudoscience, which would win assent from a majority of philosophers" (Laudan, *Beyond Positivism and Relativism,* 210).
9. Laudan, "The Demise of the Demarcation Problem."
10. Gould, "Evolution and the Triumph of Homology, or Why History Matters," 60–69.
11. For example, Pennock testified, "Intelligent design needs to have for it to be a science a way of offering a specific hypothesis that one could test in an ordinary way. They failed to do that, and so they really don't get off the ground with regard to science" (*Kitzmiller v. Dover* testimony, September 28, 2005, 39).
12. See my book, *Signature in the Cell: DNA and the Evidence for Intelligent Design* (San Francisco: HarperOne, 2009).
13. Behe, *Darwin's Black Box.*
14. Meyer, et al., "The Cambrian Explosion."
15. See the essays by William Craig and Bruce Gordon in this volume; also Robin Collins' essay "Evidence for Fine-Tuning," in Neil Manson, ed. *God and Design: The Teleological Argument and Modern Science.* New York: Routledge, 2003, 178–99.
16. See the essay by Guillermo Gonzalez in this volume; also Gonzalez and Richards, *The Privileged Planet.*
17. See Nelson and Wells, "Homology in Biology."
18. Dembski, *The Design Inference,* 36.
19. Dembski, *The Design Inference,* 36–66. See also Dembski's article "Specification: The Pattern that Signifies Intelligence" (http://www.designinference.com/documents/2005.06.Specification.pdf) for a more recent and nuanced treatment.
20. See Dembski's and Marks' essay "Life's Conservation Law: Why Darwinian Evolution Cannot Create Biological Information" in Part III of this volume.

21. Considerations of causal *existence* also play a role in the evaluation—and testing—of historical scientific theories. Indeed, historical scientific theories can fail by being unable to meet this critical test as well.

22. We discuss a number of such predictions below; see also the appendix to my book *Signature in the Cell* (2009).

23. Though historical scientists focus *primarily* on questions about the past, they clearly also have a secondary interest in questions about the present operation and cause-and-effect structure of the world. Indeed, the uniformitarian method requires that historical scientists use knowledge of the present cause-and-effect structure of the world to reconstruct what happened in the past. Nevertheless, that the historical sciences address questions about the past at all distinguishes them from many sciences that focus wholly on questions about how nature *generally* operates.

24. Gould, "Evolution and the Triumph of Homology," 61.

25. Meyer, "DNA and the Origin of Life"; Meyer, et al., "The Cambrian Explosion"; Behe, *Darwin's Black Box;* Gonzalez and Richards, *The Privileged Planet;* Craig, "Barrow and Tipler on the Anthropic Principle vs. Divine Design," 389.

26. This is not to deny that laws or process theories may play roles in support of causal explanation, as even opponents of the covering-law model, such as Scriven, admit. Scriven notes that laws and other types of general-process theories may play an important role in justifying the causal status of an explanatory antecedent and may provide the means of inferring plausible causal antecedents from observed consequents. Nevertheless, as both Scriven and I have argued elsewhere, laws are not necessary to the explanation of particular events or facts; and even when laws are present, antecedent events function as the primary causal or explanatory entity in historical explanations. See Scriven, "Truisms as the Grounds," 448–50; "Explanation and Prediction," 480; "Causes, Connections and Conditions," 249–50; Meyer, "Of Clues and Causes," 18–24, 36–72, 84–92.

27. Dawkins, *The Blind Watchmaker,* 1.

28. Ayala, "Darwin's Revolution."

29. Nagel, "Public Education and Intelligent Design."

30. For example, Judge Jones asserted in his decision: "We find that ID is not science and cannot be adjudged a valid, accepted scientific theory as it has failed to publish in peer-reviewed journals" (*Kitzmiller et al. v. Dover Area School District,* 400 F.Supp.2d).

31. Forrest, quoted in Vergano and Toppo, "'Call to Arms' on Evolution."

32. Forrest, Expert Witness Report.

33. For example, Judge Jones asserted: (1) "We find that ID is not science and cannot be adjudged a valid, accepted scientific theory as it has failed to publish in peer-reviewed journals, engage in research and testing, and gain acceptance in the scientific community"; (2) "A final indicator of how ID has failed to demonstrate scientific warrant is the complete absence of peer-reviewed publications supporting the theory"; and (3) "The evidence presented in this case demonstrates that ID is not supported by any peer-reviewed research, data or publications" (*Kitzmiller et al. v. Dover Area School District,* 400 F.Supp.2d).

34. Holden, "Random Samples"; Giles, "Peer-Reviewed Paper Defends Theory of Intelligent Design"; Klinghoffer, "The Branding of a Heretic"; Powell, "Editor Explains Reasons for 'Intelligent Design' Article."

35. See Discovery Institute, Brief of Amicus Curiae (Revised).

36. E.g., Behe, *Darwin's Black Box;* Gonzalez and Richards, *The Privileged Planet;* Thaxton, Bradley, and Olsen, *The Mystery of Life's Origin.*

37. E.g., Dembski, *The Design Inference;* Campbell and Meyer, eds., *Darwinism, Design and Public Education;* Dembski and Ruse, eds., *Debating Design.*

38. See the five articles advancing the case for the theory of intelligent design published in Campbell and Meyer, eds., *Darwinism, Design and Public Education,* and the four articles published in Dembski and Ruse, eds., *Debating Design.*

39. Minnich and Meyer, "Genetic Analysis of Coordinate Flagellar and Type III Regulatory Circuits"; Dembski, ed., *Mere Creation*.

40. Craig, "God, Creation and Mr. Davies," 163; "Barrow and Tipler on the Anthropic Principle vs. Divine Design"; Behe, "Self-Organization and Irreducibly Complex Systems."

41. Wells, "Do Centrioles Generate a Polar Ejection Force?"; Behe and Snoke, "Simulating Evolution by Gene Duplication"; Dembski and Marks, "The Conservation of Information: Measuring the Information Cost of a Successful Search"; Voie, "Biological Function and the Genetic Code Are Interdependent"; Davison, "A Prescribed Evolutionary Hypothesis." See also Lönnig and Saedler, "Chromosomal Rearrangements and Transposable Elements."

42. In 2005 my colleagues and I at the Discovery Institute actually opposed the policy of the Dover school board and urged the school board to withdraw it. Though we think there is nothing unconstitutional about teaching students about the theory of intelligent design, we feared that politicizing the issue would result in reprisals against ID proponents in university science departments. We also objected to the way some members of the Dover board attempted to justify their policy by invoking religious authority as a reason to vote for the policy. This justification guaranteed the policy would run afoul the establishment clause in the courts. From our point of view, this justification was entirely gratuitous—and incongruous—since the theory of intelligent design is not based upon a religious authority, but upon empirical evidence and standard scientific methods of reasoning. Unfortunately, the school board did not heed our advice. It lost the case, just as we predicted, ultimately causing trouble for ID proponents at universities around the country, just as we, alas, also predicted.

43. "The Council [of the Biological Society of Washington, which oversees the publication of the *Proceedings of the Biological Society of Washington*] endorses a resolution on ID published by the American Association for the Advancement of Science (www.aaas.org/news/releases/2002/1106id2.shtml), which observes that there is no credible scientific evidence supporting ID as a testable hypothesis to explain the origin of organic diversity. Accordingly, the Meyer paper does not meet the scientific standards of the Proceedings." See http://www.ncseweb.org/resources/news/2004/US/294_bsw_strengthens_statement_repu_10_4_2004.asp.

44. These charges were clearly misplaced. The president of the council that oversaw the publication of the *Proceedings* admitted to the editor in an e-mail: "I have seen the review file and comments from 3 reviewers on the Meyer paper. All three with some differences among the comments recommended or suggested publication. I was surprised but concluded that there was not inappropriate behavior vs a vis [*sic*] the review process" (Roy McDiarmid, "Re: Request for information," January 28, 2005, 2:25 P.M., to Hans Sues, Congressional Staff Report, "Intolerance and the Politicization of Science at the Smithsonian" [December 2006]: 26, at http://www.discovery.org/scripts/viewDB/ filesDB-download.php?command=download&id=1489). See also Rick Sternberg, "Statement of Facts/Response to Misinformation," at http://www.richardsternberg.org/smithsonian.php?page=statement.

45. As Stephen Jay Gould wrote with other scientists and historians of science in a brief to the U.S. Supreme Court in 1993: "Judgments based on scientific evidence, whether made in a laboratory or a courtroom, are undermined by a categorical refusal even to consider research or views that contradict someone's notion of the prevailing 'consensus' of scientific opinion. . . . Automatically rejecting dissenting views that challenge the conventional wisdom is a dangerous fallacy, for almost every generally accepted view was once deemed eccentric or heretical. Perpetuating the reign of a supposed scientific orthodoxy in this way, whether in a research laboratory or in a courtroom, is profoundly inimical to the search for truth. . . . The quality of a scientific approach or opinion depends on the strength of its factual premises and on the depth and consistency of its reasoning, not on its appearance in a particular journal or on its popularity among other scientists." Brief Amici Curiae of Physicians, Scientists, and Historians of Science in Support of Petitioners, *Daubert v. Merrell Dow Pharmaceuticals, Inc.*, 509 U.S. 579 (1993).

46. For a comprehensive annotated bibliography of ID publications, see http://discovery.org/a/2640. Some critics of design have attempted to discredit the theory because its advocates have published their case primarily in books rather than in scientific articles. But this argument ignores the important role that books have played in the history of science in establishing new scientific ideas. Anyone who understands the role that technical journals play in science will understand why this is so. Science journals are a highly specialized and conservative genre. They publish research designed to fill out an established scientific research program. They are part of what philosopher of science Thomas Kuhn calls "normal science" (*The Structure of Scientific Revolutions*). New and revolutionary ideas in science are unlikely to appear first in their pages. If the history of science is any indication, then we should expect most of the initial work in any fundamentally new scientific perspective to appear first in books. And this is precisely the pattern of publication we see in the case of intelligent design. In the last decade or so, new evidence-based arguments for the theory have made their initial appearance in books. More recently, scientific articles have begun to appear, elucidating the theory in more detail.

47. Ruse, "Darwinism," 21.

48. Ruse, "Darwinism," 23.

49. Ruse, "A Philosopher's Day in Court"; "Witness Testimony Sheet," 301; "Darwinism," 21–26.

50. Meyer, "Laws, Causes and Facts."

51. A law of nature typically describes a general relationship between two or more different types of events, entities, or properties. Many have the form, "If A occurs, then B will always follow under conditions C," such as the laws, "Pure water at sea level heated to 100 degrees C will boil" and "All unsuspended bodies will fall." Other laws describe mathematical relationships between different entities or properties that apply universally, such as the law, "Force equals mass times acceleration," or the law, "The pressure of a gas is proportional to its volume times its temperature." In any case, laws are not events; they describe *relationships* (causal, logical, or mathematical) between different types of events, entities, or properties.

52. Scriven, "Causation as Explanation," 14; Lipton, *Inference to the Best Explanation,* 47–81.

53. In the *On the Origin of Species,* Darwin proposed both a mechanism (natural selection) and a historical theory—the theory of universal common descent. Evolutionary biologists debate whether natural selection can be formulated as a law. But within Darwin's argument, the theory of common descent had its own explanatory power. Yet it did not explain by reference to a law of nature. Instead, the theory of common descent explains by postulating a hypothetical pattern of events (as depicted in Darwin's famous tree of life). The theory of common descent makes a claim about what happened in the past—namely, that a series of unobserved transitional organisms existed, forming a genealogical bridge between presently existing life-forms—to account for a variety of presently observed evidence (such as the similarity of anatomical structures in different organisms or the pattern of fossil progression). Darwin himself referred to common descent as the *vera causa* (i.e., the actual cause) of a diverse set of biological observations (*On the Origin of Species,* 195, 481–82). And in the theory of common descent, a pattern of events, not a law, does what I call the "primary explanatory work."

54. See Dembski's and Marks' essay in this volume.

55. Ruse, *Darwinism Defended,* 59.

56. Grinnell, "Radical Intersubjectivity"; Scott, "Keep Science Free from Creationism."

57. Miller, *Only a Theory,* 87; Skoog, "View from the Past"; Sober, "What Is Wrong with Intelligent Design?"

58. Ruse, "Darwinism."

59. Miller, *Only a Theory,* 87.

60. *Kitzmiller v. Dover School District* 04 cv 2688 (December 20, 2005), 22, 77; Riordan, "Stringing Physics Along," 38.

61. *Kitzmiller v. Dover School District* 04 cv 2688 (December 20, 2005), 81; Jack Krebs, "A Summary of Objections to 'Intelligent Design,'" *Kansas Citizens for Science,* June 30, 2001 (http://www.sunflower.com/~jkrebs/ JCCC/04%20Summary_Objections.html).

62. Michigan Science Teachers Association, "Evolution Education and the Nature of Science," February 3, 2007 (http://www.msta-mich.org/downloads/about/2007-02-03.doc).

63. See www.signatureinthecell.com and, specifically, "The Scientific Status of Intelligent Design: The Methodological Equivalence of Naturalistic and Non-Naturalistic Origins Theories" and "The Demarcation of Science and Religion." See also Philip Kitcher's book, *Living with Darwin*, 9–14. Kitcher is a leading philosopher of science who rejects intelligent design, but thinks that the attempts to refute it using demarcation arguments fail. Kitcher argues that the problem with intelligent design is not that it isn't scientific, but that it is "discarded" or "dead science." He argues this largely because he thinks that intelligent design cannot explain the accumulation of "junk DNA." He does not, however, address the empirical argument for intelligent design based upon the information-bearing properties of DNA or the evidence showing that non-protein-coding DNA plays a crucial functional role in the cell.

64. Meyer, "A Scopes Trial for the '90s"; "Open Debate on Life's Origin."

65. Scott, "Keep Science Free from Creationism."

66. Grinnell, "Radical Intersubjectivity."

67. Meyer, "Of Clues and Causes," 120; Darwin, *On the Origin of Species*, 398; Hull, *Darwin and His Critics*, 45.

68. Oddly, some of these same people also claim that the theory has been tested and found wanting. For example, in 1999 the U.S. National Academy of Sciences (NAS) issued a statement against intelligent design that claimed that ID is "not science because [it is] not testable by the methods of science." Yet the NAS simultaneously stated, "Molecular data counter a recent proposition called 'intelligent design theory'" and asserted that "scientists have considered the hypotheses" of intelligent design and "rejected them because of a lack of evidence." National Academy of Sciences, *Science and Creationism*, 21, ix. Similarly, Gerald Skoog argues that "the claim that life is the result of a design created by an intelligent cause cannot be tested and is not within the realm of science." Then in the next paragraph he states, "Observations of the natural world also make these dicta [concerning the theory of intelligent design] suspect" ("View from the Past"). Clearly something cannot be both untestable in principle and rendered suspect by empirical observations.

69. Pennock, Expert Witness Report. Emphasis added.

70. Scott, "Keep Science Free from Creationism."

71. Judson, *The Eighth Day of Creation*, 157–90.

72. Francis Darwin, ed., *Life and Letters*, 1: 437, emphasis added.

73. Miller, *Only a Theory*, 87.

74. Meyer, "The Return of the God Hypothesis."

75. Miller, *Finding Darwin's God*, 238.

76. For an extensive list of such predictions, see Appendix A in my book *Signature in the Cell* (2009). Note too that neither type of theory describes events that will necessarily occur *repeatedly*, but both use uniform and *repeated* experience of cause and effect to make inferences about the most likely cause of various singular happenings.

77. Shermer, *Why Darwin Matters*, 75.

78. Shermer, *Why Darwin Matters*, 75.

79. See also Kitcher, *Living with Darwin*, 57.

80. Some advocates of intelligent design think that an intelligent cause is directly responsible for only the information present in the first living organisms; other ID advocates think intelligent design is responsible for the information necessary to produce subsequent forms of life as well. Those who hold the latter view predict that the non-protein-coding DNA in both eukaryotes and prokaryotes should perform functional roles. Those who hold the former view predict that only the noncoding DNA in prokaryotes should perform functional roles. The discovery that noncoding DNA plays an important functional role in both prokaryotic

and eukaryotic organisms confirms the prediction of the more expansive ID hypothesis. I hold this latter view. See Meyer, "The Origin of Biological Information." In this essay and in my book *Signature in the Cell*, however, I have argued only for intelligent design as the best explanation of the origin of the information necessary to build the first living cell.

81. Dembski, "Intelligent Science and Design." Here's what Dembski writes about junk DNA: "[Intelligent] design is not a science stopper. Indeed, design can foster inquiry where traditional evolutionary approaches obstruct it. Consider the term 'junk DNA.' Implicit in this term is the view that because the genome of an organism has been cobbled together through a long, undirected evolutionary process, the genome is a patchwork of which only limited portions are essential to the organism. Thus on an evolutionary view we expect a lot of useless DNA. If, on the other hand, organisms are designed, we expect DNA, as much as possible, to exhibit function."

82. ENCODE Project Consortium, "Identification and Analysis of Functional Elements."

83. Von Sternberg and Shapiro, "How Repeated Retroelements Format Genome Function."

84. Han, Szak, and Boeke, "Transcriptional Disruption by the L1 Retrotransposon"; Bethany Janowski, et al., "Inhibiting Gene Expression at Transcription Start Sites"; Goodrich and Kugel, "Non-coding-RNA Regulators of RNA Polymerase II Transcription"; Li, et al., "Small dsRNAs Induce Transcriptional Activation in Human Cells"; Pagano, et al., "New Small Nuclear RNA Gene-like Transcriptional Units"; Van de Lagemaat, et al., "Transposable Elements in Mammals"; Donnelly, Hawkins, and Moss, "A Conserved Nuclear Element"; Dunn, Medstrand, and Mager, "An Endogenous Retroviral Long Terminal Repeat"; Burgess-Beusse, et al., "The Insulation of Genes"; Medstrand, Landry, and Mager, "Long Terminal Repeats Are Used as Alternative Promoters"; Mariño-Ramírez, et al., "Transposable Elements Donate Lineage-Specific Regulatory Sequences to Host Genomes."

85. Green, "The Role of Translocation and Selection"; Figueiredo, et al., "A Central Role for *Plasmodium Falciparum* Subtelomeric Regions."

86. Henikoff, Ahmad, and Malik, "The Centromere Paradox"; Bell, West, and Felsenfeld, "Insulators and Boundaries"; Pardue and DeBaryshe, "Drosophila Telomeres"; Henikoff, "Heterochromatin Function in Complex Genomes"; Figueiredo, et al., "A Central Role for *Plasmodium Falciparum*"; Schueler, et al., "Genomic and Genetic Definition of a Functional Human Centromere."

87. Jordan, et al., "Origin of a Substantial Fraction"; Henikoff, Ahmad, and Malik, "The Centromere Paradox"; Schueler, et al., "Genomic and Genetic Definition of a Functional Human Centromere."

88. Chen, DeCerbo, and Carmichael, "Alu Element-Mediated Gene Silencing"; Jurka, "Evolutionary Impact of Human Alu Repetitive Elements."; Lev-Maor, et al., "The Birth of an Alternatively Spliced Exon"; Kondo-Iida, et al., "Novel Mutations and Genotype–Phenotype Relationships"; Mattick and Makunin, "Non-coding RNA."

89. McKenzie and Brennan, "The Two Small Introns of the Drosophila Affinidisjuncta Adh Gene"; Arnaud, et al., "SINE Retroposons Can Be Used In Vivo"; Rubin, Kimura, and Schmid, "Selective Stimulation of Translational Expression"; Bartel, "MicroRNAs"; Mattick and Makunin, "Small Regulatory RNAs in Mammals."

90. Dunlap, et al., "Endogenous Retroviruses"; Hyslop, et al., "Downregulation of NANOG Induces Differentiation"; Peaston, et al., "Retrotransposons Regulate Host Genes."

91. Morrish, et al., "DNA Repair Mediated"; Tremblay, Jasin, and Chartrand, "A Double-Strand Break in a Chromosomal LINE Element"; Grawunder, *et al.* "Activity of DNA Ligase IV"; Wilson, Grawunder, and Liebe, "Yeast DNA Ligase IV."

92. Mura, et al., "Late Viral Interference Induced"; Kandouz, et al., "Connexin43 Pseudogene Is Expressed."

93. Goh, et al., "A Newly Discovered Human Alpha Globin Gene"; Kandouz, et al., "Connexin43 Pseudogene Is Expressed"; Tam, et al., "Pseudogene-Derived Small Interfering RNAs"; Watanabe, et al., "Endogenous

siRNAs from Naturally Formed dsRNAs"; Piehler, et al., "The Human ABC Transporter Pseudogene Family."

94. Mattick and Gagen, "The Evolution of Controlled Multitasked Gene Networks"; Von Sternberg and Shapiro, "How Repeated Retroelements Format Genome Function."

95. In 1994, pro-ID scientist and writer Forrest M. Mims III submitted a letter to the journal *Science* (which was rejected) predicting function for junk DNA: "DNA that molecular biologists refer to as 'junk' don't necessarily appear so useless to those of us who have designed and written code for digital controllers. They have always reminded me of strings of NOP (No OPeration) instructions. A do-nothing string of NOPs might appear as 'junk code' to the uninitiated, but, when inserted in a program loop, a string of NOPs can be used to achieve a precise time delay. Perhaps the 'junk DNA' puzzle would be solved more rapidly if a few more computer scientists would make the switch to molecular biology" ("Rejected Publications"). See also Dembski's prediction already cited. In 2004, Jonathan Wells argued that the theory of intelligent design provides a fruitful heuristic (guide to discovery) for genomic research precisely because it predicts that noncoding DNA should have latent function. As he explained: "The fact that 'junk DNA' is not junk has emerged not because of evolutionary theory but in spite of it. On the other hand, people asking research questions in an ID framework would presumably have been looking for the functions of non-coding regions of DNA all along, and we might now know considerably more about them" ("Using Intelligent Design Theory to Guide Scientific Research"). Other scientists have noted how materialistic evolutionary theories have impeded scientific progress in the study of the genome. In 2002, Richard von Sternberg reported extensive evidence for functional junk-DNA, noting that "neo-Darwinian 'narratives' have been the primary obstacle to elucidating the effects of these enigmatic components of chromosomes" and concluding that "the selfish DNA narrative and allied frameworks must join the other 'icons' of neo-Darwinian evolutionary theory that, despite their variance with empirical evidence, nevertheless persist in the literature" ("On the Roles of Repetitive DNA Elements").

96. Meyer, *Signature in the Cell: DNA and the Evidence for Intelligent Design*, 481–95.

97. Jones, *Kitzmiller et al. v. Dover Area School Board*. As Judge Jones explained, "This self-imposed convention of science, which limits inquiry to testable, natural explanations about the natural world, is referred to by philosophers as 'methodological naturalism.'"

98. Murphy, "Phillip Johnson on Trial," 33. Nancey Murphy is a philosopher and seminary professor who strongly affirms methodological naturalism. Here's what she says in full: "Science *qua* science seeks naturalistic explanations for all natural processes. Christians and atheists alike must pursue scientific questions in our era without invoking a Creator. . . . Anyone who attributes the characteristics of living things to creative intelligence has by definition stepped into the arena of either metaphysics or theology."

99. Some might object to my description of methodological naturalism as a principle that excludes all intelligent causes from science. They could point out, correctly, that some scholars construe the principle of methodological naturalism (MN) as forbidding, invoking only *supernatural* intelligent causes, not intelligent causes in general, within science. Nevertheless, nothing follows from this objection. Interpreting the principle of MN in this more limited way doesn't justify disqualifying intelligent design from consideration as a scientific theory. If methodological naturalism merely forbids reference to supernatural causes in science, then the theory of intelligent design should qualify as a scientific theory. Why? Because the theory itself claims to do no more than establish an intelligent cause, not a supernatural intelligent cause, as the best explanation for the origin of biological information.

Now, clearly, some advocates of the theory of intelligent design think (as I do) that the designing intelligence responsible for life is most likely to be a supernatural deity. But again, not much follows from this about the scientific status of the theory itself. Some advocates of the theory of intelligent design do indeed think that the theory has theistic *implications*. So what? Even if one concedes a definition of science that forbids reference to supernatural entities, all that follows is that the putative *implications* of intelligent design, not the claims of the theory itself, are unscientific. Some scientists may even offer arguments in support of their contention that the designing intelligence responsible for life is most likely to be a supernatural deity. A definition of science that forbids reference to supernatural entities would then require classifying those arguments as unscientific extrapolations from the evidence. But this would not exclude the evidence *itself,* since intelligent design theory only allows an inference to the activity of an intelligent cause (not a supernatural intelligent cause) as the best explanation of it. Even so, mere classification of such argumentative extensions as unscientific or metaphysical or religious would not show them to be invalid; science has never had the only corner on truth, and judging the merit of such arguments requires intellectual engagement, not a disdainful anti-intellectualism.

100. Some scientists attempt to justify methodological naturalism by using another demarcation argument against intelligent design: ID does not cite a mechanism as the cause of the origin of biological form or information. But this demarcation argument assumes without justification that all scientifically acceptable causes are *mechanistic* causes. To insist that all causal explanations in science must be mechanistic is to insist that all causal theories must refer only to material entities. A "mechanism" is, after all, just another word for a material cause. Yet this requirement is merely another expression of the principle of methodological naturalism, for which there is no noncircular justification. Furthermore, as Bruce Gordon discusses in his essay on quantum theory in this volume, on pain of empirical contradiction, nonlocal quantum statistical phenomena have *no* material explanation; in short, the mathematical descriptions of the quintessential expression of modern science *preclude* the universality of material mechanism as an explanatory principle for natural phenomena. Strangely enough, however, no one seems to be objecting to quantum mechanics on the grounds that it violates methodological naturalism!

As I have argued throughout this essay, scientists have tried to justify a categorical exclusion of intelligent causes from science (i.e., methodological naturalism) by reference to specific demarcation criteria such as "testability," "observability," or "must explain by natural law." But, as I have shown, these arguments have failed to justify the exclusion of intelligent causes from science and, thus, the principle of methodological naturalism as a rule for science. The failure of demarcation arguments against intelligent design (or in favor of methodological naturalism) has left scientists without any justification for treating methodological naturalism as a normative rule of scientific practice. Attempts to justify this convention simply restate it in another form: "Scientific theories must cite mechanisms." Thus, it provides no grounds for excluding consideration of intelligent causes in scientific theories, even if the intelligence in question is ultimately immaterial.

This demarcation argument against intelligent design clearly assumes the point at issue, which is whether there are independent and metaphysically neutral reasons for preferring exclusively materialistic causal explanations of origins over explanations that invoke entities such as creative intelligence, conscious agency, mental action, intelligent design or mind—entities that may not ultimately be reducible to matter alone. Since demarcation arguments have failed to provide such reasons, and since we know from first-person experience that our choices and actions as conscious intelligent agents do cause certain kinds of events, structures, and systems to arise—that minds have real causal powers—there does not seem to be any reason for prohibiting scientists from considering this type of cause as a possible explanation for certain kinds of effects. Instead, there is every reason to consider intelligent causes as explanations for effects that are known to arise only from intelligent activity.

101. Some have argued that the theory of intelligent design is unscientific because it doesn't cite a mechanism to explain how the designing intelligence responsible for life arranged the constituent parts of life. But this is also true in our own experience. We do not know how our minds influence the material substrate of our brains, the actions of our bodies, or through them the material world around us. Nevertheless, we have good reason to think *that* our conscious thoughts and decisions do influence the material world. Moreover, we can often know or infer *that* intelligent thought played a role in the arrangement of matter or outcome of events

without knowing exactly *how* mind influences matter. It's hard to see how this limitation in our understanding makes the theory of intelligent design unscientific. Many scientific theories do not explain events, evidence, or phenomena by reference to any cause whatsoever, let alone a mechanistic one. Newton's universal law of gravitation was no less a scientific theory because Newton failed—indeed, refused—to postulate a mechanism or cause for the regular pattern of attraction his law describes. As we noted in the last footnote, quantum mechanics provides another sterling example of an acausal non-mechanistic theory in modern science. In addition, many historical theories about *what* happened in the past stand on their own without any mechanistic or causal theory about *how* the events to which such theories attest could have occurred. The theory of universal common descent is generally regarded as a scientific theory even though some scientists do not think there is currently an adequate mechanism to explain how transmutation between lines of descent was achieved. In the same way, there seems little justification for asserting that the geological theory of continental drift became scientific only after the advent of plate tectonics. Although the mechanism provided by plate tectonics certainly helped render continental drift a more persuasive theory, it was nevertheless not strictly necessary to know the mechanism by which continental drift *occurs* (1) to know or theorize that drift *had occurred* or (2) to regard a theory of continental drift as scientific. In a similar way, advocates of design can affirm and detect *that* intelligence played a causal role in the design of life without knowing exactly *how* mind exerts its influence over matter. All that follows from this admission is that intelligent design does not provide an answer to every question, not that it is an unscientific (or unjustified) explanation in relation to the question it *does* answer.

102. Jones, *Kitzmiller et al. v. Dover Area School Board*.

103. There are many types of scientific inquiry in which the convention of methodological naturalism does no damage to the intellectual freedom of scientists. Consider a scientist investigating the question, "How does atmospheric pressure affect crystal growth?" The answer "Crystals were designed by a creative intelligence" (or, for that matter, "Crystals evolved via undirected natural processes") entirely misses the point of the question. The question demands an answer expressed as a general relationship describing the interaction of two physical entities: gases and crystals. A scientific law expressing such a relationship necessarily describes entirely materialistic entities.

Methodological naturalism does not limit the freedom of scientists to theorize in this case. Instead, the question motivating their inquiry ensures that scientists will only consider certain kinds of answers and that these answers will necessarily describe materialistic entities. In this case, methodological naturalism does no harm, though neither is it necessary to guide scientists to appropriate theories. Instead, the implicit question motivating the inquiry will do this on its own. Methodological naturalism and its prohibition against inferring creative intelligence does inhibit inquiry in historical sciences such as archaeology, forensics, paleobiology, cosmology, anthropology, and origin-of-life studies, however, because in the historical sciences researchers are addressing different kinds of questions. They are asking about the causal histories of particular events, events in which the purposeful design of intelligent agents might have played a role. Scientists investigating the origin of life, for example, are motivated by the question, "What happened to cause life to arise on Earth?" Since, conceivably, an intelligent designer could have played a causal role in the origin of life, any rule that prevents scientists from considering that possibility prevents scientists from considering a possibly true hypothesis.

104. Bridgman, *Reflections of a Physicist*, 535.

105. I argue this point at length in my other essay in this volume, "DNA: The Signature in the Cell."

References

Arnaud, Phillipe, Chantal Goubely, Thierry Pe'Lissier, and Jean-Marc Deragon. "SINE Retroposons Can Be Used In Vivo as Nucleation Centers for De Novo Methylation." *Molecular and Cellular Biology* 20 (2000): 3434–41.

Ayala, Francisco. "Darwin's Revolution." In *Creative Evolution?!* edited by J. Campbell and J. Schopf, 4–5. Boston: Jones and Bartlett, 1994.

Bartel, David. "MicroRNAs: Genomics, Biogenesis, Mechanism, and Function." *Cell* 116 (2004): 281–97.

Behe, Michael. Darwin's Black Box: The Biochemical Challenge to Evolution. New York: Free Press, 1996.

———. "Self-Organization and Irreducibly Complex Systems: A Reply to Shanks and Joplin." *Philosophy of Science* 67 (2000): 155–62.

Behe, Michael, and D. W. Snoke. "Simulating Evolution by Gene Duplication of Protein Features That Require Multiple Amino Acid Residues." *Protein Science* 13 (2004): 2651–64.

Bell, C., A. G. West, and G. Felsenfeld. "Insulators and Boundaries: Versatile Regulatory Elements in the Eukaryotic Genome." *Science* 291 (2001): 447–50.

Bridgman, Percy W. *Reflections of a Physicist.* 2nd ed. New York: Philosophical Library, 1955.

Burgess-Beusse, B., C. Farrell, M. Gaszner, M. Litt, V. Mutskov, F. Recillas-Targa, M. Simpson, A. West, and G. Felsenfeld. "The Insulation of Genes from External Enhancers and Silencing Chromatin." *Proceedings of the National Academy of Sciences USA* 99 (2002): 16433–37.

Campbell, John Angus, and Stephen C. Meyer, eds. *Darwinism, Design and Public Education.* East Lansing: Michigan State University Press, 2003.

Chen, Ling-Ling, Joshua N. DeCerbo, and Gordon G. Carmichael. "Alu Element-Mediated Gene Silencing." *EMBO Journal* (2008): 1–12.

Collins, Robin. "Evidence for Fine-Tuning," in Neil Manson, ed. *God and Design: The Teleological Argument and Modern Science.* New York: Routledge, 2003.

Craig, William Lane. "Barrow and Tipler on the Anthropic Principle vs. Divine Design." *British Journal for the Philosophy of Science* 38 (1988): 389–95.

———. "God, Creation and Mr. Davies." *British Journal for the Philosophy of Science* 37 (1986): 163–75.

Darwin, Charles. *On the Origin of Species by Means of Natural Selection.* A facsimile of the first edition, published by John Murray, London, 1859. Reprint, Cambridge, MA: Harvard University Press, 1964.

———. *Life and Letters of Charles Darwin, Including an Autobiographical Chapter.* Edited by Francis Darwin. 2 vols. New York: Appleton, 1898.

Davison, John A. "A Prescribed Evolutionary Hypothesis," *Rivista di Biologia/Biology Forum* 98 (2005): 155–66.

Dawkins, Richard. The Blind Watchmaker: Why the Evidence Reveals a Universe Without Design. New York: Norton, 1987.

Dembski, William. "Intelligent Science and Design." *First Things* 86 (1998): 21–27.

———. The Design Inference: Eliminating Chance Through Small Probabilities. Cambridge: Cambridge University Press, 1998.

———. ed. *Mere Creation: Science, Faith and Intelligent Design.* Downers Grove, IL: InterVarsity, 1998.

———. "Specification: The Pattern That Signifies Intelligence." *Philosophia Christi* (2005): 299–343. http://www.designinference.com/documents/2005.06.Specification.pdf.

Dembski, William A., and Robert J. Marks II. "The Conservation of Information: Measuring the Information Cost of Successful Search." *Transactions on Systems, Man and Cybernetics, Part A. http://ieeexplore.ieee.org/xpl/RecentIssue.jsp?puNumber=3468/.* Forthcoming.

———. "The Search for a Search: Measuring the Information Cost of Higher Level Search." *The International Journal of Information Technology and Intelligent Computing* (2008). http://itic.wshe.lodz.pl/.

Dembski, William A., and Michael Ruse, eds. *Debating Design: From Darwin to DNA.* Cambridge: Cambridge University Press, 2004.

DeWolf, David K., John West, and Casey Luskin. "Intelligent Design Will Survive *Kitzmiller v. Dover.*" *Montana Law Review* 68 (2007): 7–57.

DeWolf, David K., John G. West, Casey Luskin, and Jonathan Witt. *Traipsing into Evolution: Intelligent Design and the* Kitzmiller vs. Dover *Decision.* Seattle, WA: Discovery Institute Press, 2006.

Discovery Institute. Brief of Amicus Curiae (Revised). In *Kitzmiller v. Dover*, 400 F.Supp. 707 (M.D.Pa. 2005), 17. http://www.discovery.org/scripts/viewDB/filesDB-download.php?command=download&id=646. Appendix A, http://www.discovery.org/scripts/viewDB/filesDB-download.php?command=download&id=647.

Donnelly, S. R., T. E. Hawkins, and S. E. Moss. "A Conserved Nuclear Element with a Role in Mammalian Gene Regulation." *Human Molecular Genetics* 8 (1999): 1723–28.

Dunlap, K. A., M. Palmarini, M. Varela, R. C. Burghardt, K. Hayashi, J. L. Farmer, and T. E. Spencer. "Endogenous Retroviruses Regulate Periimplantation Placental Growth and Differentiation." *Proceedings of the National Academy of Sciences USA* 103 (2006): 14390–395.

Dunn, C. A., P. Medstrand, and D. L. Mager. "An Endogenous Retroviral Long Terminal Repeat Is the Dominant Promoter for Human B1,3-galactosyltransferase 5 in the Colon." *Proceedings of the National Academy of Sciences USA* 100 (2003):12841–846.

ENCODE Project Consortium. "Identification and Analysis of Functional Elements in 1% of the Human Genome by the ENCODE Pilot Project." *Nature* 447 (2007): 799–816.

Figueiredo, L. M., L. H. Freitas-Junior, E. Bottius, J.-C. Olivo-Marin, and A. Scherf. "A Central Role for *Plasmodium Falciparum* Subtelomeric Regions in Spatial Positioning and Telomere Length Regulation." *EMBO Journal* 21 (2002): L815–24.

Forrest, Barbara. Expert Witness Report. *Kitzmiller v. Dover,* 45. http://www2.ncseweb.org/kvd/experts/2005_04_01_Forrest_expert_report_P.pdf.

Forrest, Barbara, and Paul R. Gross. *Creationism's Trojan Horse: The Wedge of Intelligent Design.* New York: Oxford University Press, 2004.

Giles, Jim. "Peer-Reviewed Paper Defends Theory of Intelligent Design." *Nature* 431 (2004): 114.

Goh, S.-H., Y. T. Lee, N. V. Bhanu, M. C. Cam, R. Desper, B. M. Martin, R. Moharram, R. B. Gherman, and J. L. Miller. "A Newly Discovered Human Alpha Globin Gene." *Blood* DOI 10.1182/blood-2005–03–0948.

Gonzalez, Guillermo, and Jay W. Richards. The Privileged Planet: How Our Place in the Cosmos Was Designed for Discovery. Washington, DC: Regnery, 2004.

Goodrich, J. A., and J. F. Kugel. "Non-coding-RNA Regulators of RNA Polymerase II Transcription." *Nature Reviews Molecular and Cell Biology* 7 (2006): 612–16.

Gould, Stephen J. "Evolution and the Triumph of Homology: Or, Why History Matters." *American Scientist* 74 (1986): 60–69.

Grawunder, U., M. Wilm, X. Wu, P. Kulesza, T. E. Wilson, M. Mann, and M. R. Lieber. "Activity of DNA Ligase IV Stimulated by Complex Formation with XRCC4 Protein in Mammalian Cells." *Nature* 388 (1997): 492–95.

Green, David G. "The Role of Translocation and Selection in the Emergence of Genetic Clusters and Modules." *Artificial Life* 13 (2007): 249–58.

Grinnell, Fred. "Radical Intersubjectivity: Why Naturalism Is an Assumption Necessary for Doing Science." In *Darwinism: Science or Philosophy?* edited by Jon Buell and Virginia Hearn, 99–105. Richardson, TX: Foundation for Thought and Ethics, 1994.

Han, Jeffrey S., Suzanne T. Szak, and Jef D. Boeke. "Transcriptional Disruption by the L1 Retrotransposon and Implications for Mammalian Transcriptomes." *Nature* 429 (2004): 268–74.

Henikoff, Steven. "Heterochromatin Function in Complex Genomes." *Biochimica et Biophysica Acta* 1470 (2000): 01–08.

Henikoff, Steven, Kami Ahmad, and Harmit S. Malik. "The Centromere Paradox: Stable Inheritance with Rapidly Evolving DNA." *Science* 293 (2001): 1098–1102.

Holden, Constance. "Random Samples." *Science* 305 (2004): 1709.

Hull, David L. *Darwin and His Critics*. Chicago: University of Chicago Press, 1973.

Hyslop, L., M. Stojkovic, L. Armstrong, T. Walter, P. Stojkovic, S. Przyborski, M. Herbert, A. Murdoch, T. Strachan, and M. Lakoa. "Downregulation of NANOG Induces Differentiation of Human Embryonic Stem Cells to Extraembryonic Lineages." *Stem Cells* 23 (2005): 1035–43.

Janowski, Bethany, A., K. E. Huffman, J. C. Schwartz, R. Ram, D. Hardy, D. S. Shames, J. D. Minna, and D. R. Corey. "Inhibiting Gene Expression at Transcription Start Sites in Chromosomal DNA with Antigene RNAs." *Nature Chemical Biology* 1 (2005): 216–22.

Jones, Judge John E., III. Decision in *Kitzmiller et al. v. Dover Area School Board*. N0.04cv2688, 2005 WL 2465563, *66 (M.D.Pa. Dec. 20, 2005). http://www.pamd.uscourts.gov/kitzmiller/kitzmiller_342.pdf.

———. Kitzmiller et al. v. Dover Area School District. 400 F.Supp.2d 707, 745 (M.D.Pa. 2005).

Jordan, I. K., I. B. Rogozin, G. V. Glazko, and E. V. Koonin. "Origin of a Substantial Fraction of Human Regulatory Sequences from Transposable Elements." *Trends in Genetics* 19 (2003): 68–72.

Judson, Horace Freeland. *The Eighth Day of Creation: Makers of the Revolution in Biology*. Exp. ed. Plainview, NY: Cold Spring Harbor Laboratory Press, 1996.

Jurka, Jerzy. "Evolutionary Impact of Human Alu Repetitive Elements." *Current Opinion in Genetics and Development* 14 (2004): 603–8.

Kandouz, M., A. Bier, G. D. Carystinos, M. A. Alaoui-Jamali, and G. Batist. "Connexin43 Pseudogene Is Expressed in Tumor Cells and Inhibits Growth." *Oncogene* 23 (2004): 4763–70.

Kitcher, Philip. *Living with Darwin: Evolution, Design, and the Future of Faith*. New York: Oxford University Press, 2007.

Klinghoffer, David. "The Branding of a Heretic." *Wall Street Journal,* January 28, 2005, national edition, W11.

Kondo-Iida, E., K. Kobayashi, M. Watanabe, J. Sasaki, T. Kumagai, H. Koide, K. Saito, M. Osawa, Y. Nakamura, and T. Toda. "Novel Mutations and Genotype–Phenotype Relationships in 107 Families with Fukuyama-Type Congenital Muscular Dystrophy (FCMD)." *Human Molecular Genetics* 8 (1999): 2303–9.

Laudan, Larry. "The Demise of the Demarcation Problem." In *But Is It Science?* edited by Michael Ruse, 337–50. Buffalo, NY: Prometheus, 1988.

———. Beyond Positivism and Relativism: Theory, Method, and Evidence. Boulder, CO: Westview, 1996.

Lev-Maor, G., et al. "The Birth of an Alternatively Spliced Exon: 3′ Splice-Site Selection in Alu Exons." *Science* 300 (2003): 1288–91.

Li, Long-Cheng, S. T. Okino, H. Zhao, H., D. Pookot, R. F. Place, S. Urakami, H. Enokida, and R. Dahiya. "Small dsRNAs Induce Transcriptional Activation in Human Cells." *Proceedings of the National Academy of Sciences USA* 103 (2006): 17337–342.

Lipton, Peter. *Inference to the Best Explanation*. London and New York: Routledge, 1991.

Lönnig, Wolf-Ekkehard, and Heinz Saedler. "Chromosome Rearrangements and Transposable Elements." *Annual Review of Genetics* 36 (2002): 389–410.

Luskin, Casey. "Will Americans United Retract Their Demonstrably False Claims?" http://www.opposingviews.com/counters/will-americans-united-retract-theirdemonstrably-false-claims.

Mariño-Ramírez, L., K. C. Lewis, D. Landsman, and I. K. Jordan. "Transposable Elements Donate Lineage-Specific Regulatory Sequences to Host Genomes." *Cytogenetic and Genome Research* 110 (2005): 333–41.

Mattick, John S., and Michael J. Gagen. "The Evolution of Controlled Multitasked Gene Networks: The Role of Introns and Other Noncoding RNAs in the Development of Complex Organisms." *Molecular Biology and Evolution* 18 (2001): 1611–30.

Mattick, John S., and I. V. Makunin. "Small Regulatory RNAs in Mammals." *Human Molecular Genetics* 14 (2005): R121–32.

———. "Non-coding RNA." *Human Molecular Genetics* 15 (2006): R17–29.

McKenzie, Richard W., and Mark D. Brennan. "The Two Small Introns of the Drosophila Affinidisjuncta Adh Gene Are Required for Normal Transcription." *Nucleic Acids Research* 24 (1996): 3635–42.

Medstrand, P., J. R. Landry, and D. L. Mager. "Long Terminal Repeats Are Used as Alternative Promoters for the Endothelin B Receptor and Apolipoprotein C-I Genes in Humans." *Journal of Biological Chemistry* 276 (2001): 1896–1903.

Meyer, Stephen C. *Of Clues and Causes: A Methodological Interpretation of Origin of Life Studies.* Ph.D. dissertation. Cambridge University, 1990.

———. "A Scopes Trial for the '90s." *Wall Street Journal,* December 6, 1993, A14.

———. "Open Debate on Life's Origin." *Insight,* February 21, 1994, 27–29.

———. "Laws, Causes and Facts: A Response to Michael Ruse." In *Darwinism: Science or Philosophy?* edited by Jon Buell and Virginia Hearn, 29–40. Richardson, TX: Foundation for Thought and Ethics, 1994.

———. "The Return of the God Hypothesis." *Journal of Interdisciplinary Studies* 11 (1999): 1–38.

———. "DNA and the Origin of Life: Information, Specification and Explanation." In *Darwinism, Design and Public Education,* edited by John Angus Campbell and Stephen C. Meyer, 223–85. East Lansing: Michigan State University Press, 2003.

———. "The Cambrian Information Explosion: Evidence for Intelligent Design." In *Debating Design: From Darwin to DNA,* edited by W. Dembski and M. Ruse, 371–91. Cambridge: Cambridge University Press, 2004.

———. "The Origin of Biological Information and the Higher Taxonomic Categories." *Proceedings of the Biological Society of Washington* 117 (2004): 213–39.

———. *Signature in the Cell: DNA and the Evidence for Intelligent Design.* San Francisco: HarperOne, 2009.

Miller, Kenneth R. *Finding Darwin's God: A Scientist's Search for Common Ground Between God and Evolution.* New York: HarperCollins, 1999.

———. *Only a Theory: Evolution and the Battle for America's Soul.* New York: Viking, 2008.

Minnich, Scott A., and Stephen C. Meyer. "Genetic Analysis of Coordinate Flagellar and Type III Regulatory Circuits in Pathogenic Bacteria." In *Design and Nature II: Comparing Design in Nature with Science and Engineering,* edited by M. W. Collins and C. A. Brebbia, 295–304. Southampton: Wessex Institute of Technology, 2004.

Morrish, Tammy A., Nicolas Gilbert, Jeremy S. Myers, Bethaney J. Vincent, Thomas D. Stamato, Guillermo E. Taccioli, Mark A. Batzer, and John V. Moran. "DNA Repair Mediated by Endonuclease-Independent LINE-1 Retrotransposition." *Nature Genetics* 31 (2002): 159–65.

Mura, M., P. Murcia, M. Caporale, T. E. Spencer, K. Nagashima, A. Rein, and M. Palmarini. "Late Viral Interference Induced by Transdominant Gag of an Endogenous Retrovirus." *Proceedings of the National Academy of Sciences USA* 101 (2004): 11117–122.

Murphy, Nancey. "Phillip Johnson on Trial: A Critique of His Critique of Darwin." *Perspectives on Science and Christian Faith* 45 (1993): 26–36.

Nagel, Thomas. "Public Education and Intelligent Design." *Philosophy and Public Affairs* 36 (2008): 187–205.

National Academy of Sciences. *Science and Creationism: A View from the National Academy of Sciences.* 2nd ed. Washington, DC: National Academy Press, 1999.

Nelson, Paul, and Jonathan Wells. "Homology in Biology: Problem for Naturalistic Science and Prospect for Intelligent Design." In *Darwinism, Design and Public Education,* edited by John Angus Campbell and Stephen C. Meyer, 303–22. East Lansing: Michigan State University Press, 2003.

Pagano, A. M. Castelnuovo, F. Tortelli, R. Ferrari, G. Dieci, and R. Cancedda. "New Small Nuclear RNA Gene-like Transcriptional Units as Sources of Regulatory Transcripts." *PLoS Genetics* 3 (2007): e1.

Pardue, M.-L., and P. G. DeBaryshe. "Drosophila Telomeres: Two Transposable Elements with Important Roles in Chromosomes." *Genetica* 107 (1999): 189–96.

Peaston, E., A. V. Evsikov, J. H. Graber, W. N. de Vries, A. E. Holbrook, D. Solter, and B. B. Knowles. "Retrotransposons Regulate Host Genes in Mouse Oocytes and Preimplantation Embryos." *Developmental Cell* 7 (2004): 597–606.

Pennock, Robert. Expert Witness Report. *Kitzmiller v. Dover.* 400 F.Supp.2d 707 (M.D.Pa. 2005), 20. https://www.msu.edu/~pennock5/research/papers/Pennock_DoverExptRpt.pdf.

Piehler, A. P., M. Hellum, J. J. Wenzel, E. Kaminski, K. B. Haug, P. Kierulf, and W. E. Kaminski. "The Human ABC Transporter Pseudogene Family: Evidence for Transcription and Gene-Pseudogene Interference." *BMC Genomics* 9 (2008): 165.

Powell, Michael. "Editor Explains Reasons for 'Intelligent Design' Article." *Washington Post,* August 19, 2005, district edition. http://www.washingtonpost.com/wp-dyn/content/article/2005/08/18/AR2005081801680_pf.html.

Ratliff, Evan. "The Crusade Against Evolution." *Wired,* October 2004. http://www.wired.com/wired/archive/12.10/evolution.html.

Riordan, Michael. "Stringing Physics Along." *Physics World* 20 (2007): 38–39.

Rubin, C. M., R. H. Kimura, and C. W. Schmid. "Selective Stimulation of Translational Expression by Alu RNA." *Nucleic Acids Research* 30 (2002): 3253–61.

Ruse, Michael. *Darwinism Defended: A Guide to the Evolution Controversies.* London: Addison-Wesley, 1982.

———. "A Philosopher's Day in Court." In *But Is It Science?* edited by Michael Ruse, 13–38. Buffalo, NY: Prometheus, 1988.

———. "Witness Testimony Sheet: *McLean* v. *Arkansas.*" In *But Is It Science?* edited by Michael Ruse, 287–306. Buffalo, NY: Prometheus, 1988.

———. "Darwinism: Philosophical Preference, Scientific Inference and Good Research Strategy." In *Darwinism: Science or Philosophy?* edited by Jon Buell and Virginia Hearn, 21–28. Richardson, TX: Foundation for Thought and Ethics, 1994.

Schueler, Mary G., Anne W. Higgins, M. Katharine Rudd, Karen Gustashaw, Huntington F. Willard, "Genomic and Genetic Definition of a Functional Human Centromere." *Science,* 294 (2001): 109–15.

Scott, Eugenie C. "Keep Science Free from Creationism." *Insight,* February 21, 1994, 30.

Scriven, Michael. "Truisms as the Grounds for Historical Explanations." In *Theories of History,* edited by P. Gardiner. Glencoe, IL: Free Press, 1959.

———. "Explanation and Prediction in Evolutionary Theory." *Science* 130 (1959): 477–82.

———. "Causes, Connections, and Conditions in History." In *Philosophical Analysis and History,* edited by W. Dray, 238–64. New York: Harper & Row, 1966.

———. "Causation as Explanation." *Nous* 9 (1975): 3–15.

Shermer, Michael. Why Darwin Matters: The Case Against Intelligent Design. New York: Times/Holt, 2006

Skoog, Gerald. "A View from the Past." *Bookwatch Reviews* 2 (1989): 1–2.

Sober, Elliott "What Is Wrong with Intelligent Design?" *Quarterly Review of Biology,* 82 (2007): 3–8.

Tam, O. H., A. A. Aravin, P. Stein, A. Girard, E. P. Murchison, S. Cheloufi, E. Hodges, M. Anger, R. Sachidanandam, R. M. Schultz, and G. J. Hannon. "Pseudogene-Derived Small Interfering RNAs Regulate Gene Expression in Mouse Oocytes." *Nature* 453 (2008): 534–38.

Thaxton, Charles, Walter L. Bradley, and Roger L. Olsen. *The Mystery of Life's Origin: Reassessing Current Theories.* New York: Philosophical Library, 1984.

Tremblay, A., M. Jasin, and P. Chartrand. "A Double-Strand Break in a Chromosomal LINE Element Can Be Repaired by Gene Conversion with Various Endogenous LINE Elements in Mouse Cells." *Molecular and Cellular Biology* 20 (2000): 54–60.

Van de Lagemaat, L. N., J. R. Landry, D. L. Mager, and P. Medstrand. "Transposable Elements in Mammals Promote Regulatory Variation and Diversification of Genes with Specialized Functions." *Trends in Genetics* 19 (2003): 530–36.

Vergano, Dan, and Greg Toppo. "'Call to Arms' on Evolution." *USA Today,* March 23, 2005. http://www.usatoday.com/news/education/2005-03-23-evolution_x.htm.

Voie, Albert. "Biological Function and the Genetic Code Are Interdependent," *Chaos, Solutions and Fractals* 28 (2006): 1000–4.

Von Sternberg, Richard. "On the Roles of Repetitive DNA Elements in the Context of a Unified Genomic-Epigenetic System." *Annals of the New York Academy of Sciences* 981 (2002): 154–88.

Von Sternberg, Richard, and James A. Shapiro. "How Repeated Retroelements Format Genome Function." *Cytogenetic and Genome Research* 110 (2005): 108–16.

Wallis, Claudia. "The Evolution Wars." *Time,* August 7, 2005. http://www.time.com/time/magazine/article/0,9171,1090909,00.html.

Watanabe T., Y. Totoki, A. Toyoda, M. Kaneda, S. Kuramochi-Miyagawa, Y. Obata, H. Chiba, Y. Kohara, T. Kono, T. Nakano, M.A. Surani, Y. Sakaki, and H. Sasaki. "Endogenous siRNAs from Naturally Formed dsRNAs Regulate Transcripts in Mouse Oocytes." *Nature* 453 (2008): 539–43.

Wells, Jonathan. "Using Intelligent Design Theory to Guide Scientific Research." *Progress in Complexity, Information, and Design* (November 2004): 3.1.2.

———. "Do Centrioles Generate a Polar Ejection Force?" *Rivista di Biologia/Biology Forum* 98 (2005): 37–62.

Wilgoren, Jodi. "Politicized Scholars Put Evolution on the Defensive." *New York Times,* August 21, 2005, national edition. http://www.nytimes.com/2005/08/21/national/21evolve.html.

Worden, Amy. "Bad Frog Beer to 'Intelligent Design': The Controversial Ex-Pa. Liquor Board Chief Is Now U.S. Judge in the Closely Watched Trial." *Philadelphia Inquirer,* October 16, 2005.

West, John G., and David K. DeWolf. "A Comparison of Judge Jones's Opinion in *Kitzmiller v. Dover* with Plaintiffs' Proposed 'Findings of Fact and Conclusions of Law.'"

Wilson, Edward O. "Intelligent Evolution: The Consequences of Charles Darwin's 'One Long Argument.'" *Harvard Magazine* (November-December 2005): 29–33.

Wilson, T. E., U. Grawunder, and M. R. Liebe. "Yeast DNA Ligase IV Mediates Non-Homologous DNA End Joining." *Nature* 388 (1997): 495–98.

PART II

THE EPISTEMOLOGICAL AND ONTOLOGICAL FOUNDATIONS OF NATURALISM

Introduction

There are many varieties of naturalism, as well as many varieties of realism, antirealism, and nonrealism. The essays in Part II explore three foundational questions in relation to carefully defined variants of these philosophical positions: (1) If naturalism were true, what would the implications be for epistemology in general and the status of scientific knowledge in particular? (2) If we take seriously what fundamental physics reveals about the physical world, is naturalism ontologically credible? (3) Regardless of whether naturalism is true, must conceptual schemes always interpose themselves between the human subject and the world in such a way as to require antirealism or nonrealism? The contributors to this discussion explore these themes from a variety of philosophical convictions that lead them to defend different conclusions.

Alvin Plantinga and **William Talbott** provide updated arguments on either side of a question Plantinga first raised in the early 1990s. In "Evolution versus Naturalism," Plantinga argues that the conjunction of naturalism with current evolutionary theory is self-defeating: the probability that our cognitive faculties are reliable, given naturalism (N) and the proposition that our faculties have arisen by way of the mechanisms suggested by current evolutionary theory (E), is either low or inscrutable. Both alternatives give one who accepts N&E a defeater for the proposition that his cognitive faculties are reliable, which in turn provides a defeater for anything else he believes, including N&E itself. The epistemological problem here, as Plantinga argues, is generated by naturalism, not evolutionary theory per se. Talbott demurs, contending in "More on the Illusion of Defeat" that such is not the case and that Plantinga's argument is unsuccessful: the "defeat" of evolutionary naturalism is an illusion because Plantinga has mischaracterized the role of undercutting defeaters in the reasoning process. The polemics continue with another iteration of this exchange in which Plantinga replies to Talbott's criticisms ("It's Not an Illusion!") and Talbott offers a final rejoinder ("The End of an Illusion?"). The reader is left to decide which author has the upper hand, and why.

The next four essays argue for differing views of the relationship between naturalism and scientific knowledge. Beginning with the constraints quantum theory places on physical reality, **Bruce Gordon** contends in "A Quantum-Theoretic Argument against Naturalism" that what we know from fundamental physical theory is inconsistent with metaphysical naturalism:

the ontological significance of nonlocality, nonlocalizability, and other quantum phenomena, when properly understood, requires reconceptualizing the fundamental constituents of material reality, physical law, and causality in a way that renders naturalism untenable. Returning to epistemological questions, **Robert Koons** argues in "The Incompatibility of Naturalism and Scientific Realism" that philosophical naturalism *precludes* the possibility of scientific realism, and that this reflects poorly on the viability of naturalism. In "Truth and Realism," **Alvin Goldman** maintains to the contrary that realism is perfectly compatible with naturalism, but that nonrealism and antirealism, on the other hand, are philosophically untenable. Finally, rounding out the defense of possible viewpoints, **Michael Williams** rejects the positions of both Koons and Goldman, contending in "Must Naturalists Be Realists?" that philosophical naturalists *should not* in fact be realists and that this state of affairs allows science to function just fine.

Drawing Part II to a close, **Nicholas Wolterstorff**'s essay "The Role of Concepts in Our Access to Reality" engages the antirealist and nonrealist heritage in science through the lens of Kantian epistemology, asking whether conceptualization always interposes itself between human beings and the world in such a way as to make knowledge of the world-in-itself unachievable. His answer is that it does not, and that it is high time we dispensed with this portion of Kant's legacy, which, he argues, is the product of misinterpreting Kant, and not itself any part of Kant's own view.

5

Evolution versus Naturalism[1]

Alvin C. Plantinga

Take *philosophical naturalism* to be the belief that there is no such person as God or anything like God. Naturalism, so taken, is stronger than atheism; it is possible to be an atheist without rising to the heights (or sinking to the depths) of naturalism. A follower of Hegel could be an atheist, but because of his belief in the Absolute would not qualify for naturalism; similarly for someone who believed in the Stoic's *Nous,* or Plato's Idea of the Good or Aristotle's Prime Mover who thinks only about himself. This definition of naturalism, of course, is a bit vague: how much like God must an entity be so that endorsing it disqualifies one from naturalism? Perhaps the definition will be serviceable nonetheless; clear examples of naturalists would be the Bertrand Russell of "A Free Man's Worship," Daniel Dennett (*Darwin's Dangerous Idea*), Richard Dawkins (*The Blind Watchmaker*), the late Stephen Jay Gould, David Armstrong, and the many others that are sometimes said to endorse "The Scientific Worldview."[2]

Now naturalism is not, as it stands, a religion. Nevertheless it performs one of the most important functions of a religion: it provides its adherents with a worldview. It tells us what the world is fundamentally like, what is most important and deep in the world, how we are related to other creatures in the world, what our place in the world is, what (if anything) we can expect after death, and the like. A religion typically does that and more; it also involves worship and ritual. These latter are ordinarily (but not always) absent from naturalism; naturalism, we could therefore say, performs the cognitive functions of a religion. Naturalism has a creed, nicely stated by Peter van Inwagen (himself no naturalist):

> There is no God. There is, in fact, nothing besides the physical cosmos that science investigates. Human beings, since they are a part of this cosmos, are physical things and therefore do not survive death. . . . Like other animals, they are a product of uncaring and unconscious physical processes that did not have them, or anything else, in mind. . . . Human beings, however, have an unfortunate tendency to wish to deny these facts and to believe comforting myths according to which they have an eternal purpose. This irrational component in the psyches of most human beings—it is the great good fortune of the species that there are a few strong-minded progressives who can see through the comforting myths—encourages the confidence game called religion. . . .[3]

Since naturalism performs the cognitive or worldview function of religion, we can perhaps promote it to the status of an honorary religion. What I propose to argue is that naturalism and contemporary evolutionary theory are at serious odds with one another—and this despite the fact that the latter is ordinarily thought to be one of the main supporting members in the edifice of the former.[4] Accordingly, there is a religion/science conflict in the neighborhood of evolution—not, however, a conflict between evolution and Christian belief, but between naturalism and evolution. (Of course I am *not* attacking the theory of evolution, or the claim that human beings have evolved from simian ancestors, or anything in that neighborhood; I am instead attacking the conjunction of *naturalism* with the view that human beings have evolved in that way. I see no similar problems with the conjunction of *theism* and the idea that human beings have evolved in the way contemporary evolutionary science suggests.)

More particularly, I'll argue that the conjunction of naturalism with the belief that human beings have evolved in conformity with current evolutionary doctrine—"evolution" for short—is in a certain interesting way *self-defeating* or self-referentially incoherent. Still more particularly, I'll argue that naturalism and evolution—"N&E" for short—furnishes one who accepts it with a *defeater* for the belief that our cognitive faculties are reliable—a defeater that can't be defeated. But then this conjunction also furnishes a defeater for any belief produced by our cognitive faculties, including, in the case of one who accepts it, N&E itself: hence its self-defeating character.

1. P(R/N&E) is Low

So much for a quick overview of the argument. More specifically EAAN begins from certain doubts about the *reliability* of our cognitive faculties, where, roughly,[5] a cognitive faculty—memory, perception, reason—is reliable if the great bulk of its deliverances are true. These doubts are connected with the *origin* of our cognitive faculties. According to current evolutionary theory, we human beings, like other forms of life, have developed from aboriginal unicellular life by way of such mechanisms as natural selection and genetic drift working on sources of genetic variation: the most popular is random genetic mutation. Natural selection discards most of these mutations (they prove deleterious to the organism in which they appear), but some of the remainder turn out to have survival value and to enhance fitness; they spread through the population and thus persist. As this story has it, it is by way of these mechanisms, or mechanisms very much like them, that all the vast variety of contemporary organic life has developed; and it is by way of these same mechanisms that our cognitive faculties have arisen.

Now according to traditional Christian (and Jewish and perhaps Muslim) thought, we human beings have been created in the image of God. This means, among other things, that he created us with the capacity for achieving *knowledge*—knowledge of our environment by way of perception, of other people by way of something like what Thomas Reid calls *sympathy*, of the past by memory and testimony, of mathematics and logic by reason, of morality, our own mental life, God himself, and much more.[6] And of course most of us are inclined to think that our cognitive faculties, our belief-producing processes, are for the most part reliable. True, they may not be reliable at the upper limits of our powers, as in some of the more speculative areas of physics; and the proper function of our faculties can be skewed by envy, hate, lust, mother love, greed, and so on. But over a broad area of their operation, we think the purpose of our cognitive faculties is to furnish us with true beliefs, and that when they function properly, they do exactly that.

The above evolutionary account of our origins, clearly enough, is compatible with the theistic view that God has created us in his image.[7] So evolutionary theory taken by itself (without the patina of philosophical naturalism that often accompanies expositions of it) is not as such in tension with the idea that God has created us and our cognitive faculties in such a way that the latter are reliable, that (as the medievals liked to say) there is an adequation of intellect to reality. But if *naturalism* is true, there is no God, and hence no God (or anyone else) overseeing our development and orchestrating the course of our evolution. And this leads directly to the question whether it is at all likely that our cognitive faculties, given naturalism and given their evolutionary origin, would have developed in such a way as to be reliable, to furnish us with mostly true beliefs. The problem, from this point of view, is that the ultimate purpose or function of our cognitive faculties, if they have one, is not to produce true beliefs, but to promote reproductive fitness. As evolutionary psychologist David Sloan Wilson puts it, "the well-adapted mind is ultimately an organ of survival and reproduction."[8] What our minds are *for* is not the production of true beliefs, but the production of adaptive behavior. That our species has survived and evolved at most guarantees that our behavior is adaptive; it does not guarantee or even suggest that our belief-producing processes are for the most part reliable, or that our beliefs are for the most part true. That is because, obviously, our behavior could be adaptive, but our beliefs mainly false. Darwin himself expressed this doubt: "With me," he said, "the horrid doubt always arises whether the convictions of man's mind, which has been developed from the mind of the lower animals, are of any value or at all trustworthy. Would any one trust in the convictions of a monkey's mind, if there are any convictions in such a mind?"[9]

The same thought is put more explicitly by Patricia Churchland. She insists that the most important thing about the human brain is that it has evolved; this means, she says, that its principal function is to enable the organism to *move* appropriately:

> Boiled down to essentials, a nervous system enables the organism to succeed in the four F's: feeding, fleeing, fighting and reproducing. The principle chore of nervous systems is to get the body parts where they should be in order that the organism may survive. . . . . Improvements in sensorimotor control confer an evolutionary advantage: a fancier style of representing is advantageous *so long as it is geared to the organism's way of life and enhances the organism's chances of survival* [Churchland's emphasis]. Truth, whatever that is, definitely takes the hindmost.[10]

What Churchland means, I think, is that evolution is directly interested (so to speak) only in adaptive *behavior* (in a broad sense including physical functioning) not in true belief. Natural selection doesn't care what you *believe;* it is interested only in how you *behave*. It selects for certain kinds of behavior: those that enhance fitness, which is a measure of the chances that one's genes are widely represented in the next and subsequent generations. It doesn't select for belief, except insofar as the latter is appropriately related to behavior. But then the fact that we have evolved guarantees at most that we *behave* in certain ways—ways that contribute to our (or our ancestors') surviving and reproducing in the environment in which we have developed. Churchland's claim, I think, can perhaps be understood as the suggestion that the objective[11] probability that our cognitive faculties are reliable, given naturalism and given that we have been cobbled together by the processes to which contemporary evolutionary theory calls our attention, is low. Of course she doesn't explicitly mention naturalism, but it certainly seems that she is taking it for granted. For if theism were true, God might be directing and orchestrating the process in such a way as to produce, in the long run, beings created in his image and thus capable of knowledge; but then it wouldn't be the case that truth takes the hindmost.

We can put Churchland's claim as

$$P(R/N\&E) \text{ is low,}$$

where "R" is the proposition that our cognitive faculties are reliable, "N" the proposition that naturalism is true, and "E" the proposition that we have evolved according to the suggestions of contemporary evolutionary theory. I believe this thought—the thought that $P(R/N\&E)$ is low—is also what worries Darwin in the above quotation: I shall therefore call it "Darwin's Doubt."

Are Darwin and Churchland right? Is this probability low? Well, they are certainly right in thinking that natural selection is directly interested only in behavior, not belief, and that it is interested in belief, if at all, only indirectly, by virtue of the relation between behavior and belief. If adaptive behavior guarantees or makes probable reliable faculties, then perhaps $P(R/N\&E)$ will be rather high: we, or rather our ancestors, engaged in at least reasonably adaptive behavior, so it must be that our cognitive faculties are at least reasonably reliable, in which case it is likely that most of our beliefs are true—at least for those beliefs connected with survival and reproduction. (This likelihood, therefore, doesn't extend to philosophical beliefs, or highly abstract scientific or mathematical beliefs; it is only the occasional assistant professor of logic who needs to know how to prove, say, Gödel's theorems in order to survive and reproduce.) On the other hand, if our having reliable faculties *isn't* guaranteed by or even particularly probable with respect to adaptive behavior, then presumably $P(R/N\&E)$ will be rather low. If, for example, behavior isn't caused or governed by belief, the latter would be, so to speak, invisible to natural selection; in that case it would be unlikely that the great preponderance of true belief over false required by reliability would be forthcoming.[12] So the question of the value of $P(R/N\&E)$ really turns on the relationship between belief and behavior. Our having evolved and survived makes it likely that our cognitive faculties are reliable and our beliefs are for the most part true, only if it would be impossible or unlikely that creatures more or less like us should behave in fitness-enhancing ways but nonetheless hold mostly false beliefs.[13]

Is this impossible or unlikely? That depends upon the relation between belief and behavior. What would or could that relation be? To try to guard against interspecific chauvinism, I suggest that we think, not about ourselves and our behavior, but about a population of creatures a lot like us on a planet a lot like Earth (Darwin suggested we think about monkeys in this connection). These creatures are *rational:* that is, they form beliefs, reason, change beliefs, and the like. We imagine furthermore that they and their cognitive systems have evolved by way of the mechanisms to which contemporary evolutionary theory directs our attention, unguided by the hand of God or anyone else. Now what is $P(R/N\&E)$, specified not to us, but to them? To answer, we must think about the relationship between their beliefs and their behavior. There are four mutually exclusive and jointly exhaustive possibilities.

(1) One possibility is *epiphenomenalism:*[14] their behavior is not caused by their beliefs. On this possibility, their movement and behavior would be caused by something or other—perhaps neural impulses—which would be caused by other organic conditions including sensory stimulation: but belief would not have a place in this causal chain leading to behavior. This view of the relation between behavior and belief (and other mental phenomena such as feeling, sensation, and desire) is currently rather popular, especially among those strongly influenced

by biological science. *Time* (December 1992) reports that J. M. Smith, a well-known biologist, wrote "that he had never understood why organisms have feelings. After all, orthodox biologists believe that behavior, however complex, is governed entirely by biochemistry and that the attendant sensations—fear, pain, wonder, love—are just shadows cast by that biochemistry, not themselves vital to the organism's behavior. . . ." He could have added that (according to biological orthodoxy) the same goes for beliefs—at least if beliefs are not themselves just biochemical phenomena. If this way of thinking is right with respect to our hypothetical creatures, their beliefs would be *invisible* to evolution; and then the fact that their belief-forming mechanisms arose during their evolutionary history would confer little or no probability on the idea that their beliefs are mostly true, or mostly nearly true. Indeed, the probability of those beliefs being for the most part true would have to be rated fairly low (or inscrutable). On N&E and this first possibility, therefore, the probability of R will be rather low.

(2) A second possibility is *semantic* epiphenomenalism: it could be that their beliefs do indeed have causal efficacy with respect to behavior, but not by virtue of their *content*. Note first that nearly all naturalists are also *materialists* about human persons; we human beings are not immaterial Cartesian substances, and do not contain an immaterial part—a soul, for example. Here I'll assimilate materialism (about human beings) to naturalism; since all or nearly all naturalists are materialists, there will be little if any loss of generality. And from this point of view, a belief would have to be something like a long-term event or structure in the nervous system—perhaps a group of neurons connected and related in a certain way. This neural structure will have *neurophysiological* properties ("NP properties" for short): properties specifying the number of neurons involved in the structure, the way in which the neurons involved are connected with each other and with other structures (with muscles, sense organs, other neuronal events, etc.), the average rate and intensity of neuronal firing in various parts of this event and the ways in which this changes over time and in response to input from other areas. Now of course it is easy to see how *these* properties of this neuronal event should have causal influence on the behavior of the organism. A given belief is neurally connected both with other beliefs and with muscles; we can see how electrical impulses coming from the belief can negotiate the usual neuronal channels and ultimately cause muscular contraction and behavior.

But if this belief is really a *belief*, then it will also have *other* properties, properties in addition to its neurophysiological properties. In particular, it will have *content*; it will be the belief that p, for some proposition p—perhaps the proposition *naturalism is all the rage these days*. Having such and such a content would be a *semantic* property of the belief. Beliefs, therefore, have two very different kinds of properties: neurophysiological properties on the one hand, and semantic properties on the other. And now the question is this: does a belief—a neural structure—cause behavior, that is, enter into the causal chain leading to behavior, *by virtue of its content?* According to semantic epiphenomenalism, the answer is no. Belief is indeed causally efficacious with respect to behavior, but by virtue of the *syntactic* properties of a belief, not its semantic properties. If the first possibility is widely popular among those influenced by biological science, this possibility is widely popular among contemporary philosophers of mind; indeed, Robert Cummins goes so far as to call it the "received view."[15]

On this second possibility, as on the first, P(R/N&E) (specified to those creatures) will be low. The reason is that truth and falsehood, of course, are among the *semantic* properties of a belief, not its syntactic properties. But if the former aren't involved in the causal chain leading to behavior, then once more beliefs—or rather, their semantic properties, including truth and falsehood—will be invisible to natural selection. But then it will be unlikely that their beliefs are mostly true and hence unlikely that their cognitive faculties are reliable. The probability of R on N&E together with this possibility (as with the last), therefore, will be relatively low.

(3) It could be that beliefs are causally efficacious—by virtue of content as well as neurophysiological properties—but *maladaptive:* from the point of view of fitness these creatures would be better off without them. The probability of R on N&E together with this possibility, as with the last two, would also seem to be relatively low.

(4) Finally, it could be that the beliefs of our hypothetical creatures are indeed both causally connected with their behavior and also adaptive. (I suppose this is the commonsense view of the connection between behavior and belief in our own case.) What is the probability (on this assumption—call it C—together with N&E) that their cognitive faculties are reliable; and what is the probability that a belief produced by those faculties will be true? In WPF I argued that this probability isn't nearly as high as one is initially inclined to think. The reason is that if behavior is caused by *belief,* it is also caused by *desire* (and other factors—suspicion, doubt, approval and disapproval, fear—that we can here ignore). For any given adaptive action, there will be many belief-desire combinations that could produce that action; and very many of those belief-desire combinations will be such that the belief involved is false.

So suppose Paul is a prehistoric hominid; a hungry tiger approaches. Fleeing is perhaps the most appropriate behavior: I pointed out that this behavior could be produced by a large number of different belief-desire pairs. To quote myself:

> Perhaps Paul very much *likes* the idea of being eaten, but when he sees a tiger, always runs off looking for a better prospect, because he thinks it unlikely that the tiger he sees will eat him. This will get his body parts in the right place so far as survival is concerned, without involving much by way of true belief. . . . . Or perhaps he thinks the tiger is a large, friendly, cuddly pussycat and wants to pet it; but he also believes that the best way to pet it is to run away from it . . . or perhaps he thinks the tiger is a regularly recurring illusion, and, hoping to keep his weight down, has formed the resolution to run a mile at top speed whenever presented with such an illusion; or perhaps he thinks he is about to take part in a 1600 meter race, wants to win, and believes the appearance of the tiger is the starting signal; or perhaps. . . . Clearly there are any number of belief-cum-desire systems that equally fit a given bit of behavior. (WPF 225–26)

Accordingly, there are many belief-desire combinations that will lead to the adaptive action; in many of these combinations, the beliefs are false. Without further knowledge of these creatures, therefore, we could hardly estimate the probability of R on N&E and this final possibility as high.

A problem with the argument as thus presented is this: It is easy to see, for just *one* of Paul's actions, that there are many different belief-desire combinations that yield it; it is less easy to see how it could be that most of all of his beliefs could be false but nonetheless adaptive or fitness enhancing. Could Paul's beliefs really be mainly false, but still lead to adaptive action? Yes indeed; perhaps the simplest way to see how is by thinking of systematic ways in which his beliefs could be false but still adaptive. Perhaps Paul is a sort of early Leibnizian and thinks everything is conscious (and suppose that is false); furthermore, his ways of referring to things all involve definite descriptions that entail consciousness, so that all of his beliefs are of the form *That so-and-so conscious being is such-and-such.* Perhaps he is an animist and thinks everything is alive. Perhaps he thinks all the plants and animals in his vicinity are witches, and his ways of referring to them all involve definite descriptions entailing witchhood. But this would be entirely compatible with his belief's being adaptive; so it is clear, I think, that there would be many ways in which Paul's beliefs could be for the most part false, but adaptive nonetheless. Religious belief is nearly universal across the world; according to naturalists it is false, but nonetheless adaptive. Our question is really about the proportion of true beliefs among adaptive beliefs. For every true adaptive belief, however, it seems that one can easily think of a false belief adaptive in the same circumstances. The fact that my behavior (or that of my ancestors) has been adaptive, therefore, is at best a third-rate reason for thinking my beliefs mostly true and my cognitive faculties reliable—and that is true even given the commonsense view of the relation of belief to behavior. So we can't sensibly argue from the fact that our behavior (or that of our ancestors) has been adaptive to the conclusion that our beliefs are mostly true and our cognitive faculties reliable. It is therefore hard to see that P(R/N&E&C) is very high. To concede as much as possible to the opposition, however, let's say that this probability is either inscrutable or in the neighborhood of 0.9.

What we have seen so far is that there are four mutually exclusive and jointly exhaustive possibilities with respect to that hypothetical population: epiphenomenalism simpliciter, semantic epiphenomenalism, the possibility that their beliefs are causally efficacious with respect to their behavior but maladaptive, and the possibility that their beliefs are both causally efficacious with respect to behavior and adaptive. P(R/N&E) will be the weighted average of P(R/N&E&P_i) for each of the four possibilities P_i—weighted by the probabilities, on N&E, of those possibilities. The probability calculus gives us a formula here:

(1) P(R/N&E) = (P(R/N&E&P_1) x P(P_1/N&E)) + (P(R/N&E&P_2) x P(P_2/N&E))
      + (P(R/N&E&P_3) x P(P_3/N&E)) + (P(R/N&E&P_4) x P(P_4/N&E)).

Of course the very idea of a calculation (suggesting, as it does, the assignment of specific real numbers to these various probabilities) is laughable: the best we can do are vague estimates. Still, that will suffice for the argument. Now let's agree that P_3—the proposition that belief enters the causal chain leading to behavior both by virtue of neurophysiological properties and by virtue of semantic properties, but is nevertheless maladaptive—is very unlikely; it is then clear from (1) that its contribution to P(R/N&E) can safely be ignored. Note further that epiphenomenalism and semantic epiphenomenalism unite in declaring or implying that the *content* of belief lacks causal efficacy with respect to behavior; the content of belief does not get involved in the causal chain leading to behavior. So we can reduce these two possibilities to one: the possibility that the content of belief has no causal efficacy. Call this possibility "-C." What we have so far seen is that the probability of R on N&E&-C is low or inscrutable, and that the probability of R on N&E&C is also inscrutable or at best moderately high. We can therefore simplify (1) to

(2) P(R/N&E) = [P(R/N&E&C) x P(C/N&E)] + [P(R/N&E&-C) x P(-C/N&E)],

i.e., the probability of R on N&E is the weighted average of the probabilities of R on N&E&C and N&E&-C—weighted by the probabilities of C and -C on N&E.

We have already noted that the left-hand term of the first of the two products on the right side of the equality is either moderately high or inscrutable; the second is either low or inscrutable. What remains is to evaluate the weights, the right-hand terms of the two products. So what is the probability of -C, given N&E; what is the probability of semantic epiphenomenalism on N&E? Note that, according to Robert Cummins, semantic epiphenomenalism is in fact the received view as to the relation between belief and behavior.[16] That is because it is extremely hard to envisage a way, given materialism, in which the content of a belief *could* get causally involved in behavior. According to materialism, a belief just is a neural structure of some kind—a structure that somehow possesses content. How can its content get involved in the causal chain leading to behavior? Had a given such structure had a *different* content, one thinks, its causal contribution to behavior would be the same. What causes the muscular contractions involved in behavior are physiological states of the nervous system, including physiological properties of those structures that constitute beliefs; the content of those beliefs appears to be causally irrelevant. It is therefore exceedingly hard to see how semantic epiphenomenalism can be avoided, given N&E. (There have been some valiant efforts,[17] but things don't look hopeful.) So it looks as if P(-C/N&E) will have to be estimated as relatively high; let's say (for definiteness) 0.7, in which case P(C/N&E) will be 0.3. Of course we could easily be wrong; we don't really have a solid way of telling; so perhaps the conservative position here is that this probability too is inscrutable: one simply can't tell what it is. Given current knowledge, therefore, P(-C/N&E) is either high or inscrutable. And if P(-C/N&E) is inscrutable, then the same goes, naturally enough, for P(C/N&E). What does that mean for the sum of these two products, i.e., P(R/N&E)?

Well, we have several possibilities. Suppose we think first about the matter from the point of view of someone who doesn't find any of the probabilities involved inscrutable. Then P(C/N&E) will be in the neighborhood of 0.3, P(-C/N&E) in the neighborhood of 0.7, and P(R/N&E&-C) perhaps in the neighborhood of 0.2. This leaves P(R/N&E&C), the probability that R is true given ordinary naturalism together with the commonsense or folk-theoretical view as to the relation between belief and behavior. Given that this probability is not inscrutable, let's say that it is in the neighborhood of 0.9. And given these estimates, P(R/N&E) will be in the neighborhood of 0.41.[18] Suppose, on the other hand, we think the probabilities involved are inscrutable: then we will have to say the same for P(R/N&E). P(R/N&E), therefore, is either low—less than .5, at any rate—or inscrutable.

2. A Defeater for R

Returning to Darwin's doubt, observe that if this is the sensible attitude to take to P(R/N&E) specified to that hypothetical population, then it will also be the sensible attitude towards P(R/N&E) specified to us. We are relevantly like them in that *our* cognitive faculties have the same kind of origin and provenance as *theirs* are hypothesized to have. And the next step in the argument is to point out that each of these attitudes—the view that P(R/N&E) is low and the view that this probability is inscrutable—gives the naturalist-evolutionist a *defeater* for R. It gives him a reason to doubt it, a reason not to affirm it. I argued this by analogy. Among the crucially important facts, with respect to the question of the reliability of a group of cognitive faculties, are facts about their *origin*. Suppose I believe that I have been created by an evil Cartesian demon who takes delight in fashioning creatures who have mainly false beliefs (but think of themselves as paradigms of cognitive excellence): then I have a defeater for my natural belief that my faculties are reliable. Turn instead to the contemporary version of this scenario, and suppose I come to believe that I have been captured by Alpha-Centaurian super-scientists who have made me the subject of a cognitive experiment in which the subject is given mostly false beliefs: then, again, I have a defeater for R. But to have a defeater for R it isn't necessary that I believe that in fact I *have* been created by a Cartesian demon or been captured by those Alpha-Centaurian superscientists. It suffices for me to have such a defeater if I have considered those scenarios, and the probability that one of those scenarios is true, is inscrutable for me—if I can't make any estimate of it, or do not have an opinion as to what that probability is. It suffices if I have considered those scenarios, and *for all I know or believe* one of them is true. In these cases too I have a reason for doubting, a reason for withholding[19] my natural belief that my cognitive faculties are in fact reliable.

Perhaps it seems harder to see that one has a defeater for R in the case where the relevant probability is inscrutable, than in the case where it is low. But suppose you buy a thermometer; then you learn that the owner of the factory where it was constructed is a Luddite who aims to do what he can to disrupt contemporary technology, and he therefore makes at least some instruments that are unreliable. You can't say what the probability is of this thermometer's being reliable, given that it was made in that factory; that probability is inscrutable for you. But would you trust the thermometer? It's outside your window, and reads 30 degrees F; if you have no other source of information about the temperature outside, would you believe it is 30 degrees F?

One more analogy: you embark on a voyage of space exploration and land on a planet revolving about a distant sun, a planet that seems to have a favorable atmosphere. You crack the hatch, step out, and immediately find what appears to be an instrument that looks a lot like a terrestrial radio; you twiddle the dials, and after a couple of squawks it begins to emit strings of sounds that, oddly enough, form English sentences. These sentences express propositions only about topics of which you have no knowledge: what the weather is like in Beijing at the moment, whether Caesar had eggs on toast on the morning he crossed the Rubicon, whether the first human being to cross the Bering Strait was left-handed, and the like. Unduly impressed, indeed awed, by your find, you initially form the opinion that this instrument speaks the truth, that the propositions expressed (in English) by those sentences are true. But then you recall that you have no idea at all as to who or what constructed the instrument, what it is for, whether it has a purpose at all. You see that the probability of its being reliable, given what you know about it, is inscrutable. Then you have a defeater for your initial belief that the thing does in fact speak the truth.

3. AN OBJECTION AND A REPLY

But here an objection rears its ugly head. In trying to assess P(R/N&E), I suggested that semantic epiphenomenalism was very likely, given materialism, because a neural structure would have caused the same behavior if it had had *different* content but the *same* NP properties. But, says the objector, it *couldn't* have had the same NP properties but different content; having a given content *just is* having a certain set of NP properties. This is a sensible objection. Given materialism, there is a way of looking at the relation between content (as well as other mental properties) and NP properties according to which the objector is clearly right. We must therefore look a bit more deeply into that relation. Here there are fundamentally two positions: *reductionism* or *reductive materialism* on the one hand, and *non-reductive materialism* on the other. Consider the property of having as content the proposition *naturalism is all the rage these days,* and call this property "C." According to *reductive* materialism, C *just is* a certain combination of NP properties.[20] It might be a disjunction of such properties; more likely a complex Boolean construction on NP properties, perhaps something like

$(P_1 \& P_7 \& P_{28} \ldots) \vee (P_3 \& P_{34} \& P_{17} \& \ldots) \vee (P_8 \& P_{83} \& P_{107} \& \ldots) \vee \ldots$
(where the P_i are NP properties).

Now take any belief B you like: what is the probability that B is true given N&E and reductive materialism? What we know is that B has a certain content, that having that content just is having a certain combination of NP properties, and (we may assume) that having that combination of NP properties (in the circumstances in which the organism finds itself) is adaptive. What, then, is the probability that the content of B is *true*? Well, it doesn't *matter* whether it is true; if it is true, the NP properties constituting that content will be adaptive, but if it is false, those properties will be equally adaptive, since in each case they make the same causal contribution to behavior. That combination of NP properties is the property of having a certain content; it is the property of being associated with a certain proposition *p* in such a way that *p* is the content of the belief. Having that combination of NP properties is adaptive; hence having that belief is adaptive; but that combination of NP properties will be equally adaptive whether *p* is true or false. In this case (reductionism) content does enter into the causal chain leading to behavior, because NP properties do, and having a certain content just is displaying a certain set of NP properties. But those properties will be adaptive, whether or not the content the having of which they constitute is true. Content enters in, all right, but not, we might say, *as* content. Better, content enters the causal chain leading to behavior, but not in such a way that its truth or falsehood bears on the adaptive character of the belief.

But, someone might object, given that the belief is adaptive, isn't there a greater probability of its being true than of its being false? Why so? Because, the objector continues, the belief's being adaptive means that having this belief, in these or similar circumstances, helped the creature's ancestors to survive and reproduce; having this belief contributed to reproductive fitness. And wouldn't the best explanation for this contribution be that the belief accurately represented their circumstances, i.e., *was true*? So, probably, the belief was adaptive for the creature's ancestors because it was true. So, probably, the belief is adaptive for this creature in *its* circumstances because it is true.[21]

This objection, beguiling as it sounds, is mistaken. The proper explanation of this belief's being adaptive is that having the NP properties that constitute the content of the belief causes adaptive behavior, not that the belief is true. And, of course, having those NP properties can cause adaptive behavior whether or not the content they constitute is true. At a certain level of complexity of NP properties, the neural structure that displays those properties also acquires a certain content C. That is because having that particular complex of NP properties just is what it is to have C. Having those NP properties, presumably, is adaptive; but whether the content arising in this way is true or false makes no difference to that adaptivity. What explains the adaptivity is just that having these NP properties, this content, causes adaptive behavior.[22]

So consider again a belief B with its content C: what, given that having that belief is adaptive, is the probability that C is true, is a true proposition? Well, since truth of content doesn't make a difference to the adaptivity of the belief, the belief could be true, but it could equally likely be false. We'd have to estimate the probability that it is true as about 0.5. But then if the creature has 1,000 independent beliefs, the probability that, say, 3/4 of them are true (and at least this would be required for reliability) will be very low—less than 10^{-58}.[23] So on naturalism and reductionism, the probability of R appears to be very low.

That's how things go given reductive materialism; according to *non-reductive materialism*, the other possibility, a mental property is not an NP property or any Boolean construction on NP properties, but a new sort of property that gets instantiated when a neural structure attains a certain degree of complexity, that is, when it displays a certain complex set of NP properties. (We might call it an "emergent" property.) On this way of looking at the matter, mental properties are not identical with or reducible to NP properties; they supervene on them. But again, take any particular belief B: what is the probability, on N&E & non-reductive materialism, that B is true? What we know is that B has a content, that this content arises when the structure has a certain complex set of NP properties, and that having that set of NP properties is adaptive. But once again, it doesn't matter for adaptivity whether the content associated with those NP properties is true or false; so once again, the probability that the content is true will have to be estimated as about .5; but then once again the probability that their faculties are reliable is low. Either way, therefore, the probability is low, so that P(R/N&E) is also low—or, as we could add, if we like, inscrutable. But then either way one who accepts N&E and grasps the above argument has a defeater for R.

4. A Defeater-Defeater?

Now of course defeaters can be themselves defeated. For example, I know that you are a lifeguard and believe on that ground that you are an excellent swimmer. But then I learn that 45 percent of Frisian lifeguards are poor swimmers, and I know that you are Frisian: this gives me a defeater for the belief that you are a fine swimmer. But then I learn still further that you graduated from the Department of Lifeguarding at the University of Leeuwarden and that one of the requirements for graduation is being an excellent swimmer: that gives me a defeater for the defeater of my original belief: a defeater-defeater as we might put it.[24] But (to return to our argument) can't the defeater the naturalist has for R be in turn defeated? Could she perhaps get a defeater-defeater by doing some science—for example, determining by scientific means that her faculties really are reliable? Couldn't she go to the MIT cognitive-reliability laboratory for a check-up?

That won't help; clearly that course would *presuppose* that her faculties are reliable; she'd be relying on the accuracy of her faculties in believing that there is such a thing as MIT, that she has in fact consulted its scientists, that they have given her a clean bill of cognitive health, and the like. Thomas Reid (*Essays on the Intellectual Powers of Man*) put it like this: "If a man's honesty were called into question, it would be ridiculous to refer to the man's own word, whether he be honest or not. The same absurdity there is in attempting to prove, by any kind of reasoning, probable or demonstrative, that our reason is not fallacious, since the very point in question is, whether reasoning may be trusted" (276).

Is there any sensible way at all in which she can argue for R? It seems not. Any argument she might produce will have premises; and these premises, she claims, give her good reason to believe R. But of course she has the very same defeater for each of those premises that she has for R, as well as for the belief that if the premises of that argument are true, then so is the conclusion. So this defeater can't be defeated. We could also put it like this: any argument she offers for R is epistemically circular (to use William Alston's terminology). Naturalistic evolution gives its adherents a reason for doubting that our beliefs are mostly true; perhaps they are mostly mistaken. But then it won't help to *argue* that they can't be mostly mistaken; for the very reason for mistrusting our cognitive faculties *generally* will be a reason for mistrusting the faculties that produce belief in the goodness of the argument.

5. N&E Defeated

He who accepts N&E, therefore, has a defeater for R—a defeater, furthermore, that can't be defeated. But if you have a defeater for R, the belief that your cognitive faculties are reliable, then you have a defeater for any belief you take to be produced by your cognitive faculties. Naturally enough, that would be all of your beliefs. Once more, this defeater cannot be defeated; any proposed defeater would be an argument, and any such argument would be epistemically circular. Hence the devotee of N&E has a defeater for just any belief he holds— a defeater that is ultimately undefeated. But this means, obviously, that he has a defeater for N&E itself. N&E, therefore, cannot rationally be accepted—at any rate by someone who is apprised of this argument and sees the connections between N&E and R.

And that has two important consequences. The first is that there is a conflict between naturalism and science: one can't rationally accept the conjunction of naturalism with evolution; but evolution is one of the main pillars of contemporary science. Hence there is a religion/science conflict in this area, all right, but it isn't between Christian belief and science: it is between naturalism and science. Second, if naturalism is true, then so, in all probability, is evolution; evolution is the only game in town, for the naturalist, with respect to the question how all this variety of flora and fauna has arisen. If *that* is so, finally, then naturalism *simpliciter* is self-defeating and cannot rationally be accepted—at any rate by someone who sees the connections between N&E and R.

Notes

1. A version of "An Evolutionary Argument against Naturalism" (hereafter "EAAN") first appeared in *Logos* 12: 27–49, 1991 and in *Warrant and Proper Function* (New York: Oxford University Press, 1993; hereafter "WPF"), ch. 12. These statements of the argument contain an error: at a certain point the statement of the "preliminary argument" (WPF 228–29) confuses the unconditional objective or logical probability of R with its probability conditional on our background knowledge. (Here I was helped by Brandon Fitelson and Elliot Sober: see their paper "Plantinga's Probability Arguments against Evolutionary Naturalism" in *Pacific Philosophical Quarterly* 79 [1998], 115–29). For details and repair, see my *Warranted Christian Belief* (New York: Oxford University Press, 2000), 229ff. The main argument, happily, is unaffected. See also "Probability and Defeaters," *Pacific Philosophical Quarterly* 84:3 (Sept. 2003), my reply to Sober and Fitelson. For further critical comment on EAAN (together with my replies) see *Naturalism Defeated?* ed. James Beilby (Ithaca, NY: Cornell University Press, 2002).

2. Erroneously, in my opinion. There is no inner connection between science and naturalism; indeed, as I'm about to argue, naturalism clashes with science.

3. "*Quam Dialectica*" in *God and the Philosophers,* ed. Thomas Morris (New York: Oxford University Press, 1994), 49

4. Thus Richard Dawkins: "Although atheism might have been logically tenable before Darwin, Darwin made it possible to be an intellectually fulfilled atheist." *The Blind Watchmaker* (New York: Norton, 1986), 6–7.

5. *Very* roughly: a thermometer stuck on 72 degrees F. isn't reliable even if it is located somewhere (San Diego?) where it is 72 F. degrees nearly all of the time. What the thermometer (and our cognitive faculties) would do if things were different in certain (hard to specify) respects is also relevant. Again, if our aim were to analyze *reliability* much more would have to be said. Note that for reliability thus construed, it is not enough that the beliefs produced be fitness enhancing.

6. Thus Thomas Aquinas: "Since human beings are said to be in the image of God in virtue of their having a nature that includes an intellect, such a nature is most in the image of God in virtue of being most able to imitate God" (ST Ia q. 93 a. 4); *and* "Only in rational creatures is there found a likeness of God which counts as an image. . . . As far as a likeness of the divine nature is concerned, rational creatures seem somehow to attain a representation of [that] type in virtue of imitating God not only in this, that he is and lives, but especially in this, that he understands" (ST Ia Q.93 a.6).

7. You might think not: if our origin involves *random* genetic variation, then we and our cognitive faculties would have developed by way of *chance* rather than by way of design, as would be required by our having been created by God in his image. But this is to import far too much into the biologist's term "random." Those random variations are random in the sense that they don't arise out of the organism's design plan and don't ordinarily play a role in its viability; perhaps they are also random in the sense that they are not predictable. But of course it doesn't follow that they are random in the much stronger sense of not being caused, orchestrated, or arranged by anything at all (including God). And suppose the biologists, or others, *did* intend this stronger sense of "random": then their theory (call it "T") would indeed entail that human beings have not been designed by God. But T would not be more probable than not with respect to the evidence. For there would be an empirically equivalent theory (the theory that results from T by taking the weaker sense of "random" and adding that God has orchestrated the mutations) that is inconsistent with T but is as well supported by the evidence; if so, T is not more probable than not with respect to the relevant evidence.

8. *Darwin's Cathedral: Evolution, Religion and the Nature of Society* (Chicago: University of Chicago Press, 2002), 228.

9. Letter to William Graham, Down, July 3, 1881. In *The Life and Letters of Charles Darwin Including an Autobiographical Chapter*, ed. Francis Darwin (London: John Murray, Albermarle Street, 1887), vol. 1, 315–16. Evan Fales and Omar Mirza point out that Darwin probably had in mind, here, not everyday beliefs such as that the teapot is in the cupboard, but religious and philosophical convictions—naturalism, for example.

10. *Journal of Philosophy* 84 (Oct. 1987), 548.

11. For an account of objective probability, see my *Warrant and Proper Function* (New York: Oxford University Press, 1993), 161ff.

12. Alternatively, we might say that the probability here is inscrutable, such that we can't make an estimate of it. It is unlikely that a large set of beliefs (comparable in size to the number of beliefs a human being has) should contain mainly truths; that gives us a reason for regarding the probability in question as low. On the other hand, we know something further about the relevant set of propositions, namely, that it is a set each member of which is believed by someone. How does this affect the probability? Here perhaps we don't know what to say, and should conclude that the probability in question is inscrutable. (Here I am indebted to John Hare.)

13. Must we concur with Donald Davidson, who thinks it is "impossible correctly to hold that anyone could be mostly wrong about how things are"? ("A Coherence Theory of Truth and Knowledge" in *Kant oder Hegel?* ed. Dieter Henrich [Stuttgart: Klett-Cotta Buchhandlung, 1983], 535.) No; what Davidson shows (if anything) is that it isn't possible for me to *understand* another creature, unless I suppose that she holds mainly true beliefs. That may (or more likely, may not) be so; but it doesn't follow that there couldn't be creatures with mainly false beliefs, and *a fortiori* it doesn't follow that my own beliefs are mainly true. Davidson went on to argue that an *omniscient* interpreter would have to use the same methods we have to use and would therefore have to suppose her interlocutor held mostly true beliefs; given the omniscient interpreter's omniscience, he concluded that her interlocutor would in fact have mostly true beliefs. In so concluding, however, he apparently employs the premise that any proposition that would be believed by any omniscient being is true; this premise directly yields the conclusion that there *is* an omniscient being (since any omniscient being worth its salt will believe that there is an omniscient being), a conclusion to which Davidson might not have wished to commit himself. See WPF 80–81.

14. First so-called by T. H. Huxley, ("Darwin's bulldog"): "It may be be assumed . . . that molecular changes in the brain are the causes of all the states of consciousness. . . . [But is] there any evidence that these stages of consciousness may, conversely, cause . . . molecular changes [in the brain] which give rise to muscular motion? I see no such evidence . . . [Consciousness appears] to be . . . completely without any power of modifying [the] working of the body, just as the steam whistle . . . of a locomotive engine is without influence upon its machinery." ("On the Hypothesis that Animals are Automata and its History" [1874], chapter 5 of his *Method and Results* [London: Macmillan, 1893], 239–40.) Later in the essay: "To the best of my judgment, the argumentation which applies to brutes holds equally good of men; and therefore . . . all states of consciousness in us, as in them, are immediately caused by molecular changes of the brain-substance. It seems to me that in men, as in brutes, there is no proof that any state of consciousness is the cause of change in the motion of the matter of the organism. . . . We are conscious automata . . ." (243–44). (Note the occurrence here of that widely endorsed form of argument, "I know of no proof that not-*p;* therefore there is no proof that not-*p;* therefore *p.*") However, I am here using the term to denote *any* view according to which belief isn't involved in the causal chain leading to behavior, whether or not that view involves the dualism that seems to be part of Huxley's version.

15. *Meaning and Mental Representation* (Cambridge, MA: MIT Press, 1989), 130. In *Explaining Behavior* (Cambridge, MA: MIT Press, 1988) Fred Dretske makes a valiant (but in my opinion unsuccessful) effort to explain how, given materialism about human beings, it could be that beliefs (and other representations) play a causal role in the production of behavior by virtue of their content or semantics.

16. *Meaning and Mental Representation* (Cambridge, MA: MIT Press, 1989), 130.

17. Fred Dretske's *Explaining Behavior* (Cambridge, MA: MIT Press, 1988) is surely one of the most promising of these attempts; it fails, however (as I see it), among other things, because it implies that there are no distinct but logically equivalent beliefs, and indeed no distinct but *causally* equivalent beliefs.

18. Of course these figures are the merest approximations; others might make the estimates somewhat differently; but they can be significantly altered without significantly altering the final result. For example, perhaps you think the $P(R/N\&C)$ is higher, perhaps even 1; then (retaining the other assignments) $P(R/N)$ will be in the neighborhood of 0.44. Or perhaps you reject the thought that $P(-C/N)$ is more probable than $P(C/N)$, thinking them about equal. Then (again, retaining the other assignments) $P(R/N)$ will be in the neighborhood of 0.55.

19. I shall use this term to mean "failing to believe," so that I withhold p if either I believe its denial or I believe neither it nor its denial.

20. Or (to accommodate the thought that meaning "ain't in the head") a combination of NP properties with environmental properties. I'll assume but not mention this qualification in what follows.

21. Here I am indebted to Tom Crisp.

22. In this connection, consider dream beliefs. Take a given dream belief with its content C: Having the NP properties that constitute the property of having C is adaptive; but it makes no difference whether or not that content is true.

23. As calculated by the eminent mathematician Paul Zwier. This is the probability that the whole battery of cognitive faculties is reliable; the probability that a given faculty is reliable will be larger, but still small; if its output is, say, 100 beliefs, the probability that 3/4 of them are true will be no more than 0.000001.

24. As in fact John Pollock *does* put it.

6

More on the Illusion of Defeat

William J. Talbott

In chapter 12 of *Warrant and Proper Function*, Plantinga stated for the first time his evolutionary argument against naturalism (EAAN).[1] Later, Plantinga made some revisions to the argument.[2] In this essay, I explain why even the revised argument fails.

1. Understanding the Dialectic of the EAAN

If successful, the EAAN would lead us to reverse our ordinary common sense judgment about the relation between evolutionary theory and naturalism. The common sense view is that by providing a naturalistic explanation of potentially all biological phenomena, the development of evolutionary theory made naturalism much more plausible than it had been previously. Of course, there are many cases in which we have good reason to question common sense. But, by going against common sense in the EAAN, Plantinga assumes a substantial burden of argument.[3]

The crucial claim in the EAAN is that the evolutionary naturalist (i.e., the naturalist who accepts evolutionary theory) is in the position of being able to reason to an ultimately undefeated defeater for those very beliefs.[4] This defeater is the proposition:

(Low Probability Thesis) $P(R/N\&E)$ is low or inscrutable.[5]

The EAAN is most plausible if it is assumed that it is rational for the evolutionary naturalist to believe that $P(R/N\&E)$ is low (not merely inscrutable). Call this the *Low Probability Thesis*. I believe that EAAN fails even if it is assumed that it is rational for the evolutionary naturalist to believe the Low Probability Thesis. Therefore, for the purpose of focusing on later steps in the EAAN, I grant the assumption.[6]

There are two ways of proceeding at this point. One is to attempt to formulate precise principles of reasoning that would explain why the Low Probability Thesis is a defeater for N or would explain why it is not. Call this the *top-down approach*. The main issue for the top-down approach concerns reasoning with conditional probability. Even if $P(R/N\&E)$ is low, there are typically other propositions that the naturalist accepts (let "O" stand for their conjunction), which, when conjoined to N&E are such that:

More on the Illusion of Defeat

(High Conditional Probability Claim) $P(R/N\&E\&O)$ is high.

It is a fact about conditional probabilities that it can be rational to believe both the Low Probability Thesis and the High Conditional Probability Claim and that they can both be true. If so, then it is appropriate to ask, which attitude toward R should a naturalist take—that it is improbable, because of the Low Probability Thesis, or that it is probable because of the High Conditional Probability Claim? Plantinga in 2002 called this the problem of conditionalization (223). The top-down approach would require a principled solution to that problem.

Plantinga acknowledged that he had no principled solution to the problem (223–25). Instead, he pursued the second strategy, an *inductivist* or *particularist* strategy of focusing not on principles, but on particular cases (239–40).

Plantinga's particularist approach involved an argument by analogy (11, 239–240; also, *supra*). He presented a number of examples that he believed to be analogous in the relevant respects to the EAAN. He expected us to agree that in the analogous cases that he presented, the analogue of the low probability thesis does constitute a defeater for the beliefs of the relevant kind. Call these cases *positive cases*. I give some examples shortly. Plantinga sought to persuade us that the analogy between the positive cases he presents and the EAAN is strong enough that we should accept that in the EAAN also the Low Probability Thesis constitutes a defeater for the evolutionary naturalist's belief N, as well as for all other beliefs of the relevant kinds.

In "The Illusion of Defeat" (2002), I presented a series of what I will refer to as *negative cases*—that is, cases that are analogous to EAAN, but cases in which the analogue of the Low Probability Thesis does not constitute a defeater for the beliefs of the relevant kind. In that article, I ordered the negative cases so as to culminate in a case that would be as closely analogous as possible to the EAAN. In his reply, Plantinga discussed only the initial cases, where the analogy was weakest. Even on those cases, I believe that Plantinga is mistaken, for reasons that I explain shortly. But it is important not to become so focused on the initial cases that we overlook the cases where the analogy is strongest.

Plantinga's argument from analogy was based on a great variety of cases and my earlier response to it introduced even more of them. It seems to me that the relevant cases can be usefully grouped into four categories. I discuss each category separately.

Evaluations of the Reliability of Outside Sources of Information

The easiest way to raise the issue that Plantinga wants to raise is to consider our attitude toward the reliability of outside sources of information. Plantinga (2002) considered examples involving testimony, textual sources (e.g., the Bible), and measuring instruments (e.g., sphygmomanometers). All these examples can be used to illustrate the point that if you acquire evidence that leads you to judge that an outside source of information is probably unreliable (i.e., the analogue to the Low Probability Thesis), it is not rational to use evidence from that very source to establish its reliability. For example, Plantinga quoted with approval Reid's argument that if a man's honesty is questioned, it would not be rational to propose to settle the issue by asking him (242; also, *supra*). Similarly, if the accuracy of a text or of a measuring instrument is called into question, it is irrational to rely on that very text or that very measuring instrument to resolve the question. On this point Plantinga and I are in agreement. Thus, we can see that Plantinga has found at least some positive cases to get the argument from analogy going.

The problem with these cases is that none of them involve evaluations of the reliability of one's own cognitive faculties. Plantinga strengthened the analogy with the next group of examples.

Evaluations of the Local Reliability of Separable Cognitive Faculties

Ultimately, Plantinga will try to argue that the evolutionary naturalist has reason to doubt the reliability of all of her cognitive faculties. I refer to evaluations of reliability (or unreliability) that apply to all of one's cognitive faculties as evaluations of *global reliability*. However, Plantinga also discussed examples involving the evaluation of a single cognitive faculty, which I refer to as an evaluation of *local reliability*. There is one main example in this category discussed by Plantinga, the *example of Freudian wish-fulfillment* (e.g., WPF 229–31; IAR 207–10, 241).[7] In this example, we are to imagine that a theist obtains evidence E that convinces her that her belief in God is the product of wish-fulfillment and also is such that, given E, it is very improbable that her faculty of wish-fulfillment is reliable (the analogue of the Low Probability Thesis). Plantinga correctly claimed that it would be irrational for her to argue that there is other evidence (O) on which the reliability of her faculty of wish-fulfillment is quite probable, if O is itself the product of her faculty of wish-fulfillment.

At this point in the argument, Plantinga introduced an important distinction. He wanted to argue that the theist has a rationality defeater for belief in God. But he realized that there is a sense in which she has a defeater and a sense in which she may not. To understand the distinction, suppose that the theist's belief in God plays a crucial role in enabling her to live a productive life in a world that would otherwise seem so cold and indifferent as to render her dysfunctional. It is easy to imagine that the wish-fulfillment faculty might be part of her cognitive design, aimed at making her able to function productively in the world. In that case, her cognitive design might be such as to specify that she continue to believe the deliverances of her wish-fulfillment faculty even if she has evidence that it is unreliable. In such a case, because her cognitive faculties would be functioning according to their design plan, there would be no *proper-function rationality defeater* for her belief in God (IAR 208). But Plantinga thinks that she has a different kind of defeater for it, a *purely alethic rationality defeater* (IAR 209). Without worrying about the details, we can say that a purely alethic rationality defeater is a belief that would operate as a defeater if only our truth-aimed processes were at work (IAR 208). Plantinga insisted that, in this example, the theist would have a purely alethic rationality defeater for her belief in God, because if the relevant part of her design plan were aimed at truth, she would cease believing in God.[8]

I believe that Plantinga is correct that these examples are positive cases for his argument. In addition, they are stronger cases than those in the first category, because they are cases that involve the evaluation of the reliability of one's own cognitive faculties. However, there is still an important disanalogy. Note that in the case of Freudian wish-fulfillment, the believer was able to obtain evidence of the source of her belief in God (wish-fulfillment) and the reliability of the source without using that very source at all. I need a term for cognitive faculties that one need not rely on when evaluating their reliability. I refer to them as *separable* cognitive faculties. I believe that this second category of examples shows that in evaluating the local reliability of separable cognitive faculties, it is irrational to rely on beliefs from the very faculty whose reliability is being evaluated.

EVALUATIONS OF THE LOCAL RELIABILITY OF NON-SEPARABLE COGNITIVE FACULTIES

Are all individual cognitive faculties separable in this sense? It is clear that they are not. The clearest case of a non-separable cognitive faculty is memory. I could never hope to evaluate the reliability of my memory without relying on at least some memories. I believe that perception and reasoning are also non-separable.[9] In "The Illusion of Defeat" I referred to the cognitive faculties involved in scientific theorizing and in metaphysics as the *abstract faculties,* simply to give a name to the cognitive faculties that take us beyond perception and memory in ordinary cases of theorizing. Clearly, evaluations of the reliability of our abstract faculties would require theorizing that takes us beyond the deliverances of perception and memory. Thus, I assume that the abstract faculties are also non-separable.

If this is correct, then when the evolutionary naturalist evaluates her abstract faculties, she will be evaluating non-separable cognitive faculties. Because the situation of the evolutionary naturalist involves the evaluation of the reliability of non-separable faculties, it is important to consider whether the conclusions that Plantinga reached concerning the local evaluation of separable cognitive faculties extend to the local evaluation of non-separable cognitive faculties. In IOD, I argued that they do not. Because Plantinga had not considered any examples of the evaluation of the local reliability of non-separable cognitive faculties, I introduced some of my own.

The first example in IOD—the genetic hallucinogen blocker example—was simply intended to establish that there are some negative cases. It involved the evaluation of the local reliability of one's perceptions. Here is a brief recap. BT has overwhelming evidence that hallucinogen H produces severe hallucinations in 95 percent of subjects within one hour of taking it. BT took H one hour ago. If we let PFR = my perception faculties are reliable, then BT rationally accepts evidence E and rationally accepts the analogue of the low probability thesis (i.e., P(PFR/E) is low). However, BT also has other evidence (O) that 5 percent of the population has a gene that blocks the effects of H and that he himself has that H-blocking gene. In this case, it is also rational for BT to accept the analogue of the High Conditional Probability Claim (P(PFR/E&O) is high), and the High Conditional Probability Claim explains why it would be irrational for BT to doubt the reliability of his perceptual faculties.

In his reply, Plantinga agreed with me on this example (IAR 226). There was no reason for Plantinga to find the example threatening, because there was at best only a very weak analogy between the genetic hallucinogen blocker example and the EAAN. This first example simply established that both Plantinga and I agreed on some negative as well as positive cases.

However, Plantinga did not agree with me on my next example, which I referred to as *the first example of the tardy revelation.* This example is just like the previous example, except that at the time that BT takes the drug H, he has no idea whether or not he has the hallucinogen blocking gene. A little over an hour after taking H, BT has noticed no change in his perceptions. So, I claim, he already has some evidence that he has the hallucinogen blocking gene. Then he gets a call from his doctor telling him that he does in fact have the hallucinogen blocking gene. I claimed that the combination of apparent lack of change in his perceptions with the experience of receiving a call from his doctor informing him that he has the hallucinogen blocking gene makes it rational for BT to believe that his perceptual faculties are reliable.

Plantinga disagreed with me on this example (IAR 227), but it is not completely clear what he disagreed with. Rather than address my analysis of this example directly, he argued that there is an important disanalogy between this example and the EAAN, because the

EAAN involves a judgment of global unreliability, but in the example of the tardy revelation, I assumed that BT's memory was reliable and that BT could properly rely on it (227). I take up the issue of judgments of global reliability shortly. First, I want to say something more about Plantinga's argument that there is an important disanalogy between the example of the tardy revelation and the EAAN.

Dialectically, it would seem that by arguing that there is an important disanalogy, Plantinga must be thinking that the example of the tardy revelation as I originally described it is truly a negative case. That would make it important to distinguish it from the EAAN. Obviously, if the example were a positive case, there would be no need to argue that there is an important disanalogy between the two cases. So one question for Plantinga is whether he agrees that the example of the tardy revelation is a negative case.

Is the first example of the tardy revelation disanalogous to the EAAN in the way that Plantinga claims? In IOD, I tried to make the EAAN more plausible by interpreting it as an argument about local reliability (the reliability of one's abstract faculties) rather than as an argument about global reliability, because it seemed implausible to me that evolution would have made our perceptions and memories unreliable. Let us distinguish between a *local EAAN* (which only challenges the reliability of some cognitive faculties (e.g., the abstract faculties) not all cognitive faculties, at least not perception and memory) and a *global EAAN* (which challenges the reliability of all our cognitive faculties, including perception and memory). In his reply to me, Plantinga suggested that in focusing on the local EAAN rather than the global EAAN, I had offered him a "Trojan horse" (IAR 227). So it is important to point out that it was Plantinga himself who suggested that a local EAAN is more plausible than a global EAAN (IAR 231–32).

It seems that at one time Plantinga agreed with me that a local EAAN is more plausible than a global EAAN. Has he changed his mind? The answer to this question is important, because, if Plantinga dismisses the first example of the tardy revelation as irrelevant on the grounds that it does not involve global reliability, then it would seem that he should also dismiss as irrelevant all of the examples in the first two categories above, examples involving the reliability of outside sources of information and examples involving the evaluation of the local reliability of separable cognitive faculties.

However, nothing crucial hinges on whether Plantinga now thinks that a local EAAN is plausible, because his defense of a global EAAN fails, and the reasons that it fails are general reasons that also apply to any plausible local EAAN. To see why, we need to discuss the fourth and final category of cases.

Evaluations of Global Reliability

Plantinga in IAR discussed three cases involving evaluations of global reliability. I begin with the one that is closest to my first example of the tardy revelation, the *XX example,* and discuss the other two later. The XX example is Plantinga's variation on my hallucinogenic drug examples, in which the drug is assumed to compromise global reliability. Here is what Plantinga said about the XX example:

> Suppose therefore I take a good dose of XX, which induces not merely perceptual unreliability but global cognitive unreliability. I correctly believe that 95 percent of those in this condition are no longer reliable; I also believe that 5 percent of the population has

the blocking gene; but I have no belief as to whether I myself have that gene. I then have a defeater, so I say, for R [that my cognitive faculties are reliable]. Now suppose I come to believe that my physician has telephoned me and told me that I am among the lucky 5 percent whose reliability is unimpaired by ingesting XX. Do I have a defeater-defeater? (IAR 227)

Plantinga answered this question using the distinction he had introduced earlier. He agreed that either there is no proper-function defeater in this case, or if there is, there is also a proper-function defeater defeater, because it is part of our design plan that we continue to believe we are reliable in such circumstances (IAR 228). But Plantinga claimed that we do have a purely alethic defeater (and no purely alethic defeater defeater), because "if only our truth-aimed processes were at work in these situations, [we] would have a defeater" (208).

To see that Plantinga is mistaken about cases of this kind, let me add a few more details to the present example, thereby turning it into the *augmented XX example*. Suppose that twenty scientists who have been investigating the properties of XX have ingested a dose of it. Suppose that when they ingested it, each of them knew that one of them had the drug-blocking gene, but they did not know who. Suppose on the basis of prior studies they also knew that the drug XX, if they did not have the drug-blocking gene, would cause them to hallucinate pink elephants dancing and on that basis to believe that they were seeing pink elephants dance (and to believe that that their perceptual faculties were reliable); would cause massive memory distortions, so that, for example, they would *not* realize that they were scientists but rather would believe that they were pink-elephant trainers for a circus and all their memories would cohere with those beliefs (and they would believe that their memory is reliable); and would cause massive distortions in their scientific and other theoretical beliefs, so that, for example, they would no longer accept the chemical theory that explains the action of XX and indeed would have no beliefs at all about XX (but they would continue to believe that their theoretical beliefs were reliable); and similarly for all their other cognitive faculties.

In such a circumstance, I claim that each scientist would have a simple positive test for finding out that one possesses the drug-blocking gene (though there is no negative test for finding out that one lacks it). Suppose one of the scientists, Helen, has the following experiences: The other nineteen scientists start behaving as though they are perceiving pink elephants and seem to think they are training them to dance, but Helen does not herself have any experience of pink elephants. The other scientists behave as if they have lost all knowledge of chemical theory, but Helen continues to believe that she is a scientist who is investigating the effects of XX and can still call up her chemical knowledge of XX and wonder whether she has the gene that blocks its effects. In such a case, I believe it is clear that Helen would not even need a call from her physician telling her that she has the drug-blocking gene to make it rational for her to believe that she had it and, thus, that her cognitive faculties were reliable. However, if she did get such a call from her physician, that would give her even more reason to believe that she had the drug-blocking gene and that her cognitive faculties were reliable. I hope it is clear that it is possible to continue to add details to the example to make the evidence supporting Helen's belief that she has the drug-blocking gene even stronger. It seems to me that in this augmented XX example it is clearly rational for Helen to believe that she has the drug-blocking gene, and thus that it is rational for her to believe that her cognitive faculties are reliable.

There is one preliminary objection that I need to address and set aside. It is raised by two different ways of describing the XX example. At one point, Plantinga described it this way:

"[T]he drug also prevents its victim from detecting her unreliability" (IAR 206). This is a correct description of the example. In describing the example, I made the same point by saying that, in Helen's situation, there is no negative test for finding out that one lacks the gene. However, Plantinga has another way of describing the example that is not so uncontroversial. Referring to the person (Helen in the above example) who has the drug-blocking gene, he said: "[S]he knows of nothing that distinguishes her case from theirs" (IAR 206).[10] This is true at the beginning of the experiment. But after she has taken XX, to assume that it is true is to assume precisely what is at issue, because the issue is whether or not the continuity and coherence of Helen's memories and experience, including the experience of her doctor calling to tell her that she has the drug-blocking gene, provides her with a positive test that determines that she does have the gene.[11] That is the issue that needs to be discussed, not stipulated away.

Note that there are numerous disanalogies between this example and the EAAN. Nonetheless, this example holds some important lessons that can be applied to the EAAN. First, Plantinga would agree that Helen has no proper-function rationality defeater for her belief that her cognitive faculties are reliable (R), but he would insist that she has a purely alethic rationality defeater for R.

Why would he think that she has a purely alethic rationality defeater for R? One of the most curious features of Plantinga's position on purely alethic rationality defeaters is that he says almost nothing about how best to design cognitive faculties so that they best realize the goal of truth or reliability. Let us refer to cognitive design motivated solely by the goal of truth or reliability as *alethic design*.

The question is whether Helen's reasoning would be recommended as good alethic design. Surprisingly, most of Plantinga's discussion of the issue addresses it logically, and his objections to it seem largely *a priori*. He says that reasoning such as Helen's is pragmatically circular (WPF 234) or epistemically circular (IAR 226). In his discussion of the kind of circularity involved, he repeatedly gives examples from Hume (IAR 210–11, 269). Perhaps Hume is a good guide to our *a priori* convictions about good reasoning. Surely, no one would cite Hume as an authority on good alethic design. Hume's *a priori* constraints on good reasoning led him to dismiss as irrational all beliefs about external objects, all beliefs about causes and effects, all inductive generalizations, and all beliefs about the future. Though Hume himself did not extend the arguments further, precisely parallel arguments would brand as irrational all beliefs about the past. Surely Plantinga would not recommend a Humean cognitive design as one that is good for discovering truths.

Perhaps Plantinga thinks it is obvious that this kind of pragmatic or epistemic circularity would be bad cognitive design. But there is a problem with almost all of the examples he uses to support that conclusion. The examples are almost all from the first two categories above—that is, either examples concerning the evaluation of the reliability of outside sources of information or examples concerning the evaluation of the local reliability of separable cognitive faculties. In all cases of that kind, epistemic circularity would be bad alethic design, because it would be better for the subject to use other faculties to evaluate the reliability of outside sources of information or of separable cognitive faculties.

But in cases concerning local evaluations of the reliability of non-separable cognitive faculties or cases concerning evaluations of global reliability, if determinations of reliability are to be made at all, there is no alternative to using the faculties themselves to help determine their own reliability. Plantinga does admit that it may be appropriate to use perception and memory to evaluate the reliability of perception (242). For the same reasons, it is clear that it is appropriate to use perception and memory to evaluate the reliability of memory. It is also clear that Plantinga thinks there is an important disanalogy between these examples and the XX example (and presumably, also, the augmented XX example). However, his insistence seems to be entirely *a priori,* because he never actually considers the alethic design advantages of reasoning of the kind that Helen uses to conclude that her cognitive faculties are reliable in the augmented XX example. So let's consider them.

First, Helen's reasoning does produce true beliefs in the augmented XX example. I suspect that Plantinga is worried that endorsing Helen's reasoning would require also endorsing the reasoning of the other nineteen scientists in that example, who also believe that their cognitive faculties are reliable, and thus that it would lead to false beliefs 95 percent of the time. But this is a faulty analysis. In the case of the other nineteen scientists, their cognitive design is not responsible for their false beliefs. Their false beliefs are due to cognitive malfunction. Although good alethic design may involve including some modules for recovering from certain kinds of cognitive malfunction, no design can protect against the wholesale cognitive malfunction involved in the augmented XX example or in other examples involving global unreliability.

Of course, this is just one case, and a very hypothetical one at that. There are other cases that are not nearly so hypothetical. At one point, Plantinga mentions mad cow disease (IAR 240). Suppose mad cow disease produced complete insanity in 95 percent of cases. No cognitive design will prevent the false beliefs of those who go insane, because they will be subject to massive cognitive malfunction. But what about the 5 percent who do not go insane? Good alethic design would require them to use their cognitive faculties to determine whether or not they have gone insane. As in the case of Helen, it would be good alethic design if they used the continuity and coherence of current perceptions and memories and other evidence of this kind to conclude that they were not insane, and therefore that their cognitive faculties were reliable.

Even if the insanity rate for mad cow disease were 100 percent, it would still be possible that a cure would be found. In good alethic design, those with the disease would be able to use evidence that a cure has been found and evidence that it was administered to them in combination with evidence of the continuity and coherence of current perceptions and memories to conclude that they are not insane and that their cognitive faculties were reliable. There is currently no cure for mad cow disease, but there is a cure for syphilis. Before there was a cure, syphilis typically caused insanity. Good alethic design would make sure that those who developed syphilis before there was a cure but later found out about a cure and took it would not be trapped into thinking that they were doomed to become insane and their cognitive faculties were doomed to become unreliable.

So there are some cases where this sort of reasoning is good for getting at the truth. More such cases can easily be imagined, and I introduce another shortly. Plantinga's argument thus depends on his finding enough other cases where this sort of reasoning is bad at getting at the truth that he can convince us that, on balance, it is bad alethic design. This brings us to his final category of positive cases.

2. Plantinga's Positive Cases Involving Evaluations of Global Reliability

Plantinga (in IAR) provided two fairly equivalent cases that he relied on repeatedly. These are the only two remaining examples that play a significant role in his argument by analogy. I refer to the first as the *malicious demon example,* because it goes beyond Descartes' evil demon to suppose that the demon not only gives us experiences that lead to massively unreliable beliefs, but that the demon also gives us good evidence that that is what he is doing. Plantinga did not say what this evidence is, but we can imagine that every so often the demon interrupts the course of our experience to tell us what he is doing and to give us evidence that he has total control of our experience, perhaps by giving us experiences that seem so miraculous that the best explanation of them is that someone is completely controlling the course of our experience. Plantinga's second example is the *brain in a vat example,* where we suppose that not only are we brains in vats being given experiences by psychologists on Alpha Centauri that make our beliefs massively unreliable, but again that those psychologists have given us good evidence that that is what they are doing.

I believe that these two examples are importantly different from the XX examples[12] and from the EAAN. In the XX examples and in the EAAN, Plantinga would hold that the subject has a purely alethic, but not a proper-function, defeater for her belief in her own reliability. In contrast, I have argued that in these cases she has *neither* kind of defeater.

In the malicious demon example and the brain in the vat example, Plantinga also believes that the subject has a purely alethic but not a proper-function defeater for almost all of her beliefs, including her belief in her reliability. But I think that in these cases she has *both* kinds of defeater. So if I am correct, it is a mistake to think that the malicious demon example and the brain in the vat example are analogous to the XX examples or the EAAN in the relevant respects. In Plantinga's two examples, the subject has neither kind of defeater; in the XX examples and the EAAN, the subject has both kinds of defeater. If this is correct, Plantinga's argument from analogy fails.

Let's discuss the brain in the vat example. A parallel analysis applies to the malicious demon example. Why does it seem so obvious to Plantinga that my cognitive design requires me to continue to maintain most of my beliefs, including my belief in my own reliability, after I become convinced that I am a brain in a vat being experimented on by scientists on Alpha Centauri? Plantinga simply asserts that in such a case, "I would carry on as before"(IAR 206). I cannot speak for Plantinga here, so let me speak for myself. What would I do if I became convinced that I was a brain in a vat being stimulated to believe that I was interacting with an external world populated by other human beings? It is true that I would use external world language to describe my experience. I might say that I am seeing a dog when having the appropriate perceptions, but I would not believe it. I might be cautious around the apparent dog, to avoid being bitten, because the experience of being bitten would be painful. But I would have a very different attitude toward the world.

For example, I would no longer contribute to charities for disaster relief, unless there were some negative repercussions on my experience from my not contributing. Indeed, I would not regard myself as under any moral constraints toward those I seemed to be interacting with and would only comply with moral constraints when the negative repercussions to me of violating them would be worse than the results of complying with them. For example, if I decided to do medical research, I would have no compunction about doing experiments on other human beings, if I could get away with it. In general, there would be no prohibition on using other people in my experience as mere means, and I would think it truly crazy to regard them as ends-in-themselves. To take another example, my career decision would have been very different. Convinced that colleagues and students were imaginary, I would never have decided to become a philosopher.

In such a scenario, I would regard my life as I would regard playing a very realistic video game. Although it is possible to become so involved in a video game that one tends to forget that it is a game, most of the time one is quite aware of it. So most of the time, I would not regard my cognitive faculties as reliable, but simply as coherently producing the impression of an imaginary world. It is very likely that before long I would tire of the imaginary world, as I tire of video games. Suicide would not seem a great loss, and indeed it might be an opportunity—the opportunity to end the experiment and perhaps motivate the Alpha Centaurian psychologists to install me in a body and allow me to live a real life.

In all of this I would, I believe, be operating in accord with my design plan. So it seems to me that finding out that I was a brain in a vat would give me a proper-function rationality defeater for my prior beliefs about the external world and that I would no longer believe that they were really true. And I would certainly not believe that my cognitive faculties were reliable.

The same can be said of the malicious demon case. I do agree with Plantinga that reasoning in this way in the brain in the vat case (or the malicious demon case) would be recommended by good alethic design. I only insist that it is also a part of my actual design.

In any case, my suggestion is that, of the alternatives under consideration, the best alethic design is one that endorses reasoning that one's cognitive faculties are reliable in the XX examples and does not endorse maintaining beliefs about the external world and reasoning that the cognitive faculties producing those beliefs are reliable in the malicious demon and brain in the vat examples.

We know that it is possible to design cognitive faculties that implement that design, because mine do. It disturbs me to think that Plantinga's cognitive faculties are so poorly designed (whether judged on the basis of alethic or non-alethic goals) that he would continue to believe in external objects and in the reliability of the cognitive faculties responsible for those beliefs in the malicious demon example and the brain in the vat example. Would he really think that he had obligations not to harm the imaginary people in his imaginary world? It also disturbs me to think that in the XX examples, though Plantinga would believe that his cognitive faculties were reliable, he would be tortured with Humean doubts that that belief was produced by a design plan not aimed at the truth. It would be sad to think that his cognitive design produced a divergence between proper-function and purely alethic rationality in any of these examples.[13]

3. A Negative Case More Directly Analogous to the EAAN

If I am right about these cases, then there are important disanalogies between the XX examples and the examples of the malicious demon and of the brain in the vat. The most important disanalogy is that the subject has a rationality defeater of both kinds (proper-function and purely alethic) for R in the malicious demon and brain in the vat examples, but does not have a rationality defeater for R of either kind in the XX examples. Which kind of case more closely resembles the EAAN in the relevant respects? Here is an example, slightly modified from my previous article, that can help us to answer that question:

ZZ example. Suppose there is a chemical compound ZZ that entered the food chain years ago. All human fetuses throughout the world were exposed. Were it not for a fortuitous occurrence, ZZ would have caused all the children in the world to develop ZZ disease, which would have made their cognitive faculties globally unreliable, though they would have continued to believe that their cognitive faculties were reliable. The fortuitous occurrence was a random mutation in a benign strain of bacteria that enabled it to digest compound ZZ. Before the mutation, its objective chance of occurring was very low. After the random mutation, the benign bacteria were cultured and a small dose was given to every child in the world. No child actually developed ZZ disease from exposure to compound ZZ, even though the objective chance was high, when they absorbed compound ZZ, that they would develop it. On his twenty-first birthday, BT is told that he was exposed to ZZ in utero and is informed of the lucky mutation that saved him from insanity. After he digests this information, does he have a rationality defeater for all of his beliefs, including his belief in the reliability of his cognitive faculties?

Plantinga would say that BT has a purely alethic rationality defeater, but not a proper-function rationality defeater. I find this answer incredible. I agree that in the ZZ example, BT's believing his cognitive faculties are reliable is rational in the proper-function sense. But I insist that it is rational in the purely alethic sense, also. Any reasonable alethic design for cognitive faculties would require BT to conclude on the basis of the continuity and coherence of his memories and perceptions that he had avoided the ZZ disease and that his cognitive faculties were reliable. This would be good alethic design, because in cases like BT's it would lead to true beliefs and in cases like the ZZ example – except that no cure is developed for the disease so that all children become insane – the design plan of their cognitive faculties is irrelevant, because their beliefs are not produced in accordance with the design plan but rather as a result of a cognitive malfunction, and one that not even a good alethic design plan could be expected to correct for.

So I conclude that the ZZ example is in the relevant respects similar to the XX examples and different from the malicious demon example and the brain in the vat example. However, unlike the XX examples, the ZZ example is directly analogous to the EAAN. Both examples involve global reliability. In the EAAN, we assume, for the sake of argument, that there is a time in the past when the probability of future generations of human beings' developing reliable cognitive faculties was low. In the ZZ example, we also assume that there is a time in the past when the probability of future generations of human beings' developing reliable cognitive faculties was low. But in both cases, there is evidence provided by the coherence and continuity of memory and experience that makes it rational to continue to believe in the reliability of one's own cognitive faculties. The analogy between the EAAN and the ZZ example makes it clear why the evolutionary naturalist has no defeater of either kind for his belief in naturalism, or for any of his other beliefs either.

4. How the EAAN Might Have Succeeded

In my previous article, I considered what would be necessary to devise an XX example that was truly analogous to Plantinga's positive examples (i.e., the malicious demon and brain in a vat examples) in the relevant respects. Here I add to the augmented XX example.

Suppose it were known that XX produced the memories and beliefs typical of a scientist investigating the properties of XX. And suppose that it also was known to produce the illusion of receiving a telephone call from one's doctor calling with the information that one has the drug-blocking gene. Then someone like Harold, an undergraduate philosophy major and not a scientist investigating the properties of XX, would have no trouble telling he had the drug-blocking gene an hour after taking the drug. If he still believed he was a philosophy major, he would know he had the drug-blocking gene, even before his doctor called to confirm that fact.

Helen's situation, however, would be more problematic. An hour after taking the drug, she would not be able to tell whether she had the drug-blocking gene or not. And even if she got a call from her doctor telling her she had the drug-blocking gene, it would not help, because she would have both kinds of rationality defeater for the belief that her doctor had called because she would have both kinds of defeater for the belief that her cognitive faculties were reliable. Of course, in such a situation, if her doctor wanted to tell her that she had the drug-blocking gene, he would presumably find another way to convey the information to her than calling her on the telephone.

Once you see that it is possible to construct an XX example analogous to Plantinga's other positive examples in the relevant respects, it is clear that there are many ways to do so. For example, suppose it were known that the drug XX caused those who ingested it to believe that XX caused people who ingested it to believe they were research scientists working on XX, when really XX caused them to believe that they were undergraduate philosophy majors. If it also caused them to hallucinate receiving a call from their doctor telling them that they had the drug-blocking gene, then Harold rather than Helen would be the one whose reliability was questionable. If Harold's doctor called to tell him that he had the drug-blocking gene, Harold would have both kinds of defeater for the belief that the doctor had called, because Harold would have both kinds of defeater for the belief that the relevant cognitive faculties were reliable.

Once we see how the XX example could be made analogous in the relevant respects to Plantinga's positive examples, it is easy to see how the situation of the evolutionary naturalist could have been analogous to them in the relevant respects. If in addition to accepting the Low Probability Thesis, the evolutionary naturalist also believed that it was likely that, even if evolution gave him cognitive faculties that were globally unreliable, he would have pretty much the same experiences and memories that he has, then he would have a rationality defeater of both kinds for all (or almost all) of his beliefs, including his belief that his cognitive faculties are reliable.[14] Of course, it is not necessary that he believe that he would have exactly the same memories and experiences. If, in addition to accepting the Low Probability Thesis, the evolutionary naturalist believed that his memories and experience would be equally continuous and coherent, regardless of whether his cognitive faculties were globally reliable or whether they were globally unreliable, that would be enough to give him a rationality defeater of both kinds for all (or almost all) of his beliefs, including the belief that his cognitive faculties were reliable. Hume is one of the few naturalists I can think of who would have accepted such an extreme claim. So it seems to me that it is appropriate for Plantinga to refer to a defeater of this kind as a *Humean rationality defeater*.

Plantinga's EAAN will provide a defeater for all or almost all naturalists who have what we might call "*Humean*" *rationality* (which seems to me not any kind of rationality at all). "Humean" rationality itself provides a defeater for beliefs about the external world, the future, the past, etc., so Plantinga's EAAN will not make much practical difference to a naturalist with "Humean" rationality. To any naturalist who thinks that the coherence and continuity of experience and memory are more probable on the hypothesis that our cognitive faculties are reliable than on the hypothesis that they are not, Plantinga's EAAN will not be successful, because the analogy with the malicious demon and the brain in the vat examples fails.

5. Good Alethic Design

In my previous article I tried to explain why Plantinga's account of purely alethic rationality defeaters is problematic. If one's goal is to design cognitive faculties that are aimed at truth, Plantinga's design is unacceptably rigid. On his design, our cognitive faculties could be trapped into judging themselves to be unreliable, because they refused to admit evidence that they were not unreliable. I think he is correct that, in judging the reliability of separable cognitive faculties, it is not rational to rely on those very faculties. But there is an important disanalogy between those cases and cases involving non-separable faculties or evaluations of global reliability. In these latter cases, there are no other cognitive faculties that can be used to make reliability evaluations. A well-designed cognitive system will make use of those faculties and its determinations will, when the system is functioning properly (i.e., not when it is malfunctioning) be reasonably reliable. Fortunately for us, we actually have such a system. It would have been disastrous if Hume had convinced everyone that making inductions was a bad way of getting at the truth. It would also be disastrous, though less so, if Plantinga were able to convince us that, given our current evidence, believing in the conjunction of naturalism and evolution is not a good way to get at the truth.

When Plantinga presents the EAAN, he typically begins with a thought experiment involving a hypothetical community with a completely different evolutionary background from us (WPF 222–23; IAR 5; *supra*). Given only that the Low Probability Thesis was true at some point in their evolutionary past, he argues that we would have to expect their cognitive faculties to be unreliable. He then asks us to consider the possibility that we are that community. However, as he acknowledged in another context (IAR 249–50), there is an important disanalogy between the two cases. If we were able to observe the other community and obtain evidence that their cognitive faculties were reliable, then, even though we accepted the Low Probability Thesis as applied to the other community, it would no longer be rational for us to believe that their cognitive faculties were unreliable. For Plantinga, the important disanalogy, then, is that, as he sees it, even if we have the same kind of evidence of reliability about ourselves that we have about the reliability of the other community, it is not alethically rational for us to draw the parallel conclusion about ourselves. In this essay, I have tried to explain why it would be a poor alethic design that did not permit us, on the basis of the same kind of evidence about ourselves, to draw the parallel conclusion about ourselves. Thus, the EAAN fails.

More on the Illusion of Defeat

Notes

1. New York: Oxford University Press, 1993. Hereafter "WPF."
2. Alvin Plantinga, "Introduction and Replies," in James Beilby, ed., *Naturalism Defeated?* Ithaca, NY: Cornell University Press, 2002, 1–12 and 204–75. Hereafter "IAR."
3. Readers who, like me, believe that throughout history, epistemologists' challenges to common sense (e.g., challenging knowledge or warranted belief in external objects, theoretical objects, other minds, the future, the past, etc.) have almost always been mistaken will be even more skeptical of Plantinga's challenge. I return to this issue below.
4. For simplicity, where there is no potential for confusion, I will use "defeater" as a success term—that is, to refer to ultimately undefeated defeaters.
5. Where "N" stands for a statement of naturalism and "E" stands for an account of the evolution of human beings, including our cognitive faculties. "R" stands for a proposition stating of some or all of our cognitive faculties that they are reliable. I will say more about different substitutions for "R" shortly.
6. In "The Illusion of Defeat," in Beilby, ed., *Naturalism Defeated?* (153–64; hereafter IOD), I argued that, even given the Low Probability Thesis, the evolutionary naturalist would not have a rationality defeater for belief in N&E. This seems to have given Plantinga the mistaken impression that I agree with him that the Low Probability Thesis is true ("Introduction and Replies," 230 and especially 256). This is a mistake. In fact, I believe that we have good reason to believe that the Probability Thesis (and *a fortiori,* the Low Probability Thesis) is false. However, to discuss that issue would take me beyond the scope of this paper.
7. Other examples in this category include what he refers to as the "optimistic overrider" and what in the psychological literature is referred to as the Lake Wobegone Effect.
8. Plantinga goes on to argue that, at times of reflection (e.g., Hume in his study) a purely alethic rationality defeater becomes a proper-function rationality defeater, which he refers to as a *Humean rationality defeater*. In the text, I will simply refer to them as purely alethic rationality defeaters, though I will have more to say about "Humean" rationality below.
9. Plantinga does seem to acknowledge that it can be rational to use perception and memory to determine their own reliability (IAR 242). I discuss this further below.
10. At another place he says "relevantly distinguishes" (IAR 236).
11. Note, also, that examples of this kind pose a challenge to accounts of knowledge such as Robert Nozick's account (*Philosophical Explanations* [Cambridge, MA: Belknap Press, 1981]) in terms of tracking the truth, because it seems that Helen knows that R (that her faculties are globally reliable) even though, had -R been true (i.e, had she lacked the drug-blocking gene), she would still have believed R.
12. For ease of exposition, I refer to the original XX example and to my augmented XX example as "the XX examples."
13. I do believe there are examples in which the two conceptions of rationality diverge—for example, in Plantinga's example of the "optimistic overrider" (IAR 207–8). All the examples in which they diverge seem to involve separable cognitive faculties. In any case, as I explain in the text, the two conceptions do not diverge on the EAAN or on any of other cases more directly relevant to it.
14. I say "all or almost all," because I want to allow that some beliefs, such as "I exist," may avoid defeat.

7

Evolutionary Naturalism: Epistemically Unseated or Illusorily Defeated?

Alvin C. Plantinga and William J. Talbott

A. It's No Illusion!

Alvin C. Plantinga

First, thanks to Bill Talbott for his characteristically acute and insightful contribution. I learned something significant from his original comments on the EAAN ("The Illusion of Defeat"); now he generously offers "More on the Illusion of Defeat" (hereafter "MID"). I say it's no illusion.

Talbott begins by conceding, if only for purposes of argument, that

(1) It is rational for the naturalist to believe P(R/N&E) is low.

What I argued, of course, is not merely this, but that in fact P(R/N&E) *is* low—though naturally I do also believe that it is rational for the naturalist to believe this. What Talbott denies is the second step of the EAAN, namely that

(2) The naturalist who sees (believes) that P(R/N&E) is low has a defeater for R in her own case.

I argued for (2) by analogy, giving several relevantly similar cases where, as I saw it, the analogue of the naturalist does indeed get a defeater for R. Talbott categorizes these (together with some others he presents in his original, 2002 article) in a useful way and then adds another he thinks maximally similar in relevant ways to the EAAN situation. This crucial case is an extension of my emendation of his "first example of the tardy revelation" (I know, I know, this is starting to sound like the sort of thing that gives analytic philosophy a bad name).

My emendation of that first example of the tardy revelation went as follows. At *t*, I come to believe I've taken XX,[1] a drug that (globally) destroys cognitive reliability in 95 percent of those who take it, compromising memory, perception, reasoning, and the rest. At *t* I also believe 5 percent of the population has a blocking gene that renders XX harmless. Presumably, at *t* I have a *defeater* for R (a purely alethic defeater—see MID, 154). At *t+n* it seems to me that I receive a call from my physician, telling me that I am one of the lucky 5 percent that have that blocking gene. And the question is this: what would the proper function of my cognitive faculties and processes—my *truth-aimed* cognitive faculties and processes—require in this situation? I thought they would continue to require that I refrain from believing R. That this is correct, it seems to me, is easily seen if we reflect on what we should think in the analogous third-person case. We learn that Sam has taken XX; we also learn that after he took it, it seemed to him that he received such a call from his physician; surely proper function requires continuing to withhold the belief that Sam's faculties are reliable.

Talbott thinks I am wrong in this evaluation, but he doesn't explain why. What he does instead is to elaborate on the example (the "augmented XX example"). For the details of the elaboration, see MID. Fundamentally, the idea is this: Each of twenty scientists takes XX, a drug they believe destroys cognitive reliability; each believes that the other nineteen have also taken it; each also believes that one of the twenty has the blocking gene, but doesn't know which; and each also believes that the experiences and memories of all twenty will be coherent after taking the drug.

Under these circumstances, says Talbott, each of the scientists has an easy positive test for finding out that she has the blocking drug: noting that (a) it seems to her that the other scientists start behaving as if they are perceiving pink elephants, and as if they have lost all knowledge of chemical theory, but (b) that she herself has no experiences as of pink elephants, and also believes that she is a scientist investigating the effects of XX; it also seems to her that she has the relevant chemical knowledge; and she wonders whether she has the blocking gene. In a word (or two) it seems to her that her memories and current beliefs and experience are coherent. As Talbott puts it, "the issue is precisely whether or not the continuity and coherence of Helen's memories and experience, including the experience of her doctor calling to tell her that she has the drug blocking gene, does provide her with a positive test that determines that she does have the gene" (MID, 158). "That" he says, "is the issue that needs to be discussed, not stipulated away."

Right; that is certainly an issue that needs to be discussed, and far be it from me to stipulate it away. I'd like to discuss this issue first from a relatively abstract point of view, and then in terms of the specific example Talbott proposes. Abstractly, then, Talbott's idea seems to be this: Suppose I come to believe something (something relevant) with respect to which it is unlikely that my cognitive faculties are reliable. Perhaps I believe that I've taken XX and that 95 percent of those who do suffer massive cognitive distortions; or perhaps I believe N&E and then come to see that P(R/N&E) is low. This may initially give me a defeater for R. But I can still acquire a defeater-defeater for that defeater: I can note that my experience and memories are coherent. If I do note this, then I no longer have a defeater for R. More exactly, perhaps the idea is that if I believe this all along, so to speak, I don't get a defeater for R in the first place; my belief that my memories and experience are coherent is a "defeater-deflector."[2]

This seems to me mistaken. First, the notion of coherence; here there are several questions. (1) Talbott speaks repeatedly of coherence of my *experience* with my memories (and presumably other beliefs). Coherence is not an easy notion,[3] and much about it is unclear; it is particularly unclear, however, that or how *experience* could be coherent with *beliefs*. I take it experience is here being thought of, roughly, as or as similar to *ways of being appeared to:* but how could my being appeared to in such and such a way be coherent (or incoherent) with a belief or other proposition? Of course *the proposition* that I am being appeared to in such and such a way can be coherent with other propositions; so let's suppose this is how to understand Talbott here. Coherence requires that there be no conflict between my belief that I am being appeared to in such and such a way and the rest of my beliefs.

There is more to coherence of experience with belief. What more? Well, (2) experience often *inclines us to believe* something or other; on the basis of experience, it often seems that things are a certain way. Thus, on the basis of my present sensory experience it seems to me that I'm seated in front of a computer, that birds are singing outside my door, that the trees have fully leafed out, etc.; and this inclines me to believe those things. Sometimes experience and belief can be at odds, as when confronted with what one knows is an illusion of some sort. I know those two lines in the Muller-Lyre illusion are the same length; but one *seems* longer than the other. I cross my index and middle finger and put a marble between them; it seems as if (feels as if) there are two marbles; but I believe there is only one. Each of the propositions involved in the deduction of *non-self-exemplification both does and does not exemplify itself* seems true; but I believe at least one of them must be false. Accordingly, there can be incoherence, not just between a belief and the proposition that I am appeared to thus and so, but also between a belief and the proposition experience inclines me to believe. So coherence would also require, not that there be no conflict at all between belief and how things seem (the propositions experience inclines me to accept), but that there be minimal conflict—no doubt a weasely thing to say here, but perhaps the best that we can do.

(3) Coherence also requires an appropriate relationship among the propositions I believe. What relation? That's not an easy question. Like most of us, I am such that there is no possible world in which all of my beliefs are true, and quite properly so. That is because I believe that I believe at least one false proposition. So let B be the set of my first-order beliefs—beliefs that are not about my beliefs: I believe that B contains at least one false belief. But then there is no possible world in which all of my beliefs—second-order as well as first-order—are true. What's required here isn't logical consistency but something else extremely hard to state—but including, presumably, believing no explicit contradictions (propositions of the form P&-P), and perhaps believing no proposition P and also believing its denial, -P. Perhaps it's also required that I believe no obviously inconsistent triad of propositions, and maybe the same for obviously inconsistent quartets—beyond that, it's hard to say.

Clearly, this isn't sufficient for coherence in the relevant sense. Also required would be (4) a certain hard to characterize regularity in my experience, and a certain consonance between experience and belief. Perhaps this is required just for the coherence of my experience in itself; but at any rate it is surely required for the coherence of my experience with the beliefs I have about how the world ordinarily goes. This regularity would be violated if, e.g., at one moment it seems for all the world as if there is a house twenty-five feet in front of me, and then at the next moment it seems that I haven't moved, but now there is instead a two-acre lake, and then at the next moment a small mountain, and then a flat jungle, and then an opera house, etc. Or if at one moment it seems that my left arm is about the usual two and a half feet long, at the next moment it seems that it's fifty feet long, at the next an inch and a half, etc. Or if at one moment Helen seems to be five years old, but at the next sixty-five years old, or if I believe she is a scientist working on XX but she acts like she thinks she's an elephant trainer, or if my son Harry suddenly seems to turn into a small horse. In these cases there would be no logical inconsistency, but still a lack of coherence in some obvious sense.

So it's monumentally difficult to say just what coherence is, in the relevant sense, the sense in which Talbott no doubt intends it. And the next question: why think my believing that my experience and beliefs are coherent in this sense (call it "T-coherence") is a defeater-defeater or -deflector? Note first that it is obviously possible to be mistaken about whether one's experience and beliefs are T-coherent. Perhaps (and perhaps not) it is impossible to be mistaken about whether one is being appeared to redly; beliefs of that sort are conceivably incorrigible.[4] But the same certainly does not go for coherence, as Frege (as well as most any philosopher you pick) learned to his sorrow. I can easily believe, falsely, that some set of my beliefs is coherent. Furthermore, T-coherence involves a kind of continuity in experience; therefore, judgments of coherence involve reliance upon memory. I have to rely on memory to know that, e.g., it's not the case that at one moment it seems there is a house before me, at the next moment a lake, at the next a snowcapped volcano, etc.; and of course memories can be mistaken. So suppose I believe I've taken XX (and that 95 percent of those who do suffer from massive cognitive unreliability); then I can't sensibly take it for granted that my beliefs and experiences are T-coherent, just because it seems to me that they are. Under these conditions I can't take it for granted that there is a horse in front of me, just because it seems to me that I see a horse there; the same goes for my experience's being T-coherent.[5] The best I can do, along these lines, is to take it for granted that it *seems to me* that my beliefs and experience are T-coherent.

Hence the basic question is this: why think that if my experience and beliefs seem T-coherent, then (probably) my faculties are functioning reliably? What is P(R/XX & STC) (where "XX" denotes the proposition that I've taken XX and nineteen out of twenty people who do so are rendered unreliable, and "STC" denotes the proposition that my beliefs and experience seem T-coherent)? Is this probability high, as it must be if STC can serve as a defeater-defeater or defeater-deflector? I think not. We can see this most clearly by considering a third-person case. Suppose we initially assume R_{Sam}: that Sam's faculties function reliably; we then come to believe XX_{Sam}: that he has ingested some XX and that 95 percent of those who take it (all but that lucky 5 percent who have the blocking gene) suffer from massive cognitive dysfunction. In believing XX_{Sam} we have a defeater for R_{Sam}. Next, suppose we learn STC_{Sam}: that it seems to Sam that his experience and beliefs are T-coherent. Do we then have a defeater-defeater? If we had believed STC_{Sam} all along, would we have had a defeater-deflector? One has only to ask the question to see the answer: clearly not. Ninety-five percent of those who take XX become massively unreliable; Sam has taken XX; $P(R_{Sam})/(STC_{Sam})$ is .05; adding STC_{Sam} to the condition doesn't appreciably change that probability. If so, that proposition is neither a defeater-defeater nor a defeater-deflector.

Now surely the same goes in my own case. P(R/XX) is low, around .05; adding STC to the condition doesn't appreciably raise that probability; hence that belief is neither a defeater-defeater nor a defeater-deflector in my own case any more than in Sam's.

I said we should consider this question both from the more abstract and from the more concrete point of view: so let's turn to Talbott's example. There's no need to repeat the details; what's relevant here is (1) Helen believes the experimental set-up is as Talbott says it is, (2) she has no experiences as of pink elephants, (3) it seems to her that the others are behaving as if they have lost their knowledge of chemistry and believe that they are training pink elephants and (4) it seems to her that she has not lost her knowledge of chemistry. In a word, her experiences are T-coherent. Our question is: does the fact that her beliefs and experiences seem T-coherent give her a defeater-defeater or -deflector with respect to R?

Consider again the third-person perspective. We initially assume R with respect to Helen; we then learn that she has taken XX, and that 95 percent of those who do become massively unreliable. Thus we acquire a defeater for our initial assumption of R with respect to her. We then learn that it seems to Helen that her beliefs and experience are T-coherent. Does this give us a defeater-defeater—or, if we suppose we learned this at the very beginning of the story—a defeater-deflector? Surely not. True, she seems to remember the story about the blocking gene, how those without it would act as if they were elephant trainers, etc., while the person with it would have no elephant experiences, etc. But of course she's taken XX; so how can we credit these apparent memories of hers? $P(R_{Helen}/XX_{Helen}) = .05$; we add to the condition XX_{Helen} the proposition that Helen believes the story, and also believes that her present beliefs and experiences are T-coherent. Clearly $P(R_{Helen}/XX_{Helen}\ \&\ TC_{Helen})$ is not significantly greater than $P(R_{Helen}/TC_{Helen})$. We therefore don't have a defeater-deflector or defeater-defeater in the belief that Helen believes the story, and that her present beliefs and experiences are coherent with that story.

Surely the same goes for Helen herself. She believes that she's taken XX and that nineteen out of twenty people who take it develop massive unreliability. This gives her a defeater for R in her own case—that is, this gives her a purely alethic rationality defeater. No doubt proper function would call for her to continue to believe or anyway assume that her faculties are functioning reliably. What else could she do? But, so I say, if only the truth-aimed cognitive faculties were working in her, she would have a proper-function defeater and cease believing or assuming R. And the fact that it seems to her that her beliefs and experience were T-coherent would make no difference.

So here Talbott and I seem to be at loggerheads: I say Helen would have a purely alethic defeater; Talbott says she would not. Is this as far as we can go?

Maybe not. First, note that the third-person perspective gives us a bit of insight into the difference between a proper-function defeater here and a purely alethic defeater. Truth-aimed processes may be compromised or overridden in one's own case. There is this enormously powerful inclination to assume R in one's own case, and of course it is easy to see the utility of this inclination; failure to believe or assume R can make a shambles of one's entire noetic structure. Of course, one doesn't have nearly as strong an inclination to believe R in the case of someone else. So a sensible way to proceed here is to consider the analogous third-person situation, as I did above. It is clear, I think, that in the third-person analogues to the case in question, one doesn't get a defeater-defeater or defeater-deflector for the defeater for R.

A second consideration: Talbott complains that "one curious feature of Plantinga's position on purely alethic rationality defeaters is that he says almost nothing about how best to design cognitive faculties that best realize the goal of truth or reliability" (158); he goes on to say that "The question is whether Helen's reasoning would be recommended as good alethic design" (158); and he adds later that "In any case, my suggestion is that, of the alternatives under consideration, the best alethic design is one that endorses reasoning that one's cognitive faculties are reliable in the XX examples and does not endorse maintaining beliefs about the external world and reasoning that the cognitive faculties producing those beliefs are reliable in the malicious demon and brain in the vat examples" (161). He goes on to make interesting observations about good cognitive design.

The question," he says, "is whether Helen's reasoning would be recommended as good alethic design": but why is *that* the question? Our question is whether the devotee of N&E who knows or believes that P(R/N&E) is low has an alethic defeater for R; but *that* question has to do with what *our* truth-aimed faculties or belief-producing processes require. It is certainly possible that there be better cognitive designs than ours (perhaps enjoyed by angels); and no doubt God could design a cognitive system that always and automatically came up with only true beliefs across a wide variety of topics; still, our question is really about the truth-aimed portion of *our own* cognitive design. Talbott's thought is that a really good cognitive design plan would be such that in XX type situations, one's experience and beliefs seeming coherent would furnish a defeater-defeater or defeater-deflector. I'm inclined to doubt that; in any event our design plan doesn't enjoy that feature. We can easily see this as follows. Suppose I believe that nineteen out of twenty who take XX suffer massive cognitive distortion and are also such that they believe their beliefs and experience are T-coherent. I then come to believe I've taken XX; I am therefore in an XX situation. But now add that my experience and beliefs seem to me to be T-coherent. Don't I nevertheless still have an alethic defeater for R? As I would in the corresponding third-person case?

Be that as it may (and no doubt it will), Talbott then produces still another example, one he thinks most like the actual case involving the EAAN: the ZZ example. Here BT is told (and presumably believes) a story about ZZ, a substance producing global unreliability, and a cure. According to the story, he was exposed to ZZ but also (along with all the other children in the world) given the cure. Talbott suggests that BT doesn't, under those conditions, have a purely alethic defeater for R. I'm inclined to agree. Certainly, in that circumstance, I would continue to believe that my cognitive faculties are reliable, and I can't see that I get a defeater in being told (and believing) the story. It is as if I learn in one breath both that there was once a very high probability that my faculties would become massively unreliable, and that this eventuality was forestalled by my receiving the cure. Under those conditions I don't get a defeater for R. In learning these two things at the same time, perhaps I have a defeater-deflector, or at any rate a limiting case of a defeater-deflector—"limiting," in that in the usual cases, the defeater-deflector is a belief I have before I acquire the otherwise defeating belief.

But the ZZ case is not analogous to the EAAN situation. The reason is simple: in the EAAN situation, there is nothing analogous to the bacterium-induced cure in the ZZ story. The believer in N&E doesn't learn of any bacterial cure, or any cure of any other sort; and it doesn't seem likely or perhaps even possible that such an element could sensibly be added to the story. He who accepts N&E and sees that P(R/N&E) is low has no knowledge of any cure for the looming unreliability. He has nothing but the grim realization that the probability—given what is crucially important here, namely, the origin and provenance of his cognitive faculties—is low that his cognitive faculties are reliable.

One final consideration: perhaps Talbott thinks that what plays the role of the bacterial cure, for the evolutionary naturalist, is just the fact that it seems to her that her faculties are T-coherent; perhaps he thinks that this belief is a defeater-deflector or a defeater-defeater. Perhaps the idea is that while P(R/N&E) is low, P(R/N&E&C) (where C is the proposition that her faculties seem to her to be T-coherent) is high; and perhaps the idea is that one can simply tell, whether reliable or not, whether one's memories and beliefs seem to one to be reliable.

He also says:

If, in addition to accepting the Low Probability Thesis, the evolutionary naturalist believed that his memories and experience would be equally continuous and coherent, regardless of whether his cognitive faculties were globally reliable or whether they were globally unreliable, that would be enough to give him a rationality defeater of both kinds for all (or almost all) of his beliefs, including the belief that his cognitive faculties were reliable. Hume is one of the few naturalists I can think of who would have accepted such an extreme claim (163).

The idea, I think, is that $P(C/R) > P(C/-R)$, which means that C is evidence for R.

First, we must remember that the evolutionary naturalist can't sensibly propose, as a defeater-deflector, that his beliefs and experiences are in fact coherent. That would be just to assume that the faculties involved in the production of the belief that they are coherent are in fact reliable; and that is precisely part of what is in question. But we conceded, at least for purposes of argument, that he can't make a mistake about whether they *seem* to him to be coherent; and presumably they do. So the question is really this: what is $P(SC/R)$ and $P(SC/-R)$? What is the probability of his faculties seeming coherent, given, respectively, R and -R? Talbott apparently thinks there is a big difference here: the first probability is high, while the second is low. I can't see the slightest reason for thinking this. It's a truism that seriously deluded people often have coherent sets of beliefs; it is even more likely that their beliefs and experience will *seem* to them to be coherent. Just as there is evolutionary advantage in believing R, whether or not it is true, so there may well be evolutionary advantage in being such that one's experience and beliefs seem coherent, whether or not R is true. We'd expect that if Sam were reliable, his experiences and beliefs would seem coherent; we don't suppose that if he were not reliable, they wouldn't seem that way. Maybe they would and maybe they wouldn't; it's hard to say what that probability is; either it's inscrutable or it isn't far from .5. Neither of these will be of use to Talbott.

B. The End of an Illusion?

William J. Talbott

This is the final installment of a conversation that Al Plantinga and I have been engaged in for almost ten years. I want to thank Al for taking the time to think these issues through with me. By offering me the opportunity to have the last word here, he has shown the same generosity and goodwill that have characterized all of his contributions to our debate.

I begin this response with a contrast between Plantinga's Humean conception of rationality and my non-Humean conception. I explain why these different conceptions of rationality seem to have led Plantinga to misunderstand my use of coherence in explaining what it is rational to believe in many of the examples that I have discussed. Then I make some comments on the examples. These comments will be brief, because I think Plantinga and I have both made as strong a case as we can for our way of understanding the examples. I conclude with some suggestions for finding one's way out of a Humean conception and for moving to a more adequate conception of rationality.

1. WHAT IS RATIONALITY?

Plantinga has a conception of rationality that, in some respects, is not so different from Hume's. On Plantinga's conception, as on Hume's, rationality is understood in terms of reasoning, and reasoning is understood on the model of an argument that begins from propositions as premises and ends with propositions as conclusions, and where the rationality of the argument depends on some sort of logical relation among the propositions. Call this a *Humean* conception of rationality. Such a conception of rationality is highly intellectualistic. It is not an account on which young children or nonhuman animals could have rational beliefs. In addition, on a Humean conception of rationality, certain kinds of beliefs could never be rational, because any reasoning for them would be question-begging. Hume famously argued that it could not be rational to believe the results of induction, because any argument that induction is reliable would have to rely on the results of induction itself.

When a cognitive process is such that determinations of its reliability depend on the results of the process itself, I refer to it as *non-separable*. The non-separable processes include perception, memory, and reasoning itself, including what Hume called "induction." Consider memory. On a Humean conception of rationality, it cannot be rational for me to have beliefs about how reliable my memory is, because any such belief would have to be based, in part, on evidence supplied by memory. And yet I believe that when I was in my twenties my memory was very reliable and it was rational for me to believe that it was, and that, sadly, now that I am in my fifties, my memory is much less reliable than it was when I was in my twenties, and that it is rational for me to believe that, too.

How could these beliefs about the reliability of my memory be rational? I believe that the answer to that question requires us to find an alternative to the intellectualistic Humean conceptions of rationality. One of the benefits of my long-running debate with Plantinga has been that it has given me opportunity and the motivation to struggle with the problem of articulating such an alternative.

Because Plantinga and I are working with different conceptions of rationality, we sometimes talk past each other. For example, in my first essay in this volume, I presented a number of examples in which I claimed that there is evidence provided by the coherence and continuity of memory and experience that makes it rational to believe in the reliability of one's own cognitive faculties. When Plantinga reads this sort of claim, he is inclined to translate it into the Humean framework. First, he seems to assume that the evidence must be propositional, so he assumes that it is the belief that one's memories and experience are coherent that makes belief in the reliability of one's own cognitive faculties rational. Second, he seems to assume that coherence must be understood as a kind of logical relation among propositions. So that means that memory and experience must be given propositional content. This impression is confirmed by his reference to BonJour's coherence theory of justification, which is as intellectualistic as a theory of epistemic justification can be.

But I meant something different. I think my notion of coherence is the more ordinary, commonsense one, because it is the one that Hume appealed to when he wanted to give an example of causes of beliefs that were not truly reasons. Here is what Hume said:

> But here 'tis observable, that even in these changes they preserve a *coherence,* and have a regular dependence on each other; which is the foundation of a kind of reasoning from causation, and produces the opinion of their continu'd existence. When I return to my chamber after an hour's absence, I find not my fire in the same situation, in which I left it: But then I am accustom'd in other instances to see a like alteration produc'd in a like time, whether I am present or absent, near or remote. This coherence, therefore, in their changes is one of the characteristics of external objects, as well as their constancy.[1]

In this passage, Hume meant to be explaining why people believe in the continued existence of objects by "a kind of reasoning," even though, according to Hume, they really have no good reason to. His explanation appealed to coherence in experience and coherence between experience and memory, and his example made it easy for his readers to recognize what he meant. The main difference between Hume and me is that I think that the coherence of experience and of experience with memory is not only a cause of our beliefs, but one of the contributors to making them rational. Coherent memories and experience help make our beliefs rational without our having to have any beliefs about their coherence, and they do so without the need for any kind of logical relations that bridge experience, memory, and beliefs. So, for example, I believe that many nonhuman animals and young children have beliefs made rational by the coherence of their experience and memory, though they surely have no belief about that coherence. The upshot is that most of Plantinga's discussion of coherence is based on presuppositions about rationality that I don't share. This leads him to misunderstand the role that the coherence of experience and memory plays in my explanation of the examples.

2. More on the Examples

I focus my comments primarily on two of the examples. Consider the augmented XX case. Twenty scientists have taken the drug XX. They all know that one of them has a gene that blocks its usual effect of making it irresistible to believe that there are pink elephants dancing. Helen is one of the scientists. After a short time, she has the experience of seeing her nineteen colleagues exclaiming over the dance steps of pink elephants. So Helen concludes that the other nineteen scientists are hallucinating and that she is not because she has the genetic XX blocker. I insist that her belief is rational. Plantinga disagrees.

To explain why he disagrees, Plantinga invites us to consider what third-person beliefs would be rational. As Plantinga states the test, the third person would have the information that Helen believes her experience and memory to be coherent. He wants us to realize that that very belief might also be made unreliable by XX. Here my discussion of coherence in the previous section becomes relevant. I believe that it is the coherence of Helen's experience and memory that make the relevant beliefs rational, not her belief that they are coherent. The third-person perspective leaves out exactly what it is that makes the relevant beliefs rational, her experience and memory.

Since Plantinga and I both agree that it is part of our cognitive design for Helen to continue to believe that she is not hallucinating, the issue between us is whether it is *rational* for Helen to believe it—that is, whether she would believe it if only her truth-oriented cognitive faculties were operating. Plantinga thinks that the process that keeps Helen believing in her reliability is something like wish-fulfillment, which no one would think was a truth-oriented process. In my first essay, I tried to explain why this analogy is misleading. I gave my reasons for thinking that the process maintaining Helen's belief in her own reliability would be good alethic design—that is, would generally produce true beliefs.

In response, Plantinga agrees that it might be good alethic design in the abstract, but it is not part of *our* alethic design. Since I do not have any direct access to our alethic design, the only way I have of figuring out which of our cognitive faculties and processes are truth-oriented is to investigate which ones tend to lead to true beliefs. In the first essay, I argued that the relevant processes operative in Helen's case generally lead to true beliefs. That seems to me to be the best kind of evidence we can have that, unlike wishful thinking, they are truth-oriented. In reply, Plantinga says we can recognize that they are not truth-oriented by considering the third-person case. So I think that the third-person case misleads Plantinga on this issue also.

In the first essay, I used a series of examples to cast doubt on the EAAN. The idea was to finally reach an example that was so closely analogous to the EAAN that if Plantinga's analysis yielded the wrong result in that example, it would be reasonable to think that it yielded the wrong result in the EAAN. The last example in my series was the example of ZZ. In that example, BT finds out on his twenty-first birthday that he and everyone else on earth was exposed to compound ZZ *in utero* and, had it not been for a fortuitous and objectively improbable genetic mutation, compound ZZ would have made him and everyone else globally unreliable. So BT realizes that there was a time *in utero* when it was almost certain that he would become globally unreliable. Yet, I claimed, having had twenty-one years of evidence of his reliability, it would be irrational for BT to give up believing that his cognitive faculties were globally reliable just because there was a time in the past when it was objectively highly probable that they would not be.

Plantinga surprises me by agreeing with my conclusion on the ZZ example—that is, that BT's belief in his own reliability is rational. If he is not to give up EAAN, Plantinga must find a disanalogy between the ZZ example and EAAN. He does find one. In the ZZ example, BT is given an explanation of why his faculties are not globally unreliable at the same time that he gets the information that there was a time when it was objectively probable that they would be. But the naturalist who accepts evolution accepts (we have supposed for the sake of the argument) that there was a time in the past when it was objectively probable that human cognitive processes would not be reliable and, unlike BT in the ZZ example, he has no explanation of how human beings developed reliable cognitive faculties.

Plantinga is correct to point out that this is a disanalogy between the two cases. What puzzles me is why he thinks it is a relevant disanalogy. To test its relevance, let's consider a *modified* ZZ example that is just like the original except that BT never finds out about the genetic mutation that blocked the effect of ZZ. I believe that this modified ZZ example is not relevantly different from the original. At twenty-one years of age, BT, like anyone else with normally functioning cognitive processes, would have such coherent memories and experience that it would be irrational for him to believe that he was globally unreliable. Although BT might be curious about what happened to block the activity of ZZ, I don't see how the rationality of his belief in the reliability of his cognitive processes depends on his being able to answer that question.

Plantinga's discussion of the ZZ example suggests that there may be a conditional conclusion that he and I can agree on: If BT's belief in the modified ZZ example is rational, then the EAAN fails. Even if we don't agree on that conclusion, our discussion of that example has greatly clarified the issues raised by the EAAN.

One more thing: Plantinga's goal is to find disanalogies between my negative examples and EAAN. In my essay in this volume, I pointed to disanalogies between his positive examples and EAAN. He does not discuss them in his reply, so let me just reiterate that his position on those examples leads to strange results. Consider, for example, the malicious demon example. In that example, we are to suppose that Plantinga is convinced that his perceptual evidence is produced by a malicious demon whose intent is not only to give Plantinga massively false beliefs but also to flaunt his power by giving Plantinga plenty of evidence of exactly what he is doing. Plantinga asserts that in such a case he would still hold on to all his beliefs about the external world. For example, he would still believe that there was a New Orleans that had been devastated by Hurricane Katrina. I find this hard to believe. For example, I wonder if he would really make a monetary contribution to help the hurricane victims if he believed that all of his evidence of the disaster had been produced by a malicious demon.

3. Finding an Alternative to Humean Conceptions of Rationality

The Humean picture of rationality exerts a powerful influence on philosophical thinking about rational belief. Because it can seem to be evident *a priori* that it would be objectionably question-begging for us to rely on a cognitive process to determine its reliability, the only hope I have of helping you find a way out of the Humean picture is to rely on your being able to recognize rationality in particular cases and then show you that many of the judgments about particular cases don't fit the Humean picture.

Plantinga's EAAN is just one more example of the way that the Humean picture of rationality generates skeptical puzzles. Freed from the Humean picture, one can see the EAAN to be a kind of cognitive illusion. The move from the Humean picture of rationality to the realization that the EAAN is an illusion can be divided into three steps. The first, which is also the largest, is to recognize that we are able to use our non-separable faculties such as memory to make rational determinations of the reliability of those very faculties. Because Plantinga now apparently agrees that we can do this, his accusation that the evolutionary naturalist is involved in epistemic circularity loses its bite. All of us who use our non-separable cognitive faculties to determine their reliability are involved in a similar kind of circularity. We need a new non-Humean conception of rationality to explain how this sort of "circularity" can be rational.

The second step is to realize that there are cases in which the information that there was a time in the past at which it was objectively probable that our cognitive faculties would be globally unreliable is not a conclusive defeater for our belief that they are now globally reliable. Plantinga now concedes that, in the original ZZ example, BT's belief in his global reliability is rational. Recall that, in that example, at the same time that BT finds out about his exposure to ZZ he also finds out about the genetic mutation that prevented the chemical ZZ from making him globally unreliable. So Plantinga has also taken the second step.

The third step is the shortest. It simply involves acknowledging that BT's belief in his global reliability is rational in the modified ZZ example, where BT has no information about how the effects of the chemical ZZ were blocked. Plantinga balks at this last step.

To help him take the last step, I can divide it into many shorter sub-steps. To define the sub-steps, consider a large number of intermediate cases: First, suppose BT finds out about ZZ, but does not get any information about the sanity-saving genetic mutation until one second later. Second, imagine the first example, except that BT does not get any information about the genetic mutation until one minute later. Third, imagine the first example, except that BT does not get any information about the genetic mutation until one hour later. . . . Next-to-last, imagine the first example, except that BT does not get any information about the genetic mutation until he is on his deathbed and is moments from dying. Last, imagine the first example, but BT does not ever get any information about the genetic mutation. How many sub-steps is Plantinga willing to take? If he is willing to take one, I don't see why he would not take them all. But to refuse to take the first, he has to explain why a one-second delay in BT's getting the information about the genetic mutation would be so epistemically important.

Although the debate between Plantinga and me has taken us through several iterations of various farfetched examples, I hope it is clear that the issues are much more important than our verdicts on those examples or even than our verdict on the validity of Plantinga's EAAN. Hume showed us that what continues to be the dominant conception of rationality in epistemology leads to deep skeptical puzzles. We need a better conception of rationality. Plantinga has offered us one. He has developed his conception with great subtlety and sophistication. However, on examination we find that it is too entwined with Hume's conception to be adequate. There is more work to be done.

Notes

It's No Illusion! (Alvin C. Plantinga)

1. Not to be confused with the Mexican beer of the same name.
2. See my "Reply to Beilby's Cohorts," in James Beilby, ed., *Naturalism Defeated?* Ithaca, NY: Cornell University Press, 2002, 244.
3. See Laurence BonJour, *The Structure of Empirical Knowledge*. Cambridge: Harvard University Press, 1985, 93ff.
4. A better candidate for incorrigibility might be "I'm being appeared to like *that*."
5. In dreams it usually seems that one's experience and beliefs are T-coherent; waking sometimes reveals that they were not.

The End of an Illusion? (William J. Talbott)

1. Hume, David. (1740 [1978]) *A Treatise of Human Nature* (2nd edition, edited by P H. Nidditch. Oxford: Oxford University Press), I.4.2.

8

A Quantum-Theoretic Argument against Naturalism

Bruce L. Gordon

> Das Wirkliche is uns nicht gegeben sondern aufgegeben
> (nach Art eines Rätsel).[1]
> —Albert Einstein

A common misunderstanding, one of which we need to disabuse ourselves, is that quantum theory, while possessing astounding predictive power, actually explains the phenomena it describes. It does not. Quantum theory offers mathematical descriptions of measurable phenomena with great facility and accuracy, but it provides absolutely no understanding of why any particular quantum outcome is observed. The concepts of description, prediction, and explanation are conceptually distinct, and we must always keep this fact in mind. Mathematical descriptions, if they are accurate, tell us what mathematical relationships hold among phenomena, but not why they hold. Empirical predictions, if they are correct, tell us what we will or might observe under certain experimental conditions, but not necessarily why these things will happen. It is the province of genuine explanations to tell us how things actually work—that is, why such descriptions hold and why such predictions are true. The failure to appreciate these differences has given rise to a lot of confusion about what the impressive edifice of modern physical theory has, and has not, achieved. Quantum theory is long on the what, both mathematically and observationally, but almost completely silent on the how and the why. What is even more interesting is that, in some sense, this state of affairs seems to be a necessary consequence of the empirical adequacy of quantum descriptions. One of the most noteworthy achievements of quantum theory, I dare suggest, is the accurate prediction of phenomena that, on pain of experimental contradiction, have no physical explanation. This is perhaps a startling way to state the matter, but no less true because of it. It is such phenomena, and arguments concerning their significance, that will occupy us in this chapter.

In view of the challenge it poses to the philosophical hegemony of physical explanation, it is not surprising that quantum theory poses a problem for naturalism. Ontological naturalism, while exhibiting various niceties of definition, centrally maintains that the sum and substance of everything that exists is exhausted by physical objects and processes and whatever is causally dependent upon them. In other words, the philosophical naturalist insists on the causal closure of the material realm. A corollary of this viewpoint is that there is no such being as God or anything remotely resembling him; rather, according to the naturalist, the spatio-temporal universe of our experience, in which we exist as strictly material beings, is causally self-sufficient. The explanatory resources of this naturalistic metaphysical closure are restricted, therefore, to material objects, causes, events, and processes and their causally emergent properties.[2]

Some discussions of the role of naturalism in science have sought to characterize it instead in terms of an attitude rather than a strict adherence to certain metaphysical tenets.[3] This modification comports well with what both Arthur Fine and Bas van Fraassen have been saying for many years, and is the outworking of a distaste for metaphysics in conjunction with differing degrees of deference that both possess toward accepting the deliverances of science in respect of what sorts of things there are, and toward accepting the approximate completeness of scientific explanations.[4] As van Fraassen describes it, though not uncritically, under this conception "the apparent knowledge of what is and what is not material among newly hypothesized entities is mere appearance. The ability to adjust the content of the thesis that all is matter again and again is then explained instead by a knowing-how to retrench which derives from invariant attitudes."[5] As he goes on to note, however, it is common for the materialist to conflate the theory thereby constructed with the attitudes that generated it, thus generating a false consciousness that perhaps accounts for the conviction that science requires a presumptive materialism.[6] But regardless of whether naturalism or materialism or physicalism consist in certain tenets or are comprised by general attitudes combined with a certain know-how (albeit disingenuous) in respect of retrenchment, I want to argue that the phenomena of quantum theory pose an insuperable problem because they show that materialistic tenets, at root, are false, and that attempts at retrenchment are, at best, an exercise in self-deception.

In light of all this, it is interesting to note that both Fine and van Fraassen have chosen to argue that nonlocal quantum correlations do not need an explanation.[7] This would seem to be the only polemical path around the anti-naturalistic metaphysical conclusion quantum phenomena naturally suggest. Nonetheless, given the aversion both have expressed to any kind of metaphysics, albeit in different ways, I suspect the "no explanation needed" strategy may be their way of saying "a pox on both your houses" to materialist and anti-materialist metaphysicians alike. Regardless, I will argue that this is the wrong response, because such phenomena genuinely do require an explanation, and the correct explanation is manifestly anti-materialist. Before I fill in the details, however, let me set forth in broad outline the argument to be made.

1. The Argument in a Nutshell

Among the distinguishing characteristics of quantum phenomena are nonlocality and nonlocalizability. When quantum systems interact, their existence can become "entangled" in such a way that what happens to one of them instantaneously affects all others, no matter how far apart they have separated. Since local effects obey the constraints of special relativity and propagate at speeds less than or equal to that of light, such instantaneous correlations are called nonlocal, and the quantum systems manifesting them are said to exhibit nonlocality. A result in mathematical physics called Bell's theorem—after the Irish physicist who proved it—shows that no hidden (dynamically irrelevant) variables can be added to the description of quantum systems exhibiting nonlocal behavior that would succeed in explaining these instantaneous correlations on the basis of local considerations.

A Quantum-Theoretic Argument against Naturalism

When additional variables are introduced for this purpose, the predictions of the modified theory differ from those of quantum mechanics. A series of experiments beginning with those conducted by Alain Aspect at the University of Paris in the 1980s has demonstrated quite conclusively that quantum theory, not some theory modified by local hidden parameters, generates the correct predictions. The physical world, therefore, is fundamentally nonlocal and permeated with instantaneous connections and correlations. Nonlocalizability is a related phenomenon in relativistic quantum mechanics and quantum field theory in which it is impossible to isolate an unobserved quantum entity, such as an electron, in a bounded region of space. As we shall see, nonlocality and nonlocalizability present intractable problems for the materialist.

The ground has now been laid to summarize an argument showing not only that quantum theory does not support materialism but also that it is incompatible with materialism. The argument can be formulated in terms of the following four premises and conclusion:

P1: Naturalism is the view that the sum and substance of everything that exists is exhausted by physical objects and processes and whatever is causally dependent upon them.

P2: The explanatory resources of naturalism are therefore restricted to material objects, causes, events and processes.

P3: Neither nonlocal quantum correlations nor (in light of nonlocalizability) the nature of the fundamental constituents of material reality can be explained or understood if the explanatory constraints of naturalism are preserved.

P4: These quantum phenomena require an explanation.

C: Therefore, naturalism (materialism, physicalism) is irremediably deficient as a worldview, and consequently should be rejected not just as inadequate, but as fundamentally false.

The first two premises of this argument are uncontroversial: the first is just a definition, and the second is a consequence of this definition. The key premises of the argument are therefore the third and fourth; once these are established, the conclusion follows directly. As we shall see, the failure of material identity/individuality in the quantum realm not only undermines the ontology of naturalism, it also renders necessitarian theories of natural law untenable. This leads to the conclusion that the empirical regularities of quantum theory are mere regularities unsupported by any natural nomological structure. The presence of (near) universal regularities in nature that lack a physical explanation demonstrates the falsity of a purely naturalistic nomology, creating a second insuperable problem for naturalistic metaphysics. Our efforts therefore will be focused on justifying the claims in premises three and four. Some definitions are in order before we begin.

2. A Definitional Excursus

2.1 Criterion of Material Individuality (CMI):

All material individuals *I* are such that for every property *P* having a well-defined value or range of values, and all times *t* during which *I* may be said to exist, either *I* exemplifies a definite value (or a definite range of values) of *P* at *t* or *I* does not possess any value of *P* at *t* (i.e., *I* does not possess *P* at *t* at all). We include in the scope of such properties attributes like "being at spatiotemporal location (x,y,z,t) in reference frame *R*."

If this criterion is not met, it would be a mistake to think that we were dealing with a material individual at all, since there is no primitive substantial thisness (material haecceity) in view, no spatiotemporal location in question, and no identity-conferring properties to which we have recourse. In the absence of any individuality, all labels, names, or indices attaching to the purported material entities must be regarded as fictions: no material thing is named if the intended referent has no substantial thisness, no location, and no uniquely identifying properties. If a catchphrase is desired, we could do little better for present purposes than to adopt (*mutatis mutandis*) Quine's dictum: there is no entity without identity.

2.2 Intrinsic Properties:

Intrinsic properties (such as mass and charge) are essential properties of particle kinds in quantum physics, with particles of the same kind possessing the same values of their intrinsic properties. These properties are not individuative, however, because they do not serve to uniquely distinguish particles of the same kind.

2.3 State-Dependent Properties:

State-dependent properties (such as position, momentum, energy, and spin-direction) are contingent properties that depend on the quantum state of the particle in question. They are the only candidates for individuative properties of particles of the same kind, but they could only serve this purpose during those times when the particle is not observed if they could be regarded as the objective possession of the particle in and of itself, apart from observation.

2.4 The Precise Value Principle (PVP):

Whatever the state of a quantum system or the ensemble containing it, each observable has a precise value for the individual system.

When we briefly consider stochastic hidden variable theories later on, we will relax this assumption so that observables only need to possess an objective dispositional property (propensity) given by a definite probability distribution. Either way, the principle proves quantum-mechanically untenable.

A Quantum-Theoretic Argument against Naturalism

2.5 COMMON CAUSE (CC):

Suppose we have two factors, call them A and B, that are statistically correlated in the sense that P(A|B) ≠ P(A). If neither A nor B has probability zero, then this is a symmetric relationship that can be expressed by denying their statistical independence, i.e., P(A&B) ≠ P(A)P(B). In such case, factor C functions as a common cause for the correlation between A and B if and only if:

(i)   C precedes A and B in time;
(ii) P(A|C) > P(A|~C) and P(B|C) > P(B|~C); and
(iii) P(A&B|C) = P(A|C)P(B|C).

Note that C not only raises the probabilities of A and B but it screens them off from each other, rendering them statistically independent. Note further that specifying that C precedes A and B precludes the possibility of rendering the explanation trivial simply by setting C = (A&B). As we will note in section 3.2 below, EPR-type correlations cannot be given a local explanation in terms of common causes.

2.6 SPIN (INTRINSIC ANGULAR MOMENTUM):

In quantum mechanics, spin is the intrinsic angular momentum of a subatomic particle, nucleus, atom, or molecule, which continues to exist even when the particle comes to rest. A particle in a specific energy state may have a particular spin, just as it has a particular electric charge and mass (it may also be in a superposition of spin states). According to quantum theory, the spin is restricted to discrete and indivisible values, specified by a spin quantum number. Because of its spin, a charged particle acts like a small magnet, and is affected by magnetic fields. The direction of the spin of a spin-½ particle is a bivalent property, that is, it can be measured to be in the "up" (+) direction or the "down" (−) direction. We will consider a system of spin-½ particles (in our case, electrons) when we look at the EPR Paradox and Bell's theorems. For a spin-1 system, there are three possible values of the spin: +1, 0, and −1.

3. LOCAL COUNTERFACTUAL DEFINITENESS AND ITS ATTENDANT DIFFICULTIES

Let's get into the details of why, in principle, there is no physical explanation for quantum correlations. In 1967, Kochen and Specker constructed a geometrical argument using spin-1 systems to show that the Precise Value Principle could not be satisfied. There was a loophole in their result, however, in that it assumed that altering the experimental arrangement by changing the direction in which the spin was measured did not affect the experimental outcome. Closing this loophole requires considering the possibility that the hidden variables governing experimental outcomes might be affected by the experimental context. This can be accomplished by considering a different sort of argument, which originated with John Bell's seminal papers, in which it is assumed that the quantum mechanical observables have definite values independently of measurement, but which values they possess are contextually dependent on the experimental arrangement. Abner Shimony (1984) discusses a nice proof of a no-go theorem for local stochastic (indeterministic) contextual hidden variable theories of the environmental sort, that is, a hidden variable account in which the context is the state of the surrounding physical environment, inclusive of the experimental apparatus, with which the quantum system interacts. I note that such a proof can be given, because for the sake of simplicity of presentation, I'm not going to give it here. Rest assured that, *mutatis mutandis*, the proof I'm going to give could be extended to cover the case of local environmental stochastic hidden variables.

3.1 Wigner's Classification and the Derivation of a Bell Inequality for Local Hidden Variable Theories

We consider an electron spin experiment. Spin measurements in any direction for spin-½ systems can have one of two values, which we will refer to as either "up" (+) or "down" (−). For the purpose of this experiment, electrons are produced in pairs at the source in what is called the "spin-singlet" state. This means that they are linked together in such a way that while the probability of either one of them having a specific spin value, say "up," in a given direction is one-half, if one of them is measured to have spin up in that specific direction, it is known immediately that the other electron has spin down on that axis. No matter what axis of measurement is chosen, the spin values are anti-correlated in this way. Now, we can choose to measure the spin of electron 1 in any direction at station A in the experiment, and to measure the spin of electron 2 in any direction at station B. The values obtained will bear definite probabilistic correlations to each other in the quantum mechanical description.

Given this setup, consider the following argument, based on an assumption of what may be called "local counterfactual definiteness." We begin with two definitions:

Locality (LOC):

All the physical causes of an event lie within the past light cone of that event (have timelike separation from it); i.e., in accordance with special relativity, there is no physically causal influence between events with spacelike separation.

Local Counterfactual Definiteness (LCD):

For each (spin) measurement that could be performed on a quantum system there is a definite value of the measured observable such that, if the (spin) measurement were performed, the result would be that value independent of any other (spin) measurement performed (or not performed) at another location with spacelike separation.

The LCD assumption for state-dependent quantum properties can be used to generate an eightfold particle classification scheme akin to the one used by Eugene Wigner in his derivation of a Bell Inequality.[8]

Suppose we are performing spin-correlation measurements on composite spin-singlet systems (a system in the spin-singlet state has a total spin of zero). Consider three unit vectors $\hat{a}$, $\hat{b}$, and $\hat{c}$, which are not, in general, mutually perpendicular. Since we are dealing with spin-singlet systems, a perfect anti-correlation between the particles ensures zero total intrinsic angular momentum. Even though we cannot measure the spin in more than one direction simultaneously, we make the LCD assumption that there is a fact of the matter as to what the measured value (up or down) of the spin would be in any direction we might choose to measure it, and that it would have this definite value independent of any spin measurements performed (or not) at other locations with spacelike separation.

Having made this assumption, it follows that there is a list of types of experimental outcomes, and a fact of the matter as to which type will occur in any measurement that we might make, even if we never make it. For example, a measurement in one wing of the experiment belongs to the type ($\hat{a}$ +, $\hat{b}$ −, $\hat{c}$ +) just in case if we were to measure $\mathbf{S} \cdot \hat{a}$ (where $\mathbf{S}$ is the spin operator) we would obtain a spin up (+) outcome, if we were to measure $\mathbf{S} \cdot \hat{b}$ we would obtain a spin down (−) outcome, and if we were to measure $\mathbf{S} \cdot \hat{c}$ we would obtain a spin up (+) outcome. Because of the perfect anti-correlation of the spin-singlet state, the measurement in the other wing necessarily belongs to the type ($\hat{a}$ −, $\hat{b}$ +, $\hat{c}$ −). In any given measurement situation, the measurement outcome for the particle pair would therefore belong to one of eight mutually exclusive types. If we represent the populations of each type by N_i, $1 \leq i \leq 8$, we can catalogue the eight possibilities as in **Table 1**.

If we now suppose that the measurement performed on one particle of the spin-singlet system does not affect a measurement performed on the other particle (for good measure, let them have a spacelike separation), we may reason as follows: suppose that the experimenter at station **A** measures $\mathbf{S} \cdot \hat{a}$ and discovers the first particle to have spin up (+), and when the experimenter at station **B** measures $\mathbf{S} \cdot \hat{b}$ he finds it to have spin up (+) also. From **Table 1** we can easily see that the spin singlet pair belongs to either type 3 or type 4, and that the number of particle pairs for which this condition is satisfied is $N_3 + N_4$. Since the population in each type is greater than or equal to zero, inequality relations like

(1) $$N_3 + N_4 \leq (N_2 + N_4) + (N_3 + N_7)$$

must hold. We designate $P(\hat{a}+; \hat{b}+)$ as the probability that in a random trial, observer **A** measures $\mathbf{S} \cdot \hat{a}$ to be + and observer **B** measures $\mathbf{S} \cdot \hat{b}$ to be +, and likewise for other paired possibilities. We have straightforwardly that:

(2) $$P(\hat{a}+; \hat{b}+) = \frac{(N_3 + N_4)}{\sum_{i=1}^{8} N_i}.$$

Population	Particle 1 (Station A)	Particle 2 (Station B)
N_1	($\hat{a}$+, $\hat{b}$+, $\hat{c}$+)	($\hat{a}$−, $\hat{b}$−, $\hat{c}$−)
N_2	($\hat{a}$+, $\hat{b}$+, $\hat{c}$−)	($\hat{a}$−, $\hat{b}$−, $\hat{c}$+)
N_3	($\hat{a}$+, $\hat{b}$−, $\hat{c}$+)	($\hat{a}$−, $\hat{b}$+, $\hat{c}$−)
N_4	($\hat{a}$+, $\hat{b}$−, $\hat{c}$−)	($\hat{a}$−, $\hat{b}$+, $\hat{c}$+)
N_5	($\hat{a}$−, $\hat{b}$+, $\hat{c}$+)	($\hat{a}$+, $\hat{b}$−, $\hat{c}$−)
N_6	($\hat{a}$−, $\hat{b}$+, $\hat{c}$−)	($\hat{a}$+, $\hat{b}$−, $\hat{c}$+)
N_7	($\hat{a}$−, $\hat{b}$−, $\hat{c}$+)	($\hat{a}$+, $\hat{b}$+, $\hat{c}$−)
N_8	($\hat{a}$−, $\hat{b}$−, $\hat{c}$−)	($\hat{a}$+, $\hat{b}$+, $\hat{c}$+)

Table 1. Spin Component Matching Under the Assumption of Local Counterfactual Definiteness[9]

Similarly, we obtain that

(3) $$P(\hat{a}+;\hat{c}+) = \frac{(N_2+N_4)}{\sum_{i=1}^{8} N_i} \text{ and } P(\hat{c}+;\hat{b}+) = \frac{(N_3+N_7)}{\sum_{i=1}^{8} N_i}.$$

Incorporating (2) and (3) into the inequality (1), we obtain a Bell Inequality:

(4) $$P(\hat{a}+;\hat{b}+) \leq P(\hat{a}+;\hat{c}+) + P(\hat{c}+;\hat{b}+).$$

This inequality resulted from two major assumptions: (1) counterfactual definiteness; and (2) the separability of the measurements performed by experimenter **A** and experimenter **B** due to the locality of physical correlations. Taken together, these two conditions constitute **LCD**. What may not be clear is that there are some additional assumptions presupposed by this argument that are hidden from view. Specifically, our reasoning presupposes that every measurement has a result (**EMR**), and that the law of counterfactual conditional excluded middle holds. Counterfactual conditional excluded middle (**CCEM**) is defined as follows:

It is either the case that if A were true then B would be true, or it is the case that if A were true then *not*-B would be true, that is: (A□→B) ∨ (A□→~B).

We therefore have the following logical relationship:

LCD → EMR & CCEM.

It should be obvious that in order for every measurement that could be performed to have a definite result, we must assume that every measurement has a result (**EMR**). In fact, not every quantum mechanical experiment does have a result, and Arthur Fine exploited this inefficiency loophole to create local models for quantum correlations in which state-dependent properties can have definite values.[10] These so-called "prism models" have the drawback of a somewhat *ad hoc* feel to them, but even more importantly, they require that a certain less-than-perfect level of experimental efficiency not be exceeded. Since the time of Fine's construction of these models, more efficient experiments closing this loophole have been conducted by Mary Rowe and others at NIST.[11] As things now stand, the locality loophole exploiting the possibility of subluminal signals and the detection loophole exploiting experimental inefficiency have both been closed definitively in separate experiments, but never simultaneously in a single experiment. Nonetheless, it is unreasonable to think that a simultaneous test might still yield results disconfirming quantum predictions.

As for assuming the validity of **CCEM**, we can see that this poses no problem in the present context. **CCEM** can be stated symbolically as (A□→B) ∨ (A□→~B). To arrive at the Bell inequalities, we need to use **CCEM** to reason about the Wigner classification as follows: since the eight types of singlet systems are mutually exclusive, counterfactual definiteness necessitates either that if the measurement were performed, then a certain result would be realized, or that if the measurement were performed, that result would not be realized. Now one might worry, as van Fraassen does, that if the counterfactual reasoning we are employing presupposes a form of conditional excluded middle, it is already questionable because of the controversial status of that principle.[12] He urges that we should avail ourselves of a counterfactual logic (like David Lewis's) that does not presuppose conditional excluded middle, thereby doing an end run around questions of counterfactual definiteness and dubious assertions about how facts are related to modality.[13]

Contrary to van Fraassen's hesitancy, however, the employment of **CCEM** in this context is licensed by the fact that one of the things we are testing is whether quantum mechanics is compatible with there being a fact of the matter in respect of the outcome of quantum experiments that have not been performed. The mere assumption that there is a fact of the matter, and hence determinate outcomes that could be known, already entails the truth of **CCEM**. The reason for this is that, under the assumption that there is a fact of the matter to be known, statements having the form of **CCEM** reduce to a disjunction of strict counterfactual conditionals and **CCEM** holds for these. A strict counterfactual conditional is a counterfactual conditional preceded by the necessity operator:

$$\Box(\varphi \Box\!\!\to \psi).$$

In order to deny **CCEM** under the supposition that there is a fact of the matter about quantum measurement outcomes, we would have to be able to say both (1) that it is not necessarily the case that if we were to perform a spin measurement in a specified direction the result would be spin-up, and (2) that it is not necessarily the case that if we were to perform the measurement in a specified direction it would be spin-down (not spin-up). But assuming that there is a fact of the matter about this situation, and given the bivalent nature of the spin-property, one of these things has to be true! So one of these two denials must necessarily be false and the other necessarily true. If there is a local fact of the matter about quantum events, therefore, the employment of **CCEM** is both uncontroversial and unavoidable.

Before we forget, we've not yet considered quantum mechanical predictions for the measurements performed on this system. Quantum mechanics does not divide the singlet states into certain fractions of particle pairs belonging to specific types; rather it characterizes all singlet spin systems by the same ket:

(5) $$|\text{ spin singlet }\rangle = \frac{1}{\sqrt{2}}(|\hat{z}+;\hat{z}-\rangle - |\hat{z}-;\hat{z}+\rangle),$$

where the quantization direction has been made explicit, and, e.g., $|\hat{z}+;\hat{z}-\rangle$ indicates that the first electron is in the spin up state, and the second is in the spin down state.

Using this ket and the rules of quantum mechanics, we can evaluate all three of the terms in (4). Consider $P(\hat{a}+; \hat{b}+)$ first. Suppose that observer A measuring at the first station finds $S_1 \cdot \hat{a}$ to have spin up. Because of the perfect anti-correlation of the singlet state, we know that observer B at the second station will find $S_2 \cdot \hat{a}$ to have spin-down. To evaluate $P(\hat{a}+; \hat{b}+)$ a new quantization axis $\hat{b}$ which makes an angle θ_{ab} with axis $\hat{a}$ has to be introduced. The rules of quantum mechanics dictate that when the second particle is known to be in an eigenstate of $S_2 \cdot \hat{a}$ with negative eigenvalue, the probability that an $S_2 \cdot \hat{b}$ measurement will yield a positive eigenvalue (spin-up) is:

(6) $$\cos^2\left[\frac{(\pi - \theta_{ab})}{2}\right] = \sin^2\left(\frac{\theta_{ab}}{2}\right).$$

Since the probability of initially observing $\mathbf{S}_1 \cdot \hat{\mathbf{a}}$ to have spin up is $\frac{1}{2}$, we obtain that

(7)
$$P(\hat{\mathbf{a}}+;\hat{\mathbf{b}}+) = \frac{1}{2}\sin^2\left(\frac{\theta_{ab}}{2}\right).$$

This result and its extension to the other two terms in (4) lead to our writing the Bell Inequality as:

(8)
$$\sin^2\left(\frac{\theta_{ab}}{2}\right) \leq \sin^2\left(\frac{\theta_{ac}}{2}\right) + \sin^2\left(\frac{\theta_{cb}}{2}\right).$$

This inequality is not always geometrically possible, however. For example, if $\hat{\mathbf{a}}$, $\hat{\mathbf{b}}$, and $\hat{\mathbf{c}}$ are all chosen to lie in the same plane, with $\hat{\mathbf{c}}$ bisecting the angle made by $\hat{\mathbf{a}}$ and $\hat{\mathbf{b}}$ so that we can write $\theta_{ab} = 2\theta$, and $\theta_{ac} = \theta_{cb} = \theta$, then (8) is violated for $0 \leq \theta \leq \pi/2$. In particular, we find a maximal violation for $\theta = \pi/4$ which gives us the inequality $0.5 \leq 0.292$.

Since the Bell Inequality is violated and its derivation rests on two assumptions—the separability of quantum systems due to the locality of measurement outcomes and counterfactual definiteness—we may infer that at least one and possibly both assumptions fail. If separability/locality actually held, the proper conclusion would be that there is no fact of the matter about what quantum measurements would be apart from their actual occurrence. On the other hand, if counterfactual definiteness held, the facts in question would have to be nonlocal in character, and this would lead us to infer the existence of some sort of nonlocal quantum measurement outcome dependence, whether deterministic or stochastic. In this respect, more needs to be said about deterministic versus stochastic models, faithful measurement, and nonlocality.

As posed, the question of local counterfactual definiteness can be seen to float free of issues related to local deterministic or local indeterministic hidden variables, in fact not even requiring that the correlation be induced by a hidden variable. To see that the proof is independent of the issue of determinism, consider what local determinism entails. If it were applicable, then any two possible worlds (locally) identical in every respect (including all natural regularities) up until the time of measurement would have identical measurement results. But this need not be the case with the assumptions in the situation under examination. All that is required is that (1) there be a measurement result in each of these possible worlds; and (2) there be a fact of the matter as to what the result would be if any of these worlds were actual. Similarly, we need not assume the existence of a local stochastic hidden variable (some sort of propensity or probabilified disposition) that induces the experimental outcome. The proof does not require that there be a fact of the matter regarding objective quantum dispositions prior to a counterfactual measurement, only that there be a counterfactual truth about the outcome if the measurement had been performed. We may further note that when attempts to model quantum correlations locally using hidden variables are made, issues of deterministic versus stochastic models are irrelevant, since Arthur Fine[14] has shown that quantum phenomena have a local stochastic model just in case they have a deterministic one. The proper conclusion, therefore, is that they have neither, since a Bell inequality is generated that certain experimental arrangements violate in each case. Finally, the principle of faithful measurement, which states that the measurement process reveals a value for the measured observable actually possessed by the system prior to being measured, need not hold in this case. It need not hold, first, since it is not necessary to the proof for the system to possess any value of the measured observable prior to measurement. Second, it need not hold because even if the system did possess a value for the observable prior to measurement, it need not be the one revealed by the measurement.

3.2 Local Hidden Variables in Terms of Common Causes

I will not go through the derivation of the Bell Inequality in this case, but if it is assumed that there is a common cause (in the sense of definition 2.5 above) that explains EPR-type correlations in quantum mechanics, then it is again possible to derive Wigner's eightfold classification and thereby a Bell Inequality that is violated by the quantum system.[15] Those interested in this derivation can find it in van Fraassen's book on quantum theory, along with a discussion of the various assumptions (e.g., local determinism) feeding into the proof.[16] The quantum mechanical violation of the Bell Inequalities in this case shows that, if the world is either locally deterministic or locally stochastic, there is no physical cause and hence no naturalistic explanation for the correlations.

4. Nonlocality and Nonlocalizability: The Demise of Physical Causality and Material Objecthood

This leaves us to consider the issues of nonlocality and nonlocalizability. What if we drop the locality constraint and countenance the existence of nonlocal correlations in the experiments that have no local physical explanation, as most physicists and philosophers of science think that we must? In mathematical description, entangled quantum states, such as the spin-singlet system of the EPR experiment just discussed, exhibit nonlocal correlations that prohibit joint probabilities for the outcomes of the components in each pair of measurement events from being factored into individual probabilities. The question that now confronts us is whether such nonlocality defies all attempts at physical explanation. This mandates critical scrutiny of the nonlocal interpretive options of which there are basically two: the deterministic de Broglie-Bohm theory, and the indeterministic relational holist (dynamic emergence) model. We turn to this task now, saving a consideration of nonlocalizability in the context of relativistic quantum theory and algebraic quantum field theory until section 4.4 below.

4.1 Nonlocality and the de Broglie-Bohm Theory

The nonlocal deterministic option is represented by the de Broglie-Bohm or pilot-wave theory. In the non-relativistic case, this approach privileges the position representation and makes use of either a second-order quantum potential or first-order guidance equation controlling the dynamic behavior of structureless point particles.[17] It is generally acknowledged that this approach solves the quantum measurement problem in the non-relativistic case, though I contend (and will argue briefly in the discussion of quantum theory and physical law in section 4.6 below) that even non-relativistic Bohmian mechanics fails to rescue a viable notion of material objecthood[18] and necessitarian (deterministic causal) nomology.

Even more tellingly, pilot-wave theory seems to be fraught with insurmountable technical problems in the relativistic context such as: (1) casting it in a viable Lorentz-invariant form,[19] given the related fact that relativistic pilot-wave bosons can travel at superluminal speeds and reverse their direction in time;[20] (2) the fact that fermionic relativistic pilot-wave theory cannot account for particle-number variability under strong external potential couplings, nor the related existence of anti-matter, which is wedded to negative energy states in standard relativistic quantum mechanics;[21] and (3) the fact that when the pilot-wave approach is extended to field theory, as it must be, the appropriation of fields (however represented) as fundamental "beables" undoes the sole remaining virtue of non-relativistic Bohmian mechanics by rendering the measurement problem unsolvable.[22] So any way you look at it, salvation for philosophical naturalism is not to be found in the de Broglie-Bohm theory; neither, as we shall see, is it to be found in relational holist ontology.

4.2 Nonlocality and Relational Holism (Dynamic Emergence)

The nonlocality of quantum phenomena has served as the basis for metaphysical proposals concerning the "emergence" of the macroscopic from the microscopic realm within the context of a naturalistic metaphysics; such is the case with the treatment of indeterministic nonlocality under what Paul Teller calls "relational holism,"[23] and Fred Kronz and Justin Tiehen call "dynamic emergence." This is the form of emergence or methodological holism that will be our primary concern. Whatever may be said of the descriptive utility of this idea in other contexts, its application to nonlocal quantum phenomena is not just explanatorily vacuous, it also leads to ontological contradictions.

Emergence as limit behavior: descriptively true but metaphysically unhelpful

There is, of course, a useful sense of "emergence" appropriate to quantum physics in which classical (Maxwell-Boltzmann) statistical behavior can be understood to emerge from quantum (bosonic and fermionic) statistics in what physicists call the "classical limit." While these limits are useful in understanding how quantum descriptions can give rise to classical appearances, they are metaphysically unenlightening where relevant, and irrelevant in the case of nonlocal behavior. Let me briefly explain.[24]

With the standard definitions of the Poisson and commutator brackets, the *classical mechanical limit* (CM limit) of a quantum system is defined to be

$$\lim_{\hbar \to 0} \frac{1}{i\hbar}[\hat{A},\hat{B}] = \{A,B\}.$$

This limit is fictional, of course, because ℏ is a physical constant. The limit represents the transition between the quantum and classical *descriptions* of a system; classical behavior "emerges" when quantum effects are dampened to the point of negligibility. It is important to note, however, that there are still residual effects (dependent on Planck's constant) even after the classical mechanical limit is taken and the underlying reality is still quantum-mechanical in character.

Statistical mechanics mathematically relates the thermodynamic properties of macroscopic objects to the motion of their microscopic constituents. Since the microscopic constituents obey quantum dynamics, the correct description must lie in principle within the domain of quantum statistical mechanics. Under thermodynamic conditions of high temperature (T) and low density (n), however, classical statistical mechanics serves as a useful approximation. With this in mind, the *classical statistical limit* (CS limit) may be defined as the situation represented by:

$$T \to \infty \text{ and } n \to 0.$$

These are the same conditions as those governing the applicability of the ideal gas law ($pV = nRT$), so the CS limit could equally well be called the ideal gas limit. Unlike the CM limit, the conditions governing the CS limit are subject to experimental control. In respect of quantum statistical behavior, both the CM and the CS limits are continuous, so the quantum indistinguishability arising from permutation symmetry is not removed, even though it is dampened in the limit. Quantum "particles" retain their indistinguishability even when their aggregate behavior can be approximated by a Maxwell-Boltzmann distribution.

These reflections lay the ground for understanding why any emergentist account of the dependence or supervenience of the macroscopic realm on the the microscopic realm, while perhaps descriptively interesting, will be unenlightening as a metaphysical explanation. It is environmental decoherence (essentially, statistical damping through wave-function orthogonalization) that gives quantum-mechanical ephemera a cloak of macroscopic stability, but decoherence is not a real solution to the measurement problem. The apparent solidity of the world of our experience is a mere epiphenomenon of quantum statistics; the underlying phenomena retain their quantum-theoretic essence while sustaining classical appearances.

Emergence and Supervenience

The essence of emergentism is a layered view of nature. The world is divided into ontological strata beginning with fundamental physics and ascending through chemistry, biology, neuropsychology, and sociology. The levels correspond to successive organizational complexities of matter, and at each successive level there is a special science dealing with the complex structures possessing the distinguishing causal characteristics of that level. Higher-level causal patterns necessarily supervene on (are dependent upon) lower-level causal interactions, but are not reducible to them. The picture, then, is of emergent nomological structures irreducible to lower-level laws, with emergent features that not only affect the level at which they appear, but also exercise "downward causation" on lower-level phenomena.

Moving beyond hand-waving declarations of the "lawful" character of emergence requires giving an account of the relationship between basal physical conditions and emergent properties. McLaughlin and Kim have both attempted articulations of emergence in terms of what O'Connor and Wong term "synchronic strong supervenience":[25] given basal conditions C at time t, an emergent property P strongly supervening on conditions C will appear at time t. McLaughlin defines such emergent properties in terms of strong supervenience as follows:

> If P is a property of w, then P is emergent if and only if (1) P supervenes with nomological necessity, but not with logical necessity, on properties the parts of w have taken separately or in other combinations; and (2) some of the supervenience principles linking properties of the parts of w with w's having P are fundamental laws.[26]

McLaughlin defines a fundamental law as one that is not metaphysically necessitated by any other laws, even together with initial conditions. While Kim also understands emergence as a form of strong synchronic supervenience,[27] it is important to note that he regards emergent properties as epiphenomenal and challenges the tenability of non-reductive physicalism on this basis (he is a physical reductionist). These arguments need not concern us here but have received responses from Loewer and Shoemaker.[28] The property-fusion account of emergence developed by Humphreys circumvents Kim's objections because it is not synchronic and because emergent properties are fusions of the basal properties, which then cease to exist.[29]

The supervenience account of emergence will not suffice in the quantum context for two reasons. The first is that nonlocal phenomena quite evidently do not supervene on the properties of the various subsystems taken separately or in other combinations (the relevant joint probabilities are not factorizable), so supervenience is the wrong conception to be using here. The second is that any viable account of nomological necessity in the quantum realm would have to connect objective properties of the system immediately prior to measurement with the measurement results obtained. We have seen that such a restriction leads to empirically false consequences for both local deterministic and local stochastic models, and have remarked that, quite apart from the insurmountable technical obstacles confronting the necessary extension of de Broglie-Bohm theory to the relativistic context, the structureless point particles of non-relativistic Bohmian mechanics are incapable of supporting a sufficiently robust conception of material objecthood to ground necessitarian nomology. As we shall see momentarily, however, a non-supervenient description of quantum emergence suffers from a sort of explanatory vacuity, and it also founders on ontological contradictions arising from the postulation of nonlocal wholes (unless there is a privileged reference frame).

Property Fusion as an Account of Emergent Ontological Hierarchies

Paul Humphreys[30] has developed a concept of emergence in terms of "property fusion" that he suggests can be used to describe entangled states in quantum theory. His account assumes the existence of a hierarchy of distinct ontological levels, which he expresses in the form of a "level-assumption" (**L**):

> (**L**) There is a hierarchy of levels of properties, $L_0, L_1, \ldots, L_n, \ldots$ of which at least one distinct level is associated with the subject matter of each special science, and L_j cannot be reduced to L_i for any $i < j$.

A property P^i is then defined to be an "i-level property" just in case i is the lowest level at which instances of the property occur. A set of properties $\{P_1^i, P_2^i, \ldots, P_m^i, \ldots\}$ is associated with each level i, where P_m^i denotes the mth property at the i-level. A parallel hierarchy of entities is postulated: x^i is an i-level entity just in case i is the *lowest* level at which it exists and x_m^i denotes the mth entity at the i-level.

In order to characterize the property-fusion operation, Humphreys uses the notation $P_m^i(x_r^i)(t)$, which denotes an instantiation of property P_m^i by entity x_r^i at time t, because he regards property instances as being more fundamental than properties. We will suppress references to specific individuals and times to simplify the notation. The fusion operation [.*.] is defined by Humphreys as a process that combines two i-level properties P_m^i and P_n^i to form an $(i+1)$-level property $[P_m^i * P_n^i]$; this fusion could equally well be represented by the notation $P_{m,n}^{i+1}$. Once the basal properties have fused in this manner, they cease to exist and the new emergent property is all that remains.

Humphreys argues that entangled (or nonseparable) states in quantum mechanics lend themselves to description in terms of property fusion, maintaining that the emergent entangled state will remain intact so long as nonseparability persists. He thinks that this can be the case even after the interaction ceases, whereas Kronz and Tiehen (2002) adopt Humphreys's conception of property fusion but argue that persistence of the interaction is necessary for continued emergence. The arguments for this difference need not concern us. The more pressing concern is whether this technical account of emergence is explanatorily useful and metaphysically tenable in relation to nonlocal phenomena.

The Kronz-Tiehen Taxonomy for Quantum Mereology

On the basis of their discussion of fusion in the context of quantum chemistry, Kronz and Tiehen suggest that there are at least three ways that philosophers could develop a metaphysical account of emergence in mereological terms; they advocate the last of the three.[31] Since it is instructive to do so, we will briefly consider all three options.

Before examining these accounts, however, we need definitions of two background ideas employed by Kronz and Tiehen: independent characterizations of entities and contemporaneous parts. A characterization of an entity is an exhaustive list of the properties that are instantiated by that entity, and this characterization is said to be independent just in case the elements on the list of its properties make no essential reference to some other entity. Second, an entity is said to be a contemporaneous part of some whole just in case that part exists while the whole does (in relativistic contexts, Kronz and Tiehen make this relation reference-frame dependent in order to preserve standard interpretations of Lorentz invariance in terms of the relativity of simultaneity). So armed, they define three conceptions of emergence:

Prototypical Emergence

The idea here is that every whole consists of contemporaneous parts that have independent characterizations, but there is some criterion for distinguishing between part-whole relationships that are emergent from those which are merely resultant. The British emergentists take this line and use additivity as the relevant characterization of a resultant as opposed to an emergent property.[32] The difficulty with this view is that it seems to trivialize the notion of emergence when quantum mechanics is brought into view, either rendering every part of the universe emergent because it is entangled through past interactions with everything else in the universe, or nothing emergent, because the universe is an undivided whole that has no parts with independent characterizations. A proper interpretation of quantum theory would seem to require grasping the second horn of this dilemma.

Radical Emergence

The idea behind radical emergence is that only resultant wholes have contemporaneous parts, emergent wholes do not. Kronz and Tiehen interpret this as Humphreys's view. Emergent wholes are produced by a fusion of entities that can be likened to parts, but these parts cease to exist upon fusion, only existing when the whole does not, and vice-versa. An example of this sort of thing presumably would be a nonseparable quantum state. Prior to interaction, quantum "particles" might be taken to have independent existence, but after they interact and their wave-functions become entangled, they cease to exist as "parts," and a new entity at the next level in the ontological hierarchy comes into being. Again, it is hard to see on this view why there is not only one quantum entity: the universe itself.

Dynamic Emergence (Relational Holism)

Kronz and Tiehen proclaim themselves advocates of what they call "dynamic emergence," which seems to me a reinvention of Paul Teller's idea of relational holism. Teller defines a relationally holistic property as one in which the relevant property of the whole does not supervene on the non-relational properties of the relata,[33] as, for example, the tallness of Wilt Chamberlain relative to Mickey Rooney supervenes on the nonrelational height of each. In Kronz and Tiehen's reformulation, emergent wholes have contemporaneous parts, but these parts cannot be characterized independently of their respective wholes. These wholes are produced by an essential, ongoing interaction of their parts. Ultimately, of course, quantum theory is going to imply that every contemporaneous part of the universe, at least in its "material" respects, cannot in the final analysis be characterized independently of the whole universe, though for all practical purposes we can often treat subsystems of the universe as proximately independent.

Relational holism and quantum nonlocality: a very holey story

Granted that relational holism (to use Teller's term) seems the most reasonable description of quantum ontology, what more can be said? As Kronz and Tiehen have noted, speaking of contemporaneous parts for nonlocal wholes requires, in view of the relativity of simultaneity, a relativization of contemporaneousness to reference-frames. Though they do not discuss how this is to be done, the most plausible candidate is Gordon Fleming's theory of hyperplane dependence,[34] in which judgments of simultaneity are relativized to hyperplanes constituted by three-dimensional temporal slices of space-time; this is the solution appropriated by Teller.[35] The difficulty here is that the properties of a nonlocal quantum system can be different depending on which hyperplane is in view. In some hyperplanes, for example, the wave-function of the system may have collapsed, while in others this will not yet have happened. But there are an infinite number of such hyperplanes, some of which intersect, and it will be the case at the point of intersection that ontologically inconsistent properties are attributed to the quantum system—for example, that it has both collapsed and not collapsed. I take this situation to be suggestive of two things. Read one way, it could be a harbinger of the nonlocalizability of so-called particles, pointing to the fact that particle ontologies are not ultimately tenable because the relata don't have intrinsic states: that is, they don't really exist. This reading could be mitigated by the existence of an undetectable privileged reference frame, but as we will see in section 4.4 below, the nonlocalizability of unobserved "particles" still holds if the assumption that there is no privileged reference frame is dropped, so the conclusion that particle ontologies are untenable is secure. Taken another way, the ontological contradiction to which hyperplane dependence gives rise suggests the metaphysical necessity of an undetectable privileged reference frame that resolves the issue. A metaphysician committed to presentism rather than eternalism is forced to this position regardless, but there are indicators on the frontiers of research in quantum gravity that at least some of the physics community may also be moving this way in the effort to reconcile quantum field theory with general relativity.

As a characterization of quantum nonlocality, however, while relational holism or dynamic emergence may be descriptively accurate and revelatory of the challenge to ontological interpretation that quantum theory poses, it is explanatorily vacuous. One might protest that the "individual" described by quantum theory must ultimately be the quantum system itself, with its Hilbert Space of states, with the ontological difference between particle and field a

A Quantum-Theoretic Argument against Naturalism

mere matter of representation for a selected set of states, all of which are allowed and used by quantum field theory. But this remark will not suffice to deflate the pressing question of ontology (see section 4.3 below), nor will it obviate the fact of systematic, predictable correlation without causation: instantaneous adjustment of nonlocal relational wholes to local systemic changes, whether called "emergence" or some other term of art, remains a flagrant violation of relativistic causality that lacks a physical explanation and is present, if anything, to an even greater extent in quantum field theory than quantum mechanics.[36] Invoking "emergence" in such contexts seems little more than a terminological gambit to obscure things for which no adequate physical explanation currently exists and which arguably will not yield to the kind of explanation being sought. Furthermore, while it pays lip service to a variety of ontological levels, emergentist metaphysics, at least as we have considered it in this section, is still a species of philosophical naturalism, since it only recognizes physical properties and things that are ontologically dependent on them. Its explanatory vacuity further reinforces the untenability of ontological naturalism as a metaphysical stance.

4.3 To Be or Not To Be . . . Maybe: The Myth of Ontological Deflation

Before we move on, we need to countenance an objection based on the mathematical "equivalence" of first-quantized particle theories and second-quantized (Fock Space) field theories, since this has figured in discussions of the ontological significance of quantum statistics. In particular, Van Fraassen has suggested that the moral of this equivalence is that the whole issue of ontological interpretation can be dispensed with, because non-relativistic quantum field theory can be given either a particle interpretation or a particleless interpretation; the choice of ontology here is thus a matter of convention, not metaphysics.[37] I have dealt with this misrepresentation more thoroughly elsewhere,[38] but let me handle the matter expeditiously by making two points, one about the residual inequivalence underlying this supposed "equivalence," and the other regarding the untenability of "indexed particle" quantum field theory, since it has been proffered as evidence for the conventionality of ontology.

Are First and Second Quantized Theories Equivalent?

First and second quantized theories are equivalent only in the sense that the solution of the (second quantized) Fock Space Schrödinger equation

(9) $$[i\hbar(\partial/\partial t) - H]|\Psi_n\rangle = 0$$

can be put in the form

(10) $$|\Psi_n\rangle = (n!)^{-\frac{1}{2}} \int d^3x_1 \ldots \int d^3x_n \psi^\dagger(\mathbf{x}_n)\ldots\psi^\dagger(\mathbf{x}_1)|0\rangle \times \Psi_n(\mathbf{x}_1,\ldots,\mathbf{x}_n),$$

with the *n*-particle wavefunctions satisfying the many-particle Schrödinger equation

(11) $$[i\hbar(\partial/\partial t) - H_n]\Psi_n(\mathbf{x}_1,\ldots,\mathbf{x}_n) = 0.$$

But they are inequivalent in the important sense that *not every solution has this form*, rather just ones that that are simultaneous eigenstates of the total number operator N defined by

$$\text{(12)} \quad N \equiv \int d^3x\, \psi^\dagger(\mathbf{x})\psi(\mathbf{x}).$$

In this respect the Fock Space formalism is more general than that of many particle quantum mechanics, because it includes states that are superpositions of particle number, whereas many-particle quantum mechanics obviously does not. On the other hand, not all solutions of the wave equation **(11)** have the form

$$\text{(13)} \quad \Psi_n(\mathbf{x}_1,\ldots,\mathbf{x}_n) = (n!)^{-1/2}\langle 0 \mid \psi(\mathbf{x}_1),\ldots,\psi(\mathbf{x}_n) \mid \Psi_n \rangle,$$

with $|\Psi_n\rangle$ satifying the Fock Space equation **(9)**. The only ones that do are those satisfying the symmetry condition:

$$\Psi_n(\mathbf{x}_1,\ldots,\mathbf{x}_n) = \pm\, \Psi_n(\mathbf{x}_1,\ldots,\mathbf{x}_{i-1},\mathbf{x}_n,\mathbf{x}_{i+1},\ldots,\mathbf{x}_{n-1},\mathbf{x}_i).$$

Thus, in this regard, the wave equation is more general than the Fock Space equation because it includes the case of *n* non-identical particles by allowing for unsymmetrized wave-functions. *So the representations are equivalent only for Fock Space states that are eigenstates of N, and only for wave-functions that are either symmetric or antisymmetric.*

It is also instructive to note that total particle number is conserved in every system having the Fock Space Hamiltonian Operator H in **(9)**, because in this case the total number operator commutes with the Hamiltonian, i.e., $[N,H] = 0$. But not all Hamiltonians commute with the total number operator. In quantum field theory it is possible to have a situation when two or more fields are interacting and the interaction term does not commute with the number operator for one of the fields. This highlights another aspect of the difference between non-relativistic quantum field theory and many-particle quantum mechanics. The "equivalence" between the two representations is therefore anything but complete, and it certainly does not bear the weight of the ontological deflation that van Fraassen places on it. Many-particle quantum mechanics predicts the existence of nonsymmetric states, whereas Fock Space does not. This shows not only that the two representations are logically inequivalent, but also that the first quantized formalism is empirically deficient because the nonsymmetric states it predicts do not exist. Furthermore, if the two representations were equivalent in the sense required for ontological deflation, the Fock Space representation would need to have an empirically adequate indexed particle model, and as we shall see, it does not.

"Indexed Particle" Quantum Field Theory?

Since van Fraassen's attempted ontological deflation also relies on Willem de Muynck's construction of an "indexed particle" version of Fock Space, we need to make a brief excursion into this topic as well. De Muynck begins his discussion with the well-worn distinction, due to Jauch,[39] between the intrinsic and extrinsic properties of quanta. We alluded to this distinction earlier: intrinsic properties are defined as those independent of the state of the quantum system, whereas extrinsic properties are those dependent on the state of the system. Quanta are "identical" when they have all of the same intrinsic properties. De Muynck's suggestion is that labels (indices) might be regarded as intrinsic properties of quanta, because they are independent of the state of the system, i.e., not supposed to have dynamical consequences. This proposal motivates the attempt to construct an indexed quantum field theory that allows for the conceptual distinguishability of individual quanta despite their observational indistinguishability.

The central problem that de Muynck confronts in the context of non-relativistic quantum fields is the construction of a formalism permitting the creation and annihilation of indexed quanta. He takes as his starting point the Fock Space description of non-indexed quanta and the "equivalence" to many-particle quantum mechanics that we discussed in the last section. An indexed theory cannot get by with a single field operator, however. Rather, if all of the quanta are indexed, a different field operator $\psi_i(x)$ has to be associated with *each* quantum. The vacuum state $|0\rangle$ in this context is the direct product of the vacuum states $|0_i\rangle$ of all of the quanta in the system (indexed by $i \in I$), and defined as is customary by

(14) $$\psi_i(x)|0\rangle, \qquad \text{for all } i \in I.$$

By analogy with (10), the state vector corresponding to a system of n quanta with different indices and wavefunction $\Psi_n(x_1,...x_n)$ is defined by

(15) $$|\Psi_1,...,_n\rangle = \int dx_1 ... \int dx_n \Psi_n(\mathbf{x}_1,...,\mathbf{x}_n)\psi_n^\dagger(\mathbf{x}_n)\cdots\psi_1^\dagger(\mathbf{x}_1)|0\rangle,$$

where (cf. (13)) the wave-function is related to the state vector by

(16) $$\Psi(\mathbf{x}_1,...,\mathbf{x}_n) = \langle 0|\psi_1(\mathbf{x}_1)\cdots\psi_n(\mathbf{x}_n)|\Psi_1,...,_n\rangle.$$

De Muynck then goes on to impose as restrictions on the individual field operators only those relations which are equally valid for both bosons and fermions, deriving a number of results that are independent of the "statistics" of the quanta and therefore hold for uncorrelated quanta as well. With no symmetry requirements imposed on (15) and (16), what we get isn't ultimately that interesting because it is not an indexed version of Fock Space yielding quantum statistics, but rather a theory with no application. If symmetry considerations are introduced, the indexed theory will have to be permutation invariant in the requisite sense if it is going to produce the same results as non-relativistic quantum field theory. De Muynck protests that the idea of permuting quanta requires an interaction in order to make physical sense, and suggests that an indexed theory creates a new possibility—an interaction that exchanges just the quantal indices.[40] From a *de re* perspective, where the indices are intended to be rigid designators for the quanta in question, the idea of index swapping is a metaphysical impossibility. De Muynck seems to recognize as much, since he remarks:

> [W]hen index exchanging interactions are present it is no longer possible to use this index for distinguishing purposes. As a matter of fact precisely the presence of this kind of interaction would give the index the status of a dynamical variable. So a theory of distinguishable particles is possible only when the interactions are index preserving.[41]

In short, if the indexed theory were capable of reproducing the experimental predictions of Fock Space, the indices would have no *de re* significance.

Be this as it may, de Muynck's purpose is to develop an indexed theory as far as he can, and he pushes on to present a theory of indexed boson operators.[42] Presenting the technical details in full is not relevant for our purposes. Suffice it to say that de Muynck succeeds in developing a formalism involving annihilation and creation operators for indexed bosons, reproducing to a limited degree the correlations of symmetric bosonic statistics. These operators are not, however, simply interpretable as creating or destroying a particle with a given index in a single particle state, because the single particle states have a restricted meaning in light of the quantum correlations. For example, although the indexed creation operator adds a quantum with a specific index and single-particle state to the initial state of the system, due to (potentially nonlocal) interaction correlations, the quantum may be in a different single-particle state at the end of its interaction with the system.[43] The indexed annihilation and creation operators also have the undesirable property of being defined outside the Fock Space of symmetric states, where they have no physical meaning.[44] Furthermore, the dynamical description of a system of indexed bosons using the indexed annihilation and creation operators diverges from the Fock Space description in significant ways, not least of which the Hamiltonian sometimes has a different energy.[45] Also, in the indexed theory, the order in which particles are created or annihilated is dynamically relevant, but this is not the case in Fock Space. For this reason, the probability amplitudes associated with the indexed and non-indexed theories are different when the initial and final states are coherent superpositions of states with different numbers of particles.[46]

What we see, then, is that an indexed theory is not capable of reproducing the experimental predictions of the Fock Space description, and to the extent that it is empirically feasible, the quantal indices have no *de re* significance, i.e., they are fictions. This, along with the realization that the indexed theory of "bosons" that de Muynck develops retains the nonlocal correlations and quantal nonlocalizability characteristic of the standard formalism, confirms that quantal individuality cannot gain a foothold in the context of non-relativistic quantum fields by way of an empirically deficient theory of indexed quanta.

4.4 Nonlocalizability and Algebraic Quantum Field Theory

Let's focus on the nature of material individuality. In order for a particle to be a material individual it must possess one or more well-defined and uniquely identifying properties. A prime candidate for such a property is spatiotemporal location. In order for a material simple to exist as an individual material object, it must uniquely occupy a certain volume of space at a certain time.[47] If it does not, then whatever it is—if it's anything at all—the manner of its existence is not as a material object. The problem with this (apart from the superposition principle in quantum theory) is that the particles of relativistic quantum mechanics are not so localizable, nor do Newton-Wigner position operators in relativistic quantum field theory either localize quanta or behave so as to eliminate nonlocal correlations at spacelike separations. We can demonstrate these claims, respectively, through the consideration of Hegerfeldt nonlocalizability in relativistic quantum mechanics and the consequences of the Reeh-Schlieder theorem in algebraic quantum field theory.

The impossibility of particle interpretations of QFT

Hegerfeldt and Malament have shown that subject to the relativistic constraints that (1) a particle cannot be two places at once and that (2) the operators representing observables associated with disjoint spatial sets that have spacelike separation must commute, if one also makes the physically realistic assumption that (3) an individual particle cannot serve as an infinite source of energy, then it can be shown that such a particle has zero probability of being found in any bounded spatial region, no matter how large: measurements cannot be localized in principle.[48] In short, the supposed "particle" doesn't exist anywhere in space, and therefore, to be honest, it doesn't really exist at all. Halvorson and Clifton have extended this proof and closed some loopholes by showing that the Hegerfeldt-Malament result still holds if the assumption that there is no preferred reference frame is dropped, and if the assumption of sharply localized particles is dropped.[49] They have also shown that the necessary conditions for a particle interpretation of localized field observables cannot be satisfied in relativistic quantum field theory. In short, once relativity is taken into account, there can be no intelligible notion of microscopic material objects. Particle talk has a pragmatic utility in relation to macroscopic appearances, but it has no basis in microphysical reality.

Algebraic Quantum Field Theory (AQFT)

The realization that particle talk has no basis in microphysical reality is strengthened by a consideration of algebraic quantum field theory. AQFT originated with the efforts of Rudolf Haag in the 1960s, building on the work in axiomatic quantum field theory begun by Arthur Wightman in the late 1950s. It grows out of two convictions: that the algebraic structure of the set of quantum-theoretic observables should be given priority, and that field values must be localized in a way that makes sense from an operational perspective. What algebraic QFT does is to single out sets of axioms that apply quite generally to quantum field models that are "physically reasonable," and then use these postulates as the basis for a extended structural explorations.[50]

In the usual Hilbert Space formulation of quantum mechanics, observables are represented by self-adjoint (Hermitian) operators on Hilbert Space, and quantum states are represented by one-dimensional subspaces of Hilbert Space. Because of the priority given to algebraic structure, AQFT sets the background Hilbert Space to one side and focuses on the operator algebra associated with the observables instead.

To get the basic idea here, we need some background definitions. Let $\mathcal{B}(\mathcal{H})$ be the set of all bounded linear operators on $\mathcal{H}$ and let $A \subset \mathcal{B}(\mathcal{H})$. If $A, B \in \mathcal{A}$ and $\alpha, \beta \in \mathbb{C}$, then $\mathcal{A}$ is called an *algebra* just in case $\alpha A + \beta B \in \mathcal{A}$ and $AB \in \mathcal{A}$. Furthermore, if for every operator $A \in \mathcal{A}$, its adjoint $A^* \in \mathcal{A}$, then $\mathcal{A}$ is called a **-algebra*. In general, the operator algebras of observables in AQFT are *-algebras. We now need the following definition: a *linear form* ω over $\mathcal{A}$ is a mapping $\omega: \mathcal{A} \to \mathbb{C}$ with $\omega(\alpha A + \beta B) = \alpha \omega(A) + \beta \omega(B)$. This linear form is called *real* just in case $\omega(A^*) = \overline{\omega}(A)$, where the bar denotes the complex conjugate; it is called *positive* just in case $\omega(A^*A) \geq 0$; and it is called *normalized* just in case $\|\omega\| = 1$. The physical states of AQFT can now be identified with all the normalized positive linear forms ω over $\mathcal{A}$. It can then be shown that each ω defines a Hilbert Space $\mathcal{H}_\omega$ and a representation π_ω of $\mathcal{A}$ by linear operators acting on $\mathcal{H}_\omega$; this is known as the Gelfand-Naimark-Segal (GNS) construction, and it allows one to obtain the Hilbert Space representation from a given *-algebra. From a mathematical standpoint, then, the canonical and the algebraic approaches to QFT are obtainable from each other, and in this loose sense, equivalent. From a physical standpoint, however, the algebraic approach, by taking the *observables* as primitives, seems more directly relevant to the task of empirical and ontological interpretation.

We now need to consider the concept of locality embodied in AQFT. It would be nice if a field value $\psi(\mathbf{x})$ could be assigned to any spacetime point $\mathbf{x}$, but this is not realistic from an operational standpoint because it's always the case that we only have access to some finite spatiotemporal region $\mathcal{O}$. The approach taken in AQFT, therefore, is to use some smooth test function f of compact support (that is, it vanishes outside of $\mathcal{O}$) which spreads the field thereby localized over a corresponding space. The collection of all such smeared fields generates an algebra $\mathcal{A}(\mathcal{O})$, thus replacing $\mathbf{x} \to \psi(\mathbf{x})$ with the alternate correspondence $\mathcal{O} \to \mathcal{A}(\mathcal{O})$. Another definition is needed at this point: a *-algebra that is closed (contains all the limits of uniformly converging Cauchy sequences) in the topology (neighborhood) induced by the operator norm

$$\| A \| = \sup_{\Psi \in H} [\| A\Psi \| / \| \Psi \|],$$

and where the involution * and the norm $\|\cdot\|$ are related by $\| A * A \| = \| A^2 \|$ is called a *C*-algebra*. Provided suitable physical conditions obtain, the algebra $\mathcal{A}(\mathcal{O})$ may be treated as the C*-algebra of all bounded operators associated with $\mathcal{O}$, so the notion of locality embodied in AQFT is that of local operators representing observables in some finite spatiotemporal region $\mathcal{O}$.

Let's expand on this idea a bit. In the formalism of AQFT, the totality of spacetime can be covered by a net of local algebras

(17) $$\mathcal{A}_{loc} = \cup_{\mathcal{O}} \mathcal{A}(\mathcal{O})$$

that obeys the isotony condition $\mathcal{A}(\mathcal{O}_2) \supset \mathcal{A}(\mathcal{O}_1)$ if $\mathcal{O}_2 \supset \mathcal{O}_1$. This means that the total algebra of all observables is the union taken over all bounded regions. The notion of local operators in AQFT is in conformity with microcausality, namely the restriction that no physical effect can propagate faster than light, and expressed by the condition:

(18) $\quad [A_1, A_2] = 0$ if $A_i \in \mathcal{A}_{loc}(\mathcal{O}_i)$ for $i = 1, 2$, and $\mathcal{O}_1$ and $\mathcal{O}_2$ are *spacelike* separated.

This limns the elementary mathematical structure of AQFT. One extremely important result in AQFT is the *Reeh-Schlieder theorem*, which states, roughly speaking, that any state can be created from the vacuum. It is here that all of the familiar quantum nonlocalities reassert themselves, despite AQFT's strenuous attempt to enforce locality. Nonetheless, the abstract approach taken by AQFT allows a wide variety of formal proofs that have interesting ontological implications, as well as providing a tool that has proven very useful in statistical mechanics.

AQFT: Reeh-Schlieder trumps Newton-Wigner

For spacelike separations, relativistic causality demands that physical effects respect the speed of light as the limiting propagation velocity, which means that all field operators with spacelike separation must commute. This is known as the *microcausality* requirement. But problems arise when we try to construct the requisite Lorentz invariant theory of localized quanta. Mathematically, what a localization scheme does is define a correspondence between real linear subspaces of a one-particle Hilbert Space (which are associated with operators representing observables) and regions in physical space. In the standard localization scheme it can be shown that perfectly localized eigenstates of the local number operator do not exist because the local number operators associated with two arbitrary volumes, overlapping or not, do not commute.

But this microcausality misdemeanor is small in comparison to the implications of the Reeh-Schlieder theorem, which entails that local operations applied to the vacuum state can produce any state of the entire field, a flagrant violation of microcausality. Much as Newton-

Wigner states in the heuristic formulation of quantum field theory were introduced in an attempt to construct states of perfect localization for a Lorentz invariant theory, Irving Segal and quite recently Gordon Fleming have suggested an alternative "Newton-Wigner" localization scheme in algebraic quantum field theory that aims to avoid the Reeh-Schlieder theorem by reworking the correspondence between spatial regions and subalgebras of observables.[51] For the free Klein-Gordon (bosonic) field in the heuristic formulation, it has been shown that the Newton-Wigner localization is the best that can be done under the condition of Lorentz invariance, in the sense that any other position localization scheme would be even more badly behaved. In respect of Segal's and Fleming's efforts in algebraic quantum field theory, however, Hans Halvorson has shown that their alternative localization scheme only avoids the Reeh-Schlieder theorem in a trivial sense.[52] In particular, it remains the case that "NW-local fields allow the possibility of arbitrary spacelike distant effects from actions localized in an arbitrarily small region of space over an arbitrarily short period of time" rather than instantaneously, so it is still the case that "NW-local operators fail to commute at spacelike separation."[53]

The significance of all this, of course, is that "NW-local" position operators are not, in fact, localized, and when microcausality is egregiously violated in this fashion, nonlocal phenomena that have no physical explanation are manifested. Since the Newton-Wigner scheme is the best that can be done, the proper conclusion is that relativistic quantum field theory describes natural phenomena for which we have no physical explanation. More specifically, there are measurement-outcome correlations in nature that require a causal explanation but for which no physical explanation is in principle possible, and the nonlocalizability of field quanta entails that they fail the criterion of material individuality. So the most fundamental constituents of the so called "material world" are not material substances, and, as we have seen, their mereological fusion through environmental decoherence does not generate—and hence does not explain—macroscopic material *substances*, only macroscopic material *appearances*.

One might be inclined to wonder whether a field ontology could substitute for a particle ontology at this juncture and rescue philosophical naturalism in the process. Quite aside from the impossibility of constructing metaphysically coherent identity conditions and a viable notion of physical substance out of acausally fluctuating quantum fields, David Baker has suggested that the very considerations rendering a particle interpretation of quantum field theory untenable also preclude a field ontology.[54] More specifically, Baker argues that the regnant candidate for a field ontology, which relies on the notion of a wave-functional space, falters because it is unitarily equivalent to the Fock Space occupation number representation. In light of this equivalence, he contends that the most powerful arguments against particle ontologies count equally against field ontologies.

The general solution that Baker adopts for this difficulty—interpreting QFT ontology in terms of some suitable algebra of observables—is not new; a critical discussion of the standard options in this regard has been given by Laura Ruetsche.[55] But the crucial point to be made here is that by switching to an ontology constituted by an algebra of *observables*, we are moving away from material substances altogether to an ontology of mathematically limned phenomenological structures *without* substance; and need it be said that phenomenological structure *sans* substance is not something physical? Naturalistic metaphysics is a nonstarter at the most fundamental level of physical theory, pure and simple. Since there must be some explanation for the world of our experience, the correct explanation will therefore have to be one that is not physical and so transcends the explanatory resources of naturalism, whatever that might mean.

4.5 Structural Realism Comes (Partially) to the Rescue

To begin work on this conundrum, let's start with the eminently reasonable assumption that there is a way that the world is, that we can get it right or wrong, and that science is a useful tool in helping us to get it right. In particular, when physical theory backed by experiment demonstrates that the world must satisfy certain formal structural constraints—for example, quantizability, nonlocality as encapsulated in the Bell theorems, nonlocalizability as indicated by the Hegerfeldt-Malament and Reeh-Schlieder theorems, Lorentz symmetries in spacetime, internal symmetries like isospin, various conserved quantities as implied by Noether's theorem, and so on—then this formal feature of the world may be taken as strong evidence for a certain metaphysical state of affairs. At a minimum, such states of affairs entail that the structural constraints empirically observed to hold and represented by a given theory will be preserved (though perhaps in a different representation) by any future theoretical development; thus far structural realism.

Whether this structural realism has further ontological consequences pertaining to the actual furniture of the world (entity realism) is a matter of debate among structural realists. The epistemic structural realist believes that there are epistemically inaccessible material objects forever hidden behind the structures of physical theory and that all we can know are the structures.[56] The ontic structural realist eliminates material objects completely—it is not just that we only know structures, but rather that all that exists to be known are the structures.[57] Both these versions of structural realism are deficient, though in different ways.

We have seen that quantum theory is incompatible with the existence of material substances, even those of a relationally holistic sort. Given that this is the case, the epistemic structural realist is just wrong that there is a world of inaccessible material individuals hidden behind the structures that quantum theory imposes upon the world. The situation would therefore seem to default to ontic structural realism. But while the ontic structural realist is correct that there are no material objects behind the structures, his position is deficient too because there can be no structures simpliciter without an underlying reality that is enstructured; we cannot build castles in the air. It would seem, then, that we're in a sort of Catch-22 situation. The challenge to making sense of quantum physics is to give an account of what the world is like when it has an objective structure that does not depend on material substances. What investigations of the completeness of quantum theory have taught us, therefore, is rather than quantum theory being incomplete, it is material reality (so-called) that is incomplete. The realm that we call the "physical" or "material" or "natural" is not self-sufficient but dependent upon a more basic reality that is not physical, a reality that remedies its causal incompleteness and explains its insubstantiality, and on which its continued existence depends.

In light of this realization, the rather startling picture that begins to seem plausible is that preserving and explaining the objective structure of appearances in light of quantum theory requires reviving a type of phenomenalism in which our perception of the physical universe is constituted by sense-data conforming to certain structural constraints, but in which there is *no material reality giving rise to (causing) these sensory perceptions*. What we are left with is an ontology of (ultimately immaterial) minds experiencing and generating mental events and processes that, when sensory in nature, have a formal character limned by the fundamental symmetries and structures revealed in physical theory. The fact that these sensory perceptions are not mostly of our own individual or collective human making points to the falsity of any solipsistic or social constructivist conclusion, but it also implies the need for a transcendent source and ground of our experience. Although I will not explore the hypothesis at length in this context, I contend that there is one quite reasonable way to ground this ontology and obviate any puzzlement: metaphysical objectivity and epistemic intersubjectivity are maintained in the context of an occasionalistic theistic metaphysics that looks a lot like the immaterialism defended by George Berkeley and Jonathan Edwards and in which the only true causation is agent causation. The difference in the present case is that this explanatory hypothesis is grounded by ontological deduction from fundamental physical theory and experiment, rather than by epistemological analysis (Berkeley) or philosophico-theological argument (Edwards).

4.6 Quantum Theory and Physical Law

This may seem a bridge too far for many, but let's work our way to it along another path by considering the implications of quantum theory for physical law. There are various conceptions of natural laws that try to give an account of them as natural necessities of one variety or another. These nomological theories are called necessitarian, for obvious reasons. The causal power account sees laws of nature as grounded in the essential natures of things, that nature ultimately inhering in their material substance. Laws of interaction among material things depend upon the essential natures of the things interacting with each other and on the forces or fields mediating these interactions. Spelling this out often involves some notion of a causal power essentially possessed by an object, the possession of which follows from some other property the object has in virtue of being an instance of some natural kind.[58] Another necessitarian approach characterizes laws of nature in terms of universally quantified counterfactual conditionals of the form "*All* things of type T, were they subjected to conditions C, *would* manifest property P."[59] For example: all pure water at sea level, were it heated above 100 degrees Celsius, would boil. Note that the necessity here is implicitly embodied in the inviolable universality of the phenomenon. Lastly, there is a species of necessitarianism that explains laws of nature in terms of relations among universals. On this last account, natural laws are correlations among the properties or behavior patterns of different things in the world, the necessity of these correlations being explained by the existence of "necessary" second-order relationships among universal categories, and the behavior of individual things mirroring the necessity of the relationships among the universal categories to which they belong.[60]

All these necessitarian accounts, without exception, fail to work in the quantum context. The essential causal powers account, the counterfactual account, and the relation among universals account all require that physical systems and material objects objectively possess properties that are capable of being connected together in a law-like fashion. At a minimum, necessitarian and/or counterfactual physical law theorists have to maintain that quantum systems, or their components, objectively possess properties prior to measurement, whether these properties are determinate or indeterminate (probabilified dispositions), and that it is the *objective possession* of these properties that necessitates (or renders probable) their specific behavior. Bell's theorem demonstrates that this assumption leads to empirically false consequences in the case of local deterministic and local stochastic models.[61] As we have also seen, this assumption either leads to an ontological contradiction in the nonlocal stochastic case embodied by relational holism (dynamic emergence), or if an undetectable privileged reference frame is invoked, succumbs to the nonlocalizability and insubstantiality of the intended possessors of the requisite properties.

Furthermore, if we pursue the one remaining option for a purely naturalistic interpretation, the nonlocal deterministic model associated with the de Broglie-Bohm theory, we find it is fraught with insurmountable technical difficulties, and even if this were not the case, its ability to restore a straightforward ontology of material objects grounding a necessitarian nomology would be suspect. More specifically, even if we grant the non-relativistic Bohmian mechanical ontology of nonlocally correlated structureless point particles for the sake of argument, their lack of essential intrinsic properties[62] makes it impossible to generate the natural kinds required by necessitarian and, in particular, causal power accounts of natural laws.

What we are left with, therefore, is a situation in which there are no objective physical properties in which to ground necessitarian/counterfactual relations. So necessitarian theories of natural law cannot gain a purchase point in fundamental physical theory and must be set aside. All that remains is the so-called *regularist* account of natural laws, which asserts that while there are regularities present in the phenomenology of the world on a universal scale, there are no real *laws* of nature, that is, there is *no necessity* that inheres in the natural relationships among things or in the natural processes involving them. In short, nature behaves in ways we can count on, but it does so for no discernible physical reason. This state of affairs requires an explanation.

4.7 Humean Supervenience, Quantum Coincidence, and Explanatory Demand

In regard to this explanatory requirement a brief comment on the idea of "Humean supervenience" is in order, because the attitude it represents denies any demand for an explanation grounding the regularities present in nature, resting content with their brute factuality. As David Lewis, late of the Princeton University philosophy department, states the matter, Humean supervenience maintains that in a world like ours, the fundamental relations are spatiotemporal: distance relations that are both spacelike and timelike, and perhaps occupancy relations between point-sized things and spacetime points.[63] Furthermore, the fundamental properties are local qualities: perfectly natural intrinsic properties of points, or of point-sized occupants of points. Everything else supervenes on the spatiotemporal arrangement of local qualities throughout all of history—past, present, and future. The conception of physical law associated with this ontology is the descendant of a proposal by Frank Ramsey (and of John Stuart Mill before him, and of David Hume, of course, before him). Lewis again: take all axiomatic deductive systems whose theorems are true; the best system is the one that strikes the optimal balance between simplicity and strength (informativeness). A natural regularity is then a law just in case it is a *theorem* of the *best system* (which is postulated to exist whether we know anything about it or not).

Aside from the peculiarity of this view and puzzlement over why anyone would wish to hold it, the picture it offers obviously needs to be tweaked in order to deal with chance, and it needs substantial revision if it is going to be able to handle quantum nonlocality and the undoing of the causal metric of spacetime in quantum gravity. I am skeptical whether this needed tweaking is doable—in fact, I firmly believe that it is not—but for the sake of argument, let's suppose that the position is tenable. What does it amount to? Lewis claims that his account of laws should not be understood as epistemological. Rather he insists that Humean supervenience is an account of how nature—which, he asserts, consists in the Humean distribution of qualities—determines what is true about laws and chances, quite independently of what we humans (not to be confused with Humeans, though the latter are presumably a peculiar qualitative subset of the former) believe about the world.

A Quantum-Theoretic Argument against Naturalism

Taking Lewis at his word, I conclude that what we are left with is utter mystery and befuddlement. Quantum correlations, while nonlocally coincident, are understood in terms of local properties, requiring that we postulate random devices in harmony at spacelike separation without any deeper ontological explanation. Perhaps I can engender the requisite sense of puzzlement in the following way: how could anyone accept the plausibility of Humean supervenience in this context and still accuse two students of cheating on an exam when they sat next to each other and all their essay answers were word-for-word the same? The quantum situation, given its ubiquity, is staggeringly more improbable, with the added wrinkle that no one gets to peer over someone else's shoulder, because there is no physical signal that can pass instantaneously from one location to the other, and there's no possibility of a common text in the background that would explain the coincidence. To cling to brute factuality is to embrace irrationality, and to say that irrationality is rationally unjustifiable (though perhaps psychologically explainable) is redundant: it is definitionally so. A deeper explanation is required here, and no physical explanation is possible. Incredulity is not just the proper response to Humean supervenience, it is the necessary response. When its implications are grasped, Humean supervenience serves as a reductio of itself.[64]

5. Epilogue

If we consider carefully the progress of fundamental physics throughout the twentieth century, we find that the harder we have looked, the more ephemeral material reality has gotten, until finally it looks as though nothing is there. Yet our perceptions of the world remain and they are quite evidently not all of our own making. We have seen that neither nonlocal quantum correlations nor (in light of nonlocalizability) the nature of the fundamental constituents of material reality (so-called) can be explained or properly understood if the explanatory constraints of naturalism are preserved. Moreover, we have also seen that, short of dispensing with rationality itself, quantum phenomena such as these require an explanation. The conclusion we seek therefore follows directly: naturalism (materialism, physicalism) is irremediably deficient as a worldview, and must be rejected not just as inadequate, but as fundamentally false. The argument that has guided our discussion from the start has been vindicated.

Notes

1. "What is real is not given to us, but rather set as a task, by way of a riddle."—Albert Einstein
2. Throughout this essay, I will take the nouns "naturalism," "physicalism," and "materialism," as well as their adjectival forms, to be roughly synonymous in accordance with the definition of ontological naturalism just offered, and I will use them interchangeably. The fact remains, however, that "naturalism" is a bit of a weasel word, by which I mean that it is employed in a highly malleable way designed to insulate its fundamental thesis—the causal closure of the material realm—from disconfirming evidence. So it is that various "non-reductive naturalists" attempt to combine their monism about entities with a pluralism of properties that allows for "supervenient" or "emergent" material or non-material properties involving nonlocal correlations, normativity, intentionality, consciousness, and a variety of other things that pose problems for a naturalistic worldview. The literature on supervenience and emergence in this regard is voluminous. We will touch on concepts of supervenience, emergence, and holism in our discussion of nonlocality below, taking particular care to note that their descriptive utility, such as it is, hides their explanatory *vacuity* under a patina of technical sophistication.
3. Fine 1986a, 1986b; van Fraassen 1996.
4. Van Fraassen does, however, make explicit room for a voluntarist epistemology, in a neo-Pascalian or Jamesian vein, that allows belief to transcend the empiricist sensibilities he enforces in science (see van Fraassen 2002: 81–90, 179ff, *et passim*). Fine's epistemological preference is for a contextualist pragmatism (1986a, 1986b) that eschews any notion of deep metaphysical reality in favor of "natural ontological attitudes" arising in multiple scientific contexts, without any overarching concern for inter-contextual contradictions or desire for rapprochement and reconciliation. In giving up on the traditional project of knowledge in this way, both viewpoints devolve into the rationalizations of irrationalism characteristic of "post-modernity." We can do better than this, but I contend that doing so requires eschewing naturalism, not metaphysical realism. Both Plantinga's and Koons' essays in this volume make this point quite effectively.
5. Van Fraassen 1996: 170.
6. Ibid.
7. Fine 1989b and van Fraassen 1989a.
8. Wigner 1970; cf. also Sakurai 1985: 223–32, *et passim*.
9. Adapted from Sakurai 1985: 229.
10. Fine 1982c, 1986a, 1989a, 1991; cf. Maudlin 2002: 175–87.
11. See Rowe *et al.* 2001.
12. Van Fraassen 1991: 123–25.
13. Ibid., 124–25.
14. Fine 1982a, 1982b, 1986a.
15. Wigner 1970.
16. Van Fraassen 1991: 81–93.
17. See Holland 1993; Cushing, Fine, and Goldstein, eds. 1996; Gordon 1998: 337–406.
18. See also Bedard 1999 and Dickson 2001 in this regard.
19. See Goldstein *et al.* 1996 and Goldstein *et al.* 1999 for efforts in this regard.
20. See Kyprianidis 1985; Saunders 1999: 74–75.
21. See Saunders 1999: 75–78.
22. See Saunders 1999: 80–86.
23. Teller 1986, 1989.
24. See Gordon 2002: 402–7 for a more extensive discussion of related issues.
25. McLaughlin 1997; Kim 1999, 2006; O'Connor and Wong 2005.

26. McLauglin 1997: 39.
27. Kim 1999.
28. Loewer 2001 and Shoemaker 2002.
29. Humphreys 1997.
30. Ibid.
31. Kronz and Tiehen 2002: 344ff.
32. According to the British emergentists (see McLaughlin 1992), resultant properties are additive, like force in Newtonian mechanics, whereas emergent properties are not. This *via negativa* is taken as the definition of an emergent property and seems to be motivated by regarding forces as fundamental, then constructing a metaphysical view of emergence by analogy with the way that forces behave.
33. Teller 1986, 1989, and elsewhere.
34. Fleming 1965, 1966, 1988, 1989, 1996, 2003, 2004; Fleming and Bennett 1989.
35. Teller 1995 and elsewhere.
36. In this regard, see especially Clifton, Feldman, Halvorson, Redhead, and Wilce 1998; Halvorson and Clifton 2000.
37. Van Fraassen 1991: 434–82.
38. See Gordon 2003.
39. Jauch 1968.
40. De Muynck 1975: 340.
41. Ibid.
42. De Muynck 1975: 340–45.
43. Ibid., 342.
44. Ibid., 341.
45. Ibid., 343.
46. Ibid., 344–45.
47. Something is a material simple just in case it lacks any parts other than itself, that is to say, it has no proper parts. The condition of uniquely occupying a certain volume of space at a certain time is equivalent to saying that a material simple is impenetrable.
48. Hegerfeldt 1974, 1985, 1989, 1998a, 1998b, and Malament 1996.
49. Halvorson and Clifton 2002.
50. Haag 1996; Doplicher et al. eds. 1997; Horuzhy 1990; Clifton 1999; Kuhlmann, Lyre, and Wayne eds. 2002; Halvorson and Mueger 2007.
51. Irving Segal (1964), and quite recently Gordon Fleming (2000),
52. Halvorson 2001.
53. Halvorson 2001: 128.
54. Baker 2008.
55. Ruetsche 2002.
56. Worrall 1989; Redhead 1995; Cao 1997, 2003a, 2003b, 2003c.
57. Ladyman 1998; French 1999, 2000, 2003a, 2003b, 2005; French and Ladyman 2003a, 2003b; French and Krause 2006; Ladyman and Ross 2007.
58. See Bigelow and Pargetter 1990; Harré and Madden 1975.
59. Cf. Stalnaker 1968; Lewis 1973.
60. Armstrong 1983, 1997; Dretske 1977; Tooley 1987.
61. Bell 1987: 1–21; Redhead 1987: 71–118; Fine 1981, 1982a, 1982b; Cushing and McMullin 1989; Clifton, Feldman, Halvorson, Redhead and Wilce 1998.
62. Bedard 1999; Dickson 2001.

63. Lewis 1999: 224–47.

64. A more rigorous articulation of the argument against Humean supervenience requires a defense of the "Principle of Sufficient Reason" namely, the common-sense belief that all contingent facts have explanations. A thorough defense of this principle needs a full-length treatise and, handily, just such a treatise has been written. I am happy to recommend to the reader Alexander Pruss's book *The Principle of Sufficient Reason: A Reassessment* (Cambridge: Cambridge University Press, 2006).

References

Amstrong, David M. (1983) *What is a Law of Nature?* Cambridge: Cambridge University Press.

———. (1997) *A World of States of Affairs.* Cambridge: Cambridge University Press.

Ashtekar, A., and J. Stachel. (1991) *Conceptual Problems of Quantum Gravity.* Boston: Birkhauser.

Baker, David. (2009) "Against Field Interpretations of Quantum Field Theory," forthcoming in *British Journal for the Philosophy of Science* (http://philsci-archive.pitt.edu/archive/00004132/01/AgainstFields.pdf).

Bedard, K. (1999) "Material Objects in Bohm's Interpretation." *Philosophy of Science* 66: 221–42.

Bell, John. (1975) "The Theory of Local Beables," in J.S. Bell, *Speakable and Unspeakable in Quantum Mechanics.* Cambridge: Cambridge University Press, 1987, 52–62.

———. (1987) *Speakable and Unspeakable in Quantum Mechanics.* Cambridge: Cambridge University Press.

Belot, Gordon, and John Earman. (2001) "Pre-Socratic Quantum Gravity," in C. Callender and N. Huggett, eds., *Philosophy Meets Physics at the Planck Scale: Contemporary Theories in Quantum Gravity.* Cambridge: Cambridge University Press, 213–55.

Belousek, Darrin W. (2003) "Non-separability, Non-supervenience, and Quantum Ontology." *Philosophy of Science* 70: 791–811.

Berkeley, George. (1975 [1710–1732]) *Philosophical Works, Including the Works on Vision.* London: Everyman's Library (J.M. Dent).

Berlin, Isaiah. (1939) "The Refutation of Phenomenalism," previously unpublished essay, *The Isaiah Berlin Literary Trust,* 2004.

Bigelow, J., and R. Pargetter. (1990) *Science and Necessity.* Cambridge: Cambridge University Press.

Butterfield, Jeremy and Christopher Isham. (2001) "Spacetime and the Philosophical Challenge of Quantum Gravity," in C. Callender and N. Huggett, eds., *Philosophy Meets Physics at the Planck Scale: Contemporary Theories in Quantum Gravity.* Cambridge: Cambridge University Press, 33–89.

Butterfield. J., and C. Pagonis, eds. (1999) *From Physics to Philosophy.* Cambridge: Cambridge University Press.

Callender, Craig, and Nick Huggett, eds. (2001) *Philosophy Meets Physics at the Planck Scale: Contemporary Theories in Quantum Gravity.* Cambridge: Cambridge University Press.

Cao, Tian Yu. (1997) *Conceptual Developments of 20th Century Field Theories.* Cambridge: Cambridge University Press.

———. (2003a) "Structural Realism and the Interpretation of Quantum Field Theory." *Synthese* 136: 3–24.

———. (2003b) "Appendix: Ontological Relativity and Fundamentality—Is QFT the Fundamental Theory?" *Synthese* 136: 25–30.

———. (2003c) "Can We Dissolve Physical Entities into Mathematical Structures?" *Synthese* 136: 57–71.

Castellani, Elena, ed. (1998) *Interpreting Bodies: Classical and Quantum Objects in Modern Physics.* Princeton: Princeton University Press.

Clatterbaugh, Kenneth. (1999) *The Causation Debate in Modern Philosophy 1637—1739.* London: Routledge.

Clifton, Rob. (1999) "Beables in algebraic quantum theory," in J. Butterfield and C. Pagonis, eds., *From Physics to Philosophy.* Cambridge: Cambridge University Press, 12–44.

Clifton, R., ed. (1996) *Perspectives on Quantum Reality: Non-Relativistic, Relativistic, and Field-Theoretic.* Dordrecht: Kluwer Academic Publishers.

Clifton, R., D. V. Feldman, H. Halvorson, M. L. G. Redhead, and A. Wilce. (1998) "Superentangled states." *Physical Review A* 58(1): 135–45.

Cushing, James T., Arthur Fine and Sheldon Goldstein, eds. (1996) *Bohmian Mechanics and Quantum Theory: An Appraisal.* Dordrecht: Kluwer Academic Publishers.

Cushing, James T., and Ernan McMullin, eds. (1989) *Philosophical Consequences of Quantum Theory: Reflections on Bell's Theorem.* Notre Dame: University of Notre Dame Press.

Dembski, William A. (1998) *The Design Inference: Eliminating Chance through Small Probabilities.* Cambridge: Cambridge University Press.

———. (2002) *No Free Lunch: Why Specified Complexity Cannot Be Purchased without Intelligence.* Lanham: Rowman & Littlefield Publishers, Inc.

De Muynck, Willem. (1975) "Distinguishable and Indistinguishable Particle Descriptions of Systems of Identical Particles." *The International Journal of Theoretical Physics* 14: 327–46.

Dickson, Michael (2001) "Are There Material Objects in Bohm's Theory?" arXiv: quant-ph/0003102.

Doplicher, S., R. Longo, J.E. Roberts, and L. Zsido, eds. (1997) *Operator Algebras and Quantum Field Theory.* Cambridge: International Press.

Downing, Lisa. (2005) "Berkeley's natural philosophy and philosophy of science," in K. P. Winkler, ed., *The Cambridge Companion to Berkeley.* Cambridge: Cambridge University Press, 231–65.

Dretske, Fred. (1977) "Laws of Nature." *Philosophy of Science* 44: 248–68.

Earman, John. (1986) *A Primer on Determinism.* Dordrecht: Reidel Publishers.

Edwards, Jonathan. (1957- [1721, 1722, 1723, 1758]) *The Works of Jonathan Edwards,* vols. 3, 6. John E. Smith, General Editor. New Haven: Yale University Press.

Fine, Arthur I. (1981) "Correlations and Physical Locality," in Asquith, P., and R. Giere, eds., *PSA 1980,* vol. 2. East Lansing: Philosophy of Science Association, 535–62.

———. (1982a) "Hidden variables, joint probability, and the Bell inequalities." *Physical Review Letters* 48: 291–95.

———. (1982b) "Joint distributions, quantum correlations, and commuting observables." *Journal of Mathematical Physics* 23: 1306–10.

———. (1982c) "Some Local Models for Correlation Experiments." *Synthese* 29: 279–94.

———. (1986a) *The Shaky Game: Einstein, Realism and the Quantum Theory.* Chicago: University of Chicago Press.

———. (1986b) "Unnatural Attitudes: Realist and Instrumentalist Attachments to Science." *Mind* 95: 149–79.

———. (1989a) "Correlations and Efficiency: Testing the Bell Inequalities." *Foundations of Physics* 19: 453–78.

———. (1989b) "Do Correlations Need To Be Explained?" in Cushing, J.T. and E. McMullin, eds., *Philosophical Consequences of Quantum Theory: Reflections on Bell's Theorem.* Notre Dame: University of Notre Dame Press, 175–94.

———. (1991) "Inequalities for Non-Ideal Correlation Experiments." *Foundations of Physics* 21: 365–78.

Fleming, Gordon. (1965) "Nonlocal Properties of Stable Particles." *Physical Review B* 139: 963–68.

———. (1966) "A Manifestly Covariant Description of Arbitrary Dynamical Variables in Relativistic Quantum Mechanics." *Journal of Mathematical Physics* 7: 1959–81.

———. (1988) "Hyperplane Dependent Quantized Fields and Lorentz Invariance," in Brown, H. and R. Harré, eds., *Philosophical Foundations of Quantum Field Theory.* Oxford: Clarendon Press, 93–116.

———. (1989) "Lorentz Invariant State Reduction and Localization," In Fine, A., and J. Leplin, eds., *PSA 1988,* vol. 2. East Lansing: Philosophy of Science Association, 112–26.

———. (1996) "Just How Radical is Hyperplane Dependence?" in R. Clifton, ed., *Perspectives on Quantum Reality: Non-Relativistic, Relativistic, and Field-Theoretic.* Dordrecht: Kluwer Academic Publishers, 11–28.

———. (2003) "Observations on Hyperplanes: I State Reduction and Unitary Evolution," http://philsci-archive.pitt.edu/archive/00001533/.

———. (2004) "Observations on Hyperplanes: II Dynamical Variables and Localization Observables," http://philsci-archive.pitt.edu/archive/00002085/.

Fleming, G. and H. Bennett. (1989) "Hyperplane Dependence in Relativity and Quantum Mechanics." *Foundations of Physics* 19: 231–67.

Freddosso, Alfred. (1988) "Medieval Aristotelianism and the Case Against Secondary Causation in Nature," in Thomas V. Morris, ed., *Divine and Human Action: Essays in the Metaphysics of Theism*. Ithaca, NY: Cornell University Press, 74–118.

French, Steven. (1984) *Identity and Individuality in Classical and Quantum Physics*. Ph.D. Dissertation, University of London.

———. (1989a) "Identity and Individuality in Classical and Quantum Physics."*Australasian Journal of Philosophy* 67: 432–46.

———. (1989b) "Individuality, Supervenience, and Bell's Theorem." *Philosophical Studies* 55: 1–22.

———. (1989c) "Why the principle of the identity of indiscernibles is not contingently true either." *Synthese* 78: 141–66.

———. (1998) "On the Withering Away of Physical Objects," in E. Castellani, ed., *Interpreting Bodies: Classical and Quantum Objects in Modern Physics*. Princeton: Princeton University Press, 93–113.

———. (1999) "Models and Mathematics in Physics: The Role of Group Theory," in J. Butterfield and C. Pagonis, eds., *From Physics to Philosophy*. Cambridge: Cambridge University Press, 187–207.

———. (2000) "The Reasonable Effectiveness of Mathematics: Partial Structures and the Application of Group Theory to Physics." *Synthese* 125: 103–20.

———. (2003a) "A Model-Theoretic Account of Representation (Or, I Don't Know Much About Art . . . But I Know It Involves Isomorphism." *Philosophy of Science* 70: 1472–83.

———. (2003b) "Scribbling on the blank sheet: Eddington's structuralist conception of objects." *Studies in History and Philosophy of Modern Physics* 34: 227–59.

———. (2005) "Symmetry, Structure and the Constitution of Objects," http://philsci-archive.pitt.edu/archive/00000327/00/Symmetry&Objects_doc.pdf.

French, Steven, and Decio Krause. (2006) *Identity* in *Physics: A Historical, Philosophical,* and *Formal Analysis*. Oxford: Clarendon Press.

French, Steven, and James Ladyman. (2003a) "The Dissolution of Objects: Between Platonism and Phenomenalism." *Synthese* 136: 73–77.

———. (2003b) "Remodeling Structural Realism: Quantum Physics and the Metaphysics of Structure." *Synthese* 136: 31–56.

French, Steven and Redhead, Michael. (1988) "Quantum Physics and the Identity of Indiscernibles." *British Journal for the Philosophy of Science* 39: 233–46.

Fujita, S. (1991) "On the Indistinguishability of Classical Particles." *Foundations of Physics* 21: 439–57.

Fulling, S.A. (1989) *Aspects of Quantum Field Theory in Curved Space Time*. Cambridge: Cambridge University Press.

Goldstein, S., K. Berndl, D. Dürr, and N. Zanghí. (1996) "Nonlocality, Lorentz Invariance, and Bohmian Quantum Theory." *Physical Review A* 53: 2062–73.

Goldstein, S., D. Dürr, K. Münch-Berndl, and N. Zanghí (1999) "Hypersurface Bohm-Dirac Models." *Physical Review A* 60: 2729–36.

Gordon, Bruce L. (1998) *Quantum Statistical Mechanics and the Ghosts of Modality*. Ph.D. Dissertation, Northwestern University, Evanston, Illinois. Ann Arbor: UMI.

———. (2002) "Maxwell-Boltzmann Statistics and the Metaphysics of Modality." *Synthese* 133: 393–417.

———. (2003) "Ontology *Schmontology?* Identity, Individuation, and Fock Space." *Philosophy of Science* 70: 1343–56.

———. (2005) "Quantum Field Theory and Process Thought: An Unnecessary and Problematic Union." Presented at APA Western Division spring meeting, San Francisco.

Griffiths, R.B. (1984) "Consistent Histories and the Interpretation of Quantum Mechanics." *Journal of Statistical Physics* 36: 219–72.

———. (1993) "The Consistency of Consistent Histories." *Foundations of Physics* 23: 1601–10.

Haag, Rudolf. (1996) *Local Quantum Physics: Fields, Particles, Algebras* (2nd edition, revised). Berlin: Springer-Verlag.

Halvorson, Hans. (2001) "Reeh-Schlieder defeats Newton-Wigner: On alternative localization schemes in relativistic quantum field theory." *Philosophy of Science* 68: 111–33.

Halvorson, H. and Clifton, R. (2000) "Generic Bell correlation between arbitrary local algebras in quantum field theory." *Journal of Mathematical Physics* 41: 1711–17.

———. (2002) "No place for particles in relativistic quantum theories?" *Philosophy of Science* 69: 1–28 (arXiv:quant-ph/0103041).

Halvorson, H. and Mueger, M. (2007) "Algebraic Quantum Field Theory," in J. Butterfield and J. Earman, eds., *Philosophy of Physics (Part A)*. Amsterdam: North-Holland, 731–922.

Harré, R., and E.H. Madden. (1975) *Causal Powers: A Theory of Natural Necessity*. Oxford: Basil Blackwell.

Hegerfeldt, G.C. (1974) "Remark on Causality and Particle Localization." *Physical Review D*, 10: 3320–21.

———. (1985) "Violation of Causality in Relativistic Quantum Theory?" *Physical Review Letters* 54: 2395–98.

———. (1989) "Difficulties with Causality in Particle Localization." *Nuclear Physics B* 6: 231–37.

———. (1998a) "Causality, particle localization and positivity of the energy," in A. Bohm *et al.*, eds., *Irreversibility and Causality*. New York: Springer-Verlag, 238–45.

———. (1998b) "Instantaneous spreading and Einstein causality in quantum theory." *Annalen der Physik* 7: 716–25.

Holland, Peter R. (1993) *The Quantum Theory of Motion: An Account of the de Broglie-Bohm Causal Interpretation of Quantum Mechanics*. Cambridge: Cambridge University Press.

Horuzhy, S.S. (1990) *Introduction to Algebraic Quantum Field Theory*. Dordrecht: Kluwer Academic Publishers.

Humphreys, Paul. (1997) "How Properties Emerge." *Philosophy of Science* 64: 1–17.

Jarrett, Jon. (1984) "On the physical significance of the locality conditions in the Bell arguments." *Nous* 12: 569–89.

Jauch, J. M. (1968) *Foundations of Quantum Mechanics*. New York: Addison-Wesley.

Kim, Jaegwon. (1999) "Making Sense of Emergence." *Philosophical Studies* 95: 3–36.

———. (2006) "Being Realistic About Emergence," in P. Clayton and P. Davies, eds. *The Re-Emergence of Emergence: The Emergentist Hypothesis from Science to Religion*. Oxford: Oxford University Press, 189–202.

Kochen, S., and E. Specker (1967) "The Problem of Hidden Variables in Quantum Mechanics." *Journal of Mathematics and Mechanics* 17: 59–87.

Kronz, F., and J. Tiehen. (2002) "Emergence and Quantum Mechanics." *Philosophy of Science* 69: 324–47.

Kuhlmanm, M., H. Lyre, and A. Wayne, eds. (2002) *Ontological Aspects of Quantum Field Theory*. Singapore: World Scientific.

Kyprianidis, A. (1985) "Particle Trajectories in Relativistic Quantum Mechanics." *Physics Letters A* 111(3): 111–16.

Ladyman, James and Don Ross. (2007) *Everything Must Go: Metaphysics Naturalized*. New York: Oxford University Press.

Lewis, David K. (1973) *Counterfactuals*. Cambridge, MA: Harvard University Press.

———. (1986) *Philosophical Papers,* volume II. Oxford: Oxford University Press.

———. (1999) *Papers in Metaphysics and Epistemology*. Cambridge: Cambridge University Press.

Lo, T. K. and A. Shimony. (1981) "Proposed molecular test of local hidden variable theories." *Physical Review A* 23: 3003–12.

Loewer, Barry. (2001) "Review of J. Kim, *Mind in a Physical World*." *Journal of Philosophy* 98(6): 315–24.

Malament, David. (1996) "In Defense of Dogma: Why there cannot be a relativistic quantum mechanics of (localizable) particles," in R. Clifton, ed., *Perspectives on Quantum Reality: Non-Relativistic, Relativistic, and Field-Theoretic.* Dordrecht: Kluwer Academic Publishers, 1–9.

Maudlin, Tim. (1998) "Part and Whole in Quantum Mechanics," in Elena Castellani, ed., *Interpreting Bodies: Classical and Quantum Objects in Modern Physics*. Princeton: Princeton University Press, 46–60.

———. (2002) *Quantum Non-Locality and Relativity* (2nd edition). Oxford: Blackwell Publishers.

———. (2007) *The Metaphysics Within Physics*. Oxford: Oxford University Press.

McCracken, C.J., and I.C. Tipton, eds. (2000) *Berkeley's Principles and Dialogues: Background Source Materials*. Cambridge: Cambridge University Press.

McLaughlin, Brian P. (1992) "The Rise and Fall of British Emergentism," in A. Beckermann, J. Kim, and H. Flohr, eds. *Emergence or Reduction?* Berlin: Walter deGruyter, 49–93.

———. (1997) "Emergence and Supervenience." *Intellectica* 2: 25–43.

Morris, Thomas, and Christopher Menzel. (1986) "Absolute Creation." *American Philosophical Quarterly* 23: 353–62.

O'Connor, Timothy and Wong, Hong Yu. (2005) "The Metaphysics of Emergence." *Nous* 39: 658–78.

Omnès, Roland. (1994) *The Interpretation of Quantum Mechanics*. Princeton, NJ: Princeton University Press.

Plantinga, Alvin C. (1993) *Warrant and Proper Function*. New York: Oxford University Press.

Pruss, Alexander R. (2006) *The Principle of Sufficient Reason: A Reassessment*. Cambridge: Cambridge University Press.

Psillos, Stathis. (1999) *Scientific Realism: How Science Tracks Truth*. London: Routledge.

———. (2001) "Is Structural Realism Possible?" *Philosophy of Science* 68: 13–24.

Rea, Michael. (2002) *World Without Design: The Ontological Consequences of Naturalism*. Oxford: Oxford Univ. Press.

Rea, Michael, ed. (1998) *Material Constitution: A Reader*. Lanham: Rowman & Littlefield Publishers, Inc.

Redhead, Michael. (1987) *Incompleteness, Nonlocality and Realism: A Prolegomenon to the Philosophy of Quantum Mechanics*. Oxford: Clarendon Press.

———. (1995) *From Physics to Metaphysics*. Cambridge: Cambridge University Press.

Redhead, M. and Teller, P. (1991) "Particles, Particle Labels, and Quanta: The Toll of Unacknowledged Metaphysics." *Foundations of Physics* 21: 43–62.

———. (1992) "Particle Labels and the Theory of Indistinguishable Particles in Quantum Mechanics." *British Journal for the Philosophy of Science* 43: 201–18.

Reeh, H. and S. Schlieder (1961) "Bemerkungen zur Unitäräquivalenz von Lorentzinvarianten Feldern." *Nuovo Cimento* 22: 1051.

Robertson, Baldwin. (1973) "Introduction to Field Operators in Quantum Mechanics."*American Journal of Physics* 41: 678–90.

Rovelli, Carlo. (2001) "Quantum spacetime: What do we know?" in Callender and Huggett, 101–22.

Rovelli, C., and L. Smolin. (1995a) "Discreteness of Area and Volume in Quantum Gravity." *Nuclear Physics B* 442: 593–622.

———. (1995b) "Spin Networks and Quantum Gravity." *Physical Review D* 53: 5743.

Rowe, M. A., D. Kielpinski, V. Meyer, C. A. Sackett, W. M. Itano, C. Monroe and D. J. Wineland, (2001) "Experimental violation of a Bell's inequality with efficient detection." *Nature* 409: 791–94.

Ruetsche, Laura. (2002) "Interpreting Quantum Field Theory." *Philosophy of Science* 69: 348–78.
Sakurai, J. J. (1985) *Modern Quantum Mechanics*. Reading, MA: Addison-Wesley Publishing Company.
Salmon, Wesley. (1998) *Causality and Explanation*. New York: Oxford University Press.
Saunders, Simon. (1999) "The 'beables' of relativistic pilot-wave theory," in J. Butterfield and C. Pagonis, eds., *From Physics to Philosophy*. Cambridge: Cambridge University Press, 71–89.
Savellos, Elias, and Ümit Yalçin, eds. (1995) *Supervenience: New Essays*. Cambridge: Cambridge University Press.
Shimony, Abner. (1984) "Contextual hidden variables theories and Bell's Inequalities." *British Journal for the Philosophy of Science* 35: 25–45 (reprinted with additional comments in Shimony 1993, 104–29).
———. (1993) *Search for a Naturalistic World View, Volume II: Natural Science and Metaphysics*. Cambridge: Cambridge University Press.
Shoemaker, Sydney. (2002). "Kim on Emergence." *Philosophical Studies* 108: 53–63.
Smolin, Lee. (2001) *Three Roads to Quantum Gravity*. New York: Basic Books.
Stalnaker, Robert. (1968) "A Theory of Conditionals," in N. Rescher, ed., *Studies in Logical Theory*. Oxford: Basil Blackwell, 98–112.
Teller, Paul. (1986) "Relational Holism and Quantum Mechanics." *British Journal for Philosophy of Science* 37: 71–81.
———. (1989) "Relativity, Relational Holism, and the Bell Inequalities," in J. Cushing and E. McMullin, eds., *Philosophical Consequences of Quantum Theory: Reflections on Bell's Theorem*. Notre Dame: University of Notre Dame Press, 208–23.
———. (1995) *An Interpretive Introduction to Quantum Field Theory*. Princeton: Princeton University Press.
———. (1998) "Quantum Mechanics and Haecceities," in E. Castellani, E., ed., *Interpreting Bodies: Classical and Quantum Objects in Modern Physics*. Princeton: Princeton University Press, 114–41.
Tooley, Michael. (1987) *Causation: A Realist Approach*. Oxford: Clarendon Press.
Van Fraassen, Bas C. (1989a) "The Charybdis of Realism: Epistemological Implications of Bell's Inequality," in J. Cushing and E. McMullin, eds., *Philosophical Consequences of Quantum Theory: Reflections on Bell's Theorem*. Notre Dame: University of Notre Dame Press, 97–113.
———. (1989b) *Laws and Symmetry*. Oxford: Clarendon Press.
———. (1991) *Quantum Mechanics: An Empiricist View*. Oxford: Clarendon Press.
———. (1996) "Science, Materialism, and False Consciousness," in J. L. Kvanvig, ed. *Warrant in Contemporary Epistemology: Essays in Honor of Plantinga's Theory of Knowledge*. Lanham: Rowman & Littlefield Publishers, Inc., 149–81.
———. (2002) *The Empirical Stance*. New Haven: Yale University Press.
———. (2006) "Structure: Its Shadow and Substance," in *British Journal for the Philosophy of Science* 57: 275–307.
Votsis, Ioannis. (2003) "Is Structure Not Enough?" *Philosophy of Science* 70: 879–90.
Wigner, Eugene. (1970) "On Hidden Variables and Quantum Mechanical Probabilities." *American Journal of Physics* 38: 1005–09.
Wilczek, Frank (2008) *The Lightness of Being: Mass, Ether, and the Unification of Forces*. New York: Basic Books.
Worrall, John. (1989) "Structural Realism: The Best of Both Worlds?" *Dialectica* 43: 99–124.

9

The Incompatibilty of Naturalism and Scientific Realism

Robert C. Koons

1. Introduction

Whenever philosophers bother to offer a defense for philosophical naturalism, they typically appeal to the authority of natural science. Science is supposed to provide us with a picture of the world so much more reliable and well supported than that provided by any nonscientific source of information that we are entitled, perhaps even obliged, to withhold belief in anything that is not an intrinsic part of our best scientific picture of the world. This scientism is taken to support philosophical naturalism, since, at present, our best scientific picture of the world is an essentially materialistic one, with no reference to causal agencies other than those which can be located within space and time.

This defense of naturalism presupposes a version of scientific realism: unless science provides us with objective truth about reality, it has no authority to dictate to us the form that our philosophical ontology and metaphysics must take. Science construed as a mere instrument for manipulating experience, or merely as an autonomous construction of our society, without reference to our reality, tells us nothing about what kinds of things really exist and act.

In this essay, I will argue, somewhat paradoxically, that scientific realism can provide no support to philosophical naturalism. In fact, the situation is precisely the reverse: naturalism and scientific realism are incompatible.

Specifically, I will argue that (in the presence of certain well-established facts about scientific practice) the following three theses are mutually inconsistent:

1. Scientific realism.

2. Ontological naturalism (the world of space and time is causally closed).

3. There exists a correct naturalistic account of knowledge and intentionality (representational naturalism).

By scientific realism, I intend a thesis that includes both a semantic and an epistemological component. Roughly speaking, scientific realism is the conjunction of the following two claims:

1. Our scientific theories and models are theories and models of the real world.

2. Scientific methods tend, in the long run, to increase our stock of real knowledge.

Ontological naturalism is the thesis that nothing can have any influence on events and conditions in space and time except other events and conditions in space and time. According to the ontological naturalist, there are no causal influences from things "outside" space: either there are no such things, or they have nothing to do with us and our world.

Representational naturalism is the proposition that human knowledge and intentionality are parts of nature to be explained entirely in terms of scientifically understandable causal connections between brain states and the world. *Intentionality* is that feature of our thoughts and words that makes them *about* things, that gives them the capability of being true or false of the world. I take philosophical naturalism to be the conjunction of ontological and representational naturalism. The two theses are logically independent: it is possible to be an ontological naturalist without being a representational naturalist, and vice-versa. For example, eliminativists like the Churchlands, Stich, and (possibly) Dennett are ontological naturalists who avoid being representational naturalists by failing to accept the reality of knowledge and intentionality. Conversely, a Platonist might accept that knowledge and intentionality are to be understood entirely in terms of causal relations, including, perhaps, causal connections to the Forms, without being an ontological naturalist. I will argue that it is only the conjunction of the two naturalistic theses that is incompatible with scientific realism.

In sum, many philosophers believe that scientific realism gives us good reason to believe both ontological naturalism and representational naturalism. I will argue that scientific realism entails that either ontological naturalism or representational naturalism (or both) are false. I will argue that nature is comprehensible scientifically *only if* nature is *not* a causally closed system—only if nature is shaped by supernatural forces (forces beyond the scope of physical space and time).

My argument requires two critical assumptions:

PS: A preference for simplicity (elegance, symmetries, invariances) is a pervasive feature of scientific practice.

ER: Reliability is an essential component of knowledge and intentionality on any naturalistic account of these.

2. The Pervasiveness of Simplicity

Philosophers and historians of science have long recognized that quasi-aesthetic considerations, such as simplicity, symmetry, and elegance have played a pervasive and indispensable role in theory choice. For instance, Copernicus's heliocentric model replaced the Ptolemaic system long before it had achieved a better fit with the data because of its far greater simplicity. Similarly, Newton's and Einstein's theories of gravitation won early acceptance due to their extraordinary degree of symmetry and elegance.

The Incompatibility of Naturalism and Scientific Realism

In his book *Dreams of a Final Theory,* physicist Steven Weinberg included a chapter titled "Beautiful Theories," in which he detailed the indispensable role of simplicity in the recent history of physics. According to Weinberg, physicists use aesthetic qualities both as a way of suggesting theories and, even more importantly, as a sine qua non of viable theories. Weinberg argues that this developing sense of the aesthetics of nature has proved to be a reliable indicator of theoretical truth.

> The physicist's sense of beauty is . . . supposed to serve a purpose—it is supposed to help the physicist select ideas that help us explain nature.[1]

> [W]e demand a simplicity and rigidity in our principles before we are willing to to take them seriously.[2]

For example, Weinberg points out that general relativity is attractive, not just for its symmetry, but for the fact that the symmetry between different frames of reference requires the existence of gravitation. The symmetry built into Einstein's theory is so powerful and exacting that concrete physical consequences, such as the inverse square law of gravity, follow inexorably. Similarly, Weinberg explains that the electroweak theory is grounded in an internal symmetry between the roles of electrons and neutrinos.

The simplicity that physicists discover in nature plays a critical heuristic role in the discovery of new laws. As Weinberg explains,

> Weirdly, although the beauty of physical theories is embodied in rigid, mathematical structures based on simple underlying principles, the structures that have this sort of beauty tend to survive even when the underlying principles are found to be wrong. . . . We are led to beautiful structures by physical principles, but the beauty sometimes survives when the principles themselves do not.[3]

For instance, Dirac's 1928 theory of the electron involved an elegant formalism. Dirac's theory led to the discovery of the positron, and the mathematics of Dirac's theory has survived as an essential part of quantum field theory, despite the fact that Dirac's approach to reconciling quantum mechanics and relativity was wrong.[4] Similarly, mathematicians' pursuit of elegant mathematical theories has regularly anticipated the needs of theoretical physicists. The theory of curved space was developed by Gauss and Riemann before it was needed by Einstein, and group theory antedated its use in the theory of internal symmetry principles in particle physics.[5]

Weinberg notes that the simplicity that plays this central role in theoretical physics is "not the mechanical sort that can be measured by counting equations or symbols."[6] The recognition of this form of beauty requires an act of quasi-aesthetic judgment. As Weinberg observes,

> There is no logical formula that establishes a sharp dividing line between a beautiful explanatory theory and a mere list of data, but we know the difference when we see it.[7]

In claiming that an aesthetic form of simplicity plays a pervasive and indispensable role in scientific theory choice, I am not claiming that the aesthetic sense involved is innate or *a priori.* I am inclined to agree with Weinberg in thinking that "the universe . . . acts on us as a random, inefficient, and yet in the long-run effective, teaching machine. . . ."[8] We have become attuned to the aesthetic deep structure of the universe by a long process of trial and error, a kind of natural selection of aesthetic judgments. As Weinberg puts it,

> Through countless false starts, we have gotten it beaten into us that nature is a certain way, and we have grown to look at that way that nature is as beautiful. . . . Evidently we have been changed by the universe acting as a teaching machine and imposing on us a sense of beauty with which our species was not born. Even mathematicians live in the real universe, and respond to its lessons.[9]

Nonetheless, even though we have no reason to think that the origin of our aesthetic attunement to the structure of the universe is mysteriously prior to experience, there remains the fact that experience has attuned us to *something*, and this something runs throughout the most fundamental laws of nature. Behind the blooming and buzzing confusion of data, we have discovered a *consistent* aesthetic behind the various fundamental laws. As Weinberg concludes,

> It is when we study truly fundamental problems that we expect to find beautiful answers. We believe that, if we ask why the world is the way it is and then ask why that answer is the way it is, at the end of this chain of explanations we shall find a few simple principles of compelling beauty. We think this in part because our historical experience teaches us that as we look beneath the surface of things, we find more and more beauty. Plato and the neo-Platonists taught that the beauty we see in nature is a reflection of the beauty of the ultimate, the *nous*. For us, too, the beauty of present theories is an anticipation, a premonition, of the beauty of the final theory. And, in any case, we would not accept any theory as final unless it were beautiful.[10]

This capacity for "premonition" of the final theory is possible only because the fundamental principles of physics share a common bias toward a specific, learnable form of simplicity.

3. The Centrality of Reliability to Representational Naturalism

The representational naturalist holds that knowledge and intentionality are entirely natural phenomena, explicable in terms of causal relations between brain states and the represented conditions. In the case of knowledge, representational naturalism must make use of some form of reliability. The distinction between true belief and knowledge turns on epistemic norms of some kind. Unlike Platonists, representational naturalists cannot locate the basis of such norms in any transcendent realm. Consequently, the sort of *rightness* that qualifies a belief as knowledge must consist in some relation between the actual processes by which the belief is formed and the state of the represented conditions. Since knowledge is a form of success, this relation must involve a form of reliability, an objective tendency for beliefs formed in similar ways to represent the world accurately.

A representational naturalist might make use, as do Dretske, Papineau, and Millikan, of teleological properties, so long as these are taken to consist in a set of causal and historical relations. Knowledge could then be identified with true beliefs formed by processes whose proper functions are fulfilled in normal circumstances. However, this teleological account also connects knowledge with reliability, since the proper function of belief-forming processes is to form true beliefs, so the sort of process which fulfills this proper function must be a reliable one.

Thus, if representational naturalism is combined with epistemic realism about scientific theories, the conjunction of the two theses entails that our processes of scientific research and theory choice must reliably converge upon the truth.

A naturalistic account of intentionality must also employ some notion of reliability. The association between belief-states and their truth-conditions must, for the representational naturalist, be a matter of some sort of natural causal relation between the two. This association must consist in some sort of regular correlation between the belief-state and its truth-condition under certain conditions (the "normal" circumstances for the belief-state).

For example, according to Papineau, beliefs have teleological purposes, and these purposes fix their truth conditions, since "beliefs are true when they fulfill their purpose of co-varying with the relevant circumstances."[11] This covariation of representation and represented condition is what gives the capacity for belief its biological value. "According to the natural-selection story, it is the fact that a belief-type 'typically' obtains in certain circumstances that will explain our having it in our repertoire. . . ."[12] This regular association of belief-type and truth-conditions, and the biological purposes that the association serves, provide exactly the kind of naturalistic explication of intentionality that the representational naturalist requires.

This regular association is a form of reliability. As Fodor observed:

> [W]e shall still have this connection between the etiology of representations and their truth values: representations generated in teleologically normal circumstances must be *true*.[13]

This reliability is only a conditional reliability: reliability under teleologically *normal* circumstances. This condition provides the basis for a distinction between knowledge and true belief: an act of knowledge that *p* is formed by processes that reliably track the fact that *p* in the *actual* circumstances, whereas a belief that *p* is formed by processes that would reliably track *p* in normal circumstances.

It is possible for our reliability to be lost. Conditions can change in such a way that teleologically normal circumstances are no longer possible. In such cases, our beliefs about certain subjects may become totally unreliable:

> It is the *past* predominance of true belief over false that is required. . . . [This] leaves it open that the statistical norm from now on might be falsity rather than truth. One obvious way in which this might come about is through a change in the environment.[14]

In addition, there may be specifiable conditions that occur with some regularity in which our belief-forming processes are unreliable.

> [T]his link is easily disrupted. Most obviously, there is the point that our natural inclinations to form beliefs will have been fostered by a limited range of environments, with the result that, if we move to new environments, those inclinations may tend systematically to give us false beliefs. To take a simple example, humans are notoriously inefficient at judging sizes under water.[15]

Finally, the reliability involved may not involve a high degree of probability. The correlation of belief-type and represented condition does not have to be close to 1. As Millikan has observed, "it is conceivable that the devices that fix human beliefs fix true ones not on average, but just often enough."[16] For example, skittish animals may form the belief that a predator is near on the basis of very slight evidence. This belief will be true only rarely, but it must have a better-than-chance probability of truth under normal circumstances if it is to have a representational function at all.

Thus, despite these qualifications, it remains the case that a circumscribed form of reliable association is essential to the naturalistic account of intentionality. The reliability is conditional, holding only under normal circumstances, and it may be minimal, involving a barely greater-than-chance correlation. Nonetheless, the representational naturalist is committed to the existence of a real, objective association of the belief-state with its corresponding condition.

4. Proof of the Incompatibility

I claim that the triad of scientific realism (SR), representational naturalism (RN), and ontological naturalism (ON) is inconsistent, given the theses of the pervasiveness of the simplicity criterion in our scientific practices (PS) and the essentiality of reliability as a component of naturalistic accounts of knowledge and intentionality. The argument for the inconsistency proceeds as follows:

1. SR, RN, and ER entail that scientific methods are reliable sources of truth about the world.

As I have argued, a representational naturalist must attribute some form of reliability to our knowledge and belief-forming practices. A scientific realist holds that scientific theories have objective truth-conditions, and that our scientific practices generate knowledge. Hence, the combination of scientific realism and representational naturalism entails the reliability of our scientific practices.

2. From PS, it follows that simplicity is a reliable indicator of the truth about natural laws.

Since the criterion of simplicity as a *sine qua non* of viable theories is a pervasive feature of our scientific practices, thesis 1 entails that simplicity is a reliable indicator of the truth (at the very least, a better-than-chance indicator of the truth in normal circumstances).

3. Mere correlation between simplicity and the laws of nature is not good enough: reliability requires that there be some causal mechanism connecting simplicity and the actual laws of nature.

Reliability means that the association between simplicity and truth cannot be coincidental. A regular, objective association must be grounded in some form of causal connection. Something must be causally responsible for the bias toward simplicity exhibited by the theoretically illuminated structure of nature.

4. Since the laws of nature pervade space and time, any such causal mechanism must exist outside spacetime.

By definition, the laws and fundamental structure of nature pervade nature. Anything that causes these laws to be simple, anything that imposes a consistent aesthetic upon them, must be supernatural.

5. Consequently, ON is false.

The existence of a supernatural cause of the simplicity of the laws of nature is obviously inconsistent with ontological naturalism. Hence, one cannot consistently embrace naturalism and scientific realism.

5. Papineau and Millikan on Scientific Realism

David Papineau and Ruth Garrett Millikan are two thoroughgoing naturalists who have explicitly embraced scientific realism. If the preceding argument is correct, this inconsistency should show itself somehow in their analyses of science. This expectation is indeed fulfilled. For example, Papineau recognizes the importance of simplicity in guiding the choice of fundamental scientific theories. He also recognizes that his account of intentionality entails that a scientific realist must affirm the reliability of simplicity as a sign of the truth. Nonetheless, he fails to see the incompatibility of this conclusion with his ontological naturalism. Here is the relevant passage:

> [I]t is plausible that at this level the inductive strategy used by physicists is to ignore any theories that lack a certain kind of *physical simplicity*. If this is right, then this inductive strategy, when applied to the question of the general constitution of the universe, will inevitably lead to the conclusion that the universe is composed of constituents which display the relevant kind of physical simplicity. And then, once we have reached this conclusion, we can use it to explain why this inductive strategy is reliable. For if the constituents of the world are indeed characterized by the relevant kind of physical simplicity, then a methodology which uses observations to decide between alternatives with this kind of simplicity will *for that reason* be a reliable route to the truth.[17]

In other words, so long as we are convinced that the laws of nature *just happen to be* simple in the appropriate way, we are entitled to conclude that our simplicity-preferring methods were *reliable* guides to the truth. However, it seems clear that such a retrospective analysis would instead reveal that we succeeded by sheer, dumb luck.

By way of analogy, suppose that I falsely believed that a certain coin was two-headed. I therefore guess that all of the first six flips of the coin will turn out to be heads. In fact, the coin is a fair one, and, by coincidence, five of the first six flips did land heads. Would we say in this case that my assumption was a reliable guide to the truth about these coin flips? Should we say that its reliability was 5/6? To the contrary, we should say that my assumption led to very unreliable predictions, and the degree of success that I achieved was due to good luck, and nothing more.

Analogously, if it is a mere coincidence that the laws of nature share a certain form of aesthetic beauty, then our reliance upon aesthetic criteria in theory choice is not in any sense reliable, not even minimally reliable, not even reliable in ideal circumstances. When we use the fact that we have discovered a form of "physical simplicity" in law *A* as a reason for preferring theories of law *B* that have the same kind of simplicity, then our method is reliable only if there is some causal explanation of the repetition of this form of simplicity in nature. And this repetition necessitates a supernatural cause.

Papineau recognizes that we do rely on such an assumption of the repetition of simplicity:

> The account depends on the existence of certain general features which characterize the true answers to questions of fundamental physical theory. Far from being knowable *a priori,* these features may well be counterintuitive to the scientifically untrained.[18]

Through scientific experience, we are "trained" to recognize the simplicity shared by the fundamental laws, and we use this knowledge to anticipate the form of unknown laws. This projection of experience from one law to the next is reliable only if there is some common cause of the observed simplicity.

Similarly, Millikan believes that nature trained into us (by trial-and-error learning) certain "principles of generalization and discrimination"[19] that provided us with a solution to the problem of theoretical knowledge that was "elegant, supremely general, and powerful, indeed, I believe it was a solution that cut to the very bone of the ontological structure of the world."[20] However, Millikan seems unaware of just how deep this incision must go. A powerful and supremely general solution to the problem of theory choice must reach a ground of the common form of the laws of nature, and this ground must lie outside the bounds of nature.

Papineau and Millikan might try to salvage the reliability of a simplicity bias on the grounds that the laws of nature are—although uncaused, brute facts—*necessarily* what they are. If they share, coincidentally, a form of simplicity and do so non-contingently, then a scientific method biased toward the appropriate form of simplicity will be, under the circumstances, a reliable guide to the truth.

There are two compelling responses to this line of defense. First, there is no reason to suppose that the laws of nature are necessary. Cosmologists often explore the consequences of models of the universe in which the counterfactual laws hold.

Second, an unexplained coincidence, even if that coincidence is a brute-fact necessity, cannot ground the reliability of a method of inquiry. A method is reliable only when there is a causal mechanism that explains its reliability. By way of illustration, suppose that we grant the necessity of the past: given the present moment, all the actual events of the past are necessary. Next, suppose that a particular astrological method generates by chance the exact birth date of the first president of the United States. Since that date is now necessary, there is no possibility of the astrological method's failing to give the correct answer. However, if there is no causal mechanism explaining the connection between the method's working and the particular facts involved in Washington's birth, then it would be Pickwickian to count the astrological method as *reliable* in investigating this particular event.

Analogously, if the various laws of nature just happen, as a matter of brute, inexplicable fact, to share a form of simplicity, then, even if this sharing is a matter of necessity, using simplicity as a guide in theory choice should not count as reliable.

6. The Forster-Sober Account of Simplicity

An interesting paper by Malcolm Forster and Elliott Sober offers a justification of the scientific preference for simplicity that seems to be compatible with scientific realism and yet which does not acknowledge any sense in which simplicity is a reliable indicator of the truth.[21] If the Forster-Sober account provides an adequate explanation of the role of simplicity without any such reliable connection between simplicity and truth, then it would provide a serious challenge to the argument of the previous section. As Forster and Sober put it,

> In the past, the curve fitting problem has posed a dilemma: Either accept a realist interpretation of science at the price of viewing simplicity as an irreducible and *a prioristic* sign of truth and thereby eschew empiricism, or embrace some form of antirealism. Akaike's solution to the curve fitting problem dismantles the dilemma. It is now possible to be a realist and an empiricist at the same time.[22]

The issue for Forster and Sober is realism vs. empiricism, whereas for us it is realism vs. naturalism, but it would seem that analogous claims could be made on behalf of Akaike's solution. This solution is supposed to give the realist some reason for preferring simpler hypotheses that is independent of any supposed correlation between simplicity and truth.

The Akaike solution goes something like this. First, we must assume that all of our observations involve a certain amount of noise—that random observational error regularly occurs and the error values are normally distributed. We divide the possible hypotheses into a finite sequence of families based on the degree of simplicity (measured by the number of parameters that are allowed to vary within the family). Instead of selecting the hypothesis that best fits the actual data, we instead look for a family of hypotheses with the best combination of goodness-of-fit and simplicity, and choose the best-fitting hypothesis within that set.

The rationale for the Akaike criterion is the avoidance of *overfitting*. Since the actual data includes some unknown observational error, the curve that best fits the data is unlikely to be the true one. It will tend to fit the actual data better than the true curve, which is called the "overfitting" of the hypothesis to the data. Balancing goodness-of-fit with simplicity is supposed to mitigate this overfitting error. Consequently, the realist is given some reason to employ simplicity as a desideratum of theory choice without assuming any correlation between simplicity and truth.

Simpler, low-dimensional families are much smaller than more complex, high-dimensional families. There are therefore two reasons why the more complex families are more likely to contain the hypothesis that best fits the data:

(a) Larger families generally contain curves closer to the truth than smaller families.

(b) *Overfitting:* The higher the number of adjustable parameters, the more prone the family is to fit to noise in the data.[23]

According to Forster and Sober, we want to favor a family of hypotheses if it contains a good fit to the data because of reason (a), but not if it contains one because of reason (b). What is needed is an estimate of the expected degree of overfitting associated with each family, given the actual data. Akaike demonstrated that, under certain special conditions, we can find an *unbiased estimator* of this special form of error. By subtracting the number of parameters that are allowed to vary within a family from a measure of the degree-of-fit of the best-fitting curve within that family (this measure is one of log-likelihood or, in special cases, the sum of squares), we can arrive at a *corrected* estimate of the degree of fit of the family to the truth, which Forster and Sober call the "expected predictive accuracy" of the family.[24] The Akaike criterion tells us to choose the best-fitting hypothesis within the family with the greatest expected predictive accuracy. In this way, we have both a definite rule for trading-off goodness-of-fit for simplicity, and a plausible rationale for making the trade-off.

There are several points to be made in response to this solution. First, it is not at all clear that the role of simplicity in the kind of curve-fitting practices Forster and Sober discuss is analogous to the role simplicity plays in our choice of fundamental physical theories. As Weinberg observed, the kind of *simplicity* that guides our choice of fundamental theories is not easily defined. It does not correspond directly to what Forster and Sober mean by the *simplicity* of a family of hypotheses, viz., the number of variable parameters in the corresponding equations.

Second, the technical results upon which Forster and Sober rely are quite limited in their scope of application, as I. A. Kieseppä has demonstrated.[25] The Aikake estimator of predictive accuracy is valid only when the space of hypotheses is carefully circumscribed. For example, it is valid when the space of hypotheses includes only polynomial equations, but invalid when it includes periodic functions, like the sine wave function.[26]

Third, the rationale for the Akaike criterion is incompatible with the reliabilist implications of combining scientific realism with representational naturalism. The sort of "scientific realism" that Forster and Sober have in mind is much less specific, implying only a concern with the truth of our scientific theories. Forster and Sober make no effort to demonstrate that reliance on the Akaike criterion leads reliably to the truth. Instead, they provide only a rationale that might reasonably motivate a realist to prefer simpler theories.

Finally, it is far from clear that even this rationale provides a basis for preferring simplicity that is genuinely independent of the reliability of simplicity as a sign of the truth. As has been pointed out by Kieseppä,[27] Scott De Vito,[28] and Andre Kukla,[29] the Akaike solution presupposes that a determinate conception of simplicity is a given. There is no objective language- and representation-independent way of "counting the parameters" associated with a given curve. A linear curve is *naturally* thought of as having a single parameter, but this can easily be altered by redescribing the curve or altering the coordinate system. Sorting hypotheses into families by simplicity as we perceive it reflects a prior and unjustified preference for some hypotheses over others.

Forster and Sober might insist that the sorting of hypotheses into a hierarchy of families is entirely arbitrary or random. As they present the argument for the Akaike criterion, all that matters is that the hypotheses be sorted into a sequence of families in which the size of the families increases exponentially, and that this sorting *not* be done in an ad hoc fashion, in response to the actual data observed. Then, when we observe a relatively small family F with a hypothesis h showing a surprisingly good degree of fit to the data (surprising, that is, in light of the smallness of F), we are supposed to have good reason to believe that F has a high degree of predictive accuracy, and, therefore, that we have reason to prefer h over other hypotheses with better fit that happen to belong to much larger families. However, if it was entirely a matter of chance or caprice that h ended up in a small family, and its better-fitting competitors ended up in larger families, it is hard to see how h's good fortune provides us with any rational ground for preferring it.

To the contrary, the plausibility of the Akaike solution depends on our prior conviction that simpler hypotheses (as measured by mathematical conventions that have proved reliable at this very task) are disproportionately probable. What Forster and Sober give us is a principled way of weighing the two competing desiderata of simplicity and goodness of fit, but they do not provide us with a rationale for treating simplicity as a desideratum in the first place.

Consequently, Forster and Sober do not provide us with a way of escaping the conclusion that a reliabilist conception of scientific realism entails the reliability of simplicity as an indicator of the truth.

7. Pragmatic Accounts of the Simplicity Criterion

A popular strategy for explaining the role of simplicity in scientific theorizing has been to appeal to a variety of pragmatic considerations. For example, Reichenbach argued that we favor simpler hypotheses because they are easier to represent, to make deductions from, and to use in calculations.[30] More recently, Peter Turney has argued that simpler hypotheses are more likely (given the presence of random observational error) to be repeatedly confirmed.[31]

However, these pragmatic justifications again sidestep the central issue, that of *reliability*. If our reliance on simplicity is unreliable, resulting in a bias toward simplicity that is not reflected in the constitution of nature, then we cannot combine scientific realism with representational naturalism.

A pragmatic justification of our scientific practice, when combined with representational naturalism, yields the conclusion that scientific theories must be interpreted non-representationally, either as mere instruments for generating empirical predictions, or as conventional constructs valid only for a local culture. Pragmatism, by eschewing any commitment to the objective reliability of scientific methods, cannot be combined with a naturalistic version of scientific realism.

8. Conclusion

Philosophical naturalism, then, can draw no legitimate support from the deliverances of natural science, realistically construed, since scientific realism entails the falsity of naturalism. If scientific theories are construed nonrealistically, it seems that the status of ontology cannot be affected by the successes of natural science, nor by the form that successful theories in the natural sciences happen to take. If scientific antirealism is correct, then the "manifest image" of the scientific worldview must not be taken as authoritative. Instead, that image is merely a useful fiction, and metaphysics is left exactly as it was before the advent of science.

Of course, naturalism as a metaphysical program existed before the development of modern science (Democritus, Epicurus, Lucretius), and presumably it would survive the downfall of scientific realism. However, modern naturalists owe the rest of us a rational basis for their preferences that is independent of science.

In fact, the situation for the naturalist is even worse than I have described it. To the extent that the success of natural science provides support for scientific realism (in both its semantic and epistemic versions), to that extent it provides grounds for rejecting philosophical naturalism. Thus, conventional wisdom has the relationship between natural science and naturalism exactly backwards. In fact, the more successes natural science accumulates, the less plausible philosophical naturalism becomes.

There is a third thesis that is often included (especially since Quine) in the definition of naturalism: the continuity between the methods of philosophy and those of natural science, which we might call "meta-philosophical naturalism." Scientific antirealism, when combined with meta-philosophical naturalism, leads to the conclusion of philosophical antirealism, since philosophical theories are, according to meta-philosophical naturalism, merely a species of scientific theories.

This means that full-orbed naturalism (ontological + representational + meta-philosophical) is a self-defeating position. Full-orbed naturalism is a philosophical theory, and yet it entails philosophical antirealism, which means that such theories cannot be known, and do not even purport to represent the world. Full-orbed naturalism cannot be true, since if it were true, it would entail that no philosophical theory (itself included) could be true.

Notes

1. Steven Weinberg, *Dreams of a Final Theory: The Scientist's Search for the Ultimate Laws of Nature* (New York: Vintage Books, 1993), 133.
2. Weinberg, *Dreams of a Final Theory,* 148–49.
3. Weinberg, *Dreams of a Final Theory,* 151–52.
4. Weinberg, *Dreams of a Final Theory,* 151.
5. Weinberg, *Dreams of a Final Theory,* 153–57.
6. Weinberg, *Dreams of a Final Theory,* 134.
7. Weinberg, *Dreams of a Final Theory,* 148–49.
8. Weinberg, *Dreams of a Final Theory,* 158.
9. Weinberg, *Dreams of a Final Theory,* 158–59.
10. Weinberg, *Dreams of a Final Theory,* 165.
11. David Papineau, *Philosophical Naturalism* (Oxford: Blackwell, 1993), 177.
12. David Papineau, "Representation and Explanation," *Philosophy of Science* 51 (1984): 558.
13. Jerry A. Fodor, "Semantics, Wisconsin Style," *Synthese* 59 (1984): 247.
14. Papineau, "Representation and Explanation," 558.
15. Papineau, *Philosophical Naturalism,* 100.
16. Ruth Garrett Millikan, "Biosemantics," *Journal of Philosophy* 86 (1989): 289.
17. Papineau, *Philosophical Naturalism,* 166.
18. Papineau, *Philosophical Naturalism,* 166.
19. Millikan, "Biosemantics," 292.
20. Millikan, "Biosemantics," 294.
21. Malcolm Forster and Elliott Sober, "How to Tell when Simpler, More Unified, or Less *Ad Hoc* Theories will Provide More Accurate Predictions," *British Journal for the Philosophy of Science* 45 (1994): 1–35.
22. Forster and Sober, "How to Tell," 28.
23. Forster and Sober, "How to Tell," 8.
24. Forster and Sober, "How to Tell," 10.
25. I. A. Kieseppä, "Akaike Information Criterion, Curve-fitting and the Philosophical Problem of Simplicity," *British Journal for the Philosophy of Science* 48 (1997): 21–48.
26. Kieseppä, "Akaike Information Criterion," 34–37.
27. Kieseppä, "Akaike Information Criterion," 21–48.
28. Scott De Vito, "A Gruesome Problem for the Curve-Fitting Solution," *British Journal for the Philosophy of Science* 48 (1997): 391–96.
29. Andre Kukla, "Forster and Sober and the Curve-Fitting Problem," *British Journal for the Philosophy of Science* 46 (1995): 248–52.
30. Hans Reichenbach, "The Pragmatic Justification of Induction," in *Readings in Philosophical Analysis,* ed. H. Feigl and W. Sellars (New York: Appleton-Century-Crofts, 1949), 305–27.
31. Peter Turney, "The Curve Fitting Problem—A Solution," *British Journal for the Philosophy of Science* 41 (1990): 509–30.

10

TRUTH AND REALISM[1]

ALVIN I. GOLDMAN

1. INTRODUCTION: [EPISTEMOLOGY NATURALIZED]

Epistemology, as I conceive it, divides into two parts: individual epistemology and social epistemology. Individual epistemology—at least *primary* individual epistemology—needs help from the cognitive sciences. Cognitive science tries to delineate the architecture of the human mind-brain, and an understanding of this architecture is essential for primary epistemology. Social epistemology needs help from various of the social sciences and humanities, which jointly provide models, facts, and insights into social systems of science, learning, and culture. . . .

Along with the dominant tradition, I regard epistemology as an evaluative, or normative, field, not a purely descriptive one. This makes it far from obvious how *positive* science can have inputs into epistemology. How, exactly, do *facts* of cognition or social intercourse bear on epistemic evaluations or norms?

A few other recent characterizations of epistemology also link it with psychology. But these characterizations depict the field as a descriptive one. On a purely descriptive conception, it is not surprising that epistemology should be indebted to psychology—should even reduce to it. Thus, on W. V. Quine's naturalistic conception, the epistemologist would study how the human subject responds to certain input; how, in response to various stimulus patterns, the subject delivers a description of the external world and its history. In studying the relation between this "meager input" and "torrential output," epistemology "simply falls into place as a chapter of psychology and hence of natural science."[2] Similarly, Donald Campbell advances a conception of the field that he calls "evolutionary epistemology." On this conception epistemology takes cognizance of "man's status as a product of biological and social evolution"[3] Campbell explicitly characterizes his conception as descriptive: descriptive of man as knower.

If epistemology is a branch of psychology, or evolutionary theory, the field's empirical status needs no clarification. But this approach, though perfectly tenable, neglects the evaluative strain pervading most of historical epistemology. Epistemologists have traditionally been interested whether beliefs about the world are justified or warranted; where we are rationally entitled to these beliefs. Epistemologists seek to discover or invent proper methods of inquiry and investigation, often dismissing established procedures as irrational. Clearly, "justified," "warranted," and "rational" are evaluative terms; and the advocacy of particular methods is a normative activity. So traditional epistemology has a strong evaluative-normative strain....

My epistemological framework prominently features an objectivist standard or set of standards. The central epistemological concepts of appraisal, I argue, invoke *true belief* as their ultimate aim. So the evaluation of epistemological procedures, methods, processes, or arrangements must appeal to truth-conduciveness, an objective standard of assessment. While this emphasis on truth is hardly startling, the framework contrasts with many studies of science and opinion that explore properties of social systems and institutions. Specifically, studies in the history and sociology of science characteristically shy away from considerations of truth.

Truth-linked standards may seem useless because of circularity or vacuousness. To decide whether such a standard is satisfied, we have to employ our present beliefs about the truth. Isn't this circular? Or doesn't it imply automatic endorsement of current procedures, by which our present beliefs have been formed? Wouldn't it preclude criticism and revision, to which normative epistemology ought to be committed?

None of these objections is warranted. To be sure, application of a standard requires recourse to present beliefs. In Otto Neurath's metaphor we can only rebuild our intellectual ship while floating upon it at sea.[4] But the same point holds for any standard, truth-linked or not. So there is no objectionable circularity here. Second, criticism and revision are not precluded. We can criticize (at least some of) our belief-forming processes even with beliefs created by those very processes....

[As I see it,] epistemology is not (primarily) interested in inferences construed as argument forms. Rather, it is interested in inferences as processes of belief formation or belief revision, as sequences of psychological states. So psychological processes are certainly a point of concern, even in the matter of inference. Furthermore, additional psychological processes are of equal epistemic signficance: processes of perception, memory, problem solving, and the like.

Why is epistemology interested in these processes? One reason is its interest in epistemic justification. The notion of justification is directed, principally, at beliefs. But evaluation of beliefs, I contend, derives from evaluations of belief-forming processes. Which processes are suitable cannot be certified by logic alone. Ultimately, justificational status depends (at least in part) on properties of our basic cognitive equipment. Hence, epistemology needs to examine this equipment, to see whether it satisfies standards of justifiedness....

While few intellectual feats are achieved without any "helps" for the mind, the intrinsic properties of the mind still hold significance for epistemology. First, unless the mind has a suitable structure, it cannot *use* tools properly or successfully. Second, the invention, acquisition, and selective retention of intellectual tools must ultimately rest with native cognitive mechanisms, with the understanding "considered by itself." It is fully appropriate, then, for epistemology to inquire into cognitive architecture, to assess its strengths and weaknesses. At a minimum, this is a job for primary epistemology. Within primary epistemology, then, the objects of epistemic evaluation are cognitive processes, structures, and mechanisms....

In linking epistemology with cognitive science, I use the term "cognitive science" (or "cognitive psychology") neutrally. Some use the phrase in an "ideological" way, to advocate a particular brand or style of theorizing. For example, it may designate a computational approach, or an approach that focuses on a certain level of abstraction, say, functional as opposed to neural. In this usage cognitive science contrasts with neural and biological approaches to cognition. But my usage is not ideological in any of these ways. It includes any scientific approach to cognition....

2. Varieties of Realism

The notion of truth is important for almost every epistemology, but it is central to my epistemology for special reasons. I [invoke] truth not only, as is commonly done, as a condition for knowing, but also as a critical element in two dimensions of epistemic appraisal, namely, "justification" and "intelligence." The roles to which truth is assigned in my epistemology may be a source of queasiness to some . . . [S]ome philosophers think they discern a circularity in using "true" in an account of the epistemic term "justified," because the notion of truth is itself epistemic. Other philosophers might hold that it is misguided to assign truth, or even the pursuit of truth, a central role in problem solving, because at least one central domain of problem-solving—the domain of scientific theory—involves no pursuit of truth at all, but only predictive power, calculational convenience, or the like. For these and other reasons, a discussion of truth is [important]. It must be shown that the truth concept can perform the functions to which my epistemology assigns it, and that it has the "realistic" properties that [I attribute] to it.

As indicated, I wish to defend some kind of realistic view of truth. But not everything called "realism," not even every variety of realism about truth, is a doctrine I mean to endorse. So I need to disentangle different conceptions of realism in order to keep our bearings.

[Much] influential work on truth and realism has been done by Hilary Putnam and Michael Dummett.[5] [A lot] of my discussion will therefore relate to theses that they have advanced or explored. Dummett conceives of realism as a family of doctrines about different subject matters; you can be a realist about one subject matter and not about another. He prefers to say that realism is a view about a certain class of statements, for instance, physical object statements, mathematical statements, ethical statements, or statements in the future tense. Here is a formulation of realism he offers: "The very minimum that realism can be held to involve is that statements in the given class relate to some reality that exists independently of our knowledge of it, in such a way that that reality renders each statement in the class determinately true or false, again independently of whether we know, or are even able to discover, its truth-value."[6]

There are two components in this characterization of realism. First, there is the principle of bivalence: every statement in the class is determinately either true or false. Second, there is the principle of verification-transcendent truth: a statement is true or false independently of our knowledge, or verification, of it (or even our ability to verify it). The second of these theses will figure prominently in my discussion. But the first, the bivalence principle, is tangential.

It is not clear just how firm a criterion of realism bivalence should be. Dummett regards it as necessary for realism, though not sufficient.[7] But it is not even clear that it should be accepted as necessary. Why can't a philosopher be a realist about a given class even if he takes the Frege-Strawson line that a singular statement containing an empty proper name—such as, "The present king of France is bald"—is neither true nor false? Why can't he be a realist even if he acknowledges vague statements within the given class, which are neither true nor false?[8]

Setting these points aside, suppose we grant Dummett the label "antirealist" for philosophers who deny the principle of bivalence for a given class of statements. Is it important for my purposes to defend the corresponding kind of realism, as a general claim? Not at all. There is nothing in my epistemology that requires defense of the principle of bivalence for all classes of statements. I could readily admit the failure of this principle for any number of classes of statements: statements about the future, subjunctive conditional statements, even statements of mathematics. What is critical is that when any such statement is true (or false), what makes it true (or false) is independent of our knowledge or verification. Thus, the second of the two realist theses explained above, verification-transcendence, is critical. To put it another way, truth must not be an *epistemic* matter. But the issue of bivalence does not affect my epistemology.

Admittedly, the two theses are interconnected in some philosophies. Some denials of bivalence are consequences of a denial of verification-transcendence. The constructivist in mathematics denies verification-transcendence, saying that a mathematical statement is true only if there exists a proof of it, and is false only if there exists a proof of its negation. The denial of bivalence readily follows. However, people may also deny bivalence for other reasons. Nothing in my epistemology requires any quarrel with bivalence per se.

My concern with realism, then, is a concern with truth; with what makes a statement, or belief, true, if it is true. So I shall primarily be concerned—in much of [this chapter], at any rate—with the theory of truth.

In calling my principal topic the theory of truth, two clarifications should be made. First, I am interested (roughly) in the "meaning" of truth, not in procedures or marks for telling which propositions are true. Classical defenders of the coherence theory of truth probably failed to observe this distinction. They ran together coherence as a *test* of truth and coherence as a *definition* of truth. As a test of truth, coherence has some attractions; as a definition, it has no plausibility at all. I, of course, am only interested (here) in the definition of truth. A theory of tests for truth is better understood as a theory of justified belief.

The second clarification is that I am not concerned here with the formal theory of truth, with an account of truth that resolves the liar paradox, for example. Rather, I am only concerned with the metaphysical questions surrounding the notion of truth. There are many philosophers who doubt whether there are any interesting metaphysical questions, whether there is anything informative that can be said about these questions. They maintain that Alfred Tarski's work on the semantic conception of truth, or the idea of truth as disquotation, resolves all the legitimate questions that can be raised about truth. I believe that Putnam has given decisive arguments for rejecting this claim: the Tarskian, or disquotational, theory does not give us an adequate analysis, or understanding, of the notion of truth.[9] Substantial, classical questions remain, questions that divide realists from antirealists. These are the ones that will engage my attention.

I have isolated one central dispute about truth that I shall address: whether truth is verification-transcendent or not, whether it is a purely nonepistemic notion or in some sense epistemic. This will be the topic of the next section. But there is another important dispute about truth that standardly divides realists from antirealists; that is, whether truth consists in "correspondence" with reality. This is another way of thinking about realism versus antirealism. Whether a realist account of truth needs to be committed to correspondence, and if so, what kind of correspondence, will have to be explored. This will be the topic of section four.

3. Truth as Nonepistemic

Putnam writes of two conceptions of truth: an *externalist* versus an *internalist* conception,[10] which he earlier dubbed "metaphysical realism" and "internal realism," respectively.[11] Metaphysical realism sees truth as "radically nonepistemic."[12] Internal realism, by contrast, finds an epistemic component in truth. I do not find any of Putnam's terminology optimal. In particular, his "internal realism" is best regarded as a form of antirealism. So I shall count the nonepistemic approach to truth as realism *sans phrase,* and the epistemic approach to truth will be counted as (a species of) antirealism.

How, exactly, would an epistemic account of truth go? As we will see, the proponent of an epistemic approach is caught on the horns of a dilemma. If a strong epistemic component is introduced, the resulting account is liable to be *unnecessary* for truth. If he weakens the epistemic component, the resulting conditions are liable to be *insufficient* for truth. Before considering some sample accounts, let us consider some dimensions along which epistemic theories of truth might vary.

First, there is a choice between actual verification and merely possible verification. For example, constructivism is depicted as maintaining that a mathematical statement is true only if we are (actually) in possession of a proof of it, and false only if we are (actually) in possession of a refutation.[13] This can be contrasted with a weaker possible view, namely, that a mathematical statement is true only if it is *possible* that we should possess a proof of it.

Second, there is a choice between *present* verification and verification at some time or other, for example, at some future time. C. S. Peirce defined truth as "the opinion which is fated to be ultimately agreed to by all who investigate."[14] This is clearly a weaker requirement for truth than one which requires, for p to be true, that p is agreed to *now*.

Third, there is a choice between conclusive verification and inconclusive, or prima facie, verification. A very strong requirement for the truth of p is that it be conclusively verified. A weaker requirement is that, for p to be true, it must be supported to some degree. Of course, either of these requirements can be weakened by taking the possibility option in the first choice: p is true if it is merely possible for p to be conclusively verified.

The fourth choice concerns the person or persons for whom the proposition must be verified. Dummett regularly speaks with an indefinite first person plural: "we." In the constructivist's approach, cited above, a mathematical statement is true only if we possess a proof of it. But who, exactly, are "we"? *All* human beings? One or more human beings? Or should the restriction to human beings be lifted entirely?

Clearly, many permutations of the epistemic approach to truth can be generated from these options. Let's quickly look at some sample accounts and the difficulties they encounter.

(*a*)   All of us now have conclusive evidence for p.

(*b*)   Somebody now has conclusive evidence for p.

(*c*)   Somebody will some day have conclusive evidence for p.

(*d*)   Somebody now has prima facie evidence for p.

(*e*)   Somebody will some day have prima facie evidence for p.

(*f*)   It is possible for somebody to have prima facie evidence for p.

(*g*)   It is possible for somebody to have conclusive evidence for p.

Considered as necessary and sufficient conditions for "p is true," each of these suffers from palpable defects. Each is either too weak or too strong; some of them are both.

Let us examine them as conditions for the truth of ordinary physical object propositions. Clearly, (a) is much too strong. It could certainly be true that Julius Caesar had a mole on the nape of his neck although it is false that all of us now have conclusive evidence for this. Similarly for (b): this could be true even though nobody now has conclusive evidence for p. Even (c) is still too strong. This minor historical fact about Julius Caesar may be one for which nobody will ever have conclusive evidence. Even (d) is too stringent. Perhaps nobody now has even prima facie evidence for this fact. This would not keep the proposition from being true.

Not only is (d) too strong, it is also too weak. Suppose some historian does have prima facie evidence for this proposition. That surely does not entail that the proposition is true. For example, suppose a Roman text of the period attributes a remark to Brutus that seems to indicate the presence of a mole on Caesar's neck. That evidence is not sufficient for the truth of the proposition. The same twin problems arise for (e), for similar reasons. It is both too strong and too weak.

Turning to the "possibility" formulations, (f) is too weak for the same reasons as (d) and (e). The mere possibility of prima facie evidence for p does not guarantee p's truth. The difficulties for (g) are slightly harder to assess. They depend on how "conclusive evidence" is interpreted. If it is interpreted very strictly—so that conclusive evidence entails truth—then it is doubtful whether it is possible for somebody to have conclusive evidence for any proposition, at least any physical object proposition. So (g) would be too strong. If "conclusive evidence" is interpreted less strictly, so that it does not entail truth, then (g) is too weak.

There is another plausible objection against (g), if it is interpreted strictly. On the strict interpretation we are to understand "p is conclusively verified (for some person S)" to entail (by virtue of its meaning) "p is true." But if this is right, truth is covertly contained in the analysis of "p is conclusively verified." And then it would be circular to employ the latter locution in the analysis of "p is true"!

This point is worth underlining because it should also figure in the assessment of the epistemic account of truth endorsed by Putnam. Putnam writes:

> Truth is an *idealization* of rational acceptability. We speak as if there were such things as epistemically ideal conditions, and we call a statement "true" if it would be justified under such conditions.[15]

To clarify this proposal, let me reformulate it as follows.

(h) If someone were in epistemically ideal conditions vis-à-vis the question of p versus not-p, then he would be justified in believing p.

The correctness of this proposal depends on how "epistemically ideal conditions" is defined. Is there a way to define it so that the equivalence is correct? There is a definition, I think, in which the equivalence could succeed. Suppose we define "S is in epistemically ideal conditions vis-à-vis p" as "S is so situated and so constructed that he would believe p if and only if p is true." Being in epistemically ideal conditions vis-à-vis p might consist, for example, in (1) being in a perfect position for observing p, (2) having perfect detector organs vis-à-vis states of affairs like p, and (3) having belief-forming processes perfectly coordinated with these organs' outputs. On a suitable theory of justifiedness, then, proposal (h) might be right.

But this stratagem obviously incurs the sort of problem we posed for (g) under the strict interpretation. It makes the equivalence work only by making it *circular*. The problem, of course, is that the proposed definition of "epistemically ideal conditions" reintroduces the notion of truth!

Putnam obviously cannot use this way of defining "epistemically ideal conditions." He hints at this himself when he restricts "verified" to an "operational" sense.[16] Is there an alternative way of defining "epistemically ideal conditions" so that (1) no nonepistemic notion of truth is presupposed, and yet (2) (*h*) turns out to be equivalent to "p is true"? That has yet to be shown. I doubt that there is any such way. In short, there is no reason to think that Putnam's epistemic equivalence can succeed unless its crucial epistemic notion is cashed out circularly, in terms of the (realist) notion of truth.

The difficulty facing Putnam's proposal is a specimen of a difficulty facing any epistemic approach to truth. How are the epistemic notions appearing in any such account themselves to be understood? How are "justified," "verified," "rationally acceptable," or other such terms to be analyzed? Putnam and Dummett seem to take these terms as primitives. But I find this highly counterintuitive. Epistemic notions strike me as far more in need of explication than truth. Furthermore, given the plausibility of my sort of theory of justification—a truth-linked theory—the onus is on Putnam, Dummett, or like-minded theorists to show that these notions have non-truth-linked explications.

Let me not rest this contention on the details of my theory of justifiedness. I can argue more broadly for the necessity of having recourse to truth in explicating epistemic standards. First, as I have noted, many epistemic notions countenance degrees of strength, for example, degrees of confirmation, weight of evidence, and the like. What accounts for these degrees? What is the continuum on which these gradations occur? The only plausible answers, I think, would appeal to truth. My own theory of justifiedness appeals to true-belief-conducive psychological processes (and methods). But other theories might appeal to an evidential link between evidence and hypothesis. One has stronger evidence for p to the extent that the evidence makes p *more likely*. But "more likely" must mean "more likely to be true." Of course, we could define a technical notion of confirmation that has no commitment to truth, where the truth of the evidential propositions has no linkage—logical, probabilistic, subjunctive, or what have you—with the truth of the hypothesis. But in that case who would suppose that this technical notion has anything to do with the intuitive notion of evidential support?

Second, take the notion of epistemic "access" to a state of affairs, such as perceptual access or memory access. Again we have the notion of better and worse epistemic situations. And again this can only be spelled out in terms of truth. Perceptual access to a situation is better just in case it is easier, in that situation, to form *true* beliefs about the objects or events in the scene. The optimal distance and viewing conditions are those that promote the greatest chance of truth and accuracy.

Third, the history of science and philosophy is filled with controversies over methodology. Contending schools have vied over the right procedures for fixing belief. It is hard to make sense of these debates save as debates over which methodology is most conducive to *true* convictions. However difficult it may be to settle such questions, they seem to presuppose the notion of truth. So truth is conceptually prior to any operational specification of epistemic methods or standards. Antirealists, who seek to define truth in epistemic terms, are proceeding precisely backwards.

We must not ignore, however, an underlying impetus behind many moves toward an epistemic account of truth. Antirealists are doubtless worried about the epistemological consequences of a realist conception of truth. If truth is definitionally detached from evidence, either actual or possible, is there really any prospect that we can have epistemic access to the truth? Isn't there a threat of skepticism? Some antirealists might be prepared to concede that the ordinary, naïve conception of truth is nonepistemic. But they may feel that *replacement* of this naïve conception is required to avert epistemic catastrophe.

Given this problem, we must determine whether candidate epistemic accounts of truth really help, whether antirealist proposals really improve our epistemological opportunities. The answer is: not necessarily. Putnam's proposal, (*h*), is a case in point. Truth is there equated with what people would be justified in believing if they were in epistemically ideal conditions. But since none of us ever is in epistemically ideal conditions with respect to any proposition, how are we supposed to tell—from our actual standpoint—what a person would be justified in believing in that hypothetical situation? To assess the truth of p according to this proposed equivalence, we would have to assess the truth-value of the indicated subjunctive conditional. But that is no mean task! If anything, it imposes an even nastier epistemic predicament than the one feared from realism.

The nonepistemic nature of truth is well described by Brian Loar.[17] Our ordinary (realist) conception of truth can best be understood, he says, in *modal* terms. The idea of a proposition's being true is the idea of a state of affairs such that it could happen (or could have happened) that it be true even though we are not in a position to verify it. Furthermore, he asserts, it is part of our theory of nature that this is generally so. Whether p concerns objects in the garden, in Antarctica, or in the Andromeda galaxy, p's verifiability by us depends upon various contingent circumstances not entailed by laws of nature. For us to verify such propositions, light, sound, or some such energy must reach our sense organs. But the regions of space through which such energy must travel might have (or might have had) distorting properties. There might be black holes in the relevant regions that "soak up" the light. Also, for us to verify such propositions (indeed, *any* propositions), our brains must be constructed in certain ways. But they might not have been suitably constructed. So it might have happened that these propositions are true, although we are not in a position to verify them.

As this way of explaining truth indicates, a sharp distinction must be drawn between a proposition's being true and the proposition's being verified. The latter, but not the former, involves processes by which the truth is detected or apprehended. (Here I have in mind both extra-cognitive processes—for example, the traveling of light from object to retina—as well as intra-cognitive processes.) Indeed, only such a sharp distinction—characteristic of realism—can make good sense of certain of our verifying procedures.

In seeing how realism makes sense of verification, I follow the suggestions of William Wimsatt, who in turn credits Donald Campbell and R. Levins.[18] Our conception of reality is the conception of something robust, an object or property that is invariant under multiple modes of detection. The use of multiple procedures, methods, or assumptions to get at the same putative object, trait, or regularity is commonplace in ordinary cognition and in science. Wimsatt lists many examples, including the following: (*a*) we use different perceptual modalities to observe the same object, or the same sense to observe the object from different perspectives and under different observational circumstances; (*b*) we use different experimental procedures to (try to) verify the same alleged empirical relationships, that is, alternative procedures are used to "validate" one another; (*c*) different assumptions, models, or axiomatizations are used to derive the same result; (*d*) we use agreement of different tests, scales, or indices as a sign of validity of a trait-construct; and (*e*) we seek matches of theoretical description of a given system at different levels of organization. In short, our verification procedures, when they are careful, seek to "triangulate" on the objects or relationships under study.

We can best make sense of a need for triangulation on the assumption that the truths, or facts, about the object or system under study are sharply distinguished from the processes of verification or detection of them. Any particular method might yield putative information partly because of its own peculiarities—its biases or distortions. This is a potential danger as long as the properties of the studied object are distinct from the process of detection. Triangulation hopes to correct for such biases on the assumption that independent methods will not have the *same* distorting characteristics (if any at all). Obviously, we cannot use no method *at all* to make observations, to "get at" the system or relationship under investigation. So we run the risk of getting mere artifacts of a measurement or observation process, rather than the real character of the object. Using multiple methods of detection minimizes this risk. In general, we try to get away from the idiosyncrasies of the verification procedure in order (better) to get at the verification-independent properties of the target object. All this makes sense on the assumption that there are truths about the object independent of this or that verification procedure.

Admittedly, this realist stance is not the only possible diagnosis of the need for triangulation. Triangulation may be viewed as a variant of coherentist methodology; and coherentism is an epistemology fully available to the antirealist. However, the antirealist must take coherentism simply as a brute fact about our epistemic procedures. He cannot offer any principled rationale for triangulation of the sort available to the realist. The realist, by contrast, can explain the fallibility of our detection procedures, and hence the need for triangulation, in terms of the distinction between the real facts and the verification processes used to get at them.

As an aside, it is noteworthy that our perceptual systems seem "designed" to mark properties of the distal stimulus as opposed to those of the organic process of detection. So-called object constancies are a case in point. A moving observer does not see the stationary stimulus as moving or varying in size, though such displacements do occur at the retinal level. The perceptual system suppresses these displacements, apparently on the presumption that they are observer-induced changes. What the system seeks are properties of the observation-independent object. When input is diagnosed as resulting from observer characteristics—in the present case, observer movement—it simply is not communicated to higher cognitive levels.

Readers of Dummett may feel that I have thus far neglected a principal reason for a verificationist account of truth. Dummett approaches the topic of realism and antirealism from the context of a theory of meaning or understanding. In that context the question arises whether it is possible for us to understand, or grasp, any verification-transcendent conception of truth. Do we have a capacity to understand statements whose truth transcends their verifiability? If a proper theory of understanding, or meaning, precludes such a capacity, then no sense can be made of a realist, verification-transcendent, account of truth.[19] And none of the foregoing arguments can carry any weight.

I cannot enter in any depth into the theory of meaning or understanding. That clearly exceeds the scope of this [chapter]. Nor can I comment in detail on Dummett's proposals concerning a "use" theory of meaning, and the requirement that uses must be "manifestable" in behavior. I am inclined to regard these proposals with some skepticism, but I will not venture into this tricky terrain.[20] Instead, I put forth the following argument.

The claim that we cannot mean or understand verification-transcendent statements can only refer to verification *by us,* that is, by human beings. But as I have already argued, in the case of many if not most physical object statements, their truth certainly appears to be possible independently of human verification. For example, it might be true that such-and-such happened in the Andromeda galaxy although no human beings were (or are) in a position to verify it. (To hold otherwise would involve an untenable form of epistemic *speciesism.*) Moreover, this modal fact seems far more certain than any (interesting) doctrine in the theory of meaning. So if we are confronted with a choice between this modal fact and a meaning doctrine that excludes verification-transcendent meaning, we are better advised to doubt, or reject, the meaning doctrine.

4. Truth and Correspondence

I have now completed the discussion of whether truth is an epistemic notion. I have defended the realist position that it is wholly nonepistemic. In this section I turn to another facet of the debate over truth: whether truth consists in "correspondence." Antirealists are often anxious to pin a correspondence doctrine on their foes, because they find grave difficulties with that doctrine. But it is not clear that a realist must be committed to correspondence; that depends on what, exactly, is meant by "correspondence." Realism is in trouble on this score only if there is an untenable version of the correspondence doctrine to which realism must adhere.

One version of a correspondence doctrine, the Tractarian version, says that the world is a totality of *facts* and that a proposition is true just in case it corresponds with a fact.[21] A familiar objection to this is that the world does not contain fact-like entities, the kinds of entities that would exactly correspond to sentences or propositions. It is implausible to suppose that there are disjunctive facts in the world, existential facts, conditional facts, universal facts, and so on. Language or thought constructs disjunctive statements, existential statements, conditional statements, and the like. But it is misguided to suppose that structures of this sort constitute the world. If the correspondence doctrine is committed to this picture, it is an untenable doctrine.

The foregoing correspondence doctrine portrays the world as being prestructured into truth-like entities. This is one thing the antirealist opposes. A closely related view, also opposed by (some) antirealists, is that the world comes prefabricated in terms of categories or kinds. This is another facet of the correspondence theory to which Putnam objects.

For an internal realist like myself . . ."objects" do not exist independently of conceptual schemes. We cut up the world into objects when we introduce one or another scheme of description . . . [For an externalist the world contains] Self-Identifying Objects, . . . the *world,* and not thinkers, sorts things into kinds.[22]

The point here is essentially a Kantian point, and one also stressed by Nelson Goodman.[23] The creation of categories, kinds, or "versions" is an activity of the mind or of language. The world itself does not come precategorized, presorted, or presliced. Rather, it is the mind's "poetic" activity, or the establishment of linguistic convention, that produces categories and categorial systems. When truth is portrayed as correspondence, as thought or language *mirroring* the world, it is implied that the world comes precategorized. But that, says the antirealist, is a fiction.

Let us use "correspondence$_1$" for the constellation of views just canvassed, namely, that the world is prestructured into truth-like entities (facts), and that truth consists in language or thought mirroring a precategorized world. Although this is one conception of correspondence, a realist need not be wedded to it. Weaker variants of correspondence are still compatible with realism. Let me try to sketch—admittedly metaphorically—a less objectionable style of correspondence theory, to be called "correspondence$_2$."

The mirror metaphor is only one possible metaphor for correspondence. A different and preferable metaphor for correspondence is *fittingness:* the sense in which clothes fit a body. The chief advantage of this metaphor is its possession of an ingredient analogous to the categorizing and statement-creating activity of the cognizer-speaker. At the same time, it captures the basic realist intuition that what makes a proposition or statement true is the way the world is.

There are indefinitely many sorts of apparel that might be designed for the human body, just as there are indefinitely many categories, principles of classification, and propositional forms that might be used to describe the world. Although the body certainly has parts, it is not presorted into units that must each be covered by a distinct garment. It is up to human custom and sartorial inventiveness to decide not only what parts to cover, but what types of garments should cover which expanses of the body, and whether those garments should be snug or loose. For many bodily parts (or groups of contiguous parts), there is a wide assortment of garment-types used to clothe them. In outer footwear, for example, there are sandals, slippers, shoes, tennis shoes, snowshoes, fishing boots, and hiking boots. In inner footwear there are anklets, calf-length, and over-the-knee stockings. Among trousers there are shorts, knickers, and full-length pants. For the torso there are short-sleeve and long-sleeve shirts, jackets, vests, tunics, robes, saris, capes, and huipils.

Despite all this variety—humanly invented variety—there is still the question, for any specified type of apparel, whether a specific token of that type fits a particular customer's body. This question of fittingness is not just a question of the style of garment. It depends specifically on that customer's body. Similarly, although the forms of mental and linguistic representation are human products, not products of the world per se, whether any given sentence, thought sign, or proposition is true depends on something extra-human, namely, the actual world itself. This is the point on which realists properly insist.

In inventing or evolving sartorial styles, people devise standards for proper fittingness. These may vary from garment to garment and from fashion to fashion. Styles specify which portions of selected bodily parts should be covered or uncovered, and whether the clothing should hug snugly or hang loosely. This is all a matter of style, or convention, which determines the *conditions of fittingness* for a given type of garment. Once such fittingness-conditions are specified, however, there remains a question of whether a given garment token of that type satisfies these conditions with respect to a particular wearer's body. Whether it fits or not does not depend solely on the fittingness-conditions; it depends on the contours of the prospective wearer as well.

The case of truth is quite analogous. Which things a cognizer-speaker chooses to think or say about the world is not determined by the world itself. That is a matter of human poetic activity, lexical resources in the speaker's language, and the like. A sentence or thought sign, in order to have any truth-value, must have an associated set of *conditions of truth*. Exactly what determines truth-conditions for a sentence or thought sign is a complex and controversial matter. But let us assume that a given utterance or thought, supplemented perhaps with certain contextual factors, determines a set of truth-conditions. The question then arises whether these conditions are satisfied or not. The satisfaction or non-satisfaction of these conditions depends upon the world. Truth and falsity, then, consists in the world's "answering" or "not answering" to whatever truth-conditions are in question. This kind of answering is what I think of as "correspondence$_2$." Notice that *which* truth-conditions must be satisfied is not determined by the world. Conditions of truth are laid down not by the world, but only by thinkers or speakers. This is the sense in which the world is not precategorized, and in which truth does not consist in mirroring of a precategorized world. Unlike correspondence$_1$, correspondence$_2$ is compatible with what is correct in the constructivist themes of Kant, Goodman, and Putnam.

While these philosophers are right to emphasize constructivism, they carry it too far. There is a strong suggestion in Goodman's writing that "versions" are all there is; there is no world in itself. Although Goodman speaks of truth as being a matter of "fit," it is not fitness of a version to the world, but fitness of a version to other versions.[24] This, of course, a realist cannot accept. To return to our sartorial metaphor, the realist insists on the existence of a wearer: clothes don't make the world.

A chief motivation for the Goodman-Putnam view is epistemological. One strand of their position holds that we can never compare a version with "unconceptualized reality," so as to determine whether the world answers to a thought or statement. Comparison of a theory with perceptual experience is not comparison with unconceptualized reality because perceptual experience is itself the product of a sorting, structuring, or categorizing process of the brain. So all we can ever do, cognitively, is compare versions to versions.[25]

Conceding this point, however, does not undermine realism. Since correspondence$_2$ does not embrace the mirroring idea—nor the idea that true thoughts must resemble the world—the epistemology of getting or determining the truth need not involve comparison.

Can the same problem still be posed for my theory? It may be re-expressed by saying that, on my theory, we can never grasp or encounter the world, so as to determine whether some thought or sentence of ours fits it. And if the realist's world is unconceptualized, how can it be grasped or encountered in a manner to determine fittingness?

I see no insuperable difficulty here. Perception is a causal transaction from the world to the perceiver, so perception does involve an encounter with the world (at least in non-hallucinatory cases). To be sure, the event at the terminus of the transaction does not *resemble* the event at the starting point. The terminus of perception is a perceptual representation, which involves figure-ground organization and other sorts of grouping and structuring. The initiating event does not have these properties. Still, the transaction as a whole does constitute an encounter with something unconceptualized. We are not cut off from the world as long as this sort of encounter takes place.

But is this sort of encounter sufficient for knowledge or other forms of epistemic access? As far as I can see, realism about truth does not preclude such knowledge. Suppose that the (unconceptualized) world is such that the proposition "There is a tree before me" fits it, that is, is true. And suppose that the perceptual process is a reliable one, both locally and globally. Then, according to my account of knowledge, I may indeed *know* that there is a tree before me. The world that I learn about is an unconceptualized world. But *what* I learn about this world is that some conceptualization (of mine) fits it. *How* I learn this is by a process that begins with the unconceptualized world but terminates in a conceptualization.

Does this (realist) theory make the world into a noumenal object, an object that cannot be known or correctly described? Not at all. On the proposed version of realism, we can know of the world that particular representations fit it. So the world is not a noumenal object.

I cannot leave Putnam's critique of realism without commenting on a central component in his antirealist argument on which I have thus far been silent. In arguing against a correspondence theory of truth, Putnam has pointed out that there are *too many* correspondences. Correspondence has traditionally been construed as a word-world relationship. Putnam contends that there are too many word-world relationships of the requisite sort. Briefly, while there may be one satisfaction relation under which a given sentence turns out true, there will also be other satisfaction relations under which it turns out not to be true. So for any word-world relation purporting to be the "intended" truth relation, there are other, equally good candidates. Since no unique word-world relation can be identified with truth, the correspondence notion of truth is untenable.[26]

Actually, the multiplicity of relations Putnam discusses are often candidates for the reference relation rather than the truth relation. But since the principal topic of interest is truth, the significance of his attack on a determinate reference relation lies in its implications for truth. What we have, then, is an indeterminacy argument against a word-world relation of truth, which is taken by Putnam to be a general critique of ("metaphysical") realism.

I believe there are two sorts of replies to this critique. First, one can challenge the indeterminacy argument by denying that Putnam has taken all relevant parameters into account. This point has been argued persuasively by Alvin Plantinga.[27] Putnam seems to suppose that the terms of our language get their meaning or extensions, somehow, by virtue (solely) of the set-theoretical models of first-order formalizations of the body of our beliefs. In other words, roughly, our terms get their meaning and extensions by virtue of a vast network of implicit definitions. But this theory is extremely dubious. There must be further constraints at work in fixing our meanings and extensions. So the multiplicity that Putnam identifies of possible assignment functions to our terms does not have the implications that he alleges.

Second, the problem posed by Putnam, if it is a serious problem, is not really a problem about truth. Rather, it is a problem about interpretation, or the establishment of truth-conditions. These must be distinguished. Questions of truth cannot arise until there is a suitable bearer of truth-value, with an established set of truth-conditions, about which it can be queried *which* truth-value it has. Sentences or thought events construed as meaningless marks or nerve impulses are not bearers of truth-values. Only when a sentence or thought event is interpreted—when it has suitable semantic properties (including reference of singular terms and sense or reference of general terms)—is it even a candidate for being true or false.

Now Putnam has worries, as do other theorists in the field, about *how,* exactly, words and thought signs get their meaning and reference. How, in other words, do they get truth-conditions attached to them? These are important issues, which I will not in any way try to resolve. But unless and until sentences and thought signs are conceded to have interpretations, or truth-conditions, the question of truth cannot even arise. However, once we concede a determinate interpretation, or set of truth-conditions, for sentences and thought signs, we have assumed that Putnam's problems can be resolved. We can then turn to the question of truth, to the possession of this or that truth-value. I see no insurmountable obstacle here to the realist theory I have sketched. *Given* truth-conditions for a sentence, or thought, what makes it true or false is surely the way the world is, or whether it fits the world.

5. Scientific Realism

The doctrine of realism discussed thus far is a doctrine about truth. It is the doctrine commonly called "metaphysical realism." But another epistemological issue concerning realism is that of scientific theories and theorizing. Whereas scientific theories are ostensibly designed to describe the world, frequently the unobserved "fine structure" of the world, some philosophers of science contend that this is an improper or injudicious portrayal of scientific theories. These philosophers—variously known as instrumentalists, fictionalists, and empiricists—deny that theories purport to give literally true stories of what the world is like. Rather, they are just calculational instruments for predicting the course of experience, devices to predict (and perhaps explain) what is empirically observed. Denying the face-value meaning of theoretical statements, philosophers of this ilk try to convey the meaning of these statements in purely observational terms, or other constructions out of empirically acceptable materials. Against such reductionist maneuvers, scientific realists hold out for a literal interpretation of theoretical statements. Acceptance of such statements is seen as an ontological commitment to entities ostensibly designated by their theoretical terms. And such ontological commitment is viewed as epistemologically conscionable.

I construe scientific realism (henceforth "SR") as having a core consisting of three interconnected theses. First, there is a *semantic* thesis: SR says that theoretical statements in science are to be interpreted literally and referentially. Their meaning should not be purged of its (ostensible) extra-empirical content. Second, there is a thesis concerning the *aim* of scientists: SR says that what theorists are (generally) trying to do is to find and believe true descriptions of the world. Third, SR advances the epistemological claim that this aim is *legitimate,* since scientists sometimes are justified in believing scientific theories so construed. Or, in a slightly weaker variant, it is a legitimate aim because it is (humanly) possible for scientists to have justified belief in scientific theories realistically construed. (I will address both variants of the third thesis, but will not always announce switches from one to the other.)[28]

Although I have supported metaphysical realism as a theory of truth, the question of SR is quite independent. One could be a realist about truth while opposing SR. One could be, say, a reductionist about theoretical statements, but go on to say that the truth of any theory, given such a reduction, still depends on how the world is, nonepistemically. For example, the theory's truth would depend on whether the world really has a propensity to obey the specified observational regularities.[29] I am inclined to be a scientific realist as well as a metaphysical realist, but these doctrines are distinct. Moreover, I am *more* committed to metaphysical realism—especially its view of the nonepistemic nature of truth. My epistemology does not hinge so much on the SR issue. But to clarify my general stance, a brief excursus into SR is appropriate.

In the writings of (early) Putnam, Richard Boyd and W. H. Newton-Smith, SR has been given a new twist, rendering it a stronger doctrine than the core view I presented above.[30] SR is depicted as holding not merely that theoretical terms purport to refer, but that they (often) succeed in referring. Not only do scientists aim to describe the world with their theories, but they succeed in doing so. Or, more cautiously, their mature theories gradually get closer to the truth. In short, SR is viewed not merely as a doctrine about scientific aims, but as one about scientific success. (By "success" I mean not merely predictive success, but success in delineating the fine structure of the world.)

This brand of SR is sometimes called "convergent realism" or "historical progressivism." What I wish to emphasize about this doctrine is that it is an *addition* to the core of SR; one can therefore embrace SR's core without embracing convergent realism. This is the posture I mean to assume. More precisely, although I feel some attraction to convergent realism, at least in a modest, scaled-down version, I recognize that the historical record, viewed from our present vantage point, is not entirely clear.[31] Moreover, the thesis of convergent realism is at a different epistemological level than the core of SR. And in the present context I want to remain neutral about that level.

Am I really entitled to accept SR's core while remaining neutral about convergent realism? As I shall indicate shortly, there is reason to hesitate on this point. To assess this matter, we need to look back at the three core theses of SR. The semantic thesis of SR certainly does not depend on historical progressivism. The meaning of theoretical sentences may exceed their empirical content *however* scientists have been faring at limning the structure of reality. The second thesis of core SR, the aim component, also does not imply convergent realism. It may be true that scientists typically *intend* their theories to get at fundamental, trans-observational truths; but they may nonetheless be falling flat, or achieving only modest success, in this endeavor.

It is only when we move to the third thesis of core SR that convergent realism may be implicated. If scientific theorizing has indeed fallen flat, historically speaking, how can it be epistemically *legitimate* for scientists to aim at getting theoretical truths? How is it possible for them to be justified in believing scientific theories so construed? This seems to be a particular problem given my theory of justifiedness. After all, if beliefs are justified only when caused by reliable processes, and if scientific methodology has a poor historical track record, the processes in question must not be reliable. But then belief in scientific theories, at least construed realistically, is not and cannot be justified. So the justifiedness of accepting scientific theories appears to depend on the truth of convergent realism, or at least a good *average* record of true theoretical beliefs, even if there is no continual forward *progress*.

The problem is crystallized in what Putnam calls the "disastrous meta-induction" of scientific theorizing.[32] Suppose scientists decide that no term in the science of more than fifty years ago referred, and no theory of that vintage was true. If this keeps happening, shouldn't we be led to the meta-induction that no theoretical term now or in the future will refer, and no present or future theory will be true? From this, conjoined with my theory of justification, won't it follow that we are not, and could not be, justified in believing any theoretical statement?

No, there are several lacunae in this line of reasoning. Let us look first at the "disastrous meta-induction" itself. Upon inspection, the meta-induction is really self-undermining. We judge past theories to be false only by the lights of our present theory. If we abandon our present theory, we are no longer in a position to judge past theories false. So if we use the meta-induction to conclude that no present scientific theory is true, we thereby eliminate all grounds for believing that past theories were false. But then we are no longer entitled to believe the premises of the meta-induction.

But, it might be replied, couldn't those premises still be *true?* And if so, wouldn't this reinstate the difficulty for any prospect of justified theoretical belief? Let us see, then, whether the meta-induction might be vulnerable on other grounds. I think it is. In deciding whether we can inductively infer the falsity of present and future theory from the falsity of believed theories of the past, we have to assess the bases of those past theoretical beliefs. Scientists in previous eras may have used different—and poorer—methodologies than we now have available. After all, there have been dramatic and continued developments in statistics and other evidential procedures. Furthermore, we now have vastly more powerful instruments of discovery and detection in many domains. Perhaps these differences in instrumental and evidential tools can outweigh past failures. Thus, our present scientific situation may not be relevantly similar to past science. Suppose you are going on a long automobile trip, and on all your previous long trips you had a flat tire. Should you expect a flat on this trip too? Not necessarily. If you began all previous trips with bald tires, but this time you start with a freshly purchased set of tires, the induction would be weak. (This response may not be conclusive, but it should give us pause.)

Setting aside the force of the meta-induction, we need to look more carefully at the implications of my theory of justification for the present question.[33] There are several reasons why a poor track record of positive belief in scientific theory need not ruin scientists' chances for justified theoretical belief according to my account. By "positive" theoretical belief let us mean a belief that a theory is true; by "negative" theoretical belief let us mean a belief that a theory is false, that is, a belief in the denial of a theory. Since the same methods are used to arrive at negative theoretical beliefs as positive theoretical beliefs, even if scientific methodology (assuming, inaccurately, that a single methodology has continuously been in use) has consistently generated false positive theoretical beliefs, it may still have generated lots of true negative theoretical beliefs. Moreover, the latter class is probably much larger than the former. Think of all the scientific theories we now reject and of the innumerable theories that past scientists rejected. When the entire class of outputs of scientific methodology is included, both negative and positive, the overall truth ratio may be rather impressive.

The next point concerns the distinction between primary justifiedness (P-justifiedness) and secondary justifiedness (S-justifiedness). P-justifiedness turns on the truth ratio of basic psychological processes; and S-justifiedness turns on properties of methods.[34] Now the basic psychological processes used in assessing scientific theories probably are no different from those used for many other tasks; only the methods, at most, are distinctive. So even if these processes have a poor track record in this very special domain, it does not follow that they have poor reliability in *general*. . . .

These points indicate that, within my *full* theory of justification,[35] a poor track record of scientific beliefs—especially positive theoretical beliefs—need not preclude the possibility of justified belief in scientific theories. That suffices to show that core SR is independent of convergent realism. However, a few further comments are in order to guarantee a fair-minded assessment of just how good or bad the track record of theoretical beliefs may be.

One point is that philosophers and historians of science have a tendency to concentrate on the most encompassing theories in science, those at the highest level of generality, which also typically postulate entities or forces that are least amenable to observation. Philosophers and historians of science are preoccupied with quantum theory, or problems of evolutionary theory. They spend little time citing more prosaic theories, such as the theory that the liver's function is to detoxify the body. It may well be true that once-accepted theories at the highest level of generality are frequently found in hindsight to be mistaken. But it is unfair to restrict consideration to these kinds of theories. Lower-level theories also occupy scientists, and these typically have much greater staying power.

We should also notice that any large-scale theory normally has numerous components, and scientists who believe the whole theory also believe its several components. Even when the theory as a whole is subsequently overthrown, the several components are not necessarily all overthrown. The general theory of the electron has certainly changed over time. But the view of the electron as a basic unit of electrical charge, not resolvable into smaller units, has been retained. That component has stayed constant. If, in this fashion, we take account of beliefs in all *components* of theories, not just entire theories, the specter of a "disastrous meta-induction" is much less forbidding. It does not look, from our present vantage point, as if science has been doing so poorly at all.[36]

One final point. I have been trying to show how, given my account of justification, it could well turn out that positive belief in scientific theories is justified. But it is not critical to my epistemology that this be so. Maybe a large-scale scientific theory is such a risky intellectual commitment, and the alternative possibilities are so numerous, that scientists are not entitled to believe such theories. This admission would not lead to general skepticism, nor to any general problem for my account of justification. Even if global skepticism were fatal to a theory of justification (which is not obvious), it is not so objectionable for a theory to imply local skepticism, that beliefs in one special arena are unjustified. There is nothing in my account of justification, then, that forces me to endorse SR, especially its third thesis. Scientific realism plays a less essential role in my view than metaphysical realism. Still, for the reasons indicated, I am inclined to accept it, and to believe in its compatibility with other tenets of my epistemology.

I draw this discussion to a close, having shown why realism about truth is a plausible and defensible doctrine. Truth is not an epistemic notion; and in a properly softened form, even a correspondence theory of truth is tenable. We have seen how core scientific realism, while distinct from realism about truth, is an attractive view, and ultimately compatible with my general epistemology.

Notes

1. This chapter is adapted and reprinted with permission from Alvin I. Goldman, *Epistemology and Cognition* (Cambridge, MA: Harvard University Press, 1986), introduction and chapter 7 (see the "Sources and Permissions" section at the end of this volume).
2. W. V. Quine, "Epistemology Naturalized," in *Ontological Relativity and Other Essays* (New York: Columbia University Press, 1969), 82.
3. Donald Campbell, "Evolutionary Epistemology," in Paul Schilpp, ed., *The Philosophy of Karl Popper,* vol. 1 (La Salle, IL: Open Court, 1974), 413.
4. Otto Neurath, "Protokollsätze," *Erkenntnis* 3 (1932): 206.
5. Putnam's principle writings in this area are: *Meaning and the Moral Sciences* (London: Routledge and Kegan Paul, 1978); *Reason, Truth and History* (Cambridge: Cambridge University Press, 1981); and several papers in *Realism and Reason* (Cambridge: Cambridge University Press, 1983). Dummett's major works in this area are: several papers in *Truth and Other Enigmas* (Cambridge, MA: Harvard University Press, 1978); "What Is a Theory of Meaning? (II)" in Gareth Evans and John McDowell, eds., *Truth and Meaning* (Oxford: Oxford University Press, 1976); *Elements of Intuitionism* (New York: Oxford University Press, 1977); and "Realism," *Synthese* 52 (1982): 55–112.
6. Dummett, "Realism," 55.
7. Ibid., 55 and elsewhere.
8. Dummett comments on the case of empty singular terms but says this *is* a kind of antirealism, the contrasting opposite of which is a realism of a Meinongian kind. I find this implausible.
9. See Hilary Putnam, "On Truth," in Leigh Cauman et al., *How Many Questions?* (Indianapolis: Hackett, 1983). Also cf. Scott Soames, "What Is a Theory of Truth?" *Journal of Philosophy* (1984): 411–29.
10. *Reason, Truth and History,* chapter 3.
11. "Realism and Reason," in *Meaning and the Moral Sciences.*
12. *Meaning and the Moral Sciences,* 125.
13. See Dummett, "Realism," 69.
14. C. S. Peirce, "How to Make Our Ideas Clear," in Justus Buchler, ed., *Philosophical Writings of Peirce* (New York: Dover, 1955), 38. Note that this is, strictly, a doxastic definition, not an epistemic one. It only invokes agreement in opinion, whether or not this is justified or rational.
15. *Reason, Truth and History,* 55.
16. *Meaning and the Moral Sciences,* 125.
17. Brian Loar, "Truth Beyond All Verification," in Barry Taylor, ed., *Essays on Michael Dummett* (The Hague: Martinus Nijhoff).
18. See William Wimsatt, "Robustness, Reliability and Multiple-Determination in Science," in M. Brewer and B. Collins, eds., *Knowing and Validating in the Social Sciences* (San Francisco: Jossey-Bass, 1981). Among several works by Campbell, he cites D. T. Campbell and D. Fiske, "Convergent and Discriminant Validation by the Multi-Trait Multi-Method Matrix," *Psychological Bulletin* 56 (1959): 81–105, and T. D. Cook and D. T. Campbell, *Quasi-Experimentation* (Chicago: Rand-McNally, 1979). The relevant works by R. Levins are "The Strategy of Model Building in Population Biology," *American Scientist* 54 (1966): 421–31, and *Evolution in Changing Environments* (Princeton, NJ: Princeton University Press, 1968).
19. See especially "What Is a Theory of Meaning? (II)," 116.
20. Some critical discussions of Dummett's proposals include Loar, "Truth Beyond All Verification"; Colin McGinn, "Realist Semantics and Content-Ascription," *Synthese* 52 (1982: 113–34; Michael Devitt, "Dummett's Anti-Realism," *Journal of Philosophy* 80 (1983): 73–99; and Gregory Currie and Peter Egenberger, "Knowing and Meaning, *Nous* 17 (1983): 267–79.

21. Ludwig Wittgenstein, *Tractatus Logico-Philosophicus,* trans. C. K. Ogden and F. P. Ramsey (London: Routledge and Kegan Paul, 1922).

22. *Reason, Truth and History,* 52–53. Note, however, that in earlier work Putnam had endorsed the notion of kinds being *in the world.* See "On Properties," in Nicholas Rescher et al., eds., *Essays in Honor of Carl G. Hempel* (Dordrecht: D. Reidel, 1970).

23. Nelson Goodman, *Ways of Worldmaking* (Indianapolis: Hackett, 1978).

24. Ibid., 138.

25. Cf. Hilary Putnam, "Reflections on Goodman's *Ways of Worldmaking,*" *Journal of Philosophy* 76 (1979): 611.

26. See Putnam, "Realism and Reason," 123–26, in *Meaning and the Moral Sciences,* and *Reason, Truth and History,* chapter 2.

27. Alvin Plantinga, "How To Be an Anti-Realist," in *Proceedings and Addresses of the American Philosophical Association* 56 (1982): 47–70.

28. A similar characterization of scientific realism, minus the third thesis, is given by Bas van Fraassen in *The Scientific Image* (Oxford: Oxford University Press, 1980): "Science aims to give us, in its theories, a literally true story of what the world is like; and acceptance of a scientific theory involves the belief that it is true" (8).

29. More generally, Michael Dummett has observed that various kinds of reductionisms can be compatible with realisms about those domains (see "Realism," 75).

30. Cf. Putnam, Lecture II, in *Meaning and the Moral Sciences,* where he draws on unpublished work by Boyd; Richard Boyd, "Scientific Realism and Naturalistic Epistemology," in P. D. Asquith and R. N. Giere, eds., *PSA 1980,* vol. 2 (East Lansing, MI: Philosophy of Science Association, 1981); and W. H. Newton-Smith, *The Rationality of Science* (London: Routledge and Kegan Paul, 1981).

31. Larry Laudan raises doubts about the historical record in "A Confutation of Convergent Realism," *Philosophy of Science* 48 (1981): 19–49.

32. *Meaning and the Moral Sciences,* 24–25, 37.

33. Parenthetically, we should be reminded that past applications of a method do not exhaust the class of applications relevant to reliability. These would include present and future cases as well as counterfactual cases. But there is no reason to suppose that historical uses of scientific methods would be unrepresentative of long-run truth properties, as long as the very same methods are in question.

34. [See Alvin I. Goldman, *Epistemology and Cognition* (Cambridge, MA: Harvard University Press, 1986), 93–95.]

35. [See Goldman (1986), especially chapters 4 and 5.]

36. This point has a connection with what has been called "verisimilitude," or "truth-likeness." However, I do not mean to commit myself to any particular definition of verisimilitude, or to the prospects for a workable definition along the lines initiated by Popper. This is a very complicated issue. For some proposed definitions and critiques of verisimilitude, see Karl Popper, *Objective Knowledge* (Oxford: Oxford University Press, 1972); Ikka Niiniluoto, "Truthlikeness in First-Order Languages," in Jaakko Hintikka et al, eds., *Essays on Mathematical and Philosophical Logic* (Dordrecht: D. Reidel, 1979); and "Scientific Progress," *Synthese* 45 (1980): 427–62; W. H. Newton-Smith, *The Rationality of Science;* Pavel Tichy, "On Popper's Definition of Verisimilitude," *British Journal for the Philosophy of Science* 25 (1974): 155–60, "Verisimilitude Revisited," *Synthese* 38 (1978): 175–96; and Laudan, "A Confutation of Convergent Realism."

11

MUST NATURALISTS BE REALISTS?

MICHAEL WILLIAMS

Many contemporary philosophers call themselves naturalists. Of these, I would guess that most also think of themselves as realists, making realistic naturalism (or naturalistic realism) the dominant form of naturalism in philosophy today. The question I want to raise is whether realistic naturalism represents the only or the best way of being a naturalist? To end the suspense, my answer is "No." I think that pragmatists can be naturalists too, at least in all the ways that matter (or ought to matter). So I shall be putting in a plea for pragmatic naturalism.

1. REALISTIC NATURALISM

I now turn to naturalism in contemporary philosophy, beginning with realistic naturalism. Realistic naturalism is a complex position involving commitments in metaphilosophy, epistemology, the philosophy of language and mind, metaphysics, and the theory of truth. (Needless to say, the position I shall be describing is something of an ideal type. It is not meant to be an accurate account of the view of any particular philosopher.)

Metaphilosophy. Philosophy has often been conceived as a discipline dealing with questions about the principles that underlie all particular forms of inquiry. This conception of philosophy receives a particularly influential articulation in the writings of Kant. According to Kant, certain principles, because they function as the necessary presuppositions of all empirical investigation, can themselves only be validated *a priori*. Philosophy, as the discipline devoted to validating such fundamental principles, is therefore an essentially *a priori* undertaking, distinct from natural science. Essentially neo-Kantian conceptions of philosophy are still with us: for example, the view that the business of philosophy is conceptual analysis. Contemporary naturalism represents a rejection of the Kantian outlook. Philosophers who think of themselves as naturalists, particularly those influenced by Quine, reject appeals to the *a priori*, typically expressing skepticism about the very distinction between *a priori* and *a posteriori* knowledge. However, with this distinction in jeopardy, it is not clear what philosophy is supposed to be. Anti-apriorism is therefore taken to blur and perhaps abolish the distinction between philosophy and natural science (or to abolish philosophy altogether, letting it be replaced by science).

Epistemology. Epistemology has been much concerned with answering skepticism. Some philosophers who call themselves naturalists take the Humean view that skepticism is not so much false as pointless. The skeptic argues that we cannot justify our belief in an external world or our commitment to certain inductive practices. But, say Humean naturalists, it is pointless to argue either for or against such fundamental commitments, which are built into human nature and so are beyond questions of justification.

This Humean naturalism is not the dominant naturalistic outlook. More popular among naturalists is a positive epistemological doctrine: reliabilism. On this view, epistemology has gone wrong (and encouraged skepticism) by an excessive interest in justification. For reliabilists, knowledge has no essential connection with justification. At the most general level, knowledge is true belief reached by some kind of reliable cognitive process. Justification—inference from evidence— is at best one instance of such a process. Much knowledge—ordinary perceptual knowledge, for example—results from the unselfconscious exercise of certain cognitive capacities. Reliabilism is thus "externalist" in that the factors in virtue of which a belief of mine counts as knowledge need not be things of which I am aware. What cognitive capacities human beings have, how they work, and how reliable they are, are matters for empirical investigation, cognitive science. This aspect of externalist reliabilism makes it the epistemology of choice for metaphilosophical naturalists.

Philosophy of language and mind. Thought and language are infused with meaning or "intentionality." They represent or are about things in the world. To understand meaning in a broadly naturalistic way, realistic naturalists generally invoke another reliabilist idea: signaling. On this view, there is no deep distinction between the way in which clouds mean rain and the way in which "dog" refers to dogs. Some theorists give this approach an evolutionary twist: our discriminative responses are *about* whatever they evolved to respond to (even if they can be triggered by other stimuli). If this view can be worked out, there is hope of understanding intentionality in broadly causal terms.[1]

Metaphysics. Ontologically, realistic naturalists tend to be physicalists, holding that the only things that ultimately exist are physical things. One reason why they favor causal or causal-evolutionary theories of reference is that they find such theories physicalistically acceptable. Approaches to reference and meaning that are not physicalistically acceptable are sometimes dismissed as "magical."

The Theory of Truth. Here we come to realism. Realistic naturalists are scientific realists, holding that scientific theories are not just instruments for predicting observations but aim to determine how things really are. The goal of science is not just empirical adequacy but truth. Since this distinction would be in danger of collapse if truth were defined in epistemic terms—for example, as some kind of ideal justification—realistic naturalists insist on a "correspondence" conception of truth, though virtually all admit that this conception stands in need of clarification. One appealing suggestion is to follow Tarski in defining truth in terms of reference, adding that reference can itself be understood causally.

These, then are the commitments of realistic naturalism: global anti-apriorism, a reliabilist conception of knowledge, a broadly causal conception of meaning, physicalist ontology and a realistic or correspondence conception of truth.

2. Pragmatism

Like realistic naturalists, pragmatists are anti-apriorist. As global fallibilists, pragmatists hold that nothing we believe is in principle immune from revision. But on other issues, pragmatism and realism are starkly opposed. Let us work backwards, beginning with the theory of truth.

The original pragmatists were tempted to explain truth in epistemic terms. James said that the true is whatever is good in the way of belief or whatever "works." Peirce identified a statement's being true with its being ideally assertible: assertible at the end of inquiry. Obviously this puts them at odds with realists, for whom truth is not an epistemic notion of any kind.

Realists are right to complain about epistemic definitions of truth. "True" certainly does not mean "verified" in the sense of "rationally assertible on the basis of the evidence currently at our disposal." Something can be assertible on current evidence but not true: that is just fallibilism. As for "assertible at the end of inquiry" or "ideally assertible," it is far from clear what such phrases are supposed to mean. We have no clear conception of what it would be for inquiry to have an end. And if "ideally assertible" means anything, it means "assertible in conditions in which all relevant sources of error are either absent or have been allowed for," which is a long-winded way of saying "assertible in conditions where there are no barriers to determining what is true." Mindful of these objections, contemporary pragmatists opt for a *deflationary* approach to truth.

From the deflationary perspective, the predicate "true," while of considerable *expressive* significance, is not the name of a property requiring deep analysis. An early deflationary account of truth is Ramsey's redundancy theory, according to which "It is true that Caesar was murdered" *means no more than* that Caesar was murdered. Any difference is entirely "stylistic": for example, we may use "It is true that . . ." to speak more emphatically. Because it stresses the use of "true" in performing such special speech-acts, this approach is sometimes also called the "performative" theory. However, more recent views, such as Quine's disquotational theory, do not claim that "p" and "It is true that p" are synonymous. Rather, what matters about "true" is given by certain logical equivalences. Thus:

> "Snow is White" is true if and only if snow is white; "France is octagonal" is true if and only if France is octagonal . . . and so on.

Appending "is true" to a quoted sentence is just like canceling the quotation marks ("disquotation").

The deflationary approach does not wholly trivialize truth-talk. To the contrary, Quine's view shows why truth-talk is important. By offering a systematic way of replacing talk about the world with logically equivalent talk about words ("semantic ascent"), the truth-predicate gives us new things to generalize over—i.e., linguistic objects, sentences—thereby enabling us to express agreement and disagreement with sentences that we cannot specify: for example because we do not know exactly what they are ("What the President said is true"), because there are too many of them ("Every sentence of the form 'P or not P' is true"), and perhaps even because we do not completely understand them ("I didn't follow everything the speaker said, but I'm sure it was all true"). This expressive gain is of considerable significance. Truth-talk allows us to make explicit, in general form, the epistemic, inferential, and methodological principles that are implicit in our practices of inquiry and argument. It is thus an essential component in what Robert Brandom calls "expressive rationality."

Because they do not define truth epistemically, deflationary theories can be thought of as minimally or negatively realistic. Deflationists about truth do not exactly deny that truth is "correspondence to reality." Rather, they say that everything there is to "correspondence" intuitions is captured by their favored version of the equivalence schema. In their view, attempts to read more than this into talk of correspondence have not been successful and have no evident point. Accordingly, pragmatists and realists part company over the question of whether the concept of truth has any useful *explanatory* work to do.

Since pragmatists find no explanatory uses for truth—truth not being the right sort of concept—they cannot take truth to be a notion of deep epistemological significance. For pragmatists, epistemology—the theory of inquiry—enjoys a certain priority over the theory of truth. Thus, pragmatists tend to raise their eyebrows when told, for example, that Truth is the Goal of Inquiry. However, we should be careful here. Pragmatists do not deny that, other things being equal, we prefer true beliefs to false. Their point is twofold: in formulating such preferences, we need only deflationary truth; and in illuminating or rationalizing them, we get no help from metaphysically inflated notions such as the correspondence theory. For pragmatists, our commitment to truth as the goal of inquiry does not rationalize or explain fallibilism. Rather, it *is* fallibilism: the willingness to revise one's beliefs and theories in the face of difficulties. To improve our views is to resolve current problems, not to get closer to some *focus imaginarius,* the Truth (with a capital "T"). Why should we want to improve them? For many reasons, not just one big reason.

I used to think that James could be understood along these lines. My thought was that James did not mean to *define* truth in epistemic terms, his point being rather that, in concrete terms, the pursuit of truth just is the attempt to keep improving our theories and methods. I now think that I was being too generous. James really did want to define truth as "what works," because he wanted to suggest an expansive account of working that would be friendly to religious belief. For James, the idea of a soulless universe didn't "work" because it was emotionally unsatisfying: his definition of truth was in the service of a defense of wishful thinking. A version of pragmatism that incorporates a deflationary view of truth does not encourage this sort of thing (though it does suggest that the real issue with James is over the all-round advisability of adopting wishful thinking as a principle of belief).

Turning to ontological issues, pragmatists distinguish between admiration for the natural sciences and reductive physicalism. In particular, they see no reason to suppose that we need a physicalistically acceptable conception of meaning. I shall say more about this in a moment.

For realistic naturalists, meaning, reference, and knowledge are all to be explained using the same basic idea: reliable discriminative responsiveness. Accordingly, they see epistemology and the philosophy of language and mind as closely related. Since, in their own way, pragmatists do too, we can treat epistemology and the philosophy of language and mind together.

From a pragmatic standpoint, mere discriminative responsiveness explains neither knowledge nor meaning. Justification is essentially connected, not simply with knowledge, but with belief itself. Wilfrid Sellars deserves much of the credit for showing why this is so; Sellars links epistemology with a theory of meaning by making two moves. First, he suggests that intentionality is a fundamentally linguistic affair, in the sense that speakers of a language are the paradigmatic concept-users. Second, he asks: What distinguishes conceptual from non-conceptual activity? In particular, how does the human being who says "That's green" differ from the parrot trained to utter the same vocables in response to the presentation of a green card? Sellars's answer is that, unlike the parrot, the human reporter has the concept "green" and so understands what he is saying. This understanding consists in his grasp of inferential connections between his observation reports and other statements. Unlike the parrot, the human speaker has a grasp (perhaps implicit and practical) of what *follows from* his reports, what is *evidence for* them, how they might be challenged, how various challenges might be met, and so on.

It follows from this inferentialist approach to meaning that beliefs, as conceptually contentful, are essentially the sorts of things that can function as reasons and for which reasons can be given. But reasons can be good or bad, weak or strong. Language use is thus an activity that is essentially subject to normative-epistemological constraint. It follows that neither knowledge nor meaning can be understood in purely causal, thus wholly non-normative terms. But as I shall now argue, this commitment should not be seen as a departure from naturalism in any sense that ought to matter.

3. The Meanings of "Natural"

"Natural" has more than one antonym. "Naturalism" and "anti-naturalism" take on different meanings, depending on which contrast we have in mind.

One contrast is between the natural and the unnatural. This distinction is tied to the Aristotelian conception of nature as an inner principle of change. Individuals with the same nature, in this sense of "nature," constitute a natural kind. Change, on this view, is the actualization of an individual's potentialities, so a thing's nature determines what it can be. An acorn has the potential to develop into an oak, but not a beech and certainly not a cat.

This conception of nature is idealizing, normative, and teleological. A thing's nature determines, not what it will be, but what it ought to be: what it will be if things go as they should. But things can go wrong: if the sun does not shine, or if the rain does not fall, the acorn will not grow. Alternatively, some inner disturbance can send an individual off the rails. Such an individual will then engage in unnatural acts: acts that are against nature in the sense of contrary to that individual's proper goal or function (the goal or function that is fulfilled in the actualization of that individual's nature).

A quite different contrast is between the natural and the supernatural. Naturalists in this sense think that the natural world can be understood without appeal to factors beyond or outside that world. Thus, the naturalistic attitude excludes invocation of the Divine, indeed the magical generally.

Modern science is built on the rejection of Aristotelian naturalism. By treating the physical world exclusively as a realm of law, modern science extrudes purpose from physical nature both locally and globally. Locally, the stone falls to the Earth because of gravitational attraction, not because it is heading towards its natural place at the center of the Universe. Globally, though the Universe as we know it might have a finite lifespan, this would be an "end" only in the sense of a terminating state, brought about by laws and initial conditions, not in the sense of a goal, the reaching of which fulfills a purpose. This extrusion of purpose is what Weber famously referred to as the disenchantment of nature. It was begun by the likes of Galileo and Newton and completed by Darwin and his heirs. It has turned out to be a good idea, proving itself in the vast superiority of modern science to premodern speculation.

The disenchantment of nature, while perhaps not entailing anti-supernaturalism, certainly encourages it by making appeals to the divine seem otiose. To conceive nature as a realm of law is to conceive it as a self-sustaining causally closed system that has no need of divine supervision and no obvious room for divine intervention. This makes Providence problematic. The modern conception contrasts here with the Aristotelian conception that, by allowing for happenings that are contrary to nature, permits hands-on divine involvement.

Contemporary naturalists, realistic and pragmatic, are at one in accepting the disenchantment of nature. They concur in rejecting both supernaturalism and a teleological conception of the physical world, both of which they see as rendered incredible by modem science. Since they concur also in their metaphilosophical anti-apriorism, we need to look further to see why they are not naturalists in quite the same way.

This brings us to our final contrast, between the natural and the non-natural. This distinction is between what the Greeks called *physis* and *nomos,* nature and convention. Stones fall to earth always and everywhere, whether we like it or not. Such facts belong to nature. But acts that are acceptable in New York may be illegal in the State of Kentucky. Such facts seem to reflect locally variable standards of acceptable behavior and are, in that sense, a matter of convention (which need not mean that conventions are always arbitrary or nonrational). If we think (as pragmatists do) that norms or standards are instituted by human attitudes, then the distinction between *physis* and *nomos* turns into the distinction between the natural and the normative. In Brandom's terms, where natural laws concern the fundamental properties of matter, laws as norms concern proprieties of behavior.

With respect to this contrast, realistic and pragmatic naturalists diverge. Whereas, for naturalistic realists, knowledge and meaning are to be understood in terms of causal relations, pragmatists regard this reductive approach as hopeless. For reasons already given, they think that knowledge and meaning can be understood only in relation to normatively constrained linguistic practices. Conceptual thought is essentially embedded in what Sellars calls "the game of giving and asking for reasons," the game in which we establish epistemic entitlement and fulfill epistemic obligations. Knowledge and meaning are non-natural because normative ("fraught with 'ought,'" to use another Sellarsian phrase).

In this sense, pragmatism is anti-naturalist. But to think of knowledge and meaning as non-natural is not to think of them as *super*natural. Pragmatism about norms says that norms are instituted by human practices; standards originate in what human beings count as correct or incorrect, acceptable or unacceptable; values exist only because there are practices of evaluation. There is nothing supernatural, or even particularly mysterious, about this.

We saw earlier that realists favor causal accounts of reference because such accounts promise to be physicalistically acceptable. Theories that are not physicalistically acceptable are said to be magical. Pragmatists see this charge as conflating the distinction between the non-natural and the supernatural. It is true that, if one is already committed to a naturalistic—in the sense of "non-normative"—approach to meaning, it is reasonable to demand that one's theory be physicalistically acceptable. But otherwise, it is not reasonable at all. Meaning is non-natural, but pragmatism about norms eliminates all suspicion of magic.

Some realists will object that pragmatists cannot avail themselves of the distinction between nature and convention. By adopting a pragmatist attitude towards all norms, epistemic norms included, while discarding truth as a non-conventional guiding norm, pragmatists make all "factual" inquiry a matter of convention. Thus, pragmatism encourages (or is a version of) postmodernism, social constructivism, or some other contemporary form of antirationalism. Pragmatists reject such charges. For them, scientific knowledge is objective because constrained both theoretically and observationally. Theoretically, by our current body of entrenched theory and by various canons of theory-choice; observationally, by the demand that theories fit data. Talk of "correspondence to reality" adds nothing significant.

With regard to empirical constraint, contemporary pragmatists see externalist reliabilism as offering a valuable insight. While discriminative responsiveness is not *sufficient* for observational knowledge, because not sufficient for a report's involving an exercise of conceptual capacities, reliable discriminative responsiveness is a necessary component in such knowledge. For pragmatists, we can have noninferential observational knowledge of whatever aspects of our surrounding we can be trained to respond reliably to. Thus empirical knowledge, indeed empirical content, *involves* causal relatedness to the world without being *reducible* to it.

In playing the game of giving and asking for reasons, then, we are responsible for the rules. We make them and can in principle revise them (as we can in principle revise anything). Logical and semantic vocabulary, such as "true," aids in the process of revision precisely because it allows us to formulate explicitly in claims rules that are implicit in practice, thus making them available for critical examination. But although we make the rules, observational constraint ensures that we don't fully control the results of playing by them. The world takes a hand.

Pragmatists do not suppose that "anything goes." What they do deny is that objectivity is illuminated by the sort of truth-talk that goes beyond all accounts of methodological, theoretical and observational constraint. They need not avoid talk of "getting at the facts" as a way of reporting the results of serious and sustained inquiry. What they deny is that such talk explains what makes inquiry serious. For them, fact-talk is just a stylistic variant of truth-talk.

What does or should matter about the naturalistic spirit? Two things: anti-apriorism and the avoidance of the supernatural. On this score, pragmatism has as much claim to be naturalistic as contemporary versions of realism. Pragmatists and realists part company over the prospects of giving reductive accounts of knowledge and meaning. But since pragmatists see norms as instituted by attitudes, their view of knowledge and meaning as essentially subject to normative constraint—and so in *that* sense not "naturalistically" intelligible—does not make a mystery of either.

4. Truth Again

I take my argument so far to show that there is nothing obviously unacceptable about pragmatic naturalism. In conclusion, I want to suggest that pragmatic naturalism is more thoroughly naturalistic than its realist rival.

From the very beginning, the religiously minded were disturbed by modern science's anti-supernaturalism. Since (especially after Newton) no one wanted to say that modern physics was false, the only option was to look for features of the entire system of natural law that demanded explanation by appeal to something outside that system.

Two strategies suggest themselves, one epistemological, the other metaphysical. The epistemological argument originates with Descartes: God guarantees the truthfulness of our clear and distinct ideas (although, so far as the physics of motion are concerned, he didn't guarantee the truth of Descartes' ideas, which are almost entirely false). The metaphysical strategy is the Argument from Design: the very law-governed character of the physical world bespeaks an origin in divine intelligence. Divine intelligence is the best explanation of the order we find in the world.[2] It is possible to combine these strategies. The argument goes like this: We have a preference for simple theories; but we also want true theories. We are thus committed to simplicity's being indicative of truth. However, there is no naturalistic reason why it should be. The best explanation for simplicity's being truth-indicative is supernatural: God arranges the Universe so as to be intelligible to creatures like us.[3]

Pragmatists are unmoved by this argument. Certainly, we take simplicity as a reason for accepting—we might as well say believing—a theory; and from a deflationary perspective there is no difference between believing a theory and believing that it is true. However, this is not to say that we take simplicity itself to be an indicator of truth, for we evidently do not. Rutherford's conception of the atom is much simpler than the modern conception. But no one thinks that is more likely to be true. Rather, we prefer simpler theories to more complicated ones because of what we want from an explanation. Good explanations are easy to work with, bring unity to a wide range of phenomena, avoid *ad hoc* complications, and so on. Other things being equal, simpler explanations are rationally preferable because they are better as explanations. But this rational preferability entails no unqualified commitment to simplicity's being an indicator of truth.

I suspect, though I am not sure I can prove, that realist naturalists are less well placed than pragmatists to resist arguments like the one from the truth-conduciveness of simplicity. Their realism pushes them towards some kind of teleological conception of inquiry: inquiry as a process with Truth, understood in some non-deflationary way, as its goal. In consequence, for them there really is a problem about saying how our epistemic criteria—if not simplicity, then other criteria—are helping us get closer to this goal.

I suggested earlier that pragmatists are suspicious of truth as the goal of inquiry. As I said, they agree that truth is *one* of our goals, in the sense that we generally prefer true beliefs to false. (Pragmatists would, however, add that we have no interest in truth *as such*. What we want are interesting truths: truths that respond to questions in which we have some interest, theoretical or practical.) But this preference for truths over falsehoods can be expressed using no more than a deflationary notion of truth. Truth as the goal of science needs to be something grander: the body of truths articulated by some ideal theory of everything—the Truth with a capital "T."

Realists move towards this more seriously teleological conception of inquiry when they suggest that pragmatists cannot really make sense of progress. Naturally, pragmatists demur. We no more need—and no more understand—the idea of Truth in science than we need or understand the idea of Utopia in politics. We measure progress by our distance from where we have been, not where we are going: by the ways in which current theories improve over their predecessors, not by their proximity to the end of inquiry. By dispensing with a teleological conception of inquiry, pragmatism not only helps us disenchant nature but to disenchant science as well.

Notes

1. Causal approaches to meaning face the difficulty of articulating a principled way of determining what stage of the causal chain leading to a signaling response should count as the intentional object. Many theorists bring evolution in at this point: a response means what it was *selected* to indicate.

2. In passing, whatever its merits, this argument doesn't help those who hanker for Providence. The Divine Watchmaker is a *deus absconditus*. One could find the Argument from Design totally compelling and accept no more than a milk-and-water deism, as indeed many Enlightenment thinkers did.

3. We might even think of this as rescuing an element of Providence. While perhaps not as hands-on as we used to think, God is still nice to us in a general way.

12

The Role of Concepts in Our Access to Reality

Nicholas Wolterstorff

1. The Kantian Legacy

In the course of talking recently to a group of students about epistemological matters, I used the table we were sitting at as an example of some point and spoke confidently about what the table was like. At a certain point in the discussion, one of the students had had enough of my confident talk and asked how I or anybody else could ever know what the table was really like. After all, I conceptualized it in one way and so believed my things about the table, a person from the Fiji Islands would conceptualize it in a different way and believe his things about the table; so who's to say what it's really like? How could we ever tell? I replied, "How about looking at it in good light; that would tell you a lot." I can still see the expression of stunned incredulity that crossed his face. It said, "How could a professor of philosophy at a major university say something so stupefyingly naïve; where's he been all these years?"

The student may well have had an unduly high opinion of university philosophy professors; I won't get into that. What is important for my purposes is the fact that his question reflected acceptance of one of the main narratives shaping intellectual life today. Let's have the narrative before us.

Immanuel Kant, so the story begins, showed that the understanding of the nature, origin, and role of concepts embraced by his philosophical predecessors was fundamentally mistaken. It was their view that our concepts, for the most part, are derived from experience by such operations as abstraction, generalization, division, and combination. Concepts are Johnny-come-latelies: experience first, concepts later. Empiricists held that this is true for all concepts, rationalists held that it is true only for the bulk of them; there are, in addition, certain concepts with which the mind is innately stocked. Either way, when it comes to concepts that apply to experience, what we do when we make and evaluate judgments is, to put it metaphorically, hold the experience in one hand, the concept in the other, and see whether the latter fits the former.

What Kant taught us, so the story goes, is that concepts are interpretive in function and that there is no experience that is not conceptually interpreted. We do not *first* have experience, and *then* derive concepts from that experience and make judgments about it; experience is always already conceptualized. To suppose otherwise is "the myth of the given." Even such

The Role of Concepts in Our Access to Reality

innate concepts as we may have are not *in addition* to experience and the concepts derived therefrom; they are, like all concepts, interpretive of experience.

It follows that of reality independent of us, assuming there is such, we have no experience; our acquaintance is only with that which has been conceptually shaped by us. Between us and independent reality there is an impenetrable veil of concepts, blocking epistemic access.

All this we are said to have learned from Kant. Where Kant went wrong, the story continues, was in his contention that though there is much that is purely contingent about the concepts we have, not everything is contingent. Kant argued that certain concept-types are such that it is impossible that we should lack them—impossible that we should lack concepts of substances, for example. In addition, certain specific concepts are such that it is impossible that we should lack them—examples being the concepts of causality and of existence. These necessary-for-us concept-types, and these necessary-for-us concepts, are concepts purely of the understanding—"pure concepts of the understanding," Kant calls them. From experience we learn that there are elephants, say; there might not have been. But that there are enduring objects bearing qualities is not a contingent truth that we learn from experience; we necessarily interpret our experience with the concept of an enduring object bearing qualities.

The story says that Kant was mistaken in this claim that there are certain concepts and concept-types necessary to our interpretation of experience. We do not have to operate with the concept of causality; we just do. We do not have to operate with the concept of existence; we just do. We do not have to operate with concepts of substances; we just do. Or rather, the concept of a substance is but one of many concepts of types of non-predicables with which we can and do operate. Sets, events, heaps, waves, neutrons: these and many more are familiar parts of our conceptual repertoire. The repertoire might have been different in every respect from what it is; no doubt it will be different in the future from what it presently is. Though the concept of a substance is useful, it is not ineliminable. These doctrines of contingency, so the story goes, were taught us by the pragmatists. The pragmatists erased necessity from Kant's picture.

This pragmatist revision requires, in turn, an even more fundamental revision. Kant clearly thought that his picture as a whole was necessarily true: I mean, the picture of reality putting in its appearance to us on the condition that its appearances be located within time and space, the picture of us then always conceptualizing those appearances and of being aware only of those conceptualized appearances, and so forth. This whole large picture has to be seen as itself just one way of conceptualizing things; and it has to go. Kant still assumed that there were things true of independent reality in itself, not just of reality as it appears to us; and that what is true of independent reality in itself is not dependent on our concepts. His claim was merely that of independent reality as it is in itself we have no knowledge—though let it be added that we do have available to us some convictions well grounded in morality. Kant's assumption of a ready-made independent reality has to go. Entities exist, and propositions are true, relative to a conceptual scheme; and that's the end of the matter. Beyond the interpreted, there's nothing.

In summary: Kant taught us that we have no epistemic access to ready-made reality as it is in itself; conceptualized experience is always in the way. The pragmatist appropriation of Kant taught us that there is no independent ready-made reality; there is only conceptually interpreted reality.

In this essay I want to discuss this story which, so I claim, shapes a great deal of intellectual life today. It is my judgment that the story is mostly fiction, not history; mostly myth, not gospel. There is a ready-made reality; and to that ready-made reality we have epistemic access.

The contrary has not been shown. In this chapter, however, I have space to discuss only the first episode of the story, the Kantian episode; I will have to neglect the pragmatist episode.[1]

2. Interpreting Kant

My goal is systematic rather than historical; I will argue that concepts do not do what the narrative says they do. But given the canonical role of Kant in the narrative, I propose that we start off with some Kant interpretation. It turns out that the narrative seriously misinterprets Kant. My main text will be the section of Kant's *Prolegomena to any Future Metaphysics* that he calls "Second Part of the Main Transcendental Problem," subtitled "How is Pure Science of Nature Possible?" I find Kant more lucid and forthcoming on the issues at hand there than anywhere else. I will supplement Kant's discussion there with a few passages from his first critique.

Central to Kant's discussion in this part of the *Prolegomena* is his distinction between what he calls "judgments of perception" and "judgments of experience." Strictly speaking, of course, this is the English translation of Kant's terminology. It is important that we not let the ordinary use and connotations of the terms "perception" and "experience" shape our interpretation of what Kant is getting at; we have to treat them as terms of art. Let me add that, in my judgment, the term "acquaintance," understood as Russell understood it when he made his famous distinction between knowledge by description and knowledge by acquaintance, serves much better for what Kant was getting at than does the term "perception." But I will usually follow the translation I am using.[2]

Start, as Kant himself does, with judgments of perception; for these, as Kant sees the matter, are basic. He remarks that "All our judgments are at first merely judgments of perception . . ." (46). In judgments of perception one does no more than judge how certain of one's intuitions are related to each other or to oneself—always with a given time being understood. Thus, as Kant says, "they hold good only for us (that is, for our subject)" (46). The reason one's judgments of perception hold good only for oneself is that in making such judgments one is only making a *claim* about oneself and the intuitional content of one's mind at a certain time; one is not making a claim about other subjects, nor about objects. I think Kant's examples of judgments of perception are both helpful and infelicitous; let me quote the passage in which he gives the examples and then point out what is infelicitous in them.

> To illustrate the matter: when we say, "The room is warm, sugar sweet, and wormwood bitter," we have only subjectively valid judgments. I do not at all expect that I or any other person shall always find it as I now do; each of these sentences only expresses a relation of two sensations to the same subject, that is, myself, and that only in my present state of perception; consequently they are not valid of the object. Such are judgments of perception. (47)

A judgment, to be a judgment of perception, must lack all note of objectivation. Kant's examples fail to fill that bill in two ways. He should have said "feels warm to me" rather than "is warm," "tastes sweet to me" rather than "is sweet," and "tastes bitter to me" rather than "is bitter." Second, he should not have spoken of *the room, sugar,* and *wormwood,* since these are objects. What should he have said instead? That's not easy to say, since our language is so massively designed for making judgments of experience. "I feel warm" will do, of course, as an example of a judgment of perception; and maybe "I am having the intuitions characteristic of being in a room" will do. If that last will do, in spite of its borrowing from the language of objects, then the following will presumably also do: "I am simultaneously feeling warm and having the sensations characteristic of being in a room."

Judgments of experience are what I call *objectivated* judgments; they attribute qualities and relations to objects and events. "This wormwood tastes bitter to me" is thus not a judgment of experience—or at least not a pure one. It makes reference to an object, all right; but tasting bitter to me is not a quality of the object. (Kant does not take note, in his discussion, of the fact that some judgments, like the one just cited, are mixed.) The corresponding pure judgment of experience would be, "This wormwood is bitter." In this judgment, the talk is entirely about an object and its qualities. As a consequence, all relation to any particular subject has been dropped. If true, it is true for everybody; it has, says Kant, *universal* validity. Or as he also puts it: it is true for a consciousness in general—for the view from nowhere. Kant regarded these features, objectivation, universal validity, and impersonality, as implying each other. Let me quote him:

> [W]hen a judgment agrees with an object, all judgments concerning the same object must likewise agree among themselves, and thus the objective validity of the judgment of experience signifies nothing else than its necessary universal validity. And conversely when we have ground for considering a judgment as necessarily having universal validity . . . , we must consider that it is objective also—that is, that it expresses not merely a reference of our perception to a subject, but a characteristic of the object. For there would be no reason for the judgments of other men necessarily agreeing with mine if it were not the unity of the object to which they all refer and with which they accord; hence they must all agree with one another. Therefore objective validity and necessary universality (for everybody) are equivalent terms. . . . (46)

With the distinction in hand between judgments of perception and judgments of experience, let us now take note of some of the things Kant says about the nature and role of concepts in these two very different sorts of judgments. The question that shapes Kant's discussion is this: what has to be added to intuitions, and to judgments of perception grounded on those intuitions, to make it possible for us to form judgments of experience? He is insistent that though judgments of perception are basic, something must indeed be added if there are to be judgments of experience; the objectivation, impersonality, and universal validity definitive of judgments of experience are simply not to be found in judgments of perception. He says that "it does not, as is commonly imagined, suffice for experience that perceptions are compared and connected in consciousness through judgment; thence arises no universal validity and necessity by virtue of which alone consciousness [judgment??] can be objectively valid, that is, can be called experience" (48).

What has to be added, Kant argues, is additional concepts—specifically, the pure concepts of the understanding, the categories. He says, "Quite another judgment . . . is required before perception can become experience. The given intuition must be subsumed under a concept which determines the form of judging in general relatively to the intuition, connects empirical consciousness of intuition in consciousness in general, and thereby procures universal validity for empirical judgments. A concept of this nature is a pure *a priori* concept of the understanding . . ." (48).[3]

Notice two things about this answer. In the first place, the phenomenon of objectivation is located by Kant, as it was by Thomas Reid, in the fact that we make judgments of a certain sort—specifically, on Kant's view, the judgments that he calls judgments of experience. Not only am I aware of various states of myself; I experience myself as located among objects bearing qualities and relations to each other. The objectivation embedded in this phenomenon does not consist, on Kant's view, in my having acquaintance with external objects; it consists in my making judgments of experience that are, in their own way, based on the intuitions describable by judgments of perception. Second, an implication of this is that what has to be added to intuitions, and to judgments of perception about those intuitions, so as to make judgments of experience possible, is not additional intuitions. We do not have, in addition to the intuitions that we attribute to the self in judgments of perception, acquaintance with enduring external objects. The mind has no intuitional content beyond that which can be described with judgments of perception. A pure concept of the understanding "does nothing but determine for an intuition the general way in which it can be used for judgments [of experience]." That is to say, we have no intuitions that are to be identified, say, as causal intuitions; the concept of cause "is totally disparate from all possible perception . . ." (*ibid.*). The function of the category of causality is that it enables us to take our intuitions, about which we have already made, or could already have made, judgments of perception, and make "a universally valid [causal] judgment" (*ibid.*).

As we all know, the great bulk of Kant's attention is devoted to the nature, origin, and role of those pure concepts of the understanding that account for objectivation by making it possible for us to form judgments of experience. For my purposes, I want instead to concentrate on how Kant understood the nature, origin, and role of the concepts we use in making judgments of perception.

In the Kantian lexicon, intuitions and concepts are both *representations*. There are, on Kant's view, three defining characteristics of those representations which are concepts. In the first place, concepts, unlike intuitions, are always representations *of* representations; they represent representations. Second, they are all in principle *general* in that each is, or could be, of many representations. And third, they are the product of the spontaneity of thought, whereas intuitions are all the outcome of our receptivity to the action of reality upon us. In addition to these defining characteristics, Kant holds that though there are concepts of concepts, and so forth, the concepts at the bottom of any such tier will always be concepts of intuitions. As to the role of concepts: they enable us to make judgments. They are, in fact, the predicative component of judgments—what Russell called propositional functions.[4]

So what am I doing when I make the judgment of perception that I feel warm (now)? I am relating to myself the intuition I am presently having, and also relating it to all those other intuitions that satisfy the concept *feeling warm*. As to the latter, I do this on the basis of telltale marks or characteristics of this present intuition—what Kant calls *Merkmale*.[5] That is how it has to be, of course. Unless this present intuition has the *Merkmale* characteristic of feeling-warm intuitions in general, and unless I am aware that it does, how could I ever decide between the judgment that I am feeling warm and the judgment, say, that I am feeling hungry?

Lastly, how do we come by the concepts that we use in making judgments of perception? We know that they, like all concepts, are to be attributed to spontaneity rather than receptivity; but to which aspect, or dimension, of spontaneity? Well, we know that such a concept as *feeling warm* is not one of those pure concepts of the understanding whose origin lies wholly and entirely in the understanding, "totally disparate from all possible perception" and serving only "to make a universally valid judgment possible." So what then is its origin; how does the understanding produce the concepts used in judgments of perception?

The Role of Concepts in Our Access to Reality

Kant's view is that such concepts are *derived from intuitions by abstraction!* That is what he says in the following passage, in which he is comparing concepts begotten solely from the understanding with those which we use in making judgments of perception: judgments of experience "would be impossible were not a pure concept of the understanding superadded to the concepts abstracted from intuition, under which concept these latter are subsumed . . ." (49). The point is stated so unemphatically by Kant, given his preoccupation with pure concepts, that it bears repeating: the concepts that we use in making judgments of perception are, with the exception of the concept of *oneself,* derived from our intuitions by abstraction. Other passages make clear that if Kant were speaking here with full accuracy, he would have said: *by abstraction and comparison.*

The conclusion is unavoidable that Kant has no significant disagreement with the empiricists when it comes to judgments of perception and the origin, nature, and role of the concepts that we use in making such judgments. We are aware of our intuitions—not aware of them under concepts, but just aware of them, directly acquainted. Nowhere, to the best of my knowledge, does Kant speak of being aware of entities under concepts. The only role of concepts that he recognizes is their role in the making of judgments, concepts simply being the predicative component of judgments: "the only use which the understanding can make of . . . concepts is to judge by means of them," he says (*CPR* A68). In Kant there is no notion of conceptualized intuitions; and consequently, none of intuitions as always already conceptualized. Concepts enter the picture not with the intuitions themselves but with the judgments that we make about our intuitions.

The most basic operation that we perform on our intuitions is that of deriving concepts from them by abstraction and comparison. Such concepts are indeed Johnny-come-latelies; first intuitions, then concepts. Kant would of course agree that we do not each have to derive these concepts all by ourselves; a person can learn such a concept from someone else who has derived it from intuitions. The point is that that is the *ultimate* origin of them all. Once we have the requisite concepts in our conceptual repertoire, we can then use them to make judgments about our intuitions—judgments of perception. We determine the truth or falsehood of these judgments by attending to the *Merkmale* of the intuitions. We do exactly what the story I narrated says Kant showed we cannot do: we hold the intuition in one hand, the concept in the other, and we compare. If the given is a myth, then Kant was a purveyor of the myth.

The concepts used in judgments of perception do not have an interpretive function; they have a purely comparative-descriptive function. In judging that I feel warm, I relate a certain intuition to myself and tacitly make a comparison: I tacitly claim that this intuition resembles in the relevant respect all those intuitions which fall under the concept of *feeling warm*. And that tacit claim might be correct; this intuition might in fact resemble those others in that respect. The concept does not somehow stand between me and the intuitions given me, blocking epistemic access to them; instead it enables me to recognize in thought features of my intuitions that they do really possess.

There has been a good deal of discussion in recent years as to whether it is possible to have unconceptualized experience. The counterpart to this question in Kant's system is this: could we have intuitions that, by reason of their richness, their complexity, or whatever, go beyond our capacity at the time to capture those dimensions of our intuitions in our judgments? So far as I know, Kant never raises the question. That is not surprising. There is no need for him to commit himself one way or the other; it makes no difference. So far as Kant's system is concerned, our intuitions might well go beyond our capacity to describe them in judgments.

3. McDowell's Misinterpretation of Kant: A Critique

Let us now take a moment to probe some of the implications of what Kant did not say, but which the narrative I am discussing takes him as having said. One finds those who embrace the narrative speaking in two quite different ways. Sometimes they say that Kant taught us that it is never intuitions as such but always and only *conceptualized intuitions* that are the objects of acquaintance. Sometimes they say instead that acquaintance with something is always acquaintance under concepts. The ideas are different. The former way of speaking suggests that the work of conceptualization is to be discerned in the *objects* of mental acquaintance; the latter suggests that it is instead to be discerned in the character of the acquaintance itself. Up to this point I have done my best to speak in a way that is neutral as between these two very different interpretations. Let me now consider them separately, in the order of mention.

The former idea is dominant in John McDowell's book *Mind and World*. This is what he says in one place:

> The original Kantian thought was that empirical knowledge results from a co-operation between receptivity and spontaneity. . . . [We must] achieve a firm grip on this thought: receptivity does not make an even notionally separable contribution to the co-operation.
>
> The relevant conceptual capacities are drawn on *in* receptivity. . . . It is not that they are exercised on an extra-conceptual deliverance of receptivity. We should understand what Kant calls "intuition"—experiential intake—not as a bare getting of an extra-conceptual Given, but as a kind of occurrence or state that already has conceptual content. . . .
>
> In the view I am urging, the conceptual contents that sit closest to the impact of external reality *on* one's sensibility are not already, *qua* conceptual, some distance away from that impact. They are not the results of a first step within the space of reasons. . . . This supposed first step would be a move from an impression, conceived as the bare reception of a bit of the Given, to a judgment justified by the impression. But it is not like that: the conceptual contents that are most basic in this sense are already possessed by impressions themselves, impingements by the world on our sensibility. . . .
>
> Experiences already have conceptual content. . . .[6]

We have seen that what McDowell here calls "the original Kantian thought" was not Kant's thought at all, original or otherwise. But I want now to go beyond hermeneutics to a consideration of this view in its own right. The thought is that the object of acquaintance is always something to which the mind has already made its contribution of conceptualization. It is never simply what the mind receives but always a product of the mind's receptivity plus the mind's activity of conceptualizing what it receives.

Suppose we follow Kant and call the product of the mind's receptivity an intuition. Then one way to think of this suggestion would be to think of it as postulating three entities: the received intuition, the concept, and that new creation that is the conceptualized intuition, an amalgam of intuition and concept, this last being the sole object of acquaintance. That is not how McDowell, and those who share with him this line of thought, think of the ontology of the situation. There are not both intuitions and conceptualized intuitions; there are only conceptualized intuitions—these having an intuitional aspect and a conceptual aspect. The concept cannot be stripped from the intuition so as to yield a conceptually naked intuition. Though the intuitional aspect can perhaps be joined with another concept into a different conceptualized intuition, it cannot exist without being joined with some conceptual aspect. And only such entities, so it is said, can be present to the mind—or conversely, only with such entities can the mind be acquainted.

Now if it is true that the object of acquaintance is always and necessarily something already conceptualized, then it follows that we are never acquainted with a ready-made reality external to the mind. It is not absurd on the face of it to say that the mind is presented with amalgams in which intuition is indissolubly joined with concept; whether this is true is another matter. But it surely is absurd to suggest that the mind could be presented with an amalgam of some concept and an external object, say, a duck. For what would a conceptualized duck be—that is, not a duck of which we have a concept, but a new creation, an amalgam, neither concept nor duck but a conceptualized duck? On the view that the object of acquaintance is always a conceptualized entity, the conclusion seems inescapable that the objects of acquaintance are not external objects but interfaces between us and external reality.

And these interfaces are doubly removed from the reality that puts in its appearance to us. The intuitional aspect of the conceptualized intuition is itself not the external reality but the input of the action of that reality upon us; and secondly, these inputs never exist by themselves but only as aspects of conceptualized intuitions. Concepts are indeed impenetrable barriers between us and ready-made reality, assuming there is such; or strictly speaking, it is not concepts as such that are the barriers but conceptualized intuitions.

The picture requires that there be some entities with which we are directly acquainted, without interface—namely, our conceptualized intuitions. Once conceptualized intuitions have been produced, then the barriers to direct acquaintance with an entity are removed; if that were not the case, we would be off on an infinite regress.

Once we notice this feature of the view, then immediately and forcibly there comes to mind this question: if it has to be conceded that at this late point in the process the mind is directly acquainted with an entity, then why could that not have happened earlier? Why can there not be unconceptualized intuitions with which we are acquainted? Why, indeed, can we not be directly acquainted with external reality, with no interface?

The answer forthcoming is that only if an entity is fully mental is it of the right sort to be a direct object of acquaintance. Ducks are not candidates for acquaintance because they are not mental at all; and unconceptualized intuitions are not candidates for acquaintance because, being solely the product of receptivity, not at all of the understanding's spontaneity, they are insufficiently infused with mentality. In fact they have not been *infused* with mentality at all; they have just been received. We infuse them with mentality by conceptualizing them. Conceptualized intuitions are of the mind's own flesh and blood; that is why the mind can be directly acquainted with them.

This is the picture. Is it to be accepted? The first question to be considered is what exactly are these conceptualized intuitions? McDowell is more forthcoming on this point than most writers. He remarks that "In experience one takes in, for instance sees, *that things are thus and so*" (9). Let us leave perception out of the picture for the moment, since we are still focused on alternatives to Kant's analysis of inner experience, and work with an example of introspection. *Being aware that I am feeling a hunger pang*—for our purposes this will do nicely as an example. McDowell's thought is that the objects of acquaintance are *facts*—though he himself is rather chary of calling them that. And his ontology of facts—at least of the facts with which we have acquaintance—is that facts are the product of receptivity plus conceptualization. Conceptualized intuitions are facts.

Why should we accept this idealist ontology of facts? McDowell's argument is interesting. Experience not only evokes beliefs; some of them it *justifies*. McDowell likes to put the point metaphorically: experience has a place in the space of reasons. Not only can a belief serve as a reason for a belief; experience can also serve as a reason for a belief.

How can that be? McDowell asks. His answer is that it can only be because experience, in some deep way, is allied with belief rather than being an alien intruder. Now we know that concepts are essential components of judgments; in particular, the predicative element of a judgment is always a concept. McDowell draws the conclusion that a condition of experience justifying a judgment is that the very same concept that is the predicate of the judgment must also be present in the object of acquaintance. Of course it is not present as a predicate of a judgment; it is present instead as the conceptual aspect of the object of acquaintance. The concept of feeling a hunger pang is at one and the same time the conceptual component of what I am currently feeling and the predicate of my judgment *that I am feeling a hunger pang*. It is because concepts play this dual role, as components of the objects of experience and as predicates of judgments, that experience can justify beliefs and judgments.

If we are to understand how experience can justify beliefs and judgments, says McDowell, one more consideration must be brought into the picture. When I judge that I am feeling a hunger pang, the content of my judgment is *that I am feeling a hunger pang;* when I experience that I am feeling a hunger pang, the object of my experience is that I am feeling a hunger pang. What I judge is what I experience; that is what makes it possible for my experience to justify my judgment. Let me quote once more a sentence already quoted from McDowell, and then continue with the passage: "In experience one takes in, for instance sees, *that things are thus and so*. That is the sort of thing one can also, for instance, judge. . . . Minimally, it must be possible to decide whether or not to judge that things are as one's experience represents them to be" (9, 11).

In summary: my judgment that I have a hunger pang is justified by my experience of the fact that I have a hunger pang; and the fact that I have a hunger pang has embedded within it the very same concept that functions as predicate of the judgment, namely, the concept of having *a hunger pang*. This two-way functioning of concepts is what brings experience into the space of reasons. Concepts link experience with judgment.

An aesthetically elegant picture. But is it also compelling and true? Let me focus on the claimed identities: identity of content of experience with content of judgment; and within that, identity of the concept shaping the object of experience with the concept functioning as the predicate of the judgment. There is a serious problem with the former identification; McDowell is aware of the problem, but resolves to treat it as a puzzle to be addressed later rather than as an objection to the theory itself. All is apparently well for those experiential judgments of ours that are true. I judge that I am feeling a hunger pang and I experience that I am feeling a hunger pang: the content of my judgment appears to be identical with the content of my experience. And since the former indubitably has a concept as component, the latter must also have that same concept as component—or it would not be identical with it. The content of my experience is a fact; that is to say, it is a fact that I am feeling a hunger pang. So the content of my judgment must also be a fact.

But judgments, including experiential judgments, are often false. What is the content of a false judgment? Obviously not a fact. So the content of false judgments is of a distinct ontological type from the content of true judgments—though of course the content of both is expressed in a that-clause. This is surprising: that the contents of false judgments are of a distinct ontological type from the contents of true judgments, these latter being facts. What we want is an account of what those contents of false judgments are—and of why it is that those same entities are not contents of true judgments. McDowell refrains from discussing the issue.[7]

Rather than myself exploring what options, if any, are available to McDowell at this point, let me propose an alternative way of thinking of the whole situation; it is in fact an utterly traditional way of thinking. McDowell sets before us two phenomena of which we must take account: experience has a location in the space of reasons; and not only is the content of judgments expressed in that-clauses, but the object of much experience also can be expressed in that-clauses. In addition to these, we must take account of the phenomenon that McDowell decides not to deal with: the phenomenon of false judgments.

Here, then, is the alternative way of thinking. Fundamental to reality is factuality—or if you prefer, relationality. Factuality has to be basic. This desk of mine and the property of being brown could both exist without this desk being brown; for the desk to be brown, as it is, there must be the fact that the desk is brown. McDowell clearly thinks there is something out there that activates our receptivity so as to produce, once conceptualization has also chipped in, conceptualized intuitions. But there can be something out there only if it is factually structured. Thus, even on McDowell's account, not all facts are the product of conceptualization.

So take the fact that I am feeling a hunger pang. If it is not essential to something's being a fact that concepts enter into its ontological constitution, why suppose that they enter into the ontological constitution of this one? Why can there not be, between me and this hunger pang, the relationship of my feeling the pang whether or not I have the concept of a hunger pang? Alternatively expressed, why can there not be the fact that I am feeling this hunger pang even though I lack, or do not in this case employ, the concept of a hunger pang? Admittedly, I cannot judge that I have a hunger pang without having the concept of a hunger pang. But that's not the question. The question is why it should be supposed that the *fact* of my feeling a hunger pang has a concept as its ontological constituent—especially when McDowell's own picture requires that there be lots of facts that do not have concepts as ontological constituents. I am assuming throughout that McDowell is not a closet theist; if he were, then it would be open to him to say that the facts of the world have *God's* concepts as their ontological constituents.

And now let us suppose that between concepts and entities there is the relationship of the entity satisfying the concept—or to look at it from the other side, the relationship of the concept applying to that entity. McDowell tries working with a different idea—that of the identity of a concept embedded in an experienced fact with the concept embedded in a judgment. But he needs this other idea as well, the idea of an entity satisfying a concept. For on his account, the conceptualized intuition *that I am feeling a hunger pang* satisfies the concept of a conceptualized intuition even though that concept, the concept of a conceptualized intuition, does not itself enter into the composition of *this* fact.

Let us add to the above point that there is the phenomenon of entities satisfying concepts, the thesis that often we can tell whether an entity satisfies a concept by gaining acquaintance with the entity. McDowell takes this thesis for granted, as do all of us; he assumes throughout that he can tell which items in his mental life satisfy the concept of a conceptualized intuition (i.e., a fact), even though the concept of a conceptualized intuition does not enter into the constitution of the conceptualized intuition (the fact).

So here is how experiential judgments get justified, here is how they enter into the space of reasons—or at least, here is one way. I wonder whether to affirm that I am feeling a hunger pang. I have learned the characteristics, the *Merkmale,* whose presence is the condition for the proper application of the concept of feeling a hunger pang. If, upon noting the presence of those *Merkmale,* I then judge that I am feeling hunger pangs—which amounts to judging that the concept applies to my experience—then my judgment is justified. By contrast, if I should make the judgment in the absence of the *Merkmale* or without bothering to take note whether they are present or absent, then I am not justified. It was along these lines that Kant was thinking when it came to judgments of perception, not along McDowell's lines.

And what about falsehood? Well, suppose that the contents of judgments are not facts but Fregean *Gedanken* propositions. My judgment is true in case the proposition that is the content of my judgment has a fact corresponding to it; otherwise it is false. Propositions are not identical with facts. Rather, when a proposition is true, there is a fact corresponding to the proposition; that is what makes it true. My judgment that I am feeling a hunger pang is true because there is a fact corresponding to the propositional content of my judgment—true just in case there is the fact that I am feeling a hunger pang. The content of a false judgment is of the same ontological type as the content of a true judgment; in either case, it is a proposition.

McDowell assumes that either one builds concepts into facts and thus explains how experience justifies judgments, or one resists doing so and concedes that experience evokes judgments without justifying them. What I have just now sketched out, ever so briefly, is a way between the horns.

In short, McDowell's interpretation of Kant is not only a misinterpretation of Kant; it falls far short of establishing that the object of acquaintance is always something already conceptualized. We have seen no reason to conclude that concepts function in such a way as to make ready-made reality epistemically inaccessible to us.

4. Objects of Acquaintance and Acts of Acquaintance

I mentioned that some of those who embrace the narrative I recited claim not that the object of acquaintance is always some entity into whose ontological constitution some concept enters, but rather that acquaintance is always acquaintance under concepts. Acquaintance with something is never *mere* acquaintance with that entity but always acquaintance with that entity *as so-and-so*. Concepts play their role in the *act* rather than in the *object* of acquaintance. What unites this view with the other is that, on this view too, concepts play a bridging role between experience and judgments: we are both acquainted with entities *as so-and-so* and we judge that entities are *so-and-so*.

For our purposes here, this view can be dealt with briskly. There is nothing in this view that yields, or even threatens to yield, the conclusion that concepts function in such a way as to construct a barrier between us and ready-made reality.

Whether or not it is true that all acquaintance with something is acquaintance under a concept is an issue we need not enter, nor need we analyze what exactly it is to be acquainted with something under a concept. Take a case in which it does seem right to say that I am acquainted with something as so-and-so. Suppose I perceive what is before me as a duck. It might in fact be a duck; nothing in this view suggests otherwise. And if it is in fact a duck, then I perceive it as what it is. My perceiving it as a duck does not somehow make the object I am perceiving epistemically inaccessible to me, since it is agreed that I am perceiving what is before me, and that that might well be a duck. The functioning of the concept in my perception does not erect a cognitive barrier between me and reality; quite to the contrary, it enables cognitive access to reality. Without the concept, I could not perceive it as what it is in this respect.

But what if you, perhaps lacking the concept of *duck,* perceive what is before me not as a duck but as a fat bird; does that not create problems for the view that perception provides us with epistemic access to a ready-made world? Not at all. It is both a duck and a fat bird. I perceive it as what it is, and you perceive it as what it is—even though the 'what it is' is different in the two cases. What's the problem, given that it is both a duck and a fat bird?

But suppose you perceive it as a duck decoy, when it is not a duck decoy; it is a duck. Does that not create problems? Not at all. We both perceive the same thing, that's agreed, that thing being a duck. But whereas I perceive it as what it is, you perceive it as something that it is not. What's the problem? It happens all the time. What if I perceive it as a duck and you perceive it as a duck decoy—who's to tell what it really is? Get closer, make some noise, poke it; if it flies off, it's not a duck decoy.

5. The Kantian Letdown

In conclusion, I want briefly to return to Kant interpretation. Recent commentators on Kant have noted that in his revisions of the first edition of the *Critique of Pure Reason,* Kant removed almost all of the language suggesting that, on his view, that which is directly given to the mind is intuitions and that these are interfaces between the mind and independent reality. His new language suggests that what is given is mind-independent reality, though indeed not as it is in itself. On this account, an intuition is not an object functioning as an interface between the mind and reality, but simply the mind's intuiting of reality. This makes Kant an epistemological realist; intuition is direct acquaintance.

Kant's thought in the *Prolegomena* is still that of the first critique—or that which the first critique strongly suggests. Kant himself claimed that his revisions did not correct but only clarified the first critique. I have been working with the thought of the *Prolegomena* and that which the first critique strongly suggests because that is the interpretation of Kant that the narrative I have been engaging works with. On the realist interpretation, concepts, whatever they do, do not hinder direct epistemic access to reality independent of us.

On the traditional interpretation of Kant, the interpretation that the narrative works with, a great deal of Kant's thought in the *Critique of Pure Reason* was driven by the conclusions he arrived at in his reflections on the role of concepts. On the interpretation I have offered, those conclusions were much too close to the empiricists to have had such grand implications. But if so, what then was driving Kant's overall line of thought? If Kant's way of thinking about the role of concepts in what he calls "perception" was not significantly different from that of his empiricist predecessors, what then accounts for the fact that the first critique is far from being one more document in the history of empiricism?

I suggest that what drove Kant's thought was conclusions he drew from the fact that we have knowledge of synthetic, necessarily true, propositions—the knowledge that necessarily the intuitional component of one's mental life is temporal, the knowledge that necessarily there will be in one's experience enduring objects bearing qualities, and so forth. How could we possibly explain our knowledge of such necessarily true but synthetic propositions, Kant asks, if the necessities in question were necessities of the structure of independent reality? The only possible way of explaining our knowledge of such propositions is to suppose that the necessities are necessities of human perception and experience—these being a subset of the necessities of human nature. Human nature is such that it is impossible that one's mental life not transpire in time, impossible that one's experience be devoid of substances bearing qualities, and the like.

It is these claims about necessity, and the conclusions he draws from them, which account in good measure for the anti-empiricist structure of Kant's thought in general.

Engaging this part of Kant's thought would require an essay of its own—or a book! Let me confine myself to noting a peculiar feature of Kant's answer to his question. The question driving Kant's thought, to say it once again, is how we are to explain the fact that we have knowledge of synthetic necessarily true propositions. Experience can never ground knowledge of such necessities, or any others; on the other hand, if the necessities pertain to independent reality as it is in itself, how could one possibly know them, given that we have no acquaintance with independent reality as it is in itself. I know these propositions *a priori*. How can that be? We are in a dilemma; even if experience could ground knowledge of necessities, we have no acquaintance with independent reality as it is in itself.

The solution, says Kant, is to hold that the synthetic necessities of which we have knowledge are necessities of human nature and only of human nature. Thus, with one swipe, Kant dismissed the long tradition that held that we have knowledge of the natures of all sorts of things, and of the *de re* necessities grounded in those natures—knowledge of the nature of time, of water, of horses, etc.

Kant's argument remains curiously incomplete, however. Suppose we concede, for the sake of the argument, that the only nature of which we can have knowledge is human nature. How do we account for our knowledge of human nature? Without an answer to that question, Kant has not answered his own overarching question; he has given only a long, baroque preparation for an answer. For it's not as if the answer is so obvious as to be not worth mentioning. If the thought that one could know in *a priori* fashion synthetic necessary truths about entities other than oneself is so problematic that one ought to dismiss the thought, why then is the thought that one could know in *a priori* fashion synthetic necessary truths about human beings not equally problematic? What is the difference between human nature and other natures that makes knowledge of the latter impossible and knowledge of the former not only possible but actual? The nature I know when I know human nature is my own nature. Yes indeed. But how exactly does that make a difference?

There is one fascinating passage in which Kant is on the verge of recognizing the point. It occurs in the same part of the *Prolegomena* that I have made the center of my interpretation.

> How is nature possible in the formal sense, as the totality of the rules under which all appearances must come in order to be thought as connected in experience? The answer must be this: It is only possible by means of the constitution of our understanding, according to which all the above representations of the sensibility are necessarily referred to a consciousness, and by which the peculiar way in which we think (namely, by rules) and hence experience also are possible, but must be clearly distinguished from an insight into the object in themselves. . . .
>
> But how this peculiar property of our sensibility itself is possible, or that of our understanding and of the apperception which is necessarily its basis and that of all thinking, cannot be further analyzed or answered, because it is of them that we are in need for all our answers and for all our thinking about objects (65).

We have no explanation, says Kant, for why human nature is the way he has argued it is. Indeed we *can* have no such explanation, since we would have to use that part of our nature in explaining that part.

The Role of Concepts in Our Access to Reality

Kant is here sounding a note that Thomas Reid sounded much more powerfully than Kant ever did: in accounting for the epistemological workings of ourselves, we come to a point where further explanations are not only factually impossible but in principle impossible. In Reid, however, the note comes as the culmination of his argument; in Kant, it comes as a letdown. The structure of Kant's argument leads us to expect from him an explanation of how it is that we have *a priori* knowledge of synthetic necessary truths concerning the nature of the human knowing self. Why is it that, in speaking about our epistemic selves, we are not confined to reporting how it has always been and forming a few inductive beliefs about how it is likely to be in the future? How can we ever get beyond such beliefs to knowledge of the necessities of our nature? Kant observes in the passage quoted that we cannot *account* for human nature. Fair enough. But the question was not how to account for human nature; the question was how to account for our knowledge of human nature. Nothing in his argument has led us to expect from Kant an account of human nature; everything in his argument has led us to expect an account of how it is that we know the necessary structure of human nature. But when we open the box, there's nothing there.

Notes

1. I discussed the second part of the story, the part that talks about the pragmatist contribution, in my essay "Are Concept-Users World-Makers" in *Philosophical Perspectives, I: Metaphysics, 1987,* ed. by J. Tomberlin (Atascadero, CA: Ridgeview Publishing Co.; 1987), 233–67.

2. The Mahaffy-Carus translation, revised by Lewis White Beck. My page references will be to the edition by The Liberal Arts Press (New York, 1950).

3. *Cf.* 45–46: Judgments of perception "require no pure concept of the understanding, but only the logical connection of perception in a thinking subject. But the former always require, besides the representation of the sensuous intuition, special *concepts originally begotten in the understanding.* . . ."

4. *Cf. Critique of Pure Reason,* A68–69 (Norman Kemp-Smith translation): "Whereas all intuitions, as sensible, rest on affections, concepts rest on functions. By 'function' I mean the unity of the act of bringing various representations under one common representation. Concepts are based on the spontaneity of thought, sensible intuitions on the receptivity of impressions. Now the only use which the understanding can make of these concepts is to judge by means of them. Since no representation, save when it is an intuition, is in immediate relation to an object, no concept is ever related to an object immediately, but to some other representation of it, be that other representation an intuition, or itself a concept. Judgment is therefore the mediate knowledge of an object, that is, the representation of a representation of it. In every judgment there is a concept which holds of many representations, and among them of a given representation that is immediately related to an object. . . . Accordingly, all judgments are functions of unity among our representations; instead of an immediate representation, a *higher* representation, which comprises the immediate representation and various others, is used in knowing the object. . . . Thought is knowledge by means of concepts. But concepts, as predicates of possible judgments, relate to some representation of a not *yet* determined object. Thus the concept of body means something, for instance, metal, which can be known by means of that concept. It is therefore a concept solely in virtue of its comprehending other representations, by means of which it can relate to objects. It is therefore the predicate of a possible judgment. . . ."

5. "Objects are *given* to us by means of sensibility, and it alone yields us *intuitions;* they are *thought* through the understanding, and from the understanding arise *concepts.* But all thought must, directly or indirectly, by way of certain characters [*Merkmale*] relate ultimately to intuitions . . ." (*Critique of Pure Reason* A19). Hereafter, in the text,, "*CPR.*"

6. John McDowell, *Mind and World.* Cambridge: Harvard University Press, 1996, 9–10.

7. There is a possibility worth exploring: take the concept of states of affairs, facts being those states of affairs which obtain. Perhaps the content of a judgment is a state of affairs. Then, in the case of a true judgment pertaining to experience, the content of the judgment will be identical with the fact that is the object of experience; in the case of a false judgment, the content of the judgment is a state of affairs that proves not to be a fact. The critical points I make in the text above are as relevant to this way of thinking as they are to the option I work with, though they would of course have to be restated somewhat.

Part III

The Origin of Biological Information and The Emergence of Biological Complexity

Introduction

Part III contains nine essays, each focused on different aspects of the effort to explain the origin or development of biological information, and each exemplifying either skepticism or cautious optimism in respect of different explanatory strategies. Of all the areas of science discussed in this volume, biology is the most contentious, almost certainly because of the role that disputes over Darwinian evolution play in the wider culture. But these broader cultural considerations will not figure in the discussions here, which deal with technical issues and the explanatory resources and adequacy of different research programs.

The examination opens with **David Berlinski's** essay "On the Origins of Life," which takes a rather dismal view of abiogenesis research, arguing that the proposals currently on the table have either little chance of success or (as yet) no evidence in their favor, and that the question may very well be unanswerable by science. **Stephen Meyer** seeks a way out of this impasse in his essay "DNA: The Signature in the Cell" by arguing that DNA exhibits informational specified complexity, the origin of which cannot be explained adequately by undirected evolutionary mechanisms. He begins with a careful definition of what needs to be explained, dissects a variety of methodologically naturalistic proposals for the origin of functionally specified biological information, and concludes with a strong defense of intelligent design as the only causally sufficient explanation for the specified complexity evident in the basis of life. In contrast, Nobel Laureate **Christian de Duve** argues in "Mysteries of Life: Is There 'Something Else'?" that natural science *requires* methodological naturalism; to set aside this principle, he contends, is to give up on scientific research. Invoking the intervention of "something else" is an explanatory strategy of last resort, a measure to be used only after all the possibilities of naturalistic explanation have been exhausted. Even then, he adds, it would be very hard to say when that point had been reached—and it has certainly not been reached in regard to explanations of the origin of life. In support of this claim, de Duve adduces as evidence the fact that investigations into the mechanisms that *support* life have yielded no justification for believing that living organisms are made of matter "animated" by something else; why should the origin of life be any different? What is more, he contends, modern biological knowledge has arrived at the conclusion that all known living organisms are descended from a single ancestral form of life. This continuity of historical development shows that, once life was present in this single ancestral form, its evolutionary diversification has a natural explanation; this too gives confidence that so does its origin. De Duve then discusses various scientific strategies for research into life's origin and critiques Michael Behe's argument for intelligent design on the basis of irreducible complexity. After considering a variety of heterodox approaches to macroevolution (the origin of new body plans) he argues that paleontological evidence of continuity and the discovery of regulatory genes give us ample hope that detailed naturalistic explanations—even of possible discontinuous "jumps" in biological complexity—are out there to be found. In conclusion, he suggests that even though the idea that there is "something else" has retreated under the advance of modern science, we may still find meaning in the recognition that we are part of an immense cosmic pattern that we are only beginning to understand.

In "Life's Conservation Law: Why Darwinian Evolution Cannot Create Biological Information" **William Dembski** and **Robert Marks** take a hard look at computational biology and efforts to demonstrate the possibility of neo-Darwinian evolution *in silico*. They argue that there is a Law of Conservation of Information (LCI) that can be demonstrated mathematically to hold in the natural realm. This law characterizes the information costs that searches incur in outperforming a blind search. While searches that operate by Darwinian selection often significantly outperform a blind search, they do so because they exploit information supplied by a fitness function that is unavailable to blind searches. Searches that have a greater probability of success than a blind search do not just magically materialize. They form by some process. According to LCI, any such search-forming process must build into the search at least as much information as the search displays in raising the probability of success; in short, information obeys strict accounting principles. Dembski and Marks prove three different conservation-of-information results: one function-theoretic, one measure-theoretic, and one fitness-theoretic theorem. These three results are representative of conservation-of-information theorems in general and provide the theoretical underpinnings for the Law of Conservation of Information. The authors conclude that insofar as Darwinian evolution plays a substantial role in the history of life, the Law of Conservation of Information shows that it must be inherently teleological, and that this teleology can be measured in precise information-theoretic terms.

Reasserting the contrary viewpoint, distinguished molecular biochemist **Mark Ptashne**, focusing on genomic regulatory systems, argues in "Regulated Recruitment and Cooperativity in the Design of Biological Regulatory Systems" that complicated systems are constructed by reiterative uses of simple (and rather crude) molecular interactions, and are thus highly evolvable. What distinguishes a man from a mouse is not so much different proteins, but rather the appearance of common proteins, etc., at different times and positions in developing organisms. Thus, as specific genes are transcribed or repressed, as proteins are degraded or stabilized, as RNA transcripts are spliced one way or another, and so on, we have examples of different "regulatory" decisions taking place. A rather simple mechanism—called "regulated recruitment"—lies at the heart of many of these regulatory decisions and, Ptashne argues, serves as the mindless molecular strategy that nature has used widely in evolving biological complexity.

Before one has regulatory systems controlling the *expression* of proteins, however, one needs to have biologically functional proteins. Molecular biologist **Douglas Axe** therefore focuses on a technical problem earlier in the development of life. In his essay "The Nature of Protein Folds: Quantifying the Difficulty of an Unguided Search through Protein Sequence Space," Axe explains how current methods for DNA manipulation allow us to explore protein sequence space by direct experimentation in ways that were previously impossible. He discusses how these experimental tools have been used to answer one of the most important questions in

Introduction

molecular evolution: how common, among all the possibilities, are amino-acid sequences that form folded structures of the kind that life requires? The answer, he contends, poses a startling challenge to the claim that new protein folds can be found by unguided searches. Reinforcing these considerations in his essay "The Limits of Non-Intelligent Explanations in Molecular Biology," **Michael Behe** examines recent data involving extraordinarily large populations of microorganisms under intense selective pressure to assess the ability of undirected processes to build complex biochemical systems of the type that fill the cell. He concludes that none are to be found.

Michael Shermer is much more sanguine about the power of natural selection, and he defends the radical contingency thesis in his essay "The Chain of Accidents and the Rule of Law: The Role of Contingency and Necessity in Evolution." Shermer argues, in the tradition of Stephen Jay Gould, that if the evolutionary clock were rewound and let run forward again, the variety of species would be vastly different from what we now observe; almost certainly humanity would not be here. Finally, in opposition to this contention, biochemist **Fazale Rana** argues that even though natural selection supervening on random genetic mutations can indeed be seen to operate in microevolutionary developments as Shermer would suggest, what we in fact observe on the macroevolutionary scale is a wide range of convergences—at both the molecular and morphological levels—in what are taken to be independent evolutionary lineages. Rana argues that this situation is at complete odds with what we *should* expect if neo-Darwinian mechanisms were the correct explanation of macroevolutionary development. After presenting a wide range of evidence for convergence at the molecular level, he focuses on the independent origins of DNA replication in bacteria and in archaea and eukaryotes, and discusses just how remarkable this molecular convergence is in light of the integrated functional specificity it requires. He then illustrates the historical contingency of microevolution by discussing the results of the "Long Term Evolution Experiment" with *Escherichia coli* conducted by Richard Lenski's group at Michigan State, and concludes by arguing that the fundamental difference between microevolutionary contingency and macroevolutionary convergence is found in the differential background causes for each: undirected evolution for the former and intelligent design for the latter.

13

ON THE ORIGINS OF LIFE[1]

DAVID BERLINSKI

> For those who are studying aspects of the origin of life, the question no longer seems to be whether life could have originated by chemical processes involving non-biological components but, rather, what pathway might have been followed.
> —National Academy of Sciences (1996)

1. BEGINNINGS

It is 1828, a year that encompassed the death of Shaka, the Zulu king, the passage in the United States of the Tariff of Abominations, and the battle of Las Piedras in South America. It is, as well, the year in which the German chemist Friedrich Wöhler announced the synthesis of urea from cyanic acid and ammonia.

Discovered by H. M. Roulle in 1773, urea is the chief constituent of urine. Until 1828, chemists had assumed that urea could be produced only by a living organism. Wöhler provided the most convincing refutation imaginable of this thesis. His synthesis of urea was noteworthy, he observed with some understatement, because "it furnishes an example of the artificial production of an organic, indeed a so-called animal substance, from inorganic materials."

Wöhler's work initiated a revolution in chemistry; but it also initiated a revolution in thought. To the extent that living systems are chemical in their nature, it became possible to imagine that they might be chemical in their origin; and if chemical in their origin, then plainly physical in their nature, and hence a part of the universe that can be explained in terms of "the model for what science should be."[2]

In a letter written to his friend Sir Joseph Hooker several decades after Wöhler's announcement, Charles Darwin allowed himself to speculate. Invoking "a warm little pond" bubbling up in the dim inaccessible past, Darwin imagined that given "ammonia and phosphoric salts, light, heat, electricity, etc. present," the spontaneous generation of a "protein compound" might follow, with this compound "ready to undergo still more complex changes" and so begin Darwinian evolution itself.

Time must now be allowed to pass. Shall we say sixty years or so? Working independently, J. B. S. Haldane in England and A. I. Oparin in the Soviet Union published influential studies concerning the origin of life. Before the era of biological evolution, they conjectured, there must have been an era of *chemical* evolution taking place in something like a pre-biotic soup. A reducing atmosphere prevailed, dominated by methane and ammonia, in which hydrogen atoms, by donating their electrons (and so "reducing" their number), promoted various chemical reactions. Energy was at hand in the form of electrical discharges, and thereafter complex hydrocarbons appeared on the surface of the sea.

The publication of Stanley Miller's paper, "A Production of Amino Acids Under Possible Primitive Earth Conditions," in the May 1953 issue of *Science* completed the inferential arc initiated by Friedrich Wöhler 125 years earlier. Miller, a graduate student, did his work at the instruction of Harold Urey. Because he did not contribute directly to the experiment, Urey insisted that his name not be listed on the paper itself. But their work is now universally known as the Miller-Urey experiment, providing evidence that a good deed can be its own reward.

By drawing inferences about pre-biotic evolution from ordinary chemistry, Haldane and Oparin had opened an imaginary door. Miller and Urey barged right through. Within the confines of two beakers, they re-created a simple pre-biotic environment. One beaker held water; the other, connected to the first by a closed system of glass tubes, held hydrogen cyanide, water, methane, and ammonia. The two beakers were thus assumed to simulate the pre-biotic ocean and its atmosphere. Water in the first could pass by evaporation to the gases in the second, with vapor returning to the original alembic by means of condensation.

Then Miller and Urey allowed an electrical spark to pass continually through the mixture of gases in the second beaker, the gods of chemistry controlling the reactions that followed with very little or no human help. A week after they had begun their experiment, Miller and Urey discovered that in addition to a tarry residue—its most notable product—their potent little planet had yielded a number of the amino acids found in living systems.

The effect among biologists (and the public) was electrifying—all the more so because of the experiment's methodological genius. Miller and Urey had done nothing. Nature had done everything. The experiment alone had parted the cloud of unknowing.

2. The Double Helix

In April 1953, just four weeks before Miller and Urey would report their results in *Science,* James Watson and Francis Crick published a short letter in *Nature* entitled "A Structure for Deoxyribose Nucleic Acid." The letter is now famous, if only because the exuberant Crick, at least, was persuaded that he and Watson had discovered the secret of life. In this he was mistaken: the secret of life, along with its meaning, remains hidden. But in deducing the structure of deoxyribose nucleic acid (DNA) from X-ray diffraction patterns and various chemical details, Watson and Crick *had* discovered the way in which life at the molecular level replicates itself.

Formed as a double helix, DNA, Watson and Crick argued, consists of two twisted strings facing each other and bound together by struts. Each string comprises a series of four nitrogenous bases: adenine (A), guanine (G), thymine (T), and cytosine (C). The bases are nitrogenous because their chemical activity is determined by the electrons of the nitrogen atom, and they are bases because they are one of two great chemical clans—the other being the acids, with which they combine to form salts.

Within each strand of DNA, the nitrogenous bases are bound to a sugar, deoxyribose. Sugar molecules are in turn linked to each other by a phosphate group. When nucleotides (A,

G, T, or C) are connected in a sugar-phosphate chain, they form a polynucleotide. In living DNA, two such chains face each other, their bases touching fingers, A matched to T and C to G. The coincidence between bases is known now as Watson-Crick base pairing.

"It has not escaped our notice," Watson and Crick observed, "that the specific pairings we have postulated immediately suggests a possible *copying mechanism* for the genetic material" (emphasis added). Replication proceeds, that is, when a molecule of DNA is unzipped along its internal axis, dividing the hydrogen bonds between the bases. Base pairing then works to prompt both strands of a separated double helix to form a double helix anew.

So Watson and Crick conjectured, and so it has proved.

3. THE SYNTHESIS OF PROTEIN

Together with Francis Crick and Maurice Wilkins, James Watson received the Nobel Prize for medicine in 1962. In his acceptance speech in Stockholm before the king of Sweden, Watson had occasion to explain his original research goals. The first was to account for genetic replication. This, he and Crick had done. The second was to describe the "way in which genes control protein synthesis." This, he was in the course of doing.

DNA is a large, long, and stable molecule. As molecules go, it is relatively inert. It is the proteins, rather, that handle the day-to-day affairs of the cell. Acting as enzymes, and so as agents of change, proteins make possible the rapid metabolism characteristic of modern organisms.

Proteins are formed from the alpha-amino acids, of which there are twenty in living systems. The prefix "alpha" designates the position of the crucial carbon atom in the amino acid, indicating that it lies adjacent to (and is bound up with) a carboxyl group comprising carbon, oxygen, again oxygen, and hydrogen. And the proteins are polymers: like DNA, their amino-acid constituents are formed into molecular chains.

But just how does the cell manage to link amino acids to form specific proteins? This was the problem to which Watson alluded as the king of Sweden, lost in a fog of admiration, nodded amiably.

The success of Watson-Crick base pairing had persuaded a number of molecular biologists that DNA undertook protein synthesis by the same process—the formation of symmetrical patterns or "templates"—that governed its replication. After all, molecular replication proceeded by the divinely simple separation-and-recombination of matching (or symmetrical) molecules, with each strand of DNA serving as the template for another. So it seemed altogether plausible that DNA would likewise serve a template function for the amino acids.

It was Francis Crick who in 1957 first observed that this was most unlikely. In a note circulated privately, Crick wrote that "if one considers the physico-chemical nature of the amino-acid side chains, we do not find complementary features on the nucleic acids. Where are the knobby hydrophobic . . . surfaces to distinguish valine from leucine and isoleucine? Where are the charged groups, in specific positions, to go with acidic and basic amino acids?"

Should anyone have missed his point, Crick made it again: "I don't think that anyone looking at DNA or RNA [ribonucleic acid] would think of them as templates for amino acids."

Had these observations been made by anyone but Francis Crick, they might have been regarded as the work of a lunatic; but in looking at any textbook in molecular biology today, it is clear that Crick was simply noticing what was under his nose. Just where *are* those "knobby hydrophobic surfaces"? To imagine that the nucleic acids form a template or pattern for the amino acids is a little like trying to imagine a glove fitting over a centipede. But if the nucleic acids did not form a template for the amino acids, then the information they contained—all of the ancient wisdom of the species, after all—could only be expressed by an indirect form of transmission: a *code* of some sort.

The idea was hardly new. The physicist Erwin Schrödinger had predicted in 1945 that living systems would contain what he called a "code script"; and his short, elegant book *What Is Life?* had exerted a compelling influence on every molecular biologist who read it. Ten years later, the ubiquitous Crick invoked the phrase "sequence hypothesis" to characterize the double idea that DNA sequences spell a message *and* that a code is required to express it. What remained obscure was both the spelling of the message and the mechanism by which it was conveyed.

The mechanism emerged first. During the late 1950s, François Jacob and Jacques Monod advanced the thesis that RNA acts as the first in a chain of intermediates leading from DNA to the amino acids.

Single- rather than double-stranded, RNA is a nucleic acid: a chip from the original DNA block. Instead of thymine (T), it contains the base uracil (U), and the sugar that it employs along its backbone features an atom of oxygen missing from deoxyribose. But RNA, Jacob and Monod argued, was more than a mere molecule: it was a messenger, an instrument of conveyance, "transcribing" in one medium a message first expressed in another. Among the many forms of RNA loitering in the modern cell, the RNA bound for duties of transcription became known, for obvious reasons, as "messenger" RNA.

In transcription, molecular biologists had discovered a second fundamental process, a companion in arms to replication. Almost immediately thereafter, details of the code employed by the messenger appeared. In 1961, Marshall Nirenberg and J. Heinrich Matthaei announced that they had discovered a specific point of contact between RNA and the amino acids. And then, in short order, the full genetic code emerged. RNA (like DNA) is organized into triplets, so that adjacent sequences of three bases are mapped to a single amino acid. Sixty-four triplets (or codons) govern twenty amino acids. The scheme is universal, or almost so.

The elaboration of the genetic code made possible a remarkably elegant model of the modern cell as a system in which sequences of codons within the nucleic acids act at a distance to determine sequences of amino acids within the proteins: commands issued, responses undertaken. A third fundamental biological process thus acquired molecular incarnation. If replication served to divide and then to duplicate the cell's ancestral message, and transcription to re-express it in messenger RNA, "translation" acted to convey that message from messenger RNA to the amino acids.

For all the boldness and power of this thesis, the details remained on the level of what bookkeepers call general accounting procedures. No one had established a direct—a *physical*—connection between RNA and the amino acids.

Having noted the problem, Crick also indicated the shape of its solution. "I therefore proposed a theory," he would write retrospectively, "in which there were twenty adaptors (one for each amino acid), together with twenty special enzymes. Each enzyme would join one particular amino acid to its own special adaptor."

In early 1969, at roughly the same time that a somber Lyndon Johnson was departing the White House to return to the Pedernales, the adaptors whose existence Crick had predicted came into view. There were twenty, just as he had suggested. They were short in length; they were specific in their action; and they were nucleic acids. Collectively, they are now designated "transfer" RNA (tRNA).

Folded like a cloverleaf, transfer RNA serves physically as a bridge between messenger RNA and an amino acid. One arm of the cloverleaf is called the anti-coding region. The three nucleotide bases that it contains are curved around the arm's bulb-end; they are matched by Watson-Crick base pairing to bases on the messenger RNA. The other end of the cloverleaf is an acceptor region. It is here that an amino acid must go, with the structure of tRNA suggesting a complicated female socket waiting to be charged by an appropriate male amino acid.

The adaptors whose existence Crick had predicted served dramatically to confirm his hypothesis that such adaptors were needed. But although they brought about a physical connection between the nucleic and the amino acids, the fact that they were themselves nucleic acids raised a question: in the unfolding molecular chain, just what acted to adapt the adaptors to the amino acids? And this, too, was a problem Crick both envisaged and solved: his original suggestion mentioned both adaptors (nucleic acids) and their *enzymes* (proteins).

And so again it proved. The act of matching adaptors to amino acids is carried out by a family of enzymes, and thus by a family of proteins: the aminoacyl-tRNA synthetases. There are as many such enzymes as there are adaptors. The prefix "aminoacyl" indicates a class of chemical reactions, and it is in aminoacylation that the cargo of a carboxyl group is bonded to a molecule of transfer RNA.

Collectively, the enzymes known as synthetases have the power both to recognize specific codons and to select their appropriate amino acid under the universal genetic code. Recognition and selection are ordinarily thought to be cognitive acts. In psychology, they are poorly understood, but within the cell they have been accounted for in chemical terms and so in terms of "the model for what science should be."

With tRNA appropriately charged, the molecule is conveyed to the ribosome, where the task of assembling sequences of amino acids is then undertaken by still another nucleic acid, ribosomal RNA (rRNA). By these means, the modern cell is at last subordinated to a rich narrative drama. To repeat:

Replication duplicates the genetic message in DNA.
Transcription copies the genetic message from DNA to RNA.
Translation conveys the genetic message from RNA to the amino acids—whereupon, in a fourth and final step, the amino acids are assembled into proteins.

4. THE CENTRAL DOGMA

It was once again Francis Crick, with his remarkable gift for impressing his authority over an entire discipline, who elaborated these facts into what he called the central dogma of molecular biology. The cell, Crick affirmed, is a divided kingdom. Acting as the cell's administrators, the nucleic acids embody all of the requisite wisdom—where to go, what to do, how to manage—in the specific sequence of their nucleotide bases. Administration then proceeds by the transmission of information *from* the nucleic acids *to* the proteins.

The central dogma thus depicts an arrow moving one way, from the nucleic acids to the proteins, and never the other way around. But is anything ever routinely returned, arrow-like, from its target? This is not a question that Crick considered, although in one sense the answer is plainly no. Given the modern genetic code, which maps four nucleotides onto twenty amino acids, there can be no inverse code going in the opposite direction; an inverse mapping is mathematically impossible.

But there is another sense in which Crick's central dogma does engender its own reversal. If the nucleic acids are the cell's administrators, the proteins are its chemical executives: both the staff and the stuff of life. The molecular arrow goes one way with respect to information, but it goes the other way with respect to chemistry.

Replication, transcription, and translation represent the grand unfolding of the central dogma as it proceeds in one direction. The chemical activities initiated by the enzymes represent the grand unfolding of the central dogma as it goes in the other. Within the cell, the two halves of the central dogma combine to reveal a *system of coded chemistry*, an exquisitely intricate but remarkably coherent temporal tableau suggesting a great army in action.

From these considerations a familiar figure now emerges: the figure of a chicken and its egg. Replication, transcription, and translation are all under the control of various enzymes. But enzymes are proteins, and these particular proteins are specified by the cell's nucleic acids. DNA requires the enzymes in order to undertake the work of replication, transcription, and translation; the enzymes require DNA in order to initiate it. The nucleic acids and the proteins are thus profoundly coordinated, each depending upon the other. Without amino-acyl-tRNA synthetase, there is no translation from RNA; but without DNA, there is no synthesis of aminoacyl-tRNA synthetase.

If the nucleic acids and their enzymes simply chased each other forever around the same cell, the result would be a vicious circle. But life has elegantly resolved the circle in the form of a spiral. The aminoacyl-tRNA synthetase that is required to complete molecular translation enters a given cell from its progenitor or "maternal" cell, where it is specified by that cell's DNA. The enzymes required to make the maternal cell's DNA do its work enter that cell from *its* maternal line. And so forth.

On the level of intuition and experience, these facts suggest nothing more mysterious than the longstanding truism that life comes only from life. *Omnia viva ex vivo*, as Latin writers said. It is only when they are embedded in various theories about the *origins* of life that the facts engender a paradox, or at least a question: in the receding molecular spiral, which came first—the chicken in the form of DNA, or its egg in the form of various proteins? And if neither came first, how could life have begun?

5. The RNA World

It is 1967, the year of the Six-Day war in the Middle East, the discovery of the electroweak forces in particle physics, and the completion of a twenty-year research program devoted to the effects of fluoridation on dental caries in Evanston, Illinois. It is also the year in which Carl Woese, Leslie Orgel, and Francis Crick introduced the hypothesis that "evolution based on RNA replication *preceded* the appearance of protein synthesis" (emphasis added).

By this time, it had become abundantly clear that the structure of the modern cell was not only more complex than other physical structures but complex in poorly understood ways. And yet no matter how far back biologists traveled into the tunnel of time, certain features of the modern cell were still there, a message sent into the future by the last universal common ancestor. Summarizing his own perplexity in retrospect, Crick would later observe that "an honest man, armed with all the knowledge available to us now, could only state that, in some sense, the origin of life appears at the moment to be almost a miracle." Very wisely, Crick would thereupon determine never to write another paper on the subject—although he did affirm his commitment to the theory of "directed panspermia," according to which life originated in some other portion of the universe and, for reasons that Crick could never specify, was simply sent here.

But that was later. In 1967, the argument presented by Woesel, Orgel, and Crick was simple. Given those chickens and their eggs, *something* must have come first. Two possibilities were struck off by a process of elimination. DNA? Too stable and, in some odd sense, too perfect. The proteins? Incapable of dividing themselves, and so, like molecular eunuchs, useful without being fecund. That left RNA. While it was not obviously the right choice for a primordial molecule, it was not obviously the wrong choice, either.

The hypothesis having been advanced—if with no very great sense of intellectual confidence—biologists differed in its interpretation. But they did concur on three general principles. First: that at some time in the distant past, RNA rather than DNA controlled genetic replication. Second: that Watson-Crick base pairing governed ancestral RNA. And third: that RNA once carried on chemical activities of the sort that are now entrusted to the proteins. The paradox of the chicken and the egg was thus resolved by the hypothesis that the chicken *was* the egg.

The independent discovery in 1981 of the ribozyme—a ribonucleic enzyme—by Thomas Cech and Sidney Altman endowed the RNA hypothesis with the force of a scientific conjecture. Studying the ciliated protozoan *Tetrahymena thermophila*, Cech discovered to his astonishment a form of RNA capable of inducing cleavage. Where an enzyme might have been busy pulling a strand of RNA apart, there was a ribozyme doing the work instead. That busy little molecule served not only to give instructions: apparently it took them as well, and in any case it did what biochemists had since the 1920s assumed could only be done by an enzyme and hence by a protein.

In 1986, the biochemist Walter Gilbert was moved to assert the existence of an entire RNA "world," an ancestral state promoted by the magic of this designation to what a great many biologists would affirm as fact. Thus, when the molecular biologist Harry Noller discovered that protein synthesis within the contemporary ribosome is catalyzed by ribosomal RNA (rRNA), and not by any of the familiar, old-fashioned enzymes, it appeared "almost certain" to Leslie Orgel that "there once *was* an RNA world" (emphasis added).

6. From Molecular Biology to the Origins of Life

It is perfectly true that every part of the modern cell carries some faint traces of the past. But these molecular traces are only hints. By contrast, to everyone who has studied it the ribozyme has appeared to be an authentic relic, a solid and palpable souvenir from the pre-biotic past. Its discovery prompted even Francis Crick to the admission that he, too, wished he had been clever enough to look for such relics before they became known.

Thanks to the ribozyme, a great many scientists have become convinced that the "model for what science should be" is achingly close to encompassing the origins of life itself. "My expectation," remarks David Liu, professor of chemistry and chemical biology at Harvard, "is that we will be able to reduce this to a very simple series of logical events." Although often overstated, this optimism is by no means irrational. Looking at the modern cell, biologists propose to reconstruct in time the structures that are now plainly there in space.

Research into the origins of life has thus been subordinated to a rational three-part sequence, beginning in the very distant past. First, the constituents of the cell were formed and assembled. These included the nucleotide bases, the amino acids, and the sugars. There followed next the emergence of the ribozyme, endowed somehow with powers of self-replication. With the stage set, a system of coded chemistry then emerged, making possible what the molecular biologist Paul Schimmel has called "the theater of the proteins." Thus did matters proceed from the pre-biotic past to the very threshold of the last universal common ancestor, whereupon, with inimitable gusto, life began to diversify itself by means of Darwinian principles.

This account is no longer fantasy. But it is not yet fact. That is one reason why retracing its steps is such an interesting exercise, to which we now turn.

7. Miller Time

It is perhaps four billion years ago. The first of the great eras in the formation of life has commenced. The laws of chemistry are completely in control of things—what else is there? It is Miller Time, the period marking the transition from inorganic to organic chemistry.

According to the impression generally conveyed in both the popular and the scientific literature, the success of the original Miller-Urey experiment was both absolute and unqualified. This, however, is something of an exaggeration. Shortly after Miller and Urey published their results, a number of experienced geochemists expressed reservations. Miller and Urey had assumed that the pre-biotic atmosphere was one in which hydrogen atoms gave up (reduced) their electrons in order to promote chemical activity. Not so, the geochemists contended. The pre-biotic atmosphere was far more nearly neutral than reductive, with little or no methane and a good deal of carbon dioxide.

Nothing in the intervening years has suggested that these sour geochemists were far wrong. Writing in the 1999 issue of *Peptides,* B. M. Rode observed blandly that "modern geochemistry assumes that the secondary atmosphere of the primitive Earth (i.e., after diffusion of hydrogen and helium into space) . . . consisted mainly of carbon dioxide, nitrogen, water, sulfur dioxide, and even small amounts of oxygen." This is not an environment calculated to induce excitement.

Until recently, the chemically unforthcoming nature of the early atmosphere remained an embarrassing secret among evolutionary biologists, like an uncle known privately to dress in women's underwear; if biologists were disposed in public to acknowledge the facts, they did so by remarking that every family has one. This has now changed. The issue has come to seem troubling. A recent paper in *Science* has suggested that previous conjectures about the pre-biotic atmosphere were seriously in error. A few researchers have argued that a reducing atmosphere is not, after all, quite so important to pre-biotic synthesis as previously imagined.

In all this, Miller himself has maintained a far more unyielding and honest perspective. "Either you have a reducing atmosphere," he has written bluntly, "or you're not going to have the organic compounds required for life."

If the composition of the pre-biotic atmosphere remains a matter of controversy, this can hardly be considered surprising: geochemists are attempting to revisit an era that lies four billion years in the past. The synthesis of pre-biotic chemicals is another matter. Questions about them come under the discipline of laboratory experiments.

Among the questions is one concerning the nitrogenous base cytosine (C). Not a trace of the stuff has been found in any meteor. Nothing in comets, either, so far as anyone can tell. It is not buried in the Antarctic. Nor can it be produced by any of the common experiments in pre-biotic chemistry. Beyond the living cell, it has not been found at all.

When, therefore, M. P. Robertson and Stanley Miller announced in *Nature* in 1995 that they had specified a plausible route for the pre-biotic synthesis of cytosine from cyanoacetaldehyde and urea, the feeling of gratification was very considerable. But it has also been short-lived. In a lengthy and influential review published in the 1999 *Proceedings of the National Academy of Science,* the New York University chemist Robert Shapiro observed that the reaction on which Robertson and Miller had pinned their hopes, although active enough, ultimately went nowhere. All too quickly, the cytosine that they had synthesized transformed itself into the RNA base uracil (U) by a chemical reaction known as deamination, which is nothing more mysterious than the process of getting rid of one molecule by sending it somewhere else.

The difficulty, as Shapiro wrote, was that "the formation of cytosine and the subsequent deamination of the product to uracil occur[ed] at about the same rate." Robertson and Miller had themselves reported that after 120 hours, half of their precious cytosine was gone—and it went faster when their reactions took place in saturated urea. In Shapiro's words, "It is clear that the yield of cytosine would fall to 0 percent if the reaction were extended."

If the central chemical reaction favored by Robertson and Miller was self-defeating, it was also contingent on circumstances that were unlikely. *Concentrated* urea was needed to prompt their reaction; an outhouse whiff would not do. For this same reason, however, the pre-biotic sea, where concentrates disappear too quickly, was hardly the place to begin—as anyone who has safely relieved himself in a swimming pool might confirm with guilty satisfaction. Aware of this, Robertson and Miller posited a different set of circumstances: in place of the pre-biotic soup, drying lagoons. In a fine polemical passage, their critic Shapiro stipulated what would thereby be required:

> An isolated lagoon or other body of seawater would have to undergo extreme concentration.
>
> It would further be necessary that the residual liquid be held in an impermeable vessel [in order to prevent cross-reactions].
>
> The concentration process would have to be interrupted for some decades . . . to allow the reaction to occur.
>
> At this point, the reaction would require quenching (perhaps by evaporation to dryness) to prevent loss by deamination.
>
> At the end, one would have a batch of urea in solid form, containing some cytosine (and urea).

Such a scenario, Shapiro remarked, "cannot be excluded as a rare event on early Earth, but it cannot be termed plausible."

Like cytosine, sugar must also make an appearance in Miller Time, and, like cytosine, it too is difficult to synthesize under plausible pre-biotic conditions.

In 1861, the German chemist Alexander Bulterow created a sugar-like substance from a mixture of formaldehyde and lime. Subsequently refined by a long line of organic chemists, Bulterow's so-called formose reaction has been an inspiration to origins-of-life researchers ever since.

On the Origins of Life

The reaction is today initiated by an alkalizing agent, such as thallium or lead hydroxide. There follows a long induction period, with a number of intermediates bubbling up. The formose reaction is autocatalytic in the sense that it keeps on going: the carbohydrates that it generates serve to prime the reaction in an exponentially growing feedback loop until the initial stock of formaldehyde is exhausted. With the induction over, the formose reaction yields a number of complex sugars.

Nonetheless, it is not sugars in general that are wanted from Miller Time but a particular form of sugar, namely, ribose—and not simply ribose but dextro ribose. Compounds of carbon are naturally right-handed or left-handed, depending on how they polarize light. The ribose in living systems is right-handed, hence the prefix "dextro." But the sugars exiting the formose reaction are racemic, that is, both left- and right-handed, and the yield of usable ribose is negligible.

While nothing has as yet changed the fundamental fact that it is very hard to get the right kind of sugar from any sort of experiment, in 1990 the Swiss chemist Albert Eschenmoser was able to change substantially the way in which the sugars appeared. Reaching with the hand of a master into the formose reaction itself, Eschenmoser altered two molecules by adding a phosphate group to them. This slight change prevented the formation of the alien sugars that cluttered the classical formose reaction. The products, Eschenmoser reported, included among other things a mixture of ribose-2,4,-diphosphate. Although the mixture was racemic, it did contain a molecule close to the ribose needed by living systems. With a few chemical adjustments, Eschenmoser could plausibly claim, the pre-biotic route to the synthesis of sugar would lie open.

It remained for skeptics to observe that Eschenmoser's ribose reactions were critically contingent on Eschenmoser himself, and at two points: the first when he attached phosphate groups to a number of intermediates in the formose reaction, and the second when he removed them.

What had given the original Miller-Urey experiment its power to excite the imagination was the sense that, having set the stage, Miller and Urey exited the theater. By contrast, Eschenmoser remained at center stage, giving directions and in general proving himself indispensable to the whole scene.

Events occurring in Miller Time would thus appear to depend on the large assumption, still unproved, that the early atmosphere was reductive, while two of the era's chemical triumphs, cytosine and sugar, remain for the moment beyond the powers of contemporary pre-biotic chemistry.

8. From Miller Time to Self-Replicating RNA

In the grand progression by which life arose from inorganic matter, Miller Time has been concluded. It is now 3.8 billion years ago. The chemical precursors to life have been formed. A limpid pool of nucleotides is somewhere in existence. A new era is about to commence.

The historical task assigned to this era is a double one: forming chains of nucleic acids from nucleotides, and discovering among them those capable of reproducing themselves. Without the first, there is no RNA; and without the second, there is no life.

In living systems, polymerization or chain-formation proceeds by means of the cell's invaluable enzymes. But in the grim inhospitable pre-biotic, no enzymes were available. And so chemists have assigned their task to various inorganic catalysts. J. P. Ferris and G. Ertem, for instance, have reported that activated nucleotides bond covalently when embedded on the surface of montmorillonite, a kind of clay. This example, combining technical complexity with general inconclusiveness, may stand for many others.

In any event, polymerization having been concluded—by whatever means—the result was (in the words of Gerald Joyce and Leslie Orgel) "a random ensemble of polynucleotide sequences": long molecules emerging from short ones, like fronds on the surface of a pond. Among these fronds, nature is said to have discovered a self-replicating molecule. But how?

Darwinian evolution is plainly unavailing in this exercise or that era, since Darwinian evolution *begins* with self-replication, and self-replication is precisely what needs to be explained. But if Darwinian evolution is unavailing, so, too, is chemistry. The fronds comprise "a *random* ensemble of polynucleotide sequences" (emphasis added); but no principle of organic chemistry suggests that aimless encounters among nucleic acids must lead to a chain capable of self-replication.

If chemistry is unavailing and Darwin indisposed, what is left as a mechanism? The evolutionary biologist's finest friend: sheer dumb luck.

Was nature lucky? It depends on the payoff and the odds. The payoff is clear: an ancestral form of RNA capable of replication. Without that payoff, there is no life, and obviously, at some point, the payoff paid off. The question is the odds.

For the moment, no one knows how precisely to compute those odds, if only because within the laboratory, no one has conducted an experiment leading to a self-replicating ribozyme. But the minimum length or "sequence" that is needed for a contemporary ribozyme to undertake what the distinguished geochemist Gustaf Arrhenius calls "demonstrated ligase activity" *is* known. It is roughly 100 nucleotides.

Whereupon, just as one might expect, things blow up very quickly. As Arrhenius notes, there are 4^{100} or roughly 10^{60} nucleotide sequences that are 100 nucleotides in length. This is an unfathomably large number. It exceeds the number of atoms contained in the universe, as well as the age of the universe in seconds. If the odds in favor of self-replication are 1 in 10^{60}, no betting man would take them, no matter how attractive the payoff, and neither presumably would nature.

"Solace from the tyranny of nucleotide combinatorials," Arrhenius remarks in discussing this very point, "is sought in the feeling that strict sequence specificity may not be required through all the domains of a functional oligmer, thus making a large number of library items eligible for participation in the construction of the ultimate functional entity." Allow me to translate: why assume that self-replicating sequences are apt to be rare just because they are long? They might have been quite common.

They might well have been. And yet all experience is against it. Why should self-replicating RNA molecules have been common 3.6 billion years ago when they are impossible to discern under laboratory conditions today? No one, for that matter, has ever seen a ribozyme capable of *any* form of catalytic action that is not very specific in its sequence and thus unlike even closely related sequences. No one has ever seen a ribozyme able to undertake chemical action without a suite of enzymes in attendance. No one has ever seen anything like it.

The odds, then, are daunting; and when considered realistically, they are even worse than this already alarming account might suggest. The discovery of a single molecule with the power to initiate replication would hardly be sufficient to establish replication. What template would it replicate *against?* We need, in other words, at least two, causing the odds of their joint discovery to increase from 1 in 10^{60} to 1 in 10^{120}. Those two sequences would have been needed in roughly the same place. And at the same time. And organized in such a way as to favor base pairing. And somehow held in place. And buffered against competing reactions. And productive enough so that their duplicates would not at once vanish in the soundless sea.

In contemplating the discovery by chance of two RNA sequences a mere forty nucleotides in length, Joyce and Orgel concluded that the requisite "library" would require 10^{48} possible sequences. Given the weight of RNA, they observed gloomily, the relevant sample space would exceed the mass of the Earth. And this is the same Leslie Orgel, it will be remembered, who observed that "it was almost certain that there once was an RNA world."

To the accumulating agenda of assumptions, then, let us add two more: that without enzymes, nucleotides were somehow formed into chains, and that by means we cannot duplicate in the laboratory, a pre-biotic molecule discovered how to reproduce itself.

9. From Self-Replicating RNA to Coded Chemistry

A new era is now in prospect, one that begins with a self-replicating form of RNA and ends with the system of coded chemistry characteristic of the modern cell. The *modern* cell—meaning one that divides its labors by assigning to the nucleic acids the management of information and to the proteins the execution of chemical activity. It is 3.6 billion years ago.

It is with the advent of this era that distinctively conceptual problems emerge. The gods of chemistry may now be seen receding into the distance. The cell's system of coded chemistry is determined by two discrete combinatorial objects: the nucleic acids and the amino acids. These objects are discrete because, just as there are no fractional sentences containing three and a half words, there are no fractional nucleotide sequences containing three and a half nucleotides, or fractional proteins containing three and a half amino acids. They are combinatorial because both the nucleic acids and the amino acids are combined by the cell into larger structures.

But if information management and its administration within the modern cell are determined by a discrete combinatorial system, the *work* of the cell is part of a markedly different enterprise. The periodic table notwithstanding, chemical reactions are not combinatorial, and they are not discrete. The chemical bond, as Linus Pauling demonstrated in the 1930s, is based squarely on quantum mechanics. And to the extent that chemistry is explained in terms of physics, it is encompassed not only by "the model for what science should be" but by the system of differential equations that play so conspicuous a role in every one of the great theories of mathematical physics.

What serves to coordinate the cell's two big shots of information management and chemical activity, and so to coordinate two fundamentally different structures, is the universal genetic code. To capture the remarkable nature of the facts in play here, it is useful to stress the word *code*.

By itself, a code is familiar enough: an arbitrary mapping or a system of linkages between two discrete combinatorial objects. The Morse code, to take a familiar example, coordinates dashes and dots with letters of the alphabet. To note that codes are arbitrary is to note the distinction between a code and a purely physical connection between two objects. To note that codes embody mappings is to embed the concept of a code in mathematical language. To note that codes reflect a linkage of some sort is to return the concept of a code to its human uses.

In every normal circumstance, the linkage comes first and represents a human achievement, something arising from a point beyond the coding system. (The coordination of dot-dot-dot-dash-dash-dash-dot-dot-dot with the distress signal S-O-S is again a familiar example.) Just as no word explains its own meaning, no code establishes its own nature.

The conceptual question now follows. Can the origins of a system of coded chemistry be explained in a way that makes no appeal whatsoever to the kinds of facts that we otherwise invoke to explain codes and languages, systems of communication, the impress of ordinary words on the world of matter?

In this regard, it is worth recalling that, as Hubert Yockey observes in *Information Theory, Evolution, and the Origin of Life* (2005), "there is no trace in physics or chemistry of the control of chemical reactions by a sequence of any sort or of a code between sequences."

Writing in the 2001 issue of the journal *RNA*, the microbiologist Carl Woese referred ominously to the "dark side of molecular biology." DNA replication, Woese wrote, is the extraordinarily elegant expression of the structural properties of a single molecule: zip down, divide, zip up. The transcription into RNA follows suit: copy and conserve. In each of these two cases, structure leads to function. But where is the coordinating link between the chemical structure of DNA and the third step, namely, translation? When it comes to translation, the apparatus is baroque: it is incredibly elaborate, and it does not reflect the structure of any molecule.

These reflections prompted Woese to a somber conclusion: if "the nucleic acids cannot in any way recognize the amino acids," then there is no "fundamental *physical* principle" at work in translation (emphasis added).

But Woese's diagnosis of disorder is far too partial; the symptoms he regards as singular are in fact widespread. What holds for translation holds as well for replication and transcription. The nucleic acids cannot directly recognize the amino acids (and vice-versa), but they cannot *directly* replicate or transcribe themselves, either. Both replication and translation are enzymatically driven, and without those enzymes, a molecule of DNA or RNA would do nothing whatsoever. Contrary to what Woese imagines, no fundamental physical principles appear directly at work *anywhere* in the modern cell.

The most difficult and challenging problem associated with the origins of life is now in view. One half of the modern system of coded chemistry—the genetic code and the sequences it conveys—is, from a chemical perspective, arbitrary. The other half of the system of coded chemistry—the activity of the proteins—is, from a chemical perspective, necessary. In life, the two halves are coordinated. The problem follows: how did *that*—the whole system—get here?

The prevailing opinion among molecular biologists is that questions about molecular-biological systems can only be answered by molecular-biological *experiments*. The distinguished molecular biologist Horoaki Suga has recently demonstrated the strengths and the limitations of the experimental method when confronted by difficult conceptual questions like the one I have just posed.

The goal of Suga's experiment was to show that a set of RNA catalysts (or ribozymes) *could* well have played the role now played in the modern cell by the protein family of aminoacyl synthetases. Until his work, Suga reports, there had been no convincing demonstration that a ribozyme was able to perform the double function of a synthetase—that is, recognizing both a form of transfer RNA and an amino acid. But in Suga's laboratory, just such a molecule made a now-celebrated appearance. With an amino acid attached to its tail, the ribozyme managed to cleave itself and, like a snake, affix its amino-acid cargo onto its head. What is more, it could conduct this exercise backward, shifting the amino acid from its head to its tail again. The chemical reactions involved acylation: precisely the reactions undertaken by synthetases in the modern cell.

Horoaki Suga's experiment was both interesting and ingenious, prompting a reaction perhaps best expressed as, "Well, would you look at that!" It has altered the terms of debate by placing a number of new facts on the table. And yet, as so often happens in experimental prebiotic chemistry, it is by no means clear what interpretation the facts will sustain.

Do Suga's results really establish the existence of a primitive form of coded chemistry? Although unexpected in context, the coordination he achieved between an amino acid and a form of transfer RNA was never at issue in principle. The question is whether what was accomplished in establishing a chemical connection between these two molecules was anything like establishing the existence of a *code*. If so, then organic chemistry itself could properly be described as the study of codes, thereby erasing the meaning of a code as an arbitrary mapping between discrete combinatorial objects.

Suga, in summarizing the results of his research, captures rhetorically the inconclusiveness of his achievement. "Our demonstration indicates," he writes, "that catalytic precursor tRNA's *could have provided* the foundation of the genetic coding system." But if the association at issue is not a code, however primitive, it could no more be the "foundation" of a code than a feather could be the foundation of a building. And if it is the foundation of a code, then what has been accomplished has been accomplished by the wrong agent.

In Suga's experiment, there was no sign that the execution of chemical routines fell under the control of a molecular administration, and no sign, either, that the missing molecular administration had anything to do with executive chemical routines. The missing molecular administrator was, in fact, Suga himself, as his own account reveals. The relevant features of the experiment, he writes, "allow[ed] *us* to select active RNA molecules with selectivity toward a *desired* amino acid" (emphasis added). Thereafter, it was Suga and his collaborators who "applied *stringent* conditions" to the experiment, undertook "*selective amplification* of the self-modifying RNA molecules," and "*screened*" vigorously for "self-aminoacylation activity" (emphasis added throughout).

If nothing else, the advent of a system of coded chemistry satisfied the most urgent of imperatives: it was needed and it was found. It was needed because once a system of chemical reactions reaches a certain threshold of complexity, nothing less than a system of coded chemistry can possibly master the ensuing chaos. It was found because, after all, we are here.

Precisely these circumstances have persuaded many molecular biologists that the explanation for the emergence of a system of coded chemistry must in the end lie with Darwin's theory of evolution. As one critic has observed in commenting on Suga's experiments, "If a certain result can be achieved by direction in a laboratory by a Suga, surely it can also be achieved by chance in a vast universe."

A self-replicating ribozyme meets the first condition required for Darwinian evolution to gain purchase. It is by definition capable of replication. And it meets the second condition as well, for, by means of mistakes in replication, it introduces the possibility of variety into the biological world. On the assumption that subsequent changes to the system follow a law of increasing marginal utility, one can then envisage the eventual emergence of a system of coded chemistry—a system that can be explained in terms of "the model for what science should be."

It was no doubt out of considerations like these that, in coming up against what he called the "dark side of molecular biology," Carl Woese was concerned to urge upon the biological community the benefits of "an all-out Darwinian perspective." But the difficulty with "an all-out Darwinian perspective" is that it entails an all-out Darwinian impediment: notably, the assignment of a degree of foresight to a Darwinian process that the process could not possibly possess.

The hypothesis of an RNA world trades brilliantly on the idea that a divided modern system had its roots in some form of molecular symmetry that was then broken by the contingencies of life. At some point in the transition to the modern system, an ancestral form of RNA must have assigned some of its catalytic properties to an emerging family of proteins. This would have taken place at a given historical moment; it is not an artifact of the imagination. Similarly, at some point in the transition to a modern system, an ancestral form of RNA must have acquired the ability to code for the catalytic powers it was discarding. And this, too, must have taken place at a particular historical moment.

The question, of course, is which of the two steps came first. Without life acquiring some degree of foresight, neither step can be plausibly fixed in place by means of any schedule of selective advantages. How could an ancestral form of RNA have acquired the ability to code for various amino acids before coding was useful? But then again, why should "ribozymes in an RNA world," as the molecular biologists Paul Schimmel and Shana O. Kelley ask, "have expedited their own obsolescence?"

Could the two steps have taken place simultaneously? If so, there would appear to be very little difference between a Darwinian explanation and the frank admission that a miracle was at work. If no miracles are at work, we are returned to the place from which we started, with the chicken-and-egg pattern that is visible when life is traced backward now appearing when it is traced forward.

It is thus unsurprising that writings embodying Woese's "all-out Darwinian perspective" are dominated by references to a number of unspecified but mysteriously potent forces and obscure conditional circumstances. I quote without attribution because the citations are almost generic (emphasis added throughout):

- The aminoacylation of RNA initially *must* have provided some selective advantage.
- The products of this reaction *must* have conferred some selective advantage.
- However, the development of a crude mechanism for controlling the diversity of possible peptides *would* have been advantageous.
- [P]rogressive refinement of that mechanism *would* have provided *further* selective advantage.

And so forth—ending, one imagines, in reduction to the all-purpose imperative of Darwinian theory, which is simply that what was must have been.

10. Now It Is Now

At the conclusion of a long essay, it is customary to summarize what has been learned. In the present case, I suspect it would be more prudent to recall how much has been *assumed:*

First, that the pre-biotic atmosphere was chemically reductive; second, that nature found a way to synthesize cytosine; third, that nature also found a way to synthesize ribose; fourth, that nature found the means to assemble nucleotides into polynucleotides; fifth, that nature discovered a self-replicating molecule; and sixth, that having done all that, nature promoted a self-replicating molecule into a full system of coded chemistry.

These assumptions are not only vexing but progressively so, ending in a serious impediment to thought. That, indeed, may be why a number of biologists have lately reported a weakening of their commitment to the RNA world altogether, and a desire to look elsewhere for an explanation of the emergence of life on Earth. "It's part of a quiet paradigm revolution going on in biology," the biophysicist Harold Morowitz put it in an interview in *New Scientist,* "in which the radical randomness of Darwinism is being replaced by a much more scientific law-regulated emergence of life."

Morowitz is not a man inclined to wait for the details to accumulate before reorganizing the vista of modern biology. In a series of articles, he has argued for a global vision based on the biochemistry of living systems rather than on their molecular biology or on Darwinian adaptations. His vision treats the living system as more fundamental than its particular species, claiming to represent the "universal and deterministic features of *any* system of chemical interactions based on a water-covered but rocky planet such as ours."

This view of things—metabolism first, as it is often called—is not only intriguing in itself but is enhanced by a firm commitment to chemistry and to "the model for what science should be." It has been argued with great vigor by Morowitz and others. It represents an alternative to the RNA world. It is a work in progress, and it may well be right. Nonetheless, it suffers from one outstanding defect. There is as yet no evidence that it is true.

It is now more than 175 years since Friedrich Wöhler announced the synthesis of urea. It would be the height of folly to doubt that our understanding of life's origins has been immeasurably improved. But whether it has been immeasurably improved in a way that vigorously confirms the daring idea that living systems are chemical in their origin and so physical in their nature—that is another question entirely . . .

[I]n contemplating the origins of life, much—in fact, more—can be learned by studying the issue from the perspective of coded chemistry. In both cases, however, what seems to lie beyond the reach of "the model for what science should be" is any success beyond the local. All questions about the global origins of these strange and baffling systems seem to demand answers that the model itself cannot by its nature provide.

It goes without saying that this is a tentative judgment, perhaps only a hunch. But let us suppose that questions about the origins of the mind and the origins of life do lie beyond the grasp of "the model for what science should be." In that case, we must either content ourselves with its limitations or revise the model. If a revision also lies beyond our powers, then we may well have to say that the mind and life have appeared in the universe for no very good reason that we can discern.

Worse things have happened. In the end, these are matters that can only be resolved in the way that all such questions are resolved. We must wait and see.

Notes

1. The first version of this essay appeared in *Commentary* 121, no. 2, February 2006, 22–33. It is reprinted with permission.
2. I borrowed this phrase from the mathematicians J. H. Hubbard and B. H. West. The idea that science must conform to a certain model of inquiry is familiar. Hubbard and West identify that model with differential equations, the canonical instruments throughout physics and chemistry.

14

DNA:
THE SIGNATURE IN THE CELL

STEPHEN C. MEYER

Theories about the origin of life necessarily presuppose knowledge of the attributes of living cells. As historian of biology Harmke Kamminga has observed, "At the heart of the problem of the origin of life lies a fundamental question: What is it exactly that we are trying to explain the origin of?"[1] Or as the pioneering chemical evolutionary theorist Aleksandr Oparin put it, "The problem of the nature of life and the problem of its origin have become inseparable."[2] Origin-of-life researchers want to explain the origin of the first and presumably simplest—or, at least, minimally complex—living cell. As a result, developments in fields that explicate the nature of unicellular life have historically defined the questions that origin-of-life scenarios must answer.

Since the late 1950s and 1960s, origin-of-life researchers have increasingly recognized the complex and specific nature of unicellular life and the bio-macromolecules on which such systems depend. Further, molecular biologists and origin-of-life researchers have characterized this complexity and specificity in informational terms. Molecular biologists routinely refer to DNA, RNA, and proteins as carriers or repositories of "information."[3] Many origin-of-life researchers now regard the origin of the information in these bio-macromolecules as the central question facing their research. As Bernd-Olaf Küppers has stated, "The problem of the origin of life is clearly basically equivalent to the problem of the origin of biological information."[4]

This essay will evaluate competing explanations for the origin of the information necessary to build the first living cell. To do so will require determining what biologists have meant by the term information as it has been applied to bio-macromolecules. As many have noted, "information" can denote several theoretically distinct concepts. This essay will attempt to eliminate this ambiguity and to determine precisely what type of information origin-of-life researchers must explain "the origin of." What follows will first seek to characterize the information in DNA, RNA, and proteins as an *explanandum* (a fact in need of explanation), and second, to evaluate the efficacy of competing classes of explanation for the origin of biological information (that is, the competing *explanans*).

Part 1 will seek to show that molecular biologists have used the term "information" consistently to refer to the joint properties of complexity and functional specificity or specification. Biological usage of the term will be contrasted with its classical information-theoretic usage to show that "biological information" entails a richer sense of information than the classical mathematical theory of Shannon and Wiener. *Part 1* will also argue against attempts to treat biological "information" as a metaphor lacking empirical content and/or ontological status.[5] It will show that the term biological information refers to two real features of living systems, complexity and specificity, features that jointly do require explanation.

Part 2 will evaluate competing types of explanation for the origin of the specified biological information necessary to produce the first living system. The categories of "chance" and "necessity" will provide a helpful heuristic for understanding the recent history of origin-of-life research. From the 1920s to the mid-1960s, origin-of-life researchers relied heavily on theories emphasizing the creative role of random events—"chance"—often in tandem with some form of prebiotic natural selection. Since the late 1960s, theorists have instead emphasized deterministic self-organizational laws or properties—that is, physical-chemical "necessity." I will critique the causal adequacy of chemical evolutionary theories based on "chance," "necessity," and the combination of the two. Finally, Part 3 will suggest that the phenomenon of information understood as specified complexity requires a radically different explanatory approach. In particular, I will argue that our present knowledge of causal powers suggests intelligent design as a better, more causally adequate explanation for the origin of the specified complexity (the information so defined) present in large biomolecules such as DNA, RNA, and proteins.

1. The Nature of Biological Information

Simple to Complex: Defining the Biological Explanandum

After Darwin published *The Origin of Species* in 1859, many scientists began to think about a problem that he had not addressed.[6] Although Darwin's theory purported to explain how life could have grown gradually more complex starting from "one or a few simple forms," it did not explain or attempt to explain how life had first originated. Yet in the 1870s and 1880s, evolutionary biologists like Ernst Haeckel and Thomas Huxley assumed that devising an explanation for the origin of life would be fairly easy, in large part because they assumed that life was, in its essence, a chemically simple substance called "protoplasm," which could easily be constructed by combining and recombining simple chemicals such as carbon dioxide, oxygen, and nitrogen.

Over the next sixty years, biologists and biochemists gradually revised their view of the nature of life. During the 1860s and 1870s, biologists tended to see the cell, in Haeckel's words, as an undifferentiated and "homogeneous globule of plasm." By the 1930s, however, most biologists had come to see the cell as a complex metabolic system.[7] Origin-of-life theories reflected this increasing appreciation of cellular complexity. Whereas nineteenth-century theories of abiogenesis envisioned life arising almost instantaneously via a one- or two-step process of chemical "autogeny," early twentieth-century theories, such as Oparin's theory of evolutionary abiogenesis, envisioned a multibillion-year process of transformation from simple chemicals to a complex metabolic system.[8] Even so, most scientists during the 1920s and 1930s still vastly underestimated the complexity and specificity of the cell and its key functional components—as developments in molecular biology would soon make clear.

The Complexity and Specificity of Proteins

During the first half of the twentieth century, biochemists had come to recognize the centrality of proteins to the maintenance of life. Although many mistakenly believed that proteins also contained the source of hereditary information, biologists repeatedly underestimated the complexity of proteins. For example, during the 1930s, English X-ray crystallographer William Astbury elucidated the molecular structure of certain fibrous proteins, such as keratin, the key structural protein in hair and skin.[9] Keratin exhibits a relatively simple, repetitive structure, and Astbury was convinced that all proteins, including the mysterious globular proteins so important to life, represented variations on the same primal and regular pattern. Similarly, in 1937 biochemists Max Bergmann and Carl Niemann of the Rockefeller Institute argued that the amino acids in proteins occurred in regular, mathematically expressible proportions. Other biologists imagined that insulin and hemoglobin proteins, for example, "consisted of bundles of parallel rods."[10]

Beginning in the 1950s, however, a series of discoveries caused this simplistic view of proteins to change. From 1949 to 1955, biochemist Fred Sanger determined the structure of the protein molecule insulin. Sanger showed that insulin consisted of a long and irregular sequence of the various amino acids, rather like a string of differently colored beads arranged without any discernible pattern. His work showed for a single protein what subsequent work in molecular biology would establish as a norm: The amino acid sequence in functional proteins generally defies expression by any simple rule and is characterized instead by aperiodicity or complexity.[11] Later in the 1950s, work by John Kendrew on the structure of the protein myoglobin showed that proteins also exhibit a surprising three-dimensional complexity. Far from the simple structures that biologists had imagined earlier, an extraordinarily complex and irregular three-dimensional shape was revealed: a twisting, turning tangle of amino acids. As Kendrew explained in 1958, "The big surprise was that it was so irregular.... [T]he arrangement seems to be almost totally lacking in the kind of regularity one instinctively anticipates, and it is more complicated than has been predicted by any theory of protein structure."[12]

By the mid-1950s, biochemists recognized that proteins possess another remarkable property. In addition to their complexity, proteins also exhibit specificity, both as one-dimensional arrays and three-dimensional structures. Whereas proteins are built from rather simple amino acid "building blocks," their function (whether as enzymes, signal transducers, or structural components in the cell) depends crucially on a complex but specific arrangement of those building blocks.[13] In particular, the specific sequence of amino acids in a chain and the resultant chemical interactions between amino acids largely determine the specific three-dimensional structure that the chain as a whole will adopt. Those structures or shapes in turn determine what function, if any, the amino acid chain can perform in the cell.

For a functioning protein, its three-dimensional shape gives it a hand-in-glove fit with other molecules, enabling it to catalyze specific chemical reactions or to build specific structures within the cell. Because of its three-dimensional specificity, one protein can usually no more substitute for another than one tool can substitute for another. A topoisomerase can no more perform the job of a polymerase than a hatchet can function as a soldering iron. Instead, proteins perform functions only by virtue of their three-dimensional specificity of fit, either with other equally specified and complex molecules or with simpler substrates within the cell. Moreover, the three-dimensional specificity derives in large part from the one-dimensional sequence specificity in the arrangement of the amino acids that form proteins. Even slight alterations in sequence often result in the loss of protein function.

The Complexity and Sequence Specificity of DNA

During the early part of the twentieth century, researchers also vastly underestimated the complexity (and significance) of nucleic acids such as DNA and RNA. By then, scientists knew the chemical composition of DNA. Biologists and chemists knew that, in addition to sugars (and later phosphates), DNA was composed of four different nucleotide bases: adenine, thymine, cytosine, and guanine. In 1909, chemist P. A. Levene showed (incorrectly, as it later turned out) that the four different nucleotide bases always occurred in equal quantities within the DNA molecule.[14] He formulated what he called the "tetranucleotide hypothesis" to account for that putative fact. According to that hypothesis, the four nucleotide bases in DNA linked together in repeating sequences of the same four chemicals in the same sequential order. Since Levene envisioned those sequential arrangements of nucleotides as repetitive and invariant, their potential for expressing any genetic diversity seemed inherently limited. To account for the heritable differences between species, biologists needed to discover some source of variable or irregular specificity, some source of information, within the germ lines of different organisms. Yet insofar as DNA was seen as an uninterestingly repetitive molecule, many biologists assumed that DNA could play little if any role in the transmission of heredity.

That view began to change in the mid-1940s, for several reasons. First, Oswald Avery's famous experiments on virulent and nonvirulent strains of *Pneumococcus* identified DNA as the key factor in accounting for heritable differences between different bacterial strains.[15] Second, work by Erwin Chargaff of Columbia University in the late 1940s undermined the "tetranucleotide hypothesis." Contradicting Levene's earlier work, Chargaff showed that nucleotide frequencies actually do differ between species, even if they often hold constant within the same species or within the same organs or tissues of a single organism.[16] More importantly, Chargaff recognized that even for nucleic acids of exactly "the same analytical composition"—meaning those with the same relative proportions of the four bases (abbreviated A, T, C, and G)—"enormous" numbers of variations in sequence were possible. As he put it, different DNA molecules or parts of DNA molecules might "differ from each other . . . in the sequence, [though] not the proportion, of their constituents." As he realized, for a nucleic acid consisting of 2,500 nucleotides (roughly the length of a long gene) the number of sequences "exhibiting the same molar proportions of individual purines [A, G] and pyrimidines [T, C] . . . is not far from 10^{1500}."[17] Thus, Chargaff showed that, contrary to the tetranucleotide hypothesis, base sequencing in DNA might well display the high degree of variability and aperiodicity required by any potential carrier of heredity.

Third, elucidation of the three-dimensional structure of DNA by Watson and Crick in 1953 made it clear that DNA could function as a carrier of hereditary information.[18] The model proposed by Watson and Crick envisioned a double-helix structure to explain the Maltese-cross pattern derived from X-ray crystallographic studies of DNA by Franklin, Wilkins, and Bragg in the early 1950s. According to the now well-known Watson and Crick model, the two strands of the helix were made of sugar and phosphate molecules linked by phosphodiester bonds. Nucleotide bases were linked horizontally to the sugars on each strand of the helix and to a complementary base on the other strand to form an internal "rung" on a twisting "ladder." For geometric reasons, their model required the pairing (across the helix) of adenine with thymine and cytosine with guanine. That complementary pairing helped to explain a significant regularity in composition ratios discovered by Chargaff. Though Chargaff had shown that none of the four nucleotide bases appears with the same frequency as all the other three, he did discover that the molar proportions of adenine and thymine, on the one hand, and cytosine and guanine, on the other, do consistently equal each other.[19] Watson and Crick's model explained the regularity Chargaff had expressed in his famous "ratios."

The Watson-Crick model made clear that DNA might possess an impressive chemical and structural complexity. The double-helix structure for DNA presupposed an extremely long and high molecular-weight structure, possessing an impressive potential for variability and complexity in sequence. As Watson and Crick explained, "The sugar-phosphate backbone in our model is completely regular but any sequence of base pairs can fit into the structure. It follows that in a long molecule many different permutations are possible, and it, therefore, seems likely that the precise sequence of bases is the code which carries genetic information."[20]

As with proteins, subsequent discoveries soon showed that DNA sequences were not only complex, but also highly specific relative to the requirements of biological function. Discovery of the complexity and specificity of proteins had led researchers to suspect a functionally specific role for DNA. Molecular biologists, working in the wake of Sanger's results, assumed that proteins were much too complex (and yet also functionally specific) to arise by chance in vivo. Moreover, given their irregularity, it seemed unlikely that a general chemical law or regularity could explain their assembly. Instead, as Jacques Monod has recalled, molecular biologists began to look for some source of information or "specificity" within the cell that could direct the construction of such highly specific and complex structures. To explain the presence of the specificity and complexity in the protein, as Monod would later insist, "you absolutely needed a code."[21]

The structure of DNA as elucidated by Watson and Crick suggested a means by which information or "specificity" might be encoded along the spine of DNA's sugar-phosphate backbone.[22] Their model suggested that variations in sequence of the nucleotide bases might find expression in the sequence of the amino acids that form proteins. In 1955, Crick proposed this idea as the so-called "sequence hypothesis." According to Crick's hypothesis, the specificity of arrangement of amino acids in proteins derives from the specific arrangement of the nucleotide bases on the DNA molecule.[23] The sequence hypothesis suggested that the nucleotide bases in DNA functioned like letters in an alphabet or characters in a machine code. Just as alphabetic letters in a written language may perform a communication function depending on their sequence, so too might the nucleotide bases in DNA result in the production of a functional protein molecule depending on their precise sequential arrangement. In both cases, function depends crucially on sequence. The sequence hypothesis implied not only the complexity but also the functional specificity of DNA base sequences.

By the early 1960s, a series of experiments had confirmed that DNA base sequences play a critical role in determining amino acid sequence during protein synthesis.[24] By that time, the processes and mechanisms by which DNA sequences determine key stages of the process were known (at least in outline). Protein synthesis, or "gene expression," proceeds as long chains of nucleotide bases are first copied during a process known as transcription. The resulting copy, a "transcript" made of single-stranded "messenger RNA," now contains a sequence of RNA bases precisely reflecting the sequence of bases on the original DNA strand. The transcript is then transported to a complex organelle called a "ribosome." At the ribosome, the transcript is "translated" with the aid of highly specific adaptor molecules (called transfer-RNAs) and specific enzymes (called amino-acyl tRNA synthetases) to produce a growing amino acid chain (Figure 1).[25]

Whereas the function of the protein molecule derives from the specific arrangement of twenty different types of amino acids, the function of DNA depends on the arrangement of just four kinds of bases. This lack of a one-to-one correspondence means that a group of three DNA nucleotides (a triplet) is needed to specify a single amino acid. In any case, the sequential arrangement of the nucleotide bases determines (in large part) the one-dimensional sequential arrangement of amino acids during protein synthesis.[26] Since protein function depends critically on amino acid sequence and amino acid sequence depends critically on DNA base sequence, the sequences in the coding regions of DNA themselves possess a high degree of specificity relative to the requirements of protein (and cellular) function.

INFORMATION THEORY AND MOLECULAR BIOLOGY

From the beginning of the molecular biological revolution, biologists have ascribed information-bearing properties to DNA, RNA, and proteins. In the parlance of molecular biology, DNA base sequences contain the "genetic information," or the "assembly instructions" necessary to direct protein synthesis. Yet the term "information" can denote several theoretically

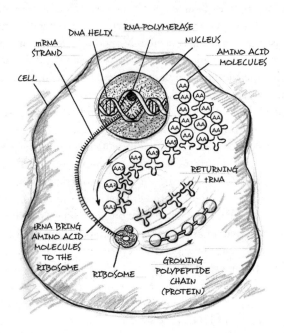

Figure 1. The intricate machinery of protein synthesis. The genetic messages encoded on the DNA molecule are copied and then transported by messenger RNA to the ribosome complex. There the genetic message is "read" and translated with the aid of other large biomolecules (transfer-RNA and specific enzyme) to produce a growing amino acid chain. (Courtesy of I. L. Cohen of New Research Publications.)

distinct concepts. Thus, one must ask which sense of "information" applies to these large bio-macromolecules. We shall see that molecular biologists employ both a stronger conception of information than that of mathematicians and information theorists and a slightly weaker conception of the term than that of linguists and ordinary users.

During the 1940s, Claude Shannon at Bell Laboratories developed a mathematical theory of information.[27] His theory equated the amount of information transmitted with the amount of uncertainty reduced or eliminated by a series of symbols or characters.[28] For example, before one rolls a six-sided die, there are six possible outcomes. Before one flips a coin, there are two. Rolling a die will thus eliminate more uncertainty, and, on Shannon's theory, will convey more information than flipping a coin. Equating information with the reduction of uncertainty implied a mathematical relationship between information and probability (or its inverse, complexity). Note that for a die each possible outcome has only a one-in-six chance of occurring, compared to a one-in-two chance for each side of the coin. Thus, in Shannon's theory, the occurrence of the more improbable event conveys more information. Shannon generalized this relationship by stating that the amount of information conveyed by an event is inversely proportional to the prior probability of its occurrence. The greater the number of possibilities, the greater the improbability of any one being actualized, and thus more information is transmitted when a particular possibility occurs.

Moreover, information increases as improbabilities multiply. The probability of getting four heads in a row when flipping a fair coin is ½ × ½ × ½ × ½, or $(½)^4$. Thus, the probability of attaining a specific sequence of heads and/or tails decreases exponentially as the number of trials increases. The quantity of information increases correspondingly. Even so, information theorists found it convenient to measure information additively rather than multiplicatively. Thus, the common mathematical expression ($I = -\log_2 p$) for calculating information converts probability values into informational measures through a negative logarithmic function, where the negative sign expresses an inverse relationship between information and probability.[29]

Shannon's theory applies most easily to sequences of alphabetic symbols or characters that function as such. Within any given alphabet of x possible characters, the placement of a specific character eliminates $x-1$ other possibilities and thus a corresponding amount of uncertainty. Or put differently, within any given alphabet or ensemble of x possible characters (where each character has an equally probable chance of occurring), the probability of any one character occurring is $1/x$. The larger the value of x, the greater the amount of information that is conveyed by the occurrence of a specific character in a sequence. In systems where the value of x can be known (or estimated), as in a code or language, mathematicians can easily generate quantitative estimates of information-carrying capacity. The greater the number of possible characters at each site and the longer the sequence of characters, the greater is the information-carrying capacity—or "Shannon information"—associated with the sequence.

The essentially digital character of the nucleotide bases in DNA, and of the amino acid residues in proteins, enabled molecular biologists to calculate the information-carrying capacity (or syntactic information) of those molecules using the new formalism of Shannon's theory. Because at every site in a growing amino acid chain, for example, the chain may receive any one of twenty amino acids, placement of a single amino acid in the chain eliminates a quantifiable amount of uncertainty and increases the Shannon or syntactic information of a polypeptide by a corresponding amount. Similarly, since at any given site along the DNA backbone any one of four nucleotide bases may occur with equal probability, the p value for the occurrence of a specific nucleotide at that site equals 1/4, or 0.25.[30] The information-carrying capacity of a sequence of a specific length n can then be calculated using Shannon's familiar expression ($I = -\log_2 p$) once one computes a p value for the occurrence of a particular sequence n nucleotides long where $p = (1/4)^n$. The p value thus yields a corresponding measure of information-carrying capacity or syntactic information for a sequence of n nucleotide bases.[31]

Complexity, Specificity, and Biological Information

Though Shannon's theory and equations provided a powerful way to measure the amount of information that could be transmitted across a communication channel, it had important limits. In particular, it did not and could not distinguish merely improbable sequences of symbols from those that conveyed a message. As Warren Weaver made clear in 1949, "The word information in this theory is used in a special mathematical sense that must not be confused with its ordinary usage. In particular, information must not be confused with meaning."[32] Information theory could measure the information-carrying capacity or the syntactic information of a given sequence of symbols, but could not distinguish the presence of a meaningful or functional arrangement of symbols from a random sequence (e.g., "we hold these truths to be self-evident" versus "ntnyhiznlhteqkhgdsjh"). Thus, Shannon information theory could quantify the amount of functional or meaningful information that *might be present* in a given sequence of symbols or characters, but it could not distinguish the status of a functional or message-bearing text from random gibberish. Thus, paradoxically, random sequences of letters often have more syntactic information (or information-carrying capacity), as measured by classical information theory, than do meaningful or functional sequences that happen to contain a certain amount of intentional redundancy or repetition.

In essence, therefore, Shannon's theory remains silent on the important question of whether a sequence of symbols is functionally specific or meaningful. Nevertheless, in its application to molecular biology, Shannon information theory did succeed in rendering rough quantitative measures of the information-carrying capacity or syntactic information (where those terms correspond to measures of brute complexity).[33] As such, information theory did help to refine biologists' understanding of one important feature of the crucial biomolecular components on which life depends: DNA and proteins are highly complex, and quantifiably so. Yet the theory by itself could not establish whether base sequences in DNA or amino acid sequences in proteins possessed the property of functional specificity. Information theory helped establish that DNA and proteins could carry large amounts of functional information; it could not establish whether they did.

The ease with which information theory applied to molecular biology (to measure information-carrying capacity) has created considerable confusion about the sense in which DNA and proteins contain "information." Information theory strongly suggested that such molecules possess vast information carrying capacities or large amounts of syntactic information, as defined by Shannon's theory. When molecular biologists have described DNA as the carrier of hereditary information, however, they have meant much more than the technically limited term information. Instead, as Sahotra Sarkar points out, leading molecular biologists defined biological information so as to incorporate the notion of specificity of function (as well as complexity) as early as 1958.[34] Molecular biologists such as Monod and Crick understood biological information—the information stored in DNA and proteins—as something more than mere complexity (or improbability). Their notion of information did associate both biochemical contingency and combinatorial complexity with DNA sequences (allowing DNA's carrying capacity to be calculated), but they also recognized that sequences of nucleotides and amino acids in functioning bio-macromolecules possessed a high degree of specificity relative to the maintenance of cellular function. As Crick explained in 1958, "By information I mean the specification of the amino acid sequence in protein. . . . Information means here the precise determination of sequence, either of bases in the nucleic acid or on amino acid residues in the protein."[35]

Since the late 1950s, biologists have equated the "precise determination of sequence" with the extra-information-theoretic property of specificity or specification. Biologists have defined specificity tacitly as "necessary to achieve or maintain function." They have determined that DNA base sequences, for example, are specified not by applying information theory, but by making experimental assessments of the function of those sequences within the overall apparatus of gene expression.[36] Similar experimental considerations established the functional specificity of proteins.

Further, developments in complexity theory have now made possible a fully general theoretical account of specification, one that applies readily to biological systems. In particular, recent work by mathematician William Dembski has employed the notion of a rejection region from statistics to provide a formal complexity-theoretic account of specification. According to Dembski, a specification occurs when an event or object (a) falls within an independently given pattern or domain, (b) "matches" or exemplifies a conditionally independent pattern, or (c) meets a conditionally independent set of functional requirements.[37]

To illustrate Dembski's notion of specification, consider these two strings of characters:

iuinsdysk]idfawqnzkl,mfdifhs

Time and tide wait for no man.

Given the number of possible ways of arranging the letters and punctuation marks of the English language for sequences of this length, both of these two sequences constitute highly improbable arrangements of characters. Thus, both have a considerable and quantifiable information-carrying capacity. Nevertheless, only the second of the two sequences exhibits a specification on Dembski's account. To see why, consider the following. Within the set of combinatorially possible sequences, only a very few will convey meaning. This smaller set of meaningful sequences, therefore, delimits a domain or pattern within the larger set of the totality of possibilities. Moreover, this set constitutes a "conditionally independent" pattern. Roughly speaking, a conditionally independent pattern corresponds to a preexisting pattern or set of functional requirements, not one contrived after the fact of observing the event in question—specifically, in this case, the event of observing the two sequences above.[38] Since the smaller domain distinguishes functional from nonfunctional English sequences and the functionality of alphabetic sequences depends on the preexisting or independently given conventions of English vocabulary and grammar, the smaller set or domain qualifies as a conditionally independent pattern.[39] Since the second string of characters ("Time and tide wait . . .") falls within this smaller conditionally independent domain (or "matches" one of the possible meaningful sentences that fall within it), the second sequence exhibits a specification according to Dembski's complexity-theoretic account. That sequence therefore exhibits the joint properties of complexity and specification and possesses not just information-carrying capacity, but both "specified" and, in this case, "semantic" information.

Biological organisms also exhibit specifications, though not necessarily semantic or subjectively "meaningful" ones. The nucleotide base sequences in the coding regions of DNA are highly specific relative to the independent functional requirements of protein function, protein synthesis, and cellular life. To maintain viability, the cell must regulate its metabolism, pass materials back and forth across its membranes, destroy waste materials, and do many other specific tasks. Each of these functional requirements in turn necessitates specific molecular constituents, machines, or systems (usually made of proteins) to accomplish these tasks.

Building these proteins with their specific three-dimensional shapes requires specific arrangements of nucleotide bases on the DNA molecule.

Since the chemical properties of DNA allow a vast ensemble of combinatorially possible arrangements of nucleotide bases, any particular sequence will necessarily be highly improbable and rich in Shannon information or information-carrying capacity. Yet within that set of possible sequences a very few will, given the multimolecular system of gene expression within the cell, produce functional proteins.[40] Those that do are thus not only improbable but also functionally "specified" or "specific," as molecular biologists use the terms. Indeed, the smaller set of functionally efficacious sequences again delimits a domain or pattern within a larger set of combinatorial possibilities. Moreover, this smaller domain constitutes a conditionally independent pattern, since (as with the English sequences above) it distinguishes functional from nonfunctional sequences, and the functionality of nucleotide base sequences depends on the independent requirements of protein function. Thus, any actual nucleotide sequence that falls within this domain (or "matches" one of the possible functional sequences that fall within it) exhibits a specification. Put differently, any nucleotide base sequence that produces a functional protein clearly meets certain independent functional requirements—in particular, those of protein function. Thus, any sequence that meets such requirements (or "falls within the smaller subset of functional sequences") is again not only highly improbable, but also specified relative to that independent pattern or domain. Thus, the nucleotide sequences in the coding regions of DNA possess both syntactic information and "specified" information.

A note of definitional clarity must be offered about the relationship between "specified" information and "semantic" information. Though natural languages and DNA base sequences are both specified, only natural language conveys meaning. If one defines "semantic information" as "subjectively meaningful information that is conveyed syntactically (as a string of phonemes or characters) and is understood by a conscious agent," then clearly the information in DNA does not qualify as semantic. Unlike a written or spoken natural language, DNA does not convey "meaning" to a conscious agent.

Rather, the coding regions of DNA function in much the same way as a software program or machine code, directing operations within a complex material system via highly complex yet specified sequences of characters. As Richard Dawkins has noted, "The machine code of the genes is uncannily computer-like."[41] Or as software developer Bill Gates noted, "DNA is like a computer program, but far, far more advanced than any software we've ever created."[42] Just as the specific arrangement of two symbols (0 and 1) in a software program can perform a function within a machine environment, so too can the precise sequencing of the four nucleotide bases in DNA perform a function within the cell.

Though DNA sequences do not convey "meaning," they do exhibit specificity or specification. Moreover, as in a machine code, the sequence specificity of DNA occurs within a syntactic (or functionally alphabetic) domain. Thus, DNA possesses both syntactic and specified information. In any case, since the late 1950s, the concept of information as employed by molecular biologists has conjoined the notions of complexity (or improbability) and specificity of function. The crucial biomolecular constituents of living organisms possess not only Shannon or syntactic information but also "specified information" or "specified complexity."[43] Biological information so defined, therefore, constitutes a salient feature of living systems that any origin-of-life scenario must explain "the origin of." Further, as we will see below, all naturalistic chemical evolutionary theories have encountered difficulty explaining the origin of such functionally "specified" biological information.

Information as Metaphor: Nothing to Explain?

Though most molecular biologists would see nothing controversial in characterizing DNA and proteins as "information-bearing" molecules, some historians and philosophers of biology have recently challenged that description. Before evaluating competing types of explanation for the origin of biological information, this challenge must be addressed. In 2000, the late historian of science Lily Kay characterized the application of information theory to biology as a failure, in particular because classical information theory could not capture the idea of meaning. She suggested, therefore, that the term "information" as used in biology constitutes nothing more than a metaphor. Since, in Kay's view, the term does not designate anything real, it follows that the origin of "biological information" does not require explanation. Instead, only the origin of the use of the term information within biology requires explanation. As a social constructivist, Kay explained this usage as the result of various social forces operating within the "Cold War Technoculture."[44] In a different but related vein, Sarkar has argued that the concept of information has little theoretical significance in biology because it lacks predictive or explanatory power.[45] He, like Kay, seems to regard the concept of information as a superfluous metaphor lacking empirical reference and ontological status.

Of course, insofar as the term information connotes semantic meaning, it does function as a metaphor within biology. That does not mean, however, that the term functions only metaphorically or that origin-of-life biologists have nothing to explain. Though information theory has a limited application in describing biological systems, it has succeeded in rendering quantitative assessments of the complexity of bio-macromolecules. Further, experimental work established the functional specificity of the sequences of monomers in DNA and proteins. Thus, the term information as used in biology does refer to two real and contingent properties of living systems: complexity and specificity. Indeed, since scientists began to think seriously about what would be required to explain the phenomenon of heredity, they have recognized the need for some feature or substance in living organisms possessing precisely these two properties together. Thus, Schrödinger envisioned an "aperiodic crystal"; Chargaff perceived DNA's capacity for "complex sequencing"; Watson and Crick equated complex sequences with "information" (which Crick in turn equated with "specificity"); Monod equated irregular specificity in proteins with the need for "a code"; and Orgel characterized life as a "specified complexity."[46] Further, Davies has recently argued that the "specific randomness" of DNA base sequences constitutes the central mystery surrounding the origin of life.[47] Whatever the terminology, scientists have recognized the need for—and now know the location of—a source of complex specificity in the cell in order to transmit heredity and maintain biological function. The incorrigibility of these descriptive concepts suggests that complexity and specificity constitute real properties of bio-macromolecules—indeed, properties that could be otherwise but only to the detriment of cellular life. As Orgel noted: "Living organisms are distinguished by their specified complexity. Crystals . . . fail to qualify as living because they lack complexity; mixtures of random polymers fail to qualify because they lack specificity."[48]

The origin of specificity and complexity (in combination), to which the term "information" in biology commonly refers, therefore does require explanation, even if the concept of information connotes only complexity in classical information theory and even if it has no explanatory or predictive value in itself. Instead, as a descriptive (rather than explanatory or predictive) concept, the term information helps to define (either in conjunction with the notion of "specificity" or by subsuming it) the effect that origin-of-life researchers must explain the origin of. Thus, only where information connotes subjective meaning does it function as a metaphor in biology. Where it refers to an analog of meaning (namely, functional specificity), it defines an essential feature of living systems.

2. NATURALISTIC EXPLANATIONS FOR THE ORIGIN OF SPECIFIED BIOLOGICAL INFORMATION

The discoveries of molecular biologists during the 1950s and 1960s raised the question of the ultimate origin of the specified complexity or specified information in both DNA and proteins. Since at least the mid-1960s, many scientists have regarded the origin of information (so defined) as the central question facing origin-of-life biology.[49] Accordingly, origin-of-life researchers have proposed three broad types of naturalistic explanation to explain the origin of specified genetic information: those emphasizing chance, necessity, or the combination of the two.

BEYOND THE REACH OF CHANCE

Perhaps the most common popular naturalistic view about the origin of life is that it happened exclusively by chance. A few serious scientists have voiced support for this view, at least, at various points in their careers. For example, in 1954 biochemist George Wald argued for the causal efficacy of chance in conjunction with vast expanses of time. As he explained, "Time is in fact the hero of the plot. . . . Given so much time, the impossible becomes possible, the possible probable, and the probable virtually certain."[50] Later, in 1968, Francis Crick would suggest that the origin of the genetic code—that is, the translation system—might be a "frozen accident."[51] Other theories have invoked chance as an explanation for the origin of genetic information, though often in conjunction with pre-biotic natural selection (see the discussion below).

Almost all serious origin-of-life researchers now consider "chance" an inadequate causal explanation for the origin of biological information.[52] Since molecular biologists began to appreciate the sequence specificity of proteins and nucleic acids in the 1950s and 1960s, many calculations have been made to determine the probability of formulating functional proteins and nucleic acids at random. Various methods of calculating probabilities have been offered by Morowitz, Hoyle and Wickramasinghe, Cairns-Smith, Prigogine, Yockey, and more recently, Robert Sauer.[53] For the sake of argument, these calculations have often assumed extremely favorable pre-biotic conditions (whether realistic or not), much more time than was actually available on the early Earth, and theoretically maximal reaction rates among constituent monomers (that is, the constituent parts of proteins, DNA, or RNA). Such calculations have invariably shown that the probability of obtaining functionally sequenced bio-macromolecules at random is, in Prigogine's words, "vanishingly small[,] . . . even on the scale of . . . billions of years."[54] As Cairns-Smith wrote in 1971: "Blind chance . . . is very limited. [It can produce] low-levels of cooperation . . . exceedingly easily (the equivalent of letters and small words), but [it] becomes very quickly incompetent as the amount of organization increases. Very soon indeed long waiting periods and massive material resources become irrelevant."[55]

Consider the probabilistic hurdles that must be overcome to construct even one short protein molecule of 150 amino acids in length. (A typical protein consists of about 300 amino acid residues, and many crucial proteins are much longer.)

First, all amino acids must form a chemical bond known as a peptide bond when joining with other amino acids in the protein chain. If the amino acids do not link up with one another via a peptide bond, the resulting molecule will not fold into a protein. In nature, many other types of chemical bonds are possible between amino acids. In fact, when amino acid mixtures are allowed to react in a test tube, they form peptide and nonpeptide bonds with roughly equal probability. Thus, with each amino acid addition, the probability of it forming a peptide bond is roughly 1/2. Once four amino acids have become linked, the likelihood that they are joined exclusively by peptide bonds is roughly (1/2 x 1/2 x 1/2) = 1/8 or (1/2)^3. The probability of building a chain of 150 amino acids in which all linkages are peptide linkages is (1/2)^{149} or roughly one chance in 10^{45}.

Second, in nature every amino acid found in proteins (with one exception) has a distinct mirror image of itself, one left-handed version (or L-form) and one right-handed version (or D-form). These mirror-image forms are called "optical isomers." Functioning proteins tolerate only left-handed amino acids, yet in abiotic amino acid production, the right-handed and left-handed isomers are produced with roughly equal frequency. Taking this into consideration compounds the improbability of attaining a biologically functioning protein. The probability of attaining at random only L-amino acids in a hypothetical peptide chain 150 amino acids long is (1/2)^{150}, or again, roughly one chance in 10^{45}. Starting from mixtures of D-forms and L-forms, the probability of building a 150 amino acid length chain at random in which all bonds are peptide bonds and all amino acids are L-form is therefore roughly one chance in 10^{90}.

Functioning proteins have a third independent requirement, the most important of all: their amino acids must link up in functionally specified sequential arrangements just as the letters in meaningful sentences must. In some cases, changing even one amino acid at a given site results in the loss of protein function. Moreover, because there are twenty biologically occurring amino acids, the probability of getting a specific amino acid at a given site is small—1/20. (Actually the probability is even lower because, in nature, there are also many nonprotein forming amino acids.) On the assumption that each site in a protein chain requires a particular amino acid, the probability of attaining a particular protein 150 amino acids long would be (1/20)^{150}, or roughly one chance in 10^{195}.

Molecular biologists have known for a while that most sites along the chain can tolerate several of the different twenty amino acids commonly found in proteins without destroying the function of the protein, though some cannot. This raised an important question: How rare, or common, are the *functional* sequences of amino acids among all the possible sequences of amino acids in a chain of any given length? In the late 1980s, several important studies were conducted in the laboratory of MIT biochemist Robert Sauer in order to investigate this question. His research team used a sampling technique known as "cassette mutagenesis" to determine how much variance among amino acids can be tolerated at any given site in several proteins. This technique would help answer an important question. If proteins can tolerate a lot of variance, then that would increase the probability that a random search through the space of possibilities would find a functional sequence. If proteins were more finicky—if the requirements of functionality imposed more rigid constraints on sequencing—then that would decrease the probability of a random process successfully producing a functional protein. Thus, whatever Sauer's team discovered would be extremely significant. The results of their experiments could help determine the probability that a functional protein would arise by chance from a pre-biotic soup.

So what did they find? Their most clear-cut experiments[56] seemed to indicate that, even taking the possibility of variance into account, the probability of achieving a functional sequence of amino acids in several known (roughly 100 residue) proteins at random is still "exceedingly small," about one chance in 10^{63} (to put this in perspective, there are 10^{65} atoms in our galaxy).[57] Using a variety of mutagenesis techniques, they and other scientists showed that proteins (and thus the genes that produce them) are highly specified relative to biological function.[58] Earlier studies had shown that amino acid residues at many sites can not vary without functional loss.[59] Now, Sauer and others had shown that even for sites that do admit some variance, not just any amino acid will do. Instead, they showed that functional requirements place significant constraints on sequencing at sites where some variance is allowed. By quantifying that allowable variance, they made it possible to calculate the probability of finding a protein with a functional sequence among the larger ensemble of combinatorial possibilities.

I first learned about the work of Robert Sauer and its relevance to assessing the chance hypothesis in 1992 from a postdoctoral researcher at Cambridge University named Douglas Axe. Axe had come to Cambridge to perform mutagenesis experiments that were similar to Sauer's. During his Ph.D. work at Caltech, Axe learned about the structure of proteins and the intricate folds they need in order to perform their functions. He began to wonder how difficult it was to produce these folds by random mutations or random molecular interactions. He began to ask a very similar question to that which had motivated Sauer: How rare, or common, are the amino acid sequences that produce the stable folds that make it possible for proteins to perform their biological functions?

As Axe began to examine Sauer's experimental method, he asked whether Sauer might have underestimated how protein sequences can vary and still maintain function. To test this possibility, he developed a more rigorous method of estimating this allowable variability in order to eliminate possible estimation error. Axe liked what Sauer had done, but wanted to produce a more definitive answer. His interest in the subject eventually led him to the Cambridge University laboratory of Alan Fersht, and then to the Center for Protein Engineering at the famous Medical Research Council Centre in Cambridge. Between 1996 and 2004, Axe's work was published in a series of papers in the *Journal of Molecular Biology, Biochemistry,* and *The Proceedings of the National Academy of Sciences.*

The results of a paper he published in a 2004 were particularly telling.[60] Axe performed a mutagenesis experiment using his refined method on a functionally significant 150-amino-acid section of a protein called beta-lactamase, an enzyme that confers antibiotic resistance upon bacteria. On the basis of his experiments, Axe was able to make a careful estimate of the ratio of (a) the number of 150-amino-acid sequences that can perform that particular function to (b) the whole set of possible amino acids sequences of this length. Based on his experiments, Axe estimated this ratio to be 1 over 10^{77}.

This was a staggering number, and it suggested that a random process would have great difficulty generating a protein with that particular function by chance. But origin-of-life researchers didn't just want to know the likelihood of finding a protein with a particular function within a space of combinatorial possibilities. They wanted to know the odds of finding *any* functional protein whatsoever within such a space. That number would make it possible to evaluate chance-based origin-of-life scenarios, by assessing the probability that a single protein—*any working protein*—would have arisen by chance on the early Earth.

Fortunately, Axe's work provided this number as well. Axe knew that in nature proteins perform many specific functions. He also knew that in order to perform these functions their amino acids chains must first fold into stable three-dimensional structures. Thus, before he estimated the frequency of sequences performing a specific (beta-lactamase) function, he first performed experiments that enabled him to estimate the frequency of sequences that will produce stable folds. On the basis of his experimental results, he calculated the ratio of (a) the number of 150-amino acid long sequences capable of folding into stable "function-ready" structures to (b) the whole set of possible amino acids sequences of that length. He determined that ratio to be 1 in 10^{74}.

Since proteins can't perform functions unless they first fold into stable structures, Axe's measure of the frequency of folded sequences within sequence space also provided a measure of the frequency of functional proteins—*any* functional proteins—within that space of possibilities. Indeed, by taking what he knew about protein folding into account, Axe estimated the ratio of (a) the number of 150-amino acid sequences that produce *any functional protein whatsoever* to (b) the whole set of possible amino acids sequences of that length. Axe's estimated ratio of 1 in 10^{74} implied that the probability of producing any properly sequenced 150 amino acid protein at random is also about 1 in 10^{74}. In other words, a random process that produced amino acid chains of this length would stumble onto a functional protein only about once in every 10^{74} attempts.

When one considers that Robert Sauer was working on a shorter protein of 100 amino acids, Axe's number might seem a bit less prohibitively improbable. Nevertheless, it still represents a startlingly small probability. Moreover, Axe's improved estimate of how rare functional proteins are within "sequence space" has now made it possible to calculate the probability that a 150-amino-acid compound assembled by random interactions in a pre-biotic soup would be a functional protein. This calculation can be made by multiplying the three independent probabilities by one another: the probability of incorporating only peptide bonds (1 in 10^{45}), the probability of incorporating only left-handed amino acids (1 in 10^{45}) and the probability of achieving correct amino acid sequencing (using Axe's 1 in 10^{74} estimate). Making that calculation (multiplying the separate probabilities by adding their exponents: $10^{45+45+74}$) gives a dramatic answer. The odds of getting a functional protein of modest length (150 amino acids) by drawing a compound of that size from a pre-biotic soup is no better than one chance in 10^{164}. In other words, the probability of constructing a rather short functional protein at random becomes so small (no more than one chance in 10^{164}) as to appear absurd on the chance hypothesis.

As I have investigated the question of whether biological information might have arisen by chance, it has become abundantly clear to me that the probability of the necessary events is exceedingly small. Nevertheless, I realized that the probability of an event by itself does not alone determine whether the event could be reasonably explained by chance. The probabilities, as small as they were, were not by themselves conclusive. I remembered that I also had to consider the number of opportunities that the event in question might have had to occur. I had to take into account what Dembski called the *"probabilistic resources."*

But what were those resources—how many opportunities did the necessary proteins or genes have to arise by chance? The advocates of the chance hypothesis envisioned amino acids—or nucleotide bases, phosphates, and sugars—knocking into each other in an ocean-sized soup until the correct arrangements of these building blocks arose by chance somewhere. Surely such an environment would have generated many opportunities for the assembly of functional proteins and DNA molecules. But how many? And were there enough such opportunities to render these otherwise exceedingly improbable events probable?

Here again Dembski's work gave me a way to answer this question. Dembski had calculated the maximum number of events that could actually have taken place during the history of the observable universe.[61] He did this to establish an upper bound on the probabilistic resources that might be available to produce any event by chance.[62]

Dembski's calculation was elegantly simple and yet made a powerful point. He noted that there were about 10^{80} elementary particles in the observable universe.[63] (Because there is an upper limit on the speed of light, only those parts of the universe observable to us can affect events on Earth. Thus, the observable universe is the only part of the universe with probabilistic resources relevant to explaining events on Earth). Dembski also noted that there had been roughly 10^{16} seconds since the Big Bang.

He then introduced another parameter that enabled him to calculate the maximum number of opportunities that any particular event would have to take place since the origin of the universe. Due to the properties of gravity, matter, and electromagnetic radiation, physicists have determined that there is a limit to the number of physical transitions that can occur from one state to another within a given unit of time. According to physicists, a physical transition from one state to another cannot take place faster than light can traverse the smallest physically significant unit of distance (an indivisible "quantum" of space). That unit of distance is the so-called Planck length of 10^{-33} centimeters. Therefore, the time it takes light to traverse this smallest distance determines the shortest time in which any physical effect can occur. This unit of time is the Planck time of 10^{-43} seconds.

Knowing this, Dembski was able to calculate the largest number of opportunities that any material event had to occur in the observable universe since the Big Bang. Physically speaking, an event occurs when an elementary particle does something, or interacts with other elementary particles. But since elementary particles can only interact with each other so many times per second (at most 10^{43} times), and since there are a limited number (10^{80}) of elementary particles, and since there has been a limited amount of time since the Big Bang (10^{16} seconds), there are a limited number of opportunities for any given event to occur in the entire history of the universe.

Dembski was able to calculate this number by simply multiplying the three relevant factors together: the number of elementary particles (10^{80}) times the number of seconds since the Big Bang (10^{16}) times the number of possible interactions per second (10^{43}). His calculation fixed the total number of events that could have taken place in the observable universe since the origin of the universe at 10^{139}. This then provided a measure of the probabilistic resources of the entire observable universe.[64]

Other mathematicians and scientists have made similar calculations.[65] During the 1930s, the French mathematician Emile Borel made a much less conservative estimate of the probabilistic resources of the universe, which he set at 10^{50}.[66] More recently, University of Pittsburgh physicist Bret Van de Sande calculated the probabilistic resources of the universe at a more restrictive 2.6×10^{92}.[67] MIT computer scientist Seth Lloyd has calculated that the most bit operations the universe could have performed in its entire history (assuming the entire universe were given over to this single-minded task) is 10^{120}, meaning that a specified physical computation with an improbability significantly greater than one chance in 10^{120} will likely never occur by chance.[68] None of these probabilistic resources is sufficient to render the chance hypothesis plausible. Dembski's calculation is the most conservative and gives chance its "best chance" to succeed. But even his calculation confirms the implausibility of the chance hypothesis, whether chance is invoked to explain the information necessary to build a single protein, or the information necessary to build the suite of proteins needed to service a minimally complex cell.

Recall that the probability of producing a single 150-amino-acid functional protein by chance stands at about 1 in 10^{164}. Thus, for each functional sequence of 150 amino acids, there are 10^{164} other nonfunctional sequences of the same length. Therefore, to have a good (i.e., better than fifty-fifty) chance of producing a single functional protein of this length by chance, a random process would have to generate (or sample) more than one half of the 10^{164} nonfunctional sequences corresponding to each functional sequence of that length. Unfortunately, that number vastly exceeds the most optimistic estimate of the probabilistic resources of the universe—that is, the number of events that could have occurred since the beginning of the universe. To see this, notice again that to have a good (better than fifty-fifty) chance of generating a functional protein by chance more than half of the 10^{164} sequences would have to be produced. Now compare that number (call it 0.5×10^{164}) to the maximum number of opportunities—10^{139}—for that event to occur in the history of the universe. Notice that the first number (0.5×10^{164}) exceeds the second (10^{139}) by more than twenty-four orders of magnitude, by more than a trillion trillion times.

What does this mean? It means that if every elementary particle in the universe over its entire history were devoted to producing combinations of amino acids of the correct length in a pre-biotic soup (an extravagantly generous and even absurd assumption), the number of combinations thus produced would still represent a tiny fraction—less than one out of a trillion trillion—of the total number of events needed to have a 50 percent chance of generating a functional protein—*any* functional protein by chance.

In other words, even if the theoretically maximum number (10^{139}) of amino acid sequences possible were generated, the number of candidates generated that way would still represent a minuscule portion of the total possible sequences (corresponding to each functional protein). For this reason, it would be vastly more probable than not that a functional protein of modest length would *not* have arisen by chance—simply too few of the possible sequences would have been sampled to provide a realistic opportunity for this to occur. In other words, it is extremely unlikely that even a single protein would have arisen *by chance* on the early Earth even taking the probabilistic resources of the entire universe into account.

More realistic calculations (taking into account the probable presence of non-proteinous amino acids, the need for much longer proteins to perform specific functions such as polymerization, and the need for hundreds of proteins working in coordination to produce a functioning cell) only compound these improbabilities, almost beyond computability. For example, recent theoretical and experimental work on the so-called minimal complexity required to sustain the simplest possible living organism suggests a lower bound of some 250 to 400 genes and their corresponding proteins.[69] The nucleotide sequence-space corresponding to such a system of proteins exceeds $4^{300,000}$. The improbability corresponding to this measure of molecular complexity again vastly exceeds one chance in 10^{139} and thus the "probabilistic resources" of the entire universe.[70] When one considers the full complement of functional biomolecules required to maintain minimal cell function and vitality, one can see why chance-based theories of the origin of life have been abandoned. What Mora said in 1963 still holds: "Statistical considerations, probability, complexity, etc., followed to their logical implications suggest that the origin and continuance of life is not controlled by such principles. An admission of this is the use of a period of practically infinite time to obtain the derived result. Using such logic, however, we can prove anything."[71]

Though the probability of assembling a functioning biomolecule or cell by chance alone is exceedingly small, it is important to emphasize that scientists have not generally rejected the chance hypothesis merely because of the vast improbabilities associated with such events or even because of the vast improbability of the events that remain after taking all probabilistic resources into account. Instead, there is one other factor involved in the elimination of chance as the best explanation.

Very improbable things do occur by chance. Any hand of cards or any series of rolled dice will represent a highly improbable occurrence. Observers often justifiably attribute such events to chance alone. What justifies the elimination of chance is not just the occurrence of a highly improbable event but also the occurrence of an improbable event that also conforms to a discernible pattern (that is, to a conditionally independent pattern; see above, "Complexity, Specificity, and Biological Information"). If someone repeatedly rolls two dice and turns up a sequence such as 9, 4, 11, 2, 6, 8, 5, 12, 9, 2, 6, 8, 9, 3, 7, 10, 11, 4, 8, and 4, no one will suspect anything but the interplay of random forces, though this sequence does represent a very improbable event given the number of combinatorial possibilities that correspond to a sequence of this length. Yet rolling twenty (or certainly two hundred) consecutive sevens will justifiably arouse suspicion that something more than chance is in play. Statisticians have long used method for determining when to eliminate the chance hypothesis; the method requires pre-specifying a pattern or "rejection region."[72] In the dice example above, one could pre-specify the repeated occurrence of seven as such a pattern in order to detect the use of loaded dice, for example. Dembski has generalized this method to show how the presence of any conditionally independent pattern, whether temporally prior to the observation of an event or not, can help (in conjunction with a small probability event) to justify rejecting the chance hypothesis.[73]

Origin-of-life researchers have tacitly, and sometimes explicitly, employed this kind of statistical reasoning to justify the elimination of scenarios relying heavily on chance. Christian de Duve, for example, has made the logic explicit in order to explain why chance fails as an explanation for the origin of life: "A single, freak, highly improbable event can conceivably happen. Many highly improbable events—drawing a winning lottery number or the distribution of playing cards in a hand of bridge—happen all the time. But a string of improbable events—drawing the same lottery number twice, or the same bridge hand twice in a row—does not happen naturally."[74]

De Duve and other origin-of-life researchers have long recognized that the cell represents not only a highly improbable but also a functionally specified system. For this reason, by the mid-1960s most researchers had eliminated chance as a plausible explanation for the origin of the specified information necessary to build a cell.[75] Many have instead sought other types of naturalistic explanations.

Pre-biotic Natural Selection: A Contradiction in Terms

Of course, even many early theories of chemical evolution did not rely exclusively on chance as a causal mechanism. For example, Oparin's original theory of evolutionary abiogenesis, first published in the 1920s and 1930s, invoked pre-biotic natural selection as a complement to chance interactions. Oparin's theory envisioned a series of chemical reactions that he thought would enable a complex cell to assemble itself gradually and naturalistically from simple chemical precursors.

For the first stage of chemical evolution, Oparin proposed that simple gases such as ammonia (NH_3), methane (CH_4), water vapor (H_2O), carbon dioxide (CO_2), and hydrogen (H_2) would have existed in contact with the early oceans and with metallic compounds extruded from the core of the Earth.[76] With the aid of ultraviolet radiation from the Sun, the ensuing reactions would have produced energy-rich hydrocarbon compounds. They in turn would have combined and recombined with various other compounds to make amino acids, sugars, and other "building blocks" of complex molecules such as proteins necessary to living cells. These constituents would eventually arrange themselves by chance into primitive metabolic systems within simple cell-like enclosures that Oparin called "coacervates." Oparin then proposed a kind of Darwinian competition for survival among his coacervates. Those that, by chance, developed increasingly complex molecules and metabolic processes would have survived to grow more complex and efficient. Those that did not would have dissolved.[77] Thus, Oparin invoked differential survival or natural selection as a mechanism for preserving complexity-increasing events, thus allegedly helping to overcome the difficulties attendant to pure-chance hypotheses.

Developments in molecular biology during the 1950s cast doubt on Oparin's scenario. Oparin originally invoked natural selection to explain how cells refined primitive metabolism once it had arisen. His scenario relied heavily on chance to explain the initial formation of the constituent bio-macromolecules on which even primitive cellular metabolism would depend. During the 1950s, discovery of the extreme complexity and specificity of such molecules undermined the plausibility of his claim. For that and other reasons, in 1968 Oparin published a revised version of his theory that envisioned a role for natural selection earlier in the process of abiogenesis. His new theory claimed that natural selection acted on random polymers as they formed and changed within his coacervate protocells.[78] As more complex and efficient molecules accumulated, they would have survived and reproduced more prolifically.

Even so, Oparin's concept of pre-biotic natural selection acting on initially unspecified bio-macromolecules remained problematic. For one thing, it seemed to presuppose a preexisting mechanism of self-replication. Yet self-replication in all extant cells depends on functional and therefore (to a high degree) sequence-specific proteins and nucleic acids. Yet the origin of specificity in these molecules is precisely what Oparin needed to explain. As Christian de Duve has stated, theories of pre-biotic natural selection "need information which implies they have to presuppose what is to be explained in the first place."[79] Oparin attempted to circumvent the problem by claiming that the first polymers need not have been highly sequence-specific. But that claim raised doubts about whether an accurate mechanism of self-replication (and thus natural selection) could have functioned at all. Oparin's latter scenario did not reckon on a phenomenon known as error catastrophe, in which small errors, or deviations from functionally necessary sequences, are quickly amplified in successive replications.[80]

Thus, the need to explain the origin of specified information created an intractable dilemma for Oparin. On the one hand, if he invoked natural selection late in his scenario, he would need to rely on chance alone to produce the highly complex and specified biomolecules necessary to self-replication. On the other hand, if Oparin invoked natural selection earlier in the process of chemical evolution, before functional specificity in bio-macromolecules would have arisen, he could give no account of how such pre-biotic natural selection could even function (given the phenomenon of error catastrophe). Natural selection presupposes a self-replication system, but self-replication requires functioning nucleic acids and proteins (or molecules approaching their complexity)—the very entities that Oparin needed to explain. Thus, Dobzhansky would insist that "prebiological natural selection is a contradiction in terms."[81]

Although some rejected the hypothesis of pre-biotic natural selection as question-begging, others dismissed it as indistinguishable from the implausible chance-based hypotheses.[82] The work of mathematician John von Neumann supported that judgment. Von Neumann showed that any system capable of self-replication would require subsystems that were functionally equivalent to the information storage, replicating, and processing systems found in extant cells.[83] His calculations established a very high minimal threshold of biological function, as would later experimental work.[84] These minimal-complexity requirements pose a fundamental difficulty for natural selection. Natural selection selects for functional advantage. It can play no role, therefore, until random variations produce some biologically advantageous arrangement of matter. Yet von Neumann's calculations—and similar ones by Wigner, Landsberg, and Morowitz—showed that in all probability (to understate the case) random fluctuations of molecules would not produce the minimal complexity needed for even a primitive replication system.[85] As noted above, the improbability of developing a functionally integrated replication system vastly exceeds the improbability of developing the protein or DNA components of such a system. Given the huge improbability and the high functional threshold it implies, many origin-of-life researchers came to regard pre-biotic natural selection as both inadequate and essentially indistinguishable from appeals to chance.

Nevertheless, during the 1980s, Richard Dawkins and Bernd-Olaf Küppers attempted to resuscitate pre-biotic natural selection as an explanation for the origin of biological information.[86] Both accepted the futility of naked appeals to chance and invoke what Küppers called a "Darwinian optimization principle." Both used computers to demonstrate the efficacy of pre-biotic natural selection. Each selected a target sequence to represent a desired functional polymer. After creating a crop of randomly constructed sequences and generating variations among them at random, their computers selected those sequences that matched the target sequence most closely. The computers then amplified the production of those sequences, eliminated the others (to simulate differential reproduction), and repeated the process. As Küppers put it, "Every mutant sequence that agrees one bit better with the meaningful or reference sequence . . . will be allowed to reproduce more rapidly."[87] In his case, after a mere thirty-five generations, his computer succeeded in spelling his target sequence, "NATURAL SELECTION."

Despite superficially impressive results, such "simulations" conceal an obvious flaw: Molecules *in situ* do not have a target sequence "in mind." Nor will they confer any selective advantage on a cell, and thus differentially reproduce, until they combine in a functionally advantageous arrangement. Thus, nothing in nature corresponds to the role that the computer plays in selecting functionally nonadvantageous sequences that happen to agree "one bit better" than others with a target sequence. The sequence "NORMAL ELECTION" may agree more with "NATURAL SELECTION" than does the sequence "MISTRESS DEFECTION," but neither of the two yields any advantage in communication over the other in trying to communicate something about "NATURAL SELECTION." If that is the goal, both are equally ineffectual. Even more to the point, a completely nonfunctional polypeptide would confer no selective advantage on a hypothetical protocell, even if its sequence happened to agree "one bit better" with an unrealized target protein than some other nonfunctional polypeptide.

Both Küppers's and Dawkins's published results of their simulations show the early generations of variant phrases awash in nonfunctional gibberish.[88] In Dawkins's simulation, not a single functional English word appears until after the tenth iteration (unlike the more generous example above that starts with actual, albeit incorrect, words). Yet to make distinctions on the basis of function among sequences that have no function is entirely unrealistic. Such determinations can be made only if considerations of *proximity to possible future function* are allowed, but that requires foresight, which natural selection does not have. A computer, programmed by a human being, can perform such functions. To imply that molecules can do so as well illicitly personifies nature. Thus, if these computer simulations demonstrate anything, they subtly demonstrate the need for intelligent agents to elect some options and exclude others—that is, to create information. (In my recent book, *Signature in the Cell: DNA and the Evidence of Intelligent Design,* I show that other more recent genetic algorithms such as *Ev* and *Avida* demonstrate this same need.)

SELF-ORGANIZATIONAL SCENARIOS

Because of the difficulties with chance-based theories, including those relying on pre-biotic natural selection, most origin-of-life theorists after the mid-1960s attempted to address the problem of the origin of biological information in a completely different way. Researchers began to look for self-organizational laws and properties of chemical attraction that might explain the origin of the specified information in DNA and proteins. Rather than invoking chance, such theories invoked necessity. If neither chance nor pre-biotic natural selection acting on chance explains the origin of specified biological information, then those committed to finding a naturalistic explanation for the origin of life must necessarily rely on physical or chemical necessity. Given a limited number of broad explanatory categories, the inadequacy of chance (with or without pre-biotic natural selection) has, in the minds of many researchers, left only one option. Christian de Duve articulated the logic: "[A] string of improbable events—drawing the same lottery number twice, or the same bridge hand twice in a row—does not happen naturally. All of which lead me to conclude that life is an obligatory manifestation of matter, bound to arise where conditions are appropriate."[89]

When origin-of-life biologists began considering the self-organizational perspective that de Duve describes, several researchers proposed that deterministic forces (stereochemical "necessity") made the origin of life not just probable but inevitable. Some suggested that simple chemicals possessed "self-ordering properties" capable of organizing the constituent parts of proteins, DNA, and RNA into the specific arrangements they now possess.[90] Steinman and Cole, for example, suggested that differential bonding affinities or forces of chemical attraction between certain amino acids might account for the origin of the sequence specificity of proteins.[91] Just as electrostatic forces draw sodium (Na+) and chloride (Cl-) ions together into highly ordered patterns within a crystal of salt (NaCl), so, too, might amino acids with special affinities for each other arrange themselves to form proteins. In 1969, Kenyon and Steinman developed that idea in a book entitled *Biochemical Predestination*. They argued that life might have been "biochemically predestined" by the properties of attraction existing between its constituent chemical parts, particularly among the amino acids in proteins.[92]

In 1977, another self-organizational theory was proposed by Prigogine and Nicolis, based on a thermodynamic characterization of living organisms. In *Self Organization in Non-Equilibrium Systems,* Prigogine and Nicolis classified living organisms as open, nonequilibrium systems capable of "dissipating" large quantities of energy and matter into the environment.[93] They observed that open systems driven far from equilibrium often display self-ordering tendencies. For example, gravitational energy will produce highly ordered vortices in a draining bathtub; thermal energy flowing through a heat sink will generate distinctive convection currents or "spiral wave activity." Prigogine and Nicolis argued that the organized structures observed in living systems might have similarly "self-originated" with the aid of an energy source. In essence, they conceded the improbability of simple building blocks arranging themselves into highly ordered structures under normal equilibrium conditions. But they suggested that, under nonequilibrium conditions, where an external source of energy is supplied, biochemical building blocks might arrange themselves into highly ordered patterns.

More recently, Kauffman and de Duve have proposed self-organizational theories with somewhat less specificity, at least with regard to the problem of the origin of specified genetic information.[94] Kauffman invokes so-called autocatalytic properties to generate metabolism directly from simple molecules. He envisions such autocatalysis occurring once very particular configurations of molecules have arisen in a rich "chemical minestrone." De Duve also envisions protometabolism emerging first with genetic information arising later as a byproduct of simple metabolic activity.

Order versus Information

For many current origin-of-life scientists, self-organizational models now seem to offer the most promising approach to explaining the origin of specified biological information. Nevertheless, critics have called into question both the plausibility and the relevance of self-organizational models. Ironically, a prominent early advocate of self-organization, Dean Kenyon, has now explicitly repudiated such theories as both incompatible with empirical findings and theoretically incoherent.[95]

First, empirical studies have shown that some differential affinities do exist between various amino acids—that is, certain amino acids do form linkages more readily with some amino acids than with others.[96] Nevertheless, such differences do not correlate to actual sequences in large classes of known proteins.[97] In short, differing chemical affinities do not explain the multiplicity of amino acid sequences existing in naturally occurring proteins or the sequential arrangement of amino acids in any particular protein.

In the case of DNA, this point can be made more dramatically. Figure 2 shows that the structure of DNA depends on several chemical bonds. There are bonds, for example, between the sugar and the phosphate molecules forming the two twisting backbones of the DNA molecule. There are bonds fixing individual (nucleotide) bases to the sugar-phosphate backbones on each side of the molecule. There are also hydrogen bonds stretching horizontally across the molecule between nucleotide bases, making so-called complementary pairs. The individually weak hydrogen bonds, which in concert hold two complementary copies of the DNA message text together, make replication of the genetic instructions possible. It is important to note, however, that there are no chemical bonds between the bases along the longitudinal axis in the center of the helix. Yet it is precisely along this axis of the DNA molecule that the genetic information is stored.

Further, just as magnetic letters can be combined and recombined in any way to form various sequences on a metal surface, so too can each of the four bases (A, T, G, and C) attach to any site on the DNA backbone with equal facility, making all sequences equally probable (or improbable). Indeed, there are no significant differential affinities between any of the four bases and the binding sites along the sugar-phosphate backbone. The same type of N-glycosidic bond occurs between the base and the backbone, regardless of which base attaches. All four bases are acceptable; none is chemically favored. As Küppers has noted, "The properties of nucleic acids indicate that all the combinatorially possible nucleotide patterns of a DNA are, from a chemical point of view, equivalent."[98] Thus, "self-organizing" bonding affinities cannot explain the sequentially specific arrangement of nucleotide bases in DNA because (1) there are no bonds between bases along the information-bearing axis of the molecule, and (2) there are no differential affinities between the backbone and the specific bases that could account for variations in sequence. Because the same holds for

DNA: The Signature in the Cell

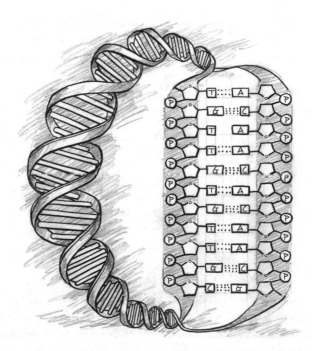

Figure 2. The bonding relationship between the chemical constituents of the DNA molecule. Sugars (designated by the pentagons) and phosphates (designated by the circled P's) are linked chemically. Nucleotide bases (A's, T's, G's and C's) are bonded to the sugar-phosphate backbones. Nucleotide bases are linked by hydrogen bonds (designated by dotted double or triple lines) across the double helix. But no chemical bonds exist between the nucleotide bases along the message-bearing spine of the helix. (Courtesy of Fred Heeren, Day Star publications.)

RNA molecules, researchers who speculate that life began in an RNA world have also failed to solve the sequence specificity problem—that is, the problem of explaining how information in functioning RNA molecules could have arisen in the first place.

For those who want to explain the origin of life as the result of self-organizing properties intrinsic to the material constituents of living systems, these rather elementary facts of molecular biology have decisive implications. The most obvious place to look for self-organizing properties to explain the origin of genetic information is in the constituent parts of the molecules that carry that information. But biochemistry and molecular biology make clear that forces of attraction between the constituents in DNA, RNA, and proteins do not explain the sequence specificity of these large, information-bearing biomolecules.

The properties of the monomers constituting nucleic acids and proteins simply do not make a particular gene, let alone life as we know it, inevitable. (We know this, in addition to the reasons already stated, because of the many variant polypeptides and gene sequences that exist in nature and that have been synthesized in the laboratory.) Yet if self-organizational scenarios for the origin of biological information are to have any theoretical import, they must claim just the opposite. And that claim is often made, albeit without much specificity. As de Duve has put it, the "processes that generated life" were "highly deterministic," making life as we know it "inevitable," given "the conditions that existed on the pre-biotic Earth."[99] Yet imagine the most favorable pre-biotic conditions. Imagine a pool of all four DNA bases and all necessary sugars and phosphates; would any particular genetic sequence inevitably arise? Given all necessary monomers, would any particular functional protein or gene, let alone a specific genetic code, replication system, or signal transduction circuitry, inevitably arise? Clearly not.

In the parlance of origin-of-life research, monomers are "building blocks," and building blocks can be arranged and rearranged in innumerable ways. The properties of stone blocks do not determine their own arrangement in the construction of buildings. Similarly, the properties of biological building blocks do not determine the arrangement of functional polymers. Rather, the chemical properties of the monomers allow a vast ensemble of possible configurations, the overwhelming majority of which have no biological function whatsoever. Functional genes or proteins are no more inevitable, given the properties of their "building blocks," than, for example, the Palace of Versailles was inevitable, given the properties of the stone blocks that were used to construct it. To anthropomorphize, neither bricks and stone, nor letters in a written text, nor nucleotide bases "care" how they are arranged. In each case, the properties of the constituents remain largely indifferent to the many specific configurations or sequences they may adopt, nor do they make any specific structures "inevitable," as self-organizationalists must claim.

Significantly, information theory makes clear that there is a good reason for this. If chemical affinities between the constituents in the DNA determined the arrangement of the bases, such affinities would dramatically diminish the capacity of DNA to carry information. Recall that classical information theory equates the reduction of uncertainty with the transmission of information, whether specified or unspecified. The transmission of information, therefore, requires physical-chemical contingency. As Robert Stalnaker has noted, "[Information] content requires contingency."[100] If, therefore, forces of chemical necessity completely determine the arrangement of constituents in a system, that arrangement will not exhibit complexity or convey information.

Consider, for example, what would happen if the individual nucleotide bases (A, C, G, and T) in the DNA molecule did interact by chemical necessity (along the information-bearing axis of DNA). Suppose that every time adenine (A) occurred in a growing genetic sequence, it attracted cytosine (C) to it.[101] Suppose every time guanine (G) appeared, thymine (T) followed. If this were the case, the longitudinal axis of DNA would be peppered with repetitive sequences in which A followed C and T followed G. Rather than a genetic molecule capable of virtually unlimited novelty and characterized by unpredictable and aperiodic sequences, DNA would contain sequences awash in repetition or redundancy—much like the arrangement of atoms in crystals. In a crystal, the forces of mutual chemical attraction do determine, to a very considerable extent, the sequential arrangement of its constituent parts. Hence, sequencing in crystals is highly ordered and repetitive, but neither complex nor informative. In DNA, however, where any nucleotide can follow any other, a vast array of novel sequences is possible, corresponding to a multiplicity of possible amino acid sequences and protein functions.

The forces of chemical necessity produce redundancy (roughly, law- or rule-generated repetition) or monotonous order but reduce the capacity to convey information and express novelty. Thus, as chemist Michael Polanyi noted:

> Suppose that the actual structure of a DNA molecule were due to the fact that the bindings of its bases were much stronger than the bindings would be for any other distribution of bases, then such a DNA molecule would have no information content. Its code-like character would be effaced by an overwhelming redundancy. . . . Whatever may be the origin of a DNA configuration, it can function as a code only if its order is not due to the

forces of potential energy. It *must be* as physically indeterminate as the sequence of words is on a printed page [emphasis added].[102]

In other words, if chemists had found that bonding affinities between the nucleotides in DNA produced nucleotide sequencing, they also would have found that they had been mistaken about DNA's information-bearing properties. Or to put the point quantitatively, to the extent that forces of attraction between constituents in a sequence determine the arrangement of the sequence, to that extent will the information-carrying capacity of the system be diminished or effaced by redundancy.[103] As Dretske has explained:

As ψ [the probability of a condition or state of affairs] approaches 1, the amount of information associated with the occurrence of ψ goes to 0. In the limiting case when the probability of a condition or state of affairs is unity [$\psi = 1$], no information is associated with, or generated by, the occurrence of ψ. This is merely another way to say that no information is generated by the occurrence of events for which there are no possible alternatives.[104]

Bonding affinities, to the extent they exist, inhibit the maximization of information because they determine that specific outcomes will follow specific conditions with high probability.[105] Yet information-carrying capacity is maximized when just the opposite situation obtains—namely, when antecedent conditions allow many improbable outcomes. Of course, as noted above ("Information Theory and Molecular Biology"), the base sequences in DNA do more than possess information-carrying capacity (or syntactic information) as measured by classical Shannon information theory. These sequences store functionally specified information—that is, they are specified as well as complex. Clearly, however, a sequence cannot be both specified and complex if it is not at least complex. Therefore, self-organizational forces of chemical necessity, which produce redundant order and preclude complexity, also preclude the generation of specified complexity (or specified information) as well. Chemical affinities do not generate complex sequences. Thus, they cannot be invoked to explain the origin of information, whether specified or otherwise.

A tendency to conflate the qualitative distinctions between "order" and "complexity" has characterized self-organizational scenarios—whether those that invoke internal properties of chemical attraction or an external organizing force or source of energy. That tendency calls into question the relevance of these scenarios for the origin of life. As Yockey has argued, the accumulation of structural or chemical order does not explain the origin of biological complexity or genetic information. He conceded that energy flowing through a system may produce highly ordered patterns. Strong winds form swirling tornados and the "eyes" of hurricanes; Prigogine's thermal baths do develop interesting convection currents; and chemical elements do coalesce to form crystals. Self-organizational theorists explain well what does not need explaining. What needs explaining in biology is not the origin of order (defined as symmetry or repetition) but the specified information—the highly complex, aperiodic, and specified sequences that make biological function possible. As Yockey warned: "Attempts to relate the idea of order . . . with biological organization or specificity must be regarded as a play on words that cannot stand careful scrutiny. Informational macromolecules can code genetic messages and therefore can carry information because the sequence of bases or residues is affected very little, if at all, by [self-organizing] physicochemical factors."[106]

In the face of these difficulties, some self-organizational theorists have claimed that we must await the discovery of new natural laws to explain the origin of biological information. As Manfred Eigen has argued, "Our task is to find an algorithm, a natural law, that leads to the origin of information."[107] Such a suggestion betrays confusion on two counts. First, scientific laws don't generally produce or cause natural phenomena, they describe them. For example, Newton's law of gravitation described, but did not cause or explain, the attraction between planetary bodies. Second, laws necessarily describe highly deterministic or predictable relationships between antecedent conditions and consequent events. Laws describe highly repetitive patterns in which the probability of each successive event (given the previous event) approaches unity. Yet information sequences are complex, not repetitive—information mounts as improbabilities multiply. Thus, to say that scientific laws can produce information is essentially a contradiction in terms. Rather, scientific laws describe (almost by definition) highly predictable and regular phenomena—that is, redundant order, not complexity (whether specified or otherwise).

Though the patterns that natural laws describe display a high degree of regularity, and thus lack the complexity that characterizes information rich systems, one could argue that we might someday discover a very particular configuration of initial conditions that routinely generates high informational states. Thus, while we cannot hope to find a law that describes an information-rich relationship between antecedent and consequent variables, we might find a law that describes how a very particular set of initial conditions routinely generates a high information state. Yet even the statement of this hypothetical seems itself to beg the question of the ultimate origin of information, since "a very particular set of initial conditions" sounds precisely like an information-rich—that is, a highly complex and specified—state. In any case, everything we know experientially suggests that the amount of specified information present in a set of antecedent conditions necessarily equals or exceeds that of any system produced from those conditions.

OTHER SCENARIOS AND THE DISPLACEMENT OF THE INFORMATION PROBLEM

In addition to the general categories of explanation already examined, origin-of-life researchers have proposed many more specific scenarios, each emphasizing random variations (chance), self-organizational laws (necessity), or both. Some of those scenarios purport to address the information problem; others attempt to bypass it altogether. Yet on closer examination, even scenarios that appear to alleviate the problem of the origin of specified biological information merely shift the problem elsewhere. Genetic algorithms can "solve" the information problem, but only if programmers provide informative target sequences and selection criteria. Simulation experiments can produce biologically relevant precursors and sequences, but only if experimentalists manipulate initial conditions or select and guide outcomes—that is, only if they add information themselves. Origin-of-life theories can leapfrog the problem altogether, but only by presupposing the presence of information in some other preexisting form.

Any number of theoretical models for the origin of life have fallen prey to this difficulty. For example, in 1964, Henry Quastler, an early pioneer in the application of information theory to molecular biology, proposed a DNA-first model for the origin of life. He envisioned the initial emergence of a system of unspecified polynucleotides capable of primitive self-replication via the mechanisms of complementary base-pairing. On Quastler's account, the polymers in his system would have initially lacked specificity (which he equated with information).[108] Only later, when his system of polynucleotides had come into association with a fully functional set of proteins and ribosomes, would the specific nucleotide sequences in the polymers take on any functional significance. He likened that process to the random selection of a combination for a lock in which the combination would only later acquire functional significance once particular tumblers had been set to allow the combination to open the lock. In both the biological and the mechanical case, the surrounding context would confer functional specificity on an initially unspecified sequence. Thus, Quastler characterized the origin of information in polynucleotides as an "accidental choice remembered."

Although Quastler's way of conceiving of the origin of specified biological information did allow a "chain of nucleotides [to] become a [functional] system of genes without necessarily suffering any change in structure," it did have an overriding difficulty. It did not account for the origin of the complexity and specificity of the system of molecules whose association with the initial sequence gave the initial sequence functional significance. In Quastler's combination-lock example, conscious agents chose the tumbler settings that made the initial combination functionally significant. Yet Quastler expressly precluded conscious design as a possibility for explaining the origin of life.[109] Instead, he seemed to suggest that the origin of the biological context—that is, the complete set of functionally specific proteins (and the translation system) necessary to create a "symbiotic association" between polynucleotides and proteins—would arise by chance. He even offered some rough calculations to show that the origin of such a multimolecular context, though improbable, would have been probable enough to expect it to occur by chance in the pre-biotic soup. Quastler's calculations now seem extremely implausible in light of the discussion of minimal complexity (see above, "Beyond the Reach of Chance").[110] More significantly, Quastler "solved" the problem of the origin of complex specificity in nucleic acids only by transferring the problem to an equally complex and specified system of proteins and ribosomes. Whereas, admittedly, any polynucleotide sequence would suffice initially, the subsequent proteins and ribosomal material constituting the translation system would have to possess an extreme specificity relative to the initial polynucleotide sequence and relative to any protocellular functional requirements. Thus, Quastler's attempt to bypass the sequence specificity problem merely shifted it elsewhere.

Self-organizational models have encountered similar difficulties. For example, chemist J. C. Walton has argued (echoing earlier articles by Mora) that the self-organizational patterns produced in Prigogine-style convection currents do not exceed the organization or structural information represented by the experimental apparatus used to create the currents.[111] Similarly, Maynard Smith, Dyson, and Shapiro have shown that Eigen's so-called hypercycle model for generating biological information actually shows how information tends to degrade over time.[112] Eigen's hypercycles presuppose a large initial contribution of information in the form of a long RNA molecule and some forty specific proteins and thus do not attempt to explain the ultimate origin of biological information. Moreover, because hypercycles lack an error-free mechanism of self-replication, the proposed mechanism succumbs to various "error catastrophes" that ultimately diminish, not increase, the (specified) information content of the system over time.

Stuart Kauffman's self-organizational theory also subtly transfers the problem of the origin of information. In *The Origins of Order*, Kauffman attempted to leapfrog the sequence-specificity problem by proposing a means by which a self-reproducing and metabolic system might emerge directly from a set of "low specificity" catalytic peptides and RNA molecules in a pre-biotic soup or "chemical minestrone." Kauffman envisioned, as Iris Frey put it, "a set of catalytic polymers in which no single molecule reproduces itself, but the system as a whole does."[113] Kauffman argued that once a sufficiently diverse set of catalytic molecules had assembled (in which the different peptides performed enough different catalytic functions), the ensemble of individual molecules would spontaneously undergo a kind of phase transition resulting in a self-reproducing metabolic system. Thus, Kauffman argued that metabolism can arise directly without genetic information encoded in DNA.[114]

Nevertheless, Kauffman's scenario does not solve, or bypass, the problem of the origin of biological information. Instead, it either presupposes the existence of unexplained sequence-specificity, or it transfers such needed specificity out of view. Kauffman claimed that an ensemble of relatively short and low specificity catalytic peptides and RNA molecules would suffice jointly to establish a metabolic system. He defended the biochemical plausibility of his scenario on the grounds that some proteins can perform enzymic functions with low specificity and complexity. He cited proteases such as trypsin that cleave peptide bonds at single amino acid sites and proteins in the clotting cascade that "cleave essentially single target polypeptides" to support his claim.[115]

Yet Kauffman's argument has two problems. First, it does not follow, nor is it the case biochemically, that just because some enzymes might function with low specificity, that all the catalytic peptides (or enzymes) needed to establish a self-reproducing metabolic cycle could function with similarly low levels of specificity and complexity. Instead, modern biochemistry shows that at least some, and probably many, of the molecules in a closed interdependent system of the type that Kauffman envisioned would require high complexity and specificity proteins. Enzymatic catalysis (which his scenario would surely necessitate) invariably requires molecules long enough (at least 50-mers) to form tertiary structures (whether in polynucleotides or polypeptides). Further, these long polymers invariably require very specific three-dimensional geometries (which can in turn derive from sequence-specific arrangements of monomers) in order to catalyze necessary reactions. How do these molecules acquire their specificity of sequencing? Kauffman did not address this question because his illustration incorrectly suggested that he need not do so.

Secondly, it turns out that even the allegedly low-specificity molecules that Kauffman cited to illustrate the plausibility of his scenario do not themselves manifest low complexity and specificity. Instead, Kauffman has confused the specificity and complexity of the parts of the polypeptides upon which the proteases act with the specificity and complexity of the proteins (the proteases) that do the enzymatic acting. Though trypsin, for example, acts upon (cleaves) peptide bonds at a relatively simple target (the carboxyl end of two separate amino acids, arginine and lysine), trypsin itself is a highly complex and specifically sequenced molecule. Indeed, trypsin is a non-repeating 200+ residue protein that possesses significant sequence-specificity as a condition of its function.[116] Further, it has to manifest significant three-dimensional (geometric) specificity to recognize the specific amino acids arginine and lysine—sites at which it cleaves peptide bonds. By equivocating in his discussion of specificity, Kauffman obscured from view the considerable specificity and complexity requirement of even the proteases he cites to justify his claim that low specificity catalytic peptides will suffice to establish a metabolic cycle. Thus, Kauffman's own illustration properly understood (that is, without equivocating about the relevant locus of specificity) shows that for his scenario to have biochemical plausibility it must presuppose the existence of many high complexity and specificity polypeptides and polynucleotides. Where does this information in these molecules come from? Kauffman, again, does not say.

Further, Kauffman must acknowledge (as he seems to in places),[117] that for autocatalysis (for which there is as yet no experimental evidence) to occur, the molecules in the "chemical minestrone" must be held in a very specific spatial-temporal relationship to one another. In other words, for the direct autocatalysis of integrated metabolic complexity to occur, a system of catalytic peptide molecules must first achieve a very specific molecular configuration, or a low configurational entropy state.[118] Yet this requirement is isomorphic with the requirement that the system must start with a highly specified complexity. Thus, to explain the origin of specified biological complexity at the systems level, Kauffman must presuppose the existence of highly specific and complex (i.e., information-rich) molecules as well as a highly specific arrangement of those molecules at the molecular level. Therefore, his work—if it has any relevance to the actual behavior of molecules—presupposes or transfers, rather than explains, the ultimate origin of specified complexity or information.

Others have claimed that the RNA-world scenario offers a promising approach to the origin-of-life problem, and with it, presumably, the problem of the origin of the first genetic information. The RNA world was proposed as an explanation for the origin of the interdependence of nucleic acids and proteins in the cell's information-processing system. In extant cells, building proteins requires genetic information from DNA, but information on DNA cannot be processed without many specific proteins and protein complexes. This poses a chicken-or-egg problem. The discovery that RNA (a nucleic acid) possesses some limited catalytic properties similar to those of proteins suggested a way to solve that problem. "RNA-first" advocates proposed an early state in which RNA performed both the enzymatic functions of modern proteins and the information-storage function of modern DNA, thus allegedly making the interdependence of DNA and proteins unnecessary in the earliest living system.

Nevertheless, many fundamental difficulties with the RNA-world scenario have emerged. First, synthesizing (and/or maintaining) many essential building blocks of RNA molecules under realistic conditions has proven either difficult or impossible.[119] Further, the chemical conditions required for the synthesis of ribose sugars are decidedly incompatible with the conditions required for synthesizing nucleotide bases.[120] Yet both are necessary constituents of RNA. Second, naturally occurring RNA possesses very few of the specific enzymatic properties of proteins necessary to extant cells. Indeed, RNA catalysts do not function as true enzyme catalysts. Enzymes are capable of coupling energetically favorable and unfavorable reactions together. RNA catalysts, so-called "ribozymes," are not. Third, RNA-world advocates offer no plausible explanation for how primitive RNA replicators might have evolved into modern cells that do rely almost exclusively on proteins to process and translate genetic information and regulate metabolism.[121] Fourth, attempts to enhance the limited catalytic properties of RNA molecules in so-called ribozyme engineering experiments have inevitably required extensive investigator manipulation, thus simulating, if anything, the need for intelligent design, not the efficacy of an undirected chemical evolutionary process.[122]

Most important for our present considerations, the RNA-world hypothesis presupposes, but does not explain, the origin of sequence specificity or information in the original functional RNA molecules. Indeed, the RNA-world scenario was proposed as an explanation for the functional interdependence problem, not the information problem. Even so, some RNA-world advocates seem to envision leapfrogging the sequence-specificity problem. They imagine oligomers of RNA arising by chance on the pre-biotic Earth and then later acquiring an ability to polymerize copies of themselves—that is, to self-replicate. In such a scenario, the capacity to self-replicate would favor the survival of those RNA molecules that could do so and would thus favor the specific sequences that the first self-replicating molecules happened to have. Thus, sequences that originally arose by chance would subsequently acquire a functional significance as an "accidental choice remembered."

Like Quastler's DNA-first model, however, this suggestion merely shifts the specificity problem out of view. First, for strands of RNA to perform enzymatic functions (including enzymatically mediated self-replication), they must, like proteins, have very specific arrangements of constituent building blocks (nucleotides in the RNA case). Further, the strands must be long enough to fold into complex three-dimensional shapes (to form so-called tertiary structures). Thus, any RNA molecule capable of enzymatic function must have the properties of complexity and specificity exhibited by DNA and proteins. Hence, such molecules must possess considerable (specified) information content. And yet, explaining how the building blocks of RNA might have arranged themselves into functionally specified sequences has proven no easier than explaining how the constituent parts of DNA might have done so, especially given the high probability of destructive cross-reactions between desirable and undesirable molecules in any realistic pre-biotic soup. As de Duve noted in a critique of the RNA-world hypothesis, "Hitching the components together in the right manner raises additional problems of such magnitude that no one has yet attempted to do so in a pre-biotic context."[123]

Second, for a single-stranded RNA catalyst to self-replicate (the only function that could be selected in a pre-biotic environment), it must find another catalytic RNA molecule in close vicinity to function as a template, since a single-stranded RNA cannot function as both enzyme and template. Thus, even if an originally unspecified RNA sequence might later acquire functional significance by chance, it could perform a function only if another RNA molecule—that is, one with a highly specific sequence relative to the original—arose in close vicinity to it. Thus, the attempt to bypass the need for specific sequencing in an original catalytic RNA only shifts the specificity problem elsewhere, namely, to a second and necessarily highly specific RNA sequence. Put differently, in addition to the specificity required to give the first RNA molecule self-replicating capability, a second RNA molecule with an extremely specific sequence—one with essentially the same sequence as the original—would also have to arise. Yet RNA-world theorists do not explain the origin of the requisite specificity in either the original molecule or its twin. Joyce and Orgel have calculated that to have a reasonable chance of finding two identical RNA molecules of a length sufficient to perform enzymatic functions would require an RNA library of some 10^{54} RNA molecules.[124] The mass of such a library vastly exceeds the mass of the Earth, suggesting the extreme implausibility of the chance origin of a primitive replicator system. Yet one cannot invoke natural selection to explain the origin of such primitive replicators, since natural selection only ensues once self-replication has arisen. Further, RNA bases, like DNA bases, do not manifest self-organizational bonding affinities that could explain their specific sequencing. In short, the same kind of evidentiary and theoretical problems emerge whether one proposes that genetic information arose first in RNA or DNA molecules. The attempt to leapfrog the sequencing problem by starting with RNA replicators only shifts the problem to the specific sequences that would make such replication possible.

3. The Return of the Design Hypothesis

If attempts to solve the information problem only relocate it, and if neither chance nor physical-chemical necessity (nor the two acting in combination) explains the ultimate origin of specified biological information, what does? Do we know of any entity that has the causal powers to create large amounts of specified information? We do. As Henry Quastler recognized, the "creation of new information is habitually associated with conscious activity."[125]

Experience affirms that specified complexity or information (defined as specified complexity) routinely arises from the activity of intelligent agents. A computer user who traces the information on a screen back to its source invariably comes to a mind, that of a software engineer or programmer. Similarly, the information in a book or newspaper column ultimately derives from a writer—from a mental, rather than a strictly material, cause.

Further, our experience-based knowledge of information flow confirms that systems with large amounts of specified complexity or information (especially codes and languages) invariably originate from an intelligent source—that is, from a mind or a personal agent.[126] Moreover, this generalization holds not only for the semantically specified information present in natural languages but also for other forms of information or specified complexity whether present in machine codes, machines, or works of art. Like the letters in a section of meaningful text, the parts in a working engine represent a highly improbable yet functionally specified configuration. Similarly, the highly improbable shapes in the rock on Mount Rushmore conform to an independently given pattern: the faces of American presidents known from books and paintings. Thus, both systems have a large amount of specified complexity or information so defined. Not coincidentally, they also originated by intelligent design, not by chance and/or physico-chemical necessity.

This generalization—that intelligence is the only known cause of specified complexity or information (at least, starting from a nonbiological source)—has received support from origin-of-life research itself. During the last fifty years, every naturalistic model proposed has failed to explain the origin of the specified genetic information required to build a living cell.[127] Thus, mind or intelligence, or what philosophers call "agent causation," now stands as the only cause known to be capable of generating large amounts of information starting from a nonliving state.[128] As a result, the presence of specified information-rich sequences in even the simplest living systems would seem to imply intelligent design.[129]

A Distinct Form of Inference

But could this intuitive connection between information and the prior activity of a designing intelligence be justified in the form of a rigorous scientific argument?

I first began to investigate this possibility during my Ph.D. research at Cambridge University in the late 1980s. To determine whether the case for intelligent design could be developed as a rigorous scientific argument, I began to examine first how scientists investigating origins events developed their arguments. Specifically, I examined the method and reasoning that historical scientists use to identify causes responsible for events in the remote past.

In the process of my investigations, I discovered that historical scientists often make inferences with a distinctive logical form. Paleontologists, evolutionary biologists, and other historical scientists often infer a past cause from present clues. For example, visit the Royal Tyrrell Museum in Alberta, Canada, and you will find there a beautiful reconstruction of the Cambrian seafloor with its stunning assemblage of exotic organisms. Or read the fourth chapter of Simon Conway Morris's book on the Burgess Shale and you will be taken on a vivid guided tour of that long-ago place. But, in both cases, what Conway Morris[130] and the museum scientists did was to reconstruct the ancient Cambrian site from an assemblage of present-day fossils. As Stephen Gould put it, historical scientists "infer history from its results."[131]

This type of inference is called an "abductive inference," or "abductive reasoning." It was first described by the American philosopher and logician C. S. Peirce. He noted that, unlike inductive reasoning, in which a universal law or principle is established from repeated observations of the same phenomena, and unlike deductive reasoning, in which a particular fact is deduced by applying a general law or rule to another particular fact or case, abductive reasoning infers unseen facts, events, or causes in the past from clues or facts in the present.

As Peirce himself showed, however, there is a problem with abductive reasoning. Consider the following syllogism:

If it rains, the streets will get wet.
The streets are wet.
Therefore, it rained.

This syllogism affirms a past condition (i.e., that it rained) but it commits a logical fallacy known as *"affirming the consequent."* Given that the street is wet (and without additional evidence to decide the matter), one can only conclude that *perhaps* it rained. Why? Because there are many other possible ways by which the street may have gotten wet. Rain may have caused the streets to get wet; a street cleaning machine might have caused them to get wet; or an uncapped fire hydrant might have done so. It can be difficult to infer the past from the present because there are many possible causes of a given effect.

Peirce's question was this: How is it that, despite the logical problem of affirming the consequent, we nevertheless frequently make reliable abductive inferences about the past? He noted, for example, that no one doubts the existence of Napoleon. Yet we use abductive reasoning to infer Napoleon's existence. That is, we must infer his past existence from present effects. But despite our dependence on abductive reasoning to make this inference, no sane or educated person would doubt that Napoleon Bonaparte actually lived. How could this be if the problem of affirming the consequent bedevils our attempts to reason abductively? Peirce's answer was revealing: "Though we have not seen the man [Napoleon], yet we cannot explain what we have seen without" the hypothesis of his existence.[132] Peirce's words imply that a particular abductive hypothesis can be strengthened if it can be shown to explain a result in a way that other hypotheses do not, and that it can be reasonably believed (in practice) if it explains in a way that no other hypotheses do. In other words, an abductive inference can be enhanced if it can be shown that it represents the best or the only adequate explanation of the effects in question.

The problem with abductive reasoning is that there is often more than one cause that can explain the same effect. To address this problem, pioneering geologist Thomas Chamberlain delineated a method of reasoning he called the "method of multiple working hypotheses."[133] Geologists and other historical scientists use this method when there is more than one possible cause or hypothesis to explain the same evidence. In such cases, historical scientists carefully weigh the evidence and what they know about various possible causes to determine which best explains the clues before them. In modern times, contemporary philosophers of science have called this the "method of *inference to the best explanation.*"

That is, when trying to explain the origin of an event or structure in the past, historical scientists compare various hypotheses to see which, if true, would best explain it. They then provisionally affirm that hypothesis that best explains the data as the most likely to be true.

But what constitutes the best explanation for the historical scientist? My research showed that among historical scientists it's generally agreed that "*best*" doesn't mean "*ideologically satisfying*" or "*mainstream*;" instead, "*best*" generally has been taken to mean, first and foremost, most "*causally adequate.*" In other words, historical scientists try to identify causes that are known to produce the effect in question. In making such determinations, historical scientists evaluate hypotheses against their present knowledge of cause and effect; causes that are known to produce the effect in question are judged to be better causes than those that are not. For instance, a volcanic eruption is a better explanation for an ash layer in the Earth than an earthquake because eruptions have been observed to produce ash layers, whereas earthquakes have not. Or consider another example: An earthquake and a bomb could explain the destruction of the building, but only a bomb could explain the presence of charring and shrapnel at the scene of the rubble. Earthquakes do not produce shrapnel, nor do they cause charring, at least not on their own. Thus, the bomb best explains the pattern of destruction at the building site. In this way, knowledge of the present causal powers of various entities and processes enables historical scientists to make inferences about possible causes in the past. Entities, conditions, or processes that have the capability (or causal powers) to produce the evidence in question constitute better explanations of that evidence than those that do not.

This principle was first articulated in the work of the great geologist Charles Lyell, a figure who exerted a tremendous influence on nineteenth-century historical science generally and on Charles Darwin specifically. Darwin read Lyell's magnum opus, *The Principles of Geology*, on the voyage of the *Beagle*, and later appealed to its uniformitarian principles to argue that observed micro-evolutionary processes of change could be used to explain the origin of new forms of life. The subtitle of Lyell's *Principles* summarized the geologist's central methodological principle: "*Being an Attempt to Explain the Former Changes of the Earth's Surface*, by Reference to Causes Now in Operation." Lyell argued that when historical scientists are seeking to explain events in the past, they should not invoke unknown or exotic causes, the effects of which we do not know, but instead they should cite causes that are known from our uniform experience to have the power to produce the effect in question (i.e., "causes now in operation").

Darwin subscribed to this methodological principle. His term for a "presently acting cause" was a *vera causa*—that is, a true or actual cause. In other words, when explaining the past, historical scientists should seek to identify established causes—causes known to produce the effect in question. For example, Darwin tried to show that the process of descent with modification was the *vera causa* of certain kinds of patterns found among living organisms. He noted that diverse organisms share many common features. He called these "*homologies*," and noted that we know from experience that descendents, although they differ from their ancestors, also resemble them in many ways, usually more closely than others who are more distantly related. So he proposed descent with modification as a *vera causa* for homologous structures. That is, he argued that our uniform experience shows that the process of descent with modification from a common ancestor is "causally adequate," or capable of producing homologous features.

Contemporary philosophers agree that causal adequacy is the key criteria by which competing hypotheses are adjudicated, but they also show that this process leads to secure inferences only where it can be shown that there is just one known cause for the evidence in question. Philosophers of science Michael Scriven and Elliot Sober, for example, point out that historical scientists can make inferences about the past with confidence when they discover evidence or artifacts for which there is only one cause known to be sufficient. When historical scientists infer to a *uniquely* plausible cause, they avoid the fallacy of affirming the consequent and the error of ignoring other possible causes with the power to produce the same effect.

It follows that the process of determining the best explanation often involves generating a list of possible hypotheses, comparing their known or theoretically plausible causal powers with respect to the relevant data, and then like a detective attempting to identify the murderer, progressively eliminating potential but inadequate explanations until, finally, one remaining causally adequate explanation can be identified as the best. As Scriven explains, such abductive reasoning (or what he calls "reconstructive causal analysis") "proceeds by the elimination of possible causes," a process that is essential if historical scientists are to overcome the logical limitations of abductive reasoning.[134] When a thorough study of various possible causes turns up only a single adequate cause for a given effect, historical or forensic scientists can make definitive inferences about the past.[135]

The matter can be framed in terms of formal logic. As C. S. Peirce noted, arguments of the form

> If X, then Y
> Y
> Therefore X

commit the fallacy of affirming the consequent. Nevertheless, as M. Scriven, E. Sober, W. P. Alston, and W. B. Gallie have observed,[136] such arguments can be restated in a logically acceptable form if it can be shown that Y has only one known cause (i.e., X) or that X is a necessary condition (or cause) of Y. Thus, arguments of the form

> X is antecedently necessary to Y,
> Y exists,
> Therefore, X existed

are accepted as logically valid by philosophers and persuasive by historical and forensic scientists. Scriven especially emphasized this point: If scientists can discover an effect for which there is only one plausible cause, they can infer the presence or action of that cause in the past with great confidence.

The Martian landscape, for example, displays erosional features—trenches and rills—that resemble those produced on Earth by moving water. Though Mars at present has no significant liquid water on its surface, some planetary scientists have nevertheless inferred that Mars did have a significant amount of water on its surface in the past. Why? Geologists and planetologists have not observed any cause other than moving water that can produce the kind of erosional features that we observe on Mars today. Since, in our experience, water alone produces erosional trenches and rills, the presence of these features on Mars allows planetologists to infer the past action of water on the surface of the red planet.

Or consider another example. Several years ago, one of the forensic pathologists from the original Warren Commission that investigated the assassination of John F. Kennedy spoke out to quash rumors about a second gunman firing from in front of the presidential motorcade. The bullet hole in the back of President Kennedy's skull apparently evidenced a distinctive beveling pattern that clearly indicated that it had entered his skull from the rear. The pathologist called the beveling pattern a "distinctive diagnostic" because the pattern indicated a single possible direction of entry. Since a rear entry was necessary to cause the beveling pattern in the back of the president's skull, the pattern allowed the forensic pathologists to diagnose the trajectory of the bullet.[137]

In many cases, of course, the investigator will have to work his way to a unique cause one painstaking step at a time. For instance, both wind shear and compressor blade failure could explain an airline crash, but the forensic investigator will want to know specifically which one did, or if the true cause lies elsewhere. Ideally, the investigator will be able to discover some crucial piece of evidence or suite of evidences for which there is only one known cause, allowing him to distinguish between competing explanations and eliminate every explanation but the correct one.

In my study of the methods of the historical sciences, I found that historical scientists, like detectives and forensic experts, routinely employ this type of abductive and eliminative reasoning in their attempts to infer the best explanation.[138] In fact, Darwin himself employed this method in *The Origin of Species*. There he argued for his theory of universal common descent, not because it could predict future outcomes under controlled experimental conditions, but because it could explain already known facts better than rival hypotheses. As he explained in a letter to Asa Gray:

> I . . . test this hypothesis [universal common descent] by comparison with as many general and pretty well-established propositions as I can find—in geographical distribution, geological history, affinities &c., &c. And it seems to me that, supposing that such a hypothesis were to explain such general propositions, we ought, in accordance with the common way of following all sciences, to admit it till some better hypothesis be found out.[139]

What does this investigation into the nature of historical scientific reasoning have to do with intelligent design, the origin of biological information, and the mystery of life's origin? For me, it was critically important to deciding whether the design hypothesis could be formulated as a rigorous scientific explanation as opposed to just an intriguing intuition. I knew from my study of origin-of-life research that the central question facing scientists trying to explain the origin of the first life was this: How did the sequence-specific digital information (stored in DNA and RNA) necessary to building the first cell arise?

My study of the methodology of the historical sciences then led me to ask a series of questions: What is the presently acting cause of the origin of specified information? What is the *vera causa* of such information? Or, what is the "only known cause" of this effect? Whether I used Lyell's, Darwin's, or Scriven's terminology, the question was the same: What type of cause has demonstrated the power to generate specified information? Based upon both common experience and my knowledge of the many failed attempts to solve the problem with "unguided" pre-biotic simulation experiments and computer simulations, I concluded that there is only one sufficient or "presently acting" cause of the origin of such functionally specified information. And that cause is intelligence. In other words, I concluded, based on our experience-based understanding of the cause-and-effect structure of the world, that intelligent design is the best explanation for the origin of the information necessary to build the first cell. Ironically, I discovered that if one applies Lyell's uniformitarian method—a practice much maligned by young Earth creationists—to the question of the origin of biological information, the evidence from molecular biology supports a new and rigorous scientific argument for (and inference to) intelligent design.

The logical calculus underlying the inference to intelligent design from the presence of specified information in DNA follows a valid and well-established method used in all historical and forensic sciences. Logically, one can infer a cause from its effect (or an antecedent from a consequent) when the cause (or antecedent) is known to be necessary to produce the effect in question. If it's true that "where there's smoke there's fire," then the presence of smoke billowing over a hillside allows us to infer a fire beyond our view. Inferences based on knowledge of empirically necessary conditions or causes ("distinctive diagnostics") are common in historical and forensic sciences, and often lead to the detection of intelligent as well as natural causes and events. Since criminal X's fingers are the only known cause of criminal X's fingerprints, X's prints on the murder weapon incriminate him with a high degree of certainty. Similarly, since intelligent design is the only known cause of large amounts of specified complexity or information, the presence of such information implies an intelligent source.

Indeed, since experience affirms mind or intelligent design as a necessary condition (and necessary cause) of information, one can detect (or "retrodict") the past action of a designing intelligence from an information-rich effect—even if the cause itself cannot be directly observed.[140] Thus, the pattern of flowers spelling "Welcome to Victoria" in Victoria Harbor in Canada allows visitors to infer the activity of intelligent agents even if they did not see the flowers planted or arranged. Similarly, the specified and complex arrangement of nucleotide sequences—the information—in DNA implies the past action of intelligence, even if such mental activity cannot be directly observed.

Scientists in many fields recognize the connection between intelligence and information, and make inferences accordingly. Archaeologists assume that a scribe produced the inscriptions on the Rosetta Stone; evolutionary anthropologists establish the intelligence of early hominids from chipped flints that are too improbably specified in form (and function) to have been produced by natural causes; NASA's search for extraterrestrial intelligence (SETI) presupposes that any information embedded in electromagnetic signals coming from space would indicate an intelligent source.[141] As yet, however, radio-astronomers have not found any such information-bearing signals. But closer to home, molecular biologists have identified information-rich sequences and systems in the cell, suggesting, by the same logic, an intelligent cause for those effects.

Argument from Ignorance? Or an Inference to the Best Explanation?

Some would object that this design argument constitutes an argument from ignorance. Objectors charge that design advocates use our present ignorance of any sufficient materialistic cause of specified information as the sole basis for inferring an intelligent cause of the information present in the cell. Since we don't yet know how specified biological information could have arisen, we invoke the mysterious notion of intelligent design. On this view, intelligent design functions not as an explanation, but as a placeholder for ignorance.

Although the inference to design from the presence of information in DNA does not qualify as a deductively certain proof of intelligent design (empirically based arguments in science rarely do), it does not constitute a fallacious argument from ignorance. Arguments from ignorance occur when evidence against a proposition X is offered as the sole (and conclusive) grounds for accepting some alternative proposition Y.

The inference to design as sketched above (see "A Distinct Form of Inference") does not commit this fallacy. True, the previous section of this essay (see Part 2, above) argued that at present all types of natural causes and mechanisms fail to account for the origin of biological information from a pre-biotic state. And clearly, this lack of knowledge of any adequate natural cause does provide *part* of the grounds for inferring design from information in the cell. (Though one could just as easily argue that even this "absence of knowledge" actually constitutes "knowledge of absence.") In any case, our "ignorance" of any sufficient natural cause is only part of the basis for inferring design. We also *know* that intelligent agents can and do produce information-rich systems: we have positive experience-based knowledge of an alternative cause that is sufficient—namely, intelligence or "conscious activity."

For this reason, the design inference defended here does not constitute an argument from ignorance but an inference to the best explanation.[142] Inferences to the best explanation do not assert the adequacy of one causal explanation merely on the basis of the inadequacy of some other causal explanation. Instead, they compare the explanatory power of many competing hypotheses to determine which hypothesis would, if true, provide the best explanation for some set of relevant data based upon our *knowledge* of the causal powers of competing explanatory entities.[143]

This essay has followed precisely this method to make a case for intelligent design as the best explanation for the origin of biological information. It has evaluated and compared the causal efficacy of four broad categories of explanation—chance, necessity, the combination of those two, and intelligent design—with respect to their ability to produce large amounts of specified complexity or information. As we have seen, neither scenarios based on chance nor those based on necessity (nor those that combine the two) can explain the origin of specified biological information in a pre-biotic context. That result comports with our uniform human experience. Natural processes do not produce information-rich structures starting from purely physical or chemical antecedents. Nor does matter, whether acting at random or under the force of physical-chemical necessity, arrange itself into complex, information-rich sequences.

Nevertheless, it is not correct to say that we do not know how information arises. We know from experience that conscious intelligent agents can create informational sequences and systems. To quote Quastler again, the "creation of new information is habitually associated with conscious activity."[144] Further, experience teaches that whenever large amounts of specified complexity or information are present in an artifact or entity whose causal story is known, invariably creative intelligence—intelligent design—played a causal role in the origin of that entity. Thus, when we encounter such information in the bio-macromolecules necessary to life, we may infer—based on our *knowledge* (not our ignorance) of established cause-effect relationships—that an intelligent cause operated in the past to produce the specified complexity or information necessary to the origin of life.

Moreover, as formulated, the argument to design from the presence of information in DNA also adheres to the standard uniformitarian canons of method employed within the historical sciences. The principle of uniformitarianism states that "the present is the key to the past." In particular, the principle specifies that our knowledge of present cause-effect relationships should govern our assessment of the plausibility of the inferences that we make about the remote causal past. Yet it is precisely such knowledge of cause-effect relationships that informs the inference to intelligent design. Since we know that intelligent agents do produce large amounts of information, and since all known natural processes do not (or cannot), we can infer design as the best explanation of the origin of information in the cell.

Insofar as the inference to design depends on present knowledge of the demonstrated causal powers of natural entities and intelligent agency, it no more constitutes an argument from ignorance than any other well-grounded inference in geology, archaeology, or paleontology—where present knowledge of cause-effect relationships guides the inferences that scientists make about the causal past.

Objectors might still deny the legitimacy of inferring intelligent design (even as a best explanation) because we are ignorant of what future inquiry may uncover about the causal powers of other materialistic entities or processes. Some would characterize the design inference presented here as invalid or unscientific because it depends on a negative generalization—that is, "purely physical and chemical causes do not generate large amounts of specified information"—which future discoveries may later falsify. They say we should "never say never."

Yet science often says "never," even if it can't say so for sure. Negative or proscriptive generalizations often play an important role in science. As many scientists and philosophers of science have pointed out, scientific laws often tell us not only what does happen, but also what does not happen.[145] The conservation laws in thermodynamics, for example, proscribe certain outcomes. The first law tells us that energy is never created or destroyed. The second tells us that the entropy of a closed system will never decrease over time. Those who claim that such "proscriptive laws" do not constitute knowledge, because they are based on past but not future experience, will not get very far if they try to use their skepticism to justify funding for research on, say, perpetual motion machines.

Further, without proscriptive generalizations, without knowledge about what various possible causes cannot or do not produce, historical scientists could not make determinations about the past. Reconstructing the past requires making abductive inferences from present effects back to past causal events.[146] Making such inferences requires a progressive elimination of competing causal hypotheses. Deciding which causes can be eliminated from consideration requires knowing what effects a given cause can—and cannot—produce. If historical scientists could never say that particular entities lack particular causal powers, they could never eliminate them, even provisionally, from consideration. Thus, they could never infer that a specific cause had acted in the past. Yet historical and forensic scientists make such inferences all the time.

Moreover, William Dembski's examples of design inferences—from fields such as archaeology, cryptography, fraud-detection, and criminal forensics—show that we often infer the past activity of an intelligent cause and do so, evidently, without worrying about committing fallacious arguments from ignorance. And we do so for good reason. A vast amount of human experience shows that intelligent agents have unique causal powers that matter (especially non-living matter) does not. When we observe features or effects that we know from experience only agents produce, we rightly infer the prior activity of intelligence.

To determine the best explanation, scientists do not need to say "never" with absolute certainty. They need only say that a postulated cause is best, given what we know at present about the demonstrated causal powers of competing entities or agencies. That cause C can produce effect E makes it a better explanation of E than some cause D that has never produced E (especially if D seems incapable of doing so on theoretical grounds), even if D might later demonstrate causal powers of which we are presently ignorant.[147]

Thus, the objection that the design inference constitutes an argument from ignorance reduces in essence to a restatement of the problem of induction. Yet one could make the same objection against any scientific law or explanation or against any historical inference that takes present, but not future, knowledge of natural laws and causal powers into account. As Barrow and Tipler have noted, to criticize design arguments, as Hume did, simply because they assume the uniformity and (normative character) of natural law cuts just as deeply against "the rational basis of any form of scientific inquiry."[148] Our knowledge of what can and cannot produce large amounts of specified information may later have to be revised, but so might the laws of thermodynamics. Inferences to design may later prove incorrect, as may other inferences implicating various natural causes. Such possibilities do not stop scientists from making generalizations about the causal powers of various entities or from using those generalizations to identify probable or most plausible causes in particular cases.

Inferences based on past and present experience constitute knowledge (albeit provisional), not ignorance. Those who object to such inferences object to science as much as they object to a particular science-based hypothesis of design.

BUT IS IT SCIENCE?

Of course, many simply refuse to consider the design hypothesis on grounds that it does not qualify as "scientific." Such critics affirm an extra-evidential principle known as methodological naturalism.[149] Methodological naturalism asserts that, as a matter of definition, for a hypothesis, theory, or explanation to qualify as "scientific," it must invoke only naturalistic or materialistic entities. On that definition, critics say, the intelligent design hypothesis does not qualify. Yet, even if one grants this definition, it does not follow that some nonscientific (as defined by methodological naturalism) or metaphysical hypothesis may not constitute a better, more causally adequate, explanation. This essay has argued that, whatever its classification, the design hypothesis does constitute a better explanation than its materialistic or naturalistic rivals for the origin of specified biological information. Surely, simply classifying an argument as metaphysical does not refute it.

In any case, methodological naturalism now lacks justification as a normative definition of science. First, attempts to justify methodological naturalism by reference to metaphysically neutral (that is, non-question-begging) demarcation criteria have failed.[150] Second, to assert methodological naturalism as a normative principle for all of science has a negative effect on the practice of certain scientific disciplines, especially the historical sciences. In origin-of-life research, for example, methodological naturalism artificially restricts inquiry and prevents scientists from seeking some hypotheses that might provide the best, most causally adequate explanations. To be a truth-seeking endeavor, the question that origin-of-life research must address is not, "Which materialistic scenario seems most adequate?" but rather, "What actually caused life to arise on Earth?" Clearly, one possible answer to that latter question is this one: "Life was designed by an intelligent agent that existed before the advent of humans." If one accepts methodological naturalism as normative, however, scientists may never consider the design hypothesis as possibly true. Such an exclusionary logic diminishes the significance of any claim of theoretical superiority for any remaining hypothesis and raises the possibility that the best "scientific" explanation (as defined by methodological naturalism) may not be the best in fact.

As many historians and philosophers of science now recognize, scientific theory-evaluation is an inherently comparative enterprise. Theories that gain acceptance in artificially constrained competitions can claim to be neither "most probably true" nor "most empirically adequate." At best, such theories can be considered the "most probably true or adequate among an artificially limited set of options." Openness to the design hypothesis would seem necessary, therefore, to any fully rational historical biology—that is, to one that seeks the truth, no holds barred.[151] A historical biology committed to following the evidence wherever it leads will not exclude hypotheses *a priori* on metaphysical grounds. Instead, it will employ only metaphysically neutral criteria—such as explanatory power and causal adequacy—to evaluate competing hypotheses. Yet this more open (and seemingly rational) approach to scientific theory evaluation would now suggest the theory of intelligent design as the best, most causally adequate, explanation for the origin of the information necessary to build the first living organism.

Notes

1. Harmke Kamminga, "Protoplasm and the Gene," in *Clay Minerals and the Origin of Life,* ed. A. G. Cairns-Smith and H. Hartman (Cambridge: Cambridge University Press, 1986), 1.
2. Aleksandr Oparin, *Genesis and Evolutionary Development of Life* (New York: Academic Press, 1968), 7.
3. F. Crick and J. Watson, "A Structure for Deoxyribose Nucleic Acid," *Nature* 171 (1953): 737–38; F. Crick and J. Watson, "Genetical Implications of the Structure of Deoxyribose Nucleic Acid," *Nature* 171 (1953): 964–67, esp. 964; T. D. Schneider, "Information Content of Individual Genetic Sequences," *Journal of Theoretical Biology* 189 (1997): 427–41; W. R. Loewenstein, *The Touchstone of Life: Molecular Information, Cell Communication, and the Foundations of Life* (New York: Oxford University Press, 1999).
4. Bernd-Olaf Küppers, *Information and the Origin of Life* (Cambridge: MIT Press, 1990), 170–72.
5. L. E. Kay, "Who Wrote the Book of Life? Information and the Transformation of Molecular Biology," *Science in Context* 8 (1994): 601–34; L. E. Kay, "Cybernetics, Information, Life: The Emergence of Scriptural Representations of Heredity," *Configurations* 5 (1999): 23–91; L. E. Kay, *Who Wrote the Book of Life?* (Stanford, CA: Stanford University Press, 2000), xv–xix.
6. Darwin's only speculation on the origin of life is found in an unpublished 1871 letter to Joseph Hooker. In it, he sketched the outlines of the chemical evolutionary idea, namely, that life could have first evolved from a series of chemical reactions. As he envisioned it, "if (and oh! what a big if!) we could conceive in some warm little pond, with all sorts of ammonia and phosphoric salts, light, heat, electricity, etc., that a proteine compound was chemically formed ready to undergo still more complex changes." Cambridge University Library, Manuscripts Room, Darwin Archives, courtesy Peter Gautrey.
7. E. Haeckel, *The Wonders of Life,* trans. J. McCabe (London: Watts, 1905), 111; T. H. Huxley, "On the Physical Basis of Life," *Fortnightly Review* 5 (1869): 129–45.
8. A. I. Oparin, *The Origin of Life,* trans. S. Morgulis (New York: Macmillan, 1938); S. C. Meyer, *Of Clues and Causes: A Methodological Interpretation of Origin of Life Studies* (Ph.D. diss., Cambridge University, 1991).
9. W. T. Astbury and A. Street, "X-Ray Studies of the Structure of Hair, Wool and Related Fibers," *Philosophical Transactions of the Royal Society of London A* 230 (1932): 75–101; H. Judson, *Eighth Day of Creation* (New York: Simon and Schuster, 1979), 80; R. Olby, *The Path to the Double Helix* (London: Macmillan, 1974), 63.
10. Olby, *Path to the Double Helix*, 7, 265.
11. Judson, *Eighth Day,* 213, 229–35, 255–61, 304, 334–35, 562–63; F. Sanger and E. O. P. Thompson, "The Amino Acid Sequence in the Glycyl Chain of Insulin," parts 1 and 2, *Biochemical Journal* 53 (1953): 353–66, 366–74.
12. Judson, *Eighth Day,* 562–63; J. C. Kendrew, G. Bodo, H. M. Dintzis, R. G. Parrish, and H. Wyckoff, "A Three-Dimensional Model of the Myoglobin Molecule Obtained by X-Ray Analysis," *Nature* 181 (1958): 662–66, esp. 664.
13. B. Alberts, D. Bray, J. Lewis, M. Raff, K. Roberts, and J. D. Watson, *Molecular Biology of the Cell* (New York: Garland, 1983), 111–12, 127–31.
14. Judson, *Eighth Day,* 30.
15. Ibid., 30–31, 33–41, 609–10; Oswald T. Avery, C. M. MacCleod, and M. McCarty, "Induction of Transformation by a Deoxyribonucleic Acid Fraction Isolated from *Pneumococcus* Type III," *Journal of Experimental Medicine* 79 (1944): 137–58.
16. Judson, *Eighth Day,* 95–96; E. Chargaff, *Essays on Nucleic Acids* (Amsterdam: Elsevier, 1963), 21.
17. Chargaff, *Essays,* 21.
18. Crick and Watson, "Structure."
19. Judson, *Eighth Day,* 96.
20. Crick and Watson, "Genetical Implications," 964–67.

21. Judson, *Eighth Day,* 611.
22. Crick and Watson, "Structure"; Crick and Watson, "Genetical Implications."
23. Judson, *Eighth Day,* 245–46, 335–36.
24. Ibid., 470–89; J. H. Matthei and M. W. Nirenberg, "Characteristics and Stabilization of DNAase-Sensitive Protein Synthesis in *E. coli* Extracts," *Proceedings of the National Academy of Sciences, USA* 47 (1961): 1580–88; J. H. Matthei and M. W. Nirenberg, "The Dependence of Cell-Free Protein Synthesis in *E. coli* upon Naturally Occurring or Synthetic Polyribonucleotides," *Proceedings of the National Academy of Sciences, USA* 47 (1961): 1588–1602.
25. Alberts et al., *Molecular Biology,* 106–8; S. L. Wolfe, *Molecular and Cellular Biology* (Belmont, Calif.: Wadsworth, 1993), 639–48.
26. We now know, of course, that in addition to the process of gene expression, specific enzymes must often modify amino acid chains after translation in order to achieve the precise sequencing necessary to allow correct folding into a functional protein. The amino acid chains produced by gene expression may also undergo further modification in sequence at the endoplasmic reticulum. Finally, even well-modified amino acid chains may require preexisting protein "chaperones" to help them fold into a functional three-dimensional configuration. All these factors make it impossible to predict a protein's final sequence from its corresponding gene sequence alone. See S. Sarkar, "Biological Information: A Skeptical Look at Some Central Dogmas of Molecular Biology," in *The Philosophy and History of Molecular Biology: New Perspectives,* ed. S. Sarkar (Dordrecht, Netherlands: Boston Studies in Philosophy of Science, 1996), 196, 199–202. Nevertheless, this unpredictability in no way undermines the claim that DNA exhibits the property of "sequence specificity," or the isomorphic claim that it contains "specified information." Sarkar argues, for example, that the absence of such predictability renders the concept of information theoretically superfluous for molecular biology. Instead, this unpredictability shows that the sequence specificity of DNA base sequences constitutes a necessary, though not sufficient, condition of attaining proper protein folding—that is, DNA does contain specified information, but not enough to determine protein folding by itself. Instead, the presence of both post-translation processes of modification and pretranscriptional genomic editing (through exonucleases, endonucleases, spliceosomes, and other editing enzymes) only underscores the need for other preexisting, information-rich biomolecules in order to process genomic information in the cell. The presence of a complex and functionally integrated information-processing system does suggest that the information on the DNA molecule is insufficient to produce proteins. It does not show that such information is unnecessary to produce proteins, nor does it invalidate the claim that DNA, therefore, stores and transmits specified genetic information.
27. C. Shannon, "A Mathematical Theory of Communication," *Bell System Technical Journal* 27 (1948): 379–423, 623–56.
28. F. Dretske, *Knowledge and the Flow of Information* (Cambridge: MIT Press, 1981), 6–10.
29. Ibid.; Shannon, "A Mathematical Theory."
30. B. Küppers, "On the Prior Probability of the Existence of Life," in *The Probabilistic Revolution,* ed. Lorenz Kruger et al. (Cambridge: MIT Press, 1987), 355–69.
31. Schneider, "Information Content"; see also H. P. Yockey, *Information Theory and Molecular Biology* (Cambridge: Cambridge University Press, 1992), 246–58, for important refinements in the method of calculating the information-carrying capacity of proteins and DNA.
32. C. Shannon and W. Weaver, *The Mathematical Theory of Communication* (Urbana: University of Illinois Press, 1949), 8.
33. Schneider, "Information Content," 58–177; Yockey, *Information Theory,* 58–177.
34. See note 26. Sarkar, "Biological Information," 199–202, esp. 196; F. Crick, "On Protein Synthesis," *Symposium for the Society of Experimental Biology* 12 (1958): 138–63, esp. 144, 153.

35. Crick, "On Protein Synthesis," 144, 153.

36. Recall that the determination of the genetic code depended, for example, on observed correlations between changes in nucleotide base sequences and amino acid production in "cell-free systems." See Judson, *Eighth Day,* 470–87.

37. W. A. Dembski, *The Design Inference: Eliminating Chance Through Small Probabilities* (Cambridge: Cambridge University Press, 1998), 1–35, 136–74.

38. Ibid., 136–74.

39. Of the two sequences, only the second meets an independent set of functional requirements. To convey meaning in English one must employ preexisting (or independent) conventions of vocabulary (associations of symbol sequences with particular objects, concepts, or ideas) and existing conventions of syntax and grammar (such as "every sentence requires a subject and a verb"). When arrangements of symbols "match" or utilize these vocabulary and grammatical conventions (that is, functional requirements), meaningful communication can occur in English. The second sequence ("Time and tide wait for no man") clearly exhibits such a match between itself and preexisting requirements of vocabulary and grammar. The second sequence has employed these conventions to express a meaningful idea. It also, therefore, falls within the smaller (and conditionally independent) pattern delimiting the domain of all meaningful sentences in English and thus, again, exhibits a "specification."

40. J. Bowie and R. Sauer, "Identifying Determinants of Folding and Activity for a Protein of Unknown Sequences: Tolerance to Amino Acid Substitution," *Proceedings of the National Academy of Sciences, USA* 86 (1989): 2152–56; J. Reidhaar-Olson and R. Sauer, "Functionally Acceptable Solutions in Two Alpha-Helical Regions of Lambda Repressor," *Proteins, Structure, Function, and Genetics* 7 (1990): 306–10.

41. R. Dawkins, *River out of Eden* (New York: Basic Books, 1995), 11.

42. B. Gates, *The Road Ahead* (Boulder, CO: Blue Penguin, 1996), 228.

43. L. E. Orgel, *The Origins of Life on Earth* (New York: John Wiley, 1973), 189.

44. See note 5. Kay, "Who Wrote the Book of Life?" 611–12, 629; Kay, "Cybernetics"; Kay, *Who Wrote the Book of Life?*

45. Sarkar, "Biological Information," 199–202.

46. E. Schrödinger, *What Is Life?* and *Mind and Matter* (Cambridge: Cambridge University Press, 1967), 82; Alberts et al., *Molecular Biology,* 21; Crick and Watson, "A Structure"; Crick and Watson, "Genetical Implications"; Crick, "On Protein"; Judson, *Eighth Day,* 611; Orgel, *Origins of Life,* 189.

47. P. Davies, *The Fifth Miracle* (New York: Simon and Schuster, 1998), 120.

48. Orgel, *Origins of Life,* 189.

49. Loewenstein, *Touchstone;* Davies, *Fifth Miracle;* Schneider, "Information Content"; C. Thaxton and W. Bradley, "Information and the Origin of Life," in The *Creation Hypothesis: Scientific Evidence for an Intelligent Designer,* ed. J. P. Moreland (Downers Grove, IL: InterVarsity Press, 1994), 173–210, esp. 190; S. Kauffman, *The Origins of Order* (Oxford: Oxford University Press, 1993), 287–340; Yockey, *Information Theory,* 178–293; Kuppers, *Information and Origin,* 170–72; F. Crick, *Life Itself* (New York: Simon and Schuster, 1981), 59–60, 88; J. Monod, *Chance and Necessity* (New York: Vintage Books, 1971), 97–98, 143; Orgel, *Origins,* 189; D. Kenyon and G. Steinman, *Biochemical Predestination* (New York: McGraw-Hill, 1969), 199–211, 263–66; Oparin, *Genesis,* 146–47; H. Quastler, *The Emergence of Biological Organization* (New Haven, CT: Yale University Press, 1964).

50. G. Wald, "The Origin of Life," *Scientific American* 191 (August 1954): 44–53; R. Shapiro, *Origins: A Skeptic's Guide to the Creation of Life on Earth* (New York: Summit Books, 1986), 121.

51. F. Crick, "The Origin of the Genetic Code," *Journal of Molecular Biology* 38 (1968): 367–79; H. Kamminga, *Studies in the History of Ideas on the Origin of Life* (Ph.D. diss., University of London, 1980), 303–4.

52. C. de Duve, "The Constraints of Chance," *Scientific American* (Jan. 1996): 112; Crick, *Life Itself,* 89–93; Quastler, *Emergence,* 7.

53. H. J. Morowitz, *Energy Flow in Biology* (New York: Academic Press, 1968), 5–12; F. Hoyle and C. Wickramasinghe, *Evolution from Space* (London: J. M. Dent, 1981), 24–27; A. G. Cairns-Smith, *The Life Puzzle* (Edinburgh: Oliver and Boyd, 1971), 91–96; I. Prigogine, G. Nicolis, and A. Babloyantz, "Thermodynamics of Evolution," *Physics Today* 23 (Nov. 1972); Yockey, *Information Theory,* 246–58; H. P. Yockey, "Self- Organization, Origin of Life Scenarios and Information Theory," *Journal of Theoretical Biology* 91 (1981): 13–31; Bowie and Sauer, "Identifying Determinants"; Reidhaar-Olson *et al., Proteins;* Shapiro, *Origins,* 117–31.

54. Prigogine, "Thermodynamics."

55. Cairns-Smith, *The Life Puzzle,* 95.

56. John Reidhaar-Olson and Robert Sauer, "Functionally Acceptable substitutions in two alpha-helical regions of lambda repressor," *Proteins: Structure, Function and Genetics,* 306–316; James Bowie and Robert Sauer, "Identifying the determinants of folding and activity for a protein of unknown structure," *Proceedings of the National Academy of Sciences USA,* 2152–56.

57. Interestingly, their descriptions of their own results often downplay the rarity of functional sequences within sequence space. Instead, they often emphasized the tolerance for different amino acids that is allowable at each site. For example, the abstract of the paper reporting the figure of 1 in 10^{63} makes no mention of that figure or its potential significance, stating instead that their results "reveal the high level of degeneracy in the information that specifies a particular protein fold." John Reidhaar-Olson and Robert Sauer, "Functionally Acceptable substitutions in two alpha-helical regions of lambda repressor," *Proteins: Structure, Function and Genetics,* 306–16.

58. James Bowie and Robert Sauer, "Identifying the determinants of folding and activity for a protein of unknown structure," *Proceedings of the National Academy of Sciences USA,* 2152–56; John Reidhaar-Olson and Robert Sauer, "Functionally Acceptable Substitutions in two Alpha-Helical Regions of Lambda Repressor," *Proteins: Structure, Function and Genetics,* 306–16; Cyrus Chothia, Israel Gelfand, & Alexander Kister, "Structural determinants in the sequences of immunoglobulin variable *Acid* domain," *Journal of Molecular Biology,* 457–79. Douglas Axe, "Extreme Functional Sensitivity to Conservative Amino Acid Changes on Enzyme Exteriors" *J. Mol. Biol.,* 585–95; Taylor, Sean V., Kai U. Walter, Peter Kast, & Donald Hilvert *et al.* "Searching sequence space for protein catalysts." *Proceedings of the National Academy of Sciences USA,* 10596–601.

59. See, for example, Max F. Perutz, Max & Hermann Lehmann, "Molecular pathology of human hemoglobin," *Nature,* 902–9.

60. Douglas Axe, "Estimating the Prevalence of Protein Sequences Adopting Functional Enzyme Folds" *J. Mol. Biol.,* 1295–15.

61. The number of possible ways to combine elementary particles (and thus the number of combinatorial possible events) is actually much greater than the number of different events that could have taken place in the history of the universe. Why? Because the occurrence of each individual event precludes the occurrence of many other possible events within the larger combinatorial space. The number of combinatorial possible events represents the number of different events that might have occurred before the universe actually unfolded in the way that it did. Dembski correctly identifies the maximum number of events that could *actually* occur in any given history of the universe as the number that determines the probabilistic resources of the universe. This smaller number determines how many opportunities the universe has to produce a particular outcome by chance. As Dembski explains, it is not the total number of combinatorial possible events (or elementary particles) in the universe which determines the available probabilistic resources, but how many opportunities there are to "individuate" actual events. See William A. Dembski, *The Design Inference: Eliminating Chance through Small Probabilities,* Chapter 6. See also note 15, page 209.

62. For Dembski's treatment of probabilistic resources at the scale of the known universe, see William A. Dembski, *The Design Inference: Eliminating Chance through Small Probabilities,* Chapter 6.

63. The elementary particles enumerated in this calculation include only protons, neutrons and electrons (fermions), because only these particles have what physicists call "half-integral spin" which allows them to form material structures. This calculation does not count bosons which cannot form material structures, but instead only transmit energy. Nor does this calculation count the quarks out of which protons and neutrons are made because quarks are necessarily bound together within these particles. Even if quarks were counted, however, the total number of elementary particles would change by less than one order of magnitude because there are only three quarks per proton or neutron.

64. To be safe, Dembski rounded the number that he had calculated up a few orders of magnitude to 10^{150}, though without any physical or mathematical justification. Since he didn't need to do this, I decided to use his more accurate, if less round number as the actual measure of the probabilistic resources of the universe in my evaluations of the chance hypothesis.

65. Cryptographers, for instance, have established 1 chance in 10^{94} as a universal limit. They interpret that improbability to mean that if it requires more than 10^{94} computational steps to decrypt a cryptosystem, then it is more likely than not that the system won't be cracked because of the limited probabilistic resources of the universe itself.

66. Emile Borel, *Probabilities and Life,* trans. M. Baudin, 28–30.

67. Bret Van de Sande, "Measuring Complexity in Dynamical Systems" presented to *RAPID II* (Research and Progress in Intelligent Design) conference," Biola University, May, 2006.

68. Kenneth W. Dam and Herbert S. Lin, eds., *Cryptography's Role in Securing the Information Society,* 380, note 17; Seth Lloyd, "Computational Capacity of the Universe," *Physical Review Letters,* 7901–4; see also Stuart Kauffman, *Investigations,* 144.

69. E. Pennisi, "Seeking Life's Bare Genetic Necessities," *Science* 272 (1996): 1098–99; A. Mushegian and E. Koonin, "A Minimal Gene Set for Cellular Life Derived by Comparison of Complete Bacterial Genomes," *Proceedings of the National Academy of Sciences, USA* 93 (1996): 10268–73; C. Bult *et al.*, "Complete Genome Sequence of the Methanogenic Archaeon, *Methanococcus jannaschi,*" *Science* 273 (1996): 1058–72.

70. Dembski, *Design Inference,* 67–91, 175–223, 209–10.

71. P. T. Mora, "Urge and Molecular Biology," *Nature* 199 (1963): 212–19.

72. I. Hacking, *The Logic of Statistical Inference* (Cambridge: Cambridge University Press, 1965), 74–75.

73. Dembski, *Design Inference,* 47–55.

74. C. de Duve, "The Beginnings of Life on Earth," *American Scientist* 83 (1995): 437.

75. Quastler, *Emergence,* 7.

76. Oparin, *Origin of Life,* 64–103; Meyer, *Of Clues,* 174–79, 194–98, 211–12.

77. Oparin, *Origin of Life,* 107–8, 133–35, 148–59, 195–96.

78. Oparin, *Genesis,* 146–47.

79. C. de Duve, *Blueprint for a Cell: The Nature and Origin of Life* (Burlington, NC: Neil Patterson, 1991), 187.

80. G. Joyce and L. Orgel, "Prospects for Understanding the Origin of the RNA World," in *RNA World,* ed. R. F. Gesteland and J. J. Atkins (Cold Spring Harbor, NY: Cold Spring Harbor Laboratory Press, 1993), 1–25, esp. 8–13.

81. T. Dobzhansky, "Discussion of G. Schramm's Paper," in *The Origins of Prebiological Systems and of Their Molecular Matrices,* ed. S. W. Fox (New York: Academic Press, 1965), 310; H. H. Pattee, "The Problem of Biological Hierarchy," in *Toward a Theoretical Biology,* ed. C. H. Waddington, vol. 3 (Edinburgh: Edinburgh University Press, 1970), 123.

82. P. T. Mora, "The Folly of Probability," in Fox, *Origins,* 311–12; L. V. Bertalanffy, *Robots, Men and Minds* (New York: George Braziller, 1967), 82.

83. J. Von Neumann, *Theory of Self-reproducing Automata,* completed and edited by A. Berks (Urbana: University of Illinois Press, 1966).

84. Pennisi, "Seeking"; Mushegian and Koonin, "Minimal Gene Set"; Bult *et al.*, "Complete Genome Sequence."

85. E. Wigner, "The Probability of the Existence of a Self-reproducing Unit," in *The Logic of Personal Knowledge,* ed. E. Shils (London: Kegan and Paul, 1961), 231–35; P. T. Landsberg, "Does Quantum Mechanics Exclude Life?" *Nature* 203 (1964): 928–30; H. J. Morowitz, "The Minimum Size of the Cell," in *Principles of Biomolecular Organization,* ed. M. O'Connor and G. E. W. Wolstenholme (London: J. A. Churchill, 1966), 446–59; Morowitz, *Energy Flow,* 10–11.

86. Dawkins, *Blind Watchmaker,* 47–49; Kuppers, "On the Prior Probability."

87. Küppers, "On the Prior Probability," 366.

88. Dawkins, *Blind Watchmaker,* 47–49; P. Nelson, "Anatomy of a Still-Born Analogy," *Origins and Design* 17 (3) (1996): 12.

89. C. de Duve, "Beginnings of Life," 437.

90. Morowitz, *Energy Flow,* 5–12.

91. G. Steinman and M. N. Cole, "Synthesis of Biologically Pertinent Peptides Under Possible Primordial Conditions," *Proceedings of the National Academy of Sciences, USA* 58 (1967): 735–41; G. Steinman, "Sequence Generation in Prebiological Peptide Synthesis," *Archives of Biochemistry and Biophysics* 121 (1967): 533–39; R. A. Kok, J. A. Taylor, and W. L. Bradley, "A Statistical Examination of Self-Ordering of Amino Acids in Proteins," *Origins of Life and Evolution of the Biosphere* 18 (1988): 135–42.

92. Kenyon and Steinman, *Biochemical Predestination,* 199–211, 263–66.

93. G. Nicolis and I. Prigogine, *Self-Organization in Non-Equilibrium Systems* (New York: John Wiley, 1977), 339–53, 429–47.

94. Kauffman, *Origins of Order,* 285–341; de Duve, "Beginnings of Life"; C. de Duve, *Vital Dust: Life as a Cosmic Imperative* (New York: Basic Books, 1995).

95. C. Thaxton, W. Bradley, and R. Olsen, *The Mystery of Life's Origin: Reassessing Current Theories* (Dallas: Lewis and Stanley, 1992), v–viii; D. Kenyon and G. Mills, "The RNA World: A Critique," *Origins and Design* 17, no. 1 (1996): 9–16; Davis, Percival and Dean Kenyon. *Of Pandas and People: The Central Question of Biological Origins,* Richardson, TX: Foundation for Thought and Ethics, 1993. S. C. Meyer, "A Scopes Trial for the '90's," *Wall Street Journal,* 6 Dec. 1993; Kok *et al.*, "Statistical Examination."

96. Steinman and Cole, "Synthesis"; Steinman, "Sequence Generation."

97. Kok *et al.*, "Statistical Examination"; B.J. Strait and G. T. Dewey, "The Shannon Information Entropy of Biologically Pertinent Peptides," *Biophysical Journal* 71: 148–55.

98. Küppers, "On the Prior Probability," 64.

99. C. de Duve, "Beginnings of Life," 437.

100. R. Stalnaker, *Inquiry* (Cambridge: MIT Press, 1984), 85.

101. This, in fact, happens where adenine and thymine do interact chemically in the complementary base-pairing across the information-bearing axis of the DNA molecule. Along the message bearing axis, however, there are no chemical bonds or differential bonding affinities that determine sequencing.

102. M. Polanyi, "Life's Irreducible Structure," *Science* 160 (1968): 1308–12, esp. 1309.

103. As noted in part I, the information-carrying capacity of any symbol in a sequence is inversely related to the probability of its occurrence. The informational capacity of a sequence as a whole is inversely proportional to the product of the individual probabilities of each member in the sequence. Since chemical affinities between constituents ("symbols") increase the probability of the occurrence of one, given another (i.e., necessity increases probability), such affinities decrease the information-carrying capacity of a system in proportion to the strength and relative frequency of such affinities within the system.

104. Dretske, *Knowledge and the Flow of Information*, 12.
105. Yockey, "Self-Organization," 18.
106. H. P. Yockey, "A Calculation of the Probability of Spontaneous Biogenesis by Information Theory," *Journal of Theoretical Biology* 67 (1977): 377–98, esp. 380.
107. M. Eigen, *Steps Toward Life* (Oxford: Oxford University Press, 1992), 12.
108. Quastler, *Emergence*, ix.
109. Ibid., 1, 47.
110. Yockey, Information Theory, 247.
111. J. C. Walton, "Organization and the Origin of Life," *Origins* 4 (1977): 16–35.
112. J. M. Smith, "Hypercycles and the Origin of Life," *Nature* 280 (1979): 445–46; F. Dyson, *Origins of Life* (Cambridge: Cambridge University Press, 1985), 9–11, 35–39, 65–66, 78; Shapiro, *Origins,* 161.
113. Iris Frey, *The Emergence of Life on Earth* (New Brunswick, NJ: Rutgers University Press, 2000), 158.
114. Kauffman, *Origins of Order,* 285–341.
115. Ibid., 299.
116. See Protein Databank at http://www.rcsb.org/pdb.
117. Kauffman, *Origins of Order,* 298.
118. Thaxton, et al., *Mystery of Life's Origin,* 127–43.
119. R. Shapiro, "Prebiotic Cytosine Synthesis: A Critical Analysis and Implications for the Origin of Life," *Proceedings of the National Academy of Sciences, USA* 96 (1999): 4396–401; M. M. Waldrop, "Did Life Really Start Out in an RNA World?" *Science* 246 (1989): 1248–49.
120. R. Shapiro, "Prebiotic Ribose Synthesis: A Critical Analysis," *Origins of Life and Evolution of the Biosphere* 18 (1988): 71–85; Kenyon and Mills, "RNA World."
121. G. F. Joyce, "RNA Evolution and the Origins of Life," *Nature* 338 (1989): 217–24. Yuri I. Wolf and Eugene V. Koonin, "On the Origin of the Translation System and the Genetic Code in the RNA World by means of Natural Selection, Exaptation, and Subfunctionalization," *Biology Direct* 2(2007): 1–25.
122. A. J. Hager, J. D. Polland Jr., and J. W. Szostak, "Ribozymes: Aiming at RNA Replication and Protein Synthesis," *Chemistry and Biology* 3 (1996): 717–25.
123. C. de Duve, *Vital Dust,* 23.
124. Joyce and Orgel, "Prospects for Understanding," 1–25, esp. 11.
125. Quastler, *Emergence,* 16.
126. A possible exception to this generalization might occur in biological evolution. If the Darwinian mechanism of natural selection acting on random variation can account for the emergence of all complex life, then a mechanism does exist that can produce large amounts of information—assuming, of course, a large amount of preexisting biological information in a self-replicating living system. Thus, even if one assumes that the selection/variation mechanism can produce all the information required for the macroevolution of complex life from simpler life, that mechanism will not suffice to account for the origin of the information necessary to produce life from nonliving chemicals. As we have seen, appeals to prebiotic natural selection only beg the question of the origin of specified information. Thus, based on experience, we can affirm the following generalization: "for all nonbiological systems, large amounts [note 128 next page] of specified complexity or information originate only from mental agency, conscious activity, or intelligent design." Strictly speaking, experience may even affirm a less qualified generalization (such as "large amounts of specified complexity invariably originate from an intelligent source"), since the claim that natural selection acting on random mutations can produce large amounts of novel genetic information depends on debatable theoretical arguments and extrapolation from observations of small scale micro-evolutionary changes that do not themselves manifest large gains in biological information. I have argued elsewhere (see S. C. Meyer, "The origin of biological information and the higher taxonomic categories," in *Proceedings of the Biological Society of*

Washington 117 (2), 2004: 213–39) that neither the neo-Darwinian mechanism nor any other current naturalistic mechanism adequately accounts for the origin of the information required to build the novel proteins and body plans that arise in the Cambrian explosion. In any case, the more qualified empirical generalization (stated above in this endnote) is sufficient to support the argument presented here, since this essay seeks only to establish intelligent design as the best explanation for origin of the specified information necessary to the origin of the first life.

127. K. Dose, "The Origin of Life: More Questions Than Answers," *Interdisciplinary Science Reviews* 13 (1988): 348–56; Yockey, *Information Theory,* 259–93; Thaxton *et al., Mystery,* 42–172; Thaxton and Bradley, "Information and the Origin," 193–97; Shapiro, *Origins.*

128. Of course, the phrase "large amounts of specified information" again begs a quantitative question, namely, "How much specified information or complexity would the minimally complex cell have to have before it implied design?" Recall that Dembski has calculated a universal probability bound of $1/10^{150}$ corresponding to the probabilistic/specificational resources of the known universe. Recall further that probability is inversely related to information by a logarithmic function. Thus, the universal small probability bound of $1/10^{150}$ translates into roughly 500 bits of information. Chance alone, therefore, does not constitute a sufficient explanation for the *de novo* origin of any specified sequence or system containing more than 500 bits of (specified) information. Further, since systems characterized by complexity (a lack of redundant order) defy explanation by self-organizational laws and since appeals to prebiotic natural selection presuppose but do not explain the origin of the specified information necessary to a minimally complex self-replicating system, intelligent design best explains the origin of the more than 500 bits of specified information required to produce the first minimally complex living system. Thus, assuming a nonbiological starting point (note 126 above), the *de novo* emergence of 500 or more bits of specified information will reliably indicate design.

129. Again, this claim applies at least in cases where the competing causal entities or conditions are nonbiological—or where the mechanism of natural selection can be safely eliminated as an inadequate means of producing requisite specified information.

130. Morris, 1998: 63–115

131. Gould, 1986: 61.

132. Peirce, 1932, vol. 2: 375.

133. Chamberlain, 1890: 92–96.

134. Scriven, 1966: 250.

135. Ibid.; E. Sober, *Reconstructing the Past* (Cambridge: MIT Press, 1988), 4–5; M. Scriven, "Causes, Connections, and Conditions in History," in *Philosophical Analysis and History,* ed. W. Dray (New York: Harper and Row, 1966), 238–64, esp. 249–50.

136. Scriven, 1959: 480; Sober, 1988: 1–5; Alston, 1971: 23; Gallie, 1959: 392.

137. *McNeil-Lehrer News Hour,* Transcript 19 May 1992.

138. Gian Capretti (1983: 143) has developed the implications of Peircian abduction. Capretti and others explore the use of abductive reasoning by Sherlock Holmes in detective fiction of Sir Arthur Conan Doyle. Capretti attributes the success of Holmesian abductive "reconstructions" to a willingness to employ a method of "progressively eliminating hypotheses."

139. Darwin, 1898, vol. 1: 437.

140. Meyer, *Of Clues,* 77–140.

141. Less exotic (and more successful) design-detection occurs routinely in both science and industry. Fraud-detection, forensic science, and cryptography all depend on the application of probabilistic or information theoretic criteria of intelligent design. Dembski, *Design Inference,* 1–35. Many would admit that we may justifiably infer a past human intelligence operating (within the scope of human history) from an information-rich artifact or event, but only because we already know that human minds exist. But, they argue, since we do

not know whether an intelligent agent(s) existed prior to humans, inferring the action of a designing agent that antedates humans cannot be justified, even if we observe an information-rich effect. Note, however, that SETI scientists do not already know whether an extraterrestrial intelligence exists. Yet they assume that the presence of a large amount of specified information (such as the first 100 prime numbers in sequence) would definitively establish the existence of one. Indeed, SETI seeks precisely to establish the existence of other intelligences in an unknown domain. Similarly, anthropologists have often revised their estimates for the beginning of human history or civilization because they discovered information-rich artifacts dating from times that antedate their previous estimates. Most inferences to design establish the existence or activity of a mental agent operating in a time or place where the presence of such agency was previously unknown. Thus, to infer the activity of a designing intelligence from a time prior to the advent of humans on Earth does not have a qualitatively different epistemological status than other design inferences that critics already accept as legitimate. T. R. McDonough, *The Search for Extraterrestrial Intelligence: Listening for Life in the Cosmos* (New York: Wiley, 1987).

142. P. Lipton, *Inference to the Best Explanation* (New York: Routledge, 1991), 32–88.

143. Ibid.; S. C. Meyer, "The Scientific Status of Intelligent Design: The Methodological Equivalence of Naturalistic and Non-Naturalistic Origins Theories," in *Science and Evidence for Design in the Universe, The Proceedings of the Wethersfield Institute, vol. 9* (San Francisco: Ignatius Press, 2000), 151–212; Meyer, "The Demarcation of Science and Religion," in *The History of Science and Religion in the Western Tradition: An Encyclopedia,* ed. G. B. Ferngren (New York: Garland, 2000), 17–23; E. Sober, *The Philosophy of Biology* (San Francisco: Westview Press, 1993); Meyer, *Of Clues,* 77–140.

144. Quastler, *Emergence,* 16.

145. Oparin, *Origin of Life,* 28; M. Rothman, *The Science Gap* (Buffalo, NY: Prometheus, 1992), 65–92; K. Popper, *Conjectures and Refutations: The Growth of Scientific Knowledge* (London: Routledge and Kegan Paul, 1962), 35–37.

146. Meyer, *Of Clues,* 77–140; Sober, *Reconstructing the Past,* 4–5; de Duve, "Beginnings of Life," 249–50.

147. R. Harré and E. H. Madden, *Causal Powers* (London: Basil Blackwell, 1975).

148. J. Barrow and F. Tipler, *The Anthropic Cosmological Principle* (Oxford: Oxford University Press, 1986), 69.

149. M. Ruse, "McLean v. Arkansas: Witness Testimony Sheet," in *But Is It Science?* ed. M. Ruse (Amherst, N.Y.: Prometheus Books, 1988), 103; Meyer, "Scientific Status"; Meyer, "Demarcation."

150. Meyer, "Scientific Status"; Meyer, "Demarcation"; L. Laudan, "The Demise of the Demarcation Problem," in Ruse, *But Is It Science?* 337–50; L. Laudan, "Science at the Bar—Causes for Concern," in Ruse, *But Is It Science?* 351–55; A. Plantinga, "Methodological Naturalism?" *Origins and Design* 18, no. 1 (1997): 18–26; A. Plantinga, "Methodological Naturalism?" *Origins and Design* 18, no. 2 (1997): 22–34.

151. Bridgman, *Reflections of a Physicist,* 2nd ed. (New York: Philosophical Library, 1955), 535.

References

Alberts, Bruce D., Dennis Bray, Julian Lewis, Martin Raff, Keith Roberts, and James D. Watson. *Molecular Biology of the Cell*. New York: Garland, 1983.

Alson, William P. "The Place of the Explanation of Particular Facts in Science." *Philosophy of Science* 38 (1971): 13–34.

Astbury, William T., and A. Street. "X-Ray Studies of the Structure of Hair, Wool and Related Fibres." *Philosophical Transactions of the Royal Society of London* 230 (1932): 75–101.

Avery, Oswald T., Colin M. MacCleod, and Maclyn McCarty. "Induction of Transformation by a Deoxyribonucleic Acid Fraction Isolated from Pneumococcus Type III." *Journal of Experimental Medicine* 79 (1944): 137–58.

Axe, Douglas. "Extreme Functional Sensitivity to Conservative Amino Acid Changes on Enzyme Exteriors." *Journal of Molecular Biology* 301 (2000): 585–95.

———. "Estimating the Prevalence of Protein Sequences Adopting Functional Enzyme Folds." *Journal of Molecular Biology* 341 (2004): 1295–315.

Barrow, John D., and Frank J. Tipler. *The Anthropic Cosmological Principle*. Oxford: Oxford University Press, 1986.

Bertalanffy, Ludwig von. *Robots, Men and Minds*. New York: Braziller, 1967.

Borel, Emile. *Probabilities and Life*. Translated by M. Baudin. New York: Dover, 1962.

Bowie, James, and Robert Sauer. "Identifying Determinants of Folding and Activity for a Protein of Unknown Structure." *Proceedings of the National Academy of Sciences USA* 86 (1989): 2152–56

Bridgman, Percy W. *Reflections of a Physicist*. 2nd ed. New York: Philosophical Library, 1955.

Bult, Carol, et al. "Complete Genome Sequence of the Methanogenic Archaeon *Methanococcus jannaschii*." *Science* 273 (1996): 1058–73.

Cairns-Smith, Alexander G. *The Life Puzzle*. Edinburgh: Oliver and Boyd, 1971.

Cairns-Smith, A. G. and H. Hartman, *Clay Minerals and the Origin of Life,* ed. Cambridge: Cambridge University Press, 1986.

Capretti, Gian. *"Peirce, Holmes, Popper."* In *The Sign of Three,* edited by U. Eco and T. Sebeok, Bloomington: Indiana University Press, 1983, 135–53.

Chamberlain, Thomas. "The Method of Multiple Working Hypotheses." *Science* (old series) 15 (1890): 92–96.

Chothia, Cyrus, Israel Gelfand, and Alexander Kister. "Structural Determinants in the Sequences of Immunoglobulin Variable Domain." *Journal of Molecular Biology* 278 (1998): 457–79.

Crick, Francis. "On Protein Synthesis." *Symposium for the Society of Experimental Biology* 12 (1958): 138–63.

———. "The Origin of the Genetic Code." *Journal of Molecular Biology* 38 (1968): 367–79.

———. *Life Itself.* New York: Simon & Schuster, 1981.

Crick, Francis and James Watson, "A Structure for Deoxyribose Nucleic Acid," *Nature* 171 (1953): 737–38.

Crick, Francis and James Watson, "Genetical Implications of the Structure of Deoxyribose Nucleic Acid," *Nature* 171 (1953): 964–67.

Dam, Kenneth W., and Herbert S. Lin, eds. *Cryptography's Role in Securing the Information Society*. Washington, DC: National Academy Press, 1996.

Darwin, Charles. Letter to Hooker. 1871. Courtesy of Mr. Peter Gautrey. Cambridge University Library, Darwin Archives, Manuscripts Room.

Darwin, Charles. *Life and Letters of Charles Darwin, Including an Autobiographical Chapter*. Edited by Francis Darwin. 2 Vols. New York: Appleton, 1898.

Davies, Paul. *The Fifth Miracle*. New York: Simon & Schuster, 1999.

Davis, Percival and Dean Kenyon. *Of Pandas and People: The Central Question of Biological Origins.* Richardson, TX: Foundation for Thought and Ethics, 1993.

Dawkins, Richard. *The Blind Watchmaker: Why the Evidence Reveals a Universe Without Design.* New York: Norton, 1987.

———. *River Out of Eden: A Darwinian View of Life.* New York: Basic Books, 1995.

De Duve, Christian. *Blueprint for a Cell: The Nature and Origin of Life.* Burlington, NC: Neil Patterson, 1991.

———. "The Beginnings of Life on Earth." *American Scientist* 83 (1995): 249–50, 428–37.

———. *Vital Dust: Life as a Cosmic Imperative.* New York: Basic Books, 1995.

———. "The Constraints of Chance," *Scientific American* 271(1996): 112.

———. *Singularities: Landmarks on the Pathways of Life.* Cambridge: Cambridge University Press, 2005.

Dembski, William. *The Design Inference: Eliminating Chance Through Small Probabilities.* Cambridge: Cambridge University Press, 1998.

Dobzhansky, Theodosius. "Discussion of G. Schramm's Paper." In *The Origins of Prebiological Systems and of Their Molecular Matrices,* edited by Sidney W. Fox, New York: Academic, 1965, 309–15.

Dose, Klaus. "The Origin of Life: More Questions Than Answers." *Interdisciplinary Science Review* 13 (1988): 348–56.

Dretske, Fred I. *Knowledge and the Flow of Information.* Cambridge, MA: MIT Press, 1981.

Dyson, Freeman F. *Origins of Life.* Cambridge: Cambridge University Press, 1985.

Eigen, Manfred. *Steps Toward Life: A Perspective on Evolution.* Translated by P. Woolley. Oxford: Oxford University Press, 1992.

Fry, Iris. *The Emergence of Life on Earth: A Historical and Scientifi c Overview.* New Brunswick, NJ: Rutgers University Press, 2000.

Gallie, Walter Bryce. "Explanations in History and the Genetic Sciences." In *Theories of History*, edited by P. Gardiner. Glencoe, IL: Free Press, 1959: 386–402.

Gates, Bill. *The Road Ahead.* Rev. ed. New York: Viking, Penguin Group, 1996.

Gould, Stephen Jay. "Evolution and the Triumph of Homology: Or, Why History Matters." *American Scientist* 74 (1986): 60–69.

Hacking, Ian. *The Logic of Statistical Inference.* Cambridge: Cambridge University Press, 1965.

Haeckel, Ernest. *The Wonders of Life.* Translated by J. McCabe. London: Harper, 1905.

Hager, Alicia J., Jack D. Polland Jr., and Jack W. Szostak. "Ribozymes: Aiming at RNA Replication and Protein Synthesis." *Chemistry and Biology* 3 (1996): 717–25.

Harre, Rom, and Edward H. Madden. *Causal Powers.* London: Blackwell, 1975.

Hoyle, Fred, and Chandra Wickramasinghe. *Evolution from Space.* London: Dent, 1981.

Huxley, Thomas H. "On the Physical Basis of Life." *Fortnightly Review* 5 (1869): 129–45.

Joyce, Gerald F. "RNA Evolution and the Origin of Life." *Nature* 338 (1989): 217–24.

Joyce, Gerald F., and Leslie Orgel. "Prospects for Understanding the Origin of the RNA World." In *The RNA World,* edited by Raymond F. Gesteland and John J. Atkins, Cold Spring Harbor, NY: Cold Spring Harbor Laboratory Press, 1993, 1–25.

Judson, Horace. *Eighth Day of Creation* New York: Simon and Schuster, 1979.

Kamminga, Harmke. "Studies in the History of Ideas on the Origin of Life." Ph.D. dissertation. University of London, 1980.

———. "Protoplasm and the Gene." In *Clay Minerals and the Origin of Life,* edited by A. G. Cairns-Smith and H. Hartman, Cambridge: Cambridge University Press, 1986, 1–10.

Kauffman, Stuart A. *The Origins of Order: Self-Organization and Selection in Evolution.* Oxford: Oxford University Press, 1993.

———. *Investigations.* New York: Oxford University Press, 2000.
Kay, Lily E. "Who Wrote the Book of Life? Information and the Transformation of Molecular Biology." *Science in Context* 8 (1994): 601–34.
———. "Cybernetics, Information, Life: The Emergence of Scriptural Representations of Heredity." *Configurations* 5 (1999): 23–91.
———. *Who Wrote the Book of Life?* Stanford, CA: Stanford University Press, 2000.
Kendrew, John C., G. Bodo, Howard M. Dintzis, R. G. Parrish, and H. Wyckoff. "A Three-Dimensional Model of the Myoglobin Molecule Obtained by X-Ray Analysis. *Nature* 181 (1958): 662–66.
Kenyon, Dean, and Gordon Mills. "The RNA World: A Critique." *Origins and Design* 17 (1996): 9–16. http://www.arn.org/docs/odesign/od171/rnaworld171.htm (last accessed September 30, 2008).
Kenyon, Dean, and Gary Steinman. *Biochemical Predestination.* New York: McGraw-Hill, 1969.
Kok, Randall A., John A. Taylor, and Walter L. Bradley. "A Statistical Examination of Self-Ordering of Amino Acids in Proteins." *Origins of Life and Evolution of the Biosphere* 18 (1988): 135–42.
Küppers, Bernd-Olaf. "On the Prior Probability of the Existence of Life." In *The Probabilistic Revolution,* vol. 2, edited by Lorenz Krüger, Gerg Gigerenzer, and Mary S. Morgan, Cambridge, MA: MIT Press, 1987, 355–69.
Küppers, Bernd-Olaf. *Information and the Origin of Life.* Cambridge, MA: MIT Press, 1990.
Landsberg, Peter T. "Does Quantum Mechanics Exclude Life?" *Nature* 203 (1964): 928–30.
Laudan, Larry. "The Demise of the Demarcation Problem." In *But Is It Science?* edited by Michael Ruse,. Buffalo, NY: Prometheus, 1988, 337–50.
———. "Science at the Bar—Causes for Concern." In *But Is It Science?* edited by Michael Ruse, Buffalo, NY: Prometheus, 1988, 351–55.
Lloyd, Seth. "Computational Capacity of the Universe." *Physical Review Letters* 88 (2002): 7901–4.
Loewenstein, W. R. *The Touchstone of Life: Molecular Information, Cell Communication, and the Foundations of Life.* New York: Oxford University Press, 1999.
Matthaei, J. Heinrich, and Marshall W. Nirenberg. "Characteristics and Stabilization of DNAase-Sensitive Protein Synthesis in *E. coli* Extracts." *Proceedings of the National Academy of Sciences USA* 47 (1961): 1580–88.
Matthei, J. Heinrich and M. W. Nirenberg, "The Dependence of Cell-Free Protein Synthesis in *E. coli* upon Naturally Occurring or Synthetic Polyribonucleotides," *Proceedings of the National Academy of Sciences, USA* 47 (1961): 1588–1602.
McDonough, Thomas R. *The Search for Extraterrestrial Intelligence: Listening for Life in the Cosmos.* New York: Wiley, 1988.
McNeil-Lehrer News Hour, Transcript 19 (May 1992).
Meyer, Stephen C. *Of Clues and Causes: A Methodological Interpretation of Origin of Life Studies.* Ph.D. dissertation. Cambridge University, 1990.
———. "A Scopes Trial for the '90s." *Wall Street Journal,* December 6, 1993, A14.
———. "The Scientific Status of Intelligent Design: The Methodological Equivalence of Naturalistic and Non-Naturalistic Origins Theories." In *Science and Evidence for Design in the Universe,* by Michael Behe, William Dembski, and Stephen C. Meyer. Proceedings of the Wethersfield Institute, vol. 9, San Francisco: Ignatius, 2000, 151–211.
———. "The Demarcation of Science and Religion." In *The History of Science and Religion in the Western Tradition: An Encyclopedia,* edited by G. Ferngren, New York: Garland, 2000, 17–23.
———. "The Origin of Biological Information and the Higher Taxonomic Categories." *Proceedings of the Biological Society of Washington* 117 (2004): 213–39.
———. *Signature in the Cell: DNA and the Evidence for Intelligent Design.* San Francisco: HarperOne, 2009.
Monod, Jacques. *Chance and Necessity: An Essay on the Natural Philosophy of Modern Biology.* New York: Vintage, 1972.

Mora, P. T. "Urge and Molecular Biology." *Nature* 199 (1963): 212–19.

———. "The Folly of Probability." In *The Origins of Prebiological Systems and of Their Molecular Matrices,* edited by Sidney W. Fox, New York: Academic, 1965, 39–52.

Morowitz, Harold J. "The Minimum Size of the Cell." In *Principles of Biomolecular Organization,* edited by Gordon E. W. Wolstenholme and Maeve O'Connor, London: Lehmann, 1966, 446–59.

———. *Energy Flow in Biology: Biological Organization as a Problem in Thermal Physics.* New York: Academic, 1968.

Mushegian, Arcady, and Eugene Koonin. "A Minimal Gene Set for Cellular Life Derived by Comparison of Complete Bacterial Genomes." *Proceedings of the National Academy of Sciences USA* 93 (1996): 10268–273.

Nelson, Paul. "Anatomy of a Still-Born Analogy." *Origins and Design* 17 (1996): 12.

Nicolis, Grégoire, and Ilya Prigogine. *Self-Organization in Nonequilibrium Systems.* New York: Wiley, 1977.

Olby, Roger. *The Path to the Double Helix.* London: Macmillan, 1974.

Oparin, Aleksandr. *The Origin of Life.* Translated by S. Morgulis. New York: Macmillan, 1938.

———. *Genesis and Evolutionary Development of Life.* New York: Academic, 1968.

Orgel, Leslie E. *The Origins of Life.* New York: Wiley, 1973.

Pattee, Howard H. "The Problem of Biological Hierarchy." In *Towards a Theoretical Biology,* vol. 3, edited by Conrad H. Waddington, Edinburgh: Edinburgh University Press, 1970, 117–36.

Peirce, Charles S. *Collected Papers,* Vol. 2. Edited by Charles Hartshorne adn Paul Weiss. Cambridge: Harvard University Press, 1932.

Pennisi, Elizabeth. "Seeking Life's Bare (Genetic) Necessities." *Science* 272 (1996): 1098–99.

Perutz, Max F., and Hermann Lehmann. "Molecular Pathology of Human Hemoglobin." *Nature* 219 (1968): 902–9.

Plantinga, Alvin. "Methodological Naturalism?" Part 1. *Origins and Design* 18.1 (1986): 18–26.

———. "Methodological Naturalism?" Part 2. *Origins and Design* 18.2 (1986): 22–34.

Polanyi, Michael. "Life's Irreducible Structure" *Science* 160 (1968): 1308–12.

Popper, Karl. *Conjectures and Refutations: The Growth of Scientific Knowledge.* London: Routledge and Kegan Paul, 1962.

Prigogine, Ilya, Grégoire Nicolis, and Agnessa Babloyantz. "Thermodynamics of Evolution." *Physics Today* 25 (1972): 23–31.

Protein Databank at http://www.rcsb.org/pdb.

Quastler, Henry. *The Emergence of Biological Organization.* New Haven, CT: Yale University Press, 1964.

Reidhaar-Olson, John, and Robert Sauer. "Functionally Acceptable Solutions in Two Alpha-Helical Regions of Lambda Repressor." *Proteins: Structure, Function, and Genetics* 7 (1990): 306–16.

Rothman, Milton A. *The Science Gap.* Buffalo, NY: Prometheus, 1992.

Ruse, Michael. "Witness Testimony Sheet: *McLean* v. *Arkansas.*" In *But Is It Science?* edited by Michael Ruse, Buffalo, NY: Prometheus, 1988, 287–306.

Sarkar, Sahotra. "Biological Information: A Skeptical Look at Some Central Dogmas of Molecular Biology." In *The Philosophy and History of Molecular Biology: New Perspectives,* edited by S. Sarkar, Dordrecht: Kluwer Academic, 1996, 187–233.

Sanger, Frederick, and E. O. P. Thompson. "The Amino Acid Sequence in the Glycyl Chain of Insulin." *Biochemical Journal* 53 (1953): 353–74.

Schneider, T. D. "Information Content of Individual Genetic Sequences," *Journal of Theoretical Biology* 189 (1997): 427–41.

Scriven, Michael. "Causes, Connections, and Conditions in History." In *Philosophical Analysis and History,* edited by W. Dray, New York: Harper & Row, 1966, 238–64.

Schrödinger, Erwin. *What Is Life? Mind and Matter.* Cambridge: Cambridge University Press, 1967.

Shannon, Claude E. "A Mathematical Theory of Communication." *Bell System Technical Journal* 27 (1948): 379–423, 623–56.

Shannon, Claude E., and Warren Weaver. *The Mathematical Theory of Communication*. Urbana: University of Illinois Press, 1949.

Shapiro, Robert. *Origins: A Skeptic's Guide to the Creation of Life on Earth*. New York: Summit, 1986.

———. "Prebiotic Ribose Synthesis: A Critical Analysis." *Origins of Life and Evolution of the Biosphere* 18 (1988): 71–85.

———. "Prebiotic Cytosine Synthesis: A Critical Analysis and Implications for the Origin of Life." *Proceedings of the National Academy of Sciences, USA* 96 (1999): 4396–401.

Smith, John Maynard. "Hypercycles and the Origin of Life." *Nature* 280 (1979): 445–46.

Sober, Elliott. *Reconstructing the Past*. Cambridge, MA: MIT Press, 1988.

———. *The Philosophy of Biology*. San Francisco: Westview, 1993.

Stalnaker, Robert C. *Inquiry*. Cambridge, MA: MIT Press, 1984.

Steinman, Gary. "Sequence Generation in Prebiological Peptide Synthesis." *Archives of Biochemistry and Biophysics* 121 (1967): 533–39.

Steinman, Gary, and Marian N. Cole. "Synthesis of Biologically Pertinent Peptides Under Possible Primordial Conditions." *Proceedings of the National Academy of Sciences USA* 58 (1967): 735–41.

Strait, B.J. and G. T. Dewey, "The Shannon Information Entropy of Biologically Pertinent Peptides," *Biophysical Journal* 71 (1996): 148–55.

Taylor, Sean V., Kai U. Walter, Peter Kast, and Donald Hilvert. "Searching Sequence Space for Protein Catalysts. *Proceedings of the National Academy of Sciences USA* 98 (2001): 10596–601.

Thaxton, Charles B., and Walter L. Bradley. "Information and the Origin of Life." In *The Creation Hypothesis: Scientific Evidence for an Intelligent Designer,* edited by J. P. Moreland, Downers Grove, IL: InterVarsity, 1994, 193–97.

Thaxton, Charles, Walter L. Bradley, and Roger L. Olsen. *The Mystery of Life's Origin: Reassessing Current Theories*. New York: Philosophical Library, 1984.

Van de Sande, Bret. "Measuring Complexity in Dynamical Systems." Paper presented to RAPID II (Research and Progress in Intelligent Design) Conference, Biola University, May 2006.

Von Neumann, John. *Theory of Self-Reproducing Automata*. Completed and edited by A. Burks. Urbana: University of Illinois Press, 1966.

Wald, George. "The Origin of Life." *Scientific American* 191 (1954): 44–53.

Waldrop, M. Mitchell. "Did Life Really Start Out in an RNA World?" *Science* 246 (1989): 1248–49.

Walton, J. C. "Organization and the Origin of Life." *Origins* 4 (1977): 16–35.

Wigner, Eugene. "The Probability of the Existence of a Self-Reproducing Unit." In *The Logic of Personal Knowledge: Essays Presented to Michael Polanyi,* edited by Edward Shils, London: Routledge and Kegan Paul, 1961, 231–35.

Wolf, Yuri I., and Eugene V. Koonin. "On the Origin of the Translation System and the Genetic Code in the RNA World by Means of Natural Selection, Exaptation, and Subfunctionalization." *Biology Direct* 2 (2007): 1–25.

Wolfe, Stephen L. *Molecular and Cellular Biology*. Belmont, CA: Wadsworth, 1993.

Yockey, Hubert P. "A Calculation of the Probability of Spontaneous Biogenesis by Information Theory." *Journal of Theoretical Biology* 67 (1977): 377–98.

———. "Self-Organization Origin of Life Scenarios and Information Theory." *Journal of Theoretical Biology* 91 (1981): 13–31.

———. *Information Theory and Molecular Biology*. Cambridge: Cambridge University Press, 1992.

15

Mysteries of Life: Is There "Something Else"?

Christian de Duve

1. Introduction[1]

Science is based on naturalism, the notion that all manifestations in the universe are explainable in terms of the known laws of physics and chemistry. This notion represents the cornerstone of the scientific enterprise. Unless we subscribe to it, we might as well close our laboratories. If we start from the assumption that what we are investigating is not explainable, we rule out scientific research.

Contrary to the view expressed by some scientists, this logical necessity does not imply that naturalism is to be accepted as an *a priori* philosophical stand, a doctrine or belief. As used in science, it is a postulate, a working hypothesis, therefore qualified as methodological naturalism by philosophers, which we should be ready to abandon if faced with facts or events that defy every attempt at a naturalistic explanation. But only then should we accept the intervention of "something else," a last resort after all possibilities of explaining a given phenomenon in naturalistic terms have been exhausted. Should we reach such a point, assuming it can be recognized, we may still have to distinguish between two alternatives: Is the "something else" an unknown law of nature now disclosed by our investigations, as has happened several times in the past? Or is it a truly supernatural agency?

Traditionally, life, with all its wonders and mysteries, has been a favorite ground for the belief in "something else." Largely muted by the spectacular advances of biology in the last century, this position has been brought back into prominence by a small but vocal minority of scientists, whose opinions have been widely relayed in various philosophical and religious circles. In this essay, I wish to examine briefly whether certain biological phenomena indeed exist, as is claimed, that truly defy every attempt at a naturalistic explanation and make it necessary to invoke the intervention of "something else." Has a stage been reached where all scientific avenues have been exhausted? Should such be the case, are we to enlarge our notion of what is natural? Or have we met the authentically supernatural?

In examining these questions, I have assumed that readers are familiar with at least the basic elements of modern biology. Also, references have been strictly limited. For additional information and a more detailed treatment of the topics addressed, readers are referred to earlier works.[2]

2. The Nature of Life

Research into the mechanisms that support life represents one of the most spectacular successes of the naturalistic approach. I can bear personal witness to this astonishing feat. When I was first exposed to biochemistry, only a number of small biological building blocks, such as sugars, amino acids, purines, pyrimidines, fatty acids, and a few others, had been identified. How these molecules were made by living organisms was largely unknown. Not a single macromolecule arising from their combination had been characterized. Of metabolism only a few central pathways, such as glycolysis and the tricarboxylic acid cycle, had been painstakingly unravelled. Enzymology was still in its infancy. So was bioenergetics, which at that time boiled down to the recent discovery of ATP and some hints of the role of this substance as a universal purveyor of biological energy. As to genetic information transfers, our ignorance was complete. We did not even know the function of DNA, let alone its structure.

We were hardly disheartened by the puniness of these achievements. On the contrary, we saw them as tremendous triumphs. They opened wide vistas, testified to the fact that life could be approached with the tools of biochemistry, and strongly encouraged us to further research. But the unsolved problems loomed huge on the horizon and their solution appeared remote. I, for my part, never imagined in my wildest dreams and I don't think any of my contemporaries did that I would live to see them elucidated. Yet, that is what has happened. We now know the structures of all major classes of macromolecules, and we have the means to isolate and analyze any single such molecule we choose. We understand in detailed fashion most metabolic pathways, including, in particular, all major biosynthetic processes. We also have an intimate understanding of the mechanisms whereby living organisms retrieve energy from the environment and convert it into various forms of work. Most impressive of all, we know how biological information is coded, stored, replicated, and expressed. It is hardly an exaggeration to say that we have come to understand the fundamental mechanisms of life. Many gaps in this knowledge remain to be filled. For all we know, some surprises may still await us—remember the discovery of split genes. But, on the whole, the basic blueprint of life is known, to the point that we are now capable of manipulating life knowingly and purposefully in unprecedented fashion.

The lesson of these remarkable accomplishments is that life is explainable in naturalistic terms. To be true, new principles have been uncovered that rule the behavior of complex molecules such as proteins and nucleic acids, or govern the properties of multimolecular complexes such as membranes, ribosomes, or multi-enzyme systems. But nothing so far has been revealed that is not explainable by the laws of physics and chemistry. These disciplines have merely been greatly enriched by the new knowledge. There is no justification for the view that living organisms are made of matter "animated" by "something else."

3. The History of Life

Modern biological knowledge has revealed another capital piece of information: all known living organisms are descendants from a single ancestral form of life. Already suspected by the early evolutionists, this view has been further bolstered by the close similarities that have been detected at the cellular and molecular levels among all analyzed living organisms, whatever their apparent diversity. Whether we look at bacteria, protists, plants, fungi, or animals, including humans, we invariably find the basic blueprint mentioned above. There are differences, of course. Otherwise, all organisms would be identical. But the differences are clearly recognized as variations on the same central theme. Life is one.

This fact is now incontrovertibly established by the sequence similarities that have been found to exist among RNA or protein molecules that accomplish the same functions in different organisms or among the DNA genes that code for these molecules. The number of examples of this sort now counts in the many hundreds and is continually growing. It is utterly impossible that molecules with such closely similar sequences could have arisen independently in two or more biological lines, unless one assumes a degree of determinism in the development of life that even the most enthusiastic supporter of this view would refuse to consider.

These sequence similarities not only prove the kinship among all the organisms that have been subjected to this kind of analysis; they can even serve to establish phylogenetic relationships on the basis of the hypothesis, subject to many refinements, that sequence differences are all the more numerous the longer the time the owners of the molecules have had to evolve separately that is, the longer the time that has elapsed since they diverged from their last common ancestor. This method, which is now widely applied, has confirmed and strengthened many of the phylogenies previously derived from the fossil record; and it has, in addition, allowed such reconstructions to be extended to the many species that have left no identifiable fossil remains.

It is fair to state that the early hopes raised by this powerful new technology have become somewhat tempered. Underlying assumptions have been attacked as oversimplified. Different competing algorithms are advocated as dealing best with the diversity of genetic changes that have to be taken into account. Most importantly, horizontal gene transfer that is, the transfer of genes between distinct species, as opposed to their vertical transfer from generation to generation has been recognized as a major complication when attempting to use molecular data to reconstruct the tree of life, especially its early ramifications. These difficulties, however, affect only the shape of the tree, not its reality. Biological evolution is an undisputable fact.

4. THE ORIGIN OF LIFE

Where, when, and especially how did life start? We don't have answers to these questions. But at least we are no longer completely in the dark about them. We now know, from unmistakable fossil traces, that elaborate forms of bacterial life, with shapes reminiscent of photosynthetic organisms known as cyanobacteria, were present on Earth at least 3.55 billion years ago. More primitive organisms must have existed before that date, perhaps even as early as 3.8 billion years ago if ^{12}C-enriched carbon deposits found in Greenland are evidence of biological activity. Considering that the Earth most likely was physically unable to harbor life during at least the first half-billion years after its birth some 4.55 billion years ago, it appears that our planet started bearing living organisms at the latest 250 million years after it became capable of doing so.

It is not known whether the first forms of Earth life arose locally or came from elsewhere. Since there is at present no compelling reason or evidence supporting an extraterrestrial origin, most investigators accept the simpler assumption of a local origin, which has the advantage of allowing the problem to be defined within the framework of the physical-chemical conditions, revealed by geological data, that may have prevailed on Earth at the time life appeared. Note that the old argument that not enough time was available on Earth for the natural development of something as complex as even the most primitive living organism is no longer considered valid. It is generally agreed that if life originated naturally, it can only have done so in a relatively short time, probably to be counted in millennia rather than in millions of years.

Whatever the pathways followed, chemical reaction rates must have been appreciable in order for fragile intermediates to reach concentrations sufficient to permit the next step to occur. There may thus have been many opportunities on the pre-biotic Earth for life to appear and disappear before it finally took root.

Another key piece of evidence has become available in the last thirty years. It has been learned, from the chemical analysis of comets and meteorites and from the spectral analysis of the radiation coming from other parts of the solar system and from outer space, that organic substances, including amino acids and other potential building blocks of life, are widely present in the cosmos. These compounds are most likely products of spontanous chemical processes, not of biological activity. Organic syntheses thus not only do not require living organisms; they can even proceed without human help, and do so on a large scale. Organic chemistry is wrongly named. It is simply carbon chemistry, which happens to be the most banal and widespread chemistry in the universe, while being extraordinarily rich thanks to the unique associative properties of the carbon atom. It seems reasonable to assume that the products of this cosmic chemistry provided the raw materials for the formation of the first living organisms.

The question is: how? Ever since the Russian pioneer Alexander Oparin first tackled the problem[3] in 1924 and especially after Stanley Miller's historic 1953 experiments,[4] some of the best organic chemists in the world have struggled with this question in the laboratory, adopting as basic premise the naturalistic postulate, the hypothesis that the origin of life can be explained in terms of physics and chemistry. Using the same premise, many bystanders, like myself, have speculated on the matter or proposed models. A number of interesting facts have been uncovered, while a much larger number of suggestive ideas or "worlds"—the RNA world, the pyrophosphate world, the iron-sulfur world, the thioester world are examples—have been bandied about. Tens of books, hundreds of scientific papers, a journal specially created for the purpose, a society devoted exclusively to the topic, regular congresses and symposia, all attest to the vitality of this new scientific discipline.

Opinions are divided on what has been accomplished by all this activity. While much has been learned, it is clear that we are still nowhere near explaining the origin of life. This is hardly surprising, considering the immense complexity of the problem. But must the naturalistic postulate be abandoned for that reason? Have we reached a stage where all attempts at a naturalistic explanation have failed and the involvement of "something else" must be envisaged?

Anybody acquainted with the field is bound to answer this question with an emphatic "No." The surface has hardly been scratched. Pronouncing the origin of life unexplainable in naturalistic terms at the present stage can only be based on an *a priori* surrender to what the American biochemist Michael Behe, a prominent defender of this thesis, calls "irreducible complexity," which he defines in his *Darwin's Black Box,*[5] as the state of "a single system composed of several well-matched interacting parts that contribute to the basic function, wherein the removal of any one of the parts causes the system to effectively cease functioning." As a simple example of an irreducibly complex system, Behe offers the "humble mousetrap," of which each part can have been made only by somebody who had the whole contraption in mind. So it is, according to Behe, with many complex biochemical systems, such as cilia or the blood-clotting cascade.

This argument from design is not new. It was made two centuries ago by the English theologian William Paley, who used the image of a watch to prove the existence of a divine watchmaker. What is new, as well as surprising, is the use of modern biochemical knowledge in support of the "intelligent design" of life, a discovery that Behe believes "rivals those of Einstein, Lavoisier and Schrödinger, Pasteur and Darwin." This is a strange claim for a "discovery" that, instead of solving a problem, removes it from the realm of scientific inquiry.

Mechanical analogies, be they to watches or mousetraps, are poor images of biochemical complexity. Proteins, the main components of biochemical systems, have none of the rigidity of mechanical parts. Their name, derived from the Greek *prôtos* (first) by Gerardus Johannes Mulder, the Dutch chemist who invented it in 1838, could just as well have been derived from the name of the god *Prôteus,* famous for his ability to take almost any shape. As is well known, replacing a single amino acid by another may completely alter the properties of a protein. Even without any sequence alteration, the shape of a protein may change simply by contact with a modified template, as evidenced by the agents of diseases such as "mad cow disease" and its human equivalent, Creutzfeldt-Jakob disease. Called prions, these infectious agents consist of abnormally shaped proteins that, according to their discoverer, the American medical scientist Stanley Prusiner, multiply by conferring their abnormal shape to their normal counterparts in the body.[6] To affirm, as is implicit in Behe's theory, that a protein playing a given role cannot be derived from a molecule that fulfilled a different function in an earlier system is a gratuitous assertion, which flies in the face of evidence. Many examples are indeed known of proteins that have changed function in the course of evolution. Crystallins, the lens proteins, are a case in point.

The time element is what is missing in Behe's reasoning. It is no doubt true that present-day biochemical systems exhibit what he calls "irreducible complexity." Let a single element of the blood coagulation system be absent, and a major loss of functional efficiency does indeed occur, as evidenced by hemophilia and other similar disorders. But what the argument ignores is that the system has behind it hundreds of millions of years of evolution, in the course of which its slow progressive assembly most likely took place by a long succession of steps each of which could be explainable in naturalistic terms. The fact that the details of this long history have not yet been unravelled is hardly proof that it could not have happened.

The molecular history of proteins also needs to be taken into account. Today's proteins are the products of almost four billion years of evolution, during which an enormous amount of innovative diversification and adaptation has taken place. Admittedly, even their remote ancestors must already have been of considerable complexity to support the kind of bacterial life revealed by early microfossils. Had those ancestral proteins arisen fully developed, in a single shot, one would indeed be entitled to invoke "irreducible complexity" explainable only by "intelligent design." But all that we know indicates that this is not what happened.

Proteins bear unmistakable evidence of modular construction. They consist of a number of small domains, or motifs, many of which are present in various combinations in a number of different protein molecules, indicating strongly that they have served as building blocks in some kind of combinatorial assembly process. This fact suggests that precursors of the modules at one time existed as independent peptides that carried out, in some primitive protocells, the rudimentary equivalents of the structural and catalytic functions devolved to proteins in present-day cells. This hypothesis is consistent with theoretical calculations, by the German chemist Manfred Eigen, showing that the first genes must have been very short in order for their information to survive the many errors that must have beset primitive replication systems.[7]

Granted this simple assumption, one can visualize a primitive stage in the development of life supported by short peptides and subject to a Darwinian kind of evolution in which a progressively improved set of peptides would slowly emerge by natural selection. Let one outcome of this process be enhanced fidelity of replication, so that doubling of gene size becomes possible without adverse effect. Then, random combination of existing genes could progressively generate a new set of more efficient peptides twice the length of the previous ones. Darwinian competition would once again allow progressive improvement of this set, eventually ushering in a new round of the same kind at the next level of complexity, and so on.

This stepwise model provides an answer to an objection, often associated with a call for a guiding agency and recently revived, and refined, by the American mathematician William Dembski, a leader in the modern "intelligent design" movement.[8] The objection points to the fact that life uses only an infinitesimally small part of what is known as the sequence space, that is, the number of possible sequences. Take proteins, for example. These molecules are long chains consisting of a large number of amino acids strung together. Consider such a string one hundred amino acids long, which is a small size for a protein. Since proteins are made with twenty different species of amino acids, this molecule is one among 20^{100}, or 10^{130} (one followed by 130 zeros), possible proteins of the same size. This number is so unimaginably immense that even trillions of trillions of universes could accommodate only an infinitesimally small fraction of the possible molecules. It is thus totally impossible that life could have arrived at the sequences it uses by some sort of random exploration of the sequence space, that is, by means of randomly assembled sequences subject to natural selection. Hence the claim that the choice was directed by some supernatural agency, which somehow "knew" what it was heading for. With the proposed stepwise model of protein genesis (by way of nucleic acids), which is strongly supported by what is known of protein structure, this claim becomes unnecessary. According to this model, the pathway to present-day proteins went by a succession of stages, each of which involved a sequence space that had been previously whittled down by selection to a size compatible with extensive, if not exhaustive, exploration.

Like all conjectures, the proposed model is of value only as a guide for investigations designed to test its validity. One such line of research, which I believe is being followed in some laboratories, consists in looking for enzyme-like catalytic activities in mixtures of small randomly generated peptides. Whether the model is to be rejected or whether it may continue to be entertained, possibly in amended form, will depend on the results of those and other experiments. This issue is irrelevant to the present discussion. The fact that a plausible model can be proposed and that its experimental testing can be contemplated suffices to show that we have not yet reached a stage where all attempts at explaining the genesis of complex biochemical systems in naturalistic terms have failed.

The example given illustrates the importance of biochemical knowledge for origin-of-life research. Far from discouraging such research, as Behe would have it, our newly gained understanding of the chemical complexities of life can open up valuable avenues for fruitful investigation. Today's organisms actually accomplish, by way of reactions that are now well understood and explained in naturalistic terms, what is believed to have taken place four billion years ago. They turn small organic building blocks, or even inorganic ones, into fully functional living cells. To be true, they do this within the context of existing living cells, with the help of thousands of specific enzymes that were not available on the pre-biotic Earth. But this fact does not necessarily impose the opinion, voiced by many origin-of-life experts, that today's metabolic pathways are totally different from those by which life first arose. On the contrary, there are strong reasons for believing that present-day metabolism arose from early pre-biotic chemistry in congruent fashion, that is, by way of a direct chemical filiation. It is thus quite possible, even likely in my opinion, that today's metabolic pathways contain many recognizable traces of the reactions whereby life originated. It remains for future generations of researchers to decipher this history and conduct appropriate experiments.

5. Evolution

A majority of biologists subscribe in one form or another to the main tenets of the theory, first proposed by Charles Darwin, that biological evolution is the outcome of accidentally arising genetic variations passively screened by natural selection according to the ability of the variants to survive and produce progeny under prevailing environmental conditions. What to Darwin was largely the product of a genial intuition, backed only by observation, has received powerful support from modern biology, which has fleshed out in clear molecular terms the vague notions of hereditary continuity and variability available to Darwin. In particular, convincing proof has been obtained that naturally occurring mutations are induced by causes that are entirely unrelated, except in purely fortuitous fashion, to their evolutionary consequences. Just as Darwin postulated, blind variation comes first, with no possible eye to the future. Selection follows, enforced just as blindly by the sum total of environmental pressures, including those exerted by other living organisms.

There are a few dissenters. I don't include here the creationists and other ideologues who not only reject natural selection but deny the very occurrence of biological evolution. Neither do I have in mind the ongoing, often tedious, sometimes acrimonious debates among evolutionist coteries on such issues as gradualism, saltation, punctuated equilibrium, genetic drift, speciation, population dynamics, and other specialized aspects of the Darwinian message. The dissenters I am referring to accept evolution but reject a purely naturalistic explanation of the process.

This stand has a long and distinguished past. Especially in the French tradition, going back to the philosopher Henri Bergson, author of *L'évolution créatrice* (1907) and father of the concept of *élan vital* (vital upsurge), many biologists have embraced a teleological view of evolution, seen as directed by a special agency that somehow induced changes according to a preconceived plan. Often associated with vitalism, finalism concurrently fell into disregard with the growing successes of biochemistry and molecular biology. In recent years, the finalistic doctrine has taken advantage of the squabbles in the Darwinian camp to stage a comeback. It enjoys something of a revival in certain French circles, often with religious overtones related to the attempt by the celebrated Jesuit Pierre Teilhard de Chardin to reconcile biological science with the Catholic faith. In the Anglo-Saxon world, it lurks, in some nebulous, so-called "holistic" form, behind writings such as those of Lynn Margulis,[9] world-renowned for her early and correct championship of the endosymbiotic origin of certain cell organelles and, more recently, for her advocacy of James Lovelock's Gaia concept and of the "autopoiesis" theory of the recently deceased Chilean biologist-cum-mystic Francisco Varela.

In a different vein, the American theoretical biologist Stuart Kauffman, an expert in computer-simulated "artificial life," also expresses dissatisfaction with classical Darwinian theory. He believes that biological systems, in addition to obeying natural selection, possess a powerful intrinsic ability to "self-organize," creating "order for free."[10] The British-Australian physicist Paul Davies, best-selling author of books with titles such as *The Mind of God* and *The Fifth Miracle*,[11] while declaring himself committed to naturalistic explanations, does not hesitate to invoke a "new type of physical law," to account for the ability of life to "circumvent what is chemically and thermodynamically 'natural.'" More explicitly finalistic views of life's origin and history are defended in two recent books, Behe's *Darwin's Black Box,* already mentioned, and *Nature's Destiny,* characteristically subtitled *How the Laws of Biology Reveal Purpose in the Universe,* by the New Zealand biologist Michael Denton.[12]

The dissenters do not reject Darwinian explanations outright. They accept such explanations for many events in what is known as microevolution, or horizontal evolution in my terminology, the process whereby diversity is generated within a fundamentally unchanged body plan. They would be willing, for example, to ascribe to natural selection the diversification of the famous Galapagos finches, which Darwin found to have differently shaped beaks, adapted on each island to the kind of food available. But they deny that an unaided mechanism could possibly account for the main steps of macroevolution, or vertical evolution, which involve major changes in body plan. The transformation of a dinosaur into a bird or any other of the apparent "jumps" revealed by the fossil record belong in this category. In such cases, so the argument goes, so many important modifications had to take place simultaneously that no natural pathway involving viable intermediates is conceivable. To illustrate this point, Behe mentions the conversion of a bicycle into a motorcycle. The latter can be derived from the former conceptually, but not physically. However, as we have already seen, mechanical contraptions are poor models for living systems.

As with the origin of life, the issue must remain undecided, considering that the details of evolutionary pathways are unknown and unlikely to be elucidated in the near future. The problem may even be more intractable than the origin of life, since evolution is a historical process that has proceeded over several billion years, leaving very few traces. But have we reached a stage in our analysis of this process where naturalistic explanations must be abandoned and an appeal to "something else" has become mandatory? This is obviously not so. Ever since the days of Darwin, the dearth of "missing links" has been brandished as an argument against his theory. But scarcity is no proof of absence, especially in an area where the "luck of the find" plays a major role. But for *Archaeopteryx* and the almost miraculously preserved imprint of its feathers in a Bavarian rock, discovered in 1864, no inkling as to the pathway from reptiles to birds would have been available until recently, when some findings of a similar nature were made in China. Nobody can predict what the future will yield.

Here also, modern science, far from spotlighting "irreducible complexity," may, instead, disclose unexpected ways of reducing the complexity. A dramatic breakthrough of this kind has recently occurred with the discovery of homeogenes. These are master genes that control the expression of a very large number of genes up to 2,500, according to their discoverer, the Swiss biologist Walter Gehring, who has reviewed this fascinating field.[13] To give just one example, the so-called *eyeless* gene of the fruitfly has the power of inducing, by itself, the whole cascade of events needed for the development of a fully functional eye. Remarkably, the same gene is found in a wide variety of invertebrates and vertebrates, where it carries out the same function, even though the resulting eyes may be as different as the single-lens eyes of mammals, the similarly single-lens but differently constructed eyes of cephalopods, the multifaceted eyes of arthropods, and the primitive eyes of flatworms. Even more astonishing, the mouse gene is perfectly active when inserted into a fruitfly; but the product of this action is a typical fruitfly eye, not a mouse eye. The switch is the same; but the battery of genes that are turned on is different in the two species, resulting in the formation of entirely different eyes. Such discoveries are not only relevant to embryological development; they also illuminate evolutionary processes by showing how huge changes in phenotype can be produced by single mutations. The mysterious "jumps" may turn out to have a naturalistic explanation after all. They may even become open to experimental study.

6. The Elusive "Something Else"

Calling on "something else" is not only heuristically sterile, stifling research; it is conceptually awkward, at least in its modern formulation. In the days when finalism and vitalism were blended into a single, all-encompassing theory, the philosophical position had the merit of being internally consistent. A mysterious "vital force" was seen as guiding living organisms in all their manifestations. Once the stand becomes selective, accepting a naturalistic explanation for some events and denying it for others, one is faced with the almost impossible task of defining the borderline between the two. The lesson of history is that this boundary has kept on shifting, as more of the unexplained, deemed unexplainable by some, came to be explained. Many biologists are willing to extrapolate this historical course to a point in the future when all will be explained. Lest we be accused of prejudging the issue, we need not be as sanguine, simply stating that the only intellectually defensible position is to accept naturalism as a working hypothesis for the design of appropriate experiments and pointing to past successes as strongly justifying such a stand.

Behe does not accept this view. To him, the boundary is fixed, set by the limits of "irreducible complexity." On one side of the boundary, events are ruled by natural laws. On the other, there is intervention by an entity he does not hesitate to identify as God. This intervention does not stop at setting life on course and then letting it run on its own steam, so to speak. Behe's God accompanies life throughout evolution, to provide the necessary nudge whenever one is needed for some probability barrier to be overcome. Behe offers no suggestion as to the molecular nature or target of this nudge. One remains perplexed by this picture of a divine engineer creating the universe with its whole set of natural laws and then occasionally breaking the laws of his creation to achieve a special objective, which, presumably, includes the advent of humankind. Why not imagine a God who, from the start, created a world capable of giving rise to life in all its manifestations, including the human mind, by the sole exercise of natural laws?

This is what Denton postulates, but in a form perilously close to Behe's view of direct interference, in spite of its being presented as "entirely consistent with the basic naturalistic assumption of modern science." Denton not only embraces the anthropic concept of a physically fine-tuned universe, uniquely fit for life, but he does so in a "hyperanthropocentric" way. He imagines the whole evolutionary script, with humankind as the crowning achievement, as written at the start into the fine print of the original DNA molecules. And he marshalls molecular arguments that purportedly support such a notion, leaving almost no room for chance events in the unfolding of evolution. He goes so far as to envisage "directed mutations," "preprogrammed genetic rearrangements," and "self-directed evolution." He even admits the possibility that certain evolutionary changes may have been sustained, not by their immediate, but by their future benefits. In discussing the development of the characteristic lung of birds, he finds it "hard not to be inclined to see an element of foresight in the evolution of the avian lung, which may well have developed in primitive birds before its full utility could be exploited." Denton finds this notion "perfectly consistent with a directed model of evolution," but offers no suggestion as to how the direction could be exerted, contenting himself with the vague concept of a Creator who has "gifted organisms . . . with a limited degree of genuine creativity."

7. A Balanced View

Naturalism has not reached the limits of its explanatory power. On the contrary, everything that has been accomplished so far encourages the belief that the origin and evolution of life are, just as are life's fundamental mechanisms, explainable in naturalistic terms. Research guided by this assumption remains a valid and promising approach to these problems. Many scientists extrapolate this scientific attitude into a philosophical worldview that denies any sort of cosmic significance to the existence of life, including its most complex manifestation to date, the human brain. Such affirmations need to be greeted with as much caution as those that claim the intervention of "something else."

Irrespective of any *a priori* position we may wish to entertain, one fact stands out as incontrovertible: we belong to a universe capable of giving rise to life and mind. Reversing a famous saying by the French biologist Jacques Monod, "the universe was pregnant with life; and the biosphere with man."[14] As shown by the exponents of the anthropic principle and reiterated by Denton,[15] this fact implies a considerable degree of physical "fine-tuning." Even minor changes in any of the physical constants would upset the material balance to such an extent that either there would be no universe or the existing universe would be such that, for one reason or another, life could not arise or subsist in it.

To Denton, this fact means that the universe is "designed" to harbor life. Others, however, are content with leaving it all to chance. The British chemist Peter Atkins,[16] a militant defender of science-based atheism, calls on a "frozen fluctuation" of "extreme improbability" to account for the birth of our universe. And he concludes: "The universe can emerge out of nothing, without intervention. By chance." According to the British cosmologist Martin Rees,[17] our universe is just one knowable because fit for life in a huge collection of universes, a "multiverse," produced by chance. The American physicist Lee Smolin[18] has adopted the same idea, but in an evolutionary context. He assumes that new universes with slightly different physical constants can arise from existing universes by way of black holes. Ability to form black holes thus appears as a measure of cosmic prolificity and serves as a selective factor in this evolutionary process. The special properties of our universe obtain because they happen to be associated with the production of a particularly high number of black holes.

Such speculations are fascinating but irrelevant to the main issue. Whether it arose by design, chance, or evolution, whether it is unique or one of a huge number, our universe did give rise to life and, through a long evolutionary process, to a living form endowed with the ability to apprehend, by science and also by other approaches, such as literature, art, music, philosophy, or religion, glimpses of the mysterious "ultimate reality" that hides behind the appearances accessible to our senses.[19] This fact seems to me supremely important.

In this respect, I accept the anthropic principle, but in its factual, not its teleological connotation. The universe is "such that," not necessarily "made for." I also reject the strongly anthropocentric bias given to the principle by Denton and other defenders of finalism. Humankind did not exist three million years ago, in contrast to life, which has been around more than one thousand times longer. There is no reason for believing that our species represents an evolutionary summit and will persist unchanged for the next billion years, which is the minimum time cosmologists tell us the Earth will stay fit for life. On the contrary, it seems highly probable that humankind, like all other living species, will continue to evolve, perhaps with human assistance. It is quite possible, even likely in my opinion, that forms of life endowed with mental faculties greatly superior to ours will one day arise, whether from the human twig or from some other branch of the evolutionary tree. To those beings, our rationalizations will look as rudimentary as the mental processes that guided the making of tools by the first hominids would look to us.

Another philosophical view often presented as irrefutably imposed by modern science is what I call the "gospel of contingency," the notion according to which biological evolution, including the advent of humankind, is the product of countless chance events not likely to be repeated anywhere, any time, and therefore devoid of any meaning. In the words of the American paleontologist Stephen Jay Gould, the most vocal and, because of his talent as a writer, persuasive advocate of this creed, "biology's most profound insight into human nature, status, and potential lies in the simple phrase, the embodiment of contingency."[20] I disagree with this affirmation.

What biology tells us is that humankind, like every other form of life, is the product of almost four billion years of evolutionary ramification and, put in schematic terms, that each new branch in this process is the consequence of an accidental genetic change that proved beneficial to the survival and proliferation of the individual concerned under the prevailing environmental conditions. The flaw in the contingency argument is to equate fortuitousness with improbability. Events may happen strictly by chance and still be obligatory. All that is needed is to provide them with enough opportunities to take place, relative to their probability. Toss a coin once, and the probability of its falling on a given side is 50 percent. Toss it ten times, and the odds are 99.9 percent that it will fall at least once on each side. Even a seven-digit lottery number is ensured a 99.9 percent chance of coming out if 69 million drawings are held. Lotteries don't work that way, of course; but the evolutionary game does.

The evolutionary tree has not, as is often assumed, spread out by a kind of random walk in an infinite space of variation. The space is limited by a number of inner constraints enforced by the sizes and structures of genomes and by outer constraints imposed by the environment. The numbers of participating individuals and the times available are such that widespread exploration of the mutational space has often been possible, leaving the main selective decision to environmental conditions.

In this context, contingency still plays an important role, but more so in horizontal evolution (microevolution) than in vertical evolution (macroevolution). Without leaves to provide a potential shield, no insect would have taken the weird evolutionary pathway that led it to look like a leaf—an occurrence, incidentally, that attests to the richness of the mutational field. On the other hand, if the particular historical circumstances that favored the conversion of a lungfish into a primitive amphibian had not taken place, it seems likely that vertebrates eventually would have invaded the lands on some other occasion, so rich were the selective bounties to be reaped by such a move. Nearer home, if our primate forebears had not been isolated some six million years ago perhaps by the opening of the Great Rift Valley in East Africa in a savannah where traits such as an erect position and manual skills became greatly advantageous, the odds are that somewhere, sooner or later, some ape group would have started on the road toward humankind. The lightning speed in terms of evolutionary time of hominization certainly shows that the process, once initiated, was very strongly favored by natural selection. This is not surprising, since a better brain is bound to be an asset in any environment.

Earlier in this paper, I pointed out how, by a series of successive steps of sequence lengthening and selection, emerging life could have explored substantial parts of the sequence spaces available to it at each stage. The subsequent hierarchization of the genes into an increasing number of levels—remember homeogenes—has produced something of a similar situation in later evolution, allowing widespread exploration of the relevant mutational spaces at each level of complexity. For this and other reasons, I defend the position that the vertical growth of the tree of life toward increasing complexity is strongly favored by the purely naturalistic factors that are believed to determine biological evolution. On the other hand, the horizontal growth of the tree toward greater diversity at each level of complexity most probably has been largely ruled by the vagaries of environmental conditions.

8. Conclusion

In conclusion, modern science, while increasingly doing away with the need for "something else" to explain our presence on Earth, does not by the same token enforce the view that we are no more than a wildly improbable and meaningless outcome of chance events. We are entitled to see ourselves as part of a cosmic pattern that is only beginning to reveal itself. Perhaps some day, in the distant future, better brains than ours will see the pattern more clearly. In the meantime, the stage we have reached, albeit still rudimentary, represents a true watershed in that, for the first time in the history of life on Earth, a species has arisen that is capable of understanding nature sufficiently to consciously, deliberately, and responsibly direct its future. It is to be hoped that humankind will be up to the challenge.

Notes

1. In preparing this essay, I have benefited from the valuable advice and editorial assistance of my friend Neil Patterson.
2. De Duve 1995, 1998, 2002.
3. Oparin 1957 (first published in Russian in 1924).
4. Miller 1953.
5. Behe 1996.
6. Prusiner 1998.
7. Eigen and Schuster 1977.
8. Dembski 1998.
9. Margulis and Sagan 1995.
10. Kauffman 1993, 1995.
11. Davies 1992, 1998.
12. Behe 1996 ; Denton 1998.
13. Gehring 1998.
14. Monod 1971.
15. Denton 1998.
16. Atkins 1981.
17. Rees 1997.
18. Smolin 1997.
19. De Duve 2002.
20. Gould 1989.

References

Atkins, Peter W. (1981) *The Creation*. Oxford and New York: Freeman & Co.

Behe, Michael J. (1996) *Darwin's Black Box. The Biochemical Challenge to Evolution*. New York: The Free Press.

Davies, Paul. (1992) *The Mind of God*. New York: Simon & Schuster.

Davies, Paul. (1998) *The Fifth Miracle. The Search for the Origin of Life*. London: Allen Lane. The Penguin Press, 1998.

de Duve, Christian. (1995) *Vital Dust. Life as a Cosmic Imperative*. New York: BasicBooks.

———. (1998) "Constraints on the Origin and Evolution of Life." *Proceedings of the. American Philosophical Society* 142: 525–32.

———. (2002) *Life Evolving: Molecules, Mind, and Meaning*. Oxford University Press.

Dembski, William A. (1998) *The Design Inference: Eliminating Chance through Small Probabilities*. Cambridge University Press.

Denton, Michael J. (1998) *Nature's Destiny. How the Laws of Biology Reveal Purpose in the Universe*. New York: The Free Press.

Eigen, M., and Schuster, P. (1977) "The Hypercycle: a Principle of Self-organization. Part A: Emergence of the Hypercycle," *Naturwissenschaften* 64: 541–65.

Gehring, W. J. (1998) *Master Control Genes in Development and Evolution: The Homeobox Story*. New Haven: Yale University Press.

Gould, Stephen J. (1989) *Wonderful Life*. New York: Norton.

Kauffman, Stuart A. (1993) *The Origins of Order*. Oxford University Press.

———. (1995) *At Home in the Universe*. Oxford University Press.

Margulis, L., and Sagan, D. (1995) *What is Life?* New York: Simon & Schuster.

Miller, Stanley L. (1953) "A Production of Amino Acids under Possible Primitive Earth Conditions." *Science* 117: 528–29.

Monod, Jacques. (1971) *Chance and Necessity*. New York: Knopf.

Oparin, Aleksandr I. (1957 [1924]) *The Origin of Life on the Earth,* 3rd Ed. New York: Academic Press.

Prusiner, S. B. (1998) "Prions." *Les Prix Nobel 1997*. Stockholm: Norstedt Tryckeri, 268–323.

Rees, Martin. (1997) *Before the Beginning*. Reading, MA: Perseus Books.

Smolin, Lee. (1997) *The Life of the Cosmos*. Oxford : Oxford University Press.

16

Life's Conservation Law:
Why Darwinian Evolution Cannot
Create Biological Information

William A. Dembski and Robert J. Marks II

1. The Creation of Information

Any act of intelligence requires searching a space of possibilities to create information. The preceding sentence illustrates this very point. In formulating it, we searched a space of letter sequences. Most such sequences are gibberish (unpronounceable arrangements of random letters). Of those which are meaningful, most are not in English. Of those which are in English, most have nothing to do with information theory or intelligent design. Only by successively eliminating vast swatches of this search space did we succeed in locating the first sentence of this paper, thereby creating a unique sentence in the history of the English language. Each such reduction of possibilities constitutes an information-generating act. Together, these reductions constitute a search that identifies one possibility (the first sentence of this paper) to the exclusion of others.

Information identifies possibilities to the exclusion of others. Unless possibilities are excluded, no information can be conveyed. To say "it's raining or it's not raining" is uninformative because it excludes no possibilities. On the other hand, to say "it's raining" excludes the possibility "it's not raining" and therefore conveys information. Tautologies, because they are true regardless, can convey no information. We don't need to be informed of them because we can figure them out on our own.

Information presupposes multiple live possibilities. Robert Stalnaker puts it this way: "Content requires contingency. To learn something, to acquire information, is to rule out possibilities. To understand the information conveyed in a communication is to know what possibilities would be excluded by its truth."[1] Fred Dretske elaborates: "Information theory identifies the amount of information associated with, or generated by, the occurrence of an event (or the realization of a state of affairs) with the reduction in uncertainty, the elimination of possibilities, represented by that event or state of affairs."[2]

According to Douglas Robertson, the defining feature of intelligent agents is their ability, as an act of free will, to create information.[3] G. K. Chesterton elaborated on this insight: "Every act of will is an act of self-limitation. To desire action is to desire limitation. In that sense every act is an act of self-sacrifice. When you choose anything, you reject everything else. . . . Every act is an irrevocable selection and exclusion. Just as when you marry one woman you give up all the others, so when you take one course of action you give up all the other courses."[4] Intelligence creates information.

But is intelligence the only causal power capable of creating information? Darwin's main claim to fame is that he is supposed to have provided a mechanism that could create information without the need for intelligence. Interestingly, he referred to this mechanism as "natural selection." Selection, as understood before Darwin, had been an activity confined to intelligent agents. Darwin's great coup was to attribute the power of selection to nature—hence "natural selection."

Nature, as conceived by Darwin and his followers, acts without purpose—it is non-teleological and therefore unintelligent. As evolutionary geneticist Jerry Coyne puts it in opposing intelligent design, "If we're to defend evolutionary biology, we must defend it as a science: a *nonteleological* theory in which the panoply of life results from the action of natural selection and genetic drift acting on random mutations."[5] But why do Coyne and fellow Darwinists insist that evolutionary biology, to count as science, must be non-teleological?[6] Where did that rule come from? The wedding of teleology with the natural sciences is itself a well-established science—it's called engineering. Intelligent design, properly conceived, belongs to the engineering sciences.

But to return to the point at hand, does nature really possess the power to select and thereby create information? To answer this question, we need to turn to the relation between matter and information. The matter–information distinction is old and was understood by the ancient Greeks. For them, there was matter, passive or inert stuff waiting to be arranged; and there was information, active stuff that did the arranging.[7] This distinction provides a useful way of carving up experience and making sense of the world. Left here, it is uncontroversial. Nonetheless, it becomes controversial once we add another dimension to it, that of nature and design:

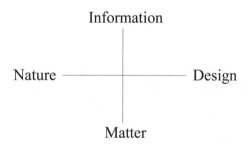

Designing intelligences are not the only agents capable of putting information into matter. Nature, too, has this capacity. Consider the difference between raw pieces of wood and an acorn. Raw pieces of wood do not have the power to assemble themselves into a ship. For raw pieces of wood to form a ship requires a designer to draw up a blueprint and then take the pieces of wood and, in line with the blueprint, fashion them into a ship. But where is the designer that causes an acorn to develop into a full-grown oak tree? The acorn possesses the power to transform itself into an oak tree.

Nature and design therefore represent two different ways of producing information. Nature produces information internally. The acorn assumes the form it does through powers it contains within itself—the acorn is a seed programmed to produce an oak tree. Accordingly, the acorn does not create its information from scratch. Rather, it expresses already existing information, which it derived from previous generations of oak trees. By contrast, a ship assumes the form it does through powers external to it—a designing intelligence needs to intentionally arrange the pieces of wood to form a ship.

Not only did the ancient Greeks know about the distinction between information and matter, but they also knew about the distinction between design and nature. Thus, Aristotle characterized design as consisting of capacities external to objects that are needed to bring about their form. On the other hand, he saw nature as consisting of capacities internal to objects that enable them to transform themselves without outside help. In Book XII of the *Metaphysics,* Aristotle wrote, "[Design] is a principle of movement in something other than the thing moved; nature is a principle in the thing itself."[8] In Book II of the *Physics,* Aristotle referred to design as completing "what nature cannot bring to a finish."[9]

The Greek word here translated "design" is *techne,* from which we get our word technology. In translations of Aristotle's work, the English word most commonly used to translate *techne* is "art" (in the sense of "artifact"). Design, art, and *techne* are synonyms. The essential idea behind these terms is that information is conferred on an object from outside it, and that the materials that make up the object lack the power to organize themselves into it. Thus, apart from that outside information, the object cannot assume the form it does. For instance, raw pieces of wood do not by themselves have the power to form a ship.

This contrasts with nature, which does have the power within itself to express information. Thus, in Book II of the *Physics,* Aristotle wrote, "If the ship-building art were in the wood, it would produce the same results by nature."[10] In other words, if raw pieces of wood had the capacity to form ships, we would say that ships come about by nature. The Greek word here translated "nature" is *phusis,* from which we get our word physics. The Indo-European root meaning behind *phusis* is growth and development. Nature produces information not by imposing it from outside but by growing or developing informationally rich structures from the capacities inherent in a thing. Consider again the acorn. Unlike wood that needs to be fashioned by a designer to form a ship, acorns produce oak trees naturally—the acorn simply needs a suitable environment in which to grow.

Aristotle claimed that the art of ship-building is not in the wood that makes up a ship. In the same vein, the art of sentence-making is not in the letters of the alphabet out of which sentences are constructed. Likewise, the art of statue-making is not in the stone out of which statues are made. Each of these cases requires a designer. So too, the theory of intelligent design contends that the art of life-making is not in the physical stuff that constitutes life. Instead, life-making requires a designer.

The central issue in the scientific debate over intelligent design and biological evolution can therefore be stated as follows: Is nature complete in the sense of possessing all the resources it needs to bring about the information-rich biological structures we see around us, or does nature also require some contribution of design to bring about those structures? Darwinian naturalism argues that nature is able to create all its own information and is therefore complete. Intelligent design, by contrast, argues that nature is merely able to re-express existing information and is therefore incomplete.

To argue for nature's incompleteness seems to presuppose a fundamental distinction between intelligent and natural causes. Can't intelligent causes be fully natural? As the scientific community understands them, natural causes operate according to deterministic and nondeterministic laws and can be characterized in terms of chance, necessity, and their combination.[11] To be sure, if one is more liberal about what one means by natural causes and includes among them end-directed (teleological) processes that are not reducible to chance and necessity (as Aristotle and the ancient Stoics did by endowing nature with immanent teleology), then our claim that natural causes are incomplete dissolves.

But for many in the scientific community, natural causes are at heart non-teleological and therefore unintelligent. Natural causes, when suitably coordinated, may exhibit intelligence. Thus, animals might be viewed as purely natural objects that act as intelligent causes to achieve ends. But the teleology they exhibit is, from a naturalistic perspective, the result of a long and blind evolutionary process whose nuts-and-bolts causal processes are non-teleological. Given naturalism, natural causes bring about intelligent causes but are not themselves intelligent. On this view, intelligent causes are always reducible to non-teleological natural causes, ultimately to the motions and interactions of particles governed by forces of attraction and repulsion.

The challenge of intelligent design, and of this paper in particular, is to show that when natural systems exhibit intelligence by producing information, they have in fact not created it from scratch but merely shuffled around existing information. Nature is a matrix for expressing already existent information. But the ultimate source of that information resides in an intelligence not reducible to nature. The Law of Conservation of Information, which we explain and justify in this paper, demonstrates that this is the case. Though not denying Darwinian evolution or even limiting its role as an immediate efficient cause in the history of life, this law shows that Darwinian evolution is deeply teleological. Moreover, it shows that the teleology inherent in Darwinian evolution is scientifically ascertainable—it is not merely an article of faith.

2. Biology's Information Problem

Caltech president and Nobel Prize–winning biologist David Baltimore, in describing the significance of the Human Genome Project, stated, "Modern biology is a science of information."[12] Biologists now recognize the crucial importance of information to understanding life and, especially, its origin. Origin-of-life researcher and Nobel Prize winner Manfred Eigen equated the problem of life's origin with uncovering "the origin of information."[13] Biologists John Maynard Smith and Eörs Szathmáry have explicitly placed information at the center of developmental and evolutionary biology: "A central idea in contemporary biology is that of information. Developmental biology can be seen as the study of how information in the genome is translated into adult structure, and evolutionary biology of how the information came to be there in the first place."[14]

Given the importance of information to biology, the obvious question is, How does biological information arise? For matter to be alive it must be suitably structured. A living organism is not a mere lump of matter. Life is special, and what makes it special is how its matter is arranged into very specific forms. In other words, what makes life special is information. Where did the information necessary for life come from?

The emergence of life constitutes a revolution in the history of matter. A vast gulf separates the organic from the inorganic worlds, and that gulf is properly characterized in terms of information. The bricks in your house and the cells in your body are both made of matter. Nevertheless, the arrangement of that matter—the information—differs vastly in these two cases. Biology's information problem is therefore to determine whether (and, if so, how) purely natural forces are able to bridge the gulf between the organic and inorganic worlds as well as the gulfs between different levels of complexity within the organic world. Conversely, biology's information problem is to determine whether (and, if so, how) design is needed to complement purely natural forces in the origin and subsequent development of life.

But how can we determine whether nature has what it takes to produce life? The sheer commonness of life on Earth tends to lull us into complacency. We look around and see life everywhere. But there was a time when the Earth contained no multicelled organisms like us. Before that, the Earth contained no life at all, not even single-celled forms. And earlier still, there was no Earth at all, no Sun or Moon or sister planets. Indeed, if physicists are right, there was a time when there were no stars or galaxies but only elementary particles like quarks densely packed at incredibly hot temperatures. That would coincide with the moment just after the Big Bang.

Suppose we go back to that moment. Given the history of the universe since then, we could say—in retrospect—that all the possibilities for complex living forms like us were in some sense present at that earlier moment in time (much as many possible statues are in some sense present in a block of marble). From that early state of the universe, galaxies and stars eventually formed, then planet Earth, then single-celled life forms, and finally life forms as complicated as us. But that still doesn't tell us how we got here or whether nature had sufficient creative power to produce us apart from design. Nature provides a necessary condition for our existence. The question is whether it also provides a sufficient condition.

As philosopher Holmes Rolston points out, humans are not invisibly present in primitive single-celled organisms in the same way that an oak tree is secretly present in an acorn. The oak tree unfolds in a lawlike or programmatic way from an acorn. But the same cannot be said for the grand evolutionary story that places single-celled organisms at one end and humans at the other ("monad to man" evolution). There's no sense in which human beings or any other multicelled organisms are latent in single-celled organisms, much less in nonliving chemicals. For Rolston, the claim that life is somehow already present or lurking in nonliving chemicals, or that complex biological systems are already present or lurking in simple biological systems, is "an act of speculative faith."[15]

Speculative faith is not science. It is unscientific simply to assert that nature possesses the causal powers capable of producing living forms. Rather, science has the job of demonstrating whether nature actually does possess such causal powers. Moreover, we do not have the luxury, like Aristotle and many ancient philosophers, of thinking that life and the universe have always been here. Advances in astrophysics and geology show that the Earth has not been here forever and that the early Earth was a tempestuous cauldron that rendered it uninhabitable. Yet somehow, from fairly simple inorganic compounds on the early lifeless Earth, life forms requiring the precisely coordinated activity of information-rich biomacromolecules emerged. How did that happen? How could it have happened?

3. The Darwinian Solution

Karl Marx is said to have joked that the twenty-six letters of the alphabet were his soldiers for conquering the world.[16] Yet to be successful, these soldiers required Marx's supervision. Left to themselves, the letters of the alphabet are inert. Mere alphabetic characters do not have the power to transform themselves into Marx's *Das Kapital*. Marx plus the letters of the alphabet, however, do have that power. But that raises the question how Marx himself came to be.

From a Darwinian perspective, any designing intelligence, such as Marx, results from a long and blind evolutionary process. Nature, without the need of any intelligence, starts off from a lifeless Earth and, over the course of natural history, produces life forms that eventually evolve into human beings like Karl Marx, who then write economic treatises like *Das Kapital*. Within Darwinian naturalism, only natural forces, principally natural selection acting on random variations, control the evolutionary process. Designing intelligences may evolve out of that process but play no role in guiding or controlling it.

Theistic evolutionists attempt to make room for God within this Darwinian scheme by claiming that God created the universe so that Darwinian processes would produce living forms. Accordingly, God brings about the natural forces that produce living things but makes sure not to interfere with those forces once they are in operation. Though logically possible, theistic evolution offers no compelling reason for thinking that nature is a divine creation. As Francisco Ayala put it, "It was Darwin's greatest accomplishment to show that the directive organization of living beings can be explained as the result of a natural process, natural selection, without any need to resort to a Creator or other external agent."[17] Darwinian evolution, far from requiring a creator God, is perfectly happy with a natural world that is entirely self-sufficient.

Theistic evolutionists think Darwin got nature right and then adapt their theology to suit Darwinian science. Proponents of intelligent design, by contrast, ask the logically prior question whether Darwin did in fact get nature right. Indeed, why should we think that natural forces, apart from intelligent guidance, have the power to create biological information? Darwin attempted to resolve this question by providing a naturalistic mechanism (natural selection acting on random variations) that could effectively account for the production of biological information.

Some of Darwin's followers consider this mechanism so effective that it makes unnecessary any intelligence behind biology. Richard Dawkins even went so far as to state, "Darwin made it possible to be an intellectually fulfilled atheist."[18] Others, like Francis Collins, think that an intelligence (for Collins, the Christian God) set up the conditions that make it possible for the Darwinian mechanism to bring about biological information. This view, according to Collins, "makes it possible for the scientist-believer to be intellectually fulfilled and spiritually alive."[19] Yet for Collins and other theistic evolutionists, God's hand is nowhere evident in the evolutionary process. Atheistic and theistic evolution join hands in proclaiming that purposeful design in biology is *scientifically undetectable*.[20]

How does the Darwinian mechanism replace intelligence in its account of biological information? Richard Dawkins uses an evolutionary algorithm to illustrate how the Darwinian mechanism is supposed to create information apart from intelligence.[21] For convenience, we refer to this example as Dawkins's WEASEL. He starts with the following target sequence:

METHINKS•IT•IS•LIKE•A•WEASEL

Note that he considers only capital Roman letters and spaces, spaces represented here by bullets—thus, twenty-seven possibilities at each location in a symbol string that is twenty-eight characters in length.

If we tried to attain this target sequence by pure chance (for example, by randomly shaking out Scrabble pieces consisting solely of letters and spaces), the probability of getting it on a given try would be around 1 in 10^{40}, and, correspondingly, it would take on average about 10^{40} tries to stand a better than even chance of getting it.[22] In other words, if we depended on pure chance to attain this target sequence, we would in all likelihood be unsuccessful. To generate the linguistic information in Dawkins's target sequence, pure chance is simply not up to the task.

Now the beauty of the Darwinian mechanism (it is a deceptive beauty) is that it is supposed to circumvent the limitations of pure chance. Thus, in place of pure chance, Dawkins considers the following evolutionary algorithm: (i) Start out with a randomly selected sequence of twenty-eight capital Roman letters and spaces, e.g.,

WDL•MNLT•DTJBKWIRZREZLMQCO•P

(ii) randomly alter individual letters and spaces in the current sequence; (iii) keep those newly formed sequences that match more letters in the target sequence, discarding the rest. This algorithm exemplifies the Darwinian mechanism: step (ii) corresponds to the random variations that supply the raw potential for the evolutionary process; step (iii) corresponds to the selection of those variants that are better fitted to their environment (in this case, those letter sequences which match the target sequence more closely).

In very short order this algorithm converges to Dawkins's target sequence. In *The Blind Watchmaker,* he summarizes a particular run of the algorithm that in a mere forty-three steps converged to the target sequence:[23]

(1) WDL•MNLT•DTJBKWIRZREZLMQCO•P

(2) WDLTMNLT•DTJBSWIRZREZLMQCO•P
         . . .
(10) MDLDMNLS•ITJISWHRZREZ•MECS•P
         . . .
(20) MELDINLS•IT•ISWPRKE•Z•WECSEL
         . . .
(30) METHINGS•IT•ISWLIKE•B•WECSEL
         . . .
(40) METHINKS•IT•IS•LIKE•I•WEASEL
         . . .
(43) METHINKS•IT•IS•LIKE•A•WEASEL

Thus, in place of 10^{40} tries on average for pure chance to produce the target sequence, by employing the Darwinian mechanism it now takes on average less than one hundred tries to produce it. In short, a search effectively impossible for pure chance becomes eminently feasible for the Darwinian mechanism.

So does Dawkins's evolutionary algorithm demonstrate the power of the Darwinian mechanism to create biological information? No. Clearly, the algorithm was stacked to produce the outcome Dawkins was after. Indeed, because the algorithm was constantly gauging the degree of difference between the current sequence from the target sequence, the very thing that the algorithm was supposed to create (i.e., the target sequence METHINKS•IT•IS•LIKE•A•WEASEL) was in fact smuggled into the algorithm from the start. The Darwinian mechanism, if it is to possess the power to create biological information, cannot merely veil and then unveil existing information. Rather, it must create novel information from scratch. Clearly, Dawkins's algorithm does nothing of the sort.

Ironically, though Dawkins uses a targeted search to illustrate the power of the Darwinian mechanism, he denies that this mechanism, as it operates in biological evolution (and thus outside a computer simulation), constitutes a targeted search. Thus, after giving his METHINKS•IT•IS•LIKE•A•WEASEL illustration, he immediately adds: "Life isn't like that. Evolution has no long-term goal. There is no long-distant target, no final perfection to serve as a criterion for selection."[24] Dawkins here fails to distinguish two equally valid and relevant ways of understanding targets: (i) targets as humanly constructed patterns that we arbitrarily impose on things in light of our needs and interests, and (ii) targets as patterns that exist independently of us and therefore regardless of our needs and interests. In other words, targets can be extrinsic (i.e., imposed on things from outside) or intrinsic (i.e., inherent in things as such).

In the field of evolutionary computing (to which Dawkins's METHINKS•IT•IS•LIKE•A•WEASEL example belongs), targets are given extrinsically by programmers who attempt to solve problems of their choice and preference. Yet in biology, living forms have come about without our choice or preference. No human has imposed biological targets on nature. But the fact that things can be alive and functional in only certain ways and not in others indicates that nature sets her own targets. The targets of biology, we might say, are "natural kinds" (to borrow a term from philosophy). There are only so many ways that matter can be configured to be alive and, once alive, only so many ways it can be configured to serve different biological functions. Most of the ways open to evolution (chemical as well as biological evolution) are dead ends. Evolution may therefore be characterized as the search for alternative "live ends." In other words, viability and functionality, by facilitating survival and reproduction, set the targets of evolutionary biology. Evolution, despite Dawkins's denials, is therefore a targeted search after all.

4. COMPUTATIONAL VS. BIOLOGICAL EVOLUTION

In the known physical universe, the number of elements that can be sampled (or queried) from a search space is always strictly limited. At the time of this writing, the fastest computer is the $133-million IBM Roadrunner supercomputer at Los Alamos that operates at 1.059 petaflops (i.e., just over a thousand trillion floating point operations per second).[25] If we imagine each floating point operation as able to take a sample of size 1, then this computer, even when run over the duration of the physical universe (circa 12 billion years), would be able to sample at most $m = 10^{34}$ elements from the search space. It is estimated that the total number of organisms, both single-celled and multicelled, that have existed on the Earth over its duration (circa 4.5 billion years) is $m = 10^{40}$.[26] Thus, it would take a million Roadrunner supercomputers running the duration of the universe to sample as many "life events" as have occurred on the Earth.

Throughout this chapter, we treat m as the upper limit on the number of elements that a given search can sample or query. Is there an upper limit to such upper limits? From examining the computational capacity of the universe at large, quantum computational theorist Seth Lloyd has shown that 10^{120} is the maximal number of bit operations that the known, observable universe could have performed throughout its entire multibillion-year history.[27] Thus, as the upper limit on the number of elements that a given search can sample, m cannot exceed 10^{120}, which sets an absolute limit on the sample size of any search.

Most search spaces that come up in the formation of biological complexity are far too large to be searched exhaustively. Take the search for a very modest protein, one that is, say, one hundred amino acids in length (most proteins are several hundreds of amino acids in length). The space of all possible protein sequences that are one hundred amino acids in length has size 20^{100}, or approximately 1.27×10^{130}, which exceeds Lloyd's limit. For this space, finding a particular protein via blind search corresponds to a 1 in 10^{130} improbability. Exhaustively or blindly searching a space this size to find a target this small is utterly beyond not only present computational capacities but also the computational capacities of the universe as we know it.

Biochemist Robert Sauer has used a technique known as *cassette mutagenesis* to determine how much variation proteins can tolerate in their amino acids without disrupting function. His results show that taking this variation into account raises the probability of forming a 100-subunit functional protein to 1 in 10^{65}. But given 10^{65} atoms in our galaxy, this probability is still vanishingly small. Add to this that most proteins are not 100 but 250 to 300 amino acids in length, and also that most proteins exist and operate in complexes requiring multiple proteins, and any prospect for blind search effectively exploring biological configuration space disappears.[28]

Fortunately for evolutionary theory, Darwinian processes operating by natural selection are available to take the place of blind search and, so the theory contends, able to overcome its limitations. Darwinian search is thus supposed to counteract the vast improbabilities that at first blush seem to undermine the formation of biological complexity. Yet evolution, even when conceived as a Darwinian search, seems a terribly inefficient search strategy. All the significant innovations in biological form and complexity attributable to evolution are supposed to have taken thousands or even millions of years. Direct experimental verification of the ability of biological evolution to produce large-scale organismal change therefore seems effectively impossible.

To accelerate the poky pace of biological evolution, early computer scientists recommended replacing the test-tube with the computer. Pioneers of evolutionary computing in the 1960s proposed that computer simulations could overcome the difficulty of demonstrating Darwinian evolution in the biology lab. "The Darwinian idea that evolution takes place by random hereditary changes and selection," wrote Nils Barricelli in 1962, "has from the beginning been handicapped by the fact that no proper test has been found to decide whether such evolution was possible and how it would develop under controlled conditions."[29]

Whereas biological evolution occurred in deep time and therefore could not be observed, computers could model evolutionary processes in real time and thus render their behavior observable and open to experimental control. As J. L. Crosby put it back in the mid 1960s, "In general, it is usually impossible or impracticable to test hypotheses about evolution in a particular species by the deliberate setting up of controlled experiments with living organisms of that species. We can attempt to partially get around this difficulty by constructing [computer] models representing the evolutionary system we wish to study, and use these to test at least the theoretical validity of our ideas."[30] Or as Heinz Pagels summarized the matter two decades later, "The only way to see evolution in action is to make computer models," because "in real time these changes take aeons, and experiment is impossible."[31]

In the last two decades, however, confidence that computational evolution elucidates biological evolution has waned. Why is that? The short answer is that programmers can cook computer simulations to achieve any evolutionary result they want. Take intelligent design critic Robert Pennock's work on the computer program AVIDA. AVIDA, written by Christoph Adami, uses selection to evolve certain types of logic functions, which are viewed as virtual organisms. From the performance of this program, Pennock infers that evolutionary processes operating in nature can produce complex biological functions.[32] Yet other computer programs, such as MESA, suggest that natural selection will have difficulty evolving features that need to form simultaneously for selective advantage. From MESA, Pennock might just as well have inferred that certain types of biological complexity (such as Michael Behe's irreducibly complex molecular machines) may be unevolvable by Darwinian means.[33] So which program gives the better insight into biological evolution, AVIDA, which seems to confirm it, or MESA, which seems to disconfirm it?

It's in large measure because computer programs can be manipulated to prove any evolutionary result one wants that ICAM was started. ICAM stands for the Institute for Complex Adaptive Matter.[34] Its mission is to understand how real-world material systems (as opposed to silicon-world virtual systems) become complex and adaptive. Talk to most working biologists, and they will tell you that computer simulations do not shed much light on actual biological evolution. Richard Lenski, Pennock's collaborator on AVIDA, can appreciate this point.

Lenski is best known not for his work on computer programs that simulate biological evolution, but rather for his work as a conventional biologist trying to evolve populations of bacteria in the lab. For many years, Lenski has cultured bacteria and placed them under selection pressure. He ran one experiment for 20,000 generations (if we think of the average generation time for humans as twenty years, then his experiment on bacteria corresponds to 400,000 years of human evolution, which is significant even on evolutionary time scales).[35] What did Lenski find in his experiments with these bacteria (i.e., with real, as opposed to virtual, organisms)? Did he find that his bacteria evolved novel irreducibly complex molecular machines of the sort that Michael Behe regards as indicating intelligent design?[36] Not at all. Lenski observed some small-scale changes, but nothing remarkable. And yet, when Lenski turned to computer simulations, he found that virtual organisms are much easier to evolve than real ones, requiring only a few hundred generations to form novel complex structures.[37]

Our ability to manipulate computer simulations of evolution has bred skepticism of the whole enterprise. Back around 1990, when artificial life was the most widely discussed form of computational evolution, John Maynard Smith called it "fact-free science."[38] For David Berlinski, that skepticism has since turned to cynicism:

> Computer simulations of Darwinian evolution fail when they are honest and succeed only when they are not. Thomas Ray has for years been conducting computer experiments in an artificial environment that he designated Tierra. Within this world, a shifting population of computer organisms meet, mate, mutate, and reproduce. Sandra Blakeslee, writing for the *New York Times,* reported the results under the headline "Computer 'Life Form' Mutates in an Evolution Experiment: Natural Selection is Found at Work in a Digital World." Natural selection found at work? I suppose so, for as Blakeslee observes with solemn incomprehension, "the creatures mutated but showed only modest increases in complexity." Which is to say, they showed nothing of interest at all. This is natural selection at work, but it is hardly work that has worked to intended effect.[39]

Berlinski raises here an important question: what does it mean for a computer simulation of Darwinian evolution to succeed? For proponents of Darwinian evolution, success means that the simulations produce the types of structures and exhibit the types of capacities that biological evolution is supposed to have displayed in the course of natural history. Christoph Adami's AVIDA, Thomas Ray's Tierra, and, Thomas Schneider's *ev* (Schneider's program attempts to model the evolution of nucleotide binding sites) are thus supposed to constitute successes.[40] Indeed, as proponents of Darwinian evolution, Adami, Ray, and Schneider have a stake in seeing their simulations confirm standard evolutionary theory.

But critics of Darwinian evolution also write computer simulations. We've mentioned MESA (by William Dembski, John Bracht, and Micah Sparacio), but there are also others, for instance, Mendel's Accountant (by John Sanford and John Baumgardner) as well as the tongue-in-cheek MutationWorks (posted anonymously online).[41] For such programs, success means showing that the simulation, despite efforts by the programmers to faithfully model biological evolution, produces no novel information or, worse yet, degrades existing information (as with Mendel's Accountant). Proponents of Darwinian evolution dismiss such simulations, claiming that their failure to evolve biologically relevant information merely reflects a failure of the programs to capture biological reality. Of course, critics of Darwinian evolution turn this criticism around, charging that only by cooking the books do Darwinists get their programs to produce biologically relevant information. Hence Berlinski's claim that these programs fail when they are honest.

Although it may seem as though we have reached an impasse, there is a way forward in this debate. Both sides have accepted a presupposition in common, namely, that it is possible to model Darwinian evolutionary processes mathematically. Indeed, it had better be possible to model evolution mathematically if it is to constitute an exact science.[42] Yet the common presupposition that Darwinian processes can be modeled mathematically raises a logically prior question. We have considered whether particular mathematical models of Darwinian evolution can produce the sort of information we find in biological systems. Some appear capable of producing it, others not. The resulting debate centers on whether these models were tendentiously manipulated to achieve a desired result. But the logically prior question is whether, and in what sense, mathematical models of Darwinian processes allow for the production of biologically relevant information at all. The remainder of this paper takes up this question.

5. Active Information

Kenneth Miller, in reviewing Thomas Schneider's work on the computer simulation *ev*, attempts to account for the apparent increase in information that results from that algorithm. "What's needed to drive this increase?" he asks. "Just three things: selection, replication, and mutation." He continues,

> Where's the new information coming from? Perhaps the investigator is sneaking it into the system at the start? No chance of that, since the starting sequences are completely randomized. Maybe there's hidden information in the program itself? Not likely. Schneider has made the source code of his program open for inspection, and there isn't even a hint of such nonsense. Did Schneider rig the parameters of the program to get the results he wanted? Not at all. In fact, changing the program in just about any way still results in an increase in measurable information, so long as we keep those three elements—selection,

replication, and mutation—intact. Where the information "comes from" is, in fact, from the selective process itself.[43]

This passage is remarkable for conceding what it seems to deny. The details of Schneider's simulation are not important here. We have argued elsewhere that *ev* is not nearly as free of investigator interference as Miller (or Schneider) makes out.[44] But let's grant, for the sake of argument, that Miller is right in denying investigator interference with the operation of the program. His claim that the information comes from the selective process is then correct but, in context, misleading. Miller suggests that *ev,* and evolution in general, outputs more information than it inputs. In fact, selective processes input as much information from the start as they output at the end. In Schneider's *ev,* for instance, the selective process inputted prior information in the form of a precisely specified error-counting function that served as a fitness measure.[45] Thus, instead of producing information in the sense of generating it from scratch, evolutionary processes produce it in the much weaker sense of merely shuffling around pre-existing information.

The view that algorithms cannot create information but only shuffle it around is long-standing and well-established. Over fifty years ago, Leon Brillouin, a pioneer in information theory, made that very point: "The [computing] machine does not create any new information, but it performs a very valuable transformation of known information."[46] Nobel prize-winning biologist Peter Medawar made the same point in the 1980s: "No process of logical reasoning—no mere act of mind or computer-programmable operation—can enlarge the information content of the axioms and premises or observation statements from which it proceeds."[47]

To see that Darwinian processes produce information only in the weaker sense of shuffling pre-existing information rather than in the stronger sense of generating it from scratch, we need some way of measuring the information outputted by evolutionary processes. This we have provided in a series of technical papers in the computational intelligence literature.[48] Yet the basic idea is straightforward. Consider again Dawkins's WEASEL. What allowed his evolutionary algorithm to converge so quickly on the target phrase METHINKS•IT•IS•LIKE•A•WEASEL is that a fitness function gauging distance from that phrase was embedded in the algorithm (indeed, the very target phrase was itself stored in the algorithm). But in that case, fitness functions gauging distance from any other string of letters and spaces could just as well have been substituted for the one Dawkins used; and with those other fitness functions, the algorithm could have converged on any sequence whatsoever.

So the target sequence METHINKS•IT•IS•LIKE•A•WEASEL initially had very small probability p (roughly 1 in 10^{40}) of arising by pure chance from a single query; and it has probability q (close to 1) of arising from Dawkins's evolutionary algorithm in a few dozen queries. But that algorithm requires a precisely specified fitness function that gauges distance from a target sequence, and such a fitness function can be built on any sequence of twenty-eight letters and spaces (and not just on METHINKS•IT•IS•LIKE•A•WEASEL). So how many such fitness functions exist? Roughly 10^{40}. And what's the probability of finding Dawkins's fitness function (which gauges distance from METHINKS•IT•IS•LIKE•A•WEASEL) among all these other possible fitness functions? Roughly 1 in 10^{40}.

Indeed, without grammatico-semantic structures on this space of fitness functions (and if such structures exist in the background and constrain the choice of fitness functions, where did that information come from?), the probability distribution on this space is uniform and the probability itself is p, or roughly 1 in 10^{40}. So the gain in probability associated with Dawkins's evolutionary algorithm readily finding the target sequence (i.e., q in place of p) is offset by the improbability of finding the fitness function that gauges distance from that sequence (i.e., p). Dawkins's algorithm, far from explaining how METHINKS•IT•IS•LIKE•A•WEASEL could be produced with high probability, simply raises the new problem of how one overcomes the low probability of finding the right fitness function for his algorithm. Dawkins has thus filled one hole by digging another.[49]

Simulations such as Dawkins's WEASEL, Adami's AVIDA, Ray's Tierra, and Schneider's *ev* appear to support Darwinian evolution, but only for lack of clear accounting practices that track the information smuggled into them. These programs capitalize on ignorance of how information works. The information hidden in them can be uncovered through a quantity we call *active information*. Active information is to informational accounting what the balance sheet is to financial accounting. Just as the balance sheet keeps track of credits and debits, so active information keeps track of inputs and outputs of information, making sure that they receive their proper due. Information does not magically materialize. It can be created by intelligence or it can be shunted around by natural forces. But natural forces, and Darwinian processes in particular, do not create information. Active information enables us to see why this is the case.

Active information tracks the difference in information between a baseline blind search, which we call the *null search,* and a search that does better at finding the target, which we call the *alternative search*. Consider therefore a search for a target T in a search space Ω (assume for simplicity that Ω is finite). The search for T begins without any special structural knowledge about the search space that could facilitate locating T. Bernoulli's principle of insufficient reason therefore applies, and we are in our epistemic rights to assume that the probability distribution on Ω is uniform, with probability of T equal to $p = |T|/|\Omega|$, where $|*|$ is the cardinality of $*$.[50] We assume that p is so small that a blind or null search over Ω for T (i.e., a search for T by uniform random sampling of Ω) is extremely unlikely to succeed. Success demands that in place of a blind search, an alternative search S be implemented that succeeds with a probability q that is considerably larger than p.

Whereas p gauges the inherent difficulty of locating the target T via a blind search, q gauges the difficulty of locating T via the alternative search S. The question then naturally arises how the blind or null search that locates T with probability p gave way to the alternative search S that locates T with probability q. In WEASEL, for instance, Dawkins starts with a blind search whose probability of success in one query is roughly 1 in 10^{40}. This is p. He then implements an alternative search (his evolutionary algorithm) whose probability of success in a few dozen queries is close to 1. This is q.

Dawkins leaves the discussion hanging, as though having furnished an evolutionary algorithm that locates the target phrase with high probability (which we are calling S), he has demonstrated the power of Darwinian processes. But in fact all he has done is shifted the problem of locating the target elsewhere, for as we showed earlier in this section, the fitness function he used for his evolutionary algorithm had to be carefully chosen and constituted 1 of 10^{40} (i.e., p) such possible fitness functions. Thus, in furnishing an alternative search whose probability of success is q, he incurred a probability cost p of finding the right fitness function, which coincides (not coincidentally) with the original improbability of the null search finding the target. The information problem that Dawkins purported to solve is therefore left completely unresolved!

In such discussions, it helps to transform probabilities to information measures (note that all logarithms in the sequel are to the base 2). We therefore define the *endogenous information* I_Ω as $-\log(p)$, which measures the inherent difficulty of a blind or null search in exploring the underlying search space Ω to locate the target T. We then define the *exogenous information* I_S as $-\log(q)$, which measures the difficulty of the alternative search S in locating the target T. And finally we define the *active information* I_+ as the difference between the endogenous and exogenous information: $I_+ = I_\Omega - I_S = \log(q/p)$. Active information therefore measures the information that must be *added* (hence the plus sign in I_+) on top of a null search to raise an alternative search's probability of success by a factor of q/p.

Dawkins's WEASEL, Adami's AVIDA, Ray's Tierra, and Schneider's *ev* are alternative searches. As such, they improve on a null search by increasing the probability of successfully locating targets. In information-theoretic terms, these simulations replace I_Ω with I_S. The endogenous information I_Ω is large, indicating the extreme difficulty of finding the target with the blind or null search. The exogenous information I_S, by contrast, is much smaller (possibly zero), indicating the relative ease of finding the target with an alternative search S. In replacing I_Ω with I_S, these simulations fail to account for the difference in these quantities. In other words, they fail to account for the active information I_+.

6. Three Conservation of Information Theorems

Active information, though easy to define mathematically, captures a profound truth. We can begin to see this with the following example. Consider an extremely large space Ω that we must search in order to find an extremely tiny target T. In other words, this is a classic needle-in-the-haystack problem. The probability of T with respect to the null search is therefore extremely small, call it p. We might picture Ω as all the dry land on Earth and T as a treasure chest buried two feet underground. Next, consider an alternative search S for T conducted in the subspace Ω' of Ω (i.e., $T \subset \Omega' \subset \Omega$) for which the probability of successfully searching for T within Ω' is q (which we assume to be much larger than p). Thus, we might picture Ω' as some small Pacific island (say Bora Bora) on which the treasure T is buried. In this case, $I_\Omega = -\log(p) = -\log \mathbf{P}(T|\Omega)$ and $I_S = -\log(q) = -\log \mathbf{P}(T|\Omega')$. (Note that the conditional probability $\mathbf{P}(T|\Omega)$ is by definition $\mathbf{P}(T \cap \Omega)/\mathbf{P}(\Omega)$, which, since T is contained in Ω and $\mathbf{P}(\Omega) = 1$, is $\mathbf{P}(T) = p$; likewise, because T is contained in Ω', $\mathbf{P}(T|\Omega') = \mathbf{P}(T \cap \Omega')/\mathbf{P}(\Omega') = \mathbf{P}(T)/\mathbf{P}(\Omega') = q$.)

The search has now become much easier, reduced from all the dry land on Earth to Bora Bora. But what enabled the search to become easier? Simply listing the endogenous and exogenous information leaves a crucial question unexplained. Indeed, it is not enough to know that the null search has the very high difficulty level $I_\Omega = -\log(p)$, but that by choosing an appropriate subspace we switch to an alternative search S whose difficulty level $I_S = -\log(q)$ is much lower. The key question that needs to be answered is how we knew to switch the search for T from Ω to the subspace Ω'. In other words, how did we know that of all places on Earth where the treasure might be hidden, we needed to look on Bora Bora?

Within the larger space Ω, the subspace Ω' has probability $\mathbf{P}(\Omega'|\Omega) = p/q$ (this follows from $T \subset \Omega' \subset \Omega$, $\mathbf{P}(T|\Omega) = p$, and $\mathbf{P}(T|\Omega') = q$). So the information associated with this subspace is $-\log \mathbf{P}(\Omega'|\Omega) = \log(q/p) = I_+$. Accordingly, when an alternative search improves performance over a null search by reducing the original search space, that improved performance must be paid for by the active information associated with knowing which reduction of the original space (i.e., which subspace) contains the target and therefore should be chosen. Indeed, what prompted us to come up with Ω' in the first place, and how did we know that it contains T? Active information assigns a cost to any knowledge that enables us to answer this question.

The preceding example is a special case of the following theorem.

Conservation of Information Theorem (function-theoretic version). Let T be a target in Ω. Assume that Ω is finite and nonempty, and that $p = |T|/|\Omega|$ (which we take to be extremely small). The endogenous information is therefore $I_\Omega = -\log(p)$. Next, let Ω' be another nonempty finite space, φ be a function that maps Ω' to Ω, and $T' = \{y \in \Omega' \mid \varphi(y) \in T\}$. Or, in standard set-theoretic notation, $T' = \varphi^{-1}(T)$. Define $q = |T'|/|\Omega'|$ (which we take to be considerably bigger than p). Given a null search for T' in Ω', φ induces an alternative search S for T in Ω. The exogenous information is therefore $I_S = -\log(q)$. Next, define $\mathcal{F}$ as the set of all functions from Ω' to Ω and $\mathcal{T}$ as the set of all functions ψ in $\mathcal{F}$ such that $|\psi^{-1}(T)|/|\Omega'| \geq q$ (i.e., each such ψ maps at least as many elements of Ω' to T as φ). Then $|\mathcal{T}|/|\mathcal{F}| \leq p/q$, or equivalently the (higher-order) endogenous information associated with finding $\mathcal{T}$ in F, i.e.,$-\log(|\mathcal{T}|/|\mathcal{F}|)$, is bounded below by the (lower-order) active information $I_+ = \log(q/p)$.

Remarks. To see that this theorem includes the preceding example as a special case, let Ω' be a subset of Ω and let φ be the embedding function that takes each element in the subset to itself in the superset.

This theorem illustrates that the inherent difficulty of a search never goes away. The original (null) search on Ω for T is difficult, characterized by the small probability p of finding the target. We might then imagine that if only we could conduct an easier null search on a space Ω' for a target T' (the probability of success being $q \gg p$) and if only there were some straightforward way to translate target elements of this alternative search to target elements of the original search, the difficulty of the original search would dissolve. But the translation scheme that connects the two searches (in this case, the function φ) resides in its own higher-order search space (the space of functions between the two search spaces), which includes lots of other translation schemes. And how do we search among them?

According to this theorem, finding a translation scheme that maintains the same level of performance as the function φ requires an expenditure of information at least that of the active information. The null search has an inherent difficulty $I_\Omega = -\log(p)$. The alternative search, by utilizing φ, has a significantly reduced difficulty $I_S = -\log(q)$. But finding the φ which enables this reduced difficulty has itself difficulty no less than $I_\Omega - I_S = I_+ = \log(q/p)$. So constructing the alternative search does nothing to make the original problem easier and quite possibly makes matters worse. It's as though one can never fool the original search.

The significance of this theorem, and of conservation of information theorems generally, is that they track the information that was externally applied to augment the probability of successfully searching for T and show that this information is bounded below by the active information. In other words, these theorems show that the improved ease of search, as represented by I_S superseding I_Ω, can only be purchased, in informational terms, at the cost of $I_+ = I_\Omega - I_S$. Conservation of information therefore shows that any improvement in the effectiveness of an alternative search over a null search is not a free lunch. Moreover, payment can never dip below the active information. In fact, *active information represents the optimal price for improving a search.*

Finally, a technical remark needs to be made about how we are representing searches. Right now, given our statement of this theorem, it appears that any search consists of exactly one query. But, in general, a search consists of up to m queries, where m is the maximum number of queries that are practicable (recall section 4). Yet, by redefining the search so that its space consists of the m-fold Cartesian product of the original search space and by redefining the target as the set of all m-tuples from this m-fold product for which at least one coordinate is in the original target, it's possible to reformulate any m-query search so that it is mathematically equivalent to a single-query search on this Cartesian product. In consequence, single-query searches such as appear in this theorem entail no loss of generality.

Proof. Let $\Omega = \{x_1, x_2, \ldots, x_K, x_{K+1}, \ldots, x_M\}$ so that $T = \{x_1, x_2, \ldots, x_K\}$, and let $\Omega' = \{y_1, y_2, \ldots, y_L, y_{L+1}, \ldots, y_N\}$ so that $T' = \{y_1, y_2, \ldots, y_L\}$. Then $p = K/M$ and $q = L/N$ and $|\mathcal{F}| = M^N$. From the binomial theorem it then follows that the number of functions in $\mathcal{F}$ that map L elements of Ω' into T and the remaining elements of Ω' into $\Omega \setminus T$ is

$$\binom{N}{L} K^L (M-K)^{N-L}.$$

From this it follows in turn that the number of functions in $\mathcal{F}$ that map L or more elements of Ω' into T and the remaining elements of Ω' into $\Omega \setminus T$ is

$$\sum_{i \geq L} \binom{N}{i} K^i (M-K)^{N-i}.$$

Now, if we divide this number by the total number of elements in $\mathcal{F}$, i.e., by M^N, we get

$$\sum_{i \geq L} \binom{N}{i} p^i (1-p)^{N-i},$$

which is a cumulative distribution for a binomial random variable with parameters N and p. It is also the probability of T. Since the mean for such a random variable is Np and since $q = L/N$, it follows that

$$\begin{aligned}
\sum_{i \geq L} \binom{N}{i} p^i (1-p)^{N-i} &= \frac{L}{L} \sum_{i \geq L} \binom{N}{i} p^i (1-p)^{N-i} \\
&= \frac{1}{L} \sum_{i \geq L} L \binom{N}{i} p^i (1-p)^{N-i} \\
&\leq \frac{1}{L} \sum_{i \geq 0} i \binom{N}{i} p^i (1-p)^{N-i} \\
&= \frac{Np}{L} \\
&= \frac{p}{q}.
\end{aligned}$$

It follows that $-\log(|T|/|\mathcal{F}|)$ is bounded below by the active information $I_+ = \log(q/p)$. This proves the theorem. $\square$

The conservation of information theorem just proved is perhaps the most basic of all the conservation of information theorems. It shows how, in constructing an alternative search that improves on the average performance over a null search, one must pay for the improvement with an amount not below the active information. All conservation of information theorems take this form. Still, one might wonder whether less expensive ways exist for improving search performance, ways that circumvent conservation of information.

Consider, for instance, Steven Pinker's proposal that the mind is a coordinated assembly of computational modules.[51] We might therefore represent the activity of the mind in performing an information-producing act as a search through a Cartesian product of several (finite) search spaces: $\Omega_1 \times \Omega_2 \times \cdots \times \Omega_n$, which we denote by Ω. The target then, let us say, consists of some set $T = T_1 \times T_2 \times \cdots \times T_n$ where each T_i is a nonempty subset of Ω_i. We assume that T has extremely small uniform probability in Ω, which we denote by p. This is the probability of a successful null search. Note that the uniform probability on the Cartesian product is the product of the uniform probabilities, and therefore that the uniform probability of T is the product of the uniform probabilities of the T_is.

Now Pinker, as a materialist and computational reductionist, would refuse to explain the mind's success in locating T as some fluke that managed, by sheer dumb luck, to find a needle in the haystack. The chance of the gaps is as scientifically unacceptable as the god of the gaps. Instead, he would want to attribute any such success to the coordination of computational modules where the computational modules and their coordination are the result of a Darwinian evolutionary process. Think of each of these modules, denoted by M_i, as working on its respective space Ω_i to find a target element in T_i. In other words, each M_i delivers an element t_i in T_i and together, as a result of their coordination, these modules induce an alternative search S that delivers the n-tuple $(t_1, t_2, \ldots, t_n)$ in the target T. In this way, computational modules representing basic mental functions can be seen to successfully search Ω for T. Pinker's modular theory of mind would therefore seem well on the road to vindication.

In fact, such an approach raises far more difficulties than it resolves. What, exactly, are these modules and how, specifically, are they coordinated? Pinker never says. But even if he did or could say, the information problem that these modules are supposed to resolve remains as unresolved as ever. To see this, consider that the success of these modules in locating the target depends on their increasing the probability of success well beyond the minuscule probability p for success of the null search, characterized by a uniform probability **U** on Ω. Pinker's modules, therefore, induce an alternative search S whose probability of success, call it q, is much bigger than p. Thus, his modules replace the uniform probability **U** with a new probability distribution on Ω, call it μ, that assigns probability q to T.

But where did this μ come from? Did it just magically materialize? No, it resides in the space of probability measures on Ω, and in that space it has to be found and identified. But how probable is it that we can find, in this space of probability measures, a probability measure (more often called a "probability distribution") at least as effective as μ at locating T? As the next theorem demonstrates, the (higher-order) probability of finding a probability distribution ν that's at least as effective as μ at locating T is less than or equal to p/q.

Conservation of Information Theorem (measure-theoretic version). Let T be a target in Ω. Assume Ω is finite and T is nonempty. Let **U** denote the uniform probability distribution on Ω and let $p = |T|/|\Omega| = \mathbf{U}(T)$ (which we take to be extremely small). The endogenous information is therefore $I_\Omega = -\log(p)$. Next, let μ be a probability distribution on Ω such that $q = \mu(T)$ (which we take to be considerably bigger than p). Suppose that μ characterizes the probabilistic behavior of an alternative search S. The exogenous information is therefore $I_S = -\log(q)$.

Next, let $\mathcal{M}$ denote the set of all probability distributions on Ω and $\mathcal{T}$ be the set of probability distributions ν in $\mathcal{M}$ such that $\nu(T) \geq q$ (i.e., each such ν assigns at least as much probability to T as μ—each such ν therefore represents a search that's at least as effective at locating T as μ). Then the (higher-order) uniform probability of $\mathcal{T}$ in $\mathcal{M}$, which may be denoted by $\mathbf{U}^*(\mathcal{T})$, is less than or equal to p/q. Equivalently, the (higher-order) endogenous information associated with finding $\mathcal{T}$ in $\mathcal{M}$, i.e., $-\log(\mathbf{U}^*(\mathcal{T}))$, is bounded below by the (lower-order) active information $I_+ = -\log(\mathbf{U}(T)) + \log(\mu(T)) = \log(q/p)$.

Proof. Let $\Omega = \{x_1, x_2, \ldots, x_K, x_{K+1}, \ldots, x_N\}$ so that $T = \{x_1, x_2, \ldots, x_K\}$. Then $p = K/N$. Next, given that μ is a probability distribution on Ω, it follows that μ has the form

$$\sum_{i=1}^{N} a_i \delta_{x_i},$$

where each a_i is nonnegative, the a_is together sum to 1, and each δ is a point mass (assigning probability 1 to the corresponding x_i). Indeed, each element of $\mathcal{M}$ has this form. It follows that $\mathcal{M}$ has the geometric structure of an $(N-1)$-dimensional simplex consisting of all convex combinations of N nonnegative real numbers. Moreover, its uniform probability is given by a normalized Lebesgue measure.

Since $\mu(T) = q$, it follows that

$$\sum_{i=1}^{K} a_i = q.$$

Moreover, any distribution ν in $\mathcal{T}$ of the form

$$\sum_{i=1}^{N} b_i \delta_{x_i}$$

satisfies

$$\sum_{i=1}^{K} b_i \geq q.$$

From these facts it now follows that the uniform probability $\mathbf{U}^*$ of $\mathcal{T}$ in $\mathcal{M}$ is given by the following expression:[52]

$$\frac{\Gamma(N)}{\Gamma(N(1-p))\Gamma(Np)} \int_0^{1-q} t^{N(1-p)-1}(1-t)^{Np-1} dt.$$

This last expression describes a cumulative beta distribution with first parameter $r = N(1-p)$ and second parameter $s = Np$. Integration by substitution shows that this expression can be rewritten as

$$\frac{\Gamma(N)}{\Gamma(Np)\Gamma(N(1-p))} \int_q^1 t^{Np-1}(1-t)^{N(1-p)-1} dt,$$

which describes a cumulative beta distribution with first parameter $r = Np$ and second parameter $s = N(1-p)$. It is well known that the mean for this distribution is $r/(r+s)$.[53] In consequence,

$$\begin{aligned}
\frac{\Gamma(N)}{\Gamma(Np)\Gamma(N(1-p))}\int_q^1 t^{Np-1}(1-t)^{N(1-p)-1}dt &= \frac{\Gamma(N)}{\Gamma(Np)\Gamma(N(1-p))}\int_q^1 \frac{q}{q}\cdot t^{Np-1}(1-t)^{N(1-p)-1}dt \\
&= \frac{1}{q}\cdot\frac{\Gamma(N)}{\Gamma(Np)\Gamma(N(1-p))}\int_q^1 q\cdot t^{Np-1}(1-t)^{N(1-p)-1}dt \\
&\leq \frac{1}{q}\cdot\frac{\Gamma(N)}{\Gamma(Np)\Gamma(N(1-p))}\int_0^1 t\cdot t^{Np-1}(1-t)^{N(1-p)-1}dt \\
&= \frac{1}{q}\cdot\frac{Np}{Np + N(1-p)} \\
&= \frac{p}{q}.
\end{aligned}$$

It follows that $-\log(\mathbf{U}^*(\mathcal{T}))$ is bounded below by the active information $I_+ = \log(q/p)$. This proves the theorem. □

Conservation of information is also implicit in a family of mathematical results known as NFL theorems (i.e., no free lunch theorems). Think of an evolutionary search as treating fitness functions as a variable, with each fitness function providing a different representation of selective advantage for the environment. A typical NFL theorem then states that the average performance of an evolutionary search across fitness functions does not exceed blind search. Conservation of information's connection to NFL now becomes evident as soon as one inquires into what enables certain fitness functions to induce alternative (evolutionary) searches that are so much better than blind search, especially given that (by NFL) averaging across fitness functions does no better than blind search. According to conservation of information, active information is required to locate the fitness functions that render evolutionary search effective. The following theorem underscores the connection between conservation of information and no free lunch.

Conservation of Information Theorem (fitness-theoretic version). Let T be a target in Ω. Assume Ω is finite and T is nonempty. Let $\mathbf{U}$ denote the uniform probability distribution on Ω and let $p = |T|/|\Omega| = \mathbf{U}(T)$ (which we take to be extremely small). The endogenous information is therefore $I_\Omega = -\log(p)$. Next, let $\mathcal{F}$ denote a finite collection of fitness functions on Ω and let $\mathbf{S}_\Omega$ denote the symmetric group on Ω (i.e., all permutations of this set). Without loss of generality, assume that any f in $\mathcal{F}$ only takes values in $\{0, 1/M, 2/M, \ldots, (M-1)/M, 1\}$ for some large fixed M and that $\mathcal{F}$ includes all such f. $\mathcal{F}$ is therefore closed under the symmetric group $\mathbf{S}_\Omega$, i.e., for any f in $\mathcal{F}$ and any σ in $\mathbf{S}_\Omega$, $f\circ\sigma$ is also in $\mathcal{F}$. Suppose further that any f in $\mathcal{F}$ induces a probability distribution $\mathbf{P}_f$ on Ω (corresponding to an alternative search). Assume that each such $\mathbf{P}_f$ satisfies the following invariance property: for any σ in $\mathbf{S}_\Omega$ and $A \subset \Omega$, $\mathbf{P}_{f\circ\sigma}(\sigma^{-1}A) = \mathbf{P}_f(A)$. An NFL result then follows:

$$\frac{1}{|\mathcal{F}|}\sum_{f\in\mathcal{F}}\mathbf{P}_f(T) = \mathbf{U}(T).$$

Assume next that g in $\mathcal{F}$ is such that $q = \mathbf{P}_g(T)$ (which we take to be considerably bigger than p) and that g induces an alternative search S for which the exogenous information is $I_S = -\log(q)$. Let $\mathcal{T}$ denote the set of all fitness functions h in $\mathcal{F}$ such that $\mathbf{P}_h(T) \geq q$ (i.e., each such h induces a probability distribution that assigns at least as much probability to T as $\mathbf{P}_g$; equivalently, each such h induces an alternative [or evolutionary] search at least as effective at locating T as S). Then, the (higher-order) uniform probability of $\mathcal{T}$ in $\mathcal{F}$, i.e., $|\mathcal{T}|/|\mathcal{F}|$, which may be denoted by $\mathbf{U}^*(\mathcal{T})$, is less than or equal to p/q. Equivalently, the (higher-order) endogenous information associated with finding $\mathcal{T}$ in $\mathcal{F}$, i.e., $-\log(\mathbf{U}^*(\mathcal{T}))$, is bounded below by the (lower-order) active information $I_+ = -\log(\mathbf{U}(T)) + \log(\mathbf{P}_g(T)) = \log(q/p)$.

Remarks. This theorem attempts to characterize the informational properties of evolutionary algorithms that use fitness functions f in $\mathcal{F}$ to search for possible targets in Ω (the target of interest being T). The assumption that the probability distributions $\mathbf{P}_f$ induced by these fitness functions are invariant under a group action (in this case, the symmetric group $\mathbf{S}_\Omega$) is customary with such theorems: it ensures that fitness is not correlated with prior knowledge of the target. As Joseph Culberson puts it: "Evolutionary algorithms (EAs) are often touted as 'no prior knowledge' algorithms. This means that we expect EAs to perform without special information from the environment. Similar claims are often made for other adaptive algorithms."[54] If f and $f \circ \sigma$ were not equally effective at locating the targets T and $\sigma^{-1}T$ respectively, presumably special information from the environment (and thus prior knowledge not available to the fitness function f) would account for the difference.

This theorem includes a no free lunch result as well as a conservation of information result (whose proof depends on the no free lunch result). The formula

$$\frac{1}{|\mathcal{F}|}\sum_{f \in \mathcal{F}} \mathbf{P}_f(T) = \mathbf{U}(T)$$

expresses a classic instance of NFL: it shows that an evolutionary search, when averaged across fitness functions, does no better at locating a target T than blind or null search (in fact, it shows the two are identical). Note that because of the invariance property of the probability distributions $\mathbf{P}_f$ under the symmetric group $\mathbf{S}_\Omega$, any constant function c over Ω that's in $\mathcal{F}$ induces a probability distribution $\mathbf{P}_c$ that is identical to the uniform probability $\mathbf{U}$. This follows because $c \circ \sigma = c$ for all σ in $\mathbf{S}_\Omega$.

Proof. Let $\Omega = \{x_1, x_2, \ldots, x_K, x_{K+1}, \ldots, x_N\}$ so that $T = \{x_1, x_2, \ldots, x_K\}$. Then $p = K/N$. Without loss of generality, assume that N is divisible by K so that $K \times L = N$ for some L. If not, simply expand Ω with enough finite elements so that T has probability p and K does divide N. Next, find a set of elements $\sigma_1, \sigma_2, \ldots, \sigma_L$ in the symmetric group $\mathbf{S}_\Omega$ such that $T_i = \sigma_i^{-1}(T)$ and the T_is partition Ω with $T_1 = T$. In other words, the T_is are disjoint, include T, and their union equals all of Ω.

Because the probability distributions induced by the fitness functions f in $\mathcal{F}$ are invariant under the symmetric group $\mathbf{S}_\Omega$, it follows that for any σ in $\mathbf{S}_\Omega$,

$$\sum_{f \in \mathcal{F}} \mathbf{P}_{f \circ \sigma}(\sigma^{-1}T) = \sum_{f \in \mathcal{F}} \mathbf{P}_f(T).$$

But since as f runs through all the fitness functions in $\mathcal{F}$, $f \circ \sigma$ likewise runs through all the fitness functions in $\mathcal{F}$ (that's because $\mathcal{F}$ is closed under composition with elements from $\mathbf{S}_\Omega$), it follows that

$$\sum_{f \in \mathcal{F}} \mathbf{P}_{f \cdot \sigma}(\sigma^{-1}T) = \sum_{f \in \mathcal{F}} \mathbf{P}_f(\sigma^{-1}T).$$

In consequence, summing all the fitness-induced probability measures over $\mathcal{F}$ is also invariant under $\mathbf{S}_\Omega$. And this in turn implies that for each i ($1 \leq i \leq L$)

$$\frac{1}{|\mathcal{F}|}\sum_{f \in \mathcal{F}} \mathbf{P}_f(\sigma_i^{-1}T) = \frac{1}{|\mathcal{F}|}\sum_{f \in \mathcal{F}} \mathbf{P}_f(T_i) = \frac{1}{|\mathcal{F}|}\sum_{f \in \mathcal{F}} \mathbf{P}_f(T)$$

is constant. But since the T_is partition Ω with $T_1 = T$, it follows that

$$\sum_{i=1}^{L} \frac{1}{|\mathcal{F}|}\sum_{f \in \mathcal{F}} \mathbf{P}_f(T_i) = 1$$

and therefore that

$$\frac{1}{|\mathcal{F}|}\sum_{f \in \mathcal{F}} \mathbf{P}_f(T) = \frac{1}{L} = \frac{K}{K \cdot L} = \frac{K}{N} = p = \mathbf{U}(T).$$

This establishes the first main claim of the theorem.

As for $\mathbf{U}^*(\mathcal{T}) = |\mathcal{T}|/|\mathcal{F}|$ being bounded above by p/q, suppose, for the sake of a contradiction, that $|\mathcal{T}| > (p/q)|\mathcal{F}|$. That would mean that in the sum

$$\sum_{f \in \mathcal{F}} \mathbf{P}_f(T),$$

more than p/q of the summation elements are greater than or equal to $\mathbf{P}_g(T) = q$. But that in turn would mean that the normalized sum

$$\frac{1}{|\mathcal{F}|}\sum_{f \in \mathcal{F}} \mathbf{P}_f(T)$$

would have to be strictly greater than p. But we just established that it is equal to p.

From this contradiction it follows that $\mathbf{U}^*(\mathcal{T}) = |\mathcal{T}|/|\mathcal{F}|$ is indeed less than or equal to p/q and therefore that $-\log(|\mathcal{T}|/|\mathcal{F}|)$ is bounded below by the active information $I_+ = \log(q/p)$. This proves the theorem. □

This fitness-theoretic conservation of information theorem is more significant than it might at first appear. One might think that its applicability is limited because it was formulated in such general and seemingly unrealistic terms. What search space, for instance, allows for all possible permutations? Most don't. Yet, insofar as they don't, it's because they exhibit structures that constrain the permissible permutations. Such constraints, however, bespeak the addition of active information. Consider, for instance, that most evolutionary algorithms are used to search, not a completely unstructured space Ω, but an m-fold Cartesian product space Ω^m, each factor Ω' signifying a single query in an m-query search.

In that case, permissible permutations must not scramble query-order or vary from one factor to the next but rather must act the same way on Ω' across factors (if a permutation moves a point from one place to another in one factor, then it must do the same in the other factors). Thus, in place of the full symmetric group on Ω'^m, whose cardinality is the factorial of $|\Omega'|^m$ (i.e., $|\Omega'|^m!$), the relevant group action would be the symmetric group on Ω', whose cardinality is the factorial of $|\Omega'|$ (i.e., $|\Omega'|!$). But in resulting ratio, $|\Omega'|!/|\Omega'|^m!$, the denominator completely overwhelms the numerator. It therefore represents a huge input of active information; indeed, unless these spaces are very, very small, this ratio will be much less than p/q.

In a similar vein, one might want to constrain the fitness functions. Thus, one might think that fitness ought to vary gradually with some underlying metric structure on the search space. But where does such a metric structure come from? And how much does it reduce the full space of fitness functions $\mathcal{F}$? If it reduced $\mathcal{F}$ to some smaller space of fitness functions $\mathcal{F}'$, then $|\mathcal{F}'|/|\mathcal{F}|$ represents a further addition of active information, as does the metric structure on the underlying search space Ω (how many other metric or topological structures were possible and what led to this one taking effect rather than the others?).

Christian Igel and Marc Toussaint have, for instance, argued that NFL theorems are unrealistic because they focus on fitness functions closed under permutation. They suggest that in realistic problems the focus should instead be on classes of fitness functions that are not closed under permutation.[55] All such focusing and constraining, however, imparts active information. Moreover, once such constraints on permissible fitness functions or permissible permutations are fleshed out to the point where we can calculate how much active information was imparted, we find invariably that conservation of information is preserved.

Indeed, all the conservation theorems listed in this section (and those we know of otherwise) give active information as the extreme lower bound on the amount of information that must be imparted to an alternative search to improve it by a given amount over a null search. Take the measure-theoretic conservation of information theorem proved earlier in this section. We have proved a more precise result showing that in the search for probability distributions that represent successful (lower-order) searches, the information cost need not simply be bounded below by the active information $I_+ = \log(q/p)$ (as we showed in the theorem proved in this section) but can grow exponentially in I_+, indicating runaway informational inflation as one attempts to account for the information required to render an alternative search successful.[56]

7. The Law of Conservation of Information

Laws of science are supposed to be universal in scope, hold with unfailing regularity, and find support from a wide array of facts and observations. We submit that conservation of information is such a law. It may be formulated as follows:

> **The Law of Conservation of Information (LCI).** Any search that proportionately raises the probability of locating a target by q/p with respect to blind search requires in its formation an amount of information not less than the active information $I_+ = \log(q/p)$.

In short, raise the probability of successful search by a factor of q/p, incur an information cost of $\log(q/p)$. The rest of this section consists of bold-titled points devoted to elucidating this law.

General Setup. The general setup for LCI is as follows. A null search B, which is blind, sets a probabilistic baseline p for searching the space Ω for a target T. Think of B as an Ω-valued random variable that induces a uniform probability distribution on Ω. Regardless of what we know or don't know about T, we can always implement B and therefore do at least as good as B in searching Ω for T. The question is how much better can we do than B. In practice, the probability p of B locating T is so small that B stands no reasonable chance of success. We therefore look to an alternative search S whose probability q of locating T is significantly larger than p. But where did S come from? S did not magically materialize. Rather, it had to be brought into existence through some formative process. LCI states that the formation of S required an investment of information not less than $I_+ = \log(q/p)$.

A Family of Theorems. LCI receives support from conservation theorems of the sort proved in the last section. Such theorems confirm that the information needed to improve a search so that its probability of successfully locating a target increases by a factor of q/p is at least $I_+ = \log(q/p)$. Even so, LCI is not itself a mathematical theorem. It says that in any circumstance where a null search gives way to an improved alternative search, at least I_+ needs to be imparted to the alternative search. But the specific forms by which null and alternative searches can be instantiated is so endlessly varied that no single mathematical theorem can cover all contingencies.

That's why, in the previous section, we offered three substantially different types of conservation of information theorems. Instead of LCI falling under one general mathematical theorem, LCI says that for any mathematically specific situation in which a blind search gives way to an improved alternative search, a mathematical theorem exists showing that the alternative search required at least I_+ to be formed. Thus, instead of LCI constituting a theorem, it characterizes situations in which we may legitimately expect to prove a conservation of information theorem. LCI might therefore be viewed as a family of theorems sharing certain common features.

No Free Lunch Theorems. In the 1990s, William Macready and David Wolpert proved several theorems to which they gave the descriptive tag "No Free Lunch" or "NFL."[57] These theorems showed how evolutionary searches, when averaged across fitness functions, did no better at locating targets than blind (or null) search. Their work had precedent. Cullen Schaffer's Law of Conservation of Generalization Performance, proved slightly earlier, compared a learner who can consistently achieve "at least mildly better-than-chance performance" to "a perpetual motion machine."[58] Schaffer's point was that, just as perpetual motion machines constitute a thermodynamic impossibility, a learner who consistently outperforms chance (even if only mildly so) constitutes an algorithmic impossibility. He elaborated,

> An essential property of the problem of inductive generalization is that it admits no general solution. An algorithm that is good for learning certain sets of concepts must necessarily be bad for learning others. Moreover, no algorithm strictly dominates any other. If two learners differ in generalization performance, there must be problems for which each is superior to the other. As a consequence, every algorithm is appropriate in some sphere of application and each is thus, in a sense, worthy of study.[59]

After Macready and Wolpert's initial work in this area, other researchers proved additional NFL theorems. Yu-Chi Ho and David Pepyne offered a simplified approach to these theorems.[60] Thomas English formulated an information-theoretic NFL theorem, which he described explicitly in terms of "conservation of information," calling it a "law" as well as a "principle" and relating it to the work of Schaffer. English's approach underscored the futility of attempting to design a generally superior optimizer without problem-specific information about a search.

Work on NFL theorems, in focusing on average performance of search algorithms, leads to a natural follow-up question, namely, what are the informational costs associated with finding algorithms that in practice are better than others? NFL is a great leveler of search algorithms. It is counterintuitive because we know that some search algorithms are much better than others *at specific tasks*. The Law of Conservation of Information, as we develop it here, presupposes the NFL theorems and then quantifies the informational costs that make some algorithms better than others at performing specific tasks. According to LCI, improved performance of an algorithm at performing a specific task is purchased at the cost of active information. Simply put, NFL says there is a cost for effective search; LCI calculates the cost.

Church-Turing Analogue. The Law of Conservation of Information, in both scope and function, parallels the Church-Turing Thesis of theoretical computer science.[61] The Church-Turing Thesis is a deep claim in mathematical logic about the nature of computation. LCI functions within the theory of intelligent design as the Church-Turing Thesis functions within the theory of computation. According to this thesis, given a procedure that is informally computable (i.e., that is finitely specifiable and rule-based), it can be coded as an algorithm running on a Turing machine. According to LCI, given an alternative search that does better than blind search, the alternative search was purchased at an information cost no less than I_+. The task of the design theorist in that case is to "follow the information trail" and show where the information that this search outputs in locating a target was first inputted (much as the task of the computer scientist is to show how some procedure that is informally computable can be explicitly formulated as an algorithm capable of being run on a Turing machine).

It follows that there is no—and indeed can be no—strict mathematical proof of either the Church-Turing Thesis or LCI. Nonetheless, the two are subject to independent verification. With the Church-Turing Thesis, the challenge is to show that any operation that is informally computable is also, upon a close mathematical analysis, formally computable (to date, this has always been confirmed). Likewise with LCI, the challenge is to find where the information that improves an alternative search over a blind search was inputted and then to calculate the amount of this information, showing that it does not fall below $I_+ = \log(q/p)$ (to date, this has also always been confirmed).

Probability Distributions vs. Probabilistic Outcomes. Much of the power of LCI comes from its focus not on particular probabilistic outcomes but on probability distributions. LCI is not saying that certain unlikely outcomes or events do not, or cannot, happen. Rather, it is saying that certain types of probability distributions cannot obtain without the addition of information. LCI begins with the observation that different probability distributions are associated with different searches (distributions which in many instances can be given precise mathematical characterizations based on theoretical or empirical considerations). These distributions can assign very different numbers to the probability of successfully concluding a search. According to LCI, any such differences between an alternative and a null search must be accounted for in terms of the information put into forming the alternative search.

Not Computational (or Kolmogorov) Complexity. Given an alternative search that improves on a null search, we might think that simply by inspecting a representation of the alternative search, it would be possible to calculate the information cost that was incurred in its formation. And we would be wrong. When a search is represented computationally, the complexity of its representation may be irrelevant to the information required for its formation. With computational representations of searches, their complexity often depends on idiosyncrasies of the underlying computational environment. Imagine searches whose targets are sequences 1,000 bits in length. These searches might use fitness functions that gauge distance from the target by counting the number of bits that fail to match up (i.e., they might use the Hamming distance).

Now imagine two possible searches in this space of 1,000-bit sequences. In the first, the fitness function gauges distance from a 1,000-bit sequence of high Kolmogorov complexity (i.e., the sequence is highly incompressible computationally, as we would expect with a random sequence of coin tosses). In the other, the fitness function gauges distance from a 1,000-bit sequence of low Kolmogorov complexity (i.e., the sequence is highly compressible, for instance, a sequence of 1,000 zeros). The first fitness function, and thus the search associated with it, will require considerable complexity for its representation. The second one, by contrast, will be much simpler to represent (the main evaluation requiring a line of code that says "add number of bits differing from 0"). And yet both searches, when identified with fitness functions, will require the same amount of information to be located and extracted from the underlying space of fitness functions (compare the fitness-theoretic conservation of information theorem of the last section).

The Search for a Search. In characterizing the information cost that alternative searches incur, LCI treats searches as residing in higher-order search spaces (i.e., spaces each of whose elements is a search). Notably, the information imparted to improve a search picks out one class of searches (those that with probability q or better locate T) to the exclusion of others (those that with probability less than q locate T). LCI says that searching the original search space Ω for a target T with probability q of success is never more difficult, and possibly much easier, than searching a higher-order search space for a search that, when applied to the lower-order search space Ω, finds T with that same probability.

To see what's at stake in such a "search for a search," imagine that you are on an island with buried treasure. The island is so large that a blind (null) search is highly unlikely to succeed in finding the treasure. Fortunately, you have a treasure map that will guide you to it. But where did you find the treasure map? Treasure maps reside in a library of possible treasure maps. The vast majority of these will not lead to the treasure. How, then, did you happen to find the right map among all these possible treasure maps? What special information did you need to find it? Conservation of information says that the information required to pick out the right map is never less than the information required to locate the treasure directly.

From the vantage of LCI, *searches are as real as the objects being searched.* Just as the existence and formation of those objects must be explained, so too the existence and formation of the searches that locate those objects must be explained. We might say that searches, by residing in a space of searches, are themselves objects to be searched. This implies a hierarchy of searches: the original search, the search for that search, the search for the search for that search, etc. LCI says that as we regress up this search hierarchy, the search problem never becomes easier and may in fact become more difficult. We clarify this point next.

The LCI Regress. Consider again our general setup: A null search B, which is blind, sets a probabilistic baseline p for searching the space Ω for a target T. Because p is so small that B

stands no practical possibility of locating T, success in searching for T requires an alternative search S that has probability q of locating T, where q is much larger than p. But that raises the question, how did we find S? S resides in a higher-order search space, call it $\Omega^{(2)}$, and specifically in a higher-order target $T^{(2)}$ consisting of all searches that have probability at least q of locating the original target T (for consistency let $\Omega^{(1)} = \Omega$ and let $T^{(1)} = T$). But how easy is it to find $T^{(2)}$ in $\Omega^{(2)}$? LCI tells us that doing so requires at least $I_+ = \log(q/p)$. Moreover, once we've found the alternative search S in $T^{(2)}$, we still have to use it to search for the original target T in Ω. This lower-order search has probability q of success, which corresponds to the exogenous information $I_S = -\log(q)$.

Thus, the information required to successfully locate the original target T by first searching a higher-order search space for S and then using S to search for T requires at least $I_+ + I_S = \log(q/p) - \log(q) = -\log(p) = I_\Omega$. This shows that the endogenous information (i.e., inherent difficulty) of using a higher-order search to locate the original target T is at least as great as the original endogenous information. We represent this fact by the inequality

$$I_{\Omega^{(1)}} \leq I_{\Omega^{(2)}},$$

where the first term denotes the original lower-order endogenous information of finding T directly (i.e., I_Ω since $\Omega = \Omega^{(1)}$) and the second denotes the higher-order endogenous information of finding T indirectly by first searching $\Omega^{(2)}$ for a higher-order target $T^{(2)}$.

Given LCI, this inequality implies (by mathematical induction) that as we move up the search hierarchy to search for higher-order targets $T^{(3)}$, $T^{(4)}$, etc. within higher-order search spaces $\Omega^{(3)}$, $\Omega^{(4)}$, etc., higher-order endogenous information associated with locating the original target T will never diminish and may well increase. We call this the LCI Regress and write

$$I_{\Omega^{(1)}} \leq I_{\Omega^{(2)}} \leq I_{\Omega^{(3)}} \leq \cdots.$$

Here $I_{\Omega^{(3)}}$ sums the information needed to locate $T^{(3)}$ plus the information needed to locate $T^{(2)}$ using a search in $T^{(3)}$ plus the information needed to locate the original T using a search in $T^{(2)}$.

So, we may ask, if an alternative search achieves a high probability of locating a target T that is extremely improbable with respect to a null (or blind) search, where did the information that enables the alternative search to be successful come from? From a higher-order search? But, as the LCI Regress shows, a higher-order search requires at least as much information to locate T as any lower-order search. Borrowing from Peter to pay Paul at best maintains, and may even intensify, the debt now owed to Peter. Likewise, borrowing from Andrew to pay Peter maintains or intensifies the debt still further. Indeed, borrowing from one lender to pay another does nothing to redress a debt. Where, then, does the LCI Regress end? In fact, it may not end, implying that the information that enables an alternative search to succeed in locating T was always present. Alternatively, it may end because an external information source added the information needed to locate T. One option suggests front-loading of information, the other direct input. Both options evoke intelligent design.

Entropy. The LCI Regress suggests a deep connection between the Law of Conservation of Information and the Second Law of Thermodynamics. In *Logic and Information*, Keith Devlin considers the thermodynamic significance of information:

Perhaps information should be regarded as (or maybe is) a basic property of the universe, alongside matter and energy (and being ultimately interconvertible with them). In such a theory (or suggestion for a theory, to be more precise), information would be an intrinsic measure of the structure and order in parts or all of the universe, being closely related to entropy (and in some sense its inverse).[62]

According to the LCI Regress, the information needed for effective search tends to increase as time moves backward (that's because searches proceed in time and the search for a given search necessarily precedes the given search). Yet, according to the Second Law of Thermodynamics, the unavailability of energy for conducting useful work in an isolated non-equilibrium system, as measured by entropy, tends to increase as time moves forward. It seems, then, that information as characterized by the Law of Conservation of Information may be regarded as inverse to entropy: increased information indicates an increased capacity for conducting a successful search, whereas increased entropy indicates a decreased capacity for doing the work necessary to conduct a search.

Given the paradox of Maxwell's Demon,[63] which indicates the ability of an information source to decrease entropy over time, the Law of Conservation of Information might actually prove more basic than the Second Law of Thermodynamics. The title of Leo Szilard's celebrated 1929 paper on Maxwell's Demon is worth recalling here: "On the Decrease of Entropy in a Thermodynamic System by the Intervention of Intelligent Beings."[64] The information source that, for Szilard, reverses entropy is intelligent. Likewise, the LCI Regress, as noted in the last bullet point, suggests that intelligence is ultimately the source of the information that accounts for successful search.

8. Applying LCI to Biology

Biology's reception of Darwinism might have been less favorable had scientists heeded Darwin's contemporary John Stuart Mill. In 1843, sixteen years before the publication of Darwin's *Origin of Species,* Mill published the first edition of his *System of Logic* (which by the 1880s had gone through eight editions).[65] In that work, Mill lays out various methods of induction. The one that interests us here is his *method of difference.* Mill described this method as follows:

> If an instance in which the phenomenon under investigation occurs, and an instance in which it does not occur, have every circumstance in common save one, that one occurring only in the former; the circumstance in which alone the two instances differ is the effect, or the cause, or an indispensable part of the cause, of the phenomenon.[66]

Essentially this method says that to discover which of a set of circumstances is responsible for an observed difference in outcomes requires finding a difference in the circumstances associated with each outcome. An immediate corollary of this method is that common circumstances cannot explain a difference in outcomes.

Suppose you and a friend have been watching TV, eating popcorn, and lounging on a vibrating couch. Yet your friend is now staggering about bleary-eyed whereas you are moving and seeing just fine. Precisely because the TV, popcorn, and vibrating couch are experiences held in common, they do not explain why your friend is having difficulties and you are doing fine. To explain the difference, you need to find not what's common to your circumstances but what's different. When you remember that your friend also consumed an inordinate amount of alcohol whereas you simply drank water, you've applied Mill's method to explain the difference.

This method, so widely used in everyday life, is crucially relevant to biological evolution. Indeed, it helps bring some sense of proportion and reality to the inflated claims so frequently made on behalf of Darwinian processes. We've already cited (in section 5) Kenneth Miller's overselling of Darwinism, where he claims that "what's needed to drive" increases in biological information is "just three things: selection, replication, and mutation."[67] Mill's method of difference gives the lie to Miller's claim. It's easy to write computer simulations that feature selection, replication, and mutation—*and that go absolutely nowhere.* It's also possible to write such simulations that solve interesting problems and produce salient patterns. But because selection, replication, and mutation are common to both such simulations, they can't, as Mill's method makes clear, account for the difference.

Nils Barricelli, writing about computer simulations of evolution as far back as 1961, understood this point:

> The *selection* principle of Darwin's theory is not sufficient to explain the evolution of living organisms if one starts with entities having only the property to *reproduce* and *mutate.* At least one more theoretical principle is needed, a principle which would explain how self-reproducing entities could give rise to organisms with the variability and evolutionary possibilities which characterize living organisms.[68]

Barricelli's point here is exactly correct: Miller's holy trinity of selection, replication, and mutation are "not sufficient"; "at least one more theoretical principle is needed." Unfortunately, Barricelli's subsequent proposal for the missing theoretical principle is mistaken. What was his proposal? Symbiogenesis, the idea that organisms or simple replicators can combine to form new, more complex organisms or replicators.[69]

These days, symbiogenesis is most closely associated with the research of biologist Lynn Margulis, who has been working on this topic for several decades.[70] Although she describes many interesting cases of symbiogenesis, she hasn't shown how this process constitutes a general solution to biology's information problem. Symbiogenesis, whether operating in real or in virtual biology, can at best mix existing traits; it cannot create fundamentally new ones. For instance, when two organisms merge in symbiogenesis, the genetic complement of the newly formed organism is simply the sum of the genes from the original two organisms—no new genes are created. Genuine novelty is therefore beyond symbiogenesis's reach. And yet, genuine novelty is precisely what the steady increase of biological information over the course of natural history exhibits.

Neither Barricelli's nor Margulis's proposals have won the day. The reason they haven't is that symbiogenesis, like genetic mutation, is simply another undirected way of producing raw variation in biological structures. To resolve biology's information problem requires harnessing that variation. For most evolutionary biologists, that's the job of natural selection. Thus, Margulis, though an outspoken critic of neo-Darwinism (which locates the source of evolutionary variation in genetic mutation), will admit, "I am definitely a Darwinist." Yet with no apparent sense of irony she immediately adds, "I think we are *missing important information* about the origins of variation."[71] The missing ingredient in her account of evolution is in fact staring her in the face: *information!* Yet rather than focus on the role of information in guiding evolution, Margulis continues to focus on undirected sources of variation. Supplementing Miller's "mutation" with other non-teleological sources of variation such as symbiogenesis does nothing to meet the challenge raised by Mill's method of difference.

The failure of selection, replication, and mutation (or undirected variation more generally—include here symbiogenesis, genetic drift, lateral gene transfer, etc.) to drive evolution is evident not just in computer simulations but also in actual biological experiments. Consider, for instance, Sol Spiegelman's work on the evolution of polynucleotides in a replicase environment. To evolve his polynucleotides, Spiegelman inserted information: the replicase protein was supplied by the investigator from a viral genome, as were the activated mononucleotides needed to feed polynucleotide synthesis. Yet even without such investigator interference, which has no analogue in a Darwinian conception of natural history, a deeper problem remains. According to Miller, selection, replication, and mutation (or variation) increase information. Yet Spiegelman demonstrated that even with all these factors at play, information steadily *decreased* over the course of his experiment. Brian Goodwin, in his summary of Spiegelman's work, underscores this point:

> In a classic experiment, Spiegelman in 1967 showed what happens to a molecular replicating system in a test tube, without any cellular organization around it. The replicating molecules (the nucleic acid templates) require an energy source, building blocks (i.e., nucleotide bases), and an enzyme to help the polymerization process that is involved in self-copying of the templates. Then away it goes, making more copies of the specific nucleotide sequences that define the initial templates. But the interesting result was that these initial templates did not stay the same; they were not accurately copied. They got shorter and shorter until they reached the minimal size compatible with the sequence retaining self-copying properties. And as they got shorter, the copying process went faster. So what happened with natural selection in a test tube: the shorter templates that copied themselves faster became more numerous, while the larger ones were gradually eliminated. This looks like Darwinian evolution in a test tube. But the interesting result was that this evolution went one way: toward greater simplicity.[72]

The problem that Spiegelman identified here is merely the tip of the iceberg. Yes, evolution had better be complexity-increasing if it is to deserve all the attention it receives. But complexity also needs to be going somewhere. In fact, in the history of life, increasing complexity has been in the service of building magnificent structures of incredible sophistication and elegance. How could evolution accomplish all this, especially given NFL and LCI? Complexity theorist Stuart Kauffman understands the challenge:

> The no-free-lunch theorem says that, averaged over all possible fitness landscapes, no search procedure outperforms any other. . . . In the absence of any knowledge, or constraint, [read "information"] on the fitness landscape, on average, any search procedure is as good as any other. But life uses mutation, recombination, and selection. These search procedures seem to be working quite well. Your typical bat or butterfly has managed to get itself evolved and seems a rather impressive entity. The no-free-lunch theorem brings into high relief the puzzle. If mutation, recombination, and selection only work well on certain kinds of fitness landscapes, yet most organisms are sexual, and hence use recombination, and all organisms use mutation as a search mechanism, where did these well-wrought

fitness landscapes come from, such that evolution manages to produce the fancy stuff around us?[73]

According to Kauffman, "No one knows."[74]

Let's be clear where our argument is headed. We are not here challenging common descent, the claim that all organisms trace their lineage to a universal common ancestor. Nor are we challenging evolutionary gradualism, that organisms have evolved gradually over time. Nor are we even challenging that natural selection may be the principal mechanism by which organisms have evolved. Rather, we are challenging the claim that evolution can create information from scratch where previously it did not exist. The conclusion we are after is that natural selection, even if it is the mechanism by which organisms evolved, achieves its successes by incorporating and using existing information.

Mechanisms are never self-explanatory. For instance, your Chevy Impala may be the principal mechanism by which you travel to and from work. Yet explaining how that mechanism gets you from home to work and back again does not explain the information required to build it. Likewise, if natural selection, as operating in conjunction with replication, mutation, and other sources of variation, constitutes the primary mechanism responsible for the evolution of life, the information required to originate this mechanism must still be explained. Moreover, by the Law of Conservation of Information, that information cannot be less than the mechanism gives out in searching for and successfully finding biological form and function.

It follows that Dawkins's characterization of evolution as a mechanism for building up complexity from simplicity fails. For Dawkins, proper scientific explanation is "hierarchically reductionistic," by which he means that "a complex entity at any particular level in the hierarchy of organization" must be explained "in terms of entities only one level down the hierarchy."[75] Thus, according to Dawkins, "the one thing that makes evolution such a neat theory is that it explains how organized complexity can arise out of primeval simplicity."[76] This is also why Dawkins regards intelligent design as unacceptable:

> To explain the origin of the DNA/protein machine by invoking a supernatural [*sic*] Designer is to explain precisely nothing, for it leaves unexplained the origin of the Designer. You have to say something like "God was always there," and if you allow yourself that kind of lazy way out, you might as well just say "DNA was always there," or "Life was always there," and be done with it.[77]

Conservation of information shows that Dawkins's primeval simplicity is not as nearly simple as he makes out. Indeed, what Dawkins regards as intelligent design's predicament of failing to explain complexity in terms of simplicity now confronts materialist theories of evolution as well. In *Climbing Mount Improbable,* Dawkins argues that biological structures that at first blush seem vastly improbable with respect to a blind search become quite probable once the appropriate evolutionary mechanism is factored in to revise the probabilities.[78] But this revision of probabilities just means that a null search has given way to an alternative search. And the information that enables the alternative search to be successful now needs itself to be explained. Moreover, by the Law of Conservation of Information, that information is no less than the information that the evolutionary mechanism exhibits in outperforming blind search. The preceding quotation, which was intended as a refutation of intelligent design, could therefore, with small modifications, be turned against natural selection as well:

To explain the origin of the DNA/protein machine by invoking *natural selection* is to explain precisely nothing, for it leaves unexplained the origin of *the information that natural selection requires to execute evolutionary searches*. You have to say something like "*the information* was always there," and if you allow yourself that kind of lazy way out, you might as well just say "DNA was always there," or "Life was always there," and be done with it.[79]

Conservation of information therefore points to an information source behind evolution that imparts at least as much information to the evolutionary process as this process in turn is capable of expressing by producing biological form and function. As a consequence, such an information source has three remarkable properties: (1) it cannot be reduced to purely material or natural causes; (2) it shows that we live in an informationally porous universe; and (3) it may rightly be regarded as intelligent. The Law of Conservation of Information therefore counts as a positive reason to accept intelligent design. In particular, it establishes ID's scientific bona fides.

Several attempts have been made to block NFL- and LCI-inspired design inferences as they apply to biology. Olle Häggström, a Swedish mathematician, thinks that the no free lunch theorems have been, as he puts it, "hype[d]." According to him, "any realistic model for a fitness landscape will have to exhibit a considerable amount of what" he calls "*clustering,* meaning that similar DNA sequences will tend to produce similar fitness values much more often than could be expected under [a model that allows all possible fitness landscapes]."[80] For Häggström, "realistic models" of fitness presuppose "geographical structures," "link structures," search space "clustering," and smooth surfaces conducive to "hill climbing."[81] All such structures, however, merely reinforce the teleological conclusion we are drawing, which is that the success of evolutionary search depends on the front-loading or environmental contribution of active information. Simply put, if a realistic model of evolutionary processes employs less than the full complement of fitness functions, that's because active information was employed to constrain their permissible range.

Building on Häggström's criticism, statistician Ronald Meester also questions the applicability of NFL (and by implication LCI) to biology. Like Häggström, Meester sees the NFL theorems as dealing "with an extreme situation: averaging over all fitness functions."[82] A less "extreme," or more "realistic," model would focus on fewer fitness functions. But, as already noted, any such model, by focusing on fewer fitness functions, needs to constrain the total space of fitness functions, and any such constraint entails an imposition of active information. Yet Meester also extends Häggström's argument:

> In a search algorithm as defined in the present article, a sufficiently special target set can only be reached when the search algorithm is very carefully tailored around the fitness function. This conclusion is a direct consequence of our discussion of the NFL theorems and beyond. This implies that this special target can only be reached by programming with insight into the future. Since Darwinian evolution cannot look into the future, this forces us to conclude that simulations cannot be used for the purpose of explaining how complex features arise into the universe.[83]

But how does Meester know that Darwinian evolution cannot look into the future? Certainly it is part of the popular mythology associated with Darwinism that it is a non-teleological theory.

We quoted Jerry Coyne to that effect in section 1. Such quotes appear across the Darwinian literature. But how do we know that evolution is non-teleological or that any teleology in it must be scientifically unascertainable? Imagine you are on an ancient ship and observe a steersman at the helm. The ship traverses difficult waters and reaches port. You conclude that the vessel's trajectory at sea was teleological. Why? Two things: you see a steersman controlling the ship's rudder who, on independent grounds, you know to be a teleological agent; also, you witness the goal-directed behavior of the ship in finding its way home.

Now imagine a variation on this story. An ancient sailor comes on board a twenty-first century ship that is completely automated so that a computer directly controls the rudder and guides the vessel to port. No humans are on board other than this sailor. Being technologically challenged, he will have no direct evidence of a teleological agent guiding the ship—no steersman of the sort that he is used to will be evident. And yet, by seeing the ship traverse difficult channels and find its way home by exactly the same routes he took with ancient ships guided by human steersmen, he will be within his rights to conclude that a purpose is guiding the ship, even if he cannot uncover direct empirical evidence of an embodied teleological agent at the helm.[84]

Now, the Law of Conservation of Information gives this conclusion extra quantitative teeth. According to LCI, any search process that exhibits information by successfully locating a target must have been programmed with no less than what we defined as the active information. Thus, armed with LCI, our ancient steersman, however technologically challenged otherwise, could reasonably infer that a teleological agent had put the necessary active information into the ship (the ship, after all, is not eternal and thus its information could not have resided in it forever). Like the ancient sailor, we are not in a position to, as it were, open the hood of the universe and see precisely how the information that runs evolution was programmed (any more than the sailor can peer into the ship's computers and see how it was programmed). But LCI guarantees that the programming that inserts the necessary information is nonetheless there in both instances.

Stricter Darwinists may resist this conclusion, but consider the alternative. When we run search algorithms in evolutionary computing, we find that these searches are inherently teleological (Meester readily concedes this point). So, we may ask, do such mathematical models adequately represent biological evolution? In these models, careful tailoring of fitness functions that assist in locating targets is always present and clearly teleological. If these models adequately represent biological evolution, then this teleological feature of fitness ought to be preserved in nature, implying that Darwinian evolution is itself teleological.

To avoid this conclusion, we must therefore hold that these models somehow fail to capture something fundamental about the inherent non-teleological character of nature. But on what basis can we hold this? It's only in virtue of such models that we can be said to have a scientific theory of evolution at all. But no non-teleological mathematical models of Darwinian evolution are known. All of them readily submit to the Law of Conservation of Information. Thus, to deny that these models adequately represent biological evolution is to deny that we have an adequate model of evolution at all. But in that case, we have no scientific basis for rejecting teleology in evolution. Without a clear mathematical underpinning, evolution degenerates into speculation and mystery-mongering.

Meester claims that these models are irrelevant to biology because Darwinian evolution is inherently non-teleological. But he simply begs the question. Darwinian evolution, as it plays out in real life, could potentially look into the future (and thus be teleological) if the fitness it employed were, as Meester puts it, "programmed with insight into the future." And how do we know that it isn't? The search algorithms in evolutionary computing give rampant evidence of teleology—from their construction to their execution to the very problems they solve. So too, when we turn to evolutionary biology, we find clear evidence of teleology: despite Dawkins's denials, biological evolution is locating targets. Indeed, function and viability determine evolution's targets (recall section 3), and evolution seems to be doing a terrific job finding them. Moreover, given that Darwinian evolution is able to locate such targets, LCI underwrites the conclusion that Darwinian evolution is teleologically programmed with active information.

Häggström and Meester, in mistakenly criticizing the applicability of NFL- and LCI-inspired design inferences to biology, are at least asking the right questions. Biologists Allen Orr and Richard Dawkins, in criticizing such inferences, display conceptual and technical confusion. Orr's criticism centers on the capacity of fitness to change as populations evolve. Fitness coevolves with an evolving population, and this, for Orr, allows natural selection to work wonders where otherwise it might be hampered. In reviewing William Dembski's *No Free Lunch,* Orr writes,

> Consider fitness functions that are as unsmooth as you like, i.e., rugged ones, having lots of peaks and few long paths up high hills. (These are the best studied of all fitness landscapes.) Now drop many geographically separate populations on these landscapes and let them evolve independently. Each will quickly get stuck atop a nearby peak. You might think then that Dembski's right; we don't get much that's interesting. But now change the environment. This shifts the landscape's topography: a sequence's fitness isn't cast in stone but depends on the environment it finds itself in. Each population may now find it's no longer at the best sequence and so can evolve somewhat even if the new landscape is still rugged. Different populations will go to different sequences as they live in different environments. Now repeat this for 3.5 billion years. Will this process yield interesting products? Will we get different looking beasts, living different kinds of lives? My guess is yes.[85]

Guessing aside, a straightforward mathematical analysis settles the matter.[86] It is quite true that fitness can change over time. But nothing in our mathematical development of LCI requires static, time-independent fitness functions. Given an m-query search of a space Ω, it can be represented as a 1-query search of the m-fold Cartesian product Ω^m. Fitness on this product may well change from factor to factor. But this observation poses no challenge to LCI. Mathematics is well able to accommodate Orr's coevolving fitness functions. Coevolving fitness functions leave the Law of Conservation of Information intact. Note that Wolpert and Macready, when they first stated the NFL theorems, proved them for time-independent as well as for time-dependent fitness functions.[87] Both NFL and LCI apply to coevolving fitness functions.

Finally, we consider a criticism by Richard Dawkins. According to him, design inferences in biology are rendered superfluous because of the presumed ability of selection to cumulate biological function and complexity by small incremental steps (each of which is quite probable). Dawkins's WEASEL is his best known illustration of cumulative selection. We analyzed the WEASEL program in sections 3 and 5, showing how active information was inserted into it to ensure that it located the target phrase METHINKS•IT•IS•LIKE•A•WEASEL. Dawkins has asserted the cumulative power of natural selection for over twenty years, going back to his 1986 book *The Blind Watchmaker,* repeating the claim in his 1996 book *Climbing Mount Improbable,* and repeating it yet again in his most recent 2006 book *The God Delusion,* where he writes,

> What is it that makes natural selection succeed as a solution to the problem of improbability, where chance and design both fail at the starting gate? The answer is that natural selection is a cumulative process, which breaks the improbability up into small pieces. Each of the small pieces is slightly improbable, but not prohibitively so.[88]

This claim cannot be sustained in the face of LCI. Dawkins here describes an evolutionary process of, let us say, m steps (or "pieces"), each of which is sufficiently probable that it could reasonably happen by chance. But even m highly probable events, if occurring independently, can have a joint probability that's extremely low. Dawkins himself makes this point: "When large numbers of these slightly improbable events are stacked up in series, the end product of the accumulation is very very improbable indeed, improbable enough to be far beyond the reach of chance."[89] Dawkins here tacitly presupposes an evolutionary search space that consists of an m-fold Cartesian product. Moreover, he rightly notes that uniform probability on this space (which is the product of uniform probabilities on the individual factors—this is elementary probability theory) assigns a very low probability (high endogenous information) to such evolutionary events.

In consequence, for natural selection to be a powerful cumulative force, it's not enough that the individual steps in Dawkins's evolutionary process be reasonably probable (or only "slightly improbable"); rather, all these reasonably probable steps must, *when considered jointly*, also be reasonably probable. And this just means the Dawkins has, with zero justification, substituted an alternative search for a null search. But whence this alternative search? Darwinian theory has no answer to this question. To paraphrase Dawkins, "You have to say something like 'the alternative search just magically appeared or was always there,' and if you allow yourself that kind of lazy way out, you might as well just say 'DNA was always there,' or 'Life was always there,' and be done with it."[90] Cumulative selection, as Dawkins characterizes it, does nothing to explain the source of evolution's creative potential. LCI, by contrast, shows that evolution's creative potential lies in its incorporation and use of active information.

9. Conclusion: "A Plan for Experimental Verification"

In a 2002 address to the American Scientific Affiliation, Francis Collins, then director of the NIH's National Human Genome Research Institute, posed the following challenge to intelligent design: "A major problem with the intelligent design theory is its lack of a plan for experimental verification."[91] We submit that the Law of Conservation of Information makes such a plan feasible.

The Law of Conservation of Information states that active information, like money or energy, is a commodity that obeys strict accounting principles. Just as corporations require money to power their enterprises and machines require energy to power their motions, so searches require active information to power their success. Moreover, just as corporations need to balance their books and machines cannot output more energy than they take in, so searches, in successfully locating a target, cannot give out more information than they take in.

It follows from the Law of Conservation of Information that active information cannot be gotten on the cheap but must always be paid for in kind. As such, this law has far-reaching implications for evolutionary theory, pointing out that the success of evolutionary processes in exploring biological search spaces always depends on pre-existing active information. In particular, evolutionary processes cannot create from scratch the active information that they require for successful search.

The Law of Conservation of Information, however, is not merely an accounting tool. Under its aegis, intelligent design merges theories of evolution and information, thereby wedding the natural, engineering, and mathematical sciences. On this view (and there are other views of intelligent design), its main focus becomes how evolving systems incorporate, transform, and export information.[92] Moreover, a principal theme of its research becomes teasing apart the respective roles of internally produced and externally applied information in the performance of evolving systems.

Evolving systems require active information. How much? Where do they get it? And what does this information enable them to accomplish? Tracking and measuring active information in line with the Law of Conservation of Information is the plan we propose for experimentally verifying intelligent design and thus meeting Collins's challenge. Evolution is a theory of process. It connects dots by telling us that one thing changes into another and then specifies the resources by which the transformation is (supposed to be) effected. According to the Law of Conservation of Information, active information will always constitute a crucial resource in driving targeted evolutionary change (and much of that change in biology, we have argued, is indeed targeted—exceptions include cyclic variation and genetic drift).

Tracking and measuring active information to verify intelligent design is readily achieved experimentally. Consider, for instance, that whenever origin-of-life researchers use chemicals from a chemical supply house, they take for granted information-intensive processes that isolate and purify chemicals. These processes typically have no analogue in realistic pre-biotic conditions. Moreover, the amount of information these processes (implemented by smart chemists) impart to the chemicals can be calculated. This is especially true for polymers, whose sequential arrangement of certain molecular bases parallels the coded information that is the focus of Shannon's theory of communication.[93] In such experiments, a target invariably exists (e.g., a simple self-replicating molecule, an autocatalytic set, or a lipid membrane).[94]

Just as information needs to be imparted to a golf ball to land it in a hole, so information needs to be imparted to chemicals to render them useful in origin-of-life research. This information can be tracked and measured. Insofar as it obeys the Law of Conservation of Information, it confirms intelligent design, showing that the information problem either intensifies as we track material causes back in time or terminates in an intelligent information source. Insofar as this information seems to be created for free, LCI calls for closer scrutiny of just where the information that was given out was in fact put in.

In such information-tracking experiments, the opponent of intelligent design hopes to discover a free lunch. The proponent of intelligent design, by contrast, attempts to track down hidden information costs and thereby confirm that the Law of Conservation of Information was preserved. There is no great mystery in any of this. Nor do such experiments to confirm intelligent design merely apply to the origin of life. Insofar as evolution (whether chemical or biological) is an exact experimental science, it will exhibit certain informational properties. Are those properties more akin to alchemy, where more information comes out than was put in? Or are they more akin to accounting, where no more information comes out than was put in? A systematic attempt to resolve such questions constitutes a plan for experimentally verifying intelligent design.

Notes

1. Robert Stalnaker, *Inquiry* (Cambridge, MA: MIT Press, 1984), 85.
2. Fred Dretske, *Knowledge and the Flow of Information* (Cambridge, MA: MIT Press, 1981), 4.
3. Douglas Robertson, "Algorithmic Information theory, Free Will, and the Turing Test," *Complexity* 4(3), 1999: 25–34.
4. G. K. Chesterton, *Orthodoxy*, in *Collected Works of G. K. Chesterton*, vol. 1 (San Francisco: Ignatius, 1986), 243.
5. Jerry Coyne, "Truckling to the Faithful: A Spoonful of Jesus Makes Darwin Go Down," posted on his blog Why Evolution Is True on April 22, 2009 at http://whyevolutionistrue.wordpress.com (last accessed April 27, 2009). Emphasis added. "Genetic drift" here refers to random changes in population gene frequencies. It too is nonteleological.
6. Darwinist philosopher of biology David Hull explicitly confirms this point: "He [Darwin] dismissed it [design] not because it was an incorrect scientific explanation, but because it was not a proper scientific explanation at all." David Hull, *Darwin and His Critics: The Reception of Darwin's Theory of Evolution by the Scientific Community* (Cambridge, MA: Harvard University Press, 1973), 26.
7. See, for instance, F. H. Sandbach, *The Stoics*, 2nd ed. (Indianapolis: Hackett, 1989), 72–75.
8. Aristotle, *Metaphysics*, trans. W. D. Ross, XII.3 (1070a, 5–10), in Richard McKeon, ed., *The Basic Works of Aristotle* (New York: Random House, 1941), 874.
9. Aristotle, *Physics*, trans. R. P. Hardie and R. K. Gaye, II.8 (199a, 15–20), in Richard McKeon, ed., *The Basic Works of Aristotle* (New York: Random House, 1941), 250.
10. Ibid., II.8 (199b, 25–30), 251.
11. Compare Jacques Monod, *Chance and Necessity* (New York: Vintage, 1972).
12. David Baltimore, "DNA Is a Reality beyond Metaphor," *Caltech and the Human Genome Project* (2000): available online at http://pr.caltech.edu:16080/events/dna /dnabalt2.html (last accessed April 23, 2007).
13. Manfred Eigen, *Steps Towards Life: A Perspective on Evolution*, trans. Paul Woolley (Oxford: Oxford University Press, 1992), 12.
14. Eörs Szathmáry and John Maynard Smith, "The Major Evolutionary Transitions," *Nature* 374 (1995): 227–32.
15. Holmes Rolston III, *Genes, Genesis and God: Values and Their Origins in Natural and Human History* (Cambridge: Cambridge University Press, 1999), 352.
16. "Give me twenty-six lead soldiers and I will conquer the world." The "lead soldiers" are the typefaces used in printing. This quote has variously been attributed to Karl Marx, William Caxton, and Benjamin Franklin. Reliable references appear to be lacking.
17. Francisco J. Ayala, "Darwin's Revolution," in *Creative Evolution?!*, eds. J. H. Campbell and J. W. Schopf (Boston: Jones and Bartlett, 1994), 4. The subsection from which this quote is taken is titled "Darwin's Discovery: Design without Designer."
18. Richard Dawkins, *The Blind Watchmaker: Why the Evidence of Evolution Reveals a Universe without Design* (New York: Norton, 1987), 6.
19. Francis S. Collins, *The Language of God: A Scientist Presents Evidence for Belief* (New York: Free Press, 2006), 201.
20. Theistic evolutionist Kenneth Miller justifies the scientific undetectability of divine action as follows: "The indeterminate nature of quantum events would allow a clever and subtle God to influence events in ways that are profound, but *scientifically undetectable* to us. Those events could include the appearance of mutations, the activation of individual neurons in the brain, and even the survival of individual cells and organisms affected by the chance processes of radioactive decay." Kenneth R. Miller, *Finding Darwin's God: A Scientist's Search for Common Ground between God and Evolution* (New York: Harper, 1999), 241. Emphasis added.

21. Richard Dawkins, *The Blind Watchmaker* (New York: Norton, 1986), 47–48.
22. For an event of probability p to occur at least once in k independent trials has probability $1-(1-p)^k$. See Geoffrey Grimmett and David Stirzaker, *Probability and Random Processes* (Oxford: Clarendon, 1982), 38. If p is small and $k = 1/p$, then this probability is greater than $1/2$. But if k is much smaller than $1/p$, this probability will be quite small (i.e., close to 0).
23. Richard Dawkins, *The Blind Watchmaker* (New York: Norton, 1987), 48.
24. Dawkins, *The Blind Watchmaker*, 50.
25. See http://www.top500.0rg/lists/2008/11 (last accessed April 20, 2009).
26. "Workers at the University of Georgia estimate that 10^{30} single-celled organisms are produced every year; over the billion-year-plus history of the Earth, the total number of cells that have existed may be close to 10^{40}." Michael J. Behe, *The Edge of Evolution: The Search for the Limits of Darwinism* (New York: Free Press, 2007), 153.
27. Seth Lloyd, "Computational Capacity of the Universe," *Physical Review Letters* 88(23) (2002): 7901–4.
28. J. Bowie and R. Sauer, "Identifying Determinants of Folding and Activity for a Protein of Unknown Sequences: Tolerance to Amino Acid Substitution," *Proceedings of the National Academy of Sciences* 86 (1989): 2152–56. J. Bowie, J. Reidhaar-Olson, W. Lim, and R. Sauer, "Deciphering the Message in Protein Sequences: Tolerance to Amino Acid Substitution," *Science* 247 (1990): 1306–10. J. Reidhaar-Olson and R. Sauer, "Functionally Acceptable Solutions in Two Alpha-Helical Regions of Lambda Repressor," *Proteins, Structure, Function, and Genetics* 7 (1990): 306–10. See also Michael Behe, "Experimental Support for Regarding Functional Classes of Proteins to be Highly Isolated from Each Other," in *Darwinism: Science or Philosophy?*, eds. J. Buell, and G. Hearn (Dallas: Foundation for Thought and Ethics, 1994), 60–71; and Hubert Yockey, *Information Theory and Molecular Biology* (Cambridge: Cambridge University Press, 1992), 246–58.
29. Nils Aall Barricelli, "Numerical Testing of Evolution Theories, Part I: Theoretical Introduction and Basic Tests," *Acta Biotheoretica* 16(1–2) (1962): 69–98. Reprinted in David B. Fogel, ed., *Evolutionary Computation: The Fossil Record* (Piscataway, NJ: IEEE Press, 1998), 166.
30. J. L. Crosby, "Computers in the Study of Evolution," *Science Progress, Oxford* 55 (1967): 279–92. Reprinted in Fogel, *Evolutionary Computation*, 95.
31. Heinz R. Pagels, *The Dreams of Reason: The Computer and the Rise of Sciences of Complexity* (New York: Simon and Schuster, 1989), 104.
32. Robert T. Pennock, "DNA by Design?" in W. A. Dembski and M. Ruse, eds., *Debating Design: From Darwin to DNA* (Cambridge: Cambridge University Press, 2004), 141. Pennock's work on AVIDA is summarized in Richard Lenski, Charles Ofria, Robert T. Pennock, and Christoph Adami, "The Evolutionary Origin of Complex Features," *Nature* 423 (May 8, 2003): 139–44.
33. MESA, which stands for Monotonic Evolutionary Simulation Algorithm, is available at http://www.iscid.org/mesa (last accessed April 6, 2009).
34. See http://www.i2cam.org (last accessed April 6, 2009).
35. Richard E. Lenski, "Phenotypic and Genomic Evolution during a 20,000-Generation Experiment with the Bacterium *Escherichia coli*," *Plant Breeding Reviews* 24 (2004): 225–65.
36. Michael Behe, *Darwin's Black Box: The Biochemical Challenge to Evolution* (New York: Free Press, 1996).
37. Lenski et al., "The Evolutionary Origin of Complex Features."
38. Quoted in John Horgan, "From Complexity to Perplexity: Can Science Achieve a Unified Theory of Complex Systems?" *Scientific American* (June 2005): 104–9.
39. David Berlinski, *The Devil's Delusion: Atheism and Its Scientific Pretensions* (New York: Crown Forum, 2008), 190.
40. For Adami's AVIDA, visit his Digital Life Lab at Caltech at http://dllab.caltech.edu. For Ray's Tierra, go to http://life.ou.edu/tierra. For Schneider's *ev*, go to http://www-lmmb.ncifcrf.gov/~toms/paper/ev.

Each of these sites was last accessed April 6, 2009. See also Thomas D. Schneider, "Evolution of Biological Information," *Nucleic Acids Research* 28(14) (2000): 2794–99.

41. These computer programs suggest sharp limits to the power of the Darwinian selection mechanism. For Mendel's Accountant visit http://mendelsaccount.sourceforge.net and for MutationWorks visit http://www.mutationworks.com. Both these sites were last accessed April 6, 2009.

42. The alternative is for evolution to constitute an inexact historical science: "To obtain its answers, particularly in cases in which experiments are inappropriate, evolutionary biology has developed its own methodology, that of *historical narratives* (tentative scenarios)." Quoted from Ernst Mayr, *What Makes Biology Unique? Considerations on the Autonomy of a Scientific Discipline* (Cambridge: Cambridge University Press, 2004), 24–25. Emphasis in the original.

43. Kenneth R. Miller, *Only a Theory: Evolution and the Battle for America's Soul* (New York: Viking, 2008), 77–78.

44. See the Evolutionary Informatic Lab's "EV Ware: Dissection of a Digital Organism" at http://www.evoinfo.org/Resources/EvWare/index.html (last accessed April 6, 2009).

45. For the details, see William A. Dembski, *No Free Lunch: Why Specified Complexity Cannot Be Purchased without Intelligence* (Lanham, MD: Rowman and Littlefield, 2002), sec. 4.9.

46. Leon Brillouin, *Science and Information Theory,* 2nd ed. (New York: Academic Press, 1962), 269.

47. Peter Medawar, *The Limits of Science* (Oxford: Oxford University Press, 1984), 79.

48. William A. Dembski and Robert J. Marks II, "The Search for a Search: Measuring the Information Cost of Higher Level Search," *Journal of Advanced Computational Intelligence and Intelligent Informatics* 14(5) (2010): forthcoming. William A. Dembski and Robert J. Marks II, "The Conservation of Information: Measuring the Information Cost of Successful Search," *IEEE Transactions on Systems, Man, and Cybernetics, Part A,* 5(5) (2009): 1051–61. For these and additional papers on tracking the active information in intelligent computing see the publications page at http://www.evoinfo.org.

49. Actually, the problem of identifying an appropriate fitness function for locating Dawkins's target phrase is much worse than sketched here. We focused entirely on the subset of single-hill fitness functions, which could have its peak at any of 10^{40} places. But there are many more fitness functions than those gauging distance from a target sequence. Indeed, the total space of fitness functions is exponential in the underlying sequence space (see the fitness-theoretic version of the conservation of information theorem in section 6) and, properly speaking, would have to be searched. In limiting ourselves to fitness functions based on the Hamming distance from a target sequence, we've already incurred a heavy informational cost.

50. Jakob Bernoulli, *Ars Conjectandi* (1713; reprinted Ann Arbor, MI: University of Michigan Library, 2006). For the applicability of Bernoulli's principle of insufficient reason in the information-theoretic context, see William A. Dembski and Robert J. Marks II, "Bernoulli's *Principle of Insufficient Reason* and Conservation of Information in Computer Search," available at http://www.evoinfo.org.

51. Steven Pinker, *How the Mind Works* (New York: Norton, 1999), 90–92.

52. This result is proved in Appendix C of Dembski and Marks, "The Search for a Search." A proof is also available online in William A. Dembski, "Searching Large Spaces: Displacement and the No Free Lunch Regress," (2005): http://www.designinference.com/documents/2005.03.Searching_Large_Spaces.pdf.

53. See Robert J. Marks II, *Handbook of Fourier Analysis and Its Applications* (Oxford: Oxford University Press, 2009), 165.

54. Joseph C. Culberson "On the Futility of Blind Search: An Algorithmic View of 'No Free Lunch,'" *Evolutionary Computation* 6(2) (1998): 109–127.

55. Christian Igel and Marc Toussaint, "On Classes of Functions for Which No Free Lunch Results Hold," *Information Processing Letters* 86 (2003): 317–321. An earlier version of this paper is available online at http://www.marc-toussaint.net/publications/igel-toussaint-01.pdf (last accessed April 13, 2009).

56. See the information bounds that come up in the Vertical No Free Lunch Theorem that is proved in Dembski and Marks, "The Search for a Search."

57. David H. Wolpert and William G. Macready, "No Free Lunch Theorems for Optimization," *IEEE Transactions on Evolutionary Computation* 1(1) (1997): 67–82.

58. Cullen Schaffer, "A Conservation Law for Generalization Performance," *Machine Learning: Proceedings of the Eleventh International Conference,* eds. H. Hirsh and W. W. Cohen, 259–65 (San Francisco: Morgan Kaufmann, 1994).

59. Cullen Schaffer, "Conservation of Generalization: A Case Study," typescript (1995): available online at http://citeseerx.ist.psu.edu/viewdoc/summary?doi=10.1.1.41.672 (last accessed April 23, 2009). Emphasis in the original.

60. Yu-Chi Ho and David L. Pepyne, "Simple Explanation of the No Free Lunch Theorem," *Proceedings of the 40th IEEE Conference on Decision and Control,* Orlando, FL (2001): 4409–14.

61. For the Church-Turing Thesis, often simply called Church's Thesis, see Klaus Weihrauch, *Computability* (Berlin: Springer-Verlag, 1987), sec. 1.7.

62. Keith Devlin, *Logic and Information* (Cambridge: Cambridge University Press, 1991), 2.

63. "Maxwell's demon inhabits a divided box and operates a small door connecting the two chambers of the box. When he sees a fast molecule heading toward the door from the far side, he opens the door and lets it into his side. When he sees a slow molecule heading toward the door from his side he lets it through. He keeps the slow molecules from entering his side and the fast molecules from leaving his side. Soon, the gas in his side is made up of fast molecules. It is hot, while the gas on the other side is made up of slow molecules and it is cool. Maxwell's demon makes heat flow from the cool chamber to the hot chamber." John R. Pierce, *An Introduction to Information Theory: Symbols, Signals and Noise,* 2nd rev. ed. (New York: Dover, 1980), 199.

64. Leo Szilard, "Über die Entropieverminderung in einem thermodynamischen System bei Eingriff intelligenter Wesen," *Zeitschrift für Physik* 53 (1929): 840–56. For the English translation, see John A. Wheeler and Wojciech H. Zurek, eds., *Quantum Theory and Measurement* (Princeton: Princeton University Press, 1983), 539–548.

65. John Stuart Mill, *A System of Logic: Ratiocinative and Inductive,* 8th ed. (1882; reprinted London: Longmans, Green, and Co., 1906).

66. Ibid., 256.

67. Miller, *Only a Theory,* 77.

68. Barricelli, "Numerical Testing of Evolution," 170–171. Emphasis added to underscore that Barricelli cites exactly the same three evolutionary mechanisms as Miller: selection, replication (or reproduction), and mutation.

69. Ibid., see especially secs. 3–6.

70. See Lynn Margulis and Dorion Sagan, *Acquiring Genomes: A Theory of the Origins of Species* (New York: Basic Books, 2002). Margulis's work on symbiogenesis goes back now over thirty years.

71. Quoted in Michael Shermer, "The Woodstock of Evolution," *Scientific American* (June 27, 2005): available online at http://www.sciam.com/article.cfm?id=the-woodstock-of-evolutio&print=true (last accessed April 27, 2009). Emphasis added. The occasion for Margulis's remark was her award of an honorary doctorate at the World Summit on Evolution, Galapagos Islands, June 8–12, 2005.

72. Brian Goodwin, *How the Leopard Changed Its Spots: The Evolution of Complexity* (New York: Scribner's, 1994), 35–36.

73. Stuart Kauffman, *Investigations* (New York: Oxford University Press, 2000), 19.

74. Ibid., 18.

75. Dawkins, *Blind Watchmaker,* 13.

76. Ibid., 316.

77. Ibid., 141. Note that intelligent design is not committed to supernaturalism. Stoic and Aristotelian philosophy eschewed the supernatural but were clearly in the ID camp.

78. Richard Dawkins, *Climbing Mount Improbable* (New York: Norton, 1996).

79. For the original, unedited passage, see Dawkins, *Blind Watchmaker,* 141. The words in italics substitute for words in the original. The words in italics are about natural selection and information; the words in the original that they replace are about God and design.

80. Olle Häggström, "Intelligent Design and the NFL Theorem," *Biology and Philosophy* 22 (2007): 226, 228.

81. Ibid., 220, 228.

82. Ronald Meester, "Simulation of Biological Evolution and the NFL Theorems," *Biology and Philosophy* 24 (2009): 461–72.

83. Ibid.

84. Our ancient sailor is, we might say, undergoing a Turing Test in which he must distinguish the unmediated teleology in a ship's course due to the direct activity of an embodied teleological agent versus the mediated teleology in a ship's course due to its having been programmed by a teleological agent. Without actually seeing the steersman, the sailor would be unable to distinguish the two forms of teleology. Such indistinguishability, according to Alan Turing, would suggest that the teleology is as real in the one case as in the other. For the Turing Test, see Alan Turing, "Computing Machinery and Intelligence," *Mind* 59 (1950): 434–60.

85. H. Allen Orr, "Book Review: *No Free Lunch,*" *Boston Review* (Summer 2002): available online at http://bostonreview.net/BR27.3/orr.html (last accessed April 28, 2009). Compare Dembski's response to Orr, "Evolution's Logic of Credulity," (2002): available online at http://www.designinference.com/documents/2002.12.Unfettered_Resp_to_Orr.htm (last accessed April 28, 2009).

86. Such an analysis was offered in section 4.10 of Dembski's *No Free Lunch,* a section Orr's review failed to address.

87. Wolpert and Macready, "No Free Lunch Theorems for Optimization." They prove an NFL theorem for "static cost functions" (i.e., time-independent fitness functions) as well as for "time-dependent cost functions" (i.e., time-dependent fitness functions).

88. Richard Dawkins, *The God Delusion* (London: Bantam Press, 2006), 121.

89. Ibid. For instance, it's highly probable that any particular person will avoid death in an automobile accident today. But given that the number of drivers m is large enough, it becomes highly improbable that none of these drivers will be killed.

90. For the original, unedited passage, see Dawkins, *Blind Watchmaker,* 141.

91. Francis S. Collins, "Faith and the Human Genome," *Perspectives on Science and Christian Faith* 55(3) (2003): 151.

92. The form of intelligent design that we are describing here falls under what we have dubbed "evolutionary informatics" (see http://www.evoinfo.org).

93. Claude Shannon and Warren Weaver, *The Mathematical Theory of Communication* (Urbana, Ill.: University of Illinois Press, 1949).

94. See William A. Dembski and Jonathan Wells, *How to Be an Intellectually Fulfilled Atheist (Or Not)* (Wilmington, DE: ISI Books, 2008), ch. 19, titled "The Medium and the Message." This book critically analyzes origin-of-life theories.

17

REGULATED RECRUITMENT AND COOPERATIVITY IN THE DESIGN OF BIOLOGICAL REGULATORY SYSTEMS[1]

MARK PTASHNE

1. INTRODUCTION

[A reasonable goal of evolutionary explanation is] to identify simple rules, or strategies, that can be used to develop complex systems. I describe here a simple molecular strategy that nature has used widely in evolving biological complexity. This description is part of an extended argument put forth in the recently published book *Genes and Signals* (2002).[2] The biological complexity I refer to has two aspects, one having to do with the development of a complex organism from a fertilized egg, and the other with the evolution of different forms of life. A similar theme is sounded in the two cases: limited sets of gene products (proteins, mainly, and RNA molecules) are used, in different combinations, to generate diversity.

For example, proteins that generate human hands also generate human feet, but the different appendages form as a result of different patterns of appearance of these proteins as the embryo develops. And, speaking somewhat more informally, a common set of gene products can be used to generate a human, a mouse, and perhaps even a fly, by appropriately regulating their appearances as each of these organisms develops. There is more to it than this, of course, but these statements suffice to pose the following problems.

How is the appearance of proteins regulated during development of any one organism? What kinds of molecular changes to the regulatory apparatus are required to modify those patterns as evolution proceeds? To return to our original theme, how is biological complexity generated from a limited set of common elements?

2. PROTEIN REGULATION

There are many ways to regulate the appearance of a protein in a cell at a given time and place, but for the purposes of illustration we consider just three: a gene can be transcribed (or not); a RNA molecule can be spliced, sometimes one way or another; and a protein can be destroyed (by proteolysis). In each case the regulation can be either/or, or it can be graded: a gene can be transcribed more or less, and so on. My general point is that a common molecular strategy has been used in the evolution of regulatory systems that effect each of these three kinds of control

Regulated Recruitment

(and many others as well). The mechanism is called "regulated recruitment." To understand it, we need first to consider the ubiquitous phenomenon of cooperative binding, as illustrated in Figure 1. I will then illustrate how the principle is used in transcription, discuss some of the implications, and then return, near the end, to RNA splicing and proteolysis.

Figure 1 shows three macromolecules; we will call A and B "proteins," and the rod "DNA," but the following characteristics apply more generally. Given specified affinities of A and B for their sites, and assuming that A and B are present at approximately some specified concentrations (as would be found in a cell, for example), neither A nor B is bound efficiently to its site when the other protein is absent. When both proteins are present, however, both proteins are bound, an effect called cooperative binding of proteins to DNA. The word "cooperativity" is laden with historical associations, and so we have to be clear what is required in this case: the proteins must simply "touch" one another when bound to DNA. Conformational changes, though they may occur, are not necessary for the effect, nor is energy in the form of adenosine triphosphate (ATP) (for example) used. Rather, one protein helps the other bind simply by providing binding energy in the form of an "adhesive" interaction between the proteins.

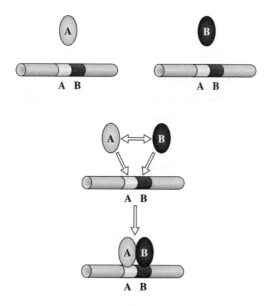

Figure 1. Cooperative binding. Interactions between proteins binding to DNA (for example) are typically weak, involving, say, 1–2 kcal of interaction energy. The resulting effect on binding, say 10- to 100-fold, can be physiologically important. Reproduced with permission from Mark Ptashne and Alexander Gann, *Genes and Signals* (Cold Spring Harbor, NY: Cold Spring Harbor Laboratory Press, 2002).

To see how the reaction of Figure 1 applies to gene regulation, consider Figure 2, which shows a bacterial transcriptional activator in action. The gene is the famous lacZ (beta-galactosidase) gene of *Escherichia coli*. (For reasons that become evident upon considering the entire argument of *Genes and Signals*,[3] analysis of gene activation, rather than gene repression—the usual entry into the subject of gene regulation—is the more revealing approach.)

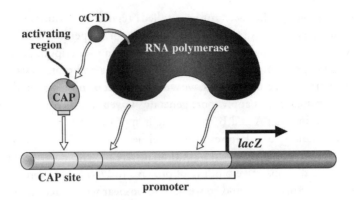

Figure 2. *Gene activation effected by cooperative binding.* The rod is a DNA fragment containing the lacZ gene, a site that binds the activator CAP, and a more extended sequence that binds the RNA polymerase, called the promoter. The "activator" CAP binds DNA only when complexed with the small molecule cyclic AMP. Glucose in the medium depletes intracellular cyAMP, and so CAP only works when glucose is absent from the medium. Reproduced with permission from Mark Ptashne and Alexander Gann, *Genes and Signals* (Cold Spring Harbor, NY: Cold Spring Harbor Laboratory Press, 2002).

The catabolite activator protein (CAP), binds specifically to its site on DNA, and RNA polymerase binds to its site, called the promoter. In the absence of CAP—and also in the absence of another regulatory protein, the lac repressor (see below)—RNA polymerase binds to the promoter, but only infrequently. CAP binds cooperatively with the bacterial RNA polymerase as shown, and thereby "activates" the gene. Note that the term "activation" as used here can be misleading. Neither the enzyme nor the gene has in any meaningful sense been "activated"; rather, the activator merely increases the frequency (by cooperative binding) with which the constitutively active polymerase encounters the gene. We say the activator "recruits" the polymerase to the gene.

How do we know that the mechanism of Figure 2 accurately describes activation of the lac and other bacterial genes? And how do we know that simple recruitment can account for activation of a typical eukaryotic gene? The latter question seems particularly difficult in view of the fact that over sixty-five proteins are required to transcribe a typical eukaryotic gene, and these proteins are found to be distributed in various complexes in the cell (see Figure 4). Part of the proof is as follows.

The mechanism of Figure 2 makes an array of experimental predictions, all of which have been verified for certain genes in bacteria. Figure 2 implies the following:

(i) The enzyme (RNA polymerase) is not pre-bound to the gene prior to activation.

(ii) DNA binding of the activator is absolutely required for gene activation. Even when expressed at high concentrations, an activating region that is not tethered to DNA cannot "activate" (because it cannot recruit).

(iii) The detailed mode of DNA binding is irrelevant. That is, many different detailed modes of DNA binding of the activator will suffice to position it so that it can contact RNA polymerase.

(iv) Two well-separated interaction surfaces are found on each protein. Thus, the polymerase has a DNA (promoter)-binding surface and, in addition, a site that contacts the activator. The activator has one site that binds DNA and another, called its "activating region," that contacts polymerase.

(v) A unique activating region–enzyme interaction is not required. That is, any of an array of positions on the polymerase can be contacted by an activator to effect cooperative binding and gene activation.

(vi) Activator-bypass experiments work. An example of such an experiment is shown in Figure 3. In place of an ordinary activator we have a fusion protein bearing a DNA-binding

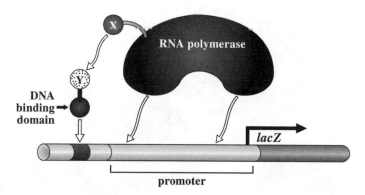

Figure 3. *An activator bypass experiment in E. coli.* This experiment has been successfully performed with different interacting pairs of proteins. In each case one is fused to a DNA-binding domain, the other to polymerase. In the case depicted here, a carboxyl terminal domain of the largest subunit of polymerase has been replaced by protein X. Reproduced with permission from Mark Ptashne and Alexander Gann, *Genes and Signals* (Cold Spring Harbor, NY: Cold Spring Harbor Laboratory Press, 2002).

domain and an attached protein called Y. The polymerase has been modified so as to bear protein X, and X and Y are known to interact. Strong transcription ensues: a result showing that even a totally heterologous protein–protein interaction suffices for gene activation (provided one of the interacting components is tethered to DNA).

The argument is given particular power thanks to the fact that there are two modes of gene activation found in bacteria that are different from the simple reaction shown here. Each of the diagnostic predictions just listed does *not* hold for one or the other or both of these alternative forms of gene activation in bacteria. (The bacterial genes activated by alternative mechanisms include a few involved in resistance to heavy metals and a group involved in nitrogen metabolism. For an explanation of how activation works in these cases, see *Genes and Signals*.[4]

3. Activation in Yeast

All of these diagnostic experimental predictions have also been realized in yeast, some in a particularly striking fashion. For example, the yeast activator Gal4 can be split into two parts, neither of which activates on its own: one binds DNA and the other bears an activating region. The activating region will work when attached to a heterologous DNA-binding domain, even a bacterial binding domain (e.g., that of the *E. coli* repressor protein LexA), provided the test gene bears the appropriate DNA-binding site. The activating region itself resembles "glue" or Velcro®: any of a wide array of sequences (particularly sequences containing an excess of hydrophobic and acidic residues) will work, and certain of those sequences have been shown to work with an efficiency approximately proportional to their lengths. The "two hybrid" system, widely used in yeast to discover interacting protein pairs, depends on the facts that activating regions must be tethered to DNA to work, and that the mode of attachment to DNA is irrelevant. One modification required as we move from bacteria to yeast is that, in the latter case, the activator evidently does not contact the polymerase itself. Rather, it is believed, other components of the complex machinery of Figure 4, some of which themselves bind polymerase, are recruited directly by the activator.

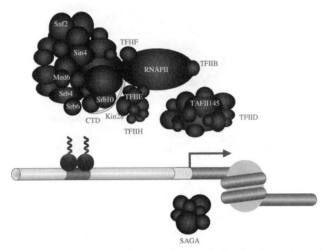

Figure 4. *Gene activation in eukaryotes.* The typical transcriptional activator [round, with curly tail] is shown bound to DNA [represented as a rod]: it comprises a DNA-binding domain fused to a peptide that works as an activating region. The activating region recruits (and assembles) the large complex required to transcribe a gene. Many components of the transcriptional machinery are pre-assembled into subcomplexes (SAGA, TFIIH, etc.), as [labeled above]. A segment of DNA is shown wrapped around histones [lower right] to form a nucleosome.

Having made the jump from bacteria to yeast, it is easy to see that many genes in higher eukaryotes must also be activated by recruitment. Thus, for example, Ga14 and other yeast activators can work in a wide array of higher organisms. Moreover, the overall nature of the transcriptional machinery, comprising the many proteins required for transcription of the typical gene, is essentially conserved between yeast and higher eukaryotes. It is therefore not surprising that (to take just one example) no higher eukaryotic gene, to my knowledge, can be activated by expressing, or even over-expressing, an activator that cannot bind DNA and hence cannot recruit. How a eukaryotic activator assembles the large complex required for transcription is currently under investigation. It is likely that some of the components are directly touched and recruited by the activator, others then binding cooperatively with the directly recruited components.

The question now arises as to why evolution exploited regulated recruitment so widely for gene regulation. What might be features of that mechanism that make it particularly "evolvable" (a term the meaning of which becomes clearer as we discuss the matter)? What are the advantages over the other mechanisms found in bacteria that we have alluded to but not discussed explicitly here? Once again, I will not discuss those alternative mechanisms here, but rather will make a few general remarks about the evolvability of regulated recruitment.

4. Genetic Regulation

It is easy to see how, in principle, nature can start with a simple system and then, in steps, make it ever more sophisticated. Figure 5 shows the imagined evolution of regulation of the lac genes of *E. coli*. Starting with just the gene and the polymerase (that is, in the absence of any regulation) the gene is transcribed at a low but biologically significant level. The bacterium constitutively synthesizes beta-galactosidase and can use lactose as a carbon source. The first modification introduced is the activator, CAP, which, as we have seen, binds cooperatively with polymerase and thereby "activates" transcription. The interaction of CAP with polymerase is small in terms of interaction energy (some 1–2 kcal) but that interaction increases transcription significantly, some 10- to 100-fold. The second modification is the addition of the repressor: this protein, by binding to a site within the promoter, excludes binding of polymerase, and its effect overrides the action of CAP. These two additions turn an unregulated system into a rather sophisticated one: the gene is on if and only if lactose is present in the medium, and it is transcribed fully only if glucose (a better carbon source than lactose) is absent (see captions to Figures 2 and 5 for more details).

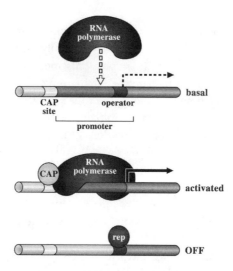

Figure 5. *Hypothetical steps in evolution of regulation of the lac genes.* In the top line, there is no regulation, and the polymerase spontaneously binds and transcribes the gene at a low level. In the second line a CAP site has been introduced upstream of the promoter, and CAP activates as in Figure 1. In the third line, a lac-repressor binding site, called an operator, has been introduced. Now the gene can be transcribed only if lactose is present, because a metabolic product of lactose binds to repressor and prevents it from binding DNA.

The system is also highly evolvable in another way. Because the specificity of CAP action depends only upon the position of its DNA-binding site (i.e., which gene it is near), it would seem quite straightforward to add that site to other genes. Indeed some 200 genes in *E. coli* are activated by CAP. Moving the CAP site from one gene to another, or adding it to new genes, effects a change, or an extension, of the "meaning" of a signal. CAP responds to glucose in the medium (see Figure 2), and so changing where CAP acts (i.e. at which gene) means changing the meaning of the signal (which is glucose in this case).

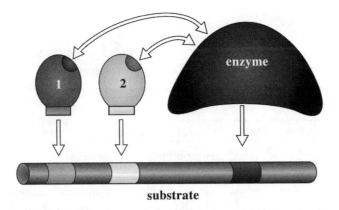

Figure 6. *Synergistic activation.* Two activators that do not interact with each other are shown simultaneously binding to and recruiting polymerase. The two activators work "synergistically": the level of transcription elicited by the two activators working together is more than the sum of the two activators working separately.

The system is highly modular. This modularity is found at many levels (including at the structure of the activator). For many genes and activators, any activator will work on any gene, precisely because what activators do is so simple. This mechanism—recruiting a common enzyme to whatever gene one wishes to express—should be contrasted to a hypothetical world in which a new enzymatic machinery would be built for each gene.[5]

Building complex organisms from a rather "small" set of genes requires signal integration: genes must be regulated in response to combinations of regulators. For example, a given gene might be activated only when worked on by several activators. Regulated recruitment lends itself to signal integration in an obvious way. Thus, any two (for example) activators, binding near a gene will work synergistically, because each will contribute to the binding reaction as shown in Figure 6. This effect is easily observed in yeast: in general, any two activators placed near a gene work synergistically, even if those activators are unrelated and do not interact directly with each other.

As we move to higher eukaryotes we have ever more signal integration in regulating genes. Figure 7 shows an extension of the ideas we have discussed: in this case three separate activators bind cooperatively to form an "enhancesome" that activates the human interferon-beta gene. Only if all three activators are working does the gene become activated. These (and many other) eukaryotic activators are found working in different combinations with different partners. This "combinatorial control" expands even further the ways that signals can be integrated.

5. LAMBDA SWITCH

The interactions I have been discussing—between enzyme, activator and DNA, and between different activators as they bind DNA—are instantiations of the cooperative binding mechanism illustrated in Figure 1. The simple "adhesive" interactions required for cooperative binding can be used to generate an "on–off" switch that responds in an all-or-nothing manner to an extra-cellular switch. The familiar example, the bacteriophage lambda switch, is shown in Figure 8. The lambda repressor (despite its name) activates transcription of its own gene as

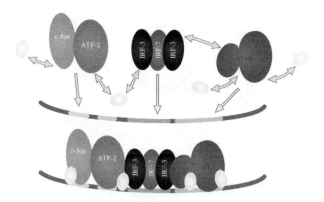

Figure 7. *The human interferon-beta enhancer.* Three activators (called Jun/ATF, IRF3/7 and NfkB) are shown binding cooperatively (with small auxiliary proteins) to form an "enhancesome." The enhancesome forms, and the gene is activated, only when all three activators are present and capable of binding DNA. Modified with permission from Mark Ptashne and Alexander Gann, "Imposing specificity by localization: mechanism and evolvability," *Current Biology* 8 (1998): 812–22.

it represses transcription of an adjacent gene. Repressor monomers are in concentration-dependent equilibrium with dimers, as shown, and those dimers bind cooperatively to two adjacent sites on DNA. This autocatalytic system depends (speaking roughly) on the fourth power of the monomer concentration. An inducing signal (e.g., UV light), which destroys repressor, has little effect until the repressor concentration is reduced to a critical point, and from there on induction occurs essentially instantaneously. Each of the reactions depicted in the figure—between repressor monomers, between dimers and between repressor and polymerase—is merely adhesive. And so a rather sophisticated switch has been produced by molecules that interact like Lego® toys.

But there is another reason I bring up the lambda switch. I wish to illustrate a very general problem that arises from the use of cooperativity (regulated recruitment) to control genes. Going back to our basic description of Figure 1, we see that, for the binding of proteins A and B to be dependent upon one another, their concentrations must be below a specified level. The typical interaction between cooperatively binding proteins is weak: some 1–2 kcal, as we have already noted for a particular case (see lac genes above). That means that the system is highly sensitive to the concentrations of the interacting proteins. In the lambda case, for example, if the repressor concentration rises (say 5- to 10-fold) above the level found in cells, the cooperative effects on repressor binding to DNA are lost. For example, the binding of one dimer no longer depends on binding of the other. And so the "fourth power" dependence of the switch, and hence its on–off character, would be lost.

Lambda has evolved a seemingly elaborate mechanism for controlling the level of repressor in the cell. The principle is simple: at higher concentrations, the repressor binds to a DNA site called Or3 (see figure 8) and turns off its own synthesis. To effect this negative feedback, the repressor binds cooperatively not only to sites in and around its promoter, but simultaneously with sites located some 3 kb (3000 base pairs) away. A large looped DNA structure is thereby formed, the sole purpose of which is to maintain the concentration of repressor below a specified level in the cell.

All systems controlled by regulated recruitment must keep the concentration of key components below specified levels, and nature has found many ways to do this. For example,

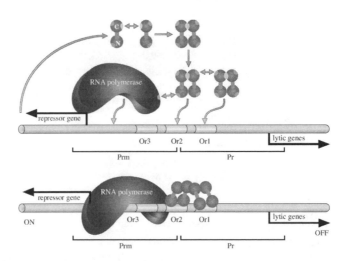

Figure 8. *Autoregulation of the lambda repressor gene.* The small dumbbell-shaped protein is the repressor. It forms dimers, two of which bind cooperatively to sites (called Or1 and Or2) in the repressor gene's promoter, as shown. The repressor recruits the polymerase and activates its own gene (at promoter Prm) as it excludes polymerase from the adjacent promoter (Pr). The start sites and directions of transcription of the regulated genes are indicated by the arrows. At higher concentrations repressor negatively regulates transcription of its own gene by binding to the third site shown (Or3) (see text). Other features of the lambda system reinforce the on–off nature of the switch [see Ptashne and Gann, *Genes and Signals,* and Mark Ptashne, *A Genetic Switch: Phage Lambda Revisited,* 3rd edition (Cold Spring Harbor: Cold Spring Harbor Laboratory Press, 2004)].

eukaryotic transcriptional activators are typically held outside the nucleus unless they are signalled to work, thus keeping their concentrations in the nucleus low. Some of the bewildering complexities of regulatory systems reflect nature's various ways of controlling the concentrations of active forms of various regulators.

Now I can return to the two other examples I mentioned at the start: RNA splicing and proteolysis. Recall that regulation of these two processes, like regulation of transcription, can determine when, where, and to what extent any given protein might appear in a developing embryo. Figure 9 shows that a large splicing machine (comprising some 145 proteins) is recruited to a specific site on RNA, where it then works spontaneously. Just as with RNA polymerase in *E. coli* and the more complex polymerase and associated proteins in eukaryotes, the enzyme depicted here works spontaneously at a low level on many different splice sites. Only when recruited to a specific site does it work with high efficiency.

Figure 10 shows how regulated recruitment controls proteolysis. A large protein complex adds ubiquitin molecules to a protein destined to be destroyed by another large complex called the proteosome. The figure illustrates how a protein is chosen to be ubiquitylated: an adaptor (or "activator") protein called an F-box protein binds cooperatively with the target protein and with the ubiquitylating complex, and the ubiquitination reaction then proceeds spontaneously.

6. Conclusion

I will end with a slightly different formulation of what I have stated before. We might recast the problem as one of *enzyme specificity.* The classical enzymes we all studied many years ago

Regulated Recruitment

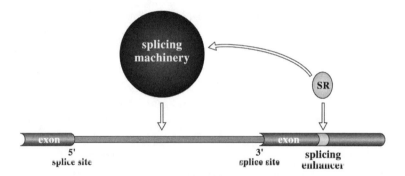

Figure 9. *Control of splicing by regulated recruitment.* The SR protein works analogously to the transcriptional activator of figure 1: it recruits an enzyme, in this case the splicing machinery, to a potential splice site in the RNA. Different SR proteins, working with various cooperatively binding partners, bind specific splicing enhancer sites in RNA and thereby promote one or another splicing event. Modified with permission from B. R. Graveley, K. J. Hertel, and T. Maniatis, "SR proteins are 'locators' of the RNA splicing machinery." *Current Biology* 9 (1998): R6–R7.

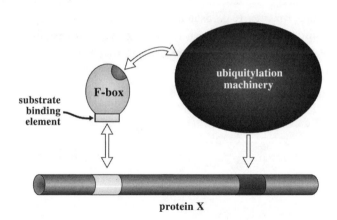

Figure 10. *Control of ubiquitylation by regulated recruitment.* The F-box protein brings a specific substrate to the ubiquitylating enzyme. If the substrate specificity of the F-box protein is altered so that it recognizes a different protein, that protein will then be recruited and ubiquitylated.

(e.g. beta-galactosidase) have highly specific active sites that typically recognize one and only one substrate. The term "regulation" as applied to these enzymes (where it applies) usually refers to allosteric inhibitory or activating effects of substrates and other small molecules. But many of the enzymes studied by molecular biologists, including those we have discussed here, can work on many different substrates: *E. coli* RNA polymerase can transcribe any of 3,000 genes, for example; the ubiquitinating machinery can work on thousands of proteins in the cell; and so on. The regulatory problem thus reduces to one of substrate choice: which gene will be transcribed, which protein will be ubiquitylated, which RNA site will be spliced, and so on.

When viewed in this way we see that nature has hit upon a generally applicable trick for imparting specificity: the use of compact protein domains, which can be inserted at many places on the surfaces of proteins, and which impart specificity for binding to another macromolecule. We have, for example, families of DNA-binding domains (e.g., helix–turn–helix and zinc-finger motifs) that direct transcriptional regulatory proteins to one or another site on DNA; families of protein binding domains (e.g., Sh2 and Sh3) that direct one protein to another; other domains that direct proteins to specific sites on RNA molecules or on membranes; and so on. The specificity of each domain is limited, and we usually find two or more of these locator domains working cooperatively. "New" genes that are found in higher eukaryotes often encode the same enzyme found in a lower eukaryote but attached to arrays of "locator" domains that give the enzyme a new specificity in the sense I have been discussing.

The general principle extends even further than this brief description: "locator" functions can be provided by molecules other than proteins. For example, in the process called RNA interference, a bit of double-stranded RNA recruits, to a specific mRNA site, a machinery that destroys that mRNA. RNA works as a recruiter in several other instances as well (directing telomerase to specific sequences at the ends of chromosomes, for example), and other examples are likely to be discovered.

I have stressed, by way of illustration, how enzymes that are sometimes found in large complexes can be directed to one or another substrate by regulated recruitment, and have only mentioned in passing that such regulatory events often occur in response to extra-cellular signals. A more complete discussion would reveal that the pathways which convey signals to the regulatory apparatus, so-called "signal transduction pathways," are themselves made up of essentially common enzymes (e.g., protein kinases) that are given specificity by the kinds of simple binding interactions I have been discussing.[6] Pawson describes in more detail the plethora of "locator" domains found in nature and how they are used in various ways to create these signal transduction pathways.[7]

Notes

1. Adapted with permission from *Philosophical Transactions of the Royal Society of London* A 361 (2003): 1223–34.
2. M. Ptashne and A. Gann, *Genes and Signals* (Cold Spring Harbor, NY: Cold Spring Harbor Laboratory Press, 2002).
3. Ibid.
4. Ibid.
5. See M. Ptashne and A. Gann, "Imposing Specificity by Localization: Mechanism and Evolvability." *Current Biology* 8 (1998): 812–22.
6. See Ptashne and Gann, *Genes and Signals*.
7. T. Pawson, "Organization of Cell-Regulatory Systems through Modular-Protein-Interaction Domains." *Philosophical Transactions of the Royal Society of London* A361 (2003): 1251–62.

18

THE NATURE OF PROTEIN FOLDS: QUANTIFYING THE DIFFICULTY OF AN UNGUIDED SEARCH THROUGH PROTEIN SEQUENCE SPACE

Douglas D. Axe

1. Introduction

The elucidation of the genetic code in the late 1960s provided a precise framework for understanding the effects of genetic mutations on protein sequences. And because proteins perform most of the molecular tasks needed for life, solving the code also opened the possibility of understanding the connection between genotype and phenotype on a scale that was not previously possible—the fine scale of nucleotide bases rather than the coarse scale of whole genes.

Among other benefits, this promised unprecedented insight into the inner workings of the neo-Darwinian process at the molecular level. Along with this benefit, however, came a potential problem. The code had made it clear that the vast set of *possible* proteins, each of which could conceivably be constructed by genetic mutations, is far too large to have actually been sampled to any significant extent in the history of life. Yet how could highly incomplete sampling have been so successful? How could it have located the impressive array of protein functions required for life in all its forms, or the comparably impressive array of protein structures that perform those functions? This concern was raised repeatedly in the early days of the genetic code,[1] but it received little attention from the biological community.

One possible reason for the lack of serious attention was the simplicity of the analyses being offered and their reliance on guesswork to fill in for missing data. With today's unprecedented catalogue of information on whole genomes and proteomes, we find ourselves in a considerably better situation. The information now at hand should enable us to reevaluate the issue that was raised four decades ago much more conclusively.

2. Biological Proteins and the Sampling Problem They Pose

Proteins are natural polymers—large molecules made by connecting smaller building blocks to form unbranched chains. In the general terminology of polymer chemistry, the building blocks are called monomers. Amino acids, which come in twenty different kinds, are the monomers used to construct biological proteins. The twenty amino acids differ not in the way they connect to form the main chain, but in their chemically distinct appendages, called *side chains,*

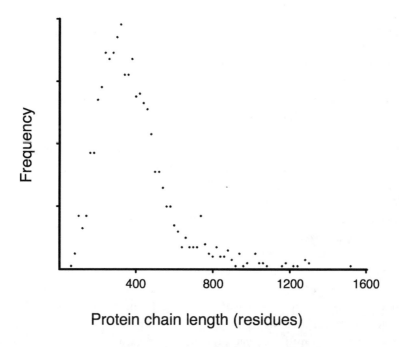

Figure 1. Approximate length distribution for 1,326 proteins known to be enzymes or enzyme components in *E. coli*. The mean and median lengths are 389 residues and 350 residues, respectively. Molecular weights of protein chains, obtained from EcoCyc version 9.0 (*proteins.dat* data file), were converted to approximate chain lengths by using an average per-residue molecular weight of 110 g/mol. Lengths were binned in 20-residue increments, the most occupied bin containing 78 protein chains.

that protrude from the main chain. Since there are n^l possible ways to build a polymer chain of length *l* with *n* distinct monomer types, protein chains a mere twelve residues long (*residue* being the term for an amino acid monomer that has been incorporated into a protein polymer) can be built in four quadrillion ways ($20^{12} = 4 \times 10^{15}$).

Because the number of distinct sequence possibilities grows very rapidly as the chain length is increased, we begin by asking how long biological proteins tend to be. It should be emphasized, though, that this is only a starting point. Several other aspects of proteins will need to be examined before we can decide whether their size complicates evolutionary explanations of their origins.

The simple relationship between gene sequences and protein sequences in bacteria allows protein sizes to be determined directly from genomic data. This, in combination with abundant data on protein structures and functions, makes the well-studied gut bacterium *Escherichia coli* an excellent model system for examining a simple proteome.[2] A clear picture of the size of *E. coli* proteins with known functions can be had by analyzing the data files provided by EcoCyc,[3] a comprehensive database of information on this organism. Figure 1 shows the distribution of protein chain lengths for all proteins known to be involved in enzymatic functions in *E. coli*, either alone or in combination with other proteins. The mode of the distribution shows that the most common length of these proteins is around 300 amino acid residues, with the higher mean and median lengths reflecting the existence of numerous protein chains that are much longer than this.

If we take 300 residues as a typical chain length for functional proteins, then the corresponding set of sequence possibilities numbers 20^{300} (= 10^{390}). How or whether this number should figure into our assessment of origins scenarios will be examined in the following sections. Here the point is simply that biological protein sequences are indeed members of astoundingly large sets of sequence possibilities. And by "large" we mean much more numerous than any evolutionary events we might postulate as having successfully searched through them. Consider, for example, Dembski's generous estimate of the maximum number of distinct physical events that have occurred within the visible universe, which he places at 10^{150}.[4] Since only a minute fraction of these events have any bearing on protein evolution, we can assert with confidence that there is a vast disparity between the number of distinct protein sequences of normal length that are conceptually *possible*, on the one hand, and the number that might have become *actual* on the other. In other words, real events have provided only an *exceedingly* sparse sampling of the whole set of sequence possibilities.

We will refer to this as the *sampling problem*, with the intent of deciding whether it is in fact a problem for unguided evolution. At the very least it raises the important question of how such sparse sampling would uncover so many highly functional protein sequences. To picture the difficulty, imagine being informed that a valuable gemstone was lost somewhere in the Sahara Desert. Without more specific information, any proposal for finding the missing gem would have to come to terms with the vastness of this desert. If only an infinitesimal fraction of the expanse can feasibly be searched, we would judge the odds of success to be infinitesimally small.

Evolutionary searches for functional proteins might seem less hopeless in some respects, though. For one, there is a highly many-to-one mapping of protein sequences onto protein functions. Consequently it could be argued that vast numbers of comparably valuable targets (protein sequences that are comparably suitable for any particular function) are there to be found. Therefore, while it is effectively impossible to stumble upon a particular 1-in-10^{390} protein sequence by chance, the likelihood of stumbling upon a particular protein *function* by chance will be m-fold higher, where m represents the multiplicity of sequences capable of performing that function.

There are good reasons to be cautious about this, however. Natural proteins would have to be *greatly* oversized and/or *highly* indifferent to the specifics of their amino acid sequences in order for m to be large enough to resolve the problem. We can imagine a different world where, for example, the planetary surface has rich deposits of abiotic amino acids, and cells indiscriminately incorporate these amino acids into long polypeptide chains, and these chains somehow benefit the cells without performing complex functions. In that world the problem we address here would not exist. But in our world things are strikingly different. Here we see a planet loaded with amino acids of strictly biological origin, and we see cells going to extraordinary lengths to manufacture, use, recycle, and scavenge all twenty of them. We see elaborate error-checking mechanisms that minimize the chances of confusing any one amino acid for any other during protein synthesis, and (as already noted) we see that the products of this tightly controlled process are *long* proteins. Lastly, we see that these long proteins perform an impressive variety of functions with equally impressive specificity and efficiency.

In the face of this, either we accept that proteins are what they seem to be—long amino acid chains that need to meet stringent sequence requirements in order to work—or we suppose that they are really much simpler than they seem to be. It has to be said that the second option arouses immediate suspicion by disregarding matters of plain fact—both the actual properties of proteins and the cellular processes that only make sense if the first option is correct. This is admittedly more a suggestion than a proof, and yet it does clearly add to the burden of justifying the second option. More conclusive arguments will require a closer look at the data.

3. THE CONNECTION BETWEEN PROTEIN SIZE AND PROTEIN FUNCTION

Enzymes are proteins or protein complexes that perform chemical transformations in a highly efficient and specific way. When students first encounter them, one of the things they may find puzzling is that they tend to be quite large in comparison to their active sites—the portion that actually binds the reactants (or *substrates,* as they are known) and converts them into products. Catalase, for example, is an enzyme consisting of four identical protein chains with individual molecular weights of around 80,000 atomic mass units (amu). Each chain forms an active site that functions as an extremely efficient converter of individual hydrogen peroxide (H_2O_2) molecules into water and oxygen. At 34 amu, though, this tiny substrate molecule has less than 1/2000th the mass of the protein that works on it. Mass ratios differ widely from one enzyme to the next, but as a rule, small-molecule metabolism employs enzymes that are very large in comparison to their substrates.

Why *are* these enzymes so much larger than the things they manipulate? Although we are some way from a complete answer to this, several aspects of the relationship between enzyme structures and their functions provide at least partial answers. On the most basic level, it has become clear that protein chains have to be of a certain length in order to fold into stable three-dimensional structures. At the low end, this requires something like forty amino acid residues, with more complex structures requiring much longer chains. In addition to this minimal requirement of stability, most folded protein chains perform their functions in physical association with other folded chains. The complexes formed by these associations may have symmetrical structures made by combining identical proteins or asymmetrical ones made by combining different proteins. In either case the associations involve specific inter-protein contacts with extensive interfaces (typically 1000 $Å^2$ or more of contact area). The need to stabilize these contacts between proteins therefore adds to their size, over and above the need to stabilize the structures of the individual folded chains.

Beyond these general principles, we have specific understanding of the need for structural complexity with respect to many particular protein functions. In catalase, for example, the active sites are deeply buried within the enzyme, such that the H_2O_2 molecules must pass through a long channel before they are catalytically converted. By replacing some of the amino acids in the enzyme, it has been shown that an electrical potential gradient within the channel makes an important contribution to the catalytic process.[5] So in this case, as in many others, the enzyme has important interactions with the substrate some distance away from the place where the actual chemical conversion occurs. We see in such examples that enzymes may guide reactants and/or products through a process that is more extensive than mere catalysis, and this processing requires a structure than extends well beyond the active site.

Another common functional constraint with implications for protein size is the need to achieve direct coupling of processes occurring at different sites on the same enzyme. This is distinct from the more general indirect coupling achieved by diffusion of shared metabolites within the cell. Direct coupling, unlike simple diffusion, has to be mediated by structural connections between the sites being coupled, and this requires more extensive protein structures.

Three examples will illustrate the importance of coupling in biology. The first is carbamoyl phosphate synthetase (CPS), which has been aptly described as "a remarkably complex enzyme."[6] It uses bicarbonate, glutamine, and ATP to make carbamoyl phosphate, which is required for the biosynthesis of both arginine and pyrimidine ribonucleotides. In order to couple the reactions occurring at its three active sites, this enzyme uses internal molecular tunnels for efficient transfer of reactants. To achieve this channel-coupled multisite architecture, CPS uses two protein chains with a combined length of over 1,400 amino acid residues (Figure 2). Thoden and coworkers describe the design rationale for this complexity as follows:

> From extensive biochemical data, it is now known that a fully coupled CPS requires the hydrolysis of one glutamine and the utilization of two molecules of MgATP for every molecule of carbamoyl phosphate formed. The three active sites of the enzyme must therefore be synchronized with one another in order to maintain the overall stoichiometry of the reaction without wasteful hydrolysis of glutamine and/or MgATP.[7]

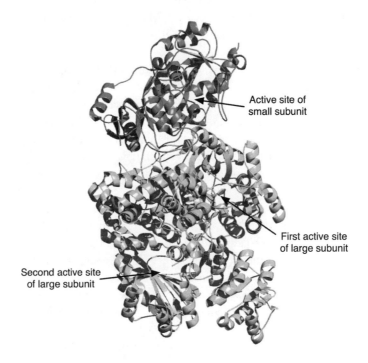

Figure 2. Structure of carbamoyl phosphate synthetase from *E. coli*, based on coordinate file 1T36 in the Protein Data Bank (PDB).[8] In this common way of representing protein structure, alpha helices are shown as coiled ribbons and beta strands are shown as ribbons with arrowheads. The two proteins chains are differentiated by shade.

So again we see an enzyme having to coordinate a complex process rather than a simple reaction, and having to be large in the sense of the sampling problem in order to achieve this.

As a second example of direct coupling, consider the following representations of cellular processes:

$$H^+_{ext} \to H^+_{int}, \text{ and}$$
$$ADP + P_i \to ATP + H_2O.$$

The Nature of Protein Folds

The first describes the flow of protons down an electrochemical potential gradient from the exterior of a membrane-enclosed compartment to the interior, and the second describes the generation of ATP from ADP and inorganic phosphate. These two processes are entirely unrelated from a purely physicochemical perspective. That is, there is no general principle of physics or chemistry by which ATP synthesis and proton fluxes have anything to do with each other. From an engineering perspective, however, it is often possible and desirable to design devices that *force* a relation upon otherwise unrelated processes. Of particular interest in this regard are devices that harness energy from an available source in order to accomplish useful tasks that require energy.

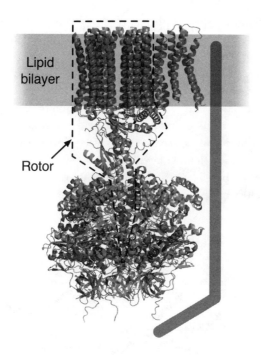

Figure 3. Partial structure of the proton translocating ATP synthase, based on PDB entries 1C17 and 1E79. The energy liberated by proton flow causes rapid rotation of the rotor, which is an assembly of many protein chains. The other protein chains form the stator (the stationary part of the motor). The bent grey bar shows the approximate location of a portion of the stator for which the structure is not fully known. The different proteins used to construct the *E. coli* synthase (some used multiple times) have 79, 139, 156, 177, 271, 287, 460, and 513 amino acid residues, for a combined length (non-redundant) of 2,082 residues.

Life likewise crucially depends on many such devices, one of which provides highly efficient energetic coupling of the above two processes. This coupler, the proton-translocating ATP synthase, is a rotary engine built from eight or more protein types, some of which are used multiple times to form symmetric substructures (Figure 3). Various versions of this ingenious device are found in all forms of life. The mitochondrial version couples the processes in the direction shown above, with an energetically favorable proton flux driving the energetically unfavorable (but biologically crucial) synthesis of ATP. The mechanism by which it operates is fascinating, but for the present purposes the key point is that the crucial function it performs absolutely requires an overall structure that is very large with respect to the sampling problem described in the previous section. The ATP synthase has to manage the flow of protons through the lipid bilayer in such a way as to produce the rotation of a molecular rotor, which in turn forces a conformational shift in the portion of the stator that contacts the other end of the rotor, which in turn causes cyclic binding and phosphorylation of ADP, followed by release of the desired product—ATP. Clearly only a substantial protein structure could possibly orchestrate a process of this physical and spatial complexity. Indeed, in view of what it accomplishes, what amazes us is how *small* it is.

For the purposes of this discussion, the ATP synthase provides an opportunity to refine the connection between protein size and the sampling problem. Returning to the lost gemstone metaphor, the gem is a new beneficial function that can be provided by a protein or a set of proteins working together, and the desert is the space of sequence possibilities within which successful solutions exist. Although some of the component proteins that form the ATP synthase are at the small end of the distribution shown in Figure 1 (see Figure 3 legend), none of these performs a useful function in itself. Rather, the function of ATP production requires the whole suite of protein components acting in a properly assembled complex. Consequently, the desert is most precisely thought of as a space of DNA sequences of sufficient length to encode that full suite. For our purposes, though, it will suffice to picture the space of protein sequences of a length equaling the combined length of the different protein types used to form the working complex (around 2,000 residues for the ATP synthase; see Figure 3 legend). This takes into

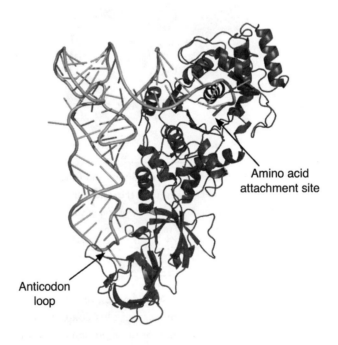

Figure 4. Structure of *E. coli* glutaminyl-tRNA synthetase, based on PDB entry 1GTR. RNA bases are here rendered as sticks protruding from the RNA backbone. The enzyme is a single protein chain of 554 amino acid residues.

account both the need for multiple non-identical chains in many working structures and the sequence redundancy that exists when multiple identical chains are used.

It also dramatically expands the size of the search space in the common case where a protein chain of one kind is useful only in combination with other kinds. Possible ways that undirected searches might succeed in spaces of this magnitude will be considered in the next section. The point here is simply that many biological functions require proteins or sets of proteins with combined lengths that are very large with respect to the sampling problem.

As a final example of the role of coupling in biology, we return to the connection between protein sequences and the DNA sequences that encode them. When expressed as an abstract mapping of codons to amino acids, this is the familiar genetic code often represented in table form. The physical embodiment of that code, though, is the set of aminoacyl-tRNAs—large RNA derivatives incorporating both the anticodon loops that "recognize" their respective codons on mRNA and the amino acids that these codons specify. Indeed, the genetic code only has its law-like status because aminoacyl-tRNAs are synthesized with anticodons paired very reliably with their corresponding amino acids. And because the anticodon loops in tRNAs are spatially distant from the amino acid attachment sites, the enzymes that accomplish this reliable pairing have to be large in order to attach the amino acids while simultaneously "verifying" that they are the correct ones (see Figure 4).

Many more examples could be given, but these adequately make the point that cellular functions often require proteins that are themselves large with respect to the sampling prob-

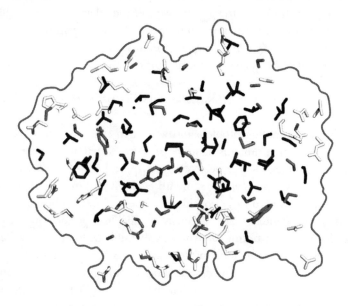

Figure 5. A slice through the TEM-1 beta-lactamase showing amino acid side chains shaded according to their affinity for water, based on PDB entry 1ERM. Histidine, lysine, arginine, asparagine, aspartic acid, glutamine, and glutamic acid side chains (shown white) have a high affinity for water. Leucine, isoleucine, methionine, valine, and phenylalanine (shown black) have a low affinity for water. The remaining amino acids (shown grey) have an intermediate affinity for water. There is a pronounced tendency for white side chains to be near the surface (shown as outline) and for black side chains to be in the interior.

lem. When we consider the sets of distinct proteins that commonly provide these functions, the sampling problem becomes even more challenging. And since many of those functions are essential to life, it seems inescapable that this challenge goes all the way back to the origin of protein-catalyzed metabolism.

Moreover, that conclusion is supported by a different line of reasoning based on structure rather than on function. In essence, it appears to be physically implausible for the large protein structures we see in biology to have been built up from tiny ancestral structures in a way that: 1) employed only simple mutation events, and 2) progressed from one well-formed structure to another. Simply put, the reason for this is that folded protein structures consist of discrete multi-residue units in hierarchical arrangements that cannot be built through continuous accretion. The material on the outer surface of an accretive structure, such as a stalagmite, is converted to interior material as successive layers are added. For structures of that kind the distinction between exterior and interior is one of time-dependent geometry rather than one of substance. By contrast, the process by which proteins fold involves a substantive distinction between interior and exterior that is evident in the final folded form (Figure 5). Since an evolutionary progression from tiny protein structures to large globular ones would have to repeatedly convert exterior surface to interior interface, this means that any such progression would have to coordinate the addition of appropriate new residues with the simultaneous conversion of existing ones. Considering that these structural additions and conversions would both involve many residues, it seems inescapable that one or the other of the above two conditions would be violated. Furthermore, on top of these conditions is the one of primary consideration in this section—that of function. Bringing this back into the picture, we conclude that it appears highly implausible for the protein structures we see in biology to have been built up from tiny ancestral structures in a way that: 1) employed only simple mutation events, 2) progressed from one well-formed structure to another, and 3) adequately performed the essential tasks of biology at each step.

4. THE EXTREME RARITY OF FUNCTIONAL PROTEIN FOLDS IN SEQUENCE SPACE

Having shown that the sampling problem introduced in the first section is *real*—meaning that cellular functions require proteins or suites of proteins that are of necessity far too large for the sequence possibilities to have been sampled appreciably—we now turn to the question of whether it is really a *problem* for standard evolutionary models. Two possibilities for circumventing the problem need to be considered. One of these has been mentioned already. It is the possibility that the multiplicity of sequences capable of performing the requisite functions, m, is large enough that working sequences can feasibly be found by undirected searches. The second possibility is that functional protein sequences bear a relationship to one another that greatly facilitates the search. In the desert metaphor, imagine all the different gems being together in close proximity, or perhaps at regular spacing along lines of longitude and latitude. Either way, finding the first gem would greatly facilitate finding the others because of the relationship their positions bear to one another.

We will complete our examination of the first of these possibilities before moving to the second. As noted previously, for m to be large enough to compensate for the size of the search spaces, modern proteins would have to be either much larger than necessary or constructed with much higher sequence fidelity than necessary. I pointed out above that both of these assumptions, superfluous size and superfluous sequence fidelity, make it hard to explain the sophisticated cellular systems that are devoted to producing the twenty amino acids and precisely incorporating them into long protein chains. Subsequently, I presented the evidential case for rejecting the notion of superfluous size. What remains is to consider the specific evidence pertaining to the possibility of superfluous fidelity.

Superfluous fidelity implies that protein synthesis is much more fastidious about amino acid identities than protein function is. Consequently, we can reframe this possibility in terms of functional constraints. Namely, for m to be large enough to resolve the sampling problem, it would have to be the case that protein functions place very loose constraints on amino acid sequences. A number of studies have in fact been construed as supporting this.[9] Rather than recount and analyze each of these, it makes sense to ask what would need to be established in order for us to conclude that the sampling problem has in fact been resolved.

To answer this in quantitative terms, we begin with a generous estimate of the number of independent opportunities for spontaneous mutation events to have produced a new protein that successfully provides a particular new function. Success here refers to evolutionary success, which is achieved by fixation of the new trait in a population through natural selection. Bacterial species are most conducive to this because of their large effective population sizes.[10] So let us assume, generously, that an ancient bacterial population sustained an effective size of 10^{10} individuals[11] while passing through 10^4 generations per year. Let us also assume that this population tolerated a mutation rate of one change per cell (roughly 300-fold higher than current rates),[12] and that most of these mutations occurred in portions of the genome that are not constrained by existing functions, making them free to evolve new ones. This presupposes a much higher tolerance of nonfunctional ("junk") DNA than modern bacteria exhibit. After five billion years, such a population would have had at most 5×10^{23} ($=5 \times 10^9 \cdot 10^4 \cdot 10^{10}$) opportunities for mutations to craft a protein that performs the function in question successfully.

Let us suppose for a moment, then, that the prevalence of functional sequences is above this cutoff. Since most necessary cellular functions require proteins or suites of proteins with 300 or more total residues,[13] we are effectively supposing that the multiplicity factor m introduced in the previous section can be as large as $20^{300}/5 \times 10^{23} \approx 4 \times 10^{366}$. In other words, we are supposing that particular functions requiring a 300-residue structure are realizable through something like 10^{366} distinct amino acid sequences. If that were so, what degree of sequence degeneracy

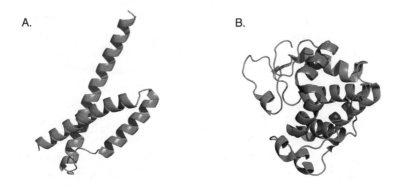

Figure 6. Structures of protein chains used for experimental measurement of the rarity of working sequences, based on PDB entries 1ECM and 1ERM. A) Structure of a single chain (93 residues) from the AroQ-type chorismate mutase examined by Taylor *et al.*[14] B) Structure of the 153-residue portion of the beta-lactamase examined by Axe.[15]

would be implied? More specifically, if 1 in 5×10^{23} full-length sequences are supposed capable of performing the function in question, then what proportion of the twenty amino acids would have to be suitable on average at any given position? The answer is calculated as the 300th root of $(5\times10^{23})^{-1}$, which amounts to about 83 percent, or 17 of the 20 amino acids. That is, by the current assumption, proteins would provide the function in question by merely *avoiding* three or so unacceptable amino acids at each position along their lengths.

No study of real protein functions suggests anything like this degree of indifference to sequence. In evaluating this, keep in mind that the indifference referred to here would have to characterize the whole protein. Natural proteins commonly tolerate some sequence change without complete loss of function. But this merely implies that completely inactivating mutations are relatively rare when the starting point is a highly functional wild-type sequence (e.g., 5 percent of the total in one study).[16] It does not imply that most mutations are harmless. Rather, the fact that several amino acid changes may be tolerated at certain positions within most proteins is readily explained by the capacity of well-formed structures to sustain moderate damage without complete loss of function.[17] Conditional tolerance of that kind does not extend to whole proteins, though, for the simple reason that there are strict limits to the amount of damage that can be sustained.

A study of the cumulative effects of replacing amino acid residues with chemically similar alternatives has demonstrated this.[18] Two unrelated bacterial enzymes, a ribonuclease and a beta-lactamase, were both found to suffer complete loss of function *in vivo* at or near the point of 10 percent substitution, despite the conservative nature of the changes. Since most substitutions would be more disruptive than these conservative ones, it is clear that these protein functions place much more stringent demands on amino acid sequences than the above supposition requires.

Two experimental studies provide reliable data for estimating the proportion of protein sequences that perform specified functions. One study focused on the AroQ-type chorismate mutase, which is formed by the symmetrical association of two identical 93-residue chains.[19] These relatively small chains form a very simple folded structure (Figure 6A). The other study examined a 153-residue section of a 263-residue beta-lactamase.[20] That section forms a compact structural component known as a *domain* within the folded structure of the whole beta-lactamase (Figure 6B). Compared to the chorismate mutase, this beta-lactamase domain has both larger size and a more complex *fold* (this being the term for the overall chain structure).

In both studies, large sets of extensively mutated genes were produced and tested. By placing suitable restrictions on the allowed mutations and counting the proportion of working genes that result, it is possible to estimate the expected prevalence of working sequences for the hypothetical case where those restrictions are lifted. In this way, prevalence values that are far too low to be measured directly can be estimated with reasonable confidence.

The results allow the average fraction of sampled amino acid substitutions that are functionally acceptable at a single amino acid position to be calculated. By raising this fraction to the power l, it is possible to estimate the overall fraction of working sequences expected when l positions are simultaneously substituted (see reference 17 for details). Applying this approach to the data from the chorismate mutase and the beta-lactamase experiments gives a range of values (bracketed by the two cases) for the prevalence of protein sequences that perform a specified function. The reported range[21] is one in 10^{77} (based on the beta-lactamase data: $l = 153$) to one in 10^{53} (based on the chorismate mutase data adjusted to the same length: $l = 153$). This shows that constraints vary from protein to protein and suggests that structural complexity may be an

The Nature of Protein Folds

important determinant of the severity of these constraints. More importantly, though, it shows that even very simple protein folds can place very severe constraints on sequence.

Rescaling the figures to reflect a more typical chain length of 300 residues gives a prevalence range of one in 10^{151} to one in 10^{104}. On the one hand, this range confirms the very highly many-to-one mapping of sequences to functions. The corresponding range of m values is 10^{239} (=$20^{300}/10^{151}$) to 10^{286} (=$20^{300}/10^{104}$), meaning that countless viable sequence possibilities exist for each protein function. But on the other hand it appears that these functional sequences are nowhere near as common as they would have to be in order for the sampling problem to be dismissed. The shortfall is itself a staggering figure—some 80 to 127 orders of magnitude (comparing the above prevalence range to the cutoff value of 1 in 5×10^{23}). So it appears that even when m is taken into account, protein sequences that perform particular functions are far too rare to be found by random sampling.

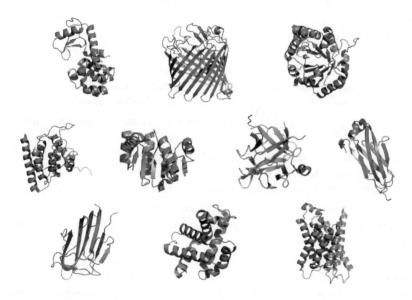

Figure 7. A sample of the structural variety of protein domain folds. Shown in the top row (left to right): bacteriophage T4 lysozyme (PDB: 167L), ompF porin from *E. coli* (PDB: 2OMF), human triosephosphate isomerase (PDB: 1HTI); middle row: run domain of mouse rap2-interacting protein X (PDB: 2CXF), a transport protein from *B. subtilis* (PDB: 1LSU), human interleukin 1-beta (PDB: 1L2H), mouse antibody light chain (PDB: 1UM5); bottom row: a fragment of human collagen (PDB: 1GR3), human hemoglobin alpha chain (PDB: 1IRD), sheep aquaporin (PDB: 1SOR).

The possibility yet to be examined is that functional protein sequences might bear a relationship to one another that allows natural mutations to discover new protein functions much more readily than wholly random sampling would. The most straightforward way for this to happen would be if functional sequences were generally much more similar to each other than random sequences are. The effect of such a general correlation between sequence and function would be to concentrate all the useful protein sequences within a tiny region of sequence space, making searches that start in that region much more likely to succeed.

Localized searches of this kind are known to work in some cases. Many enzymes, for example, can be made to perform their catalytic conversions on different substrates by changing just one or two amino acids within their active sites. Bacteria often evolve resistance to antibiotic variants in this way, by an existing resistance enzyme acquiring a new specificity. The evolutionary search for resistance to the new antibiotic works in these cases because the original enzyme needs only slight adjustment in order to perform the new task. Consequently a local search of point-mutation variants has a reasonably good chance of succeeding.

The problem comes when we attempt to generalize this local phenomenon. Although there are definite correlations between the various kinds of functions that proteins perform and the respective fold structures used to perform them, and these structural correlations often imply sequence correlations as well, it is simply not the case that *all* functional folds or sequences are substantially alike. Consequently, while local searches may explain certain local functional transitions, we are left with the bigger problem of explaining how so many fundamentally new protein structures and functions first appeared.

To get an idea of the scale of this problem, consider that the SCOP classification of protein structures currently lists 1,777 different structural forms for protein domains, the basic units of folded protein structure.[22] But since that figure is based upon known protein structures, it is certainly an underestimate. Each new genome project reveals numerous protein sequences with no significant similarity to any previously known sequence, which implies that the actual number of fundamentally distinct protein domains is much higher.[23] Whatever the true figure turns out to be, it is clearly large enough that no model of protein origins can be considered satisfactory without accounting for the origin of this great variety of domain forms.

In fact, although the sampling problem has here been framed in terms of protein chains, it could equally be framed in terms of domains. Since domains are presumed to be the fundamental units of conserved structure in protein evolution,[24] the question of whether functional sequences are confined to a small patch of sequence space is best addressed at this level of structure. And it turns out that domain sequences are not confined in this way. When structurally unrelated protein domain sequences are aligned optimally, the resulting alignment scores are very similar to the expected scores for randomized sequences with the same amino acid composition.[25] Since random sequences produced in this way are widely scattered through sequence space, this means that dissimilar natural sequences are as well. In fact, because amino acid composition correlates with structural class,[26] we would expect random sequences with average compositions to align somewhat better than dissimilar natural sequences do. Indeed, such wide dispersion of natural domain sequences throughout sequence space is not surprising considering the great variety of domain structures that these sequences form (Figure 7).

On average, protein domains tend to be about half the size of complete protein chains (compare Figure 8 to Figure 1), implying that a typical chain forms two domains. Of course, this means that the space of sequence possibilities for an average domain, while vast, is nowhere near as vast as the space for an average chain. But as discussed above, the relevant sequence space for evolutionary searches is determined by the combined length of *all* the new domains needed to produce a new beneficial phenotype.

As a rough way of gauging how many new domains are typically required for new adaptive phenotypes, the SUPERFAMILY database[27] can be used to estimate the number of different protein domains employed in individual bacterial species, and the EcoCyc database[28] can be used to estimate the number of metabolic processes served by these domains. Based on analysis of the genomes of 447 bacterial species,[29] the projected number of different domain structures per species averages 991.[30] Comparing this to the number of pathways by which metabolic processes are carried out, which is around 263 for *E. coli*,[31] provides a rough figure of three or four

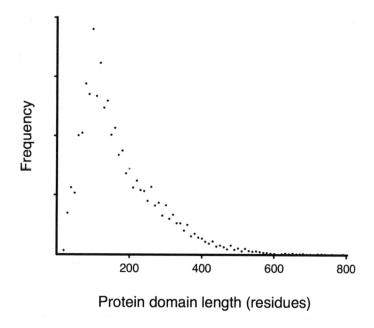

Figure 8. Length distribution for 9,535 SCOP-defined protein domains, based on a non-redundant set (less than 40 percent pairwise sequence identity) obtained from ASTRAL SCOP version 1.73.[32] The mean and median lengths are 178 residues and 145 residues, respectively. Lengths were binned in 10-residue increments, the most occupied bin containing 665 protein domains.

new domain folds having arisen, on average, for every new metabolic pathway.[33] Consequently, a successful evolutionary mechanism would need to be capable of locating sequences that amount to anything from one in 10^{159} to one in 10^{308} possibilities,[34] something the standard model falls short of by well over a hundred orders of magnitude.

5. Conclusion

What was raised decades ago as an apparent limitation to the evolution of new proteins has here been dubbed the *sampling problem*—the impossibility of any evolutionary process sampling anything but a minuscule fraction of the possible protein sequences. At that time, several missing pieces of information made it difficult to conclude with certainty whether this limitation presented a serious challenge to neo-Darwinian accounts of the origin of new proteins. With those pieces now in place, it has become clear both that the sampling problem is real and that it actually does present a serious challenge.

We have used a picture of gems hidden in a vast desert at various points in our discussion in order to illustrate the challenge. Now that we have estimated the relevant fractions it may be helpful to return to this picture. Imagine that the search for gems is conducted by specifying sample points as mathematically exact geographic coordinate pairs (longitude and latitude). Sampling then consists of determining whether a gemstone rests at any of these specified points. A target the size of a grain of sand amounts to about one part in 10^{20} of a search space the size of the Sahara, which is above the feasibility threshold of one part in 5×10^{23}. So under favorable circumstances a Darwinian search would be capable of locating a sand-grain-sized gemstone in a Sahara-sized search space. As mentioned above, the ability to accomplish a search on this scale is clearly of some practical significance.

But as a generator of molecular complexity, it turns out not to be very significant. Extending our desert picture, imagine that the top surface of every grain of sand in the Sahara has a miniature desert of its own resting upon it—one in which the entire Sahara is replicated in minute detail. We may call the sub-microscopic sand in these miniature deserts *level-1* sand, referring to the fact that it is one level removed from the real world (where we find *level-0* sand). This terminology can be applied to arbitrarily small targets by invoking a succession of levels (along the lines of De Morgan's memorable recursion of fleas[35]). In terms of this picture, the sampling problem stems from the fact that the targets for locating new protein folds are *much* smaller than a grain of level-0 sand. For example, the target that must be hit in order to discover one new functional domain fold of typical size covers (in total) not more than one ten-trillionth of the surface of a single grain of level-1 sand.[36] Under favorable circumstances a Darwinian search will eventually sample the grain of level-0 sand on which the right grain of level-1 sand rests, but even then the odds of sampling that level-1 grain are negligible, to say nothing of the target region on that grain.[37] And the situation rapidly deteriorates when we consider the relevant target—not just a new domain, but a beneficial new phenotype that employs (typically) several new protein structures. In the end, a search mechanism that cannot locate a small patch on a grain of level-14 sand is not apt to provide the explanation of fold origins that we seek.[38]

But consider this closing thought. The e-mail messages that we routinely compose reside in comparably small patches within another kind of sequence space. Consequently the claim that intelligence can successfully navigate to such small targets is indisputable. On the other hand, the claim that intelligence has probably played a decisive role in the origin of protein folds is vigorously disputed. Yet it is not entirely clear that there is a scientific or rational basis for this dispute. What *is* clear is that any such basis would require a convincing alternative to intelligence, which starts with a frank recognition of the difficulties these alternatives face.

The Nature of Protein Folds

Notes

1. M. Eden, "Inadequacies of Neo-Darwinian Evolution as a Scientific Theory." In P. S. Moorhead and M. M. Kaplan, eds., *Mathematical Challanges to the Neo-Darwinian Interpretation of Evolution* (Philadelphia: Wistar Institute Press, 1967), 109–11. L. M. Spetner, "Information Transmission in Evolution," *IEEE Transactions on Information Theory* IT-14 (1968): 3–6. F. B. Salisbury, "Natural Selection and the Complexity of the Gene," *Nature* 224 (1969): 342–43. L. M. Spetner, "Natural Selection versus Gene Uniqueness," *Nature* 226 (1970): 948–49.

2. The term *proteome* refers to the complete set of proteins that operate in an organism or cell type.

3. http://ecocyc.org

4. W. A. Dembski, *The Design Inference* (Cambridge: Cambridge University Press, 1998), 209.

5. P. Chelikani, X. Carpena, I. Fita, and P. C. Loewen. "An Electrical Potential in the Access Channel of Catalases Enhances Catalysis," *Journal of Biological Chemistry* 278 (2003): 31290–296.

6. J. B. Thoden, X. Huang, J. Kim, F. M. Raushel, and H. M. Holden, "Long-Range Allosteric Transitions in Carbamoyl Phosphate Synthetase," *Protein Science* 13 (2004): 2398–405.

7. Ibid.

8. http://www.rcsb.org/pdb/home/home.do

9. See, for example: A. R. Davidson, K. J. Lumb, and R. T. Sauer, "Cooperatively Folded Proteins in Random Sequence Libraries," *Nature Structural Biology* 2 (1995): 856–63. A. D. Keefe and J. W. Szostak, "Functional Proteins from a Random Sequence Library," *Nature* 410 (2001): 715–18. A. Yamouchi, T. Nakashima, N. Tokuriki, M. Hosokawa, H. Nogamai, S. Arioka, I. Urabe, and T. Yomo, "Evolvability of Random Polypeptides through Functional Selection within a Small Library," *Protein Engineering* 15 (2002): 619–26. Y. Hayashi, H. Sakata, Y. Makino, I. Urabe, and T. Yomo, "Can an Arbitrary Sequence Evolve towards Acquiring a Biological Function?" *Journal of Molecular Evolution* 56 (2003): 162–68.

10. Stochastic aspects of survival having nothing to do with fitness make it unlikely that any particular instance of beneficial mutation will lead to fixation of the resulting genotype. Roughly speaking, the effective size of a real population is the size of a hypothetical population lacking these stochastic influences that is as conducive to fixation of new genotypes as the real one is.

11. M. Lynch and J. S. Conery, "The Origins of Genome Complexity," *Science* 302 (2003): 1401–4.

12. J. W. Drake, B. Charlesworth, D. Charlesworth, and J. F. Crow, "Rates of Spontaneous Mutation," *Genetics* 148 (1998): 1667–86.

13. We will refine this number below.

14. S. V. Taylor, K. U. Walter, P. Kast, and D. Hilvert, "Searching sequence space for protein catalysts." *Proceedings of the National Academy of Sciences (USA)* 98 (2001): 10596–601.

15. D. D. Axe, "Estimating the prevalence of protein sequences adopting functional enzyme folds." *Journal of Molecular Biology* 341 (2004): 1295–315.

16. D. D. Axe, N. W. Foster, and A. R. Fersht, "A Search for Single Substitutions That Eliminate Enzymatic Function in a Bacterial Ribonuclease," *Biochemistry* 37 (1998): 7157–66.

17. This is the *buffering* effect described in D. D. Axe, "Estimating the Prevalence of Protein Sequences Adopting Functional Enzyme Folds," *Journal of Molecular Biology* 341 (2004): 1295–315.

18. D.D. Axe, "Extreme functional sensitivity to conservative amino acid changes on enzyme exteriors." *Journal of Molecular Biology* 301 (2000): 585–95.

19. S.V. Taylor, K.U. Walter, P. Kast, and D. Hilvert, "Searching sequence space for protein catalysts." *Proceedings of the National Academy of Sciences (USA)* 98 (2001): 10596–601.

20. D.D. Axe, "Estimating the prevalence of protein sequences adopting functional enzyme folds." *Journal of Molecular Biology* 341 (2004): 1295–315.

21. Ibid.

22. SCOP version 1.73 (http://scop.mrc-lmb.cam.ac.uk/scop/) organizes domain structures into 1,777 categories at the 'superfamily' level, based on "structures and, in many cases, functional features [that] suggest a common evolutionary origin" (A.G. Murzin, S.E. Brenner, T. Hubbard, and C. Chothia, "SCOP: A structural classification of proteins database for the investigation of sequences and structures." *Journal of Molecular Biology* (1995): 536–40).

23. N. Siew and D. Fischer, "Twenty thousand ORFan microbial protein families for the biologist?" *Structure* 11 (2003): 7–9.

24. M. Bashton and C. Chothia, "The generation of new protein functions by the combination of domains." *Structure* 15 (2007): 85–99.

25. The Z-score of an alignment compares the raw alignment score to the raw scores of optimally aligned randomized versions of the initial pair of sequences. Z-scores plotted for dissimilar domain sequences are distributed around zero, meaning that the actual alignments tend to be comparable to randomized alignments (see the Z-scores plotted in C. Webber and G. J. Barton "Estimation of P-values for global alignments of protein sequences." *Bioinformatics* 17 (2001): 1158–167).

26. K. C. Chou, "Prediction of protein folding types from amino acid composition by correlation angles." *Amino Acids* 6 (1994): 231–46.

27. http://supfam.mrc-lmb.cam.ac.uk/SUPERFAMILY/.

28. http://ecocyc.org.

29. From SUPERFAMILY 1.69 release (http://supfam.mrc-lmb.cam.ac.uk/SUPERFAMILY/).

30. Calculated by dividing the number of superfamilies detected in each species by the fraction of that species' genome assigned to any superfamily, then taking the mean of this ratio.

31. This is the number of metabolic pathways in the 12.5 release of EcoCyc, excluding "superpathways" to avoid redundancy.

32. http://astral.berkeley.edu/.

33. Consistent with this is the tendency for domains of similar structure (as classified at the superfamily level in SCOP [http://scop.mrc-lmb.cam.ac.uk/scop/]) to appear in only one or two metabolic pathways (see M. A. Saqi and M. J. Sternberg "A structural census of metabolic networks for *E. coli*." *Journal of Molecular Biology* 313 (2001): 1195–206).

34. The higher prevalence is based on the chorismate mutase data in Taylor *et al.* (2001) with l = 153 for each of three domains; the lower is based on the beta-lactamase data in Axe (2004) with l = 153 for each of four domains.

35. "Great fleas have little fleas upon their backs to bite 'em, And little fleas have lesser fleas, and so *ad infinitum*. . . ." (Augustus De Morgan).

36. Based on the chorismate mutase data in Taylor *et al.* (2001) and a domain of average size (l = 153).

37. The target might actually be fragmented into dots that appear on many different grains. Nonetheless it is the total target size in comparison to the total search space that determines the difficulty of the search.

38. Based on the one in 10^{308} figure, from the beta-lactamase data [Axe (2004)] with l = 153 for each of four domains.

19

THE LIMITS OF NON-INTELLIGENT EXPLANATIONS IN MOLECULAR BIOLOGY

MICHAEL J. BEHE

I. INTRODUCTION

In 1859, Charles Darwin published his seminal book *On The Origin of Species by Means of Natural Selection,* which purported to explain how life on Earth changed over eons of time by simple, entirely natural means. Until Darwin's theory appeared, the reason why biota changed with time was a matter of speculation, but for the most part naturalists envisioned that there was an external driving force for what came to be called evolution. Perhaps God periodically replaced animal life on Earth, or perhaps he implanted an inner drive in life to guide life along the path he intended it to take. Although the reason was unknown, the great variety and complexity of life was thought to be the result of an external teleological agency, as William Paley argued in *Natural Theology*.

Then came Darwin. The importance of Darwin's theory, in both a philosophical and scientific sense, was not that it hypothesized the idea of common descent—that modern life forms derived over time from earlier life forms. Rather, its shock value lay in its proposal that no teleology undergirded the process. The life forms of the past gave rise to different life forms in the present, not because there was a plan for that to happen or a guiding hand to shepherd the process along a path. Instead, life meandered along a rugged evolutionary landscape governed solely by the blind forces of nature, both animate and inanimate. The path of life was determined in the same way that, say, the path of a boulder tumbling down a mountainside is determined: by happenstance and unyielding laws of nature, but not by guidance.

Darwin's *Origin of Species* was "one long argument" in the sense that, while he argued for his ideas, he could not prove them with data that were available in his lifetime. Although he brought many facts to the table, Darwin was unable to point to a single example of the transmutation of species in nature; the best he could do with the data available to him was to point to examples of domestic breeding, which selected traits useful to the breeder, and claim that nature must go through an analogous selection process when she weeds out the weak and preserves the strong in the process he called natural selection.

Over the next few decades teleology was swept away from the study of biology. Interestingly, however, Darwin's specific mechanism of evolution was not well regarded, and other, non-teleological ones were proposed by other scientists. The idea of Darwin's that was widely adopted by biologists was not the process of natural selection working on random variation. Rather it was the presupposition that life could be explained without appealing to teleology. As the eminent biologist, historian of biology, and leader of the mid-twentieth-century neo-Darwinian revolution Ernst Mayr wrote in *One Long Argument*,[1] although almost all late-nineteenth-century biologists accepted common descent, almost none accepted Darwin's mechanism (see Table 1).

Theorist	Common Descent	Multiplication of Species	Gradualism	Natural Selection
Lamarck	No	No	Yes	No
Darwin	Yes	Yes	Yes	Yes
Haeckel	Yes	?	Yes	In part
Neo-Lamarckians	Yes	Yes	Yes	No
T. H. Huxley	Yes	No	No	(No)[a]
de Vries	Yes	No	No	No
T. H. Morgan	Yes	No	(No)[a]	Unimportant

a. Parentheses indicate ambivalence or contradiction

Table 1.[2] The composition of the evolutionary theories of various evolutionists[3]

This may surprise modern readers. Yet the reason why Darwin's proposal met with mixed success in the late nineteenth century is that his theory is actually an amalgam of multiple hypotheses, some much more firmly supported than others. In both scholarly and popular literature one frequently finds references to "Darwin's theory of evolution," as though it were a unitary entity. In reality, Darwin's "theory" of evolution was a whole bundle of theories, and it is impossible to discuss Darwin's evolutionary thought constructively if one does not distinguish its various components. The current literature can easily leave one perplexed over the disagreements and outright contradictions among Darwin specialists, until one realizes that to a large extent these differences of opinion are due to a failure of some of these students of Darwin to appreciate the complexity of his paradigm.[4]

In a chapter of *One Long Argument* titled "Ideological Opposition to Darwin's Five Theories," Ernst Mayr stressed that what is commonly called "Darwin's Theory" actually contains at least five distinct claims. Mayr lists these five separate claims as:

(1) *Evolution as such.* This is the theory that the world is not constant nor recently created nor perpetually cycling but rather is steadily changing and that organisms are transformed in time.

(2) *Common descent.* This is the theory that every group of organisms descended from a common ancestor, and that all groups of organisms—including animals, plants, and microorganisms—ultimately go back to a single origin of life on Earth.

(3) *Multiplication of species.* This theory explains the origin of the enormous organic diversity. It postulates that species multiply, either by splitting into daughter species or by "budding," that is, by the establishment of geographically isolated founder populations that evolve into new species.

The Limits of Non-Intelligent Explanations in Molecular Biology

(4) *Gradualism*. According to this theory, evolutionary change takes place through the gradual change of populations and not by the sudden (saltational) production of new individuals that represent a new type.

(5) *Natural selection*. According to this theory, evolutionary change comes about through the abundant production of genetic variation in every generation. The relatively few individuals who survive, owing to a particularly well-adapted combination of inheritable characters, give rise to the next generation.

The strength of evidence showing that, say, "the world is not constant" (the first claim) is disputed by virtually no one. On the other hand, the strength of evidence showing that natural selection is the sole or principle mechanism of that change (the fifth claim) is much weaker. Mayr wrote that in the years after Darwin published his theory, although most scientists accepted change in the world and common descent, few scientists thought that the mechanism of random variation and natural selection was convincing, as is shown in Table 1, which is reproduced from Mayr's book.

Not only was Darwin unable to test his theory, no other scientist of the time could either. It was not until the 1950s, nearly a century after the publication of *The Origin,* that the first experiment was conducted that was widely credited as testing Darwin's hypothesis of natural selection. In a series of trials the British naturalist Bernard Kettlewell obtained data that seemed to show that a white variety of the moth *Biston betularia* was less prone to predation by birds when resting on light-colored tree trunks, while a darker variety of the same species was less prone to predation on dark tree trunks. *Scientific American* hailed the work as the first test of Darwin's theory of natural selection. Later work disputed Kettlewell's results, but in the meantime many other experiments on animals, plants, and microorganisms firmly established the principle of natural selection. In the early twenty-first century, the idea of natural selection is about as well established as most other facets of Darwin's theory: change over time; common descent; and multiplication of species.

In fact, only one facet of Darwin's theory remains unproven today—a facet that, although critical, wasn't specifically mentioned by either Darwin or Ernst Mayr. And that is the efficacy of random mutation, when acted upon by natural selection, to build the complex structures of life. Darwin neglected random mutation for the excellent reason that the basis of heredity was unknown in his time—he did not know about the properties of DNA. He assumed that there would be a supply of random "variation" in nature that would be sufficient for his theory, but he knew not what caused the variation. Mayr seems to have folded randomness into Darwin's idea of natural selection, in the presumption that there is "abundant production of genetic variation in every generation."

It is critical to realize that the assumption that random variation (random mutation in modern understanding) is adequate for its assigned role in Darwin's scheme had up to the time of Ernst Mayr's writing never been adequately tested. Nonetheless, it is the "randomness" in Darwin's theory—the assumption that no guiding hand or teleological principle affected the history of life—which was, and still is, by far its most important scientific and philosophical contention. It is the adequacy of random mutation which this chapter will address.

2. Why Not Randomness?

It is clear that random events do occur in biology, as elsewhere in the world. Birth defects, genetic mutations, and other haphazard phenomena certainly happen, and they provide variation upon which natural selection can act. So why should we question whether random variation/natural selection can perform the task that Darwin assigned to it?

There are two main reasons. The first reason that random mutation looks like a poor choice for supplying appropriate variation to natural selection is that we now know that life is much more complex than was known in Darwin's day. In the mid-nineteenth century the cell—which we now know to be the foundation of life—was thought by such prominent scientists as Ernst Haeckel and Thomas Huxley to be little more than a glob of protoplasm, essentially a microscopic piece of Jell-O®. With the foundation of life thought to be so simple, it was easy to imagine that life just might appear spontaneously. In fact, Huxley and Haeckel famously supposed that some odd-looking mud dug up by an English exploring vessel was cellular life, and so concluded that cells could be generated spontaneously from sea mud.[5] And if the foundation of life were relatively simple, then shaping and transforming it by random changes and natural selection might not be all that difficult, like shaping clay into pottery.

Yet we now know that Huxley and Haeckel and other scientists of Darwin's day were spectacularly wrong. The cell, the foundation of life, far from being a simple glob of protoplasm, is much better described as an automated nanoscale factory, containing technology that we humans can only gape at.[6] As the very basis of life grows exponentially in complexity, the theory that seemed it might work with imaginary "protoplasm" appears less and less adequate to real life.

The second reason to doubt the power of randomness is that, at least in some circumstances, it is known to be inadequate to solve complex problems. For example, Richard Watson, a computer scientist and lecturer at the University of Southampton, considers the poor record of success of an analog of random mutation/natural selection in solving computer problems:[7]

> In computer science we recognize the algorithmic principle described by Darwin—the linear accumulation of small changes through random variation and selection—as *hill climbing*, more specifically *random mutation hill climbing*. However, we also recognize that hill climbing is the simplest possible form of optimization and is known to work well only on a limited class of problems.

Those problems include very simple ones that can be solved by changing just one or a few variables—changes in size, color, antibiotic resistance, or other traits that can be affected by changes in single genes. "Darwin's masterful contribution was to show that there was *some* principle of optimization that could be implemented in biological systems" (his emphasis), writes Watson—apparently, however, not the right one for complex systems. To rescue Darwin, Watson proposes his idea of compositional evolution, where several biological modules join together for the benefit of all. However, Watson's proposal remains substantially untested.

The Limits of Non-Intelligent Explanations in Molecular Biology

3. Results from Randomness in Nature

Malaria Studies

Human genetic responses to malaria. In the past few decades, new results from the study of evolution in nature and in the lab have become available. These results, I argue, strongly circumscribe the role of randomness, demonstrating that, although it can supply the variation for modest changes in biological systems, as Richard Watson thought, it is ineffective for building more complex systems.

The first category of results I will consider is the human genetic response to malaria (Table 2). Although we in the developed world may give it little thought, for most people on this planet, malaria remains a scourge and a killer. About a million humans die each year from malaria, mostly children under the age of five. For ten thousand years humans and the fiercest species of malaria, *Plasmodium falciparum,* have coexisted in a classic evolutionary struggle. The single-celled parasite enters the human bloodstream with the bite of an infected mosquito, travels to the liver where it undergoes multiplication and development, and then several hundred thousand parasites break out of the liver and enter the victim's bloodstream. Through a complex process, an individual parasite enters a human red blood cell, feeds off the hemoglobin, and proceeds to make a score of copies of itself. The hemoglobin consumed, the malarial parasites burst the blood cell and then enter other red cells. In the second round, four hundred are produced. Multiplying exponentially, the parasites can consume a large fraction of a person's blood in days.

Over the past ten millennia, humans have acquired a handful of genetic mutations that confer a measure of resistance to malaria. Let's consider them.[8]

Perhaps the best known anti-malarial genetic change in humans is the sickle-cell gene. A single change at the sixth position of the beta chain from a glutamic acid residue to a valine residue grants to heterozygotes (whose other globin allele is the normal adult form) resistance to the lethal affects of malaria. Unfortunately, as is widely known, if a child inherits two copies of the sickle gene (homozygosity) she acquires the genetic disease sickle-cell anemia, which is generally lethal before the age of twenty in areas without modern medical care.

Some carriers of the sickle gene have also acquired a second mutation in a separate area of the genome. The control region upstream of the gamma chain of fetal hemoglobin is altered so that, rather than shutting off the fetal gene at birth—as happens in normal individuals—the fetal form of hemoglobin is manufactured into adult life. This arrangement apparently is helpful for homozygotes of the sickle gene; it seems to dilute the sickle hemoglobin in the red blood cell with fetal hemoglobin, which attenuates red cell sickling and the crippling symptoms that result.

Besides the well-known sickle mutation, in the past ten thousand years a handful of other mutations have appeared in the human genome that confer some resistance to malaria. These are listed in Table 2. Several other specific point mutations in hemoglobin (HbC and HbE) appear to confer some resistance to malaria. So also do the thalassemias, where a broken copy (by dint of point mutation or deletion) of either the gene for the alpha or beta chain of hemoglobin is inherited. The imbalance in the number of functioning alpha and beta hemoglobin genes in the thalassemia patient somehow makes him more resistant to malaria.

Other mutations that help in the struggle with the malarial parasite are ones to the glucose-6-phosphate dehydrogenase gene. G6PD is normally found in the red blood cell and is a housekeeping gene. The beneficial mutations either decrease or eliminate G6PD function. Also, if the gene for Duffy antigen is switched off in the red blood cell by a specific point mutation, the patient is immune to a species of malaria called *Plasmodium vivax,* which ordinarily enters the red blood cell by attaching to that antigen. The gene for Band 3 protein normally produces a membrane protein which is an ion exchanger. Heterozygotes in which one of the copies of the gene has been deleted have some resistance to malaria. No homozygotes for the condition have been found, presumably because the total lack of the gene kills a child before birth.

Gene	Mutation	Adverse effects (clinical / molecular)
Hemoglobin, beta chain	HbS: specific point mutation	anemia; usually lethal in two copies / increased fragility of red blood cell
	HbC: specific point mutation	slight anemia / increased fragility of red blood cell
	HbE: specific point mutation	none apparent
	thalassemia: various point mutations, deletions	anemia / broken gene
Hemoglobin, alpha chain	thalassemia: deletion	anemia / broken gene
Hemoglobin, gamma chain	HPFH: various deletions, point mutations in control regions	none apparent / broken genetic controls
G6PD	various point mutations, deletions	anemia / decrease or loss of G6PD function
Band 3 protein	deletion	lethal in two copies / broken gene
Duffy antigen	specific point mutation in control region	none apparent / protein no longer expressed in red blood cells

Table 2. Human genetic effects of malaria trench warfare[9]

Significance of malaria results. What is the significance of our understanding of the genetic changes in the human genome as a result of our evolutionary battle with malaria over the past ten thousand years? The first thing to remark is that this is one of our best examples of evolution in action. Because we are concerned with human health, we have studied the genetics of our interaction with malaria perhaps more than any other evolutionary phenomenon. What's more, because we have relatively ready access to human subjects and their families, we can obtain good reliable data, more so than on other creatures. Finally, because of the large human population that is affected by malaria, and the length of time that we have been battling it, the data is more precise than for other studies. So what lessons can we glean?

The first lesson is that, while all the mutations assist survival in malarious regions, almost all the mutations are degradative. In thalassemias, genes are deleted or rendered non-functional. The same goes for mutations in G6PD and Duffy antigen. In the case of the control region of the fetal hemolobin gene in hereditary persistence of fetal hemoglobin, it is rendered nonfunctional, so that its normal function of turning off fetal hemoglobin production at birth is destroyed. Even several of the substitution mutations (HbS, C) seem to have mild deleterious consequences such as anemia.

It has been learned over the past few decades that by far most mutations which have an effect on an organism are deleterious. The results from the study of human genetic changes with regard to malaria show that even within the small fraction of mutations that are beneficial, the great majority of them are degradatory, either deleting genes completely or degrading their function. Thus we now know that the fraction of mutations that are both beneficial and constructive—helping to build a complex structure of systems—is at best an infinitesimal portion of possible mutations. Worse, however, is that beneficial-yet-degradatory mutations will be selected by natural selection and become fixed in a population. Over time, then, genomes will degrade in response to natural selection acting on random mutation.

Another factor working against Darwin's theory should not be overlooked here: It is much easier and more rapid to degrade a gene than to make a specific change in a gene. The reason is easy to see in the case of thalassemias. To produce a thalassemic individual, any of hundreds of different mutations will suffice. Many changes in the gene for a globin chain will insert a stop codon, cause a frame shift, substitute a deleterious amino acid, and so forth. On the other hand, to get the sickle-cell gene, a single nucleotide has to be altered. Therefore, thalassemias will on average pop up in a population at a rate hundreds of times faster than the sickle mutation.

Extrapolating this lesson from malaria shows us why a Darwinian process of random mutation and natural selection is such a poor candidate to build complex, integrated biochemical machinery. The reason is that such machinery is not made by breaking genes, but by integrating genes into a coherent system. At the very least it requires the components of a molecular machine to develop specific binding sites for their correct partners. However, from what we have learned by our study of malaria, such a process is doomed to failure.

Suppose there existed an organism in some environment that was under a certain selective pressure. Suppose further that, theoretically, the organism could counter the selective pressure by evolving a new molecular machine that is built by assembling molecular machinery from five proteins that previously had not worked together. At the least, that assembly would require mutating specific residues in the proteins so that they could bind each other and interact. In the meantime, however, suppose the organism could also respond to the selective pressure by deleting or disabling a specific gene. The second route, disabling the gene, would occur at a much faster evolutionary rate than assembling components of a new molecular machine. That would obviate the selective pressure, so the net result would be an organism that can survive in its environment but with fewer components than it had before. Over time, the organism would degrade to a minimum.

A final lesson from our malaria data is that only a very limited number of genes were able to help. Even though a very large fraction of humanity (hundreds of millions to billions of people) have been under relentless life-and-death selective pressure from malaria for over ten thousand years, the number of genes or genetic control regions that have been useful to mutate is only a handful: alpha and beta globin, fetal globin transcription controls, and the other elements listed in Table 2. The other twenty thousand genes or so could not be mutated to be helpful, even in such dire circumstances.

Laboratory Studies of *E. Coli* Evolution

Results of long-term studies. Although studies on human genetic changes in response to the selective pressure of malaria are very informative, they have some limitations. First, although the population size of humans is large as far as large animals go, it is quite small when compared to the number of organisms that have existed on this Earth since its formation. Since the more chances that random mutation is given, the more likely it is to stumble across a fortunate change, then everything else being equal, studies that include greater numbers of organisms will be more reliable and informative about the power of random mutation. Second, although humans have battled malaria for ten thousand years, that is roughly only four hundred generations. Again, other things being equal, the more time and generations given to random mutation and natural selection, the more likely it is to come up with a beneficial result. To increase the number of organisms and generations in a study, one can turn to microorganisms. In several such studies, results reinforce the lessons learned from human genetic changes in response to malaria.

In the early 1990s, the Michigan State microbiologist Richard Lenski set out to study the long-term evolution of cultures of the bacterium *Escherichia coli* in his laboratory.[10] Every day he grew small flasks of the bacterium such that they underwent six to seven generations until the culture reached saturation. Each day he would transfer a potion of the previous day's culture to a new flask and repeat the process. At present more than forty thousand generations of *E. coli* have lived and died in the Michigan lab. That is about a hundred times the number of generations of humans that have battled malaria in the past one hundred centuries. What's more, each flask cultured in the lab contained about a hundred million *E. coli* cells, which is about a hundred times the number of humans in a typical generation over the past ten thousand years (only relatively recently has the population of humans mushroomed). The bottom line is that the laboratory study gave random mutation and natural selection orders of magnitude more opportunities to come up with clever changes to *E. coli* than humans have had in their malaria battle.

Unlike humans, the *E. coli* in Lenski's lab usually had no specific threat to their wellbeing, such as malaria, to contend with. However, that hardly means that evolution took a holiday. One of the most important evolutionary factors in an organism's environment is the presence of other organisms. Just as modern humans apparently competed successfully against Neanderthals, an *E. coli* in a flask in Lenski's lab that fortuitously acquired a beneficial mutation would out-compete its fellow organisms and eventually dominate the population. What's more, it is not tethered to a single task, such as trying to deal with a parasite such as malaria. Any inventive strategy that an *E. coli* luckily acquired through random mutation of any of its molecular systems could rocket it ahead of its competitors, and it would inherit the flask. Repetition of the beneficial mutation/selection process potentially could build some new interactive system.

That didn't happen. Instead, the *E. coli* in Lenski's experiment behaved much like humans in their battle with malaria. Degrading a handful of genes gave some bacteria an advantage over others in the flask.

As Lenski began and proceeded with his experiment, he quickly determined that some *E. coli* had acquired beneficial mutations that allowed them to dominate the population. Unfortunately, determining exactly which mutations had occurred, of the many millions of possible ones, was a daunting task. Nonetheless, after more than a decade of labor, Lenski was able to track down a handful of mutations that contributed to the dominance of the winning bacteria. Like human mutations in response to malaria, they all appear to be degradative.[11] This is indicated from the fact that the same genes in separate experimental flasks acquired mutations that benefitted the bacterium, but the specific changes to the genes were different. Apparently, much like the case of thalassemia in humans, where any of hundreds of mutations that switch off or degrade the globin gene are useful in malarious regions, switching off or degrading certain *E. coli* genes was helpful to the bacteria in the Lenski laboratory. As with the human genetic changes, there were no mutations in the bacterial studies that appeared to be interacting to build a complex system, such as those that fill a typical cell.

Significance of* E. coli *results. Comparing the results of Lenski's study with human changes in response to malaria allows us to draw an important conclusion: the general type of results—degradation of various genes but no constructive genetic changes—is seen in both situations, even though many specific factors are different. For example, *E. coli* are single-celled, asexual bacteria while humans are multicelled, sexual animals. The number of generations and population sizes differed by orders of magnitude but made no apparent difference to the evolutionary outcome of the studies. In both cases only a handful of genes changed. The similar results over vast phylogenetic distances, population sizes, and generations indicates that the results are insensitive to those factors, so we can confidently extrapolate to even greater distances. The take-home lesson from the work on human genetic changes in response to malaria is that random changes much more easily break or degrade genes—and occasionally that can be helpful to an organism. Constructing new genes or features of genes is much more difficult.

STUDIES OF TWO TYPES OF MICROORGANISMS IN THE WILD

Work on humans and *E. coli* is very informative, since it shows similar evolutionary results for two very disparate systems. Yet if it does reflect the real limitations of random mutation and natural selection, we should find substantially similar results when we look at other systems in nature. Two well-studied systems that bear closely on our questions are those of the microorganisms *Plasmodium falciparum* (malaria) and the human immunodeficiency virus (HIV). Because they are so well studied, and because evolutionary changes in the microorganisms are tracked closely, a wealth of data is available for both. Furthermore, both attain astronomical population sizes. Consider that the total number of *E. coli* cells produced in the years of work by Richard Lenski is on the order of $10^{13.}$ Yet the number of malarial cells produced in the world each year is on the order of 10^{20}.[12] Thus, random mutation has had hundreds of millions to billions of times more opportunities to alter *P. falciparum* than in the *E. coli* laboratory studies.

Nonetheless, results with malaria have shown similarities to results with *E. coli* and humans—in an evolutionary challenge, only a few genes can mutate to help, and the changes are simple ones to pre-existing genes, not the start of a new, complex system such as the machinery that fills the cell. A specific challenge malaria has faced in the past century has come from human medicines and drugs used to treat the infection. The drug of choice from the 1950s to the 1970s was chloroquine, a small organic molecule that killed the malarial parasite but had few detrimental side effects for the human patient. After several decades, however, a strain of the parasite developed immunity to chloroquine, and over succeeding years this strain spread widely around the world.[13]

It has only been in the past few years, as genomic sequencing tools improved, that the specific mutations conferring chloroquine resistance have been elucidated. It turns out that the principal gene conferring resistance is *pfcrt,* which codes for a protein called the *Plasmodium falciparum* chloroquine resistance trait, or PFCRT. Even more recent genetic studies have shown that multiple resistant and nonresistant strains seem to differ in variability in only the region of their genomes coding for PFCRT and one other region.[14] Thus, the genes that can help malaria in the teeth of chloroquine poisoning are very few, despite the fact that an enormous number of opportunities are available. What's more, the change to PFCRT was restricted to amino acid substitutions at a handful of residues (different resistant strains from different parts of the world exhibited some differences in the substitutions). In other words, the pattern shown by evolutionary studies of humans and *E. coli* continued with malaria, despite the huge numbers.

Let's examine results for another microorganism in the wild, HIV. The number of human immunodeficiency viruses produced over the past few decades is about the same number as malaria produced in a year, around 10^{20}. Yet what the virus lacks in numbers compared to *P. falciparum*, it makes up in mutation rate. HIV is a retrovirus whose mutation rate is roughly ten thousand times greater than that of cellular organisms such as malaria. That means multiple mutations will occur much more frequently in a single virus than in a cellular organism. For example, the mutation rate for nucleotide substitutions in cells is on the order of 10^{-8}, but for HIV it is 10^{-4}. That means it would require only 10^8 viruses (which are present in the body of a single infected individual) to produce a particular double-point mutation, where two specific nucleotides change simultaneously in the same virus. However, it would require 10^{16} cells to produce a particular double-point mutation. To produce five simultaneous mutations would require 10^{20} viruses, which actually have been produced in the past few decades while we have studied the AIDS epidemic. To achieve the same feat with cells would require the astounding number of 10^{40}. There have probably only been 10^{40} cells on the Earth since its formation roughly four billion years ago. So in the past few decades HIV has undergone more of some kinds of mutations than all the cells in the history of the world.

Nonetheless, the general evolutionary results with HIV have been the same as with other organisms. Only minor functional changes to the virus have popped up. (Of course, the virus only has a handful of genes, which restricts the number of genes that might change.) In response to human antibiotics that bind to and interfere with the functioning of a protein, the virus responds the way other organisms' genes do: by mutating amino acid residues near the binding site of the drug to make it less complementary.[15] In the course of the decades, the nucleotide sequence of the virus and amino acid sequence of its proteins have changed a lot, but with not a lot to show for it.

One protein in particular, Vpu, has acquired the ability to cause the infected cell's membranes to leak cations (probably by degrading the integrity of the membrane), which helps viruses to keep from being bound to the membrane.[16] This is a good reminder that random mutation and natural selection can accomplish some amazing feats. But it also shows the limits of processes underwritten by random events such as the Darwinian mechanism. If this is the best substantive change the virus can accomplish with as many mutational chances as cells have had since the beginning of the world, then it dramatically demonstrates the feebleness of the process.

To sum up, in four very disparate organisms (humans, *E. coli*, malaria, and HIV) representing four very different branches of life (multicell eukaryote, bacterium, single-cell eukaryote, and virus) over enormous ranges of population size and generations, the same paltry evolutionary results are obtained. This allows us to be confident that the role of randomness in constructive molecular evolution is very limited indeed.

The Limits of Non-Intelligent Explanations in Molecular Biology

Not Just Darwin's Mechanism

Although Darwin's mechanism is favored by the majority of biologists as the driving force behind evolution, there are other views as well, including self-organization, nonadaptive processes, and so on. It is important to realize, however, that the data cited above count against *all* evolutionary mechanisms that depend critically on random events, not just against Darwinian random mutation and natural selection. There were no artificial restrictions placed on any of the organisms described above. They could have mutated or evolved by any process that was available in nature and able to do them some good. However, after an astronomical number of chances, little was accomplished. Thus, we can conclude that no process based on randomness—not just Darwinism—is effective at building coherent biological systems.

4. Response to Criticisms of My Argument

In my recently published book, *The Edge of Evolution*, I use the data discussed above on the limits of randomness, in conjunction with experimental work that has been published on the specificity of protein-protein interactions, to make a further argument. I contend that forming a new protein-protein interaction is a difficult evolutionary problem, and that forming two protein-protein interactions that have to work together to confer a new selectable function on an organism is an extremely difficult evolutionary problem if the mechanism of evolution is based on random biological events. In fact, it would be so rare that it would be expected to occur only once in the history of life on Earth. But since most molecular machines in the cell function as complexes of a half dozen or more proteins, each of which is bound to at least one other protein in the complex, then the great majority of molecular machinery is beyond any process that involves randomness at its core.

The book has been widely reviewed by leading Darwinian evolutionary biologists who have spared no opportunity to curl their lips at it. However, virtually all of the critical Darwinian fire has been concentrated on the argument concerning the development of protein-protein interactions.[17] Little has been said about the basic data the book brings to wide attention to show that random mutation leads to incoherent changes in the genome of organisms. For example, reviewing the book for *Science*, the University of Wisconsin biologist Sean Carroll writes:[18]

> Very simple calculations indicate how easily such motifs evolve at random. If one assumes an average length of 400 amino acids for proteins and equal abundance of all amino acids, any given two-amino acid motif is likely to occur at random in every protein in a cell. (There are 399 dipeptide motifs in a 400-amino acid protein and 20 x 20 = 400 possible dipeptide motifs.) Any specific three-amino acid motif will occur once at random in every 20 proteins and any four-amino acid motif will occur once in every 400 proteins. That means that, without any new mutations or natural selection, many sequences that are identical or close matches to many interaction motifs already exist. New motifs can arise readily at random, and any weak interaction can easily evolve, via random mutation and natural selection, to become a strong interaction. Furthermore, any pair of interacting proteins can readily recruit a third protein, and so forth, to form larger complexes.

So Carroll argues against my contention that protein-protein binding sites are difficult to develop. Yet whether his counterpoints are valid or not (they aren't—I answer Carroll's and other reviewers' criticisms on my Amazon.com blog), he and other reviewers fail to address the foundational observational data showing that given huge population sizes and tens of thousands of generations, little of evolutionary significance was produced in very disparate organisms, few genes were ever able to help in any circumstance, and the changes that did occur to aid against a selectional challenge were incoherent—disconnected from each other and not forming integrated systems such as the kind that pack cells.

A fundamental feature of randomness, of course, is that it can't plan ahead. Any given evolutionary step taken by a population of organisms is therefore likely to be disconnected from the previous step. That is, it is likely to be incoherent, like a drunk staggering on a road. If the foundation of life were simple, as Darwin's contemporaries thought, then perhaps a random mechanism might work, as Richard Watson acknowledges it does for simple computer problems. This can be visualized as what has been called an "evolutionary fitness landscape." The simple landscape shown in Figure 1A is the kind in which a blind Darwinian mechanism would easily find the peak by following a simple rule such as "always climb higher" (which mimics the Darwinian evolutionary imperative to always improve fitness). However, the foundation of life is about as far from simple as it could be, so a realistic evolutionary landscape does not resemble this picture. Rather, it much more likely resembles that of Figure 1B, a "rugged" evolutionary fitness landscape.[19] In such a landscape, random processes might bring some system's features to a local peak, but local peaks are disconnected from each other, and there is no coherent way to traverse the landscape by a blind process.

5. Conclusion

Recent evolutionary data at the genetic level on astronomical numbers of organisms in species from widely varying branches of life have shown that any process based on randomness cannot explain the complex, coherent, integrated features of the cell, such as molecular machinery. Therefore, the evolution of life was not a random process.

The Limits of Non-Intelligent Explanations in Molecular Biology

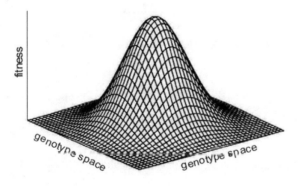

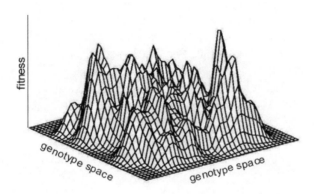

Figure 1. Evolutionary fitness landscapes. A) The top figure represents a simplistic evolutionary landscape, where only one or a few traits can vary, and fitness can increase smoothly. Ordinary Darwinian processes would easily drive a species to the single pinnacle. B) The bottom figure represents a more realistic, rugged evolutionary landscape, where many traits can vary. Here random mutation and natural selection would drive a species to some local peak, where it would remain stuck. Natural selection would actively inhibit a species from traversing such a landscape. If a limited scientific study focuses on just one peak of a rugged landscape, the results can misleadingly seem to match the smooth peak.[20]

Notes

1. Ernst Mayr, *One Long Argument: Charles Darwin and the Genesis of Modern Evolutionary Thought* (Cambridge, MA: Harvard University Press, 1991).
2. Mayr, *One Long Argument,* 37.
3. All these authors accepted a fifth component—that of evolution as opposed to a constant, unchanging world.
4. Mayr, *One Long Argument,* 35.
5. J. Farley, *The Spontaneous Generation Controversy from Descartes to Oparin* (Baltimore: Johns Hopkins University Press, 1977).
6. Special issue on molecular machines, *Bioessays* 25, no. 12, 2003.
7. R. A. Watson, *Compositional Evolution: The Impact of Sex, Symbiosis, and Modularity on the Gradualist Framework of Evolution* (Cambridge: MIT Press, 2006), 272.
8. R. Carter and K. N. Mendis, "Evolutionary and Historical Aspects of the Burden of Malaria," *Clinical Microbiology Reviews* 15 (2002): 564–94.
9. "This burden is composed not only of the direct effects of malaria but also of the great legacy of debilitating, and sometimes lethal, inherited diseases that have been selected under its impact in the past" (Carter and Mendis, 2002, 589).
10. R. E. Lenski, "Phenotypic and Genomic Evolution during a 20,000-Generation Experiment with the Bacterium *Escherichia coli.*" *Plant Breeding Reviews* 24 (2004): 225–65.
11. Woods, R., Schneider, D., Winkworth, C. L., Riley, M. A., and Lenski, R. E. "Tests of parallel molecular evolution in a long-term experiment with *Escherichia coli.*" *Proc. Natl. Acad. Sci. U.S.A.* 103 (2006): 9107–112.
12. N. J. White, "Antimalarial Drug Resistance," *Journal of Clinical Investigation* 113 (2004): 1084–92.
13. I. M. Hastings, P. G. Bray, and S. A. Ward, "Parasitology: A Requiem for Chloroquine," *Science* 298 (2002): 74–75.
14. D.C. Jeffares, A. Pain, A. Berry, A.V. Cox, J. Stalker, C.E. Ingle, A. Thomas, M.A. Quail, K. Siebenthall, A.C. Uhlemann, S. Kyes, S. Krishna, C. Newbold, E.T. Dermitzakis, and M. Berriman, "Genome variation and evolution of the malaria parasite Plasmodium falciparum." *Nature Genetics* 39 (2007): 120–25; J. Mu, P. Awadalla, J. Duan, K.M. McGee, J. Keebler, K. Seydel, G.A. McVean, and X.Z. Su, "Genome-wide variation and identification of vaccine targets in the Plasmodium falciparum genome." *Nature Genetics* 39 (2007): 126–30; S.K. Volkman, P.C. Sabeti, D. DeCaprio, D.E. Neafsey, S.F. Schaffner, D.A. Milner Jr., J.P. Daily, O. Sarr, D. Ndiaye, O. Ndir, S. Mboup, M.T. Duraisingh, A. Lukens, A. Derr, N. Stange-Thomann, S. Waggoner, R. Onofrio, L. Ziaugra, E. Mauceli, S. Gnerre, D.B. Jaffe, J. Zainoun, R.C. Wiegand, B.W. Birren, D.L. Hartl, J.E. Galagan, E.S. Lander, and D.F. Wirth, "A genome-wide map of diversity in Plasmodium falciparum." *Nat. Genet.* 39 (2007): 113–19.
15. For example: S.G. Sarafianos. K. Das, S.H. Hughes, and E. Arnold, "Taking aim at a moving target: designing drugs to inhibit drug-resistant HIV-1 reverse transcriptases." *Current Opinion in Structural Biology* 14 (2004: 716–30; Y. Tie, P.I. Boross, Y.F. Wang, L. Gaddis, F. Liu, X. Chen, J. Tozser, R.W. Harrison, I.T Weber, "Molecular basis for substrate recognition and drug resistance from 1.1 to 1.6 angstroms resolution crystal structures of HIV-1 protease mutants with substrate analogs." *Federation of European Biochemical Societies (FEBS) Journal* 272 (2005): 5265–77.
16. M. E. Gonzalez and L. Carrasco, "Viroporins," *FEBS Letters* 552 (2003): 28–34.
17. I respond to critics on my Amazon.com blog.
18. S. B. Carroll, "God as Genetic Engineer," *Science* 316 (2007): 1427–28.
19. S. Gavrilets, *Fitness Landscapes and the Origin of Species* (Princeton, NJ: Princeton University Press, 2004).
20. From Gavrilets, *Fitness Landscapes and the Origin of Species.* Reprinted courtesy of Sergey Gavrilets.

20

THE CHAIN OF ACCIDENTS AND THE RULE OF LAW: THE ROLE OF CONTINGENCY AND NECESSITY IN EVOLUTION

MICHAEL SHERMER

Humans are pattern-seeking, storytelling animals. We look for and find patterns in our world and in our lives, then weave narratives around those patterns to bring them to life and give them meaning. Such is the stuff of which myth, religion, history, and science are made. Sometimes the patterns we find represent reality—DNA as the basis of heredity or the fossil record as the history of life. But sometimes the patterns are imposed by our minds—the face on Mars or the image of the Virgin Mary on a window. The rub lies in distinguishing which patterns are true and which are false, and the essential tension pits skepticism against credulity as we struggle to determine the nature of reality.

That tension is at the forefront of the debate over how likely or unlikely the evolution of intelligent life is, particularly a culture-generating, language-producing, technology-using form of intelligent life like ours. Evolutionary theorists are interested in the question because it taps into the relative roles of chance and law in nature and natural history. Astronomers and astrobiologists are interested in the question because it cuts to the heart of how likely or unlikely it is that we will find life on other planets or make contact with an extraterrestrial intelligence.

Does the pattern of life's history and nature's laws indicate that our existence was preordained from the beginning, or does the pattern indicate that we are the product of chance events and random occurrences? Theists have an easy answer in their proclamation that a deity created us, and thus we were preordained by divine fiat. But non-theists can make a similar argument, substituting the laws of nature for God. Either way, the temptation to conclude that we are therefore special is overwhelming. The physicist Steven Weinberg well captured this conceit in his 1977 book, *The First Three Minutes:* "It is almost irresistible for humans to believe that we have some special relation to the universe, that human life is not just a more-or-less farcical outcome of a chain of accidents reaching back to the first three minutes, but that we were somehow built in from the beginning."[1]

Is our existence a *necessity*—it could not have been otherwise? Or is our existence a *contingency*—it need not have been? A coarse-grained look at the question finds scientists roughly divided between astrobiologists and SETI astronomers who tend to be fairly optimistic, estimating a relatively high probability of intelligent life evolving in the cosmos (or re-evolving in an earth-bound thought experiment), and biologists and evolutionary theorists who tend to be fairly pessimistic, estimating a relatively low probability of intelligent life evolving elsewhere (or re-evolving here). Since the question cannot be answered in a laboratory experiment, we must turn to those sciences that attempt to answer it indirectly, such as those employed by SETI (Search for Extra-Terrestrial Intelligence) scientists and evolutionary theorists.

1. The SETI Optimists

Astrobiologists and SETI astronomers base their optimism on numbers that they plug into the infamous Drake equation, proposed in 1961 by the radio astronomer Frank Drake for estimating the number of technological civilizations that reside in our galaxy:

$$N = R f_p n_e f_l f_i f_c L,$$

where N = the number of communicative civilizations, R = the rate of formation of suitable stars, f_p = the fraction of those stars with planets, n_e = the number of earth-like planets per solar system, f_l = the fraction of planets with life, f_i = the fraction of planets with intelligent life, f_c = the fraction of planets with communicating technology, and L = the lifetime of communicating civilizations.[2]

Although we have a fairly good idea of the rate of stellar formation, and we are confident that a significant number of these stars have planets, it is too soon to estimate the rate of formation of Earth-like planets because the technology is not yet available to detect planets smaller than Jupiter-size behemoths. In the SETI literature, a 10 percent figure is often used, where, say, in a galaxy of 100 billion stars, there will be 10 billion sun-like stars, one billion Earth-like planets, 100 million planets with life, 10 million planets with intelligent life, and one million planets with intelligent life capable of radio technology.[3]

Although most SETI astronomers are realistic about the limitations of such estimates, I was puzzled to encounter numerous caveats about L, the lifetime of technological civilizations, such as this one from SETI Institute astronomer Seth Shostak: "The lack of precision in determining these parameters pales in comparison to our ignorance of L."[4] Similarly, the Mars Society President and space-exploration visionary Robert Zubrin says that "the biggest uncertainty revolves around the value of L; we have very little data to estimate this number and the value we pick for it strongly influences the results of the calculation."[5] Estimates of L by astronomers reflect this uncertainty, ranging from 10 years to 10 million years, with a mean of about 50,000 years.

Using a conservative Drake equation calculation, where L = 50,000 years (and R = 10, f_p = .5, n_e = .2, f_l = .2, f_i = .2, f_c = .2), N = 400 civilizations, or one per 4,300 light years. Applying Robert Zubrin's optimistic (and modified) Drake equation, where L = 50,000 years, N = 5,000,000 galactic civilizations, or one per 185 light years. (Zubrin's calculation assumes the Milky Way galaxy has 400 billion stars, of which 10 percent are suitable G and K type stars not part of multiple star systems, with almost all having planets, and 10 percent of these containing an active biosphere, and 50 percent of those as old as Earth.) Estimates of N range wildly in between these figures of 400 and five million, from Planetary Society SETI scientist Thomas R. McDonough's 4,000 to Carl Sagan's one million galactic civilizations.[6]

I find this inconsistency in the estimation of L perplexing because it is the one component in the Drake equation for which we have copious empirical data from the history of civilization on earth. To compute my own value of L, I compiled the lengths of 60 civilizations (the number of years from inception to demise), including: Sumeria, Mesopotamia, Babylonia, the eight dynasties of Egypt, the six civilizations of Greece, the Roman Republic and Empire, and others in the ancient world, plus various civilizations since the fall of Rome, including the nine dynasties (and two Republics) of China, four in Africa, three in India, two in Japan, six in Central and South America, and six modern states of Europe and America. For all 60 civilizations in my database, there was a total of 25,276 years, or $L = 421.27$ years. For more modern and technological societies, L became shorter, with the 28 civilizations since the fall of Rome averaging only 306.04 years. Plugging these figures into the Drake equation goes a long way toward explaining why ET has yet to visit or call. Where $L = 421.27$ years, $N = 3.37$ civilizations in our galaxy; where $L = 306.04$ years, $N = 2.49$ civilizations in our galaxy.[7]

Despite my skepticism, I am an unalloyed enthusiast for the SETI program because the importance and impact of making contact far outweigh the relatively minuscule costs to conduct a search. Nevertheless, I am not optimistic that contact will be made in my lifetime, or that of my descendents for the next several thousand years, because of the number of evolutionary contingencies (which are not even represented in the Drake equation) necessary to drive an animal lineage toward technological intelligence. And once they get there, they may off themselves with their advanced technologies, as we have seen happen to nearly every civilization that ever flourished on Earth.

2. The Evolutionary Pessimists

Ever since Darwin, biologists and evolutionary theorists have tended to be skeptical of the possibility of extraterrestrial intelligence (and thus of intelligent life re-evolving here again). Their reasons have tended to focus on the fact that there are simply too many contingent steps along the way from bacteria to big brains where the sequence could have bifurcated down some other equally plausible path. Alfred Russel Wallace, Darwin's contemporary who co-discovered natural selection, well expressed this conclusion in his 1903 book *Man's Place in the Universe:*

> The ultimate development of man has, therefore roughly speaking, depended on something like a million distinct modifications, each of a special type and dependent on some precedent changes in the organic and inorganic environments, or in both. The chances against such an enormously long series of definite modifications having occurred twice over . . . are almost infinite.[8]

And Wallace did not know what we know about the length and breadth of hominid evolution.

Shortly after the SETI program was launched in the early 1960s, the evolutionary biologist George Gaylord Simpson outlined his skepticism in a 1964 *Science* essay with a title befitting his conclusion, "On the Non-Prevalence of Humanoids."[9] Evolutionary theorist Ernst Mayr concurred, presenting his skepticism in a debate on the likelihood of success of the SETI program with the astronomer Carl Sagan.[10] "Life originated on Earth about 3.8 billion years ago, but high intelligence did not develop until about half a million years ago." That is, for 3.75 of the 3.8 billion years of the history of life on earth, there was no advanced intelligence. For the two billion years after life began there were only simple prokaryote cells. Eukaryote cells with a nucleus and other modern characteristics evolved about 1.8 billion years ago, followed by three groups of multicellular organisms: fungi, plants, and animals. Of the sixty to eighty phyla of animals, only one, the chordates, led to intelligence; and only one of these, the vertebrates, developed intelligence; and of all the vertebrates—including fishes, amphibians, reptiles, birds, and mammals—only mammals developed intelligence; and of the twenty-four orders of mammals only one—us—has intelligence. Mayr concludes definitively, "Nothing demonstrates the improbability of the origin of high intelligence better than the millions of phyletic lineages that failed to achieve it."

Given these brute facts about evolution, why would anyone bother to listen for ET's signal? Carl Sagan provided an answer. First, we don't need "humans" per se, just

> creatures able to build and operate radio telescopes. They may live on the land or in the sea or air. They may have unimaginable chemistries, shapes, sizes, colors, appendages and opinions. We are not requiring that they follow the particular route that led to the evolution of humans. There may be many different evolutionary pathways, each unlikely, but the sum of the number of pathways to intelligence may nevertheless be quite substantial.

Sagan notes that there are trends in evolution that lead to SETI optimism: "other things being equal, it is better to be smart than to be stupid, and an overall trend toward intelligence can be perceived in the fossil record. On some worlds, the selection pressure for intelligence may be higher; on others, lower." Our sun is not a first-generation star in our galaxy; there are stars (and thus solar systems) 10 billion years old. Sagan asks us to imagine two curves:

> The first is the probable timescale to the evolution of technical intelligence. It starts out very low; by a few billion years it may have a noticeable value; by 5 billion years, it's something like 50 percent; by 10 billion years, maybe it's approaching 100 percent. The second curve is the ages of Sun-like stars, some of which are very young—they're being born right now—some of which are as old as the Sun, some of which are 10 billion years old. If we convolve these two curves, we find there's a chance of technical civilizations on planets of stars of many different ages—not much in the very young ones, more and more for the older ones. The most likely case is that we will hear from a civilization considerably more advanced than ours.[11]

I have noticed an interesting difference between the SETI optimists and evolutionary pessimists (we would consider ourselves realists), and that is that astronomers traffic in the rule of nature's laws, which are repetitive, reliable, and necessary, whereas biologists traffic in the chain of accidents, which are quirky, chancy, and contingent. Among biologists, no one has emphasized the role of contingency more than Stephen Jay Gould, most notably in his 1989 book, *Wonderful Life*,[12] which has become something of a watershed for those who study complexity, especially applied to organisms, societies, and history. Walter Fontana and Leo Buss, for example, ask, "What Would be Conserved if 'The Tape Were Played Twice'?"[13] This is a direct reference to Gould's suggestion in *Wonderful Life* that if the tape of life were rewound to the time of the organisms found in the Canadian outcrop known as the Burgess Shale, dated to about 530 million years ago, and replayed with a few contingencies tweaked here and there, humans would most likely never have evolved. So powerful are the effects of contingency that a small change in the early stages of a sequence can produce large effects in the later stages. Yet, Fontana and Buss contend that plenty would be conserved if the tape were rerun again. Stuart Kauffman, in his 1995 book, *The Origins of Order,* references Gould and asks about the Cambrian explosion of life: "Was it Darwinian chance and selection alone . . . or did principles of self-organization mingle with chance and necessity?"[14] His answer is that the necessitating laws of self-organization defy contingency. Mathematicians Jack Cohen and Ian Stewart responded to Gould in a feature article in *Nonlinear Science Today,* pointing out: "Nowhere in *Wonderful Life* does Gould give an adequate treatment of the possible existence of evolutionary mechanisms, convergences, universal constants, that might constrain the effects of contingency."[15] *Wired* magazine's Kevin Kelly actually ran Gould's thought experiment in a sandbox, with contrary results: "First thing you notice as you repeat the experiment over and over again, as I have, is that the landscape formations are a very limited subset of all possible forms."[16]

Philosophers also got in on the debate. Murdo William McRae published a critique titled "Stephen Jay Gould and the Contingent Nature of History," concluding: "Gould's argument for contingency ultimately returns to the notions of progress and predictability it set out to challenge."[17] Daniel Dennett devoted a long chapter in his book *Darwin's Dangerous Idea* to Gould and contingency, concluding that the question depends on what is meant by "us" in the thought experiment:

> There is a sliding scale on which Gould neglects to locate his claim about rewinding the tape. If by 'us' he meant something very particular—Steve Gould and Dan Dennett, let's say—then we wouldn't need the hypothesis of mass extinction to persuade us how lucky we are to be alive. . . . If, at the other extreme, by "us" Gould meant something very general, such as 'air-breathing, land-inhabiting vertebrates,' he would probably be wrong.[18]

Dennett's point is well made. Surely the history of life is not completely contingent. Rerun the tape and certainly some things would reappear, such as vision, flight, locomotion, and thought.

That is precisely the point made by the Cambridge University paleontologist Simon Conway Morris in *Life's Solutions,* in which he argues that humans were an inevitable product of the necessitating laws of nature.[19] (Paradoxically, he simultaneously claims that another intelligent species anywhere in the cosmos is so highly improbable that we are alone.) Conway Morris began his anti-contingency campaign in his earlier book, *The Crucible of Creation,* a full-frontal attack on Gould's interpretation (as well as his own earlier explication) of the Burgess Shale fauna, and an indirect assault on non-anthropocentric models of evolution.[20] Conway Morris's *bête noire* is contingency, which he counters with the phenomenon of evolutionary convergence, in which necessitating physical laws and biological constraints dictate a limited number of solutions to life's problems: eggs, eyes, wings, brains, echolocation, sociality, and the like. Convergence is a well-known phenomenon in evolutionary theory, but in the last couple of decades biologists have focused more on biological uniqueness, noting that convergence is the exception, not the rule. The classification system of cladistics, in fact, is based on evolutionary novelties and uniqueness as the criteria of distinguishing species from one another.

The paleontologist Donald Prothero notes that Conway Morris has missed Gould's central point about contingency: "Once groups of organisms are established and develop a body plan and set of niches, biological constraints are such that convergence and parallelism can be expected. But the issue of who gets this head start in the first place may be more a matter of luck and contingency that has nothing to do with adaptation."[21] Consider the Burgess Shale chordate *Pikaia,* which led to the evolution of vertebrates and is thus our earliest ancestor.

What if *Pikaia* had gone extinct like most of the rest of the Burgess fauna? Prothero notes that "there would have been no vertebrates, and half the examples of convergence that Conway Morris details could not have occurred, because they are peculiar to the basic vertebrate body plan, and not shared in any other phylum." Maybe some other body plan would have evolved intelligence. Not likely, Prothero counters.

> The jointed-legged phylum *Arthropoda* is the most successful, diverse, and numerous organisms on the land, sea, and air today. This phylum includes the huge numbers and varieties of insects, spiders, scorpions, mites, ticks, millipedes, centipedes, and crustaceans. As numerous and diverse as they are, their habit of molting their exoskeletons when they grow means they can never exceed a certain body size—they would fall apart like a blob of jelly during their soft stage after molting without any skeletal support.

Mollusks are also supremely successful as clams, snails, and squids, "but their body plan yields few convergences with vertebrates, let alone humans." Likewise, such phyla as the *Echinodermata* (sea stars, sea urchins, sea cucumbers, brittle stars, and sea lilies) are flourishing, "yet they are even less likely to produce a humanoid with their peculiar specializations, such as a radial symmetry with no head or tail, spiny calcite plates, and lack of eyes. In addition, they lack any sort of circulatory, respiratory, or excretory system, so they are forever bound to marine waters of normal salinity." Running down the list of Cambrian survivors to test Conway Morris's hypothesis, Prothero concludes that "the role of contingency as to which body plans survived becomes more and more obvious—and the significance of the convergence that Conway Morris details in so many examples is moot."[22]

In *Wonderful Life*, Gould wrote: "Replay the tape a million times from a Burgess beginning, and I doubt that anything like *Homo sapiens* would ever evolve again."[23] Per Dennett's critique of Gould, the debate boils down to what is meant by "anything like *Homo sapiens*." If one means the species itself, then of course a rerun of the tape would produce a world sans *Homo sapiens*. But if by "anything like" we mean a brainy land-based organism capable of tool use, culture, and symbolic communication, then that is another question, a hypothesis that I contend is testable in two ways: SETI and history.

In 1982, Sagan published a petition in *Science* urging the scientific respectability of SETI, for which he garnered many distinguished signatories, including such biologists as David Baltimore, Melvin Calvin, Francis Crick, Manfred Eigen, Thomas Eisner, Stephen Jay Gould, Matthew Meselson, Linus Pauling, David Raup, and E. O. Wilson. The petition proposed that in addition to debating the issue, the hypothesis should be tested by running the experiment: "We are unanimous in our conviction that the only significant test of the existence of extraterrestrial intelligence is an experimental one. No *a priori* arguments on this subject can be compelling or should be used as a substitute for an observational program."[24] The SETI search continues and we all await its outcome. The other test is history itself, which I will conduct in the final section of this paper, once we define with semantic precision what is meant by contingency and necessity, and consider a model of how they interact.

3. Contingent-Necessity

The issue of contingency and necessity remains one of the great issues of our time because it touches so deeply on history and our place in the cosmos. Yet many of those who oppose the idea of a predominantly contingent universe have misread contingency to mean accidental or random. Cohen and Stewart, for example, wrote, "The survivors, who produced us, did so by contingency, by sheerest accident"; "Gould's argument that contingency—randomness—plays a major role in the results of evolution . . ."[25] and Gould "sees the evolution of humanity as being accidental, purely contingent."[26] Yet Gould was clear in *Wonderful Life* about what he meant by contingency:

> I am not speaking of randomness (for E had to arise, as a consequence of A through D), but of the central principle of all history—contingency. A historical explanation does not rest on direct deductions from laws of nature, but on an unpredictable sequence of antecedent states, where any major change in any step of the sequence would have altered the final result. This final result is therefore dependent, or contingent, upon everything that came before—the unerasable and determining signature of history.[27]

Contingency is an *unpredictable sequence of antecedent states,* not randomness, chance, or accident. Nor does contingency exclude necessity. Murdo William McRae, for example, writes:

> In spite of all his dedication to contingency and its attendant questioning of progress and predictability, Gould equivocates often enough to cast doubt upon the depth of his revolutionary convictions. . . . At times he insists that altering any antecedent event, no matter how supposedly insignificant, diverts the course of history; at other times he suggests that such antecedents must be significant ones.[28]

The reason for the apparent prevarication is that Gould sometimes emphasizes contingency over necessity to make a particular point. Yet, in an earlier work on contingency, he noted that, "incumbency also reinforces the stability of a pathway once the little quirks of early flexibility push a sequence into a firm channel. Stasis is the norm for complex systems; change, when it happens at all, is usually rapid and episodic."[29]

Because of the confusion over semantics in this debate, I asked Gould directly what he meant. He responded:

> My argument in *Wonderful Life* is that there is a domain of law and a domain of contingency, and our struggle is to find the line between them. The reason why the domain of contingency is so vast, and much vaster than most people thought, is not because there isn't a law-like domain. It is because we are primarily interested in ourselves and we have posited various universal laws of nature. It is because we want to see ourselves as results of law-like predictability and sensible products of the universe in that sense.[30]

That "domain of law and domain of contingency" may be more formally described in a model I developed to clarify this issue. In this model *contingency* is taken to mean: *a conjuncture of events occurring without design* and *necessity* to mean *constraining circumstances compelling a certain course of action*. Contingencies are the sometimes small, apparently insignificant, and usually unexpected events of life—the kingdom hangs in the balance awaiting the horseshoe nail. Necessities are the large and powerful laws of nature and trends of history—once the kingdom has collapsed, the arrival of 100,000 horseshoe nails will not save it. The past is composed of both contingencies and necessities. Therefore, it is useful to combine the two into one term that expresses this interrelationship—*contingent-necessity*—taken to mean: *a conjuncture of events compelling a certain course of action by constraining prior conditions.*

There is in this system a rich matrix of interactions between contingencies and necessities, varying over time, in what I call the *model of contingent-necessity,* which states: *In the development of any historical sequence, the role of contingencies in the construction of necessities is accentuated in the early stages and attenuated in the later.* There are corollaries that encompass six aspects of the model, including:

> *Corollary 1*: The earlier in the development of any historical sequence, the more chaotic the actions of the individual elements of that sequence; and the less predictable are future actions and necessities.
>
> *Corollary 2*: The later in the development of any historical sequence, the more ordered the actions of the individual elements of that sequence, and the more predictable are future actions and necessities.
>
> *Corollary 3*: The actions of the individual elements of any historical sequence are generally postdictable but not specifically predictable, as regulated by Corollaries 1 and 2.
>
> *Corollary 4*: Change in historical sequences from chaotic to ordered is common, gradual, followed by relative stasis, and tends to occur at points where poorly established necessities give way to dominant ones so that a contingency will have little effect in altering the direction of the sequence.
>
> *Corollary 5*: Change in historical sequences from ordered to chaotic is rare, sudden, followed by relative non-stasis, and tends to occur at points where previously well-established necessities have been challenged by others so that a contingency may push the sequence in one direction or the other.
>
> *Corollary 6*: Between origin and bifurcation, sequences self-organize through the interaction of contingencies and necessities in a feedback loop driven by the rate of information exchange.

At the beginning of a historical sequence, actions of the individual elements (atoms, molecules, organisms, people) are chaotic, unpredictable, and have a powerful influence on the future development of that sequence. But as the sequence slowly but ineluctably evolves, and the pathways become more worn, the chaotic system self-organizes into an orderly one. The individual elements sort themselves, and are sorted into their allotted positions, as dictated by what came before, with the conjuncture of events compelling a certain course of action by constraining prior conditions.

In the language of contingent-necessity, a bifurcation, or "trigger of change," is any stimulus that causes a shift from the dominance of necessity and order to the dominance of contingency and chaos in a historical sequence, which might be anything from an asteroid impact to famine, disease, and war. A trigger of change, however, will not cause a shift at just any point in the sequence. Corollary 5 states that it will be most effective when well-established necessities have been challenged by others so that a contingency may push the sequence in one direction or the other. Similarly for the butterfly effect, described in Corollaries 1 and 2, where the power of the trigger depends on when in the chronological sequence it enters. The flap of the butterfly's wings in Brazil may indeed set off a tornado in Texas, but only when the system has started anew or is precariously hanging in the balance. Once the storm is well under way, the flap of a million butterfly wings would not alter the outcome for the tornado-leery Texans. The potency of the sequence grows over time.

Elsewhere, I have provided numerous historical examples of the model of contingent-necessity.[31] Here I would like to conduct a historical experiment to test the hypothesis that contingency played such a significant role in evolution that an intelligent technological intelligence would not likely evolve again on Earth, and is probably extremely rare in the galaxy.

4. Testing the Contingency Hypothesis

I first thought of a way to test the contingency hypothesis when I wrote a review of Robert Wright's book, *Nonzero: The Logic of Human Destiny*[32] for the *Los Angeles Times*.[33] Wright's thesis emphasizes the power of necessity in evolution through the force of cooperation: over billions of years of natural history, and over thousands of years of human history, there has been an increasing tendency toward the playing of cooperative "nonzero" games between organisms. This tendency has allowed more nonzero gamers to survive. In zero-sum games, the margin of victory for one is the margin of defeat for the other. If the Yankees beat the Red Sox 8–2, the Red Sox lose 2–8—where the margin of victory is +6 and the margin of defeat -6, summing to zero. In non-zero-sum games, both players gain, as in an economic exchange where I win by purchasing your product and you win by receiving my money.

Although competition between individuals and groups was common in both biological evolution and cultural history, Wright argues that symbiosis among organisms and cooperation among people has gradually displaced competition as the dominant form of interaction. Why? Because those who cooperated by playing nonzero games were more likely to survive and pass on their genes for cooperative behavior. And this process has been ongoing, Wright says, "from the primordial soup to the World Wide Web." From the Paleolithic to the present, human groups have evolved from bands of hundreds, to tribes of thousands, to chiefdoms of tens of thousands, to states of hundreds of thousands, to nations of millions. This could not have happened through zero-sum exchanges alone, Wright claims. The hallmarks of humanity—language, tools, hunting, gathering, farming, writing, art, music, science, and technology—could not have come about through the actions of isolated zero-sum gamers.

Wright's deeper theme is directionality in evolution, from which we can derive meaning: "the evolutionary process is subordinate to a larger purpose—a 'higher' purpose, you might even say."[34] That purpose is to be found, says Wright, in the fact that our existence was necessary and inevitable. Replay the time line of life over and over and "we" would appear again and again, "we" being an intelligent social species that carries its "social organization to planetary breadth." Therefore, he concludes,

> Globalization, it seems to me, has been in the cards not just since the invention of the telegraph or the steamship, or even the written word or the wheel, but since the invention of life. All along, the relentless logic of non-zero-sumness has been pointing toward this age in which relations among nations are growing more non-zero-sum year by year.[35]

Is social globalization an inevitable necessity of the evolutionary process? If *Homo sapiens* had not filled this ineluctable position of global dominance, would one of the other hominids or great apes have done so? This is a counterfactual question, and counterfactual ("what if") history is a legitimate form of thought experiment that provides an opportunity for understanding cause and effect relationships in the past. Wright asks, counterfactually: "If they [*Homo sapiens*] had died out, would they have been the last?"[36] No, Wright answers. "If our own ancestors had died out around that time, it probably would have been at the hands of the Neanderthals, who could have then continued on their co-evolutionary ascent, unmolested by the likes of us." What if Neanderthals had also gone extinct? "I'd put my money on chimps. In fact, I suspect that they are already feeling some co-evolutionary push; if they're not quite on the escalator, they're in the vicinity." What if all the great apes had gone extinct? "Well, monkeys, though more distant human relatives than any apes, can be pretty impressive. Baboons are cleverly coalitional, and macaques are quite creative."[37] Wright continues in the counterfactual mode:

> What if somehow the entire primate branch had been nipped in the bud? Or suppose that the whole mammalian lineage had never truly flourished? For example, if the dinosaurs hadn't met their untimely death, mightn't all mammals still be rat-sized pests scurrying around underfoot? Actually, I doubt it, but as long as we're playing "What if," let's suppose the answer is yes. So what? Toward the end of the age of dinosaurs—just before they ran into their epoch-ending piece of bad luck—a number of advanced species had appeared, with brain-to-body ratios as high as those of some modern mammals. It now looks as if some of the smarter dinosaurs could stand up and use grasping forepaws. And some may have been warm-blooded and nurtured their young. Who knows? Give them another 100 million years and their offspring might be riding on jumbo jets.[38]

Maybe, but I doubt it. Let's consider just one relatively recent counterfactual: what if Neanderthals won and we lost? In our time line, Neanderthals went extinct between 40,000 and 30,000 years ago. What if we ran an alternate time line where *Homo sapiens* went extinct and Neanderthals continued flourishing in Europe, Asia, and the Levant? Would some big brow-ridged, stooped-shouldered, hirsute hominid be communicating on a World Wide Web and conducting radio searches for extraterrestrial intelligences?

Consider the facts. Neanderthals split off from the common ancestor shared with us between 690,000 and 550,000 years ago, and they were in Europe at least 242,000 (and perhaps 300,000) years ago, giving them free reign there for a quarter of a million years. They had a cranial capacity just as large as ours (ranging from 1,245 to 1,740 cc, with an average of 1,520 cc compared to our average of 1,560 cc), were physically more robust than us—with barrel chests and heavy muscles—and they sported a reasonably complex toolkit of about sixty different tools. On paper, it certainly seems reasonable to argue that Neanderthals had a good shot at "becoming us."

But if we dig deeper we see that there is almost no evidence that Neanderthals would have ever "advanced" beyond where they were when they disappeared 30,000 years ago. Even though paleoanthropologists disagree about a great many things, there is near total agreement in the literature that Neanderthals were not on their way to becoming "us." They were perfectly well-adapted organisms for their environments.

This progressivist bias, in fact, is pervasive in nearly all evolutionary accounts and is directly challenged by counterfactual thinking. I once explained to my young daughter that polar bears are a good example of a transitional species between land and marine mammals, since they are well adapted for both land and marine environments. But this is not correct. Polar bears are not "becoming" marine mammals. They are not becoming anything. They are perfectly well adapted for doing just what they do. They may become marine mammals should, say, global warming melt the polar ice caps. Then again, they may just go extinct. In either case, there is no long-term drive for polar bears to progress to anything, since evolution only creates immediate adaptations for local environments. The same applies to our hominid ancestors.

Paleoanthropologist Richard Klein, in his authoritative work *The Human Career*, concludes that "the archeological record shows that in virtually every detectable aspect—artifacts, site modification, ability to adapt to extreme environments, subsistence, and so forth—the Neanderthals were behaviorally inferior to their modern successors, and to judge from their distinctive morphology, this behavioral inferiority may have been rooted in their biological makeup."[39] Neanderthals had Europe to themselves for at least 200,000 years, unrestrained by the presence of other hominids, yet their tools and culture are not only simpler than those of *Homo sapiens*, they show almost no sign of change at all, let alone progress toward social globalization. Paleoanthropologist Richard Leakey notes that Neanderthal tools "remained unchanged for more than 200,000 years—a technological stasis that seems to deny the workings of the fully human mind. Only when the Upper Paleolithic cultures burst onto the scene 35,000 years ago did innovation and arbitrary order become pervasive."[40]

Likewise, Neanderthal art objects are comparatively crude and there is much controversy over whether many of them were not the product of natural causes instead of artificial manipulation.[41] The most striking exception to this is the famous Neanderthal bone flute dated from between 40,000 to 80,000 BC, which some archaeologists speculate means that the maker was musical. Yet even Christopher Wills, a rare dissenting voice who rejects the inferiority of the Neanderthals, admits that it is entirely possible that the holes were naturally created by an animal gnawing on the bone, not by some Paleolithic Ian Anderson. And even though Wills argues that "Recent important discoveries suggest that toward the end of their career, the Neanderthals might have progressed considerably in their technology," he has to confess that "it is not yet clear whether this happened because of contact with the Cro-Magnons and other more advanced peoples or whether they accomplished these advances without outside help."[42]

Probably the most dramatic claim for the Neanderthals' "humanity" is the burial of their dead that often included flowers strewn over carefully laid-out bodies in a fetal position. I even used this example in my book *How We Believe* on the origins of religion.[43] But new research is challenging this interpretation. Klein notes that graves "may have been dug simply to remove corpses from habitation areas" and that in six of twenty of the best documented burial sites "the bodies were tightly flexed (in near fetal position), which could imply a burial ritual or simply a desire to dig the smallest possible burial trench."[44] Paleoanthropologist Ian Tattersall agrees: "Even the occasional Neanderthal practice of burying the dead may have been simply a way of discouraging hyena incursions into their living spaces, or have a similar mundane explanation, for Neanderthal burials lack the 'grave goods' that would attest to ritual and belief in an afterlife."[45]

Much has been made about the possibility of Neanderthal language—that quintessential component of modern cognition. This is inferential science at best, since soft brain tissue and vocal box structures do not fossilize. Inferences can be drawn from the hyoid bone that is part of the vocal box structure, as well as the shape of the basicranium, or the bottom of the skull. But the discovery of part of an apparent Neanderthal hyoid bone is inconclusive, says

Tattersall: "However the hyoid argument works out, however, when you put the skull-base evidence together with what the archaeological record suggests about the capacities of the Neanderthals and their precursors, it's hard to avoid the conclusion that articulate language, as we recognize it today, is the sole province of fully modern humans."[46] As for the cranial structure, in mammals the bottom of the cranium is flat, but in humans it is arched (related to how high up in the throat the larynx is located). In ancestral hominids the basicranium shows no arching in *Australopithecines,* some in *Homo erectus,* and even more in archaic *Homo sapiens.* In Neanderthals, however, the arching largely disappears, evidence that does not bode well for theories about Neanderthal language, as Leakey concludes: "Judging by their basicrania, the Neanderthals had poorer verbal skills than other archaic sapiens that lived several hundred thousands years earlier. Basicranial flexion in Neanderthals was less advanced even than in *Homo erectus.*"[47]

Leakey then speculates, counterfactually, what might have happened had our ancestors survived: "I conjecture that if, by some freak of nature, populations of *Homo habilis* and *Homo erectus* still existed, we would see in them gradations of referential language. The gap between us and the rest of nature would therefore be closed, by our own ancestors."[48] That "freak of nature" is the contingency in our time line that allowed us to survive while no other hominids did, and thus Leakey concludes, "*Homo sapiens* did eventually evolve as a descendant of the first humans, but there was nothing inevitable about it."[49] Ian Tattersall also reasons in the contingent mode: "If you'd been around at any earlier stage of human evolution, with some knowledge of the past, you might have been able to predict with reasonable accuracy what might be coming up next. *Homo sapiens,* however, is emphatically not an organism that does what its predecessors did, only a little better; it's something very—and potentially very dangerously—different. Something extraordinary, if totally fortuitous, happened with the birth of our species."[50]

Had Neanderthals won and we lost, there is every reason to believe that they would still be living in a stone-age culture of hunting, fishing, and gathering, roaming the hinterlands of Europe in small bands of a couple of dozen individuals, surviving in a world without towns and cities, without music and art, without science and technology . . . a world so different from our own that it is almost inconceivable.

As for the great apes or monkeys succeeding, had both humans and Neanderthals gone extinct, apes have never shown any inclination toward progressive cultural evolution, now or in the fossil record, and monkeys proliferated throughout Asia and the new world for tens of millions of years without any interference from hominids, yet they didn't take step one toward developing a complex culture.

The fossil record, while still fragmented and desultory, is complete enough now to show us that over the past thirty million years we can conservatively estimate that hundreds of primate species have lived out their lives in the nooks and crannies of rain forests around the world; over the past ten million years dozens of great ape species have forged specialized niches on the planet; and over the last six million years—since the hominid split from the common ancestor of gorillas, chimps, and orangutans—dozens of bipedal, tool-using hominid species have struggled for survival. If these hominids were so necessitated by the laws of evolutionary progress, why is it that only a handful of those myriad pongids and hominids survived? If braininess is such an inevitable product of necessitating trends of nature, then why has only one hominid species managed to survive long enough to ask the question? What happened to those bipedal, tool-using *Australopithecines: anamensis, afarensis, africanus, aethiopicus, robus-*

The Chain of Accidents and the Rule of Law

tus, boisei, and *garhi?* What happened to those big-brained culture-generating *Homos: habilis, rudolfensis, ergaster, erectus, heidelbergensis,* and *neanderthalensis?* If big brains are so great, why did all but one of their owners go extinct?

Historical experiment after experiment reveals the same answer: we are a fluke of nature, a quirk of evolution, a glorious contingency. It is tempting to fall into the oldest trap of all pattern-seeking, storytelling animals: writing yourself into the story as the central pattern in order to find purpose and meaning in this gloriously contingent cosmos. But skeptical alarms should toll whenever anyone claims that science has discovered that our deepest desires and oldest myths are true after all. If there is a necessity in this story, it is that a purpose-seeking animal will find itself as the purpose of nature.

Notes

1. S. Weinberg, *The First Three Minutes: A Modern View of the Origins of the Universe* (New York: Basic Books, 1977).
2. The Drake equation is referenced throughout SETI literature, and at the SETI Institute web page (http://www.seti.org/site/pp.asp?c=ktJ2J9MMIsE&b=179074) you can plug in your own estimates to generate different Ns.
3. For estimates of L by SETI scientists, see the summary on page 441 of Steven Dick's book, *The Biological Universe* (Cambridge: Cambridge University Press, 1996). For a thorough overview that samples many different positions on the debate see: S. Webb, *If the Universe Is Teeming with Aliens . . . Where Is Everybody? Fifty Solutions to Fermi's Paradox and the Problem of Extraterrestrial Life* (New York: Springer, 2003); R. D. Ekers (ed.), *SETI 2020: A Roadmap for the Search for Extraterrestrial Intelligence* (Mountain View: SETI Press, 2002); J. Heidmann, *Extraterrestrial Intelligence* (New York: Cambridge University Press, 1995); P. Davies, *Are We Alone?* (New York: Basic Books, 1995); S. J. Dick, *Life on Other Worlds: The 20th-Century Extraterrestrial Life Debate* (New York: Cambridge University Press, 1998).

   On the optimistic side of SETI see: D. Darling, *Life Everywhere: The Maverick Science of Astrobiology* (New York: Perseus, 2002); A. D. Aczel, *Probability 1: Why There Must Be Intelligent Life in the Universe* (New York: Harcourt, 1998); C. Sagan and J. Agel, *Carl Sagan's Cosmic Connection: An Extraterrestrial Perspective* (New York: Cambridge University Press, 2000); C. Sagan, *The Cosmic Connection: An Extraterrestrial Perspective* (New York: Dell Books, 1973).

   On the pessimistic side of SETI see: P. D. Ward and D. Brownlee, *Rare Earth: Why Complex Life Is Uncommon in the Universe* (New York: Copernicus Books, 2000); M. Shermer, "Why ET Hasn't Called," *Scientific American*, August 2002, 34.

   On the SETI scientists themselves, see: D. Swift, *SETI Pioneers: Scientists Talk About Their Search for Extraterrestrial Intelligence* (Tucson, AZ: University of Arizona Press, 1990); M. Shermer, "The Exquisite Balance: Carl Sagan and the Difference Between Orthodoxy and Heresy in Science," in *The Borderlands of Science: Where Sense Meets Nonsense* (New York: Oxford University Press, 2001).
4. S. Shostak, *Cosmic Company: The Search for Life in the Universe* (New York: Cambridge University Press, 2003), 124.
5. R. Zubrin, "Galactic Society," *Analog*, April 2002, 33.
6. McDonough's estimate comes from a personal correspondence, May 7, 2002: "I assumed L would be 1/1000th the age of the home world. I took the latter to be 10 billion years, so got a lifetime of 10 million years. I got $N \sim 4{,}000$." McDonough summarized the SETI estimates as follows: "Most SETI people think the lifetime is large, although some pessimists in the past (during the Cold War) thought it might be decades, when WWIII would end it. The reason for the large figure among optimists is that we don't really care how long Greece or Rome lasted. The idea is that, if we don't destroy ourselves through war or pollution, then radio or laser technology should be around forever, or at least until the sun cooks the earth before it dies. So even if American civilization decays and is replaced by Chinese civilization, which is replaced by Costa Rican civilization, etc., they should all have access to the technology of the past." See also: T. McDonough, *The Search for Extraterrestrial Intelligence* (New York: Wiley, 1987). Carl Sagan's estimate comes from I. S. Shklovskii and C. Sagan, *Intelligent Life in the Universe* (New York: Holden-Day, 1966, reissued in 1998 by Emerson-Adams Press), 413: "As an average for all technical civilizations, both short-lived and long-lived, I adopt $L \sim 10^7$ [10 million] years." This yields $N \sim 1$ million.
7. For my calculation of L, I used the civilization chronology compiled by Northpark University historian David W. Koeller (http://campus.northpark.edu/history/WebChron/WestCiv/WestCiv.html):

The Chain of Accidents and the Rule of Law

Sumeria 2800–1900 BC: 900 years
Mesopotamia 2800–1200 BC: 1,600 years
Babylonia 1900–1100 BC: 800 years
Hittite 1600–717 BC: 883 years
Israel 1000 BC–AD 70: 1,070 years (from David to the destruction of Herod's temple)
Persia 550–323 BC: 227 years
Parthia 250 BC–AD 225: 475 years
Maurya 321–185 BC: 136 years
"Axial Age" 600–400 BC: 200 years
Egypt (3100–30 BC: 3,070 years)
Early Dynastic 3100–2686 BC: 414 years
Old Kingdom 2686–2181 BC: 505 years
First Intermediate 2181–2040 BC: 141 years
Middle Kingdom 2133–1786 BC: 347 years
Second Intermediate 1786–1567 BC: 219 years
New Kingdom 1567–1085 BC: 482 years
Late Dynastic 1085–341 BC: 744 years
Ptolemaic 332–30 BC: 302 years
Greece (2900–146 BC: 2,754 years)
Minoan 2900–1150 BC: 1,750 years
Mycenean 1600–1150 BC: 450 years
Dark Ages 1100–750 BC: 350 years
Archaic 750–500 BC: 250 years
Hellenic 479–323 BC: 156 years
Hellenistic 323–146 BC: 177 years
Rome (509 BC–AD 312: 817 years)
Republic 509–31 BC: 478 years
Empire 31 BC–AD 312: 343 years
Byzantine Empire 312–1453: 1,141 years
Sassanid Persia 226–642: 416 years
Ottoman 1350–1918: 568 years

Africa
Axum 300–700: 400 years
Ghana 900–1100: 200 years
Mali 1200–1450: 250 years
Songhai 1460–1591: 131 years

India
Indus Valley/Harappan 3000–1900 BC: 1,100 years
Mauryan Empire 332–185 BC: 147 years
Mughal Empire 1483–1757: 274 years

China
Xia dynasty 2205–1818: 387 years
Shang dynasty 1523–1027: 496 years

Zhou dynasty 1027–771: 256 years
Han dynasty 206 BC–AD 220: 426 years
Sui dynasty 581–618: 37 years
Tang dynasty 618–907: 289 years
Yuan dynasty 1280–1367: 87 years
Ming dynasty 1368–1644: 276 years
Qing dynasty 1644–1911: 267 years
Republic of China 1911–1949: 38 years
People's Republic 1949–2002: 53 years

Japan
Shogunate 1338–1867: 529 years
Meiji Restoration 1868–1937: 69 years

Central/South America
Olmec civilization 800–300 BC: 500 years
Toltec empire 300–600: 300 years
Mayan civilization 300–1300: 1,000 years
Aztec empire 1400–1519: 119 years
Inca empire 1438–1538: 100 years
Spanish empire 1519–1810: 291 years

Modern States
United States of America 1776–2009: 233 years
Germany 1871–2009: 138 years
England 1066–2009: 943 years
France 1789–2009: 220 years
Italy 1870–2009: 139 years
Israel 1948–2009: 61 years
BC/AD total: 25,276 years ÷ 60 civilizations = 421.27 years, average length of a civilization.
AD only total: 8,569 years ÷ 28 civilizations = 306.04 years, average length of a civilization.

8. A. R. Wallace, *Man's Place in the Universe: A Study of the Results of Scientific Research in Relation to the Unity or Plurality of Worlds* (London: Chapman and Hall, 1903), 73.
9. G. G. Simpson, "On the Non-Prevalence of Humanoids." *Science* 143 (1964): 769–75.
10. E. Mayr and C. Sagan, "Can SETI Succeed: Carl Sagan and Ernst Mayr Debate." *Bioastronomy News* 7 (1995), no. 4. Also available at http://www.planetary.org/html/UPDATES/seti/Contact/debate/default.html.
11. Ibid.
12. S. J. Gould, *Wonderful Life: The Burgess Shale and the Nature of History* (New York: W. W. Norton, 1989).
13. W. Fontana and L. Buss, "What Would be Conserved if 'The Tape Were Played Twice'?" In G. Cowan, D. Pines, and D. Meltzer, eds., *Complexity* (Reading, MA: Addison Wesley, 1994).
14. S. A. Kauffman, *The Origins of Order: Self-Organization and Selection in Evolution* (New York: Oxford University Press, 1993), 13.
15. J. Cohen, and I. Stewart, "Chaos, Contingency, and Convergence," *Nonlinear Science Today.* 1 (1991), no. 2: 13.

16. K. Kelly, *Out of Control* (Reading, MA: Addison Wesley, 1994), 410.

17. M. W. McRae, "Stephen Jay Gould and the Contingent Nature of History," *Clio* 22 (1993), no. 3: 241.

18. D. C. Dennett, *Darwin's Dangerous Idea: Evolution and the Meanings of Life* (New York: Simon and Schuster, 1995), 310.

19. S. Conway Morris, *Life's Solution: Inevitable Humans in a Lonely Universe* (New York: Cambridge University Press, 2004).

20. S. Conway Morris, *The Crucible of Creation: The Burgess Shale and the Rise of Animals* (New York: Oxford University Press, 1998).

21. D. Prothero, "Inevitable Humans or Hidden Agendas: A Review of *Life's Solution*," *Skeptic* 10 (2003), no. 3: 54–57.

22. Ibid., 57.

23. Gould, *Wonderful Life*, 289.

24. Mayr and Sagan, "Can SETI Succeed?"

25. Cohen and Stewart, "Chaos, Contingency, and Convergence," 9.

26. J. Cohen and I. Stewart, *The Collapse of Chaos* (New York: Viking, 1994), 131–32.

27. Gould, *Wonderful Life*, 283.

28. McRae, "Stephen Jay Gould and the Contingent Nature of History," 244.

29. S. J. Gould, "The Panda's Thumb of Technology," *Natural History*, January 1987: 22.

30. M. Shermer, "Gould's Dangerous Idea: Contingency, Necessity, and the Nature of History." *Skeptic* 4 (1996), no. 1: 91–95, 88.

31. M. Shermer, "The Chaos of History: On a Chaotic Model That Represents the Role of Contingency and Necessity in Historical Sequences," *Nonlinear Science Today* 2 (1993), no. 4: 1–13; M. Shermer, "Exorcising LaPlace's Demon: Chaos and Antichaos, History and Metahistory," *History and Theory* 34 (1995), no. 1: 59–83; M. Shermer, "Cycles and Curves," *Skeptic* 3 (1995), no. 3: 58–61; M. Shermer, "The Crooked Timber of History," *Complexity* 2 (1997), no. 6: 23–29.

32. R. Wright, *Nonzero: The Logic of Human Destiny* (New York: Pantheon, 2000).

33. M. Shermer, "We Are the World: A Review of *Nonzero* by Robert Wright," *Los Angeles Times Book Review*, Feb. 6, 2000.

34. Wright, *Nonzero*, 317.

35. Ibid., 7.

36. Ibid., 291.

37. Ibid., 292.

38. Ibid., 292–93.

39. R. G. Klein, *The Human Career* (Chicago: University of Chicago Press, 1999), 367–493.

40. R. Leakey, *The Origin of Humankind* (New York: Basic Books, 1994), 134.

41. Klein, *The Human Career*, 441–42.

42. C. Wills, *Children of Prometheus* (Reading, MA: Perseus Books, 1998), 143–45.

43. M. Shermer, *How We Believe: The Search for God in an Age of Science* (New York: W. H. Freeman, 1999).

44. Klein, *The Human Career*, 469.

45. I. Tattersall, "Once We Were Not Alone," *Scientific American*, January 2000: 56–62.

46. I. Tattersall, *The Fossil Trail* (New York: Oxford University Press, 1995), 212.

47. Leakey, *The Origin of Humankind*, 132.

48. Ibid., 138.

49. Ibid., 20.

50. Tattersall, *The Fossil Trail*, 246.

21

MOLECULAR CONVERGENCE: REPEATED EVOLUTION OR REPEATED DESIGNS?[1]

FAZALE R. RANA

1. INTRODUCTION: RECURRING DESIGNS

Even the uninitiated can recognize art by Pablo Picasso, partly because of his worldwide fame, but also because distinct styles and recurring themes characterize his work. Perhaps the most easily identifiable paintings come from his cubist period.[2] More sophisticated patrons of the arts may recognize paintings from other stages of Picasso's career,[3] and even distinct phases within his cubist period. While in his analytical cubist stage, he analyzed objects by taking them apart; during his synthetic cubist stage, he incorporated collages into his paintings.[4]

Picasso is not unique in this regard. All artists identify with particular schools of art and media of expression. They use colors and materials in characteristic ways and typically gravitate toward certain objects and themes. That is what makes it possible to identify a piece of art as the work of a particular artist. Each artist has his own style.

Artists are not the only ones who create in characteristic ways; other human designers do as well. Engineers, inventors, and architects typically produce works that reflect their own signature styles. This was certainly the case for Frank Lloyd Wright (1867–1959). Trained as a civil engineer, Wright is considered among America's greatest architects. Known for radical innovations, Wright's houses are characterized by open plans that eliminate walls between rooms.[5]

Just as artists generally gravitate toward the same themes, architects, inventors, and engineers also reuse the same techniques and technologies. Wright did. It's much more prudent and efficient for an architect to reapply a successful strategy (even when it may be unconventional) than to invent a new approach, particularly when confronted with complicated problems that already have solutions.

The tendency of artists and other human designers to revisit the same themes and reuse the same designs provides insight into the process of intelligent design. If human craftsmen reuse the same techniques and technologies, it's reasonable to expect that any intelligent agent would do the same. So, if life were the product of intelligent design, it also would be reasonable to expect the same designs to appear repeatedly throughout nature.

2. Identical Accidents?

While repeated occurrences of a design naturally point to intelligent causation, that is not the case for independent evolutionary processes. If biological systems are the product of evolution, then the same designs would *not* be expected to recur independently throughout nature.

Random variation governs biological evolution at its most fundamental level. Evolutionary pathways consist of a historical sequence of chance genetic changes operated on by natural selection, which in turn depends on radically contingent features of the environment. The consequences of this are profound. If evolutionary events could be repeated, the outcome should be dramatically different every time. The extraordinary improbability of separate evolutionary processes tracing the *same* path makes it highly unlikely that the same biological designs would be repeated in different evolutionary lineages.

Darwin was perhaps the first to recognize this feature of unguided evolution. In a later edition of *On the Origin of Species* he wrote:

> It is incredible that the descendants of two organisms, which had originally differed in a marked manner, should ever afterwards converge so closely as to lead to a near approach to identity throughout their whole organisation. If this had occurred, we should meet with the same form, independently of genetic connection, recurring in widely separated geological formations; and the balance of evidence is opposed to any such an admission.

Given the vagaries of evolutionary mechanisms, Darwin couldn't fathom how radically different organisms might evolve toward the same biological form. In other words, he didn't think evolutionary processes could converge on the same design.

Evolutionary biologist Stephen J. Gould popularized Darwin's initial intuition using the notion of historical contingency. This concept is the theme of his book, *Wonderful Life*. According to Gould:

> No finale can be specified at the start, none would ever occur a second time in the same way, because any pathway proceeds through thousands of improbable stages. Alter any early event, ever so slightly, and without apparent importance at the time, and evolution cascades into a radically different channel.[6]

To help clarify the idea of historical contingency, Gould used the metaphor of "replaying life's tape." If one could push the rewind button and erase life's history, then let the tape run again, the results would be completely different each time.[7] The very essence of the evolutionary process renders its outcomes nonrepeatable.

3. Putting the Facts to the Test: Contingency or Convergence?

Most scientists argue that the design so prevalent in biological systems is not true design. It only appears that way as an artifact of evolutionary processes. Accordingly, this apparent design stems from natural selection operating repeatedly on random genetic changes over vast periods of time to fine-tune biological systems.

Does contingency account for the patterns observed in the biological realm?[8] The idea of historical contingency suggests a way to discriminate between the "appearance of design" and intelligent design. If life results exclusively from evolutionary processes, then scientists should expect to see few, if any, cases in which evolution has repeated itself. However, if life is the product of intelligent design, then it would be natural for the same structures to appear repeatedly in living systems.

4. MOLECULAR CONVERGENCE

In his day, Darwin saw little evidence for convergence. Today such evidence abounds. In his book, *Life's Solution,* paleontologist Simon Conway Morris describes numerous examples of morphological and behavioral convergence at the organismal level.[9] But convergence isn't only confined to macroscopic systems. Over the last decade or so, scientists exploring the origin of biochemical systems have made a series of remarkable discoveries. When viewed from an evolutionary perspective, a number of life's molecules and processes, though virtually identical, appear to have originated independently, multiple times.[10] Evolutionary biologists refer to this independent origin of identical biomolecules and biochemical systems as molecular convergence. According to this concept, these molecules and processes arose *separately* when different evolutionary pathways converged on the same structure or system.

When molecular biologists first began studying biochemical origins, they expected to find few, if any, instances of molecular convergence.[11] One of the first examples was recognized in 1943 when two distinct forms of the enzyme fructose 1,6-bisphosphate aldolase were discovered in yeast and also in rabbit muscles. From an evolutionary perspective, it appears as if these two enzymes had separate evolutionary histories.[12]

At the time, this result was viewed as an evolutionary oddity. In the past decade, however, the advent of genomics—which now makes it possible to sequence, analyze, and compare the genomes of organisms—has made it evident that molecular convergence is a recurring pattern in nature rather than an exception to the rule. Contrary to expectations, biochemists are uncovering a mounting number of repeated independent biochemical origin events.

Evolutionary biologists recognize five different types of molecular convergence:[13]

1. *Functional convergence* describes the independent origin of biochemical functionality on more than one occasion.

2. *Mechanistic convergence* refers to the multiple independent emergences of biochemical processes that use the same chemical mechanisms.

3. *Structural convergence* results when two or more biomolecules independently adopt the same three-dimensional structure.

4. *Sequence convergence* occurs when either proteins or regions of DNA arise separately, but have identical amino acid or nucleotide sequences, respectively.

5. *Systemic convergence* is the most remarkable of all. This type of molecular convergence describes the independent emergence of identical biochemical systems.

Table 1 lists one hundred recently discovered examples of molecular convergence. This table is neither comprehensive nor exhaustive; it simply calls attention to the pervasiveness of molecular convergence.

Table 1
Examples of Molecular Convergence

Example	Reference
RNA	
Small nucleolar RNAs in eukaryotes and archaea	Omer, Arina D., et al. "Homologs of Small Nuclear RNAs in Archaea." *Science* 288 (April 21, 2000): 517–22.
Hammerhead Ribozyme	Salehi-Ashtiani, Kourosh, and Jack W. Szostak. "*In Vitro* Evolution Suggests Multiple Origins for the Hammerhead Ribosome." *Nature* 414 (November 1, 2001): 82–84.
DNA and Genes	
Gene structure of lamprin, elastins, and insect structural proteins	Robson, Paul, et al. "The Structure and Organization of Lamprin Genes: Multiple-Copy Genes with Alternative Splicing and Convergent Evolution with Insect Structural Proteins." *Molecular Biology and Evolution* 17 (November 2000): 1739–52.
Major histocompatibility complex DRB gene sequences in humans and Old and New World monkeys	Kriener, K., et al. "Convergent Evolution of Major Histocompatibility Complex Molecules in Humans and New World Monkeys." *Immunogenetics* 51 (March 2000): 169–78.
Structure and expression of the Ω–crystallin gene in vertebrates and invertebrates	Carosa, Eleonora, et al. "Structure and Expression of the Scallop Ω–Crystallin Gene. Evidence for Convergent Evolution of Promoter Sequences." *Journal of Biological Chemistry* 277 (January 4, 2002): 656–64.
Group I introns in mitochondria and chloroplasts, and hyperthermophilic bacteria	Nesbo, Camillia L., and W. Ford Doolittle. "Active Self-Splicing Group I Introns in 23S rRNA Genes of Hyperthermophilic Bacteria, Derived from Introns in Eukaryotic Organelles." *Proceedings of the National Academy of Sciences, USA* 100 (September 16, 2003): 10806–11.
Flanking sequences to microsatellite DNA in the human genome	Vowles, Edward J., and William Amos. "Evidence for Widespread Convergent Evolution around Human Microsatellites." *PLoS Biology* 2 (August 17, 2004): e199.

Proteins and Enzymes	
Immunoglobin G-binding proteins in bacteria	Frick, Inga-Maria, et al. "Convergent Evolution among Immunoglobulin G-Binding Bacterial Proteins." *Proceedings of the National Academy of Sciences, USA* 89 (September 15, 1992): 8532–36.
The α/β hydrolase fold of hydrolytic enzymes	Ollis, David L., et al. "The α/β Hydrolase Fold." *Protein Engineering* 5 (April 1992): 197–211.
Peptidases	Rawlings, Neil D., and Alan J. Barrett. "Evolutionary Families of Peptidases." *Biochemical Journal* 290 (February 15, 1993): 205–18.
Myoglobins in humans and gastropods	Suzuki, Tomohiko, H. Yuasa, and Kiyohiro Imai. "Convergent Evolution. The Gene Structure of Sulculus 41 kDa Myoglobin Is Homologous with that of Human Indoleamine Dioxygenase." *Biochimica Biophysica Acta* 1308 (July 31, 1996): 41–48.
Tubulin in eukaryotes and FtsZ in bacteria	Desai, Arshad, and Timothy J. Mitchison. "Tubulin and FtsZ Structures: Functional and Therapeutic Implications." *Bioessays* 20 (July 1998): 523–27.
D-alanine:D-alanine ligase and cAMP-dependent protein kinase	Denessiouk, K. A., et al. "Two 'Unrelated' Families of ATP-Dependent Enzymes Share Extensive Structural Similarities about Their Cofactor Binding Sites." *Protein Science* 7 (May 1998): 1136–46.
Cytokines in vertebrates and invertebrates	Beschin, Alain, et al. "Convergent Evolution of Cytokines." *Nature* 400 (August 12, 1999): 627–28.
Zinc peptidases	Makarova, Kira S., and Nick V. Grishin. "The Zn-Peptidase Superfamily: Functional Convergence after Evolutionary Divergence." *Journal of Molecular Biology* 292 (September 10, 1999): 11–17.
Redox regulation of glucose 6-phosphate dehydrogenase in plants and cyanobacteria	Wendt, Urte K., et ak. "Evidence for Functional Convergence of Redox Regulation in G6PDH Isoforms of Cyanobacteria and Higher Plants." *Plant Molecular Biology* 40 (June 1999): 487–94.
MDR ethanol dehydrogenase / acetaldehyde reductase in vertebrates and *Escherichia coli*	Shafqat, Jawed, et al. "An Ethanol-Inducible MDR Ethanol Dehydrogenase / Acetaldehyde Reductase in *Escherichia coli*." *European Journal of Biochemistry* 263 (July 1999): 305–11.

Carbonic anhydrase in bacteria and archaea	Smith, Kerry S., et al. "Carbonic Anhydrase Is an Ancient Enzyme Widespread in Prokaryotes." *Proceedings of the National Academy of Sciences, USA* 96 (December 21, 1999): 15184–89.
Inositol-1-phosphate synthase in eukaryotes, bacteria, and archaea	Bachhawat, N. N., and S. C. Mande. "Complex Evolution of the Inositol-1-Phosphate Synthase Gene among Archaea and Eubacteria." *Trends in Genetics* 16 (March 2000): 111–13.
Myoglobins in eukaryotes and myoglobin-like, heme-containing protein in Archaea	Hou, Shaobin, Randy W. Larsen, Dmitri Boudko, Charles W. Riley, et al. "Myoglobin-Like Aerotaxis Transducers in Archaea and Bacteria." *Nature* 403 (February 3, 2000): 540–44.
Calmodulins in vertebrates and cephalochordates	Karabinos, Anton, and Debashish Bhattacharya. "Molecular Evolution of Calmodulin and Calmodulin-Like Genes in the Cephalochordate Branchiostoma." *Journal of Molecular Evolution* 51 (August 2000): 141–48.
Pheromone binding proteins in moths	Willett, Christopher S. "Do Pheromone Binding Proteins Converge in Amino Acid Sequence When Pheromones Converge?" *Journal of Molecular Evolution* 50 (February 2000): 175–83.
DNA Holliday junction resolvases in bacteria and eukaryotic viruses	Garcia, Alonzo D., et al. "Bacterial-Type DNA Holliday Junction Resolvases in Eukaryotic Viruses." *Proceedings of the National Academy of Sciences, USA* 97 (August 1, 2000): 8926–31.
The DNA replication protein, RepA, in Gram-negative bacteria plasmids, archaea, and eukaryotes	Giraldo, Rafael, and Ramón Díaz-Orejas. "Similarities between the DNA Replication Initiators of Gram-Negative Bacteria Plasmids (RepA) and Eukaryotes (Orc4p) / Archaea (Cdc6p)." *Proceedings of the National Academy of Sciences, USA* 98 (April 24, 2001): 4938–43.
Spider silk fibroin sequences	Gatesy, John, et al. "Extreme Diversity, Conservation, and Convergence of Spider Silk Fibroin Sequences." *Science* 291 (March 30, 2001): 2603–5.
Alcohol dehydrogenase in *Drosophila* and Medfly, Olive Fly, and Flesh Fly	Brogna, Saverio, et al. "The *Drosophila* Alcohol Dehydrogenase Gene May Have Evolved Independently of the Functionally Homologous Medfly, Olive Fly, and Flesh Fly Genes." *Molecular Biology and Evolution* 18 (March 2001): 322–29.

Type II restriction enzymes	Bujnicki, Janusz, Monika Radlinska, and Leszek Rychlewski. "Polyphyletic Evolution of Type II Restriction Enzymes Revisited: Two Independent Sources of Second-Hand Folds Revealed." *Trends in Biochemical Sciences* 26 (January 2001): 9–11.
Heavy metal binding domains of copper chaperones and copper-transporting ATPases	Jordan, I. King, et al. "Independent Evolution of Heavy Metal-Associated Domains in Copper Chaperones and Copper-Transporting ATPases." *Journal of Molecular Evolution* 53 (December 2001): 622–33.
Opsin in vertebrates and invertebrates	Zakon, Harold H. "Convergent Evolution on the Molecular Level." *Brain, Behavior and Evolution* 59, no. 5–6 (2002): 250–61.
Ionotropic and metabotropic neurotransmitter receptors	Zakon, Harold H. "Convergent Evolution on the Molecular Level." *Brain, Behavior and Evolution* 59, no. 5–6 (2002): 250–61.
Gap junction proteins in invertebrates and vertebrates	Zakon, Harold H. "Convergent Evolution on the Molecular Level." *Brain, Behavior and Evolution* 59, no. 5–6 (2002): 250–61.
Neurotoxins in invertebrates and vertebrates	Zakon, Harold H. "Convergent Evolution on the Molecular Level." *Brain, Behavior and Evolution* 59, no. 5–6 (2002): 250–61.
Anti-β-elimination mechanism in 1 and 10 polysaccharide lyases	Charnock, Simon J., et al. "Convergent Evolution Sheds Light on the Anti-β-Elimination Mechanism in 1 and 10 Polysaccharide Lyases." *Proceedings of the National Academy of Sciences, USA* 99 (September 17, 2002): 12067–72.
α1,4,-Fucosyltransferase activity in primates	Dupuy, Fabrice, et al. "α1,4,-Fucosyltransferase Activity: A Significant Function in the Primate Lineage Has Appeared Twice Independently." *Molecular Biology and Evolution* 19 (June 2002): 815–24.
Aldehyde oxidase into xanthine dehydrogenase two separate times	Rodriguez-Trelles, Francisco, Rosa Tarrio, and Francisco J. Ayala. "Convergent Neofunctionalization by Positive Darwinian Selection after Ancient Recurrent Duplications of the *Xanthine Dehydrogenase* Gene." *Proceedings of the National Academy of Sciences, USA* 100 (November 11, 2003): 13413–17.
RuBisCO-like protein of *Bacillus* in non-photosynthetic bacteria and archaea and photosynthetic RuBisCo in photosynthetic bacteria	Ashida, Hiroki, et al. "A Functional Link between RuBisCO-Like Protein of Bacillus and Photosynthetic RuBisCO." *Science* 302 (October 10, 2003): 286–90.

Active site of creatinine amidohydrolase of *Pseudomonas putida* and hydantoinase-like cyclic amidohydrolases	Beuth, B., K. Niefind, and D. Schomburg. "Crystal Structure of Creatininase from *Pseudomonas putida*: A Novel Fold and a Case of Convergent Evolution." *Journal of Molecular Biology* 332 (September 5, 2003): 287–301.
The insect flight muscle protein arthrin in Diptera and Hemiptera	Schmitz, Stephan, et al. "Molecular Evolutionary Convergence of the Flight Muscle Protein Arthrin in Diptera and Hemiptera." *Molecular Biology and Evolution* 20 (December 2003): 2019 33.
The enzyme, tRNA(m1G37) methyltransferase, in bacteria and archaea	Christian, Thomas, et al. "Distinct Origins of tRNA(m1G37) Methyltransferase." *Journal of Molecular Biology* 339 (June 11, 2004): 707–19.
β-lactam-hydrolyzing function of the B1 + B2 and B3 subclasses of metallo-β-lactamases	Hall, Barry, Stephen Salipante, and Miriam Barlow. "Independent Origins of Subgroup B1 + B2 and Subgroup B3 Metallo-β-Lactamases." *Journal of Molecular Evolution* 59 (July 2004): 133–41.
Catabolic enzymes for galactitol, and D-tagatose in enteric bacteria	Shakeri-Garakani, et al. "The Genes and Enzymes for the Catabolism of Galactitol, D-Tagatose, and Related Carbohydrates in *Klebsiella oxytoca* M5a1 and Other Enteric Bacteria Display Convergent Evolution." *Molecular Genetics and Genomics* 271 (July 2004): 717–28.
Lipases and GDSL esterases/lipases	Akoh, C. C., et al. "GDSL Family of Serine Esterases/Lipases." *Progress in Lipid Research* 43 (November 2004): 534–52.
Chitosanases	Adachi, Wataru, et al. "Crystal Structure of Family GH-8 Chitosanase with Subclass II Specificity from Bacillus sp. K17." *Journal of Molecular Biology* 343 (October 22, 2004): 785–95.
Plant and cyanobacterial phytochromes	Lamparter, T. "Evolution of Cyanobacterial Plant Phytochromes." *FEBS Letters* 573 (August 27, 2004): 1–5.
The outer membrane protein, OmpA, in Enterobacteriaceae	Gophna, U., et al. "OmpA of a Septicemic *Escherichia coli* O78—Secretion and Convergent Evolution." *International Journal of Medical Microbiology* 294 (October 2004): 373–81.
Cardiovascular risk factor, LPA in hedgehogs and primates	Boffelli, D., J. F. Cheng, and E. M. Rubin. "Convergent Evolution in Primates and an Insectivore." *Genomics* 83 (January 2004): 19–23.

Lectin-like activity of cytokines in vertebrates and invertebrates	Beschin, A., et al. "Functional Convergence of Invertebrate and Vertebrate Cytokine-Like Molecules Based on a Similar Lectin-Like Activity." *Progress in Molecular and Subcellular Biology* 34 (2004): 145–63.
Temperature adaptation of A4-lactate dehydrogenases of Pacific damsel fishes	Johns, Glenn C., and George N. Somero. "Evolutionary Convergence in Adaptation of Proteins to Temperature: A_4-Lactate Dehydrogenases of Pacific Damselfishes (*Chromis* spp.)." *Molecular Biology and Evolution* 21 (February 2004): 314–20.
Scorpion and sea anemone toxins that bind to voltage-gated potassium ion channels	Gasparini, S., B. Gilquin, and A. Menez. "Comparison of Sea Anemone and Scorpian Toxins Binding to Kv1 Channels: An Example of Convergent Evolution." *Toxicon* 43 (June 15, 2004): 901–8.
Feruloyl esterase A in microorganisms	Hermoso, J. A., et al. "The Crystal Structure of Feruloyl Esterase A from *Aspergillus niger* Suggests Evolutive Functional Convergence in Feruloyl Esterase Family." *Journal of Molecular Biology* 338 (April 30, 2004): 495–506.
The proofreading domain of the enzyme threonyl-tRNA synthetase in archaea and bacteria	Korencic, Dragana, et al. "A Freestanding Proofreading Domain Is Required for Protein Synthesis Quality Control Archaea." *Proceedings of the National Academy of Sciences, USA* 101 (July 13, 2004): 10260–65.
Protein inhibitors of proteases	Otlewski, Jacek, et al. "The Many Faces of Protease-Protein Inhibitor Interaction." *EMBO Journal* 24 (April 6, 2005): 1303–10.
Alginate lyases	Osawa, Takuo, et al. "Crystal Structure of the Alginate (Poly α-L-Guluronate) Lyase from Corynebacterium sp. At 1.2 Å Resolution." *Journal of Molecular Biology* 345 (February 4, 2005): 1111–18.
Defensins from insects and mollusks and ABF proteins in nematodes	Froy, Oren. "Convergent Evolution of Invertebrate Defensins and Nematode Antibacterial Factors." *Trends in Microbiology* 13 (July 2005): 314–19.
Blue and red light photoreceptors in diatoms	Falciatore, Angela, and Chris Bowler. "The Evolution and Function of Blue and Red Light Photoreceptors." *Current Topics in Developmental Biology* 68 (2005): 317–50.

Red light photoreceptors in ferns and green algae	Suetsugu, Noriyuki, et al. "A Chimeric Photoreceptor Gene, NEOCHROME, Has Arisen Twice during Plant Evolution." *Proceedings of the National Academy of Sciences, USA* 102 (September 20, 2005): 13705–9.
Xanthine oxidation in fungus	Cultrone, Antonietta, et al. "Convergent Evolution of Hydroxylation Mechanisms in the Fungal Kingdom: Molybdenum Cofactor Independent Hydroxylation of Xanthine via α-Ketoglutarate - Dependent Dioxygenases." *Molecular Microbiology* 57 (July 2005): 276–90.
The muscle protein troponin C in various insect orders	Herranz, Raúl, Jesús Mateos, and Roberto Marco. "Diversification and Independent Evolution of Troponin C Genes in Insects." *Journal of Molecular Evolution* 60 (January 2005): 31–44.
Structure of immunoglobulin and C type lectin receptors	Feng, Jianwen, et al. "Convergence on a Distinctive Assembly Mechanism by Unrelated Families of Activating Immune Receptors." *Immunity* 22 (April 2005): 427–38.
The placental development syncytin family of proteins in primates and muridae from separate endogenous retrovirus infections	Dupressoir, Anne, et al. "Syncytin-A and Syncytin-B, Two Fusogenic Placenta-Specific Murine Envelop Genes of Retroviral Origin Conserved in Muridae." *Proceedings of the National Academy of Sciences, USA* 102 (January 18, 2005): 725–30.
Structure and function of S-adenosylmethionine-binding proteins	Kozbial, Piotr Z., and Arcady R. Mushegian. "Natural History of S-Adenosylmethionine-Binding Proteins." *BMC Structural Biology* 14 (October 14, 2005): 5–19.
2-methylbutyryl-CoA dehydrogenase in potato and short/branched-chain acyl-CoA dehydrogenase in humans	Goetzman, Eric S., et al. "Convergent Evolution of a 2-Methylbutyryl-CoA Dehydrogenase from Isovaleryl-CoA Dehydrogenase in *Solanum tuberosum*." *Journal of Biological Chemistry* 280 (February 11, 2005): 4873–79.
Dynamin-mediated endocytosis in multicellular animals and ciliates	Elde, Nels C., et al. "Elucidation of Clathrin-Mediated Endocytosis in *Tetrahymena* Reveals an Evolutionarily Convergent Recruitment of Dynamin." *PLoS Genetics* 1 (November 4, 2005): e:52.

The animal glycan-recognizing proteins—lectins and sulfated glycosaminoglycan binding proteins—in animals	Varki, Ajit, and Takashi Angata. "Siglecs—The Major Subfamily of I-Type Lectins." *Glycobiology* 16 (January 2006): 1R–27R.
Sodium channel in the electric organ of the mormyriform and gymnitoform electric fishes	Zakon, Harold H., et al. "Sodium Channel Genes and the Evolution of Diversity in Communication Signals of Electric Fishes: Convergent Molecular Evolution." *Proceedings of the National Academy of Sciences, USA* 103 (March 7, 2006): 3675–80.
Clathrin heavy and light chain isoforms in chordates	Wakeham, Diane E., et al. "Clathrin Heavy and Light Chain Isoforms Originated by Independent Mechanisms of Gene Duplication during Chordate Evolution." *Proceedings of the National Academy of Sciences, USA* 102 (May 17, 2005): 7209–14.
Protein-binding receptor that readies proteins for import into the mitochondria of animals and plants	Perry, Andrew J., et al. "Convergent Evolution of Receptors for Protein Import into Mitochondria." *Current Biology* 16 (February 7, 2006): 221–29.
ICP C1 cysteine peptidase inhibitors	Smith, Brian O., et al. "The Structure of *Leishmania mexicana* ICP Provides Evidence for Convergent Evolution of Cysteine Peptidase Inhibitors." *Journal of Biological Evolution* 281 (March 3, 2006): 5821–28.
TRIM5 anti-retroviral resistance factor in primates and bovines	Si, Zhihai, et al. "Evolution of a Cytoplasmic Tripartite Motif (TRIM) Protein in Cows that Restricts Retroviral Infection." *Proceedings of the National Academy of Sciences, USA* 103 (May 9, 2006): 7454–59.
Protein receptors that bind bitter compounds in humans and chimpanzees	Wooding, Stephen, et al. "Independent Evolution of Bitter-Taste Sensitivity in Humans and Chimpanzees." *Nature* 440 (April 13, 2006): 930–34.
Protoporphyrin (IX) ferrochelatase in prokaryotes and eukaryotes	Shepherd, Mark, Tamara A. Dailey, and Harry A. Dailey. "A New Class of [2Fe-2S]-Cluster-Containing Protoporphyrin (IX) Ferrochelatases." *Biochemical Journal* 397 (July 1, 2006): 47–52.
SPFH (stomatin-prohibitin-flotillin-HflC/K)-like proteins	Rivera-Milla, E., C. A. Stuermer, and E. Malaga-Trillo. "Ancient Origin of Reggie (Flotillin), Reggie-Like, and Other Lipid-Raft Proteins: Convergent Evolution of the SPFH Domain." *Cellular and Molecular Life Sciences* 63 (February 2006): 343–57.
Disulfide-rich protein domains	Cheek, S., S. S. Krishna, and N. V. Grishin. "Structural Classification of Small, Disulfide-Rich Protein Domains." *Journal of Molecular Biology* 359 (May 26, 2006): 215–37.

Fatty acid synthases in fungi and animals	Maier, Timm, Simon Jenni, and Nenad Ban. "Architecture of Mammalian Fatty Acid Synthase at 4.5 Å Resolution." *Science* 311 (March 3, 2006): 1258–62; Jenni, Simon, Marc Leibundgut, Timm Maier, and Nanad Ban. "Architecture of a Fungal Fatty Acid Synthase at 5 Å Resolution." *Science* 311 (March 3, 2006): 1263–67.
NAD(P)H:quinone oxidoreductase (NQO)	Vasiliou, Vasilis, David Ross, and Daniel W. Nebert. "Update of the NAD(P)H:Quinone Oxidoreductase *(NQO)* Gene Family." *Human Genomics* 2 (March 2006): 329–35.
Adenylation activity in BirA, lipoate protein ligase and class II tRNA synthetases	Wood, Zachary A., et al. "Co-Repressor Induced Order and Biotin Repressor Dimerization: A Case for Divergent Followed by Convergent Evolution." *Journal of Molecular Biology* 357 (March 24, 2006): 509–23.
D7 and lipocalin salivary proteins in insects	Calvo, Eric, et al. "Function and Evolution of a Mosquito Salivary Protein Family." *Journal of Biological Chemistry* 281 (January 27, 2006): 1935–42.
Cold shock domain of cold shock proteins in bacteria and higher plants	Nakaminami, Kentaro, Dale T. Karlson, and Ryozo Imai. "Functional Conservation of Cold Shock Domains in Bacteria and Higher Plants." *Proceedings of the National Academy of Sciences, USA* 103 (June 27, 2006): 10122–27.

Biochemical Systems

Bioluminescent systems	Hastings, J. W. "Biological Diversity, Chemical Mechanisms, and the Evolutionary Origins of Bioluminescent Systems." *Journal of Molecular Evolution* 19 (September 1983): 309–21.
Chlorocatechol catabolic pathway in *Rhodococcus opacus* and proteobacteria	Eulberg, Dirk, et al. "Evolutionary Relationship between Chlorocatechol Catabolic Enzymes from *Rhodococcus opacus* 1CP and Their Counterparts in Proteobacteria: Sequence Divergence and Functional Convergence." *Journal of Bacteriology* 180 (March 1998): 1082–94.
Nucleotide excision DNA repair in humans and *Escherichia coli*	Petit, C., and A. Sancar. "Nucleotide Excision Repair: From *E. Coli* to Man." *Biochimie* 81 (January–February 1999): 15–25.
DNA replication in bacteria and archaea	Leipe, Detlef D., L. Aravind, and Eugene V. Koonin. "Did DNA Replication Evolve Twice Independently?" *Nucleic Acids Research* 27 (September 27, 1999): 3389–3401.

DNA repair proteins	Aravind, L., D. Ronald Walker, and Eugene V. Koonin. "Conserved Domains in DNA Repair Proteins and Evolution Systems." *Nucleic Acids Research* 27 (March 1, 1999): 1223–42.
Toxin resistance	Zakon, Harold H. "Convergent Evolution on the Molecular Level." *Brain, Behavior and Evolution* 59, no. 5–6 (2002): 250–61.
Biosurfactants in archaea and bacteria	Maier, Raina M. "Biosurfactants: Evolution and Diversity in Bacteria." *Advances in Applied Microbiology* 52 (2003): 101–21.
Glycolytic pathways in archaea and bacteria	Verhees, Corné H., et al. "The Unique Features of Glycolytic Pathways in Archaea." *Biochemical Journal* 375 (October 15, 2003): 231–46.
Type III and Type IV secretion systems of gram-negative and gram-positive bacteria	Blocker, Ariel, Kaoru Komoriya, and Shin-Ichi Aizawa. "Type III Secretion Systems and Bacterial Flagella: Insights into Their Function from Structural Similarities." *Proceedings of the National Academy of Sciences, USA* 100 (March 18, 2003): 3027–30.
Phosphopantothenate biosynthesis in archaea and bacteria	Genschel, Ulrich. "Coenzyme A Biosynthesis: Reconstruction of the Pathway in Archaea and an Evolutionary Scenario Based on Comparative Genomics." *Molecular Biology and Evolution* 21 (July 2004): 1242–51.
Crassulacean acid metabolism, a specialized form of photosynthesis in the Bromeliaceae family of plants	Crayn, Darren M., Klaus Winter, and J. Andrew C. Smith. "Multiple Origins of Crassulacean Acid Metabolism and the Epiphytic Habit in the Neotropical Family Bromeliaceae." *Proceedings of the National Academy of Sciences, USA* 101 (March 9, 2004): 3703–8.
Alternate splicing of tandem exons in ion-channel genes in humans and *Drosophila melanogaster*	Copley, Richard R. "Evolutionary Convergence of Alternative Splicing in Ion Channels." *Trends in Genetics* 20 (April 2004): 171–76.
Viral capsid structure of viruses that infect archaea, bacteria, and eukarya	Rice, George, et al. "The Structure of a Thermophilic Archaeal Virus Shows a Double-Stranded DNA Capsid Type that Spans All Domains of Life." *Proceedings of the National Academy of Sciences, USA* 101 (May 18, 2004): 7716–20.
Hub-based design of gene regulatory networks	Amoutzias, Gregory D., et al. "Convergent Evolution of Gene Networks by Single-Gene Duplications in Higher Eukaryotes." *EMBO Reports* 5 (March 2004): 274–79.

The two Mg^{2+} metal ion mechanism in protein phosphoryltransferases and RNA phosphoryltransferases	Stahley, Mary R., and Scott A. Strobel. "Structural Evidence for a Two-Metal-Ion Mechanism of Group I Intron Splicing." *Science* 309 (September 2, 2005): 1587–90.
Halophilic biochemical adaptations in bacteria and archaea	Mongodin, E. F., et al. "The Genome of *Salinibacter ruber*: Convergence and Gene Exchange Among Hyperhalophilic Bacteria and Archaea." *Proceedings of the National Academy of Sciences*, USA 102 (December 13, 2005): 18147–52.
Regulatory network linking DNA synthesis to cell cycle in yeast and bacteria	Brazhnik, Paul, and John J. Tyson. "Cell Cycle Control in Bacteria and Yeast: A Case of Convergent Evolution?" *Cell Cycle* 5 (March 2006): 522–29.
Biochemical mechanisms for ion regulation in invertebrates	Zanotto, Flavia Pinheiro, and Michele G. Wheatly. "Ion Regulation in Invertebrates: Molecular and Integrative Aspects." *Physiological and Biochemical Zoology* 79 (March / April 2006): 357–62.
Apoptosis and immune response defenses in large nuclear and cytoplasmic DNA viruses of eukaryotes	Iyer, Lakshminarayan M., et al. "Evolutionary Genomics of Nucleo-Cytoplasmic Large DNA Viruses." *Virus Research* 117 (April 2006): 156–84.
Endocannabinoid system	McPartland, John M., et al. "Evolutionary Origins of the Endocannabinoid System." *Gene* 370 (March 29, 2006): 64–74.
Non-universal codon usage in the genetic code of arthropod mitochondrial genomes	Abascal, Federico, et al. "Parallel Evolution of the Genetic Code in Arthropod Mitochondrial Genomes." *PLoS Biology* 4 no. 5 (April 25, 2006): e127.
Xist RNA gene-mediated X chromosome inactivation in eutherian and marsupial mammals	Duret, Laurent, et al. "The *Xist* RNA Gene Evolved in Eutherians by Pseudogenization of a Protein-Coding Gene." *Science* 312 (June 16, 2006): 1653–55.

Currently recognized examples of molecular convergence are likely just the tip of the iceberg. For instance, researchers from Cambridge University (United Kingdom) examined the amino acid sequences of over six hundred peptidase enzymes.[14] When viewed from an evolutionary standpoint, these workers discovered that there appear to have been over sixty separate origin events for peptidases. This result stands in sharp contrast to what the researchers expected to find: a handful of peptidase families with separate origins. In many cases, the peptidases appeared to converge on the same enzyme mechanisms and reaction specificities.[15]

Researchers from the National Institutes of Health (NIH) recently made a similar discovery. These scientists systematically examined protein sequences from 1,709 EC (enzyme commission) classes and discovered that 105 of them consisted of proteins that catalyzed the same reaction, but must have had separate evolutionary origins.[16]

In a separate study, this same team discovered that in at least twelve clear-cut cases the same essential cellular functions were carried out by unrelated enzymes (from an evolutionary vantage point) when the genomes of the bacteria *Mycoplasma genitalium* and *Haemophilus influenzae* were compared.[17] The researchers noted that the genomes of these two microbes are small, close to the size of the minimal gene set. It's quite likely a greater number of convergent systems would be identified if the genomes of more complex organisms were compared.

The explosion in the number of examples of molecular convergence is unexpected if life results from historical sequences of chance evolutionary events. Yet, if life is the result of intelligent design, it's reasonable to expect the repeated appearance of the same designs. These artifacts would give the appearance of multiple independent origin events when viewed from an evolutionary vantage point.

It is beyond the scope of this essay to detail each example of molecular convergence. Rather, a discussion of one of the most arresting examples of molecular convergence, the independent origins of DNA replication in bacteria and archaea/eukaryotes, illustrates how remarkable molecular convergence is from an evolutionary standpoint and why it is preferable to view the repeated independent origins of a biochemical system as a product of intelligent causation.

5. The Origin of DNA Replication

The process of generating two "daughter" molecules identical to the "parent" DNA molecule—DNA replication—is essential for life. This duplication plays a central role in reproduction, inaugurating the cell division process. Once replicated, a complex ensemble of enzymes distributes the two newly made DNA molecules between the emerging daughter cells.

Because of its extremely complex nature (described below), most biochemists previously thought DNA replication arose once, prior to the origin of LUCA (the last universal common ancestor). Figure 1 shows the relationship between this supposed "organism" and the evolutionary tree of life.

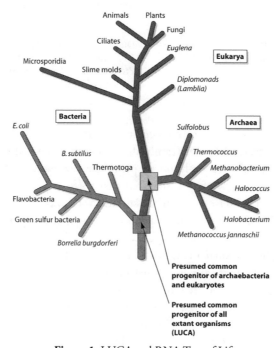

Figure 1. LUCA and RNA Tree of Life

Molecular Convergence

Many biochemists have long regarded the close functional similarity of DNA replication, observed in all life, as evidence for the single origin of DNA replication prior to the emergence of LUCA. The common features of DNA replication include:

1. semi-conservative replication;

2. initiation at a defined origin by an origin-replication complex;

3. bidirectional movement of the replication fork;

4. continuous (leading strand) replication for one DNA strand and discontinuous (lagging strand) replication for the other DNA strand;

5. use of RNA primers; and

6. the use of nucleases, polymerases, and ligases to replace RNA primer with DNA (see DNA replication discussion below).

Surprisingly, in 1999, researchers from NIH demonstrated that the core enzymes in the DNA replication machinery of bacteria and archaea/eukaryotes (the two major trunks of the evolutionary tree of life) did not share a common evolutionary origin. From an evolutionary perspective, it appears as if two identical DNA replication systems emerged independently in bacteria and archaea—after these two evolutionary lineages supposedly diverged from the last universal common ancestor.[18] (If evolutionary processes explain the origin of DNA replication, then two *different* forms of DNA replication should exist in archaea and bacteria.)

6. DNA Replication—A Signature Style[19]

DNA consists of two polynucleotide chains aligned in antiparallel fashion. (The two strands are arranged parallel to one another with the starting point of one strand in the polynucleotide duplex located next to the ending point of the other strand and vice-versa.) The paired polynucleotide chains twist around each other forming the well-known DNA double helix. The polynucleotide chains are generated using four different nucleotides: adenosine (A); guanosine (G); cytidine (C); and thymidine (T).

A special relationship exists between the nucleotide sequences of the two DNA strands. These sequences are considered complementary. When the DNA strands align, the A side chains of one strand always pair with T side chains from the other strand. Likewise, the G side chains from one DNA strand always pair with C side chains from the other strand. Biochemists refer to these relationships as Watson-Crick base-pairing rules.

As a result of this base pairing, if biochemists know the sequence of one DNA strand, they can readily determine the sequence of the other strand. Base pairing plays a critical role in DNA replication.

Following a Pattern. The nucleotide sequences of the parent DNA molecule function as a template directing the assembly of the DNA strands of the two daughter molecules. It is a semi-conservative process because after replication, each daughter DNA molecule contains one newly formed DNA strand and one strand from the parent molecule (see Figure 2).

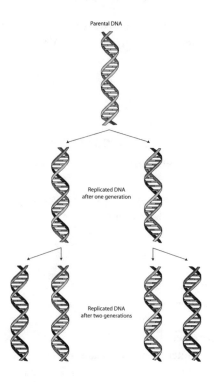

Figure 2. Semi-conservative DNA Replication

Conceptually, template-directed, semi-conservative DNA replication entails the separation of the parent DNA double helix into two single strands. According to the base-pairing rules, each strand serves as a template for the cell's machinery to follow as it forms a new DNA strand with a nucleotide sequence complementary to the parent strand. Because each strand of the parent DNA molecule directs the production of a new DNA strand, two daughter molecules result. Each possesses an original strand from the parent molecule and a newly formed DNA strand produced by a template-directed synthetic process.

The Start of It All. DNA replication begins at specific sites along the DNA double helix. Typically, prokaryotic cells have only a single origin of replication. More complex eukaryotic cells have multiple origins.

The DNA double helix unwinds locally at the origin of replication to produce a replication bubble (see Figure 3). The bubble expands in both directions from the origin during the course of replication. Once the individual strands of the DNA double helix unwind and are exposed within the replication bubble, they are available to direct the production of the daughter strand. The site where the double helix continuously unwinds is the replication fork. Because DNA replication proceeds in both directions away from the origin, each bubble contains two replication forks.

Moving on. DNA replication can proceed only in a single direction, from the top of the DNA strand to the bottom (5' to 3'). Because the strands that form the DNA double helix align in an antiparallel fashion with the 5' end of one strand juxtaposed to the 3' end of the other strand, only one strand at each replication fork has the proper orientation (bottom-to-top) to direct the assembly of a new strand in the top-to-bottom direction. For this leading strand, DNA replication proceeds rapidly and continuously in the direction of the advancing replication fork (see Figure 3).

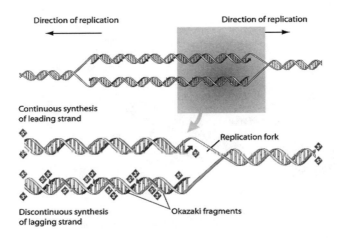

Figure 3. DNA Replication Bubble

DNA replication can't proceed along the strand with the top-to-bottom orientation until the replication bubble expands enough to expose a sizeable stretch of DNA. When this happens, DNA replication moves away from the advancing replication fork. It can proceed only a short distance along the top-to-bottom oriented strand before the replication process has to stop and wait for more of the parent DNA strand to be unwound. After a sufficient length of the parent DNA template is exposed the second time, DNA replication can proceed again, but only briefly before it has to stop and wait for more DNA.

The process of discontinuous DNA replication takes place repeatedly until the entire strand is replicated. Each time DNA replication starts and stops, a small fragment of DNA is produced. These pieces of DNA (that eventually comprise the daughter strand) are called Okazaki fragments, after the biochemist who discovered them. The discontinuously produced strand is the one that lags, because DNA replication for this strand falls behind the more rapidly, continuously produced leading strand (see Figure 3).

One additional point: the leading strand at one replication fork is the lagging strand at the other replication fork because the replication diverges as the two ends of the replication bubble advance in opposite directions.

The Protein Palette. An ensemble of proteins is needed to carry out DNA replication (see Figure 4). Once the origin recognition complex (which consists of several different proteins) identifies the replication origin, a protein called helicase unwinds the DNA double helix to form the replication fork. The process of unwinding introduces torsional stress in the DNA helix downstream from the replication fork. Another protein, gyrase, relieves the stress, thus preventing the DNA molecule from supercoiling.

Single-strand binding proteins bind to the DNA strands exposed by the unwinding process. This association keeps the fragile DNA strands from breaking apart.

Once the replication fork is established and stabilized, DNA replication can begin. Before the newly formed daughter strands can be produced, a small RNA primer must be made. The protein that synthesizes new DNA by reading the parent DNA template strand (DNA polymerase) cannot start from scratch. It must be primed. A massive protein complex, the primosome, which consists of over fifteen different proteins, produces the RNA primer needed by DNA polymerase.

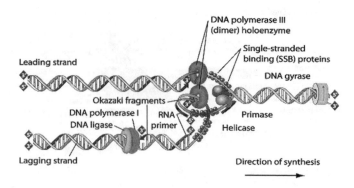

Figure 4. The Proteins of DNA Replication

Primed and ready to go. Once primed, DNA polymerase will continuously produce DNA along the leading strand. However, for the lagging strand, DNA polymerase can only generate DNA in spurts to produce Okazaki fragments. Each time DNA polymerase generates an Okazaki fragment, the primosome complex must produce a new RNA primer.

After DNA replication is completed, the RNA primers are removed from the continuous DNA of the leading strand and the Okazaki fragments that make up the lagging strand. A protein called a 3′–5′ exonuclease removes the RNA primers. A different DNA polymerase fills in the gaps created by the removal of the RNA primers. Finally, a ligase protein connects all the Okazaki fragments together to form a continuous piece of DNA out of the lagging strand.

This cursory description of DNA replication clearly illustrates its complexity and intricacies (many details were omitted). It is phenomenal to think this biochemical system evolved a single time, let alone twice. There is no obvious reason for DNA replication to take place by a semi-conservative, RNA primer-dependent, bidirectional process that depends on leading and lagging strands to produce DNA daughter molecules. Even if DNA replication could have evolved independently on two separate occasions, it is reasonable to expect that functionally distinct processes would emerge for bacteria and archaea/eukaryotes, given their idiosyncrasies. But this is not what happened.

7. A Challenge to Historical Contingency

Considering the complexity of life's chemical systems, pervasive molecular convergence fits uncomfortably within an evolutionary framework. Paleontologist J. William Schopf, one of the world's leading authorities on Earth's early life, says,

> Because biochemical systems comprise many intricately interlinked pieces, any particular full-blown system can only arise once . . . Since any complete biochemical system is far too elaborate to have evolved more than once in the history of life, it is safe to assume that microbes of the primal LCA [last common ancestor] cell line had the same traits that characterize all its present-day descendants. [20]

But the pattern expected by Schopf and other evolutionary biologists is simply not observed at the biochemical level. An inordinate number of independent examples of molecular convergence have already been discovered. In all likelihood, many more will be identified in the future. Each new instance of molecular convergence makes an evolutionary explanation for life less likely.

Yet some evolutionary biologists express skepticism about historical contingency. Simon Conway Morris is one of these. In his book, *Life's Solution,* Conway Morris argues that the pervasiveness of biological and biochemical convergence reveals something fundamental about the evolutionary process. Though the pathways that evolution takes may be historically contingent, the process always finds its way to the same endpoints under the influence of natural selection. In other words, evolution does *indeed* repeat itself, with contingency left to the minor details.

Given that large-scale evolutionary processes can't be observed, how is it possible to resolve this controversy? New work by Richard Lenski's group at Michigan State University provides significant headway toward this end.[21] They provided the first real-time scientific test for historical contingency within the framework of their Long Term Evolution Experiment (LTEE).

Long Term Evolution Experiment. Designed to monitor evolutionary changes in *Escherichia coli,* this study was inaugurated in 1988. The LTEE began with a single cell of *E. coli* that was used to generate twelve genetically identical lines of cells. The twelve clones of the parent *E. coli* cell were then inoculated into a minimal growth medium that contained low levels of glucose as the only carbon source. After growing overnight, an aliquot of the cells was transferred into fresh growth media. This process was repeated every day for about twenty years.

Throughout the experiment, aliquots of cells were frozen every 500 generations. These frozen cells represent a "fossil record" of sorts that can be thawed out and compared to current and other past generations of cells. Furthermore, throughout the experiment the forces of natural selection were carefully controlled. The temperature, pH, nutrients, and oxygen exposure were held constant for the last twenty years. Starvation was the primary challenge facing these cells.

Lenski and coworkers noted evolutionary changes in the cells, some of which have occurred in parallel. For example, all of the populations evolved to increase cell size, grow more efficiently on glucose, and grow more rapidly when transferred to fresh media. These changes make sense given the near starvation conditions of the cells.

***Evolution of the* Cit+ *Strain*.** One evolutionary change that should have occurred, but until recently did not, involves the use of citrate as a carbon source. *E. coli* grown under aerobic conditions can't use this compound as a food stuff. This bacterium has the biochemical machinery to metabolize citrate; it just can't transport the compound across its cell envelope.

Lenski and his team took advantage of *E. coli's* deficiency to monitor for contamination in their experiment. They added citrate at relatively high levels to the growth media. Since other microbes can typically make use of citrate, any contaminating microbe accidently introduced during the transfer steps will grow to greater cell densities than *E. coli,* thus causing the media to turn cloudy. Cloudy media means something is using citrate. Whenever the media turned cloudy, Lenski's collaborators attempted to confirm the contamination by identifying the unwanted microbial intruder.

Surprisingly, around 32,000 generations into the LTEE, *E. coli* started utilizing citrate as a carbon source. Lenski and his coworkers reasoned that this newly acquired ability was due either to a single, rare genetic change, or a sequence of mutations that formed a pathway culminating in the ability to use citrate.

The *E. coli* genome consists of about 4.6 million base pairs. The cells that are part of the LTEE collectively experienced billions of mutations over the course of the experiment; enough to cause changes in each position of the *E. coli* genome several times over. Additionally, the mutation rate producing the citrate-using strain of *E. coli* is extremely low compared to the rate that mutations, in general, and beneficial mutations, specifically, occur in the *E. coli* genome. Taken together, these two observations imply that the evolution of citrate-eating cells must have involved a *sequence* of mutations, not a single rare genetic change.

A Test of Historical Contingency. The evolved ability to use citrate presented Lenski's team with a rare opportunity to test the notion of historical contingency. To conduct this test, Lenski and his fellow researchers took frozen samples of ancestral *E. coli* cells, grew them up, and transferred them to fresh media every day monitoring them for about 4,000 generations to see if any acquired the ability to use citrate.

They did not observe the changes they hoped to see for cells sampled prior to about 27,000 generations. It was only the cells taken after that point in "history" that developed the ability to use citrate, and only rarely did this change occur.

This result indicates that a potentiating series of mutational changes made it possible for *E. coli* to grow on citrate. Cells that did not experience that same pathway of changes could not evolve citrate-utilizing capabilities. In other words, historical contingency characterized the evolution of the citrate-growing strains.

Citrate utilization did evolve repeatedly for several cells taken after about 27,000 generations. These cells required only a *single* change to make use of citrate. This result means that, given enough opportunities, evolution can repeat if it involves a single mutational event. But when *several* mutations have to be sequenced in the right order, evolution cannot repeat. Evolutionary processes are historically contingent.

In a sense, the LTEE represents the best possible opportunity for repeated evolution, since the growth conditions and the forces of selection (near starvation) have been constant throughout. The high levels of citrate have also been constant. By incorporating such high levels of this carbon source into the experiment, Lenski's group was "daring" *E. coli* to evolve citrate-utilizing capabilities.

In the real world—outside of the controlled conditions of the laboratory—the growth conditions, the forces of selection, and the opportunities for evolutionary advance are variable. This variability adds to the contingency of the evolutionary process.

One Final Point. Does the evolution of *E. coli* in the LTEE validate the evolutionary paradigm? Not really. The evolutionary changes experienced by this microbe are equivalent to microevolutionary changes observed in complex multicellular organisms.

The evolution of citrate-utilizing capabilities appears to involve changes in a transport protein associated with the cell envelope. To evolve a transport protein with the capability to move citrate across the cell envelope under aerobic conditions appears to require several changes that work in concert to alter the properties of the transporter. Given the large number of cells and the large number of generations (30,000), it is not surprising that evolutionary changes like the ability to utilize citrate take place.

But the experimental demonstration of historical contingency in *E. coli* raises significant questions about the validity of evolutionary explanations for life's origin and diversification. Even though evolution *shouldn't* repeat, it appears as if it *has* numerous times at a biochemical level and an organismal level.

Molecular Convergence

Biological convergence not only raises questions about the validity of biological evolution, it points to intelligent causation. Designers and engineers frequently reapply successful strategies when they face closely related problems. Why reinvent the wheel? It's much more prudent and efficient for an inventor to reuse the same good designs as much as possible, particularly when confronted with a problem he or she has already solved.

The tendency of engineers and designers to reuse the same designs provides insight into the way that intelligence might have played a role in biological history. If human engineers reutilize the same techniques and technologies when they invent, it's reasonable to expect that *any* intelligent agent would do the same. If life stems from the work of an intelligent cause, then it's reasonable to expect that the same designs would repeatedly appear throughout nature. Use of good, effective designs over and over again would reflect prudence and efficiency.

Notes

1. This essay has been adapted by the author from pages 203–24 of *The Cell's Design* by Fazale Rana (Grand Rapids: Baker Books, 2008). It is used by permission of Baker Books, a division of Baker Publishing Group.
2. Picasso and his friend George Braque invented this school of art around 1910. The cubists fragmented three-dimensional objects and redefined them as a series of interlocking planes.
3. For example, before inventing cubism, he went through two periods characterized by the use of particular colors. During his Blue Period, Picasso produced blue-tinted paintings depicting acrobats, harlequins, prostitutes, beggars, and artists. Orange and pink colors defined Picasso's Rose Period.
4. See http://www.answers.com/cubism for more on cubism and Picasso. A summary of Picasso's life and the various artistic movements he catalyzed can be found at http://en.wikipedia.org/wiki/Pablo_Picasso.
5. More about Frank Lloyd Wright can be found in *The Columbia Encyclopedia* article "Wright, Frank Lloyd"; see http://education.yahoo.com/reference/encyclopedia/entry/Wright-FL.
6. Stephen J. Gould, *Wonderful Life: The Burgess Shale and the Nature of History* (New York: Norton, 1989), 51.
7. Ibid., 48.
8. John Cafferky, *Evolution's Hand: Searching for the Creator in Contemporary Science* (Toronto: Eastendbooks, 1997), 66–69.
9. Simon Conway Morris, *Life's Solution: Inevitable Humans in a Lonely Universe* (New York: Cambridge University Press, 2003).
10. See, for example, Russell F. Doolittle, "Convergent Evolution: The Need to Be Explicit," *Trends in Biochemical Sciences* 19 (January 1994): 15–18; Eugene V. Koonin, L. Aravind, and Alexy S. Kondrashov, "The Impact of Comparative Genomics on Our Understanding of Evolution," *Cell* 101 (June 9, 2000): 573–76; Harold H. Zakon, "Convergent Evolution on the Molecular Level," *Brain, Behavior and Evolution* 59, no. 5–6 (2002): 250–61.
11. Michael Y. Galperin, D. Roland Walker, and Eugene V. Koonin, "Analogous Enzymes: Independent Inventions in Enzyme Evolution," *Genome Research* 8 (August 1998): 779–90.
12. Ibid.
13. Russell F. Doolittle, "Convergent Evolution," 15–18.
14. Peptidases are proteins that break down other proteins by breaking bonds between amino acids.
15. Neil D. Rawlings and Alan J. Barrett, "Evolutionary Families of Peptidases," *Biochemical Journal* 290 (February 15, 1993): 205–18.
16. Galperin, Walker, and Koonin, "Analogous Enzymes," 779–90.
17. Eugene V. Koonin, Arcady R. Mushegian, and Peer Bork, "Non-Orthologous Gene Displacement," *Trends in Genetics* 12 (September 1996): 334–36.
18. Detlef D. Leipe, L. Aravind, and Eugene V. Koonin, "Did DNA Replication Evolve Twice Independently?" *Nucleic Acids Research* 27 (September 1, 1999): 3389–401.
19. The material for this section was adapted from Alan G. Atherly, Jack R. Girton, and John F. McDonald, *The Science of Genetics* (Fort Worth: Saunders College Publishing, 1999), 256–77.
20. J. William Schopf, "When Did Life Begin?" in *Life's Origin: The Beginnings of Biological Evolution*, ed. J. William Schopf (Berkeley: University of California Press, 2002), 163.
21. Zachary D. Blount, Christina Z. Borland, and Richard E. Lenski, "Historical Contingency and the Evolution of a Key Innovation in an Experimental Population of *Escherichia coli*," *Proceedings of the National Academy Sciences USA* 105 (2008): 7899–906.

Part IV

Cosmological Origins and Fine-Tuning

Introduction

The viability of naturalistic cosmological explanations of universal origins and the fine-tuning of the laws and constants of nature is the concern of the essays in Part IV. **Alan Guth**, the inventor of inflationary cosmology, opens the discussion in "Eternal Inflation and Its Implications" by giving a state-of-the-art account of cosmological inflation and the evidence for it, outlining the inflationary model of what happened in the first few milliseconds after the big bang and how it explains the observed homogeneity and isotropy of the universe, among other things. He defends the infinity of bubble universes brought about by eternal inflation on the basis of the consequences that inflation has for the way predictions are extracted from theoretical models, and goes on to discuss some of the ambiguities and paradoxes of probability that arise in eternally inflating spacetimes, including string landscape models. He concludes with a discussion of his proof that inflationary spacetimes are past-incomplete, arguing that this requires "a different physics" at the past-boundary of the inflating region, possibly some "quantum creation event" of presumably naturalistic provenance.

William Lane Craig's paper "Naturalism and the Origin of the Universe" is structured around an examination of the Leibnizian conundrum: why does the universe exist rather than nothing? He covers the history of modern cosmology from its inception in general relativity, through various inflationary and quantum gravitational proposals, to a consideration of string-theoretic "ekpyrotic" models. The ground of examination then shifts to the cosmological constant question and the thermodynamics of the universe. His argument culminates by pressing the point that, since modern cosmology continues to bear evidence that the universe had a beginning, its *ex nihilo* origin generates an explanatory requirement that naturalism lacks the resources to satisfy.

Howard Van Till attempts to change the terms of the debate in his essay "Cosmic Evolution, Naturalism, and Divine Creativity." He argues that not all forms of naturalism are inherently materialistic, and that two forms of "naturalistic theism" provide fruitful alternatives to the usual stark choice between materialistic atheism and supernatural theism. Naturalistic theism therefore serves as a compromise intended to quell the disputes between theists and naturalists.

Nobel Laureate **Steven Weinberg** defends the multiverse as an explanation for cosmological fine-tuning in his essay "Living in the Multiverse." Beginning with a discussion of how advances in the history of science have at certain points involved discoveries about science itself and led to changes in what we regard as an acceptable theory, he argues that the discovery of a vast number of solutions to the string-theoretic equations may place us at just such a turning point in regard to what we should accept as a legitimate foundation for physical theory. Weinberg defends the legitimacy of anthropic cosmological considerations in the string landscape by suggesting ways that they can gain empirical traction, then outlines four different senses in which the "universes" in the string landscape could be imagined to exist. He closes by defending the possibility that, while anthropic reasoning may represent a retreat from what we had hoped that physical theory would do for us, it nonetheless provides enough testable consequences that, if the tests go well, we might convince ourselves that the theory is right.

In "Balloons on a String: A Critique of Multiverse Cosmology," philosopher of physics **Bruce Gordon** takes up the challenges raised by Guth and Weinberg, subjecting quantum, inflationary, and string cosmology to critical examination. After explaining how inflationary scenarios are grafted onto Big Bang cosmology and emphasizing the significance of the proof that all inflationary spacetimes are past-incomplete, Gordon offers a substantial critique of quantum cosmology in its role as the "different physics" intended to obviate the implications of an initial singularity at the past-boundary of our universe. He then proceeds to enumerate the deficiencies of eternal inflation and string-theoretic multiverse explanations, focusing especially on three proposals that have received much popular attention: the Steinhardt-Turok cyclic ekpyrotic model, the Gasperini-Veneziano pre–Big Bang Inflationary scenario, and the inflationary string landscape hypothesis advanced by Susskind, Polchinski, Bousso, and Linde. Gordon argues that all these models suffer from fine-tuning problems of their own—thus merely displacing the complex specified information that ultimately needs to be explained—and are beset with a variety of technical difficulties, implausibilities, and self-destructive consequences that render them inadequate as vehicles for the tasks they are intended to perform. The essay then revisits the question of whether matter or mind should be seen as the ultimate explanatory principle in cosmology, and concludes that intelligent design offers the best explanation, both ontologically and epistemically, for cosmological origins and fine-tuning.

The final essay dealing with issues in physics and cosmology, "Habitable Zones and Fine-Tuning," is written by astronomer and astrobiologist **Guillermo Gonzalez**. He takes fine-tuning analysis in a different direction, showing that constraints on circumstellar habitable zones (CHZ), galactic habitable zones (GHZ), and cosmic habitable ages (CHA) create a confluence of fine-tuning factors for life that point in the direction of the uniqueness and improbability of our terrestrial habitat, lending further support to the design hypothesis.

22

Eternal Inflation and Its Implications[1]

Alan Guth

1. Introduction: The Successes of Inflation

Since the proposal of the inflationary model some twenty-five years ago,[2] inflation has been remarkably successful in explaining many important qualitative and quantitative properties of the universe. In this [essay] I will summarize the key successes, and then discuss a number of issues associated with the eternal nature of inflation. In my opinion, the evidence that our universe is the result of some form of inflation is very solid. Since the term *inflation* encompasses a wide range of detailed theories, it is hard to imagine any reasonable alternative. The basic arguments are as follows:

(i) *The universe is big.* First of all, we know that the universe is incredibly large: the visible part of the universe contains about 10^{90} particles. Since we have all grown up in a large universe, it is easy to take this fact for granted: of course the universe is big, it's the whole universe! In "standard" FRW cosmology, without inflation, one simply postulates that about 10^{90} or more particles were here from the start. However, in the context of present-day cosmology, many of us hope that even the creation of the universe can be described in scientific terms. Thus, we are led to at least think about a theory that might explain how the universe got to be so big. Whatever that theory is, it has to somehow explain the number of particles, 10^{90} or more. However, it is hard to imagine such a number arising from a calculation in which the input consists only of geometrical quantities, quantities associated with simple dynamics, and factors of 2 or π. The easiest way by far to get a huge number, with only modest numbers as input, is for the calculation to involve an exponential. The exponential expansion of inflation reduces the problem of explaining 10^{90} particles to the problem of explaining 60 or 70 e-foldings of inflation. In fact, it is easy to construct underlying particle theories that will give far more than 70 e-foldings of inflation. Inflationary cosmology therefore suggests that, even though the observed universe is incredibly large, it is only an infinitesimal fraction of the entire universe.

(ii) *The Hubble expansion.* The Hubble expansion is also easy to take for granted, since we have all known about it from our earliest readings in cosmology. In standard FRW cosmology, the Hubble expansion is part of the list of postulates that define the initial conditions. But inflation actually offers the possibility of explaining how the Hubble expansion began. The repulsive gravity associated with the false vacuum is just what Hubble ordered. It is exactly the kind of force needed to propel the universe into a pattern of motion in which any two particles are moving apart with a velocity proportional to their separation.

(iii) *Homogeneity and isotropy.* The degree of uniformity in the universe is startling. The intensity of the cosmic background radiation is the same in all directions, after it is corrected for the motion of the Earth, to the incredible precision of one part in 100,000. To get some feeling for how high this precision is, we can imagine a marble that is spherical to one part in 100,000. The surface of the marble would have to be shaped to an accuracy of about 1,000 angstroms, a quarter of the wavelength of light.

Although modern technology makes it possible to grind lenses to quarter-wavelength accuracy, we would nonetheless be shocked if we unearthed a stone, produced by natural processes, that was round to an accuracy of 1,000 angstroms. If we try to imagine that such a stone were found, I am sure that no one would accept an explanation of its origin that simply proposed that the stone started out perfectly round. Similarly, I do not think it makes sense to consider any theory of cosmogenesis that cannot offer some explanation of how the universe became so incredibly isotropic.

The cosmic background radiation was released about 300,000 years after the Big Bang, after the universe cooled enough so that the opaque plasma neutralized into a transparent gas. The cosmic background radiation photons have mostly been traveling on straight lines since then, so they provide an image of what the universe looked like at 300,000 years after the Big Bang. The observed uniformity of the radiation therefore implies that the observed universe had become uniform in temperature by that time. In standard FRW cosmology, a simple calculation shows that the uniformity could be established so quickly only if signals could propagate at 100 times the speed of light, a proposition clearly contradicting the known laws of physics. In inflationary cosmology, however, the uniformity is easily explained. The uniformity is created initially on microscopic scales, by normal thermal-equilibrium processes, and then inflation takes over and stretches the regions of uniformity to become large enough to encompass the observed universe.

(iv) *The flatness problem.* I find the flatness problem particularly impressive, because of the extraordinary numbers that it involves. The problem concerns the value of the ratio

$$(1) \qquad \Omega_{tot} \equiv \frac{\rho_{tot}}{\rho_c},$$

where ρ_{tot} is the average total mass density of the universe and $\rho_c = 3H^2/8\pi G$ is the critical density, the density that would make the universe spatially flat. (In the definition of "total mass density," I am including the vacuum energy $\rho_{vac} = \Lambda/8\pi G$ associated with the cosmological constant Λ, if it is nonzero.)

By combining data from the Wilkinson Microwave Anisotropy Probe (WMAP), the Sloan Digital Sky Survey (SDSS), and observations of type Ia supernovae, Tegmark *et al* deduced that the present value of Ω_{tot} is equal to one within a few percent (Ω_{tot} = 1.012, +0.018 or

−0.022).³ Although this value is very close to one, the really stringent constraint comes from extrapolating Ω_{tot} to early times, since $\Omega_{tot} = 1$ is an unstable equilibrium point of the standard model evolution. Thus, if Ω_{tot} was ever *exactly* equal to one, it would remain exactly one forever. However, if Ω_{tot} differed slightly from one in the early universe, that difference—whether positive or negative—would be amplified with time. In particular, it can be shown that $\Omega_{tot} - 1$ grows as

$$\text{(2)} \qquad \Omega_{tot} - 1 \propto \begin{cases} t & \text{(during the radiation dominated era)} \\ t^{2/3} & \text{(during the matter dominated era).} \end{cases}$$

Dicke and Peebles⁴ pointed out that at $t = 1$ second, for example, when the processes of Big Bang nucleosynthesis were just beginning, Ω_{tot} must have equaled one to an accuracy of one part in 10^{15}. Classical cosmology provides no explanation for this fact—it is simply assumed as part of the initial conditions. In the context of modern particle theory, where we try to push things all the way back to the Planck time, 10^{-43} s, the problem becomes even more extreme. If one specifies the value of Ω_{tot} at the Planck time, it has to equal one to 59 decimal places in order to be in the allowed range today.

While this extraordinary flatness of the early universe has no explanation in classical FRW cosmology, it is a natural prediction for inflationary cosmology. During the inflationary period,

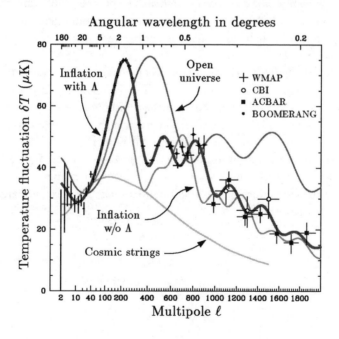

Figure 1. Comparison of the latest observational measurements of the temperature fluctuations in the CMB with several theoretical models, as described in the text. The temperature pattern on the sky is expanded in multipoles (i.e., spherical harmonics), and the intensity is plotted as a function of the multipole number ℓ. Roughly speaking, each multipole ℓ corresponds to ripples with an angular wavelength of $360°/\ell$.

instead of Ω_{tot} being driven away from one as described by Eq. (2), Ω_{tot} is driven towards one, with exponential swiftness:

$$(3) \qquad \Omega_{tot} - 1 \propto e^{-2H_{inf} t},$$

where H_{inf} is the Hubble parameter during inflation. Thus, as long as there is a long enough period of inflation, Ω_{tot} can start at almost any value, and it will be driven to unity by the exponential expansion.

(v) *Absence of magnetic monopoles.* All grand unified theories predict that there should be, in the spectrum of possible particles, extremely massive particles carrying a net magnetic charge. By combining grand unified theories with classical cosmology without inflation, Preskill found[5] that magnetic monopoles would be produced so copiously that they would outweigh everything else in the universe by a factor of about 10^{12}. A mass density this large would cause the inferred age of the universe to drop to about 30,000 years! Inflation is certainly the simplest known mechanism to eliminate monopoles from the visible universe, even though they are still in the spectrum of possible particles. The monopoles are eliminated simply by arranging the parameters so that inflation takes place after (or during) monopole production, so the monopole density is diluted to a completely negligible level.

(vi) *Anisotropy of the cosmic background radiation.* The process of inflation smooths the universe essentially completely, but density fluctuations are quantum fluctuations of the inflaton field.[6] Generically these are adiabatic Gaussians generated as inflation ends by the fluctuations with a nearly scale-invariant spectrum.

Until recently, astronomers were aware of several cosmological models that were consistent with the known data: an open universe, with $\Omega \approx 0.3$; an inflationary universe with considerable dark energy (Λ); an inflationary universe without Λ; and a universe in which the primordial perturbations arose from topological defects such as cosmic strings. Each of these models leads to a distinctive pattern of resonant oscillations in the early universe, which can be probed today through its imprint on the CMB. As can be seen in Figure 1,[7] three of the models are now definitively ruled out. The full class of inflationary models can make a variety of predictions, but the predictions of the simplest inflationary models with large Λ, shown on the graph, fit the data beautifully.

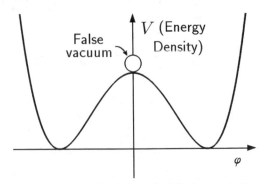

Figure 2. Evolution of the inflaton field during new inflation.

2. Eternal Inflation: Mechanisms

The remainder of this article will discuss eternal inflation—the questions that it can answer, and the questions that it raises. In this section, I discuss the mechanisms that make eternal inflation possible, leaving the other issues for the following sections. I will discuss eternal inflation first in the context of new inflation, and then in the context of chaotic inflation, where it is more subtle.

Eternal New Inflation. The eternal nature of new inflation was first discovered by Steinhardt, and later that year Vilenkin showed that new inflationary models are generically eternal.[8] Although the false vacuum is a metastable state, the decay of the false vacuum is an exponential process, very much like the decay of any radioactive or unstable substance. The probability of finding the inflaton field at the top of the plateau in its potential energy diagram, Figure 2, does not fall sharply to zero, but instead trails off exponentially with time.[9] However, unlike a normal radioactive substance, the false vacuum exponentially expands at the same time that it decays. In fact, in any successful inflationary model the rate of exponential expansion is always much faster than the rate of exponential decay. Therefore, even though the false vacuum is decaying, it never disappears, and in fact the total volume of the false vacuum, once inflation starts, continues to grow exponentially with time, ad infinitum.

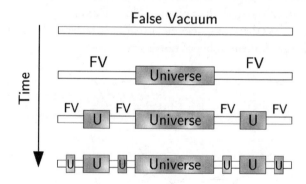

Figure 3. A schematic illustration of eternal inflation.

Figure 3 shows a schematic diagram of an eternally inflating universe. The top bar indicates a region of false vacuum. The evolution of this region is shown by the successive bars moving downward, except that the expansion could not be shown while still fitting all the bars on the page. So the region is shown as having a fixed size in comoving coordinates, while the scale factor, which is not shown, increases from each bar to the next. As a concrete example, suppose that the scale factor for each bar is three times larger than for the previous bar. If we follow the region of false vacuum as it evolves from the situation shown in the top bar to the situation shown in the second bar, in about one-third of the region the scalar field rolls down the hill of the potential energy diagram, precipitating a local big bang that will evolve into something that will eventually appear to its inhabitants as a universe. This local big-bang region is shown in gray and labeled "Universe." Meanwhile, however, the space has expanded so much that each of the two remaining regions of false vacuum is the same size as the starting region. Thus, if we follow the region for another time interval of the same duration, each of these regions of false vacuum will break up, with about one-third of each evolving into a local universe, as shown on the third bar from the top. Now there are four remaining regions of false vacuum, and again each is as large as the starting region. This process will repeat itself literally forever, producing a kind of a fractal structure to the universe, resulting in an infinite number of the local universes shown in gray. There is no standard name for these local universes, but they are often called bubble universes. I prefer, however, to call them pocket universes, to avoid the suggestion that they are round. While bubbles formed in first-order phase transitions are round,[10] the local universes formed in eternal new inflation are generally very irregular, as can be seen for example in the two-dimensional simulation by Vanchurin, Vilenkin, and Winitzki.[11]

The diagram in Figure 3 is, of course, an idealization. The real universe is three-dimensional, while the diagram illustrates a schematic one-dimensional universe. It is also important that the decay of the false vacuum is really a random process, while the diagram was constructed to show a very systematic decay, because it is easier to draw and to think about. When these inaccuracies are corrected, we are still left with a scenario in which inflation leads asymptotically to a fractal structure[12] in which the universe as a whole is populated by pocket universes on arbitrarily small comoving scales. Of course this fractal structure is entirely on distance scales much too large to be observed, so we cannot expect astronomers to see it. Nonetheless, one does have to think about the fractal structure if one wants to understand the very large scale structure of the spacetime produced by inflation.

Most important of all is the simple statement that once inflation happens, it produces not just one universe, but an infinite number of universes.

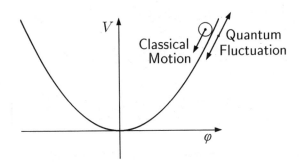

Figure 4. Evolution of the inflaton field during eternal chaotic inflation.

Eternal Chaotic Inflation. The eternal nature of new inflation depends crucially on the scalar field lingering at the top of the plateau of Figure 2. Since the potential function for chaotic inflation, Figure 4, does not have a plateau, it is not obvious how eternal inflation can happen in this context. Nonetheless, Andrei Linde showed in 1986 that chaotic inflation can also be eternal.[13]

In this case inflation occurs as the scalar field rolls down a hill of the potential energy diagram, as in Figure 4, starting high on the hill. As the field rolls down the hill, quantum fluctuations will be superimposed on top of the classical motion. The best way to think about this is to ask what happens during one time interval of duration $\Delta t = H^{-1}$ (one Hubble time), in a region of one Hubble volume H^{-3}. Suppose that φ_0 is the average value of φ in this region, at the start of the time interval. By the definition of a Hubble time, we know how much expansion is going to occur during the time interval: exactly a factor of e. (This is the only exact number in [this essay], so I wanted to emphasize the point.) That means the volume will expand by a factor of e^3. One of the deep truths that one learns by working on inflation is that e^3 is about equal to 20, so the volume will expand by a factor of 20. Since correlations typically extend over about a Hubble length, by the end of one Hubble time the initial Hubble-sized region grows and breaks up into 20 independent Hubble-sized regions.

As the scalar field is classically rolling down the hill, the change in the field $\Delta\varphi$ during the time interval Δt is going to be modified by quantum fluctuations $\Delta\varphi_{qu}$, which can drive the field upward or downward relative to the classical trajectory. For any one of the 20 regions at the end of the time interval, we can describe the change in φ during the interval by

$$(4) \qquad \Delta\varphi = \Delta\varphi_{cl} + \Delta\varphi_{qu} ,$$

where $\Delta\varphi_{cl}$ is the classical value of $\Delta\varphi$. In lowest order perturbation theory the fluctuations are calculated using free quantum field, which implies that $\Delta\varphi_{qu}$, the quantum fluctuation averaged over one of the 20 Hubble volumes at the end, will have a Gaussian probability distribution, with a width of order $H/2\pi$.[14] There is then always some probability that the sum of the two terms on the right-hand side will be positive—that the scalar field will fluctuate up and not down. As long as that probability is bigger than 1 in 20, then the number of inflating regions with $\varphi \geq \varphi_0$ will be larger at the end of the time interval Δt than it was at the beginning. This process will then go on forever, so inflation will never end.

Thus, the criterion for eternal inflation is that the probability for the scalar field to go up must be bigger than $1/e^3 \approx 1/20$. For a Gaussian probability distribution, this condition will be met provided that the standard deviation for $\Delta\varphi_{qu}$ is bigger than $0.61|\Delta\varphi_{cl}|$. Using $\Delta\varphi_{cl} \approx (d\varphi_{cl}/dt)H^{-1}$, the criterion becomes

$$(5) \qquad \Delta\varphi_{qu} \approx \frac{H}{2\pi} > 0.61|\dot\varphi_{cl}|H^{-1} \Leftrightarrow \frac{H^2}{|\dot\varphi_{cl}|} > 3.8$$

We have not discussed the calculation of density perturbations in detail, but the condition (5) for eternal inflation is equivalent to the condition that $d\rho/\rho$ on ultra-long length scales is bigger than a number of order unity.

The probability that $\Delta\varphi$ is positive tends to increase as one considers larger and larger values of φ, so sooner or later one reaches the point at which inflation becomes eternal. If one takes, for example, a scalar field with a potential

$$(6) \qquad V(\varphi) = \tfrac{1}{4}(\lambda\varphi^4) ,$$

then the de Sitter space equation of motion in flat Robertson-Walker coordinates takes the form

$$(7) \qquad \ddot\varphi + 3H\dot\varphi = -\lambda\varphi^3$$

where spatial derivatives have been neglected. In the "slow-roll" approximation one also neglects the $d^2\varphi/dt$ term, so $d\varphi/dt \approx -\lambda\varphi^3/(3H)$, where the Hubble constant H is related to the energy density by

$$(8) \qquad H^2 = \frac{8\pi}{3}G\rho = \frac{2\pi}{3}\frac{\lambda\varphi^4}{M_p^2}.$$

Putting these relations together, one finds that the criterion for eternal inflation, Eq. (5), becomes

$$(9) \qquad \varphi > 0.75\,\lambda^{-1/6}\,M_p.$$

Since λ must be taken very small, on the order of 10^{-12}, for the density perturbations to have the right magnitude, this value for the field is generally well above the Planck scale. The corresponding energy density, however, is given by

$$(10) \qquad V(\varphi) = (\tfrac{1}{4})\lambda\varphi^4 = (.079)\lambda^{1/3}M_p^4,$$

which is actually far below the Planck scale.

So, for these reasons, we think inflation is almost always eternal. I think the inevitability of eternal inflation in the context of new inflation is really unassailable—I do not see how it could possibly be avoided, assuming that the rolling of the scalar field off the top of the hill is slow enough to allow inflation to be successful. The argument in the case of chaotic inflation is less rigorous, but I still feel confident that it is essentially correct. For eternal inflation to set in, all one needs is that the probability for the field to increase in a given Hubble-sized volume during a Hubble time interval is larger than 1/20.

Thus, once inflation happens, it produces not just one universe, but an infinite number of universes.

3. Implications for the Landscape of String Theory

Until recently, the idea of eternal inflation was viewed by most physicists as an oddity, of interest only to a small subset of cosmologists who were afraid to deal with concepts that make real contact with observation. The role of eternal inflation in scientific thinking, however, was greatly boosted by the realization that string theory has no preferred vacuum, but instead has perhaps 10^{1000} metastable vacuum-like states.[15] Eternal inflation then has potentially a direct impact on fundamental physics, since it can provide a mechanism to populate the landscape of string vacua. While all of these vacua are described by the same fundamental string theory, the apparent laws of physics at low energies could differ dramatically from one vacuum to another. In particular, the value of the cosmological constant (e.g., the vacuum energy density) would be expected to have different values for different vacua.

The combination of the string landscape with eternal inflation has in turn led to a markedly increased interest in anthropic reasoning, since we now have a respectable set of theoretical ideas that provide a setting for such reasoning. To many physicists, the new setting for anthropic reasoning is a welcome opportunity: in the multiverse, life will evolve only in very rare regions where the local laws of physics just happen to have the properties needed for life, giving a simple explanation for why the observed universe appears to have just the right properties for the evolution of life. The incredibly small value of the cosmological constant is a telling example of a feature that seems to be needed for life, but for which an explanation from fundamental physics is painfully lacking. Anthropic reasoning can give the illusion of intelligent design without the need for any intelligent intervention.[16]

On the other hand, many other physicists have an abhorrence of anthropic reasoning. To this group, anthropic reasoning means the end of the hope that precise and unique predictions can be made on the basis of logical deduction.[17] Since this hope should not be given up lightly, many physicists are still trying to find some mechanism to pick out a unique vacuum from string theory. So far, there is no discernible progress.

It seems sensible, to me, to consider anthropic reasoning to be the explanation of last resort. That is, in the absence of any detailed understanding of the multiverse, life, or the evolution of either, anthropic arguments become plausible only when we cannot find any other explanation. That said, I find it difficult to know whether the cosmological constant problem is severe enough to justify the explanation of last resort.

Inflation can conceivably help in the search for a nonanthropic explanation of vacuum selection, since it offers the possibility that only a small minority of vacua are populated. Inflation is, after all, a complicated mechanism that involves exponentially large factors in its basic description, so it is possible that it populates some states overwhelmingly more than others. In particular, one might expect that those states that lead to the fastest exponential expansion rates would be favored. Then these fastest expanding states—and their decay products—could dominate the multiverse.

But so far, unfortunately, this is only wishful thinking. As I will discuss in the next section, we do not even know how to define probabilities in eternally inflating multiverses. Furthermore, it does not seem likely that any principle that favors a rapid rate of exponential inflation will favor a vacuum of the type that we live in. The key problem, as one might expect, is the value of the cosmological constant. The cosmological constant Λ in our universe is extremely small, i.e., $\Lambda < 10^{-120}$ in Planck units. If inflation singles out the state with the fastest exponential expansion rate, the energy density of that state would be expected to be of order Planck scale or larger. To explain why our vacuum has such a small energy density, we would need to find some reason why this very high energy density state should decay preferentially to a state with an exceptionally small energy density.[18]

There has been some effort to find relaxation methods that might pick out the vacuum,[19] and perhaps this is the best hope for a nonanthropic explanation of the cosmological constant. So far, however, the landscape of nonanthropic solutions to this problem seems bleak.

4. Difficulties in Calculating Probabilities

In an eternally inflating universe, anything that can happen will happen; in fact, it will happen an infinite number of times. Thus, the question of what is possible becomes trivial—anything is possible, unless it violates some absolute conservation law. To extract predictions from the theory, we must therefore learn to distinguish the probable from the improbable.

However, as soon as one attempts to define probabilities in an eternally inflating spacetime, one discovers ambiguities. The problem is that the sample space is infinite, in that an eternally inflating universe produces an infinite number of pocket universes. The fraction of universes with any particular property is therefore equal to infinity divided by infinity—a meaningless ratio. To obtain a well-defined answer, one needs to invoke some method of regularization.

To understand the nature of the problem, it is useful to think about the integers as a model system with an infinite number of entities. We can ask, for example, what fraction of the integers are odd. Most people would presumably say that the answer is ½, since the integers alternate between odd and even. That is, if the string of integers is truncated after the Nth, then the fraction of odd integers in the string is exactly ½ if N is even, and is $(N + 1)/2N$ if N is odd. In any case, the fraction approaches ½ as N approaches infinity.

However, the ambiguity of the answer can be seen if one imagines other orderings for the integers. One could, if one wished, order the integers as

$$(11) \qquad 1, 3, 2, 5, 7, 4, 9, 11, 6, \ldots,$$

always writing two odd integers followed by one even integer. This series includes each integer exactly once, just like the usual sequence $(1, 2, 3, 4, \ldots)$. The integers are just arranged in an unusual order. However, if we truncate the sequence shown in Eq. (11) after the Nth entry, and then take the limit $N \to \infty$, we would conclude that two-thirds of the integers are odd. Thus, we find that the definition of probability on an infinite set requires some method of truncation, and that the answer can depend nontrivially on the method that is used.

In the case of eternally inflating spacetimes, the natural choice of truncation might be to order the pocket universes in the sequence in which they form. However, we must remember that each pocket universe fills its own future light cone, so no pocket universe forms in the future light cone of another. Any two pocket universes are spacelike separated from each other, so some observers will see one as forming first, while other observers will see the opposite. One can arbitrarily choose equal-time surfaces that foliate the spacetime, and then truncate at some value of t, but this recipe is not unique. In practice, different ways of choosing equal-time surfaces give different results.

5. The Youngness Paradox

If one chooses a truncation in the most naïve way, one is led to a set of very peculiar results which I call the *youngness paradox*.

Specifically, suppose that one constructs a Robertson-Walker coordinate system while the model universe is still in the false vacuum (de Sitter) phase, before any pocket universes have formed. One can then propagate this coordinate system forward with a synchronous gauge condition[20] and one can define probabilities by truncating at a fixed value t_f of the synchronous time coordinate t. That is, the probability of any particular property can be taken to be proportional to the volume on the $t = t_f$ hypersurface which has that property. This method of defining probabilities was studied in detail by Linde, Linde, and Mezhlumian, in a paper with the memorable title "Do We Live in the Center of the World?"[21] I will refer to probabilities defined in this way as synchronous gauge probabilities.

The youngness paradox is caused by the fact that the volume of false vacuum is growing exponentially with time with an extraordinary time constant, in the vicinity of 10^{-37} s. Since the rate at which pocket universes form is proportional to the volume of false vacuum, this rate is increasing exponentially with the same time constant. That means that in each second the number of pocket universes that exist is multiplied by a factor of $\exp\{10^{37}\}$. At any given time, therefore, almost all of the pocket universes that exist are universes that formed very, very recently, within the last several time constants. The population of pocket universes is therefore an incredibly youth-dominated society, in which the mature universes are vastly outnumbered by universes that have just barely begun to evolve. Although the mature universes have a larger volume, this multiplicative factor is of little importance, since in synchronous coordinates the volume no longer grows exponentially once the pocket universe forms.

Probability calculations in this youth-dominated ensemble lead to peculiar results, as discussed in [the article by Linde, Linde, and Mezhlumian just mentioned]. These authors considered the expected behavior of the mass density in our vicinity, concluding that we should find ourselves very near the center of a spherical low-density region. Here I would like to discuss a less physical but simpler question, just to illustrate the paradoxes associated with synchronous gauge probabilities. Specifically, I will consider the question: "Are there any other civilizations in the visible universe that are more advanced than ours?" Intuitively, I would not expect inflation to make any predictions about this question, but I will argue that the synchronous gauge probability distribution strongly implies that there is no civilization in the visible universe more advanced than us.

Suppose that we have reached some level of advancement, and suppose that t_{min} represents the minimum amount of time needed for a civilization as advanced as we are to evolve, starting from the moment of the decay of the false vacuum—the start of the Big Bang. The reader might object on the grounds that there are many possible measures of advancement, but I would respond by inviting the reader to pick any measure she chooses; the argument that I am about to give should apply to all of them. The reader might alternatively claim that there is no sharp minimum t_{min}, but instead we should describe the problem in terms of a function which gives the probability that, for any given pocket universe, a civilization as advanced as we are would develop by time t. I believe, however, that the introduction of such a probability distribution would merely complicate the argument without changing the result. So, for simplicity of discussion, I will assume that there is some sharply defined minimum time t_{min} required for a civilization as advanced as ours to develop.

Since we exist, our pocket universe must have an age t_0 satisfying

(12) $$t_0 \geq t_{min}.$$

Suppose, however, that there is some civilization in our pocket universe that is more advanced than we are, let us say by 1 second. In that case Eq. (12) is not sufficient, but instead the age of our pocket universe would have to satisfy

(13) $$t_0 \geq t_{min} + 1 \text{ second.}$$

However, in the synchronous gauge probability distribution, universes that satisfy Eq. (13) are outnumbered by universes that satisfy Eq. (12) by a factor of approximately $\exp\{10^{37}\}$. Thus, if we know only that we are living in a pocket universe that satisfies Eq. (12), it is extremely improbable that it also satisfies Eq. (13). We would conclude, therefore, that it is extraordinarily improbable that there is a civilization in our pocket universe that is at least 1 second more advanced than we are.

Perhaps this argument explains why SETI has not found any signals from alien civilizations, but I find it more plausible that it is merely a symptom that the synchronous gauge probability distribution is not the right one.

Although the problem of defining probabilities in an eternally inflating universe has not been solved, a great deal of progress has been made in exploring options and understanding their properties. For many years, Vilenkin and his collaborators were almost the only cosmologists working on this issue,[22] but now the field is growing rapidly.[23]

6. Is Inflation Eternal into the Past?

If the universe can be eternal into the future, is it possible that it is also eternal into the past? Here I will describe a recent theorem that shows,[24] under plausible assumptions, that the answer to this question is no.[25]

The theorem is based on the well-known fact that the momentum of an object traveling on a geodesic through an expanding universe is redshifted, just as the momentum of a photon is redshifted. Suppose, therefore, we consider a timelike or null geodesic extended backwards, into the past. In an expanding universe such a geodesic will be blueshifted. The theorem shows that under some circumstances the blueshift reaches infinite rapidity (i.e., the speed of light) in a finite amount of proper time (or affine parameter) along the trajectory, showing that such a trajectory is (geodesically) incomplete.

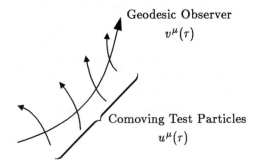

Figure 5. An observer measures the velocity of passing test particles to infer the Hubble parameter.

To describe the theorem in detail, we need to quantify what we mean by an expanding universe. We imagine an observer whom we follow backwards in time along a timelike or null geodesic. The goal is to define a local Hubble parameter along this geodesic, which must be well defined even if the spacetime is neither homogeneous nor isotropic. Call the velocity of the geodesic observer $v^\mu(\tau)$, where τ is the proper time in the case of a timelike observer, or an affine parameter in the case of a null observer. (Although we are imagining that we are following the trajectory backwards in time, τ is defined to increase in the future timelike direction, as usual.) To define H, we must imagine that the vicinity of the observer is filled with "comoving test particles," so that there is a test particle velocity $u^\mu(\tau)$ assigned to each point τ along the geodesic trajectory, as shown in Figure 5. These particles need not be real—all that will be necessary is that the worldlines can be defined, and that each worldline should have zero proper acceleration at the instant it intercepts the geodesic observer.

To define the Hubble parameter that the observer measures at time τ, the observer focuses on two particles, one that he passes at time τ, and one at $\tau + \Delta\tau$, where in the end he takes the limit $\Delta\tau \to 0$. The Hubble parameter is defined by

$$(14) \qquad H \equiv \Delta v_{\text{radial}} / \Delta r,$$

where Δv_{radial} is the radial component of the relative velocity between the two particles, and Δr is their distance, where both quantities are computed in the rest frame of one of the test particles, not in the rest frame of the observer. Note that this definition reduces to the usual one if it is applied to a homogeneous isotropic universe.

The relative velocity between the observer and the test particles can be measured by the invariant dot product,

$$(15) \qquad \gamma \equiv u_\mu v^\mu,$$

which for the case of a timelike observer is equal to the usual special relativity Lorentz factor

$$(16) \qquad \gamma = \frac{1}{\sqrt{1 - v_{\text{rel}}^2}}.$$

If H is positive we would expect γ to decrease with τ, since we expect the observer's momentum relative to the test particles to redshift. It turns out, however, that the relationship between H and changes in γ can be made precise. If one defines

$$(17) \qquad F(\gamma) \equiv \begin{cases} 1/\gamma & \text{for null observers} \\ \text{arctanh}(1/\gamma) & \text{for timelike observers,} \end{cases}$$

then

$$(18) \qquad H = dF(\gamma)/d\tau.$$

I like to call $F(\gamma)$ the "slowness" of the geodesic observer, because it increases as the observer slows down, relative to the test particles. The slowness decreases as we follow the geodesic backwards in time, but it is positive definite, and therefore cannot decrease below zero. $F(\gamma) = 0$ corresponds to $\gamma = \infty$, or a relative velocity equal to that of light. This bound allows us to place a rigorous limit on the integral of Eq. (18). For timelike geodesics,

$$(19) \qquad \int^{\tau_f} H \, d\tau \;\leq\; \operatorname{arctanh}\!\left(\frac{1}{\gamma_f}\right) \;=\; \operatorname{arctanh}\!\left(\sqrt{1 - v_{rel}^2}\right),$$

where γ_f is the value of γ at the final time $\tau = \tau_f$. For null observers, if we normalize the affine parameter τ by $d\tau/dt = 1$ at the final time τ_f, then

$$(20) \qquad \int^{\tau_f} H \, d\tau \;\leq\; 1,$$

Thus, if we assume an *averaged expansion condition*, i.e., that the average value of the Hubble parameter H_{av} along the geodesic is positive, then the proper length (or affine length for null trajectories) of the backwards-going geodesic is bounded. Thus, the region for which $H_{av} > 0$ is past-incomplete.

It is difficult to apply this theorem to general inflationary models, since there is no accepted definition of what exactly defines this class. However, in standard eternally inflating models, the future of any point in the inflating region can be described by a stochastic model[26] for inflaton evolution, valid until the end of inflation. Except for extremely rare large quantum fluctuations, $H \geq \sqrt{(8\pi/3)G\rho_f}$, where ρ_f is the energy density of the false vacuum driving the inflation. The past for an arbitrary model is less certain, but we consider eternal models for which the past is like the future. In that case H would be positive almost everywhere in the past-inflating region. If, however, $H_{av} > 0$ when averaged over a past-directed geodesic, our theorem implies that the geodesic is incomplete.

There is, of course, no conclusion that an eternally inflating model must have a unique beginning, and no conclusion that there is an upper bound on the length of all backwards-going geodesics from a given point. There may be models with regions of contraction embedded within the expanding region that could evade our theorem. Aguirre and Gratton have proposed a model that evades our theorem, in which the arrow of time reverses at the $t = -\infty$ hypersurface, so the universe "expands" in both halves of the full de Sitter space.[27]

The theorem does show, however, that an eternally inflating model of the type usually assumed, which would lead to $H_{av} > 0$ for past-directed geodesics, cannot be complete. Some new physics (i.e., not inflation) would be needed to describe the past boundary of the inflating region. One possibility would be some kind of quantum creation event.

One particular application of the theory is the cyclic ekpyrotic model of Steinhardt and Turok.[28] This model has $H_{av} > 0$ for null geodesics for a single cycle, and since every cycle is identical, $H_{av} > 0$ when averaged over all cycles. The cyclic model is therefore past-incomplete, and requires a boundary condition in the past.

7. Conclusion

In this paper I have summarized the arguments that strongly suggest that our universe is the product of inflation. I argued that inflation can explain the size, the Hubble expansion, the homogeneity, the isotropy, and the flatness of our universe, as well as the absence of magnetic monopoles, and even the characteristics of the nonuniformities. The detailed observations of the cosmic background radiation anisotropies continue to fall in line with inflationary expectations, and the evidence for an accelerating universe fits beautifully with the inflationary preference for a flat universe. Our current picture of the universe seems strange, with 95 percent of the energy in forms of matter that we do not understand, but nonetheless the picture fits together extraordinarily well.

Next I turned to the question of eternal inflation, claiming that essentially all inflationary models are eternal. In my opinion this makes inflation very robust: if it starts anywhere, at any time in all of eternity, it produces an infinite number of pocket universes. A crucial issue in our understanding of fundamental physics is the selection of the vacuum, which, according to current ideas in string theory, could be any one of a colossal number of possibilities. Eternal inflation offers at least a hope that a small set of vacua might be strongly favored. For that reason it is important for us to learn more about the evolution of the multiverse during eternal inflation. But so far it is only wishful thinking to suppose that eternal inflation will allow us to determine the vacuum in which we should expect to find ourselves.

I then discussed the past of eternally inflating models, concluding that, under mild assumptions, the inflating region must have a past boundary, and that new physics (other than inflation) is needed to describe what happens at this boundary.

Although eternal inflation has fascinating consequences, our understanding of it remains incomplete. In particular, we still do not understand how to define probabilities in an eternally inflating spacetime.

We should keep in mind, however, that observations in the past few years have vastly improved our knowledge of the early universe, and that these new observations have been generally consistent with the simplest inflationary models. It is the success of these predictions that justifies spending time on the more speculative aspects of inflationary cosmology.[29]

Notes

1. This article is reprinted with permission from *Journal of Physics A: Mathematical and Theoretical* 40 (2007): 6811–6826, doi: 10.1088/1751–8113/40/25/S25.
2. A.H. Guth, "The inflationary universe: A possible solution to the horizon and flatness problems." *Physical Review* D 23 (1981): 347–356; A.D. Linde, "A new inflationary universe scenario: a possible solution of the horizon, flatness, homogeneity, isotropy and primordial monopole problems." *Physics Letters* B 108 (1982): 389–93; A. Albrecht and P.J. Steinhardt, "Cosmology for grand unified theories with radiatively induced symmetry breaking." *Physical Review Letters* 48 (1982): 1220–3. For an earlier example of an inflationary model with a completely different motivation, see A.A. Starobinsky, *Journal of Experimental and Theoretical Physics Letters* 30 (1979): 682; A.A. Starobinsky "A new type of isotropic cosmological models without singularity." *Phys. Lett.* B 91 (1980): 99–102.
3. M. Tegmark *et al.* "Cosmological parameters from SDSS and WMAP," *Physical Review* D 69 (2004): 103501 (arXiv:astro-ph/0310723).
4. R.H. Dicke and P.J.E. Peebles, (1979) in S. W. Hawking and W. Israel, eds. *General Relativity: An Einstein Centenary Survey* (Cambridge: Cambridge University Press, 1979).
5. J.P. Preskill, "Cosmological production of superheavy magnetic monopoles," *Phys. Rev. Lett.* 43 (1979): 1365–8.
6. The history of this subject has become a bit controversial, so I'll describe my best understanding of the situation. The idea that quantum fluctuations could be responsible for the large scale structure of the universe goes back at least as far as Sakharov's 1966 paper [A. D. Sakharov, "The initial stage of an expanding universe and the appearance of a nonuniform distribution of matter," *Journal of Experimental and Theoretical Physics Letters* 22 (1966): 241–49], and it was re-introduced in the modern context by Mukhanov and Chibisov [V.F. Mukhanov and G.V. Chibisov, "Quantum fluctuations and a nonsingular universe." *Journal of Experimental and Theoretical Physics Letters* 33 (1981): 532–35; V.F. Mukhanov and G.V. Chibisov, (1982) "Vacuum energy and large-scale structure of the universe," *Journal of Experimental and Theoretical Physics Letters* 56 (1982): 258–65], who considered the density perturbations arising during inflation of the Starobinsky type [A. A. Starobinsky, *Journal of Experimental and Theoretical Physics Letters* 30 (1979): 682; A.A. Starobinsky "A new type of isotropic cosmological models without singularity." *Phys. Lett.* B 91 (1980): 99–102]. The calculations for "new" inflation, including a description of the evolution of the perturbations through "horizon exit," reheating, and "horizon reentry," were first carried out in a series of papers [Starobinsky, A.A. "Dynamics of phase transition in the new inflationary universe scenario and generation of perturbations." *Physics Letters* B 117 (1982): 175–8; A.H. Guth and S-Y. Pi, "Fluctuations in the new inflationary universe." *Physics Review Letters* 49 (1982): 1110–3; S.W. Hawking, "The development of irregularities in a single bubble inflationary universe." *Physics Letters* B 115 (1982): 295–7; and J.M. Bardeen, P.J. Steinhardt and M.S. Turner, "Spontaneous creation of almost scale-free density perturbations in an inflationary universe." *Physical Review* D 28 (1983): 679–93] arising from the Nuffield Workshop in Cambridge, U.K., in 1982. For Starobinsky inflation, the evolution of the conformally flat perturbations during inflation (as described in V.F. Mukhanov and G.V. Chibisov, (1982) "Vacuum energy and large-scale structure of the universe," *Journal of Experimental and Theoretical Physics Letters* 56 (1982): 258–65) into the post-inflation nonconformal perturbations was calculated, for example, in A.A. Starobinsky, "The perturbation spectrum evolving from a nonsingular, initially de Sitter cosmology, and the microwave background anisotropy." *Soviet Astronomy Letters* 9 (1983): 302–4, and V.F. Mukhanov, "Quantum theory of cosmological perturbations in R^2 gravity." *Physics Letters* B 218 (1989): 17–20. For a different perspective, the reader should see V.F. Mukhanov, "CMB, quantum fluc-

tuations and the predictive power of inflation" (2003): arXiv:astro-ph/03030779. For modern reviews, see for example S. Dodelson, *Modern Cosmology* (San Diego, CA: Academic Press, 2003); A.R. Liddle and D.H. Lyth, *Cosmological Inflation and Large-Scale Structure* (Cambridge: Cambridge University Press, 2000); and V.F. Mukhanov, H.A. Feldman, and R.H. Brandenberger, "Theory of cosmological perturbations." *Physics Report* 215 (1992): 203–333.

7. I thank Max Tegmark for providing this graph, an earlier version of which appeared in A.H. Guth and D.I. Kaiser, "Inflationary cosmology: Exploring the universe from the smallest to the largest scales." *Science* 307 (2005): 884–90 (arXiv:astro-ph/0502328). The graph shows the most precise data points for each range of l from recent observations as summarized in M. Tegmark *et al.* "Cosmological parameters from SDSS and WMAP," *Physical Review* D 69 (2004): 103501 (arXiv:astro-ph/0310723) and D.N. Spergel *et al.* "Wilkinson Microwave Anisotropy Probe (WMAP) three year results: Implications for cosmology" (2006): arXiv:astro-ph/0603449. The cosmic string prediction is taken from U.-L. Pen, U. Seljak, and N. Turok "Power spectra in global defect theories of cosmic structure formation." *Physics Review Letters* 79 (1997): 1611 (arXiv:astro-ph/9704165), and the "Inflation with Λ" curve was calculated from the best-fit parameters to the WMAP 3-year data from Table 5 [in the paper by Spergel *et al.* just mentioned]. The other curves were both calculated for $n_s = 1$, $\Omega_{baryon} = 0.05$, and $H = 70$ km s^{-1} Mpc^{-1}, with the remaining parameters fixed as follows. "Inflation without Λ": $\Omega_{DM} = 0.95$, $\Omega_\Lambda = 0$, $\tau = 0.06$; "Open universe": $\Omega_{DM} = 0.25$, $\Omega_\Lambda = 0$, $\tau = 0.06$. With our current ignorance of the underlying physics, none of these theories predicts the overall amplitude of the fluctuations; the "Inflation with Λ" curve was normalized for a best fit, and the others were normalized arbitrarily.

8. P.J. Steinhardt, "Natural inflation," in G.W. Gibbons, S.W. Hawking, and S.T.C. Siklos, eds., *The Very Early Universe,* Proceedings of the Nuffield Workshop, Cambridge, 21 June–9 July, 1982 (Cambridge: Cambridge University Press, 1983, 251–66); A. Vilenkin, "The birth of inflationary universes." *Physical Review* D 27 (1983): 2848–55.

9. A.H. Guth and S-Y. Pi, "Quantum mechanics of the scalar field in the new inflationary universe." *Physical Review* D 32 (1985): 1899–1920.

10. S. Coleman and F. De Luccia, "Gravitational effects on and of vacuum decay." *Physical Review* D 21 (1980): 3305–15.

11. See Figure 2 in V. Vanchurin, A. Vilenkin, and S. Winitzki, "Predictability crisis in inflationary cosmology and its resolution." *Physical Review* D 61 (2000): 083507 (arXiv:gr-qc/9905097).

12. M. Aryal and A. Vilenkin, "The fractal dimension of inflationary universe." *Physics Letters* B 199 (1987): 351–7.

13. A.D. Linde, "Eternal chaotic inflation." *Modren Physics Letters* A 1 (1986): 81–85; A.D. Linde, "Eternally existing self-reproducing chaotic inflationary universe." *Phys. Lett.* B 175 (1986): 395–400; A.S. Goncharov, A.D. Linde and V.F. Mukhanov, "The global structure of the inflationary universe." *International Journal of Modern Physics* A 2 (1987): 561–91.

14. A.A. Starobinsky, "Dynamics of phase transition in the new inflationary universe scenario and generation of perturbations." *Physics Letters* B 117 (1982): 175–78; A. Vilenkin and L.H. Ford, "Gravitational effects upon cosmological phase transitions." *Physical Review* D 26 (1982): 1231–41; A.D. Linde, "Scalar field fluctuations in expanding universe and the new inflationary universe scenario." *Physics Letters* B 116 (1982): 335; and A.A. Starobinsky, in H.J. de Vega and N. Sánchez, eds., *Field Theory, Quantum Gravity and Strings: Lecture Notes in Physics,* vol. 246 (Berlin: Springer-Verlag, 1986, 107–26).

15. R. Bousso and J. Polchinski, "Quantization of four form fluxes and dynamical neutralization of the cosmological constant." *Journal of High Energy Physics* 06 (2000): 006 (arXiv:hep-th/0004134); L. Susskind, "The anthropic landscape of string theory" (2003): arXiv:hep-th/0302219.

16. L. Susskind, *The Cosmic Landscape: String Theory and the Illusion of Intelligent Design.* (New York: Little, Brown and Company, 2006).

17. See, for example, D. J. Gross, "Where do we stand in fundamental string theory," *Physica Scripta* T117 (2005): 102–5; D. Gross, "The future of physics." *International Journal of Modern Physics* A 20 (2005): 5897–909.

18. I thank Joseph Polchinski for convincing me of this point.

19. See, for example, L.F. Abbott "A mechanism for reducing the value of the cosmological constant," *Physics Letters* B 150 (1985): 427; J. L. Feng, J. March-Russell, S. Sethi, and F. Wilczek, "Saltatory relaxation of the cosmological constant." *Nuclear Physics* B 602 (2001): 307–28 (arXiv:hep-th/0005276); and P.J. Steinhardt and N. Turok, "Why the cosmological constant is small and positive." *Science* 312 (2006): 1180–182 (arXiv:astro-ph/0605173) and references therein.

20. By a synchronous gauge condition, I mean that each equal-time hypersurface is obtained by propagating every point on the previous hypersurface by a fixed infinitesimal time interval Δt in the direction normal to the hypersurface.

21. A. D. Linde, D. Linde, and A. Mezhlumian, "Do we live in the center of the world?" *Physics Letters* B 345 (1995): 203–10 (arXiv:hep-th/9411111).

22. V. Vanchurin, A. Vilenkin, and S. Winitzki, "Predictability crisis in inflationary cosmology and its resolution." *Physical Review* D 61 (2000): 083507 (arXiv:gr-qc/9905097); A. Vilenkin, "Unambiguous probabilities in an eternally inflating universe." *Physical Review Letters* 81 (1998): 5501–504 (arXiv:hep-th/9806185); J. Garriga and A. Vilenkin "A prescription for probabilities in eternal inflation." *Physical Review* D 64 (2001): 023507 (arXiv:gr qc/0102090); J. Garriga, D. Schwartz-Perlov, A. Vilenkin, and S. Winitzki, "Probabilities in the inflationary multiverse." *Journal of Cosmology and Astroparticle Physics* JCAP 01 (2006): 017 (arXiv:hep-th/0509184).

23. M. Tegmark, "What does inflation really predict?" *Journal of Cosmology and Astroparticle Physics* JCAP 04 (2005): 001 (arXiv:astro-ph/0410281); R. Easther, E. A. Lim, and M. R. Martin, "Counting pockets with world lines in eternal inflation." *Journal of Cosmology and Astroparticle Physics* JCAP 03 (2006): 016 (arXiv:astro-ph/0511233); R. Bousso, B. Freivogel, and M. Lippert, "Probabilities in the landscape: The decay of nearly flat space." *Physical Review* D 74 (2006): 046008 (arXiv:hep-th/0603105); R. Bousso, "Holographic probabilities in eternal inflation." *Physical Review Letters* 97 (2006): 191302 (arXiv:hep-th/0605263); and A. Aguirre, S. Gratton, and M.C. Johnson, "Measures on transitions for cosmology in the landscape." (2006): arXiv:hep-th/0612195.

24. A. Borde, A.H. Guth, and A. Vilenkin, "Inflationary spacetimes are not past complete," *Physical Review Letters* 90 (2003): 151301 (arXiv:gr-qc/0110012).

25. There were also earlier theorems about this issue by Borde and Vilenkin [A. Borde and A. Vilenkin, "Eternal inflation and the initial singularity." *Physical Review Letters* 72 (1994): 3305–9 (arXiv:gr-qc/9312022), and A. Borde and A. Vilenkin, "Singularities in inflationary cosmology: A review." Talk given at 6th Quantum Gravity Seminar, Moscow, Russia, 6–11 Jun 1996, *International Journal of Modern Physics* D 5 (1996): 813–24 (arXiv:gr-qc/9612036)], and by Borde [A. Borde, "Open and closed universes, initial singularities and inflation," *Physical Review* D 50 (1994): 3692–702 (arXiv:gr-qc/9403049)], but these theorems relied on the weak energy condition, which for a perfect fluid is equivalent to the condition $\rho + p \geq 0$. This condition holds classically for forms of matter that are known or commonly discussed as theoretical proposals. It can, however, be violated by quantum fluctuations [see A. Borde and A.Vilenkin, "Violations of the weak energy condition in inflating spacetimes." *Physical Review* D 56 (1997): 717–23 (arXiv:gr-qc/9702019)], and so the applicability of these theorems is questionable.

26. A.S. Goncharov, A.D. Linde, and V.F. Mukhanov, "The global structure of the inflationary universe." *International Journal of Modern Physics* A 2 (1987): 561–91.

27. A. Aguirre and S. Gratton, "Steady state eternal inflation." *Physical Review* D 65 (2002): 083507 (arXiv:astro-ph/0111191), and A. Aguirre and S. Gratton, "Inflation without a beginning: A null boundary proposal." *Physical Review* D 67 (2003): 083515 (arXiv:gr-qc/0301042).

28. P. J. Steinhardt and N. G. Turok, "Cosmic evolution in a cyclic universe." *Physical Review* D 65 (2002): 126003 (arXiv:hep-th/0111098).

29. This work is supported in part by funds provided by the U.S. Department of Energy (D.O.E.) under grant #DF-FC02–94ER40818. The author would particularly like to thank Joan Sola and his group at the University of Barcelona, who made the IRGAC- 2006 conference so valuable and so enjoyable.

23

Naturalism and the Origin of the Universe[1]

William Lane Craig

1. The Fundamental Question

From time immemorial men have turned their gaze toward the heavens and *wondered*. Both cosmology and philosophy trace their roots to the wonder felt by the ancient Greeks as they contemplated the cosmos. According to Aristotle,

> it is owing to their wonder that men both now begin and at first began to philosophize; they wondered originally at the obvious difficulties, then advanced little by little and stated difficulties about the greater matters, e.g., about the phenomena of the moon and those of the sun and the stars, and about the origin of the universe.[2]

The question of why the universe exists remains the ultimate mystery. Derek Parfit declares that "No question is more sublime than why there is a Universe: why there is anything rather than nothing."[3]

This question led the great German mathematician and philosopher Gottfried Wilhelm Leibniz to posit the existence of a metaphysically necessary being that carries within itself the sufficient reason for its own existence and constitutes the sufficient reason for the existence of everything else in the world.[4] Leibniz identified this being as God. Leibniz's critics, on the other hand, claimed that the space-time universe is itself at least factually necessary—that is to say, eternal, uncaused, incorruptible, and indestructible[5]—while dismissing the demand for a logically necessary being. Thus, the Scottish skeptic David Hume queried, "Why may not the material universe be the necessarily existent Being . . . ?" Indeed, "How can anything, that exists from eternity, have a cause, since that relation implies a priority in time and a beginning of existence?"[6] There is no warrant for going beyond the universe to posit a supernatural ground of its existence, said the skeptics. As Bertrand Russell put it so succinctly in his BBC radio debate with Frederick Copleston, "The universe is just there, and that's all."[7]

Naturalism and the Origin of the Universe

2. The Expansion of the Universe

It was thus the presumed past eternity of the universe that allowed naturalistic minds to rest comfortably in the face of the mystery of existence. Tremors of the impending earthquake that would demolish this comfortable cosmology were first felt in 1917, when Albert Einstein made a cosmological application of his newly discovered gravitational theory, the General Theory of Relativity (GR).[8] In so doing, he assumed that the universe is homogeneous and isotropic and that it exists in a steady state, with a constant mean mass density and a constant curvature of space. To his chagrin, however, he found that GR would not permit such a model of the universe unless he introduced into his gravitational field equations a certain "fudge factor" Λ in order to counterbalance the gravitational effect of matter and so ensure a static universe. Einstein's universe was balanced on a razor's edge, however, and the least perturbation—even the transport of matter from one part of the universe to another—would cause the universe either to implode or to expand. By taking this feature of Einstein's model seriously, the Russian mathematician Alexander Friedman and the Belgian astronomer Georges Lemaître were able to formulate independently in the 1920s solutions to the field equations that predicted an expanding universe.[9]

The monumental significance of the Friedman-Lemaître model lay in its historization of the universe. As one commentator has remarked, up to this time, the idea of the expansion of the universe "was absolutely beyond comprehension. Throughout all of human history the universe was regarded as fixed and immutable and the idea that it might actually be changing was inconceivable."[10] But if the Friedman-Lemaître model were correct, the universe could no longer be adequately treated as a static entity existing, in effect, timelessly. Rather the universe has a history, and time will not be a matter of indifference for our investigation of the cosmos.

In 1929 Edwin Hubble showed that the red shift in the optical spectra of light from distant galaxies was a common feature of all measured galaxies and was proportional to their distance from us.[11] This red-shift, first observed by Vesto Slipher at the Lowell Observatory,[12] was taken to be a Doppler effect indicative of the recessional motion of the light source in the line of sight. Incredibly, what Hubble had discovered was the isotropic expansion of the universe predicted by Friedman and Lemaître on the basis of Einstein's GR. It was a veritable turning point in the history of science. "Of all the great predictions that science has ever made over the centuries," exclaims John Wheeler, "was there ever one greater than this, to predict, and predict correctly, and predict against all expectation a phenomenon so fantastic as the expansion of the universe?"[13]

The Standard Big Bang Model

According to the Friedman-Lemaître model, as time proceeds, the distances separating the ideal particles of the cosmological fluid constituted by the matter and energy of the universe become greater. It is important to appreciate that, as a GR-based theory, the model does not describe the expansion of the material content of the universe into a preexisting, empty, Newtonian space, but rather the expansion of space itself. The ideal particles of the cosmological fluid are conceived to be at rest with respect to space but to recede progressively from one another as space itself expands or stretches, just as buttons glued to the surface of a balloon would recede from one another as the balloon inflates. As the universe expands, it becomes less and less dense. This has the astonishing implication that as one reverses the expansion and extrapolates back in time, the universe becomes progressively denser until one arrives at a state of infinite density at some point in the finite past. This state represents a singularity at which space-time curvature, along with temperature, pressure, and density, becomes infinite. It therefore constitutes an edge or boundary to space-time itself. P. C. W. Davies comments,

> If we extrapolate this prediction to its extreme, we reach a point when all distances in the universe have shrunk to zero. An initial cosmological singularity therefore forms a past temporal extremity to the universe. We cannot continue physical reasoning, or even the concept of spacetime, through such an extremity. For this reason most cosmologists think of the initial singularity as the beginning of the universe. On this view the big bang represents the creation event; the creation not only of all the matter and energy in the universe, but also of spacetime itself.[14]

The term "Big Bang," originally a derisive expression coined by Fred Hoyle to characterize the beginning of the universe predicted by the Friedman-Lemaître model, is thus potentially misleading, since the expansion cannot be visualized from the outside (there being no "outside," just as there is no "before" with respect to the Big Bang).[15]

The standard Big Bang model, as the Friedman-Lemaître model came to be called, thus describes a universe that is not eternal in the past, but came into being a finite time ago. Moreover—and this deserves underscoring—the origin it posits is an absolute origin *ex nihilo*. For not only all matter and energy, but space and time themselves come into being at the initial cosmological singularity. As Barrow and Tipler emphasize, "At this singularity, space and time came into existence; literally nothing existed before the singularity, so, if the Universe originated at such a singularity, we would truly have a creation *ex nihilo*."[16] Thus, we may graphically represent space-time as a cone (Figure 1).

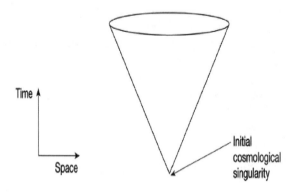

Figure 1. Conical Representation of Standard Model Space-Time. Space and time begin at the initial cosmological singularity, before which literally nothing exists.

On such a model, the universe originates *ex nihilo* in the sense that at the initial singularity it is true that *there is no earlier space-time point* or it is false that *something existed prior to the singularity*.

Naturalism and the Origin of the Universe

Now such a conclusion is profoundly disturbing for anyone who ponders it. For the question cannot be suppressed: *Why does the universe exist rather than nothing?* Sir Arthur Eddington, contemplating the beginning of the universe, opined that the expansion of the universe was so preposterous and incredible that "I feel almost an indignation that anyone should believe in it—except myself."[17] He finally felt forced to conclude, "The beginning seems to present insuperable difficulties unless we agree to look on it as frankly supernatural."[18] The problem of the origin of the universe, in the words of one astrophysical team, thus "involves a certain metaphysical aspect which may be either appealing or revolting."[19]

The Steady State Model

Revolted by the stark metaphysical alternatives presented by an absolute beginning of the universe, certain theorists have been understandably eager to subvert the Standard Model and restore an eternal universe. Sir Fred Hoyle, for example, could countenance neither an uncaused nor a supernaturally caused origin of the universe. With respect to the singularity, he wrote, "This most peculiar situation is taken by many astronomers to represent *the origin of the universe.* The universe is supposed to have begun at this particular time. From where? The usual answer, surely an unsatisfactory one, is: from nothing!"[20] Equally unsatisfactory was the postulation of a supernatural cause. Noting that some accept happily the universe's absolute beginning, Hoyle complained,

> To many people this thought process seems highly satisfactory because a "something" outside physics can then be introduced at $\tau = 0$. By a semantic manoeuvre, the word "something" is then replaced by "god," except that the first letter becomes a capital, God, in order to warn us that we must not carry the enquiry any further.[21]

To Hoyle's credit, he did carry the inquiry further by helping to formulate in 1948 the first competitor to the Standard Model, namely, the Steady State Model of the universe.[22] According to this theory, the universe is in a state of isotropic cosmic expansion, but as the galaxies recede, new matter is drawn into being *ex nihilo* in the interstices of space created by the galactic recession (Figure 2).

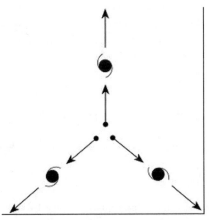

Figure 2. Steady State Model. As the galaxies mutually recede, new matter comes into existence to replace them. The universe thus constantly renews itself and so never began to exist.

If one extrapolates the expansion of the universe back in time, the density of the universe never increases because the matter and energy simply vanish as the galaxies mutually approach!

The Steady State theory never secured a single piece of experimental verification; its appeal was purely metaphysical.[23] The discovery of progressively more radio galaxies at ever greater distances undermined the theory by showing that in the past the universe was significantly different than it is today, thus contradicting the notion of a steady state of the universe. Instead it became increasingly evident that the universe had an evolutionary history. But the decisive refutation of the Steady State Model came with two discoveries that constituted, in addition to the galactic red shift, the most significant evidence for the Big Bang theory: the cosmogonic nucleosynthesis of the light elements and the microwave background radiation. Although the heavy elements were synthesized in the stellar furnaces, stellar nucleosynthesis could not manufacture the abundant light elements such as helium and deuterium. These could only have been created in the extreme conditions present in the first moment of the Big Bang. In 1965 a serendipitous discovery revealed the existence of a cosmic background radiation predicted in the 1940s by George Gamow on the basis of the Standard Model. This radiation, now shifted into the microwave region of the spectrum, consists of photons emitted during a very hot and dense phase of the universe. In the minds of almost all cosmologists, the cosmic background radiation decisively discredited the Steady State Model.

Oscillating Models

The Standard Model was based on the assumptions of homogeneity and isotropy. In the 1960s and 1970s some cosmologists suggested that by denying homogeneity and isotropy, one might be able to craft an Oscillating Model of the universe and thus avert the absolute beginning predicted by the Standard Model.[24] If the internal gravitational pull of the mass of the universe were able to overcome the force of its expansion, then the expansion could be reversed into a cosmic contraction, a Big Crunch. If the universe were not homogeneous and isotropic, then the collapsing universe might not coalesce at a point, but the material contents of the universe might pass by one another, so that the universe would appear to bounce back from the contraction into a new expansion phase. If this process could be repeated indefinitely, then an absolute beginning of the universe might be avoided (Figure 3).

Such a theory is extraordinarily speculative, but again there were metaphysical motivations for adopting this model.[25] The prospects of the Oscillating Model were severely dimmed in 1970, however, by Roger Penrose and Stephen Hawking's formulation of the Singularity

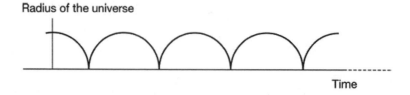

Figure 3. Oscillating Model. Each expansion phase is preceded and succeeded by a contraction phase, so that the universe in concertina-like fashion exists beginninglessly and endlessly.

Theorems that bear their names.[26] The theorems disclosed that under very generalized conditions an initial cosmological singularity is inevitable, even for inhomogeneous and non-isotropic universes. Reflecting on the impact of this discovery, Hawking notes that the Hawking-Penrose Singularity Theorems "led to the abandonment of attempts (mainly by the Russians) to argue that there was a previous contracting phase and a non-singular bounce into expansion. Instead almost everyone now believes that the universe, and time itself, had a beginning at the big bang."[27]

Despite the fact that no space-time trajectory can be extended through a singularity, the Oscillating Model exhibited a stubborn persistence. Two further strikes were lodged against it. First, there is no known physics that would cause a collapsing universe to bounce back to a new expansion. If, in defiance of the Hawking-Penrose Singularity Theorems, the universe rebounds, this is predicated upon a physics that is completely unknown. Physics predicts that a universe in a state of gravitational self-collapse will not rebound like a basketball dropped to the floor, but rather land like a lump of clay.[28] Second, attempts by observational astronomers to discover the mass density sufficient to generate the gravitational attraction required to halt and reverse the expansion continually came up short. In January 1998 astronomical teams from Princeton, Yale, the Lawrence Berkeley National Laboratory, and the Harvard-Smithsonian Astrophysics Institute reported at the American Astronomical Society meeting that their various tests all show that "the universe will expand forever."[29] A spokesman for the Harvard-Smithsonian team stated that they were now at least 95 percent certain that "the density of matter is insufficient to halt the expansion of the universe."[30]

At the same time, however, observations of the red shifts of 1a-type supernovae yielded unexpected results that have thrown the debate over the universe's fate into a wholly new arena and served to render questions of its density and geometrical openness or closure irrelevant. The red-shift data gathered from the distant supernovae indicate that, far from decelerating, the cosmic expansion is actually accelerating.[31] There is some sort of mysterious "dark energy" in the form of either a variable energy field (called "quintessence") or, more probably, a positive cosmological constant or vacuum energy that at a certain point in the evolution of the cosmos kicks the expansion into a higher gear, causing the expansion to proceed more rapidly.[32] The universe's internal gravitational attraction is a function of its matter density and pressure, which for ordinary matter are positively correlated. But the dark energy, while contributing to the mass density of the universe, is inversely correlated with the pressure, so that as its contribution to the density increases, the universe's internal gravitational pull declines due to the decrease in pressure. Consequently, even universes with a sufficient mass density to warp space into a positively curved, finite geometrical hypersurface may expand forever; a potentially infinite future is no longer the privileged prerogative of spatially open universes. Highly accurate recent measurements of the anisotropies in the cosmic microwave background radiation by the Wilkinson Microwave Anisotropy Probe (WMAP) indicate that the universe is composed of around 4 percent ordinary luminous matter, 23 percent non-luminous matter, and a startling 73 percent invisible dark energy, yielding a value of 1.02 ± 0.02 for the density parameter Ω.[33] Though the universe may therefore well possess a closed geometry, that fact is not determinative for its fate in view of the acceleration effect wrought by the negative pressure of its abundant dark energy. Indeed, if that dark energy does represent the existence of a positive cosmological constant, as the evidence increasingly suggests, then the universe will expand forever. According to the WMAP website, "For the theory that fits our data, the Universe will expand forever."[34]

Although the above difficulties have always been well known, proponents of the Oscillating Model tenaciously clung to it until a new alternative to the Standard Model emerged during the 1970s.[35] Looking back, quantum cosmologist Christopher Isham muses,

> Perhaps the best argument in favor of the thesis that the Big Bang supports theism is the obvious unease with which it is greeted by some atheist physicists. At times this has led to scientific ideas, such as continuous creation or an oscillating universe, being advanced with a tenacity which so exceeds their intrinsic worth that one can only suspect the operation of psychological forces lying very much deeper than the usual academic desire of a theorist to support his/her theory.[36]

The Oscillating Model drew its life from its avoidance of an absolute beginning of the universe; but once other models became available claiming to offer the same benefit, the Oscillating Model sank into oblivion under the weight of its own deficiencies.

Vacuum Fluctuation Models

It was realized that a physical description of the universe prior to the Planck time (10^{-43} second after the Big Bang singularity) would require the introduction of quantum physics in addition to GR. On the quantum level, so-called virtual particles are thought to arise due to fluctuations in the energy locked up in the vacuum, particles that the Heisenberg Indeterminacy Principle allows to exist for a fleeting moment before dissolving back into the vacuum. In 1973 Edward Tryon speculated whether the universe might not be a long-lived virtual particle, whose total energy is zero, born out of the primordial vacuum.[37] This seemingly bizarre speculation gave rise to a new generation of cosmogonic theories that we may call Vacuum Fluctuation Models. These models were closely related to an adjustment to the Standard Model known as Inflation. In an attempt to explain the astonishing large-scale homogeneity and isotropy of the universe, certain theorists proposed that between 10^{-35} and 10^{-33} sec after the Big Bang singularity, the universe underwent a phase of super-rapid, or inflationary, expansion that served to push the inhomogeneities out beyond our event horizon.[38] Prior to the Inflationary Era the universe was merely empty space, or a vacuum, and the material universe was born when the vacuum energy was converted into matter via a quantum mechanical phase transition. In most inflationary models, as one extrapolates backward in time beyond the Planck time, the universe continues to shrink down to the initial singularity. But in Vacuum Fluctuation Models, it is hypothesized that, prior to inflation, the Universe-as-a-whole was not expanding. This Universe-as-a-whole is a primordial vacuum that exists eternally in a steady state. Throughout this vacuum subatomic energy fluctuations constantly occur, by means of which matter is created and mini-universes are born (Figure 4).

Our expanding universe is but one of an indefinite number of mini-universes conceived within the womb of the greater Universe-as-a-whole. Thus, the beginning of our universe does not represent an absolute beginning, but merely a change in the eternal, uncaused Universe-as-a-whole.

Vacuum Fluctuation Models did not outlive the decade of the 1980s. Not only were there theoretical problems with the production mechanisms of matter, but these models faced a deep internal incoherence.[39] According to such models, it is impossible to specify precisely when and where a fluctuation will occur in the primordial vacuum that will then grow into a universe.

Naturalism and the Origin of the Universe

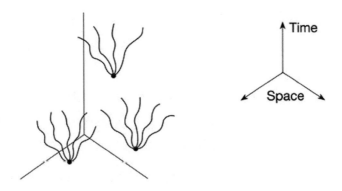

Figure 4. Vacuum Fluctuation Models. Within the vacuum of the wider universe, fluctuations occur which grow into mini-universes. Ours is but one of these and its relative beginning does not imply a beginning for the Universe-as-a-whole.

Within any finite interval of time, there is a positive probability of such a fluctuation occurring at any point in space. Thus, given infinite past time, universes will eventually be spawned at *every* point in the primordial vacuum, and, as they expand, they will begin to collide and coalesce with one another. Thus, given infinite past time, we should by now be observing an infinitely old universe, not a relatively young one. One theorist tried to avoid this problem by stipulating that fluctuations in the primordial vacuum only occur infinitely far apart, so that each mini-universe has infinite room in which to expand.[40] Not only is such a scenario unacceptably *ad hoc*, but it does not even solve the problem. For given infinite past time, each of the infinite regions of the vacuum will have spawned an open universe which by now will have entirely filled that region, with the result that all of the individual mini-universes would have coalesced.

Isham has called this problem "fairly lethal" to Vacuum Fluctuation Models and says that therefore they "have not found wide acceptance."[41] About the only way to avert the problem would be to postulate an expansion of the primordial vacuum itself; but then we are right back to the absolute origin implied by the Standard Model. According to Isham these models were therefore jettisoned long ago and "nothing much" has been done with them since.[42]

Chaotic Inflationary Model

Inflation also forms the context for the next alternative we shall consider: the Chaotic Inflationary Model. Inflationary theory, though criticized by some as unduly "metaphysical," has been widely accepted among cosmologists. One of the most fertile of the inflation theorists has been the Russian cosmologist Andrei Linde, who has championed his Chaotic Inflationary Model.[43] In Linde's model inflation *never* ends: each inflating domain of the universe, when it reaches a certain volume, gives rise via inflation to another domain, and so on, *ad infinitum* (Figure 5).

Linde's model thus has an infinite future. But Linde is troubled at the prospect of an absolute beginning. He writes, "The most difficult aspect of this problem is not the existence of the singularity itself, but the question of what was *before* the singularity. . . . This problem lies somewhere at the boundary between physics and metaphysics."[44] Linde therefore proposed

Figure 5. Chaotic Inflationary Model. The wider universe produces via inflation separate domains that continue to recede from one another as the wider space expands.

that chaotic inflation is not only endless, but beginningless. Every domain in the universe is the product of inflation in another domain, so that the singularity is averted and with it as well the question of what came before (or, more accurately, what caused it).

In 1994, however, Arvind Borde and Alexander Vilenkin showed that a universe eternally inflating toward the future cannot be geodesically complete in the past; that is to say, there must have existed at some point in the indefinite past an initial singularity. They write,

> A model in which the inflationary phase has no end . . . naturally leads to this question: Can this model also be extended to the infinite past, avoiding in this way the problem of the initial singularity?
>
> . . . this is in fact not possible in future-eternal inflationary spacetimes as long as they obey some reasonable physical conditions: such models must necessarily possess initial singularities.
>
> . . . the fact that inflationary spacetimes are past incomplete forces one to address the question of what, if anything, came before.[45]

In response, Linde concurred with the conclusion of Borde and Vilenkin: there must have been a Big Bang singularity at some point in the past.[46]

In 2001, Borde and Vilenkin, in cooperation with Alan Guth, were able to strengthen their theorem by crafting a new theorem independent of the assumption of the so-called "weak energy condition," which partisans of past-eternal inflation might have denied in an effort to save their theory.[47] The new theorem, in Vilenkin's words, "appears to close that door completely."[48] Inflationary models, like their predecessors, thus failed to avert the beginning predicted by the Standard Model.

Quantum Gravity Models

At the close of their analysis of Linde's Chaotic Inflationary Model, Borde and Vilenkin say with respect to Linde's metaphysical question, "The most promising way to deal with this problem is probably to treat the Universe quantum mechanically and describe it by a wave function rather than by a classical spacetime."[49] For "it follows from the theorem that the inflating region has a boundary in the past, and some new physics (other than inflation) is necessary to determine the conditions of that boundary. Quantum cosmology is the prime candidate for this role."[50] They thereby bring us to the next class of models that we shall consider, namely, Quantum Gravity Models. Vilenkin and, more famously, James Hartle and Stephen Hawking have proposed models of the universe that Vilenkin candidly calls exercises in "metaphysical cosmology."[51] In his best-selling popularization of his theory, Hawking even reveals an explicitly theological orientation. He concedes that on the Standard Model one could legitimately identify the Big Bang singularity as the instant at which God created the universe.[52] Indeed, he thinks that a number of attempts to avoid the Big Bang were probably motivated by the feeling that a beginning of time "smacks of divine intervention."[53] He sees his own model as preferable to the Standard Model because there would be no edge of space-time at which one "would have to appeal to God or some new law."[54] As we shall see, he is not at all reluctant to draw theological conclusions on the basis of his model.

Both the Hartle-Hawking and the Vilenkin models eliminate the initial singularity by transforming the conical hypersurface of classical space-time into a smooth, curved hypersurface having no edge (Figure 6).

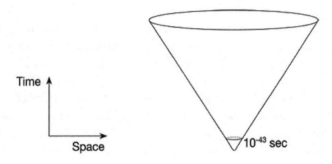

Figure 6. Quantum Gravity Model. In the Hartle-Hawking version, space-time is "rounded off" prior to the Planck time, so that although the past is finite, there is no edge or beginning point.

This is accomplished by the introduction of imaginary numbers for the time variable in Einstein's gravitational equations, which effectively eliminates the singularity. Hawking sees profound theological implications in the model:

> The idea that space and time may form a closed surface without boundary . . . has profound implications for the role of God in the affairs of the universe. . . . So long as the universe had a beginning, we could suppose it had a creator. But if the universe is really completely self-contained, having no boundary or edge, it would have neither beginning nor end. What place, then, for a creator?[55]

Hawking does not deny the existence of God, but he does think his model eliminates the need for a creator of the universe.

The key to assessing this theological claim is the physical interpretation of Quantum Gravity Models. By positing a finite (imaginary) time on a closed surface prior to the Planck time rather than an infinite time on an open surface, such models actually seem to support, rather than undercut, the fact that time and the universe had a beginning. Such theories, if successful, would enable us to model the beginning of the universe without an initial singularity involving infinite density, temperature, pressure, and so on. As Barrow points out, "This type of quantum universe has not always existed; it comes into being just as the classical cosmologies could, but it does not start at a Big Bang where physical quantities are infinite. . . ."[56] Barrow points out that such models are "often described as giving a picture of 'creation out of nothing,'" the only caveat being that in this case "there is no definite . . . point of creation."[57] Hartle-Hawking themselves construe their model as giving "the amplitude for the Universe to appear from nothing," and Hawking has asserted that, according to the model, the universe "would quite literally be created out of nothing: not just out of the vacuum, but out of absolutely nothing at all, because there is nothing outside the universe."[58] Similarly, Vilenkin claims that his model describes the creation of the universe "from literally *nothing*."[59] Taken at face value, these statements entail the beginning of the universe. According to Vilenkin, "The picture presented by quantum cosmology is that the universe starts as a small, closed 3-geometry and immediately enters the regime of eternal inflation, with new thermalized regions being constantly formed. In this picture, the universe has a beginning but no end."[60]

Hence, Hawking's statement quoted above concerning the theological implications of his model must therefore be understood to mean that on such a model there are no beginning or ending *points*. But having a beginning does not entail having a beginning point. Even in the Standard Model, theorists sometimes "cut out" the initial singular point without thinking that therefore space-time no longer begins to exist and the problem of the origin of the universe is thereby resolved. Time begins to exist just in case for any finite temporal interval, there are only a finite number of equal temporal intervals earlier than it. That condition is fulfilled for Quantum Gravity Models as well as for the Standard Model. Nor should we think that by giving the amplitude for the universe to appear from nothing, quantum cosmologists have eliminated the need for a creator, for that probability is conditional upon several choices that only the creator could make (such as selecting the wave function of the universe) and is dubiously applied to absolute nothingness.[61] Thus, Quantum Gravity Models, like the Standard Model, imply the beginning of the universe.

Perhaps it will be said that such an interpretation of Quantum Gravity Models fails to take seriously the notion of "imaginary time." Introducing imaginary numbers for the time variable in Einstein's equation has the peculiar effect of making the time dimension indistinguishable from space. But in that case, the imaginary time regime prior to the Planck time is not a space-time at all, but a Euclidean four-dimensional space. Construed realistically, such a four-space would be evacuated of all temporal becoming and would simply exist timelessly. Thus, Vilenkin characterizes this regime as a "state in which all our basic notions of space, time, energy, entropy, etc. lose their meaning."[62] Hawking describes it as "completely self-contained and not affected by anything outside itself. It would be neither created nor destroyed. It would just BE."[63]

The question that arises for this construal of the model is whether such an interpretation is meant to be taken realistically or instrumentally. On this score, there can be little doubt that the use of imaginary quantities for time is a mere mathematical device without ontological significance. For, first, there is no intelligible physical interpretation of imaginary time on offer. What, for example, would it mean to speak of the lapse of an imaginary second or of a physical object's enduring through two imaginary minutes? Second, time is metaphysically distinct from space, its moments being ordered by an *earlier than* relation that does not similarly order points in space. But this essential difference is obscured by imaginary time. Thus, "imaginary time" is most plausibly construed as a mathematical *Hilfsmittel*. Barrow observes, physicists have often carried out this "change time into space" procedure as a useful trick for doing certain problems in ordinary quantum mechanics, although they did not imagine that time was *really* like space. At the end of the calculation, they just swap back into the usual interpretation of there being one dimension of time and three . . . dimensions of . . . space.[64]

In his model, Hawking simply declines to re-convert to real numbers. If we do, then the singularity reappears. Hawking admits, "Only if we could picture the universe in terms of imaginary time would there be no singularities. . . . When one goes back to the real time in which we live, however, there will still appear to be singularities."[65] Hawking's model is thus a way of redescribing a universe with a singular beginning point in such a way that that singularity is transformed away; but such a redescription is not realist in character.

Hawking has since stated explicitly that he interprets the Hartle-Hawking model non-realistically. He confesses, "I'm a positivist. . . . I don't demand that a theory correspond to reality because I don't know what it is."[66] Still more extreme, "I take the positivist viewpoint that a physical theory is just a mathematical model and that it is meaningless to ask whether it corresponds to reality."[67] In assessing the worth of a theory, "All I'm concerned with is that the theory should predict the results of measurements."[68] The clearest example of Hawking's instrumentalism is his analysis of particle pair creation in terms of an electron quantum tunneling in Euclidean space (with time being imaginary) and an electron/positron pair accelerating away from each other in Minkowski space-time.[69] This analysis is directly analogous to the Hartle-Hawking cosmological model; and yet no one would construe particle pair creation as literally the result of an electron's transitioning out of a timelessly existing four-space into our classical space-time. It is just an alternative description employing imaginary numbers rather than real numbers.

Significantly, the use of imaginary quantities for time is an inherent feature of *all* Quantum Gravity Models.[70] This precludes their being construed realistically as accounts of the origin of the space-time universe in a timelessly existing four-space. Rather, they are ways of modeling the real beginning of the universe *ex nihilo* in such a way as to not involve a singularity. What brought the universe into being remains unexplained on such accounts.

Moreover, we are not without positive reasons for affirming the reality of the singular origin of space-time postulated by the Standard Model. John Barrow has rightly cautioned that "one should be wary of the fact that many of the studies of quantum cosmology are motivated by the desire to avoid an initial singularity of infinite density, so they tend to focus on quantum cosmologies that avoid a singularity at the expense of those that might contain one."[71] As we shall see, the initial cosmological singularity may be a virtual thermodynamical necessity. But whether it was at a singular point or not, the fact that the universe began to exist remains a prediction of any realistic interpretation of Quantum Gravity Models.

String Scenarios

We come finally to the extreme edge of cosmological speculation: string cosmology. These scenarios are based on an alternative to the standard quark model of elementary particle physics. So-called string theory (or M-theory) conceives of the fundamental building blocks of matter to be, not particles like quarks, but tiny, vibrating, 1-dimensional strings of energy. String theory is so complicated and embryonic in its development that all its equations have not yet even been stated, much less solved. But that has not deterred some cosmologists from trying to envision cosmological scenarios based on concepts of string theory to try to avert the beginning predicted by customary Big Bang cosmology.

Two sorts of scenarios have been proposed. The first of these is the Pre–Big Bang Scenario championed by the Italian physicists Gabriele Veneziano and Maurizio Gasperini.[72] They conceive of the Big Bang as the transitional event between a contraction phase chronologically prior to the Big Bang and the observed expansion phase after it. Such a rebound is postulated on the basis of limits set by the size and symmetries of strings to the increase in quantities like spacetime curvature, density, temperature, and so forth. Prior to the Big Bang a black hole formed in the eternally pre-existing, static, string perturbative vacuum space and collapsed to the maximum allowed values of such quantities before rebounding in the current expansion observed today (Figure 7).

The scenario differs from the old oscillating models in that the prior contraction is conceived to take place within a wider, static space and to be the mirror image of the current expansion. If that expansion will go on forever, then the contraction has gone on forever.

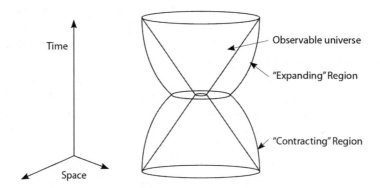

Figure 7. Pre-Big Bang Scenario. Our observable universe results from the rebound of the collapse of a black hole in a wider preexisting vacuum space. The pre-Big Bang contraction is the mirror image of the post-Big Bang expansion.

The further one regresses into the infinite past, the less dense the universe becomes, as one approaches a limit in the infinite past of a nearly empty universe consisting of an ultra-thin gas of radiation and matter. As one moves forward in time, the material contents of various regions of space begin to collapse into black holes. But rather than collapsing to singularities, these black holes reach a maximum of spacetime curvature, density, and so on, before rebounding into expansion phases. Our universe is just one of these collapsing and rebounding regions within the wider universe. Thus, an absolute beginning of the universe is averted.

Apart from the fact that the scenario is based on a nonexistent theory and so is pointedly speculative, the Pre-Big Bang Scenario is dogged with problems concerning how to join the pre- and post-Big Bang phases together and to smooth the transition to our matter-dominated universe.[73] These purely physical problems, however, pale in comparison to the deep conceptual difficulties attending such a scenario. Like the old Vacuum Fluctuation Models, the Pre-Big Bang Scenario postulates a wider, static space in which our observable universe originates via a

Naturalism and the Origin of the Universe

Big Bang event a finite time ago. But since there is a nonzero probability of a black hole forming in any patch of pre-existing space, such an event would have happened infinitely long ago, which is inconsistent with the finite age of our observable universe. Moreover, all the pre-Big Bang black holes should in infinite time have coalesced into one massive black hole coextensive with the universe, so that the post-Big Bang universe ought to be observed as infinitely old. Similarly, such a static wider universe should, given infinite past time, have already arrived at a state of thermodynamic equilibrium, in contradiction to the observed disequilibrium (more on this in the sequel).[74] Andrei Linde has pointed out the related difficulty that, to match observation, the black hole that spawned our universe would have to have had an unusually large size, much larger than the length scale of string theory. Veneziano responds to Linde's difficulty by appealing to the Anthropic Principle: "An answer to this objection is that the equations predict black holes of all possible sizes. Our universe just happened to form inside a sufficiently large one."[75] This response fails to appreciate the force of Linde's objection. The point is that on the Pre-Big Bang Scenario we should in all probability expect to be observing length scales much smaller than those observed. The fact that we do not therefore disconfirms that scenario (cf. the Boltzmann hypothesis discussed below). In their efforts to explain the origin of the observable universe from a pre-Big Bang condition, Gasperini and Veneziano have been singularly inattentive to the problematic issues arising from their supposition of a wider, eternally pre-existing state.

The more celebrated of the string scenarios has been the so-called Ekpyrotic Scenario championed by Paul Steinhardt.[76] In the most recent revision, the Cyclic Ekpyrotic Scenario, we are asked to envision two three-dimensional membranes (or "branes" for short) existing in a five-dimensional space-time (Figure 8). One of these branes is our universe. These two branes are said to be in an eternal cycle in which they approach each other, collide, and retreat again from each other. It is the collision of the other brane with ours that causes the expansion of our universe. With each collision, the expansion is renewed. Ripples in the branes are said to account for the large-scale structure of our three-dimensional universe. Thus, even though our universe is expanding, it never had a beginning.

Now, apart from its speculative nature, the Ekpyrotic Scenario is plagued with problems.[77] For example, the Horava-Witten version of string theory on which the scenario is based requires that the brane on which we live have a positive tension. But in the ekpyrotic scenario it has a

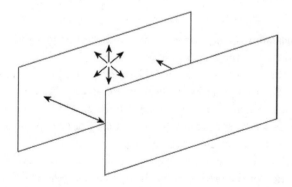

Figure 8. Cyclic Ekpyrotic Scenario. Two three-dimensional membranes in an eternal cycle of approach, collision, and retreat. With each collision the expansion of our universe is reinvigorated.

negative tension in contradiction to the theory. Attempts to rectify this have been unsuccessful. Second, the model requires an extraordinary amount of ad hoc fine turning. For example, the two branes have to be so perfectly aligned that even at a distance of 10^{30} greater than the space between them, they cannot deviate from being parallel by more than 10^{-60}. There is no explanation at all for this extraordinary setup. Third, the collapsing and retreating branes are the equivalent of a four-dimensional universe that goes through an eternal cycle of contractions and expansions. In this sense, the Cyclic Ekpyrotic Model is just the old Oscillating Model writ large in five dimensions. As such, it faces exactly the same problem as the original: there is no way for the universe to pass through a singularity at the end of each cycle to begin a new cycle and no physics to cause a nonsingular bounce. Finally, even if the branes could bounce back, there is no means of the physical information in one cycle being carried through to the next cycle, so that the Ekpyrotic Scenario has been unable to deliver on its promises to explain the large-scale structure of the observable universe. These are just some of the problems that have been alleged to afflict the model. It's no wonder that Andrei Linde has complained that while the Cyclic Ekpyrotic Scenario is "very popular among journalists," it has remained "rather unpopular among scientists."[78]

But let all that pass. Perhaps all these problems can be somehow solved.[79] The more important point is that it turns out that, like the Chaotic Inflationary Model, the Cyclic Ekpyrotic Scenario cannot be eternal in the past. With the formulation of their stronger theorem, Borde, Vilenkin, and Guth were able in 2001 to generalize their earlier results on inflationary models in such a way as to extend their conclusion to other models. Indeed, the new theorem implies that any universe that has on average been globally expanding at a positive rate is geodesically incomplete in the past and therefore has a past boundary. Specifically, they note, "Our argument can be straightforwardly extended to cosmology in higher dimensions," specifically brane-cosmology like Steinhardt's.[80] According to Vilenkin, "It follows from our theorem that the cyclic universe is past-incomplete,"[81] that is to say, the need for an initial singularity has not been eliminated. Therefore, such a universe cannot be past-eternal. Steinhardt has himself come to recognize this implication of the theorem for ekpyrotic scenarios and so now acknowledges that on his scenario the universe has a past boundary at some point in the metrically finite past.[82]

Summary

With each successive failure of alternative cosmogonic theories to avoid the absolute beginning of the universe predicted by the Standard Model, that prediction has been corroborated. It can be confidently said that no cosmogonic model has been as repeatedly verified in its predictions and as corroborated by attempts at its falsification, or as concordant with empirical discoveries and as philosophically coherent, as the Standard Big Bang Model. Of course, in view of the metaphysical issues raised by the prospect of a beginning of the universe, we may be confident that the quest to avert the absolute beginning predicted by the Standard Model will continue unabated. Such efforts are to be encouraged, and we have no reason to think that such attempts at falsification of the prediction of the Standard Model will result in anything other than further corroboration of its prediction of a beginning. While scientific evidence is always provisional, there can be little doubt in this case where the evidence points.

3. The Thermodynamics of the Universe

Traditionally, the study of the Last Judgment and the final state, eschatology is no longer exclusively the province of theology but has, in the last quarter century or so, emerged as a new branch of cosmology, being a sort of mirror image of cosmogony, that branch of cosmology that studies the origin of the universe. Not that the future of the universe will resemble its past; far from it. But just as physical cosmogony looks back in time to retrodict the history of the cosmos based on traces of the past and the laws of nature, so physical eschatology looks forward in time to predict the future of the cosmos based on present conditions and the laws of nature.

The key to physical eschatology is the Second Law of Thermodynamics. According to the Second Law, processes taking place in a closed system tend toward a state of equilibrium. Now the cosmologist's interest in the law concerns what it predicts when it is applied to the universe as a whole. For the universe is, on a naturalistic view at least, a gigantic closed system, since it is everything there is and there is nothing outside it. Already in the nineteenth century, scientists realized that the application of the Second Law to the universe as a whole implied a grim eschatological conclusion: given sufficient time, the universe will eventually come to a state of equilibrium and suffer "heat death." Once the universe reaches this state, no further change is possible. The universe thus faces inevitable extinction.

But this apparently firm projection raised an even deeper question: if, given sufficient time, the universe will suffer heat death, then why, if it has existed forever, is it not now in a state of heat death? If in a finite amount of time the universe will inevitably come to equilibrium, from which no significant further change is physically possible, then it should already be at equilibrium by now, if it has existed for infinite time. Like a ticking clock, it should by now have run down. Since it has not yet run down, this implies, in the words of Richard Schlegel, "In some way the universe must have been *wound up*."[83]

The German physicist Ludwig Boltzmann offered a daring proposal in order to explain why we do not find the universe in a state of heat death or thermodynamic equilibrium.[84] Boltzmann hypothesized that the universe as a whole *does,* in fact, exist in an equilibrium state, but that over time fluctuations in the energy level occur here and there throughout the universe, so that by chance alone there will be isolated regions where disequilibrium exists. Boltzmann referred to these isolated regions as "worlds." We should not be surprised to see our world in a highly improbable disequilibrium state, he maintained, since in the ensemble of all worlds there must exist by chance alone certain worlds in disequilibrium, and ours just happens to be one of these.[85]

The problem with Boltzmann's daring Many Worlds Hypothesis is that if our world were merely a fluctuation in a sea of diffuse energy, then it is overwhelmingly more probable that we should be observing a much tinier region of disequilibrium than we do. In order for us to exist, a smaller fluctuation, even one that produced our world instantaneously by an enormous accident, is inestimably more probable than a progressive decline in entropy over billions of years to fashion the world we see. In fact, Boltzmann's hypothesis, if adopted, would force us to regard the past as illusory, everything having the mere appearance of age, and the stars and planets as illusory, mere "pictures" as it were, since that sort of world is vastly more probable given a state of overall equilibrium than a world with genuine, temporally and spatially distant events. Therefore, Boltzmann's Many Worlds Hypothesis has been universally rejected by the scientific community, and the present disequilibrium is usually taken to be just a result of the initial low entropy condition mysteriously obtaining at the beginning of the universe.

The advent of relativity theory and its application to cosmology altered the shape of the eschatological scenario predicted on the basis of the Second Law of Thermodynamics but did not materially affect the fundamental question. Assuming that there is no positive cosmological constant fueling the expansion of the universe, that expansion will decelerate over time. Two radically different eschatological scenarios then present themselves. If the density of the universe exceeds a certain critical value, then the internal pull of the universe's own gravity will eventually overcome the force of the expansion and the universe will collapse in upon itself in a fiery Big Crunch. Beatrice Tinsley described such a scenario:

> If the average density of matter in the universe is great enough, the mutual gravitational attraction between bodies will eventually slow the expansion to a halt. The universe will then contract and collapse into a hot fireball. There is no known physical mechanism that could reverse a catastrophic big crunch. Apparently, if the universe becomes dense enough, it is in for a hot death.[86]

If the universe is fated to re-contraction, then, as it contracts, the stars gain energy, causing them to burn more rapidly so that they finally explode or evaporate. As everything in the universe grows closer together, the black holes begin to gobble up everything around them, and eventually begin themselves to coalesce. In time, "All the black holes finally coalesce into one large black hole that is coextensive with the universe," from which the universe will never reemerge.[87] There is no known physics that would permit the universe to bounce back to a new expansion prior to a final singularity or to pass through the singularity into a subsequent state.

On the other hand, if the density of the universe is equal to or less than the critical value, then gravity will not overcome the force of the expansion and the universe will expand forever at a progressively slower rate. Tinsley described the fate of this universe:

> If the universe has a low density, its death will be cold. It will expand forever at a slower and slower rate. Galaxies will turn all of their gas into stars, and the stars will burn out. Our own sun will become a cold, dead remnant, floating among the corpses of other stars in an increasingly isolated Milky Way.[88]

At 10^{30} years the universe will consist of 90 percent dead stars, 9 percent supermassive black holes formed by the collapse of galaxies, and 1 percent atomic matter, mainly hydrogen. Elementary particle physics suggests that thereafter protons will decay into electrons and positrons, so that space will be filled with a rarefied gas so thin that the distance between an electron and a positron will be about the size of the present galaxy. At 10^{100} years, some scientists believe that the black holes themselves will dissipate by a strange effect predicted by quantum mechanics. The mass and energy associated with a black hole so warp space that they are said to create a "tunnel" or "worm-hole" through which the mass and energy are ejected in another region of space. As the mass of a black hole decreases, its energy loss accelerates, so that it is eventually dissipated into radiation and elementary particles. Eventually all black holes will completely evaporate and all the matter in the ever-expanding universe will be reduced to a thin gas of elementary particles and radiation. Because the volume of space constantly increases, the universe will never actually arrive at equilibrium, since there is always more room for entropy production. Nonetheless, the universe will become increasingly cold, dark, dilute, and dead.

Very recent discoveries provide strong evidence that there is effectively a positive cosmological constant which causes the cosmic expansion to accelerate rather than decelerate. Paradoxically, since the volume of space increases exponentially, allowing greater room for further entropy production, the universe actually grows farther and farther from an equilibrium state as time proceeds. But the acceleration only hastens the cosmos's disintegration into increasingly isolated material particles no longer causally connected with similarly marooned remnants of the expanding universe.

Thus, the same pointed question raised by classical physics persists: why, if the universe has existed forever, is it not now in a cold, dark, dilute, and lifeless state? In contrast to their nineteenth-century forbears, contemporary physicists have come to question the implicit assumption that the universe is past eternal. Davies reports,

> Today, few cosmologists doubt that the universe, at least as we know it, did have an origin at a finite moment in the past. The alternative—that the universe has always existed in one form or another—runs into a rather basic paradox. The sun and stars cannot keep burning forever: sooner or later they will run out of fuel and die.

The same is true of all irreversible physical processes; the stock of energy available in the universe to drive them is finite, and cannot last for eternity. This is an example of the so-called Second Law of Thermodynamics, which, applied to the entire cosmos, predicts that it is stuck on a one-way slide of degeneration and decay towards a final state of maximum entropy, or disorder. As this final state has not yet been reached, it follows that the universe cannot have existed for an infinite time.[89]

Davies concludes, "The universe can't have existed forever. We know there must have been an absolute beginning a finite time ago."[90]

Some theorists have tried to escape this conclusion by adopting an oscillating model of the universe that never reaches a final state of equilibrium (recall Figure 3). But wholly apart from the aforementioned physical and observational difficulties confronting such a model, the thermodynamic properties of this model imply the very beginning of the universe that its proponents sought to avoid. For entropy is conserved from cycle to cycle in such a model, which has the effect of generating larger and longer oscillations with each successive cycle (Figure 9).

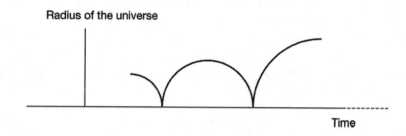

Figure 9. Oscillating Model with Entropy Increase. Due to the conservation of entropy each successive oscillation has a larger radius and longer expansion time.

As one scientific team explains, "The effect of entropy production will be to enlarge the cosmic scale, from cycle to cycle. . . . Thus, looking back in time, each cycle generated less entropy, had a smaller cycle time, and had a smaller cycle expansion factor then [sic] the cycle that followed it."[91] Thus, as one traces the oscillations back in time, they become progressively smaller until one reaches a first and smallest oscillation. Zeldovich and Novikov therefore conclude, "The multicycle model has an infinite future, but only a finite past."[92] In fact, astronomer Joseph Silk estimates on the basis of current entropy levels that the universe cannot have gone through more than 100 previous oscillations.[93]

Even if this difficulty were avoided, a universe oscillating from eternity past would require an infinitely precise tuning of initial conditions in order to perdure through an infinite number of successive bounces. A universe rebounding from a single, infinitely long contraction is, if entropy increases during the contracting phase, thermodynamically untenable and incompatible with the initial low entropy condition of our expanding phase. Postulating an entropy decrease during the contracting phase in order to escape this problem would require us to postulate inexplicably special low entropy conditions at the time of the bounce in the life of an infinitely evolving universe. Such a low entropy condition at the beginning of the expansion is more plausibly accounted for by the presence of a singularity or some sort of quantum creation event.[94] In either case, such a universe involves a radical fine-tuning of a very special sort, since the initial conditions are set at $-\infty$.[95]

If one is to avoid the inference that the universe has not existed forever, then one must find some scientifically plausible way to overturn the findings of physical eschatology so as to permit the universe to return to its youthful condition. But no realistic and plausible scenario is forthcoming. Linde once proposed, as we have seen, that a future-eternal inflationary model might be extended infinitely into the past, with the result that the beginning of the universe was averted. Perhaps one could thereby account for the appearance of youth in the observable universe. But we have seen that future-eternal inflationary space-times cannot be past-eternal: they must involve initial boundaries and so an absolute beginning of the universe. In any case, even if such an extension into the past were feasible, there is no reason to think that further inflation would subvert the implications of the Second Law or serve to restore the universe's youth. On the contrary, the effect of further inflation would be to increasingly isolate stars and galaxies in the universe until they cease to be in causal contact with one another and so disappear from view.

It has been suggested that the universe might in the future undergo quantum tunneling into a radically new state. For example, if the universe were currently in a false vacuum state, then it would eventually tunnel into a lower-energy vacuum state (Figure 10).

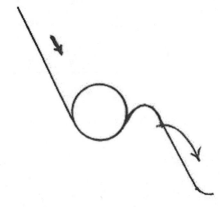

Figure 10. Tunneling from a False Vacuum State. If the universe is currently caught in a false vacuum state, eventually it will tunnel to the lower energy state of the true vacuum, resulting in a metamorphosis of nature.

Naturalism and the Origin of the Universe

In going through such a phase transition, all the physical constants' values would change and a totally new universe would emerge. Perhaps one could hypothesize that such a transition took place at some point in the finite past after an infinite lapse of time. But even if such a transition were to take place, the probability that the values of all the constants would fall into the unimaginably narrow life-permitting range is vanishingly small (a staple of discussions of cosmic fine-tuning).[96] Hence, it is highly improbable that our present life-permitting constellation of physical constants is the chance result of such a phase transition from a higher-level vacuum state about 13 billion years ago. Worse, if there is *any* nonzero probability that such a meta-stable state would tunnel to a true vacuum state, then given infinite past time it should have already occurred infinitely long ago, not just some 13.5 billion years ago. But then it again becomes inexplicable why the universe is not already dead.

Speculations about our universe begetting future "baby universes" have also been floated in eschatological discussions. It has been conjectured that black holes may be portals of wormholes through which bubbles of false vacuum energy can tunnel to spawn new expanding baby universes, whose umbilical cords to our universe may eventually snap as the wormholes close up, leaving the baby universe an independently existing spacetime (Figure 11).

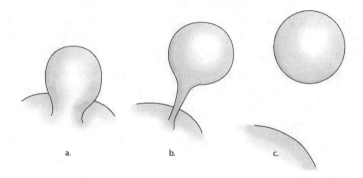

Figure 11. Birth of a Baby Universe. A baby universe spawned from its mother universe eventually becomes a disconnected and causally isolated space-time.

Perhaps we might imagine that our observable universe is just one of the newly birthed offspring of an infinitely old, pre-existing universe.

The conjecture of our universe's spawning future offspring by such a mechanism was the subject of a bet between Stephen Hawking and James Preskill, which Hawking in 2004 finally admitted, in an event much publicized in the press, that he had lost.[97] The conjecture would require that information locked up in a black hole could be utterly lost forever by escaping to another universe. One of the last holdouts, Hawking finally came to agree that quantum theory requires that information is preserved in black hole formation and evaporation. The implications? "There is no baby universe branching off, as I once thought. The information remains firmly in our universe. I'm sorry to disappoint science fiction fans, but if information is preserved, there is no possibility of using black holes to travel to other universes."[98] But suppose Hawking is wrong about this. Could such an eschatological scenario be in any case successfully extrapolated into the past, such that our universe is one of the baby universes spawned by the mother universe or by an infinite series of ancestors? It seems not, for while such baby universes appear as black holes to observers in the mother universe, an observer in the baby universe itself will see the Big Bang as a white hole spewing out energy. But this is in sharp contrast to our observation of the Big Bang as a low-entropy event with a highly constrained geometrical structure (see below). And again, what rescues the infinite sequence of cosmic descendants from the consequences of the Second Law of Thermodynamics is unclear.

Since such speculative conjectures fail to elude the problem, we seem left with the conclusion that the universe is not past eternal. The Big Bang represents the absolute beginning of the universe, just as it does in the Standard Big Bang Model; and the low entropy condition was simply an initial condition.

Indeed, thermodynamics may provide good reasons for affirming the reality of the singular origin of space-time postulated by the Standard Model. Roger Penrose states, "I have gradually come around to the view that it is actually misguided to ask that the space-time singularities of classical relativity should disappear when standard techniques of quantum (field) theory are applied to them."[99] For if the initial cosmological singularity is removed, then "we should have lost what seems to me to be the best chance we have of explaining the mystery of the second law of thermodynamics."[100] What Penrose has in mind is the remarkable fact that as one goes back in time the entropy of the universe steadily decreases. Just how unusual this is can be demonstrated by means of the Bekenstein-Hawking formula for the entropy of a stationary black hole. The total observed entropy of the universe is 10^{88}. Since there are around 10^{80} baryons in the universe, the observed entropy per baryon must be regarded as extremely small. By contrast, in a collapsing universe, the entropy would be 10^{123} near the end. Comparison of these two numbers reveals how absurdly small 10^{88} is compared to what it might have been. Thus, the structure of the Big Bang must have been severely constrained in order that thermodynamics as we know it should have arisen. So how is this special initial condition to be explained? According to Penrose, we need the initial cosmological singularity, conjoined with the Weyl Curvature Hypothesis, according to which initial singularities (as opposed to final singularities) must have vanishing Weyl curvature.[101] In standard models, the Big Bang does possess vanishing Weyl curvature. The geometrical constraints on the initial geometry have the effect of producing a state of very low entropy. So the entropy in the gravitational field starts at zero at the Big Bang and gradually increases through gravitational clumping. The Weyl Curvature Hypothesis thus has the time asymmetric character necessary to explain the Second Law. By contrast, on a time symmetrical theory like Hawking's, we should have white holes spewing out material, in contradiction to the Weyl Curvature Hypothesis, the Second Law of Thermodynamics, and probably also observation.[102] Penrose illustrates the difference using the picture (Figure 12) we have reproduced on the next page. If we remove the initial cosmological singularity, we render the Weyl Curvature Hypothesis irrelevant and "we should be back where we were in our attempts to understand the origin of the second law."[103]

Could the special initial geometry have arisen sheerly by chance in the absence of a cosmic singularity? Penrose's answer is decisive: "Had there not been any constraining principles (such as the Weyl curvature hypothesis) the Bekenstein-Hawking formula would tell us that the probability of such a 'special' geometry arising by chance is at least as small as about one part in $10^{1000B(3/2)}$ where B is the present baryon number of the universe [~10^{80}]."[104] Thus Penrose calculates that, aiming at a phase space whose regions represent the likelihood of various possible configurations of the universe, "the accuracy of the Creator's aim" would have to have been one part in $10^{10\exp(123)}$ in order for our universe to exist.[105] He comments, "I cannot even recall seeing anything else in physics whose accuracy is known to approach, even remotely, a

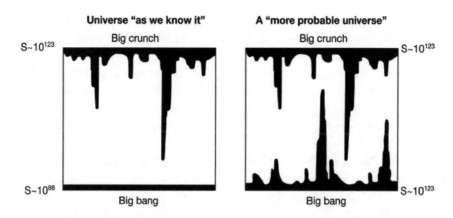

Figure 12. Contrast between the universe as we know it (assumed for convenience to be closed) with a more probable universe. In both cases the Big Crunch is a high-entropy (~10^{123}), complicated, unconstrained singularity. For the left-hand picture the Big Bang is a low-entropy (<10^{88}), highly constrained, initial singularity, while for the right-hand picture it is an unconstrained, much more probable Big Bang. The "stalactites" represent singularities of black holes, while the "stalagmites" represent singularities of white holes.

figure like one part in $10^{10^{\exp(123)}}$."[106] Thus, the initial cosmological singularity may be a virtual thermodynamic necessity.

Summary

Whether, then, one adopts a re-contracting model, an ever-expanding model, or an oscillating model, thermodynamics implies that the universe had a beginning. In a certain respect, this evidence of thermodynamics is even more impressive than the evidence afforded by the expansion of the universe. For while an accurate physical description of the universe prior to the Planck time remains and perhaps always will remain unknown, thereby affording room for speculations aimed at averting the origin of time and space implied by the expanding cosmos, no such uncertainty attends the laws of thermodynamics and their application. Indeed, thermodynamics is so well established that this field is virtually a closed science. Even though we may not like it, concludes Davies, we must say on the basis of the thermodynamic properties of the universe that the universe's energy was somehow simply "put in" at the creation as an initial condition.[107] Prior to the creation, says Davies, the universe simply did not exist.

4. Conclusion

This beginning of the universe, of space and time themselves, reveals the contingency of the universe. The universe is evidently not necessarily existent, as Hume suggested, since it is not eternal, and therefore its existence does cry out for explanation.[108] It is no longer sufficient to dismiss this problem with a shrug and a slogan, "The universe is just there, and that's all." Given its origin *ex nihilo,* the demand why the universe exists rather than nothing presses insistently upon us.[109]

Notes

1. Precursors of this paper were presented at the 22nd International Wittgenstein Symposium of the Austrian Ludwig Wittgenstein Society, Kirchberg am Wechsel, Austria, August 15–21, 1999; at the international conference "Towards a New Millennium in Galaxy Morphology," Eskom Conference Centre, Midrand, South Africa, September 13–18, 1999; and at the international meeting "Analytic Philosophy without Naturalism," Catholic University of Milan, June 11–13, 2003; and published as "The Ultimate Question of Origins: God and the Beginning of the Universe," *Astrophysics and Space Science* 269–270 (1999): 723–40; "Naturalism and Cosmology," in *Naturalism: a Critical Analysis,* ed. Wm. L. Craig and J. P. Moreland, Routledge Studies in Twentieth-Century Philosophy (London: Routledge, 2000), 215–52; and "Naturalism and Cosmology," in *Analytic Philosophy without Naturalism,* ed. A. Corradini, S. Galvan, and J. Lowe (London: Routledge, 2006), forthcoming.
2. *Metaphysics* A. 2. 982^{b}10–15.
3. Derek Parfit, "Why Anything? Why This?" *London Review of Books* 20/2 (January 22, 1998), 24.
4. Gottfried Wilhelm Leibniz, "The Principles of Nature and of Grace, Founded on Reason," in *The Monadology and Other Philosophical Writings,* trans. Robert Latta (London: Oxford University Press, 1951), 415; idem, "The Monadology," in *Monadology and Other Philosophical Writings,* 237–39.
5. For this analysis of so-called factual necessity, see John Hick, "God as Necessary Being," *Journal of Philosophy* 57 (1960): 733–4.
6. David Hume, *Dialogues concerning Natural Religion,* ed. with an Introduction by Norman Kemp Smith, Library of Liberal Arts (Indianapolis: Bobbs-Merrill, 1947), pt. IX, 190.
7. Bertrand Russell and F. C. Copleston, "The Existence of God," in *The Existence of God,* ed. with an Introduction by John Hick, Problems of Philosophy Series (New York: Macmillan, 1964), 175.
8. A. Einstein, "Cosmological Considerations on the General Theory of Relativity," in *The Principle of relativity,* by A. Einstein, *et. al.,* with Notes by A. Sommerfeld, trans. W. Perrett and J. B. Jefferey (rep. ed.: New York: Dover Publications, 1952), 177–88.
9. A. Friedman, "Über die Krümmung des Raumes," *Zeitschrift für Physik* 10 (1922): 377–86; G. Lemaître, "Un univers homogène de masse constante et de rayon croissant, rendant compte de la vitesse radiale des nébuleuses extragalactiques," *Annales de la Société scientifique de Bruxelles* 47 (1927): 49–59.
10. Gregory L. Naber, *Spacetime and Singularities: an Introduction* (Cambridge: Cambridge University Press, 1988), 126–27.
11. E. Hubble, "A Relation between Distance and Radial Velocity among Extra-galactic Nebulae," *Proceedings of the National Academy of Sciences* 15 (1929): 168–73.
12. Slipher's early papers are now available on-line at <http://www.roe.ac.uk/~jap/slipher/>.
13. John A. Wheeler, "Beyond the Hole," in *Some Strangeness in the Proportion,* ed. Harry Woolf (Reading, Mass.: Addison-Wesley, 1980), 354.
14. P. C. W. Davies, "Spacetime Singularities in Cosmology," in *The Study of Time III,* ed. J. T. Fraser (Berlin: Springer Verlag), 78–79.
15. As Gott, Gunn, Schramm, and Tinsley write,

> "The universe began from a state of infinite density about one Hubble time ago. Space and time were created in that event and so was all the matter in the universe. It is not meaningful to ask what happened before the big bang; it is somewhat like asking what is north of the North Pole. Similarly, it is not sensible to ask where the big bang took place. The point-universe was not an object isolated in space; it was the entire universe, and so the only answer can be that the big bang happened everywhere" (J. Richard Gott III, James E. Gunn, David N. Schramm, and Beatrice M. Tinsley, "Will the Universe Expand Forever?" *Scientific American* [March 1976], 65).

Naturalism and the Origin of the Universe

The Hubble time is the time since the singularity if the rate of expansion has been constant. The singularity is a point only in the sense that the distance between any two points in the singularity is zero. Anyone who thinks that there must be a place in the universe where the Big Bang occurred still has not grasped that it is space itself which is expanding; it is the two-dimensional *surface* of an inflating balloon which is analogous to three-dimensional space. The spherical surface has no center and so no location where the expansion begins. The analogy of the North Pole with the beginning of time should not be pressed, since the North Pole is not an edge to the surface of the globe; the beginning of time is more like the apex of a cone. But the idea is that just as one cannot go further north than the North Pole, so one cannot go earlier than the initial singularity.

16. John Barrow and Frank Tipler, *The Anthropic Cosmological Principle* (Oxford: Clarendon Press, 1986), 442.
17. Arthur Eddington, *The Expanding Universe* (New York: Macmillan, 1933), 124.
18. Ibid., 178.
19. Hubert Reeves, Jean Audouze, William A. Fowler, and David N. Schramm, "On the Origin of Light Elements," *Astrophysical Journal* 179 (1973): 912.
20. Fred Hoyle, *Astronomy Today* (London: Heinemann, 1975), 165.
21. Fred Hoyle, *Astronomy and Cosmology: A Modern Course* (San Francisco: W. H. Freeman, 1975), 658.
22. H. Bondi and T. Gold, "The Steady State Theory of the Expanding Universe," *Monthly Notices of the Royal Astronomical Society* 108 (1948): 252–70; F. Hoyle, "A New Model for the Expanding Universe," *Monthly Notices of the Royal Astronomical Society* 108 (1948): 372–82.
23. As Jaki points out, Hoyle and his colleagues were inspired by "openly anti-theological, or rather anti-Christian motivations" (Stanley L. Jaki, *Science and Creation* [Edinburgh: Scottish Academic Press, 1974], 347). Martin Rees recalls his mentor Dennis Sciama's dogged commitment to the Steady State Model:

> "For him, as for its inventors, it had a deep philosophical appeal—the universe existed, from everlasting to everlasting, in a uniquely self-consistent state. When conflicting evidence emerged, Sciama therefore sought a loophole (even an unlikely seeming one) rather as a defense lawyer clutches at any argument to rebut the prosecution case" (Martin Rees, *Before the Beginning*, with a Foreword by Stephen Hawking [Reading, Mass.: Addison-Wesley, 1997], 41). The phrase "from everlasting to everlasting" is the Psalmist's description of God (Ps. 90.2). Rees gives a good account of the discoveries leading to the demise of the Steady State model.

24. See, *e.g.*, E. M. Lifschitz and I. M Khalatnikov, "Investigations in Relativist Cosmology," *Advances in Physics* 12 (1963): 207.
25. As evident from the sentiments expressed by John Gribbin:

> "The biggest problem with the Big Bang theory of the origin of the universe is philosophical—perhaps even theological—what was there before the bang? This problem alone was sufficient to give a great initial impetus to the Steady State theory; but with that theory now sadly in conflict with the observations, the best way round this initial difficulty is provided by a model in which the universe expands from a singularity, collapses back again, and repeats the cycle indefinitely" (John Gribbin, "Oscillating Universe Bounces Back," *Nature* 259 [1976]: 15).

Scientists not infrequently misexpress the difficulty posed by the beginning of the universe as the question of what existed before the Big Bang (which invites the easy response that there was no "before"). The real question concerns the causal conditions of this event, why the universe exists at all.

26. R. Penrose, "Gravitational Collapse and Space-Time Singularities," *Physical Review Letters* 14 (1965): 57–59; S. W. Hawking and R. Penrose, in *The Large-Scale Structure of Space-Time*, ed. S. W. Hawking and G. F. R. Ellis (Cambridge: Cambridge University Press, 1973), 266.

27. Stephen Hawking and Roger Penrose, *The Nature of Space and Time,* The Isaac Newton Institute Series of Lectures (Princeton, NJ: Princeton University Press, 1996), 20.

28. Alan Guth and Mark Sher, "The Impossibility of a Bouncing Universe," *Nature* 302 (1983): 505–6; Sidney A. Bludman, "Thermodynamics and the End of a Closed Universe," *Nature* 308 (1984): 319–22.

29. Associated Press News Release, 9 January 1998.

30. Ibid.

31. Adam G. Riess, *et al.*, "Observational Evidence from Supernovae for an Accelerating Universe and a Cosmological Constant," *Astronomical Journal* 116 (1998): 1009–38; S. Perlmutter, *et al.*, "Measurements of Omega and Lambda from 42 High-Redshift Supernovae," *Astrophysical Journal,* 517:2 (1999): 565–86.

32. Ryan Scranton, *et al.*, "Physical Evidence for Dark Energy," http://arXiv.org/abs/astro-ph/0307335 (20 July 2003).

33. See D. N. Spergel, *et al.*, "First Year Wilkinson Microwave Anisotropy Probe (WMAP) Observation: Determination of Cosmological Parameters," http://arXiv:astro-ph/0302209v3 (17 June 2003) and the WMAP website http://map.gsfc.nasa.gov/gov/index.html; J-. P. Uzan, U. Kirchner, and George F. R. Ellis, "WMAP Data and the Curvature of Space," arXiv:astro-ph/0302597v2 (10 March 2003).

34. See http://map.gsfc.nasa.gov/m_mm/mr_limits.html.

35. One thinks, for example, of the late Carl Sagan on his *Cosmos* television series propounding this model and reading from Hindu scriptures about cyclical Brahman years in order to illustrate the oscillating universe, but with nary a hint to his viewers about the difficulties attending this model.

36. Christopher Isham, "Creation of the Universe as a Quantum Process," in *Physics, Philosophy and Theology: a Common Quest for Understanding,* ed. R. J. Russell, W. R. Stoeger, and G. V. Coyne (Vatican City: Vatican Observatory, 1988), 378.

37. Edward Tryon, "Is the Universe a Vacuum Fluctuation?" *Nature* 246 (1973): 396–7.

38. A. Guth, "Inflationary Universe: A Possible Solution to the Horizon and Flatness Problems," *Physical Review D* 23 (1981): 247–56.

39. See Isham, "Creation of the Universe," 385–87.

40. J. R. Gott III, "Creation of Open Universes from de Sitter Space," *Nature* 295 (1982): 304–7.

41. Christopher Isham, "Space, Time, and Quantum Cosmology," paper presented at the conference "God, Time, and Modern Physics," March 1990.

42. Christopher Isham, "Quantum Cosmology and the Origin of the Universe," lecture presented at the conference "Cosmos and Creation," Cambridge University, 14 July 1994.

43. See, *e.g.,* A. D. Linde, "The Inflationary Universe," *Reports on Progress in Physics* 47 (1984): 925–86; idem, "Chaotic Inflation," *Physics Letters* 1298 (1983): 177–81. For a critical review of inflationary scenarios, including Linde's, see John Earman and Jesus Mosterin, "A Critical Look at Inflationary Cosmology," *Philosophy of Science* 66 (1999): 1–49.

44. Linde, "Inflationary Universe," 976.

45. A. Borde and A. Vilenkin, "Eternal Inflation and the Initial Singularity," *Physical Review Letters* 72 (1994): 3305, 3307.

46. Andrei Linde, Dmitri Linde, and Arthur Mezhlumian, "From the Big Bang Theory to the Theory of a Stationary Universe," *Physical Review D* 49 (1994): 1783–1826.

47. Arvind Borde, Alan Guth, and Alexander Vilenkin, "Inflation Is Not Past-Eternal," http://arXiv:gr-qc/0110012v1 (1 Oct 2001): 4. The article was updated in January 2003.

48. Alexander Vilenkin, "Quantum Cosmology and Eternal Inflation," http://arXiv:gr-qc/0204061v1 (18 April 2002): 10.

49. Borde and Vilenkin, "Eternal Inflation," 3307.

50. Vilenkin, "Quantum Cosmology and Eternal Inflation," 11.

51. A. Vilenkin, "Birth of Inflationary Universes," *Physical Review D* 27 (1983): 2854. See J. Hartle and S. Hawking, "Wave Function of the Universe," *Physical Review D* 28 (1983): 2960–75; A. Vilenkin, "Creation of the Universe from Nothing," *Physical Letters* 117B (1982): 25–8.
52. Stephen Hawking, *A Brief History of Time* (New York: Bantam Books, 1988), 9.
53. Ibid., 46.
54. Ibid., 136.
55. Ibid., 140–41.
56. John D. Barrow, *Theories of Everything* (Oxford: Clarendon Press, 1991), 68.
57. Ibid., 67–68.
58. Hartle and Hawking, "Wave Function of the Universe," 2961; Hawking and Penrose, *Nature of Space and Time*, 85.
59. Vilenkin, "Creation of the Universe," 26.
60. Alexander Vilenkin, "Quantum Cosmology and Eternal Inflation,"11.
61. See my "Hartle-Hawking Cosmology and Atheism," *Analysis* 57 (1997): 291–5. With respect to determining the wave function of the universe Bryce DeWitt says, "Here the physicist must play God" (B. DeWitt, "Quantum Gravity," *Scientific American* 249 [1983]: 120).
62. Vilenkin, "Birth of Inflationary Universes," 2851.
63. Hawking, *Brief History of Time*, 136.
64. Barrow, *Theories of Everything*, 66–67.
65. Hawking, *Brief History of Time*, 138–39. In fact, according to Frank Tipler, there still remains a quantum singularity in such quantum gravity models, so that the classical singularity is merely replaced by a singularity of a different sort at the beginning of time (Frank Tipler, "The Mind of God," *The Times Higher Education Supplement* [10 October 1988], 23). See further Frank Tipler, "The Structure of the World from Pure Numbers," *Reports on Progress in Physics* 68 (2005): 897–964.
66. Hawking and Penrose, *Nature of Space and Time*, 121.
67. Ibid., 3–4. Cf. his comment, "I . . . am a positivist who believes that physical theories are just mathematical models we construct, and that it is meaningless to ask if they correspond to reality, just whether they predict observations" (Stephen Hawking, "The Objections of an Unashamed Positivist," in *The Large, the Small, and the Human,* by Roger Penrose [Cambridge: Cambridge University Press, 1997], 169).
68. Hawking and Penrose, *Nature of Space and Time*, 121; cf. 4, 53–55.
69. Ibid., 53–55.
70. As pointed out by Christopher Isham, "Quantum Theories of the Creation of the Universe," in *Quantum Cosmology and the Laws of Nature*, ed. R. J. Russell, N. Murphy, and C. J. Isham (Vatican City: Vatican Observatory, 1993), 56.
71. John D. Barrow, *The Origin of the Universe* (New York: Basic Books, Harper Collins, 1994), 113.
72. Gabriele Veneziano, "A Simple/Short Introduction to Pre-Big Bang Physics/Cosmology," http://arXiv:hep-th/9802057v2 (2 March 1998); M. Gasperini, "Looking Back in Time beyond the Big Bang," *Modern Physics Letters A* 14/16 (1999): 1059–1066; M. Gasperini, "Inflation and Initial Conditions in the Pre-Big Bang Scenario," *Physics Review D* 61 (2000): 87301–87305; M. Gasperini and Veneziano, "The Pre-Big Bang Scenario in String Cosmology," http://arXiv:hep-th/0207130v1 (12 July 2002).
73. T. Damour and M. Henneaux, "Chaos in Superstring Cosmology," Physical Review Letters 85 (2000): 920–923 (http://arXiv: hep-th/0003139); T. Damour, "String Cosmology and Chaos," Invited lecture at the International Conference on Theoretical Physics, Paris, France, 22–27 July 2002. Veneziano calls the Pre-Big Bang Scenario and the Ekpyrotic Scenario (to be discussed below) two "guesses" about the pre-big bang state and concedes that both scenarios share "the difficult (and unresolved) problem of making the transition between the pre- and post-Big Bang phase" (Gabriele Veneziano, "The Myth of the Beginning of Time,"

Scientific American [May 2004], 63–64).

74. Of course, if in an attempt to avoid such problems one postulates the expansion of the pre-existing vacuum space itself, then one has merely pushed the question of the origin back a notch.

75. Veneziano, "Myth of the Beginning of Time,"

76. See http://feynman.princeton.edu/~steinh/.

77. For these criticisms see especially Gary Felder, Andrei Frolov, Lev Kaufman, and Andrei Linde, "Cosmology with Negative Potentials," http://arXiv:hep-th/0202017v2 (16 February 2002) and the therein cited literature, particularly the studies by David Lyth.

78. "Cyclic Universe Runs into Criticism," *Physics World* (June 2002): 8.

79. Steinhardt maintains that some of them, such as the alleged fine-tuning, have been solved.

80. Borde, Guth, and Vilenkin, "Inflation Is Not Past-Eternal," 4. See also Alexander Vilenkin, "Quantum Cosmology and Eternal Inflation," 11.

81. Alexander Vilenkin, personal communication.

82. See www.phy.princeton.edu/~steinh/ under "Answers to Frequently Asked Questions: Has the cyclic model been cycling forever?" Steinhardt seeks to mollify the impact of the Borde-Guth-Vilenkin theorem by maintaining that clocks run progressively faster as one approaches the past boundary, so that elapsed time becomes what he calls "semi-infinite." This trick does nothing to abrogate the finitude of the past or the beginning of the universe.

83. Richard Schlegel, "Time and Thermodynamics," in *The Voices of Time,* ed. J. T. Fraser (London: Penguin, 1968), 511.

84. Ludwig Boltzmann, *Lectures on Gas Theory,* trans. Stephen G. Brush (Berkeley: University of California Press, 1964), §90 (446–48).

85. For a fascinating contemporary reprise of Boltzmann's hypothesis and a discussion of its central weakness, see Lin Dyson, Matthew Kleban, and Leonard Susskind, "Disturbing Implications of a Cosmological Constant," http://arXiv.org/abs/hep-th/0208013v3 (14 November 2002). Their point of departure is Henri Poincaré's argument that in a closed box of randomly moving particles every configuration of particles, no matter how improbable, will eventually recur, given enough time; given infinite time, every configuration will recur infinitely many times. Eschewing a global perspective in favor of a restriction to our causally connected patch of the universe, they argue for the inevitability of cosmological Poincaré recurrences, allowing the process of cosmogony to begin anew. "The question then is whether the universe can be a naturally occurring fluctuation, or must it be due to an external agent which starts the system out in a specific low entropy state?" (Ibid, 4). They recognize that the central weakness of the fluctuation hypothesis is that there are "far more probable ways of creating livable ('anthropically acceptable') environments" than those that begin in a low entropy condition. See further note 107 below.

86. Beatrice Tinsley, "From Big Bang to Eternity?" *Natural History Magazine,* October 1975, 103.

87. Duane Dicus, *et al.*, "*The* Future of the Universe," *Scientific American* (March 1983), 99.

88. Tinsley, "Big Bang," 105.

89. Paul Davies, "The Big Bang—And Before," The Thomas Aquinas College Lecture Series, Thomas Aquinas College, Santa Paula, Calif., March 2002.

90. Paul Davies, "The Big Questions: In the Beginning," ABC Science Online, interview with Phillip Adams, http://aca.mq.edu.au/pdavieshtml.

91. Duane Dicus, *et al.*, "Effects of Proton Decay on the Cosmological Future," *Astrophysical Journal* 252 (1982): 1, 8.

92. I. D. Novikov and Ya. B. Zeldovich, "Physical Processes near Cosmological Singularities," *Annual Review of Astronomy and Astrophysics* 11 (1973): 401–2.

93. Joseph Silk, *The Big Bang,* 2d ed. (San Francisco: W. H. Freeman, 1989), 311–12.

94. Thanks to Donald Page for these points.

95. George Ellis remarks,

> "The problems are related: first, initial conditions have to be set in an extremely special way at the start of the collapse phase in order that it is a Robertson-Walker universe collapsing; and these conditions have to be set in an acausal way (in the infinite past). It is possible, but a great deal of inexplicable fine tuning is taking place: how does the matter in widely separated causally disconnected places at the start of the universe know how to correlate its motions (and densities) so that they will come together correctly in a spatially homogeneous way in the future?? Secondly, if one gets that right, the collapse phase is unstable, with perturbations increasing rapidly, so only a very fine-tuned collapse phase remains close to Robertson-Walker even if it started off so, and will be able to turn around as a whole (in general many black holes will form locally and collapse to a singularity)" (G. F. R. Ellis to James Sinclair, 25 January, 2006).

Ellis then pointedly asks, "Who focused the collapse so well that it turns around nicely?"

96. See my "Design and the Anthropic Fine-Tuning of the Universe," in *God and Design: The Teleological Argument and Modern Science,* ed. Neil Manson (London: Routledge, 2003), 178–99; also the brief discussion of fine-tuning in my paper "Theism Defended."

97. For a first-hand account see James Preskill's website www.theory.caltech.edu/~preskill/jp-24ju104.html.

98. S. W. Hawking, "Information Loss in Black Holes," http://arXiv:hep-th/0507171v2 (15 September 2005): 4.

99. Roger Penrose, "Some Remarks on Gravity and Quantum Mechanics," in *Quantum Structure of Space and Time,* ed. M. J. Duff and C. J. Isham (Cambridge: Cambridge University Press, 1982), 4.

100. Ibid., 5.

101. Weyl curvature is the curvature of space-time which is not due to the presence of matter and is described by the Weyl tensor. Space-time curvature due to matter is described by the Ricci tensor. Together they make up the Riemann tensor giving the metric for space-time.

102. Hawking and Penrose, *Nature of Space and Time,* 130.

103. Penrose, "Remarks on Gravity," 5.

104. Ibid.

105. Roger Penrose, "Time-Asymmetry and Quantum Gravity," in *Quantum Gravity 2,* ed. C. J. Isham, R. Penrose, and D. W. Sciama (Oxford: Clarendon Press, 1981), 249; cf. Hawking and Penrose, *Nature of Space and Time,* 34–35.

106. Penrose, "Time-Asymmetry," 249.

107. P. C. W. Davies, *The Physics of Time Asymmetry* (London: Surrey University Press, 1974), 104. Dyson, Kleban, and Susskind (see note above) respond to such a suggestion as follows: "Another possibility is that an unknown agent intervened in the evolution and for reasons of its own restarted the universe in the state of low entropy characterizing inflation. However, even this does not rid the theory of the pesky recurrences. Only the first occurrence would evolve in a way that would be consistent with usual expectations" (Dyson, Kleban, and Susskind, "Disturbing Implications of a Cosmological Constant," 20–21). But so saying, they have misconstrued the hypothesis. The hypothesis was not of an external agent who "restarted" the universe but of "an external agent who *starts the system out* in a specific low entropy state" (Ibid., 4). On such a hypothesis "Some unknown agent initially started the inflation high up on its potential, and the rest is history" (Ibid., 2). On this hypothesis the recurrence problems do not even arise. By contrast, Dyson, Kleban, and Susskind are finally driven to suggest that "Perhaps the only reasonable conclusion is that we do not live in a world with a true cosmological constant" (Ibid., 21), a desperate hypothesis which flies in the face of the evidence.

108. My paper "Theism Defended" in this volume explores what sort of explanation it must be.

109. I am grateful to Don Page and especially James Sinclair for helpful discussion of several issues of theoretical cosmogony in connection with this paper.

24

Cosmic Evolution, Naturalism, and Divine Creativity,
or
Who Owns the Robust Formational Economy Principle?

Howard J. Van Till

From several of the natural sciences we have come to know a great deal about the Universe's formational history. Modern scientific cosmology, for instance, has crafted a remarkable account of the evolution of the Universe from a hot Big Bang beginning to a universe now populated with varied forms of energy, atoms, molecules, and star-studded galaxies. Similarly, stellar astronomy has learned how to describe the formation and evolution of the stars within galaxies—from proto-stellar globules to main sequence stars to red giants, white dwarfs, neutron stars, and black holes. Likewise, planetary astronomy and historical geology have come to understand the grand drama of planetary formation and the evolution of Earth's geological features. Finally, modern biology provides us with an account of the biological evolution of life on planet Earth over the last four billion years.

1. God and the Nature of Nature

But the positing of these scientific accounts of evolutionary development has often met with skepticism and resistance rooted in religious concerns. Chief among those religious concerns, I suspect, is the one captured by this question: If Nature has all of the resources, potentialities, and capabilities to make possible its evolutionary history, without need for any episodes of form-conferring supernatural intervention, is God superfluous? Or, to put it more casually, if cosmic evolution is possible is the Creator out of a job?

But the question then becomes: What is the Creator's job? If the job of a Creator is to impose forms on matter in order to actualize particular physical systems or particular organisms or parts of organisms that the Universe is unable (for whatever reason) to actualize, then a robustly equipped universe would seem to put a Creator out of *that* job. Form-imposing interventions would be unnecessary. But what if the role of divine creativity is something quite different from form-imposing intervention? That's a question we will consider later in this essay.

For now, however, let us go back to the question of whether or not the formational capabilities of the Universe are adequate to the task of making possible the cosmic, stellar, planetary, and biological evolution now envisioned by the natural sciences. Because of the way in which discussion on this issue is often carried out, let me break the question into two parts—one regarding inanimate physical structures and the other regarding life forms: (1) Are the formational capabilities of the Universe adequate for the task of actualizing the full array of inanimate *physical structures* (nucleons, nuclei, atoms, molecules, planets, stars, galaxies, and the like) that have appeared from time zero until now? Cosmology, physics, chemistry, astronomy, and geology have found it fruitful to assume that they are. (2) Are the formational capabilities of the Universe adequate for the task of actualizing the full array of *life forms* that have appeared in the course of Earth history? Biology has found it fruitful to assume that they are.

Before looking at some of the differing ways in which this combination of questions can be answered—along with what the answers might imply for religion—let me outline a way to restate the question in a manner that I have found helpful in highlighting what I judge to be the core issue for those who see great religious significance in the answers.

2. Restating the Question

I must begin by offering a definition of what I mean by the "formational economy" of the Universe.[1]

> Definition: The *formational economy* of the Universe is the set of all of the *resources, potentialities,* and *capabilities* of the Universe that have contributed to its formational history.

Let me provide a bit of clarification for three particular terms used in this definition. (1) By *resources* I mean such entities as the interactive fundamental units of matter/energy that occupy the Universe, along with the dynamic space-time context in which they function. Aggregate systems made of these fundamental units could, in turn, play the role of resources from which still larger structures and systems are made. (2) By *potentialities* I mean the possibilities (whether actualized or not) for functional physical structures (such as atoms, molecules, stars, or planets) and for viable living organisms (such as bacteria, fish, mammals, or primates). (3) By *capabilities* I mean to call attention especially to the Universe's abilities for organizing or transforming its resources in such a way as to actualize some of its structural and functional potentialities in the course of time. Atoms, for example, have the capabilities to act and interact in such a way as to self-organize into a huge variety of molecules.

One of the questions at the heart of the naturalism versus supernaturalism debate concerns the nature of the Universe's formational economy: Is this set of resources, potentialities, and capabilities sufficient to make possible such remarkable phenomena as cosmic evolution, stellar evolution, planetary evolution, and biological evolution as now envisioned by the natural sciences?

A positive answer to this question can be stated in the form of a general principle concerning "the nature of Nature," a principle I call the "Robust Formational Economy Principle" (RFEP):

> The RFEP: *The formational economy of the Universe is sufficiently robust to make possible—without need for occasional episodes of form-imposing intervention by any extra-natural agent—the actualization of both (a) all of the types of physical structures (nucleons, nuclei, atoms, molecules, galaxies, stars, planets, etc.) and (b) all of the life forms that have appeared in time.*

Finally, then, our fundamental question about "the nature of Nature" can be stated in this simple way: *Does the Universe satisfy the RFEP?* The answer one gets depends, of course, on who is being asked. In fact, this is a question that divides thinkers into two distinct camps that I will call the "episodic creationism camp" and the "RFEP camp" respectively.

Episodic Creationism

In its most traditional and familiar manifestation, the episodic creationist camp is formed by persons who affirm both the Judeo-Christian *doctrine* of creation and a particular *portrait* of the Universe's formational history.[2] Stated as succinctly as I am able, the traditional *doctrine* of creation affirms the idea that the Universe owes its existence to the voluntary act of the Creator who called the Creation into being from nothing. Distinct from this doctrine, the episodic creationist *portrait* of the Universe's formational history is a what-happened-and-when story that includes, as essential elements, a series of occasional episodes of form-conferring intervention by a Creator (sometimes called a Designer).

There are many variations on the episodic creationist theme, mostly variations on the number, type, and timeframe of the form-conferring interventions that are visualized (on the basis of some chosen combination of scriptural interpretation and scientific input). Included among these variant strains of episodic creationism are *young-Earth special creationism, old-earth special creationism, progressive creationism, and Intelligent Design (ID)*. Although most proponents of ID have strategically chosen to leave their models of form-conferring intervention unspecified and vague, the non-natural action that is ascribed to their unidentified Designer has essentially the same theological character as envisioned by other forms of episodic creationism.

Does the created Universe satisfy the Robust Formational Economy Principle? No way, says the episodic creationist. Why not? Because (presumably by the Creator's purposeful choice) the universe lacks the requisite formational capabilities to make either cosmic evolution or biotic evolution possible. In other words, the formational economy of the Universe is incomplete. There are gaps (the consequence of missing capabilities) in the Universe's formational economy. Specifically, there are gaps that could be bridged only by the form-imposing action of some extra-natural (from outside of nature) or supernatural (having power over nature) agent. From the episodic creationist perspective, certain structures and forms (especially biotic forms) could not have been actualized without occasional episodes of supernatural intervention.

A variant strain of episodic creationism splits the question into two parts, one having to do with the formation of inanimate physical structures, the other having to do with life forms. *Cosmic* evolution (including galaxies, stars, planets, atoms, and some molecules) from the Big Bang until now may well be possible, some say, but the molecules to mankind *biotic* evolution now envisioned by biologists is, from this perspective, just not possible. The Universe may have been provided by its Creator with a robust formational economy for the actualization of inanimate structures, but there are, it is claimed, empirically discernible gaps in the formational economy of life. Occasional episodes of form-imposing intervention by some non-natural agent are considered to be essential to actualize organisms or parts of organisms that possess *specified complexity* or *irreducible complexity*. As I understand it, most of the advocates of Intelligent Design are members of this camp.

Proponents of the RFEP

Does the Universe satisfy the RFEP? Persons in this camp would say, Yes, the formational economy of the Universe is indeed sufficiently robust to make possible the actualizing (forming, assembling) of the full diversity of both inanimate physical structures and life forms that have ever existed. There are no gaps in the formational economy of the Universe. Consequently, there has never been a need to bridge such gaps by form-imposing acts of extra-natural assembly. Contemporary cosmology and biology are judged to be on the right track of theory formulation when they presume that this Universe is equipped with a robust formational economy.

3. Implications for Religious Belief Systems

What if contemporary cosmology and biology are correct in this assessment? What if the nature of Nature is such that the RFEP is true?

Some voices, representing one extreme form of naturalism (in essence, a fully atheistic naturalism) have in effect taunted, "Well then, *if there are no gaps* in the Universe's formational economy, what need is there for a Creator?" The implied answer: "None."

In response to this taunt, some proponents of theism have, in effect, retorted, "Well, then *there must be gaps* in the Universe's formational economy and we will show you the empirical evidence that natural processes are not sufficient to accomplish the actualization of specific organisms or parts of organisms."

And the shouting match goes on.

Note carefully: this shouting match is real. There *is* a controversy. But it is not a *scientific* controversy. It is a battle between two ideologies, with the RFEP caught in the crossfire. Extreme (atheistic) naturalism and episodic creationism obviously disagree vehemently on both the reality of God and the nature of Nature. Ironically, however, they appear to agree on the basic premise (one that I shall soon dispute) that if atoms, molecules, cells, and organisms can accomplish everything that evolutionary biology envisions, then divine creative action is effectively unnecessary and atheism is favored by the empirical evidence. Or, to use the terminology just developed, they agree on the proposition that *if the RFEP is true, then atheism wins.*

In my judgment, however, this premise is radically flawed and the inadequate examination of it lies at the heart of the fruitless shouting match to which I just alluded. I can see why some proponents of atheistic naturalism might find it strategically convenient to leave the premise unchallenged (it gives them the appearance of ownership of the RFEP), but I am astounded that the majority of theists (at least the majority of the North American Christian communities with which I am familiar) are also content to let this premise stand without critical examination.

4. Who Owns the Robust Formational Economy Principle?

I ask this question about RFEP ownership as a way of testing the common presumption that if the RFEP is true, then naturalism (treated as if it were indistinguishable from atheism) would be the victor over all forms of theism. Our first steps toward performing this test require important clarifications regarding the meanings of key terms. We begin with "naturalism."

There is a broad sense in which any worldview that accepts the RFEP constitutes a form of naturalism. *However, this can be said only if one understands and respects the fact that the term* "naturalism," without further qualification or clarification, cannot honestly be equated to atheism or materialism. In what follows I shall use the term "naturalism," when unqualified, to represent neither more nor less than *the rejection of supernaturalism*. Stated positively, naturalism is committed to the belief that all events that occur within this Universe are consistent with and adequately explained by the system of natural causes. This commitment necessarily entails the additional belief that the system of natural causes is fully adequate to account for all events that transpire. Focusing on the issue of the Universe's formational economy, we can say that naturalism—as here defined—entails the RFEP.

In defining *naturalism* to be the rejection of *supernaturalism*, however, it is essential to understand that "supernaturalism" must be clearly understood as a term that can *not* glibly be treated as a synonym for "theism." Supernaturalism is but one form of theism among many. By "supernaturalism" I mean the specific form of theism (among North Americans it is perhaps the most common form) that entails a commitment to the belief that God is both able and, on occasion, willing to act *coercively* in the sense of exercising unilateral power *over nature*.

Coercive divine action is a determinative form of divine action that supersedes natural action and brings about outcomes in the world that are either beyond or contrary to the capabilities of natural action. Form-imposing interventions, of the sort discussed above, would clearly fall into the category of coercive divine action. Defining *naturalism* as *anti-supernaturalism* may be somewhat unconventional, but I have come to appreciate the fruitfulness of this approach in my own reflections on the issues at hand. Most importantly, I believe, it provides the basis for dispelling the rhetorically popular but logically mistaken notion that all forms of naturalism are enemies of all forms of theism.

Once we have limited the meaning of *naturalism* to the rejection of *supernaturalism*, it then becomes possible to make useful distinctions among several naturalistic perspectives that function either as comprehensive worldviews or as important tenets within worldviews. In my judgment, a widespread failure to make these distinctions has contributed a great deal of confusion to evaluations of evolution in the context of religious concerns. Following are six perspectives (along with the labels that I shall use to identify them) that differ from one another regarding the application of naturalism to the formational history of the Universe. Note that each of the first four perspectives accepts the RFEP, *but for very different reasons*.

First, two differing forms of naturalism:

1. By *atheistic naturalism* I mean a comprehensive worldview (one that includes tenets relevant to religion) that explicitly rejects not only supernaturalism but also the existence of any transcendent "God." It could also be called simply *atheism* (because it rejects the existence of God) or *ontological naturalism* (because it asserts that Nature is the only category of being that is real) or *materialism* (because it presumes that Nature, taken to be a purely physical/material system, is all there is). Materialism takes the RFEP as an unexplained given. In a sense, the RFEP simply must be true because there is no useful alternative. Matter is the only active agent that exists, the only agent available to do what needs to be done. Questions regarding the grand purpose of the Universe are usually set aside as meaningless. From here on we shall use the term *materialism* to designate this worldview.

2. By *religiously agnostic naturalism* I mean a point of view (not a complete worldview) that makes no commitment regarding the existence or nonexistence of a transcendent God, but nonetheless does preclude the occurrence of supernatural divine interventions. The nature

of Nature is taken to be such that the existence of a God who acts only non-coercively is possible but not essential. Religiously agnostic naturalism accepts the RFEP as descriptive of the nature of Nature largely because of the RFEP's track record of fruitfulness as a fundamental premise for scientific theorizing. For this reason I would consider it appropriate to call this perspective by the name *scientific naturalism* and will do so from here on. It is important to note the distinction between scientific naturalism (agnostic, not atheistic) and materialism (explicitly atheistic). Compared with materialism, scientific naturalism focuses its attention not on the ontology of the universe but on the manner in which it functions (without interruptions of the natural causal nexus by supernatural interventions).

Next, three versions of theism: two that entail a rejection of supernatural intervention as a means of actualizing new physical structures or new life forms, followed by the more familiar version that embraces the idea of supernatural intervention.

1. By *naturalistic theism* I mean a comprehensive theistic worldview that takes the existence and *non-coercive* action of God to be essential to the nature of Nature. This worldview sees supernatural (coercive) divine intervention as something that is precluded by the very natures of God, the World, and the God/World relationship, thereby ensuring that the World would necessarily be characterized by the RFEP. Although episodic creationism (with its coercive, form-imposing divine interventions) is categorically rejected by this worldview, non-coercive divine action is presumed to be effective—contributing to the outcome of natural events and processes—and manifest in such phenomena as the remarkable fruitfulness of the Universe's formational history.

2. By *kenotic theism* I mean a comprehensive theistic worldview that presumes the existence of a Creator-God who is capable of supernatural intervention but who nonetheless voluntarily chooses *not* to engage in supernatural intervention, at least not for the purposes of actualizing either inanimate physical structures or living organisms.[3] From this perspective, there is no need for occasional episodes of form-imposing divine intervention because the Universe has been provided by its Creator with a robust formational economy. The RFEP is accepted with the proviso that supernatural intervention could, however, play a role in events outside of the arena of the Universe's formational history. One could characterize this perspective as a *partially* naturalistic theism.

3. By *supernatural theism* I mean a comprehensive worldview that posits the existence of a transcendent God (or Creator, or Designer) who is not only able and, on occasion, willing to act supernaturally on the world, but is also presumed to have actually done so. Specifically, supernatural theism presumes that God has engaged in occasional episodes of form-imposing intervention as the only means by which certain physical structures and life forms could have been actualized in time. This worldview welcomes the episodic creationist view of the Universe's formational history and rejects the RFEP. As I see it, the Intelligent Design movement represents an attempt to create a place for the episodic creationist perspective in public science education by stripping it of all explicit references to its religious motivation and then focusing on the presentation of arguments in support of the claim that there is empirical evidence against the RFEP.

Finally, a matter of scientific methodology: The term *methodological naturalism* is often used to name the conventional practice of positing only natural causes in the formulation of scientific theories. This is not a metaphysical statement about the nature of Nature, but only a statement about the way the sciences have chosen to conduct their work. As such, methodological naturalism is *consistent* with all of the four naturalistic perspectives defined and named above (materialism, scientific naturalism, naturalistic theism, and kenotic theism), but provides no definitive means of favoring any one of them over the others. One could, however, argue that the success of the natural sciences does serve as evidence that there is no clear need for positing supernatural interventions as part of any science-like explanation.

5. Common Misunderstandings Regarding RFEP Ownership

The literature of both critics and advocates of episodic creationism is permeated with rhetoric based on the dual premise that materialism owns the RFEP and theism necessarily rejects it. In his book, *Darwin's Dangerous Idea*, for instance, philosopher Daniel C. Dennett speaks with eloquent playfulness about the long history of episodic creationism's dream of defeating materialism by finding proof that the RFEP fails. In the following brief excerpt, Dennett uses the word "skyhook" as a metaphor for what we have been calling *occasional episodes of form-imposing divine intervention*. "Cranes," in contrast, represent the accomplishment of the Universe's robust formational economy of natural capabilities.

> For over a century, skeptics have been trying to find a proof that Darwin's idea just can't work, at least not *all the way*. They have been hoping for, hunting for, praying for skyhooks, as exceptions to what they see as the bleak vision of Darwin's algorithm churning away. And time and again, they have come up with truly interesting challenges—leaps and gaps and other marvels that do seem, at first, to need skyhooks. But then along have come the cranes, discovered in many cases by the very skeptics who were hoping to find a skyhook.[4]

My own response to rhetoric of this sort would be to point out that the presence and sufficiency of "cranes," while consistent with the RFEP, constitutes neither the defeat of theism nor a definitive victory for materialism. Naturalistic theism, kenotic theism, and scientific naturalism all remain as viable alternatives.

Leaders of the ID movement, however, have argued for their perspective along quite different lines. In an exchange that we had several years ago in the religious periodical *First Things*, law professor Phillip E. Johnson expressed it this way:

> If God had created a lifeless world, even with oceans rich in amino acids and other organic molecules, and thereafter had left matters alone *[that is, had refrained from form-imposing intervention (hvt)]*, life would not have come into existence. If God had done nothing *[that is, nothing supernatural (hvt)]* but create a world of bacteria and protozoa, it would still be a world of bacteria and protozoa.
>
> Persons who believe that chemicals unassisted by intelligence *[that is, by non-natural, choice-making, form-imposing agents (hvt)]* can combine to create life, or that bacteria can evolve by natural processes *[presumed to exclude non-coercive divine action (hvt)]* into complex animals, are making an *a priori* assumption that nature has the resources to do its own creating.
>
> I call such persons metaphysical *naturalists*.[5]

Note the clear implication that accepting the RFEP (no need for occasional episodes of form-imposing divine intervention) is, by Johnson's measure, tantamount to being a "metaphysical naturalist." Throughout Johnson's writing, it is clear that what he calls a *metaphysical naturalist* is identical to what we have called a *materialist*. If I understand Johnson correctly, what I have called "naturalistic theism" and "kenotic theism" would not be counted by him as positions that could be held with either intellectual or religious integrity, and the truth of the RFEP would constitute a de facto victory of materialism over theism.

But the equating of the truth of the RFEP with the victory of materialism is perhaps articulated most colorfully by British chemist Peter Atkins, an ardent defender of materialism. Using a vocabulary drawn in part from biblical literature, Atkins announces that,

> In the beginning there was nothing. Absolute void, not merely empty space. There was no space; nor was there time, for this was before time. The Universe was without form and void.[6]
>
> What comes after this formless beginning, devoid of structure?
>
> By chance there was a fluctuation, and a set of points, emerging from nothing and taking their existence from the pattern they formed, defined a time.... From absolute nothing, absolutely without intervention, there came into being rudimentary existence. The emergence of the dust of points and their chance organization into time was the haphazard, unmotivated action that brought them into being.[7]

Notice how quickly (and without explanation) Atkins's story moved from his conjectures about what sort of events may have occurred to his declaration concerning the lack of motivation for them. Moving from the inanimate to the animate, Atkins continues his story in a way that includes no hint whatsoever that positing the truth of the RFEP should be an occasion for either candor or wonder.

> A great deal of the Universe does not need any explanation. Elephants, for instance. Once molecules have learnt to compete and to create other molecules in their own image, elephants, and things resembling elephants, will in due course be found roaming through the countryside.[8]

Having treated the advance from replicating molecules to elephants as one of those "obvious things" that we have a right to expect, the move from elephants to humans (including Mr. Atkins himself, presumably) is similarly announced as amusing, but without significance.

> Some of the things resembling elephants will be men. They are equally unimportant.... Their special but not significant function is that they are able to act as commentators on the nature, content, structure and source of the Universe and that, as a sideline, they can devise and take pleasure from communicable fantasies.[9]

Whether I find his perspective attractive or not, Atkins is, of course, free to espouse his materialist worldview. I must, however, take exception to the glibness with which his rhetoric presumes that materialism owns the RFEP and that the truth of the RFEP reduces any form of theism to being no more than one of the "communicable fantasies" that insignificant humans sometimes devise. I would assume that, as an expression of his personal sense of fairness,

Atkins would place his own story of a Universe that created itself from "absolutely nothing" in the same category, but he does not say so explicitly.

6. Theism Without Supernaturalism: The Options Seldom Offered

Kenotic Theism: The Fully-Gifted Creation Perspective

As a member of a religious community that valued careful theorizing in both Christian theology and the natural sciences, I considered it mandatory to develop a theological perspective that welcomed the work of empirical science. I worked toward an approach that was comfortable with both the historic Judeo-Christian *doctrine* of Creation (the Universe owes its existence to a Creator) and the contemporary scientific understanding of the Universe's formational history. Quite obviously, such an approach would entail a rejection of atheistic naturalism of the sort espoused by Dennett and Atkins. At the same time, however, I found it imperative also to reject all approaches that differed from materialism only by inserting occasional episodes of form-imposing divine intervention into presumed gaps in an otherwise atheistic process, approaches that I have sometimes characterized as "punctuated materialism".

In broad outline, these are basic tenets of what I have often called the "fully-gifted Creation" perspective:

1. The Universe is a Creation. That is, the Universe exists now only because its Creator has *given it being* and continues to *sustain it in being* from moment to moment. (Note that this says nothing about the particulars of its formational history. This *doctrine* of Creation does not specify any particular what-happened-and-when story for the actualization of new forms.)

2. The *being* of the Creation includes all that it *is* (resources, properties, etc.), all that it could potentially *become* (represented by a potentiality space of structures and functional systems), and all that it is capable of *doing* (for instance, to achieve some of its structural and functional potentialities).

3. Each resource, potentiality, and capability present in the Universe can be celebrated as a "gift of being" that is an expression of its Creator's *creativity* (in conceptualizing an economy of resources, potentialities, and capabilities that would accomplish the Creator's purposes) and of its Creator's *generosity* (in giving such fullness of being as the sciences have found this Creation to possess).

4. All of this applies to the Creation's capabilities for forming new structures—its capabilities for occupying new locations in its potentiality space. From this perspective, each of the Universe's formational capabilities represents a gift of being that can be celebrated as a manifestation of the Creator's creativity and generosity. Contrary to the inclination of persons who look for gaps in the Universe's formational economy (as in the Intelligent Design movement, for example), the *more* robust the Creation's formational economy is, the *more* the Creation owes to the Creator for the richness of its being.

5. In the spirit of this perspective, a person would be inclined to have high expectations regarding the wealth of formational capabilities that contribute to its formational economy. In fact, one would expect it to be complete—gapless—lacking no capabilities that would be needed in order to bring about the formation of any cosmic structure or life form in the course of time.

In other words, the *fully-gifted Creation perspective* provided a way to see the RFEP in a positive theological light. The job of the Creator was not to perform occasional form-imposing interventions to compensate for gaps (missing capabilities) in the Universe's formational economy, but rather it was to give the Creation a robust formational economy of precisely the sort that would accomplish the Creator's intentions. Although supernatural divine action was not categorically ruled out as something that was *impossible,* it was nonetheless seen as *unnecessary* for the particular purpose of actualizing new structures and forms in the course of time. As a symbol of the Creator's respect for what was given being in the first place, the Creator *voluntarily* chose to refrain from overpowering it with coercive form-imposing interventions.

Persons who, for whatever reason, choose to hold to the possibility that God is able and, on occasion, willing to perform supernatural miracles can, I believe, hold to this fully-gifted Creation perspective with integrity. Positing God's voluntary choice to refrain from form-imposing intervention as the means for actualizing new structures and forms over time does not in any way preclude the idea that God could nonetheless choose to act supernaturally in other circumstances and for other reasons.[10] The form of naturalism here maintained is both *partial* (it applies only to the actualization of new structures and life forms) and *voluntary* (freely chosen by God, not constrained by any other being or relationship). As such, *kenotic theism* is consistent with the mainstream of the Judeo-Christian heritage and could, I believe, serve well as an approach for members of the traditional Christian community who also maintain a high respect for the empirical sciences and the idea that the formational history of the Universe is evolutionary in nature.

Naturalistic theism and process theology

For most of my career I was comfortable with the fully-gifted Creation perspective and I encouraged the Christian community to give it a sympathetic hearing. At the same time, I often felt as if I were standing on a narrow ledge at the perimeter of traditional Christian supernaturalism. Standing on that ledge, I found supernatural intervention to be *unnecessary* as a means of forming new creatures, but I was not yet prepared, it seems, to exclude supernaturalism categorically. This is where I found process theologian David Ray Griffin's critical engagement of my work especially helpful.[11] To summarize it as succinctly as I can, these are the two principal challenges that he offered for my consideration:

1. *Dare to be consistent in regard to supernatural divine action.* If supernatural action was unnecessary for something as astounding as the formational history of the entire Universe, then why hold to the need, or even to the possibility, for occasional episodes of coercive supernatural action in any other arena? Griffin's concept of variable and effective, but non-coercive, divine action struck me as an attractive alternative to the traditional concept of supernatural action held by the majority of Christians today.

2. *Naturalism and theism need not be enemies.* Naturalism comes in significantly differing forms that must be carefully distinguished from one another. That recommendation is the basis for the set of distinctions set out earlier in this essay. As there noted, *materialism* does indeed rule out the existence of God and it builds its worldview on the premise that Nature (taken to be no more than a physical/material system) is all there is. Other forms of naturalism, however, require no such denial of God, no categorical rejection

of divine action, and no rejection of the RFEP. *Naturalistic theism,* for example, rejects supernaturalism but then proceeds to develop an enriched concept of natural phenomena by incorporating purposeful and effective but non-coercive divine action as an essential component of all natural processes.

Griffin's two challenges to my earlier strategy have been helpful in my continuing thoughts on the issue of cosmic evolution. I see no reason to reject the RFEP and I find Griffin's theological perspective attractive. That being the case, perhaps I should reformulate the RFEP in a way that explicitly welcomes the contribution of non-coercive divine action to all natural processes. Doing so would lead me to posit the following:

> Revised RFEP: We assume that the Universe possesses all of the right *physical resources,* all of the right *structural and functional potentialities,* and all of the right *formational capabilities* to make possible the forming—by natural processes alone—of every type of physical structure and life form that has ever existed, *without need for occasional episodes of form-imposing divine intervention. At the same time, however, we understand that natural processes and events, while they do preclude any form of coercive divine intervention, may nonetheless include non-coercive divine action as an effective factor.*

In this context, the chief empirical evidence of the non-coercive divine action of which process theology speaks would be the remarkable fruitfulness of the entire evolutionary process and the astounding richness of the World's experience. Why does the Universe satisfy the RFEP? Because God and the World are related in such a way that the World must be free and fully capable of responding to the love of God. Why is the outcome of the Universe's formational history so fruitful as to include self-conscious, rational, morally aware, and morally capable creatures? Because that is God's effective will for the World's experience and the outcome of God's (non-coercive) loving action within the World.

Much more needs to be said to flesh out the perspective of naturalistic theism and of process theology's articulation of it.[12] For the limited purposes of this essay, however, this brief reference to it should be sufficient to illustrate two important points: 1) theism can be fully naturalistic, partially naturalistic, or supernaturalistic, and each version of theism deserves to be considered on its own merits; and 2) the RFEP is not owned by materialism, and arguments that fail to recognize this will only add to the confusion that prevails in much of the naturalism versus supernaturalism debate. Choices among the diverse worldviews considered in this essay will have to be made on criteria far beyond the scope of the RFEP. If the nature of Nature is such that the Universe satisfies the RFEP, and cosmic evolution has proceeded in the manner consistent with scientific (agnostic) naturalism, questions concerning the role and manifestation of divine creativity remain on the table for open and meaningful discussion.

Notes

1. As employed here, the word "economy" should *not* be taken to denote qualities like thrift or parsimony. Compare it instead to such concepts as the global *economy* or the European *economy*—vast systems of resources and capabilities that contribute to the activities of production, commerce, development, and growth.

2. I here confine myself to the Judeo-Christian tradition because it is the religious tradition with which I am most familiar.

3. The word "kenotic" is here employed to convey the idea of voluntary divine restraint, thereby giving up a position of power out of respect for the integral nature of the creature.

4. Daniel C. Dennett, *Darwin's Dangerous Idea: Evolution and the Meanings of Life* (New York: Touchstone, 1995), 75.

5. Phillip E. Johnson, *First Things,* June/July 1993: 38.

6. P. W. Atkins, *Creation Revisited* (New York: W. H. Freeman, 1992), 149.

7. Ibid.

8. Ibid., 3, in the chapter titled "Obvious Things."

9. Ibid.

10. For a major portion of my career, I held to this option in spite of the obvious tension entailed in maintaining a naturalistic perspective in the arena of cosmic evolution while still holding to the possibility of supernatural divine action (as in miracles) in other arenas.

11. See David Ray Griffin, *Religion and Scientific Naturalism: Overcoming the Conflicts* (Albany, NY: SUNY Press, 2000), especially chapter 3.

12. In addition to the work of Griffin cited in the previous endnote, see also his book *Two Great Truths: A New Synthesis of Scientific Naturalism and Christian Faith* (Louisville: Westminster John Knox Press, 2004). The foreword to this book was written by Howard J. Van Till.

25

Living in the Multiverse[1]

Steven Weinberg

1. Introduction

We usually mark advances in the history of science by what we learn about nature, but at certain turning points we have made discoveries about science itself. These discoveries lead to changes in how we score our work, in what we consider to be an acceptable theory.

For an example, look back to a discovery made just one hundred years ago. Before 1905 there had been numerous unsuccessful efforts to detect changes in the speed of light due to the motion of the Earth through the ether. Attempts were made by Fitzgerald, Lorentz, and others to construct a mathematical model of the electron (which was then conceived to be the chief constituent of all matter) that would explain how rulers contract when moving through the ether in just the right way to keep the apparent speed of light unchanged. Einstein instead offered a symmetry principle, which stated that not just the speed of light but all the laws of nature are unaffected by a transformation to a frame of reference in uniform motion. Lorentz grumbled that Einstein was simply assuming what he and others had been trying to prove. But history was on Einstein's side. The 1905 Special Theory of Relativity was the beginning of a general acceptance of symmetry principles as a valid basis for physical theories.

This was how Special Relativity made a change in science itself. From one point of view, Special Relativity was no big thing—it just amounted to the replacement of one ten-parameter spacetime symmetry group, the Galileo group, with another ten-parameter group, the Lorentz group. But never before had a symmetry principle been taken as a legitimate hypothesis on which to base a physical theory.

As usually happens with this sort of revolution, Einstein's advance came with a retreat in another direction: The effort to construct a classical model of the electron was suspended for decades. Instead, symmetry principles increasingly became the dominant foundation for physical theories. This tendency was accelerated after the advent of quantum mechanics in the 1920s, because the survival of symmetry principles in quantum theories imposes highly restrictive consistency conditions (existence of antiparticles, connection between spin and statistics, cancellation of infinities and anomalies) on physically acceptable theories. Our present Standard Model of elementary particle interactions can be regarded as simply the consequence of certain gauge symmetries and the associated quantum mechanical consistency conditions.

The development of the Standard Model did not involve any changes in our conception of what was acceptable as a basis for physical theories. Indeed, the Standard Model can be regarded as just quantum electrodynamics writ large. Similarly, when the effort to extend the Standard Model to include gravity led to widespread interest in string theory, we expected to score the success or failure of this theory in the same way as for the Standard Model: String theory would be a success if its symmetry principles and consistency conditions led to a successful prediction of the free parameters of the Standard Model.

Now we may be at a new turning point, a radical change in what we accept as a legitimate foundation for a physical theory. The current excitement is of course a consequence of the discovery of a vast number of solutions of string theory, beginning in 2000 with the work of Bousso and Polchinski.[2] The compactified six dimensions in Type II string theories typically have a large number (tens or hundreds) of topological fixtures (3-cycles), each of which can be threaded by a variety of fluxes. The logarithm of the number of allowed sets of values of these fluxes is proportional to the number of topological fixtures. Further, for each set of fluxes one obtains a different effective field theory for the modular parameters that describe the compactified 6-manifold, and for each effective field theory the number of local minima of the potential for these parameters is again proportional to the number of topological fixtures. Each local minimum corresponds to the vacuum of a possible stable or metastable universe.

Subsequent work by Giddings, Kachru, Kallosh, Linde, Maloney, Polchinski, Silverstein, Strominger, and Trivedi (in various combinations)[3] established the existence of a large number of vacua with positive energy densities. Ashok and Douglas[4] estimated the number of these vacua to be of order 10^{100} to 10^{500}. String theorists have picked up the term "string landscape" for this multiplicity of solutions from Susskind,[5] who took the term from biochemistry, where the possible choices of orientation of each chemical bond in large molecules leads to a vast number of possible configurations. Unless one can find a reason to reject all but a few of the string theory vacua, we will have to accept that much of what we had hoped to calculate are environmental parameters, like the distance of the Earth from the Sun, whose values we will never be able to deduce from first principles.

We lose some and win some. The larger the number of possible values of physical parameters provided by the string landscape, the more string theory legitimates anthropic reasoning as a new basis for physical theories: Any scientists who study nature must live in a part of the landscape where physical parameters take values suitable for the appearance of life and its evolution into scientists.

An apparently successful example of anthropic reasoning was already at hand by the time the string landscape was discovered. For decades there seemed to be something peculiar about the value of the vacuum energy density ρ_V. Quantum fluctuations in known fields at well-understood energies (say, less than 100 GeV) give a value of ρ_V larger than observationally allowed by a factor of 10^{56}. This contribution to the vacuum energy might be cancelled by quantum fluctuations of higher energy, or by simply including a suitable cosmological constant term in the Einstein field equations, but the cancellation would have to be exact to fifty-six decimal places. No symmetry argument or adjustment mechanism could be found that would explain such a cancellation. Even if such an explanation could be found, there would be no reason to suppose that the remaining net vacuum energy would be comparable to the *present* value of the matter density, and since it is certainly not very much larger, it was natural to suppose that it is very much less, too small to be detected.

On the other hand, if ρ_V takes a broad range of values in the multiverse, then it is natural for scientists to find themselves in a subuniverse in which ρ_V takes a value suitable for the appearance of scientists. I pointed out in 1987 that this value for ρ_V can't be too large and positive, because then galaxies and stars would not form.[6] Roughly, this limit is that ρ_V should be less than the mass density of the universe at the time when galaxies first condense. Since this was in the past, when the mass density was larger than at present, the anthropic upper limit on the vacuum energy density is larger than the present mass density, but not many orders of magnitude greater.

But anthropic arguments provide not just a bound on ρ_V, they give us some idea of the value to be expected: ρ_V should be not very different from the mean of the values suitable for life. This is what Vilenkin calls the "principle of mediocrity."[7] This mean is positive, because if ρ_V were negative, it would have to be less in absolute value than the mass density of the universe during the whole time that life evolves (since otherwise the universe would collapse before any astronomers come on the scene),[8] while if ρ_V were positive, it would only have to be less than the mass density of the universe at the time when most galaxies form, giving a much broader range of possible positive than negative values. In 1997–98, Martel, Shapiro, and I carried out a detailed calculation of the probability distribution of values of ρ_V seen by astronomers throughout the multiverse,[9] under the assumption that the *a priori* probability distribution is flat in the relatively very narrow range that is anthropically allowed.[10] At that time, the value of the primordial root-mean-square (rms) fractional density fluctuation σ was not well known, since the value inferred from observations of the cosmic microwave background depended on what one assumed for ρ_V. It was therefore not possible to calculate a mean expected value of ρ_V, but for any assumed value of ρ_V we could estimate σ and use the result to calculate the fraction of astronomers that would observe a value of ρ_V as small as the assumed value. In this way, we concluded that if Ω_Λ (the dimensionless density parameter associated with ρ_V) turned out to be much less than 0.6, anthropic reasoning could not explain why it was so small. The editor of the *Astrophysical Journal* objected to publishing papers about anthropic calculations, and we had to sell our article by pointing out that we had provided a strong argument for abandoning an anthropic explanation of a small value of ρ_V, if it turned out to be too small.

Of course, it turned out that ρ_V is not too small. Soon after this work, observations of type Ia supernovae revealed that the expansion of the universe is accelerating[11] and gave the result that $\Omega_V \approx 0.7$. In other words, the ratio of the vacuum energy density to the present mass density ρ_{M0} in *our* subuniverse (which I use just as a convenient measure of density) is about 2.3, a conclusion subsequently confirmed by observations of the microwave background.[12]

This is still a bit low. Martel, Shapiro, and I had found that the probability of a vacuum energy density this small was 12 percent. I have now recalculated the probability distribution, using WMAP data and a better transfer function, with the result that the probability of a random astronomer seeing a value as small as $2.3\rho_{M0}$ is increased to 15.6 percent.[13] Now that we know σ, we can also calculate that the median vacuum energy density is $13.3\rho_{M0}$.

I should mention a complication in these calculations. The average of the product of density fluctuations at different points becomes infinite as these points approach each other, so the rms fractional density fluctuation σ is actually infinite. Fortunately, it is not σ itself that is really needed in these calculations, but the rms fractional density fluctuation averaged over a sphere of comoving radius R taken large enough so that the density fluctuation is able to hold on efficiently to the heavy elements produced in the first generation of stars. The results mentioned above were calculated for R (projected to the present) equal to 2 Mpc. These results are rather sensitive to the value of R; for $R = 1$ Mpc, the probability of finding a vacuum energy as small as $2.3\rho_{M0}$ is only 7.2 percent. The estimate of the required value of R involves complicated astrophysics and needs to be better understood.

2. Problems

Now I want to take up four problems we have to face in working out the anthropic implications of the string landscape.

What is the shape of the string landscape?

Douglas[14] and Dine and coworkers[15] have taken the first steps in finding the statistical rules governing different string vacua. I can't comment usefully on this, except to say that it wouldn't hurt in this work if we knew what string theory is.

What constants scan?

Anthropic reasoning makes sense for a given constant if the range over which the constant varies in the landscape is large compared with the anthropically allowed range of values of the constant, for then it is reasonable to assume that the *a priori* probability distribution is flat in the anthropically allowed range. We need to know what constants actually "scan" in this sense. Physicists would like to be able to calculate as much as possible, so we hope that not too many constants scan.

The most optimistic hypothesis is that the only constants that scan are the few whose dimensionality is a positive power of mass: the vacuum energy, and whatever scalar mass or masses set the scale of electroweak symmetry breaking. With all other parameters of the Standard Model fixed, the scale of electroweak symmetry breaking is bounded by about 1.4 to 2.7 times its value in our subuniverse, by the condition that the pion mass should be small enough to make the nuclear force strong enough to keep the deuteron stable against fission.[16] (The condition that the deuteron be stable against beta decay, which yields a tighter bound, does not seem to me to be necessary. Even a beta-unstable deuteron would live long enough to allow cosmological helium synthesis; helium would be burned to heavy elements in the first generation of very massive stars; and then subsequent generations could have long lifetimes burning hydrogen through the carbon cycle.) But the mere fact that the electroweak symmetry breaking scale is only a few orders of magnitude larger than the QCD scale should not in itself lead us to conclude that it must be anthropically fixed. There is always the possibility that the electroweak symmetry breaking scale is determined by the energy at which some gauge coupling constant becomes strong, and if that coupling happens to grow with decreasing energy a little faster than the QCD coupling then the electroweak breaking scale will naturally be a few orders of magnitude larger than the QCD scale.

If the electroweak symmetry breaking scale is anthropically fixed, then we can give up the decades-long search for a natural solution of the hierarchy problem. This is a very attractive prospect, because none of the "natural" solutions that have been proposed, such as technicolor or low-energy supersymmetry, were ever free of difficulties. In particular, giving up low-energy supersymmetry can restore some of the most attractive features of the non-supersymmetric standard model: automatic conservation of baryon and lepton number in interactions up to dimension 5 and 4, respectively; natural conservation of flavors in neutral currents; and a small neutron electric dipole moment. Arkani-Hamed and Dimopoulos[17] and others[18] have even shown how it is possible to keep the good features of supersymmetry, such as a more accurate convergence of the $SU(3) \times SU(2) \times U(1)$ couplings to a single value, and the presence of candidates for dark matter, WIMPs. The idea of this "split supersymmetry" is that, although supersymmetry is broken at some very high energy, the gauginos and higgsinos are kept light by a chiral symmetry. (An additional discrete symmetry is needed to prevent lepton-number violation in higgsino-lepton mixing, and to keep the lightest supersymmetric particle stable.) One of the nice things about split supersymmetry is that, unlike many of the things we talk about these days, it makes predictions that can be checked when the LHC starts operation. One expects a single neutral Higgs with a mass in the range 120 to 165 GeV, possible winos and binos but no squarks or sleptons, and a long-lived gluino. (Incidentally, a Stanford group[19] has recently used considerations of Big Bang nucleosynthesis to argue that a 1 TeV gluino must have a lifetime less than 100 seconds, indicating a supersymmetry breaking scale less than 10^{10} GeV, which might create problems for proton stability. But I wonder whether, even if the gluino has a longer lifetime and decays after nucleosynthesis, the universe might not thereby be reheated above the temperature of helium dissociation, giving Big Bang nucleosynthesis a second chance to produce the observed helium abundance.)

What about the dimensionless Yukawa couplings of the Standard Model? If these couplings are very tightly constrained anthropically, then we might reasonably suspect that they take a wide range of values in the multiverse, so that anthropic considerations can have a chance to affect the values we observe. Hogan has analyzed the anthropic constraints on these couplings,[20] with the electroweak symmetry breaking scale and the sum of the u and d Yukawa couplings held fixed, to avoid complications due to the dependence of nuclear forces on the pion mass. He imposes the following conditions: (1) $m_d - m_u - m_e > 1.2$ MeV, so that the early universe doesn't become all neutrons; (2) $m_d - m_u + m_e < 3.4$ MeV, so that the pp reaction is exothermic; and (3) $m_e > 0$. With three conditions on the two parameters $m_u - m_d$ and m_e, he naturally finds these parameters are limited to a finite region, which turns out to be quite small. At first sight, this gives the impression that the quark and lepton Yukawa couplings are subject to stringent anthropic constraints, in which case we might infer that the Yukawa couplings probably scan.

I have two reservations about this conclusion. The first reservation is that the pp reaction is not necessary for life. For one thing, the pep reaction $p + p + e^- \rightarrow d + \nu$ can keep stars burning hydrogen for a long time. For this, we do not need $m_d - m_u + m_e < 3.4$ MeV, but only the weaker condition $m_d - m_u - m_e < 3.4$ MeV. The three conditions then do not constrain $m_d - m_u$ and m_e separately to any finite region, but only constrain the single parameter $m_d - m_u - m_e$ to lie between 1.2 MeV and 3.4 MeV, not a very tight anthropic constraint. (In fact, He^4 will be stable as long as $m_d - m_u - m_e$ is less than about 13 MeV, so stellar nucleosynthesis can begin with helium burning in the heavy stars of Population III, followed by hydrogen burning in later generations of stars.) My second reservation is that the anthropic constraints on the Yukawa couplings are alleviated if we suppose (as discussed above) that the electroweak symmetry breaking scale is not fixed, but free to take whatever value is anthropically necessary. For instance, according to the results of Agrawal et al.,[21] the deuteron binding energy could be made as large as about 3.5 MeV by taking the electroweak breaking scale to be much less than it is in our universe, in which case even the condition that the pp reaction be exothermic becomes much looser.

Incidentally, I don't set much store by the famous "coincidence" emphasized by Hoyle, that there is an excited state of C^{12} with just the right energy to allow carbon production via α–Be^8 reactions in stars. We know that even–even nuclei have states that are well described as composites of α-particles. One such state is the ground state of Be^8, which is unstable against fission into two α-particles. The same α–α potential that produces that sort of unstable state in Be^8 could naturally be expected to produce an unstable state in C^{12} that is essentially a composite of three α-particles, and that therefore appears as a low-energy resonance in α–Be^8 reactions. So the existence of this state doesn't seem to me to provide any evidence of fine-tuning.

What else scans? Tegmark and Rees have raised the question whether the rms density fluctuation σ may itself scan.[22] If it does, then the anthropic constraint on the vacuum energy becomes weaker, resuscitating to some extent the problem of why ρ_V is so small. But Garriga and Vilenkin have pointed out that it is really ρ_V/σ^3 that is constrained anthropically,[23] so that even if σ does scan, the anthropic prediction of this ratio remains robust.

Arkani-Hamed, Dimopoulos, and Kachru,[24] referred to below as ADK, have offered a possible reason to suppose that most constants do not scan. If there are a large number N of decoupled modular fields, each taking a few possible values, then the probability distribution of quantities that depend on all these fields will be sharply peaked, with a width proportional to $1/\sqrt{N}$. According to Distler and Varadarajan,[25] it is not really necessary here to make arbitrary assumptions about the decoupling of the various scalar fields; it is enough to adopt the most general polynomial superpotential that is stable, in the sense that radiative corrections do not change the effective couplings for large N by amounts larger than the couplings themselves. Distler and Varadarajan emphasize cubic superpotentials, because polynomial superpotentials of order higher than cubic presumably make no physical sense. But it is not clear that even cubic superpotentials can be plausible approximations, or that peaks will occur at reasonable values in the distribution of dimensionless couplings rather than of some combinations of these couplings.[26] It also is not clear that the multiplicity of vacua in this kind of effective scalar field theory can properly represent the multiplicity of flux values in string theories,[27] but even if not, it presumably can represent the variety of minima of the potential for a given set of flux vacua.

If most constants do not effectively scan, then why should anthropic arguments work for the vacuum energy and the electroweak breaking scale? ADK point out that, even if some constant has a relatively narrow distribution, anthropic arguments will still apply if the anthropically allowed range is even narrower and near a point around which the distribution is symmetric. (ADK suppose that this point would be at zero, but this is not necessary.) This is the case, for instance, for the vacuum energy if the superpotential W is the sum of the superpotentials W_n for a large number of decoupled scalar fields, for each of which there is a separate broken R symmetry, so that the possible values of each W_n are equal and opposite. The probability distribution of the total superpotential $W = \sum_{n=1}^{N} W_n$ will then be a Gaussian peaked at $W = 0$ with a width proportional to $1/\sqrt{N}$, and the probability distribution of the supersymmetric vacuum energy $-8\pi G|W|^2$ will extend over a correspondingly narrow range of negative values, with a maximum at zero. When supersymmetry-breaking is taken into account, the probability distribution widens to include positive values of the vacuum energy, extending out to a positive value depending on the scale of supersymmetry-breaking. For any reasonable supersymmetry-breaking scale, this probability distribution, though narrow compared with the Planck scale, will be very wide compared with the very narrow anthropically allowed range around $\rho_V = 0$, so within this range, the probability distribution can be expected to be flat, and anthropic arguments should work. Similar remarks apply to the μ-term of the supersymmetric Standard Model, which sets the scale of electroweak symmetry breaking.

How should we calculate anthropically conditioned probabilities?

We would expect the anthropically conditioned probability distribution for a given value of any constant that scans to be proportional to the number of scientific civilizations that observe that value. In the calculations described above, Martel, Shapiro, and I took this number to be proportional to the fraction of baryons that find themselves in galaxies, but what if the total number of baryons itself scans? What if it is infinite?

How is the landscape populated?

There are at least four ways in which we might imagine the different "universes" described by the string landscape actually to exist:

(i) The various subuniverses may be simply different regions of space. This is most simply realized in the chaotic inflation theory.[28] The scalar fields in different inflating patches may take different values, giving rise to different values for various effective coupling constants. Indeed, Linde speculated about the application of the Big Bang anthropic principle to cosmology soon after the proposal of chaotic inflation.[29]

(ii) The subuniverses may be different eras of time in a single Big Bang. For instance, what appear to be constants of nature might actually depend on scalar fields that change very slowly as the universe expands.[30]

(iii) The subuniverses may be different regions of spacetime. This can happen if, instead of changing smoothly with time, various scalar fields on which the "constants" of nature depend change in a sequence of first-order phase transitions.[31] In these transitions metastable bubbles form within a region of higher vacuum energy; then within each bubble there form further bubbles of even lower vacuum energy; and so on. In recent years this idea has been revived in the context of the string landscape.[32] In particular, it has been suggested that in this scenario the curvature of our universe is small for anthropic reasons, and hence possibly large enough to be detected.[33]

(iv) The subuniverses could be different parts of quantum mechanical Hilbert space. In a reinterpretation of Hawking's earlier work on the wave function of the universe,[34] Coleman showed that certain topological fixtures known as wormholes in the path integral for the Euclidean wave function of the universe would lead to a superposition of wave functions in which any coupling constant not constrained by symmetry principles would take any possible value.[35] Ooguri, Vafa, and Verlinde have argued for a particular wave function of the universe,[36] but it escapes me how anyone can tell whether this or any other proposed wave function is *the* wave function of the universe.

These alternatives are by no means mutually exclusive. In particular, it seems to me that, whatever one concludes about the first three alternatives, we will still have the possibility that the wave function of the universe is a superposition of different terms representing different ways of populating the landscape in space and/or time.

3. Conclusion

In closing, I would like to comment about the impact of anthropic reasoning within and beyond the physics community. Some physicists have expressed a strong distaste for anthropic arguments. (I have heard David Gross say "I hate it.") This is understandable. Theories based on anthropic calculation certainly represent a retreat from what we had hoped for: the calculation of all fundamental parameters from first principles. It is too soon to give up on this hope, but without loving it, we may just have to resign ourselves to a retreat, just as Newton had to give up Kepler's hope of a calculation of the relative sizes of planetary orbits from first principles.

There is also a less creditable reason for hostility to the idea of a multiverse, based on the fact that we will never be able to observe any subuniverses except our own. Livio and Rees, and Tegmark have given thorough discussions of various other ingredients of accepted theories that we will never be able to observe, without our being led to reject these theories.[37] The test of a physical theory is not that everything in it should be observable and every prediction it makes should be testable, but rather that enough is observable and enough predictions are testable to give us confidence that the theory is right.

Finally, I have heard the objection that, in trying to explain why the laws of nature are so well suited for the appearance and evolution of life, anthropic arguments take on some of the flavor of religion. I think that just the opposite is the case. Just as Darwin and Wallace explained how the wonderful adaptations of living forms could arise without supernatural intervention, so the string landscape may explain how the constants of nature that we observe can take values suitable for life without being fine-tuned by a benevolent creator. I found this parallel well understood in a surprising place, a *New York Times* article by Christoph Schönborn, Cardinal Archbishop of Vienna.[38] His article concludes as follows:

> Now, at the beginning of the twenty-first century, faced with scientific claims like neo-Darwinism and the multiverse hypothesis in cosmology invented to avoid the overwhelming evidence for purpose and design found in modern science, the Catholic Church will again defend human nature by proclaiming that the immanent design evident in nature is real. Scientific theories that try to explain away the appearance of design as the result of "chance and necessity" are not scientific at all, but, as John Paul put it, an abdication of human intelligence.

It's nice to see work in cosmology get some of the attention given these days to evolution, but of course it is not religious preconceptions like these that can decide any issues in science.

It must be acknowledged that there is a big difference in the degree of confidence we can have in neo-Darwinism and in the multiverse. It is settled, as well as anything in science is ever settled, that the adaptations of living things on Earth have come into being through natural selection acting on random undirected inheritable variations. About the multiverse, it is appropriate to keep an open mind, and opinions among scientists differ widely. In the Austin airport on the way to this meeting I noticed for sale the October issue of a magazine called

Astronomy, having on the cover the headline "Why You Live in Multiple Universes." Inside I found a report of a discussion at a conference at Stanford, at which Martin Rees said that he was sufficiently confident about the multiverse to bet his dog's life on it, while Andrei Linde said he would bet his own life. As for me, I have just enough confidence about the multiverse to bet the lives of both Andrei Linde *and* Martin Rees's dog.[39]

Notes

1. Reprinted with permission from *Universe or Multiverse?* ed. Bernard Carr (Cambridge: Cambridge University Press, 2007), 29–42.
2. R. Bousso and J. Polchinski (2000) *Journal of High Energy Physics* 0006: 006. Lee Smolin had noted earlier that string theory has a large number of vacuum solutions, and explored an imaginative possible consequence of this multiplicity [L. Smolin (1997) *Life of the Cosmos*. New York: Oxford University Press]. Even earlier, in the 1980s, Duff, Nilsson and Pope had noted that $D = 11$ supergravity has an infinite number of possible compactifications, but of course it was not then known that this theory is a version of string theory. For a summary, see M. J. Duff, B. E. W. Nilsson and C. N. Pope (1986) *Physics Reports* 130: 1.
3. S. B. Giddings, S. Kachru and J. Polchinski (2002) *Physical Review* D 66: 106006; A. Strominger, A. Maloney and E. Silverstein (2003), in G. W. Gibbons, E. P. S. Shellard and S. J. Ranken, eds. *The Future of Theoretical Physics and Cosmology*, 570–91; S. Kachru, S. Kalloh, A. D. Linde and S. P. Trivedi (2003) *Physical Review* D 68: 046005.
4. S. K. Ashok and M. Douglas (2004) *Journal of High Energy Physics* 0401: 060.
5. L. Susskind (2003) "The Anthropic Landscape of String Theory" (hep-th/0302219).
6. S. Weinberg (1987) *Physical Review Letters* 59: 2607.
7. A. Vilenkin (1995) *Physical Review Letters* 74: 846.
8. J. D. Barrow and F. J. Tipler (1986) *The Anthropic Cosmological Principle*. Oxford: Clarendon Press.
9. H. Martel, P. Shapiro and S. Weinberg (1998) *Astrophysics Journal* 492: 29.
10. For earlier calculations, see G. Efstathiou (1995) *Monthly Notes of the Royal Astronomical Society* 274: L73 and S. Weinberg (1997), in N. Turok, ed. *Critical Dialogues in Cosmology*. Singapore: World Scientific Publishers.
11. A. G. Reiss, A. V. Filippenko, P. Challis *et al.* (1998) *Astronomical Journal* 116: 1009, and S. Perlmutter, G. Aldering, G. Goldhaber *et al.* (1999) *Astrophysical Journal* 517: 565.
12. D. N. Spergel, L. Verde, H. V. Peiris *et al.* (2003) *Astrophysical Journal Supplement* 148: 175.
13. This situation has improved since the release of the second and third year WMAP results. Assuming flat space, the ratio of the vacuum energy density to the matter density is now found to be about 3.2 rather than 2.3.
14. M. R. Douglas (2004) *Comptes Rendus Physique* 5: 965 (hep-ph/0401004).
15. M. Dine, D. O'Neil and Z. Sun (2005) *Journal of High Energy Physics* 0507: 014, and M. Dine and Z. Sun (2006) *Journal of High Energy Physics* 0601: 129 (hep-th/0506246).
16. V. Agrawal, S. M. Barr, J. F. Donoghue and D. Seckel (1998) *Physical Review* D 57: 5480.
17. N. Arkani-Hamed and S. Dimopoulos (2005) *Journal of High Energy Physics* 0506: 073.
18. G. F. Giudice and A. Romanino (2004) *Nuclear Physics* B 699: 65; N. Arkani-Hamed, S. Dimopoulos, G.F. Giudice and A. Romanino (2005) *Nuclear Physics* B 709: 3; and A. Delgado and G. F. Giudice (2005) *Physics Letters* B 627: 155 (hep-ph/0506217).
19. A. Arvnitaki, C. Davis, P. W. Graham, A. Peirce and J. G. Wacker (2005) *Physical Review* D 72: 075011 (hep-ph/05042100).
20. C. Hogan (2000) *Reviews of Modern Physics* 72: 1149, and C. Hogan (2004) "Quarks, electrons and atoms in closely related universes" (astro-ph/0407086).
21. V. Agrawal, S. M. Barr, J. F. Donoghue and D. Seckel (1998) *Physical Review* D 57: 5480.
22. M. Tegmark and M. J. Rees (1998) *Astrophysical Journal* 499: 526.
23. J. Garriga and A. Vilenkin (2006) *Progress in Theoretical Physics Supplement* 163: 245 (hep-th/0508005).
24. N. Arkani-Hamed, S. Dimopoulos and S. Kachru (2005), (hep-th/0501082).
25. J. Distler and U. Varadarajan (2005), (hep-th/0507090).

26. M. Douglas, private communication.

27. T. Banks (2000), (hep-th/0011255).

28. A. D. Linde (1986) *Physics Letters* B 129: 305; A. Vilenkin (1983), *Physical Review* D 27: 2848; A. D. Linde (1986) *Physics Letters* B 175: 305; A. D. Linde (1987) *Physica Scripta* T 15: 100; A. D. Linde (1988) *Physics Letters* B 202: 194.

29. A. D. Linde (1983), in G. W. Gibbons, S. W. Hawking and S. Siklos, eds., *The Very Early Universe*. Cambridge: Cambridge University Press; A. D. Linde (1984) *Reports on Progress in Physics* 47: 925.

30. T. Banks (1985) *Nuclear Physics* B 249: 332.

31. L. Abbott (1985) *Physics Letters* B 150: 427; J. D. Brown and C. Teitelboim (1987) *Physisc Letters* B 195: 177; and J. D. Brown and C. Teitelboim (1987) *Nucelar Physics* B 297: 787.

32. J. L. Feng, J. March-Russel, S. Sethi and F. Wilczek (2001) *Nuclear Physics* B 602: 307; and H. Firouzjahi, S. Sarangji and S.-H. Tye (2004) *Journal of High Energy Physics* 0409: 060.

33. B. Freivogel, M. Kleban, M. R. Martinez, and L. Suskind (2006) *Journal of High Energy Physics* 3: 39 (hepth/0505232).

34. S. W. Hawking (1984) *Nuclear Physics* B 239: 257 and S. W. Hawking (1984), in B. S. DeWitt and R. Stora, eds. *Relativity, Groups, and Topology II*, NATO Advanced Study Institute Session XL, Les Houches, 1983. Amsterdam: Elsevier, 336. Some of this work is based on an initial condition for the origin of the universe proposed by Hartle and Hawking in J. Hartle and S. W. Hawking (1983) *Physical Review* D 28: 2960.

35. S. Coleman (1988) *Nuclear Physics* B 307: 867. It has been argued by Hawking and others that the wavefunction of the universe is sharply peaked at values of the constants that yield a zero vacuum energy at late times [see S. W. Hawking (1985), in R. Jackiw et al., eds. *Shelter Island II—Proceedings of the 1983 Shelter Island Conference on Quantum Field Theory and the Fundamental Problems of Physics*. Cambridge: MIT Press; S. W. Hawking (1984) *Physics Letters* B 134: 403; E. Baum (1984) *Physics Letters* B 133: 185; and S. Coleman (1985) *Nuclear Physics* B 310: 643]. This view has been challenged in W. Fischler, I. Klebanov, J. Polchinski and L. Susskind (1989) *Nuclear Physics* B 237: 157. I am assuming here that there are no such peaks.

36. H. Ooguri, C. Vafa and E. Verlinde (2005) *Letters in Mathematical Physics* 74: 311 (hep- th/0502211).

37. M. Livio and M. J. Rees (2003) *Science* 309: 1022, and M. Tegmark (1998) *Annalen der Physik* 270: 1.

38. C. Schönborn (2005) *New York Times*, 7 July 2005, A23.

39. This material is based upon work supported by the National Science Foundation under Grants Nos. PHY-0071512 and PHY-0455649 and with support from The Robert A. Welch Foundation, Grant No. F-0014, and also grant support from the US Navy, Office of Naval Research, Grant nos. N00014–03–1–0639 and N00014–04–1–0336, Quantum Optics Initiative.`

26

BALLOONS ON A STRING:
A CRITIQUE OF MULTIVERSE COSMOLOGY[1]

BRUCE L. GORDON

We can have some confidence in the story of the evolution of the universe from the time of electron-positron annihilation to the present. . . . About earlier times, so far we can only speculate.
—Steven Weinberg (2008)

If a philosopher, deep within his study, should try to move matter, he can do with it what he wishes: nothing resists him. This is because the imagination sees whatever it wishes to, and sees nothing more. But such arbitrary hypotheses throw light on no verity; on the contrary, they retard the progress of science and become most dangerous through the errors they lead us to adopt.
—Étienne Bonnot de Condillac (1715–80)

Is the ultimate explanatory principle of the universe to be found in matter or mind? Perhaps no topic moves to the heart of this question more quickly than that of the origin of the universe and the fine-tuning of many of its physical parameters for the existence of life.[1] It is obvious that any answer to this question will be influenced by philosophical assumptions about the nature of reality, science, and legitimate explanatory principles. In this respect, rather than asking whether methodological naturalism is a necessary or desirable constraint on science,[2] I will simply argue that an *adequate* explanation of the origin of the universe and its properties cannot be had if the constraints of methodological naturalism are retained. So reserve the designation "science" as an honorific for whatever you wish, and persist in maintaining its heuristics are methodologically naturalistic if you are persuaded you must, the fact remains that the word "science" has *never* been coterminous with all that is true—indeed, historically it has encompassed a good deal that we now recognize to be false. Having set this issue aside, therefore, we are left to consider the work of various cosmologists who are about the business of fabricating a purely naturalistic explanation of cosmological origins and fine-tuning. I will argue that, by any reasonable standard of assessment, they are not succeeding in this effort, and that this lack of success is not in the least surprising, because it has a principled basis.[3]

Since we must concern ourselves with naturalistic cosmological models that claim resources sufficient to explain not just the origin, but *also* the fine-tuning of the laws and constants of nature, our focus here will be on the explanatory adequacy of various concatenations of quantum cosmology, inflationary cosmology, and the embarrassment of riches constituted by the solutions (vacua) of string theory. Our examination of universal origins and fine-tuning will begin with a discussion of inflationary scenarios grafted onto Big Bang cosmology and the proof that all inflationary spacetimes are past-incomplete. After diverting into a lengthy critical examination of the "different physics" offered by quantum cosmologists at the past-boundary of the universe, we will proceed to dissect the inadequacies of inflationary explanations and string-theoretic constructs in the context of three cosmological models that have received much attention: the Steinhardt-Turok cyclic ekpyrotic model (which does not invoke inflation), the Gasperini-Veneziano pre-Big Bang inflationary model, and the inflationary string landscape model advanced by Susskind, Polchinski, Bousso, and Linde. We will argue that none of these highly speculative string cosmologies removes the necessity of a beginning to the process of universe generation, and we will emphasize the implications of this fact. Then, since the inflationary "mechanism" only really addresses the fine-tuning of the *initial* conditions of the universe and *not* the conditions embodied in its finely tuned laws and constants, we will analyze the adequacy of the string multiverse in its three versions (cyclic ekpyrotic, pre-Big Bang, and landscape) for explaining the nomological structure and values of these precisely tuned life-compatible universal parameters. When all is said and done, it will be clear that transcendent intelligent agency is not just the *only causally sufficient* and therefore *metaphysically sound* explanation for universal origins and fine-tuning, but it is also much more parsimonious, elegant, and resonant with meaning than all of the ad hoc machinations of multiverse cosmology.[4]

1. Universal Origins and Inflationary Cosmology

As is common knowledge, the "Big Bang theory" of the origin of the universe was widely accepted on the basis of its theoretical description in general relativity and the Hawking-Penrose singularity theorems,[5] as well as its empirical confirmation in the discovery of Hubble expansion, the cosmic background radiation permeating the universe, and the massive energies required for the nucleosynthesis of deuterium, helium-3, helium-4 and lithium-7, which exceed even those found on the interior of stars. The wide acceptance of the theory did not relieve the unease with which many cosmologists received it, however. Fred Hoyle states the reason for this unease about the universe's absolute beginning rather bluntly:

> Many people are happy to accept this position. . . . The abrupt beginning is regarded as *meta*physical—i.e., *outside* physics. The physical laws are therefore considered to break down at $\tau = 0$, *and to do so inherently.* To many people this thought process seems highly satisfactory because a "something" outside physics can then be introduced at $\tau = 0$. By a semantic manoeuvre, the word "something" is then replaced by "god," except that the first letter becomes a capital, God, in order to warn us that we must not carry the inquiry any further.[6]

As William Craig remarks,[7] it seems clear that it was Hoyle's desire to avoid the potential intrusion of theism that led him to defend steady-state models well beyond the bounds of plausibility. This motivation is alive and well in cosmology today: it galvanized the (now failed) hope that inflationary processes might be regarded as past-eternal,[8] and it infuses life into fantastical mathematical constructions involving universal quantum-gravitational wavefunctions quantum-tunneling from imaginary time, eternally oscillating 3-brane collisions, everlastingly ancient string perturbative vacua, and infinitely many bubble universes with different initial conditions and countless variations of laws and constants. Since much of this recent speculation rests on the postulation of cosmic inflation as a solution to the horizon and flatness problems,[9] we begin with a consideration of how inflationary cosmology got started.

Inflationary Cosmology

One aspect of the homogeneity and isotropy of our universe is the uniformity of the cosmic microwave background radiation (CMB), which has the same temperature throughout the observable cosmos to within one part in a hundred thousand. This is regarded as a puzzle in standard Big Bang cosmology because until about 300,000 years after the Big Bang, the photons in the CMB would have been interacting with electrons in the hot plasma that filled the entire universe as it expanded. At about 300,000 years, the universe cooled enough for electrically neutral atoms to form and release the background radiation to travel unfettered, thus giving us a picture of the universe at this early stage. But the uniformity of this radiation, which has the same temperature in every direction to within a small fraction of a degree, requires the aboriginal plasma itself to be extraordinarily uniform. This in turn would require *very* precise initial conditions, since calculations in standard Big Bang cosmology tell us that radiation arriving from opposite directions in the sky at that time would have been separated by about 100 horizon distances, that is, by 100 times the distance light could have traveled since the beginning of the universe. This thermal equilibrium can only be explained in standard Big Bang cosmology by postulating an initial state of almost perfect uniformity.

Another aspect of the uniformity of the universe is its flatness. Homogeneous universes are called *flat* if they are on the borderline between eventual gravitational collapse and eternal expansion; in such a case, their geometry is precisely Euclidean. If the *actual* mass density in the universe is very close to the *critical* mass density required to gravitationally halt the expansion, that is, if their ratio is close to one, the expansion rate of the universe will asymptotically approach zero. WMAP (Wilkinson Microwave Anisotropy Probe) data reveal our universe to have an actual mass density that is extremely close to the critical mass density, with indications of an ever-so-slight positive curvature that would imply a geometrically closed universe, but other recent observations show there to be an exceedingly small *positive* cosmological constant. As a result of this positive cosmological constant our universe's expansion is accelerating, which would seem to suggest that the universe will continue to expand without gravitational collapse.[10] Nobody knows for sure. But because actual and critical mass densities are so precisely balanced, space itself has hardly *any* overall curvature. The precisely balanced character of these quantities is again surprising from the standpoint of standard Big Bang cosmology, since it also requires very precisely tuned initial conditions.

Inflationary cosmology tries to alleviate this puzzlement by proposing that the horizon and flatness problems are resolved if the observable universe underwent an exponentially rapid rate of expansion in the first fraction of a second after the Big Bang; this rapid inflation then halted abruptly and universal growth settled down to the less frenetic pace we observe today. In current models, inflation is hypothesized to have begun around 10^{-37} seconds (or so) after the Big Bang and lasted until 10^{-35} seconds (or so), during which time the space constitutive of our observable universe expanded by a factor of 10^{60} (or so). At the beginning of the inflationary epoch, the observable universe was, say, about 10^{-60} meters in size, and at the end of it, therefore, about a meter across. In this scenario, the horizon size at the start of inflation would have been 10^{-37} light-seconds, which is far *larger* than the tiny patch postulated to grow into our observable universe. There was thus plenty of time *before* inflation started for the aboriginal observable universe to thermalize uniformly, whence the inflationary process stretched this homogeneous region immensely, and the patch constituting our visible universe then continued to expand more sedately out of this inflated volume. Any residual inhomogeneities, if they exist, would therefore lie beyond the bounds of what we can see.

In respect of the flatness problem, given the assumption that the universe began as a de Sitter space that then transitioned (somehow) to a Friedmann-Lemaître-Robertson-Walker (FLRW) metric, the effect is thought to be similar. During the inflationary epoch, all the distances in the region that became our observable universe increased by a measure of 10^{60} or so, which means the radius of the observable universe increased by this factor as well. To illustrate the effect of this, suppose four-dimensional space-time prior to inflation had positive curvature, like the surface of a balloon does in three dimensions, and that its radius was a billionth of a meter (a nanometer). After inflation, its radius would be 10^{51} meters, or about 10 billion trillion trillion light years (the radius of the observable universe is on the order of 13.7 billion light years). Just as inflating a balloon to larger and larger sizes makes a small patch on its surface look flatter, so inflating the entire universe makes the patch we can see look quite flat. This, at least, is how the inflationary explanation is intended to work; as we shall see momentarily, whether it realizes the intentions of its inventors is questionable.

According to the currently dominant "chaotic eternal inflationary model,"[11] the rapid expansion of the early universe was driven (as in all inflationary models) by the false vacuum of a hypothesized scalar field called an *inflaton field*, which represents the effect of a massive *repulsive* gravitational force. After an initial phase of expansion, this field is assumed to have decayed *locally* to produce our universe. In the chaotic scenario, however, it is necessary to suppose that the inflaton field starts fairly high up in a range of energies (the "energy landscape") having *no* upper bound, and so (by quantum-mechanical description) the field continues to oscillate chaotically outside the local area for a time much greater than the inflationary doubling time. This assumption entails that inflating regions multiply faster than they decay, with the consequence that inflation continues eternally into the future and produces a boundless expansion of space into which other universes are birthed as the ever-expanding inflaton field decays at other locations. So it is that current inflationary cosmologists postulate the decay of the inflaton field as a "mechanism" by which a potentially infinite number of "bubble universes" can be created. Since the chaotic inflaton field continues to expand at a rate vastly greater than the bubble universes growing within it, none of these bubbles will ever encounter each other, so we can never empirically confirm the existence of any bubble universe save our own. The hypothesized aboriginal inflaton field therefore gives birth to endless bubble universes, a scenario Alexander Vilenkin picturesquely describes as "many worlds in one."[12]

THE BGV THEOREM AND ITS SIGNIFICANCE

One of the hopes expressed by Andrei Linde and other advocates of chaotic eternal inflation was that it could be conceived as eternal into the past as well as the future, thus obviating the implications of an absolute beginning to the universe. Results indicating the falsity of this hope emerged in the mid-1990s and were established beyond reasonable doubt in 2003. The earliest theorem demonstrating that inflationary models are past-incomplete, which depended on a weak-energy condition that allowed for exceptions,[13] has now been established instead by an argument that needs no energy condition.[14]

This stronger proof of geodesic past-incompleteness, put forward in 2003 by Arvind Borde, Alan Guth and Alexander Vilenkin (henceforth BGV), considers spacetimes satisfying the condition that the *average* Hubble expansion in the past is greater than zero, i.e., $H_{av} > 0$. It is shown that a suitable construction of the Hubble parameter, H, allows it to be defined for arbitrary (inhomogeneous, anisotropic) cosmological models in a way that reduces to its standard definition in simple models. With this generalized Hubble parameter in hand, a demonstration is given that its integral along any null or timelike geodesic is bounded, so that any backward-going null or timelike geodesic with $H_{av} > 0$ must have finite length, i.e., be past-incomplete. The class of cosmologies satisfying the assumption that the Hubble parameter has a positive value when averaged over the affine parameter of a past-directed null or non-comoving timelike geodesic also includes cosmologies of *higher* dimensions, which is why the BGV result is applicable to Steinhardt's and Turok's cyclic ekpyrotic string cosmology as well as the inflationary string landscape model. We will examine string cosmology in due course, but, for now, the importance of the BGV result is its demonstration that all inflationary spacetimes (or merely those expanding on average) have a beginning in the finite past, at which point some different kind of physics allegedly applies.[15] This "different kind of physics" is usually taken to be a universal nucleation event via some kind of quantum cosmological construct[16] that mitigates the breakdown of physical theory that classical general relativity requires at the Big Bang singularity.

Apart from theoretical and observational considerations on expanding spacetimes that imply an absolute universal origin, other arguments against an actual temporal infinity leading up to the present are exceedingly strong: given the reality of temporal progression, if reaching the present required traversing an infinite temporal past, the present would never have been reached; but the present (obviously) has been reached, therefore the temporal past is *not* infinite. Considered in terms of historical events or instants of time, an actual infinity is metaphysically nonsensical and incapable of coherent positive construction—it would require, for instance, that the number of events in a proper subset of universal history could be set in one-to-one correspondence with that history as a whole, an ontological state of affairs that is internally contradictory. David Hilbert provided the first trenchant expression of this argument,[17] which might be rendered as "an infinite cannot be actualized by any finitary algorithmic process."[18] The mathematical description of an infinite past is therefore a theoretical limit that does not correspond to any reality, an extrapolation that, quite apart from its ontological impossibility, founders on the necessary *meta*-stability of the primordial state in cosmological models generating universes, like our own, that are not static.

By the impeccable metaphysical logic of the *kalām* argument,[19] the BGV theorem implies that spacetimes expanding on average throughout their histories are *caused*: they are caused because they began to exist, and everything that begins to exist requires an ontologically and logically—though not always temporally (as demonstrated by the case of space-time itself)—prior cause.[20] What is more, this cause cannot be mathematical in character, for mathematical descriptions are both abstract and causally inert—or rather, they are causally inert *because* they are abstract. To ascribe efficient material causality to them is to commit what Whitehead has called the "fallacy of misplaced concreteness."[21] This, in part, is the problem with Max Tegmark's remarkable assertion that every consistent mathematical structure is physically instantiated.[22] His claim lacks any discernible metaphysical basis and, advanced as an "explanation" for cosmological fine-tuning, undercuts the possibility of rational explanation altogether by entailing *every* possible state of affairs and undermining *every* claim of improbability.[23, 24]

In regard to the principle that everything that begins to exist must have a cause, some have alleged that the failure of efficient material causality in quantum theory is a counterexample—no less so in the case of universal origins since any quantum cosmological constructs or inflaton fields embody *quantum* behavior. But this conceit begs the very question at issue by assuming that every efficient cause must be material and concluding that since there is no material cause for various quantum behaviors, no explanation of *any* kind is necessary. As I have argued elsewhere,[25] since the explanatory resources of naturalism are restricted to material objects, causes, events, and processes, and since neither nonlocal quantum correlations nor (in light of nonlocalizability) the nature of the fundamental constituents of "material reality" can be explained or understood when the explanatory constraints of naturalism are preserved, and since these quantum phenomena require a rational explanation, what must be rejected here is *naturalism,* not explanatory demand.

When we further reflect on the nature of the cause that brought the universe into existence, it is evident that it must be *transcendent* in nature. Space-time and mass-energy do not conceptually entail any principle of self-causation, so prior to the existence of all space, time, matter, and energy there was *no* universe to be described, and hence *no* physical laws or initial conditions that could have played a role in its genesis. Instead, space-time and mass-energy came into existence out of nothing, so a transcendent *immaterial* cause must have acted.

A "Different Physics" at the Past Boundary?

There are many in the community of cosmologists who, like Fred Hoyle, do not like the implications of an absolute beginning requiring a transcendent cause, and who therefore argue that a "different physics" is required at the past boundary. What is meant by this, for good reason, is not always clear, but usually involves some form of quantum cosmology like the Vilenkin tunneling model[26] or the Hartle and Hawking "no boundary" proposal.[27] We will start with Vilenkin's quantum origination theory, then focus on the Hartle-Hawking scenario, which has received much more attention as a consequence of the public awareness generated by Hawking's popular book *A Brief History of Time* (1988).[28]

Let's begin with a rough-and-ready definition of superspace, which is the domain of the hypothesized universal quantum-gravitational wavefunction Ψ. Superspace is the *mathematical* space, S, of all curved 3-dimensional spaces, which, when matter is present, is extended to include the set of all pairings of curved 3-spaces and matter configurations on those spaces.[29] This infinite-dimensional space plays a central role in quantum cosmological *descriptions* of universal origins.[30] The quantum-gravitational wavefunction Ψ assigns a complex number to every point in S. Each path in S describes a 4-dimensional spacetime and its matter configuration, and *any* parameterization of a path that is strictly increasing provides an admissible measure of time for the spacetime represented by a given path and is, therefore, a possible "history" of the universe.[31]

So far so good, one might suppose, but now we need to discuss quantum "origination" scenarios. These origination scenarios deviate considerably from paths representative of classical spacetimes because Ψ must oscillate rapidly in certain directions within S in order to establish as highly probable the "right" quantum-mechanical correlations between the curvature and matter variables and their velocities in S. These correlations do not generate unique classical histories of the universe but rather whole *families* of classical histories in superposition—in short, they instantiate the measurement problem on a universal scale.

Setting questions of the plausibility of this scenario momentarily aside, even if one were able to select a *unique* universal wavefunction, and even if one granted credence to it and adopted a consistent histories approach[32] to the path-integral formalism in which a stable "classical" world was a possible result, there would still be an infinite number of other solutions that were not classical at all. Furthermore, the consistent histories approach allows for a world that is classical right *now,* but was an arbitrary superposition of classical states in the past and will return to this jumbled state in the next instant.[33] Taken by itself, the consistent histories formalism implies at best that the universe as a whole is comprised of *many* different internally consistent but mutually incompatible (in the sense that they cannot be simultaneously experienced) histories. Within the formalism, each possible history is equally real and the experience of any one of these histories as actual is something that is radically context dependent. If we are trying to explain the origin of the universe and this is the representation foisted upon us, we are off to a rather poor start. Insofar as we take the formalism to be an instrumental expedient it provides no explanation, and since it has no other discernible use, we have no reason to continue indulging in the mathematical fantasy it offers; on the other hand, insofar as we suggest that the formalism be interpreted realistically, we are saddled with an infinite expansion of reality that is as untestable as it is fantastical and ontologically profligate. What such an approach needs, one might surmise, is supplementation in the form of an external cause and constraints that guarantee a *unique* real history and a future that, when it happens, *will also be unique.* I argue elsewhere that the fact that quantum physics, on pain of empirical contradiction, provides *no* such causes or constraints indicates that the basic reality it describes is not self-sufficient.[34]

Be this as it may, faced with an infinitely split reality in a universal wavefunction that is compounded by worries of its infinitely arbitrary decomposition into orthogonal states, quantum cosmologists have at least tried to make headway by inventing "natural" constraints on the boundary of superspace or a "natural" algorithm for computing Ψ that would contribute to the uniqueness of the wavefunction itself. In the tradition of Bryce DeWitt,[35] Vilenkin's approach is to invent constraints on superspace. While this procedure has analogues in ordinary physics—for example, Maxwell's equations uniquely determine the electromagnetic field inside a charge-free region of normal space once the values of the field on the boundary are specified—what is being attempted here is not ordinary physics but rather highly mathematicized *meta*physical construction. Both S and its boundary are infinite dimensional and there is no guarantee, even if we ever find halfway coherent quantum gravitational equations, that the metaphysically dubious Ψ thereby obtained will be amenable to this procedure. Furthermore, all this arbitrariness is symptomatic of the deeper, unresolvable difficulty that quantum cosmology is not an observational science and its theories will forever remain untestable.[36] In light of these realities, those interested in playing the infamous demarcation game—where *everyone* loses or *everyone* wins—might like to try their hand at questioning the status of quantum cosmology as a "science."[37]

Before discussion can proceed, we need a conception of the boundary of superspace. A point on the boundary of superspace is the limit of a sequence of points in S that converge to something that is *not* in S. Boundary points thus represent universal configurations that are singular in some respect, as, for example, where the curvature of the 3-space is infinite, or a matter variable possesses infinite density, or where the 4-dimensional spacetime associated with a curve in S is singular (as in the classical Big Bang scenario). Quantum cosmologists distinguish between *singular* and *regular* boundaries of superspace. A curve in S representing a classical universe beginning from a Big Bang would spring from one of the 4-dimensional spacetime singularities constitutive of the singular boundary of superspace, whereas a singular spatial 3-geometry that does not coincide with a 4-dimensional spacetime singularity, but rather one with finite curvature, is classified as part of the regular boundary of S.

Bryce DeWitt was the first to attempt the requisite constraining of Ψ by suggesting that a unique wavefunction might be obtained by requiring it to *disappear* on the boundary of superspace.[38] If the amplitude of Ψ were zero on the boundary, then the probability of any singular configuration would be zero and a breakdown of "physical" description avoided. This prescription turned out not to work well at all, but Alexander Vilenkin later offered a slightly better construction.[39] At the singular boundary of S, Vilenkin proposed that the suitably defined flux of Ψ (roughly, the oscillatory behavior restricted to include only the paths of a family of classical solutions of Einstein's equations) be oriented so it is moving *out* of S. This means (by the theoretician's fiat) that classical spacetimes are allowed to end in a singularity, but not to begin in one. Under these conditions, a variety of approximate calculations have been made that predict a unique Ψ, but as Christopher Isham observes:

> [T]hese approximations involve ignoring all but a small number of the infinite possible modes of the universe and it is by no means clear that uniqueness will be preserved in the full theory. The problem is compounded by the fact that the equations for Ψ that we are trying to solve are mathematically ill-defined. . . . [A]ny proper resolution of this issue must await the discovery of a fully consistent unification of general relativity and quantum theory.[40]

As Isham also notes, proponents of the competing Hartle-Hawking scenario (which we will discuss momentarily) have conceded that their approach definitely does *not* predict a unique universal wavefunction.[41]

If the universe does not originate in a singularity, however, where *does* it come from in Vilenkin's construction? It has to originate from a location within superspace, and Vilenkin's suggestion is that it emerges from the rather ill-defined boundary internal to S between regions where Ψ oscillates (and hence where an underlying classical picture is possible) and regions where it does not (which are purely quantum-mechanical in character). These regions in S where no oscillatory behavior is possible depend on the parameterization of Ψ by a complex time variable (imaginary time) and a precise choice of matter and its interactions. Real time therefore "begins" at the internal boundary between the non-oscillatory and oscillatory regions with the acausal "quantum tunneling" transition of the wavefunction from imaginary to real time, after which the flux of Ψ, constrained to include only oscillations involving classical spacetimes, then moves outward through superspace to terminate on its external (singular and regular) boundaries. Geometrically, imaginary time amounts to the *spatialization* of time in the form of a Euclidean rather than a Lorentzian metric. In the Euclidean domain, there is no special ordering of events as described in the real time of general relativity; a typical exact solution of Einstein's equations in this region is a 4-sphere, as opposed to the conic real-time solutions of general relativity. The geometry of the Hartle-Hawking origination scenario is similar, which is why it is often represented as a round bowl fused to the bottom of a cone with its point removed (a badminton shuttlecock), or a flared conic flask fused to a large bulbous bottom (an old-fashioned bicycle horn).

We turn, then, to a discussion of the famous Hartle-Hawking "no boundary" proposal. Rather than proceeding by first specifying conditions on the boundary of superspace in an effort to extract a unique universal wavefunction, the Hartle-Hawking procedure instead chooses a specific algorithm for the computation of Ψ: a Euclidean path integral summing over compact (closed and bounded) 4-geometries that interpolate between a point and a finite 3-geometry, where the "boundary conditions" correspond to a specification of the class of histories over which the sum is taken.[42] As a first approximation, this latter condition involves restricting the number of degrees of freedom of the gravitational and matter fields in superspace to a finite number and then solving the Wheeler-DeWitt equation on this "minisuperspace."[43] Since this is just an approximation and what is ultimately needed is the wavefunction on the *whole* of superspace, a further step can be taken by using a "midisuperspace" approximation in which the action is considered to all orders for a finite number of degrees of freedom, and to the second order for the remaining degrees of freedom. When this is undertaken, the picture that emerges out of the WKB approximation to the oscillating part of the wavefunction is that of a universe tunneling out of the "minimal temporal radius" in imaginary time, expanding in an inflationary and then matter-dominated fashion to a maximum radius, then *recollapsing* to a singularity.[44] While recent calculations based on WMAP data *seem* to indicate a minute positive curvature that hints of a closed universe, the simultaneous existence of a minuscule positive cosmological constant accelerating its expansion suggests that it might expand forever, even with a closed geometry. If the universe has negative curvature—or more likely is flat—and will continue to expand forever (at a rate asymptotically approaching zero in the flat case), then insofar as the Hartle-Hawking proposal and these approximations can be taken seriously, they are based on false premises and will have to be reworked with a different algorithm and boundary conditions. Regardless, the game of mathematical brinksmanship such calculations represent is just so much whistling in the dark—a case of highly malleable mathematical descriptions adaptable to the exigencies of the moment and therefore devoid of geniune empirical content and explanatory power.

Let's backtrack briefly and fill in some of the details. The whole "no boundary" approach was motivated by Hawking's earlier work on black hole radiation. In 1974, Hawking discovered that black holes would radiate particles via a quantum-mechanical process and these particles would have a thermal spectrum.[45] The re-derivation of these results using thermal Green's functions that, in standard quantum field theory, involve substituting for the time-coordinate an imaginary number inversely proportional to the temperature, inspired Hawking to take a Euclidean approach to quantum gravity in which Lorentzian metrics are replaced by Riemannian ones via a Wick rotation in which the time coordinate t is "rotated" by a complex transformation into $\tau = it$.

Specifically, Hawking proposed to study functional integrals having the form

$$(1) \qquad Z(\mathcal{M}) := \int \mathcal{D}g \, e^{\frac{1}{\hbar} \int_{\mathcal{M}} |\det g|^{1/2} R^4(g)},$$

where the integral is taken over all Riemannian metrics g on a 4-manifold $\mathcal{M}$, with $R^4(g)$ being the curvature of g, and det g the determinant of the metric.[46] This expression can be generalized to a weighted "quantum topology" where every 4-manifold M has a weight $w(\mathcal{M})$ that it contributes to expressions of the type:

(2) $$Z := \sum_{\mathcal{M}} w(\mathcal{M}) Z(\mathcal{M}).$$

If applied to a manifold with a single 3-boundary Σ, (2) will represent a functional $\Psi[h]$ as long as the functional integral is taken over all 4-metrics g on $\mathcal{M}$ that induce the requisite 3-metric h on Σ. So defined, the functional of h can be taken to satisfy the Wheeler-DeWitt equation,[47] and by including matter fields ϕ, we may write

(3) $$\Psi[h, \varphi_0, \Sigma] = \sum_{\mathcal{M}} w(\mathcal{M}) \int \mathcal{D}\varphi \, \mathcal{D}g \, e^{-\frac{1}{\hbar} I(g, \varphi, \mathcal{M})},$$

where I is the Euclidean action. Equation (3) is the basis of the Hartle-Hawking quantum cosmological "wave-function of the universe."[48] The "no boundary" proposal requires that the sum in (3) be taken just over compact (closed and bounded) manifolds $\mathcal{M}$ that have the connected 3-manifold Σ (carrying the 3-geometries of quantum geometrodynamics) as their only boundary. What is meant by suggesting that this represents the universe coming into existence by "tunneling from nothing" is summarized well by Butterfield and Isham:

> The word "nothing" just reflects (very obscurely!) the idea that $\mathcal{M}$ has only Σ as its boundary. The word "tunneling" refers to the facts that (i) moving from a Lorentzian to a Riemannian manifold corresponds, roughly, to moving from a time variable that is a real number to one that is purely imaginary (in the sense of complex numbers); and (ii) in normal quantum theory, a good approximation to the probability of tunneling through a potential barrier can be found by computing the action I for a solution to the *classical* equations of motion with an *imaginary* time; the probability amplitude in question is then proportional to exp $-I / \hbar$.[49]

The computational expedient of Euclideanization in quantum cosmology is used solely for the purpose of constructing a convergent path integral[50] and is *not* applied to the *background space* in Hawking's appropriation, but to the *individual spacetimes* constitutive of *each path* in the path integral. When this imaginary time coordinate is retained in the final answer, the singularity at $\tau = 0$ disappears, yielding mathematical representations of universes with no temporal beginning but just a "minimal temporal radius." In his popular book,[51] Hawking suggests that this procedure provides a model in which the universe has no beginning and hence is completely self-contained: without a beginning or an end, it just *is*. His remark is more than a little disingenuous, however, since a "realistic" solution using this procedure requires completing the transformation back to a Lorentzian metric and the *real time* in which we live, since none of the Riemannian universes in the path integral are capable of being models of our Lorentzian spacetime. Once this is done, however, the initial singularity reappears, and all the spacetimes in the sum again have a beginning. Hawking knows this, of course, and says as much in passing: "When one goes back to the real time in which we live, however, there will still appear to be singularities. . . . Only if [we] lived in imaginary time would [we] encounter no singularities . . . In real time, the universe has a beginning *and an end* at singularities that form a boundary to space-time and at which the laws of science break down."[52] It seems clear, therefore, that the "no boundary" proposal does not genuinely remove universal beginnings from its descriptions, but introduces the device of "imaginary time" as an intermediate computational expedient interpolating between a real-time singular beginning and (probably contrary to fact) singular end. Furthermore, even if it did remove the beginning of time, the logically and metaphysically unnecessary existence and structure of such a spacetime would still create an explanatory demand only satisfiable by something like the cosmological argument from contingency.[53]

If we pause to take stock of what has been offered, we can see clearly that any claim that quantum cosmology provides an explanation of the origin of the universe that renders intelligent agency otiose either as a cause or a constraint is false. The mathematical expressions of quantum cosmology cannot describe any real process without an external (transcendent) activation of some sort: they are *causally inert* mathematical descriptions predicated of what is falsely referred to as "nothing" so the pretense may be sustained that quantum cosmologists have explained how nothing can turn, unaided, into a universe.

Let me expand on this point, because it is critical: contrary to the zealous assertions of some who maintain that quantum cosmology succeeds in conjuring a universe out of nothing—just like the proverbial rabbit pulled out of a hat, except that there's no hat and no magician doing the pulling—the mathematical description proffered belies the claim made on its behalf. If the universe really had begun from absolutely nothing, the original "state" would not have *any* positive properties. You certainly could not predicate quantum mechanical commutation relations of it or describe it by some artfully gerrymandered universal wavefunction. Such a predication assumes the existence of positive structural content that absolute nothing, by definition, does not possess. This is why the remark of some theoretical physicists (Frank Wilczek and Michio Kaku most prominent among them) that "nothing is unstable" is the assertion of an ontological absurdity. To the contrary, there isn't anything more stable than absolute nothing, which can be expected to *do* exactly what it *is*: absolutely nothing.

Furthermore, since the original state and boundary conditions in quantum cosmology do have positive structural content, what might this imply? Even if you postulate the eternal existence of superspace and various quantum-theoretic mathematical relations as platonic forms, these abstract objects are *causally inert*: they cannot act or bring into existence concrete realities that satisfy their descriptions. To suppose otherwise would be another instance of Tegmark's temptation: the fallacy of misplaced concreteness. What this means is that the proposals of quantum cosmologists, if they describe anything at all, describe processes that are *transcendently caused*. The reason is simple: if we are going to get space, time, matter, and energy out of absolutely nothing, we require a cause that is not dependent upon any of these things and is also capable of acting. What must lie behind these scenarios, insofar as they can be taken seriously, is a transcendent, immaterial, timeless, intelligent, and hence personal cause of immense power. Without recourse to such an agent, quantum cosmology as an enterprise is a metaphysical non-starter, pure and simple. The Parmenidean dictum that nothing comes from nothing (*ex nihilo, nihil fit,* in its common Latin rendering) is as true today as when it was first stated.

That such a transcendent cause must of necessity be intelligent is evident from the finely tuned parameters we will discuss shortly. In the present context, it is also evinced by the background structure assumed. The quantum cosmological constructions being examined invoke a number of hugely restrictive intelligent choices involving geometrical, quantum-theoretic, and material configurations. The universe under these descriptions is thus neither the metaphysical nor informational "free lunch" that would-be quantum cosmological conjurors portray it to be. Rather, such universes originate under highly specified mathematico-structural constraints that are, by themselves, causally impotent descriptions. Universal ontogenesis, therefore, cannot credibly be ascribed to "nothing," nor can the constraints it requires credibly be left to "chance." The ineluctable and proper conclusion is that the universe was brought into being by a transcendent intelligent cause.

Truthfully, we could entirely bid goodbye to any pretense of a naturalistic explanation for cosmological origins and fine-tuning at this juncture, for the points just made apply to the whole edifice of cosmological research under discussion. Nonetheless, by pressing onward we can reinforce the certainty of this metaphysical conclusion.

Popping the Balloon of Cosmic Inflation[54]

The inflaton field has been offered as the panacea for any aboriginal inhomogeneity and anisotropy in the universe, but propaedeutic and evaluative questions really need to be asked. First, is inflationary cosmology free from arbitrary assumptions and gerrymandering? And second, does inflation really solve the problems it was invented out of whole cloth to address? The answer to the first question is a resounding *no*; and the answer to the second is a highly cautionary *only under very special assumptions*.

Inflation was initially proposed for the purpose of solving three problems: the horizon problem illustrated by the uniformity of the cosmic microwave background (CMB) radiation, the flatness problem constituted by the precision with which the universe's actual mass density approximates its critical mass density, and the absence of the magnetic monopoles predicted by favorably regarded grand unified theories (GUTs). Yet these three problems do not have to be addressed by the assumptions of the contemporary inflationary model.

First, the multiple scalar fields postulated by chaotic inflation are arbitrary. They constitute false vacua that bear no relation to any other known fields in physics and have properties invented solely for the purpose of making inflation work. In short, the explanation they offer is completely ad hoc.[55]

Secondly, Hawking and Page have shown that when an inflaton field is grafted onto standard FLRW cosmology, while the measure of the set of models that inflate is infinite, so is the set of models that do *not* inflate.[56] This is not an inconsequential observation. As Earman and Mosterin have observed,[57] even when inflation is restricted to the class of homogeneous and isotropic cosmologies, inflationary cosmologists have not been able to show, without invoking highly speculative hypotheses, that inflationary mechanisms actually resolve the fine-tuning issues associated with the hot Big Bang model that prompted their invention.

Thirdly, and this is related to the previous point, inflation may not be an adequate solution to the flatness problem. The matter density (ordinary and dark) in the universe is very close to the critical density that would imply a perfectly flat universe. Exceedingly precise measurements in 1998 demonstrating a very small positive vacuum energy (cosmological constant) led cosmologists to conclude that the universe will expand forever, though recent calculations based on WMAP data seem to indicate very slight positive curvature, which would imply it has a closed geometrical structure. No one knows for sure—it is that close—but the general bet is for a flat universe. Inflation is put forward as an explanation for this flatness, but so far serious attempts to calculate inflationary consequences for flatness have assumed an FLRW metric and have not addressed what would happen in the generic case. As Penrose points out,[58] expansion from a generic singularity can become whatever type of irregular universe we please, independent of whether there is an inflationary phase. As a consequence, unless a special metric and other special assumptions are in view, inflation is not an adequate explanation of flatness.

Fourthly, as Thomas Banks explains, any suggestion that inflation resolves the problems created by applying the Second Law of Thermodynamics to cosmology—primarily, the recognition that the universe had to be created in a state of very low entropy—is mistaken.[59] The initial inflationary patch would have had a very small number of degrees of freedom describable by effective field theory; most of the degrees of freedom in the observable universe are not capable of description using quantum field theory until a large number of e-folds have occurred.[60] To handle this deficiency, the standard approach to these degrees of freedom, in its most sophisticated form, begins with the assumption that they were in the ground state of some slowly varying Hamiltonian that approaches the conventional field-theoretic Hamiltonian in the inflationary background, comoving mode by mode, as the physical size of each mode crosses the Planck scale. As Banks notes, this approach involves many ad hoc assumptions, including a low-entropy initial condition that is smuggled in by assuming the system was in its ground state. Furthermore, the excited states of every known large quantum system are highly degenerate, and the adiabatic theorem[61] does not apply to generic initial conditions chosen as a linear combination of highly degenerate states. What this means is that the standard assumption in inflation of a very special state for a huge number of degrees of freedom is completely unjustified, because we do not have a reliable dynamical description of these variables. As Banks concludes, therefore, inflationary cosmology does not, in this sense, solve the problem of the homogeneity and isotropy of the early universe.[62]

Fifthly, inflation is regarded as explaining why the monopoles predicted by various favorably regarded grand unified theories have yet to be observed by effectively diluting their density in the observable universe. Invoking inflation in this context, however, is using it as an ad hoc measure to spare other favored yet unconfirmed theories from disconfirming evidence. That inflation can be used in this way is not evidence of its merit. If the GUTs do not stand the test of time and additional empirical evidence, then there will be no need to explain why magnetic monopoles have not been detected by appealing to inflation—or any other rescue strategy, for that matter.[63]

Finally, while inflationary cosmology has made some predictions about the distribution of the CMB at various wavelengths that are independent of its original motivations in terms of a "solution" to the horizon and flatness problems, and while some of these predictions seem to hold, there are also some anomalies that haven't been resolved. Inflation predicts an isotropic distribution of the CMB at all frequencies on a large scale, yet analysis of WMAP data has yielded a preferred direction for large-scale modes of the CMB that disagrees with such a prediction.[64] This issue is still being resolved and looks like it may have been mitigated,[65] but if anistropy holds up, not only will inflation's theoretical basis remain woefully insufficient, it will fail observational testing in its only real area of empirical contact.

DEFLATION: THE FINE-TUNING OF INITIAL CONDITIONS REVISITED

Given that inflation was invented as a strategy for explaining the fine-tuning of certain initial conditions in our universe, it is highly ironic that the inflaton field requires very special assumptions and exquisite fine-tuning itself. Let's look at a few telling examples.

The mechanism for bubble formation in the inflationary multiverse is Einstein's equation in general relativity, which, even though there is no intrinsic connection between the theories, is assumed to constrain the process of inflation in such a way that bubbles will form from local decay of the inflaton field while the field itself continues to expand. In the creation of these bubbles, however, the inflaton field must be shut off and "converted" to normal mass-energy. This shut-off point is delicate, operating in the first 10^{-37} to 10^{-35} seconds (or so) of the universe's existence, while causing space to expand by a factor of around 10^{60} (or so). The conversion from the inflation to the preheating era necessary to bring about particle production in an initially cold and empty universe involves a variety of highly speculative models with inflaton-preheating coupling parameters that have to be finessed to produce the right results.[66] Furthermore, depending on the inflationary model under consideration, the initial energy of the inflaton field is anywhere from 10^{53} to 10^{123} times the maximum vacuum energy consistent with our universe having the properties it does. This means that the energy decay of the inflaton field also has to be fine-tuned to at least one part in 10^{53} and possibly as much as one part in 10^{123}.[67] In short, the decay of the shut-off energy needs to be fine-tuned at a *minimum* to one part in a hundred thousand trillion trillion trillion trillion. Compared to such levels of precision, the fine-tuning of the Big Bang inherent in the so-called horizon and flatness problems, like an unruly friend, seems rather manageable.

There is another massive fine-tuning problem that turns out to be affected by inflation: the incredibly precise value of universal entropy needed at the Big Bang to produce a universe consistent with our observation. As Roger Penrose points out,[68] inflation solves the horizon problem only by exponentially increasing the fine-tuning of the already hyper-exponentially fine-tuned entropy of the Big Bang. The event that initiated our universe was precise almost beyond the point of comprehensibility.

How precise? The fine-tuning of the initial conditions for universal entropy can be calculated as follows.[69] In the observable universe there are about 10^{80} baryons (protons and neutrons). The *observed* statistical entropy per baryon in the universe can be estimated by supposing that the universe consists of galaxies populated mainly by ordinary stars, where each galaxy has a million solar-mass black hole at its center. Under such conditions, the entropy per baryon (a dimensionless number inclusive of the entropy in the cosmic background radiation) is calculated to be 10^{21}, yielding an observed universal entropy on the order of 10^{101} ($10^{80} \times 10^{21}$). If we run the clock backward to the beginning of time, thus mimicking universal collapse, the entropy per baryon near the resulting "big crunch" is calculable from the Bekenstein-Hawking formula for black-hole entropy by considering the whole universe to have formed a black hole. Performing this calculation leads to a value of 10^{43} for the entropy per baryon, yielding a total value of 10^{123} ($10^{43} \times 10^{80}$) for universal entropy. Since this number also indicates the possible entropy for a universe our size emerging from a Big Bang singularity, we can compare it with what we now observe to estimate how fine-tuned the Big Bang had to be to give us a universe compatible with the Second Law of Thermodynamics and what we now observe to be the case. Since 10^{123} is the natural logarithm of the volume of the position-momentum (phase) space associated with all of the baryons in the universe, the volume itself is given by the exponential: $V = e^{10\exp(123)}$; similarly, the observed total entropy is $W = e^{10\exp(101)}$. For numbers this size, it makes little difference if we substitute base 10 for the natural logarithm, so Penrose does that. Following his lead, the required precision in the Big Bang is therefore given by:

$$W/V = 10^{10\exp(101)}/10^{10\exp(123)} = 10\exp[10^{101} - 10^{123}] \approx 10\exp(-10^{123}).$$

In other words, to satisfy the observed entropy of our universe, the Big Bang had to be fine-tuned to *one part* in 10exp(10^{123}). This latter number is difficult to grasp; suffice it to say that with 10^{80} baryons in the observable universe, if we attached a zero to every one of them, it would take 10^{43} universes the size of ours just to write it out![70]

So how is this result affected by inflation's "resolution" of the horizon problem? Again, the fundamental strategy is for inflation to push beyond the observable universe the particle horizons that would preclude explaining the uniformity of the CMB on the basis of thermalization. But, as Penrose has observed, if thermalization serves the role of driving background temperatures to equilibrium in the inflationary context, then it represents a definite *increase* in universal entropy that requires the Big Bang to be even *more* finely tuned to account for its current observed value, because the universe exponentially inflates into a normal expansion the initial entropy of which is fine-tuned to one part in 10exp(10^{123}). In other words, if inflation explains the horizon problem through thermalization, it turns a hyper-exponentially fine-tuned initial entropy into a *hyper*-hyper-exponentially fine-tuned quantity. On the other hand, if thermalization plays *no* role in explaining the horizon problem, then inflationary cosmology is completely *irrelevant* to its solution.

2. Universal Origins and String Cosmology

Having catalogued the reasons for profound skepticism where quantum cosmology and the inflationary multiverse are concerned, we must turn to the subject of string theory and its invocation in the cosmological context. It is by conjoining the resources of quantum and inflationary cosmology with the landscape of string vacua that multiverse theorists hope to obviate the fine-tuning of the laws and constants of the universe along with that inherent in its initial conditions. As mentioned earlier, the string cosmological models that have received the most attention are the Steinhardt-Turok cyclic ekpyrotic model (which does not invoke inflation but satisfies the conditions of the BGV theorem), the Gasperini-Veneziano pre-Big Bang inflationary model (which circumvents the BGV theorem despite employing inflation, but nonetheless has a beginning in the finite past due to the meta-stability of its primordial state), and the inflationary string landscape model advanced by Susskind, Polchinski, Bousso, and Linde (which invokes inflation and is subject to the BGV result). In this section, we document the fact that none of these highly speculative string cosmologies remove the necessity of a beginning to the process of universe generation. A brief primer on string theory will help to facilitate our discussion.

A Primer on String Theory

String theory was initially proposed in the mid-1960s as a description of the strong nuclear force generating mesons and baryons, but it lapsed into obscurity after the success of quantum chromodynamics. The fundamental constituents of string theory are one-dimensional filaments existing as open strings or closed loops on the scale of the Planck length (10^{-33} cm). The theory was revived in the late 1970s when John Schwarz and other researchers discovered that the spin-2 particle that had thwarted its nuclear ambitions could be reinterpreted as the quantum of the gravitational field, producing a theory that, when the demands of quantum-theoretic consistency were satisfied, reconciled gravity with quantum mechanics in a ten-dimensional spacetime. The extra six spatial dimensions of string theory require compactification into Planck-scale Calabi-Yau manifolds to suggest any connection with reality as we know it, and this division of the spatial dimensions into three large and six small transforms some of the N=1 SUSY gravitational modes in nine large dimensions into a variety of non-gravitational bosonic and fermionic vibrations.[71]

Just what *kind* of non-gravitational forces (spin-1 bosons) and matter (fermions and supersymmetric scalar partners) are produced by this transformation depends on the size and shape of the compactified dimensions[72] and, alas, there is an *unlimited* number of ways of compactifying them. This embarrassment of riches was once regarded as a vice, but increasing appreciation of the degree to which the laws and constants of our universe are fine-tuned for life has led some to extol it as a virtue, speaking instead of the "landscape" of string solutions.[73] We will examine the string landscape hypothesis in more detail presently.

One of the most recalcitrant technical problems in the early stages of the string revival was solved in 1984 by John Schwarz and Michael Green.[74] Ten-dimensional string theory exhibited a quantum anomaly resulting from unphysical longitudinal modes that were shown to be eliminable if the strings obeyed a specific gauge symmetry, SO(32). It was subsequently shown in fairly quick order that there were actually *five* anomaly-free classes of ten-dimensional string theories characterized by different gauge symmetries: Type I, Type IIA, Type IIB, $E_8 \times E_8$ heterotic, and SO(32) heterotic, all Calabi-Yau compactifiable in the six extra spatial dimensions and each with countless numbers of models.

In the 1990s evidence began to collect that these five classes of string theories were not, in fact, independent of each other. This suspicion was given life in 1995 when Edward Witten demonstrated the equivalence of heterotic SO(32) string theories with low energy effective string field theories of Type I.[75] Subsequently, the community of string theorists found dualities expressing the equivalence of all five classes of string theories as well as eleven-dimensional supergravity. The key to these equivalences proved to be a string with finite *width* in addition to its length—in essence a two-dimensional membrane—which therefore existed in *eleven* rather than ten dimensions. Thus was born eleven-dimensional "M-theory."

The additional spatial direction in M-theory potentially plays a different role than the others. The ten-dimensional spacetime of string theory is a slice of the eleven-dimensional bulk of M-theory. Since the new spatial dimension is orthogonal everywhere to the other nine, it can be regarded as a line segment connecting two ten-dimensional string universes (9-branes), each hidden from the other and having only the gravitational force in common.[76] Since six of the extra ten spatial dimensions are compactified throughout the eleven-dimensional bulk, the effective picture is that of a five-dimensional spacetime bulk with two four-dimensional universes (3-branes) at the ends of a line segment. Since gravity would vary as the inverse cube of the distance in four spatial dimensions and the inverse square law has been tested down to 55 micrometers,[77] if this M-theoretic model had any basis in reality, this minuscule distance would provide the current maximum separation between our universe and its twin. Additional tests of this highly speculative scenario have been proposed and are being pursued.[78]

Steinhardt-Turok Cyclic Ekpyrotic Universes

M-theory also permits the possibility of freely moving universes (branes); it is this possibility that is explored in the cyclic ekpyrotic models of string cosmology.[79] In this scenario, a bulk of four spatial dimensions exists between two 3-branes. The collisions between these branes, which happen on average once every trillion years, release sufficient energy to catalyze the hot Big-Bang stage of new universes.[80] Because of Planck-scale quantum fluctuations neither 3-brane remains perfectly flat, so energy release is greatest at points of first contact. Steinhardt and Turok estimate that on each bounce cycle such brane-brane collisions have the potential to produce staggering numbers of new Big Bang regions (10^{100} to 10^{500}) that are causally isolated from each other.

Of course, each such universe has a beginning in a quantum string nucleation event induced by a brane-brane collision and so has a finite past. The original brane spacetimes were postulated to be nonsingular, however, and this served to ground the claim that the cyclic ekpyrotic scenario did *not* require initial conditions and could be past-eternal.[81] This turned out not to be the case, however, as Borde, Guth and Vilenkin made clear and Steinhardt and Turok have now acknowledged.[82] An essential feature of the ekpyrotic model, which enables it to deal with the thermodynamic objection that defeats conventional cyclic cosmologies,[83] is that the volume of the universe consisting of the aboriginal bouncing branes increases with each cycle while the energy released into the branes by each collision gets renewed by the inexhaustible resource of gravitational potential energy. This entails that on average the cyclic universe is expanding, i.e., $H_{av} > 0$, and so the BGV theorem requires its geodesic incompleteness—in short, it has a beginning in the finite past.[84]

Gasperini-Veneziano Pre-Big Bang Inflationary Scenarios

Another pre-Big Bang scenario in string cosmology that merits our attention was proposed by Maurizio Gasperini and Gabriele Veneziano.[85] Anachronistically speaking, it sidesteps the $H_{av} > 0$ condition governing the BGV result and, from a *purely* mathematical standpoint, can be geodesically extended into the infinite past. The Gasperini-Veneziano pre-Big Bang inflationary (PBBI) model proposes that the universe started its evolution from the simplest possible string-theoretic initial state, namely, its perturbative vacuum, which corresponds to a universe that, for all practical purposes, is empty, cold, and flat. This string perturbative vacuum (SPV) phase is neither expanding nor contracting in whole or in part; in this sense, it is static. Since the spacetime manifold of this state has models in which timelike and null geodesics can be past-extended for infinite values of their affine parameter, it is proposed that this phase could have been of infinite duration, which would mean that the universe did not have a beginning.

The assumption that the primordial universe was a string perturbative vacuum means that the dilaton field[86] started very large and negative, which allows the early history of the universe to be treated classically.[87] The assumption of virtual flatness also enables employment of the low-energy approximation to string theory. This means the evolution of the universe can be described using classical field equations for the low-energy effective action,[88] from which, under the assumptions of homogeneity and flatness, inflationary behavior automatically follows from the hypothesis that the primordial universe was an SPV.[89] Of course, homogeneity and flatness are fine-tuning conditions. If the assumption of homogeneity is relaxed and replaced with generic initial conditions approximating the perturbative vacuum, it can be shown that it is possible for a chaotic version of the pre-Big Bang scenario to arise through dilaton-driven inflation in patches of the primordial SPV[90] *as long as* the kinetic energy in the dilaton is a non-negligible fraction of the critical density.[91] One of the controversial features of this latter scenario is that in order to have sufficient inflation in a patch, dilaton-driven inflation has to last long enough to reach a hot big bang nucleation event. Since PBBI is limited in the past by the initial value of spatial curvature, it has to be extremely small in string units if sufficient inflation is to be achieved.[92] In other words, no matter which approach to PBBI one takes, *considerable fine-tuning is necessary.*[93]

A word is in order about a feature of PBBI model that some may find puzzling: how is it that an inflationary phase *leads* to a Big Bang rather than *following* from it? This result is a consequence of one of the peculiar features of string theory called *T-duality,* which relates small and large distance scales. T-duality implies that, at some deep level, the separation between large and small distance scales in physics is fluid. In the "inflationary phase" of the PBBI model, spatial expansion is taking place in the *string frame* coordinates while, in the classical *Einstein frame* coordinates, matter is collapsing into trapped surfaces, i.e., black holes.[94] At the conclusion of this dilaton-driven inflationary phase, a transition is supposed to take place to an FLRW phase typical of the standard hot Big Bang model—though models for *how* this happens are, to say the least, not well-understood.[95]

What shall we say, then: does the PBBI model do an end run around the BGV theorem and provide a viable picture of a universe with no beginning? Not really. While the null and timelike geodesics of the SPV phase can in theory be extended into the infinite past, asymptotically approaching exact equilibrium, the fact remains that at every point in the finite past the string perturbative vacuum is unstable.[96] Quantum fluctuations of the background fields, particularly the dilaton, move the SPV from equilibrium, so that at any given finite physical time, the system is in a *non*-equilibrium state. Since each patch of the SPV has a non-zero probability of decaying into dilaton-driven inflation, quite apart from issues of metaphysical incoherence stemming from the non-traversability of an infinite past, a realistic interpretation of the model, however implausible in itself, requires acknowledging that the SPV phase has finite duration. Since the other two phases of the model (inflationary and FLRW) are also finite in duration, the universe has a beginning. So, even in Gasperini-Veneziano PBBI scenarios, the universe begins to exist—and as we saw earlier, since it begins to exist, it must have a transcendent cause.

The String-Theoretic Landscape Hypothesis

As we observed earlier, the laws and constants of different string-theoretic universes are determined respectively by the shape and size of their compactified dimensions. If there were a mechanism for navigating around the "landscape" of these moduli, each combination describing a different solution (vacuum) of the string-theoretic equations, there would be a way to *generate* universes with different laws and constants—at least 10^{500} of them, in fact, if we restrict ourselves to versions of string theory having a positive-valued cosmological constant, as required by our own universe.

Bousso, Kachru, Kallosh, Linde, Maldecena, McAllister, Polchinski, Susskind, and Trivedi *all* contributed to devising a mechanism that might do this by finding a way to combine inflationary cosmology with the string landscape:[97] bubbles of lower energy string vacua nucleate when moduli decay at random locations throughout higher energy string vacua that continue to inflate forever, so the whole landscape (they contend) gets explored as a series of nested bubble universes. Interior bubbles inflate at a slower rate than their parent universes, and bubbles of still lower energy nucleate inside of them, while all of the vacua so created inflate eternally. According to this picture, we live in one such bubble universe.

Note that the BGV theorem applies to the string landscape hypothesis because the inflationary mechanisms on which it is premised require its overall expansion. The landscape is thus past-incomplete, so if it existed, it too would have a beginning in the finite past that required a transcendent catalyst.

An Original Requirement

So where does this leave us? Even if inflation were ultimately upheld as theoretically viable, empirically sustainable, and metaphysically and epistemologically credible—an unlikely outcome in my estimation—a beginning and hence a transcendent cause would be required in all models subject to the BGV theorem. As we have seen, this applies to every higher-dimensional cosmology, inclusive of cyclic ekpyrotic and landscape models, that involves spacetimes satisfying the condition that the average Hubble expansion in the past is greater than zero, i.e., $H_{av} > 0$. Furthermore, inflationary models such as the one proposed by Gasperini and Veneziano, to which the BGV result does not apply, do not ameliorate the need for a beginning because their realistic interpretation, however implausible, still requires the meta-stability of the primordial state, which means that the earliest phase has finite duration. Since subsequent phases must also be of finite duration, any universe satisfying this scenario will have a beginning as well.

But what if inflation is not upheld as a viable explanation? Then, assuming the universe isn't suffering from noninflationary cyclic ekpyrosis, which also requires an absolute beginning, we revert from multiverse scenarios to a single universe again. In this context, the singularity theorems of classical general relativity regain their traction—qualified, perhaps, by a different quantum cosmological physics at the past boundary that, as we have seen, does not alter the final metaphysical verdict—and lead to the conclusion that our universe has an absolute beginning in the finite past, and thus the necessity of a transcendent cause for space-time, energy, and matter.

It appears, therefore, that a beginning and a transcendent cause of the universe (or multiverse) are unavoidable.

3. Cutting the Gordian Knot of String Cosmology

We now turn to an examination of the assumptions governing string multiverse cosmologies and an evaluation of their effectiveness as explanations of cosmological fine-tuning.

Fine-Tuning and the Cyclic Ekpyrotic Model

Khoury, Steinhardt, and Turok[98] have shown that the phenomenological constraints on the scalar field potential in cyclic ekpyrotic models necessitate a degree of fine-tuning comparable to that of inflationary models—the number of degrees of freedom, the number of tunings, and the quantitative degree of tuning are similar.[99] While the claim to be just as good as inflationary models might be received with favor in some quarters, our discussion to this point has established grounds for a somewhat less sanguine attitude.

Kallosh, Kofman, Linde, and others take an even less sanguine view, arguing that the ekpyrotic model faces additional problems.[100] For instance, they point out that the Hořava-Witten version of string theory on which the ekpyrotic scenario is based requires the 3-brane of our universe to have positive tension, but the ekpyrotic model requires negative tension. To make the ekpyrotic scenario workable, therefore, they argue that the problem of the negative cosmological constant on the visible brane must be solved and the bulk brane potential fine-tuned with an accuracy of 10^{-50}. Furthermore, they contend that the mechanism for the generation of density perturbations is not brane-specific; rather, it is a particular limiting case of the mechanism of tachyonic preheating, which exponentially amplifies not only quantum fluctuations, but any initial inhomogeneities.[101] As a result, to solve the homogeneity problem the ekpyrotic scenario would require the branes to be parallel to each other with an accuracy of better than 10^{-60} on a scale 10^{30} times greater than the distance between the branes. With some gerrymandering assumptions, Steinhardt and Turok have managed to ameliorate some of these difficulties,[102] but significant technical problems and fine-tuning issues remain—in particular, Veneziano and Bozza[103] have shown that a smooth bounce cannot generate a scale-invariant density perturbation spectrum via the mode-mixing mechanism advocated by Steinhardt, Turok, and others;[104] and Kim and Hwang have argued that it is not possible to obtain the requisite near Harrison-Zel'dovich scale-invariant density spectrum through a bouncing world model as long as the seed fluctuations were generated from quantum fluctuations of the curvature perturbation in the collapsing phase—rather, the spectrum is significantly blue-shifted in comparison with what is needed.[105]

It is worthwhile considering whether the cyclic ekpyrotic scenario has the probabilistic resources to address the one in $10\exp(10^{123})$ fine-tuning of the Big Bang entropy of our universe.[106] It does not. It is *not* an inflationary model—though it does involve dark energy—so it does not invoke an unending chaotic cascade of string vacua.[107] Rather, each trillion-year cycle produces 10^{100} to 10^{500} Big Bang events with opportunities for finely tuned entropy. This means that with each new cycle there is *at best* a $\{10^{500}/10\exp(10^{123})\} = 10\exp(500 - 10^{123}) \approx 10\exp(-10^{123})$ chance that the requisite entropy condition will be met. In short, the ekpyrotic universe would have to go through a significant fraction of $10\exp(10^{123})$ trillion-year cycles for there to be any reasonable probability of getting a universe like ours. But we have already seen that such cyclic models are geodesically incomplete and, as Steinhardt and Turok admit,[108] the most likely story is that the cycling stage was preceded by a singular beginning. Furthermore, even if this picture were true, there is in principle no measurement that could be made to determine how many cycles have taken place. It would be a highly unwarranted assumption, therefore, to presume that the model has the probabilistic resources necessary to resolve the problem of universal entropy; in fact, the incomprehensibly large number of trillion-year cycles required inspires deep skepticism, especially when the logico-metaphysical necessity of a transcendent cause for the singular beginning of any ekpyrotic universe brings with it the far more plausible scenario of intelligently directed fine-tuning.

While the ekpyrotic model confronts some extraordinary fine-tuning issues of its own, we may nonetheless reasonably ask whether it resolves any. Steinhardt and Turok have recently claimed that it does,[109] most specifically that it offers a credible explanation for why the cosmological constant (vacuum energy) is small and positive. What they essentially do is engineer a "relaxation mechanism" that can be incorporated into the cyclic model that slowly decreases the value of the cosmological constant over time, while taking account of contributions to the vacuum density over all energy scales. The mechanism works by allowing the relaxation time to grow exponentially as vacuum density decreases, generating asymptotic behavior in which every volume of space spends the majority of time at a stage when the cosmological constant is small and positive—just as it is observed to be today. Again, the solution is ad hoc: a mechanism was *invented* to produce the desired behavior and then declared to be a virtue of the model simply because a way was found to make it work. Furthermore, there is no reason intrinsic to the ekpyrotic scenario, which as we have seen is subject to the BGV theorem, that explains why it must start with a vacuum energy greater than what we observe today, yet invoking a relaxation mechanism must assume that it does, since this condition is needed in order to "explain" the value it now has.[110]

As a last consideration, Alexander Vilenkin argues that the cyclic relaxation mechanism provides no explanation for the fact that the vacuum density, which is fine-tuned to 120 decimal places, is roughly twice the average energy density of matter in the universe.[111] These two densities behave very differently with cosmic expansion—the former stays constant while the latter decreases—so why do we live in an epoch when the values are close? This is known as the "cosmic coincidence" problem. The ekpyrotic model provides no answer to it, but Vilenkin contends that standard inflationary cosmology conjoined with the string landscape does: the universe on the largest scale is postulated to be in a state of high-energy expansion that is spawning lower energy bubble universes like our own, having, in virtue of the string landscape, *all* possible values for a wide variety of "universal" constants. Since galaxies and observers only exist in those rare bubbles where the vacuum energy is small and a variety of other parameters are appropriately adjusted (the anthropic principle), and since analysis reveals that during the epoch of galaxy formation—which includes *our* present time—most galaxies will form in regions where vacuum and matter densities are about the same, he contends this cosmic coincidence is thereby "explained."

Fine-Tuning and the Pre-Big Bang Inflationary Model

How does the Gasperini-Veneziano PBBI model fare in relation to issues of cosmological fine-tuning? Turner and Weinberg have shown that pre-Big Bang dilaton-driven inflation of an SPV patch has to last long enough to reach a hot Big Bang nucleation event,[112] but since the PBBI period is tightly constrained by the initial value of spatial curvature, this curvature has to be extremely small in string units if sufficient inflation is to be achieved to "solve" the flatness and horizon problems. It is not completely obvious on this account, however, just how strong this fine-tuning has to be, and others have argued that it may be possible to mitigate this conclusion if the universe is open[113] or if the pre-Big Bang conditions are restricted in just the right way.[114]

More tellingly, Kaloper, Linde, and Bousso have shown that PBB dilaton-driven inflation can address the horizon and flatness problems *only if* the primordial SPV is extremely large and homogeneous from the outset[115]—in short, the fine-tuning of our universe is "explained" by pushing *all* the fine-tuning into the SPV era. Let me elaborate. The authors show that if our universe appeared as the result of PBBI then it had to originate from a homogeneous domain of exponentially large initial size, with enormously large initial mass and entropy at the onset of inflation. Furthermore, if this PBB universe is *closed,* then at the time the SPV becomes describable by the low-energy effective action, it can be shown that it must consist of at least 10^{24} causally disconnected regions of nearly equal density. Needless to say, this is extremely improbable and is a re-expression of the horizon problem with a vengeance—one of the very problems the PBBI scenario was intended to solve.

On the other hand, if the universe in the SPV era is *open,* then in order to account for the homogeneity of our part of the universe, it must start as a Milne universe (roughly, an infinitely large patch of Minkowski space) in the distant past with an infinitesimally small and spatially homogeneous dilaton kinetic energy density of infinite extent. In order for the PBB era to be of infinite duration, it would be necessary for the SPV universe to shrink uniformly for an infinitely long time until the dilaton density grows sufficiently large to cause the scale factor to bounce and undergo super-inflation. Of course, such an SPV state is highly unstable and can be completely destroyed by quantum fluctuations of the dilaton field, which is why, as we saw earlier, the PBBI universe cannot be of infinite duration and must have a beginning. Nonetheless, even if the exquisitely fine-tuned homogeneity required of an open PBB universe were explainable, Kaloper, Linde and Bousso demonstrate that the possibility of resolving the flatness problem depends on being able to explain the unlikely existence and value of two very large dimensionless parameters on which this flatness depends: $g_0^{-2} > 10^{53}$ and $B > 10^{38} g_0^{-2} > 10^{91}$.

Finally, Kaloper, Linde and Bousso demonstrate that the dynamics of PBB cosmology preclude the possibility of self-reproduction and hence do not lead to a period of eternal inflation because quantum fluctuations during the inflationary stage are never large enough to overtake the rolling of the dilaton-field. As a consequence of this, not only is the PBBI scenario incapable of alleviating the fine-tuning of its own initial conditions, it has no resources for addressing the one in $10\exp(10^{123})$ fine-tuning of the Big Bang entropy of our universe.[116]

FINE-TUNING AND THE STRING-THEORETIC LANDSCAPE MODEL

As a final case, let's evaluate the explanatory power of the string landscape to account for cosmological fine-tuning.[117] The "landscape" of string theory is the brainchild of Leonard Susskind, Joseph Polchinski, Raphael Bousso, and Andrei Linde.[118] It aims to turn the vice of the countless moduli associated with the Calabi-Yau compactification of the higher dimensions in string/M-theory into the virtue of a probabilistic resource for anthropic explanations of cosmological fine-tuning.

As explained earlier, the idea is that bubbles of lower energy string vacua nucleate when moduli decay at random locations throughout higher energy string vacua, which continue to inflate forever. It is suggested that the whole landscape is eventually explored by this means as a series of nested bubble universes. Since only the tiniest fraction of such bubbles exemplify laws and constants hospitable to life, and since observer selection (the weak anthropic principle) places us in just such a bubble, the *anthropic principle* becomes the fundamental "explanation" for cosmological fine-tuning, that is, for why our universe has the laws and constants that it does.

Setting the debatable legitimacy of anthropic explanations aside, what should we make of the string landscape as an entity and of the proposed mechanism for exploring it? Just as with inflationary cosmology, there are some very serious reasons to doubt the tenability of string theory;[119] conjoining the two in one picture would seem to provide twice the ground for skepticism. There is no question that string theory has produced some beautiful and interesting mathematics, but there are some very good reasons to question whether it has told us anything about the universe. First of all, string theory does not make any *unique* predictions that are testable by current experiments (the hypothesis of extra dimensions to reality is separable from its string-theoretic embodiment). Secondly, if models with *non*-positive values for the cosmological constant are also included, string theory comes in an infinite number of versions. With an appreciative nod toward the cleverness of string phenomenologists who have found a set of models consistent with the minimal supersymmetric standard model (MSSM),[120] we still have no idea whether any of these match our reality, and there remain an impossibly large number of them.[121] Thirdly, it is also the case that nobody knows whether eleven-dimensional M-theory, which provides the necessary connection among the five anomaly-free classes of ten-dimensional string theory, is itself mathematically consistent, that is, whether it avoids assigning infinite values to physical quantities. Finally, since we don't really have a clue what the underlying M-theory *is,* we don't even know whether a complete and coherent framework exists that would justify calling the web of conjectures and approximations about strings a unified "theory." This assessment is not the isolated view of a few cranks; it is the considered judgment of a healthy portion of the physics community and it deserves serious consideration.

Even if we grant string theory as a working hypothesis, however, there are reasons internal to it that cast doubt on the tenability of the landscape.[122] Michael Dine argues that if a string landscape of meta-stable ground states exists, it is likely to lead to a prediction of low energy supersymmetry. But in the discretuum of the landscape, he contends, the parameters of low energy physics seem to be *random* numbers, and if this is true, the landscape is *not* a correct description of physics as we know it and so must be rejected. Alternatively, there might be some set of principles in the landscape that explain those laws of nature which do not seem to be anthropically constrained, but it is far from obvious what such principles might be, so even if the landscape were a coherent entity, we would have no key that would enable us to interpret it properly.

Susskind and Douglas think this criticism is very serious and do their best to counter it.[123] Susskind argues, somewhat weakly, that the string landscape is unexplored territory and it is possible that the gauge hierarchy does *not* favor low energy supersymmetry. Douglas's argument is stronger. Building on earlier work,[124] he argues that the vast majority of string vacua do not produce exponentially small symmetry breaking scales and that, given many supersymmetry breaking parameters, adding together the positive breaking terms will produce a distribution weighted toward high scales. It is true that models of supersymmetry breaking driven by a single parameter favor low scale breaking, but models involving more than one independently distributed parameter lead to an expectation of high scale breaking. Nonetheless, the idea of "favoring" one type of vacuum over another is not a strong result. Since we do not yet have the mathematical wherewithal to provide a definitive answer to how the SUSY-breaking scale is distributed in a complete ensemble of phenomenologically viable vacua,[125] Dine's observations remain solid, casting doubt on the intrinsic coherence and phenomenological tenability of the string-theoretic landscape.

Setting these additional doubts aside, we still need to ask whether the proposed inflationary mechanism reifying exploration of the landscape is sufficient to the task. As we know, the landscape is subject to the BGV theorem and has an absolute beginning in the finite past. It therefore has a transcendent cause. But in which string false vacuum state did it begin? Scattered throughout the landscape are at least 10^{500} relative minima constitutive of meta-stable false vacua in which the string moduli can get stuck for a very long time. There is no reason intrinsic to the landscape that *necessitates* that it began with a false vacuum energy greater than what we observe today—indeed, there is no necessity to the supposition that the universe started off in an inflationary state at all, save the convenience of such an assumption for anthropic explanations.[126] Furthermore, the quantum tunneling mechanism by which modulus decay leads to the nucleation of bubble universes with different vacuum energies is exponentially suppressed for transitions to higher energies (and can only occur in the presence of gravity), so it is vastly more likely in the landscape scenario that higher inflationary energy states cascade to lower ones.[127] The assumption of such a cascade is theoretically expedient for the purpose of anthropic explanations, but again not guaranteed; there is in principle no way of knowing whether it is true. Given the exponential suppression of transitions to higher energy states, the only way to ensure that the entirety of the landscape gets explored is either to assume it starts in its highest possible energy state, or if in a lower energy state, to assume that the first string vacuum that came into existence is exponentially older than today's Hubble time. If it started off in a state of low enough energy, however, the hypothesized landscape would have no relevance to the explanation of our finely tuned cosmological constant. So even if some version of string/M-theory were true and reification of the landscape were legitimate—an overly generous concession by any measure—there is no way, even in principle, to determine what proportion of space lands in vacua of each type and hence no reason to think that the whole landscape could or would be explored by such means.

It is worth observing too that a cosmological model that randomly varied the laws and constants of nature in the universes it generated would itself have to be subject to lawful constraints were it not to break down. Such lawful constraints, presumably, fall to string/M-theory functioning as a "meta-theory" that governs which laws and constants, and hence which vacua, are possible. In other words, the principles governing the string/M-theoretic process of variation (whatever they may be) would have to remain stable for the description to be coherent. Of course, were such an explicit meta-theoretic construction to prove consistently realizable, the carefully structured variation process it exemplified would be subject to non-negotiable meta-laws and fine-tuned meta-parameters indicative of design at this higher level—unless, of course, we entertain the absurd notion of an infinite regress of meta-theoretical constructions.[128]

We may also, briefly, raise the issue of whether assuming the existence of 10^{500} universes with different laws and constants generates enough probabilistic resources for anthropic explanation of the fine-tuning of the universe in which we live. The cosmological constant (vacuum energy) of our universe is fine-tuned to 120 decimal places.[129] Given the forty orders of magnitude difference in coupling strengths between the gravitational force and the strong force, plus the absence of any theoretical justification for the size of Newton's gravitational constant, it is reasonable to assume that it might have varied over this range. In consideration of the effects of such a variation, we may conclude that the gravitational constant is fine-tuned to one part in 10^{40} of its physically possible range.[130] Similar considerations lead to the recognition that the weak force is fine-tuned to one part in a billion.[131] The proton/neutron mass difference is fine-tuned to at least four decimal places.[132] As Spitzer notes, there are at least seventeen other independent constants and factors that are fine-tuned to a high degree of precision,[133] some of them requiring a cooperative assignment of values to achieve effects necessary for the existence of life that would be unattainable separately. The cumulative effect of all of these fine-tunings significantly erodes the probabilistic resources inherent in the landscape. A precise calculation of cumulative fine-tuning on the basis of current theory has not yet been made, though significant work continues to be done.[134]

This leaves us to consider the fine-tuning of universal entropy. Unfortunately, the one in $10\exp(10^{123})$ probability—rendered exponentially smaller by the inflationary mechanism—that would swamp the resources of 10^{500} or more string vacua is not a fine-tuning of the laws and constants characteristic of the vacua themselves, but rather a fine-tuning of the *initial conditions* of bubble nucleation. In the inflationary picture, assuming a cascade down the string landscape from an initial vacuum with an energy higher than our own that produces inflationary bubbles decaying at random locations while continuing an eternal expansion, the landscape advocate will contend that there is an unbounded number of instantiations of each string vacuum in the cascade.[135] Given an unbounded number of instantiations of the vacuum characteristic of our universe, so the argument goes, we would expect the one in $10\exp(10^{123})$ initial entropic condition it exemplifies to be instantiated an unbounded number of times. So it is that we encounter the standard but startling claim from practitioners of inflationary cosmology that there are "infinitely many" universes just like our own. A typical example is Alexander Vilenkin,[136] who contends that "[i]n the worldview that has emerged from eternal inflation, our Earth and our civilization are anything but unique. Instead, countless *identical* civilizations are scattered in the infinite expanse of the cosmos." Indeed, clones of each of us are endlessly reproduced throughout the inflationary universe, for "the existence of clones is . . . an inevitable consequence of the theory."[137]

The less sanguine among us might be inclined to remark that if it is a consequence of the theory that endless copies of ourselves exist holding every conceivable opinion and involved in every conceivable activity, then *so much the worse* for inflationary (string) cosmology: it has successfully reduced itself to an absurdity. In this regard, it is worthwhile to ask what the consequences of embracing this theory would be for science itself. A fundamental implication of the theory is that every possible event, no matter how improbable (say, one in $10\exp(10^{123})$, just to pick a number) will happen countlessly many times. Indeed, this conclusion has led to a flurry of articles by cosmologists discussing the string landscape in relation to "Boltzmann Brains" and the question of our universe's "typicality"[138]—a discussion so fantastical that it drew the incredulous attention of the *New York Times*.[139]

If, as inflation standardly assumes, the de Sitter space in which our universe began is a thermal system,[140] then a free-floating "Boltzmann Brain" (BB) can spontaneously appear in this space due to thermal fluctuations.[141] Since quantum fluctuations into large volumes are vastly more improbable than fluctuations into small ones, the overwhelmingly most probable configuration would be the smallest fluctuation compatible with our individual awareness, which is presumed to be a universe containing nothing more than a single brain with external sensations fed into it. Under standard conditions for bubble universe generation in the string landscape,[142] the problem formulated by Dyson, Kleban and Susskind[143] giving rise to the BB phenomenon becomes quite serious.[144] In fact, some calculations lead to free-floating BBs swamping the number of normal brains,[145] in which case it becomes a virtual certainty that we *ourselves* are free-floating BBs rather than persons with a history living in an orderly universe 13.7 billion years old. Not to put too fine a point on it, the BB issue suggests that the multiverse is falsified because the persons we take ourselves to be are not typical observers within it.

Inflationary cosmologists recognize the absurdity of their predicament and are trying to circumvent it, but they cannot agree on how or whether progress on the problem is being made.[146] While there is a sense in which anything with a nonzero probability of happening *will* happen in an infinite eternally inflating multiverse—and an infinite number of times at that—from the perspective of these cosmologists, a viable typicality condition would nonetheless succeed in privileging events that we take to be preconditions of our existence. One idea in this regard has been to finagle the decay time of the inflaton fields in an ad hoc manner so that bubble universes don't get large enough to make BBs more likely than ordinary observers. Of course, setting aside the complete arbitrariness and the principled impossibility of evidence for this strategy of convenience, what results at best is the *relative* typicality of observers like us in an infinite universe where the set of Boltzmann Brains and the set of normal observers have equicardinality and Hilbert's Hotel is open for business. But desperate straits require desperate measures and the fate of universal naturalistic explanation hangs in the balance.[147]

Given what has been wrought, it is perhaps unsurprising that the inflationary multiverse has recently been invoked by a prominent molecular biologist as an "explanation" for the intractably improbable origin of life:

> Despite considerable experimental and theoretical effort, no compelling scenarios currently exist for the origin of replication and translation, the key processes that together comprise the core of biological systems and the apparent pre-requisite of biological evolution.... The MWO [Many Worlds in One] version of the cosmological model of eternal inflation could suggest a way out of this conundrum because, in an infinite multiverse with a finite number of distinct macroscopic histories (each repeated an infinite number of times), emergence of even highly complex systems by chance is not just possible but inevitable.... Specifically, it becomes conceivable that the minimal requirement (the breakthrough stage) for the onset of biological evolution is a primitive coupled replication-translation system that *emerged by chance.* That this extremely rare event occurred on Earth and gave rise to life as we know it is *explained by anthropic selection alone.*... By showing that highly complex systems, actually, can emerge by chance and, moreover, are inevitable, if extremely rare, in the universe, the present model sidesteps the issue of irreducibility and leaves no room whatsoever for any form of intelligent design.[148]

It is not hard to see that anthropic explanation in an infinite multiverse, were it to become the standard default when naturalistic mechanisms have reached the end of their tether, would spell the end of science as a rational enterprise. By providing an all-too-easy explanation for anything that has happened or may happen, the multiverse ends up explaining nothing at all.

Whatever else may be said, it is clear that the string landscape hypothesis is a highly speculative construction built on assumptions that strain the limits of credulity. Even if taken seriously, the reality is that: (1) the mechanisms of the landscape will *require meta-level fine-tuning* themselves; (2) there are substantial reasons to think the landscape as a whole may be *intrinsically incoherent* and *phenomenologically untenable*; (3) the mechanism for exploring the landscape may be *unequal to the task* required of it, and beyond this—like all quantum-theoretic constructs—it lacks an immanent principle of sufficient causality, thus pointing to its metaphysical incompleteness and *need for ongoing transcendent catalyzation*;[149] (4) even if the whole landscape were capable of being explored, the number of string vacua compatible with a positive cosmological constant *may not ultimately prove sufficient* to account for the actual fine-tuning of the laws and constants of our universe; and (5) given its reliance on the equally dubious mechanism of eternal inflation, the string landscape contains the seeds for *destroying science altogether* as a rational enterprise. Such are the follies of scientism.

4. End Game: Mind over Matter

Given this sobering assessment, one wonders why the string landscape has provoked so much enthusiasm. Leonard Susskind provides a revealing answer:

> If, for some unforeseen reason, the landscape turns out to be inconsistent—maybe for mathematical reasons, or because it disagrees with observation... [then] as things stand now we will be in a very awkward position. Without any explanation of nature's fine-tunings we will be hard pressed to answer the ID critics.[150]

Indeed, and if Eugene Koonin is to be believed, this inability to avoid intelligent design will carry over into origins of life research if the inflationary multiverse fails.[151] But what gives rise to this reticence about intelligent design? Its detractors mutter about a "god of the gaps" and "arguments from ignorance," but such objections miss the mark and deflect back on their own appeals to chance, especially in contexts such as these where, in the absence of any causally sufficient story, blind luck is invoked as as *deus ex machina* for naturalistic explanations. Intelligent design, by contrast, provides an argument from what we know intelligent causes are sufficient to produce and, furthermore, only intelligent causes are known to be sufficient to produce: structures incredibly rich in complex specified information.[152] Having satisfied the conditions of causal sufficiency and causal uniqueness for the phenomenon in question, therefore, an inference to intelligent design as the best explanation for cosmological origins and fine-tuning is conspicuously warranted.

Nonetheless, reticence remains, and the evolutionary biologist and geneticist Richard Lewontin helps to put a finger on one source of it:

> Our willingness to accept scientific claims that are against common sense is the key to an understanding of the real struggle between science and the supernatural. We take the side of science in spite of the patent absurdity of some of its constructs, in spite of its failure to fulfill many of its extravagant promises of health and life, in spite of the tolerance of the scientific community for unsubstantiated just-so stories, because we have a prior commitment, a commitment to materialism. It is not that the methods and institutions of science somehow compel us to accept a material explanation of the phenomenal world, but, on the contrary, that we are forced by our *a priori* adherence to material causes to create an apparatus of investigation and a set of concepts that produce material explanations, no matter how counter-intuitive, no matter how mystifying to the uninitiated. Moreover, that materialism is absolute, for we cannot allow a Divine Foot in the door. The eminent Kant scholar Lewis Beck used to say that anyone who could believe in God could believe in anything. To appeal to an omnipotent deity is to allow at any moment the regularities of nature may be ruptured, that miracles may happen.[153]

Setting aside possible motivations arising from a desire for freedom from transcendent moral constraints and accountability, there is little doubt that Lewontin is right about the motivation and the reasoning behind scientistic fear of the miraculous, but in articulating the matter so clearly he has exposed a central irony: in their theophobic flight, scientific materialists have found it necessary to affirm a universe in which anything can happen—fully functioning brains popping out of the quantum vacuum, for instance—without a sufficient causal antecedent and for no rhyme or reason. So who believes in miracles *now*? What is more, the naturalist believes in *random* miracles. In a theistic universe, on the other hand, nothing happens without a reason, and while nature is not self-sufficient and therefore not causally closed, any miracles constituted by intelligently directed deviations from purposefully maintained regularities are also expressions of divine purpose. In the ultimate irony, therefore, what we see is that the purposes of scientific naturalism cannot survive the purposelessness they create, for out of the random void is birthed the end of scientific rationality itself.

Stephen Hawking once asked, somewhat poetically, "What is it that breathes fire into the equations and makes a universe for them to describe?"[154] He intended the question rhetorically, but it both deserves and has a genuine answer. Mathematical descriptions may have ontological implications, but they do not function as efficient causes, either metaphysically or materially. They are causally inert abstract objects. If quantum cosmology describes string vacua tunneling into existence from a highly structured faux-nothingness or from another vacuum state, or if relativistic quantum field theory describes evanescent matter scintillating in the quantum vacuum or manifesting nonlocally correlated behavior, neither mathematical construction provides an explanation, let alone an efficient cause, for these events. To believe otherwise is to be guilty of an ontological category mistake. So with all due respect to Leonard Susskind and his coterie of devout string landscape naturalists,[155] there is no landscape of mathematical possibilities that gives rise to a megaverse of actualities and provides a *mindless* solution to the problem of cosmological fine-tuning, for even an infinite arena of mathematical possibilities lacks the power to generate one solitary universe.

The mindless multiverse "solution" to the problem of fine-tuning is, quite literally, a metaphysical non-starter. What the absence of efficient material causality in fundamental physics and cosmology reveals instead is the limit of scientific explanations and the need for a deeper metaphysical understanding of the world's rationality and orderliness. That explanation has always been, and will forever be, Mind over matter. When the logical and metaphysical necessity of an efficient cause, the demonstrable absence of a material one, and the realized implication of a universe both contingent and finite in temporal duration, are all conjoined with the fact that we exist in an ordered cosmos the conditions of which are fine-tuned beyond the capacity of any credible mindless process, the scientific evidence points inexorably toward *transcendent intelligent agency* as the only sufficient cause, and thus the only reasonable explanation. In short, a clarion call to intellectual honesty and metaphysical accountability reverberates throughout the cosmos: release the strings of nihilism and let the balloons of naturalism drift unaccompanied into their endless night. If anyone has ears to hear, let him hear.

Notes

1. My thanks to Fr. Robert Spitzer, David Berlinski, Arthur Fine, Don Page, and most especially Gerald Cleaver, James Sinclair, and Robin Collins for comments on an earlier draft of this essay. I am solely responsible for any residual infelicities.
2. This question is the focus of the essays by Numbers, McMullin, and Meyer in Part I of this volume.
3. I do not mean to suggest that such research should not be pursued; something invariably is learned from the attempt, and the shortcomings of such efforts contain lessons of a more profound sort.
4. There is, of course, always the third option, viz., maintaining that some demands for explanation may well be excessive or unreasonable, and that seeking an explanation for the origin of the universe and its properties is an instance in which explanatory demand can and should be rejected. To the contrary, I would maintain—in the company of the vast majority of Western scientists and intellectuals up through the end of the nineteenth century—that the correct explanation is perfectly obvious and evidentially supported beyond reasonable doubt: *transcendent intelligent causation*. Moreover, it is clear from conceptually related historical considerations that Judeo-Christian monotheism provided the basis and impetus for the quest for order in nature by modern science, as well as the confidence that such order would be found. Indeed, a belief of this kind is justifiably regarded as the *transcendental ground* for the very possibility of science as a rational truth-conducive enterprise; evolutionary naturalism certainly does not provide grounds for science, in fact, just the opposite (see Plantinga's and Koons's contributions to this volume). Far from the Middle Ages being the "Dark Ages," therefore, they were the period during which the foundations and expectations of modern science were planted, nurtured, and eventually blossomed. This historical corrective to conventional conceits regarding the "folly" of medieval scholasticism receives ample documentation in Edward Grant's *The Foundations of Modern Science in the Middle Ages* (1996) and Rodney Stark's *For the Glory of God: How Monotheism Led to Reformations, Science, Witch-Hunts, and the End of Slavery* (2003) and *The Victory of Reason: How Christianity Led to Freedom, Capitalism, and Western Success* (2005). Insofar as obviation of the supernatural is indeed the motivation for current efforts in multiverse cosmology, the consequence, foreseen or not, is that the very foundations of rationality and morality are undermined (in the latter regard, see Dallas Willard's contribution to this volume). Much more than the simple truth is at stake, therefore, in the inference to transcendent intelligent agency as the best and only rational explanation for the data of cosmology. Since the *causal sufficiency* and *causal uniqueness* of transcendent intelligent agency as a cosmological explanation justly inspire confidence in the truth of what is independently recognized to be essential to the grounding of rationality itself, to rest content with the brute factuality of the universe is to recommend a form of skepticism without merit—both baseless and destructive. For a more extensive discussion of these matters, see my essay "The Rise of Naturalism and Its Problematic Role in Science and Culture" in Part I of this volume.
5. Einstein 1916, 1917; Friedmann 1922; Lemaître 1927; Hawking and Penrose 1970; Hawking and Ellis 1973: 256–98.
6. Fred Hoyle 1975: 684–85.
7. Craig 1993: 46.
8. Linde 1986b.
9. Guth 1981; Linde 1982.
10. It is worth noting the difference in the singularities characteristic of spatiotemporally finite versus spatiotemporally infinite universes in classical general relativity. Finite (elliptic) universes begin with a singular point of infinite density from which they expand to produce a universe that has a finite but unbounded geometry (like the surface of a sphere). Universes that may be conceived as infinite (flat or hyperbolic) are thought of as

having begun from a state of infinite density that is also *infinite in extent* and from which they expand *everywhere at once.* This is why general relativists sometimes speak of open universes as being infinite in extent: in the mathematical model, the singularity from which they began may be conceived as *already infinitely extended;* it is *not* the case, *per impossible,* that such universes grow to infinite size in finite time. If a certain air of unreality should seem to the reader to pervade such mathematical modeling, he may be credited with an admirable degree of metaphysical astuteness. It is the metaphysical absurdities and ineluctable causal incompleteness frequently exhibited by the constructs of fundamental physical theory that, among other things, pose insurmountable problems for ontological naturalism and require radical metaphysical recontextualization before sense can be made of them (see my essay "A Quantum-Theoretic Argument against Naturalism" in this volume).

11. Linde 1986a.
12. Garriga and Vilenkin 2001; Vilenkin 2006a.
13. Borde and Vilenkin 1994: 3305–309; Borde and Vilenkin 1997: 717–23.
14. Borde, Guth, and Vilenkin 2003.
15. See Alan Guth's essay in this volume.
16. Vilenkin 1982, 1988, 1994, 2002; Hartle and Hawking 1983; Hawking 1987, 1988.
17. Hilbert 1925.
18. See Spitzer 2009 (Chapter 5, Section III) for a more complete discussion.
19. Craig 1979; Craig and Smith 1993; Copan and Craig 2004: 147–266.
20. One might also argue that contingent entities, like our universe, even if there were no time $\tau = 0$ at which they began to exist, would still require an explanation of their existence in virtue of their *contingent* character. In respect of the universe as a whole especially, this explanation would have to be given in terms of something that existed *transcendently* and *necessarily* and was capable of *activity* (and hence not an abstract object). Since cosmological models addressing the issue of the fine-tuning of the initial conditions and the laws and constants of nature do not lead us in this direction, we will not pursue this train of thought any further. Those interested in a rigorous development of the cosmological argument from contingency should consult Rob Koons's 1997 essay "A New Look at the Cosmological Argument" and also consult Alexander Pruss's book *The Principle of Sufficient Reason: A Reassessment* (Cambridge: Cambridge University Press, 2006).
21. Whitehead 1925.
22. Tegmark 1998, 2003.
23. A more thorough refutation of Tegmark can be found in Robin Collins's (2009) essay "The Fine-Tuning Argument for Theism."
24. In another vein, this latter argument might also be advanced against the "many worlds" interpretation (MWI) of quantum mechanics, which is based on the idea that quantum wavefunctions never collapse; rather, every possible outcome of every quantum process is realized in actuality, but each occurs in a different "parallel" universe empirically inaccessible to our own. Aside from the ontologically profligate, completely untestable, generally unwarranted and deeply implausible character of this proposed resolution of the measurement problem, it also suffers from some intractable technical difficulties. There is a serious problem with the concept of probability in the MWI context. If I am going to perform a quantum experiment with two possible outcomes such that standard quantum mechanics predicts probability 1/3 for outcome A and 2/3 for outcome B, then, according to the MWI, *both the world with outcome A and the world with outcome B will exist.* It is then *meaningless* to ask "What is the probability that I will observe A instead of B?" because *both* events will happen and parallel versions of myself will observe each outcome in its associated world. So whence the "probabilities" of quantum theory? Furthermore, quantum theory allows *infinitely* many ways to decompose the quantum state of the whole universe into a superposition of orthogonal states. So the question arises for the many worlds interpretation: "Why choose *this* particular decomposition and not any other?" Since alternate decompositions might lead to very *different* pictures, the whole construction is *arbitrary* and *devoid of empirical content.*

25. See my article "A Quantum-Theoretic Argument against Naturalism" in this volume; also footnote 4.
26. Vilenkin 1982; 1988; 1994; 2002.
27. Hartle and Hawking 1983.
28. *A Brief History of Time* (1988).
29. Isham 1993a: 68.
30. For a short discussion of the conceptual differences among description, prediction, and explanation, see my paper "A Quantum-Theoretic Argument against Naturalism" in this volume. Mere descriptions, even when accurate, do not function as explanations of that which they describe; the description provided by the quantum-gravitational wavefunction—which, apart from an MWI interpretation (see footnote 24 above), must collapse *acausally* to give us the reality in which we live—is no exception.
31. For a discussion of the "problem of time" in quantum gravity, see Butterfield and Isham (1999; 2001) as well as many of the essays in Callender and Huggett (2001).
32. Griffiths 1984, 1993; Omnes 1994: 122–43, 268–323; Gell-Mann and Hartle 1990, 1996.
33. Dowker and Kent 1996, Gell-Mann and Hartle 1996.
34. Again, see my essay "A Quantum-Theoretic Argument against Naturalism" in this volume.
35. DeWitt 1967.
36. Vilenkin 2002: 12–13.
37. See Stephen Meyer's essay "Methodological Naturalism and the Failure of Demarcation Arguments" in this volume.
38. DeWitt 1967.
39. Vilenkin 1988.
40. Isham 1993a: 72–73.
41. Isham 1993a: 72n38.
42. Hartle and Hawking 1983; Vilenkin 1994.
43. Hawking 1987: 640–45.
44. Hawking 1987: 646–47.
45. Hawking 1975.
46. See Butterfield and Isham 1999.
47. DeWitt 1967.
48. Hartle-Hawking 1983.
49. Butterfield and Isham 1999.
50. Hawking 1987: 640.
51. Hawking 1988: 140–41.
52. Hawking 1988: 139.
53. See footnote 20 above; also Koons 1997 and Pruss 2006.
54. For further discussion of the shortcomings of inflationary cosmology see Hawking and Page 1987: 789–809; Penrose 1989b: 249–64; Rees 1997; Earman and Mosterin 1999; Martin and Brandenberger 2001; Hollands and Wald 2002: 2043–55; Holder 2004: 130–43; Penrose 2005: 746–57, 762–65; and van Elst 2008.
55. Penrose 2005: 754.
56. Hawking and Page 1987.
57. Earman and Mosterin 1999.
58. Penrose 1989b: 249–64; 2005: 746–57.
59. Banks 2007: 4.
60. Standard quantification of the inflaton field is given by the number of its "e-foldings," N, which provide a way of measuring the inflationary expansion. If standard slow-roll inflation is operative, then $N = \ln(a_f/a_i)$,

where "i" and "f" denote respectively the initial and final values of the scale factor of the universe (the global multiplier to universe size). For other cases, such as oscillating inflation, this definition must be modified. See section III of Liddle and Mazumdar (1998) for a brief discussion.

61. The adiabatic theorem states that a quantum mechanical system, subjected to *gradually* changing external conditions, can adapt its functional form; in the case of *rapidly* varying conditions, though, there is no time for the functional form of the state to adapt, so its probability density remains unchanged. See Messiah 1999: 739–50.

62. See also Martin and Brandenberger 2001.

63. Rees 1997: 185; Earman and Mosterin 1999; Holder 2004: 130–43.

64. Land and Magueijo 2005.

65. Land and Magueijo 2007. The worry is that the anisotropies are the result of insufficient subtraction of Milky Way polar-aligned contributions, since the preferred direction seems to be aligned with the Milky Way pole. Further research should shed definitive light on this issue, especially data from the Planck satellite launched in spring 2009.

66. For a *small* sample of these discussions, see Kofman 1996; Boyanovsky, Cormier, de Vega, Holman, Singh, and Srednecki 1997; Bassett 1997; Boyanovsky, Cormier, de Vega, Holman, and Kumar 1998; Bassett and Viniegra 2000; Tsujikawa, Bassett and Viniegra 2000; Felder, García-Bellido, Greene, Kofman, Linde, and Tkachev 2000; Felder, Kofman, and Linde 2001; Green and Malik 2002; Watanabe and Komatsu 2007; and Brandenberger, Frey, and Lorenz 2008.

67. Collins 2003; Cohn 1998; Sahni and Starobinsky 2000: 373–444.

68. Penrose 1989b: 249–64; Penrose 2005: 746–57, 762–65.

69. Penrose 1981: 245–72; Penrose 1989a: 423–47; Penrose 2005: 726–32, 762–65.

70. Penrose offers another quite powerful entropy-based argument against the anthropic inflationary universe. I will not discuss it here, but I commend it to the reader's consideration (Penrose 2005: 762–65).

71. Since even when compactified, bosons remain bosons and fermions remain fermions, it takes compactification of the gravitino to produce spacetime fermionic matter. I thank Gerald Cleaver for this clarification.

72. Green, Schwarz and Witten 1987.

73. Bousso and Polchinski 2000; Kachru, Kallosh, Linde and Trivedi 2003; Kachru, Kallosh, Linde, Maldecena, McAllister and Trivedi 2003; Susskind 2003; Susskind 2004; Freivogel and Susskind 2004; Bousso and Polchinski 2004: 60–69; Ashok and Douglas 2004; Douglas 2004a; Kobakhidze and Mersini-Houghton 2004: 869–73; Ooguri and Vafa 2006; Riddle and Urena-Lopez 2006; Barvinsky and Kamenshchik 2006; Susskind 2006; Vanchurin and Vilenkin 2006; Denef and Douglas 2006; Kumar 2006: 3441–472; Polchinski 2006; Cleaver 2006. Indeed, the embarrassment of riches is so extreme that, were cosmological fine-tuning not so *incredibly* stringent, one might be inclined to modify William Unruh's quip that he could fit any dog's leg you handed him with inflation (as reported in *Science,* August 30, 1996) to state "I'll fit any dog's leg you hand me with string theory."

74. Green and Schwarz 1985: 93–114; Green, Schwarz and Witten 1987.

75. Witten 1995.

76. Arkani-Hamed, Dimopoulos, and Dvali 1998: 263–72.

77. Adelberger, Heckel, and Nelson 2003: 87–100; Kapner, Cook, Adelberger, Gundlach, Heckel, Hoyle, and Swanson 2007.

78. Adelberger, Heckel, and Hoyle 2005; Giddings and Thomas 2002.

79. Khoury, Ovrut, Steinhardt and Turok 2001a; Steinhardt and Turok 2002a; Steinhardt and Turok 2002b; Khoury, Ovrut, Seiberg, Steinhardt and Turok 2002; Turok, Perry and Steinhardt 2004; McFadden, Turok and Steinhardt 2005; Steinhardt and Turok 2007.

80. Khoury, Ovrut, Steinhardt and Turok 2001a.

81. Steinhardt and Turok 2002a.

82. Borde, Guth and Vilenkin 2003; Vilenkin 2006a; Steinhardt and Turok 2005: 43–7; Steinhardt 2004.
83. Steinhardt and Turok 2007; Vilenkin 2006a; Steinhardt 2004.
84. Borde, Guth and Vilenkin 2003; Vilenkin 2006a; Steinhardt and Turok 2005; Steinhardt 2004.
85. Gasperini and G. Veneziano 2003: 1–212; Veneziano 1998; Hawking and Penrose 1970: 529–48; Veneziano 1995; Veneziano 1997: 297–303; Feinstein, Lazkoz and Vazquez-Mozo 1997; Barrow and Dabrowski 1997; Saygili 1999: 225–40; Buananno, Meissner, Ungarelli and Veneziano 1998; Barrow and Kunze 1997; Gasperini 1999; Turner and Weinberg 1997: 4604–609; Maggiore and Sturani 1997: 335–43; Kaloper, Linde and Bousso 1999; Brustein and Veneziano 1994: 429–34; Gasperini, Maharana and Veneziano 1996: 349–60; Rey 1996: 1929–32; Gasperini, Maggiore and Veneziano 1997: 315–30; Gasperini 2000; Brandenberger, Easther and Maia 1998; Foffa 2003; Gasperini 2007; Gasperini 2008.
86. In string theory, *dilatons* (radions, graviscalars) are quanta of a massless scalar field ϕ that obeys a generalized Klein-Gordon equation and is always linked with gravity. Perturbative string theories automatically contain dilatons in ten dimensions, but M-theory doesn't include them in its spectrum *unless* it's compactified. The dilatonic coupling constant is a dynamical variable in string theory. If supersymmetry is *un*broken, these scalar fields can take arbitrary values (they are *moduli* characteristic of *different* string solutions); supersymmetry breaking, however, creates a potential energy for scalar fields that localizes near a minimum value that is, at least in principle, calculable.
87. Veneziano 1998.
88. Hawking and Penrose 1970: 529–48.
89. Veneziano 1995.
90. Veneziano 1997: 297–303; Feinstein, Lazkoz and Vazquez-Mozo 1997; Barrow and Dabrowski 1998: 7204–22; Saygili 1999: 225–40.
91. Veneziano 1998; Veneziano 1997: 297–303; Feinstein, Lazkoz and Vazquez-Mozo 1997; Barrow and Dabrowski 1998: 7204–22; Saygili 1999: 225–40.
92. Turner and Weinberg 1997: 4604–609.
93. Turner and Weinberg 1997: 4604–609; Maggiore and Sturani 1997: 335–43; Kaloper, Linde and Bousso 1999; Brustein and Veneziano 1994: 429–34; Gasperini, Maharana and Veneziano 1996: 349–60; Rey 1996: 1929–32; Gasperini, Maggiore and Veneziano 1997: 315–30; Gasperini 1999: 1059–66.
94. Gasperini and Veneziano 2003: 1–212.
95. Veneziano 1998; Brustein and Veneziano 1994: 429–34; Gasperini, Maharana and Veneziano 1996: 346–60; Rey 1996: 1929–32; Gasperini, Maggiore and Veneziano 1997: 315–30; Gasperini 2000; Brandenberger, Easther and Maia 1998; Foffa 2003; Gasperini 2007.
96. Veneziano 1998; Veneziano 1995; Veneziano 1997: 297–303; Feinstein, Lazkoz and Vazquez-Mozo 1997; Barrow and Dabrowski 1997: 7204–22; Saygili 1999: 225–40; Buananno, Meissner, Ungarelli and Veneziano 1998: 2543–56; Barrow and Kunze 1998.
97. Bousso and Polchinski 2000; Kachru, Kallosh, Linde and Trivedi 2003; Kachru, Kallosh, Linde, Maldecena, McAllister and Trivedi 2003; Susskind 2003; Susskind 2004; Freivogel and Susskind 2004; Bousso and Polchinski 2004: 60–69.
98. Khoury, Steinhardt and Turok 2004.
99. Khoury, Steinhardt and Turok 2004; Khoury, Steinhardt and Turok 2003; Gratton, Khoury, Steinhardt and Turok 2004.
100. Kallosh, Kofman and Linde 2001; Linde 2001: 89–104; Felder, Frolov, Kofman and Linde 2002; Lyth 2002: 1–4; Räsänen 2002: 183–206; Heyl and Loeb 2002.
101. Felder, García-Bellido, Greene, Kofman, Linde and Tkachev 2000; Felder, Kofman and Linde 2001.
102. Turok, Perry and Steinhardt 2004; Khoury, Ovrut, Steinhardt and Turok 2001b; Donagi, Khoury, Ovrut, Steinhardt and Turok 2001; Steinhardt and Turok 2002c; Khoury, Ovrut, Steinhardt and Turok 2002; Tolley, Steinhardt and Turok 2004; Erickson, Gratton, Steinhardt and Turok 2006.

103. Bozza and Veneziano 2005a: 177–83; Bozza and Veneziano 2005b; Bozza 2005.

104. Gratton, Khoury, Steinhardt and Turok 2004; Khoury, Ovrut, Steinhardt and Turok 2002; Tolley, Steinhardt and Turok 2004; Erickson, Gratton, Steinhardt and Turok 2006.

105. Kim and Hwang 2007.

106. Not that Steinhardt and Turok would recommend this course of action, since they deplore anthropic arguments (see Steinhardt and Turok 2007: 231–36).

107. Steinhardt and Turok 2002c.

108. Steinhardt and Turok 2005: 43–47.

109. Steinhardt and Turok 2006: 1180–82.

110. For a slight qualification of this assertion, see footnote 126.

111. Vilenkin 2006b.

112. Turner and Weinberg 1997.

113. Veneziano 1998; Buananno, Meissner, Ungarelli and Veneziano 1998: 2543–56.

114. Maggiore and Sturani 1997: 335–43.

115. Kaloper, Linde, and Bousso 1999.

116. Veneziano (1999) suggests that gravitational contraction is a scale-free phenomenon in that regional patches of perturbed SPV of all different sizes will contract to create Big Bang events. He then argues that an anthropic explanation of entropic fine-tuning is possible on the basis of a multiverse created by regional contractions. There are two responses to be made here. The *first* is that the fine-tuning considerations that make the contraction of *any* SPV patch exponentially unlikely make the contraction of *multiple* patches to create a multiverse *hyper-exponentially unlikely*. The hyper-exponential unlikeliness of multiple contractions will far outrun any probabilistic resources that multiple contractions might generate to address entropic fine-tuning, especially when the meta-stability of the SPV guarantees an origin in the finite past. The *second* is that Gasperini and Veneziano often talk (and rightly so) as if they regard an infinite past to the SPV phase in the PBBI model as an idealization that has *no* real existence. If it has no real existence, however, as we have just argued, it cannot provide the resources to explain away entropic or other kinds of cosmological fine-tuning.

117. Bousso and Polchinski 2000; Kachru, Kallosh, Linde and Trivedi 2003; Kachru, Kallosh, Linde, Maldecena, McAllister and Trivedi 2003; Susskind 2003; Susskind 2004; Freivogel and Susskind 2004; Bousso and Polchinski 2004: 60–69; Ashok and Douglas 2004; Douglas 2004a; Kobakhidze and Mersini-Houghton 2004: 869–73; Ooguri and Vafa 2006; Riddle and Urena-Lopez 2006; Barvinsky and Kamenshchik 2006; Susskind 2006; Vanchurin and Vilenkin 2006; Denef and Douglas 2006; Kumar 2006: 3441–72; Polchinski 2006; Cleaver 2006.

118. Bousso and Polchinski 2000; Kachru, Kallosh, Linde and Trivedi 2003; Kachru, Kallosh, Linde, Maldecena, McAllister and Trivedi 2003; Susskind 2003; Susskind 2004; Freivogel and Susskind 2004; Bousso and Polchinski 2004: 60–69.

119. Krauss 2005; Smolin 2006; Woit 2006.

120. Cleaver, Faraggi and Nanopoulos 1999: 135–46; Cleaver 2008.

121. Kumar 2006.

122. Banks, Dine and Gorbatov 2003; Dine 2004; Robbins and Sethi 2005.

123. Susskind 2004; Douglas 2004b.

124. Denef and Douglas 2004.

125. Kumar 2006.

126. Some would suggest that it is "natural" to start off with an order 1 cosmological constant since the most straightforward calculation of the vacuum energy in quantum field theory comes to roughly one Planck mass per cubic Planck length, which exceeds the *actual* value by 120 orders of magnitude—the worst prediction in the entire history of physics! Others argue that, given a probability distribution of cosmological constants for *possible* universes over the *whole* landscape, an order 1 cosmological constant is not an unreasonable assump-

tion. My basic point, however, is that the landscape hypothesis *needs* this condition to generate anthropic explanations and there is nothing that guarantees it, rather it is an additional and unjustified postulate required for explanatory traction.

127. See Linde 2007; also Coleman and De Luccia 1980 and Guth and Weinberg 1983.

128. This point has also been made quite powerfully by both Robin Collins and John Polkinghorne.

129. Bousso and Polchinski 2004.

130. Collins 2003: 189–90.

131. Collins 2003: 188–89.

132. Collins 2003: 186–88.

133. Spitzer 2009.

134. See, for example, Collins 2003, 2007, 2009 and *The Well-Tempered Universe* (forthcoming); Carter 1967; Carr and Rees 1979; Davies 1982; Barrow and Tipler 1986; Leslie 1989; Hogan 2000; Oberhummer, Csótó and Schlattl 2000; Rees 2000; Gonzalez and Richards 2004; Carr (ed.) 2007; and Spitzer 2009.

135. As we discussed earlier and is worth reinforcing by repetition, this assumption is *not* mandated by the landscape hypothesis and, even though the exponentially suppressed transitions to higher energy states render an uphill climb possible in a timespan exponentially *longer* than today's Hubble time, a sufficiently low-energy genesis to the landscape would completely vitiate the landscape's utility for anthropic explanations. Even on the unlikely assumption that the landscape exists, therefore, we have no way of knowing (apart from our own universe) in what it consists, and there is no principled way to tell what proportion of space lands in vacua of which type, and hence no reason to think that the whole landscape could or would be explored.

136. Vilenkin 2006a: 117, 114.

137. Indeed, Vilenkin's "many worlds in one" model (Garriga and Vilenkin 2001; Vilenkin 2006a) has the consequence that *all* macroscopic sequences of events not forbidden by physical conservation laws not only occur *somewhere* in an eternally inflating universe, but occur over and over again *without limit* as inflation endlessly spawns new expanding regions of spacetime. For instance, the model suggests there is an *unlimited* number of macroscopically exact copies of the Earth and everything that exists on it, even though the probability of any given observable region of the universe containing such a copy is vanishingly small.

138. Dyson, Kleban and Susskind 2002; Albrecht and Sorbo 2004; Page 2006; Ceresole, Dall'Agata, Giryavets, Kallosh and Linde 2006; Linde 2006; Bousso and Freivogel 2006; Page 2006; Carlip 2007; Hartle and Srednecki 2007; Giddings and Maroff 2007; Page 2007a; Page 2007b.

139. Overbye 2008.

140. Dyson, Kleban and Susskind 2002.

141. Linde 2006; Bousso and Freivogel 2006; Page 2006.

142. Kachru, Kallosh, Linde and Trivedi 2003.

143. Dyson, Kleban and Susskind 2002; Albrecht and Sorbo 2004; Page 2006.

144. Bousso and Freivogel 2006.

145. Dyson, Kleban and Susskind 2002.

146. Bousso and Freivogel 2006; Page 2006; Carlip 2007; Hartle and Srednecki 2007; Giddings and Maroff 2007; Page 2007a; Page 2007b.

147. A more rigorous treatment of the Boltzmann Brain issue can be found in Collins (2009).

148. Koonin 2007. Indeed, the probabilistic difficulties attending the undirected production of life in a *life-compatible* universe would be comparable to (and in some cases even exceed) the unattended production of a universe with the finely-tuned properties of our own if the raw "mechanisms" of universe production were capable of autonomous operation. The reader is referred to Stephen Meyer's *The Signature in the Cell* (2009) for an extensive introduction to these issues in origin of life research, as well as to the essays by Meyer, Axe, Dembski and Marks, and Behe in Part III of this volume. For ongoing scientific research into this and related

issues in biological science, the reader is referred to the work of Biologic Institute (http://www.biologicinstitute.org).

149. Once more, see my paper "A Quantum-Theoretic Argument against Naturalism" in this volume.

150. Susskind, as quoted in *New Scientist* magazine, December 17, 2005.

151. Koonin 2007.

152. See Dembski (1998; 2002; 2005), Dembski and Marks (2009a, 2009b), and especially the contribution to this volume by Dembski and Marks.

153. Lewontin 1997: 31.

154. Hawking 1988: 174.

155. Susskind 2006.

References

Adelberger, E., Heckel, B. and Nelson, A. (2003) "Test of the Gravitational Inverse-Square Law" *Nuclear Physics* B672: 87–100 (hep-th/0307284).

Adelberger, E., Heckel, B. and Hoyle, C. (2005) "Testing the Gravitational Inverse-Square Law," http://physicsworld.com/cws/article/print/21822.

Albrecht, A. and Sorbo, L. (2004) "Can the universe afford inflation?" *Journal of Physics* A38: 1345–70 (hep-th/0405272).

Arkani-Hamed, N., Dimopoulos, S. and Dvali, G. (1998) "The Hierarchy Problem and New Dimensions at a Millimeter," *Physics Letters* B429: 263–272 (hep-ph/9803315).

Ashok, S. and Douglas, M. (2004) "Counting Flux Vacua," *Journal of High Energy Physics* 0401 060 (hep-th/0307049).

Banks, T. (2007) "Entropy and initial conditions in cosmology" (hep-th/0701146).

Banks, T., Dine, M. and Gorbatov, E. (2004) "Is There a String Theory Landscape?" *Journal of High Energy Physics* 0408 058 (hep-th/0309170).

Barrow, J. and Tipler, F. (1986) *The Anthropic Cosmological Principle*. New York: Oxford University Press.

Barrow, J. and Kunze, D. (1997) "Spherical Curvature Inhomogeneities in String Cosmology" (hep-th/9710018).

Barrow, J. and Dabrowski, M. (1998) "Is There Chaos in Low-Energy String Cosmology?" *Physical Review* D57: 7204–7222 (hep-th/9711049).

Barvinsky, A. and Kamenshchik, A. (2006) "Cosmological Landscape from Nothing: Some Like It Hot," *Journal of Cosmology and Astroparticle Physics* 0609 014 (hep-th/0605132).

Bassett, B. (1997) "The Preheating-Gravitational Wave Correspondence: I," *Physical Review* D56: 3439 (hep-ph/9704399).

Bassett, B., and Viniegra, F. (2000) "Massless metric preheating," *Physical Review* D62: 043507 (hep-ph/9909353).

Borde, A. and Vilenkin, A. (1994) "Eternal inflation and the initial singularity," *Physical Review Letters* 72: 3305–309 (gr-qc/9312022).

_____. (1997) "Violation of the weak energy condition in inflating spacetimes," *Physical Review* D56: 717–23 (gr-qc/9702019).

Borde, A., Guth, A., and Vilenkin, A. (2003) "Inflationary spacetimes are not past-complete," *Physical Review Letters* 90 151301 (gr-qc/0110012).

Bousso, R. and Polchinski, J. (2000) "Quantization of Four-form Fluxes and Dynamical Neutralization of the Cosmological Constant," *Journal of High Energy Physics* 0006 006, (hep-th/0004134).

_____. (2004) "The String Theory Landscape," *Scientific American* 291: 60–69.

Bousso, R. and Freivogel, B. (2006) "A paradox in the global description of the multiverse" (hep-th/0610132).

Boyanovsky, D. Cormier, D., de Vega, H., Holman, R. and Kumar, S. (1998) "Out of Equilibrium Fields in Inflationary Dynamics: Density Fluctuations" (hep-ph/9801453).

Boyanovsky, D. Cormier, D., de Vega, H., Holman, R., Singh, A., and Srednecki, M. (1997) "Scalar Field Dynamics in Friedman-Robertson-Walker Spacetimes," *Physical Review* D56: 1939–57 (hep-ph/9703327).

Bozza, V. (2005) "General solution for scalar perturbations in bouncing cosmologies," *Journal of Cosmology and Astroparticle Physics* 0602 009 (hep-th/0512066).

Bozza, V. and Veneziano, G. (2005a) "Scalar Perturbations in Regular Two-Component Bouncing Cosmologies," *Physics Letters* B625: 177–183 (hep-th/0502047).

_____. (2005b) "Regular two-component bouncing cosmologies and perturbations therein," *Journal of Cosmology and Astroparticle Physics* 0509 007 (gr-qc/0506040).

Brandenberger, R., Easther, R. and Maia, J. (1998) "Nonsingular Dilaton Cosmology," *Journal of High Energy Physics* 9808 007 (gr-qc/9806111).

Brandenberger, R., Frey, A., and Lorenz, L. (2008) "Entropy Fluctuations in Brane Inflation Models" (arXiv:0712.2178).

Brustein, R. and Veneziano, G. (1994) "The Graceful Exit Problem in String Cosmology," *Physics Letters* B329: 429–434 (hep-th/9403060).

Buananno, A., Meissner, K., Ungarelli, C. and Veneziano, G. (1998) "Classical Inhomogeneities in String Cosmology," *Physical Review* D57: 2543–2556 (hep-th/9706221).

Butterfield, J., and Isham, C. J. (1999) "On the Emergence of Time in Quantum Gravity," in J. Butterfield, ed. *The Arguments of Time.* Oxford: Oxford University Press, 111–68 (gr-qc/9901024v1).

———. (2001) "Spacetime and the philosophical challenge of quantum gravity," in C. Callender and N. Huggett, eds. *Physics Meets Philosophy at the Planck Scale: Contemporary Theories of Quantum Gravity.* Cambridge: Cambridge University Press, 33–89.

Callender, C., and Huggett, N., eds. (2001) *Physics Meets Philosophy at the Planck Scale: Contemporary Theories of Quantum Gravity.* Cambridge: Cambridge University Press.

Carlip, S. (2007) "Transient Observers and Variable Constants *or* Repelling the Invasion of the Boltzmann Brains" (hep-th/0703115).

Carr, B., ed. (2007) *Universe or Multiverse?* Cambridge: Cambridge University Press.

Carr, B. and Rees, M. (1979) "The anthropic cosmological principle and the structure of the physical world." *Nature* 278: 605–12.

Carter, B. (1967) "The significance of numerical coincidences in nature" (http://arxiv.org/abs/0710.3543).

Ceresole, A., Dall'Agata, G., Giryavets, A., Kallosh, R. and Linde, A. (2006) "Domain walls, near-BPS bubbles and probabilities in the landscape," *Physical Review* D74: 086010 (hep-th/0605266).

Cleaver, G., Faraggi, A. and Nanopoulos, D. (1999) "String Derived MSSM and M-Theory Unification," *Physics Letters* B455 (Issues 1–4): 135–46 (hep-ph/9811427).

Cleaver, G. (2006) "Before the Big Bang, String Theory, God, and the Origin of the Universe," unpublished manuscript (http://www.metanexus.net/conferences/pdf/conference2006/Cleaver.pdf).

———. (2008) "In Search of the (Minimal Supersymmetric) Standard Model String," in F. Balogh, ed. *String Theory Research Progress.* Chapter 2. Nova Science Publishers, Inc. (hep-ph/0703027).

Cohn, J. (1998) "Living with Lambda," *Astrophysical Journal Supplement* 259: 213 (astro-ph/9807128).

Coleman, S., and De Luccia, F. (1980) "Gravitational Effects On And Of Vacuum Decay," *Physical Review* D21: 3305.

Collins, R. (2003) "Evidence for Fine-Tuning," in N. Manson, ed. *God and Design: The Teleological Argument and Modern Science.* New York: Routledge, 178–99.

———. (2007) "How to Rigorously Define Fine-Tuning," (http://home.messiah.edu/~rcollins/FINETUNE/chapter%203%20how%20to %20rigorously%20define%20fine-tuning.htm#TOC2_1).

———. (2009) "The Teleological Argument: An Exploration of the Fine-Tuning of the Universe," in W. Craig and J. P. Moreland, eds. *The Blackwell Companion to Natural Theology.* Oxford: Blackwell Publishers, 202–81.

———. (forthcoming) *The Well-Tempered Universe.*

Copan, P. and Craig, W. (2004) *Creation out of Nothing: A Biblical, Philosophical and Scientific Exploration.* Grand Rapids: Baker Academic.

Craig, W. L. (1979) *The Kalām Cosmological Argument.* Eugene: Wipf and Stock Publishers.

———. (1993) "The Finitude of the Past and the Existence of God," in William Lane Craig and Quentin Smith, *Theism, Atheism and Big Bang Cosmology.* Oxford: Clarendon Press, 3–76.

Davies, P. (1982) *The Accidental Universe.* Cambridge: Cambridge University Press.

Dembski, W. A. (1998) *The Design Inference: Eliminating Chance through Small Probabilities*. Cambridge: Cambridge University Press.

———. (2002) *No Free Lunch: Why Specified Complexity Cannot Be Purchased without Intelligence*. Lanham: Rowman & Littlefield Publishers, Inc.

———. (2005) "Specification: The Pattern That Signifies Intelligence" (http://www.designinference.com/documents/2005.06.Specification.pdf).

Dembski, W. A. and Marks, R. J. (2009a) "Conservation of Information in Search: Measuring the Cost of Success," forthcoming in *Transactions on Systems, Man and Cybernetics, Part A* (see also http://marksmannet.com/ RobertMarks/REPRINTS/short/CoS.pdf).

———. (2009b) "The Search for a Search: Measuring the Information Cost of Higher Level Search," forthcoming in *The International Journal of Information Technology and Intelligent Computing* (see also http://marksmannet.com/RobertMarks/REPRINTS/short/S4S.pdf).

Denef, F. and Douglas, M. (2004) "Distributions of Flux Vacua," *Journal of High Energy Physics* 0405 072 (hep-th/ 0404116).

———. (2006) "Computational Complexity of the Landscape" (hep-th/ 0602072).

DeWitt, B. (1967) "Quantum Theory of Gravity I. The Canonical Formulism," *Physical Review* 160: 1113–48.

Dine, M. (2004) "Is There a String Theory Landscape: Some Cautionary Remarks" (hep-th/ 0402101).

Donagi, R., Khoury, J., Ovrut, B., Steinhardt, P. and Turok, N. (2001) "Visible Branes with Negative Tension in Heterotic M-Theory," *Journal of High Energy Physics* 0111 041 (hep-th/0105199).

Douglas, M. (2004a) "The statistics of string/M-theory vacua," *Journal of High Energy Physics* 0305 046 (hep-th/ 0303194).

———. (2004b) "Statistical analysis of the supersymmetry breaking scale" (hep-th/ 0405279)

Dowker, F., and Kent, A. (1996) "On the consistent histories approach to quantum mechanics." *Journal of Statistical Physics* 82: 1575–646.

Dyson, L., Kleban, M. and Susskind, L. (2002) "Disturbing Implications of a Cosmological Constant," *Journal of High Energy Physics* 0210 011 (hep-th/0208013).

Earman, J. and Mosterin, J. (1999) "A Critical Look at Inflationary Cosmology," *Philosophy of Science* 66: 1–49.

Einstein, A. (1916) "Die Grundlage der allgemeinen Relativitätstheorie," *Annalen der Physik* 49 (7): 284–339.

———. (1917) "Kosmologische Betrachtungen zur allgemeinen Relativitätstheorie," *Königlich Preussische Akademie der Wissenschaften* (Berlin). *Sitzungsberichte,* 421–45.

Erickson, J., Gratton, S., Steinhardt, P. and Turok, N. (2006) "Cosmic Perturbations through the Cyclic Ages" (hep-th/0607164).

Feinstein, A., Lazkoz, R. and Vazquez-Mozo, M. (1997) "Closed Inhomogeneous String Cosmologies," *Physical Review* D56: 5166 (hep-th9704173).

Felder, G., García-Bellido, J., Greene, P., Kofman, L., Linde, A. and Tkachev, I. (2000) "Dynamics of Symmetry Breaking and Tachyonic Preheating," *Physical Review Letters* 87: 011601 (hep-ph/0012142).

Felder, G., Kofman, L. and Linde, A. (2001) "Tachyonic Instability and Dynamics of Spontaneous Symmetry Breaking," *Physical Review* D64: 123517 (hep-th/0106179).

Felder, G., Frolov, A., Kofman, L. and Linde, A. (2002) "Cosmology with Negative Potentials," *Physical Review* D66: 023507 (hep-th/0202017).

Foffa, S. 2003. "Bouncing Pre-Big Bang on the Brane," *Physical Review* D68: 043511 (hep-th/ 0304004).

Freivogel, B. and Susskind, L. (2004) "A Framework for the Landscape," *Physical Review* D70: 126007 (hep-th/0408133).

Friedmann, A. (1922) "Über die Krümmung des Raumes," *Zeitschrift für Physik* 10(1): 377–86.

Garriga, J. and Vilenkin, A. (2001) "Many Worlds in One." *Physical Review* D64: 43511 (http://xxx.tau.ac.il/PS_cache/gr-qc/pdf/0102/0102010v2.pdf).

———. (2008) "Prediction and Explanation in the Multiverse," *Physical Review* D77: 043526 (hep-th/0711.2559).

Gasperini, M. (1999) "Looking back in time beyond the Big Bang," *Modern Physics Letters* A14: 1059–66 (gr-qc/9905062).

———. (2000) "Inflation and Initial Conditions in the pre-Big Bang Scenario," *Physical Review* D61: 087301 (gr-qc/9902060).

———. (2007) "Dilaton Cosmology and Phenomenology" (hep-th/0702166).

———. (2008) "The Pre-Big Bang Scenario Bibliography"(http://www.ba.infn.it/~gasperin/).

Gasperini, M., Maharana, J. and Veneziano, G. (1996) "Graceful exit in quantum string cosmology," *Nuclear Physics* B472: 349–60 (hep-th/9602087).

Gasperini, M., Maggiore, M. and Veneziano, G. (1997) "Towards a Non-Singular pre-Big Bang Cosmology," *Nuclear Physics* B494: 315–30 (hep-th/9611039).

Gasperini, M., and Veneziano, G. (2003) "The pre-Big Bang Scenario in String Cosmology," *Physics Reports* 373: 1–212 (hep-th/0207130).

Gell-Mann, M., and Hartle, J. (1990) "Quantum Mechanics in the Light of Quantum Cosmology," in W. Zurek, ed. *Complexity, Entropy and the Physics of Information*. New York: Addison-Wesley Publishing Company, 425–58.

———. (1996) "Equivalent sets of histories and multiple quasiclassical realms" (arXiv:gr-qc/9404013).

Giddings, S. and Thomas, S. (2002) "High Energy Colliders as Black Hole Factories: The End of Short Distance Physics," *Physical Review* D65: 056010 (hep-ph/0106219).

Giddings, S. and Maroff, D. (2007) "A global picture of quantum de Sitter space" (hep-th/0705.1178).

Gonzalez, G. and Richards, J. (2004) *The Privileged Planet: How Our Place in the Cosmos is Designed for Discovery*. Washington, DC: Regnery.

Gott, J. R. and Li, L-X. (1998) "Can the Universe Create Itself?" *Physical Review* D58: 023501 (astro-ph/9712344v1).

Grant, Edward. (1996) *The Foundations of Modern Science in the Middle Ages: Their Religious, Institutional, and Intellectual Contexts*. Cambridge: Cambridge University Press.

Gratton, S., Khoury, J., Steinhardt, P. and Turok, N. (2004) "Conditions for Generating Scale Invariant Density Perturbations," *Physical Review* D69: 103505 (astro-ph/0301395).

Green, A., and Malik, K. (2001) "Primordial black hole production due to preheating," *Physical Review* D64: 021301 (hep-ph/0008113).

Green, M., and Schwarz, J. (1985) "The Hexagon Gauge Anomaly in Type I Superstring Theory," *Nuclear Physics* B255: 93–114.

Green, M., Schwarz, J. and Witten. E. (1987) *Superstring theory* (2 volumes). Cambridge: Cambridge University Press.

Griffiths, R.B. (1984) "Consistent Histories and the Interpretation of Quantum Mechanics." *Journal of Statistical Physics* 36: 219–72.

———. (1993) "The Consistency of Consistent Histories." *Foundations of Physics* 23: 1601–10.

Guth A. H. (1981) "The Inflationary Universe: A Possible Solution to the Horizon and Flatness Problems." *Physical Review* D23: 347.

———. (1997) *The Inflationary Universe: The Quest for a New Theory of Cosmic Origins*. Reading: Perseus Books.

Guth, A. H. and Weinberg, E. (1983) "Could the Universe Have Recovered From a Slow First Order Phase Transition?" *Nuclear Physics* B212: 321.

Hartle, J. and Hawking, S. (1983) "Wave function of the universe," *Physical Review* D28: 2960–75.

Hartle, J. and Srednecki, M. (2007) "Are We Typical?" (hep-th/0704.2630).

Hawking, S. W. (1975) "Particle creation by black holes." *Communications in Mathematical Physics* 43: 199–220.

———. (1987) "Quantum cosmology," in S. W. Hawking and W. Israel, eds. *300 Years of Gravitation*. Cambridge: Cambridge University Press, 631–51.

———. (1988) *A Brief History of Time from the Big Bang to Black Holes*. New York: Bantam Books.

Hawking, S. and Ellis, G. F. R. (1973) *The large scale structure of space-time*. Cambridge: Cambridge University Press.

Hawking, S. and Page, D. (1987) "How Probable is Inflation?" *Nuclear Physics* B298: 789–809.

Hawking, S. and Penrose, R. (1970) "The singularities of gravitational collapse and cosmology," *Proceedings of the Royal Society of London* A 314: 529–48.

Heyl, J. and Loeb, A. (2002) "Vacuum Decay Constraints on a Cosmological Scalar Field," *Physical Review Letters* 88: 121302 (astro-ph/0111570).

Hilbert, D. (1925) "Über das Unendliche," *Mathematische Annalen* 95 (1926): 161–90. A translation into English by S. Bauer-Mengelberg is reproduced in J. van Heijenoort, ed. *From Frege to Gödel: A Sourcebook in Mathematical Logic, 1879–1931*. Cambridge: Harvard University Press, 1967, 367–92.

Hogan, C. (2000) "Why the Universe is Just So," *Reviews of Modern Physics* 72: 1149–61 (astro-ph/ 9909295).

Holder, R. (2004) *God, the Multiverse, and Everything*. Burlington, VT: Ashgate.

Hollands, S. and Wald, R. (2002) "An Alternative to Inflation," *General Relativity and Gravitation* 34: 2043–55 (gr-qc/0205058).

Hoyle, F. (1975) *Astronomy and Cosmology: A Modern Course*. San Francisco: W H. Freeman.

Isham, C. J. (1993a) "Quantum Theories of the Creation of the Universe," in R. J. Russell, N. Murphy and C. J. Isham, eds. *Quantum Cosmology and the Laws of Nature*. Notre Dame: University of Notre Dame Press, 51–89.

———. (1993b) "Prima Facie Questions in Quantum Gravity" (gr-qc/9310031v1).

Kachru, S., Kallosh, R. Linde, A. and Trivedi, S. (2003) "de Sitter Vacua in String Theory,"*Physical Review* D68: 046005 (hep-th/0301240).

Kachru, S., Kallosh, R., Linde, A., Maldecena, J., McAllister, L. and Trivedi, S. (2003) "Towards Inflation in String Theory," *Journal of Cosmology and Astroparticle Physics* 0310 013 (hep-th/0308055).

Kallosh, R., Kofman, L. and Linde, A. (2001) "Pyrotechnic Universe," *Physical Review* D64: 123523 (hep-th/0104073).

Kaloper, N., Linde, A. and Bousso, R. (1999) "Pre-Big Bang Requires the Universe to be Exponentially Large from the Beginning," *Physical Review* D59: 043508 (hep-th/9801073).

Kapner, D., Cook, T., Adelberger, E., Gundlach, J., Heckel, B., Hoyle, C., and Swanson, H. (2007) "Tests of the Gravitational Inverse-Square Law below the Dark-Energy Length Scale," *Physical Review Letters* 98: 021101 (hep-ph/0611184).

Khoury, J., Ovrut, B., Steinhardt, P. and Turok, N. (2001a) "The Ekpyrotic Universe: Colliding Branes and the Origin of the Hot Big Bang," *Physical Review* D64: 123522 (hep-th/0103239).

Khoury, J., Ovrut, B., Steinhardt, P. and Turok, N. (2001b) "A Brief Comment on 'The Pyrotechnic Universe'" (hep-th/0105212).

Khoury, J., Ovrut, B., Steinhardt, P. and Turok, N. (2002) "Density Perturbations in the Ekpyrotic Scenario," *Physical Review* D66: 046005 (hep-th/0109050).

Khoury, J., Ovrut, B., Seiberg, N., Steinhardt, P. and Turok, N. (2002) "From Big Crunch to Big Bang," *Physical Review* D65: 086007 (hep-th/0108187).

Khoury, J., Steinhardt, P. and Turok, N. (2003) "Inflation versus Cyclic Predictions for Spectral Tilt," *Physical Review Letters* 91: 161301 (astro-ph/0302012).

Khoury, J., Steinhardt, P. and Turok, N. (2004) "Designing Cyclic Universe Models," *Physical Review Letters* 92: 031302 (hep-th/0307132).

Kim, H. and Hwang, J. (2007) "Evolution of linear perturbations through a bouncing world model: Is the near Harrison-Zel'dovich spectrum possible via a bounce?" *Physical Review* D75: 043501 (astro-ph/0607464).

Kobakhidze, A. and Mersini-Houghton, L. (2007) "Birth of the Universe from the Landscape of String Theory," *European Physics Journal C* 49: 869–873 (hep-th/0410213).

Kofman, L. (1996) "The Origin of Matter in the Universe: Reheating after Inflation" (astro-ph/ 9605155).

Koonin, E. (2007) "The cosmological model of eternal inflation and the transition from chance to biological evolution in the history of life," *Biology Direct* (http://www.biology-direct.com/content/2/1/15).

Koons, R. C. (1997) "A New Look at the Cosmological Argument," *American Philosophical Quarterly* 34: 171–92.

Krauss, L. (2005) *Hiding in the Mirror: The Mysterious Allure of Extra Dimensions, from Plato to String Theory and Beyond*. New York: Viking.

Kumar, J. (2006) "A Review of Distributions on the String Landscape," *International Journal of Modern Physics* A21: 3441–72 (hep-th/0601053).

Land, K. and Magueijo, J. (2005) "The Axis of Evil," *Physical Review Letters* 95: 071301 (astro-ph/0502237).

———. (2007) "The Axis of Evil Revisited" (astro-ph/0611518).

Lemaître, G. (1927) "Un Univers homogène de masse constante et de rayon croissant rendant compte de la vitesse radiale des nébuleuses extra-galactiques," *Annales de la Société Scientifique de Bruxelles* A47: 49–59.

Leslie, J. (1989) *Universes*. New York: Routledge.

Lewontin, R. (1997) "Billions and billions of demons," *The New York Review of Books,* 9 January 1997: 28–32.

Liddle, A., and Mazumdar, A. (1998) "Inflation during oscillations of the inflaton," *Physical Review* D58: 083508 (astro-ph/9806127).

Linde, A. (1982) "A New Inflationary Universe Scenario: A Possible Solution of the Horizon, Flatness, Homogeneity, Isotropy and Primordial Monopole Problems," *Physics Letters* B108: 389.

———. (1986a) "Eternal Chaotic Inflation," *Modern Physics Letters* A1: 81.

———. (1986b) "Eternally existing self-reproducing chaotic inflationary universe," *Physics Letters* B175: 395.

———. (2001) "Inflation and String Cosmology," *International Journal of Modern Physics* A17S1: 89–104 (hep-th/0107176).

———. (2007) "Sinks in the Landscape, Boltzmann Brains, and the Cosmological Constant Problem," *Journal of Cosmology and Astroparticle Physics* 0701 022 (hep-th/0611043).

Lyth, D. (2002) "The primordial curvature perturbation in the ekpyrotic universe," *Physics Letters* B524: 1–4 (hep-ph/0106153).

Maggiore, M. and Sturani, R. (1997) "The fine-tuning problem in pre-Big Bang inflation," *Physics Letters* B415: 335–343 (gr-qc/9706053).

Martin, J. and Brandenberger, R. (2001) "The Trans-Planckian Problem of Inflationary Cosmology," *Physical Review* D63: 123501 (hep-th/0005209).

McFadden, P., Turok, N. and Steinhardt, P. (2007) "Solution of a Braneworld Big Crunch/Big Bang Cosmology," *Physical Review* D76: 104038 (hep-th/0512123).

Messiah, A. (1999 [1958]) *Quantum Mechanics* (Two Volumes Bound as One). Mineola, NY: Dover Publications.

Meyer, Stephen C. (2009) *The Signature in the Cell: DNA and the Evidence for Intelligent Design*. San Francisco: Harper One.

Oberhummer, H., Csótó, A. and Schlattl, H. (2000) "Fine-tuning of carbon-based life in the universe by triple-alpha process in red giants," *Science* 289 (July): 88–90.

Omnès, Roland. (1994) *The Interpretation of Quantum Mechanics*. Princeton, NJ: Princeton University Press.

Ooguri, H. and Vafa, C. (2006) "On the Geometry of the String Landscape and the Swampland" (hep-th/0605264).

Overbye, D. (2008) "Big Brain Theory: Have Cosmologists Lost Theirs?" *The New York Times,* Science Section, January 15th.

Page, D. (2000) "Is Our Universe Likely to Decay within 20 Billion Years?" (hep-th/0610079).

———. (2006) "Return of the Boltzmann Brains" (hep-th/0611158).

———. (2007a) "Typicality Defended" (arXiv:0707.4169).

———. (2007b) "Predictions and Tests of Multiverse Theories," in B. Carr, ed. *Universe or Multiverse?* Cambridge: Cambridge University Press, 411–30.

Penrose, R. (1981) "Time-asymmetry and quantum gravity," in C. Isham, R. Penrose and D. Sciama, *Quantum Gravity 2*. Oxford: Clarendon Press, 245–72.

———. (1989a) *The Emperor's New Mind: Concerning Computers, Minds and the Laws of Physics*. New York: Vintage.

———. (1989b) "Difficulties with inflationary cosmology," in E. Fergus, ed., *Proceedings of the 14th Texas Symposium on Relativistic Astrophysics, Ann. NY Acad. Sci.* 571: 249–64.

———. (2005) *The Road to Reality: A Complete Guide to the Laws of the Universe*. New York: Alfred A. Knopf.

Polchinski, J. (2006) "The Cosmological Constant and the String Landscape" (hep-th/0603249).

Pruss, Alexander R. (2006) *The Principle of Sufficient Reason: A Reassessment*. Cambridge: Cambridge University Press.

Räsänen, S. (2002) "On ekpyrotic brane collisions," *Nucl. Phys.* B626: 183–206 (hep-th/0111279).

Rees, M. (1997) *Before the Beginning: Our Universe and Others*. Reading: Addison-Wesley.

———. (2000) *Just Six Numbers: The Deep Forces that Shape the Universe*. New York: Basic Books.

Rey, S-J. (1996) "Back Reaction and Graceful Exit in String Inflationary Cosmology," *Physical Review Letters* 77: 1929–32 (hep-th/9605176).

Riddle, A. and Urena-Lopez, L. (2006) "Inflation, dark matter and dark energy in the string landscape" *Physical Review Letters* 97: 161301 (astro-ph/0605205).

Robbins, D. and Sethi, S. (2005) "A Barren Landscape?" *Phys. Rev.* D71: 046008 (hep-th/0405011).

Sahni, V. and Starobinsky, A. (2000) "The Case for a Positive Cosmological Λ-term," *International Journal of Modern Physics* D9: 373–444 (astro-ph/9904398).

Saygili, K. (1999) "Hamilton-Jacobi Approach to Pre-Big Bang Cosmology at Long Wavelengths," *International Journal of Modern Physics* A14: 225–40 (hep-th/9710070).

Smolin, L. (2006) *The Trouble with Physics: The Rise of String Theory, the Fall of a Science, and What Comes Next*. New York: Houghton Mifflin.

Spitzer, R. J. (2009) *New Proofs for the Existence of God: Contributions of Contemporary Physics and Philosophy*. Grand Rapids: Eerdmans.

Stark, R. (2003) *For the Glory of God: How Monotheism Led to Reformations, Science, Witch-Hunts, and the End of Slavery*. Princeton: Princeton University Press.

———. (2005) *The Victory of Reason: How Christianity Led to Freedom, Capitalism, and Western Success*. New York: Random House, Inc.

Steinhardt, P. (2004) "Has the cyclic model been cycling forever?" http://www.physics.princeton.edu/~steinh/cyclicFAQS/index.html#eternal

Steinhardt, P., and Turok, N. (2002a) "Cosmic Evolution in a Cyclic Universe," *Phys. Rev.* D65: 126003 (hep-th/0111098).

———. (2002b) "A Cyclic Model of the Universe" (hep-th/0111030).

———. (2002c) "Is Vacuum Decay Significant in Ekpyrotic and Cyclic Models?" *Physical Review* D66: 101302 (astro-ph/0112537).

———. (2005) "The Cyclic Model Simplified," *New Astronomy Reviews* 49: 43–7 (www.physics.princeton.edu/~steinh/dm2004.pdf).

———. (2006) "Why the cosmological constant is small and positive," *Science* 312: 1180–182 (astro-ph/0605173).

———. (2007) *Endless Universe: Beyond the Big Bang.* New York: Random House.

Susskind, L. (2003) "The Anthropic Landscape of String Theory" (hep-th/0302219).

———. (2004) "Supersymmetry Breaking in the Anthropic Landscape" (hep-th/0405189).

———. (2006) *The Cosmic Landscape: String Theory and the Illusion of Intelligent Design.* New York: Little, Brown and Company.

Tegmark, M. (1998) "Is 'The Theory of Everything' Merely the Ultimate Ensemble Theory?" *Annals of Physics* 270: 1–51.

———. (2003) "Parallel Universes," in J. D. Barrow, P. C. W. Davies and C. L. Harper, eds. *Science and Ultimate Reality: Quantum Theory, Cosmology and Complexity.* Cambridge: Cambridge University Press, 459–91 (http://arxiv.org/PS_cache/astro-ph/ pdf/0302/0302131v1.pdf).

Tolley, A., Turok, N. and Steinhardt, P. (2004) "Cosmological Perturbations in a Big Crunch/Big Bang Spacetime," *Physical Review* D69: 106005 (hep-th/0306109).

Tsujikawa, S., Bassett, B., and Viniegra, F. (2000) "Multi-field fermionic preheating," *Journal of High Energy Physics* 0008 019 (hep-ph/0006354).

Turner, M. and Weinberg, E. (1997) "Pre-Big Bang Inflation Requires Fine-Tuning," *Physical Review* D56: 4604–9 (hep-th/9705035).

Turok, N., Perry, M. and Steinhardt, P. (2004) "M Theory of a Big Crunch/Big Bang Transition," *Physical Review* D70: 106004 (hep-th/0408083).

Vanchurin, V., and Vilenkin, A. (2006) "Eternal Observers and Bubble Abundances in the Landscape," *Physical Review* D74: 043520 (hep-th/0605015).

Van Elst, H. (2008) "Inflationary Cosmological Models/Scalar Field Solutions," http://www.maths.qmul.ac.uk/~hve/ref_dir/chinfl.html (bibliography).

Veneziano, G. (1995) "String Cosmology: Basic Ideas and General Results" (hep-th/9510027).

———. (1997) "Inhomogeneous Pre-Big Bang String Cosmology," *Physics Letters* B406: 297–303 (hep-th/9703150).

———. (1998) "A Simple/Short Introduction to pre-Big Bang Physics/Cosmology" (hep-th/9802057).

———. (1999) "Pre-bangian origin of our entropy and time arrow," *Physics Letters* B454: 22–26 (hep-th/9902126).

Vilenkin, A. (1982) "Creation of universes from nothing," *Physics Letters* B117: 25–28.

———. (1988) "Quantum Cosmology and the Initial State of the Universe," *Physical Review* D37: 888–97.

———. (1994) "Approaches to Quantum Cosmology," *Physical Review* D50: 2581–2594 (gr-qc/9403010v1).

———. (2002) "Quantum cosmology and eternal inflation" (gr-qc/0204061v1).

———. (2006a) *Many Worlds in One: The Search for Other Universes.* New York: Hill and Wang.

———. (2006b) "The vacuum energy crisis," *Science* 312: 1148–49 (astro-ph/0605242).

Watanabe, Y., and Komatsu, E. (2008) "Gravitational inflaton decay and the hierarchy problem," *Physical Review* D77: 043514 (arXiv:0711.3442).

Weinberg, Steven. (2008) *Cosmology.* Cambridge: Cambridge University Press.

Whitehead, A. N. (1925) *Science and the Modern World.* New York: The Macmillan Company.

Witten, E. (1995) "Some Comments on String Dynamics" (hep-th/9507121).

Woit, P. (2006) *Not Even Wrong: The Failure of String Theory and the Search for Unity in Physical Law.* New York: Basic Books.

27

Habitable Zones and Fine-Tuning

Guillermo Gonzalez

1. Introduction

In the second half of the twentieth century physicists discovered that the values of the physical constants and cosmological initial conditions must not differ greatly from their observed values for life to be possible in the universe.[1] Small changes to the values of some constants, for example, would result in universes far too short-lived, or too simple (e.g., only hydrogen or black holes), or too chaotic for life. The conclusion from these theoretical considerations is that the universe must be "fine-tuned" for life.

In considering fine-tuning, physicists assume that the constants and initial conditions (and possibly the physical laws) could have been different. In other words, our universe is not logically necessary. Thus, the question arises as to whether the properties of our particular universe were designed and selected *for* us. Alternately, how much of what we observe was selected *by* us? The latter question falls under the category of one of the species of the *anthropic principle*. The observer self-selection "explanation" for the properties of the universe we inhabit (the *weak anthropic principle*), however, suffers from lack of independent observational evidence for other universes or domains, and theoretical motivation for them is controversial (see Bruce Gordon's contribution in the preceding chapter).

It is helpful to split fine-tuning into two distinct types, which we will call "global" and "local." Global tuning deals with the global properties of the observable universe. These include the masses of the fundamental particles, the strengths of the four fundamental forces, the initial cosmological conditions, and the cosmological constant.

In contrast, local tuning includes things that are not universal in their properties: planets, stars, and galaxies. Not only do we know that planets, stars, and galaxies do not have fixed properties, we actually observe them to vary in their properties over a broad range. We can study how life depends on the local parameters while keeping the global parameters fixed. We can also tally their numbers. For local tuning, then, we have the hope of accurately quantifying the available probabilistic resources and estimating how much of our local circumstances can be explained by observer self-selection.

Habitable Zones and Fine-Tuning

Although it is helpful to examine fine-tuning in this way, eventually we must rejoin local and global tuning if we are to determine how finely tuned our universe is for life. Discussions of global tuning do not get us very far unless we understand how the global parameter values are instantiated locally in planets, stars, and galaxies.

Historically, local tuning has been explored within the context of exobiology/astrobiology. Motivated by the desire to find other inhabited planets, astrobiologists have sought to determine the full range of environments compatible with life (i.e., habitable environments). Over the past twenty years, considerable progress has been made towards this end. In the following section I review the state of our knowledge about habitable environments.[2] In Section 3 I return to the topic of global tuning and describe how local and global tuning are linked.

2. Habitable Zones

Introduction

Since its introduction over four decades ago, the Circumstellar Habitable Zone (CHZ) concept has served to focus scientific discussions about habitability within planetary systems. Early studies simply defined the CHZ as that range of distances from the Sun that an Earth-like planet must be within to maintain liquid water on its surface. Too close, and too much water enters the atmosphere, leading to a runaway greenhouse effect. Too far, and too much water freezes, leading to runaway glaciation. Since these modest beginnings, CHZ models have become more complex and realistic, mostly due to improvements in the treatment of energy transport in planetary atmospheres and the inclusion of the carbon-silicate cycle. Along the way, Mars and Venus have served as "real-world" test cases of the CHZ boundaries.

The CHZ has been an important unifying concept in astrobiology. Research on the CHZ requires knowledge of stellar evolution, planetary dynamics, climatology, biology, and geophysics. Yet even modern CHZ models are far from complete. Many factors relating to planet formation processes and subsequent gravitational dynamics have yet to be incorporated in a formal way.

While they were not the first to discuss habitability beyond the Solar System, Gonzalez et al. were the first to introduce a unifying concept called the "Galactic Habitable Zone" (GHZ).[3] The GHZ describes habitability on the scale of the Milky Way galaxy. While the GHZ appears superficially similar to the CHZ, it is based on a very different set of physical processes, including the radial gradients of the supernova rate, gas metallicity, density of gas, and density of stars in the galactic disk. It should also be possible to define habitable zones for other galaxies and even to extend the concept to the whole universe.[4] The largest of all habitable zones can be termed the "Cosmic Habitable Age" (CHA), which describes the evolution of the habitability of the universe over time.

In the following, I review published studies relevant to the CHZ, GHZ, and CHA, but first I will review life's basic needs.

The Needs of Life

Published studies of the CHZ focus on the maintenance of minimal habitable conditions on the surface of a terrestrial planet. These conditions are constrained most fundamentally by limits on the planet's mean surface temperature, the presence of liquid water, and the composition of its atmosphere. To these we can add constraints on the temporal and spatial variations of a planet's surface temperature; a slowly rotating Earth-like planet, for example, will experience larger temperature variations than a similar but faster rotating planet with the same mean temperature.

It may seem that the requirement of liquid water is merely an assumption of convenience for defining the CHZ based on our knowledge of "life as we know it." The evidence from chemistry, however, lends support to the view that liquid water and carbon are essential for life.[5] In addition, single-celled life requires some sixteen elements and mammals require an additional ten for essential biological processes,[6] all of which must be cycled in the environment.[7]

Astrobiologists sometimes treat habitability as a binary, either-or quantity. A planet is either sterile or it is teeming with life; it either has liquid water on its surface or it doesn't. Franck et al. advanced beyond this simplistic approach,[8] quantifying the habitability of a planet in terms of its photosynthetic productivity. While photosynthesis is not the most basic form of habitability, it is one that has existed on Earth since very early times and has proven critical for the oxygenation of the atmosphere. Following Franck et al., we propose that a Basic Habitability Index (BHI) be adopted as a measure of habitability. We can additionally define a habitability index for Earthly animal life (i.e., large, oxygen-breathing, mobile metazoans); we can call it the Animal Habitability Index (AHI). According to the *Rare Earth* hypothesis,[9] the AHI would be more restrictive on the environment than the BHI. The limits on the mean surface temperature and the surface temperature variations would both be narrower for the AHI. An upper limit on the carbon dioxide partial pressure also needs to be added, as well as a lower limit on the oxygen partial pressure for the AHI. These limits can be estimated from the physiology of extant animals, the reconstructed evolution of the partial pressures of carbon dioxide and oxygen in Earth's atmosphere,[10] and the history of life. While such limits will be necessarily parochial, certain general physiological principles we have learned from Earthly life will apply universally; for example, large metazoans (e.g., you and me) require an oxygen-rich atmosphere.[11]

Particularly helpful in quantifying the AHI and BHI is knowledge of the global ecological patterns of the present Earth. Ecologists have noted that a few large-scale spatial patterns account for the distribution of biodiversity.[12] The most prominent among these are a decrease in biodiversity (quantified as species richness) with increasing latitude and altitude. More fundamentally, Allen et al. argue that biodiversity increases with increasing temperature and nutrient availability; they explain the temperature dependence in terms of basic biochemical kinetics.[13] Biodiversity also correlates positively with primary ecosystem productivity;[14] for example, Schneider and Rey-Benayas show how the diversity of vascular plants correlates with productivity.[15] Other factors that influence biodiversity and ecosystem productivity include temperature variability and mean insolation, both of which are more important at high latitudes. Finally, productivity is sensitive to essential nutrient availability. An interesting example concerns the molybdenum (Mo) concentration in the oceans. Mo is necessary for fixing nitrogen (N) in living things. Most Mo in the oceans comes from the continents, but its concentration is sensitive to the oxygen content of the atmosphere.[16]

Another possibly fruitful approach towards generalizing habitability would be to construct an "equation of state of life." For example, Méndez compiled a database of the physiological properties of several hundred genera of prokaryotes and studied statistical trends in it.[17] Prokaryotes are an important element of the primary producers, and thus, of biodiversity. He found that about 85 percent of prokaryotes have an optimum growth temperature between 295 and 315 K. This is interesting, because it implies that the biophysical limitations of prokaryotes have been more important to their distribution on Earth than adaptations. A complete equation of state for prokaryotes would include at least temperature, pressure, and water concentration as parameters.

The history of life on Earth is another important source of information on factors relevant to habitability.[18] The fossil record reveals that single-celled life appeared on Earth at least 3.5 billion years ago,[19] shortly after the end of the "late heavy bombardment." The "Cambrian explosion" occurred about 540 million years ago. Since then, there have been many extinction events with global footprints.[20] Only the K/T extinction (65 million years ago) has been securely linked to a well-dated extraterrestrial event—the Chixzulub impact structure. Once additional extinction events can be linked to individual impacts, it will be possible to produce a "kill curve," which relates the magnitude of extinction and the size of the impact crater.[21] It will probably be necessary to include some threshold impactor energy required to trigger global extinctions, given that other large impacts, such as the two that occurred 35.5 million years ago (Chesapeake and Popigai; about 100 km each), had relatively little global effect on the biosphere.

The habitability of a terrestrial planet depends sensitively on its total water content. Planets with scarce surface water, like Mars, experience larger temporal and spatial temperature variations. On the other hand, planets with much more surface water than the Earth are not necessarily more habitable. On first consideration we should expect such planets to have less variable surface temperature and therefore to be more habitable. However, reduced dry land area also means less opportunity for land-based life and less surface area for chemical weathering, an important part of the carbon-silicate cycle.[22] Marine organisms depend on nutrients and minerals washed off the continents and on the regulation of the oceanic salt content by the continents.[23] With enough water, dry land can be completely absent on a terrestrial planet. Such a "waterworld" is unlikely to be habitable. Models of planet formation, though still in their infancy, predict that terrestrial planets can vary widely in their water content.[24]

Therefore, we should define an Earth-*like* planet as a terrestrial planet with surface water, dry land, and similar geophysics to the Earth. This should be contrasted with a merely Earth-*mass* terrestrial planet.

The Circumstellar Habitable Zone

All published studies of the CHZ start with an Earth-like planet. The planet is assumed to be habitable as long as liquid water can be maintained on its surface. It is imbedded in a planetary system identical to ours, except possibly a different host star. It has the same orbital eccentricity, Moon, and planetary neighbors. Thus, all the difficult questions about the formation of a planetary system are avoided. This is the traditional definition of the CHZ.

Kasting et al. defined the boundaries of the CHZ in multiple ways.[25] One definition of the inner boundary is based on the "moist greenhouse." In this process water gets into the stratosphere, where it is dissociated by solar UV radiation and the hydrogen atoms are lost from the top of the atmosphere. A second definition for the inner boundary is based on the runaway greenhouse. They calculated the outer boundary according to the maximum possible CO_2 greenhouse or the increase of planetary albedo due to formation of CO_2 clouds. The inner and outer boundaries were also estimated from the states of Venus and Mars, respectively. Their most restrictive case has inner and outer boundaries of 0.95 and 1.37 Astronomical Units (AUs), respectively.

Franck et al. presented a new set of CHZ models based on a more realistic treatment of geophysical processes.[26] Previous studies had assumed constant continental area, metamorphic outgassing of CO_2, and weathering rate over geologic timescales. Building on the climate models of Kasting et al., and Caldeira and Kasting, and relaxing these assumptions, Franck et al. thus modeled Earth's coupled climate-geologic systems as dynamical processes.[27] Their CHZ is defined by surface temperature bounds of 0°C and 100°C and CO_2 partial pressure above 10^{-5} bar. They added the CO_2 partial pressure requirement to ensure that conditions are suitable for biological productivity via photosynthesis. It sets the inner boundary of their CHZ, while the minimum temperature requirement sets the outer boundary; their CHZ inner and outer bounds for the present Solar System are 0.95 and 1.2 AUs, respectively. Franck et al. also determined that the maximum lifespan for an Earth-like planet around a star between 0.6 and 1.1 M_{sun} (6.5 billion years) is limited by planetary geodynamics.

While CHZ models have improved steadily over the past few decades, they are still at an immature stage of development. They lack many deterministic and stochastic processes relevant to habitability, and the modelers have yet to describe how the formative processes of a planetary system set the initial conditions for their CHZ calculations. The relevant initial conditions include the locations, masses, compositions, initial volatile inventories, initial rotation periods, initial obliquities, initial orbital inclinations, presence of moons, initial eccentricities of the terrestrial planets, and the orbits and masses of the giant planets; they also include the properties of the asteroid and comet reservoirs. These have significant stochastic components, and they cannot properly be treated in isolation, as there are many complex interdependencies among them. Proper treatment of the initial conditions requires simulations that begin with a protoplanetary nebula of a given mass, composition, and environment and follow its evolution through the final stages of star and planet formation.

Lissauer identified four dynamical stages of planet formation in a protoplanetary disk: 1) condensation and growth of grains, 2) grains grow to km size either by pairwise accretion or gravitational instability of the solid disk, 3) oligarchic growth to Mars-size terrestrial bodies and giant planet runaway accretion, and 4) development of crossing orbits leading to giant impacts.[28] Numerical simulations have shown that, while stochastic processes are important, the final distributions of orbital periods, eccentricities, and masses of the terrestrial planets are significantly constrained by the initial and boundary conditions.[29]

The origin of planetary rotation is still a matter of some controversy. Simulations indicate that large impacts near the terminal stage of terrestrial planet formation may dominate any systematic preference for one spin direction over the other.[30] For example, the formation of the Moon via an impact by a Mars-size body probably imparted more angular momentum to the Earth than it had prior to that event.[31] Following the early formative phase, the rotation periods of terrestrial planets continue to evolve via tidal torques from the host star and from any orbiting moons. Whether the rotation periods increase or decrease and how fast they change depend on the details of a planet's interior, presence of oceans, and atmosphere, as well as the direction of its rotation and the rotation period in comparison to its moon's orbital period.

Planetary rotation is highly relevant to habitability. A planet's rotation period affects its day-night temperature variation, obliquity stability, and magnetic field generation.[32] Unless a terrestrial planet has a thick carbon dioxide atmosphere, slower rotation will result in larger day-night temperature differences. In addition, prolonged absence of light will be a factor for photosynthetic life on any slowly rotating terrestrial planet. For the extreme case of synchronous rotation, the complete freeze-out of water on the dark hemisphere is very likely (see below).

Habitable Zones and Fine-Tuning

The details of the origin of the atmospheres of the terrestrial planets are also uncertain. The two general classes of sources of volatiles are accretion from local material in the protoplanetary nebula and collisions with comets and bodies from the asteroid belt. Among the volatiles, most research has focused on water, given its importance in defining the CHZ. According to protoplanetary disk models, Earth could not have received its water from material formed near 1 AU, as the protoplanetary disk temperature would have been too high for it to condense. Water must have been delivered from beyond about 2.5 AU. Apparently, nearly all of Earth's water came from large bodies in the region of the outer asteroid belt.[33] Contrary to previous expectations, isotopic and dynamical data indicate that comets contributed no more than about 10 percent of Earth's crustal water.[34]

The net quantity of water and other volatiles delivered to and retained by a terrestrial planet also depends on its size and location. Smaller planets, like Mars, are subject to a much greater degree of atmospheric impact erosion.[35] Earth's gravity, however, is sufficiently large that impacts added much more to its atmosphere than they removed. Even the giant impact proposed to have formed the Moon probably removed only a modest portion of Earth's atmosphere.[36] The impact velocity depends, in part, on the impactor's original orbit and on the orbit of the target body. Comets, which originate far from the Sun, impact at higher velocity than objects from the asteroid belt. Likewise, terrestrial planets closer to their host star will encounter objects at greater velocities. Higher velocity impacts tend to erode planetary atmospheres more effectively.

Lunine argued that the delivery of volatiles to the terrestrial planets in the Solar System should be very sensitive to the location and eccentricity of Jupiter's orbit.[37] One of the critical quantities is the location of the innermost giant planet in relation to the so-called snowline. The presence of Jupiter near the snowline in the Solar System allowed it to transfer water-rich embryos efficiently from the asteroid belt into the terrestrial planet region. Recent N-body simulations of the formation of the terrestrial planets have generally confirmed this. Raymond et al. showed that increasing the eccentricity of Jupiter produces drier terrestrial planets, and moving it farther from the Sun produces more massive, water-rich planets;[38] they also find that the volatile delivery has considerable stochastic variability. In addition, terrestrial planets in the CHZ of a lower mass star tend to be drier.[39]

Today, the radial distributions of asteroid and comet perihelia peak just outside the orbit of Mars.[40] As the outermost terrestrial planet, Mars takes the brunt of asteroid and comet impacts (except that, because it is smaller than Earth, it has a smaller cross-section for collision). In fact, any planet that is the outermost terrestrial planet in a system similar to the Solar System will take the brunt of the asteroid and comet impacts.

With the discovery of the first extrasolar giant planet around a nearby Sun-like star in 1995, it became immediately obvious that other planetary systems can be very different from ours. About 10 percent of the detected systems have a giant planet within about 0.1 AU of their host stars. The remaining systems have giant planets with eccentricities that scatter nearly uniformly between 0.0 and 0.80. Several processes were proposed to account for the great variety of orbits observed. These include inward planet migration[41] and strong disk-planet and planet-planet interactions.[42] Some of these processes also result in non-coplanar orbits, which tend to produce less stable systems.[43] Veras and Armitage, assuming that the observed eccentricities are due to planet-planet scattering, determined that terrestrial planets are unlikely to form in a star's habitable zone if an eccentric giant planet has a semimajor axis between 2 and 3 AU.[44]

Giant planet migration is also important to the habitability of the terrestrial planets. For example, migration of a giant planet toward its host star will remove any terrestrial protoplanets in the CHZ and reduces the probability that more planets will form there afterward.[45] Raymond et al., however, find that the terrestrial planets that do form in the wake of a migrating giant planet are very water-rich (and thus not Earth-like).[46]

Similarly, Kuchner and Léger et al. note that an icy planet like Uranus or Neptune or something smaller, if it migrates into the CHZ, can become an "ocean planet."[47] Such a planet would have a very deep ocean of water surrounding a thick ice mantle, which would separate the deeper silicate mantle from the ocean. The pressure at the bottom of its ocean would be too high for any known life. Such a planet would also be more sensitive to tidal torques from its host star and any large moons, causing more rapid spin-down. A small influx of life-essential elements at the surface could be provided by micrometeorites, but the quantities could support at most a feeble biosphere. Finally, an ocean planet would not be able to regulate the concentration of salt dissolved in its oceans.[48]

Migrating giant planets would probably bring along at least some of their moons. How habitable would an Earth-size moon in the CHZ be? Williams et al. explored this possibility.[49] Even if such a moon could be as large as Earth,[50] it is unlikely to be as habitable for several reasons.[51] For example, a moon formed far beyond the CHZ would contain a great deal of water (a possible exception would be a large moon intermediate in composition between Io and Europa). Other relevant factors include rotational synchronization (causing longer days and nights and a weakened magnetic field), tidal-induced migration, immersion in the host planet's radiation belts, and higher frequency and energy of small-body impacts.

The giant planets also have significant influences on the obliquity variations of the terrestrial planets. Laskar et al. showed that a terrestrial planet can exhibit large and chaotic obliquity variations,[52] which are caused by resonances between its precession frequency and combinations of secular orbital frequencies of the giant planets in the system. The chaotic zones are broad, and they depend on several parameters, including the orbital period, rotation period, and mass of the terrestrial planet, the presence of a large moon, and the orbital periods and masses of the giant planets. Certain combinations of these parameters result in very small obliquity variations. Today, Earth exhibits tiny obliquity variations of ±1.3 degrees around an average value of 23.4 degrees. The lunar gravitational torque increases Earth's precession frequency by a factor of about three compared to what it would be without the Moon, taking it far from a spin-orbit resonance.[53] The Moon has a similar effect to reducing Earth's rotation period by the same factor.

Ward et al. demonstrated that the region of chaotic obliquity variation is very broad in the Solar System.[54] They calculated the amplitude of obliquity variations for Mars over a broad range of locations and rotation periods. They found stability comparable to Earth's only for distances less than 0.7 AU from the Sun and with faster rotation (but solar-induced tides would slow the rotation over a few billion years). Interestingly, if Mars had a large moon (keeping all else the same), it would still exhibit large obliquity variations, because it would be brought closer to a resonance.[55]

Atobe et al. and Atobe and Ida have conducted more general simulations of obliquity variations of terrestrial planets.[56] They include an analysis of hypothetical terrestrial planets in known extrasolar planetary systems and an exploration of the varieties of dynamical histories possible for collision-formed moons.

A moonless Earth would have exhibited a stable obliquity if its rotation period were less than about ten hours. The likelihood of such a state depends primarily on the last few large collisions it experienced near the end of its formation. Earth's initial rotation period was indeed less than ten hours, but it has since slowed to twenty-four hours, mostly by the action of the lunar tides. Ironically, Earth likely received its fast initial rotation from the impact that resulted in the Moon's formation.

Mercury is presently locked into a 3:2 spin-orbit resonance with a stable low obliquity value, but it was very likely born with a much faster rotation. Its precession frequency gradually declined via core-mantle interactions and tidal dissipation from the Sun that gradually slowed its rotation. Before reaching its present state, however, Mercury must have passed through a large chaotic zone in obliquity.

The case of Venus offers additional insights on obliquity variations. Correia and collaborators show that most initial conditions drive Venus to its present state and that this is generally true of terrestrial planets with very thick atmospheres.[57]

Touma and Wisdom studied the core-mantle, spin-orbit interactions for Earth and Venus and concluded that both planets have passed through major heat-generating core-mantle resonances.[58] They speculate that Earth's passage through such a resonance about 250 million years ago may have been responsible for generating the Siberian traps and causing the Permo-Triassic extinction, and that Venus's passage through a similar resonance caused the planet to resurface itself about 700 million years ago and generate its thick atmosphere. They find that terrestrial planets with retrograde rotation generate much more heat from such resonance passages. In addition, the Moon's tidal torque on Earth allowed it to pass quickly through its resonance, avoiding the fate of Venus. They also speculate that Venus's high surface temperature caused by its thick atmosphere maintained a magma ocean, which led to a rapid slowdown of its rotation through tidal dissipation. As a result, Touma and Wisdom argue that Venus was born with retrograde rotation; otherwise, it would have generated less heat through its core-mantle resonance passage. The case of Venus shows us how intimately linked are the geology, rotation, obliquity, orbit, and atmosphere of a terrestrial planet.

Both the value of the obliquity and the amplitude of its variation affect the habitability of a planet. Seasonal variations would be absent on a planet with a stable obliquity near zero degrees. While it would have constant surface temperatures, this benefit to life would be offset by two problems. First, weather systems would be more constant, some areas receiving steady precipitation, others receiving very little. More seriously, the polar regions would experience smaller maximum surface temperatures. Analogous to the water "cold traps" on synchronously rotating planets, polar ice would extend to lower latitudes, and it is possible that all the water would eventually freeze out at high latitudes. A thick atmosphere would be a possible way out of such a state, as would a deep ocean, but they would have other consequences for life that would have to be examined in detail.

At the other extreme, a stable obliquity near ninety degrees would result in very large surface temperature variations over most of the surface of a terrestrial planet. Most planets will have unstable obliquities over at least a few billion years, varying between small and large angles. Each case will have to be treated in detail to determine overall habitability. Williams and Pollard have explored seasonal surface temperature variations for a wide range of obliquities using general circulation climate models of Earth-like planets, confirming that high obliquity angles produce more extreme variations in surface temperatures.[59]

Low eccentricities characterize the orbits of the planets in the Solar System. The present eccentricity of Earth's orbit is 0.016, smaller than most of the other planets. While exhibiting chaos, the planets have maintained low eccentricity orbits since they formed.[60]

Numerical orbital simulations demonstrate the sensitivity of the Earth's orbit to changes in the orbit of Jupiter.[61] Earth's eccentricity increases significantly as Jupiter's orbit is made more eccentric. What is more, its orbit becomes unstable if Jupiter's eccentricity is greater than about 0.15. Increasing the mass of Jupiter reduces the eccentricity of Earth's orbit, while reducing its mass increases it. Decreasing Jupiter's semi-major axis has a significant effect on Earth's orbit only near resonances. Interestingly, Earth's orbit becomes significantly more eccentric if its mass is less than about half its present value.

Once it is known how the eccentricity of a terrestrial planet evolves, it is possible to calculate how its climate responds. There are two ways to do this for Earth. Williams and Pollard approached the problem with simulations of the climates of Earth-like planets under the assumptions of different eccentricities.[62] Not surprisingly, they found that larger eccentricities produce larger annual temperature variations; larger eccentricity can also reduce the amount of time a planet spends in the CHZ. One of the most important factors describing the response of surface temperature to insolation variation is the radiative time constant of the atmosphere. It depends on, among other quantities, the surface pressure and heat capacity of the atmosphere. Earth's atmosphere has a time constant of about one month; planets with thicker atmospheres will have longer time constants. If the radiative time constant is much smaller than the orbital period, then a planet's surface temperature will be more sensitive to eccentricity-induced insolation variations.

Study of Earth's ancient climate via proxies stored in sediments and polar ice is another way to elucidate the relationships between orbital variations and climate change. The Milankovitch cycles (obliquity, precession, and eccentricity) have been detected in several such records.[63] Even the small ranges of variation in the obliquity and eccentricity of 23.4 ± 1.3 degrees and 0.00–0.04, respectively, have been sufficient to leave their marks in the paleoclimate records. Earth's climate has been fluctuating dramatically in response to these small forcings for the past three million years.

The size of a terrestrial planet affects its habitability in diverse ways. Hart and Kasting et al. considered the effects of changing the size of a terrestrial planet on the evolution of its atmosphere.[64] Hart described how outgassing, atmospheric escape and surface chemical processes depend on the size of a terrestrial planet. Kasting et al. noted that a change in a planet's size has substantial effects on the greenhouse effect, albedo, atmospheric loss, and internal heat flow. Also, Lissauer noted that a larger terrestrial planet, all else being equal, should have a deeper ocean and higher surface pressure, due in part to the increasing importance of self-compression for terrestrial planets larger than Earth.[65]

A planet's geophysics is also sensitive to its size. A smaller planet will lose its interior heat more quickly, primarily due to its larger surface area to volume ratio. A smaller planet also has smaller pressure throughout its interior, which affects core formation.[66] O'Neill and Lenardic argue that terrestrial planets more massive than the Earth are unlikely to exhibit Earth-like plate tectonics,[67] but Valencia et al. reach a different conclusion.[68]

Mars is an important comparison case; it is half the size of Earth and lacks plate tectonics and a global magnetic field. Evidence from space missions and Martian meteorites indicates that it did once possess a global magnetic field and was volcanically active for about the first billion years,[69] confirming that its interior cooled more quickly than Earth's interior. The generation of a global magnetic field is closely linked to the operation of plate tectonics.[70] Plate tectonics produces a larger surface heat flux than a one-plate mode of tectonics, which, in turn, produces a larger temperature gradient in the deep interior and convection in the liquid portion of the core. A convecting outer core and fast planetary rotation are considered to be necessary requirements for generating a global magnetic field.

The generation of a global magnetic field also depends on the composition of the core. The presence of alloying light elements affects the melting temperature of iron (the most abundant element in the core) and can cause chemical convection. In the case of Mars, the volatile element sulfur is speculated to be the most abundant light element in the core.[71] In the early stages of its growth, a terrestrial planet's "feeding zone" spans a relatively narrow range in the protoplanetary nebula.[72] It is only during the later stages of its growth that a terrestrial planet accretes embryos from more distant regions. Thus, because of the negative radial temperature gradient in the protoplanetary disk, terrestrial planets farther from the Sun should have relatively more sulfur in their cores. For this reason, even if Mars were as massive as Earth, it would not have the same core chemistry, structure, and magnetic field evolution as Earth.

Relatively little research has been done on the biological consequences of a weak or absent magnetic field. Mars does provide some clues, however. The absence of a strong magnetic field over most of Mars's history has been implicated in the loss of a substantial fraction of its atmosphere through solar wind stripping.[73] Additional consequences include increased secondary cosmic ray particle radiation at the surface.[74]

The exchange of water between a planet's interior and its surface also depends on its geophysics.[75] Over billions of years, water on the surface of a terrestrial planet is lost to space and sequestered into its mantle.[76] Water itself is intimately linked to the operation of plate tectonics.[77] For this reason, it is likely that the present mode of Venus's geophysics depends, in part, on the loss of its water.

A planet's host star plays very important roles in the evolution of the CHZ. The host star affects the planets with its gravity and radiation. Assuming the core-accretion model for giant planet formation,[78] Laughlin et al., and Ida and Lin, find that giant planets are less likely to form around stars less massive than the Sun.[79] This theoretical result is consistent with observations.[80] Simulations also indicate that terrestrial planets forming around low mass stars in their CHZs are less massive.[81] Clearly, then, planetary systems forming around stars different than the Sun should be very different from the Solar System.

The Sun's radiation allows Earth to maintain liquid water on its surface and allows plants to produce chemical energy from photosynthesis, but it also has negative effects on life. There is evidence that small variations in the Sun's energy output on timescales from decades to millennia affect Earth's climate. They are caused by changes in the Sun's atmosphere and are not related to what goes on in its core. However, we know from stellar evolution models that the Sun has brightened by about 30 percent since its formation 4.6 gigayears (Gyrs) ago.[82] Unlike the short-term variations, this one is taken into account in modern CHZ models.

We can infer the evolution of the Sun's chromospheric activity from observations of nearby Sun-like stars spanning a broad range in age. Such a research program, called "The Sun in Time," began about twenty years ago.[83] Young Sun-like stars are observed to have shorter rotation periods, higher UV and X-ray luminosities, and more frequent flares (which produce temporary high fluxes of ionizing radiation). From satellite observations of such stars, the Sun's X-ray luminosity is inferred to have decreased by about three orders of magnitude, while its UV declined by about a factor of twenty.[84] Observations also imply that the Sun's optical variability has declined by a factor of about fifty over the same period.[85] Interestingly, Radick et al. also confirmed that the Sun's optical variability on decadal timescales is anomalously small compared to otherwise similar stars. The early higher activity is very relevant to habitability in part because the higher flux of ionizing radiation stripped a significant fraction of the terrestrial planets' early atmospheres.[86]

Chromospheric activity also correlates with the rate of flares. Flares produce X-ray and proton radiation. X-rays entering the top of the atmosphere are downgraded to UV line emission at the bottom of the atmosphere.[87] Protons can alter the chemistry in the middle atmosphere and stratosphere. In particular, the two strongest solar proton events of the past four decades (1972 and 1989) were calculated to have temporarily reduced the total ozone by 1–2 percent.[88] Even stronger flares should occur over longer timescales, and they should have been more frequent in the Sun's past.[89]

Stellar activity also varies along the main sequence. Many red dwarf stars exhibit extremely powerful and frequent flares. West et al. studied a large sample of nearby M dwarfs and found that activity peaks near spectral type M8 at about 80 percent incidence.[90] The UV flux can increase by a factor of 100 during a flare.[91] Active M dwarfs have a soft X-ray to bolometric flux ratio several orders of magnitude greater than the Sun; during flares this flux ratio can be 10^6 times the Sun's. Since the size of the CHZ is set by the bolometric flux, a planet in the CHZ of an M dwarf will be subjected to a much greater flux of ionizing radiation. Smith et al. found that the UV flux in the lower atmosphere of a planet around an M dwarf is completely dominated by redistributed energy from flare X-rays and reaches biologically significant levels.

The red color of an M dwarf star means that relatively less blue light will reach the surface of its orbiting planets compared to the Sun. Although photosynthesis doesn't require blue light, it generally becomes less effective without abundant light blueward of 6800 Å. Wolstencroft and Raven showed that Earth-like planets in the CHZ of cooler stars should be less effective at producing oxygen from photosynthesis.[92] Some bacteria can still use infrared light, but not to produce oxygen. Any marine photosynthetic organisms would have difficulty using red light as an energy source, since ocean water transmits blue-green light much better than violet-blue or red light. The precise wavelength of peak transmission will depend on the minerals dissolved in the oceans. The transmittance of pure water peaks at 4300 Å, while that of Earth's oceans peaks a few hundred Å to the red.

Planets around M dwarfs should be less habitable than Earth also because they achieve rotational synchronization quickly.[93] If a rotationally synchronized planet has a circular orbit, then one side will continuously face its host star while the other remains in darkness. In other words, the length of the year equals the length of the day. This will lead to large temperature differences between the day and night sides and to the freeze out of its water. Slower rotation would also result in a weaker magnetic field.[94] Rotational synchronization can be avoided in two ways. First, a planet can have a sufficiently eccentric orbit so that, like Mercury, it has a 3:2 spin-orbit resonance. Such a planet would avoid water freeze-out, but it would still exhibit large temperature variations over the course of its orbit and have a weak magnetic field. The second way is to have a planet-size moon orbiting around a giant planet in the habitable zone. Such a planet, if it could exist in such a configuration (which is doubtful), would suffer from many of the same difficulties already noted above.

Finally, the recent discovery of a possible link between the cloud fraction and the galactic cosmic ray (GCR) flux in Earth's atmosphere has a direct bearing on models of the CHZ. Higher GCR flux leads to enhanced production of low altitude clouds, which leads to a cooling of the Earth.[95] Svensmark et al. presented the first experimental evidence for what had been a missing part of this link—demonstration that the ionization induced by cosmic rays enhances the formation of cloud condensation nuclei (and therefore clouds).[96] The GCR flux is modulated by the solar interplanetary magnetic fields. Therefore, Earth's albedo is determined partly by the state of the magnetic fields of the Sun and the GCR flux arriving at the Solar System.[97] More generally, it is determined by the location of a terrestrial planet in the Solar System. At a greater distance from the Sun, the influence of the solar magnetic field is weaker, and therefore variations in the GCR flux entering the Solar System are relatively more important. This existence of the GCR–cloud link also implies that the carbon-silicate cycle is not as important for climate regulation as it is currently believed to be.[98]

In summary, the definition of the CHZ depends on much more than just the flux of radiation a terrestrial planet receives from its host star. A terrestrial planet's habitability also depends on its orbital eccentricity, presence of a large moon, size, initial volatile inventory, initial rotation period and its evolution, the locations and properties of any giant planets, the distributions of small bodies, and the host star's modulation of the cosmic ray flux. Much progress is being made on each of these factors, but eventually they will have to be treated together, given their sometimes strong interdependencies.

The Galactic Habitable Zone

Habitability on the scale of the Milky Way Galaxy has been discussed at least since Schklovsky and Sagan considered the perturbations to the biosphere by nearby supernovae (SNe).[99] Since then, several studies have re-examined this question. Some have considered the effects of ionizing radiation from a nearby SN on Earth's ozone layer.[100] Others have searched the paleobiological and geological records for signatures of nearby SNe. For example, Benitez et al. present evidence linking the Pliocene-Pleistocene boundary marine extinction event about two million years ago to a nearby SN,[101] and Knie et al. and Wallner et al. discovered spikes in the ^{60}Fe and ^{244}Pu concentrations, respectively, in marine sediments from about the same time, consistent with a SN explosion about forty parsecs from Earth.[102]

The Galactic environment also affects the orbits of comets in our Solar System and (presumably) others. The weakly bound Oort cloud comets are sensitive to large-scale gravitational perturbations, including the Galactic vertical (disk)[103] and radial[104] tides, Giant Molecular Clouds (GMCs),[105] and nearby star encounters.[106] Of these, the Galactic disk tide is the dominant perturber of the present outer Oort cloud comets; the disk tide is about fifteen times greater than the radial tide.[107] Such perturbations can cause "comet showers" in the inner Solar System and thus increase the comet impact rate on Earth. Matese et al. argue that the imprint of the radial Galactic tide is present in the observed distribution of long-period comet aphelia on the sky.[108] Stars closer to the Galactic center will experience more intense comet showers due to the increased radial Galactic tide and more frequent nearby star encounters.[109]

Gonzalez et al. unified the various Galactic factors by introducing the Galactic Habitable Zone (GHZ) concept (Lineweaver introduced a similar idea).[110] The inner boundary of the GHZ is set by the various threats to the biosphere (e.g., nearby SNe and comet impacts), and its outer boundary is set by the minimum metallicity required to form an Earth-*like* planet, not merely an Earth-*mass* planet. If the metallicity is too low, then it will not be possible to form an Earth-like planet, which is composed mostly of O, Mg, Si and Fe.[111] These elements are produced primarily by massive star SNe, which enrich the originally pure H and He interstellar gas with their processed ejecta. The incidence of giant planets is much higher among metal-rich stars, but the planets detected to date tend to have highly eccentric or very short period orbits.[112] Such orbits are less likely to be compatible with the presence of habitable terrestrial planets. The temporal evolution of the GHZ is determined primarily by the evolution of the metallicity of the interstellar gas, the interstellar abundances of the geologically important radioisotopes (^{40}K, ^{235}U, ^{238}U and ^{232}Th), and the rate of transient radiation events.

Lineweaver et al. further quantified the GHZ by applying Galactic chemical evolution models.[113] They only included the effects of the evolving interstellar gas metallicity and SN rate and estimated that 10 percent of all the stars that have ever existed in the Milky Way have been in the GHZ. While this is a helpful study, as we show below, several other factors need to be included to provide a more complete picture of the GHZ.

There is strong evidence for extraterrestrial influence on Earth's climate.[114] Bond et al., for example, employed ^{14}C and ^{10}Be as proxies for solar variations and marine sediments in the North Atlantic as proxies for polar ice extent.[115] They found a strong correlation between the polar climate and variations of solar activity on centennial to millennial timescales. This and many other recent studies present evidence that Earth's climate has varied in response to solar variations over the entire Holocene. Several mechanisms for solar-induced climate change have been proposed, of which the leading contender is the GCF–cloud link noted above.

The GCR flux from Galactic sources also varies over long timescales. Thus, the importance of GCR-induced low cloud formation will depend on the Galactic star formation rate and the location of the Solar System in the galaxy, as well as the location of a planet in the Solar System.

Transient radiation events important on the Galactic scale include SNe, Gamma Ray Bursts (GRBs) and Active Galactic Nucleus (AGN) outbursts. Less powerful events, such as novae and magnetar outbursts, while more frequent, are less important on average and will not be considered here. I will provide a brief review of the rates, distributions, and energetics of these three classes of Galactic transient radiation events along with their possible effects on the biosphere.

The possible threats posed by SNe to life on a planet over the history of the Milky Way Galaxy depend on their spatial distribution, temporal frequency, and typical total radiant energy. Galactic chemical evolution models are required to estimate the temporal evolution of SNe, but observations can give us helpful constraints. A simplifying starting assumption is that the average SN rate in galaxies similar to the Milky Way (Hubble type Sbc) is representative of the rate in the Milky Way.[116] This is probably a good assumption, but over relatively short time intervals the Milky Way's SN rate will sometimes deviate from the average significantly. Extragalactic SN surveys also yield reliable estimates of the average rates[117] of the various types of SNe (the main types are Type Ia (SN Ia) and Type II+Ib/c (SN II)) and their luminosity distributions.[118] Supernova rates are usually given in units of number per century per 10^{10} solar luminosities of blue light (SNu). For example, the total SN rate for elliptical galaxies is 0.18 ± 0.06 SNu, increasing to 1.21 ± 0.37 SNu for Scd-Sd Hubble types;[119] SNIa are the only type of supernovae observed in elliptical galaxies. The total SN rate in the Milky Way is 2–3 SN per century.[120]

It is helpful to know the rates for the SN Ia and SN II separately, given their different distributions in the Milky Way. SN II, which result from massive stars, are observed in the thin disk and in the spiral arms. SN Ia result from older stars and occur throughout the Milky Way. Surveys of supernova remnants in the Milky Way,[121] pulsars in the Milky Way,[122] and SNe in nearby galaxies[123] show that SNe peak at about 5 kiloparsec (kpc) from the Galactic center.

Gehrels et al. explored the effects on Earth's atmosphere of the gamma ray and cosmic ray radiation from a nearby SN II.[124] The gamma ray radiation resulting from the decay of ^{56}Co lasts a few hundred days, while the elevated cosmic ray flux can last thousands of years. They simulated twenty years of atmospheric evolution and found that a SN would have to occur within 8 pc for the UV radiation at Earth's surface to be at least doubled. They estimate a rate of about 1.5 dangerous SNe per billion years. They did not explore the biological effects of the secondary particle radiation produced in the atmosphere nor the effects of the cosmic ray flux on the cloud fraction. Thus, their calculations should underestimate the important biological effects of a nearby SN II.

GRBs are rare, very short-duration (~10s), very luminous (~10^{51} ergs s^{-1}) explosions that produce most of their luminous energy in gamma rays with energies between 100 keV and at least 1 MeV. Scalo and Wheeler, assuming GRBs are produced by massive star supernovae (SN Ib/c), calculated that the GRB rate in the Milky Way is about one per ≥ 30,000 SN Ib/c.[125] This rate depends on the assumed degree of collimation of the GRB radiation, GRB evolution with redshift, and the properties of the GRB progenitor (e.g., minimum mass star to explode as a SN Ib/c).

Still, the number of GRBs that affect Earth is not dependent on the degree of collimation, only on the observed rate (only those GRBs which we can see will affect us). The gamma-ray photons from a GRB cannot reach the ground for planets with thick atmospheres like Earth. The incoming photons are first downgraded to X-rays via Compton scattering, and then the X-rays are absorbed and generate photoelectrons. The energetic electrons then collide with oxygen and nitrogen atoms, exciting them and causing ultraviolet emission, which makes it to the surface. Scalo and Wheeler estimate that GRBs aimed at a planet with a thin atmosphere can do significant damage to the DNA of eukaryotes from as far away as 14 kpc (nearly twice the distance to the Galactic center);[126] the corresponding distance for prokaryotes is 1.4 kpc. The critical distance for significant UV production at the surface of Earth is about 11 kpc. They estimate that this occurs once every 2–4 million years. The very short duration of the photon radiation burst on the surface of a planet resulting from a GRB implies that only life on one hemisphere of the planet will suffer its direct effects.

Melott et al., Thomas et al., and Thomas and Melott also consider the possible long-term damaging effects of a GRB's energetic photons on a planet's atmosphere, including ozone destruction, global cooling, and acid rain.[127] They estimated that a GRB within 3 kpc of Earth can cause significant damage to its ozone layer, and that such an event should occur every 170 million years. They also suggested that a GRB might have caused the late Ordovician mass extinction.

GRBs should also generate collimated jets of energetic particles.[128] Dar et al. and Dar and De Rújula consider the effects on the biosphere of particle jets from GRBs impinging on Earth's atmosphere.[129] These include atmospheric muons, radioactive spallation products, and ozone destruction. The muons can penetrate deep underwater and underground. The duration of the cosmic ray irradiation is expected to be a few days, long enough to cover all longitudes (but not necessarily all latitudes). Dar and De Rújula estimate that cosmic rays from a GRB at the Galactic center aimed at Earth would produce a lethal dose of atmospheric muons to eukaryotes.[130]

The Milky Way's nucleus is presently in a relatively inactive state, but there is strong evidence that a 2.6 million solar mass black hole resides there.[131] It is among the smaller black holes detected in the nuclei of nearby large galaxies. The Milky Way's nuclear black hole has grown over time by accreting gas and disrupted stars. While it is accreting, the black hole's disk emits electromagnetic and particle radiation.

High-resolution optical observations indicate that all large galaxies have a supermassive nuclear black hole.[132] When active, such black holes are observed in the bulges of galaxies as AGNs (called Seyfert galaxies). The fraction of galaxies observed in the nearby universe with AGNs is related to the duty cycles of their black holes. Assuming the AGN–nuclear black hole paradigm is correct, then the larger the observed fraction of AGNs, the larger the average duty cycle. Large duty cycles are possible if AGNs are long-lived and/or frequent. Miller et al. studied the distribution of AGNs in the nearby universe and concluded that about 40 percent of massive galaxies have an AGN.[133] They conclude from this that the typical AGN lifetime is about 2×10^8 years, or that the typical AGN has burst forty times over the $5.7 \square 10^8$ years covered by their survey. Marconi et al. modeled the growth of nuclear black holes during AGN phases and found that duty cycles have declined over the history of the universe and that they are larger for smaller black holes.[134]

If these numbers can be applied to the recent history of the Milky Way Galaxy's nuclear black hole, then it should have been in an active state during about 40 percent of the past half billion years. The luminosity of its nucleus in an active state would be about 10^{44} ergs s^{-1} in 2-10 keV X-rays.[135] Emission in this range alone would generate the energy of a typical supernova in less than a year. Above about 5 keV, there is relatively little attenuation by the interstellar medium towards the Galactic center. The total interstellar extinction towards the Galactic center is also a sensitive function of the distance from the Galactic midplane; planets located near the midplane will be the most protected from ionizing photons produced by an AGN. At Earth, the X-ray flux would be 130 erg m^{-2} s^{-1}, assuming no intervening absorption. This is about twenty times the typical flux from the Sun in the same energy band and is comparable to the peak flux of an M-class X-ray solar flare. Including absorption would make the Galactic center X-ray flux comparable to that of the Sun's average X-ray output. Thus, the X-ray emission from an AGN outburst would probably not be very important for life on Earth, but it probably would be damaging for planets within ~ 2 kpc of the Galactic center. BL Lacertae objects and flat-spectrum radio quasars are observed to have gamma ray luminosities up to 10^{49} erg s^{-1} and 10^{50} erg s^{-1}, respectively.[136] Whether the Milky Way was ever in such a state, however, is an open question.

Clarke argued that the particle radiation from an AGN outburst would be much more important to life than its ionizing photons.[137] He calculated, assuming no propagation energy losses, that particle radiation fluxes at Earth would increase by a factor of ~100, causing significant damage to the ozone layer and increased radiation at the surface.

In order to model the GHZ, it is also important to understand the spatial gradients and temporal evolution of the disk gas metallicity in the Milky Way. The disk radial metallicity gradient is determined from a variety of objects, including open clusters, H II regions, planetary nebulae, and Cepheid variables. Recent abundance measurements of cepheids (arguably the best type of indicator of the gradient) give a gradient value of –0.07 dex kpc^{-1}.[138] Maciel et al. determined that the gradient has been flattening over the past 8 billion years and is in the range of 0.005 to 0.010 dex kpc^{-1} Gyr^{-1}.[139] Observations of thin disk G dwarfs in the solar neighborhood show that the disk gas metallicity is increasing at a rate of about 0.035 dex Gyr^{-1}.[140]

Knowledge of the metallicity gradient is critical to understanding the GHZ, as the initial gas metallicity of a cloud strongly constrains the properties of the terrestrial and giant planets that form from it. First, the incidence of Doppler-detected giant planets around nearby Sun-like stars is now known to be very sensitive to the host star's metallicity. It rises steeply from about 3 percent at solar metallicity to 25 percent for stars with twice the solar metallicity.[141]

The best explanation for the correlation between metallicity and the presence of giant planets is that giant planets are more likely to form around metal-rich stars.[142] How the Solar System fits into this picture is still unsettled, but it is beginning to appear that it is not typical.[143]

Ida and Lin have explored the metallicity dependence of giant planet formation with simulations based on the core-accretion theory.[144] They succeed in reproducing qualitatively the observed metallicity dependence on the incidence of giant planets. Since the processes in the early phases of giant planet formation also apply to terrestrial planet formation (prior to the gas accretion phase), studies like Ida and Lin's should help us understand the metallicity dependence of terrestrial planet formation as well. Gonzalez et al. suggested that the mass of a typical terrestrial planet formed in the CHZ should depend on the initial metallicity raised to the 1.5 power,[145] while Lineweaver assumed that the dependence should be linear.[146] Gonzalez explores this question but concludes that we are not yet ready to decide between these two assumptions.[147]

The major comet reservoirs in the Solar System reside beyond the orbits of Neptune and Pluto.[148] Three reservoirs are typically recognized (listed with heliocentric distances): the Kuiper belt (30 to 1,000 AU), the inner Oort cloud (1,000 to 20,000 AU) and the outer Oort cloud (20,000 to 50,000 AU). There is direct observational support for the existence of the Kuiper belt and indirect evidence for the Oort cloud.[149] The Kuiper belt has about 7×10^9 comets, while the Oort cloud has about 10^{12} comets;[150] the inner Oort cloud has about five times as many comets as the outer Oort cloud.

The properties of the Oort comet cloud around a given planetary system depend in part on the properties of its giant planets and the initial metallicity of its birth cloud. Presumably, a planetary system forming from an initially more metal-rich cloud will form a more populated Oort cloud, but this needs to be confirmed with self-consistent simulations that include the metallicity dependence of the properties of the giant planets. Given the high sensitivity of the incidence of giant planets to metallicity, it seems likely that this is a reasonable assumption. Granting this and the known Galactic metallicity disk gradient, planetary systems born in the inner Galaxy should start with more populous Oort clouds. The subsequent history of interaction between an Oort cloud and its Galactic environment is also critical to understanding the threats from comets.

Interstellar gas metallicity, stellar velocity dispersion, stellar density, GMC density, and Galactic tides all vary with location and all affect the orbits of the Oort cloud comets. To first order, the initial number of comets in the Oort cloud can be approximated as being proportional to the initial metallicity. The stellar density, GMC density, and tides are all greater in the inner Galaxy, relative to the Sun's location. Moving the Solar System half way to the Galactic center would increase the comet flux by a factor of about seven.

In addition to the threat from comets residing in the Sun's gravitational domain, there is also the threat from interstellar comets. They can be grouped into two distinct types: 1) free-floating comets lost from Oort clouds around other stars (IS1), and 2) comets gravitationally bound to other stars (IS2).

The probability that an interstellar comet collides with Earth is much greater than just the geometrical cross-section of Earth amplified by the Sun's gravity; it depends on the cross section of the orbits of the giant planets, which can capture an interstellar comet into an orbit that brings it close to the inner planets repeatedly. Taking this factor into account and assuming a local density of IS1 comets of 10^{12} pc^{-3}, Zheng and Valtonen calculated that Earth should have collided with about 100 such comets over its history, or one every 40 million years.[151]

Presently, there are no stars sufficiently close to the Sun for their bound comets to pass near enough to the Sun for us to see them. A type IS2 comet only poses a threat to us when its host star passes within about one-third of an AU of the Sun. Thus, a nearby star passage will threaten Earth both from its comets and from the perturbed comets in the Sun's domain. Taking an average heliocentric distance in the Sun's inner Oort cloud (10,000 AUs) and the present number of comets there (~10^{12}), the average number density is 2×10^{15} pc^{-3} (or, about one comet per five cubic AUs). This is at least three orders of magnitude greater than the number density of IS1 comets in the solar neighborhood.

The probability of impact from interstellar comets should vary with Galactic location. The three key factors are the initial metallicity, stellar density, and stellar velocity dispersion. The higher gas metallicity in the inner Galaxy should result in the formation of more interstellar comets. The inner Galaxy should be populated by a higher density of IS1 comets resulting from more frequent star-star and star-GMC encounters and stronger Galactic tides. On the other hand, the larger stellar velocity dispersion in the inner Galaxy will reduce the cross-section for comet captures. The more efficient stripping of comets from their Oort clouds in the inner Galaxy will reduce the importance of type IS2 comet collisions relative to those of type IS1; the accelerated stripping of comets from a star's Oort cloud in the inner Galaxy is at least partly compensated by the increased number of interstellar comets.

While a planetary system is traversing interstellar space, it will occasionally encounter a GMC (and more often, lower density clouds), and the probability of encounter is increased when it is crossing a spiral arm. Talbot and Newman calculated that the Solar System should encounter an average density GMC (~330 H atoms cm^{-3}) every 100 million years and a dense GMC (~2×10^3 H atoms cm^{-3}) every billion years.[152] The biologically significant effects of such an encounter are varied. These could include comet showers, exposure to a greater flux of cosmic ray particles, glaciations, and more nearby SNe.

Begelman and Rees first noted that passage of the Sun through interstellar clouds with densities of at least 10^2 to 10^3 H atoms cm^{-3} are sufficient to push the heliopause inside Earth's orbit.[153] This would leave Earth exposed to interstellar matter. Scherer et al. note that the shrinking of the heliopause during passage of the Solar System through an interstellar cloud will also eliminate the solar modulation of the cosmic rays at Earth and expose it to a higher cosmic ray flux,[154] possibly leading to more clouds due to the GCR-cloud link noted above. Florinski et al. determine that the GCR flux at Earth would be enhanced by a factor of 1.5–3 by a cloud with a hydrogen density of 8.5 cm^{-3}, which is only about thirty times the present local interstellar gas density.[155] In addition, the cosmic ray flux within a GMC would be much greater due both to recent SNe within it and to the longer cosmic ray diffusion time within it.

Yeghikyan and Fahr modeled the passage of the Solar System through a dense interstellar cloud (10^3 cm^{-3}), confirming that the heliopause is pushed into the region of the terrestrial planets.[156] In such a situation, the interstellar matter interacts directly with Earth's atmosphere. They found that the ozone in the upper atmosphere is depleted, and Earth is cooled by about 1 degree C, possibly causing an ice age.

Pavlov et al. noted two effects neglected in previous studies that should significantly increase the effects to the biosphere of a cloud passage.[157] First, the flux of anomalous cosmic rays (generated from interstellar neutrals) would increase much more than the GCR flux. Second, during passage through a typical cloud (lasting about 1 million years) there would be one or two magnetic field reversals, during which time Earth's atmosphere would remain unprotected from cosmic rays at all latitudes. Pavlov et al. calculated that passage through a cloud with a density of 150 H atoms cm^{-3} would decrease the ozone column by 40 percent globally and up to 80 percent near the poles. Pavlov et al. studied the possible effects of dust accumulation in Earth's atmosphere during passage through a GMC.[158] They concluded that such an event could produce global snowball glaciations and that less dense clouds could still produce moderate ice ages.

Of course, a planet closer to the outer edge of the CHZ would feel the effects of cloud passages more intensely and more often, as would a planet with a weak or absent magnetic field. Thus, among the terrestrial planets in the Solar System, Mars should have been subjected to the most severe interactions with interstellar clouds. This is another reason that the CHZ definition should be expanded beyond merely the maintenance of liquid surface water.

How would the effects from passages through GMCs vary with location in the Galaxy? First, GMC encounters should be more frequent in the inner Galaxy due to the higher density of GMCs there. Second, the encounter velocities should be greater in the inner Galaxy, making it more likely that the heliopause will be pushed back to the vicinity of the terrestrial planets, even for relatively low-density interstellar clouds. In addition, at high encounter velocity, the energy deposited by interstellar dust impacting a planet's atmosphere might become an important factor in atmospheric loss, but this needs to be explored quantitatively.

Galactic dynamics is arguably the most complex aspect of the GHZ. While the Sun's original birth orbit in the Galactic disk cannot be determined, there is little doubt that it has changed significantly. Older stars in the disk have larger velocity dispersions. Stars form in relatively circular orbits, and over time they experience gravitational perturbations that make their orbits more eccentric and send them farther from the disk midplane. Thus, older stars tend to pass through the midplane at higher velocity and traverse a greater range of radial distances from the Galactic nucleus.

Given that stars migrate in the disk over billions of years, the present metallicity of interstellar gas in a star's vicinity, corrected for the time of its formation, will not be representative of the star's metallicity. The disk radial metallicity gradient allows us to calculate the metallicity of a star at the time of its formation, but the star will later wander to a region with a higher or lower gas metallicity. As a result, the GHZ has fuzzy boundaries.

The spiral arms are important structures for determining the boundaries of the GHZ. They contain most of the GMCs and SNe II in the Milky Way. The interarm regions have less star-formation activity, but the star density there is only modestly less than that in the arms at the same Galactocentric radius. Within the framework of the density wave hypothesis, the spiral arms rotate about the nucleus of the Milky Way like a solid body, with a constant angular frequency.[159] Assuming the Milky Way's spiral arm pattern can be maintained for at least several billion years,[160] surveys of the spiral arms can be used to conduct long-term simulations of stars' motions with respect to them.

Stars at the corotation circle will orbit the Galactic center with the same period as the spiral arm pattern. Thus, the interval between spiral arm crossings is longest for a star in a circular orbit at the corotation circle. However, Lepine et al. show that a star near the corotation circle experiences resonant perturbations with the spiral density waves that cause it to wander in the radial direction by 2–3 kpc in less than a Gyr.[161] This would imply that, over Gyr timescales, spiral arm crossings are minimized, instead, at a moderate distance from the corotation circle. If the Sun is indeed very close to the corotation circle, as some recent studies indicate,[162] then it is a surprise that its orbit has such a small eccentricity.

Additional benefits for life at the corotation circle include reduced gas density and reduced star formation. It is interesting to note that on this basis alone, Marochnik concluded that there must be a habitable zone in the Milky Way not unlike the GHZ.[163]

The interarm spacing at the Sun's location is 2.8 ± 0.3 kpc.[164] The Sun is presently located 0.20 kpc inside its mean Galactic orbital radius and about 0.14 kpc from its perigalactic radius.[165] It is 0.9 kpc from the Sagittarius arm.[166] Thus, the Sun is presently safe from radial excursions into either the inner or outer arms, assuming it is near the corotation circle.

Shaviv reports having found a link between spiral arm crossings and long-term variations in Earth's climate, especially glaciations.[167] At the core of his thesis is the GCR–cloud link. He reconstructs the historical GCR flux from meteorite exposure ages and the concurrent temperature variations from ancient calcite shells. He compares these data to the calculated cosmic ray flux fluctuations resulting from spiral arm crossings (varying from about 35 percent to 140 percent of the present value). The strength of the correlation, however, depends sensitively on the still-controversial value of the Sun's position relative to the corotation circle. His data could also be interpreted in terms of a varying star-formation rate combined with close encounters with (as opposed to crossings of) spiral arms.

In summary, the evolution of the GHZ is determined by a number of astrophysical processes. In a given region of the Milky Way, Earth-like planets are unlikely to form within the CHZ until the interstellar gas metallicity reaches a value close to solar metallicity. Survival of life depends on the distribution of radiation hazards, interstellar clouds, and comets. Threats to life increase toward the center of the Galaxy and decrease with time. The greatest uncertainty about the GHZ concerns stellar dynamics and how a given star's orbit interacts with the spiral arms.

THE COSMIC HABITABLE AGE

The Steady State Theory held that the universe has always appeared as it does today.[168] The replacement of the Steady State Theory by the Big Bang theory in the 1960s, however, made it clear that such is not the case. Analysis of the Wilkinson Microwave Anisotropy Probe (WMAP) data indicates that the universe began in a hot dense state 13.7 billion years ago.[169] Analyses of distant galaxies show that the global star-formation rate has been declining for the past 5 billion years.[170] Old stars in the Milky Way are observed to be systematically deficient in metals compared to young stars, mirroring the evolution of metals in the broader universe. The universe has changed drastically since its formation, and these changes bear on the question of habitability.

Discussions of the Cosmic Habitable Age (CHA) have usually been framed in terms of the anthropic principle.[171] Given that the universe has changed so dramatically since its origin, the question naturally arises why we observe ourselves to be living during this particular time as opposed to some other time. Clearly, chemically based life is not possible in the very early universe before atoms formed or in the distant future, after all the stars burn out. Other considerations indicate that the boundaries of the CHA are much narrower than these extreme limits.

Lineweaver estimated the probability of forming Earth-like planets over the history of the universe based on the evolution of the global metallicity.[172] He assumed that habitable terrestrial planets are most probable over a narrow range of metallicity centered on the solar value. Von Bloh et al. used the results from Lineweaver to calculate the peak time of the incidence of Earth-like planets in the Milky Way; they found it to be at about the time of formation of the Earth.[173]

Habitable Zones and Fine-Tuning

Progress in refining the CHA will come primarily from improvements in our understanding of the evolution of the cosmic star-formation rate. Star formation, in turn, determines the evolution of the average supernova rate, AGN activity, and gas-phase metallicity in galaxies. Many of the same studies relevant to the GHZ can also be applied in a broader sense to the CHA.

If all galaxies were just like the Milky Way, then the GHZ could just be applied to other galaxies. But they aren't; there is great variation in their properties. Galaxies differ in their Hubble types (elliptical, spiral, or irregular), metallicities, luminosities, masses, and environments. Some of these properties correlate with each other. For example, mass correlates with luminosity. Metallicity also correlates with luminosity, in the sense that more luminous galaxies are more metal-rich.[174]

Calura and Matteucci calculated the evolution of the production of metals over the history of the universe.[175] They determined that the present mean metallicity of galaxies is close to the solar value. This is consistent with the fact that the Milky Way Galaxy is among the 1–2 percent most luminous galaxies in the local universe. The inner Galaxy experienced more metal enrichment over its history and has a higher mass density than the solar neighborhood. Thus, the mass-weighted average of the Galaxy's metallicity is larger than the solar value. An additional consideration is that luminosity-weighted metallicity averages for galaxies tend to be smaller than mass-weighted ones, because the bright red giants in a galaxy tend to be more metal-poor.

Even though the average metallicity of local Galactic matter is now close to solar, there were many metal-rich stars formed within the first 2–3 billion years after the Big Bang. The metals first built up quickly in the inner regions of (now) massive galaxies. Presumably, these metal-rich stars have been accompanied by planets. However, like the juvenile Milky Way Galaxy, these were also the regions with the most dangerous radiation hazards, including supernovae, GRBs, and intense AGN activity.

Galaxy evolution is not the same everywhere. Ellipticals are much more common in dense galaxy clusters than in small groups, like our Local Group. Spiral galaxy disks tend to be stripped of gas more efficiently in dense cluster environments, and star formation is suppressed as a result.[176] Galaxy collisions are also less frequent in sparse groups.

Galaxy collisions shape galaxies and alter their star-formation rates. Major collisions can temporarily increase the star-formation rates throughout each of the involved galaxies (becoming starburst galaxies) and possibly consume most of the gas. Most often, colliding galaxies become ellipticals following exhaustion of their gas, with a greatly reduced star-formation rate (and thus a reduced planet-formation rate). Mergers can also feed fresh gas into any supermassive black holes in their nuclei, causing AGN outbursts. Mergers were frequent in early times, but even today they continue. For example, Hammer et al. argue that about three-quarters of intermediate-size spirals have experienced major collisions within the last 8 billion years.[177]

Surveys of supernovae and their remnants tell us not only about the present supernovae distribution and rate in the Milky Way, they also reveal how these quantities vary among the various types of galaxies. Surveys of SNRs in large nearby galaxies show that their radial scale lengths are generally similar (about 30 percent their disk radii), and many have peaks at 20 percent to 40 percent of their disk radii (the Milky Way being such a case).[178] The starburst spiral galaxy NGC 6946, however, has a sharp peak of the SNR density at its nucleus. Particularly helpful are studies like Capellaro's, which catalogs the observed rates of all SN types for all types of galaxies.[179] Such studies reveal that SNe II do not occur in ellipticals, while they predominate in late-type spirals, where the overall SN rate is greater.

With the star-formation rate continuing to slow, the mean metallicity will increase ever more slowly. Many galaxies presently below solar metallicity will eventually build up enough metals to form Earth-size terrestrial planets. The time in the history of the universe when a particular region in a given galaxy reaches this critical stage is important—too early, and the radiation environment may be too harsh for life to flourish. There is also a limit at late times. As the star-formation rate declines, the production of the long-lived geologically important radioisotopes cannot keep up with their decay in the interstellar medium. Earth-size terrestrial planets forming in the future will have less radiogenic heating.[180] Of course, increasing the size of a planet can compensate for this deficit, but then all the processes dependent on planet size discussed above will need to be taken into account.

In the future, massive stars will become rarer, which means that the SN rate will continue to decline. AGN activity will also decline. Galaxies will continue to recede from each other. Thus, the universe should become safer from powerful transient radiation events. On the other hand, G dwarfs will also become rarer, and stellar galactic orbits will become hotter. Adams and Laughlin speculated on the fate of the universe in the far distant future, after nucleons decay and black holes evaporate.[181] While these are very great extrapolations from the present, there is no question that the future history of the universe will differ drastically from its present state. The changes will take it ever further from the conditions we know are compatible with life (especially complex life).

Based on the evolution of the global star-formation rate, the CHA probably encompasses the last few billion years and the next ten billion years. This is brief compared to the possible future history of the universe. Interestingly, there appears to be a convergence of several timescales that permits life to flourish on a planet within this brief window. These include: the nuclear evolutionary timescale of the Sun; the rate of decline of the Sun's activity; the mean half-life of the geologically important radioisotopes; the loss rate of Earth's volatiles; the recession rate of the Moon and the related slowdown of Earth's rotation; the evolution of the star formation rate in the Galaxy; and the expansion rate of the universe. Is this just a coincidence, or is this telling us that these timescales must match for life to be possible now?

By way of summary, some of the complexity of the many habitability factors is presented in highly schematic form in Figure 1. The many interrelationships are represented by the links between the boxes. The overall problem of habitable zones is highly nonlinear. It will require considerable computational power to perform the required Monte Carlo simulations and long temporal integrations. Its solution will require continued advancements in astrophysics, geophysics, climatology, and biology.

The multitude of interactions shown in Figure 1, both positive and negative for life, makes it unlikely that another "island of habitability" will be found in parameter space. For example, if Earth had formed 20 percent farther from the Sun, it would still be within the traditional CHZ, but it would be subjected to a different asteroid and comet impact rate, different gravitational perturbations to its orbit and rotation, and different exposure to interstellar clouds and cosmic ray flux, and it would have needed more carbon dioxide in its atmosphere to maintain liquid surface water. Similar arguments apply to the type of host star, the location within the Milky Way, and to the type of galaxy. If we change one parameter, it is not often the case that we can change another parameter to compensate for any deleterious effects to life. As astrobiologists continue to learn about the formation and long-term evolution of planetary systems, they are discovering that processes that affect planetary habitability are intertwined in a complex web that sets severe constraints.

3. Implications for Global Tuning

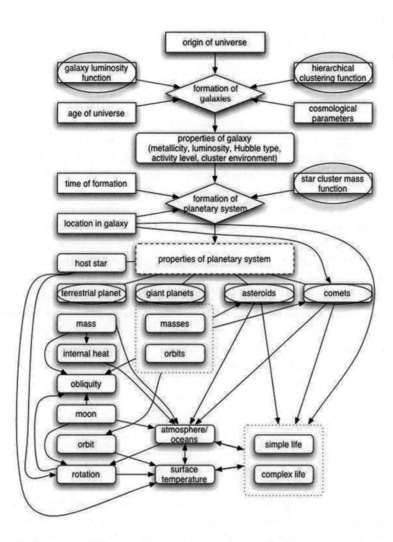

Figure 1. Highly schematic diagram showing the many interrelationships among the important habitability factors. The factors with the ellipses have significant stochastic components. The factors for terrestrial planets continue along the bottom of the figure, as they did not all fit along the left column.

Once we've established what kinds of environments are habitable, we can use this information to examine how changes in the global parameters affect habitability at the local scale. For example, if it turns out to be the case that an environment very similar to ours is necessary for complex life, as now seems likely, even relatively small changes to the global parameters will result in a lifeless universe. On the other hand, if environments significantly different from ours can be just as habitable, then relatively large changes in the global parameters are likely to find other local islands of habitability in parameter space. When adjusting a global parameter, it is necessary to follow its effects all the way down to the local scale. The physicist and cosmologist must partner with the astrobiologist. We will briefly explore one example of changes to global parameters to illustrate how the global parameters are linked to local parameters.

Probably the most famous example of fine-tuning is Fred Hoyle's discovery of the critical placement of a resonant energy level in the carbon-12 nucleus within the context of nuclear reactions inside stars. Changes in the strong nuclear force by about 0.5 percent or in the electromagnetic force by about 4 percent would significantly change the relative yields of carbon and oxygen.[182] Carbon and oxygen are needed by living things in comparable amounts, and stars are the only sources of these key elements.

The key steps in getting from the universal physical constants to the needed mix of carbon and oxygen for life are as follows. Given a set of global parameters including the cosmological initial conditions, the masses of the fundamental particles and the strengths of the four forces (or higher-level forces), how does changing one of these—say, the strength of the strong force—alter the abundances of carbon and oxygen at the local scale? The first step is to calculate the sensitivity of the production of carbon and oxygen in stars to changes in the strong force.[183] The next step is to calculate how carbon and oxygen, thus produced, are distributed in galaxies over the history of the universe.

Following the mixing of the ejected carbon and oxygen atoms into the interstellar medium, we must next consider their incorporation into star-forming clouds. As a planetary system forms out of a denser clump of gas and dust in a giant molecular cloud, carbon and oxygen react with other elements, mostly hydrogen, to generate several molecular species. The gas chemistry is highly sensitive to the carbon-to-oxygen (C/O) ratio.[184] If the ratio is greater than one, then molecules containing carbon and hydrogen dominate, while water dominates for ratios less than one. The final critical step is the formation of the solid planetary building material from the cooling gas.

The C/O ratio in condensed solids varied with location in the early protoplanetary disk. Solids formed close to the Sun were more strongly fractionated and contained less C. Solids formed farther from the Sun were more volatile-rich and had a C/O ratio closer to the initial value of the birth cloud. As a result, the C/O ratio of the bulk Earth is much smaller than the solar ratio. This follows from the low abundance of C in the Earth's core and mantle. The C/O ratio of the crust, however, is closer to the solar value. Earth's crust is believed to have formed from a late accretion of material rich in volatiles from the outer asteroid belt. Of course, it is in the crust where the ratio needs to be close to unity for life.

The C/O ratio varies over time and location in the Milky Way in a systematic way.[185] Its value at a given location depends on the star-formation history there. The C/O in matter returned to the interstellar medium depends on the mix of low- and high-mass stars. If the C/O ratio at a given location and time differs from the Solar System value, other aspects of the environment are also likely to differ. For example, the C/O ratio correlates with metallicity in the Milky Way. As I noted above, metallicity is a critical parameter in determining whether a system is habitable. A region with a different C/O ratio from the Sun will likely also have a different metallicity.

4. Multiple Global Tuning

The example above shows how a change in the strong force can alter the C/O ratio locally. However, changing the strong force has other local effects, such as nuclear reactions in stars and the stability of nuclei and thus the length of the periodic table. The other forces also display multiple sensitivities. Changing the weak force strength affects the relative amounts of hydrogen and helium produced in the first few minutes after the Big Bang, the fusion reactions inside stars, the explosion of massive stars as supernovae, and the decay of radioactive isotopes.[186] Changing the electromagnetic force changes all of chemistry and all processes involving the interaction of light with matter. If we change gravity, we change planets, stars, galaxies, and the large-scale dynamics of the universe.

When combined with the complex web of interdependent habitability factors illustrated in Figure 1, this phenomenon of "multiple global tuning" makes it even less likely that changes in global parameters will result in another island of habitability.

5. Conclusions

Astrobiology research is revealing the high specificity and interdependence of the local parameters required for a habitable environment. These two features of the universe make it unlikely that environments significantly different from ours will be as habitable. At the same time, physicists and cosmologists have discovered that a change in a global parameter can have multiple local effects. Therefore, the high specificity and interdependence of local tuning and the multiple effects of global tuning together make it unlikely that our tiny island of habitability is part of an archipelago. Our universe is a small target indeed.

Notes

1. Barrow and Tipler 1986; Rees 2000, 2003; Collins 2003.
2. The material in Section 2 is a shortened and updated version of my review paper on habitable zones (Gonzalez, 2005). The interested reader is encouraged to consult that paper for more detailed discussions of this topic.
3. Gonzalez et al. 2001a,b.
4. For example, Lineweaver 2001.
5. Barrow and Tipler, 1986; Pace, 2001; Brack, 2002; Chaplin, 2006; Ball, 2008.
6. Davies and Koch 1991.
7. Wilkinson, 2003.
8. Frank et al. 2001.
9. Ward and Brownlee 2000.
10. For example, Berner et al. 2000.
11. Catling et al. 2005.
12. Gaston 2000.
13. Allen et al. 2002.
14. Waide et al. 1999.
15. Schneider and Rey-Benayas 1994
16. Scott et al. 2008.
17. Méndez 2001, 2002.
18. Nisbet and Sleep 2001.
19. Schopf et al. 2002.
20. Sepkoski 1995.
21. Rampino 1998.
22. Kump et al. 1999.
23. Knauth 2005.
24. Raymond 2008.
25. Kasting et al. 1993.
26. Franck et al. 2001.
27. Kasting et al. 1993; Caldeira and Kasting 1992.
28. Lissauer 1993.
29. For example, Raymond et al. 2004.
30. Lissauer et al. 2000; Kokubo and Ida 2007.
31. Canup 2004.
32. Griebmeier et al. 2005.
33. Morbidelli et al. 2000.
34. Morbidelli et al.
35. Lunine et al. 2003.
36. Genda and Abe, 2003, 2004; Melosh, 2003.
37. Lunine 2001.
38. Raymond et al. 2004.
39. Raymond et al. 2007.
40. See Figure 1 of Gonzalez 2005.
41. Lin et al., 1996; originally proposed in the 1980s.
42. Chiang et al., 2002; Goldreich and Sari, 2003; Marzari and Weidenschilling, 2002.

43. Thommes and Lissauer 2003.
44. Veras and Armitage 2005.
45. Armitage 2003.
46. Raymond et al. 2006.
47. Kuchner 2003; Léger et al. 2004.
48. Knauth 2005.
49. Williams et al. 1997.
50. Actually, it probably can't; see Canup and Ward 2006.
51. Gonzalez et al. 2001a.
52. Laskar et al. 1993.
53. Laskar and Robutel 1993.
54. Ward et al. 2002.
55. See Figure 1 of Ward et al. 2002.
56. Atobe et al. 2004 and Atobe and Ida 2007.
57. Correia et al. 2003 and Correia and Laskar 2003.
58. Touma and Wisdom 2001.
59. Williams and Pollard 2003.
60. Laskar 1994.
61. Gonzalez 2005.
62. Williams and Pollard 2002.
63. Berger and Loutre 1994; Petit et al. 1999; EPICA 2004.
64. Hart 1982; Kasting et al. 1993.
65. Lissauer 1999.
66. Agee 2004.
67. O'Neill and Lenardic 2007.
68. Valencia et al. 2007.
69. Zuber 2001.
70. Nimmo and Stevenson 2000.
71. Spohn et al. 2001.
72. Lissauer 1995.
73. Jakosky and Phillips 2001.
74. Griebmeier et al. 2005.
75. Hauck and Phillips 2002; Franck et al. 2001.
76. Bounama et al. 2001.
77. Regenauer-Lieb et al. 2001.
78. Pollack et al. 1996.
79. Laughlin et al. 2004; Ida and Lin 2005.
80. Laws et al. 2003; Johnson et al. 2007.
81. Raymond et al. 2007.
82. As the reader can infer, a gigayear (Gyr) is a billion years.
83. Ribas et al. 2005.
84. Güdel, 2003; Ribas et al., 2005.
85. Radick et al. 2004.
86. Pepin 1997.
87. Smith et al. 2004.
88. Jackman et al. 2000.

89. See Smith et al. for the flare frequency–energy relation.
90. West et al. 2004.
91. Gershberg et al. 1999.
92. Wolstencroft and Raven 2002.
93. Kasting et al. 1993.
94. Griebmeier et al. 2005.
95. Marsh and Svensmark 2005.
96. Svensmark et al. 2006.
97. Svensmark 2006.
98. Shaviv 2005.
99. Schklovsky and Sagan 1966.
100. For example, Gehrels et al. 2003.
101. Benitez et al. 2002.
102. Knie et al. 2004 and Wallner et al. 2004 discovered spikes in the ^{60}Fe (but see Basu et al. 2007) and ^{244}Pu concentrations, respectively.
103. Heisler and Tremaine 1986; Matese et al. 1995.
104. Heisler and Tremaine 1986; Matese et al. 1999.
105. Hut and Tremaine 1985.
106. Matese and Lissauer 2002.
107. Heisler and Tremaine 1986.
108. Matese et al. 1999.
109. Masi et al. 2003.
110. Gonzalez et al. 2001a,b; Lineweaver 2001.
111. It might be possible to form an Earth-mass planet from metal-poor gas, but it would consist mostly of ices.
112. Marcy et al. 2005.
113. Lineweaver et al. 2004.
114. See the review by Kirkby (2007).
115. Bond et al. 2001.
116. Dragicevich et al. 1999.
117. Capellaro 2004.
118. Richardson et al. 2002.
119. Capellaro 2004.
120. Ferrière 2001.
121. Case and Bhattacharya 1998.
122. Yusifov and Küçük 2004.
123. Van den Bergh 1997.
124. Gehrels et al. 2003.
125. Scalo and Wheeler 2002.
126. Scalo and Wheeler 2003.
127. Melott et al. 2004; Thomas et al. 2005; Thomas and Melott 2006.
128. Waxman 2004; Dermer and Holmes 2005.
129. Dar et al. 1998; Dar and De Rújula 2002.
130. Dar and De Rújula 2002.
131. Morris et al. 1999.
132. Miller et al. 2003; Marconi et al. 2004.

133. Miller et al. 2003.
134. Marconi et al. 2004.
135. Gursky and Schwartz 1977.
136. Hartman et al. 1999.
137. Clarke 1981.
138. Luck et al. 2006; Lemasle et al. 2007. Dex is a logarithmic unit. The disk gradient -0.07 dex kpc^{-1} corresponds to a decline of 17 percent per kpc.
139. Maciel et al. 2005.
140. Gonzalez 1999.
141. Santos et al., 2004 2005; Fischer and Valenti 2005.
142. Gonzalez 2006.
143. Beer et al. 2004.
144. Ida and Lin 2004.
145. Gonzalez et al. 2001a.
146. Lineweaver 2001.
147. Gonzalez 2008.
148. See the review by Stern (2003).
149. Levison et al. 2001.
150. Stern 2003.
151. Zheng and Valtonen 1999.
152. Talbot and Newman 1977.
153. Begelman and Rees 1976.
154. Scherer et al. 2002.
155. Florinski et al. 2003.
156. Yeghikyan and Fahr 2004a,b.
157. Pavlov et al. 2005a.
158. Pavlov et al. 2005b.
159. Lin et al. 1969.
160. See Lepine et al. 2001.
161. Lepine et al. 2003.
162. For example, Dias and Lépine 2005.
163. Marochnik 1983.
164. Vallée 2008.
165. Sellwood and Binney 2002.
166. Vallée 2008.
167. Shaviv 2003.
168. Hoyle 1948.
169. Komatsu et al. 2008.
170. Heavens et al. 2004.
171. Dicke, 1961; Garriga et al. 2000; Rees 2000, 2003.
172. Lineweaver 2001.
173. Von Bloh et al. 2003.
174. The so-called L-Z relation (see Lee et al. 2006).
175. Calura and Matteucci 2004.
176. Vogt et al. 2004.
177. Hammer et al. 2005.

178. Matonick and Fesen 1997; Sasaki 2004.
179. Capellaro 2004.
180. Gonzalez et al. 2001a.
181. Adams and Laughlin 1997.
182. Oberhummer et al. 2000; there are other effects of such changes in other related nuclei which would make this sensitivity even greater.
183. Schlattl et al. 2004.
184. Watt 1985.
185. Carigi et al. 2005.
186. Recently, Harnik et al. (2006) argued that a universe lacking the weak force would still be habitable. Their arguments, however, are seriously flawed, as shown by Clavelli and White (2006).

References

Adams, F. and Laughlin, F. (1997) "A Dying Universe: The Long-term Fate and Evolution of Astrophysical Objects." *Reviews of Modern Physics* 69: 337–72.
Agee, C. B. (2004) "Hot Metal." *Nature* 429: 33–35.
Allen, A. P., Brown, J. H. and Gillooly, J. F. (2002) "Global Biodiversity, Biochemical Kinetics, and the Energetic-Equivalence Rule." *Science* 297: 1545–48.
Armitage, P. J. (2003) "A Reduced Efficiency of Terrestrial Planet Formation Following Giant Planet Migration." *Astrophysical Journal* 582: L47-L50.
Atobe, K. and Ida, S. (2007) "Obliquity Evolution of Extrasolar Terrestrial Planets." *Icarus* 188: 1–17.
Atobe, K., Ida, S., and Ito, T. (2004) "Obliquity Variations of Terrestrial Planets in Habitable Zones." *Icarus* 168: 223–36.
Ball, P. (2008) "Water as an Active Constituent in Cell Biology." *Chemical Review* 108: 74–108.
Barrow, J. D. and Tipler, F. J. (1986) *The Anthropic Cosmological Principle.* Oxford: Oxford University Press.
Basu, S., Stuart, F. M., Schnabel, C., and Klemm, V. (2007) "Galactic-Cosmic-Ray-Produced He-3 in a Ferromanganese Crust: Any Supernova Fe-60 Excess on Earth?" *Physical Review Letters* 98: ID 141103.
Begelman, M. C., and Rees, M. J. (1976) "Can Cosmic Clouds Cause Climatic Catastrophes?" *Nature* 261: 298–99.
Benitez, N., Maiz-Apellaniz, J. and Canelles, M. (2002) "Evidence for Nearby Supernova Explosions." *Physical Review Letters* 88: 081101.
Berger, A. and Loutre, M. F. (1994) "Astronomical Forcing Through Geological Time," in De Boer, P. L. and Smith, D. G., eds., *Orbital Forcing and Cyclic Sequences: Special Publication of the International Association of Sedimentologists.* Malden, MA: Blackwell Scientific Publications, 15–24.
Berner, R. A., Petsch, S. T., Lake, J. A., Beerling, D. J., Popp, B. N., Lane, R. S., Laws, E. A., Westley, M. B., Cassar, N., Woodward, F. I. and Quick, W. P. (2000) "Isotope Fractionation and Atmospheric Oxygen: Implications for Phanerozoic O_2 Evolution." *Science* 287: 1630–33.
Bond, G., Kromer, B., Beer, J., Muscheler, R., Evans, M. N., Showers, W., Hofmann, S., Lotti-Bond, R., Hajdas, I. and Bonani, G. (2001) "Persistent Solar Influence on North Atlantic Climate During the Holocene." *Science* 294: 2130–36.
Bounama, C., Franck, S. and von Bloh, W. (2001) "The Fate of Earth's Ocean." *Hydrology and Earth System Science* 5: 569–75.
Brack, A. (2002) "Water, Spring of Life," in Horneck, G. and Baumstark-Khan, C., eds., *Astrobiology: The Quest for the Conditions of Life.* Heidelberg: Springer-Verlag, 79–88.
Caldeira, K. and Kasting, J. F. (1992) "Life Span of the Biosphere Revisited." *Nature* 360: 721–23.
Calura, F. and Matteucci, F. (2004) "Cosmic Metal Production and the Mean Metallicity of the Universe." *Monthly Notes of the Royal Astronomical Society* 350: 351–64.
Canup, R. M. (2004) "Simulations of a late Lunar-Forming Impact." *Icarus* 168: 433–56.
Canup, R. M. and Ward, W. R. (2006) "A Common Mass Scaling for Satellite Systems of Gaseous Planets." *Nature* 441: 834–39.
Cappellaro, E. (2004) "The Evolution of the Cosmic SN Rate." *Memorie della Società Astronomica Italiana* 75: 206–13.
Carigi, L., Peimbert, M., Esteban, C., and Garcia-Rojas, J. (2005) "Carbon, Nitrogen, and Oxygen Galactic Gradients: A Solution to the Carbon Enrichment Problem." *Astrophysical Journal* 623: 213–24.
Case, G. L. and Bhattacharya, D. (1998) "A New -D Relation and its Application to the Galactic Supernova Remnant Distribution." *Astrophysical Journal* 504: 761–72.

Catling, D. C., Glein, C. R., Zahnle, K. J., and McKay, C. P. (2005) "Why O_2 is required by Complex Life on Habitable Planets and the Concept of Planetary Oxygenation Time." *Astrobiology* 5: 415–38.

Chaplin, M. (2006) "Do We Underestimate the Importance of Water in Cell Biology?" *Nature Reviews in Molecular Cell Biology* 7: 861–866.

Chiang, E. I., Fischer, D. and Thommes, E. (2002) "Excitation of Orbital Eccentricities of Extrasolar Planets by Repeated Resonance Crossings." *Astrophysical Journal* 564: L105-L109.

Clarke, J. N. (1981) "Extraterrestrial Intelligence and Galactic Nuclear Activity." *Icarus* 46: 94–96.

Clavelli, L. and White, R. E. III. (2006) "Problems in a Weakless Universe." hep-ph/0609050.

Collins, R. (2003) "Evidence for Fine-Tuning," in *God and Design,* N. A. Manson, ed. Routledge, London, 178–99.

Correia, A. C. M. and Laskar, J. (2003) "Long-Term Evolution of the Spin of Venus II. Numerical Simulations." *Icarus* 163: 24–45.

Correia, A. C. M., Laskar, J. and Neron de Surgy, O. (2003) "Long-Term Evolution of the Spin of Venus I. Theory." *Icarus* 163: 1–23.

Dar, A., Laor, A. and Shaviv, N. J. (1998) "Life Extinctions by Cosmic Ray Jets." *Physical Review Letters* 80: 5813–16.

Dar, A. and De Rújula, A. (2002) "The Threat to Life from Eta Carinae and Gamma Ray Bursts," in A. Morselli and P. Picozza, eds., *Astrophysics and Gamma Ray Physics in Space,* Frascati Physics Series, Vol. XXIV: 513–23.

Davies, R. E. and Koch, R. H. (1991) "All the Observed Universe has Contributed to Life." *Philosophical Transactions of the Royal Society* B 334: 391–403.

Dermer, C. D. and Holmes, J. M. (2005) "Cosmic Rays from Gamma-Ray Bursts in the Galaxy." *Astrophysical Journal* 628: L21-L24.

Dias, W. S. and Lépine, J. R. D. (2005) "Direct Determination of the Spiral Pattern Rotation Speed of the Galaxy." *Astrophysical Journal* 629: 825–31.

Dicke, R. H. (1961) "Dirac's Cosmology and Mach's Principle." *Nature* 192: 440–41.

Dragicevich, P. M., Blair, D. G. and Burman, R. R. (1999) "Why are Supernovae in Our Galaxy so Frequent?" *Monthly Notes of the Royal Astronomical Society* 302: 693–99.

Ferrière, K. M. (2001) "The Interstellar Environment of Our Galaxy." *Reviews of Modern Physics* 73: 1031–66.

Fischer, D. A. and Valenti, J. (2005) "The Planet-Metallicity Correlation." *Astrophysical Journal* 622: 1102.

Florinski, V., Zank, G. P. and Axford, W. I. (2003) "The Solar System in a Dense Interstellar Cloud: Implications for Cosmic-ray Fluxes at Earth and ^{10}Be Records." *Geophysical Research Letters* 30:2206, doi:10.1029/ 2003GL017566.

Franck, S., Block, A., von Bloh, W., Bounama, C., Garrido, I. and Schellnhuber, H.-J. (2001) "Planetary Habitability: Is Earth Commonplace in the Milky Way?" *Naturwissenschaften* 88: 416–26.

Garriga, J., Livio, M. and Vilenken, A. (2000) "Cosmological Constant and the Time of its Dominance." *Physical Review* D 61: 023503.

Gaston, K. J. (2000) "Global Patterns in Biodiversity." *Nature* 405: 220–27.

Gehrels, N., Laird, C. M., Jackman, C. H., Cannizo, J. K., Mattson, B. J. and Chen, W. (2003) "Ozone Depletion from Nearby Supernovae" *Astrophysical Journal* 585: 1169–76.

Genda, H. and Abe, Y. (2003) "Survival of a Proto-Atmosphere through the Stage of Giant Impacts: The Mechanical Aspects." *Icarus* 164: 149–62.

Genda, H. and Abe, Y. (2004) "Hydrodynamic Escape of a Proto-Atmosphere Just after a Giant Impact." *Lunar Planetary Science* 35: 1518.

Gershberg, R. E., Katsova, M. M., Lovkaya, M. N., Terebizh, A. V. and Shakhovskaya, N. I. (1999) "Catalogue and Bibliography of the UV Ceti-Type Flare Stars and Related Objects in the solar Vicinity." *Astronomy and Astrophysics Supplement* 139: 555–58.

Goldreich, P. and Sari, R. (2003) "Eccentricity Evolution for Planets in Gaseous Disks." *Astrophysical Journal* 585: 1024–37.

Gonzalez, G. (1999) "Are Stars with Planets Anomalous?" *Monthly Notes of the Royal Astronomical Society* 308: 447–58.

Gonzalez, G. (2005) "Habitable Zones in the Universe." *Origins of Life and Evolution of Biospheres* 35: 555–606.

Gonzalez, G. (2006) "The Chemical Compositions of Stars with Planets: A Review." *Publications of the Astronomical Society of the Pacific* 118: 1494–505.

Gonzalez, G. (2008) "Planet Formation and the Galactic Habitable Zone." *Icarus*, in press.

Gonzalez, G., Brownlee, D. and Ward, P. (2001a) "The Galactic Habitable Zone: Galactic Chemical Evolution." *Icarus* 152: 185–200.

Gonzalez, G., Brownlee, D. and Ward, P. (2001b) "The Galactic Habitable Zone." *Scientific American* 285 (October): 60–67.

Griebmeier, J. -M., Stadelmann, A., Motschmann, U., Belisheva, N. K., Lammer, H. and Biernat, H. K. (2005) "Cosmic Ray Impact on Extrasolar Earth-Like Planets in Close-in Habitable Zones." *Astrobiology* 5: 587–603.

Güdel, M. (2003) "The Sun in Time: From PMS to Main Sequence," in Arnaud, J. and Maunier, N. eds., *Magnetism and Activity of the Sun and Stars*. France: EDP Sciences, 339–49.

Gursky, H. and Schwartz, D. A. (1977) "Extragalactic X-ray Sources." *Annual Review of Astronomy and Astrophysics* 15: 541–68.

Hammer, F., Flores, H., Elbaz, D., Zheng, X. Z., Liang, Y. C. and Cesarsky, C. (2005) "Did Most Present-Day Spirals form during the last 8 Gyrs?: A Formation History with Violent Episodes Revealed by Panchromatic Observations." *Astronomy and Astrophysics* 430: 115–28.

Harnik, R., Kribs, G. D. and Perez, G. (2006) "A Universe without Weak Interactions." *Physical Review* D 74: 035006.

Hart, M. H. (1982) "The Effect of a Planet's Size on the Evolution of its Atmosphere." *Southwest Regional Conference for Astronomy and Astrophysics:* 111–26.

Hartman, R. C. *et al.* (1999) "The Third EGRET Catalog of High-Energy Gamma-Ray Sources." *Astrophysical Journal* 123: 79–202.

Hauck, S. A. and Phillips, R. J. (2002) "Thermal and Crustal Evolution of Mars." *Journal of Geophysical Research* 107 (E7): 5052.

Heavens, A., Panter, B., Jimenez, R. and Dunlop, J. (2004) "The Star-Formation History of the Universe from the Stellar Populations of Nearby Galaxies." *Nature* 428: 625–27.

Heisler, J. and Tremaine, S. (1986) "Influence of the Galactic Tidal Field on the Oort Cloud." *Icarus* 65: 13–26.

Hoyle, F. (1948) "A New Model for the Expanding Universe." *Monthly Notes of the Royal Astronomical Society*. 108: 372–82.

Hut, P. and Tremaine, S. (1985) "Have Interstellar Clouds Disrupted the Oort Comet Cloud?" *Astronomical Journal*. 90: 1548–57.

Ida, S. and Lin, D. N. C. (2004) "Toward a Deterministic Model of Planetary Formation. II. The Formation and Retention of Gas Giant Planets Around Stars with a Range of Metallicities." *Astrophysical Journal* 616: 567–72.

Ida, S. and Lin, D. N. C. (2005) "Dependence of Exoplanets on Host Stars' Metallicity and Mass, *Progress in Theoretical Physics Supplement* No. 158: 68–85.

Jackman, C. H., Fleming, E. L. and Vitt, F. M. (2000) "Influence of Extremely Large Solar Proton Events in a Changing Stratosphere." *Journal of Geophysical Research* 105: 11659–670.

Jakosky, B. M. and Phillips, R. J. (2001) "Mars' Volatile and Climate History." *Nature* 412: 237–44.

Johnson, J. A., Butler, R. P., Marcy, G. W., Fischer, D. A., Vogt, S. S., Wright, J. T. and Peek, K. M. G. (2007) "A New Planet around an M Dwarf: Revealing a Correlation Between Exoplanets and Stellar Mass." *Astrophysical Journal* 670: 833–40.

Kasting, J. F., Whitmire, D. P. and Reynolds, R. T. (1993) "Habitable Zones around Main Sequence Stars." *Icarus* 101: 108–28.

Kirkby, J. (2007) "Cosmic Rays and Climate." *Survey of Geophysics* 28: 333–75.

Knauth, L. P. (2005) "Temperature and Salinity History of the Precambrian Ocean: Implications for the Course of Microbial Evolution." *Palaeogeography, Palaeoclimatology, and Palaeoecology* 219: 53–69.

Knie, K., Korschinek, G., Faestermann, T., Dorfi, E. A., Rugel, G. and Wallner, A. (2004) "^{60}Fe Anomaly in a Deep-Sea Manganese Crust and Implications for a Nearby Supernova Source." *Physical Review Letters* 93: 171103.

Kokubo, E. and Ida, S. (2007) "Formation of Terrestrial Planets from Protoplanets II. Statistics of Planetary Spin." *Astrophysical Journal* 671: 2082–90.

Komatsu, E. et al. (2008) "Five-Year Wilkinson Microwave Anisotropy Probe (WMAP) Observations: Cosmological Interpretation." *Astrophysical Journal Supplement,* in press.

Kuchner, M. J. (2003) "Volatile-Rich Earth-Mass Planets in the Habitable Zone." *Astrophysical Journal* 596: L105-L108.

Kump, L. R., Kasting, J. F. and Crane, R. G. (1999) *The Earth System.* New Jersey: Prentice Hall, 126–51.

Laskar, J., Joutel, F. and Robutel, P. (1993) "Stabilization of the Earth's Obliquity by the Moon." *Nature* 361: 615–17.

Laskar, J. and Robutel, P. (1993) "The Chaotic Obliquity of the Planets." *Nature* 361: 608–12.

Laughlin, G., Bodenheimer, P. and Adams, F. (2004) "Core-Accretion Model Predicts Few Jovian-Mass Planets Orbiting Red Dwarfs." *Astrophysical Journal Letters* 612: L73-L76.

Laws, C., Gonzalez, G., Walker, K. M., Tyagi, S., Dodsworth, J., Snider, K., Suntzeff and N. B. (2003) "Parent Stars of Extrasolar Planets VII. New Abundance Analyses of 30 Systems." *Astronomical Journal* 125: 2664–77.

Lee, H., Skillman, E. D., Cannon, J. M., Jackson, D. C., Gerhz, R. D., Polomski, E. F., Woodward, C. E. (2006) "On Extending the Mass-Metallicity Relation of Galaxies by 2.5 Decades in Stellar Mass." *Astrophysical Journal* 647: 970–83.

Léger, A., Selsis, F., Sotin, C., Guillot, T., Despois, D., Mawet, D., Ollivier, M., Labèque, A., Valette, C., Brachet, F., Chazelas, B. and Lammer, H. (2004) "A New Family of Planets? 'Ocean-planets'" *Icarus* 169: 499–504.

Lemasle, B., Francois, P., Bono, G., Mottini, M., Primas, F., Romaniello, M. (2007) "Detailed chemical composition of Galactic cepheids. A determination of the Galactic abundance gradient in the 8–12 kpc region." *Astronomy and Astrophysics* 467: 283–94.

Lepine, J. R. D., Mishurov, Yu. N. and Dedikov, S. Yu. (2001) "A New Model for the Spiral Structure of the Galaxy: Superposition of 2- and 4-Armed Patterns." *Astrophysical Journal* 546: 234–47.

Lepine, J. R. D., Acharova, I. A. and Mishurov, Yu. N. (2003) "Corotation, Stellar Wandering, and Fine Structure of the Galactic Abundance Pattern." *Astrophysical Journal* 589: 210–16.

Levison, H. F., Dones, L. and Duncan, M. J. (2001) "The Origin of Halley-Type Comets: Probing the Inner Oort Cloud." *Astronomical Journal* 121: 2253–67.

Lin, D. N. C., Bodenheimer, P. and Richardson, D. C. (1996) "Orbital Migration of the Planetary Companion of 51 Pegasi to its Present Location." *Nature* 380: 606–7.

Lineweaver, C. H. (2001) "An Estimate of the Age Distribution of Terrestrial Planets in the Universe: Quantifying Metallicity as a Selection Effect." *Icarus* 151: 307–13.

Lineweaver, C. H., Fenner, Y. and Gibson, B. K. (2004) "The Galactic Habitable Zone and the Age Distribution of Complex Life in the Milky Way." *Science* 303: 59–62.

Lissauer, J. J. (1995) "Urey Prize Lecture: On the Diversity of Plausible Planetary Systems." *Icarus* 114: 217–36.

Lissauer, J. J. (1999) "How Common are Habitable Planets?" *Nature* 402: C11-C14.

Lissauer, J. J., Dones, L. and Ohtsuki, K. (2000) "Origin and Evolution of Terrestrial Planet Rotation," in Canup, R.M. and Righter, K. eds., *Origin of the Earth and Moon.* Tuscon and Houston: The University of Arizona Press and Lunar and Planetary Institute, 101–12.

Luck, R. E., Kovtyukh, V. V., Andrievsky, S. M. (2006) "The distribution of the elements in the Galactic disk." *Astrononical Journal* 132: 902–18.

Lunine, J. I., Chambers, J., Morbidelli, A. and Leshin, L. A. (2003) "The Origin of Water on Mars." *Icarus* 165: 1–8.

Maciel, W. J., Lago, L. G. and Costa, R. D. D. (2005) "An Estimate of the Time Variation of the Abundance Gradient from Planetary Nebulae II. Comparison with Open Clusters, Cepheids and Young Objects." *Astronomy and Astrophysics* 433: 127–35.

Marconi, A., Risaliti, G., Gilli, R., Hunt, L. K., Maiolino, R. and Salvati, M. (2004) "Local Supermassive Black Holes, Relics of Active Galactic Nuclei and the X-ray Background." *Monthly Notes of the Royal Astronomical Society* 351: 169–85.

Marcy, G., Butler, R. P., Fischer, D., Vogt, S., Wright, J. T., Tinney, C. G., and Jones, H. R. A. (2005) "Observed Properties of Exoplanets: Masses, Orbits, and Metallicites." *Progress of Theoretical Physics Supplement,* No. 158: 24–42.

Marochnik, L. S. (1983) "On the Origin of the Solar System and the Exceptional Position of the Sun in the Galaxy." *Astrophysics and Space Science* 89: 61–75.

Marsh, N. and Svensmark, H. (2005) "Solar Influence on Earth's Climate." *Space Science Reviews* 107: 317–25.

Marzari, F. and Weidenschilling, S. J. (2002) "Eccentric Extrasolar Planets: The Jumping Jupiter Model." *Icarus* 156: 570–79.

Masi, M., Secco, L., and Vanzani, V. (2003) "Dynamical effects of the Galaxy on the Oort's Cloud." *Memorie della Società Astronomica Italiana* 74: 494–95.

Matese, J. J. and Lissauer, J. J. (2002) "Characteristics and Frequency of Weak Stellar Impulses of the Oort Cloud." *Icarus* 157: 228–40.

Matese, J. J., Whitman, P. G., Innanen, K. A. and Valtonen, M. J. (1995) "Periodic Modulation of the Oort Cloud Comet Flux by the Adiabatically Changing Galactic Tide." *Icarus* 116: 255–68.

Matese, J. J., Whitman, P. G. and Whitmire, D. P. (1999) "Cometary Evidence of a Massive Body in the Outer Oort Cloud." *Icarus* 141: 354–66.

Matonick, D. M. and Fesen, R. A. (1997) "Optically Identified Supernova Remnants in the Nearby Spiral Galaxies: NGC 5204, NGC 5585, NGC 6946, M81, and M101." *Astrophysical Journal Supplement* 112: 49–107.

Melosh, H. J. (2003) "The History of Air." *Nature* 424: 22–23.

Melott, A. L., Lieberman, B. S., Laird, C. M., Martin, L. D., Medvedev, M. V., Thomas, B. C., Cannizzo, J. K., Gehrels, N. and Jackman, C. H. (2004) "Did a Gamma-Ray Burst Initiate the Late Ordovician Mass Extinction?" *International Journal of Astrobiology* 3: 55–61.

Méndez, A. (2001) "Planetary Habitable Zones: The Spatial Distribution of Life on Planetary Bodies." *Lunar and Planetary Science* 32, 2001.

Méndez, A. (2002) "Habitability of Near-Surface Environments on Mars." *Lunar and Planetary Science* 33, 1999.

Miller, C. J., Nichol, R. C., Gómez, P. L., Hopkins, A. M. and Bernardi, M. (2003) "The Environment of Active Galactic Nuclei in the Sloan Digital Sky Survey." *Astrophysical Journal* 597: 142–56.

Morbidelli, A., Chambers, J., Lunine, J. I., Petit, J. M., Robert, F., Valsecchi, G. B. and Cyr, K. E. (2000) "Source Regions and Timescales for the Delivery of Water on Earth." *Meteoritics and Planetary Science* 35: 1309–20.

Nimmo, F. and Stevenson, D. J. (2000) "Influence of Early Plate Tectonics on the Thermal Evolution and Magnetic Field on Mars." *Journal of Geophysical Research* 105 (E5): 11969–980.

Nisbet, E. G. and Sleep, N. H. (2001) "The Habitat and Nature of Early Life." *Nature* 409: 1083–91.

Oberhummer, H., Csótó, A. and Schlattl, H. (2000) "Stellar Production Rates of Carbon and its Abundance in the Universe." *Science* 289: 88–90.

O'Neill, C., Lenardic, A. (2007) "Geological Consequences of Super-sized Earths." *Geophysical Research Letters* 34: L19204.

Pace, N. R. (2001) "The Universal Nature of Biochemistry." *Proceedings of the National Academy of Sciences* 98: 805–8.

Pavlov, A. A., Pavlov, A. K., Mills, M. J., Ostryakov, V. M., Vasilyev, G. I. and Toon, O. B. (2005a) "Catastrophic Ozone Loss During Passage of the Solar System through an Interstellar Cloud." *Geophysical Research Letters* 32: L01815, doi:10.1029/2004GL021601.

Pavlov, A. A., Toon, O. B., Pavlov, A. K., Bally, J. and Pollard, D. (2005b) "Passing Through a Giant Molecular Cloud: 'Snowball' Glaciations Produced by Interstellar Dust." *Geophysical Research Letters* 32: L03705, doi:10.1029/2004GL021890.

Pepin, R. O. (1997) "Evolution of Earth's Noble Gases: Consequences of Assuming Hydrodynamic Loss Driven by Giant Impact." *Icarus* 126: 148–56.

Radick, R. R., Lockwood, G. W. Henry, G. W. and Baliunas, S. L. (2004) "The Variability of Sunlike Stars on Decadal Timescales," in Dupree, A. K. and Benz, A. O. eds., *Stars as Suns: Activity, Evolution, and Planets*. San Francisco: Astronomical Society of the Pacific, 264–68.

Rampino, M. R. (1998) "The Galactic Theory of Mass Extinctions: An Update." *Celestial Mechanics and Dynamical Astronomy* 69: 49–58.

Raymond, S. N. (2008) "Terrestrial Planet Formation in Extra-Solar Planetary Systems," astro-ph/0801.2560v1.

Raymond, S. N., Mandell, A. M. and Sigurdsson, S. (2006) "Exotic Earths: Forming Habitable Worlds with Giant Planet Migration." *Science* 313: 1413–16.

Raymond, S., Quinn, T. and Lunine, J. I. (2004) "Making other Earths: Dynamical Simulations of Terrestrial Planet Formation and Water Delivery." *Icarus* 168: 1–17.

Raymond, S., Scalo, J. and Meadows, V. S. (2007) "A Decreased Probability of Habitable Planet Formation around Low-mass Stars." *Astrophysical Journal* 669: 606–14.

Rees, M. (2000) *Just Six Numbers: The Deep Forces that Shape the Universe.* New York: Basic Books.

Rees, M. (2003) "Numerical Coincidences and 'Tuning' in Cosmology." *Astrophysics and Space Science* 285: 375–88.

Regenauer-Lieb, K., Yuen, D. A. and Branlund, J. (2001) "The Initiation of Subduction: Criticality by Addition of Water?" *Science* 294: 578–80.

Ribas, I., Guinan, E. F., Güdel, M. and Audard, M. (2005) "Evolution of the Solar Activity over Time and Effects on Planetary Atmospheres: I. High-Energy Irradiances (1–1700 Å)." *Astrophysical Journal* 622: 680–94.

Richardson, D., Branch, D., Casebeer, D., Millard, J., Thomas, R. C. and Baron, E. (2002) "A Comparative Study of the Absolute Magnitude Distributions of Supernovae." *Astronomical Journal* 123: 745–52.

Sasaki, M., Breitschwerdt, D. and Supper, R. (2004) "SNR Surface Density Distribution in Nearby Galaxies." *Astrophysics and Space Science* 289: 283–86.

Scalo, J. and Wheeler, J. C. (2002) "Astrophysical and Astrobiological Implications of Gamma-Ray Burst Properties." *Astrophysical Journal* 566: 723–737.

Scherer, K., Fichtner, H. and Stawicki, O. (2002) "Shielded by the Wind: The Influence of the Interstellar Medium on the Environment of Earth." *Journal of Atmospheric and Solar-Terrestrial Physics* 64: 795–804.

Schneider, S. M. and Rey-Benayas, J. M. (1994) "Global Patterns of Plant Diversity." *Evolutionary Ecology* 8: 331–47.

Schlattl, H., Heger, A., Oberhummer, H., Rauscher, T. and Csótó, A. (2004) "Sensitivity of the C and O Production on the 3-alpha Rate." *Astrophysics and Space Science* 291: 27–56.

Scott, C., Lyons, T. W., Bekker, A., Shen, Y., Poulton, S. W., Chu, X., and Anbar, A. D. (2008) "Tracing the Stepwise Oxygenation of the Protezoroic Ocean." *Nature* 452: 456–59.

Sepkoski, J. J. (1995) "Patterns of Phanerozoic Extinction: A Perspective from Global Databases," in Walliser, O. H. ed., *Global Events and Event Stratigraphy in the Phanerozoic*. Berlin: Springer, 35–51.

Shaviv, N. J. (2003) "The Spiral Structure of the Milky Way, Cosmic Rays, and Ice Age Epochs on Earth." *New Astronomy Reviews* 8: 39–77.

Shaviv, N. J. (2005) "On Climate Response to Changes in the Cosmic Ray Flux and Radiative Budget." *Journal of Geophysical Research* 110: A08105.

Shklovsky, J. S. and Sagan, C. (1966) *Intelligent Life in the Universe*. San Francisco: Holden-Day.

Smith, D. S., Scalo, J. and Wheeler, J. C. (2004) "Importance of Biologically Active and Aurora-Like Ultraviolet Emission: Stochastic Irradiation of Earth and Mars by Flares and Explosions." *Origins of Life and Evolution of Biospheres* 34: 513–32.

Spohn, T., Acuña, M. H., Breuer, D., Golombek, M., Greeley, R., Halliday, A., Hauber, E., Jaumann, R. and Sohl, F. (2001) "Geophysical Constraints on the Evolution of Mars." *Space Science Reviews* 96: 231–62.

Stern, S. A. (2003) "The Evolution of Comets in the Oort Cloud and Kuiper Belt." *Nature* 424: 639–42.

Svensmark, H. (2006) "Cosmic Rays and the Biosphere over 4 Billion Years." *Astronomische Nachrichten* 327: 871–75.

Svensmark, H., Pederson, J. O. P., Marsh, N. D., Enghoff, N. B. and Uggerhoj, U. I. (2006) "Experimental Evidence for the Role of Ions in Particle Nucleation Under Atmospheric Conditions." *Proceedings of the Royal Society* A, doi:10.1098.

Talbot, R. J. Jr. and Newman, M. J. (1977) "Encounters Between Stars and Dense Molecular Clouds." *Astrophysical Journal Supplement* 34: 295–308.

Thomas, B. C. *et al.* (2005) "Gamma Ray Bursts and the Earth: Exploration of Atmospheric, Biological, Climatic, and Biogeochemical Effects." *Astrophysical Journal* 634: 509–33.

Thomas, B. C. and Melott, A. L. (2006) "Gamma-ray Bursts and Terrestrial Planetary Atmospheres." *New Journal of Physics* 8, doi:10.1088/1367–2630/8/7/120.

Thommes, E. W. and Lissauer, J. J. (2003) "Resonant Inclination Excitation of Migrating Giant Planets." *Astrophysical Journal* 597: 566–80.

Touma, J. and Wisdom, J. (2001) "Nonlinear Core-Mantle Coupling." *Astronomical Journal* 122: 1030–1050.

Valencia, D., O'Connell, R. J., Sasselov, D. D. (2007) "Inevitability of Plate Tectonics on Super-Earths." *Astrophysical Journal* 670: L45-L48.

Vallée, J. P. (2008) "New Velocimetry and Revised Cartography of the Spiral Arms in the Milky Way—A Consistent Symbiosis." *Astronomical Journal* 135: 1301–10.

Van den Bergh, S. (1997) "Distribution of Supernovae in Spiral Galaxies, *Astronomical Journal* 113: 197–200.

Veras, D. and Armitage, P. J. (2005) "The Influence of Massive Planet Scattering on Nascent Terrestrial Planets." *Astrophysical Journal* 620: L111-L114.

Vogt, N. P., Haynes, M. P., Giovanelli, R. and Herter, T. (2004) "M/L, H Rotation Curves, and H I Gas Densities for 329 Nearby Cluster and Field Spirals. III. Evolution in Fundamental Galaxy Parameters, *Astronomical Journal* 127, 3325–37.

Von Bloh, W., Franck, S., Bounama, C. and Schellnhuber, H.-J. (2003) "Maximum Number of Habitable Planets at the Time of Earth's Origin: New Hints for Panspermia?" *Origins of Life and Evolution of Biospheres* 33: 219–31.

Waide, R. B., Willig, M. R., Steiner, C. F., Mittelbach, G., Gough, L., Dodson, S. I., Juday, G. P. and Parmenter, R. (1999) "The Relationship Between Productivity and Species Richness." *Annual Review of Ecology and Systematics* 30: 257–300.

Wallner, C., Faestermann, T., Gerstmann, U., Knie, K., Korschinek, G., Lierse, C. and Rugel, G. (2004) "Supernova Produced and Anthropogenic ^{244}Pu in Deep Sea Manganese Encrustations." *New Astronomy Review* 48: 145–50.

Ward, P. D. and Brownlee, D. (2000) *Rare Earth: Why Complex Life is Uncommon in the Universe.* New York: Copernicus.

Ward, W. R., Agnor, C. B. and Canup, R. M. (2002) "Obliquity Variations in Planetary Systems." *Lunar and Planetary Science* 33: 2017.

Watt, G. D. (1985) "Time-Dependent Chemistry. II—Dependence of the Chemistry on the Initial C/O Abundance Ratio." *Monthly Notes of the Royal Astronomical Society* 212: 93–103.

Waxman, E. (2004) "High-Energy Cosmic Rays from Gamma-Ray Burst Sources: A Stronger Case." *Astrophysical Journal* 606: 988–93.

West, A. A., Hawley, S. L., Walkowicz, L. M., Covey, K. R., Silvestri, N. M., Raymond, S. N., Harris, H. C., Munn, J. A., McGehee, P. M., Ivezic′, Z. and Brinkmann, J. (2004) "Spectroscopic Properties of Cool stars in the Sloan Digital Sky Survey: An Analysis of Magnetic Activity and a Search for Subdwarfs." *Astronomical Journal* 128: 426–36.

Wilkinson, D. M. (2003) "The Fundamental Processes in Ecology: A Thought Experiment on Extraterrestrial Biospheres." *Biological Reviews* 78: 171–79.

Williams, D. M., Kasting, J. F. and Wade, R. A. (1997) "Habitable Moons Around Extrasolar Giant Planets." *Nature* 385: 234–36.

Williams, D. M. and Pollard, D. (2002) "Earth-Like Planets on Eccentric Orbits: Excursions Beyond the Habitable Zone." *International Journal of Astrobiology* 1: 61–69.

Wolstencroft, R. D. and Raven, J. A. (2002) "Photosynthesis: Likelihood of Occurrence and Possibility of Detection on Earth-Like Planets." *Icarus* 157: 535–48.

Yeghikyan, A. and Fahr, H. (2004a) "Effects Induced by the Passage of the Sun through Dense Molecular Clouds I. Flows Outside of the Compressed Heliosphere." *Astronomy and Astrophysics* 415: 763–70.

Yeghikyan, A. and Fahr, H. (2004b) "Terrestrial Atmospheric Effects Induced by Counterstreaming Dense Interstellar Cloud Material." *Astronomy and Astrophysics* 425: 1113–18.

Yusifov, I. and Küçük, I. (2004) "Revisiting the Radial Distribution of Pulsars in the Galaxy." *Astronomy and Astrophysisc* 422: 545–53.

Zheng, J. Q. and Valtonen, M. J. (1999) "On the Probability that a Comet that has Escaped from another Solar System will Collide with the Earth." *Monthly Notes of the Royal Astronomical Society* 304: 579–82.

Zuber, M. T. (2001) "The Crust and Mantle of Mars." *Nature* 412: 220–27.

Part V

Mathematics

Introduction

Is it possible to account for the nature of mathematical objects and mathematical knowledge, as well as the effectiveness of mathematical techniques in the natural sciences, on a naturalistic basis? Are the presence of mathematical order and regularity in nature, and the utility of abstract mathematics as a heuristic in natural science, evidence of some sort of transcendent metaphysical or teleological constraint on the structure of the universe?

Philip Kitcher argues that neither the nature of mathematics, nor its scientific utility, is mysterious; both are intelligible within the framework of philosophical naturalism. In his essay "Mathematical Naturalism," he suggests that regarding the philosophy of mathematics as an attempt to secure the foundations of mathematical knowledge presupposes a commitment to apriorism—the view that the foundational principles of mathematics are known *a priori*. He argues instead that mathematical knowledge arises from our *experience* of the world and maintains that it can be naturalized by considering the achievements of one generation of mathematicians as resting on those of the preceding generation. Through this reconceptualization, the focus is shifted to giving an account of the rational development of mathematics and mathematical progress. This perspectival shift allow us to see that mathematical change is driven by factors external to the discipline, and that mathematical progress needs to be understood in terms of the advancement of ends that are *external* to mathematics itself. Kitcher also asserts that this viewpoint provides an alternative to Platonist accounts of mathematical truth, which can be reconceived along Peircean lines as what is attained in the long run through the application of the rational principles governing the historical development of mathematics.

Mark Steiner develops a radically different approach. His essay "Mathematics—Application and Applicability" examines various senses in which mathematics is applied to the material world. He distinguishes between canonical and noncanonical applications of mathematics, the former being those applications for which the mathematics was developed to describe. A good example of a canonical empirical application is the use of differential calculus to describe accelerated motion, which Newton developed precisely for this purpose. Steiner also distinguishes between empirical and nonempirical applications, thus leading to four different species of applications: canonical empirical, noncanonical empirical, canonical nonempirical, and noncanonical nonempirical (what Steiner calls "logical applications").

Part V

The distinctions among these categories are made clear through an extensive discussion of examples. Philosophical problems connected with each category of application are treated in a nuanced manner that suggests a picture of the relationship between mathematics and the empirical world that is utterly at odds with naturalism: a form of Pythagoreanism in which the classification of empirical phenomena is induced by the classification of mathematical structures by mathematicians.

28

MATHEMATICAL NATURALISM[1]

PHILIP KITCHER

Virtually all the discussion of the "philosophy of mathematics" in our century has been concerned with the enterprise of providing a foundation for mathematics. There is no doubt that this enterprise has often been mathematically fruitful. Indeed, the growth of logic as an important field within mathematics owes much to the pioneering work of scholars who hoped to exhibit the foundations of mathematics. Yet it should be almost equally obvious that the major foundational programs have not achieved their main goals. The mathematical results they have brought forth seem more of a piece with the rest of mathematics than first points from which the entire edifice of mathematics can be built.

Under these circumstances, it is tempting to echo a title question of Dedekind's and to ask what the philosophy of mathematics is and what it ought to be. Many practicing mathematicians and historians of mathematics will have a brusque reply to the first part of the question: a subject noted as much for its irrelevance as for its vaunted rigor, carried out with minute attention to a small number of atypical parts of mathematics and with enormous neglect of what most mathematicians spend most of the their time doing. My aim in the present essay is to offer an answer to the second half of the question, leaving others to quarrel about the proper response to the first. I shall argue that there are good reasons to reject the presuppositions of the foundationalist enterprises and that, by doing so, we obtain a picture of mathematics that raises different—and, to my mind, more interesting—philosophical problems. By trying to formulate these problems clearly, I shall try to draw an agenda for a naturalistic philosophy of mathematics that will, I hope, have a more obvious relevance for mathematicians and historians of mathematics.

1. Epistemological Commitments

Seeking a foundation for a part of mathematics can make exactly the same sense as looking for a foundation for some problematic piece of scientific theory. At some times in the history of mathematics, practitioners have self-consciously set themselves the task of clarifying concepts whose antecedent use skirted paradox or of systematizing results whose connections were previously only dimly perceived. Weierstrass's efforts with concepts of convergence and Lagrange's explanation of the successes of techniques for solving cubic and quartic equations are prime examples of both forms of activity. But the grand foundational programs move beyond these local projects of intellectual slum clearance.[2]

Foundationalist philosophies of mathematics bear a tacit commitment to apriorist epistemology. If mathematics were not taken to be *a priori*, then the foundational programs would have point only insofar as they responded to some particular difficulty internal to a field of mathematics. Subtract the apriorist commitments and there is no motivation for thinking that there must be some first mathematics, some special discipline from which all the rest must be built.

Mathematical apriorism has traditionally been popular, so popular that there has seemed little reason to articulate and define it, because it has been opposed to the most simplistic versions of empiricism. Apriorists come in two varieties. Conservative apriorists claim that there is no possibility of obtaining mathematical *knowledge* without the use of certain special procedures: one does not know a theorem unless one has carried out the appropriate procedures for gaining knowledge of the axioms (enlightenment by Platonic intuition, construction in pure intuition, stipulative fixing of the meanings of terms, or whatever) and has followed a gapless chain of inferences leading from axioms to theorem. Frege, at his most militant, is an example of a conservative apriorist. By contrast, liberals do not insist on the strict impossibility of knowing a mathematical truth without appealing to the favored procedures. Their suggestion is that any knowledge so obtained can ultimately be generated through the use of *a priori* procedures and that mathematical knowledge is ultimately improved through the production of genuine proofs.

Empiricists and naturalists[3] dissent from both versions of apriorism by questioning the existence or the power of the alleged special procedures. Insofar as we can make sense of the procedures to which apriorist epistemologies make their dim appeals, those procedures will not generate knowledge that is independent of our experience. Platonic or constructivist intuition, stipulative definition, yield knowledge—to the extent that they function at all—only against the background of a kindly experience that underwrites their deliverances. From the naturalistic perspective, apriorists have misjudged the epistemological status of the features they invoke, supposing that processes that are analogous to the heuristic arguments or *Gedankenexperimente* of the natural scientist are able to warrant belief come what may.

The contrast may be sharpened by considering Zermelo's introduction of his set-theoretic axioms.[4] For the conservative apriorist, there was no set-theoretic knowledge prior to 1905 and, in consequence, no knowledge of analysis, arithmetic, or any other part of mathematics before the pure cumulative hierarchy beamed in on Zermelo's consciousness. Liberals may relax this harsh judgment about the pre-Zermelian ignoramuses, but they remain committed to the view that mathematical knowledge underwent a dramatic transformation when the "merely empirical" justifications of the Greeks, of Fermat, Newton, Euler, Gauss, Cauchy, and Weierstrass finally gave way, in the first decade of our century, to genuine proof.

Mathematical naturalism opposes to both positions a different philosophical picture. Zermelo's knowledge of the axioms that he introduced was based on his recognition of the possibility of systematizing the prior corpus of claims about sets, claims that had been tacitly or explicitly employed in reasoning about real numbers. Zermelo proposed that these antecedently accepted claims could be derived from the principles he selected as basic. The justification is exactly analogous to that of a scientist who introduces a novel collection of theoretical principles on the grounds that they can explain the results achieved by previous workers in the field.

Strictly speaking, one might accept this picture of Zermelo's knowledge without subscribing to mathematical naturalism. Perhaps there are some who believe that Zermelo may have known the axioms of his set theory by recognizing their ability to systematize previous mathematical work but that his successor did better. Their consciousnesses have been illuminated by the beauties of the cumulative hierarchy. Mathematical naturalists take this prospect to be illusory. Not only is Zermelo's knowledge to be understood in the way that I have suggested, but—in a fashion that I shall describe in more detail below—the same type of justification is inherited by those of us who come after him.

Yet even this much may be conceded without adopting naturalism. There is a hybrid position, with which Russell and Whitehead seem briefly to have flirted,[5] according to which our knowledge of the principles of set theory is based on our knowledge of arithmetic, even though the principles of arithmetic are themselves taken to be *a priori*. Mathematical intuition, stipulative definition, or whatever, is supposed to give us a prior knowledge of the Peano postulates and of all the theorems we can deduce from them; however, we know the axioms of set theory by seeing that they suffice to systematize the Peano postulates.

Mathematical naturalism embodies a ruthless consistency. Having set its face against the procedures to which *a priori* epistemologies for mathematics have appealed, it will not allow that those procedures operate by themselves to justify any mathematical beliefs possessed by any mathematician, past or present. I have tried elsewhere to give reasons for this principled objection against the varieties of intuition and apprehension of meanings on which apriorist accounts have traditionally relied—usually without much explicit epistemological commentary.[6]

Here I shall only offer the sketch of a motivating argument. Many people will have no hesitation in supposing that mathematicians know the statements they record on blackboards and in books and articles. Perhaps there are some who think of the practice of inscribing mathematical symbols as a contentless formal game, and who feel squeamish in talking about mathematical "knowledge." Yet, even for those who adopt this position, there must be an analog to the epistemic concept of justification. Inscriptional practices can be performed well or badly, usefully or pointlessly, and our common view is that the mathematical community has some justification for proceeding in the way that it does.

Now it seems that there can be a mathematical analog of the distinction between knowledge and conjecture (or ungrounded belief). Sometimes a professional mathematician begins with a conjecture and concludes with a piece of knowledge. Sometimes completely uninitiated people guess correctly. If two friends who know barely enough of number theory to understand some outstanding problems decide to divide up the alternatives, one claiming that Goldbach's conjecture is true, the other that it is false, and so forth, then there will not be a sudden advance in mathematical knowledge. Mathematical knowledge, like knowledge generally, requires more than believing the truth. (Similar distinctions can be drawn from the perspective of the approach that takes mathematical statements to be contentless. I shall henceforth ignore this approach and leave it to the interested reader to see how the argument would be developed from the preferred point of view.)

The obvious way to distinguish mathematical knowledge from mere true belief is to suggest that a person only knows a mathematical statement when that person has evidence of the truth of the statement—typically, though not invariably, what mathematicians count as a proof.[7] But that evidence must begin somewhere and an epistemology for mathematics ought to tell us where. If we trace the evidence for the statements back to the acknowledged axioms for some part of mathematics, then we can ask how the person knows those axioms.

In almost all cases, there will be a straightforward answer to the question of how the person learned the axioms. They were displayed on a blackboard or discovered in a book, endorsed by the appropriate authorities, and committed to the learner's memory. But non-naturalistic epistemologies of mathematics deny that the axioms are known because they were acquired in this way. Apriorists offer us the picture of individuals throwing away the props that they originally used to obtain their belief in the axioms and coming to know those axioms in special ways. At this point we should ask a series of questions. What are these special ways of knowing? How do they function? Are they able to produce knowledge that is independent of the processes through which the beliefs were originally acquired? One line of naturalistic argument consists in examining the possibilities and showing that the questions cannot be answered in ways that are consistent with apriorism.[8]

There is a second, simpler way to argue for a naturalistic epistemology for mathematics. Consider the special cases, the episodes in which a new axiom or concept is introduced and accepted by the mathematical community. Naturalists regard such episodes as involving the assembly of evidence to show that the modification of mathematics through the adoption of the new axiom or concept would bring some advance in mathematical knowledge. Often the arguments involved will be complex—in the way that scientific arguments in behalf of a new theoretical idea are complex—and the pages in which the innovators argue for the merits of their proposal will not simply consist in epistemologically superfluous rhetoric. On the rival picture, the history of mathematics is punctuated by events in which individuals are illuminated by new insights *that bear no particular relation to the antecedent state of the discipline.*

To appreciate the difference, and the merits of naturalism, consider Cauchy's introduction of the algebraic concept of limit in the definitions of the concepts of convergence, continuity, and derivative. If we suppose that Cauchy's claims were backed by some special *a priori* insight, then it is appropriate to ask why this insight was unavailable to his predecessors, many of whom had considered the possibility of employing the concept of limit to reconstruct fundamental arguments in the differential calculus. We seem compelled to develop an analogy with superior powers of visual discrimination: Cauchy just had better powers of mathematical intuition, so he saw what Lacroix, Lagrange, l'Huilier, d'Alembert, Euler, Maclaurin, Leibniz, and Newton had all missed. This farfetched story is not only unnecessary, it also fails to do justice to the argumentative structure of Cauchy's work on the calculus. The *Cours d'Analyse* displays how a thorough use of the algebraic limit concept can be employed to reformulate problematic reasoning in the calculus and thus to prepare the way for the resolution of outstanding problems. Cauchy does not invite his fellow mathematicians to intuit the correctness of his new claims. He shows, at some length, how they are useful in continuing the practice of mathematics.[9]

This second line of argument anticipates a theme that I shall elaborate in the next section. Mathematical naturalism offers an account of mathematical knowledge that does justice to the historical development of mathematics. Apriorism has seemed attractive because the only available rival seemed to be the simplistic empiricism of John Stuart Mill.[10] I claim that we can transcend a pair of inadequate alternatives by recognizing that mathematical knowledge is a historical product.

2. A Role for History

Mathematical knowledge is not built from the beginning in each generation. During the course of their education, young mathematicians absorb the ideas accepted by the previous generation. If they go on to creative work in mathematics, they may alter that body of ideas in ways that are reflected in the training of their successors. Like any other part of science, mathematics builds new knowledge on what has already been achieved. For the epistemologist of mathematics, as for the epistemologist of science, a crucial task is to identify those modifications of the corpus of knowledge that can yield a new corpus of knowledge.

We can now outline a naturalist account of mathematical knowledge. Our present body of mathematical beliefs is justified in virtue of its relation to a prior body of beliefs; that prior body of beliefs is justified in virtue of its relation to a yet earlier corpus; and so it goes. Somewhere, of course, the chain must be grounded. Here, perhaps, we discover a type of mathematics about which Mill was right, a state of rudimentary mathematical knowledge in which people are justified through their perceptual experiences in situations where they manipulate their environments (for example, by shuffling small groups of objects). What naturalism has to show is that contemporary mathematical knowledge results from this primitive state through a sequence of rational transitions.[11]

A preliminary task is to replace the vague talk of "states of mathematical knowledge" with a more precise account of the units of mathematical change. The problem is formally parallel to that which confronts the philosopher concerned with the growth of scientific knowledge, and the solutions are also analogous. In both cases, I suggest, we should understand the growth of knowledge in terms of changes in a multidimensional unit, a *practice* that consists of several different components.[12] Each generation transmits to its successor its own practice. In each generation, the practice is modified by the creative workers in the field. If the result is knowledge, then the new practice emerged from the old by a rational *interpractice transition*.

As a first analysis, I propose that a mathematical practice has five components: a language employed by the mathematicians whose practice it is; a set of statements accepted by those mathematicians; a set of questions that they regard as important and as currently unsolved; a set of reasoning that they use to justify the statements they accept; and a set of mathematical views embodying their ideas about how mathematics should be done, the ordering of mathematical disciplines, and so forth. I claim that we can regard the history of mathematics as a sequence of changes in mathematical practices, that most of these changes are rational,[13] and that contemporary mathematical practice can be connected with the primitive, empirically grounded practice through a chain of interpractice transitions, all of which are rational.[14]

It is probably easier to discern the various components in a mathematical practice by looking at a community of mathematicians other than our own. As an illustration of my skeletal account, let us consider the practice of the British community of mathematicians around 1700, shortly after Newton's calculus had become publicly available.[15] This community adopted a language that not only lacks concepts belonging to contemporary mathematics but also contains concepts we no longer use: Newtonian concepts of fluxion, number, series, and function are, I think, different from anything that has survived to the present, although there are contemporary concepts that are similar to these in certain respects. The British mathematicians of the time accepted a wide variety of claims about the properties of tangents to curves, areas under curves, motions of bodies, and sums of infinite series. Many of these statements can be translated into results that we would not accept, although some of the suggestions about infinite series might give us pause. The mathematicians posed for themselves the problems of achieving analogous claims for a broader class of curves and motions; they did not endorse (as their Leibnizian contemporaries did) the general question of finding canonical algebraic representations for the integrals of arbitrary functions or for the sums of infinite series; problems in algebra arose out of, and were to be interpreted in terms of, problems in geometry and kinematics. Members of the community were prepared to justify some of their claims by offering geometric proofs—which they took to be strict synthetic demonstrations in the style of traditional geometry—but, in some cases, they were forced to rely on reasoning that appealed to infinitesimals. Because of their background metamathematical views, reasoning of this kind appeared less than ideal. On the Newtonian conception of mathematics, the fundamental mathematical disciplines are geometry and kinematics, and there is a serious foundational problem of showing how infinitesimalist justifications can either be recast as, or replaced by, reasonings in proper geometric style.

The central part of the naturalistic account of mathematical knowledge will consist in specifying the conditions under which transitions between practices preserve justification. As I have tried to show elsewhere,[16] the Newtonian practice just described can be connected with the practice of the late-nineteenth-century community of analysts by a series of transitions that are recognizably rational. But it is important to acknowledge that my previous efforts at articulating naturalism with respect to this example fall short of providing a full account of rational interpractice transitions in mathematics. We can appreciate rationality in the growth of mathematical knowledge by describing episodes in the history of mathematics in a way that makes clear their affinity with transitions in science (and in our everyday modification of our beliefs) that we count as rational. A *theory* of rationality in mathematics (or, more precisely, in the growth of mathematics) would go further by providing principles that underlie such intuitive judgments, making clear the fundamental factors on which the rationality of the various transitions rests.

By emphasizing the connection between transitions in the history of mathematics and transitions in the history of science that we view as rational, it is possible to argue for a reduction of the epistemology of mathematics to the epistemology of science. In both cases it will be important to distinguish two main types of transition. The first type justifies some change in some component (or components) of a prior practice on the basis of the state of that practice alone. I shall call such transitions *internal transitions*. The modifications of the second type, the *external transitions*, owe their rationality to something outside the prior state of mathematics, some recent experience on the part of those who make the change or some feature of another area of inquiry. It is tempting to believe that the natural sciences are altered through external transitions, and through these alone, and that mathematics grows solely through internal transitions. The temptation should be resisted. Theoretical sciences are often changed in dramatic ways through attempts to resolve tensions in the prevailing practice—think of Copernicus's original efforts to resolve the problem of the planets or of the development of population genetics in the early 1930s. By the same token, the development of mathematics is sometimes affected by the state of other sciences, or even by ordinary experience. To cite just one example, the pursuit of analysis in the early nineteenth century was profoundly modified through the study of problems in theoretical physics. I shall consider the importance of external transitions in mathematics in great detail below.

The insight that underlies the tempting—but oversimple—thesis is that there is a continuum of cases. Some areas of inquiry develop primarily through external transitions; others grow mainly through internal transitions. What is usually called "pure" mathematics lies at one end of the continuum; the applied sciences (for example, metallurgy) lie at the other. We can appreciate the fact that there are differences of degree without making the mistake of believing that there are differences in kind.

Because internal transitions are obviously important in mathematics, it is useful to consider some main patterns of mathematical change, to consider how they fit into the scheme that

I have outlined, and to see how they are exemplified in the history of mathematics. Unresolved problems that have emerged as significant frequently provide the spur for the further articulation of a branch of mathematics. New language, new statements, and new forms of reasoning are introduced to solve them. In at least some cases, the newly introduced language is initially ill understood, the workings of the new reasonings may be mysterious, and there may be legitimate doubts about the truth of the new statements. However, when things turn out well, the unclarities are ultimately resolved and the collection of answers is eventually systematized by principles that accord with prevailing metamathematical criteria. Moreover, we can easily understand prominent ways in which new questions are generated in mathematics: as new language is introduced or old problems are resolved, it becomes rational to pose new, often more general, questions.

Consider many famous examples, each of which counts as an internal interpractice transition. Descartes' introduction of the concepts of analytic geometry was motivated by the ability of the new concepts and modes of reasoning to yield answers to questions about locus problems, answers that were recognizably correct and recognizably more general than those attained by his predecessors. Mysterious though the language of the Leibnizian calculus may have been, its defenders could point out, as the Marquis de l'Hospital did, that it enabled them to solve problems that "previously no one had dared to attempt." Lagrange's use of concepts that we see as precursors of group-theoretic ideas was justified through its ability to bring order to the ramshackle collection of methods for solving cubic and quartic equations that had been assembled by earlier mathematicians.

In each case, the new extension generated new problems. After Descartes, it was rational to seek techniques for associating recalcitrant curves with algebraic equations and for computing the "functions" of curves.[17] Through their use of infinite series, Newton and Leibniz generated new families of mathematical questions: What is the series expansion of a given function? What is the sum of a given infinite series? This process of question generation illustrates a general pattern that has been repeated countless times in the history of function theory: given two schemes for representing functions, one asks how the two are coordinated, typically by seeking canonical representations in well-established terms for functions that are defined in some novel fashion. Similarly, the combination of two discoveries in the theory of equations—the proof of the insolubility of the quintic and Gauss's demonstration that the cyclotomic equation, $x^p - 1 = 0$, is soluble when p is prime—posed the question Galois answered: Under what conditions is an equation soluble in radicals?

All these episodes can be represented within the framework I have sketched, and when they are so represented we see interpractice transitions that are recognizably rational. So much suffices for the reduction of the epistemology of mathematics to the epistemology of science. But it is only the first step toward a complete naturalistic epistemology. Eventually our confident judgments of rationality should be subsumed under principles that explain the rationality of the transitions we applaud. Instead of responding to a skeptic by pointing to the details of the episode, saying "Look! Don't you see that what is going on when Galois poses this problem or when Leibniz introduces this reasoning is just like what goes on in any number of scientific situations?," we would strive to identify the crucial epistemic features of the situation and to incorporate them within a general theory of rational interpractice transitions, a theory that would cover both the mathematical and the scientific cases.

When we see the problem, and our current ability to tackle it, in this way, we should realize that there is presently no prospect of resolving hard cases. Consider the current situation in the foundations of set theory. (Set theory deserves to count as *a* subject for philosophers of mathematics to explore, even if it is not the only such subject.) There are several different proposals for extending the standard collection of set-theoretic axioms. Ideally, a naturalistic epistemology of mathematics would enable us to adjudicate the situation, to decide which, if any, of the alternatives should be incorporated into set theory.[18] If any naturalistic reduction of the philosophy of mathematics is correct, then the problem has been properly defined. But it has not been solved.[19]

Mathematical naturalism thus identifies a program for philosophical research. Apriorists will not find the program worth pursuing because they will suppose that the genuine sources of mathematical knowledge lie elsewhere. (But, with luck, the naturalistic challenge may prompt them to be more forthcoming about exactly where.) However, even those who join me in rejecting apriorism may maintain that naturalism is wrong-headed. For they may perceive a deep difficulty in understanding the rationality of mathematics—either because they are skeptical about the rationality of science or because they think that the conditions that allow the rationality of science are absent in the mathematical case. The objector speaks:

> Naturalists take the reconstruction of our mathematical knowledge to consist in a selective narration of parts of the history of mathematics, a narration designed to give us comfortable feelings about the innovative ideas of the great mathematicians. But why should we count the end product as knowledge? What is it about these transitions that makes us dignify them with the name *rational*? Even if we had a clear set of principles that subsumed the patterns of interpractice transition to which the label is attached, it would still be incumbent on the naturalist to explain why they are to be given a preferred epistemic status.

There is a tangle of problems here, among which is a cousin of Humean worries about the grounds of methodological principles. In the rest of this essay, I shall try to confront the problems, showing that they may help us to clarify the naturalistic picture and that some interesting and unanticipated conclusions emerge.

3. Varieties of Rationality

Those who offer theories of the growth of scientific knowledge must answer two main problems. The first of these, the problem of progress, requires us to specify the conditions under which fields of science make progress. The second, the problem of rationality, requires us to specify the conditions under which fields of science proceed rationally. Everyone ought to agree that the two problems are closely connected. For a field to proceed rationally, the transitions between states of the field at different times must offer those who make the transitions the best available strategies for making progress. Rationality, as countless philosophers have remarked, consists in adjustment of means to ends. Our typical judgments of rationality in discussions about scientific change or the growth of knowledge tacitly assume that the ends are epistemic. When we attribute rationality to a past community of scientists, we consider how people in their position would best act to achieve those ends which direct all inquiry, and we recognize a fit between what was actually done and our envisaged ideal.

So I start with a general thesis, one that that applies equally to mathematics and to the sciences. Interpractice transitions count as rational insofar as they maximize the chances of attaining the ends of inquiry. The notion of rationality with which I am concerned is an absolute one. I suppose that there are some goals that dominate the context of inquiry, that are not goals simply because they would serve as stepping stones to yet further ends. The assumption is tricky. What are these goals? Who (or what) is it that stands a chance of attaining them? In what ways should the chances be maximized? Finally, are there special mathematical goals, or does mathematics serve us in our attempts to achieve more general ends?

Following Kant, I take it that there are ends of rational inquiry and that to be a rational inquirer is to be a being who strives to achieve these ends.[20] The ends I identify as ultimate are the achieving of truth and the attainment of understanding. If we ask a mathematician why a certain style of proof has been widely adopted, or why a particular systematization of a body of mathematical knowledge has been proposed and accepted, then we shall expect to be told that the proof-pattern is likely to issue in the acceptance of true conclusions and that the favored axiomatization yields understanding of the body of mathematical theory that it systematizes. Similar responses are anticipated if we raise questions about the value of certain mathematical concepts or the urgency of some mathematical questions: the former help us to appreciate the interrelationships among mathematical claims, the latter signal gaps in our current understanding. But if we continue by asking why we should be interested in true conclusions or in understanding, we can only answer that these are the ends of inquiry.

The ends of rational inquiry are not our only ends, and, in consequence, they do not completely dictate the course of rational development of mathematics. Our aims also include the goals of providing for the welfare of present and future members of our species (and perhaps members of other species as well), of securing free and just social arrangements, and so forth. The relation between the growth of science and these practical ends is typically tenuous and indirect, and the connection with mathematics is even more attenuated. Nevertheless, if asked to justify the pursuit of a particular collection of mathematical problems, we may reply that these problems arise within the context of a particular scientific inquiry. That scientific inquiry, in its turn, may be motivated by the desire for rational ends of scientific inquiry—greater understanding of some facet of the universe, say of the structure of matter or of the springs of animal behavior. Or it may be justified through its contribution to some practical project, the securing of a steady supply of food for a group of people or the improvement of transmission of information or resources around the world. Finally, that practical project, in its turn, may connect directly with our ultimate practical ends.

My division between ends of rational inquiry and practical ends obviously relates to the previous distinction between internal and external rational interpractice transitions. More exactly, an interpractice transition is rational in virtue of its advancement of the ends of rational inquiry *in mathematics*. An external interpractice transition is rational in virtue of its advancement either of the ends of rational inquiry in some other branch of knowledge or of some practical ends.[21]

When we consider the rationality of mathematics and science as a problem about the adjusting of means to ends, then it becomes evident that judgments of rationality are highly ambiguous. I want to distinguish four main senses:

1. Individual epistemic rationality: an individual is epistemically rational when he or she adjusts practice so as to maximize the probability that he or she will attain his or her epistemic ends.

2. Individual overall rationality: an individual is rational overall when that individual adjusts practice so as to maximize the probability that he or she will attain his or her total set of ends.

3. Community epistemic rationality: a community is epistemically rational when the distribution of practices within it maximizes the probability that the community will ultimately attain its epistemic ends.

4. Community overall rationality: a community is rational overall when the distribution of practices within it maximizes the probability that the community will ultimately attain its total set of ends.

I claim that these notions of rationality are quite distinct and that it is important to know which we have in mind before we inquire about the rationality of mathematics or of some other science.[22]

To see the difference between the individual and community perspectives, consider the following kind of example. Workers in a scientific field seek the answer to a certain question, with respect to which various methods are available. Considering the decision for each individual scientist, we quickly reach the conclusion that, if there is one method that is recognizably more likely to lead to the solution, then each scientist ought to elect to pursue that method. However, if all the methods are roughly comparable, then this is a bad bargain from the point of view of the community. Each individual ought to prefer that he or she belongs to a community in which the full range of methods is employed. Hence, from the community perspective, a rational transition from the initial situation would be one in which the community splits into subgroups, each homogeneous with respect to the pursuit of method. The intuition is that the ends of the community can be better advanced if some of the members act against their own epistemic interests.[23]

Although the points I have been making may seem to have little to do with the growth of mathematical knowledge, they enable us to resolve an important type of disagreement among historians and philosophers of mathematics. Consider the different research strategies pursued by British and Continental mathematicians in the years after the priority dispute between Newton and Leibniz.[24] Leibnizians confidently set about using new algebraic techniques, vastly increased the set of problems in analysis, and postponed the task of attempting to provide a rigorous account of theory concepts and reasonings. Their attitude is not only made explicit in Leibniz's exhortations to his followers to extend the scope of his methods, without worrying too much about what the more mysterious algebraic maneuvers might mean, but also in the acceptance of results about infinite series sums that their successors would abandon as wrongheaded.[25] Insofar as they were concerned to articulate the foundations of the new mathematics, the Leibnizians seem to have thought that the proper way to clarify their concepts and reasonings would emerge from the vigorous pursuit of the new techniques. In retrospect, we can say that their confidence was justified.

By contrast, Newton's successors were deeply worried about the significance of the symbols that they employed in solving geometric and kinematic problems. They refused to admit into their mathematical work questions or modes of reasoning that could not be construed in geometric terms, and they lavished attention on the problem of giving clear and convincing demonstrations of elementary rules for differentiating and integrating (to speak somewhat anachronistically). Some of the ideas that emerged from their work were ultimately made central to analysis in the reconstruction offered by Cauchy.[26]

Now, we ask, "Which, if either, of these developments was rational?" Before we answer the question, it is important to be aware of the distinctions that I have drawn above. When we demand an individualistic assessment of the rationality of the rival interpractice transitions, we seem compelled to engage in a difficult cost-benefit analysis. To a first approximation, we can view both groups of mathematicians as aiming at two kinds of epistemic ends: proliferation of answers to problems and increased understanding of those answers, the concepts they contain, and the methods by which they are generated. Newtonians emphasize the need for clarity first, holding that secure results in problem solving will be found once the concepts and reasoning are well understood. Leibnizians suggest that the means of clarification will emerge once a wide variety of problems have been tackled. Each tradition is gambling. What are the expected costs and benefits?

Those who are uneasy, as I am, about attributing rationality to one tradition and denying it to the other have an obvious first response to the situation. They may propose that the alternative strategies were initially so close in terms of their objective merits that it would have been perfectly reasonable to pursue either of them. Hence there can be no condemnation of either the early Newtonians or the early Leibnizians. Neither group behaved irrationally. Yet this proposal is unsatisfactory as it stands. For it is hard to extend it to account for the continued pursuit of the Newtonian tradition once it had become apparent that the Continental approach stemming from Leibniz and the Bernoullis was achieving vast numbers of solutions to problems that the Newtonians also viewed as significant (problems that could readily be interpreted in geometric or kinematic terms, even though the techniques used to solve them could not). Moreover, historians who like to emphasize the role of social factors in the development of science will note, quite correctly, that the national pride of the British mathematicians and the legacy of the dispute over priority in the elaboration of the calculus both played an important role in the continued opposition to Continental mathematics.

I suggest that we can understand why British mathematicians doggedly persisted in offering clumsy and often opaque geometric arguments if we recognize the broader set of goals that they struggled to attain. Among these goals was that of establishing the eminence of indigenous British mathematics, and we can imagine that this end became especially important after the Hanoverian succession and after Berkeley's clever challenge to the credentials of the Newtonian calculus.[27] Moreover, before we deplore the fact that some of Newton's successors in the 1740s and 1750s may have been moved by such nonepistemic interests as national pride, we should also appreciate the possibility that the maintenance of a variety of points of view (which chauvinism may sometimes achieve) can advance the epistemic ends of the community. The goals of promoting acceptance of truth and understanding in the total mathematical community were ultimately achieved (in the nineteenth century) because both traditions were kept alive through the eighteenth century. If, as I suspect, one of the traditions was maintained because some mathematicians were motivate by nonepistemic interests, then perhaps we should envisage the possibility that the deviation from individual epistemic rationality signals the presence of an institution in the community of knowers that promotes community epistemic rationality. A rational community of knowers will find ways to exploit individual overall rationality in the interest of maximizing the chances that the community will attain its epistemic ends.[28]

Only if we restrict ourselves to the notion of individual epistemic rationality and seek to find this everywhere in the history of mathematics (or of science generally) does the search for rationality in that history commit us to a Whiggish enterprise of distributing gold stars and black marks. The general form of an interpractice transition is more complex than we might have supposed, and the historiography of mathematicians should reflect the added complexity. We are to imagine that the community of mathematics is initially divided into a number of homogeneous groups that pursue different practices. In some cases, the claims made by members of the groups will be incompatible (Newtonians versus Leibnizians, Kronecker and his disciples against Dedekind and Cantor, militant constructivists against classical mathematicians). In other cases, they will just be different. An interpractice transition may modify the particular practices, the group structure, or both. Such interpractice transitions may be viewed from the perspective of considering whether (a) they maximize the chances that the individuals who participate in them will achieve their individual epistemic ends, (b) they maximize the chances that those individuals will attain their total set of ends with some epistemic ends being sacrificed to nonepistemic ends, (c) they maximize the chances that the community will attain its epistemic ends, or (d) they maximize the chances that the community will attain its total set of ends, with epistemic ends being sacrificed to nonepistemic ends. Moreover, with respect to each case we may focus on epistemic ends that are internal to mathematics or we may look to see where the ends of other areas of inquiry are also involved and, where (b) and (c) both obtain, we may look for institutions within the total mathematical community that promote the attainment of epistemic ends by the community at cost to the individual.

I have been concerned to stress the broad variety of questions that arise for the history of mathematics once we adopt the naturalistic perspective I have outlined and once we have extended it by differentiating notions of rationality. To the best of my knowledge, these questions remain virtually unanswered for virtually all of the major transitions in the history of mathematics.

4. The Ends of Inquiry

So far, I have remained vague about the epistemic ends at which I take inquiry to be directed. The vagueness should provoke questions. Are there epistemic ends intrinsic to mathematics? If so, what are they? In other words, what counts as *mathematical* progress? How are the epistemic ends of mathematics, if there are any, to be balanced against the epistemic ends of other branches of inquiry or against nonepistemic (practical) ends? These are important issues for a naturalistic philosophy of mathematics to address. I shall try to show that they connect with questions that should concern historians of mathematics and even professional mathematicians.

Compare paradigm cases of internal and external interpractice transitions. For the former, imagine that a group of mathematicians introduces new concepts or new axioms for the sole purpose of improving their understanding of prior mathematical claims. For the latter, suppose that they decide to lavish great attention on methods for solution to complicated differential equations with important applications, or to develop broadly applicable techniques of approximation that allow practical problems to be solved with assignable error. In either case, we may applaud the transition for its contribution to some end. Yet the applause is consistent with a more global criticism. Perhaps the transition may be felt to strike the wrong balance between our epistemic and practical (or, more precisely, our extramathematical) ends. The gain in understanding was achieved by sacrificing opportunities to pursue a more significant goal, or the concern with the practical applications overshadowed a more important epistemic interest.

Mathematical Naturalism

The recent history of mathematics has shown that the criticism can run either way. Critics of post-Weierstrassian analysis (and of the education program that it influenced in the Cambridge mathematical Tripos of most decades of our century) may suggest that the emphasis on solving recondite problems involving special functions (originally introduced into mathematics because of their significance for certain problems in physics)[29] detracted from the achievement of insight into fundamental theorems. Detractors of contemporary mathematics might conclude that the habit of discerning maximum generality, and the embedding of classical results in the broadest possible context, has yielded a collection of sterile investigations that should be abandoned in favor of projects more closely connected with the projects of the sciences and with everyday life.

To formulate these issues in a clear way we need a solution to the problem of mathematical progress. We need to know the epistemic ends at which mathematics aims. Let us begin with a liberal conception of progress. According to this conception, we must distinguish between epistemic and pragmatic appraisal. We make progress so long as we add to the store of knowledge in *any* way. Perhaps some additions are more profitable in advancing us toward the totality of our goals, but that is a purely pragmatic matter.

The liberal conception suggests that our applause for both our paradigm cases should not be tempered with criticism. To see if so ecumenical an attitude can be sustained, we need to consider what theses about the ontology of mathematics can be integrated with a naturalistic approach. There is a variant of ontological Platonism that is compatible with almost all my claims about mathematical knowledge. According to this variant, mathematical knowledge begins in prehistoric times, with the apprehension of those structures which are instantiated in everyday physical phenomena. In the simplest version, we perceive the properties of small concrete sets (that is, sets whose members are physical objects).[30] Mathematics proceeds by systematically investigating the abstract realm, to which our rudimentary perceptual experiences give us initial access. That investigation is guided by further perception, by the uses of mathematics in the physical sciences, and by attempts to explain and to systematize the body of results that has so far been acquired. Platonists can simply take over my stories about rational interpractice transitions, regarding those transitions as issuing in the recognition of further aspects of the realm of abstract objects.

Mathematical naturalists who are also Platonists can readily draw the distinction between pragmatic appraisal and epistemic appraisal. The mathematician's task is to draw a map of Platonic heaven, and the acquisition of any "geographical" information constitutes progress.[31] In some cases (the embedding of classical results in an abstract and highly general context), we may learn about abstract features of that heaven that are of little practical value in steering ourselves around the earth. In other instances (the discovery of techniques for solving very special classes of differential equations), we may come to know details of considerable utility but of little general import. So long as we continue to amass truths, there is no question about making progress.

Now I think that it is possible to attack this liberal view of progress even within the *Platonistic* perspective. Those who see the natural sciences as aiming at truth do not have to suppose that *any* accumulation of truths constitutes scientific progress: insignificant truth is typically not hard to come by.[32] Thus, my point is not that a naturalistic Platonist is compelled to adopt the liberal conception of progress, but that Platonism is one way of elaborating the liberal conception. However, I want to press a more radical challenge. I claim that we should reject the Platonist's view of the ontology of mathematics, substituting for it a picture that does not allow for the distinction between pragmatic assessment and epistemic assessment on which the liberal conception builds.

Many of the reasons for worrying about Platonism are familiar. How is reference to mathematical objects to be secured? How is it that perception provides us with knowledge about such objects? Even though these questions may be blunted by claiming that we initially refer to and know about *concrete* sets, it still remains mysterious how we are ultimately able to refer to and know about *abstract* sets.[33] Other anxieties arise in the particular context of a naturalistic approach to mathematical knowledge. Like other theoretical realists, Platonists must explain why our ability to systematize a body of results provides a basis for belief in the existence of antecedently unrecognized entities. It is not easy to understand how Lagrange's insights into the possibilities of leaving certain expressions invariant through the permutation of roots should constitute recognition of the existence of hitherto unappreciated mathematical objects—to wit, groups.

The general problem of understanding the historical transitions that have occurred in mathematics as revelations of the inhabitants of Platonic heaven arises with particular force when we consider a particular type of interpractice transition that is common in mathematics: the resolution of apparent incompatibility by reinterpreting alternatives. When mathematicians discovered that non-Euclidean geometries were consistent, the traditional vocabulary of geometry was reconstrued. On the Platonist's account, new objects—non-Euclidean spaces and their constituents—were discovered. By contrast, the resolution of incompatibilities in the sciences often proceeds by dismissing previously countenanced entities. The chemical revolution did not conclude by allowing Lavoisierian and non-Lavoisierian combustibles.

Focus on this last example may prompt the conventionalist reaction that mathematics simply consists in exploring the consequences of arbitrarily selected conventions and that the history of mathematics reveals the emergence of the conventions that have happened to interest past mathematicians. This is an overreaction to the pitfalls of Platonism. Interpractice transitions typically involve a nonarbitrary modification of antecedent concepts, reasonings, problems, and statements. Thus, our task is to find an account of the ontology of mathematics that will avoid supposing that practices are altered in accordance with the whim of the moment, without lapsing into the idea that the alterations issue in the disclosure of the truths about Platonic heaven.

I have suggested elsewhere that the problem can be solved if we treat mathematics as an idealized science of human operations.[34] The ultimate subject matter of mathematics is the way in which human beings structure the world, either through performing crude physical manipulations or through operations of thought. We idealize the science of human physical and mental operations by considering all the ways in which we could collect and order the constituents of our world if we were freed from various limitations of time, energy, and ability. One way to articulate the content of the science is to conceive of mathematics as a collection of stories about the performances of an ideal subject to whom we attribute powers in the hope of illuminating the abilities we have to structure our environment.

This proposal goes beyond conventionalism in placing restrictions on the stories we tell when we make progress in mathematics. Some stories, the stories of elementary mathematics, achieve the epistemic end of directly systematizing the operations we find ourselves able to perform on physical objects (and on mental representations of such objects). Others achieve the epistemic end of answering questions that arise from stories that achieve an epistemic end, or of systematizing the results obtained within stories that achieve an epistemic end. Ultimately, there must be a link, however indirect, to operations on our environment. Nevertheless, it would be wrong to think of the entire structure of mathematics as an attempt to systematize and illuminate the elementary operations that are described in the rudimentary portions of the subject. For, in an important sense, *mathematics generates its own content.* The new forms of mathematical notation that we introduce not only enable us to systematize and extend the mathematics that has already been achieved, but also to perform new operations or to appreciate the possibility that beings released from certain physical limitations could perform such operations. Such extensions of our repertoire seem to me to have occurred with the development of notation for representing morphisms on groups and for constructing sets of sets. In both cases, the notation is a vehicle for iterating operations that we would not be able to perform without it.[35]

This proposal (which I shall call *naturalistic constructivism*) resolves the difficulties that beset Platonism. There is no question of securing reference to or gaining perceptual knowledge of anything other than unproblematic entities—to wit, operations that we perform and recognize ourselves as performing. The transitions that occur in the history of mathematics are taken at face value: they consist in introducing concepts, statements, problems, and reasonings as parts of stories that help us understand the operations we are able to perform on our environments; in some cases, the activity of elaborating these stories itself generates new kinds of operations for later mathematics to consider. Finally, the instances in which an apparent disagreement is resolved through reinterpretation are cases in which the illumination of what has so far been achieved is obtained by embedding what appeared to be a single story within a range of alternatives.

Naturalistic constructivism collapses the notions of justification and truth in an interesting way. To say that a mathematical statement is true is to make a claim about the powers that are properly attributed to the ideal subject (or, more generally, to make a claim to the effect that the statement figures in a story that is properly told). What "properly" means here is that, in the limit of the development of rational mathematical inquiry, our mathematical practice contains that statement. Truth is what rational inquiry will produce, in the long run.

We can now see why naturalistic constructivism undercuts the distinction between pragmatic and epistemic appraisal. What we mean by "the limit of the development of rational mathematical inquiry" is the state to which mathematics will tend if we allow our entire body of investigations to run their course and to be guided by procedures designed to maximize our chances of attaining our ends. Since there is no independent notion of mathematical truth, the only epistemic end in the case of mathematics is the understanding of the mathematical results so far achieved. Mathematics proceeds autonomously through attempting to systematize whatever claims have previously been made about our powers to order and collect—including those which have been made possible for us through the activity of mathematics itself—and is inevitably dependent on other sciences and on practical concerns for the material on which it will work. Alternatively, we can say that the goal of mathematics is to bring system and understanding to the physical and mental operations *we find it worth performing* on the objects of our world, so that the shape and content of mathematics are ultimately dictated by our practical interests and the epistemic goals of other sciences.

If this is correct, then epistemic appraisal in mathematics inevitably involves pragmatic considerations. It is not only legitimate to reform a body of mathematics on the grounds that it fails to answer to our pragmatic needs, but also senseless to defend the mathematics so reformed on the grounds that it is "useless knowledge." There is room for just one kind of useless knowledge in mathematics: claims that have in themselves no practical implications but serve to enhance our understanding of results that are practically significant. This does not mean that Hardy was wrong when he gloried in the uselessness of the results of number theory. But, if he was right, then there must be a chain of mathematical practices culminating in the refined abstractions of number theory and beginning with material of genuine practical significance, such that each member of the chain illuminates its predecessor.

Mathematical progress, in a nutshell, consists in constructing a systematic and idealized account of the operations that humans find it profitable to perform in organizing their experience. Some of these operations are the primitive manipulations with which elementary arithmetic and elementary geometry begin. Others are first performed by us through the development of mathematical notation that is then employed in the sciences as a vehicle for the scientific organization of some area of experience. But there is no independent notion of mathematical truth and mathematical progress that stands apart from the rational conduct of inquiry and our pursuit of non-mathematical ends, both epistemic and nonepistemic.

I draw a radical conclusion. *Epistemic* justification of a body of mathematics must show that the corpus we have obtained contributes either to the aims of science or to our practical goals. If parts can be excised without loss of understanding or of fruitfulness, then we have no *epistemic* warrant for retaining them. If there is a distinction between mathematics as art and mathematics as cognitive endeavor, it is here that it must be drawn.[36] Drawing it must wait on the development of a full theory of rational interpractice transitions, both in mathematics and in the sciences.

5. An Agenda

In *The Nature of Mathematical Knowledge,* I focused on internal transitions and on individual epistemic rationality. My project was to show that there have been important and unrecognized inferences in the history of mathematics and that similar inferences underlie our knowledge of those statements philosophers have often taken to be the foundations of our mathematical knowledge. I now think that the position I took was too conservative in several different ways.

First, external interpractice transitions need more emphasis. As I have suggested above, mathematics is dependent on other sciences and on our practical interests for the concepts that are employed in the spinning of mathematical stories. It is shortsighted to think that the systematization of a branch of mathematics can proceed in neglect of the ways in which adjacent fields are responding to external demands. Hamilton thought that generalization of claims about complex numbers would necessarily be a fruitful project. Ultimately, the field in which he labored was changed decisively through attempts to come to terms with problems in mathematical physics. Vector algebra and analysis offered a perspective from which the lengthy derivations of recondite properties of quaternions look beside the point.[37]

Second, the history of mathematics, like the history of other areas of science, needs to be approached from both the individual and the community perspectives. We should ask not only about the reasons people have for changing their minds, but also about the fashion in which the community takes advantage of our idiosyncrasies to guide us toward ends we might otherwise have missed. As I have already noted, the pursuit of more than one research program may often advance the community's epistemic projects—despite the fact that it will require of some members of the community that they act against their individual epistemic interests. We can properly ask whether there are enough incentives in mathematics for the encouragement of diversity.

Third, there are serious questions about the balance between the pursuit of epistemic ends and the pursuit of nonepistemic ends. I have been arguing that it is difficult to distinguish questions of "science policy" from questions of epistemology when the science under study is mathematics. An important part of an epistemology for mathematics ought to be the consideration of the relative importance of mathematical understanding and the articulation of methods that will promote our extramathematical projects (including our practical projects).

These conclusions are not likely to be popular, for they contradict an image of purity that has dominated much contemporary thinking about mathematics and much past philosophy of mathematics. I hold them because I take them to follow from a thoroughly naturalistic approach to mathematical knowledge, and because I believe that there is no plausible alternative to a naturalistic mathematical epistemology. These conclusions should be corrected, refined, or discarded by undertaking a pair of studies.

The first of these studies is the philosophical enterprise of giving a precise account of rationality and progress in the sciences. The discussions of the last sections need to give way to a detailed account of our epistemic ends, of the ways in which interpractice transitions contribute to these ends, and of the kinds of institutions that can play a role in shaping the community pursuit of these ends. If my claims above are correct, then the problems of epistemology of mathematics reduce to questions in the philosophy of science, questions that I have tried to formulate in a preliminary fashion here.

The second project is more historical. As I have already suggested, virtually all the major questions about the growth of mathematical knowledge remain unanswered for almost every major transition in the history of mathematics. I want to conclude by mentioning two examples that seem to me to illustrate some of the historical issues that I am recommending and that point toward distinctions that have been underemphasized in the previous discussion.

Philosophers of mathematics routinely concentrate their attention on the emergence of set theory and of modern logic in the early decades of our century. That historical episode is important, but, to my mind, it is far less significant than a contemporaneous development. *Part* (but only part) of that development begins in papers by Dedekind (specifically the supplements to Dirichlet's lectures on number theory), is pursued by Emmy Noether and her colleagues, and is completed with the publication of van der Waerden's *Moderne Algebra*. In the process, the language of mathematics was radically changed and the problems of mathematics were transformed. If we are to assess the way in which the mathematics of our century has been driven by internal changes, then there can be no better model for our study than the evolution in which the lineage begun by Dedekind is one significant strand.

The example is not only important because it focuses on a transition in which an entirely new ideal of mathematical understanding was fashioned—an ideal that sees our ability to identify particular results about numbers, spaces, functions, and so forth, as special cases of claims about very abstract structure. Insight into this transition is essential if the naturalistic approach I espouse is to be defended against a commonly heard objection. "Naturalism," say the critics, "is all very well for pre-twentieth-century mathematics, but in our own times the subject has come of age and has been transformed."[38] Now I shall assume that the historical reconstructions that I have given in the case of the emergence of the abstract group concept and (in far more detail) in the emergence of the main concepts of nineteenth-century analysis are satisfactory. If that is so, then it is possible to show how major modifications of mathematical practice were achieved on the basis of tensions within prior practice—how such tensions led to the introduction of new problems, how the new problems were solved by employing new concepts and methods of reasoning, and how ill-understood reasonings and concepts were finally systematized with axioms and definitions. There is no obvious reason why the development of contemporary mathematics should have gone differently, and it is natural to suggest that every successful generation of inquirers takes itself to have brought the discipline to maturity. Nonetheless, the critics make a telling point when they emphasize the extent of the differences between current investigations and nineteenth-century mathematics, and it is a legitimate challenge to ask whether the transition can be understood in the same way as the examples of group theory and classical analysis.

Here is a simple version of a fragment of the story. In the tenth supplement to Dirichlet's lectures, Dedekind undertakes to introduce a new, general, and perspicuous way of reformulating some of Kummer's results on "ideal numbers." Part of his strategy is to consider collections of elements in the number domain and to define an analog of multiplication when one of the multiplicands is not an element of the domain but a certain kind of collection. So we might see set-theoretic constructions as admitted into mathematics because of their ability to systematize a class of results that had already been assembled in tackling significant problems in number theory.

But the simple version is far too simple. As Harold Edwards has argued, there was a serious debate about whether or not Dedekind's method offered an explanatory systematization (and extension) of the theory of ideal numbers.[39] Kronecker offered a rival version, bereft of set-theoretic constructions but also avoiding a detour that Dedekind's reasoning was forced to undertake.[40] So why did Dedekind's proposals win adherents and, eventually, success? Here we must look at the manifold ways in which thinking about various types of abstract structures was beginning to prove useful in late-nineteenth-century mathematics. I conjecture that a more realistic version of the rational acceptance of the concept of an ideal (or, more generally, of the idea of defining analogs of operations on elements on collections of such elements) will need to consider the ways in which related notions proved useful in a number of quite different mathematical subdisciplines.[41]

This is *only* a conjecture. I offer it merely to turn back the charge that it is quite incomprehensible how the naturalistic approach that I have used in understanding the transitions that took place in analysis from Newton and Leibniz to the end of the nineteenth century could achieve similar success in understanding the emergence of contemporary mathematics. Critics who contend that it is *impossible* that twentieth-century algebra (for example) could have been the product of a sequence of interpractice transitions of the kinds I describe are mistaken. Whether any of the sequences of rational interpractice transitions that my scheme allows corresponds to the actual historical developments is another matter—a matter for detailed historical research.

Contrast this first example with cases suggested by an obvious question. There is little doubt that the practice of some areas of pre-twentieth-century mathematics was decisively affected by developments in the natural sciences (particularly in physics). Has this process continued into our own times? Three obvious possible positive instances come to mind. The first is the development of catastrophe theory, with its roots in attempts to study aspects of biological and social systems. Second is the work of Traub and his associates on identifying the reliability of error-prone algorithms. Third, and perhaps most exciting, is the investigation of periodic equilibria and "chaos," pioneered by Feigenbaum and culminating in the identification of new "fundamental constants."[42] In each of these instances, mathematics is being used in novel ways to address practical and scientific problems. The interesting issue is whether the new work provides a basis for modifying mathematical practice.

Recall a paradigm from the nineteenth-century analysis. Fourier's studies of the diffusion equation led to problems that could not be resolved within the framework of the analysis of the time. In Cauchy's treatment, the central concepts of analysis were redefined and clarified so as to provide means of answering the new questions. Conservatives may declare that the contemporary applications to which I have alluded are quite different from the paradigm. All that Thom, Traub, and Feigenbaum have done is to call to our attention hitherto unanticipated ways of developing received ideas. Others may see in these (or in different) contemporary developments the basis for introducing new concepts that will radically alter mathematical practice.

Who is right? Again, the question can only be resolved by detailed investigation. But it is important to see that there is a live issue here, an issue related to the exchanges in which pessimistic mathematicians engage over the coffeepot at numerous institutions all over the world. Is the field going to the dogs because it has lost itself in arid abstractions that serve no cognitive purpose? Or is it in decline because mathematics is no longer queen, but very much the servant, of science (or, worse, engineering)? On my view of mathematical knowledge, these common complaints pose central—but neglected—problems in the philosophy of mathematics: What is the right balance between the epistemic end of understanding the mathematics already achieved and the ends set for us by the sciences and our practical needs? Does contemporary practice strike that balance? I have tried to construct a framework within which historians, philosophers, and mathematicians can collaborate to find answers.

Notes

1. This essay is reprinted with permission from William Aspray and Philip Kitcher, eds. (1988) *History and Philosophy of Modern Mathematics*. Minneapolis: University of Minnesota Press, 293–325.
2. The transition is evident in the work of Frege, who notes in the introduction to the *Grundlagen (The Foundations of Arithmetic* [Oxford: Blackwell, 1959]) that he has "felt bound to go back rather further into the general logical foundations of our science than perhaps most mathematicians will consider necessary" (x). His stated reason is that, without the successful completion of the project he undertakes, mathematics has no more than "an empirical certainty" (ibid.). Later, he suggests that the achievements of the nineteenth century in defining the main concepts of analysis point inexorably to an analogous clarification of the concept of natural number. To those who ask why, Frege offers the same epistemological contrast, pointing out that mathematics prefers "proof, where proof is possible, to any confirmation by inductions" (1–2). For a more detailed investigation of Frege's epistemological motives and of his transformation of the philosophy of mathematics, see my "Frege's Epistomology," *Philosophical Review* 88 (1979): 235–62, and "Frege, Dedekind, and the Philosophy of Mathematics" (to appear in *Synthesizing Frege*, ed. L. Haaparanta and J. Hintikka).
3. In *The Nature of Mathematical Knowledge* (New York: Oxford University Press, 1983), I developed an antiapriorist philosophy of mathematics that I called *empiricism*. However, that position differs from most approaches that have called themselves *empiricism* in several important ways, and I now prefer the name *naturalism*, both for the view I defended in the book and the refinement of it presented in this essay. I hope that the change of labels will help some of my mathematical readers to avoid the provocations of my earlier terminology. I am grateful to Felix Browder for suggesting that the choice of that earlier terminology was "a philosophical fetish."
4. This example was used to similar ends by Hilary Putnam in "What Is Mathematical Truth?" in his *Collected Papers*, vol. 1, *Mathematics, Matter, and Method* (Cambridge: Cambridge University Press, 1976).
5. *Principia Mathematica*, vol. 1 (Cambridge: Cambridge University Press, 1910), x.
6. See *Nature of Mathematical Knowledge*, chap. 1–4. One important part of my argument is that the notion of *a priori* knowledge stands in need of preliminary clarification. Perhaps the thesis that mathematical knowledge is *a priori* obtains its total credibility from the fact that the crucial notion of apriority is usually left so vague that it is possible to mistake heuristic devices for vehicles of *a priori* knowledge.
7. Not, of course, what logicians count as a proof. Complete derivations with all steps made explicit are not available except in the case of extremely elementary parts of mathematics. Moreover, it is a serious epistemological question to ask what such derivations would do for us if we had them.
8. Again, see *Nature of Mathematical Knowledge*, chap. 1–4. I think that the best reply for the apriorist is to contend that the conditions I take to be necessary for *a priori* knowledge are too stringent, and this reply has been offered by a number of people, most lucidly by Charles Parsons (review of *The Nature of Mathematical Knowledge*, in *Philosophical Review* 95 [1986]: 129–37.) As Parsons correctly points out, my arguments against apriorism depend on a condition to the effect that the procedures that give *a priori* justificatory function no matter what our experience. So apriorism could be salvaged by dropping that condition. But it seems to me that the suggestion is vulnerable in two different ways. First, if "*a priori*" procedures could be undermined by recalcitrant experiences, then it would appear that experience is doing some positive work when the procedures actually function to justify our beliefs. Thus our "*a priori*" knowledge would be dependent on our experience. To overcome this difficulty, it would seem necessary to argue for an asymmetry: recalcitrant experience can play a negative role, but kindly experience does nothing positive. I do not at present see how such an asymmetry could be articulated and defended. Second, there seem to be many procedures in the natural sciences—the thought experiments of Galileo and Einstein, for example—that also seem to fit the notion of "*a priori* knowledge" if one abandons the condition that generates trouble for the apriorist. Hence, I do not think there is much point in defending "apriorism" by dropping that condition.

9. For an account with much more detail, see chap. 10 of *Nature of Mathematical Knowledge*. Other significant features of the episode are presented in Judith Grabiner's *The Origins of Cauchy's Rigorous Calculus* (Cambridge, MA: MIT Press, 1981).

10. However, Mill was not as muddled as his critics (notably Frege) have often made him out to be. For a more charitable assessment of his views, see Glenn Kessler's "Frege, Mill, and the Foundations of Arithmetic," *Journal of Philosophy* 77 (1980): 65–79; and my paper "Arithmetic for the Millian," *Philosophical Studies* 37 (1980): 215–36.

11. Two important points need to be noted here. First, because the chain is so long, it seems misleading to emphasize the *empirical* character of the foundation. Indeed, it seems to me to be possible that the roots of primitive mathematical knowledge may lie so deep in prehistory that our first mathematical knowledge may be coeval with our first propositional knowledge of any kind. Thus, as we envision the evolution of human thought (or of hominid thought, or of primate thought) from a state in which there is no propositional knowledge to a state in which some of our ancestors know some propositions, elements of mathematical knowledge may emerge with the first elements of the system of representation. Of course, this is extremely speculative, but it should serve as a reminder that the main thrust of the naturalistic approach to mathematical knowledge is to understand *changes* in mathematical knowledge; although a naturalist contends that mathematical knowledge originated in some kind of responses to the environment, it is eminently reasonable to propose that there are a number of possibilities and that this aspect of the naturalistic theory of knowledge is (for the moment, and perhaps permanently) less accessible to elaboration. (I am indebted to Thomas Kuhn for an illuminating discussion of this point.)

Second, it is not necessary for a naturalist to believe that *all* the transitions that have occurred in the history of mathematics were rational. There may be temporary fallacies in the arguments that the mathematicians initially give to introduce new ideas, or the reasons they present may be inadequate. Such lapses are of no account provided that good reasons are later supplied, for, in such cases, we may see the change as an episode in which mathematical practice develops for a while in an unjustified fashion before securing proper justification that then prepares the way for later transitions. Hence it is possible for the chain of rational interpractice transitions to diverge from the actual course of events. It seems to me likely that this occurred in the case of the introduction of complex numbers, where the initial reasons for extending the language of mathematics were not very strong. However, Euler's demonstration of the fruitfulness of the new language eventually provided compelling reasons for the modification, and the "rational reconstruction" favored by the naturalistic epistemologist will depart from the chronology by treating the Eulerian transition as the pertinent link in the justificatory chain and by ignoring the arguments originally proposed by Bombelli.

12. The same approach is also useful in understanding issues in the growth of scientific knowledge. I have discussed two particularly important cases in "1953 and All That: A Tale of Two Sciences," *Philosophical Review* 93 (1984: 335–73; and "Darwin's Achievement," in *Reason and Rationality in Science,* ed. N. Rescher (Washington, D.C., University Press of America, 1985), 127–89.

13. Strictly speaking, it is not necessary to assume the rationality of the majority—or even of *any*—of the major interpractice transitions that have actually occurred in the history of mathematics. All that is needed is to suppose that unjustified leaps are later made good through the provision of reasons that support the practices that emerged from the leaps. As a matter of fact, I think that modifications of practice in the history of mathematics are usually made for good reason, and hence I offer the formulation of the text. However, because the point has so frequently been misunderstood, it seems worth reemphasizing one conclusion of note 11: the naturalism I have proposed allows for a distinction between discovery and justification, and it does not commit the genetic fallacy.

14. One part of this work is done in chap. 10 of *Nature of Mathematical Knowledge.*

15. For an illuminating account of how Newton's mathematical ideas became available to his contemporaries, see R. S. Westfall's *Never at Rest* (Cambridge: Cambridge University Press, 1980).

16. *Nature of Mathematical Knowledge,* chap. 10.

17. The original notion of function is geometric. The functions of curves are such things as subtangents, subnormals, radii of curvature, and so forth.

18. It is possible that mathematics should rationally proceed by adopting more than one, so that there would be alternative set theories, as there are alternative geometries.

19. Penelope Maddy has argued that a naturalistic account of mathematical knowledge ought to enable us to resolve current disputes in the foundations of set theory (review of *Nature of Mathematical Knowledge,* in *Philosophy of Science* 52 [1985]: 312–14). Here, I think she expects too much. The initial task for naturalistic epistemology, the task undertaken in *Nature of Mathematical Knowledge,* is to integrate mathematical knowledge into the naturalistic framework. A subsequent project is to give a sufficiently detailed account of scientific methodology to allow for the resolution of hard cases. The latter is the vast problem of characterizing scientific rationality.

20. Here I draw on an interpretation of Kant that I have developed in "Kant's Philosophy of Science," in *Self and Nature in Kant's Philosophy,* ed. A. Wood (Ithaca, NY: Cornell University Press, 1984), 185–215; and "Projecting the Order of Nature" (to appear in *Kant's Philosophy of Physical Science,* ed. R. E. Butts). I should note that my interpretation is heretical in cutting away the apriorist strands in Kant's thought.

21. Of course, external transitions may create new branches of mathematics that are then subject to internal interpractice transitions. If we credit popular anecdotes about Pascal and Euler, then the fields of probability theory and topology may have originated in this way.

22. A large number of further distinctions may obviously be drawn there, for we may take very different approaches to the question of how the maximization is to be done. This is especially clear in cases where our ends admit of degrees, so that we may contrast maximizing the expected value with minimizing the risk of failing to obtain a certain value, and so forth. I ignore such niceties for the purposes of present discussion.

23. I think that it can often be shown in cases of this type that consideration of the practical interests of individuals reveals that the *community* optimum is more likely to result if the individuals are motivated by nonepistemic factors. In other words, a *sine qua non* for community epistemic rationality may be the abandonment by some individuals of individual epistemic rationality. However, it is possible that those individuals are overall rational. Something like this has been suggested by Kuhn (see, for example, his "Objectivity, Value Judgment, and the Theory Choice," in *The Essential Tension* [Chicago: University of Chicago Press, 1977], 320–39.

24. For a brief account, see chap. 10 of *Nature of Mathematical Knowledge,* and, for more detail about the Leibnizians, Ivor Grattan-Guinness, *The Development of the Foundations of Analysis from Euler to Riemann* (Cambridge, MA: MIT Press, 1970).

25. An especially clear example is furnished by the discussion among the Leibnizians of the "result" that $1-1 + 1-1 + \ldots = \frac{1}{2}$. See Leibneiz's *Mathematische Schriften,* ed. Gerhardt, 5 vols. (Halle, 1849–63), vol. 5, 382ff., and vol. 4., 388. Euler was extremely dubious about the conclusions favored by Leibniz and Varignon. Nevertheless, his own writings are full of inspired attempts to assign sums to divergent series that such later writers as Abel would find appalling.

26. Important figures in the sequence are Benjamin Robins, Colin Maclaurin, and Simon l'Huilier. The case of Maclaurin offers a clear contrast with the Continental tradition. When *Treatise on Fluxions* is compared with any volume of Euler's works in analysis, one sees two talented (though not *equally* talented mathematicians) proceeding by working on very different problems. Maclaurin turns again and again to the question of finding an explanation of the basic rules of the Newtonian calculus. Euler builds up a wealth of results about integrals, series, maximization problems, and so forth, and is almost perfunctory about the basic algorithms for differentiating and integrating.

27. In *The Analyst* (reprinted in *The Works of George Berkeley,* vol. 4, A. Luce and T. Jessop, eds. [London: Nelson, 1950]). Berkeley's challenge provoked a number of responses, some fairly inept (the essays of James

Jurin, for example), others that helped elucidate some important Newtonian ideas (the work of Maclaurin and, even more, the papers of Benjamin Robins).

28. Plainly, this is simply part of a long and complicated story. The purpose of telling it here is to show that a simplistic historiography is not forced on us by thinking about the rationality of mathematical change. (I am grateful to Lorraine Daston for some penetrating remarks that raised for me the issue of whether my ascriptions of rationality to past mathematicians commit me to Whig history. See her review of *Nature of Mathematical Knowledge*, in *Isis* 75 [1984]: 717–21).

29. See Roger Cooke, *The Mathematics of Sonya Kovaleskaya* (New York: Springer, 1984), for some beautiful examples of the influence of physical problems on late-nineteenth-century analysis.

30. See Penelope Maddy's "Perception and Mathematical Intuition," *Philosophical Review, 89,* (1980): 163–96. Related views have been elaborated by Michael Resnik in his "Mathematics as a Science of Patterns: Ontology," *Nous 15* (1981): 529–50.

31. The geographical analogy stems from Frege; see *Grundlagen*, 108.

32. This is a point that has been emphasized by Karl Popper (see, for example, *The Logic of Scientific Discovery* [London: Hutchinson, 1959], 27–145); in the Popperian tradition, it leads to the notorious problems of constructing measures of verisimilitude (for reviews, see I. Niiniluoto's "Scientific Progress," *Synthese* 45 [1980]: 427–62; and W. Newton-Smith's *The Rationality of Science* [London: Routledge and Kegan Paul, 1981], chap. 2 and 8). I believe that problem can be overcome if we break the spell of the idea that the search for the significant is always the search for the general, but this is a long story for another occasion.

33. Thus, for example, the account offered by Maddy in her "Perception and Mathematical Intuition" seems at best to reveal how we are able to refer and to know about concrete sets. It is not at all clear how this knowledge is supposed to provide us with a basis for reference to and knowledge of abstract objects, where we are no longer in causal interaction with the supposed objects. So even if we grant that our causal relation to an object provides us with a basis for knowledge about the set whose sole member is that object, it is hard to see how we obtain a similar basis when the sets under discussion do not have concrete objects as members.

34. *Nature of Mathematical Knowledge,* chap. 6. This chapter has often been misunderstood, and I have been taken to substitute one kind of abstract object (ideal agents) for another (sets). But, as I took some pains to emphasize, there are no more any ideal agents than there are such things as ideal gases. In both ideal gas theory and in mathematics, we tell stories—stories designed to highlight salient features of a messy reality. I hope that my present stress on storytelling will forestall any further misconceptions on this point.

35. See *Nature of Mathematical Knowledge,* 128–29. It is crucial to appreciate that some forms of human constructive activity consist in achieving representations of objects—as when, paradigmatically, we cluster objects in thought. My claim is that the use of various kinds of mathematical notation—designed to describe the properties of various constructions—makes possible new constructive activity. Thus we have constructive operations that are iterated, sometimes to quite dizzying complexity, through the use of notation. Mathematics does not describe the notation but does provide (idealized) descriptions of the constructive acts that we can carry out with the help of the notation.

36. Mathematicians sometimes toy with the idea that mathematics is art—or is like art. One consequence of my naturalistic epistemology for mathematics is that it enables us to see what this idea might amount to and how it might apply to various parts of mathematics.

37. Since quaternions are now coming back into fashion, it may appear that the example does not support my claims. However, what is now being done with quaternions is quite distinct from what Hamilton did. For Hamilton, quaternions were to be treated in just the ways that real and complex numbers had previously been treated. So, to cite only one example, Hamilton set himself the task of defining the logarithm of a quaternion. So far as I know, that perspective is a long way from the context of present discussion.

38. This is a common response from mathematicians who have read *Nature of Mathematical Knowledge*. As I shall argue below in the text, the complaint seems to me a very important one, and its justice can only be resolved by combining sophisticated understanding of contemporary mathematics with sophisticated understanding of the philosophical and historical issues. Here, I think, collaboration is clearly required.

39. See H. M. Edwards, "The Genesis of Ideal Theory," *Archive for the History of the Exact Sciences* 23 (1980): 321–78.

40. For Dedekind's argument, see his essay *Sur la théorie des nombres entiers algebriques,* in *Gesammelte Mathematische Werke,* vol. 3, ed. E. Noether and O. Ore (Braunschweigh: Vieweg, 1932). The crucial passage occurs on 268–69.

41. In his presentation in Minneapolis, Garrett Birkhoff stressed the intertwining of threads in "the tapestry of mathematics." It seems to me that there are numerous occasions in the history of mathematics in which one area of mathematical practice is modified in response to the state of others, and that concentration on the development of a mathematical field can blind one to the ways in which fields emerge, modify one another, and are fused. For some stimulating attempts to reveal these processes in concrete cases, see Emily Grosholz's papers, "Descartes' Unification of Algebra and Geometry," in *Descartes, Mathematics and Physics,* ed. S. Gaukroger (Hassocks: Harvester Press, 1980); and "The Unification of Logic and Topology," *British Journal for the Philosophy of Science* 36 (1985): 147–57.

42. See R. Thom, *Structural Stability and Morphogenesis,* trans. D. Fowler (New York: Benjamin, 1975); J. F. Traub and H. J. Wozniakowski, "Information and Computation," *Advances in Computers* 23 (1984): 23–92; and M. Feigenbaum, "Universal Behavior in Nonlinear Systems," *Los Alamos Science* (Summer 1980), 3–27.

29

MATHEMATICS—APPLICATION AND APPLICABILITY[1]

MARK STEINER

1. INTRODUCTION

At the beginning of a book titled *Symplectic Techniques in Physics*,[2] the authors (both mathematicians) state: "Not enough has been written about the philosophical problems involved in the application of mathematics, and particularly of group theory, to physics."[3]

While I applaud this statement, and mean this essay to address their concerns, if only partially (hoping to avoid replacing their "not enough" with "too much"), I must point out that the disregard by the philosophical community of issues of mathematical application is quite recent.

To an unappreciated degree, the history of Western philosophy is the history of attempts to understand why mathematics is applicable to Nature, despite apparently good reasons to believe that it should not be. A cursory look at the great books of philosophy bears this out.

Plato's *Republic* invokes the theory of "participation" to explain why, for instance, geometry is applicable to ballistics and the practice of war, despite the Theory of Forms, which places mathematical entities in a different (higher) realm of being than that of empirical Nature. This argument is part of Plato's general claim that theoretical learning, in the end, is more useful than "practical" pursuits.

Descartes' *Meditations* invokes no less than God to explain why the ideas of "true and immutable essences" of mathematics (triangle, circle, etc.) that we grasp with our mind must represent existing entities in nature (meaning empirical space). Thus the applicability of mathematics is co-opted by the mind–body problem. And Spinoza's monism, such as it is, is intended to solve the same problem without invoking the explanatory, or other, power of the Deity.[4]

Berkeley's problem is the same as Plato's, except in reverse: for Berkeley, standard mathematics is an obfuscation, and is even incoherent; examples of incoherent ideas in mathematics are the dimensionless point, the infinitesimal, and numbers (understood as abstract objects). His problem, then, is: How does an inconsistent theory like Newtonian calculus give the right numbers? (Note that Hartry Field's explanation in the twentieth century would not help Berkeley, because it applies only to consistent mathematical theories.) His instrumentalist explanation is ingenious: the infinitesimal calculus gives the right predictions by a sort of canceling out of errors, which reminds me of our contemporary practice of renormalization in quantum field theory.

Kant's *Critique of Pure Reason* returns mathematics to its status of synthetic *a priori* truth without a return to Plato. Kant's transcendental idealism, according to which the mind itself imposes mathematical order on empirical reality, is an answer to the question of how a synthetic *a priori* truth can be applied to empirical reality, an answer which avoids the Theory of Forms, theology, and instrumentalism. It turns out, contrary to Plato, that the *only* synthetic *a priori* truths are about empirical reality.

John Stuart Mill's account of the applicability of mathematics to nature is unique: it is the only one of the major Western philosophies that denies the major premise upon which all the above accounts are based. Mill simply asserts that mathematics itself is empirical, so there is no problem to begin with.[5]

This short sample of Western philosophy illustrates that the central philosophical doctrines of these major philosophers were conceived in great measure to explain the applicability of mathematics to Nature. What is more, we conclude that (though they didn't use the word "apply") all the doctrines presupposed the same *concept* of application: they all assumed that application is a relation (some approximation to a homomorphism) between mathematical theorems and empirical facts, a relation that can be used to "read off" empirical facts from mathematical theorems. The question they ask is: Given the nature of mathematics, why should such a homomorphism exist? And their strategy is either to provide an explanation (participation, God, etc.) or to deny the existence of applicability in the first place—like Berkeley or Mill, but to explain only why it looks as though mathematics were applicable (what my teacher, Professor Sidney Morgenbesser, called "explaining away" the phenomenon).

We must now ask a question which is not often asked: What do we *mean* when we say that mathematics is "applicable"? And before this question, another: What are we doing when we "apply" mathematics to Nature?

2. Canonical and Noncanonical Empirical Applications

The sort of applications the classical philosophers puzzled over could be called "canonical empirical applications" of mathematics, because the available empirical applications of mathematics were either canonical or reducible to canonical explanations. The classical philosophers, like their counterparts today, mostly did not realize that some of the canonical applications of (even elementary) mathematics are not empirical, nor of course could they have predicted the rise of empirical applications that are not canonical. But let's define these terms.

I call an application of a mathematical theory *canonical* if the theory was developed in the first place to describe the application. For example, suppose we have an apparent empirical regularity R of some kind; mathematics M is developed in an attempt to describe this regularity; perhaps we should say that mathematics M is developed so that R should be, at least approximately, its model. Then R is a canonical empirical application of M. Our classical philosophers want to understand this procedure and how it could work—but they *don't* need an explanation why R, *rather than* some other regularity R^*, is an application of M, because M was introduced for this purpose.

An obvious case of canonical empirical application is the use of the differential calculus to describe accelerated motion. This was developed by Newton precisely for this purpose. Any philosophical problem concerning this that we find in the philosophers would be equally a problem concerning any canonical empirical application.[6]

Ironically, it is much harder to see the role of canonical empirical applications in elementary mathematics, because in these cases, typically, the mathematics evolved together with the applications over a long period of time, rather than being invented; and there is thus a tendency to confuse the applications of elementary mathematical theories with the theories themselves.

An example of what I speak is the "application" of addition to finding the size of collections of bodies. Many people, even mathematicians, find it hard to recognize that we are speaking of application of mathematics rather than mathematics itself, but let me try to get things clear.

Consider a set S of bodies. Consider the scattered physical object S^* which is the "mereological sum" of the elements of S. (The mereological sum of a set of objects is the smallest body of which each of the objects is a part.) A mereological sum is not a set and does not have members, but only parts; a set has both members and subsets.[7] Another difference is that the sum of a sum of bodies is no different from the sum itself; yet the set containing a set of objects is different from the latter.

Suppose, however, we think of bodies as maximal irregular polyhedrons (i.e., polyhedrons not part of larger polyhedrons). Then these bodies play the role, to some extent, of set members—since the parts of these bodies are not themselves bodies. There is, therefore, a sense in which a mereological sum of bodies is a model or image of a set of those same bodies. And since this is true because of the empirical properties of those bodies (stability, for example, as well as discreteness), we can say that mereological sums of bodies are empirical applications of the set concept. Let us call the mereological sum of bodies a "collection" of those bodies, so that collections are empirical applications of sets.[8]

But this works in the opposite direction as well: we can read some of the properties of the sets from the collections for reasons which are both mathematical and empirical.

Consider "counting." This is basically an empirical process in which we point to all the bodies in a collection while reciting numerals in order. The last *numeral* recited expresses the *number* or *size* of the set of those same bodies. This is true because, as we say, the collection is a physical model of the set. But it is also true for a mathematical reason: the finite numbers are both ordinal and cardinal numbers. (For infinite sets, it matters a great deal in what order we place them—the ordinal number of the natural numbers in the standard ordering is called ω, but if we "count" the numbers starting from 1 and put 0 after the rest, we get the ordinal $\omega + 1$.) This means, for example, that it matters not in what order we count the bodies in a collection—we will get the same result. The result, invariant over order, is thus the size of the set of those bodies. Once again reversing the story, if we know the cardinal number of a set of bodies, we can predict the result of counting the collection of the bodies, an empirical prediction that has a mathematical explanation.

Next consider addition. Thought of as an operation on cardinal numbers m and n, which are the sizes of disjoint sets X and Y, $m + n$ is the size of the union of X and Y. Suppose X and Y are sets of bodies, as before, and X^* and Y^* the corresponding collections of those same bodies. Then the mereological sum of X^* and Y^* is an image or model of the union of X with Y. If we know the sum $m + n$ in canonical notation, we can then predict the result of counting the collection of X^* with Y^*. This is, of course, because of the above-mentioned identity between cardinal and ordinal finite numbers.

But the cardinal–ordinal equivalence has another implication, and a profound one. This is that cardinal addition is equivalent to ordinal addition; extensionally, they are the same operation. But ordinal addition is based on recursion (iteration), and recursion yields computational algorithms, such as the ones we learn in school. Many educators today decry the emphasis placed on these algorithms, and they have a point: these algorithms are for ordinal arithmetic, which has little to do with the more applicable cardinal arithmetic, which measures the sizes of various sets. On the other hand, the mathematicians have a point as well, since ordinal arithmetic can immediately be applied to cardinal arithmetic to make calculation a cinch. The mathematicians suffer from the lack of algorithms in infinite cardinal arithmetic, and as a result cannot make the simplest calculations, such as $2^{\aleph_0}$.

Cardinal arithmetic, then, is of great application but does not allow calculation; ordinal arithmetic has no great application, but does allow algorithmic calculation. Together we have the following story: take two distinct collections—of bodies—and count them both. We then have the ordinal number, and thus the cardinal number, of the two sets corresponding to the collections. The cardinal number of the (disjoint) union of the sets corresponds to the sum of the two cardinal numbers. This can be calculated using ordinal arithmetic (i.e., algorithms). From this we can predict the result of counting the (mereological) sum of the two collections, the smallest collection of which the two original collections are parts.

It is a considerable intellectual strain, we see, to separate pure arithmetic from its canonical empirical applications. Small wonder that the distinction is seldom made.

To sum this all up (pardon the pun), the applicability of addition algorithms to counting physical collections is based on an amazingly complicated series of facts and mathematical "accidents": (a) the empirical stability of many objects; (b) the resulting model of sets of physical objects as collections of those objects; (c) the equivalence of cardinal and ordinal finite addition, which allows (1) finding the cardinal number of a set by counting the corresponding collection and (2) the use of powerful algorithms of ordinal arithmetic to solve problems in cardinal arithmetic.

A noncanonical, yet still empirical, application of addition would be to weights. Seven unit weights and another five give twelve unit weights. This is an empirical application, so empirical that it is even (slightly) wrong—according to Einstein's General Theory of Relativity, weight is not quite linear, just as velocity is not quite linear in Special Relativity. However, for ordinary purposes we can speak of the "linearity" of weight. In that case, we can use arithmetic to predict the weight of a collection of unit weights. (Of course, for a fuller story one would discuss the application of the theory of real numbers to weights that are incommensurable.)

This is not, strictly, a canonical application, insofar as the theory of arithmetic operations was not put together to do weights, but to count. However, weight is "near" canonical, in that (given the empirical properties of the magnitude "weight") we can reduce weighing to counting. (I again stress that for simplicity I am discussing only commensurable weights.)

A more advanced empirical, yet noncanonical, application is the application of the Apollonian theory of conic sections by Kepler to planetary orbits. Obviously the Greeks had not introduced this theory to describe the orbits of the planets, which in any case were thought of as strictly circular. Indeed, as far as I know, Kepler was the first to apply the theory in physics.

An even more advanced case is the application of non-Euclidean geometry to gravity, as in Einstein's General Theory of Relativity. What is noncanonical here is not necessarily the description of space as "curved" (it is plausible that both Gauss and Riemann had this in mind), but rather the application of Riemannian geometry to space-time.

This raises the question of whether all mathematics is applicable in physics. Of any given mathematical theory, it is quite risky to say that it has no physical application.[9] Nevertheless, we can say that the beautiful theory of prime numbers, which so caught the imagination of the Pythagoreans, has to this day no certified applications in physics. Of course, arithmetic itself is applicable in physics, and arithmetic does treat prime numbers, but this is not the sense of "applicability" I have in mind, which demands that the concept "prime number" should actually appear in a physical theory. Suppose that the nth energy level of hydrogen had a certain property if and only if n is a prime number—that would be an example, but there aren't any like that.[10] If string theory is ever confirmed empirically, the role of the prime numbers in science will finally be established, because string theory actually does rest on some characteristic concepts of the theory of numbers.

The converse issue, raised by the celebrated physicist Eugene Wigner in a famous paper,[11] is whether the level and type of applicability we do see in these noncanonical applications is "unreasonable." Wigner argued that it is, because:

(1) Mathematical concepts—at least in the last 150 years—are subject primarily to criteria internal to the mathematical community. Among these criteria are aesthetic ones; and most of the criteria are not such as to make it likely that mathematical concepts should be applicable at all. In some cases, the mathematical concepts have, and had, no *canonical* empirical applications.

(2) In physics, the reliance on mathematical concepts in formulating laws of nature has led to laws of unbelievable accuracy.

Skeptics argue that Wigner suppressed the failures in applying mathematics (there are hundreds monthly). Other skeptics argue the reverse, that we should expect mathematics to be applicable because (unlike what Wigner asserts) the real and ultimate source of mathematical concepts is experience; thus, it is not at all unreasonable that mathematical concepts should return the favor.

I'm not particularly impressed by these objections (particularly because they contradict one another), since I think that each of the examples that Wigner gives is so extraordinary that it requires explanation.[12] For example, I feel that even Newton's law of gravitation, based on astronomical observations which were wildly inaccurate by today's standards, nevertheless has withstood the increase in the accuracy of our measurements to the extent that we can now say that the elementary particle known as the neutrino falls to Earth just like the Moon. There is nothing like the accuracy of the inverse square law, for example, in economic models, which retain roughly the experimental error of the data that suggested them in the first place. Kant once argued that the inverse square law is to be expected (if not *a priori*), because gravity spreads out spherically, and the area of the (surface of) the sphere is proportional to the square of its radius. But Charles Peirce already answered this by the telling rejoinder that what spreads out in space is the potential, not the force, and the gravitational potential goes as the inverse, not the inverse square, of the radius! Thus, it's not clear that we have anything that would count as an explanation for how much more we "got out" of the inverse square law than we "put in."

My objection to Wigner's thesis is completely different: it's not a thesis at all. He gives persuasive examples of successes that cry out for explanation—but he doesn't prove that they add up to one phenomenon that cries out for explanation. Each success is a story in itself, which may or may not have an explanation. Wigner does not make a case that what is unreasonably effective is *mathematics,* even though the individual examples he gives are of concepts that happen to be mathematical. In other words, Wigner may give examples of a number of applications that are "unreasonably effective"—applications of concepts that happen to be mathematical. But he doesn't show that these successes have anything to do with the fact that the concepts are mathematical. Of course, this is connected with Wigner's failure or inability to give a definition of "mathematical concept."

3. Canonical Nonempirical Applications

I would like now to speak, finally, of canonical applications of mathematics that are not empirical at all—these are applications in which one uses mathematics to "read off" results in another theory that itself is mathematical. One might think that these kinds of applications would have to be quite advanced, yet I was surprised to find that there is a very elementary example: multiplication.

Multiplication is often "defined" as repeated addition, but this, I hold, is a confusion between a definition and an application.

An intuitive way to see that something has gone wrong with the pseudo-definition "multiplication is repeated addition" is to inquire concerning the commutativity of multiplication. Why is it always the case that adding x, y times, gives the same result as adding y, x times? To be sure, this fact can be proved by mathematical induction. But the proof is far from trivial, and my experience teaching this material from textbooks like Mendelson's *Introduction to Mathemetical Logic* is that most college students are not capable of discovering the proof.[13] Morever, proofs by mathematical induction usually prove only *that,* but not *why,* a theorem is true.[14]

The "definition" of multiplication as repeated addition has its source in ordinal arithmetic (i.e., recursion). We have already seen that ordinal arithmetic is dandy as algorithm, horrible for applications. Small wonder that children cannot understand why multiplication is commutative (the noted philosopher and logician Saul Kripke recalled—in a class I attended—being amazed as a child at the commutativity of multiplication). Nor can they understand why one can "add candies to candies" but not "multiply candies by candies." What we need is a *cardinal* definition of multiplication.

Now let's look at the matter from the logician's point of view. To see that "repeated addition" cannot be a definition of multiplication, consider the formula xy, having two free variables. The "repeated addition" definition (so called) would dictate that xy means

$$y + y + y \ldots + y \ (x \text{ times}).$$

But the ellipsis here (. . .) is not defined mathematically; and the parenthetical comment (x times) means only that the letter y is repeated an unspecified number x of times. The most coherent interpretation of this "definition," then, is a schema containing infinitely many definitions of the following type:

$$1 \times y = y$$
$$2 \times y = y + y$$
$$3 \times y = y + y + y$$
$$\ldots \ldots \ldots \ldots \ldots$$

Mathematics—Application and Applicability

The numbers 1, 2, 3, . . . count the number of occurrences of the letter "y"; to put it another way, x is a metalinguistic, rather than a mathematical, variable, where y is a "true" variable.[15] To see what is wrong with this, take a specific example: $7 \times 5 = 35$, which, according to the standard pseudo-definition is to *mean* $5 + 5 + 5 + 5 + 5 + 5 + 5 = 35$. But standard rules of logic allow the inference

(E)   There is an x such that $x \times 5 = 35$

from $7 \times 5 = 35$, but not from $5 + 5 + 5 + 5 + 5 + 5 + 5 = 35$. Hence, the definition of multiplication in terms of repeated addition is wrong. The most we can say is that $5 + 5 + 5 + 5 + 5 + 5 + 5 = 35$—if we assume the axioms of Peano (which of course contain multiplication as a primitive, undefined notion)—is equivalent to $7 \times 5 = 35$ (and therefore implies E), but the equivalence is not purely logical.

A good way to look at multiplication is, in fact, set-theoretic: xy is the (cardinal) number of x disjoint sets, each of which has the cardinal number y. The role of x here is different from that of y (though it is *not* metamathematical); suppose we are calculating the number of candies we need in order to give four children five candies each—this is the same as calculating the number of elements in four disjoint sets with five candies each. Five, then, tells us the number of candies, where four tells us the number of sets. In other words, the "second-order" concept of "set of sets" is inherent in multiplication. We have already seen that though collections, in a sense, can be considered physical models for sets, there is no such model for "sets of sets." This is the deeper reason why we can't "multiply candies by candies," though we can "add candies to candies": adding candies to candies is to model addition by mereological sum, forming a larger collection from two smaller ones; for multiplication there is no such model.

At the same time, we see from this definition that each multiplication is equivalent to (though not synonymous with) a repeated addition—we can "read off" the results of a repeated addition from a product of two numbers. In this sense, it could be said that multiplication is being "applied" to addition! But this is not an empirical application at all!

Consider, now, the following table:

	Candy No. 1	Candy No. 2	Candy No. 3
Child No. 1	A	B	C
Child No. 2	D	E	F
Child No. 3	G	H	I
Child No. 4	J	K	L

Table 1

We have labeled the candies with capital letters. If we have four children and have to give each child three candies, we need 4×3 candies, the cardinal number of a disjoint union of 4 disjoint sets with 3 members each. The rows of the table, marked out as above, symbolize the sets of candies. If we permute "candy" with "child" everywhere and switch rows for columns, we get:

	Child No. 1	Child No. 2	Child No. 3
Candy No. 1	A	E	I
Candy No. 2	B	F	J
Candy No. 3	C	G	K
Candy No. 4	D	H	L

Table 2

It is clear from this that multiplication is commutative; from the abstract mathematical point of view, what we have done here is set up a one–one map between

$$\bigcup_{i=1}^{|A|} B_i ,$$

all the Bs equinumerous, and $A \times B$, the set of ordered pairs $\{<x, y>: x \in A \,\&\, y \in B\}$, B being any set equinumerous to the Bs, and then noted that, obviously,

$$|A \times B| = |B \times A|.$$

In effect, we have two set-theoretic definitions of multiplication here that are equivalent: in terms of disjoint unions, or in terms of Cartesian products. The former definition is the one that leads to applications—but distinguishes between the operands m and n in that the latter numbers sets; the former numbers sets of sets. In the latter definition, both m and n number sets.

A final comment: I maintained that "repeated addition" should be regarded as a *nonempirical*, if canonical, application of multiplication; what of *areas*, as in geometry? Is this not a canonical *empirical* application? Didn't the Egyptians and the Greeks develop this application of multiplication?

Yes and no. Though it is true that area is certainly an application of multiplication, it originally was not an application of the operation we teach today in school. The Greeks regarded multiplication as an operation on magnitudes, not numbers; when we multiply linear magnitudes, we get square magnitudes. In other words, multiplication of magnitudes was not a closed operation.

Furthermore, the use of multiplication in areas is connected with the Euclidean structure of the plane. If space is not Euclidean, then the area of a square is not actually the product of the lengths of two sides. We will still need multiplication to find the area, which is still definable (as an integral) as long as space is still locally Euclidean (the smaller the space we take, the "flatter" it gets). However, I don't think that we have here anything like a canonical application any more.

Let us leave elementary arithmetic, and look at some of the modern theories of mathematics. Indeed, let us heed the call of the mathematicians we began with, and discuss group theory.

Mathematics—Application and Applicability

Group theory is known today as the theory of symmetries—we define a symmetry as a property that is invariant under a group of transformations. The classical symmetries, which are visual symmetries, involve transformations of space, such as rotations, translations, and reflections. For example, a cube is invariant under rotations of multiples of ninety degrees around axes that go through, and are perpendicular to, its faces. And one might have thought that group theory developed in order to describe these symmetries, for example, in crystallography—and it is true that group theory can indeed be applied to "read off" results in that field. Once we know, for example, the symmetry group of a substance like salt, we can predict all the possible forms of the salt crystal, even the forms that occur "naturally" and are not regular polyhedra. We can even use group theory to explain why so few visual symmetries are seen in nature—crystals are made up of a lattice structure (atoms), which restricts the number of symmetry groups possible.[16]

Yet this is not how group theory came into existence. In fact, group theory was constructed to be applied not to empirical reality, but internally—in mathematics. Namely, group theory was introduced into mathematics by Galois, to be applied to algebra. The Galois group of an equation, for example, can be thought of as a group of certain permutations of the roots of that equation.[17]

By studying the group of an equation, mathematicians are able to extract information about the equation itself—in fact, the application of group theory in mathematics has precisely this form: we find that a certain group characterizes a much more complicated structure, and we can get needed information about that structure by studying the properties of the group. (This is how group theory is used in topology, for example.) Thus, what Galois did was to inaugurate modern mathematics, in which the main applications of mathematics are to mathematics itself. It would be fair, then, if surprising, to state that the canonical applications of group theory are mathematical, not empirical. Ironically, the origins of group theory in pure mathematics not only did not prevent physicists from applying group theory (after a period of resistance), but made it possible.

Briefly, the reason is this.[18] Already in classical mechanics it was discovered (by the mathematician Emmy Noether) that there is a mathematical connection between symmetry groups and *conservation laws*. For each continuous group G of symmetries[19] of the laws of nature, there is a magnitude P_G which is conserved. If the law has rotational symmetry, then angular momentum is conserved, for example.

In quantum mechanics, the state of a system is represented by a unit vector in a (complex) vector space V of infinite dimension. The group G must act on this vector space V, rather than directly on physical space.[20] This action is called a "representation" of the group, in which each element of the group is represented by a square matrix.[21] Suppose that a particle has a determinate value v for property P_G, which is conserved under the action of G. Then,[22] as G acts, (a) the unit vector never moves out of a subspace W of V that is associated with that very value v and (b) there is no subspace Y of W that so confines the unit vector. W, taken together with the action of G, is called an "irreducible representation" of G. Suppose W is an n dimensional subspace of V. The physical meaning of this is that a particle with value v of property P_G can assume n different states. It should not be assumed without proof that G has such an irreducible representation for every dimension n.

Consider, as an example, the electron and let P_G be the property known as "spin," so called because the electron is a little magnet analogous to a spinning charged ball in classical physics. As one might expect, rotating an electron (or, equivalently, rotating the measuring device relative to the electron) does not change the value of its spin (which remains at ½ Planck's constant). Yet rotating it 360 degrees does not bring the electron back to its original *state* (the unit vector is multiplied by −1); this requires two full turns. What is most significant is that no matter how we rotate the electron (or the coordinates), the electron is observed in only one of two states: (a) with its north pole "up" or (b) with its north pole "down." (This is of course utterly unlike a classical spinning ball, which can be observed spinning in any direction.) Rotating the electron changes only the *probability* of observing the electron in an up or down position. Thus, it has two states corresponding to its spin value.

This might suggest the following group-theoretical description of the electron (in fact, this picture was actually resisted by physicists for quite a while): the symmetry group of electronic spin is the one called SU(2). This is a continuous group that has a two-dimensional representation,[23] just as required, and is also the "double covering group" of the rotation group in ordinary space (i.e., a continuous action of this group, which brings a unit vector back to itself, is homomorphic to two rotations, again as required).

So far the group-theoretical description does not seem to add anything to the physical facts, which is why physicists referred to Hermann Weyl and his colleagues in the mathematical community as the "Gruppenpest." Yet, as Sternberg points out, one can reverse this procedure, and go from the group description to the physical facts. For example, in 1932 Heisenberg argued that the neutron and proton are actually two different states of the same particle (today called the "nucleon"), with the value ½ of some property yet to be explored, and with the same symmetry group as the electronic spin. He called the property, again by analogy, "isotopic spin," though this is nothing more than saying "the property conserved in the nucleus by the spin group SU(2)." Actually, even today, there is no known physical connection between the spin of the electron and the isotopic spin (known today as isospin) of the nucleus—nature merely utilizes the same symmetry in both places.

Having gone from physics to group theory, we now go back to physics: SU(2) also has a three-dimensional irreducible representation, which means that it is possible that there is a particle which can take on three different states (we can think of these particles as being "made out of," respectively, a neutron-neutron, a proton-proton, and a neutron-proton). The different states of the particle will look like three different particles in the laboratory, just as the nucleon looks like two different particles. Sure enough, the three *pions* were found: particles analogous to the neutron and proton in that they are *hadrons*—they exert the powerful nuclear force. The three "directions" of the pions are marked out by their charge: positive, neutral, and negative. By now, the group-theoretical method began impressing physicists. These predictions were being made with no knowledge of the properties of hadrons, except for their symmetry groups.

The next step was even more startling. The neutron and proton themselves turned out to be part of a family of eight hadrons with comparable rest mass. Aside from them, there were nine more hadrons with comparable mass (though by no means the same), greater than that of the first family. This suggested to Yuval Ne'eman and Murray Gell-Mann (independently) that there was a higher symmetry than SU(2). Mathematical analogies suggested a group known as SU(3), of which SU(2) is a subgroup, which has representations of dimensions 3, 8, 10, . . . And, indeed, experimental evidence suggested that the neutron and proton belonged to a family of eight similar hadrons. Furthermore, there was a family of nine heavier hadrons. But if SU(3) were the right symmetry group of the hadrons, the family of nine particles would require a tenth. Hence, both Gell-Mann and Ne'eman predicted the tenth, calling it in advance the omega minus particle, which was later discovered. Gell-Mann then predicted the existence of the triplet from the three-dimension representation, *none* of which had been observed—and for good reason: mathematical considerations indicated that such a triplet, called by Gell-Mann "quarks" after Joyce, would have to have fractional charge. At the same time, the quark hypothesis (if one could come up with an excuse why nobody had ever detected these ubiquitous particles) could serve as a partial explanation of the SU(3) symmetry: one could use quarks to "construct" all the families of the hadrons (as we "constructed" the pions out of nucleons). Note, however, that this explanation was ex post facto, like most explanations: the success of SU(3) symmetry led to quarks, not the other way around.

One could push back the use of group-theoretic methods even earlier than I have done, by including in this story the saga of "identical particles." This is, in fact, what Steven French does, in an illuminating attempt to account for the success of the group-theoretical method in physics.[24]

The principle of identical particles is that there is no physical difference between a state of two identical particles A and B (e.g., two photons in which A is in state X and B in state Y, or one in which A is in state Y and B in state X). The most straightforward application of this principle is the situation where the state vector of the system of two particles simply does not change on permuting A with B. Let us also consider the case in which A and B must be either in state X or state Y. Then there are three cases: both particles are in state X, both in state Y, or they are in different states (not four cases where we distinguish A in state X, B in state Y, from the converse case with A in state Y, B in state X). This means that the probability of finding the particles in the same state is 2/3, rather than 1/2, which we might have thought. Another application of this principle is that permuting A with B brings the state vector to minus itself, since there is no observable difference in quantum mechanics between the vector ψ and the vector $-\psi$. These two group actions lead to an important classification of particles: those that change sign are called "fermions," and those that do not are called "bosons."

Now there is no question that we can see this is as an application of group theory—in hindsight, though, the historical players would mostly have rejected this characterization. They did not see the category of "groups" as a natural way to categorize phenomena; hence the "success" of applying one group could not, for them, suggest the application of a completely different one (such as SU(2)).[25] The fact that we today see the permutation group and SU(2) as examples of the "same thing" and as relevant to physics does not mean that previous generations did. Our inclination to see the quantum mechanical treatment of "identical particles" as a triumph of group theory betrays our implicit belief that mathematical language is the deepest language of physics and that mathematical classification of structures is the ultimate physical classification, too. But this belief amounted to nothing less than a revolution in thinking about nature, and prominent in this revolution was Eugene Wigner, one of the greatest pioneers of the group-theoretical approach to physics. For him, it could be said, a physical object is not much more than the sum of all its symmetries, including the symmetries of the geometry of space-time.[26]

We have, then, a new kind of application of mathematics. In the past, mathematics was used to get quantitative descriptions of phenomena that could also be described qualitatively. Today mathematics gives even the qualitative descriptions, because we often have no deeper language than mathematics. Related to this is the use of mathematics to make discoveries and "predictions." Our story here has shown that pure mathematics was the basis of analogies for which there was (at least at the time) no underlying physical basis. Scientists postulated symmetries without knowing what they were symmetries *of.* "Predictions" were made in a new way: in the past, mathematical calculations were made to show what necessarily had to be the case.

In our story here, mathematics was used to show what is possible, the assumption being that what is mathematically possible is physically actual.

All of this suggests (and Ne'eman has recently written this explicitly) that modern physics has a Pythagorean streak to it. So far, at least the quality of *mass* has not been shown to be amenable to a group-theoretical treatment. If that one day happens, however, the world itself could turn out to be nothing but a mathematical structure: a reducible representation of the ultimate symmetry group,[27] a bizarre possibility we will nevertheless discuss in the concluding section of this essay.

4. Logical Applications

Another nineteenth-century development that influences our subject is the development of modern logic and the foundations of mathematics. An attempt was made to put mathematical deductions on a rigorous footing and, at the same time, to find the foundations of mathematics. This research program presupposed, or perhaps introduced, concepts of application of mathematics that could not have been formulated before. These are the logical concepts of application that we can now survey.

For example, one sense in which mathematics is applied is as an engine of deduction. Even if mathematics is not literally logic, as Frege and many others once thought, one of its functions is certainly akin to that of logic: as what some called a "juice extractor" in which we use mathematics to derive consequences of non-mathematical premises. Of course, any set of sentences can be used to derive consequences from others, but mathematical sentences seem "topic neutral" in a way other sentences are not.

But for this purpose, attention must be paid to the logical form both of pure mathematics and of "mixed" contexts, in which the same sentence contains both mathematical and non-mathematical references. An elementary example is that of the use of arithmetic to shorten or eliminate counting, as when we conclude that there are twelve fruits on the table after observing that there are seven apples and five pears, and no other fruits on the table, from the pure mathematical proposition $7 + 5 = 12$.

The question arises: What does the pure mathematical proposition, which on its face is about "natural numbers," have to do with fruits? This question has two aspects, the logical and the metaphysical. The metaphysical question arises from the ontological gap, noted already by Plato, between mathematical and empirical objects. How can truths about mathematical objects be in any way relevant to the empirical world?

The logical question arises from the difference between the numerals "seven" and "five," used as adjectives, and the numerals 7 and 5, used as nouns. This ambiguity produces an equivocation that threatens to spoil the logical validity of our deduction. Obviously the two questions are related.

A strategy that suggests itself to many observers, past and present, is to argue that there is no such thing as pure mathematics, or at least that there are two kinds of mathematics, pure and applied. Applied mathematics (whether or not it is the only mathematics) is the empirical theory of certain properties, such as "seven" and "five." There are no mathematical entities, but only mathematical properties, and these are empirical properties like any other. The seven apples are a collection with the property "seven," for example. The arithmetical statement $7 + 5 = 12$, then, is a general empirical law about any such collections. This is the position, in fact, of the great nineteenth-century British philosopher J. S. Mill in his *A System of Logic*.

However, there are so many problems with this view that, strikingly, even the empiricists have by and large rejected it—with some notable exceptions. Most of the objections have been epistemological—the empiricist account of arithmetic does not seem to account for the various peculiarities of mathematical knowledge. I would like, however, to draw attention to a logical objection: What is the status of a "collection" of which a number is a property? If a collection is a physical object (as is probably intended) rather than, say, a set, then there is no one number that characterizes it: A heap of socks could be characterized by the number of socks, the number of pairs of socks, or the number of molecules in the heap, or perhaps the number of atoms, and so on. Thus, numbers cannot be regarded as properties of empirical bodies like collections. This objection was raised by Frege in his *Grundlagen*, which contains a scathing critique of "pebble and gingerbread arithmetic," referring to Mill. Frege does not seem to be aware, however, that basically the same objection was raised by Plato in the *Theaetetus*, against the more general view that "perception is knowledge."[28]

Instead, Frege, like Plato before him, took the opposite tack: he eliminated the adjectival numerals, and with them the numerical properties entirely, leaving only the mathematical entities. Plato's solution, however, is not entirely coherent (as he himself pointed out in various places): he thought of the empirical world as *participating* in the mathematical world. Since, however, as he himself had argued, the empirical world can participate in the mathematical world in conflicting ways, he ended up arguing that the empirical world itself is subject to logical conflict (and thus is not the object of knowledge). Frege, though he gives no credit to Plato, in fact made two revisions to Plato's account: first, he replaced "participation" with something like "class membership." Second, he eliminated the logical incoherence by denying outright that arithmetic relates directly to the world of empirical objects. Rather, it is the empirical (and nonempirical) *concepts* which are members of the different numbers. Thus, the number two is the class of all *concepts* that are true of two objects.[29] The metaphysical problem (How can facts about the world of abstract entities be relevant to the empirical world?) is also dissolved: in Frege's scheme, numbers are related (directly) only to concepts.

Frege's account of (deductive) applicability comes with a price, however. It is heavily committed to the existence of non-material objects (and, of course, concepts), and thus heir to all the traditional attacks on Platonism. I cannot discuss those attacks here, except to point out that some philosophers have been influenced enough by these attacks to attempt to gain the benefits of Frege's account of logical applicability without paying the price. One of these strategies is "fictionalism," understood as the doctrine that the Platonist commitments of Frege's account are real, in the same way that Shakespeare's *The Merchant of Venice* is committed to Shylock as well as to Venice.

The version of fictionalism set forth by Field is actually a research project:[30] given a theory *MT* of mathematical physics, to find a "nominalist" theory *T* that has the same "nominalist" consequences as *MT*. Field claims to have done just that for the classical equation of the gravitational field. Success in such a replacement program shows that mathematics is theoretically not necessary and thus can be regarded as pure fiction. Mathematics, Field argues, has only an instrumental value in shortening proofs. Hence it need not be regarded as true, but only as useful.

Field's book occasioned a great deal of comment and criticism, much of it centering around his major example of classical gravity: whether his replacement for it was truly nominalist (he helps himself to the entire continuum of space-time points, whereas Aristotle had rejected even one space-time point as too Platonist for his liking); whether he had truly demonstrated that it was a replacement; whether he had demonstrated that mathematics is really not deductively necessary in physics;[31] whether he could reproduce his success in other areas of physics, particularly quantum mechanics, where, as we have already seen, the mathematical formalism is central in a way that it is not in classical mechanics. Many of these criticisms seem to me cogent; in fact, I made some of them myself,[32] although in Hebrew.[33]

I would like to point out something else in this connection that seems to have been missed: we have seen already that mathematics is not only a medium of proof; it is an engine of discovery. We have seen, that is, how the mathematical form of a theory can serve as a springboard for its development or even, paradoxically, its replacement. The great scientists relied on mathematical analogies to suggest replacements or at least generalizations of existing theories. The idea was to replace false theories by (hopefully) true ones, in which what was preserved was mathematical structure.

Consider the following example. Measuring devices, no matter how accurate, give results in rational numbers only. On the other hand, the solutions of differential equations are functions on the continuum. One could presumably replace current theories, involving the continuum, with physical theories invoking only rational numbers—with no observable difference. Differential equations, for example, can be replaced by "difference equations," which are used as algorithms to calculate the solutions of differential equations. In so doing, however, we would destroy the mathematical form of the latter; and it is the mathematical form that was used in finding generalizations or corrections of present equations. Physicists insist on the form of equations even when their content is obscure. For example, Richard Feynman introduced mathematical notation for calculations in quantum electrodynamics that he himself suspected of inconsistency and that, in any case, lacks a consistent mathematical interpretation even today. The inconsistency of this notation, however, has not prevented scientists from using it to calculate the magnetic moment of the electron correctly to twelve decimal places!

Why is this relevant to Field's program? Field regards Platonism as refuted if, for each "Platonist" physical theory A with nominalist consequences N, we can replace it with a "nominalist" theory B with the same nominalist consequences N. Since, then, B "can do everything that A can," Ockham's razor suggests that we are committed only to what B is committed to. This being a pragmatist concept of ontology, one could argue that epistemic values should be factored into this equation: there is an epistemic sense in which B *cannot* do everything that A can.[34]

5. Speculative Concluding Remarks

What philosophical problems are presented by the applicability of mathematics? Since we have seen there is more than one concept of applicability, there is more than one problem. In discussing them, I will begin with the issues raised by the *logical* application of mathematics.

We saw that Frege showed how to use pure arithmetic alone in making deductions about the number of physical bodies. He did this by standardizing the expressions of the form "There are n Fs" as "The number of Fs = n," an equation. The variable F, for Frege, takes concepts as values. Hence the number n is directly associated only with a concept when saying "The number of Fs is n." The number n is not related to the Fs themselves, thus eliminating the traditional puzzlement concerning the relevance of mathematical objects to the empirical world. In fact, they are not relevant, and need not be.

However, Frege's solution raises the specter of Platonism, since his reduction of applied to pure mathematics involves the commitment to mathematical objects at the most elementary level. One can, of course, say that Frege's accomplishment was to *reduce* the (logical) problem of applicability (of arithmetic, since his success with analysis is questionable) to that of Platonism. Of course, Platonists don't see any problem with Platonism; they feel entitled to say that Frege has solved the (logic) problem of applicability. What of the nominalists?

Earlier, we saw how Hartry Field attempts to have his cake and eat it, too: he accepts all the details of the Fregean solution, but whisks away the ontological commitment involved with the claim that one "could," in principle, dispense with numbers (i.e., "number-talk") entirely. Fregean or "Platonist" mathematics can be regarded instrumentally. Field thus claims that he can get all the benefits of Frege's program without paying any ontological price.

Although I have given some of my reasons for being skeptical about Field's "free lunch" solution, I would like to point out an underappreciated virtue of Field's "piecemeal" approach to eliminating Platonism. That Field aims to replace each physical theory individually with a nominalist counterpart is regarded as a serious flaw to his argument by writers such as Dummett,[35] who complain that what he really should do is replace all the theories simultaneously, or are suspicious that Field won't be able to carry through the program in other, more intractable cases (like quantum mechanics). Yet the very piecemeal character of the project makes it ideal for solving a completely different problem of applicability: Wigner's problem of the noncanonical empirical applications of mathematics. This is not a logical problem but a descriptive problem, involving the use of mathematics in describing nature—rather than in making deductions.

For, as I suggested above,[36] Wigner's problem is not one problem: every unreasonable success in using mathematics to describe any physical phenomenon is a separate problem. Field's reduction of classical gravitational field theory to a theory of spatial regions may not be a nominalist reduction, but it does give an account of the usefulness of analysis in physics without using analysis. The bugaboo of Platonism—and the inability of most philosophers, including perhaps Field himself—is to see that there is more than one problem of the applicability of mathematics. Wigner's problem is not Frege's. Wigner's problem, for each application of a mathematical theory M, is solved by articulating, without using M, the conditions under which M applies.

Wigner's problem—or really problems, as I believe—is to give an account of various noncanonical empirical applications. Wigner says this is difficult to do on account of the gap between the goals of mathematics and the goals of physics, so that it is unreasonable to expect a mathematical concept to describe an empirical phenomenon with as much precision as we find in physics, typically. Field's method gives an example of a solution.

We have seen, moreover, that there are also philosophical problems connected with *canonical* empirical applications. Here, of course, Wigner's problem does not arise—by definition. But the opposite problem arises—the close relationship between the mathematics and its application makes it often quite difficult to distinguish between them. We saw an example of this in the application of addition to predict the result of empirical counting. Inability to distinguish between mathematics and its *canonical* applications leads to positions like that of J. S. Mill, according to which mathematics is a crudely empirical science, a position that most philosophers regard as severely flawed.[37]

Finally, we discussed the application of mathematics within mathematics itself. For example, we use one kind of mathematical structure in order to characterize another. A striking illustration of this is the concept of group representations, in which the elements of a single group G are homomorphically "represented" by square matrices (i.e., linear transformations of a vector space). The existence of irreducible representations of various dimensions characterizes G by its possible actions.

The reason that these internal applications are so important to our subject is the existence of mathematical formalisms in physics. These are mathematical structures that, though they do not directly describe physical systems, nevertheless contain an enormous amount of information about those systems, which can be extracted from the formalism by rules whose major justification is that they work (or at least, they worked). Of course, the mathematical structure of quantum mechanics is such a formalism. In this formalism, a particle "is" (Pythagoreans will remove the scare quotes) not much more than an irreducible representation of its most comprehensive symmetry group. In that case, the theory of group representations, and the categories it works with, turn into the fundamental classification of reality, deeper than the fire/air/earth/water of the ancients, and even deeper than the periodic table of the elements. And this classification, in turn, makes it possible to draw analogies even from false theories in the process of guessing the true ones, using the mathematical hierarchies as (what Nelson Goodman called) "projectible predicates," or what were called, in the Middle Ages, natural kinds.

My claim, then, is that a good deal of what passes for applications of mathematics in physics is really the application of mathematics to itself.[38] The end result remains empirical—make no mistake—because there are rules connecting the mathematical formalism to empirical predictions that could, but don't, fail. Nevertheless, the procedure is deeply Pythagorean, because the classification of empirical phenomena is induced by the classification of mathematical structures by mathematicians.

There is a great vindication here of Galileo's vision of the "Book of Nature," written in the language of circles and triangles, except of course that the circles and the triangles themselves are replaced by their symmetry groups. The vision is still empiricist, as long as we distinguish between the Book and Nature itself. It is Nature which gives validity to the Book, but we use the Book as a map of Nature.

Yet the boundary between the Book and Nature itself has recently been becoming more and more blurred. Good old space-time is now in some quarters regarded as a mere low-energy approximation to what there really is in the way of dimensions (I have heard lectures in which over fifty dimensions were postulated).[39] We can detect by observation only four dimensions, or at least it appears to us that we do—and these appearances remain the basis for accepting the various hypotheses offered. Other dimensions may be, for example, cylindrical, so tightly rolled up that we cannot detect them. We may fundamentally misrepresent what we do detect—for example, when we think that the four dimensions of space-time are continua (they may, for all we know, be lattices). Finally, mass itself, together with gravitation, may be amenable to the group-theoretic approach. This would be the ultimate irony: Pythagoras and Democritus might turn out ultimately to have been saying exactly the same thing—of course, the world is made only of matter, but look what matter is! Materialism as a doctrine might turn to be not so much wrong as pointless—in a world in which matter, energy, and space-time turn out to be mathematical structures. What of atomism? Well, what are atoms? When we say "protons are made of quarks," all we might mean is that a certain irreducible representation of the symmetry group SU(3) is "constructed" from its basic ("atomic"), three-dimensional, irreducible representation, by means of mathematical operations such as "tensor product."[40] (Pythagoras thought the world is made of numbers, but that is too simple: it's made of matrices of numbers, and the numbers are not real, but complex!) In such a world, by the way, causality itself would be a pointless concept,[41] valid only of the world as it appears to our poor receptors (as Bertrand Russell argued almost one hundred years ago). For as the late Pythagoreans argued, Aristotle's Four Causes can themselves be reduced to mathematical properties and relations.[42]

Mathematics—Application and Applicability

Does this raise any philosophical problems? You bet. First of all, there is the question of whether Pythagoreanism is a *coherent* doctrine—I believe that it is. I am persuaded by John Locke's central insight that the world could turn out to be fundamentally unlike what it appears to be, and the framework of space and time is no exception. Of course, Locke's rigid separation of primary from secondary qualities makes it seem as though it is some kind of an *a priori* truth that the world has the geometric properties it seems to have. Yet some astute commentators have pointed out passages in the *Essay* that hint that Locke saw that even the primary qualities may not be the last word, and that if we had the ability to perceive what is going on at the atomic level, even the primary qualities might turn out to be appearances.

The ultimate problem, then, would be whether Pythagoreanism is an *acceptable* doctrine.[43] If it is, then we end this article with the greatest irony of all.

Notes

1. This essay is reprinted with permission from Stewart Shapiro, ed. (2005) *The Oxford Handbook of Philosophy of Mathematics and Logic*. Oxford: Oxford University Press, 625–50.
2. Guillemin and Sternberg 1990a.
3. I would like to thank Professor Sternberg for valuable discussions concerning the role of symmetry in physics; Stanley Ocken for his insights into advanced aspects of elementary mathematics; and Mark Colyvan for his detailed suggestions. This research was supported by the Israel Science Foundation (Grant No. 949102), and I am very grateful for the support.
4. I once heard this idea from Stuart Hampshire.
5. Colyvan points out that I could have listed Quine as also holding that mathematics is empirical. This makes sense, but Quine rejects the empirical/*a priori* dualism in the first place, so it is far from clear what the doctrine, as espoused by Quine, comes to. Also, Quine believes that mathematics is the science of objects that have no empirical properties whatever. See also note 37.
6. One caveat, nevertheless, is in order. Modern foundational achievements in analysis have turned the "calculus" almost into a logical device, analytic truths concerning accelerated motion. So it's not clear that we have here an *empirical* application, one which could be refuted by experience. In the seventeenth century, however, I believe it would be fair to say that Newton was "reading off" the characteristics of accelerated motion from his "calculus," which justifies the name. At the same time, the calculus was developed for just this purpose, justifying the label "canonical."
7. There is a sense in which sets also have parts—we can view the subsets of S as parts of S. See Lewis (1991).
8. The reader will find it instructive to compare the present approach with the thoughtful discussion of the relationship between sets and physical bodies in Maddy (1990).
9. Most physicists before 1930 would have said this of group theory—it is now, of course, the centerpiece of elementary particle physics. The famous mathematician G. H. Hardy argued that no truly mathematical theory could be applied to warfare, but this was wishful thinking, since one of his examples of a "truly mathematical theory" was $E = mc^2$.
10. As Professor Sternberg pointed out to me, however, the concept of a prime number is crucial to cryptography, and of course he's right. This is another—rather whopping—counterexample to Hardy's wishful thinking (see note 8), since a lot of research in number theory today must be classified.
11. Wigner 1967.
12. I regret that Wigner, no doubt out of modesty, left out of his article the most striking examples of the unreasonable effectiveness of mathematics: his own contributions in the field of applying group theory to quantum mechanics.
13. Mendelson 1997.
14. See Steiner (1978a) for a discussion.
15. Interestingly enough, Wittgenstein in *RFM* (Wittgenstein 1978) points this out.
16. Perhaps the best treatment of this subject is Sternberg (1994, ch. 1).
17. Given an equation E with coefficients in field K, one can construct an extension field M in which the equation "splits" (can be factored, has roots). The Galois group of the equation is the group of automorphisms of M which leave K fixed. Since the image of a root of the equation under any element of the Galois group is also a root, the Galois group can be thought of as a group of permutations of the roots.
18. For a detailed account, cf. Sternberg 1991; Sternberg also graciously offered his help in cleaning up this entire section.
19. Technically, a one-parameter group of symmetry transformations. In addition, these transformations must not change the "symplectic" geometry of the phase space (see note 20). In the case of quantum mechan-

ics, to generate a conservation law the action of the group must not change the geometry of the Hilbert space—hence it must be represented by unitary matrices (these are invertible complex square matrices M such that the inverse is equal to the adjoint, $MM^* = I$). So in both classical and quantum mechanics, conserved quantities arise from the action of continuous symmetry groups that preserve both the laws of nature and the geometry of the abstract space (not necessarily space-time) which is the mathematical "arena" of these laws.

20. Even in classical mechanics, symmetry groups *need* not act on physical space. The fact, for example, that the orbit of a planet does not precess (turn) is a conservation law which arises from the action of a group of rotations, but not rotations of three-dimensional physical space as we now describe:

A deep comparison of classical physics with quantum mechanics is through the medium of geometry. In classical mechanics, the state of a system is given as a point in "phase space." For a free particle (no constraints or forces) this will be a six-dimensional vector space with three coordinates of position and three of momentum. As soon as we introduce forces or constraints, as in a particle in a central gravitational field, however, the coordinates are no longer rectilinear (Cartesian) but curved. Only at each point (locally) can we say that phase space looks like a vector space. Phase space is thus what is called a "manifold," indeed, a symplectic manifold (the term "symplectic" refers to the particular geometry of the manifold). The rotation group, when acting on the space dimensions, yields conservation of angular momentum for the planet; when acting on another submanifold of this "phase space" (namely, the submanifold of the phase space which contains all the possibilities for planetary motion), it yields a different conservation law. For details of this, see Guillemin and Sternberg (1990b). In quantum mechanics, on the other hand, the state of a system is given by a unit vector in an infinite dimensional complex vector space, known as a Hilbert space. (The term "Hilbert space" expresses the particular geometry of the vector space.) The coordinates remain rigidly rectilinear (even if complex), and thus the restrictive theory of group representations by square matrices must apply.

21. We are, again, interested primarily in representation by single parameter groups of unitary matrices because, as stated above, unitary transformations do not change the geometry of the Hilbert space. The geometry of a Hilbert space is given by the scalar product of two vectors, analogous to the scalar product of Euclidean space, and unitary matrices preserve the scalar product.

22. The success of group theoretical methods in physics has actually made this statement into a definition: what we mean by a "particle with a fixed value of a physical magnitude" is described by definition as an irreducible representation of the group associated with the conservation of that magnitude.

23. The way SU(2) is usually defined, as a group of 2 × 2 "unitary" matrices, this statement is trivial, true by definition; but there are other ways to define the group.

24. Cf. French (2000).

25. Suppose we have a physical theory, like string theory, which postulates a 26-dimensional space. The number 26 happens to be the numeral value of the Divine Tetragrammaton in Hebrew. Should this encourage us to try other of the Hebrew Names of God?

26. Let's for the record make precise what we mean by "geometry," at least for a two-dimensional vector space. If we consider two vectors (x_1, x_2) and (y_1, y_2), then the different quadratic expressions we can make from the two define different kinds of geometries: Euclidean geometry would be defined by $x_1 y_1 + x_2 y_2$; a (two-dimensional complex) Hilbert space would be defined by $x_1 \bar{y}_1 + x_2 \bar{y}_2$, thinking of the numbers here as complex numbers; Minkowsky geometry by $x_1 y_1 - x_2 y_2$; and symplectic geometry by $x_1 y_2 - x_2 y_1$. We also have the four corresponding symmetry transformations which leave each of these quadratic forms invariant.

27. This is a point made by Steven Weinberg (1986).

28. "[Soc.] Very good; and now tell me what is the power which discerns, not only in sensible objects, but in all things, universal notions, such as those which are called being and not-being, and those others about which we were just asking—what organs will you assign for the perception of these notions? [Theaet.] You are thinking of being and not-being, likeness and unlikeness, sameness and difference, and also of unity and

other numbers which are applied to objects of sense; and you mean to ask, through what bodily organ the soul perceives odd and even numbers and other arithmetical conceptions. [Soc.] You follow me excellently, Theaetetus; that is precisely what I am asking. [Theaet.] Indeed, Socrates, I cannot answer; my only notion is, that these, unlike objects of sense, have no separate organ, but that the mind, by a power of her own, contemplates the universals in all things."

29. Though this seems circular, in fact it is not, when the whole definition is written out. The "class of all concepts true of two objects," on the other hand, is a very tricky notion, and when Frege tried to axiomatize it, he landed up in Russell's Paradox. This is not the place to discuss the (by now) hoary question, is Frege's definition of the numbers "correct"—and what, indeed, constitutes "correctness" in this context? Since his definition cannot even be given in standard set theory (ZF), mathematics students are usually deprived of the pleasure of studying Frege's interesting ideas.

30. Field 1980.

31. In this category, we can cite an ingenious paper (Shapiro 1983), which exposes a rather subtle error in Field's argument that mathematics does not add any logical power to (nominalist) physics. Field has confused the notion of logical consequence with deductive consequence, so that even if it is true that physics with mathematics does not *imply* more facts than physics without mathematics, there are nevertheless facts, so implied, which cannot be *deductively proven* without mathematics. The distinction between logical consequence and deductive consequence arises in Field's book because he does not use first-order logic (for which there is no difference between logical and deductive consequences) as his underlying logical system.

32. In the next section, however, I will point out a great virtue of Field's book. Hang on.

33. Cf. Steiner (1982).

34. After writing this, I discovered that Mark Colyvan (2001) already made this point. See chapter 4 there, which contains some nice examples.

35. Cf. Dummett (1991).

36. In Section 2. See Steiner (1998) for more argumentation on this point.

37. By "crudely" empirical, I mean that the arithmetical operations are identified with everyday operations such as gathering (for plus). Even Quine, who regards mathematical theories as part of our entire scientific doctrine, which does face the tribunal of experience, rejects such a crude interpretation of arithmetic. Arithmetical theory for him is more like the theory of quarks than the theory of pebbles and gingerbread (to borrow phrases from Frege).

38. This claim is analogous to the one I made (Steiner 1978b) concerning mathematical *explanations* in science.

39. Charles Peirce (1999) foresaw this possibility.

40. For the definition of the tensor product of representations, see Sternberg (1994, Appendix B, 320ff.). The tensor product of two irreducible representations is again irreducible, which is the basis of our Pythagorean "atomism."

41. I am referring to the pointlessness of causal *laws*. Causal *reasoning* will continue to play a role in physics, for example, in the labeling of certain solutions of a valid equation "nonphysical"—as in Einstein's controversial rejection of travel faster than the speed of light on causal grounds, in Special Relativity. Ironically enough, however, in *General* Relativity, recent studies of black holes seem to allow the kind of "time travel" that Einstein ruled out! (Thanks to Shlomo Sternberg for pointing out this work to me.) This just shows that causal reasoning (even of Einstein) is fallible, not that it doesn't exist.

42. Cf. O'Meara (1989).

43. Readers familiar with the writings of W. V. Quine will recall that Pythagoreanism (especially rampant Pythagoreanism) was a sign that something had gone seriously wrong. But the context there was the use of the "Skolem–Löwenheim Theorem" to reinterpret any consistent first-order theory (including the theory of the

continuum) in a universe consisting solely of natural numbers. There, however, the *predicates* turn out to be true of arbitrary classes of numbers (i.e., the predicates are not arithmetical). Here, everything is interpreted as usual, except it turns out that every property of the physical world is mathematical. My own view, that Pythagoreanism is ultimately an anthropocentric doctrine, is expounded in Steiner (1998).

References

Colyvan, Mark. (2001) *The Indispensability of Mathematics.* New York: Oxford University Press.

Dummett, Michael. (1991) *Frege: Philosophy of Mathematics.* Cambridge, MA: Harvard University Press.

Field, Hartry. (1980) *Science without Numbers: A Defense of Nominalism.* Princeton, NJ: Princeton University Press.

French, Steven (2000) "The Reasonable Effectiveness of Mathematics: Partial Structures and the Application of Group Theory to Physics." *Synthèse* 125: 103–20.

Guillemin, Victor, and Shlomo Sternberg. (1990a) *Symplectic Techniques in Physics.* Cambridge: Cambridge University Press.

Guillemin, Victor, and Shlomo Sternberg. (1990b) *Variations on a Theme by Kepler.* American Mathematical Society Colloquium Publications, 42. Providence, RI: American Mathematical Society.

Lewis, David. (1991) *Parts of Classes.* Oxford: Basil Blackwell.

Maddy, Penelope. (1990) *Realism in Mathematics.* Oxford: Oxford University Press.

Mendelson, Elliot. (1997) *Introduction to Mathematical Logic,* 4th ed. New York: Chapman & Hall.

O'Meara, Dominic J. (1989) *Pythagoras Revived: Mathematics and Philosophy in Late Antiquity.* Appendix; The Excerpts from Iamblichus; On Pythagoreanism V–VII in Psellus. Text, Translation, and Notes. Oxford: Clarendon Press.

Peirce, Charles. (1999) "Note on the Analytical Representation of Space as a Section of a Higher Dimensional Space," in *Writings of Charles S. Peirce: A Chronological Edition,* ed. Nathan Houser, vol. 6. 1886–90: 260–62. Indianapolis: Indiana University Press.

Shapiro, Stewart. (1983) "Conservativeness and Incompleteness." *Journal of Philosophy* 80: 521–31.

Steiner, Mark. (1978a) "Mathematical Explanation." *Philosophical Studies* 34: 135–51.

Steiner, Mark. (1978b) "Mathematics, Explanation, and Scientific Knowledge." *Noûs* 12: 17–28.

Steiner, Mark. (1982) "Review (in Hebrew) of Hartry Field, *Science without Numbers.*" *Iyyun* 31: 211–17.

Steiner, Mark. (1998) *The Applicability of Mathematics as a Philosophical Problem.* Cambridge, MA: Harvard University Press.

Sternberg, Shlomo. (1994) *Group Theory and Physics.* Cambridge: Cambridge University Press.

Weinberg, Steven. (1986) "Lecture on the Applicability of Mathematics." *Notices of the American Mathematical Society* 33: 725–28.

Wigner, Eugene. (1967) "The Unreasonable Effectiveness of Mathematics in the Natural Sciences," in *Symmetries and Reflections,* Bloomington: Indiana University Press, 222–37.

Wittgenstein, Ludwig. (1978) *Remarks on the Foundations of Mathematics,* 3rd ed., rev., G.H. von Wright, R. Rhees, and G.E.M. Anscombe, eds., Cambridge: MIT Press.

PART VI

EVOLUTIONARY PSYCHOLOGY, NEUROSCIENCE, AND CONSCIOUSNESS

Introduction

Three tremendous mysteries confront human comprehension of the cosmos. How these questions are answered have ramifications that extend far beyond their territorial significance. The first of these is why there is something rather than nothing. Why is there a universe at all? The second is the transition from nonlife to life: abiogenesis. How did matter become animated? The difference between life and nonlife is vast. And finally, how did consciousness and rationality arise? Human beings have intellect and will of an entirely different order from those of any other life on this planet; we are, in a word, exceptional. What is the fundamental nature of this kind of consciousness? It seems, if not completely dependent upon our biology, at the very least not reducible to it. How, after all, could meaning and first-person experience be nothing more than biochemical syntax? One is left with the conclusion either that the self is merely an illusion (but then who is drawing the conclusion?) or that consciousness, while correlated with neurochemistry, has a good deal more to say for itself. It is this latter mystery that will be the subject of examination in this section.

We open the discussion with evolutionary accounts of the mind. In their essay "Toward Mapping the Evolved Functional Organization of Mind and Brain," **Leda Cosmides** and **John Tooby** argue that cognitive neuroscience is a branch of evolutionary biology because the human brain is a biological system produced by the evolutionary process. On this premise, cognitive neuroscientists may benefit by learning about technical advances in modern evolutionary biology and applying them in their research. In particular, Tooby and Cosmides suggest that evolutionary biology supplies cognitive science with three investigative tools: (1) a biologically rigorous concept of function appropriate to neural and cognitive systems; (2) an expanding list of functions that the human brain evolved to perform; and (3) an ability to distinguish narrowly functional features of our neural and cognitive architecture—those responsible for its organization—from a larger set of properties that are mainly byproducts or systemic noise. By using these tools and others, Cosmides and Tooby believe that evolutionary psychologists can construct experiments to investigate a wide variety of functionally dedicated subunits of the human brain. On the hypothesis that the brain is comprised of many such subunits, they argue that the evolved architecture of the human mind can be investigated by employing evolutionarily meaningful—as opposed to arbitrary—stimuli and tasks that will elicit behavioral responses illuminating the mind's complex functional organization. Integral to this approach is the premise that the mind is a set of information-processing devices, embodied in neural tissue, that are responsible for all conscious and nonconscious mental activity, that generate all behavior, and that regulate the body. The ultimate goal of the research program Cosmides and Tooby advocate is that through the collaborative interdisciplinary effort of many researchers, human nature—the human mind and human behavior—will eventually be encapsulated in a set of precise, high-resolution models of our evolved computational architecture, models that can be cashed out genetically, at the cellular level, developmentally, physiologically, and neurally.

David Berlinski is highly skeptical of this whole research program, and in his essay "On the Origins of the Mind," he sets out to dissect systematically what he regards as the pretensions of evolutionary psychology. In particular, he identifies three "metaphors" that give rise to divisions of labor within this field—that the human mind is like a digital computer in the way it functions, that it is like any other organ of the body by way of its embryological development, and that it is like any other biological artifact in the way it evolved through random variation and natural selection—and he sets about the task of critically examining each contention. He concludes that each metaphor is deeply flawed and that evolutionary psychology is a pseudoscience.

The discussion then moves from evolutionary psychology to the nature of consciousness and its neurophysiological basis. In his essay, titled simply "Consciousness," renowned philosopher **John Searle** argues for the thesis of *biological naturalism:* that in the end, the right way to think of the problem of consciousness is as a biological problem like any other, because consciousness is a biological phenomenon in exactly the same sense as digestion or growth or photosynthesis are biological phenomena. Unlike other problems in biology, however, the problem of consciousness is surrounded by a persistent set of philosophical problems. Searle therefore undertakes to circumscribe and deal with these philosophical problems before outlining his methodology for the scientific study of phenomena associated with consciousness.

In "A Framework for Consciousness," **Christof Koch** and the late **Francis Crick** summarize their research approach to consciousness as a neurobiological problem. After outlining a general strategy, they articulate a framework that provides a coherent scheme for explaining the neural correlates of *visual* consciousness in terms of competing cellular assemblies. They then outline some general experimental approaches to the problem of consciousness sui generis, and acknowledge some relevant aspects of the brain that have been left out of their proposed framework.

Nancey Murphy's article "Supervenience and the Downward Efficacy of the Mental: Nonreductive Physicalism and the Christian Tradition" adds to the considerations intended to show that physicalists do not need to be reductionists. Making connections to theological anthropology, she argues that while Christians have an important stake in this argument, so does anyone who values the notion of human reason. That is, the coherence or intelligibility of non-reductive physicalism has been called into question, particularly with regard to the problem of mental causation: how do we account for the role that reasons play in mental/neural processes if we assume the causal closure of the neurobiological level? Since reductionism seems to preclude the intelligibility of conscious rationality, as a committed physicalist, Murphy works to rescue a non-reductive physicalist model of mental causation. In doing so, she argues not only that there is a solution to this difficulty, but also that it provides a basis for a new approach to the problem of free will (though she does not explore the connections here).

Introduction

The next two essays seek to illuminate the mystery of the human mind by exploring possible connections between quantum theory and consciousness. **Stuart Hameroff** and **Roger Penrose**'s collaborative essay, "Conscious Events as Orchestrated Space-Time Selections," contends that certain elements of quantum theory (e.g., quantum coherence), and a newly proposed physical phenomenon of quantum wavefunction self-collapse that Penrose calls "objective reduction," are essential for consciousness. Furthermore, they hypothesize that these phenomena occur in cytoskeletal microtubules and other structures within each of the brain's neurons. The particular characteristics of microtubules suitable for quantum effects include their crystal-like lattice structure, hollow inner core, organization of cell function, and capacity for information processing. Hameroff and Penrose suggest that conformational states of microtubule subunits (tubulins) are coupled to internal quantum events and cooperatively interact (compute) with other tubulins in such a way that macroscopic coherent superposition of quantum-coupled tubulin conformational states occurring throughout significant brain volumes are able to provide the global binding essential to consciousness (compare this with discussions of the "binding problem" in Searle, as well as Koch and Crick). In the picture they propose, the emergence of the microtubule quantum coherence is equated with preconscious processing which grows (for up to 500 milliseconds) until the mass-energy difference among the separated states of tubulins reaches a threshold related to *quantum gravity*. Objective reduction of the quantum state in the microtubules is a self-collapse resulting in particular patterns of microtubule–tubulin conformational states that regulate neuronal activities, including synaptic functions. By providing a connection among preconscious to conscious transitions, fundamental space-time notions, and noncomputability, as well as binding various (timescale and spatial) reductions into an instantaneous event (a "conscious now"), Penrose and Hameroff argue that orchestrated objective reduction in brain microtubules gives the most specific and plausible model for consciousness yet proposed. In a newly written addendum to the essay, Hameroff details some of the historical development of the orchestrated objective reduction model since it was first proposed, some of the technical obstacles to its proper testing, and some of the tests it has passed. He concludes with a reflection on its metaphysical and scientific significance.

Taking an entirely different approach, **Henry Stapp**, in his essay "Quantum Interactive Dualism: The Libet and Einstein-Podolsky-Rosen Causal Anomalies" argues that the orthodox approach to quantum theory converts materialist conceptions of nature to a structure that does not require the causal closure of the physical. The causal gaps in quantum laws are filled in practice by inputs from our stream of consciousness. He further argues that this gives evidence of an underlying ontology of psychophysical *events* rather than substances, and of objective tendencies for these events to occur. Such events constitute mind-brain connections that form the fundamental link between brain processes and events in our stream of consciousness. Stapp further maintains that this "quantum ontology" grants our conscious intentions the causal efficacy we take them to have and leads to a form of interactive dualism that eschews notions of substance. He then uses this ontology to reconcile causally efficacious conscious free will with causal anomalies of the kind discussed by Libet and the type exemplified in the EPR paradox.

Finally, **James P. Moreland**'s essay "Neuroscience and Substance Dualism" provides a philosophical defense of dualist models of consciousness and the mind–brain connection. He argues that once we get clear on the central first- and second-order issues in philosophy of mind, it becomes evident that stating and resolving these issues is a philosophical matter for which discoveries in the hard sciences are largely *irrelevant*. Put differently, these philosophical issues are, with rare exceptions, *autonomous from* and *authoritative with respect to* the so-called deliverances of the hard sciences. After clarifying certain preliminary notions, Moreland defends his central thesis by focusing on two paradigm cases—one regarding property dualism and another regarding substance dualism—that are representative of the actual dialectic in the literature in philosophy of mind, and then he responds to two principled objections to his position.

30

TOWARD MAPPING THE EVOLVED FUNCTIONAL ORGANIZATION OF MIND AND BRAIN

JOHN TOOBY AND LEDA COSMIDES

> Nothing in biology makes sense except in the light of evolution.
> —T. Dobzhansky
>
> It is the theory which decides what we can observe.
> —A. Einstein

1. SEEING WITH NEW EYES: TOWARD AN EVOLUTIONARILY INFORMED COGNITIVE NEUROSCIENCE

The task of cognitive neuroscience is to map the information-processing structure of the human mind and to discover how this computational organization is implemented in the physical organization of the brain. The central impediment to progress is obvious: The human brain is, by many orders of magnitude, the most complex system that humans have yet investigated. Purely as a physical system, the vast intricacy of chemical and electrical interactions among hundreds of billions of neurons and glial cells defeats any straightforward attempt to build a comprehensive model, as one might attempt to do with particle collisions, geological processes, protein folding, or host–parasite interactions. Combinatorial explosion makes the task of elucidating the brain's computational structure even more overwhelming: There are an indefinitely large number of specifiable inputs, measurable outputs, and possible relationships between them. Even worse, no one yet knows with certainty how computations are physically realized. They depend on individuated events within the detailed structure of neural microcircuitry largely beyond the capacity of current technologies to observe or resolve. Finally, the underlying logic of the system has been obscured by the torrent of recently generated data.

Historically, however, well-established theories from one discipline have functioned as organs of perception for others (e.g., statistical mechanics for thermodynamics). They allow new relationships to be observed and make visible elegant systems of organization that had previously eluded detection. It seems worth exploring whether evolutionary biology could provide a rigorous metatheoretical framework for the brain sciences, as they have recently begun to do for psychology.[1]

Cognitive neuroscience began with the recognition that the brain is an organ designed to process information and that studying it as such would offer important new insights. Cognitive neuroscientists also recognize that the brain is an evolved system, but few realize that anything follows from this second fact. Yet these two views of the brain are intimately related and, when considered jointly, can be very illuminating.

2. WHY BRAINS EXIST

The brain is an organ of computation that was built by the evolutionary process. To say that the brain is an organ of computation means that (1) its physical structure embodies a set of programs that process information, and (2) that physical structure is there *because* it embodies these programs. To say that the brain was built by the evolutionary process means that its functional components—its programs—are there *because* they solved a particular problem-type in the past. In systems designed by natural selection, function determines structure.

Among living things, there are whole kingdoms filled with organisms that lack brains (plants, Monera, fungi). The sole reason that evolution introduced brains into the designs of some organisms—the reason brains exist at all—is because brains performed computations that regulated these organisms' internal processes and external activities in ways that promoted their fitness. For a randomly generated modification in design to be selected—that is, for a mutation to be incorporated by means of a nonrandom process into a species-typical brain design—it had to improve the ability of organisms to solve adaptive problems. That is, the modification had to have a certain kind of effect: It had to improve the organisms' performance of some activity that systematically enhanced the propagation of that modification, summed across the species' range and across many generations. This means that the design of the circuits, components, systems, or modules that make up our neural architecture must reflect, to an unknown but high degree, (1) the computational task demands inherent in the performance of those ancestral activities, and (2) the evolutionarily long-enduring structure of those task environments.[2]

Activities that promoted fitness in hominid ancestral environments differ in many ways from activities that capture our attention in the modern world, and they were certainly performed under radically different circumstances. (Consider: hunting *vs.* grocery shopping; walking everywhere *vs.* driving and flying; cooperating within a social world of around two hundred relatives and friends *vs.* fifty thousand strangers in a medium-sized city). The design features of the brain were built to specifications inherent in ancestral adaptive problems and selection pressures, often resulting in talents or deficits that seem out of place or irrational in our world. A baby cries—alerting her parents—when she is left to sleep alone in the dark, not because hyenas roam her suburban household, but because her brain is designed to keep her from being eaten under the circumstances in which our species evolved.

There is no single algorithm or computational procedure that can solve every adaptive problem.[3] The human mind (it will turn out) is composed of many different programs for the same reason that a carpenter's toolbox contains many different tools: Different problems require different solutions. To reverse-engineer the brain, one needs to discover functional units that are native to its organization. To do this, it is useful to know, as specifically as possible, what the brain is for—which specific families of computations it was built to accomplish and what counted as a biologically successful outcome for each problem-type. The answers to this question must be phrased in computational terms because that is the only language that can capture or express the functions that neural properties were naturally selected to embody. They must also refer to the ancestral activities, problems, selection pressures, and environments of the species in question because jointly these define the computational problems each component was configured to solve.[4]

For these reasons, evolutionary biology, biological anthropology, and cognitive psychology (when integrated, called *evolutionary psychology*) have the potential to supply to cognitive neuroscientists what might prove to be a key missing element in their research program: a partial list of the native information-processing functions that the human brain was built to execute, as well as clues and principles about how to discover or evaluate adaptive problems that might be proposed in the future.

Just as the fields of electrical and mechanical engineering summarize our knowledge of principles that govern the design of human-built machines, the field of evolutionary biology summarizes our knowledge of the engineering principles that govern the design of organisms, which can be thought of as machines built by the evolutionary process.[5] Modem evolutionary biology constitutes, in effect, a foundational "organism design theory" whose principles can be used to fit together research findings into coherent models of specific cognitive and neural mechanisms.[6] To apply these theories to a particular species, one integrates analyses of selection pressures with models of the natural history and ancestral environments of the species. For humans, the latter are provided by hunter–gatherer studies, biological anthropology, paleoanthropology, and primatology.[7]

3. First Principles: Reproduction, Feedback, and the Antientropic Construction of Organic Design

Within an evolutionary framework, an organism can be described as a self-reproducing machine. From this perspective, the defining property of life is the presence in a system of "devices" (organized components) that cause the system to construct new and similarly reproducing systems. From this defining property—self-reproduction—the entire deductive structure of modern Darwinism logically follows.[8] Because the replication of the design of the parental machine is not always error free, randomly modified designs (i.e., mutants) are introduced into populations of reproducers. Because such machines are highly organized so that they cause the otherwise improbable outcome of constructing offspring machines, most random modifications interfere with the complex sequence of actions necessary for self-reproduction. Consequently, such modified designs will tend to remove themselves from the population—a case of negative feedback.

However, a small residual subset of design modifications will, by chance, happen to constitute improvements in the design's machinery for causing its own reproduction. Such improved designs (by definition) cause their own increasing frequency in the population a case of positive feedback. This increase continues until (usually) such modified designs out-reproduce and thereby replace all alternative designs in the population, leading to a new species-standard design. After such an event, the population of reproducing machines is different from the ancestral population: The population- or species-standard design has taken a step "uphill" toward a greater degree of functional organization for reproduction than it had previously. This spontaneous feedback process—natural selection—causes functional organization to emerge naturally, that is, without the intervention of an intelligent "designer" or supernatural forces.

Over the long run, down chains of descent, this feedback cycle pushes designs through state-space toward increasingly well-organized—and otherwise improbable—functional arrangements.[9] These arrangements are functional in a specific sense: the elements are improbably well organized to cause their own reproduction in the environment in which the species evolved. Because the reproductive fates of the inherited traits that coexist in the same organism are linked together, traits will be selected to enhance each other's functionality.[10] As design features accumulate, they will tend to sequentially fit themselves together into increasingly functionally elaborated machines for reproduction, composed of constituent mechanisms—called adaptations—that solve problems that either are necessary for reproduction or increase its likelihood.[11] Significantly, in species like humans, genetic processes ensure that complex adaptations virtually always are species-typical (unlike nonfunctional aspects of the system). This means that *functional* aspects of the architecture will tend to be universal at the genetic level, even though their expression may often be sex or age-limited, or environmentally contingent.[12]

Because design features are embodied in individual organisms, they can, generally speaking, propagate themselves in only two ways: by solving problems that increase the probability that offspring will be produced either by the organism they are situated in or by that organism's kin.[13] An individual's relatives, by virtue of having descended from a recent common ancestor, have an increased likelihood of having the same design feature as compared to other conspecifics. This means that a design modification in an individual that causes an increase in the reproductive rate of that individual's kin will, by so doing, tend to increase its own frequency in the population. Accordingly, design features that promote both direct reproduction and kin reproduction, and that make efficient trade-offs between the two, will replace those that do not. To put this in standard biological terminology, design features are selected to the extent that they promote their inclusive fitness.[14]

In addition to selection, mutations can become incorporated into species-typical designs by means of chance processes. For example, the sheer impact of many random accidents may cumulatively propel a useless mutation upward in frequency until it crowds out all alternative design features from the population. Clearly, the presence of such a trait in the architecture is not explained by the (nonexistent) functional consequences that it had over many generations on the design's reproduction; as a result, chance-injected traits will not tend to be coordinated with the rest of the organism's architecture in a functional way.

Although such chance events play a restricted role in evolution and explain the existence and distribution of many simple and trivial properties, organisms are not primarily chance agglomerations of stray properties. Reproduction is a highly improbable outcome in the absence of functional machinery designed to bring it about, and only designs that retain all the necessary machinery avoid being selected out. To be invisible to selection and, therefore, not organized by it a modification must be so minor that its effects on reproduction are negligible. As a result, chance properties do indeed drift through the standard designs of species in a random way, but they are unable to account for the complex organized design in organisms and are, correspondingly, usually peripheralized into those aspects that do not make a significant impact on the functional operation of the system.[15] Random walks do not systematically build intricate and improbably functional arrangements such as the visual system, the language faculty, face recognition programs, emotion recognition modules, food aversion circuits, cheater detection devices, or motor control systems, for the same reason that wind in a junkyard does not assemble airplanes and radar.

Toward Mapping the Evolved Functional Organization of Mind and Brain

4. Brains are Composed Primarily of Adaptive Problem-Solving Devices

In fact, natural selection is the only known cause of and explanation for complex functional design in organic systems. Hence, all naturally occurring functional organization in organisms should be ascribed to its operation, and hypotheses about function are likely to be correct only if they are the kinds of functionality that natural selection produces.

This leads to the most important point for cognitive neuroscientists to abstract from modern evolutionary biology: Although not everything in the designs of organisms is the product of selection, all complex functional organization is. Indeed, selection can only account for functionality of a very narrow kind: approximately, design features organized to promote the reproduction of an individual and his or her relatives in ancestral environments.[16] Fortunately for the modern theory of evolution, the only naturally occurring complex functionality that ever has been documented in undomesticated plants, animals, or other organisms is functionality of just this kind, along with its derivatives and byproducts.

This has several important implications for cognitive neuroscientists:

1. *Technical definition of function.* In explaining or exploring the reliably developing organization of a cognitive device, the *function* of a design refers solely to how it systematically caused its own propagation in ancestral environments. It does not validly refer to any intuitive or folk definitions of function such as "contributing to personal goals," "contributing to one's well-being," or "contributing to society." These other kinds of usefulness may or may not exist as side effects of a given evolved design, but they can play no role in explaining how such designs came into existence or why they have the organization that they do.

It is important to bear in mind that the evolutionary standard of functionality is entirely independent of any ordinary human standard of desirability, social value, morality, or health.[17]

2. *Adapted to the past.* The human brain, to the extent that it is organized to do anything functional at all, is organized to construct information, make decisions, and generate behavior that would have tended to promote inclusive fitness in the ancestral environments and behavioral contexts of Pleistocene hunter–gatherers and before. (The pre-agricultural world of hunter–gatherers is the appropriate ancestral context because natural selection operates far too slowly to have built complex information-processing adaptations to the post-hunter–gatherer world of the last few thousand years.)

3. *No evolved "reading modules."* The problems that our cognitive devices are designed to solve do not reflect the problems that our modern life experiences lead us to see as normal, such as reading, driving cars, working for large organizations, reading insurance forms, learning the oboe, or playing Go. Instead, they are the odd and seemingly esoteric problems that our hunter-gatherer ancestors encountered, generation after generation, over hominid evolution. These include such problems as foraging, kin recognition, "mind reading" (i.e., inferring beliefs, desires, and intentions from behavior), engaging in social exchange, avoiding incest, choosing mates, interpreting threats, recognizing emotions, caring for children, regulating immune function, and so on, as well as the already well-known problems involved in perception, language acquisition, and motor control.

4. *Side effects are personally important but scientifically misleading.* Although our architectures may be capable of performing tasks that are "functional" in the (nonbiological) sense that we may value them (e.g., weaving, playing piano), these are incidental side effects of selection for our Pleistocene competencies—just as a machine built to be a hair-dryer can, incidentally, dehydrate fruit or electrocute. But it will be difficult to make sense of our cognitive mecha-

nisms if one attempts to interpret them as devices designed to perform functions that were not selectively important for our hunter–gatherer ancestors, or if one fails to consider the adaptive functions these abilities are side effects of.

5. *Adaptationism provides new techniques and principles.* Whenever one finds better-than-chance functional organization built into our cognitive or neural architecture, one is looking at adaptations—devices that acquired their distinctive organization from natural selection acting on our hunter–gatherer or more distant primate ancestors.

Reciprocally, when one is searching for intelligible functional organization underlying a set of cognitive or neural phenomena, one is far more likely to discover it by using an adaptationist framework for organizing observations because adaptive organization is the only kind of functional organization that is there to be found.

Because the reliably developing mechanisms (i.e., circuits, modules, functionally isolable units, mental organs, or computational devices) that cognitive neuroscientists study are evolved adaptations, all the biological principles that apply to adaptations apply to cognitive devices. This connects cognitive neuroscience and evolutionary biology in the most direct possible way. This conclusion should be a welcome one because it is the logical doorway through which a very extensive body of new expertise and principles can be made to apply to cognitive neuroscience, stringently constraining the range of valid hypotheses about the functions and structures of cognitive mechanisms. Because cognitive neuroscientists are usually studying adaptations and their effects, they can supplement their present research methods with carefully derived adaptationist analytic tools.

6. *Ruling out and ruling in.* Evolutionary biology gives specific and rigorous content to the concept of function, imposing strict rules on its use.[18] This allows one to rule out certain hypotheses about the proposed function of a given cognitive mechanism. But the problem is not just that cognitive neuroscientists sometimes impute functions that they ought not to. An even larger problem is that many fail to impute functions that they ought to. For example, an otherwise excellent recent talk by a prominent cognitive neuroscientist began with the claim that one would not expect jealousy to be a "primary emotion"—that is, a universal, reliably developing part of the human neural architecture (in contrast to others, such as disgust or fear). Yet there is a large body of theory in evolutionary biology—sexual selection theory—that predicts that sexual jealousy will be widespread in species with substantial parental investment in offspring (particularly in males); behavioral ecologists have documented mate-guarding behavior (behavior designed to keep sexual competitors away from one's mate) in a wide variety of species, including various birds, fish, insects, and mammals;[19] male sexual jealousy exists in every documented human culture;[20] it is the major cause of spousal homicides,[21] and in experimental settings, the design features of sexual jealousy have been shown to differ between the sexes in ways that reflect the different adaptive problems faced by ancestral men and women.[22] From the standpoint of evolutionary biology and behavioral ecology, the hypothesis that sexual jealousy is a primary emotion—more specifically, the hypothesis that the human brain includes neurocognitive mechanisms whose function is to regulate the conditions under which sexual jealousy is expressed and what its cognitive and behavioral manifestations will be like—is virtually inescapable.[23] But if cognitive neuroscientists are not aware of this body of theory and evidence, they will not design experiments capable of revealing such mechanisms.

7. *Biological parsimony, not physics parsimony.* The standard of parsimony imported from physics, the traditional philosophy of science, or from habits of economical programming is inappropriate and misleading in biology, and hence, in neuroscience and cognitive science,

which study biological systems. The evolutionary process never starts with a clean work board, has no foresight, and incorporates new features solely on the basis of whether they lead to systematically enhanced propagation. Indeed, when one examines the brain, one sees an amazingly heterogeneous physical structure. A correct theory of evolved cognitive functions should be no less complex and heterogeneous than the evolved physical structure itself and should map on to the heterogeneous set of recurring adaptive tasks faced by hominid foragers over evolutionary time. Theories of engineered machinery involve theories of the subcomponents. One would not expect that a general, unified theory of robot or automotive mechanism could be accurate.

8. *Many cognitive adaptations.* Indeed, analyses of the adaptive problems humans and other animals must have regularly solved over evolutionary time suggest that the mind contains a far greater number of functional specializations than is traditionally supposed, even by cognitive scientists sympathetic to "modular" approaches. From an evolutionary perspective, the human cognitive architecture is far more likely to resemble a confederation of hundreds or thousands of functionally dedicated computers, designed to solve problems endemic to the Pleistocene, than it is to resemble a single general purpose computer equipped with a small number of domain-general procedures, such as association formation, categorization, or production rule formation.[24]

9. *Cognitive descriptions are necessary.* Understanding the neural organization of the brain depends on understanding the functional organization of its computational relationships or cognitive devices. The brain originally came into existence and accumulated its particular set of design features only because these features functionally contributed to the organism's propagation. This contribution—that is, the evolutionary function of the brain—is obviously the adaptive regulation of behavior and physiology *on the basis of information* derived from the body and from the environment. The brain performs no significant mechanical, metabolic, or chemical service for the organism—its function is purely informational, computational, and regulatory in nature. Because the function of the brain is informational in nature, its precise functional organization can only be accurately described in a language that is capable of expressing its informational functions—that is, in cognitive terms, rather than in cellular, anatomical, or chemical terms. Cognitive investigations are not some soft, optional activity that goes on only until the "real" neural analysis can be performed. Instead, the mapping of the computational adaptations of the brain is an unavoidable and indispensable step in the neuroscience research enterprise. It must proceed in tandem with neural investigations and provides one of the primary frameworks necessary for organizing the body of neuroscience results.

The reason is straightforward. Natural selection retained neural structures on the basis of their ability to create adaptively organized relationships between information and behavior (e.g., the sight of a predator activates inference procedures that cause the organism to hide or flee) or between information and physiology (e.g., the sight of a predator increases the organism's heart rate, in preparation for flight). Thus, it is the information-processing structure of the human psychological architecture that has been functionally organized by natural selection, and the neural structures and processes have been organized insofar as they physically realize this cognitive organization. Brains exist and have the structure that they do because of the computational requirements imposed by selection on our ancestors. The adaptive structure of our computational devices provides a skeleton around which a modern understanding of our neural architecture should be constructed.

5. Brain Architectures Consist of Adaptations, Byproducts, and Random Effects

To understand the human (or any living species') computational or neural architecture is a problem in reverse engineering: We have working exemplars of the design in front of us, but we need to organize our observations of these exemplars into a systematic functional and causal description of the design. One can describe and decompose brains into properties according to any of an infinite set of alternative systems, and hence there is an indefinitely large number of cognitive and neural phenomena that could be defined and measured. However, describing and investigating the architecture in terms of its adaptations is a useful place to begin, because (1) the adaptations are the cause of the system's organization (the reason for the system's existence), (2) organisms, properly described, consist largely of collections of adaptations (evolved problem-solvers), (3) an adaptationist frame of reference allows cognitive neuroscientists to apply to their research problems the formidable array of knowledge that evolutionary biologists have accumulated about adaptations, (4) all of the complex, functionally organized subsystems in the architecture are adaptations, and (5) such a frame of reference permits the construction of economical and principled models of the important features of the system, in which the wealth of varied phenomena fall into intelligible, functional, and predictable patterns. As Ernst Mayr put it, summarizing the historical record, "the adaptationist question, 'What is the function of a given structure or organ?' has been for centuries the basis for every advance in physiology."[25] It should prove no less productive for cognitive neuroscientists. Indeed, all of the inherited design features of organisms can be partitioned into three categories: (1) adaptations (often, although not always, complex); (2) the by products or concomitants of adaptations; and (3) random effects. Chance and selection, the two components of the evolutionary process, explain different types of design properties in organisms, and all aspects of design must be attributed to one of these two forces. The conspicuously distinctive cumulative impacts of chance and selection allow the development of rigorous standards of evidence for recognizing and establishing the existence of adaptations and distinguishing them from the nonadaptive aspects of organisms caused by the nonselectionist mechanisms of evolutionary change.[26]

Design Evidence. Adaptations are systems of properties ("mechanisms") crafted by natural selection to solve the specific problems posed by the regularities of the physical, chemical, developmental, ecological, demographic, social, and informational environments encountered by ancestral populations during the course of a species' or population's evolution (Table 1). Adaptations are recognizable by "evidence of special design"[27]—that is, by recognizing certain features of the evolved species—typical design of an organism "as components of some special problem-solving machinery."[28] Moreover, they are so well organized and such good engineering solutions to adaptive problems that a chance coordination between problem and solution is effectively ruled out as a counter-hypothesis. Standards for recognizing special design include whether the problem solved by the structure is an evolutionarily long-standing adaptive problem, and such factors as economy, efficiency, complexity, precision, specialization, and reliability, which, like a key fitting a lock, render the design too good a solution to a defined adaptive problem to be coincidence.[29] Like most other methods of empirical hypothesis testing, the demonstration that something is an adaptation is always, at core, a probability assessment concerning how likely a set of events is to have arisen by chance alone. Such assessments are made by investigating whether there is a highly nonrandom coordination between the recurring properties of the phenotype and the structured properties of the adaptive problem, in a

> An adaptation is:
>
> 1. A cross-generationally recurring set of characteristics of the phenotype
> 2. that is reliably manufactured over the developmental life history of the organism,
> 3. according to instructions contained in its genetic specification,
> 4. in interaction with stable and recurring features of the environment (i.e., it reliably develops normally when exposed to normal ontogenetic environments),
> 5. whose genetic basis became established and organized in the species (or population) over evolutionary time, because
> 6. the set of characteristics systematically interacted with stable and recurring features of the ancestral environment (the "adaptive problem"),
> 7. in a way that systematically promoted the propagation of the genetic basis of the set of characteristics better than the alternative designs existing in the population during the period of selection. This promotion virtually always takes place through enhancing the reproduction of the individual bearing the set of characteristics, or the reproduction of the relatives of that individual.
>
> *Adaptations.* The most fundamental analytic tool for organizing observations about a species' functional architecture is the definition of an adaptation. To function, adaptations must evolve such that their causal properties rely on and exploit these stable and enduring statistical structural regularities in the world, and in other parts of the organism. Things worth noticing include the fact that an adaptation (such as teeth or breasts) can develop at any time during the life cycle, and need not be present at birth; an adaptation can express itself differently in different environments (e.g., speaks English, speaks Tagalog); an adaptation is not just any individually beneficial trait, but one built over evolutionary time and expressed in many individuals; an adaptation may not be producing functional outcomes currently (e.g., agoraphobia), but only needed to function well in ancestral environments; finally, an adaptation (like every other aspect of the phenotype) is the product of gene–environment interaction. Unlike many other phenotypic properties, however, it is the result of the interaction of the species-standard set of genes with those aspects of the environment that were present and relevant during the species' evolution. For a more extensive definition of the concept of adaptation see Tooby and Cosmides (1990b, 1992).

Table 1
The formal properties of an adaptation

way that meshed to promote fitness (genetic propagation) in ancestral environments.[30] For example, the lens, pupil, iris, retina, visual cortex, and other parts of the eye are too well coordinated, both with each other and with features of the world, such as the properties of light, optics, geometry, and the reflectant properties of surfaces, to have co-occurred by chance. In short, like the functional aspects of any other engineered system, they are recognizable as adaptations for analyzing scenes from reflected light by their organized and functional relationships to the rest of the design and to the structure of the world.

In contrast, concomitants or byproducts of adaptations are those properties of the phenotype that do not contribute to functional design per se, but that happen to be coupled to properties that are. Consequently, they were dragged along into the species-typical architecture because of selection for the functional design features to which they are linked. For example, bones are adaptations, but the fact that they are white is an incidental byproduct. Bones were selected to include calcium because it conferred hardness and rigidity to the structure (and was dietarily available), and it simply happens that alkaline earth metals appear white in many compounds, including the insoluble calcium salts that are a constituent of bone. From the point of view of functional design, byproducts are the result of "chance," in the sense that the process that led to their incorporation into the design was blind to their consequences (assuming that they were not negative). Accordingly, such byproducts are distinguishable from adaptations by the fact that they are not complexly arranged to have improbably functional consequences (e.g., the whiteness of bone does nothing for the vertebrae).

In general, byproducts will be far less informative as a focus of study than adaptations because they are consequences and not causes of the organization of the system (and hence are functionally arbitrary, unregulated, and may, for example, vary capriciously between individuals). Unfortunately, unless researchers actively seek to study organisms in terms of their adaptations, they usually end up measuring and investigating arbitrary and random admixtures of functional and functionless aspects of organisms, a situation that hampers the discovery of the underlying organization of the biological system. We do not yet, for example, even know which exact aspects of the neuron are relevant to its function and which are byproducts, so many computational neuroscientists may be using a model of the neuron that is wildly inaccurate.

Finally, entropic effects of many types are always acting to introduce disorder into the design of organisms. Traits introduced by accident or by evolutionary random walks are recognizable by the lack of coordination that they produce within the architecture or between the architecture and the environment, as well as by the fact that they frequently cause uncalibrated variation between individuals. Examples of such entropic processes include genetic mutation, recent change in ancestrally stable environmental features, and developmentally anomalous circumstances.

6. How Well-Engineered are Adaptations?

The design of our cognitive and neural mechanisms should only reflect the structure of the adaptive problems that our ancestors faced to the extent that natural selection is an effective process. Is it one? How well or poorly engineered are adaptations? Some researchers have argued that evolution primarily produces inept designs, because selection does not produce perfect optimality.[31] In fact, evolutionary biologists since Darwin have been well aware that selection does not produce perfect designs.[32] Still, because natural selection is a hill-climbing process that tends to choose the best of the variant designs that actually appear, and because of the immense numbers of alternatives that appear over the vast expanse of evolutionary time, natural selection tends to cause the accumulation of very well-engineered functional designs.

Empirical confirmation can be gained by comparing how well evolved devices and human engineered devices perform on evolutionarily recurrent adaptive problems (as opposed to arbitrary, artificial modern tasks, such as chess). For example, the claim that language competence is a simple and poorly engineered adaptation cannot be taken seriously, given the total amount of time, engineering, and genius that has gone into the still unsuccessful effort to produce artificial systems that can remotely approach—let alone equal—human speech perception, comprehension, acquisition, and production.[33]

Even more strikingly, the visual system is composed of collections of cognitive adaptations that are well-engineered products of the evolutionary process, and although they may not be "perfect" or "optimal"—however these somewhat vague concepts may be interpreted—they are far better at vision than any human-engineered system yet developed.

Wherever the standard of biological functionality can be clearly defined—semantic induction, object recognition, color constancy, echolocation, relevant problem-solving generalization, chemical recognition (olfaction), mimicry, scene analysis, chemical synthesis—evolved adaptations are at least as good as and usually strikingly better than human engineered systems, in those rare situations in which humans can build systems that can accomplish them at all. It seems reasonable to insist that, before a system is criticized as being poorly designed, the critic ought to be able to construct a better alternative—a requirement, it need hardly be pointed out, that has never been met by anyone who has argued that adaptations are poorly designed. Thus, although adaptations are certainly suboptimal in some ultimate sense, it is an empirically demonstrable fact that the short-run constraints on selective optimization do not prevent the emergence of superlatively organized computational adaptations in brains. Indeed, aside from the exotic nature of the problems that the brain was designed to solve, it is exactly this sheer functional intricacy that makes our architecture so difficult to reverse-engineer and to understand.

7. Cognitive Adaptations Reflect the Structure of the Adaptive Problem and the Ancestral World

Four lessons emerge from the study of natural competences, such as vision and language: (1) most adaptive information-processing problems are complex; (2) the evolved solution to these problems is usually machinery that is well-engineered for the task; (3) this machinery is usually specialized to fit the particular nature of the problem; and (4) its evolved design often embodies substantial and contentful "innate knowledge" about problem-relevant aspects of the world.

Well-studied adaptations overwhelmingly achieve their functional outcomes because they display an intricately engineered coordination between their specialized design features and the detailed structure of the task and task environment. Like a code that has been torn in two and given to separate couriers, the two halves (the structure of the mechanism and the structure of the task) must be put together to be understood. To function, adaptations evolve such that their causal properties rely on and exploit these stable and enduring statistical and structural regularities in the world. Thus, to map the structures of our cognitive devices, we need to understand the structures of the problems that they solve and the problem-relevant parts of the hunter–gatherer world. If studying face recognition mechanisms, one must study the recurrent structure of faces. If studying social cognition, one must study the recurrent structure of hunter–gatherer social life. For vision, the problems are not so very different for a modern scientist and a Pleistocene hunter–gatherer, so the folk notions of function that perception researchers use are not a problem. But the more one strays from low-level perception, the more one needs to know about human behavioral ecology and the structure of the ancestral world.

8. Experimenting with Ancestrally Valid Tasks and Stimuli

Although bringing cognitive neuroscience current with modem evolutionary biology offers many new research tools,[34] we have out of necessity limited discussion to only one: an evolutionary functionalist research strategy.[35] The adoption of such an approach will modify research practice in many ways. Perhaps most significantly, researchers will no longer have to operate purely by intuition or guesswork to know which kinds of tasks and stimuli to expose subjects to. Using knowledge from evolutionary biology, behavioral ecology, animal behavior, and hunter–gatherer studies, they can construct ancestrally or adaptively valid stimuli and tasks. These are stimuli that would have had adaptive significance in ancestral environments, and tasks that resemble (at least in some ways) the adaptive problems that our ancestors would have been selected to be able to solve.

The present widespread practice of using arbitrary stimuli of no adaptive significance (e.g., lists of random words, colored geometric shapes) or abstract experimental tasks of unknown relevance to Pleistocene life has sharply limited what researchers have observed and can observe about our evolved computational devices. This is because the adaptive specializations that are expected to constitute the majority of our neural architecture are designed to remain dormant until triggered by cues of the adaptively significant situations that they were designed to handle. The Wundtian and British Empiricist methodological assumption that complex stimuli, behaviors, representations, and competences are compounded out of simple ones has been empirically falsified in scores of cases,[36] and so, restricting experimentation to such stimuli and tasks simply restricts what researchers can find to a highly impoverished and unrepresentative set of phenomena. In contrast, experimenters who use more biologically meaningful stimuli have had far better luck, as the collapse of behaviorism and its replacement by modern behavioral ecology have shown in the study of animal behavior. To take one example of its applicability to humans, effective mechanisms for Bayesian inference—undetected by 20 years of previous research using "modern" tasks and data formats—were activated by exposing subjects to information formatted in a way that hunter–gatherers would have encountered it.[37] Equally, when subjects were given ancestrally valid social inference tasks (cheater detection, threat interpretation), previously unobserved adaptive reasoning specializations were activated, guiding subjects to act in accordance with evolutionarily predicted but otherwise odd patterns.[38]

Everyone accepts that one cannot study human language specializations by exposing subjects to meaningless sounds: the acoustic stimuli must contain the subtle, precise, high level relationships that make sound language. Similarly, to move on to the study of other complex cognitive devices, subjects should be exposed to stimuli that contain the subtle, ancestrally valid relationships relevant to the diverse functions of these devices. In such an expanded research program, experimental stimuli and tasks would involve constituents such as faces, smiles, disgust expressions, foods, the depiction of socially significant situations, sexual attractiveness, habitat quality cues, animals, navigational problems, cues of kinship, rage displays, cues of contagion, motivational cues, distressed children, species-typical "body language," rigid object mechanics, plants, predators, and other functional elements that would have been part of ancestral hunter–gatherer life. Investigations would look for functional subsystems that not only deal with such low-level and broadly functional competences as perception, attention, memory, and motor control, but also with higher-level ancestrally valid competences as well—mechanisms such as eye direction detectors,[39] face recognizers,[40] food memory subsystems,[41] person-specific memory, child care motivators,[42] and sexual jealousy modules.

Although these proposals to look for scores of content-sensitive circuits and domain-specific specializations will strike many as bizarre and even preposterous, they are well grounded in modern biology. We believe that, in a decade or so, they will look tame. If cognitive neuroscience is anything like investigations in domain-specific cognitive psychology[43] and in modern animal behavior, researchers will be rewarded with the materialization of a rich array of functionally patterned phenomena that have not been observed so far because the mechanisms were never activated in the laboratory by exposure to ecologically appropriate stimuli. Although presently, the functions of most brain structures are largely unknown, pursuing such research directions may begin to populate the empty regions of our maps of the brain with circuit diagrams of discrete, functionally intelligible computational devices.

In short, because theories and principled systems of knowledge can function as organs of perception, the incorporation of a modern evolutionary framework into cognitive neuroscience

9. Conclusion

The aforementioned points indicate why cognitive neuroscience is pivotal to the progress of the brain sciences. There are an astronomical number of physical interactions and relationships in the brain, and blind empiricism rapidly drowns itself among the deluge of manic and enigmatic measurements. Through blind empiricism, one can equally drown at the cognitive level in a sea of irrelevant things that our computational devices can generate, from writing theology or dancing the mazurka to calling for the restoration of the Plantagenets to the throne of France. However, evolutionary biology, behavioral ecology, and hunter–gatherer studies can be used to identify and supply descriptions of the recurrent adaptive problems humans faced during their evolution.

Supplemented with this knowledge, cognitive research techniques can abstract out of the welter of human cognitive performance a series of maps of the functional information-processing relationships that constitute our computational devices and that evolved to solve this particular set of problems: our cognitive architecture. These computational maps can then help us abstract out of the ocean of physical relationships in the brain that exact and minute subset that implements those information-processing relationships because it is only these relationships that explain the existence and functional organization of the system. The immense number of other physical relationships in the brain are incidental byproducts of those narrow aspects that implement the functional computational architecture. Consequently, an adaptationist inventory and functional mapping of our cognitive devices can provide the essential theoretical guidance for neuroscientists that will allow them to home in on these narrow but meaningful aspects of neural organization and to distinguish them from the sea of irrelevant neural phenomena.[44]

Notes

1. Shepard 1984, 1987a, 1987b; Gallistel 1990; Cosmides and Tooby 1987; Pinker 1994, 1997; Marr 1982; Tooby and Cosmides 1992.
2. Marr 1982; Shepard 1987a; Tooby and Cosmides 1992.
3. Cosmides and Tooby 1987; Tooby and Cosmides 1990a, 1992.
4. Cosmides and Tooby 1987; Tooby and Cosmides 1990a, 1992.
5. For overviews, see Daly and Wilson (1984), Dawkins (1976; 1982; 1986), Krebs and Davies (1997).
6. Tooby and Cosmides 1992.
7. Lee and DeVore 1968.
8. Dawkins 1976; Williams 1985; Tooby and Cosmides 1990a.
9. Dawkins 1986; Williams, 1966, 1985.
10. However, see Cosmides and Tooby (1981), and Tooby and Cosmides (1990a) for the relevant genetic analysis and qualifications.
11. Darwin 1859; Dawkins 1986; Thornhill 1991; Tooby and Cosmides 1990a; Williams 1966, 1985.
12. See Tooby and Cosmides (1990b). The genes underlying complex adaptations cannot vary substantially between individuals because if they did, the obligatory genetic shuffling that takes place during sexual reproduction would break apart the complex adaptations that had existed in the parents when these are recombined in the offspring generation. All the genetic subcomponents necessary to build the complex adaptation rarely would reappear together in the same individual if they were not being supplied reliably by both parents in all matings (for a discussion of the genetics of sexual recombination, species-typical adaptive design, and individual differences, see Tooby 1982, Tooby and Cosmides 1990b).
13. Hamilton 1964; Williams and Williams 1957. However, see Cosmides and Tooby (1981) and Haig (1993) for intragenomic methods.
14. Hamilton 1964.
15. Tooby and Cosmides 1990a, 1990b, 1992.
16. Williams 1966; Dawkins 1986.
17. Cosmides and Tooby 1999.
18. Williams 1966; Dawkins 1982, 1986.
19. Krebs and Davies 1997; Wilson and Daly 1992.
20. Daly *et al.* 1982; Wilson and Daly 1992.
21. Daly and Wilson 1988.
22. Buss 1994.
23. For an evolutionary/cognitive approach to emotions, see Tooby and Cosmides (1990a, 1990b).
24. For discussion, see Cosmides and Tooby (1987, 1994), Gallistel (1990), Pinker (1997), Sperber (1994), Symons (1987), and Tooby and Cosmides (1992).
25. Mayr 1983: 32.
26. Williams 1966, 1985; Pinker and Bloom 1992; Symons 1992; Thornhill 1991; Tooby and Cosmides 1990a, 1990b, 1992; Dawkins 1986.
27. Williams 1966.
28. Williams 1985: 1.
29. Williams 1966.
30. Tooby and Cosmides 1990b, 1992.
31. Gould and Lewontin 1979.
32. Darwin 1859; Williams 1966; Dawkins 1976, 1982, 1986. For a recent convert from the position that organisms are optimally designed to the more traditional adaptationist position, see Lewontin (1967, 1979);

see Dawkins (1982) for an extensive discussion of the many processes that prevent selection from reaching perfect optimality.

33. Pinker and Bloom 1992.
34. Preuss 1995.
35. See Tooby and Cosmides (1992) for a description; for examples, see chapters in Barkow *et al.* (1992), Daly and Wilson (1995), and Gaulin (1995).
36. See, for example, Gallistel (1990).
37. Brase *et al.* 1998; Cosmides and Tooby 1996; Gigerenzer and Hoffrage 1995.
38. Cosmides 1989; Cosmides and Tooby 1992.
39. Baron-Cohen 1994.
40. For example, Johnson and Morton (1991).
41. For example, Hart *et al.* (1985) and Caramazza and Shelton (1998).
42. Daly and Wilson 1995.
43. Hirschfeld and Gelman 1994.
44. The authors gratefully acknowledge the financial support of the James S. McDonnell Foundation, the National Science Foundation (NSF grant BNS9157–449 to John Tooby), and a Research Across Disciplines grant (Evolution and the Social Mind) from the UCSB Office of Research.

References

Barkow, J., L. Cosmides, and J. Tooby, eds., (1992) *The Adapted Mind: Evolutionary Psychology and the Generation of Culture.* New York: Oxford University Press.

Baron-Cohen, S. (1994) "The eye-direction detector: A case for evolutionary psychology," in *Joint-Attention: Its Origins and Role in Development,* C. Moore and P. Dunham, eds. Hillsdale, NJ: Erlbaum.

Brase, G., L. Cosmides, and J. Tooby. (1998) "Individuation, counting, and statistical inference: The role of frequency and whole-object representations in judgment under uncertainty." *Journal of Experimental Psychology: General* 127: 3–21.

Buss, D. (1994) *The Evolution of Desire.* New York: Basic Books.

Caramazza, A., and J. Shelton. (1998) "Domain-specific knowledge systems in the brain: The animate-inanimate distinction." *Journal of Cognitive Neuroscience* 10: 1–34.

Cosmides, L. (1989) "The logic of social exchange: Has natural selection shaped how humans reason? Studies with the Wason selection task." *Cognition* 31: 187–276.

Cosmides, L., and J. Tooby. (1981) "Cytoplasmic inheritance and intragenomic conflict." *Journal of Theoretical Biology* 89: 83–129.

———. (1987) "From evolution to behavior: Evolutionary psychology as the missing link." in *The Latest on the Best: Essays on Evolution and Optimality,* J. Dupre, ed. Cambridge: MA: MIT Press, 277–306.

———. (1992) "Cognitive adaptations for social exchange," in *The Adapted Mind: Evolutionary Psychology and the Generation of Culture,* J. Barkow, L. Cosmides, and J. Tooby, eds. New York: Oxford University Press, 163–228.

———. (1994) "Beyond intuition and instinct blindness: The case for an evolutionarily rigorous cognitive science." *Cognition* 50: 41–77.

———. (1996) "Are humans good intuitive statisticians after all? Rethinking some conclusions from the literature on judgment under uncertainty." *Cognition* 58: 1–73.

———. (1999) "Toward an evolutionary taxonomy of treatable conditions," *Journal of Abnormal Psychology* 108 (3): 453–64.

Daly, M., and M. Wilson. (1984) *Sex, Evolution and Behavior,* Second Edition. Boston: Willard Grant.

———. (1988) *Homicide.* New York: Aldine.

———. (1995) "Discriminative parental solicitude and the relevance of evolution- ary models to the analysis of motivational systems," in *The Cognitive Neurosciences,* M.S. Gazzaniga, ed. Cambridge, MA: MIT Press, 1269–86.

Daly, M., M. Wilson, and S. J. Weghorst. (1982) "Male sexual jealousy." *Ethology and Sociobiology* 3: 11-27.

Darwin, C. (1859) *On the Origin of Species.* London: Murray. New edition: Cambridge, MA: Harvard University Press.

Dawkins, R. (1976) *The Selfish Gene.* New York: Oxford University Press.

———. (1982) *The Extended Phenotype.* San Francisco: W.H. Freeman.

———. (1986) *The Blind Watchmaker.* New York: Norton.

Gallistel, C. R. (1990) *The Organization of Learning.* Cambridge, MA: MIT Press.

Gaulin, S. (1995) "Does evolutionary theory predict sex differences in the brain?" in *The Cognitive Neurosciences,* M.S. Gazzaniga, ed. Cambridge, MA: MIT Press, 1211–25.

Gigerenzer, G., and U. Hoffrage. (1995) "How to improve Bayesian reasoning without instruction: Frequency formats." *Psychological Review* 102: 684–704.

Gould, S. J., and R. C. Lewontin. (1979) "The spandrels of San Marco and the Panglossian program: A critique of the adaptationist programme." *Proceedings of the Royal Society of London* 205: 281–88.

Haig, D. (1993) "Genetic conflicts in human pregnancy." *Quarterly Review of Biology* 68: 495–532.

Hamilton, W. D. (1964) "The genetical evolution of social behavior." *Journal of Theoretical Biology* 7: 1–52.

Hart, J. Jr., R. S. Berndt, and A. Caramazza (1985) "Category-specific naming deficit following cerebral infarction." *Nature* 316: 439–40.

Hirschfeld, L., and S. Gelman, eds. (1994) *Mapping the Mind: Domain Specificity in Cognition and Culture.* New York: Cambridge University Press.

Johnson, M., and J. Morton. (1991) *Biology and Cognitive Development: The Case of Face Recognition.* Oxford: Blackwell.

Krebs, J. R., and N. B. Davies. (1997) *Behavioural Ecology: An Evolutionary Approach,* 4th edition. London: Blackwell Science.

Lee, R. B., and I. DeVore. (1968) *Man the Hunter.* Chicago: Aldine.

Lewontin, R. (1967) Spoken remarks in *Mathematical Challenges to the Neo-Darwinian Interpretation of Evolution,* P. Moorhead and M. Kaplan, eds. *Wistar Institute Symposium Monograph* 5: 79.

Lewontin, R. (1979) "Sociobiology as an adaptationist program." *Behavioral Sciences* 24: 5–14.

Marr, D. (1982) *Vision: A Computational Investigation into the Human Representation and Processing of Visual Information.* San Francisco: Freeman.

Mayr, E. (1983) "How to carry out the adaptationist program." *American Naturalist* 121: 324–34.

Pinker, S. (1994) *The Language Instinct.* New York: Morrow.

———. (1997) *How the Mind Works.* New York: Norton.

Pinker, S., and P. Bloom. (1992) "Natural language and natural selection," reprinted in *The Adapted Mind: Evolutionary Psychology and the Generation of Culture,* J. Barkow, L. Cosmides, and J. Tooby, eds. New York: Oxford University Press, 451–93.

Preuss, T. (1995) "The argument from animals to humans in cognitive neuroscience," in *The Cognitive Neurosciences,* M. S. Gazzaniga, ed. Cambridge, MA: MIT Press, 1227–41.

Shepard, R. N. (1984) "Ecological constraints on internal representation: Resonant kinematics of perceiving, imagining, thinking, and dreaming." *Psychological Review* 91: 417–47.

———. (1987a) "Evolution of a mesh between principles of the mind and regularities of the world," in *The Latest on the Best: Essays on Evolution and Optimality,* J. Dupre, ed. Cambridge, MA: MIT Press, 251–75.

———. (1987b) "Towards a universal law of generalization for psychological science." *Science* 237: 1317–23.

Sperber, D. (1994) "The modularity of thought and the epidemiology of representations," in *Mapping the Mind: Domain Specificity in Cognition and Culture,* L. Hirschfeld and S. Gelman, eds. New York: Cambridge University Press, 39–67.

Symons, D. (1987) "If we're all Darwinians, what's the fuss about?" in *Sociobiology and Psychology,* C. B. Crawford, M. F. Smith, and D. L. Krebs, eds. Hillsdale, NJ: Erlbaum, 121–46.

———. (1992) "On the use and misuse of Darwinism in the study of human behavior," in *The Adapted Mind: Evolutionary Psychology and the Generation of Culture,* J. Barkow, L. Cosmides, and J. Tooby, eds. New York: Oxford University Press, 137–59.

Thornhill, R. (1991) "The study of adaptation," in *Interpretation and Explanation in the Study of Behavior,* M. Bekoff and D. Jamieson, eds. Boulder, CO: Westview Press.

Tooby, J. (1982) "Pathogens, polymorphism, and the evolution of sex." *Journal of Theoretical Biology* 97: 557–76.

Tooby, J., and L. Cosmides. (1990a) "The past explains the present: Emotional adaptations and the structure of ancestral environments." *Ethology and Sociobiology* 11: 375–424.

———. (1990b) "On the universality of human nature and the uniqueness of the individual: The role of genetics and adaptation." *Journal of Personality* 58: 17–67.

———. (1992) "The psychological foundations of culture," in *The Adapted Mind: Evolutionary Psychology and the Generation of Culture,* J. Barkow, L. Cosmides, and J. Tooby, eds. New York: Oxford University Press, 19–136.

Williams. G. C. (1966) *Adaptation and Natural Selection: A Critique of Some Current Evolutionary Thought*. Princeton, NJ: Princeton University Press.

———. (1985) "A defense of reductionism in evolutionary biology." *Oxford Surveys in Evolutionary Biology* 2: 1–27.

Williams, G. C., and D. C. Williams. (1957) "Natural selection of individually harmful social adaptations among sibs with special reference to social insects." *Evolution* 17: 249–53.

Wilson, M., and M. Daly, 1992. "The man who mistook his wife for a chattel," in J. Barkow, L. Cosmides, and J. Tooby, eds. *The Adapted Mind: Evolutionary Psychology and the Generation of Culture*. New York: Oxford University Press, 289–322.

31

On the Origins of the Mind[1]

David Berlinski

> *It's all scientific stuff; it's been proved.*
> —Tom Buchanan in *The Great Gatsby*

At some time in the history of the universe, there were no human minds, and at some time later, there were. Within the blink of a cosmic eye, a universe in which all was chaos and void came to include hunches, beliefs, sentiments, raw sensations, pains, emotions, wishes, ideas, images, inferences, the feel of rubber, *Schadenfreude,* and the taste of banana ice cream.

A sense of surprise is surely in order. How did *that* get *here?*

If the origin of the human mind is mysterious, so too is its nature. There are, Descartes argued, two substances in the universe, one physical and the other mental.

To many contemporary philosophers, this has seemed rather an embarrassment of riches. But no sooner have they ejected mental substances from their analyses than mental properties pop up to take their place, and if not mental properties then mental functions. As a conceptual category, the mental is apparently unwilling to remain expunged.

And no wonder. Although I may be struck by a thought, or moved by a memory, or distracted by a craving, these familiar descriptions suggest an effect with no obvious physical cause. Thoughts, memories, cravings—they are what? Crossing space and time effortlessly, the human mind deliberates, reckons, assesses, and totes things up; it reacts, registers, reflects, and responds. In some cases, like inattention or carelessness, it invites censure by doing nothing at all or doing something in the wrong way; in other cases, like vision, it acts unhesitatingly and without reflection; and in still other cases, the human mind manages both to slip itself into and stay aloof from the great causal stream that makes the real world boom, so that when *it* gives the go-ahead, what *I* do is, as Thomas Aquinas observed, "inclined but not compelled."

These are not properties commonly found in the physical world. They are, in fact, not found at all.

And yet, the impression remains widespread that whoever is responsible for figuring out the world's deep things seems to have figured out the human mind as well. Commenting on negative advertising in political campaigns, Kathleen Hall Jamieson, the director of the Annenberg Public Policy Center at the University of Pennsylvania, remarked that "there appears to be something hard-wired into humans that gives special attention to negative information." There followed what is by now a characteristic note: "I think it's evolutionary biology."

Negative campaign advertisements are the least of it. There is, in addition, war and male aggression, the human sensitivity to beauty, gossip, a preference for suburban landscapes, love, altruism, marriage, jealousy, adultery, road rage, religious belief, fear of snakes, disgust, night sweats, infanticide, and the fact that parents are often fond of their children. The idea that human behavior is "the product of evolution," as the *Washington Post* puts the matter, is now more than a theory: it is a popular conviction.

It is a conviction that reflects a modest consensus of opinion among otherwise disputatious philosophers and psychologists: Steven Pinker, Daniel Dennett, David Buss, Henry Plotkin, Leda Cosmides, John Tooby, Peter Gärdenfors, Gary Marcus. The consensus is constructed, as such things often are, on the basis of a great hope and a handful of similes. The great hope is that the human mind will in the end find an unobtrusive place in the larger world in which purely material causes chase purely material effects throughout the endless night. The similes are, in turn, designed to promote the hope.

Three similes are at work, each more encompassing than the one before. They give a natural division of labor to what is now called evolutionary psychology. First, the human mind is *like* a computer in the way that it works. And it is just because the mind *is* like a computer that the computer comprises a model of the mind. "My central thesis," the cognitive psychologist H. A. Simon has written, is that "conventional computers can be, and have been, programmed to represent symbol structures and carry out processes on those structures that parallel, step by step, the way the human brain does it."

Second, the individual human mind is *like* the individual human kidney, or any other organ of the body, in the way that it is created anew in every human being. "Information," Gary Marcus writes, "copied into the nucleus of every newly formed cell, guides the gradual but powerful process of successive approximation that shapes each of the body's organs." This is no less true of the "organ of thought and language" than of the organs of excretion and elimination.

Third, the universal human mind—the expression in matter of human nature—is *like* any other complicated biological artifact in the way that it arose in the human species by means of random variation and natural selection. These forces, as Steven Pinker argues, comprise "the only explanation we have of how complex life *can* evolve. . . ."

Taken together, these similes do succeed wonderfully in suggesting a coherent narrative. The ultimate origins of the human mind may be found in the property of irritability that is an aspect of living tissue itself. There is a primordial twitch, one that has been lost in time but not in memory; various descendant twitches then enlarged themselves, becoming, among the primates at least, sophisticated organs of perception, cognition, and computation. The great Era of Evolutionary Adaptation arrived in the late Paleolithic, a veritable genetic Renaissance in which the contingencies of life created, in the words of the evolutionary psychologist Leda Cosmides, "programs that [were] well-engineered for solving problems such as hunting, foraging for plant foods, courting mates, cooperating with kin, forming coalitions for mutual defense, avoiding predators, and the like." There followed the long Era in Which Nothing Happened, the modern human mind retaining in its structure and programs the mark of the time that human beings spent in the savannah or on the forest floor, hunting, gathering, and reproducing with Darwinian gusto.

Three quite separate scientific theories do much to justify this grand narrative and the three similes that support it. In the first instance, computer science; in the second, theories of

On the Origins of the Mind

biological development; in the third, Darwin's theory of evolution. At times, indeed, it must seem that only the width of a cigarette paper separates evolutionary psychology from the power and the glory of the physical sciences themselves.

1. The Model for What Science Should Be

If the claims of evolutionary psychology are ambitious, the standard against which they should be assessed must be mature, reasonable, and persuasive. If nothing else, that standard must reflect principles that have worked to brilliant success in the physical sciences themselves. This is more than a gesture of respect; it is simple common sense.

In stressing the importance of their subject, the mathematicians J. H. Hubbard and B. H. West begin their textbook on differential equations by observing that "historically, Newton's spectacular success in describing mechanics by differential equations was a *model for what science should be*" (emphasis added). Hubbard and West then add what is in any case obvious: that "all basic physical laws are stated as differential equations, whether it be Maxwell's equations for electrodynamics, Schrödinger's equation for quantum mechanics, or Einstein's equations for general relativity."

Equations do lie close to the mathematician's heart, and differential equations closer than most. On one side of such an equation, there is a variable denoting an unknown mathematical function; on the other, a description of the rate at which that unknown function is changing at every last moment down to the infinitesimal. Within the physical sciences, such changes express the forces of nature, the Moon perpetually falling because perpetually accelerated by the universal force of gravitation. The mathematician's task is to determine the overall, or global, identity of the unknown function from its local rate of change.

In describing the world by means of a differential equation, the mind thus moves from what is local to what is global. It follows that the "model for what science should be" involves an interdiction against action at a distance. "One object," the Russian mathematician Mikhael Gromov observes, "cannot influence another one removed from it without involving local agents located one next to another and making a continuous chain joining the two objects." As for what happens when the interdiction lapses, Gromov, following the French mathematician René Thom, refers to the result as *magic*. This contrast between a disciplined, differential description of a natural process and an essentially magical description is a useful way of describing a fundamental disjunction in thought.

A differential equation, it is important to stress, offers only a general prescription for change. The distance covered by a falling object is a matter of how fast it has been going and how long it has been going fast; this, an equation describes. But how *far* an object has gone depends on how high it was when it began falling, and this, the underlying equation does not specify and so cannot describe. The solutions to a differential equation answer the question, how is the process changing? The data themselves answer a quite different question: how or where does the process *start?* Such specifications comprise the initial conditions of a differential equation and represent the intrusion into the mathematical world of circumstances beyond the mathematical.

It is this which in 1902 suggested to the French mathematician Jacques Hadamard the idea of a "well-posed problem" in analysis. For a differential equation to be physically useful, Hadamard argued, it must meet three requirements. Solutions must in the first place exist. They must be unique. And they must in some reasonable sense be stable, the solutions varying continuously as the initial conditions themselves change.

With these requirements met, a well-posed differential equation achieves a coordination among continuous quantities that is determined for every last crack and crevice in the manifold of time. And is this the standard I am urging on evolutionary psychology? Yes, absolutely.

Nothing but the best.

2. That the Human Mind Is Like a Digital Computer

Although evolutionary psychologists have embraced the computational theory of mind, it is not entirely a new theory; it has been entertained, if not embraced, in other places and at other times. Gottfried Leibniz wrote of universal computing machines in the seventeenth century, and only the limitations of seventeenth-century technology prevented him from toppling into the twenty-first. As it was, he did manage to construct a multipurpose calculator, which, he claimed, could perform the four elementary operations of addition, subtraction, division, and multiplication. But when he demonstrated the device to members of the Royal Society in London, someone in the wings noticed that he was carrying numbers by hand.

I do not know whether this story is true, but it has a very queer power, and in a discussion dominated by any number of similes it constitutes a rhetorical figure—shaped as a warning—all its own.

In 1936, the British logician Alan Turing published the first of his papers on computability. Using nothing more than ink, paper, and the resources of mathematical logic, Turing managed to create an imaginary machine capable of incarnating a very smooth, very suave imitation of the human mind.

Known now as a Turing machine, the device has at its disposal a tape divided into squares and a reading head mounted over the tape. It has, as well, a finite number of physical symbols, most commonly 0's and 1's. The reading head may occupy one of a finite number of distinct physical states. And thereafter the repertoire of its action is extremely limited. A Turing machine can, in the first place, recognize symbols, one square at a time. It can, in the second place, print symbols or erase them from the square it is scanning. And it can, in the third place, change its internal state, and move to the left or to the right of the square it is scanning, one square at a time.

There is no fourth place. A Turing machine can do nothing else. In fact, considered simply as a mechanism, a Turing machine can do nothing whatsoever, the thing existing in that peculiar world—my own, and I suspect others' as well—in which everything is possible but nothing gets done.

A Turing machine gains its powers of imitation only when, by means of a program or an algorithm, it is told what to do. The requisite instructions consist of a finite series of commands, written in a stylized vocabulary precisely calibrated to take advantage of those operations which a Turing machine can perform. What gives to the program its air of cool command is the fact that its symbols function in a double sense. They are symbols by virtue of their *meaning*, and so reflect the intentions of the human mind that has created them; but they are *causes* by virtue of their structure, and so enter into the rhythms of the real world. Like the word "bark," which both expresses a human command and sets a dog to barking, the symbols do double duty.

Although imaginary at its inception, a Turing machine brilliantly anticipated its own realization in matter. Through a process of intellectual parthenogenesis, Turing's ideas gave rise to the modern digital computer. And once the sheer physical palpability of the computer was recognized—there it is, as real as the rocks, the rifts, and the rills of the physical sciences—there was nothing to stand in the way of the first controlling simile of evolutionary psychology: that the human mind is itself a computer, one embodied in the human brain.

The promotion of the computer from an imaginary to a physical object serves the additional purpose of restoring it to the world that can be understood in terms of the "model for what science should be." As a physical device, nothing more than a collection of electronic circuits, the digital computer can be represented entirely by Clerk Maxwell's theory of the electromagnetic field, with the distinction between a Turing machine and its program duplicated in the distinction between a differential equation and its initial conditions. We are returned to the continuous and infinite world studied by mathematical physics, the world in which differential equations track the evolution of material objects moving through time in response to the eternal forces of nature itself.

The intellectual maneuvers that I have recounted serve to make the computer an irresistibly compelling object. But they serve, as well, to displace attention from the human mind. The effect is to endow the simile that the human mind is like a computer with a plausibility it might not otherwise enjoy.

A certain "power to alter things," Albertus Magnus observed, "indwells in the human soul." The *existence* of this power is hardly in doubt. It is evident in every human act in which the mind imposes itself on nature by taking material objects from their accustomed place and rearranging them; and it is evident again whenever a human being interacts with a machine. Writing with characteristic concision in the *Principia,* Isaac Newton observed that "the power and use of machines consist only in this, that by diminishing the velocity *we* may augment the force, and the contrary" (emphasis added). Although Newton's analysis was restricted to mechanical forces (he knew no others), his point is nonetheless general. A machine is a material object, a *thing,* and as such, its capacity to do work is determined by the forces governing its behavior and by its initial conditions.

Those initial conditions must themselves be explained, and in the nature of things they cannot be explained by the very device that *they* serve to explain. This is precisely the problem that Newton faced in the *Principia*. The magnificent "system of the world" that he devised explained why the orbits of the planets around the Sun must be represented by a conic section; but Newton was unable to account for the initial conditions that he had himself imposed on his system. Facing an imponderable, he appealed to divine intervention. It was not until Pierre Simon Laplace introduced his nebular hypothesis in 1796 that some form of agency was removed from Newtonian mechanics.

This same pattern, along with the problem it suggests, recurs whenever machines are at issue, and it returns with a vengeance whenever computers are invoked as explanations for the human mind. A computer is simply an electromechanical device, and this is precisely why it is useful as a model of the human brain. By setting its initial conditions, a computer's program allows the machine to do work in the real world. But the normal physical processes by which a computer works are often obscured by their unfamiliarity—who among us *really* understands what a computer is and how it works? No doubt, this is why the thesis that the mind is like a computer resonates with a certain intellectual grandeur.

An abacus conveys no comparable air of mystery. It is a trifle. Made of wood, it consists of a number of wires suspended in a frame and a finite number of beads strung along the wires. Nevertheless, an idealized abacus has precisely the power of a Turing machine, and so both the abacus and the Turing machine serve as models for a working digital computer. By parity of reasoning, they also both serve as models for the human mind.

Yet the thesis that the human mind is like an abacus seems distinctly less plausible than the thesis that the human mind is like a computer, and for obvious reasons. It is precisely when things have been reduced to their essentials that the interaction between a human being and a simple machine emerges clearly. That interaction is naked, a human agent handling an abacus with the same directness of touch that he might employ in handling a lever, a pulley, or an inclined plane. The force that human beings bring to bear on simple machines is muscular and so derived from the chemistry of the human body, the causes ultimately emptying out into the great ocean of physical interactions whose energy binds and loosens the world's large molecules. But what we need to know in the example of the abacus is not the nature of the forces controlling its behavior but the circumstances by which those forces come into play.

No chain of causes known to date accommodates the inconvenient fact that, by setting the initial conditions of a simple machine, a human agent brings about a novel, an unexpected, an entirely idiosyncratic distribution of matter. Every mechanical artifact represents what the anthropologist Mary Douglas calls "matter out of place." The problem that Newton faced but could not solve in the *Principia* returns when an attempt is made to provide a description of the simplest of human acts, the trivial tap or touch that sets a polished wooden bead spinning down a wire. Tracing the causal chain backward leads only to a wilderness of causes, each of them displacing material objects from their proper settings, so that in the end the mystery is simply shoveled back until the point is reached when it can be safely ignored.

A chain of physical causes is thus not obviously useful in explaining how a human agent exhibits the capacity to "alter things." But neither does it help to invoke, as some have done, the hypothesis that another abacus is needed to fix the initial conditions of the first. If each abacus requires yet another abacus in turn, the road lies open to the madness of an infinite regress, a point observed more than seventy years ago by the logicians Kurt Gödel and Alfred Tarski in their epochal papers on incompleteness.

If we are able to explain how the human mind works neither in terms of a series of physical causes nor in terms of a series of infinitely receding mechanical devices, what then is left? There is the ordinary, very rich, infinitely moving account of mental life that without hesitation we apply to ourselves. It is an account frankly magical in its nature. The human mind registers, reacts, and responds; it forms intentions, conceives problems, and then, as Aristotle dryly noted, it *acts*. In analyzing action, we are able to say only, as Albertus Magnus said, that a certain power to alter things inheres in the human soul.

A simile that for its persuasiveness depends on the very process it is intended to explain cannot be counted a great success.

3. That the Human Mind Is Like Any Other Organ of the Body

If the computational account of the human mind cannot be brought under the control of the "model for what science should be," what of the thesis that the human mind can be comprehended by reference to the laws of biological development? Here we come to the second simile of evolutionary psychology.

"As the ruler of the soul," Ptolemy wrote in the *Tetrabiblos,* "Saturn has the power to make men sordid, petty, mean-spirited, indifferent, mean-minded, malignant, cowardly, diffident, evil-speaking, solitary, tearful, shameless, superstitious, fond of toil, unfeeling, devisors of plots against their friends, gloomy, taking no care of their body." We know the type; there is no need to drown the point in words. Some men are just rotten.

On the Origins of the Mind

The analysis that Ptolemy offers in defense of his thesis is anything but crude. "The chronological starting point of human nativities," he writes, "is naturally the very time of conception, for to the seed is given once and for all the very qualities that will mark the adult and that are expressed in growth." It is Saturn's position that affects the seed, and the seed thereafter that affects the man.

Ptolemy's sophistication notwithstanding, no one today is much minded to study the *Tetrabiblos* as a guide to human psychology. Even if a convincing correlation could be established between the position of the planets and the onset of human rottenness, persuading us that we have identified some remote cause in nature for some human effect, that cause would quite obviously violate the interdiction against action at a distance. Ptolemy himself was sensitive to the distinction between astrological knowledge and real knowledge. In trying to construct a continuous chain between the position of the planets and the advent of human rottenness, he was at as great a loss as we are. It is for this reason that the word he employs to describe the way in which heavenly objects evoke their effects is *influence*; it is a word that does not appear, and is not needed, in the *Almagest,* Ptolemy's great treatise on astronomy.

More than two thousand years have gone by since Ptolemy composed the *Tetrabiblos*. The stars have withdrawn themselves; their role in human affairs has been assigned to other objects. Under views accepted by every evolutionary psychologist, the source of human rottenness may be found either in the environment or within the human genome.

The first of these, the environment, has been the perpetual Plaintiff of Record in *Nurture* v. *Nature et al*. But for our purposes it may now be dismissed from further consideration. If some men are made bad, then they are not born that way; and if they are not born that way, an explanation of human rottenness cannot be expressed in evolutionary terms.

The question at hand is thus whether the path initiated by the human genome in development can be understood in terms of "the model for what science should be." A dynamical system is plainly at work, one that transforms what Ptolemy called "the seed" into a fully formed human being in nine months, and then into an accomplished car thief in less than twenty years. What evolutionary psychology requires is a demonstration that this process may itself be brought under control of a description meeting the standard that "one object cannot influence another one removed from it without involving local agents located one next to another and making a continuous chain joining the two objects."

Well, can it?

"Our basic paradigm," Benjamin Levin writes in his textbook on genetics, "is that genes encode proteins, which in turn are responsible for the synthesis of other structures." Levin is a careful and a conscientious writer. By "other structures" he means only the nucleic acids. But his "basic paradigm" is now a part of a great cultural myth, and by "other structures" most evolutionary psychologists mean *all* of the structures that are made from the proteins, most notably the human brain.

The myth begins solidly enough—with the large bio-molecules that make up the human genome. The analysis of the precise, unfathomably delicate steps that take place as the genome undertakes various biochemical activities has been among the glories of modern science. Unfortunately, however, the chain of causes that begins in the human genome gutters out inconclusively long before the chain can be continued to the human brain, let alone the human mind. Consider in this regard the following sequence of representative quotations, in which tight causal connections are progressively displaced in favor of an ever more extravagant series of metaphors:

(1) *Quantum chemistry:* "For a molecule, it is reasonable to split the kinetic energy into two summations—one over the electrons, and one over the nuclei."

(2) *Biochemistry:* "Initiation of prokaryotic translation requires a tRNA bearing N-formyl methionine, as well as three initiation factors (IF1,2,3), a 30S ribosomal subunit GTP," etc.

(3) *Molecular biology:* "Once the protein binds one site, it *reaches* the other by *crawling* along the DNA, thus *preserving* its *knowledge* of the orientation of the first site" (emphasis added).

(4) *Embryology:* "In the embryo, cells divide, *migrate, die, stick to each other, send out* processes, and *form* synapses" (emphasis added).

(5) and (6) *Developmental genetics:* "But genes are simply regulatory elements, molecules that *arrange* their surrounding environments into an *organism*" (emphasis added).

"Genes *prescribe* epigenetic *rules,* which are the neural *pathways* and *regularities* in *cognitive development* by which the individual *mind assembles* itself" (emphasis added).

(7) *Developmental biology:* "The *pattern* of neural connections (synapses) *enables* the human cortex to *function* as the *center* for *learning, reasoning,* and *memory,* to *develop* the *capacity* for *symbolic expression,* and to *produce voluntary responses* to interpreted stimuli" (emphasis added).

(8) and (9) *Evolutionary psychology:* "Genes, of course, do *influence* human development" (emphasis added).
"[Genes] *created* us, body and mind" (emphasis added).

Now the very sober (1) and (2) are clearly a part of "the model for what science should be." By the time we come to (3), however, very large molecular chains have acquired powers of agency: they are busy reaching, crawling, and knowing; it is by no means clear that these metaphors may be eliminated in favor of a biochemical description. Much the same is true of (4). In (5) and (6), a connection is suggested between genes, on the one hand, and organisms, on the other, but the chain of causes and their effects has become very long, the crucial connections now entirely expressed in language that simply disguises profound gaps in our understanding.

In (7) the physical connection between morphology and the mind is reduced to wind, while (8) defiantly resurrects "influence," Ptolemy's original term of choice. It is the altogether exuberant (9)—the quotation is from Richard Dawkins—that finally drowns out any last faint signal from the facts.

These literary exercises suggest that the longer the chain of causes, the weaker the links between and among them. Whether this represents nothing more than the fact that our knowledge is incomplete, or whether it points to a conceptual deficiency that we have no way of expressing, let alone addressing—these are matters that we cannot now judge.

Curiously enough, it has been evolutionary psychologists themselves who are most willing to give up in practice what they do not have in theory. For were that missing theory to exist, it would cancel—it would *annihilate*—any last lingering claim we might make on behalf of human freedom. The physical sciences, after all, do not simply trifle with determinism: it is the heart and soul of their method. Were Boron salts at liberty to discard their identity, the claims of inorganic chemistry would seem considerably less pertinent than they do.

Thus, when Steven Pinker writes that "nature does not dictate what we should accept or how we should live our lives," he is expressing a hope entirely at odds with his professional commitments. If ordinary men and women are, like the professor himself, perfectly free to tell their genes to "go jump in the lake," why then pay the slightest attention to evolutionary psychology—why pay the slightest attention to Pinker?

Irony aside, a pattern is at work here. Where (in the first simile) computational accounts of the mind are clear enough to be encompassed by the model for what science should be, they are incomplete—radically so. They embody what they should explain. Where (in the second simile) biochemical and quantum chemical accounts of development are similarly clear and compelling, they extend no farther than a few large molecules. They defer what they cannot explain. In both cases, something remains unexplained.

This is a disappointing but perhaps not unexpected conclusion. We are talking, after all, about the human mind.

4. That the Human Mind Is Like Any Other Biological Artifact

Evolutionary psychologists believe that the only force in nature adequate to the generation of biological complexity is natural selection. It is an axiom of their faith. But although natural selection is often described as a force, it is certainly not a force of *nature*. There are four such forces in all: gravitational, electromagnetic, and the strong and weak forces. Natural selection is not one of them. It appears, for the most part, as a free-floating form of agency, one whose identity can only be determined by field studies among living creatures—the ant, the field mouse, and the vole.

But field studies have proved notoriously inconclusive when it comes to natural selection. After three decades spent observing Darwin's finches in the Galapagos, P. R. and B. R. Grant were in the end able to state only that "further continuous long-term studies are needed." It is the conclusion invariably established by evolutionary field studies, and it is the only conclusion established with a high degree of reliability.

The largest story told by evolutionary psychology is therefore anecdotal. Like other such stories, it subordinates itself to the principle that we are what we are because we were what we were. Who could argue otherwise? All too often, however, this principle is itself supported by the counter-principle that we were what we were because we are what we are, a circle not calculated to engender confidence.

Thus, in tests of preference, Victor Johnson, a bio-psychologist at New Mexico State University, has reported that men throughout the world designate as attractive women with the most feminine faces. Their lips are large and lustrous, their jaws narrow, their eyes wide. On display in every magazine and on every billboard, such faces convey "accented hormonal markers." These are a guide to fertility, and it is the promise of fertility that prompts the enthusiastic male response.

There is no reason to doubt Johnson's claim that on the whole men prefer pretty young women to all the others—the result, I am sure, of research extending over a score of years. It is the connection to fertility that remains puzzling. If male standards of beauty are rooted in the late Paleolithic era, men worldwide should now be looking for stout muscular women with broad backs, sturdy legs, a high threshold to pain, and a welcome eagerness to resume foraging directly after parturition. It has not been widely documented that they do.

In any case, an analysis of human sexual preferences that goes no farther than preferences is an exercise in tiptoeing to the threshold of something important and never peering over. The promise of evolutionary psychology is nothing less than an explanation of the human *mind*. No psychological theory could possibly be considered complete or even interesting that did not ask *why* men exhibit the tastes or undertake the choices they do. When it comes to sexual "preferences," what is involved is the full apparatus of the passions—beliefs, desires, sentiments, wishes, hopes, longings, aching tenderness. To study preferences without invoking the passions is like studying lightning without ever mentioning electricity.

This is one of those instances where evolutionary psychology betrays a queer family resemblance to certain theories in philosophy and psychology that (as we have seen in the case of determinism) evolutionary psychologists are themselves eager to disown. Behaviorism in psychology, as in the work of John Watson and B. F. Skinner, came to grief because human behavior is itself a contested category, and one that lapses into irrelevance once it is enlarged to accommodate the sources of behavior in the mind itself. It may be possible to analyze the mating strategies of the vole, the subject of much current research, by means of a simple assessment of what the vole does: a single genetic switch seems sufficient to persuade an otherwise uxorious male vole to become flamboyantly promiscuous. But human beings, it goes without saying, are not voles, and what *they* do becomes intelligible to them only when it is coordinated with what they are.

Despite the palpably unreliable stories that evolutionary psychologists tell about the past, *is* there, nevertheless, a scientifically reasonable structure that may be invoked to support those stories (as fine bones may support an otherwise frivolous face)?

The underlying tissue that connects the late Paleolithic and the modern era is the gene pool. Changes to that pool reflect a dynamic process in which genes undergo change, duplicate themselves, surge into the future or shuffle off, and by means of all the contingencies of life serve in each generation the purpose of creating yet another generation. This is the province of population genetics, a discipline given a remarkably sophisticated formulation in the 1930s and '40s by Ronald Fisher, J. B. S. Haldane, and Sewall Wright. Excellent mathematicians, these men were interested in treating evolution as a process expressed by some underlying system of equations. In the 1970s and '80s, the Japanese population geneticist Motoo Kimura revived and then extended their theories.

Kimura's treatise, *The Neutral Theory of Molecular Evolution* (1983), opens with words that should prove sobering to any evolutionary psychologist: "The neutral theory asserts that the great majority of evolutionary changes at the molecular level, as revealed by comparative studies of protein and DNA sequences, are caused not by Darwinian selection but by random drift of selectively neutral or nearly neutral mutants."

If Darwin's theory is a matter of random variation *and* natural selection, it is natural selection that is demoted on Kimura's view. Random variation is paramount; chance is the driving force. This is carefully qualified: Kimura is writing about "the great majority of evolutionary changes," not all. In addition, he is willing to accept the Darwinian disjunction: either complex adaptations are the result of natural selection or they are the result of nothing at all. But the effect of his work is clear: insofar as evolution is neutral, it is not adaptive, and insofar as it is not adaptive, natural selection plays no role in life.

Like his predecessors, Kimura writes within a particular tradition, one whose confines are fixed by the "model for what science should be." Thus, in trying to describe the fate of a mutant gene, Kimura is led to a differential equation—the Fokker-Planck equation, previously used to model diffusion processes. Although complicated, the equation has a straightforward interpretation. It describes the evolution of a probability distribution, tracking the likelihood over every instant of time that a specific gene will change its representation in a population of genes. Kimura is able to provide an explicit solution for the equation, and thus to treat molecular evolution as a well-posed problem in analysis.

But if the "model for what science should be" is powerful, it is also limited. Stretching it beyond its natural limits often turns out to be an exercise in misapplied force, like a blow delivered to the empty air.

As I have noted several times, the power of a differential equation to govern the flow of time is contingent on some specification of its initial conditions. It is precisely these initial conditions that anecdotal accounts of human evolution cannot supply. We can say of those hunters and gatherers only that they hunted and they gathered, and we can say this only because it seems obvious that there was nothing else for them to do. The gene pool that they embodied cannot be directly recovered.

The question very naturally arises: might that gene pool be recovered from the differential equations of mathematical genetics, much as the original position and momentum of a system of particles moving under the influence of gravitational forces might be recovered from their present position and momentum? This is the question posed by Richard Lewontin.[2] Writing in a recent issue of the *Annual Review of Genetics,* Lewontin observes that if Kimura's equations carry "a population forward in time from some initial conditions," then what is needed is a second theory, one "that can reverse the deductions of the first theory and infer backward from a particular observed state at present."

Lewontin is correct: this is precisely what is needed. Given the trajectory described by the solution of the Fokker-Planck equation, it *is* certainly possible to track the equation backward, past the Middle Ages, well past the Roman and then the Sumerian empires, and then into the era of the hunter-gatherers. There is nothing troubling about this. Kimura's equation has an explicit solution, and seeing where it led from is like running a film backward.

But whether, in running this particular film backward, we inevitably channel the temporal stream into a *unique* set of initial conditions is not altogether clear. With questions of this sort, we are in the domain of inverse problems, in which the past is contingent on the present. The solution to an inverse problem, the Russian mathematician Oleg Alifanov remarked, "entails determining unknown causes based on observation of their effects." It is this problem that evolutionary psychology must solve if its engaging stories about the Paleolithic era are to command credibility at the molecular level.

And it is this problem that Lewontin argues cannot be solved in the context of mathematical genetics. "A dynamical theory that predicts the present state generally requires that we know not only the nature and magnitude of the forces that have operated, but also the initial conditions and how long the process has been in operation." This double requirement—*know the forces, specify the initial conditions*—cannot simultaneously be met in going backward from the present. One item of knowledge is needed for the other.

This specific argument may now be enlarged to accommodate the general case. Inverse problems arise in mathematics when the attempt is made to run various mathematical films backward, and they are by now sufficiently well understood so that something may be said about them in a rough-and-ready way. Inverse problems are *not*, in general, well posed. Observing a pot of boiling liquid, we cannot use the heat equations to determine its identity. Many liquids reach the same boiling point in roughly the same time.

With inverse problems, what is, in fact, lost is the essential sureness and power of the "model for what science should be," and we are returned to a familiar world in which things and data are messy, disorganized, and partial, and in which theories, despite our best intentions, find themselves unable to peep over the hedge of time into the future or the past.

A familiar and by now depressing shape has revealed itself beneath the third and final simile of evolutionary psychology. It succeeds in meeting the demands of "the model for what science should be," but it succeeds in meeting those demands only at an isolated point. The rest is darkness, mystery, and magic.

5. THE ORIGINS OF THE HUMAN MIND

If the chief similes of evolutionary psychology have not improved our understanding of the human mind in any appreciable sense, might we at least say that they have done something toward promoting the field's principal hope, namely, that the mind will in the end take its place as a material object existing in a world of other material objects?

This too is by no means clear. As Leda Cosmides has very sensibly observed, evolutionary psychology is more a research program than a body of specific results. As a program, it rather resembles a weekend athlete forever preparing to embark on a variety of strenuous exercises. In the literature of evolutionary psychology, there is thus no very determined effort to assess any of the classical topics in the philosophy of mind with the aim of doing more than affirming vaguely that some aspect of the mind exists because it may well have been useful. There is, in evolutionary psychology, no account of the emotions beyond the trivial, or of the sentiments, no account of action or intention, no account of the human ability to acquire mathematical or scientific knowledge, no very direct exploration of the mind's power to act at a distance by investing things with meaning—no account, that is, of any of the features of the mind whose existence prompts a question about its origins. In its great hope, as in so many other respects, evolutionary psychology has reposed its confidence on the bet that in time these things will be explained. If that is so, all that we on the outside can say is that time will tell.

Yet any essay on evolutionary psychology would be incomplete if it did not acknowledge the moving power of its chief story. For that story, involving as it does our own ancestors, suggests that the human mind that we now occupy had its source in circumstances that, although occluded by time and damaged by distance, are nonetheless familiar.

The time is the distant past. "In Babylonia," the third-century historian Eusebius writes in recounting the lost histories of Berossos the Chaldean, a large number of people "lived without discipline and without order, just like the animals." A frightening monster named Oannes then appeared to the Babylonians after clambering out of the Red Sea. "It had the whole body of a fish, but underneath and attached to the head of the fish there was another head, human, and joined to the tail of the fish, feet, like those of a man, and it had a human voice." The monster "spent his days with men, never eating anything, but teaching men the skills necessary for writing, and for doing mathematics, and for all sorts of knowledge."

Since that time, Eusebius adds regretfully, "nothing further has been discovered."

Notes

1. This essay first appeared in *Commentary*, vol. 118, no. 4, November 2004, 26–35. It is reprinted with permission.
2. I am grateful to Robert Berwick of MIT for calling my attention to this article, and for insisting on its importance.

32

Consciousness[1]

John R. Searle

1. Resistance to the Problem[2]

As recently as two decades ago there was little interest among neuroscientists, philosophers, psychologists, and cognitive scientists generally in the problem of consciousness. Reasons for the resistance to the problem varied from discipline to discipline. Philosophers had turned to the analysis of language, psychologists had become convinced that a scientific psychology must be a science of behavior, and cognitive scientists took their research program to be the discovery of the computer programs in the brain that, they thought, would explain cognition. It seemed especially puzzling that neuroscientists should be reluctant to deal with the problem of consciousness, because one of the chief functions of the brain is to cause and sustain conscious states. Studying the brain without studying consciousness would be like studying the stomach without studying digestion, or studying genetics without studying the inheritance of traits. When I first got interested in this problem seriously and tried to discuss it with brain scientists, I found that most of them were not interested in the question.

The reasons for this resistance were various but they mostly boiled down to two. First, many neuroscientists felt—and some still do—that consciousness is not a suitable subject for neuroscientific investigation. A legitimate brain science can study the microanatomy of the Purkinje cell, or attempt to discover new neurotransmitters, but consciousness seems too airy-fairy and touchy-feely to be a real scientific subject. Others did not exclude consciousness from scientific investigation, but they had a second reason: "We are not ready" to tackle the problem of consciousness. They may be right about that, but my guess is that a lot of people in the early 1950s thought we were not ready to tackle the problem of the molecular basis of life and heredity. They were wrong; and I suggest, for the current question, the best way to get ready to deal with a research problem may be to try to solve it.

There were, of course, famous earlier twentieth-century exceptions to the general reluctance to deal with consciousness, and their work has been valuable. I am thinking in particular of the work of Sir Arthur Sherrington, Roger Sperry, and Sir John Eccles.

Whatever was the case twenty years ago, today many serious researchers are attempting to tackle the problem. Among neuroscientists who have written recent books about consciousness are Cotterill, Crick, Damasio, Edelman, Freeman, Gazzaniga, Greenfield, Hobson, Libet, and Weiskrantz.[3] As far as I can tell, the race to solve the problem of consciousness is already on. My aim here is not to try to survey this literature but to characterize some of the neurobiological problems of consciousness from a philosophical point of view.

2. Consciousness as a Biological Problem

What exactly is the neurobiological problem of consciousness? The problem, in its crudest terms, is this: How exactly do brain processes cause conscious states and how exactly are those states realized in brain structures? So stated, this problem naturally breaks down into a number of smaller but still large problems: What exactly are the neurobiological correlates of conscious states (NCC), and which of those correlates are actually causally responsible for the production of consciousness? What are the principles according to which biological phenomena such as neuron firings can bring about subjective states of sentience or awareness? How do those principles relate to the already well understood principles of biology? Can we explain consciousness with the existing theoretical apparatus or do we need some revolutionary new theoretical concepts to explain it? Is consciousness localized in certain regions of the brain or is it a global phenomenon? If it is confined to certain regions, which ones? Is it correlated with specific anatomical features, such as specific types of neurons, or is it to be explained functionally with a variety of anatomical correlates? What is the right level for explaining consciousness? Is it the level of neurons and synapses, as most researchers seem to think, or do we have to go to higher functional levels such as neuronal maps,[4] or whole clouds of neurons;[5] or are all of these levels much too high so that we have to go below the level of neurons and synapses to the level of the microtubules?[6] Or do we have to think much more globally in terms of Fourier transforms and holography?[7] As stated, this cluster of problems sounds similar to any other such set of problems in biology or in the sciences in general. It sounds like the problem concerning microorganisms: How, exactly, do they cause disease symptoms, and how are those symptoms manifested in patients? Or the problem in genetics: By what mechanisms exactly does the genetic structure of the zygote produce the phenotypical traits of the mature organism? In the end I think that is the right way to think of the problem of consciousness—it is a biological problem like any other, because consciousness is a biological phenomenon in exactly the same sense as digestion, growth, or photosynthesis. But unlike other problems in biology, there is a persistent series of philosophical problems that surround the problem of consciousness, and before addressing some current research I would like to address some of these problems.

3. Identifying the Target: The Definition of Consciousness

One often hears it said that "consciousness" is frightfully hard to define. But if we are talking about a definition in common-sense terms, sufficient to identify the target of the investigation, as opposed to a precise scientific definition of the sort that typically comes at the end of a scientific investigation, then the word does not seem to me hard to define. Here is the definition: Consciousness consists of inner, qualitative, subjective states and processes of sentience or awareness. Consciousness, so defined, begins when we wake in the morning from a dreamless sleep and continues until we fall asleep again, die, go into a coma, or otherwise become "unconscious." It includes all of the enormous variety of the awareness that we think of as characteristic of our waking life. It includes everything from feeling a pain, to perceiving objects visually, to states of anxiety and depression, to working out crossword puzzles, playing chess, trying to remember your aunt's phone number, arguing about politics, or to just wishing you were somewhere else. Dreams, on this definition, are a form of consciousness, though of course they are in many respects quite different from waking consciousness.

This definition is not universally accepted, and the word consciousness is used in a variety of other ways. Some authors use the word to refer only to states of self-consciousness, that is, the consciousness that humans and some primates have of themselves as agents. Some use it to refer to the second-order mental states about other mental states; so according to this definition, a pain would not be a conscious state, but worrying about a pain would be a conscious state. Some use "consciousness" behavioristically to refer to any form of complex intelligent behavior. It is, of course, open to anyone to use any word any way he likes, and we can always redefine consciousness as a technical term. Nonetheless, there is a genuine phenomenon of consciousness in the ordinary sense, however we choose to name it; and it is that phenomenon that I am trying to identify now, because I believe it is the proper target of the investigation.

Consciousness has distinctive features that we need to explain. Because I believe that some, not all, of the problems of consciousness are going to have a neurobiological solution, what follows is a shopping list of what a neurobiological account of consciousness should explain.

4. The Essential Feature of Consciousness: The Combination of Qualitativeness, Subjectivity, and Unity

Consciousness has three aspects that make it different from other biological phenomena and, indeed, different from other phenomena in the natural world. These three aspects are qualitativeness, subjectivity, and unity. I used to think that for investigative purposes we could treat them as three distinct features, but because they are logically interrelated, I now think it best to treat them together, as different aspects of the same feature. They are not separate because the first implies the second, and the second implies the third. I discuss them in order.

Qualitativeness

Every conscious state has a certain qualitative feel to it, and you can see this clearly if you consider examples. The experience of tasting beer is very different from hearing Beethoven's Ninth Symphony, and both of those have a different qualitative character from smelling a rose or seeing a sunset. These examples illustrate the different qualitative features of conscious experiences. One way to put this point is to say that for every conscious experience there is something that it feels like, or something that it is like, to have that conscious experience. Nagel made this point over two decades ago when he pointed out that if bats are conscious, then there is something that "it is like" to be a bat.[8] This distinguishes consciousness from other features of the world, because in this sense, for a nonconscious entity such as a car or a brick, there is nothing that "it is like" to be that entity. Some philosophers describe this feature of consciousness with the word "qualia," and they say there is a special problem of qualia. I am reluctant to adopt this usage because it seems to imply that there are two separate problems, the problem of consciousness and the problem of qualia. But, as I understand these terms, "qualia" is just a plural name for conscious states. Because "consciousness" and "qualia" are coextensive, there seems no point in introducing a special term. Some people think that qualia are characteristic of only perceptual experiences, such as seeing colors and having sensations such as pains, but that there is no qualitative character to thinking. As I understand these terms, that is wrong. Even conscious thinking has a qualitative feel to it. There is something it is like to think that two plus two equals four. There is no way to describe it except by saying that it is the character of thinking consciously "two plus two equals four." But if you believe there is no qualitative character to thinking that, then try to think the same thought in a language you do not know well. If I think in French, "deux et deux fait quatre," I find that it feels quite different. Or try thinking, more painfully, "two plus two equals one hundred eighty-seven." Once again, I think you will agree that these conscious thoughts have different characters. However, the point must be trivial; that is, whether or not conscious thoughts are qualia must follow from our definition of qualia. As I am using the term, thoughts definitely are qualia.

Subjectivity

Conscious states exist only when they are experienced by some human or animal subject. In that sense, they are essentially subjective. I used to treat subjectivity and qualitativeness as distinct features, but it now seems to me that properly understood, qualitativeness implies subjectivity, because in order for there to be a qualitative feel to some event, there must be some subject that experiences the event. No subjectivity, no experience. Even if more than one subject experiences a similar phenomenon, say, two people listening to the same concert, all the same, the qualitative experience can exist only as experienced by some subject or subjects. And even if the different token experiences are qualitatively identical—that is, they all exemplify the same type—nonetheless each token experience can exist only if the subject of that experience has it. Because conscious states are subjective in this sense, they have what I call a first-person ontology, as opposed to the third-person ontology of mountains and molecules, which can exist even if no living creatures exist. Subjective conscious states have a first-person ontology ("ontology" here means mode of existence) because they exist only when they are experienced by some human or animal agent. They are experienced by some "I" that has the experience, and it is in that sense that they have a first-person ontology.

Unity

All conscious experiences at any given point in an agent's life come as part of one unified conscious field. If I am sitting at my desk looking out the window, I do not just see the sky above and the brook below shrouded by the trees, and at the same time feel the pressure of my body against the chair, the shirt against my back, and the aftertaste of coffee in my mouth. Rather, I experience all of these as part of a single, unified conscious field. This unity of any state of qualitative subjectivity has important consequences for a scientific study of consciousness. I say more about them later on. At present I just want to call attention to the fact that the unity is already implicit in subjectivity and qualitativeness for the following reason: If you try to imagine that my conscious state is broken into seventeen parts, what you imagine is not a single conscious subject with seventeen different conscious states but rather seventeen different centers of consciousness. A conscious state, in short, is by definition unified, and the unity will follow from the subjectivity and the qualitativeness, because there is no way you could have subjectivity and qualitativeness except with that particular form of unity.

There are two areas of current research where the aspect of unity is especially important. These are, first, the study of the split-brain patients by Gazzaniga and others,[9] and, second, the study of the binding problem by a number of contemporary researchers. The interest of the split-brain patients is that both the anatomical and the behavioral evidence suggest that in these patients there are two centers of consciousness that, after commissurotomy, are communicating with each other only imperfectly. They seem to have, so to speak, two conscious minds inside one skull.

The interest of the binding problem is that it looks like this problem might give us in microcosm a way of studying the nature of consciousness because, just as the visual system binds all of the different stimulus inputs into a single, unified visual percept, so the entire brain somehow unites all of the variety of our different stimulus inputs into a single unified conscious experience.

Several researchers have explored the role of synchronized neuron firings in the range of forty Hz to account for the capacity of different perceptual systems to bind the diverse stimuli of anatomically distinct neurons into a single perceptual experience.[10] For example, in the case of vision, anatomically separate neurons specialized for such things as line, angle, and color all contribute to a single, unified, conscious visual experience of an object. Crick extended the proposal for the binding problem to a general hypothesis about the NCC.[11] He put forward a tentative hypothesis that perhaps the NCC consists of synchronized neuron firings in the general range of forty Hz in various networks in the thalamocortical system, specifically in connections between the thalamus and layers four and six of the cortex.

This kind of instantaneous unity has to be distinguished from the organized unification of conscious sequences that we get from short-term or iconic memory. For non-pathological forms of consciousness, at least some memory is essential in order that the conscious sequence across time can come in an organized fashion. For example, when I speak a sentence, I have to be able to remember the beginning of the sentence at the time I get to the end if I am to produce coherent speech. Whereas instantaneous unity is essential to, and is part of, the definition of consciousness, organized unity across time is essential to the healthy functioning of the conscious organism, but it is not necessary for the very existence of conscious subjectivity.

This combined feature of qualitative, unified subjectivity is the essence of consciousness, and it, more than anything else, is what makes consciousness different from other phenomena studied by the natural sciences. The problem is to explain how brain processes, which are objective third-person biological, chemical, and electrical processes, produce subjective states of feeling and thinking. How does the brain get us over the hump, so to speak, from events in the synaptic cleft and the ion channels to conscious thoughts and feelings? If you take seriously this combined feature as the target of explanation, I believe you get a different sort of research project from what is currently the most influential. Most neurobiologists take what I call the building-block approach: Find the NCC for specific elements in the conscious field, such as the experience of color, and then construct the whole field out of such building blocks. Another approach, which I call the unified-field approach, takes the research problem to be one of explaining how the brain produces a unified field of subjectivity to start with. On the unified-field approach, there are no building blocks; rather there are just modifications of the already existing field of qualitative subjectivity. I say more about this later.

Some philosophers and neuroscientists think we can never have an explanation of subjectivity: We can never explain why warm things feel warm and red things look red. To these skeptics there is a simple answer: We know it happens. We know that brain processes cause all of our inner qualitative, subjective thoughts and feelings. Because we know that it happens, we ought to try to figure out how it happens. Perhaps in the end we will fail, but we cannot assume the impossibility of success before we try.

Many philosophers and scientists also think that the subjectivity of conscious states makes it impossible to have a strict science of consciousness. For, they argue, if science is by definition objective, and consciousness is by definition subjective, it follows that there cannot be a science of consciousness. This argument is fallacious. It commits the fallacy of ambiguity over the terms objective and subjective. Here is the ambiguity: We need to distinguish two different senses of the objective–subjective distinction. In one sense, the epistemic sense ("epistemic" here means having to do with knowledge), science is indeed objective. Scientists seek truths that are equally accessible to any competent observer and that are independent of the feelings and attitudes of the experimenters in question. An example of an epistemically objective claim would be "Bill Clinton weighs 210 pounds." An example of an epistemically subjective claim would be "Bill Clinton was a good president." The first is objective because its truth or falsity is settleable in a way that is independent of the feelings and attitudes of the investigators. The second is subjective because it is not so settleable.

But there is another sense of the objective–subjective distinction, and that is the ontological sense ("ontological" here means having to do with existence). Some entities, such as pains, tickles, and itches, have a subjective mode of existence, in the sense that they exist only as experienced by a conscious subject. Others, such as mountains, molecules, and tectonic plates, have an objective mode of existence, in the sense that their existence does not depend on any consciousness. The point of making this distinction is to call attention to the fact that the scientific requirement of epistemic objectivity does not preclude ontological subjectivity as a domain of investigation. There is no reason whatever why we cannot have an objective science of pain, even though pains only exist when they are felt by conscious agents. The ontological subjectivity of the feeling of pain does not preclude an epistemically objective science of pain. Though many philosophers and neuroscientists are reluctant to think of subjectivity as a proper domain of scientific investigation, in actual practice we work on it all the time. Any neurology textbook will contain extensive discussions of the etiology and treatment of such ontologically subjective states as pains and anxieties.

5. Some Other Features

To keep this list short, I mention some other features of consciousness only briefly.

Feature 2: Intentionality

Most important, conscious states typically have "intentionality," that property of mental states by which they are directed at or about objects and states of affairs in the world. Philosophers use the word intentionality not just for "intending" in the ordinary sense but for any mental phenomena at all that have referential content. According to this usage, beliefs, hopes, intentions, fears, desires, and perceptions all are intentional. So if I have a belief, I must have a belief about something. If I have a normal visual experience, it must seem to me that I am actually seeing something, and so forth. Not all conscious states are intentional and not all intentionality is conscious; for example, undirected anxiety lacks intentionality, and the beliefs a man has even when he is asleep lack consciousness then and there. But I think it is obvious that many of the important evolutionary functions of consciousness are intentional: For example, an animal has conscious feelings of hunger and thirst, engages in conscious perceptual discriminations, embarks on conscious intentional actions, and consciously recognizes both friend and foe. All of these are conscious intentional phenomena and all are essential for biological survival. A general neurobiological account of consciousness will explain the intentionality of conscious states. For example, an account of color vision will naturally explain the capacity of agents to make color discriminations.

Feature 3: The Distinction between the Center and the Periphery of Attention

It is a remarkable fact that within my conscious field, at any given time, I can shift my attention at will from one aspect to another. So, for example, right now I am not paying any attention to the pressure of the shoes on my feet or the feeling of the shirt on my neck. But I can shift my attention to them any time I want. There is already a fair amount of useful work done on attention.

Feature 4: All Human Conscious Experiences Are in Some Mood or Other

There is always a certain flavor to one's conscious states, always an answer to the question "How are you feeling?" The moods do not necessarily have names. Right now I am not especially elated or annoyed, not ecstatic or depressed, not even just blah. But, all the same, I will become acutely aware of my mood if there is a dramatic change, if I receive some extremely good or bad news, for example. Moods are not the same as emotions, though the mood we are in will predispose us to having certain emotions.

We are, by the way, closer to having pharmacological control of moods with such drugs as Prozac than we are to having control of other internal features of consciousness.

Feature 5: All Conscious States Come to Us in the Pleasure/Unpleasure Dimension

For any total conscious experience there is always an answer to the question of whether it was pleasant, painful, unpleasant, neutral, and so forth. The pleasure/unpleasure feature is not the same as mood, though of course some moods are more pleasant than others.

Feature 6: Gestalt Structure

The brain has a remarkable capacity to organize very degenerate perceptual stimuli into coherent conscious perceptual forms. I can, for example, recognize a face, or a car, on the basis of very limited stimuli. The best known examples of Gestalt structures come from the researches of the Gestalt psychologists.

Feature 7: Familiarity

There is, in varying degrees, a sense of familiarity that pervades our conscious experiences. Even if I see a house I have never seen before, I still recognize it as a house; it is of a form and structure that is familiar to me. Surrealist painters try to break this sense of the familiarity and ordinariness of our experiences, but, even in surrealist paintings, the drooping watch still looks like a watch, and the three-headed dog still looks like a dog.

One could continue this list, and I have done so in other writings.[12] The point now is to get a minimal shopping list of the features that we want a neurobiology of consciousness to explain. In order to look for a causal explanation, we need to know what the effects are that need explanation. Before examining some current research projects, we need to clear more of the ground.

6. The Traditional Mind–Body Problem and How to Avoid It

The confusion about objectivity and subjectivity I mentioned earlier is just the tip of the iceberg of the traditional mind–body problem. Though ideally I think scientists would be better off if they ignored this problem, the fact is that they are as much victims of the philosophical traditions as anyone else, and many scientists, like many philosophers, are still in the grip of the traditional categories of mind and body, mental and physical, dualism and materialism, etc. This is not the place for a detailed discussion of the mind–body problem, but I need to say a few words about it so that, in the discussion that follows, we can avoid the confusions it has engendered.

The simplest form of the mind–body problem is this: What exactly is the relation of consciousness to the brain? There are two parts to this problem, a philosophical part and a scientific part. I have already been assuming a simple solution to the philosophical part. The solution, I believe, is consistent with everything we know about biology and about how the world works. It is this: Consciousness and other sorts of mental phenomena are caused by neurobiological processes in the brain, and they are realized in the structure of the brain. In a word, the conscious mind is caused by brain processes and is itself a higher-level feature of the brain.

The philosophical part is relatively easy but the scientific part is much harder. How, exactly, do brain processes cause consciousness and how, exactly, is consciousness realized in the brain? I want to be very clear about the philosophical part because it is not possible to approach the scientific question intelligently if the philosophical issues are unclear. Notice two features of the philosophical solution. First, the relationship of brain mechanisms to consciousness is one of causation. Processes in the brain cause our conscious experiences. Second, this does not force us to any kind of dualism because the form of causation is bottom-up, and the resulting effect is simply a higher-level feature of the brain itself, not a separate substance. Consciousness is not like some fluid squirted out by the brain. A conscious state is rather a state that the brain is in. Just as water can be in a liquid or solid state without liquidity and solidity being separate substances, so consciousness is a state that the brain is in without consciousness being a separate substance.

Notice that I stated the philosophical solution without using any of the traditional categories of "dualism," "monism," "materialism," and all the rest of it. Frankly, I think those categories are obsolete. But if we accept those categories at face value, then we get the following picture: You have a choice between dualism and materialism. According to dualism, consciousness and other mental phenomena exist in a different ontological realm altogether from the ordinary physical world of physics, chemistry, and biology. According to materialism, consciousness, as I have described it, does not exist. Neither dualism nor materialism, as traditionally construed, allows us to get an answer to our question. Dualism says that there are two kinds of phenomena in the world, the mental and the physical; materialism says that there is only one, the material. Dualism ends up with an impossible bifurcation of reality into two separate categories and thus makes it impossible to explain the relation between the mental and the physical. But materialism ends up denying the existence of any irreducible subjective qualitative states of sentience or awareness. In short, dualism makes the problem insoluble; materialism denies the existence of any phenomenon to study, and hence of any problem.

On the view that I am proposing, we should reject those categories altogether. We know enough about how the world works to know that consciousness is a biological phenomenon caused by brain processes and realized in the structure of the brain. It is irreducible not because it is ineffable or mysterious, but because it has a first-person ontology and therefore cannot be reduced to phenomena with a third-person ontology. The traditional mistake that people have made in both science and philosophy has been to suppose that if we reject dualism, as I believe we must, then we have to embrace materialism. But on the view that I am putting forward, materialism is just as confused as dualism, because it denies the existence of ontologically subjective consciousness in the first place. Just to give it a name, the resulting view that denies both dualism and materialism I call biological naturalism.

7. How Did We Get into This Mess? A Historical Digression

For a long time I thought scientists would be better off if they ignored the history of the mind–body problem, but I now think that unless you understand something about the history, you will always be in the grip of historical categories. I discovered this when I was debating people in artificial intelligence and found that many of them were in the grip of Descartes, a philosopher many of them had not even read.

What we now think of as the natural sciences did not really begin with ancient Greece. The Greeks had almost everything, and in particular they had the wonderful idea of a "theory." The invention of the idea of a theory—a systematic set of logically related propositions that attempt to explain the phenomena of some domain—was perhaps the greatest single achievement of Greek civilization. However, they did not have the institutionalized practice of systematic observation and experiment. That came only after the Renaissance, especially in the seventeenth century. When you combine systematic experiment and testability with the idea of a theory, you get the possibility of science as we think of it today. But there was a feature of the seventeenth century that was a local accident and is still blocking our path. It is that in the seventeenth century, there was a very serious conflict between science and religion, and it seemed that science was a threat to religion. Part of the way that the apparent threat posed by science to orthodox Christianity was deflected was due to Descartes and Galileo. Descartes, in particular, argued that reality divides into two kinds, the mental and the physical, *res cogitans* and *res extensa*. Descartes made a useful division of the territory: Religion had the territory of the soul, and science could have material reality. But this gave people the mistaken concept that science could deal only with objective third-person phenomena; it could not deal with the inner qualitative subjective experiences that make up our conscious life. This was a perfectly harmless move in the seventeenth century because it kept the church authorities off the backs of the scientists. (It was only partly successful. Descartes, after all, had to leave Paris and live in Holland where there was more tolerance, and Galileo had to make his famous recantation to the church authorities of his heliocentric theory of the planetary system.) However, this history has left us with a tradition and a tendency not to think of consciousness as an appropriate subject for the natural sciences, in the way that we think of disease, digestion, or tectonic plates as subjects of the natural sciences. I urge us to overcome this reluctance, and in order to overcome it, we need to overcome the historical tradition that made it seem perfectly natural to avoid the topic of consciousness altogether in scientific investigation.

8. Summary of the Argument to This Point

I am assuming that we have established the following: Consciousness is a biological phenomenon like any other. It consists of inner qualitative subjective states of perceiving, feeling, and thinking. Its essential feature is unified, qualitative subjectivity. Conscious states are caused by neurobiological processes in the brain, and they are realized in the structure of the brain. To say this is analogous to saying that digestive processes are caused by chemical processes in the stomach and the rest of the digestive tract, and that these processes are realized in the stomach and the digestive tract. Consciousness differs from other biological phenomena in that it has a subjective or first-person ontology. But ontological subjectivity does not prevent us from having epistemic objectivity. We can still have an objective science of consciousness. We abandon the traditional categories of dualism and materialism for the same reason we abandon the categories of phlogiston and vital spirits: They have no application to the real world.

9. The Scientific Study of Consciousness

How, then, should we proceed in a scientific investigation of the phenomena involved?

Seen from the outside, it looks deceptively simple. There are three steps. First, one finds the neurobiological events that are correlated with consciousness (the NCC). Second, one tests to see that the correlation is a genuine causal relation. And third, one tries to develop a theory, ideally in the form of a set of laws, that would formalize the causal relationships.

These three steps are typical of the history of science. Think, for example, of the development of the germ theory of disease. First we find correlations between brute empirical phenomena. Then we test the correlations for causality by manipulating one variable and seeing how it affects the others. Then we develop a theory of the mechanisms involved and test the theory by further experiment. For example, Semmelweis in Vienna in the 1840s found that women obstetric patients in hospitals died more often from puerperal fever than did those who stayed at home. So he looked more closely and found that women examined by medical students who had just come from the autopsy room without washing their hands had an exceptionally high rate of puerperal fever. Here was an empirical correlation. When he made these young doctors wash their hands in chlorinated lime, the mortality rate went way down. He did not yet have the germ theory of disease, but he was moving in that direction. In the study of consciousness we appear to be in the early Semmelweis phase.

At the time of this writing we are still looking for the NCC. Suppose, for example, we found, as Crick once put forward as a tentative hypothesis, that the neurobiological correlate of consciousness was a set of neuron firings between the thalamus and the cortex layers four and six, in the range of forty Hz. That would be step one, and step two would be to manipulate the phenomena in question to see if you could show a causal relation. Ideally, we need to test for whether the NCC in question is both necessary and sufficient for the existence of consciousness. To establish necessity, we find out whether a subject who has the putative NCC removed thereby loses consciousness; to establish sufficiency, we find out whether an otherwise unconscious subject can be brought to consciousness by inducing the putative NCC.

Pure cases of causal sufficiency are rare in biology, and we usually have to understand the notion of conditions against a set of background presuppositions, that is, within a specific biological context. Thus, our sufficient conditions for consciousness would presumably operate only in a subject who was alive, had his brain functioning at a certain level of activity, at a certain appropriate temperature, and so forth. But what we are trying to establish ideally is a proof that the element is not just correlated with consciousness, but that it is both causally necessary and sufficient, other things being equal, for the presence of consciousness. Seen from the outsider's point of view, that looks like the ideal way to proceed.

Why has it not yet been done? I do not know. It turns out, for example, that it is very hard to find an exact NCC, and the current investigative tools, most notably in the form of positron emission tomography scans, CAT scans, and functional magnetic resonance imaging techniques, have not yet identified the NCC. There are interesting differences between the scans of conscious subjects and sleeping subjects with REM sleep, on the one hand, and slow-wave sleeping subjects on the other. But it is not easy to tell how much of the differences are related to consciousness. Many things are going on in both the conscious and the unconscious subjects' brains that have nothing to do with the production of consciousness. Given that a subject is already conscious, you can get parts of his or her brain to light up by getting him or her to perform various cognitive tasks, such as perception or memory. But that does not give you the difference between being conscious in general and being totally unconscious. So, to establish this first step, we still appear to be in an early state of the technology of brain research. In spite of all of the hype surrounding the development of imaging techniques, we still, as far as I know, have not found a way to image the NCC.

With all this in mind, let us turn to some actual efforts at solving the problem of consciousness.

10. The Standard Approach to Consciousness: The Building-Block Model

Most theorists tacitly adopt the building-block theory of consciousness. The idea is that any conscious field is made of its various parts: the visual experience of red, the taste of coffee, the feeling of the wind coming in through the window. It seems that if we could figure out what makes even one building block conscious, we would have the key to the whole structure. If we could, for example, crack visual consciousness, that would give us the key to all the other modalities. This view is explicit in the work of Crick and Koch.[13] Their idea is that if we could find the NCC for vision, then we could explain visual consciousness, and we would then know what to look for to find the NCC for hearing and for the other modalities, and if we put all those together, we would have the whole conscious field.

The strongest and most original statement I know of the building-block theory is by Bartels and Zeki.[14] They see the binding activity of the brain not as one that generates a conscious experience that is unified, but rather one that brings together a whole lot of already conscious experiences. As they put it, "[C]onsciousness is not a unitary faculty, but . . . it consists of many microconsciousnesses."[15] Our field of consciousness is thus made up of a lot of building blocks of microconsciousnesses. "Activity at each stage or node of a processing-perceptual system has a conscious correlate. Binding cellular activity at different nodes is therefore not a process preceding or even facilitating conscious experience, but rather bringing different conscious experiences together."[16]

There are at least three lines of research that are consistent with, and often used to support, the building-block theory.

Blindsight

Blindsight is the name given by the psychologist Weiskrantz to the phenomenon whereby certain patients with damage to V1 can report incidents occurring in their visual field even though they report no visual awareness of the stimulus. For example, in the case of DB, the earliest patient studied, if an X or an O were shown on a screen in that portion of DB's visual field where he was blind, the patient, when asked what he saw, would deny that he saw anything. But if asked to guess, he would guess correctly that it was an X or an O. His guesses were right nearly all the time. Furthermore, the subjects in these experiments are usually surprised at their results. When the experimenter asked DB in an interview after one experiment, "Did you know how well you had done?" DB answered, "No, I didn't, because I couldn't see anything. I couldn't see a darn thing."[17] This research has subsequently been carried on with a number of other patients, and blindsight is now also experimentally induced in monkeys.[18]

Some researchers suppose that we might use blindsight as the key to understanding consciousness. The argument is the following: In the case of blindsight, we have a clear difference between conscious vision and unconscious information processing. It seems that if we could discover the physiological and anatomical difference between regular sight and blindsight, we might have the key to analyzing consciousness because we would have a clear neurological distinction between the conscious and the unconscious cases.

Binocular Rivalry and Gestalt Switching

One exciting proposal for finding the NCC for vision is to study cases where the external stimulus is constant but where the internal subjective experience varies. Two examples of this are the Gestalt switch, where the same figure, such as the Necker cube, is perceived in two different ways, and binocular rivalry, where different stimuli are presented to each eye but the visual experience at any instant is of one or the other stimulus, not both. In such cases, the experimenter has a chance to isolate a specific NCC for the visual experience independently of the neurological correlates of the retinal stimulus.[19] The beauty of this research is that it seems to isolate a precise NCC for a precise conscious experience. Because the external stimulus is constant and there are (at least) two different conscious experiences, A and B, it seems there must be some point in the neural pathways where one sequence of neural events causes experience A and another point where a second sequence causes experience B. Find those two points and you have found the precise NCCs for two different building blocks of the whole conscious field.

The Neural Correlates of Vision

Perhaps the most obvious way to look for the NCC is to track the neurobiological causes of a specific perceptual modality, such as vision. In a recent article, Crick and Koch assume as a working hypothesis that only some specific types of neurons will manifest the NCC.[20] They do not think that any of the NCCs of vision are in V1.[21] The reason for thinking that V1 does not contain the NCC is that V1 does not connect to the frontal lobes in such a way that would make V1 contribute directly to the essential information-processing aspect of visual perception. Their idea is that the function of visual consciousness is to provide visual information directly to the parts of the brain that organize voluntary motor output, including speech. Thus, because the information in V1 is recoded in subsequent visual areas and does not transmit directly to the frontal cortex, they believe that V1 does not correlate directly with visual consciousness.

11. Doubts about the Building-Block Theory

The building-block theory may be right but it has some worrisome features. Most important, all the research done to identify the NCCs has been carried out with subjects who are already conscious, independently of the NCC in question. Going through the cases in order, the problem with the blindsight research as a method of discovering the NCC is that the patients in question only exhibit blindsight if they are already conscious. That is, it is only in the case of fully conscious patients that we can elicit the evidence of information processing that we get in the blindsight examples. So we cannot investigate consciousness in general by studying the difference between the blindsight patient and the normally sighted patient because both patients are fully conscious. It might turn out that what we need in our theory of consciousness is an explanation of the conscious field that is essential to both blindsight and normal vision or, for that matter, to any other sensory modality.

Similar remarks apply to the binocular rivalry experiments. All this research is immensely valuable, but it is not clear how it will give us an understanding of the exact differences between the conscious brain and the unconscious brain because, for both experiences in binocular rivalry, the brain is fully conscious.

Similarly, Crick and Koch investigated only subjects who were already conscious.[22] What one wants to know is: how is it possible for the subject to be conscious at all? Given that a subject is conscious, his consciousness will be modified by having a visual experience, but it does not follow that this consciousness is made up of various building blocks of which the visual experience is just one.

I wish to state my doubts precisely. There are (at least) two possible hypotheses:

1. *The building-block theory:* The conscious field is made up of small components that combine to form the field. To find the causal NCC for any component is to find an element that is causally necessary and sufficient for that conscious experience. Hence, to find even one is, in an important sense, to crack the problem of consciousness.

2. *The unified-field theory* (explained in more detail below): Conscious experiences come in unified fields. In order to have a visual experience, a subject has to be conscious already and the experience is a modification of the field. Neither blindsight, binocular rivalry, nor normal vision can give us a genuine causal NCC because only already conscious subjects can have these experiences.

It is important to emphasize that both hypotheses are rival empirical hypotheses to be settled by scientific research and not by philosophical argument. Why then do I prefer hypothesis 2 to hypothesis 1? The building-block theory predicts that in a totally unconscious patient, if the patient meets certain minimal physiological conditions (he is alive, the brain is functioning normally, he has the right temperature, etc.), and if you could trigger the NCC for, say, the experience of red, then the unconscious subject would suddenly have a conscious experience of red and nothing else. One building block is as good as another. Research may prove me wrong, but on the basis of what little I know about the brain, I do not believe that is possible. Only a brain that is already over the threshold of consciousness, that already has a conscious field, can have a visual experience of red.

Furthermore, on the multistage theory of Bartels and Zeki, the microconsciousnesses are all capable of a separate and independent existence.[23] It is not clear to me what this means. I know what it is like for me to experience my current conscious field, but who experiences all the tiny microconsciousnesses? And what would it be like for each of them to exist separately?

12. Basal Consciousness and a Unified-Field Theory

There is another way to look at matters that implies another research approach. Imagine that you wake from a dreamless sleep in a completely dark room. So far, you have no coherent stream of thought and almost no perceptual stimulus. Save for the pressure of your body on the bed and the sense of the covers on top of your body, you are receiving no outside sensory stimuli. All the same, there must be a difference in your brain between the state of minimal wakefulness you are now in and the state of unconsciousness you were in before. That difference is the NCC I believe we should be looking for. This state of wakefulness is basal or background consciousness.

Now you turn on the light, get up, move about, and so forth. What happens? Do you create new conscious states? Well, in one sense you obviously do, because previously you were not consciously aware of visual stimuli and now you are. But do the visual experiences stand to the whole field of consciousness in the part–whole relation? Well, that is what nearly everybody thinks and what I used to think, but here is another way of looking at it. Think of the visual experience of the table not as an object in the conscious field the way the table is an object in the room, but think of the experience as a modification of the conscious field, as a new form that the unified field takes. As Llinas and his colleagues put it, consciousness is "modulated rather than generated by the senses."[24]

I want to avoid the part–whole metaphor, but I also want to avoid the proscenium metaphor. We should not think of my new experiences as new actors on the stage of consciousness but rather as new bumps or forms or features in the unified field of consciousness. What is the difference? The proscenium metaphor gives us a constant background stage with various actors on it. I think that is wrong. There is just the unified conscious field, nothing else, and it takes different forms.

If this is the right way to look at things (and again this is a hypothesis on my part, nothing more), then we get a different sort of research project. There is no such thing as a separate visual consciousness, so looking for the NCC for vision is barking up the wrong tree. Only the already conscious subject can have visual experiences, so the introduction of visual experiences is not an introduction of consciousness but a modification of a pre-existing consciousness.

The research program that is implicit in the hypothesis of unified-field consciousness is that at some point we need to investigate the general condition of the conscious brain as opposed to the condition of the unconscious brain. We will not explain the general phenomenon of unified, qualitative subjectivity by looking for specific local NCCs. The important question is not what the NCC for visual consciousness is, but how does the visual system introduce visual experiences into an already unified conscious field, and how does the brain create that unified conscious field in the first place. The problem becomes more specific. What we are trying to find is which features of a system that is made up of a hundred billion discrete elements, neurons, connected by synapses, can produce a conscious field of the sort that I have described. There is a perfectly ordinary sense in which consciousness is unified and holistic, but the brain is not in that way unified and holistic. So what we have to look for is some massive activity of the brain capable of producing a unified holistic conscious experience.

For reasons that we now know from lesion studies, we are unlikely to find this as a global property of the brain, and we have very good reason to believe that activity in the thalamocortical system is probably the place to look for unified-field consciousness. The working hypothesis would be that consciousness is in large part localized in the thalamocortical system and that the various other systems feed information to the thalamocortical system that produces modifications corresponding to the various sensory modalities. To put it simply, I do not believe we will find visual consciousness in the visual system and auditory consciousness in the auditory system. We will find a single, unified, conscious field containing visual, auditory, and other aspects.

Notice that if this hypothesis is right, it will solve the binding problem for consciousness automatically. The production of any state of consciousness at all by the brain is the production of a unified consciousness.

We are tempted to think of our conscious field as made up of the various components—visual, tactile, auditory, stream of thought, etc. The approach whereby we think of big things as being made up of little things has proved so spectacularly successful in the rest of science that it is almost irresistible to us. Atomic theory, the cellular theory in biology, and the germ theory of disease are all examples. The urge to think of consciousness as likewise made of smaller building blocks is overwhelming. But I think it may be wrong for consciousness. Maybe we should think of consciousness holistically, and perhaps for consciousness we can make sense of the claim that "the whole is greater than the sum of the parts." Indeed, maybe it is wrong to think of consciousness as made up of parts at all. I want to suggest that if we think of consciousness holistically, then the aspects I have mentioned so far, especially our original combination of subjectivity, qualitativeness, and unity all in one feature, will seem less mysterious. Instead of thinking of my current state of consciousness as made up of the various bits—the perception of the computer screen, the sound of the brook outside, the shadows cast by the evening sun falling on the wall—we should think of all of these as modifications, forms that the underlying basal conscious field takes after my peripheral nerve endings have been assaulted by the various external stimuli. The research implication of this is that we should look for consciousness as a feature of the brain emerging from the activities of large masses of neurons, which cannot be explained by the activities of individual neurons. I am, in sum, urging that we take the unified-field approach seriously as an alternative to the more common building-block approach.

13. VARIATIONS ON THE UNIFIED-FIELD THEORY

The idea that one should investigate consciousness as a unified field is not new—it goes back at least as far as Kant's doctrine of the transcendental unity of apperception.[25] In neurobiology I have not found any contemporary authors who state a clear distinction between what I have been calling the building-block theory and the unified-field theory, but at least two lines of contemporary research are consistent with the approach urged here, the work of Llinas and his colleagues and that of Tononi, Edelman, and Sporns.[26]

On the view of Llinas and his colleagues,[27] we should not think of consciousness as produced by sensory inputs but rather as a functional state of large portions of the brain, primarily the thalamocortical system, and we should think of sensory inputs serving to modulate a pre-existing consciousness rather that creating consciousness anew. On their view, consciousness is an "intrinsic" state of the brain, not a response to sensory stimulus inputs. Dreams are of special interest to them, because in a dream the brain is conscious but unable to perceive the external world through sensory inputs. They believe the NCC is synchronized oscillatory activity in the thalamocortical system.[28]

Tononi and Edelman have advanced what they call the dynamic core hypothesis.[29] They are struck by the fact that consciousness has two remarkable properties, the unity mentioned earlier and the extreme differentiation or complexity within any conscious field. This suggests to them that we should not look for consciousness in a specific sort of neuronal type, but rather in the activities of large neuronal populations. They seek the NCC for the unity of consciousness in the rapid integration that is achieved through the reentry mechanisms of the thalamocortical system. The idea they have is that in order to account for the combination of integration and differentiation in any conscious field, they have to identify large clusters of neurons that function together and fire in a synchronized fashion. Furthermore, this cluster, which they call a functional cluster, should also show a great deal of differentiation within its component elements in order to account for the different elements of consciousness. They think that synchronous firing among cortical regions between the cortex and the thalamus is an indirect indicator of this functional clustering. Then, once such a functional cluster has been identified, they wish to investigate whether it contains different activity patterns of neuronal states. The combination of functional clustering together with differentiation they submit as the dynamic core hypothesis of consciousness. They believe a unified neural process of high complexity constitutes a "dynamic core." They also believe the dynamic core is not spread over the brain but is primarily in the thalamocortical regions, especially those involved in perceptual categorization, and contains reentry mechanisms of the sort that Edelman discussed in his earlier books.[30] In a new study, they and their colleagues claim to find direct evidence of the role of reentry mapping in the NCC.[31] Like the adherents of the building-block theory, they seek such NCCs of consciousness as one can find in the studies of binocular rivalry.

As I understand this view, it seems to combine features of both the building-block and the unified-field approach.

14. Conclusion

In my view, the most important problem in the biological sciences today is the problem of consciousness. I believe we are now at a point where we can address this problem as a biological problem like any other. For decades, research has been impeded by two mistaken views: first, that consciousness is just a special sort of computer program, a special software in the hardware of the brain; and second, that consciousness is just a matter of information processing. The right sort of information processing—or on some views any sort of information processing—would be sufficient to guarantee consciousness. I have criticized these views at length elsewhere and do not repeat these criticisms here.[32] But it is important to remind ourselves how profoundly antibiological these views are. On these views brains do not really matter. We just happen to be implemented in brains, but any hardware that could carry the program or process the information would do just as well. I believe, on the contrary, that understanding the nature of consciousness crucially requires understanding how brain processes cause and realize consciousness. Perhaps when we understand how brains do that, we can build conscious artifacts using some nonbiological materials that duplicate, and not merely simulate, the causal powers that brains have. But first we need to understand how brains do it.

Notes

1. This article is an updated version of the talk "Current Research into Consciousness," given at the *Nature of Nature* conference, April 15, 2000. This text has been reprinted, with permission, from John R. Searle, "Consciousness," in *Annual Review of Neuroscience,* vol. 23, no. 1, 2000, 557–78; it also appears as chapter 3 in John R. Searle, *Consciousness and Language.* Cambridge: Cambridge University Press, 2002, 36–60.

2. I am indebted to many people for discussion of the issues [in this essay]. None of them is responsible for any of my mistakes. I especially wish to thank Samuel Barondes, Dale Berger, Francis Crick, Gerald Edelman, Susan Greenfield, Jennifer Hudin, John Kihlstrom, Jessica Samuels, Dagmar Searle, Wolf Singer, Barry Smith, and Gunther Stent.

3. Cotterill 1998; Crick 1994; Damasio 1999; Edelman 1989, 1992; Freeman 1995; Gazzaniga 1988; Greenfield 1995; Hobson 1999; Libet 1993; and Weiskrantz 1997.

4. Edelman 1989, 1992.

5. Freeman 1995.

6. Penrose 1994; Hameroff 1998a,b.

7. Pribram 1976, 1991, 1999.

8. Nagel 1974.

9. Gazzaniga 1998; Gazzaniga *et al.* 1962, 1963.

10. Llinas 1990; Llinas and Pare 1991; Llinas and Ribary 1992, 1993; Singer 1993, 1995; Singer and Gray 1995.

11. Crick 1994.

12. Searle 1992.

13. Crick and Koch 1998 (reprinted in this volume).

14. Bartels and Zeki 1998; Zeki and Bartels 1998.

15. Bartels and Zeki 1998: 2327.

16. Bartels and Zeki 1998: 2330.

17. Weiskrantz 1986: 24.

18. Stoerig and Cowey 1997.

19. Logothetis 1998; Logothetis and Schall 1989.

20. Crick and Koch 1998.

21. Crick & Koch 1995.

22. Crick 1996; Crick and Koch 1998.

23. Bartels & Zeki 1998; Zeki and Bartels 1998.

24. Llinas *et al.* 1998: 1841.

25. Kant 1787.

26. Llinas 1990; Llinas *et al.* 1998; Tononi & Edelman 1998; Tononi *et al.* 1992, 1998.

27. Llinas *et al.* 1998.

28. Llinas *et al.* 1998: 1845.

29. Tononi and Edelman 1998.

30. Edelman 1989, 1992.

31. Srinivasan *et al.* 1999.

32. Searle 1980, 1992, 1997.

References

Bartels, A., and Zeki, S. (1998) "The theory of multistage integration in the visual brain." *Proceedings of the Royal Society of London* B 265: 2327–32.

Cotterill, R. (1998) *Enchanted Looms: Consciousness Networks in Brains and Computers.* Cambridge, UK: Cambridge University Press.

Crick, F. (1994) *The Astonishing Hypothesis: The Scientific Search for the Soul.* New York: Scribner.

Crick, F. (1996) "Visual perception: rivalry and consciousness." *Nature* 379: 485–86.

Crick, F., and Koch, C. (1995) "Are we aware of neural activity in primary visual cortex?" *Nature* 374: 121–23.

Crick, F., and Koch, C. (1998) "Consciousness and neuroscience." *Cerebral Cortex* 8: 97-107 (reprinted in this volume).

Damasio, A. (1999) *The Feeling of What Happens, Body and Emotion in the Making of Consciousness.* New York: Harcourt Brace Jovanovich.

Edelman, G. (1989) *The Remembered Present: A Biological Theory of Consciousness.* New York: Basic Books.

Edelman, G. (1992) *Bright Air, Brilliant Fire: On the Matter of the Mind.* New York: Basic Books.

Freeman, W. (1995) *Societies of Brains: A Study in the Neuroscience of Love and Hate.* Hillsdale, NJ: Erlbaum.

Gazzaniga, M. (1988) *How Mind and Brain Interact to Create Our Conscious Lives.* Boston: Houghton Mifflin; and Cambridge, MA: in association with MIT Press.

Gazzaniga, M. (1998) "The split brain revisited." *Scientific American* 279: 35–39.

Gazzaniga, M., J. Bogen, and R. Sperry. (1962) "Some functional effects of sectioning the cerebral commissures in man." *Proceedings of the National Academy of Sciences USA* 48: 1765–69.

———. (1963) "Laterality effects in somesthesis following cerebral commissurotomy in man." *Neuropsychologia* 1: 209–15.

Greenfield, S. (1995) *Journeys to the Centers of the Mind: Toward a Science of Consciousness.* New York: Freeman.

Hameroff, S. (1998a) "Funda-Mentality: Is the conscious mind subtly linked to a basic level of the universe?" *Trends in Cognitive Science* 2 (4): 119–27.

Hameroff, S. (1998b) "Quantum computation in brain microtubules? The Penrose–Hameroff 'Orch OR' model of consciousness. *Philosophical Transactions of the Royal Society of London* A 356: 1869–96.

Hobson, J. (1999) *Consciousness.* New York: Scientific American Library (Freeman).

Kant, I. (1787) *The Critique of Pure Reason.* Riga: Hartknock.

Libet, B. (1993) *Neurophysiology of Consciousness: Selected Papers and New Essays.* Boston: Birkhauser.

Llinas, R. (1990) "Intrinsic electrical properties of mammalian neurons and CNS function." *Fidea Research Foundation Neuroscience Award Lecture* 4: 1–10.

Llinas, R., and D. Pare. (1991) "Of dreaming and wakefulness." *Neuroscience* 44: 521–35

Llinas, R., and U. Ribary. (1992) "Rostrocaudal scan in human brain: a global characteristic of the 40-Hz response during sensory input," in *Induced Rhythms in the Brain,* Bullock Basar, ed. Boston: Birkhauser, 147–54.

Llinas, R., and U. Ribary. (1993) "Coherent 40-Hz oscillation characterizes dream state in humans." *Proceedings of the National Academy of Sciences USA* 90: 2078–81.

Llinas, R., U. Ribary, D. Contreras, and C. Pedroarena. (1998) "The neuronal basis for consciousness." *Philosophical Transactions of the R.oyal Society of London* B 353: 1841–49.

Logothetis, N. (1998) "Single units and conscious vision." *Philosophical Transactions of the R.oyal Society of London* B 353: 1801–18.

Logothetis, N., and J. Schall. (1989) "Neuronal correlates of subjective visual perception." *Science* 245: 761–63.

Consciousness

Nagel, T. (1974) "What is it like to be a bat?" *Philosophical Review* 83: 435–50.

Penrose, R. (1994) *Shadows of the Mind: A Search for the Missing Science of Consciousness.* New York: Oxford University Press.

Pribram, K. (1976) "Problems concerning the structure of consciousness," in *Consciousness and Brain: A Scientific and Philosophical Inquiry,* G. Globus, G. Maxwell, and I. Savodnik, eds. New York: Plenum, 297–313.

———. (1991) *Brain and Perception: Holonomy and Structure in Figural Processing.* Hillsdale, NJ: Erlbaum.

———. (1999) "Brain and the composition of conscious experience." *Journal of Consciousness Studies* 6 (5): 19–42.

Searle, J. R. (1980) "Minds, brains and programs." *Behavioral and Brain Sciences* 3: 417–57.

———. (1983) *Intentionality: An Essay in the Philosophy of Mind.* Cambridge: Cambridge University Press.

———. (1984) *Minds, Brains and Science.* Cambridge, MA: Harvard University Press.

———. (1992) *The Rediscovery of the Mind.* Cambridge, MA: MIT Press.

———. (1997) *The Mystery of Consciousness.* New York: New York Review Book.

Singer, W. (1993) "Synchronization of cortical activity and its putative role in information processing and learning." *Annual Review of Physiology* 55: 349–75.

Singer, W. (1995) "Development and plasticity of cortical processing architectures." *Science* 270: 758–64.

Singer, W., and Gray, C. (1995) "Visual feature integration and the temporal correlation hypothesis." *Annual Review of Neuroscience* 18: 555–86.

Srinivasan, R., D. Russell, G. Edelman, and G. Tononi. (1999) "Frequency tagging competing stimuli in binocular rivalry reveals increased synchronization of neuromagnetic responses during conscious perception." *Journal of Neuroscience* 19: 5435–48.

Stoerig, P., and A. Cowey. (1997) "Blindsight in man and monkey." *Brain* 12: 535–59.

Tononi, G., and G. Edelman. (1998) "Consciousness and complexity." *Science* 282: 1846–851.

Tononi, G., G. Edelman, and O. Sporns. (1998) "Complexity and coherency: integrating information in the brain." *Trends in Cognitive Science* 2:12: 474–84.

Tononi, G., O. Sporns, and G. Edelman. (1992) "Reentry and the problem of integrating multiple cortical areas: simulation of dynamic integration in the visual system." *Cerebral Cortex* 2: 310–35.

Tononi, G., R. Srinivasan, D. Russell, and G. Edelman. (1998) "Investigating neural correlates of conscious perception by frequency-tagged neuromagnetic responses." *Proceedings of the National Academy of Sciences USA* 95: 3198–203.

Weiskrantz, L. (1986) *Blindsight: A Case Study and Implications.* New York: Oxford University Press.

Weiskrantz, L. (1997) *Consciousness Lost and Found.* Oxford: Oxford University Press.

Zeki, S., and A. Bartels. (1998) "The autonomy of the visual systems and the modularity of conscious vision." *Philosophical Transactions of the Royal Society of London* B 353: 1911–14.

33

Consciousness and Neuroscience[1]

Francis Crick and Christof Koch

> "When all's said and done, more is said than done."
> —Anon.

The main purpose of this chapter is to set out for neuroscientists one possible approach to the problem of consciousness and to describe the relevant ongoing experimental work. We have not attempted an exhaustive review of other approaches.

1. Clearing the Ground

We assume that when people talk about "consciousness," there is something to be explained. While most neuroscientists acknowledge that consciousness exists, and that at present it is something of a mystery, most of them do not attempt to study it, mainly for one of two reasons:

(1) They consider it to be a philosophical problem, and so best left to philosophers.

(2) They concede that it is a scientific problem, but think it is premature to study it now.

We have taken exactly the opposite point of view. We think that most of the philosophical aspects of the problem should, for the moment, be left on one side, and that the time to start the scientific attack is now.

We can state bluntly the major question that neuroscience must first answer: It is probable that at any moment some active neuronal processes in your head correlate with consciousness, while others do not—what is the difference between them? In particular, are the neurons involved of any particular neuronal type? What is special (if anything) about their connections? And what is special (if anything) about their way of firing? The "neuronal correlates of consciousness" are often referred to as the NCC. Whenever some information is represented in the NCC, it is represented in consciousness.

In approaching the problem, we made the tentative assumption that all the different aspects of consciousness (pain, visual awareness, self-consciousness, and so on) employ a basic common mechanism, or perhaps a few such mechanisms.[2] If one could understand the mechanism for one aspect, then, we hope, we will have gone most of the way towards understanding them all.

We made the personal decision that several topics should be set aside or merely stated without further discussion, for experience had shown us that otherwise valuable time can be wasted arguing about them without coming to any conclusion.[3]

(1) Everyone has a rough idea of what is meant by being conscious. For now, it is better to avoid a precise definition of consciousness because of the dangers of premature definition. Until the problem is understood much better, any attempt at a formal definition is likely to be either misleading or overly restrictive, or both. If this seems evasive, try defining the word "gene." So much is now known about genes that any simple definition is likely to be inadequate. How much more difficult, then, to define a biological term when rather little is known about it.

(2) It is plausible that some species of animals—in particular the higher mammals—possess some of the essential features of consciousness, but not necessarily all. For this reason, appropriate experiments on such animals may be relevant to finding the mechanisms underlying consciousness. It follows that a language system (of the type found in humans) is not essential for consciousness—that is, one can have the key features of consciousness without language. (This is not to say that language does not enrich consciousness considerably.)

(3) It is not profitable at this stage to argue about whether simpler animals (such as octopi, fruit flies, nematodes), or even plants, are conscious.[4] It is probable, however, that consciousness correlates to some extent with the degree of complexity of any nervous system. When one clearly understands, both in detail and in principle, what consciousness involves in humans, then will be the time to consider the problem of consciousness in much simpler animals. For the same reason, we won't ask whether some parts of our nervous system have a special, isolated consciousness of their own. If you say, "Of course my spinal cord is conscious, but it's not telling me," we are not, at this stage, going to spend time arguing with you about it. Nor will we spend time discussing whether a digital computer could be conscious.

(4) There are many forms of consciousness, such as those associated with seeing, thinking, emotion, pain, and so on. Self-consciousness—that is, the self-referential aspect of consciousness—is probably a special case of consciousness. In our view, it is better left to one side for the moment, especially as it would be difficult to study self-consciousness in a monkey. Various rather unusual states—such as the hypnotic state, lucid dreaming, and sleepwalking—will not be considered here, since they do not seem to us to have special features that would make them experimentally advantageous.

2. Visual Consciousness

How can one approach consciousness in a scientific manner? Consciousness takes many forms, but for an initial scientific attack, it usually pays to concentrate on the form that appears easiest to study. We chose visual consciousness rather than other forms because humans are very visual animals and our visual percepts are especially vivid and rich in information. In addition, the visual input is often highly structured yet easy to control.

The visual system has another advantage. There are many experiments that, for ethical reasons, cannot be done on humans but can be done on animals. Fortunately, the visual system of primates appears fairly similar to our own,[5] and many experiments on vision have already been done on animals such as the macaque monkey.

This choice of the visual system is a personal one. Other neuroscientists might prefer one of the other sensory systems. It is, of course, important to work on alert animals. Very light anesthesia may not make much difference to the response of neurons in macaque V1, but it certainly does to neurons in cortical areas like V4 or IT (inferotemporal).

3. Why Are We Conscious?

We have suggested that the biological usefulness of visual consciousness in humans is to produce the best current interpretation of the visual scene in the light of past experience, either of ourselves or of our ancestors (embodied in our genes), and to make this interpretation directly available, for a sufficient time, to the parts of the brain that contemplate and plan voluntary motor output, of one sort or another, including speech.[6]

Philosophers, in their carefree way, have invented a creature they call a "zombie," who is supposed to act just as normal people do but to be completely *un*conscious.[7] This seems to us to be an untenable scientific idea, but there is now suggestive evidence that part of the brain does behave like a zombie. That is, in some cases, a person uses the current visual input to produce a relevant motor output, without being able to say what was seen. Milner and Goodale point out that a frog has at least two independent systems for action[8] as shown by Ingle.[9] These may well be unconscious. One is used by the frog to snap at small, prey-like objects, and the other for jumping away from large, looming discs. Why doesn't our brain consist simply of a series of such specialized zombie systems?

We suggest that such an arrangement is inefficient when very many such systems are required. Better to produce a single but complex representation and make it available for a sufficient time to the parts of the brain that make a choice among many different but possible plans for action. This, in our view, is what seeing is about. As pointed out to us by Ramachandran and Hirstein, it is sensible to have a *single* conscious interpretation of the visual scene, in order to eliminate hesitation.[10]

Milner and Goodale suggest that in primates there are two systems, which we shall call the on-line system and the seeing system.[11] The latter is conscious, while the former, acting more rapidly, is not. The general characteristics of these two systems, and some of the experimental evidence for them, are outlined below in the section on the on-line system. There is anecdotal evidence from sports. It is often stated that a trained tennis player reacting to a fast serve has no time to see the ball; the seeing comes afterwards. In a similar way, a sprinter is believed to start running before he consciously hears the starting pistol.

4. The Nature of the Visual Representation

We have argued elsewhere that, to be aware of an object or event, the brain has to construct a multilevel, explicit, symbolic interpretation of part of the visual scene.[12] By "multilevel," we mean, in psychological terms, different levels such as those that correspond, for example, to lines or eyes or faces. In neurological terms, we mean, loosely, the different levels in the visual hierarchy.[13]

The important idea is that the representation should be explicit. We have had some difficulty getting this idea across.[14] By "explicit representation," we mean a smallish group of neurons which employ coarse coding (as it is called[15]) to represent some *aspect* of the visual scene. In the case of a particular face, all of these neurons can fire to somewhat face-like objects.[16] We postulate that one set of such neurons will be all of one type (say, one type of pyramidal cell in one particular layer or sublayer of cortex), will probably be fairly close together, and will all project to roughly the same place. If all such groups of neurons—there may be several of them, stacked one above the other—were destroyed, then the person would not see a face, though he or she might be able to see the parts of a face, such as the eyes, the nose, the mouth, etc. There may be other places in the brain that explicitly represent other aspects of a face, such as the emotion the face is expressing.[17]

Notice that while the *information* needed to represent a face is contained in the firing of the ganglion cells in the retina, there is, in our terms, no explicit representation of the face there.

How many neurons are there likely to be in such a group? This is not yet known, but we would guess that the number to represent one aspect is likely to be closer to the 100 to 1000 range than the 10,000 to 1,000,000 range.

A representation of an object or an event will usually consist of representations of many of the relevant aspects of it, and these are likely to be distributed, to some degree, over different parts of the visual system. How these different representations are bound together is known as the "binding problem."[18]

Much neural activity is usually needed for the brain to construct a representation. Most of this is probably unconscious. It may prove useful to consider this unconscious activity as the computations needed to find the best interpretation, while the interpretation itself may be considered to be the *results* of these computations, only some of which we are then conscious of. To judge from our perception, the results probably have something of a winner-take-all character.

As a working hypothesis we have assumed that only some types of specific neurons will express the NCC. It is already known that the firing of many cortical cells does not correspond to what the animal is currently seeing (see the discussion under "Bistable Percepts," below). An alternative possibility is that the NCC is necessarily global.[19] In one extreme form, this would mean that, at one time or another, any neuron in the cortex and associated structures could express the NCC. At this point, we feel it more fruitful to explore the simpler hypothesis—that only particular types of neurons express the NCC—before pursuing the more global hypothesis. It would be a pity to miss the simpler one if it were true. As a rough analogy, consider a typical mammalian cell. The way its complex behavior is controlled and influenced by its genes could be considered to be largely global, but its genetic instructions are localized and coded in a relatively straightforward manner.

5. Where Is the Visual Representation?

The conscious visual representation is likely to be distributed over more than one area of the cerebral cortex and possibly over certain subcortical structures as well. We have argued that in primates, contrary to most received opinion, it is not located in cortical area V1 (also called the "striate cortex" or "area 17").[20] Some of the experimental evidence in support of this hypothesis is outlined below. This is not to say that what goes on in V1 is not important, and indeed, it may be crucial for most forms of vivid visual awareness. What we suggest is that the neural activity there is not directly correlated with what is seen.

We have also wondered whether the visual representation is largely confined to certain neurons in the lower cortical layers (layers 5 and 6).[21] This hypothesis is still very speculative.

6. What Is Essential for Visual Consciousness?

The term "visual consciousness" almost certainly covers a variety of processes. When one is actually looking at a visual scene, the experience is very vivid. This should be contrasted with the much less vivid and less detailed visual images produced by trying to remember the same scene. (A vivid recollection is usually called a hallucination.) We are concerned here mainly with the normal vivid experience. (It is possible that our dimmer visual recollections are mainly due to the back pathways in the visual hierarchy acting on the random activity in the earlier stages of the system.)

Some form of very short-term memory seems almost essential for consciousness, but this memory may be very transient, lasting for only a fraction of a second. Edelman has used the striking phrase "the remembered present" to make this point.[22] The existence of iconic memory, as it is called, is well-established experimentally.[23]

Psychophysical evidence for short-term memory suggests that if we do not pay attention to some part or aspect of the visual scene, our memory of it is very transient and can be overwritten (masked) by the visual stimuli that follow.[24] This probably explains many of our fleeting memories when we drive a car over a familiar route. If we do pay attention (e.g., a child running in front of the car) our recollection of this can be longer lasting. Our impression that at any moment we see all of a visual scene clearly and in great detail is illusory, partly due to ever-present eye movements, and partly due to our ability to use the scene itself as a readily available form of memory, since in most circumstances the scene usually changes rather little over a short span of time.[25]

Although working memory expands the time frame of consciousness,[26] it is not obvious that it is *essential* for consciousness. It seems to us that working memory is a mechanism for bringing an item—or a small sequence of items—into vivid consciousness, by speech (or silent speech), for example. In a similar way, the episodic memory enabled by the hippocampal system is not essential for consciousness, though a person without it is severely handicapped.[27] Consciousness, then, is enriched by visual attention, though attention is not essential for visual consciousness to occur.[28] Attention is broadly of two types: bottom-up, caused by the sensory input; and top-down, produced by the planning parts of the brain. This is a complicated subject, and we will not try to summarize here all the experimental and theoretical work that has been done on it.

Visual attention can be directed to either a location in the visual field or to one or more (moving) objects.[29] The exact neural mechanisms that achieve this are still being debated. In order to interpret the visual input, the brain must arrive at a *coalition* of neurons whose firing represents the best interpretation of the visual scene, often in competition with other possible but less likely interpretations; and there is evidence that attentional mechanisms appear to bias this competition.[30]

7. Recent Experimental Results

We shall not attempt to describe all the various experimental results of direct relevance to the search for the neuronal correlates of visual consciousness in detail, but rather, outline a few of them and point the reader to fuller accounts.

7.1 Action without Seeing

Classical Blindsight

This will already be familiar to most neuroscientists. It is discussed, along with other relevant topics, in an excellent book by Weiskrantz.[31] It occurs in humans (where it is rare) when there is extensive damage to cortical area V1, and has also been reproduced in monkeys.[32] In a typical case, the patient can indicate, well above chance level, the direction of movement of a spot of light over a certain range of speed, while denying that he sees anything at all. If the movement is less salient, his performance falls to chance; if more salient (that is, brighter or faster), he may report that he had some ill-defined visual percept, considerably different from the normal one. Other patients can distinguish large, simple shapes or colors.[33]

The pathways involved have not yet been established. The most likely one is from the superior colliculus to the pulvinar and from there to parts of visual cortex; several other known weak anatomical pathways from the retina and bypassing V1 are also possible. Recent functional magnetic resonance imaging of the blindsight patient G.Y. directly implicated the superior colliculus as being active *specifically* when G.Y. correctly discriminated the direction of motion of some stimulus without being aware of it at all.[34]

The On-line System

The broad properties of the two hypothetical systems—the on-line system and the seeing system—are shown in Table 1, following the account by Milner and Goodale in their book *The Visual Brain in Action*.[35] The on-line system may have multiple subsystems (e.g., for eye movements, for arm movements, for body posture adjustment, and so on). Normally, the two systems work in parallel, and indeed there is evidence that in some circumstances the seeing system can interfere with the on-line system.[36]

One striking piece of evidence for an on-line system comes from studies on patient D.F. by Milner, Perrett, and their colleagues.[37] D.F.'s brain had diffuse damage produced by carbon monoxide poisoning. She was able to see color and texture very well, but was deficient in seeing orientation and form. In spite of this, she was very good at catching a ball. She could "post" her hand or a card into a slot without difficulty, though she could not report the slot's orientation.

It is obviously important to discover the difference between the on-line system, which is unconscious, and the seeing system, which is conscious. Milner and Goodale suggest that the on-line system mainly uses the dorsal visual stream.[38] They propose that rather than being the "where" stream (as suggested by Ungerleider and Mishkin[39]), it is really the "how" stream. This might imply that all activity in the dorsal stream is unconscious. On the other hand, they consider the ventral stream to be largely conscious. An alternative suggestion, due to Steven Wise,[40] is that direct projections from parietal cortex into premotor areas are unconscious, whereas projections to them via prefrontal cortex are related to consciousness.

Our suspicion is that while these suggestions about two systems are along the right lines, they are probably overly simple. The little that is known of the neuroanatomy would suggest that there are likely to be *multiple* cortical streams, with numerous anatomical connections between them.[41] This is implied in Figure 1, a diagram often used by Fuster.[42] In short, the neuroanatomy does not suggest that the sole pathway goes up to the highest levels of the visual system, and from there to the highest levels of the prefrontal system and then down to the

The Nature of Nature

	On-line system	*Seeing system*
Visual inputs handled	must be simple	can be complex
Motor outputs produced	stereotyped responses	many possible responses
Minimum time needed for response	short	longer
Effect of a few seconds' delay	may not work	can still work
Coordinates used	egocentric	object-centered
Certain perceptual illusions	not effective	seen
Conscious	no	yes

Table 1
Comparison of the hypothetical on-line system and the seeing system[35]

motor output. There are numerous pathways from most intermediate levels of the visual system to intermediate frontal regions.

We would therefore like to suggest a general hypothesis: that the brain always tries to use the *quickest appropriate* pathway for the situation at hand. Exactly how this idea works out in detail remains to be discovered. Perhaps there is competition, and the fastest stream wins. The postulated on-line system would be the quickest of these hypothetical cortical streams. This would be the "zombie" part of you.

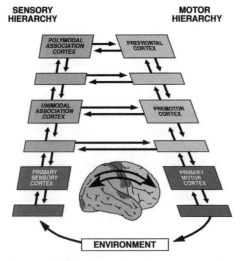

Figure 1. Fuster's figure showing the fiber connections between cortical regions participating in the perception-action cycle [adapted with permission from J. M. Fuster, "The Prefrontal Cortex—An Update: Time Is of the Essence." *Neuron* 30 (May 2001): 319–33; figure 10 on page 329]. Empty rhomboids stand for intermediate areas or subareas of the labeled regions. Notice that there are connections between the two hierarchies at several levels, not just at the top level.

7.2 Bistable Percepts

Perhaps the present most important experimental approach to finding the NCC is to study the behavior of single neurons in the monkey's brain when it is looking at something that produces a bistable percept. The visual input, apart from minor eye movements, is constant; but the subject's percept can take one of two alternative forms. This happens when one looks at a drawing of the well-known Necker cube, for example.

It is not obvious where to look in the brain for the two alternative views of the Necker cube. Allman suggested a more practical alternative: to study the responses in the visual system during binocular rivalry.[43] If the visual input into each eye is different, but perceptually overlapping, one usually sees the visual input as received by one eye alone, then by the other one, then by the first one, and so on. The input is constant, but the percept changes. Which neurons in the brain mainly follow the input, and which the percept?

This approach has been pioneered by Logothetis and his colleagues, working on the macaque visual system. They trained the monkey to report which of two rival inputs it saw. The experiments are difficult, and elaborate precautions had to be taken to make sure the monkey was not cheating. The fairly similar distribution of switching times strongly suggests that monkeys and humans perceive these bistable visual inputs in the same way.

The first set of experiments studied neurons in cortical area MT (medial temporal, also called V5), since they preferentially respond to movement.[44] The stimuli were vertically drifting horizontal gratings. Only the first response was recorded. Of the relevant neurons, only about 35 percent were modulated according to the monkey's reported percept. Surprisingly, half of these responded in the opposite direction to the one expected.

The second set of experiments used stationary gratings.[45] The orientation was chosen in each case to be optimal for the neuron studied, and orthogonal to it in the other eye. The researchers recorded how the neuron fired during several alterations of the reported percept. The neurons were in foveal V1/V2 and in V4. The fraction following the percept in V4 was similar to that in MT, but a rather smaller fraction of V1/V2 neurons followed the percept. Also, here, but not in V4, none of the cells were anticorrelated with the stimulus.

The results of the third set of experiments were especially striking.[46] In this case, the visual inputs included images of humans, monkeys, apes, wild animals, butterflies, reptiles, and various manmade objects. The rivalrous image was usually a sunburst-like pattern (see Figure 2). If a new image was flashed into one eye while the second eye was fixating on another pattern, the new stimulus was the one that was always perceived ("flash suppression"). Recordings were made in the upper and lower banks of the superior temporal sulcus (STS) and inferior temporal cortex (IT). Overall, approximately 90 percent of the recorded neurons in STS and IT were found to reliably predict the perceptual state of the animal. Moreover, many of these neurons responded in an almost all-or-none fashion, firing strongly for one percept, yet only at noise level for the alternative one.

More recently, Bradley and his colleagues have studied a different bistable percept in macaque MT, produced by showing the monkey, on a television, the 2D projection of a transparent, rotating cylinder with random dots on it, without providing any stereoscopic disparity information.[47] Human subjects exploit structure-from-motion and see a 3D cylinder rotating around its axis. Without further clues, the direction of rotation is ambiguous, and observers first report rotation in one direction, a few seconds later, rotation in the other direction, and so on. The trained monkey responds as if it saw the same alteration. In their studies on the monkey, about half the relevant MT neurons the researchers recorded followed the percept (rather than the "constant" retinal stimulus).

These are all exciting experiments, but they are still in the early stages. Just because a particular neuron follows the percept, it does not automatically imply that its firing is part of the NCC. The NCC neurons may be mainly elsewhere, such as higher up in the visual hierarchy. It is obviously important to discover, for each cortical area, *which* neurons are following the percept.[48] That is, what type of neurons are they, in which cortical layer or sublayer do they lie, in what way do they fire, and most important of all, *where do they project?* At the moment, it is technically difficult to do this, but it is essential to have this knowledge, or it will be almost impossible to understand the neural nature of consciousness.

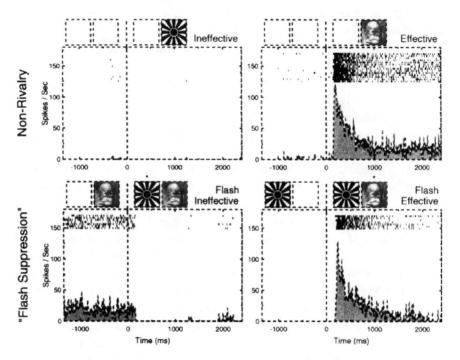

Figure 2. The activity of a single neuron in the superior temporal sulcus (STS) of a macaque monkey in response to different stimuli presented to the two eyes (taken from Sheinberg and Logothetis, 1997). In the upper left panel a sunburst pattern is presented to the right eye without evoking any firing response ("ineffective" stimulus). The same cell will fire vigorously in response to its "effective" stimulus, here the image of a monkey's face (upper right panel). When the monkey is shown the face in one eye for a while, and the sunburst pattern is flashed onto the monitor for the other eye, the monkey signals that it is "seeing" this new pattern and that the stimulus associated with the rival eye is perceptually suppressed ("flash suppression"; lower left panel). At the neuronal level, the cell shuts down in response to the ineffective yet perceptual dominant stimulus following stimulus onset (at the dotted line). Conversely, if the monkey fixates the sunburst pattern for a while, and the image of the face is flashed on, it reports that it perceives the face, and the cell will now fire strongly (lower right panel). Neurons in V4, earlier in the cortical hierarchy, are largely unaffected by perceptual changes during flash suppression.

7.3 ELECTRICAL BRAIN STIMULATION

An alternate approach, with roots going back to Penfield,[49] involves directly stimulating cortex or related structures in order to evoke a percept or behavioral act. Libet and his colleagues have used this technique to great advantage on the somatosensory system of patients.[50] They established that a stimulus, at or near threshold, delivered through an electrode placed onto the surface of somatosensory cortex or into the ventrobasal thalamus, required a minimal stimulus duration (between 0.2–0.5 seconds) in order to be consciously perceived. Shorter stimuli were not perceived, even though they could be detected with above-chance probability using a two-alternative forced choice procedure. In contrast, a skin or peripheral sensory-nerve stimulus of very short duration could be perceived. The difference appears to reside in the amount and type of neurons recruited during peripheral stimulation versus direct central stimulation. Using sensory events as a marker, Libet also established that events caused by direct cortical stimulation were back-dated to the beginning of the stimulation period.[51]

In a series of classical experiments, Newsome and colleagues studied the macaque monkey's performance in a demanding task involving visual motion discrimination.[52] They established a quantitative relationship between the performance of the monkey and the neuronal discharge of neurons in its medial temporal cortex (MT). In 50 percent of all the recorded cells, the psychometric curve—based on the behavior of the entire animal—was statistically indistinguishable from the neurometric curve—based on the averaged firing rate of a single MT cell. In a second series of experiments, cells in MT were directly stimulated via an extracellular electrode (MT cells are arranged in columnar structure for direction of motion).[53] Under these conditions, the performance of the animal shifted in a predictable manner, compatible with the idea that the small brain stimulation caused the firing of enough MT neurons, encoding for motion in a specific direction, to influence the final decision of the animal. It is not clear, however, to what extent visual consciousness for this particular task is present in these highly trained monkeys.

The V1 Hypothesis

We have argued that one is not directly conscious of the features represented by the neural activity in the primary visual cortex.[54] Activity in V1 may be necessary for vivid and veridical visual consciousness (as is activity in the retinae), but we suggest that the firing of none of the neurons in V1 directly correlates with what we consciously see.[55]

Our reasons are that, at each stage in the visual hierarchy, the explicit aspects of the representation we have postulated is always recoded. We have also assumed that any neurons expressing an aspect of the NCC must project directly, without recoding, to at least some of the parts of the brain that plan voluntary action—that is what we have argued seeing is for. We think that these plans are made in some parts of frontal cortex (see the discussion of the frontal lobe hypothesis below).

The neuroanatomy of the macaque monkey shows that V1 cells do not project directly to any part of frontal cortex.[56] Nor do they project to the caudate nucleus of the basal ganglia,[57] the intralaminar nuclei of the thalamus,[58] the claustrum,[59] nor to the brain stem, with the exception of a small projection from peripheral V1 to the pons.[60] It is plausible, but not yet established, that this lack of connectivity is also true for humans.

The strategy to verify or falsify this and related hypotheses is to relate the receptive field properties of individual neurons in V1 or elsewhere to perception in a quantitative manner.

If the structure of perception does not map to the receptive field properties of V1 cells, it is unlikely that these neurons directly give rise to consciousness. In the presence of a correlation between perceptual experience and the receptive field properties of one or more groups of V1 cells, it is unclear whether these cells just correlate with consciousness or directly give rise to it. In that case, further experiments need to be carried out to untangle the exact relationship between neurons and perception.

A possible example may make this clearer. It is well known that the color we perceive at one particular visual location is influenced by the wavelengths of the light entering the eye from surrounding regions in the visual field.[61] This form of (partial) color constancy is often called the "Land effect." It has been shown in the anesthetized monkey that neurons in V4—but *not* in V1—exhibit the Land effect.[62] As far as we know, the corresponding information is lacking for alert monkeys. If the same results could be obtained in a behaving monkey, it would follow that it would not be *directly* aware of the "color" neurons in V1.

Some Experimental Support

In the last two years, a number of psychophysical, physiological, and imaging studies have provided some support for our hypothesis, although this evidence falls short of proving it.[63] Let us briefly discuss two other cases.

When two isoluminant colors are alternated at frequencies beyond 10 Hz, humans perceive only a single fused color with a minimal sensation of brightness flicker. In spite of the perception of color fusion, color opponent cells in the primary visual cortex of two alert macaque monkeys follow high-frequency flicker well above heterochromatic fusion frequencies.[64] In other words, neuronal activity in V1 can clearly represent certain retinal stimulation, yet is not perceived. (This is supported by fMRI studies on humans by Engel and colleagues.).[65]

The study by He and his collaborators is based on a common visual after-effect (see Figure 3a).[66] If a subject stares for a fraction of a minute at a horizontal grating, and is then tested with a faint grating at the same location to decide whether it is oriented vertically or horizontally, the subject's sensitivity for detecting a horizontal grating will be reduced. This adaptation is orientation specific—the sensitivity for vertical gratings is almost unchanged—and disappears quickly. He and his colleagues projected a single patch of grating onto a computer screen some twenty-five degrees from the fixation point. It was clearly visible and their subjects showed the predictable orientation-selective adaptation effect. Adding one or more similar patches of gratings to either side of the original grating—which remained exactly as before—removed the lines of the grating from visibility; it was now "masked." Subjectively, one still sees "something" at the location of the original grating, but one is unable to make out its orientation, even when given unlimited viewing time. Yet despite this inability to "see" the adapting stimulus, the after-effect was as strong and as specific to the orientation of the "invisible" grating as when the grating was visible (see Figure 3b). What this shows—foreshadowed by earlier experiments,[67]—is that visual awareness in such cases must occur at a higher stage in the visual hierarchy than orientation-specific adaptation. This after-effect is thought to be mediated by oriented neurons in V1 and beyond, implying that at least in this case the neurons which mediate visual awareness must be located past this stage.

Our ideas regarding the absence of the NCC from V1 are not disproven by PET experiments showing that in at least some people V1 is activated during visual imagery tasks,[68] though severe damage to V1 is compatible with visual imagery in patients.[69] There is no obvi-

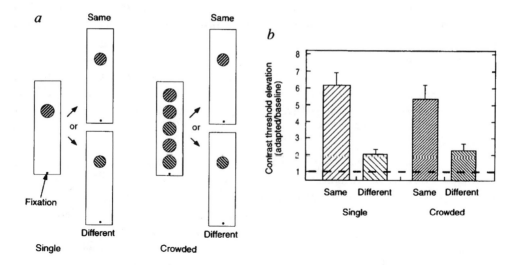

Figure 3. Psychophysical displays (schematic) and results pertaining to an orientation-dependent after-effect induced by "crowded" grating patches (reproduced with the permission of Nature Publishing Group from He, Cavanagh and Intriligator).

a. Adaptation followed by contrast threshold measurement for a single grating (left) and a crowded grating (right). In each trial, the orientation of the adapting grating was either the same or orthogonal to the orientation of the test grating. Observers fixated at a distance of approximately 25 degrees from the adapting and test gratings.

b. Threshold contrast elevation after adaptation relative to baseline threshold contrast before adaptation. Data are averaged across four subjects. The difference between same and different adapt-test orientations reflects the orientation-selective after-effect of the adapting grating. The data show that this after-effect is comparable for a crowded grating (whose orientation is not consciously perceived) and for a single grating (whose orientation is readily perceived).

ous reason why such top-down effects should not reach V1. Such V1 activity would not, by itself, prove that we are *directly* aware of it, any more than the V1 activity produced there when our eyes are open proves this. We hope that further neuroanatomical work will make our hypothesis plausible for humans, and that further neurophysiological studies will show it to be true for most primates. If correct, it would narrow the search to areas of the brain farther removed from the sensory periphery.

7.4 THE FRONTAL LOBE HYPOTHESIS

As mentioned several times, we hypothesize that the NCC must have access to explicitly encoded visual information and directly project into the planning stages of the brain, associated with the frontal lobes in general and with the prefrontal cortex in particular.[70] We would therefore predict that patients unfortunate enough to have lost their entire prefrontal cortex on both sides (including Broca's area) would not be visually conscious, although they might still have well-preserved, but unconscious, visual-motor abilities. No such patient is known to us (not even Brickner's famous patient).[71] The visual abilities of any such "frontal lobe" patient need to be carefully evaluated using a battery of appropriate psychophysical tests.

The fMRI study of the blindsight patient G.Y. provides direct evidence for our view by revealing that prefrontal areas 46 and 47 are active when G.Y. is visually aware of a moving stimulus.[72]

The recent findings of neurons in the inferior prefrontal cortex (IPC) of the macaque that respond selectively to faces—and that receive direct input from regions around the superior temporal sulcus and the inferior temporal gyrus that are well known to contain face-selective neurons—is also very encouraging in this regard.[73] This raises the question of why face cells would be represented in *both* IT and IPC. Do they differ in some important aspect? Large-scale lesion experiments carried out in monkeys suggest that the absence of frontal lobes leads to complete blindness.[74] One would hope that future monkey experiments reversibly inactivate specific prefrontal areas and demonstrate the specific loss of abilities linked to visual perception while visual-motor behaviors—mediated by the on-line system—remain intact.

It will be important to study the pattern of connections between the highest levels of the visual hierarchy—such as inferotemporal cortex—and premotor and prefrontal cortex. In particular, does the anatomy reveal any feedback loops that might sustain activity between IT and prefrontal neurons?[75] There is suggestive evidence that projections from prefrontal cortex back into IT might terminate in layer 4,[76] but these need to be studied directly.

7.5 Gamma Oscillations

Much has been made of the presence of oscillations in the gamma range (30–70 Hz) in the local-field potential and in multi-unit recordings in the visual and sensory-motor system of cats and primates.[77] The existence of such oscillations remains in doubt in higher visual cortical areas.[78] We remain agnostic with respect to the relevance of these oscillations to conscious perception; it is possible that they subserve attention or figure-ground in early visual processing.

8. Philosophical Matters

There is, at the moment, no agreed philosophical answer to the problem of consciousness, except that most living philosophers are not Cartesian dualists—they do not believe in an immaterial soul which is distinct from the body. We suspect that the majority of neuroscientists do not believe in dualism, the most notable exception being the late Sir John Eccles.[79]

We shall not describe here the various opinions of philosophers, except to say that while philosophers have, in the past, raised interesting questions and pointed to possible conceptual confusions, historically they have had a very poor record at arriving at valid scientific answers. For this reason, neuroscientists should listen to the questions philosophers raise, but should not be intimidated by their discussions. In recent years, the amount of discussion about consciousness has reached absurd proportions compared to the amount of relevant experimentation.

8.1 The Problem of Qualia

What is it that puzzles philosophers? Broadly speaking, it is *qualia*—the blueness of blue, the painfulness of pain, and so on. This is also the layman's major puzzle. How can you possibly explain the vivid scene you see before you in terms of the firing of neurons? The argument that you cannot explain consciousness by the action of the parts of the brain goes back at least as far as Leibniz.[80] But compare an analogous assertion: that you cannot explain the "livingness" of living things (such as bacteria, for example) by the action of "dead" molecules. This assertion sounds extremely hollow now, for a number of reasons. Scientists understand the enormous power of natural selection. They know the chemical nature of genes, and that inheritance is particulate, not blending. They understand the great subtlety, sophistication, and variety of protein molecules, the elaborate nature of the control mechanisms that turn genes on and off, and the complicated way that proteins interact with, and modify, other proteins. It is entirely possible that the very elaborate nature of neurons and their interactions, far more elaborate than most people imagine, is misleading us, in a similar way, about consciousness.

Some philosophers are rather fond of this analogy between "livingness" and "consciousness,"[81] and so are we. But, as Chalmers has emphasized, an analogy is only an analogy.[82] He has given philosophical reasons why he thinks it is wrong. Neuroscientists know only a few of the basics of neuroscience, such as the nature of the action potential and the chemical nature of most synapses. Most important, there is not a comprehensive, overall theory of the activities of the brain. To be shown to be correct, the analogy must be filled out by many experimental details and powerful general ideas. Most of these are still lacking.

This problem of *qualia* is what Chalmers calls "the hard problem": a full account of the manner in which subjective experience arises from cerebral processes.[83] As we see it, the hard problem can be broken down into several questions, of which the first is the major problem: How do we experience anything at all? What leads to a particular conscious experience (such as the blueness of blue)? What is the function of conscious experience? Why are some aspects of subjective experience impossible to convey to other people—in other words, why are they private?

We believe we have answers to the last two questions.[84] We have already explained in section 3 above ("Why Are We Conscious") what we think consciousness is for. The reason that visual consciousness is largely private is, we think, an inevitable consequence of the way the brain works.[85] To be conscious, we have argued, there must be an explicit representation of each aspect of visual consciousness. At each successive stage in the visual cortex, what is made explicit is recoded. To produce a motor output, such as speech, the information must be recoded again, so that what is expressed by the motor neurons is related, but not identical, to the explicit representation expressed by the firing of the neurons associated with, for example, the color experience at some level in the visual hierarchy.

It is thus not possible to convey with words the exact nature of a subjective experience. It is possible, however, to convey a *difference* between subjective experiences—to distinguish between red and orange, for example. This is possible because a difference in a high-level visual cortical area can still be associated with a difference at the motor stage. The implication is that we can never explain to other people the nature of any conscious experience, only—in some cases—its relation to other ones.

Is there any sense in asking whether the blue color you see is subjectively the same as the blue color I see? If it turns out that the neural correlate of blue is exactly the same in your brain as in mine, it would be scientifically plausible to infer that you see blue as I do. The problem lies in the word "exactly." How precise one has to be will depend on a detailed knowledge of the processes involved. If the neural correlate of blue depends, in an important way, on my past experience, and if my past experience is significantly different from yours, then it may not be possible to deduce that we both see blue in exactly the same way.[86]

Could this problem be solved by connecting two brains together in some elaborate way? It is impossible to do this at the moment, or in the easily foreseeable future. One is therefore tempted to use the philosopher's favorite tool, the thought experiment. Unfortunately, this enterprise is fraught with hazards, since it inevitably makes assumptions about how brains behave, and most of these assumptions have so little experimental support that conclusions based on them are valueless—for example, how much is a person's percept of the blue of the sky due to early visual experiences?

8.2 The Problem of Meaning

An important problem neglected by neuroscientists is the problem of meaning. Neuroscientists are apt to assume that if they can see that a neuron's firing is roughly correlated with some aspect of the visual scene, such as an oriented line, then that firing must be part of the neural correlate of the seen line. They assume that because they, as outside observers, are conscious of the correlation, the firing must be part of the NCC. This by no means follows, as we have argued for neurons in V1.

But this is not the major problem, which is rather: How do other parts of the brain know that the firing of a neuron (or of a set of similar neurons) produces the conscious percept of, say, a face? How does the brain know what the firing of those neurons represents? In other words, how is meaning generated by the brain?

This problem has two aspects. How is meaning expressed in neural terms? And how does this expression of meaning arise? We suspect that meaning derives both from the correlated firing described above and from the linkages to related representations.[87] For example, neurons related to a certain face might be connected to ones expressing the name of the person whose face it is, and to others for her voice, memories involving her, and so on, in a vast associational network, similar to a dictionary or a relational database. Exactly how this works in detail is unclear.

But how are these useful associations derived? The obvious idea is that they depend very largely on the consistency of the interactions with the environment, especially during early development. Meaning can also be acquired later in life. The usual example is a blind man with a stick: He comes to feel what the stick is touching, not merely the stick itself.[88]

9. Future Experiments

Although experiments on attention, short-term and working memory, the correlated firing of neurons, and related topics may make finding the NCC easier, at the moment the most promising experiments are those on bistable percepts. These experiments should be continued in numerous cortical and thalamic areas and need extending to cover other such percepts. It is also important to discover *which* neurons express the NCC in each case (e.g., which neuronal subtype, in what layer, and so on), how they fire (e.g., do they fire in bursts), and especially, where they project. To assist this, more detailed neuroanatomy of the connectivity will be needed. This is relatively easy to do in the macaque but difficult in humans.[89] It is also important to discover how the various on-line systems work, so that one can contrast their (unconscious) neuronal activity with the NCC.

To discover the exact role (if any) of the frontal cortex in visual perception, it would be useful to inactivate it reversibly by cooling and/or the injection of GABA agonists, perhaps using the relatively smooth cortex of an owl monkey.

Inevitably, it will be necessary to compare the studies on monkeys with similar studies on humans, using both psychophysical experiments as well as functional imaging methods such as

PET or fMRI. Conversely, functional imaging experiments on normal subjects or patients—showing, for instance, the involvement of prefrontal areas in visual perception[90]—can provide a rationale for appropriate electrophysiological studies in monkeys. It would help considerably if there were more detailed architectonic studies of cortex and thalamus, since these can be done postmortem on monkeys, apes, and humans. The extremely rapid pace of molecular biology should soon provide a wealth of new markers to help in this endeavor.

To understand a very complex nonlinear system, it is essential to be able to interfere with it both specifically and delicately. The major impact of molecular biology is likely to be the provision of methods for the inactivation of all neurons of a particular type. Ideally, this should be done reversibly on the mature animal.[91] At the moment this is only practical on mice, but in the future, one may hope for methods that can be used on mature monkeys (perhaps using a viral vector), as such methods are also needed for the medical treatment of humans.

As an example, consider the question of whether the cortical feedback pathways—originating in a higher visual area and projecting into a lower area[92]—are essential for normal visual consciousness. There are at least two distinct types of back pathways: one, from the upper cortical layers, goes back only a few steps in the visual hierarchy; the other, from the lower cortical layers, can also go back over longer distances.[93] We would like to be able to selectively inactivate these pathways, both singly and collectively, in the mature macaque. Present methods are not specific enough to do this, but new methods in molecular biology should, in time, make this possible.

It will not be enough to show that certain neurons embody the NCC in certain, limited visual situations. Rather, we need to locate the NCC for all types of visual inputs, or at least for a sufficiently large and representative sample of them. For example, when one blinks, the eyelids briefly (30–50 milliseconds) cover the eyes, yet the visual percept is scarcely interrupted (blink suppression).[94] We would therefore expect the NCC to be also unaffected by eye blinks—for example, the firing activity should not drop noticeably during the blink—but not to blanking out of the visual scene for a similar duration due to artificial means. Another example is the large number of visual illusions. For instance, under appropriate circumstances, humans clearly perceive a transient motion after-effect. On the basis of fMRI imaging, it has been found that the human equivalent of cortical area MT is activated by the motion after-effect (in the absence of any moving stimuli).[95] The time course of this illusion parallels the time course of activity as assayed using fMRI. In order to really pinpoint the NCC, one would need to identify individual cells expressing this, and similar visual after-effects. We have assumed that the visual NCC in humans is very similar to the NCC in the macaque, mainly because of the similarity of their visual systems. Ultimately, the link between neurons and perception will need to be made in humans.

The problem of meaning and how it arises is more difficult, since as yet there is not even an outline formulation of this problem in neural terms. For example, do multiple associations depend on transient priming effects? Whatever the explanation, it would be necessary to study the developing animal to show how meaning arises; in particular, how much is built in epigenetically and how much is due to experience.

In the long run, finding the NCC will not be enough. A complete theory of consciousness is required, including its functional role. With luck this might illuminate the hard problem of *qualia*. It is likely that scientists will then stop using the term consciousness except in a very loose way. After all, biologists no longer worry whether a seed or a virus is "alive." They just want to know how it evolved, how it develops, and what it can do.

10. Finale

We hope we have convinced the reader that the problem of the neural correlate of consciousness (the NCC) is now ripe for direct experimental attack. We have suggested a possible framework for thinking about the problem, but others may prefer a different approach; and of course, our own ideas are likely to change with time. We have outlined the few experiments that directly address the problem and mentioned briefly other types of experiments that might be done in the future. We hope that some of the younger neuroscientists will seriously consider working on this fascinating problem. After all, it is rather peculiar to work on the visual system and not worry about exactly what happens in our brains when we "see" something. The explanation of consciousness is one of the major unsolved problems of modern science. After several thousand years of speculation, it would be very gratifying to find an answer to it.[96]

Notes

1. Adapted with permission from *Cerebral Cortex* 8, 1998: 97–107.
2. Crick and Koch 1990.
3. Ibid.
4. Nagel 1997.
5. Tootell *et al.* 1996.
6. Crick and Koch 1995a.
7. Chalmers 1995.
8. Milner and Goodale 1995.
9. Ingle 1973.
10. Ramachandran and Hirstein 1997.
11. Milner and Goodale 1995.
12. Crick and Koch 1995a.
13. Felleman and Van Essen 1991.
14. Crick and Koch 1995a.
15. Ballard *et al.* 1983.
16. Young and Yamane 1992.
17. Adolphs *et al.* 1994.
18. Von der Malsberg 1995.
19. Greenfield 1995.
20. Crick and Koch 1995a.
21. Crick 1994.
22. Edelman 1989.
23. Coltheart 1983; Gegenfurtner and Sperling 1993.
24. Potter 1976; Subramaniam *et al.* 2000.
25. O'Regan 1992.
26. Baddeley 1992; Goldman-Rakic 1995.
27. Zola-Morgan and Squire 1993.
28. Rock *et al.* 1992; Braun and Julesz 1998.
29. Kanwisher and Driver 1992.
30. Luck *et al.* 1997.
31. Weiskrantz 1997.
32. Cowey and Stoerig 1995.
33. For Weiskrantz's comments on Gazzaniga's criticisms, see Weiskrantz 1997: 152–53; on Zeki's criticisms, see 1997: 247–48.
34. Sahraie *et al.* 1997. This paper should be consulted for further details of the areas involved.
35. Milner and Goodale 1995. The reader is referred to this book for a more extended account. For a review, see Boussaoud *et al.* (1996).
36. Rosetti 1998.
37. Milner *et al.* 1991.
38. Milner and Goodale 1995.
39. Ungerleider and Mishkin 1982.
40. Personal communication; see also Boussaoud *et al.* 1996.
41. Distler *et al.* 1993.
42. Fuster 1997: see his Figure 8.4; Fuster 2001: see his Figure 10.

43. Myerson *et al.* 1981.
44. Logothetis and Schall 1989.
45. Leopold and Logothetis 1996.
46. Sheinberg and Logothetis 1997.
47. Bradley *et al.* 1998.
48. Crick 1996.
49. Penfield 1958.
50. Libet 1993.
51. Ibid.
52. Britten *et al.* 1992.
53. Salzman *et al.* 1990.
54. Crick and Koch 1995a.
55. For a critique of our hypothesis, see Pollen (1995); our reply is given in Crick and Koch (1995b).
56. Crick and Koch 1995a.
57. Saint-Cyr *et al.* 1990.
58. L. G. Ungerleider, personal communication.
59. Sherk 1986.
60. Fries 1990.
61. Land and McCann 1971; Blackwell and Buchsbaum 1988.
62. Zeki 1980, 1983; Schein and Desimone 1990.
63. He *et al.* 1995; Cumming and Parker 1997; Kolb and Braun 1995; summarized in Koch and Braun 1996; but see also Morgan *et al.* 1997.
64. Gur and Snodderly 1997.
65. Engel *et al.* 1997.
66. He *et al.* 1996.
67. Blake and Fox 1974.
68. Kosslyn *et al.* 1995.
69. Goldenberg *et al.* 1995.
70. Fuster 1997.
71. For an extensive discussion of this, see Damasio and Anderson (1993).
72. Sahraie *et al.* 1997.
73. Scalaidhe, Wilson and Goldman-Rakic 1997.
74. Nakamura and Mishkin 1980, 1986.
75. Crick and Koch 1997.
76. Webster *et al.* 1994.
77. Singer and Gray 1995.
78. Young *et al.* 1992.
79. Eccles 1994.
80. Leibniz 1686; see the translation (1965).
81. Searle 1984; Dennett 1996.
82. Chalmers 1995.
83. Ibid.
84. Crick and Koch 1995c.
85. By "private," we mean that it is inherently impossible to communicate the exact nature of what we are conscious of.
86. Crick 1994.

87. Crick and Koch 1995c.
88. For an ingenious recent demonstration along similar lines, see Ramachandran and Hirstein (1997).
89. Crick and Jones 1993.
90. Weiskrantz 1997; Sahraie *et al.* 1997.
91. See, for example, No *et al.* (1996); Nirenberg and Meister (1997).
92. In the sense of Felleman and Van Essen (1991).
93. Salin and Bullier 1995.
94. Volkmann *et al.* 1980.
95. Tootell *et al.* 1995.
96. We thank the J.W. Kieckhefer Foundation, the National Institute of Mental Health, the Office of Naval Research, and the National Science Foundation. For helpful comments, we thank David Chalmers, Leslie Orgel, John Searle, and Larry Weiskrantz.

References

Adolphs, R., D. Tranel, H. Damasio, and A. Damasio. (1994) "Impaired recognition of emotion in facial expressions following bilateral damage to the human amygdala." *Nature* 372: 669–72.

Baddeley, A. (1992) "Working memory." *Science* 255: 556–59.

Ballard, D. H., G. E. Hinton, T. J. Sejnowski. (1983) "Parallel visual computation." *Nature* 306: 21–26.

Blackwell, K. T., Buchsbaum, G. (1988) "Quantitative studies of color constancy." *Journal of the Optical Society of America* A5: 1772–80.

Blake, R., and R. Fox. (1974) "Adaptation to invisible gratings and the site of binocular rivalry suppression." *Nature* 249: 488–90.

Boussaoud, D., G. di Pellegrino, and S. P. Wise. (1996) "Frontal lobe mechanisms subserving vision-for-action versus vision-for-perception." *Behavioral Brain Research* 72: 1–15.

Bradley, D. C., G. C. Chang, and R. A. Andersen. (1998) "Encoding of 3D structure from motion by primate area MT neurons." *Nature* 392: 714–17.

Braun. J., and B. Julesz. (1998) "Withdrawing attention at little or no cost: Detection and discrimination tasks." *Perception and Psychophysics* 60 (1): 1–23.

Britten, K. H., M. N. Shadlen, W. T. Newsome, and J. A. Movshon. (1992) "The analysis of visual motion: a comparison of neuronal and psychophysical performance." *Journal of Neuroscience* 12: 4745–65.

Chalmers, D. (1995) *The Conscious Mind: In Search of a Fundamental Theory.* Oxford: Oxford University Press.

Coltheart, M. (1983) "Iconic memory." *Philosophical Transactions of the Royal Society of London* B 302: 283–94.

Cowey, A., and P. Stoerig. (1995) "Blindsight in monkeys." *Nature* 373: 247–49.

Crick, F. (1994) *The Astonishing Hypothesis.* New York: Scribner's.

Crick, F. (1996) "Visual perception: rivalry and consciousness." *Nature* 379: 485–86.

Crick, F., and E. Jones (1993). "Backwardness of human neuroanatomy." *Nature* 361: 109–10.

Crick, F., and C. Koch. (1990) "Towards a neurobiological theory of consciousness." *Seminars in the Neurosciences* 2: 263–275.

———. (1995a) "Are we aware of neural activity in primary visual cortex?" *Nature* 375: 121–23.

———. (1995b) "Cortical areas in visual awareness—Reply." *Nature* 377: 294–95.

———. (1995c) "Why neuroscience may be able to explain consciousness." *Scientific American* 273: 84–85.

———. (1998) "Constraints on cortical and thalamic projections: The no-strong-loops hypothesis." *Nature* 391: 245–50.

Cumming, B. G., and A. J. Parker. (1997) "Responses of primary visual cortical neurons to binocular disparity without depth perception." *Nature* 389: 280–83.

Damasio, A. R., and S. W. Anderson, (1993) "The frontal lobes," in *Clinical Neuropsychology,* 3rd ed., K.M. Heilman and E. Valenstein, eds., Oxford: Oxford University Press, 409–60.

Dennett, D. (1996) *Kinds of minds: Toward an understanding of consciousness.* New York: Basic Books.

Distler C, Boussaoud D, Desimone R, Ungerleider LG (1993) Cortical connections of inferior temporal area IEO in macaque monkeys. *Journal of Computational Neurology* 334:125–50.

Eccles, J. C. (1994) *How the self controls its brain.* Berlin: Springer-Verlag.

Edelman, G. M. (1989) *The remembered present: a biological theory of consciousness.* New York: Basic Books.

Engel, S., Zhang, X., Wandell, B. (1997) "Colour tuning in human visual cortex measured with functional magnetic resonance imaging." *Nature* 388: 68–71.

Felleman, D. J., Van Essen, D. (1991) "Distributed hierarchical processing in the primate cerebral cortex." *Cerebral Cortex* 1: 1–47.

Fries, W. (1990) "Pontine projection from striate and prestriate visual cortex in the macaque monkey: an anterograde study." *Visual Neuroscience* 4: 205–16.

Fuster, J.M. (1997) *The prefrontal cortex: anatomy, physiology, and neuropsychology of the frontal lobe,* 3rd ed., Philadelphia: Lippincott-Raven.

———. (2001) "The Prefrontal Cortex—An Update: Time Is of the Essence." *Neuron* 30: 319–33.

Gegenfurtne, K. R., and G. Sperling. (1993) "Information transfer in iconic memory experiments." *Journal of Experimental Psychology: Human Perception and Performance* 19: 845–66.

Goldenberg, G., W. Müllbacher, and A. Nowak. (1995) "Imagery without perception—a case study of anosognosia for cortical blindsight." *Neuropsychologia* 33: 1373–82.

Goldman-Rakic, P.S. (1995) "Cellular basis of working memory." *Neuron* 14: 477–85.

Greenfield, S. A. (1995) *Journey to the centers of the mind.* New York: W.H. Freeman.

Gur, M., and D. M. Snodderly. (1997) "A dissociation between brain activity and perception: chromatically opponent cortical neurons signal chromatic flicker that is not perceived." *Vision Research* 37: 377–82.

He, S., P. Cavanagh, and J. Intriligator. (1996) "Attentional resolution and the locus of visual awareness." *Nature* 383: 334–37.

He, S., H. Smallman, and D. MacLeod. (1995) "Neural and cortical limits on visual resolution." *Investigative Opthalmology and Visual Science* 36: 2010.

Ingle, D. (1973) "Two visual systems in the frog." *Science* 181: 1053–55.

Kanwisher, N., Driver, J. (1992) "Objects, attributes, and visual attention: which, what, and where." *Current Directions in Psychological Science* 1: 26–31.

Koch, C., Braun, J. (1996) "On the functional anatomy of visual awareness." *Cold Spring Harbor Symposia on Quantitative Biology* 61: 49–57.

Kolb, F. C., and J. Braun. (1995) "Blindsight in normal observers." *Nature* 377: 336–39.

Kosslyn, S. M., W. L Thompson, I. J. Kim, and N. M. Alpert. (1995) "Topographical representations of mental images in primary visual cortex." *Nature* 378: 496–98.

Land, E. H., and J. J. McCann. (1971) "Lightness and retinex theory." *Journal of the Optical Society of America* 61: 1–11.

Leibniz, G. W. (1965) *Monadology and other philosophical essays,* P. Schreckerand A. M. Schrecker, translators. Indianapolis: Bobbs-Merrill.

Leopold, D. A., Logothetis, N. K. (1996) "Activity changes in early visual cortex reflect monkeys' percepts during binocular rivalry." *Nature* 379: 549–53.

Libet, B. (1993) *Neurophysiology of consciousness: selected papers and new essays by Benjamin Libet.* Boston: Birkhäuser.

Logothetis, N., and J. Schall. (1989) "Neuronal correlates of subjective visual perception." *Science* 245: 761–63.

Luck, S. J., L. Chelazzi, S.A. Hillyard, and R. Desimone. (1997) "Neural mechanisms of spatial selective attention in areas V1, V2, and V4 of macaque visual cortex." *Journal of Neurophysiology* 77: 24–42.

Milner, D., Goodale, M. (1995) *The Visual Brain in Action.* Oxford: Oxford University Press.

Milner, A. D., D. I. Perrett, R. S. Johnston, P. J. Benson, T. R. Jordan, D. W. Heeley et al. 1991) "Perception and action in 'visual form agnosia.'" *Brain* 114: 405–28.

Morgan, M. J., A. J. S. Mason, and J. A. Solomon. (1997) "Blindsight in normal subjects?" *Nature* 385: 401–2.

Myerson, J., F. Miezin, and J. Allman. (1981) "Binocular rivalry in macaque monkeys and humans: a comparative study in perception." *Behavioral Analysis Letters* 1: 149–56.

Nakamura, R. K., and M. Mishkin. (1980) "Blindness in monkeys following nonvisual cortical lesions." *Brain Research* 188: 572–7.

———. (1986) "Chronic blindness following lesions of nonvisual cortex in the monkey." *Experimental Brain Research* 62: 173–84.

Nagel, A. H. M. (1997) "Are plants conscious?" *Journal of Consciousness Studies* 4: 215–30.

Nirenberg, S., and M. Meister. (1997) "The higher response of retinal ganglion cells is truncated by a displaced amacrine circuit." *Neuron* 18: 637–50.

No, D., T. P. Yao, and R. M. Evans. (1996) "Ecdysone-inducible gene expression in mammalian cells and transgenic mice." *Proceedings of the National Academy of Sciences USA* 93: 3346–51.

O'Regan, J. K. (1992) "Solving the 'real' mysteries of visual perception: the world as an outside memory." *Canadian Journal of Psychology* 46: 461–88.

Penfield, W. (1958) *The excitable cortex in conscious man.* Liverpool: Liverpool University Press.

Pollen, D. A. (1995) "Cortical areas in visual awareness." *Nature* 377: 293–4.

Potter, M. C. (1976) "Short-term conceptual memory for pictures." *Experimental Psychology: Human Learning and Memory* 2: 509–22.

Ramachandran, V. S., and Hirstein, W. (1997) "Three laws of qualia: what neurology tells us about the biological functions of consciousness." *Journal of Consciousness Studies* 4: 4–29.

Rock, I., Linnett, C. M., Grant, P., Mack, A. (1992) "Perception without attention: results of a new method." *Cognitive Psychology* 24: 502–34.

Rossetti, Y. (1997) "Implicit perception in action: short-lived motor representations of space evidenced by brain-damaged and healthy subjects," in *Finding Consciousness in the Brain,* P.G. Grossenbacher, ed. Philadelphia: Benjamins.

Sahraie, A., L. Weiskrantz, J. L. Barbur, A. Simmons, S. C. R. Williams, and M. J. Brammer. (1997) "Pattern of neuronal activity associated with conscious and unconscious processing of visual signals." *Proceedings of the National Academy of Sciences USA* 94: 9406–11.

Saint-Cyr, J. A., L. G. Ungerleider, and R. Desimone. (1990) "Organization of visual cortex inputs to the striatum and subsequent outputs to the pallidonigral complex in the monkey." *Journal of Computational Neurology* 298: 129–56.

Salin, P. A., and J. Bullier. (1995) "Corticocortical connections in the visual system: structure and function." *Physiological Review* 75: 107–54.

Salzman, C. D., K. H. Britten, and W. T. Newsome. (1990) "Cortical microstimulation influences perceptual judgements of motion direction." *Nature* 346: 174–7.

Scalaidhe, S. P. O., F. A. W. Wilson, and P. S. Goldman-Rakic. (1997) "Areal segregation of face-processing neurons in prefrontal cortex." *Science* 278: 1135–8.

Schein, S. J., and R. Desimone. (1990) "Spectral properties of V4 neurons in the macaque." *Journal of Neuroscience* 10: 3369–89.

Sheinberg, D. L., and N. K. Logothetis. (1997) "The role of temporal cortical areas in perceptual organization." *Proceedings of the National Academy of Sciences USA* 94: 3408–13.

Sherk, H. (1986) "The claustrum and the cerebral cortex," in *Cerebral Cortex,* vol 5: *sensory-motor areas and aspects of cortical connectivity.* E. G. Jones and A. Peters, eds., New York: Plenum Press, 467–99.

Singer, W., Gray, C. M. (1995) "Visual feature integration and the temporal correlation hypothesis." *Annual Review of Neuroscience* 18: 555–86.

Subramaniam, S., I. Biederman, and S. A. Madigan. (2000) "Accurate identification but no chance recognition memory for pictures in RSVP sequences." *Visual Cognition* 7(4): 511–535.

Tootell, R. B. H., A. M. Dale, M. I. Sereno, and R. Malach, R. (1996) "New images from human visual cortex." *Trends in Neuroscience* 19:.481–9.

Tootell, R. B. H., J. B. Reppas, A. M. Dale, R. B. Look, M. I. Sereno, R. Malach, T. J. Brady, and B. R. Rosen. (1995) "Visual motion aftereffect in human cortical area MT revealed by functional magnetic resonance imaging." *Nature* 375: 139–41.

Ungerleider, L. G., and M. Mishkin. (1982) "Two cortical visual systems," in *Analysis of visual behavior,* D.J. Ingle, M. A. Goodale, and R. J. W. Mansfield, eds., Cambridge, MA: MIT Press, 549–86.

Volkmann, F. C., L. A. Riggs, and R. K. Moore. (1980) "Eye-blinks and visual suppression." *Science* 207: 900–02.

von der Malsburg, C. (1995) "Binding in models of perception and brain function." *Current Opinion in Neurobiology* 5: 520–6.

Webster, M. J., J. Bachevalier, and L. G. Ungerleider, L. G. (1994) "Connections of inferior temporal areas TEO and TE with parietal and frontal cortex in macaque monkeys." *Cerebral Cortex* 5: 470–83.

Weiskrantz, L. (1997) *Consciousness lost and found.* Oxford: Oxford University Press.

Young, M.P., K. Tanaka, and S. Yamane. (1992) "On oscillating neuronal responses in the visual cortex of the monkey." *Journal of Neurophysiology* 67: 1464–74.

Young, M.P., and S. Yamane. (1992) "Sparse population coding of faces in the inferotemporal cortex." *Science* 256: 1327–31.

Zeki, S. (1980) "The representation of colours in the cerebral cortex." *Nature* 284: 412–8.

Zeki,, S. (1983) "Colour coding in the cerebral cortex: the reaction of cells in monkey visual cortex to wavelengths and colours." *Neuroscience* 9: 741–65.

Zola-Morgan, S., and L. R. Squire. (1993) "Neuroanatomy of memory." *Annual Review of Neuroscience* 16: 547–63.

34

SUPERVENIENCE AND THE DOWNWARD EFFICACY OF THE MENTAL: NONREDUCTIVE PHYSICALISM AND THE CHRISTIAN TRADITION

NANCEY MURPHY

1. INTRODUCTION

It is a strange feature of contemporary culture that radically different theories persist regarding something so central and important as the nature of the human person. It seems especially strange that these differences have only begun rather recently to be debated in public. The major contestants are trichotomists, dualists, and physicalists or materialists. Dualists come in two sorts, body and mind or body and soul; physicalists in two sorts, reductive and non-reductive. I have done informal surveys among my audiences when I speak on these issues. Among the general population, trichotomism (the view that humans consist of body, soul, and spirit) is usually the majority view, followed by dualism. I find very few physicalists, even in audiences with a large proportion of scientists.

Motivation for both trichotomism and dualism seems to come primarily from religious commitments. This is certainly true for trichotomism, and probably true for most dualists. I know of only one philosopher who is a dualist but does not have any religious commitments.[1] So my first move in this paper will consist in a very brief rehearsal of some of the reasons why Christians do not need to argue for dualism.

The central purpose of my paper will be to contribute to arguments showing that physicalists do not need to be reductionists. This is an argument in which Christians *do* have an important stake, but, I maintain, so does anyone who values the notion of human *reason*. That is, the coherence or intelligibility of non-reductive physicalism (NRP) has been called into question, particularly with regard to the problem of mental causation—how to account for the role of reasons in mental/neural processes if we assume the causal closure of the neurobiological level. I think there is a solution, and one that will turn out also to provide a basis for a new approach to the problem of free will, but I will not have space to make the connections here.

2. WHY NOBODY NEEDS DUALISM

It is certainly the case that most Christian scholars throughout most of Christian history have held some version of dualism—earlier forms more Platonic; in the late Middle Ages, more

Aristotelian; and in the modern period strongly influenced by Descartes. However, beginning about a century ago, biblical scholars and historians of doctrine began to question whether the *original* Christian writers were dualists. By mid-century there was near consensus among mainline Christian scholars that dualism was a later development, a consequence of accommodation to cultures more Hellenized than the original Jewish Christians. How could so many Christians be wrong about their own authoritative texts? Part of the answer is that Greek philosophical conceptions were read into the language of the New Testament and equally into the Greek translations of the Hebrew Scriptures—conceptions that were then preserved in later translations.

New Testament scholar James Dunn argues that the very questions we address to the texts about the various constitutive parts of the person are foreign to biblical thought. He distinguishes between partitive and aspective understandings, the latter being the tendency of the Bible. Here one speaks of the whole person from various aspects, or in light of the person's various relationships to something else. So what appears to us as a part—for example, the body—is a term for the whole person thought of from a certain angle.[2] Biblical anthropology is concerned about relationships—to others, to the natural world, and to God.

If the scholars upon whom I depend and I are right in claiming that dualism is not a part of earliest Christian teaching, one may well ask why, in light of pressure from both science and philosophy, dualism has not been rejected by more Christians. The answer in many cases is that it is thought to be necessary in order to make sense of other Christian teachings. For example, many assume that an immortal soul is required to guarantee life after death. Yet central Christian teaching about the afterlife has always focused instead on resurrection of the body. For some who do believe in resurrection, there is the problem of the intermediate state: the teaching by John Calvin and the Catholic Church that the blessed enjoy conscious awareness of God between death and the general resurrection at the end of history. This would seem to require a soul to maintain consciousness.[3] However, the biblical evidence for this doctrine is questionable.[4]

My conclusion, then, although I cannot come close to justifying it here, is that Christians have no need of dualism (and even less, trichotomism).[5] I turn now to an explication and partial defense of an account of human nature that I claim is compatible with both Christian teaching and current science.

3. Defining Non-reductive Physicalism

There are two routes by which to arrive at a physicalist account of human beings. One is to begin with dualism, say, of a Cartesian sort, and then subtract the mind or soul. John Searle, among others, has argued persuasively against this move.[6] The other route begins with science. We recognize a certain "layered" feature of reality: subatomic particles at the lowest level combine in increasingly complex structures to give us the features of the world known to chemists, and these in turn combine into incredibly complex organizations to give us biological organisms. The version of physicalism I espouse begins with the denial of the complete reducibility of the biological level to that of chemistry and physics. It then argues that just as life appears as a result of complex organization, so too, sentience and consciousness appear as non-reducible products of biological organization.

There are a variety of benefits in approaching physicalism scientifically rather than through a reaction against Cartesianism. As Searle has pointed out, it frees one from the (apparent) necessity of attempting to deny or define away obvious facts of experience—such as the fact that we're conscious. Another benefit is this. Arguments against the reducibility of the mental to the physical can draw upon parallel arguments against reductionism in other scientific domains. So here I come to the focus of my paper: employment of resources from philosophy of biology to address the problem of mental causation.

4. Defending Non-reductive Physicalism

The most pressing problem for such an account of human nature, I believe, is a more compelling argument against the *total* reduction of the mental (and moral and spiritual)[7] levels of human functioning to the level of neurobiology. That is, I intend to answer the question, if mental events are intrinsically related to (supervene on) neural events, how can it *not* be the case that the contents of mental events are ultimately governed by the laws of neurobiology? If neurobiological determinism is true, then it would appear that there is no freedom of the will, that moral responsibility is in jeopardy, and, indeed, that our talk about the role of reasons in any intellectual discipline is misguided. Thus, the main goal of this paper will be to show why, in certain sorts of cases, complete *causal* reduction of the mental to the neurobiological *fails*. I shall attempt to do this, first, by clarifying the concept of *supervenience*, which is used by philosophers of mind to give an account of the relations between the mental and the physical. This redefinition of "supervenience" will make it clear that in many cases, supervenient properties are *functional* properties.

I then turn to Donald Campbell's account of downward causation, which also trades on the functional character of the state of affairs being explained. Then, employing accounts of cognition and neurobiology that emphasize functionality and feedback loops, I show that Campbell's account of downward causation occurs in a variety of sorts of cognitive processes. It turns out that rational and moral principles can be seen to exert a top-down effect on the formation and functioning of neural assemblies.

Thus, I hope to contribute to arguments to the effect that Christian theology has nothing to lose by substituting a non-reductive physicalist account of human nature for the various forms of body–soul dualism that have appeared in Christian history.

Supervenience

The concept of supervenience is now used extensively in philosophy of mind. Many suppose that, in contrast to mind–brain *identity* theses, it allows for a purely physicalist account of the human person without entailing the explanatory or causal reduction of the mental. In other words, it leaves room for the causal efficacy of the mental. However, it is not clear to me that typical approaches to constructing a formal definition of "supervenience" have this consequence.[8] In this section I first review some of the history of the development of this concept in philosophy. I then offer an alternative characterization of the supervenience relation and attempt to motivate this alternative by showing that it fits examples from domains in which the concept of supervenience is most often used—ethics, biology, and philosophy of mind. I claim that my definition sheds light on the question of causal reductionism, helping us indeed to account for instances of mental causation without giving up the dependence of the mental on the neurobiological.

Accounts of the development of the concept of supervenience in philosophy generally mention R. M. Hare's use in ethics and Donald Davidson's in philosophy of mind. Hare used "supervenience" as a technical term to describe the relation of evaluative judgments (including ethical judgments) to descriptive judgments. Hare says:

> [L]et us take that characteristic of "good" which has been called its supervenience. Suppose that we say "St. Francis was a good man." It is logically impossible to say this and to maintain at the same time that there might have been another man placed in precisely the same circumstances as St. Francis, and who behaved in them in exactly the same way, but who differed from St. Francis in this respect only, that he was not a good man.[9]

In 1970 Davidson used the concept to describe the relation between mental and physical characteristics. He describes the relation as follows:

> [M]ental characteristics are in some sense dependent, or supervenient, on physical characteristics. Such supervenience might be taken to mean that there cannot be two events alike in all physical respects but differing in some mental respect, or that an object cannot alter in some mental respect without altering in some physical respect. Dependence or supervenience of this kind does not entail reducibility through law or definition. . . . [10]

David Lewis characterizes the intuition that definitions of "supervenience" are meant to capture: "The idea is simple and easy: we have supervenience when [and only when] there could be no difference of one sort without difference of another sort."[11]

Jaegwon Kim has been influential in the development of formal definitions of supervenience. It is now common to distinguish three types: weak, strong, and global. Kim has defined these as follows, where A and B are two nonempty families of properties:

> A *weakly supervenes* on B if and only if necessarily for any property F in A, if an object x has F, then there exists a property G in B such that x has G, and if any y has G it has F.[12]

> A *strongly supervenes* on B just in case, necessarily, for each x and each property F in A, if x has F, then there is a property G in B such that x has G, and *necessarily* if any y has G, it has F.[13]

> A *globally supervenes* on B just in case worlds that are indiscernible with respect to B ("B-indiscernible," for short) are also A-indiscernible.[14]

In short, supervenience is now widely understood as an asymmetrical relation of property covariation, and the interesting questions of definition are taken to turn on the placement of modal operators.

However, consider again Hare's original *use* of "supervenience" (see above). I believe that the qualification "placed in precisely the same circumstances" is an important one. St. Francis's behavior (e.g., giving away all his possessions) would be evaluated quite differently were he in different circumstances (e.g., married and with children to support). If this is the case, then the standard definitions of supervenience are not only in need of qualification but are entirely wrong-headed. On the standard account the fact that G supervenes on F means that F materially implies G. But if circumstances make a difference it may well be that F implies G under circumstance c, but that F implies not-G under c . Thus, as Theo Meyering has pointed out, there is room for "multiple supervenience" as well as multiple realizability.[15]

A second reason for redefining supervenience is that property covariation is too weak a relation to capture some notions of supervenience. A number of authors argue that what supervenience is really about is a relationship between properties such that the individual has the supervening property *in virtue of* having the subvenient property or properties.

Thus, I propose the following as a more adequate characterization of supervenience:

> Property G supervenes on property F if and only if x's instantiating G is in virtue of x's instantiating F under circumstance c.[16]

A number of authors call attention to the sorts of factors that I mean to highlight by making the supervenience relation relative to circumstances. Externalists in philosophy of mind argue that relevant features of the way the world is are crucial for determining what intentional state supervenes on a given brain state. My definition makes it possible to say that mental properties supervene on brain properties and at the same time recognize that some mental properties are codetermined by the way the world is. Another relevant case is Thomas Grimes's example of the economic properties of currency.[17] To put his example in my terms, the property of being, say, a U.S. penny supervenes on its being a copper disk with Lincoln's head stamped on one side, and so forth, only under the circumstances of its having been made at a U.S. mint, and under a vast number of other, more complex, circumstances having to do with the federal government, its powers, and its economic practices.

Berent Enç claims that there is a *species* of supervenient properties that have causal efficacy that does not get fully accounted for by the causal role played by the micro base properties. "The properties I have in mind," he says,

> . . . are locally supervenient properties of an individual that are associated with certain globally supervenient properties. These globally supervenient properties will have their base restricted to a region outside the individual in which properties causally interact with the properties of the individual in question. I do not know how to give a general formula that captures all of these globally supervenient properties. But some examples will illustrate the idea. . . .
>
> 1. Properties that are defined "causally," for example, being a skin condition that is caused by excessive exposure to sun rays, that is, being a sunburn. . . .
>
> 2. Properties that are defined in terms of what distal properties they have, for example, fitness in biology. . . .
>
> 3. Properties that are defined in terms of what would have caused them under a set of specifiable conditions, like being a representation of some state of affairs.[18]

In my terminology, the property of being a sunburn supervenes on a micro-condition of the skin cells under the circumstance of its having been brought about by overexposure to the sun. Fitness supervenes on any particular configuration of biological characteristics only within certain environmental circumstances.

Paul Teller makes similar points:

Supervenience and the Downward Efficacy of the Mental

Let us restrict attention to properties that reduce in the sense of having a physical realization, as in the cases of being a calculator, having a certain temperature, and being a piece of money. Whether or not an object counts as having properties such as these will depend, not only on the physical properties of that object, but on various circumstances of the context. Intensions of relevant language users constitute a plausible candidate for relevant circumstances. In at least many cases, dependence on context arises because the property constitutes a functional property, where the relevant functional system (calculational practices, heat transfer, monetary systems) are much larger than the property-bearing object in question. These examples raise the question of whether many and perhaps all mental properties depend ineliminably on relations to things outside the organisms that have the mental properties.[19]

So the moves I am making are not unheard of in the supervenience literature. Furthermore, I claim that my definition does in fact meet the desiderata for the concept of supervenience in that it reflects both dependence and non-reducibility. However, my claim for non-reducibility is circumscribed—I claim that it gives us non-reducibility only where and in the sense in which we should want it. That is, it gives us a way of talking about the genuine dependence of human characteristics on the brain, but leaves room for the codetermination of *some* of those characteristics by the external world, especially by culture.

There seem to be at least three types of dependence or in-virtue-of relations that deserve to be counted as supervenience relations. On some accounts, the supervenience of moral on nonmoral properties involves what we may call "conceptual supervenience." St. Francis's goodness supervenes on his generosity because being generous is part of what we *mean* by goodness. Notice, though, that the kind of supervenience in question depends on one's moral theory. For example, if rule utilitarianism is true, then we need an intervening level of description between the property of being a generous act and the description of it as good—namely, the property of being a pleasure-enhancing action. The relation between the goodness and the pleasure enhancement, again, is conceptual. But the relation between generosity and pleasure-enhancement is functional.

"Supervenience" is often used to characterize the relation between the temperature of a gas and the mean kinetic energy of its molecules: the gas has the temperature it does *in virtue of* the kinetic energy of its particles. This is an instance of microdetermination.

The type of supervenience that may repay the most attention is functional. The sort of relation envisioned here is that between a supervenient functional property and its realizand. Examples are plentiful in biology: the relation between fitness and an organism's physical properties; the relation between being the gene for red eye pigmentation and being a particular sequence of base pairs in a strand of DNA. The relation between being a coin and being a metal disk with certain characteristics is an example from another domain.

Functional terms describe causal roles in larger systems. Reference to these larger systems is one way of making more specific the circumstances that I highlight in my definition, and thus shows in a vivid manner the *codetermination* of the supervenient property by the subvenient property or properties and the circumstances. For example, fitness is codetermined by the environment; being currency is codetermined by the system of economic exchange.[20]

So, one advantage of a stronger definition of supervenience in terms of the in-virtue-of relation is that the dependence of supervenient on subvenient properties is built in. However, this dependence is (usefully) moderated by the recognition that in some cases, due to circumstances of various sorts, the subvenient properties are not sufficient determinants of the supervenient property.

The second advantage of defining supervenience in terms of the in-virtue-of relation is its heuristic value. It leads us to look at actual purported cases of supervenience and ask *how it is* that properties at one level exist in virtue of properties described at another level. In the process, we learn more about the complex "layered" reality in which we live. There is a great deal of room here for further exploration—exploration of types of supervenience based on different types of dependence relations. I project that such exploration would result in a body of literature more interesting than the current one focusing on strong, weak, and global supervenience—more interesting, at least, to philosophers of science and ethicists, if not to logicians.

Non-reducibility

The supervenience relation, as I have already stated, is supposed by some to be a non-reducible relation. A number of authors complain that the question of reduction is complicated by the fact that there is no accepted account of reduction.[21] In this section, I distinguish a variety of kinds of reduction and claim that we should want a definition of supervenience that avoids only certain of these kinds. The most significant open issue, I believe, involves causal reduction. However, it is only in limited instances that we should want to argue for a definition of supervenience that entails causal non-reducibility. My primary concern in this paper is to preserve the distinction between reasons and causes. However, free will and moral responsibility are also at issue.[22] Thus, the desideratum for a definition of supervenience is that the relation turn out to be non-reducible in only the right way and for only the right kinds of cases.

Francisco Ayala has usefully distinguished three forms of reduction: methodological, ontological, and epistemological.[23] Methodological reductionism is a research strategy, seeking explanations by investigation of lower levels of complexity. Ontological reductionism, as Ayala uses it, is the denial of non-material ontological entities such as vital forces and souls. Ayala's epistemological reduction corresponds to the theoretical reduction that has been the focus in much of the supervenience literature to date; that is, translation of higher-level laws or theories into special cases of lower-level laws via bridge rules. There are at least two other possible sorts of reduction: semantic and causal. Semantic reduction, of course, is related to epistemological or theoretical reduction in that bridge laws or definitions are needed to relate the vocabulary of the reduced theory to that of the reducing theory. Causal reduction is also related to theoretical reduction if one takes law-like regularity to be the criterion for causal relations. Note that.

Let us consider, now, in the particular case of philosophy of mind, where these various reductionist theses stand. Methodological reductionism in philosophy of mind is the thesis that one should pursue explanation of mental phenomena by investigating the underlying neural mechanisms. It cannot be denied that the recent marriage between the cognitive and neurosciences has been fruitful, so methodological reductionism here seems plainly unobjectionable, so long as it is not taken to the extreme of saying that cognition can be approached only in this way.

I believe it is safe to say that most philosophers of mind and neuroscientists now hold to an ontologically reductionist view of the relation between the mental and the physical. In fact, one of the primary uses of the concept of supervenience is to enable us to say that while complex living organisms have mental properties, there is no need to postulate a substantial mind to account for those properties. So most would say that ontological reductionism in this domain is both true and desirable, and any good account of supervenience ought to help us see why it is true.

Most of the discussions of supervenience and reductionism in the literature have focused on semantic and theoretical reduction. In philosophy of mind, this amounts to the question whether the laws of psychology are practically or in principle derivable from the laws of neurobiology. This is the kind of reductionism that an understanding of supervenience is intended (by many) to thwart. I believe it is still an open question whether multiple realizability blocks semantic and therefore epistemological reduction.[24] However, I shall not attempt to contribute to this discussion, since I see the question of causal reduction to be the more interesting and important.

Some would argue that, causation being understood in terms of lawful regularity, the fate of causal reductionism hangs exactly on sorting out the issue of epistemological reduction. However, if we consider a more primitive notion of causality, that of "making something happen," then the failure of epistemological reductionism does not entail the absence of causal determination of the mental by the neurobiological.

Note that the question here is not whether the subvenient property causes the supervenient, but rather whether casual *relations* described at the subvenient level are sufficient to determine the sequence of events as described at the supervenient level. Bear in mind that the best account we can give of the causal efficacy of mental states is to assume that they supervene on physical states, and it is the physical states that are causally efficacious. That is, it is because of the fact that my thought "I should take an umbrella" is realized neurobiologically that it is capable of initiating the chain of physical events resulting in my picking up the umbrella. But if all mental states are physically realized, and if we assume causal closure at the physical level, then what are we to make of the supposed reasoned, as opposed to causal, relations among intentional states? That is, how can we avoid having to assume that what are usually taken for reasoned connections among intentional states are not in fact determined by the laws of neurobiology? A central goal of this essay is to show that the reasoned connections *do* matter.

Consider this example: let M_1 be the thought that it is cloudy and the wind is out of the southwest, and M_2 be the thought that probably it will rain today, and M_3 be the decision to take the umbrella. By hypothesis, M_1 supervenes on some brain state B_1, and M_2 on B_2. If we assume in addition that B_1 is an adequate physical cause of B_2, then how can it *not* be the case that M_2 is merely *caused* by B_1 rather than it being a reasoned judgment based on M_1?[25]

So the question can be expressed symbolically as follows: Let the dollar sign represent the supervenience relation, the solid arrow represent the assumed causal relation between brain states (which we need in order to give an account of the causal efficacy of mental states) and the dashed arrow represent the supposed reasoned connection between mental states (judgments). We can then draw the following picture:

$$M_1 \dashrightarrow M_2$$
$$\$ \quad \quad \$$$
$$B_1 \rightarrow B_2$$

The question now is the relation between the two arrows. Causal reductionism here means that the bottom arrow is the significant one; the top arrow is dependent on or determined by it. Yet we seem to need to show, in order to maintain the meaningfulness and efficacy of the mental qua mental, that the order of dependence is reversed. This is not to say that M_1 *causes* B_2 in a straightforward manner—I reject all moves to make supervenience or realization a causal relation. Rather, it is to say that the reasoned relation between M_1 and M_2 plays an indispensable role in the neurobiological processes. I shall show how this may be the case below.

I think that several attempts in the literature to argue for the causal relevance of intentional states are on the right track but not fully adequate.[26] However, I shall not take time here to comment on them; rather I shall attempt to show the value of my own account of supervenience for explaining the causal relevance of *some* intentional states (which involves the limitation of causal reductionism in certain cases).

Downward Causation

Let me emphasize that I want no general argument against the causal reducibility of the supervenient to the subvenient, or even of the mental in general to the neurobiological. If so, we would lose the benefits of methodological reductionism and would be in danger of losing ontological reductionism as well. That is, we *want* to be able to give an account in neurobiological terms of the necessary and sufficient conditions of, say, feeling pain or seeing a patch of blue. We can assume that perceptions and other low-level conscious states[27] are realized by neurobiological events or properties and that these sorts of conscious states are perfectly determined by the neurobiological (or other physical) antecedents of the neural events that realize them. For example, a pin prick is followed by a deterministic chain of physical events, one of which realizes the experience of a stabbing pain.

Where underdetermination at the neurobiological level becomes important is primarily with regard to "higher-level" mental events: deciding, judging, and reasoning. To see how it might be possible to argue against causal reductionism in certain instances I turn to Donald Campbell's account of downward causation in the evolutionary process. We shall see that Campbell's analysis involves what I have called functional supervenience and that nonreducible circumstances play an essential role.

Here is Campbell's account of downward causation:

> Consider the anatomy of the jaws of a worker termite or ant. The hinge surfaces and the muscle attachments agree with Archimedes' laws of levers, that is, with macromechanics. They are optimally designed to apply maximum force at a useful distance from the hinge. A modern engineer could make little if any improvement on their design for the uses of gnawing wood, picking up seeds, etc., given the structural materials at hand. This is a kind of conformity to physics, but a different kind than is involved in the molecular, atomic, strong and weak coupling processes underlying the formation of the particular proteins of the muscle and shell of which the system is constructed. The laws of levers are one part of the complex selective system operating at the level of whole organisms. Selection at that level has optimised viability, and has thus optimised the form of parts of organisms, for the worker termite and ant and for their solitary ancestors. We need the laws of levers, *and organism-level selection* . . . to explain the particular distribution of proteins found in the jaw and *hence* the DNA templates guiding their production. . . . Even the *hence* of the previous sentence implies a reverse-directional 'cause' in that, by natural selection, it is protein efficacy that determines which DNA templates are present, even though the immediate micro determination is from DNA to protein.[28]

This example is meant to illustrate the following set of theses:

(1) All processes at the higher levels are restrained by and act in conformity to the laws of lower levels, including the levels of subatomic physics.

(2) The teleonomic achievements at higher levels require for their implementation specific lower-level mechanisms and processes. Explanation is not complete until these micromechanisms have been specified.[29]

But in addition:

(3) (The emergentist principle) Biological evolution in its meandering exploration of segments of the universe encounters laws, operating as selective systems, which are not described by the laws of physics and inorganic chemistry, and which will not be described by the future substitutes for the present approximations of physics and inorganic chemistry.

(4) (Downward causation) Where natural selection operates through life and death at a higher level of organisation, the laws of the higher-level selective system determine in part the distribution of lower-level events and substances. Description of an intermediate-level phenomenon is not completed by describing its possibility and implementation in lower-level terms. Its presence, prevalence or distribution (all needed for a complete explanation of biological phenomena) will often require reference to laws at a higher level of organisation as well. Paraphrasing Point 1, all processes at the lower levels of a hierarchy are restrained by and act in conformity to the laws of the higher levels.[30]

Campbell uses the term "downward causation" reluctantly, he says, "because of the shambles that philosophical analysis has revealed in our common sense meanings of 'cause.'" If it is causation, he says, "it is the back-handed variety of natural selection and cybernetics, causation by a selective system which edits the products of direct physical causation."[31]

We can represent the bottom-up aspect of the causation as follows:

$$\text{jaw structure}$$
$$\$$$
$$\text{DNA} \rightarrow \text{protein structures}$$

That is, the information encoded in the DNA contributes to the production of certain proteins upon which the structure of the termite jaw supervenes.

However, to represent the top-down aspect of causation, we need a more complex diagram, as in Figure 1. Here the dashed lines represent the top-down aspects, solid lines represent bottom-up causation.

I suggest that this "back-handed" variety of causation is all that is needed for the purposes of blocking *total* causal reduction of the mental to the neurobiological.[32] The key to my argument is the fact that neuroscientists and cognitive scientists describe mental processes such as concept formation, memory, and learning in terms formally identical to the evolutionary process in biology. In short, because of the organism's actions and interaction with the envi-

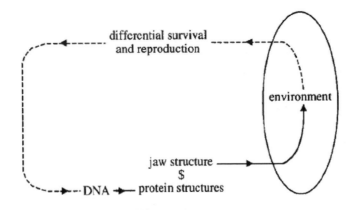

Figure 1. The top-down aspect of causation

ronment, cognitive processes can be understood to develop by means of feedback systems that suppress some responses and enhance others. Gerald Edelman describes neuron growth and the development of synaptic connections as random, rather than instructed, processes. These processes provide the somatic diversity upon which "selection" can occur. Learning is selection whereby some groups of neurons and their connections survive and thrive at the expense of others.[33] Jean-Pierre Changeaux uses the term "selective stabilization" to refer to this process.[34]

Tuning Neural Networks

The distinction I made above between lower and higher mental functions may correspond roughly to different levels of "functional validation"; this refers to differences among cognitive processes regarding the extent to which they are "wired in" for the species versus needing to be shaped in each individual by interaction with the environment. Actually, three levels are recognized here. There are cognitive processes whose character is entirely independent of the environment, even though the organism needs environmental stimulation to activate them. Other processes are partially independent of the environment, but inappropriate stimulation may change their character (for example, kittens raised in visual environments with nothing but vertical lines will later respond only to vertical, not horizontal, lines). Finally, there are cognitive processes whose development is entirely dependent upon the character of the environmental stimulation.[35] Notice that even in the case of states such as pain or color sensation only part of the explanation of why, say, hot surfaces produce this *kind* of feeling can be given by tracing the neural realization of pain back to its physical antecedents, since selection surely helps account for the fact that a burn feels *like that* rather than feeling pleasurable.

While color perception may be built into the species, learning to associate the visual experience of blue with the word "blue" has to be explained in terms of the history of the individual. In a present instance, the tendency to think "blue" in the presence of a certain visual stimulus may be causally explained by response thresholds of a given set of neurons. So, in this instance, the physical realization of the visual sensation causes, in a fairly straightforward manner, the physical realization of the thought "That's blue."[36] But the complete causal account requires, in exact parallel with Campbell's example of downward causation, reference to history—to

Supervenience and the Downward Efficacy of the Mental

the individual's having been taught the word and the concept, with appropriate associations of stimulus and response strengthened and others being eliminated. The social environment here provides the selective pressure.

For a slightly more complex example, I draw upon an account by Paul Churchland of experiments in computer modeling of recognition tasks. We can hypothesize that the processes in human learning are formally identical. His simplest example is training a target cell to detect a "T" shape when it is projected onto a "retinal" grid with nine squares (see figure on right). Each square is connected to the target cell. The target cell has been successfully trained to recognize a T when it responds maximally to the combined input from A_1, A_2, A_3, B_2, and C_2, but does not respond if it receives signals from B_1, B_3, C_1, or C_3. It may also respond weakly to patterns similar to its "preferred stimulus." Training of a receptor cell requires feedback. At first the receptor cell's firing is random with relation to the incoming stimuli. Feedback from the environment weeds out inappropriate responses and strengthens appropriate ones.[37]

A_1	A_2	A_3
B_1	B_2	B_3
C_1	C_2	C_3

I find Donald MacKay's treatment of cognitive processes particularly useful. He diagrams cognitive processes as feedback systems, as in Figure 2. This sort of feedback loop is exemplified in simple self-governing systems such as a thermostat. Here, the action of the effector system E, in the field F, is monitored by the receptor system R, which provides an indication I_f of the state of F. This indication is compared with the goal criterion I_g in the comparator C, which informs the organizing system O of any mismatch. O selects from the repertoire of E an action calculated to reduce the mismatch.[38]

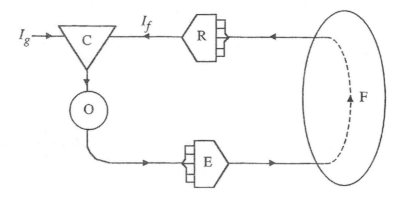

Figure 2. Cognitive processes as feedback systems

If we assume that something comparable to what Churchland describes is actually going on in the brain of a child learning to recognize letters (and to name them), then E effects the child's response to the visual stimulation and to the question "Is this a T?" R is the reception of feedback from the teacher, which is recognized by C as either matching or mismatching the goal of "getting it right." When there is a mismatch, the organizing system, in this simple case, would simply generate randomly a different response to the next set of stimuli. However, a "match" would result in strengthening the receptor cell's response to the correct pattern of input and strengthen the connection between this and the verbal representation.

Once the receptor cell is trained there is bottom-up causation whenever the child recognizes a T—the connection between the (imaginary) nine receptors in the visual field and the T-receptor cell is now "hard wired." However, top-down causation in the interaction with the social environment needs to be invoked to explain why that particular connection has come to exist. This is formally identical to the case Campbell describes, where the existence of the DNA that causes, in bottom-up fashion, the protein structures that constitute the termite jaw, needs to be explained by means of a history of feedback from the environment (top-down).

Schemas and Downward Causation

A schema, according to Michael Arbib, is a basic functional unit of action, thought, and perception. Schemas, in the first instance, are realized by means of neural processes in an individual brain. So here is an instance of *functional supervenience*. A schema *is* nothing but a pattern of neural activity, but to understand the schema qua schema it is necessary to know what it does; that is, what function it serves in cognition and action.

Arbib's account of *tuning* schemas via interaction with the environment is an account at a higher level of description that parallels and partially overlaps the foregoing account of tuning neural networks. Here, the unit upon which selection and downward causation operates is not individual neural connections but rather schemas, which are realized by medium-scale neural structures and distributed systems. Apparently some schemas are innate yet subject to restructuring; others are learned. In addition to tuning as a result of action and interaction with the environment, schemas are enriched as a result of interaction with other schemas.

Arbib's emphasis on the action-oriented character of schemas, not merely their representational functions, is crucial for my account. This creates the feedback loops that allow for downward causation from the environment, including the sociocultural environment. Another crucial ingredient is the element of competition among schemas and schema instances. "It is as a result of competition that instances which do not meet the evolving (data-guided) consensus lose activity."[39]

The Downward Efficacy of Reason

Colin McGinn points out that one of the most significant problems in philosophy of mind is to explain how a physical organism can be subject to the norms of rationality: "How, for example, does *modus ponens* get its grip on the causal transitions between mental states?"[40] Here I employ the concept of downward causation to begin to develop an account of the top-down efficacy of the *intellectual* environment. In the following examples, I intend to show that causal closure at the neurobiological level does not conflict with giving an account at the mental level in terms of reasons. To the question, "What is 5 × 7?" most of us respond automatically with "35." Let us presume that understanding the question (M_1) and thinking the answer (M_2) supervene upon (or are realized neurologically by) B_1 and B_2 respectively. Then we can ask the question whether M_2 occurs because "35" is the *correct* answer or because B_1 *caused* B_2 and M_2 supervenes on B_2. However, considering again the role of feedback loops, we can see that it is clearly both. In short, the laws of arithmetic are needed to account for the prior development of the causal link between B_1 and B_2. In this case, using again the diagram in Figure 2, the organizing system can be thought of as first producing random responses to the question, "What is 5 × 7?" The effector writes or speaks the answer. The field of operation, again, is the classroom; the receptor system receives feedback from the teacher, and the comparator registers "right" or "wrong." If wrong, the organizing system selects a different response. In time, wrong guesses are extinguished, and the causal connection from B_1 to B_2 is strengthened. Then, of course, the response "35" is caused (bottom up) by the connection between B_1 and B_2. The complete explanation requires a top-down account of interaction with the teacher. But this raises in turn the question why the teacher reinforces one answer rather than another, which can only be answered, ultimately, by an account of the *truth* of "5 × 7 = 35."

It is clear, I believe, from Arbib's work as well as that of Edelman and his colleagues that analysis of the mental in terms of selection can be extended to all higher mental functions. Higher-order consciousness involves the ability to categorize lower-order conscious functions (schemas) and to evaluate them. So, just as the concept "blue" or the connection of "5 × 7" with "35" can be learned, *schemas having to do with the evaluation of thought itself can be learned and internalized.* Thereafter, they function to select, among lower-order thought processes, those which are and are not "fit."

We can see this by turning again to the learning of arithmetic. Instruction does not aim simply at rote learning but at teaching both skills and evaluation procedures that ultimately allow for internal correction of mental operations. Consider the slightly more complex case of learning to multiply. Here a problem is posed (say, 55 × 77). The organizing system now involves not a random generator of responses but a *skill* or *operation* that produces answers.[41] The feedback system, then, not only corrects for wrong answers but also, via a supervisory system (see Figure 3), corrects flaws in the operation itself. The student learns *how* to multiply.

Figure 3 represents a slightly more complex information-flow system that involves internal evaluation of cognitive processes. This is a map of a system capable of modifying its own goals, including: SS, supervisory system; C, comparator; O, organizing system; F, field of action; E, effectors; R, Receptors; I_f, indicated state of field; and I_g, indicated goal-state or criterion of evaluation. The organizing system O organizes the repertoire of possible activity, which may

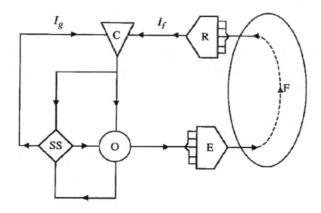

Figure 3. Information-flow system that involves internal evaluation of cognitive processes.

be prewired or built up in response to the regularities of its field of action. The supervisory system selects from O parts of the repertoire appropriate to incoming mismatch signals from the comparator and supervises the development and trial and updating of the repertoire.[42]

With this diagram we can represent cognitive processes in which the criteria for correct performance are internalized. The supervisory system now allows for another feedback loop, internal to the information-flow system, and thus allows for more sophisticated instances of downward causation.

MacKay provides a more complex information-flow map to represent both the hierarchical nature of internal evaluation of cognitive processes and also to make it possible to represent a system capable of self-organized goal-directed activity (see Figure 4). Here the supervisory system of Figure 3 is represented by two components, the meta-comparator MC, and the meta-organizing system MO. Heavy lines represent the supervisory or meta-organizing level. (FF represents a feed-forward system with feature filters.)

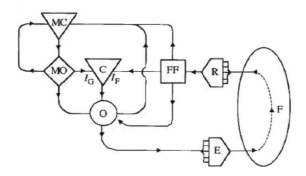

Figure 4. A representation of the hierarchical nature of internal evaluation of cognitive processes

In Figure 2, I_g, the goal state of the system, is set by some criterion outside the system. For example, the goal of getting right answers in class is (initially, at least) set by the teacher. In the system represented in Figure 4, the goal state itself is set by higher-level processes within the system. MacKay says, "we have drawn the meta-organizing system with a meta-evaluative procedure to take stock of how things are going. It adjusts and reselects the current goal in the light of its evaluation of the success of ongoing agency, and it also keeps up to date the organizing system to match the state of the world indicated by the feed forward from the receptor system. . . . So there is a meta-organizing system, which organizes the activity of the organizing system in the way that the organizing system organizes the activity in the outside world."[43]

A simple example of the role of the meta-organizing system, building again on the examples from mathematics classes, would be a student who decides to rebel and rejects the goal of getting right answers. This change in priorities completely changes the operation of the entire system, but may itself be revised if the environmental consequences make the new strategy too costly. And here we are imagining yet a higher level of evaluation, which is removed from the sphere of algorithmic cognitive processes and instead falls within the sphere of practical reasoning.

This puts us in position to consider an example of moral reasoning. The point is to show that, just as in the simpler cases of arithmetic, the reasons (here moral reasons) can have top-down efficacy despite the presumed causal closure at the neurobiological level.

The fight or flight response is prewired in humans, as in other animals. The ability to recognize threatening behavior is apparently built up easily from pre-existing perceptual capacities.[44] Thus, a typical series of events can be represented by Figure 2, involving, say, perception of the behavior of another person (R); evaluation of the behavior as threatening (C); selection of response—fleeing, fighting, conciliation—by the organizing system (O); and effecting the response (E). Feedback from the field of operation (F) will refine the actor's ability to choose responses that have the best survival value.

Now we complicate the picture by adding a moral supervisory system. Let us say that our agent is a pacifist. Now we need Figure 4 to represent the system. The meta-comparator now places a higher value on nonviolent resolution of conflict than on survival. The meta-organizing system then adjusts C's priorities accordingly. C's job will now be to evaluate threatening behavior not in terms of threats to survival, but in terms of threats to the peace of the community. A different repertoire of skills, maps, and norms will have to be developed in O. As this system develops, the FF path, which selects relevant features of sensory input, will be affected by action and reactions of the environment. G. Simon Harak points out that virtuous behavior effects changes in the agent's perceptions. As an example, he recounts an incident from a time in his life when he was practicing martial arts. He commented to a companion on the threatening demeanor of a man they had just encountered. However, the companion, a seminarian dedicated to pacifist principles, had not perceived the stranger as threatening. So patterns of action, and thus patterns of readiness to act, gradually shape one's perceptions of reality.[45] This is a particular instance of Arbib's more general claim that action shapes one's schemas and schemas in turn shape expectations and perceptions.[46]

Notice that in MacKay's diagram there is feedback from the field of operation to the meta-comparator. This represents the fact that, in the case in question, the moral principle is subject to readjustment in light of the effects produced by acting in accordance with it. For example, it is often supposed that pacifist responses increase others' aggression. The pacifist might re-evaluate her commitment to this principle if this turned out to be true. So here we have a representation of the top-down efficacy of moral principles (as well as the environment) in shaping human behavior.

The foregoing is not intended to be an adequate account of moral reasoning, but only to show the possible downward efficacy of moral principles.

5. Conclusion

Let me try now to tie all of this together. I began with a brief overview of the status of physicalism in Christian thought, claiming that it is at least as compatible with early Christian teaching as dualism. However, if the challenge of neurobiological reductionism cannot be met, then it is not an acceptable account of human nature. To begin to meet this challenge I suggested, first, that we redefine the concept of supervenience, which is used by philosophers of mind to give an account of how the mental relates to the neurobiological. I emphasized two ingredients that need to go into a definition; one is that supervenient properties be taken to obtain *in virtue of* subvenient properties. This builds into the very meaning of "supervenience" the sort of asymmetric property dependence that is wanted but not, apparently, explicable when the relation is defined in terms of material implication and modal qualifiers. Second, I suggested that *circumstances* pertaining to the supervenient level of description will often codetermine the presence or absence of the supervenient property, and thus this qualification needs to be reflected in the definition.

Concentration on the in-virtue-of relation has heuristic value in leading us to ask, of real properties in the real world, what kinds of dependence relations there are. I suggested that there are at least three that deserve attention: conceptual, microdeterminational (to coin a term), and functional. The functional sort turns out to be particularly useful, in part because it draws our attention to the additional circumstances at the supervenient or functional level that have to be taken into account—namely the rest of the causal system into which this part fits. So concern with the functional relation has also led, in this essay, to examination of cases widely recognized as instances of supervenience in which downward causation obtains.

In philosophy of mind and cognitive science, then, we can take mental events or schemas to supervene, in a functional sense, on neural processes. I used Donald MacKay's understanding of cognitive processes as information-flow feedback systems, and spelled out increasingly complex cases where we can see the top-down efficacy of epistemological (mathematical) and moral principles. That is, evaluative standards of an epistemological sort shape, by means of downward causation, the very structure of the individual's neural networks.

In her comments on an earlier draft of this paper, Leslie Brothers summarized my position so well that I end with a quotation from her remarks. She begins with an example that well illustrates my concept of supervenience and then shows how it works to account, first, for the non-reducibility of the social to the individual level, and second, for the top-down efficacy of higher-level mental events:

> Take the sound of a single musical note. (1) It can be reduced to soundwave frequencies that causally give pitch. (2) If a component of a melody, the sound of the note will also have a relational property with the notes around it, where the relations are determined by the musical system in which the melody was composed. Standard supervenience applies to the note in isolation (so-and-so many cycles per second) and Murphy-supervenience to the note as a diminished seventh, which it is in virtue both of the soundwaves and the musical context. . . .
>
> Think of persons as having two dimensions that intersect. The first is that they are loci of experience—for example, they can "have" pain. The second . . . is that they are locations in a moral-social order, an order woven together of shoulds and reasons and villains and *mensches* and so on. The actions of persons are accounted for, by themselves and others, in the context of this order. Now, if people like Crick and Koch are right, in the fullness of time we will causally reduce qualia to a feature of collective neuronal activity (personally I suspend judgment on this but let's give it to them for the sake of argument). This will be like reducing the note to its soundwave frequencies. On the other hand, what Murphy refers to as "higher level" mental events (deciding, judging, reasoning) will not be able to be extricated from the moral-social order to which they belong. Furthermore, I see no problem at all with downward neural causation here: if I choose to practice the golden rule by returning a lost wallet, I activate neural assemblies that represent the worried state of someone whom I might imagine to be like a friend or relative; I activate assemblies representing the person's joy when her wallet is returned, etc. By themselves, the neurons can only encode various states of the world, not a moral dimension. But moral behavior will select neural activity having to do with scenarios of giving something to someone, as opposed to scenarios having to do with spending the cash myself. It is a backhand, selective system of downward causation, but . . . [it] is what Murphy has in mind as "all we need or want for the purposes of blocking total causal reduction of the mental to the neurobiological."[47]

Notes

1. This is William D. Hart, who delivered a lecture titled "Unity and Dualism" at a symposium on mind and body at Westmont College, Santa Barbara, CA on February 15, 2002.
2. J. D. G. Dunn, *The Theology of the Apostle Paul* (Grand Rapids: Eerdmans, 1998), 51ff.
3. John W. Cooper, *Body, Soul, and Life Everlasting: Biblical Anthropology and the Monism-Dualism Debate* (Grand Rapids: Eerdmans, 1989); second enlarged ed., 2000.
4. Joel B. Green, "Eschatology and the Nature of Humans: A Reconsideration of Pertinent Biblical Evidence," *Science and Christian Belief* 14, no. 1 (April 2002): 33–50.
5. For a more detailed treatment, see Nancey Murphy, *Bodies and Souls, Or Spirited Bodies?* (Cambridge: Cambridge University Press, 2006), chap. 1.
6. John R. Searle, *The Rediscovery of the Mind* (Cambridge, MA: MIT Press, 1992).
7. I deal with religious experience in "Nonreductive Physicalism: Philosophical Issues," in Warren S. Brown, Nancey Murphy, and H. Newton Malony, eds., *Whatever Happened to the Soul?: Scientific and Theological Portraits of Human Nature* (Minneapolis: Fortress Press, 1998), chap. 6. I describe religious experience as supervenient on ordinary cognitive and emotional experiences.
8. Jaegwon Kim now argues that non-reductive physicalism is a myth; that is, it is an unstable position that tends toward outright eliminativism or some form of dualism (see "The Myth of Nonreductive Materialism," in Richard Warren and Tadeusz Szubka, eds., *The Mind-Body Problem* [Oxford: Blackwell, 1994], 242–60). I agree, but only so long as one sticks to Kim's own definition of "supervenience."
9. R. M. Hare, *The Language of Morals* (New York: Oxford University Press, 1966), 145. Originally published in 1952.
10. Donald Davidson, *Essays on Actions and Events* (Oxford: Clarendon Press, 1980), 214. Reprinted from *Experience and Theory,* ed. Lawrence Foster and J.W. Swanson (University of Massachusetts Press and Duckworth, 1970).
11. David Lewis, *On the Plurality of Worlds* (Oxford: Basil Blackwell, 1986), 14. Quoted by Brian McLaughlin, "Varieties of Supervenience," in Elias E. Savellos and Ümit D. Yalçin, eds., *Supervenience: New Essays* (Cambridge: Cambridge University Press, 1995), 16–59; quotation 17. McLaughlin added "and only when," claiming that it is clear from the context that this is what Lewis meant.
12. Jaegwon Kim, "Concepts of Supervenience," *Philosophy and Phenomenological Research,* xiv, no. 2 (December 1984): 153–76; quotation 163.
13. Ibid., 165.
14. Ibid., 168.
15. See Theo C. Meyering, "Physicalism and the Autonomy of the Person," in Robert J. Russell et al., eds. *Neuroscience and the Person: Scientific Perspectives on Divine Action* (Vatican City State: Vatican Observatory Press, 1999), 165–77.
16. See also my "Supervenience and the Nonreducibility of Ethics to Biology," in R. J. Russell et al., eds. *Evolutionary and Molecular Biology: Scientific Perspectives on Divine Action* (Vatican City State: Vatican Observatory Press, 1998), 463–89.
17. Thomas R. Grimes, "The Tweedledum and Tweedledee of Supervenience," in Savellos and Yalçin, eds., *Supervenience,* 110–23; 117.
18. Berent Enç, "Nonreducible Supervenient Causation," in Savellos and Yalçin, *Supervenience,* 169–80; 175.
19. Paul Teller, "Reduction," in Robert Audi, ed., *The Cambridge Dictionary of Philosophy* (Cambridge: Cambridge University Press, 1995), 679–80; 680.

20. I have argued that a speech act or illocutionary act in J. L. Austin's terms supervenes on a locutionary act (*Anglo-American Postmodernity: Philosophical Perspectives on Science, Religion, and Ethics* [Boulder, CO: Westview Press, 1997], 24–25).

21. Brian McLaughlin, "Varieties of Supervenience," 45–46; Paul K. Moser and J. D. Trout, "Physicalism, Supervenience, and Dependence," in Savellos and Yalçin, eds., *Supervenience,* 187–217; 190–91.

22. Warren Brown and I deal with these further issues in *Did My Neurons Make Me Do It? Philosophical and Neurobiological Perspectives on Moral Responsibility and Free Will* (forthcoming).

23. Francisco J. Ayala, "Introduction," in F. J. Ayala and T. Dobzhansky, eds., *Studies in the Philosophy of Biology: Reduction and Related Problems* (Berkeley and Los Angeles: University of California Press, 1974).

24. See Andrew Melnyk, "Two Cheers for Reductionism: Or, the Dim Prospects for Non-Reductive Materialism," *Philosophy of Science,* 62 (1995): 370–88; especially 379.

25. I shall in the end (try to) solve this problem by showing how M_2 can be both reasoned and caused.

26. Macdonald, "Psychophysical Supervenience"; and Enç, "Nonreducible Supervenient Causation."

27. "Low-level conscious states" cries out for explication. Roughly what I have in mind are the sorts of conscious states that we presume we share with the higher animals: the deliverances of the senses, proprioception, etc. My point here is that nothing is lost in terms of our sense of our humanness if these turn out to be strictly determined by the laws of neurobiology. However, I believe that downward causation is as operative in many of these instances as it is in reasoning, judging, etc. For example, "mental set" (i.e., expectation) affects sensory perception.

28. Donald T. Campbell, "'Downward Causation' in Hierarchically Organised Biological Systems," in Ayala and Dobzhansky, eds., *Studies in the Philosophy of Biology,* 179–86; 181.

29. Campbell, "Downward Causation," 180.

30. Ibid.

31. Campbell, "Downward Causation," 180–81.

32. It may be useful to remind the reader that my concern here is with causal reductionism, not epistemological reduction, since it is already clear that we often need the higher-level epistemological accounts for practical reasons. And the reason it is important to show the limits of causal reduction is to allow for the top-down effects of culture on the brain.

33. Gerald M. Edelman, *Bright Air, Brilliant Fire: On the Matter of the Mind* (New York: Basic Books, 1992).

34. Jean-Pierre Changeaux, *Neuronal Man: The Biology of the Mind,* tr. L. Garey (New York: Pantheon, 1985).

35. Bryan Kolb and Ian Q. Whishaw, *Fundamentals of Human Neuropsychology,* 4th ed., (New York: W. H. Freeman, 1996), 500.

36. I say *fairly* straightforward because of the complications of analyzing probabilistic causes.

37. Paul M. Churchland, *The Engine of Reason, the Seat of the Soul: A Philosophical Journey into the Brain* (Cambridge, MA: MIT Press, 1995), 36–39.

38. Donald M. MacKay, *Behind the Eye,* the Gifford Lectures, ed. Valerie MacKay (Oxford: Basil Blackwell, 1991), 43–44.

39. Michael Arbib, "Towards a Neuroscience of the Person," in Russell et al. eds., *Neuroscience and the Person,* 77–100.

40. Colin McGinn, "Consciousness and Content," in Ned Block, Owen Flanagan, and Güven Güzeldere, eds., *The Nature of Consciousness: Philosophical Debates* (Cambridge, MA: MIT Press, 1997), 295–307; quotation 305, n. 2.

41. MacKay conceives of the typical organizing system as involving norms, maps, and skills (*Behind the Eye,* 144).

42. MacKay, *Behind the Eye,* 51.

43. McKay, *Behind the Eye,* 141, 142.

44. Leslie Brothers, *Friday's Footprint: How Society Shapes the Human Mind* (Oxford and New York: Oxford University Press, 1997), 28–29.

45. G. Simon Harak, *Virtuous Passions: The Formation of Christian Character* (New York: Paulist Press, 1993), 34.
46. Arbib, "Towards a Neuroscience of the Person."
47. The major part of this essay is a slight revision of my "Supervenience and the Downward Efficacy of the Mental: A Nonreductive Physicalist Account of Human Action," in Robert J. Russell et al., eds. *Neuroscience and the Person: Scientific Perspectives on Divine Action* (Vatican City State: Vatican Observatory Press, 1999), 147–64. I thank the Vatican Observatory for permission to use it here.

35

CONSCIOUS EVENTS AS ORCHESTRATED
SPACE-TIME SELECTIONS[1]

STUART HAMEROFF AND ROGER PENROSE

1. INTRODUCTION: SELF-SELECTION IN AN EXPERIENTIAL MEDIUM?

The "hard problem" of incorporating the phenomenon of consciousness into a scientific worldview involves finding scientific explanations of qualia, or the subjective experience of mental states.[2] On this subject, reductionist science is still at sea. Why do we have an inner life, and what exactly is it?

One set of philosophical positions, addressing the hard problem, views consciousness as a fundamental component of physical reality. For example, an extreme view—"panpsychism"—is that consciousness is a quality of all matter: atoms and their subatomic components having elements of consciousness (e.g. Spinoza, Rensch).[3] "Mentalists," such as Leibniz and Whitehead,[4] contended that systems ordinarily considered to be physical are constructed in some sense from mental entities. Bertrand Russell described "neutral monism," in which a common underlying entity, neither physical nor mental, gave rise to both.[5] Recently, Stubenberg has claimed that *qualia* are that common entity.[6] In monistic idealism, matter and mind arise from consciousness—the fundamental constituent of reality (e.g. Goswami).[7] Wheeler has suggested that information is fundamental to the physics of the universe.[8] From this, Chalmers proposes a double-aspect theory in which information has both physical and experiential aspects.[9]

Among these positions, the philosophy of Alfred North Whitehead may be most directly applicable.[10] Whitehead describes the ultimate concrete entities in the cosmos as being actual "occasions of experience," each bearing a quality akin to "feeling." Whitehead construes "experience" broadly—in a manner consistent with panpsychism—so that even "temporal events in the career of an electron have a kind of 'protomentality.'" Whitehead's view may be considered to differ from panpsychism, however, in that his discrete "occasions of experience" can be taken to be related to "quantum events."[11] In the standard descriptions of quantum mechanics, randomness occurs in the events described as quantum state reductions—these being events which appear to take place when a quantum-level process gets magnified to a macroscopic scale.

Quantum state reduction (**R**) is the random procedure that is adopted by physicists in their descriptions of the quantum measurement process.[12] It is still a highly controversial matter whether **R** is to be taken as a "real" physical process, or whether it is some kind of illusion and not to be regarded as a fundamental ingredient of the behavior of nature. Our position is to take **R** to be indeed real—or rather, to regard it as a close approximation to an objectively real process, **OR** (objective reduction), which is to be a noncomputable process instead of merely a random one.[13] In almost all physical situations, **OR** would come about in situations in which the random effects of the environment dominate, so **OR** would be virtually indistinguishable from the random **R** procedure that is normally adopted by quantum theorists. However, when the quantum system under consideration remains coherent and well-isolated from its environment, it becomes possible for its state to collapse spontaneously, in accordance with the **OR** scheme we adopt, and to behave in noncomputable, rather than random, ways. Moreover, this **OR** scheme intimately involves the geometry of the physical universe at its deepest levels.

Our viewpoint is to regard experiential phenomena as also inseparable from the physical universe, and in fact to be deeply connected with the very laws which govern the physical universe. The connection is so deep, however, that we perceive only glimmerings of it in our present-day physics. One of these glimmerings, we contend, is a necessary noncomputability in conscious thought processes; and we argue that this non-computability must also be inherent in the phenomenon of quantum state *self*-reduction—the "objective reduction" (OR) referred to above. This is the main thread of argument in *Shadows of the Mind*.[14] The argument that conscious thought, whatever other attributes it may also have, is noncomputable (as follows most powerfully from certain deductions from Gödel's incompleteness theorem) grabs hold of one tiny but extremely valuable point. This means that at least some conscious states cannot be derived from previous states by an algorithmic process—a property which distinguishes human and other animal minds from computers. Noncomputability per se does not directly address the "hard problem" of the nature of experience, but it is a clue to the kind of physical activity that lies behind it; this points to **OR**, an underlying physical action of a completely different character from that which seems to underlie nonconscious activity. Following this clue with sensitivity and patience should ultimately lead to real progress towards understanding mental phenomena in their inward manifestations as well as their outward ones.

In the **OR** description, consciousness occurs if an organized quantum system is able to isolate and sustain coherent superposition until its quantum gravity threshold for space-time separation is met; it then *self*-reduces (noncomputably). For consciousness to occur, *self*-reduction is essential, as opposed to reduction triggered by the system's random environment. (In the latter case, the reduction would itself be effectively random and would lack useful noncomputability, being unsuitable for direct involvement in consciousness.) We take the *self*-reduction to be an instantaneous event—the climax of a *self*-organizing process fundamental to the structure of space-time—and apparently consistent with a Whitehead "occasion of experience."

As **OR** could, in principle, occur ubiquitously within many types of inanimate media, it may seem to imply a form of "panpsychism" (in which individual electrons, for example, possess an experiential quality). However, according to the principles of **OR**,[15] a single superposed electron would spontaneously reduce its state (assuming it could maintain isolation) only once in a period much longer than the present age of the universe. Only large collections of particles, acting coherently in a single macroscopic quantum state, could possibly sustain isolation and support coherent superposition in a time frame brief enough to be relevant to our consciousness. Thus only very special circumstances could support consciousness:

1. High degree of coherence of a quantum state—a collective mass of particles in superposition for a time period long enough to reach threshold, and brief enough to be useful in thought processes.

2. Ability for the **OR** process to be at least transiently isolated from a "noisy" environment until the spontaneous state reduction takes place. This isolation is required so that reduction is not simply random. Mass movement in the environment which entangles with the quantum state would effect a random (not noncomputable) reduction.

3. Cascades of **OR**s to give a "stream" of consciousness, and huge numbers of **OR** events taking place during the course of a lifetime.

By reaching quantum gravity threshold, each **OR** event has a fundamental bearing on space-time geometry. One could say that a cascade of **OR** events charts an actual course of physical space-time geometry selections.

It may seem surprising that quantum gravity effects could plausibly have relevance at the physical scales relevant to brain processes, for quantum gravity is normally viewed as having only absurdly tiny influences at ordinary dimensions. However, we shall show later that this is not the case, and the scales determined by basic quantum gravity principles are indeed those that are relevant for conscious brain processes.

We must ask how such an **OR** process could actually occur in the brain. How could it be coupled to neural activities at a high rate of information exchange; how could it account for preconscious to conscious transitions, have spatial and temporal binding, and both simultaneity and time flow?

We here nominate an **OR** process with the requisite characteristics occurring in cytoskeletal microtubules within the brain's neurons. In our model, microtubule-associated proteins "tune," or "orchestrate," the quantum oscillations leading to **OR**; we thus term the process "orchestrated objective reduction" (**Orch OR**).

2. Space-Time: Quantum Theory and Einstein's Gravity

Quantum theory describes the extraordinary behavior of the matter and energy which comprise our universe at a fundamental level. At the root of quantum theory is the wave/particle duality of atoms, molecules and their constituent particles. A quantum system such as an atom or subatomic particle which remains isolated from its environment behaves as a "wave of possibilities" and exists in a coherent complex-number valued "superposition" of many possible states. The behavior of such wave-like, quantum-level objects can be satisfactorily described in terms of a state vector which evolves deterministically according to the Schrödinger equation (unitary evolution), denoted by **U**.

Somehow, quantum microlevel superpositions lead to unsuperposed stable structures in our macro-world. In a transition known as wave function collapse, or reduction (**R**), the quantum wave to alternative possibilities reduces to a single macroscopic reality, an "eigenstate" of some appropriate operator. (This would be just one out of many possible alternative eigenstates relevant to the quantum operator.) This process is invoked in the description of a macroscopic measurement, when effects are magnified from the small quantum scale to the large classical scale.

Conscious Events as Orchestrated Space-Time Selections

According to conventional quantum theory (as part of the standard "Copenhagen interpretation"), each choice of eigenstate is entirely random, weighted according to a probability value that can be calculated from the previous state according to the precise procedures of quantum formalism. This probabilistic ingredient was a feature with which Einstein, among others, expressed displeasure—as he once wrote to Max Born, "You believe in a God who plays dice and I in complete law and order." Penrose has contended that, at a deeper level of description, the choices may more accurately arise as a result of some presently unknown "noncomputational" mathematical/physical (i.e., "platonic realm") theory—that is, they cannot be deduced algorithmically.[16] Penrose argues that such noncomputability is essential to consciousness, because (at least some) conscious mental activity is unattainable by computers.

It can be argued that present-day physics has no clear explanation for the cause and occurrence of wave function collapse **R**. Experimental and theoretical evidence through the 1930s led quantum physicists (such as Schrödinger, Heisenberg, Dirac, von Neumann, and others) to postulate that quantum-coherent superpositions persist indefinitely in time, and would in principle be maintained from the micro to macro levels. Or perhaps they would persist until conscious observation collapses or reduces the wave function (subjective reduction, or "**SR**"). Accordingly, even macroscopic objects, if unobserved, could remain superposed. To illustrate the apparent absurdity of this notion, Erwin Schrödinger described his now-famous "cat in a box" as simultaneously both dead and alive until the box was opened and the cat observed.[17]

As a counter to this unsettling prospect, various new physical schemes for collapse according to objective criteria (objective reduction—"**OR**") have recently been proposed. According to such a scheme, the growth and persistence of superposed states could reach a critical threshold, at which collapse, or **OR**, rapidly occurs.[18] Some such schemes are based specifically on gravitational effects mediating **OR**.[19]

Table 1 categorizes types of reduction.

Context	Cause of Collapse (Reduction)	Description	Acronym
Quantum coherent superposition	No collapse	Evolution of the wave function (Schrödinger equation)	U
Conventional quantum theory (Copenhagen interpretation)	Environmental entanglement, Measurement, Conscious observation	Reduction; Subjective reduction	R SR
New physics (Penrose 1994)	Self-collapse—quantum gravity induced (Penrose, Diósi, etc)	Objective reduction	OR
Consciousness (present paper)	Self-collapse, quantum gravity threshold in microtubules orchestrated by MAPs etc	Orchestrated objective reduction	Orch OR

Table 1. Descriptions of wave function collapse.

The physical phenomenon of gravity, described to a high degree of accuracy by Isaac Newton's mathematics in 1687, has played a key role in scientific understanding. However, in 1915, Einstein created a major revolution in our scientific worldview. According to Einstein's theory, gravity plays a unique role in physics for several reasons.[20] Most particularly, these are:

1. Gravity is the only physical quality that influences causal relationships between space-time events.

2. Gravitational force has no local reality, as it can be eliminated by a change in space-time coordinates; instead, gravitational tidal effects provide a *curvature* for the very *space-time* in which all other particles and forces are contained.

It follows from this that gravity cannot be regarded as some kind of "emergent phenomenon," secondary to other physical effects, but is a "fundamental component" of physical reality.

There are strong arguments to suggest that the appropriate union of general relativity (Einstein's theory of gravity) with quantum mechanics—a union often referred to as "quantum gravity"—will lead to a significant change in *both* quantum theory and general relativity, and when the correct theory is found, will yield a profoundly *new* understanding of physical reality.[21] And although gravitational *forces* between objects are exceedingly weak (feebler than, for example, electrical forces by some forty orders of magnitude), there are significant reasons for believing that gravity has a fundamental influence on the behavior of quantum systems as they evolve from the micro to the macro levels. The appropriate union of quantum gravity with biology, or at least with advanced biological nervous systems, may yield a profoundly new understanding of consciousness.

3. Curved Space-Time Superpositions and Objective Reduction

According to modern accepted physical pictures, reality is rooted in three-dimensional space and a one-dimensional time, combined together into a four-dimensional space-time. This space-time is slightly curved, in accordance with Einstein's general theory of relativity, in a way that encodes the gravitational fields of all distributions of mass density. Each mass density effects a space-time curvature, albeit a tiny one.

This is the standard picture according to *classical* physics. On the other hand, when *quantum* systems have been considered by physicists, this mass-induced tiny curvature in the structure of space-time has been almost invariably ignored, gravitational effects having been assumed to be totally insignificant for normal problems in which quantum theory is important. Surprisingly as it may seem, however, such tiny differences in space-time structure *can* have large effects, for they entail subtle but fundamental influences on the very rules of quantum mechanics.

Superposed quantum states for which the respective mass distributions differ significantly from one another will have space-time geometries which correspondingly differ. Thus, according to standard quantum theory, the superposed state would have to involve a quantum superposition of these differing space-times. In the absence of a coherent theory of quantum gravity, there is no accepted way of handling such a superposition. Indeed, the basic principles of Einstein's general relativity begin to come into profound conflict with those of quantum mechanics.[22] Nevertheless, various tentative procedures have been put forward in attempts to describe such a superposition. Of particular relevance to our present proposals are the suggestions of certain authors[23] that it is at this point that an objective quantum state reduction (**OR**) ought to occur, and the rate or timescale of this process can be calculated from basic quantum gravity considerations. These particular proposals differ in certain detailed respects, and for definiteness we shall follow the specific suggestions made in Penrose.[24] Accordingly, the quantum superposition of significantly differing space-times is unstable, with a lifetime given by that time scale. Such a superposed state will decay—or "reduce"—into a single universe state, which is one or the other of the space-time geometries involved in that superposition.

Whereas such an **OR** action is not a generally recognized part of the normal quantum-mechanical procedures, there is no plausible or clear-cut alternative that standard quantum theory has to offer. This **OR** procedure avoids the need for "multiple universes."[25] There is no agreement, among quantum gravity experts, about how else to address this problem. For the purposes of the present article, it will be assumed that a gravitationally induced **OR** action is indeed the correct resolution of this fundamental conundrum.

Figure 1 schematically illustrates the way in which space-time structure can be affected when two macroscopically different mass distributions take part in a quantum superposition. Each mass distribution gives rise to a separate space-time, the two differing slightly in their curvatures. So long as the two distributions remain in quantum superposition, we must consider that the two space-times remain in superposition. Since, according to the principles of general relativity, there is no natural way to identify the points of one space-time with corresponding points of the other, we have to consider the two as separated from one another in some sense, resulting in a kind of "blister" where the space-time bifurcates.

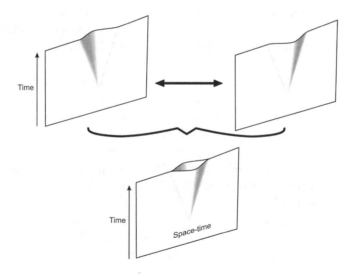

Figure 1. Quantum coherent superposition represented as a separation of space-time. In the lowest of the three diagrams, a bifurcating space-time is depicted as the union ("glued-together version") of the two alternative space-time histories that are depicted at the top of the figure. The bifurcating space-time diagram illustrates two alternative mass distributions actually in quantum superposition, whereas the top two diagrams illustrate the two individual alternatives which take part in the superposition (adapted from Penrose 1994: 338).

A bifurcating space-time is depicted in the lowest of the three diagrams, this being the union ("glued-together version") of the two alternative space-time histories that are depicted at the top of Figure 1. The initial part of each space-time is at the lower end of each individual space-time diagram. The bottom space-time diagram (the bifurcating one) illustrates two alternative mass distributions actually in quantum superposition, whereas the top two illustrate the individual alternatives that take part in the superposition. The combined space-time describes a superposition in which the alternative locations of a mass move gradually away from each other as we proceed in the upward direction in the diagram. Quantum-mechanically (so long as **OR** has not taken place), we must think of the "physical reality" of this situation as being illustrated as an actual superposition of these two slightly differing space-time manifolds, as indicated in the bottom diagram.

As soon as **OR** has occurred, one of the two individual space-times takes over, as depicted as one of the two sheets of the bifurcation. For clarity only, the bifurcating parts of these two sheets are illustrated as being one convex and the other concave. Of course, there is additional artistic license involved in drawing the space-time sheets as two-dimensional, whereas the actual space-time constituents are four-dimensional. Moreover, there is no significance to be attached to the imagined "three-dimensional space" within which the space-time sheets seem to be residing. There is no "actual" higher dimensional space there, the "intrinsic geometry" of the bifurcating space-time being all that has physical significance. When the "separation" of the two space-time sheets reaches a critical amount, one of the two sheets "dies"—in accordance with the **OR** criterion—the other being the one that persists in physical reality. The quantum state thus reduces (**OR**), by choosing between either the "concave" or "convex" space-time of Figure 1.

It should be made clear that this measure of separation is only very schematically illustrated as the "distance" between the two sheets in the lower diagram in Figure 1. As remarked above, there is no physically existing "ambient higher dimensional space" inside which the two sheets reside. The degree of separation between the space-time sheets is a more abstract mathematical thing; it would be more appropriately described in terms of a *symplectic measure* on the space of four-dimensional metrics[26]—but the details (and difficulties) of this will not be important for us here. It may be noted, however, that this separation is a space-time separation, not just a spatial one. Thus the *time* of separation contributes as well as the spatial displacement. Roughly speaking, it is the product of the temporal separation T with the spatial separation S that measures the overall degree of separation, and **OR** takes place when this overall separation reaches the critical amount.[27] Thus, for small S, the lifetime T of the superposed state will be large; on the other hand, if S is large, then T will be small. To calculate S, we compute (in the Newtonian limit of weak gravitational fields) the gravitational self-energy E of the difference between the mass distributions of the two superposed states (that is, one mass distribution counts positively and the other negatively).[28] The quantity S is then given by:

$$S = E,$$
$$\text{thus}$$
$$E = \hbar / T.$$

Schematically, since S represents three dimensions of displacement rather than the one dimension involved in T, we can imagine that this displacement is shared equally between each of these three dimensions of space—and this is what has been depicted in Figure 3 (below).

However, it should be emphasized that this depiction is for pictorial purposes only, the appropriate rule being the one given above. These two equations relate the mass distribution, time of coherence, and space-time separation for a given **OR** event. If, as some philosophers contend, experience is contained in space-time, **OR** events are *self*-organizing processes in that experiential medium, and a candidate for consciousness.

But where in the brain, and how, could coherent superposition and **OR** occur? A number of sites and various types of quantum interactions have been proposed. We strongly favor microtubules as an important ingredient; however, various organelles and biomolecular structures, including clathrins, myelin (glial cells), pre-synaptic vesicular grids,[29] and neural membrane proteins,[30] might also participate.

4. Microtubules

Properties of brain structures suitable for quantum coherent superposition and **OR** that are relevant to consciousness might include: 1) high prevalence; 2) functional importance (for example, regulating neural connectivity and synaptic function); 3) periodic, crystal-like lattice dipole structure with long-range order; 4) ability to be transiently isolated from external interaction/observation; 5) functionally coupled to quantum-level events; 6) hollow, cylindrical (possible wave guide); and 7) suitable for information-processing. Membranes, membrane proteins, synapses, DNA and other types of structures have some, but not all, of these characteristics. Cytoskeletal microtubules appear to qualify in all respects.

Interiors of living cells, including the brain's neurons, are spatially and dynamically organized by *self*-assembling protein networks: the cytoskeleton. Within neurons, the cytoskeleton establishes neuronal form, and maintains and regulates synaptic connections. Its major components are microtubules, hollow cylindrical polymers of individual proteins known as tubulin. Microtubules (MTs) are interconnected by linking proteins (microtubule-associated proteins: MAPs) to other microtubules and cell structures to form cytoskeletal lattice networks (Figure 2).

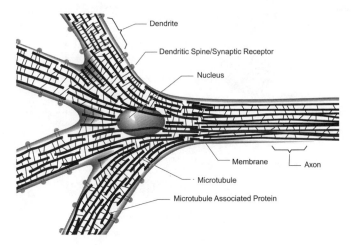

Figure 2. Schematic of central region of neuron (distal axon and dendrites not shown), showing parallel arrayed microtubules interconnected by MAPs. Microtubules in axons are lengthy and continuous, whereas in dendrites they are interrupted and of mixed polarity. Linking proteins connect microtubules to membrane proteins including receptors on dendritic spines.

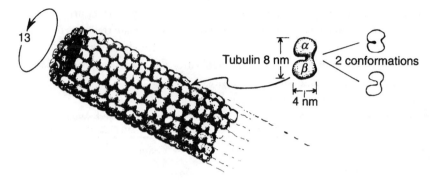

Figure 3. Microtubule structure: a hollow tube of 25 nanometers diameter, consisting of thirteen columns of tubulin dimers. Each tubulin molecule is capable of (at least) two conformations (reprinted with permission from Penrose 1994: 359).

MTs are hollow cylinders 25 nanometers (nm) in diameter, whose lengths vary and may be quite long within some nerve axons. MT cylinder walls are composed of thirteen longitudinal protofilaments, which are each a series of subunit proteins known as tubulin (Figure 3). Each tubulin subunit is a polar, 8 nm dimer which consists of two slightly different 4 nm monomers (alpha and beta tubulin: see Figure 4). Tubulin dimers are dipoles, with surplus negative charges localized toward monomers,[31] and within MTs are arranged in a hexagonal lattice that is slightly twisted, resulting in helical pathways which repeat every three, five, eight, and other numbers of rows. Traditionally viewed as the cell's "bone-like" scaffolding, microtubules and other cytoskeletal structures also appear to fill communicative and information processing roles. Numerous types of studies link the cytoskeleton to cognitive processes.[32] Theoretical models and simulations suggest how conformational states of tubulins within microtubule lat-

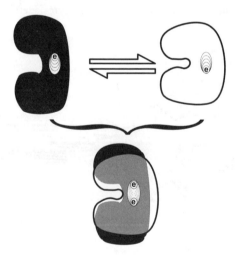

Figure 4. Top: Two states of tubulin in which a single quantum event (electron localization) within a central hydrophobic pocket is coupled to a global protein conformation. Switching between the two states can occur on the order of nanoseconds to picoseconds. Bottom: Tubulin in quantum coherent superposition of both states.

tices can interact with neighboring tubulins to represent, propagate, and process information as in molecular-level "cellular automata," or "spin glass" type computing systems (Figure 5).[33]

In an earlier essay,[34] we presented a model linking microtubules to consciousness, using quantum theory as viewed in the particular "realistic" way that is described in *Shadows of the Mind*.[35]

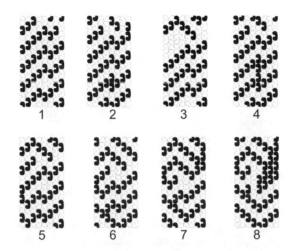

Figure 5. Microtubule automaton simulation (from Rasmussen et al. 1990). Black and white tubulins correspond to states shown in Figure 2. Eight nanosecond time steps of a segment of one microtubule are shown in "classical computing" mode in which patterns move, evolve, interact, and lead to emergence of new patterns.

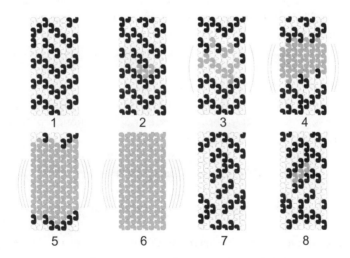

Figure 6. Microtubule automaton sequence simulation in which classical computing (step 1) leads to emergence of quantum coherent superposition (steps 2–6) in certain (gray) tubulins due to pattern resonance. Step 6 (in coherence with other microtubule tubulins) meets critical threshold related to quantum gravity for self-collapse (**Orch OR**). Consciousness (**Orch OR**) occurs in the step 6 to 7 transition. Step 7 represents the eigenstate of mass distribution of the collapse, which evolves by classical computing automata to regulate neural function. Quantum coherence begins to re-emerge in step 8.

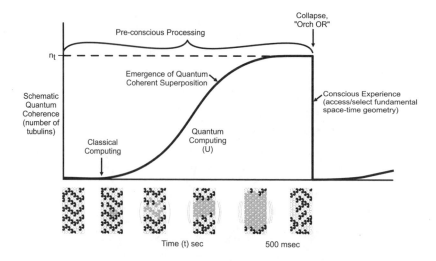

Figure 7. Schematic graph of proposed quantum coherence (number of tubulins) emerging vs time in microtubules. Five hundred milliseconds is time for preconscious processing (Libet 1979). Area under curve connects mass-energy differences with collapse time in accordance with gravitational **OR**. This degree of coherent superposition of differing space-time geometries leads to abrupt quantum-classical reduction ("*self*-collapse," or "orchestrated objective reduction": **Orch OR**).

In our model, quantum coherence emerges and is isolated in brain microtubules until the differences in mass-energy distribution among superposed tubulin states reach the threshold of instability described above, related to quantum gravity (Figure 6). The resultant *self*-collapse (**OR**), considered to be a time-irreversible process, creates an instantaneous "now" event. Sequences of such events create a flow of time and consciousness (Figures 7 and 8).

We envisage that attachments of MAPs on microtubules "tune" quantum oscillations, and "orchestrate" possible collapse outcomes (Figure 9). Thus we term the particular *self*-organizing **OR** occurring in MAP-connected microtubules, and relevant to consciousness, orchestrated objective reduction (**Orch OR**). **Orch OR** events are thus *self*-selecting processes in fundamental space-time geometry. If experience is truly a component of fundamental space-time, **Orch OR** may begin to explain the "hard problem" of consciousness.

5. Summary of the Orch OR Model for Consciousness

The picture we are putting forth involves the following ingredients: [36]

> 1. Aspects of quantum theory (e.g., quantum coherence) and of the suggested physical phenomenon of quantum wave function "*self*-collapse" (objective reduction: **OR**)[37] are essential for consciousness, and occur in cytoskeletal microtubules (MTs) and other structures within each of the brain's neurons.

> 2. Conformational states of MT subunits (tubulins) are coupled to internal quantum events, and cooperatively interact with other tubulins in both classical and quantum computation (Figures 4, 5, and 6).[38]

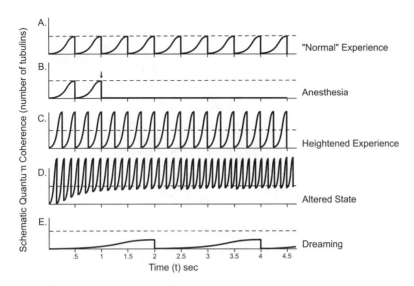

Figure 8. Quantum coherence in microtubules schematically graphed on longer time scale for five different states related to consciousness (area under each curve equivalent in all cases). A. Normal experience: as in Figure 8. B. Anesthesia: anesthetics bind in hydrophobic pockets and prevent quantum delocalizability and coherent superposition (Louria and Hameroff 1996). C. Heightened experience: increased sensory experience input (for example) increases rate of emergence of quantum coherent superposition. **Orch OR** threshold is reached faster (e.g., 250 msec) and **Orch OR** frequency is doubled. D. Altered state: even greater rate of emergence of quantum coherence due to sensory input and other factors promoting quantum state (e.g., meditation, psychedelic drug, etc.). Predisposition to quantum state results in baseline shift and only partial collapse so that conscious experience merges with normally subconscious quantum computing mode. E. Dreaming: prolonged quantum coherence time.

3. Quantum coherence occurs among tubulins in MTs, pumped by thermal and bi chemical energies.[39] Evidence for coherent excitations in proteins has been reported by Vos.[40]

It is also considered that water at MT surfaces is "ordered," dynamically coupled to the protein surface. Water ordering within the hollow MT core (acting like a quantum wave guide) may result in quantum coherent photons (as suggested by the phenomena of "super-radiance" and "*self*-induced transparency."[41] We require that coherence be sustained (protected from environmental interaction) for up to hundreds of milliseconds by isolation a) within hollow MT cores; b) within tubulin hydrophobic pockets; c) by coherently ordered water; and d) sol-gel layering.[42] The feasibility of quantum coherence in the seemingly noisy, chaotic cell environment is supported by the observation that quantum spins from biochemical radical pairs which become separated retain their correlation in cytoplasm.[43]

4. During preconscious processing, quantum coherent superposition/computation occurs in MT tubulins, and continues until the mass-distribution difference among the separated states of tubulins reaches a threshold related to quantum gravity. Self-collapse (**OR**) then occurs (Figures 6 and 7).

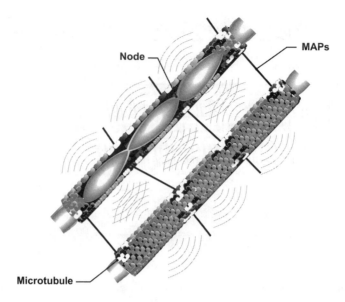

Figure 9. Quantum coherence in microtubules. Having emerged from resonance in classical automaton patterns, quantum coherence nonlocally links superpositioned tubulins (gray) within and among microtubules. Upper microtubule: cutaway view shows coherent photons generated by quantum ordering of water on tubulin surfaces, propagating in microtubule waveguide. MAP (microtubule-associated-protein) attachments breach isolation and prevent quantum coherence; MAP attachment sites thus act as "nodes" which tune and orchestrate quantum oscillations and set possibilities and probabilities for collapse outcomes (orchestrated objective reduction: **Orch OR**).

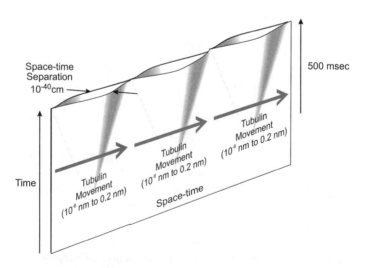

Figure 10. Schematic space-time separation illustration of three superposed tubulins. The space-time differences are very tiny in ordinary terms (10–40 nm), but relatively large mass movements (e.g., billions of tubulins, each moving from 10^{-6} nm to 0.2 nm) indeed have precisely such very tiny effects on the space-time curvature.

5. The **OR** *self*-collapse process results in classical "outcome states" of MT tubulins, which then implement neurophysiological functions. According to certain ideas for **OR**,[44] the outcome states are "noncomputable"; that is, they cannot be determined algorithmically from the tubulin states at the beginning of the quantum computation.

6. Possibilities and probabilities for post-**OR** tubulin states are influenced by factors including initial tubulin states and attachments of microtubule-associated proteins (MAPs) acting as "nodes" that tune and orchestrate the quantum oscillations (Figure 9). We thus term the *self*-tuning **OR** process in microtubules "**Orch OR**."

7. According to the arguments for **OR** put forth in Penrose,[45] superposed states each have their own space-time geometries. When the degree of coherent mass-energy difference leads to sufficient separation of space-time geometry, the system must choose and decay—reduce, collapse—to a single universe state. Thus **Orch OR** involves *self*-selections in fundamental space-time geometry (Figures 10 and 11).

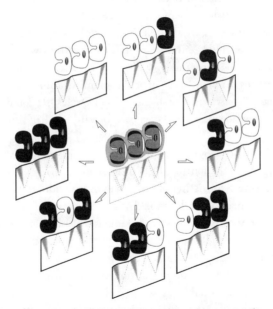

Figure 11. Center: Three superposed tubulins (cf. Figure 4) with corresponding schematic space-time separation illustrations (Figures 1 and 10). Surrounding the superposed tubulins are the eight possible post-reduction "eigenstates" for tubulin conformation, and corresponding space-time geometry.

8. To quantify the **Orch OR** process, in the case of a pair of roughly equally superposed states, each of which has a reasonably well-defined mass distribution, we calculate the gravitational *self*-energy E of the difference between these two mass distributions, and then obtain the approximate lifetime T for the superposition to decay into one state or the other by the formula $T = \hbar/E$. We call T the coherence time for the superposition (how long coherence is sustained). If we assume a coherence time $T = 500$ msec (shown by Libet[46] and others to be a relevant time for preconscious processing), we calculate E, and determine the number of MT tubulins whose coherent superposition for 500 msec will elicit **Orch OR**. This turns out to be about 10^9 tubulins.

9. A typical brain neuron has roughly 10^7 tubulins.[47] If, say, 10 percent of tubulins within each neuron are involved in the quantum coherent state, then roughly 10^3 (one thousand) neurons would be required to sustain coherence for 500 msec, at which time the quantum gravity threshold is reached and **Orch OR** occurs.

10. We consider each *self*-organized **Orch OR** as a single conscious event; cascades of such events would constitute a "stream" of consciousness. If we assume some form of excitatory input (e.g., you are threatened, or enchanted) in which quantum coherence emerges faster, then, for example, 10^{10} coherent tubulins could **Orch OR** after 50 msec (e.g., Figure 8c). Turning to see a Bengal tiger in your face might perhaps elicit 10^{11} in 5 msec, or more tubulins, faster. A slow emergence of coherence (your forgotten phone bill) may require longer times. A single electron would require more than the age of the universe to have a conscious moment.

11. Quantum states are nonlocal (because of quantum entanglement—or Einstein-Podolsky-Rosen (EPR) effects), so that the entire nonlocalized state reduces all at once. This can happen if the mass movement that induces collapse takes place in a small region encompassed by the state, or if it takes place uniformly over a large region. Thus, each instantaneous **Orch OR** could "bind" various superpositions which may have evolved in separated spatial distributions and over different time scales, but whose net displacement *self*-energy reaches threshold at a particular moment. Information is bound into an instantaneous event (a "conscious now"). Cascades of **Orch OR**s could then represent our familiar "stream of consciousness," and create a "forward" flow of time.[48]

It may be interesting to compare our considerations with subjective viewpoints that have been expressed with regard to the nature of the progression of conscious experience. For example, support for consciousness consisting of sequences of individual, discrete events is found in Buddhism, where trained meditators describe distinct "flickerings" in their experience of reality.[49] Buddhist texts portray consciousness as "momentary collections of mental phenomena," and as "distinct, unconnected and impermanent moments which perish as soon as they arise." Each conscious moment successively becomes, exists, and disappears—its existence is instantaneous, with no duration in time, as a point has no length. Our normal perceptions, of course, are seemingly continuous, presumably as we perceive "movies" as continuous despite their actual makeup being a series of frames. Some Buddhist writings even quantify the frequency of conscious moments. For example, the Sarvaastivaadins described 6,480,000 "moments" in twenty-four hours (an average of one "moment" per 13.3 msec),[50] and some Chinese Buddhists as one "thought" per 20 msec. These accounts, including variations in frequency, are consistent with our proposed **Orch OR** events. For example, a 13.3 msec preconscious interval would correspond with an **Orch OR** involving 4×10^{10} coherent tubulins; a 0.13 msec interval would correspond with 4×10^{12} coherent tubulins; and a 20 msec interval with 2.5×10^{10} coherent tubulins. Thus, Buddhist moments of experience, Whiteheadian occasions of experience, and our proposed **Orch OR** events seem to correspond tolerably well with one another.

The **Orch OR** model thus appears to accommodate some important features of consciousness:

1. control/regulation of neural action
2. preconscious to conscious transition
3. noncomputability
4. causality
5. binding of various (time scale and spatial) superpositions into instantaneous "now"
6. a "flow" of time
7. a connection to fundamental space-time geometry in which experience may be based.

6. Conclusion: What Is It Like to Be a Worm?

The **Orch OR** model has the implication that an organism able to sustain quantum coherence among, for example, 10^9 tubulins for 500 msec, might be capable of having a conscious experience. More tubulins coherent for a briefer period, or fewer for a longer period ($E = \hbar/T$), will also have conscious events. For example, human brains appear capable of 10^{11} tubulin, 5 msec "Bengal tiger experiences," but what about simpler organisms?

From an evolutionary standpoint, the introduction of a dynamically functional cytoskeleton (perhaps symbiotically from spirochetes)[51] greatly enhanced eukaryotic cells by providing cell movement, internal organization, separation of chromosomes, and numerous other functions. As cells became more specialized, with extensions like axopods and eventually neural processes, increasingly larger cytoskeletal arrays, providing transport and motility, may have developed quantum coherence, via the Fröhlich mechanism, as a byproduct of their functional coordination.

Another possible scenario for emergence of quantum coherence leading to **Orch OR** and conscious events is "cellular vision." Albrecht-Buehler has observed that single cells utilize their cytoskeletons in "cellular vision"—detection, orientation, and directional response to beams of red/infra-red light.[52] Jibu and his colleagues[53] argue that this process requires quantum coherence in microtubules and ordered water, and Hagan[54] suggests that quantum effects/cellular vision provided an evolutionary advantage for cytoskeletal arrays capable of quantum coherence. For whatever reason, quantum coherence emerged, one could then suppose that one day an organism achieved sufficient microtubule quantum coherence to elicit **Orch OR**, and had a "conscious" experience.

At what level of evolutionary development might this primitive consciousness have emerged? A single-celled organism like *Paramecium* is extremely clever, and utilizes its cytoskeleton extensively. Could a paramecium be conscious? Assuming a single paramecium contains, like each neuronal cell, 10^7 tubulins, then for a paramecium to elicit **Orch OR**, 100 percent of its tubulins would need to remain in quantum coherent superposition for nearly a minute. This seems unlikely.

Consider the nematode worm *C. elegans*. Its 302 neuron nervous system is completely mapped. Could *C. elegans* support **Orch OR**? With 3×10^9 tubulins, *C. elegans* would require one third of its tubulins to sustain quantum coherent superposition for 500 msec. This seems unlikely, but not altogether impossible. If not *C. elegans,* then perhaps *Aplysia*, with a thousand neurons, or some higher organism. **Orch OR** provides a theoretical framework to entertain such possibilities.

Would a primitive **Orch OR** experience be anything like ours? If *C. elegans* were able to *self*-collapse, what would it be like to be a worm?[55] Compared to a human **Orch OR** event at $T = 25$ msec (gamma synchrony) and $E = 2 \times 10^{11}$ tubulins, a 10^9 tubulin, 500 msec **Orch OR** event in *C. elegans* would only be 1/200 in experiential intensity. And we have many **Orch OR** events sequentially (e.g., forty per second) whereas *C. elegans* could generate, at most, two low-intensity events per second. *C. elegans* would also presumably lack extensive memory and associations, and have poor sensory data and context, but nonetheless, by our criteria, a 10^9 tubulin, 500 msec **Orch OR** in *C. elegans* would be a conscious experience: a mere smudge of known reality, the next space-time move.

Consciousness has an important place in the universe. **Orch OR** in microtubules is a model depicting consciousness as sequences of non-computable *self*-selections in fundamental space-time geometry. If experience is a quality of space-time, then **Orch OR** indeed begins to address the "hard problem" of consciousness in a serious way.[56]

Addendum (2010)

A Place for Consciousness in Nature: Perspective on "Conscious Events as Orchestrated Space-Time Selections"

Stuart Hameroff

Sir Roger Penrose and I wrote the foregoing essay for a special issue of the *Journal of Consciousness Studies* devoted to the "hard problem."[57] Coined by philosopher David Chalmers, the term "hard problem" refers to the enigmatic question of how the brain produces conscious awareness, subjective feelings, and phenomenal experience—composed of what philosophers call *qualia*.[58]

On the other hand, *nonconscious* cognitive brain functions, including sensory processing and control of habitual behaviors, *can* be explained by computation among brain neurons in which axonal firings and synaptic transmissions play the roles of "bit" states and switches. Though complex, these nonconscious brain functions are referred to as "easy problems,"[59] "zombie modes,"[60] or "auto-pilot."[61] Many neuroscientists and philosophers dispute or disregard the hard problem, and assume and assert that consciousness emerges as an epiphenomenon of neuron-based complex computation. Indeed, the idea that consciousness involves something in addition to neuronal computing has been derided as dualism, as expecting a "ghost in the machine."[62] Accordingly, proponents of "strong" artificial intelligence (AI) argue that silicon computers will inevitably attain brain equivalence, including conscious awareness.[63]

In his 1989 book, *The Emperor's New Mind*, Roger Penrose laid out an argument against strong AI through Gödel's theorem, contending that consciousness required something beyond classical computation, some "noncomputable" element that influenced or augmented the computation.[64] The missing ingredient in consciousness, he suggested, was a specific form of quantum computing in the brain.

Quantum computing utilizes strange properties of the quantum world, differing markedly from those of our everyday classical world. Quantum properties include quantum superposition in which particles exist in multiple states or locations simultaneously, e.g., as quantum bits, or "qubits," which interact/compute with other qubits by entanglement. In a quantum computer, superpositioned, entangled qubits eventually reduce ("collapse") to classical bits as the solution. Mechanisms underlying quantum state reduction—"collapse of the wave function"—in quantum computing (and nature in general) are not understood, although numerous theories exist. One view, stemming from Niels Bohr, is that quantum systems persist until consciously observed—that conscious observation causes collapse of the wave function. This pragmatic view puts consciousness outside science.

Penrose turned this view around. He introduced a new theory of *self*-collapse—quantum state reduction due to a specific objective threshold ("objective reduction" [**OR**]). Rather than being *caused* by conscious observation, each **OR** event entails a conscious perception or choice. To identify the threshold for such conscious **OR** events, Penrose first characterized superpositions as separations in fundamental space-time geometry, the level of quantum gravity at the Planck scale.[65] He further proposed that such objective reduction/conscious events could be influenced by platonic information (e.g., mathematical truth), but also perhaps ethical and aesthetic values embedded in Planck scale geometry. These platonic values could influence (noncomputably) the output of a quantum computation mediated by this type of OR. Penrose thus placed consciousness in nature, precisely on the edge between quantum and classical worlds.

Penrose was calling for a particular form of quantum computation in the brain to explain consciousness. As qubits, he suggested that neurons could exist in superposition of both firing and not firing, but worried that neurons were too large for quantum effects and left the door open for a biomolecular qubit. When I read *The Emperor's New Mind* in the early 1990s, I was struck and somewhat bewildered by the breadth and depth of Penrose's arguments leading to this seemingly strange conclusion. He was obviously brilliant, and he had a theory which portrayed consciousness as actual physical events, if as yet theoretical physical events.

I readily agreed that consciousness involved something other than computation among neurons, having worked for twenty years on the premise that consciousness extended inside neurons. In medical school I had become obsessed with the notion of molecular-scale computing in cytoskeletal structures called microtubules, which regulate synapses and axonal firings from within neurons.[66] I believed that viewing neurons as simple input-output devices was an insult to neurons, but also realized that extending computation inside neurons (and increasing brain computational capacity enormously) didn't solve the hard problem. As an anesthesiologist, I had also studied how anesthetic gases selectively erase consciousness through quantum London forces[67] and knew the work of Herbert Fröhlich[68] who had proposed biochemically driven quantum coherence in biomolecular lattices such as microtubules. I was open to new ideas, including quantum computing in the brain. Maybe tubulins were qubits, and microtubules were the quantum computers Penrose sought?

I contacted Penrose, and we soon met in his cluttered Oxford office. He was, and is, a remarkable man, both gentle and strong, with an open but critical mind. As I described microtubules, he was most impressed by their symmetry and geometry. We began collaborating toward a theoretical model.[69]

We faced difficult issues. Technological quantum computers required isolation and extreme cold to avoid environmental decoherence, so quantum computing in the warm brain seemed extremely unlikely.[70] And even if quantum computing could occur in one neuron, how could it extend globally across cell boundaries throughout the brain? And if quantum systems were isolated within the brain, how could they interact with the external world for input and output? Regarding the latter, Penrose and I described consciousness as a sequence of discrete **OR**-mediated quantum computations occurring at frequencies compatible with brain electrophysiology, such as EEG rhythms ranging from 2 Hz to 40 Hz or higher. In the classical intervals between such isolated quantum events, neurons could express outputs to the external world, and synaptic inputs could "orchestrate" **OR**-mediated quantum computing via microtubule-associated proteins (**Orch OR**).

Casting consciousness as sequences of discrete events made **Orch OR** compatible not only with brain electrophysiology and **OR** quantum physics, but also with the work of process philosopher Alfred North Whitehead.[71] Consciousness, according to Whitehead, was a series of "occasions of experience" occurring in a "wider, basic field of proto-consciousness." As quantum state reductions, **Orch OR** events qualified as "Whitehead occasions." Regarding the "wider, basic field," Penrose and I pointed to fundamental space-time geometry at the Planck scale in which **Orch OR** was suggested to occur. Penrose had placed platonic values there, and for our "hard problem" paper reproduced here, we added protoconscious *qualia,* or experience, as an irreducible component of fundamental space-time geometry, akin to other irreducible components like mass, spin, or charge. **Orch OR** became consistent with Whitehead's philosophical construct containing consciousness.

Orch OR came under heavy attack, primarily on the issue of decoherence in the warm brain.[72] But evidence in recent years has revealed quantum coherent superposition in warm biomolecules and proteins, specifically in the type of nonpolar hydrophobic pockets that mediate anesthetic effects and are foci of tubulin quantum states in **Orch OR**.[73] Fröhlich coherence has been demonstrated in microtubules at 8 MHz;[74] Penrose **OR** is being tested as an explanation for quantum state reduction.[75]

In 1998, I published a list of twenty testable predictions of **Orch OR**, some of which have been corroborated.[76] For example, to account for extension of isolated quantum states in one neuron to global neuronal assemblies spanning the brain, neurons involved in conscious processes were proposed to require linkage by gap junctions, window-like connections that fuse and synchronize neighboring neurons and glia, and through which quantum states could extend by quantum tunneling. In recent years, gap junctions between brain neuronal dendrites have been shown to be essential for gamma synchrony EEG (~40 Hz), the best measurable correlate of consciousness.[77] Considering that habitual cognitive brain functions like driving or walking could at times be nonconscious easy problems (zombie modes, autopilot), and at other times be accompanied by conscious perception and control, I proposed a "conscious pilot" model, in which gap junction-defined spatiotemporal envelopes of gamma synchrony (within which **Orch OR** can occur) move through the brain as a mobile agent (the conscious pilot) to consciously perceive and control otherwise nonconscious auto-pilot modes.[78] The **Orch OR** conscious pilot may be seen as an actual quantum "ghost in the machine."

Orch OR encompasses aspects of philosophy, neuroscience, and physics, goes out on several limbs simultaneously, and contains much to criticize. But, as yet, **Orch OR** has weathered every objection[79] and remains the most ambitious and complete theory of consciousness yet put forward. Most importantly, it characterizes consciousness not as an epiphenomenon, but as an intrinsic component of the universe, a process of actual events, of Whitehead occasions of experience occurring on the edge between quantum and classical worlds. Moreover, noncomputable platonic information in space-time geometry can influence not only conscious thoughts and behaviors, but also life and evolution through quantum-level mutations in DNA.[80]

Notes

1. Adapted with permission from *Journal of Consciousness Studies* 1996, 3(1): 36–53.
2. Chalmers 1996a-c.
3. Spinoza 1677; Rensch 1960.
4. For example, Whitehead 1929.
5. Russell 1954 (originally published 1927).
6. Stubenberg 1996.
7. Goswami 1993.
8. Wheeler 1990.
9. Chalmers 1996a-c.
10. Whitehead 1929, 1933.
11. Shimony 1993.
12. See Penrose 1989, 1994.
13. Ibid.
14. Penrose 1994.
15. As expounded in Penrose 1994, 1996.
16. Penrose 1989, 1994.
17. Schrödinger 1935.
18. For example, Pearle 1989; Ghirardi *et al.* 1986.
19. For example, Károlyházy 1986; Diósi 1989; Ghirardi *et al.* 1990; Penrose 1989, 1994; Pearle and Squires 1994; Percival 1995.
20. Cf. Penrose 1994.
21. For a number of such arguments, see Penrose 1987, 1995.
22. Cf. Penrose 1996.
23. Károlyházy 1996, 1974; Károlyházy *et al.* 1986; Kibble 1991; Diósi 1989; Ghirardi *et al.* 1990; Pearle and Squires 1995; Percival 1995; Penrose 1993, 1994, 1996.
24. Penrose 1994, 1996.
25. Cf. Everett 1957; Wheeler 1957, for example.
26. Cf. Penrose 1993.
27. This critical amount would be of the order of unity, in absolute units, for which the Planck-Dirac constant (Planck's constant over 2π), the gravitational constant G, and the velocity of light c, all take the value unity, cf. Penrose 1994: 337–339.
28. See Penrose 1994, 1995.
29. Beck and Eccles 1992.
30. Marshall 1989.
31. De Brabander 1982.
32. For review, see Hameroff and Penrose 1996.
33. For example, Hameroff and Watt 1982; Rasmussen *et al.* 1990; Tuszynski *et al.* 1995.
34. Hameroff and Penrose 1996; and in summary form, Penrose and Hameroff 1995.
35. Penrose 1994.
36. The full details of this model are given in Hameroff and Penrose 1996b.
37. Penrose 1994, 1996.
38. Hameroff *et al.* 1992; Rasmussen *et al.* 1990.
39. Perhaps in the manner proposed by Frohlich (1968, 1970, 1975).
40. Vos *et al.* 1993.

41. Jibu *et al.* 1994, 1995.
42. Hameroff and Penrose 1996b.
43. Walleczek 1995.
44. Penrose 1994.
45. Penrose 1994.
46. Libet 1979.
47. Yu and Baas 1994.
48. Aharonov and Vaidman 1990; Elitzur 1996; Tollaksen 1996.
49. Tart 1995.
50. Von Rospatt 1995.
51. For example, Margulis 1975.
52. Albrecht-Buehler 1992.
53. Jibu *et al.* 1995.
54. Hagan 1995.
55. Nagel 1974.
56. Acknowledgments: Thanks to Dave Cantrell for artwork and to Carol Ebbecke for technical support.
57. Hameroff and Penrose 1996a.
58. Chalmers 1996b.
59. Ibid.
60. Koch and Crick 2001.
61. Hodgson 2007.
62. Ryle 1949.
63. Kurzweil 2005.
64. Penrose 1989.
65. Penrose 1996.
66. Hameroff and Watt 1982; Hameroff 1987; Rasmussen et al. 1990.
67. Hameroff 1998a; Hameroff 2006b.
68. Fröhlich 1968, 1970.
69. Hameroff and Penrose R 1996a; Penrose and Hameroff 1995; Hameroff and Penrose 1996b; Hameroff 1998b; Hameroff 2007.
70. Tegmark 2000; Hagan *et al.* 2002.
71. Whitehead 1929.
72. Tegmark 2000; Hagan et al. 2002.
73. Engel et al. 2007; Collini and Scholes 2009.
74. Pokorny 2004.
75. Marshall et al. 2003.
76. Hameroff 1998b; Hameroff 2006a.
77. Galarreta and Hestrin 1999; Bennett and Zukin 2004.
78. Hameroff 2009.
79. Reimers et al. 2009; Samsonovich et al. 1992.
80. Hameroff 2008.

References

Aharonov, Y., and L. Vaidman. (1990) "Properties of a quantum system during the time interval between two measurements." *Physical Review* A 41: 11.

Albrecht-Buehler, G. (1992) "Rudimentary form of 'cellular vision'." *Cell Biology* 89: 8288–92.

Beck, F., and J.C. Eccles. (1992) "Quantum aspects of brain activity and the role of consciousness." *Proceedings of the National Academy of Science USA* 89 (23): 11357–61.

Bennett, M. V., and R. S. Zukin. (2004) "Electrical coupling and neuronal synchronization in the Mammalian brain." *Neuron* 41(4): 495–511.

Chalmers, D. (1996a) "Facing up to the problem of consiousness," in *Toward a Science of Consciousness—The First Tucson Discussions and Debates*. S. R. Hameroff, A. Kaszniak, and A. C. Scott, eds. Cambridge: MIT Press.

———. (1996b) *The Conscious Mind—In Search of a Fundamental Theory*. New York: Oxford University Press.

———. (1996c) *Toward a Theory of Consciousness*. Berlin: Springer-Verlag.

Collini, E. and G. D. Scholes. (2009) "Cogerent intrachain energy migration in conjugated polymers at room temperature." *Science* 323: 369–73.

Conze, E. (1988) "Buddhist Thought in India," (Louis de La Vallee Poussin, translator), in *Abhidharmako sabhaa.syam:* English translation by Leo M. Pruden, 4 vols. (Berkeley), 85–90.

De Brabander, M. (1982) "A model for the microtubule organizing activity of the centrosomes and kinetochores in mammalian cells." *Cell Biology International Reports* 10 (Oct. 6): 901–15.

Diósi, L. (1989) "Models for universal reduction of macroscopic quantum fluctuations." *Physical Review* A 40: 1165–74.

Elitzur, Avshalom C. (1996) "Time and consciousness: The uneasy bearing of relativity theory on the mind–body problem," in S. R. Hameroff, A. Kaszniak, and A. C. Scott, eds., *Toward a Science of Consciousness—The First Tucson Discussions and Debates*. Cambridge: MIT Press.

Engel, G. E., T. R. Calhoun, E. L. Read, T. K. Ahn, T. Mancal, Y. C. Cheng, R. E. Blankenship, and G. R. Fleming. (2007) "Evidence for wavelike energy transfer through quantum coherence in photosynthetic systems." *Nature* 446: 782–86.

Everett, H. (1957) "Relative state formulation of quantum mechanics," in J.A. Wheeler and W.H. Zurek, eds. *Quantum Theory and Measurement*. Princeton: Princeton University Press, 1983; originally in *Reviews of Modern Physics* 29: 454–62.

Frohlich, H. (1968) "Long-range coherence and energy storage in biological systems." *International Journal of Quantum Chemistry* 2: 641–9.

———. (1970) "Long range coherence and the actions of enzymes." *Nature* 228: 1093.

———. (1975) "The extraordinary dielectric properties of biological materials and the action of enzymes." *Proceedings of the National Academy of Sciences* 72: 4211–15.

Galarreta, M. and S. Hestrin. (1999) "A network of fast-spiking cells in the neocortex connected by electrical synapses." *Nature* 402: 72–5.

Ghirardi, G. C., R. Grassi, and A. Rimini. (1990) "Continuous-spontaneous reduction model involving gravity." *Physical Review* A 42: 1057–64.

Ghirardi, G. C., A. Rimini, and T. Weber. (1986) "Unified dynamics for microscopic and macroscopic systems." *Physical Review* D 34: 470.

Goswami, A. (1993) *The Self-Aware Universe: How Consciousness Creates the Material World*. New York: Tarcher/Putnam.

Hagan, S. (1995) Personal communication.

Hagan, S., S. Hameroff, and J. Tuszynski. (2002) "Quantum Computation in Brain Microtubules? Decoherence and Biological Feasibility." *Physical Reviews* E 65: 061901.

Hameroff, S. (1987) *Ultimate computing: Biomolecular consciousness and nanotechnology.* Dordrecht: Elsevier-North Holland (http://www.quantumconsciousness.org/ultimatecomputing.html).

———. (1998a) "Anesthesia, consciousness and hydrophobic pockets—A unitary quantum hypothesis of anesthetic action." *Toxicology Letters* 100/101: 31–9

———. (1998b) "Quantum computation in brain microtubules—The Penrose-Hameroff '**Orch OR**' model of consciousness." *Philosophical Transactions Royal Society London* A 356:1869–96. (http://www.quantumconsciousness.org/penrose-hameroff/quantumcomputation.html).

———. (2006a) "Consciousness, neurobiology and quantum mechanics: The case for a connection," in Jack Tuszynski, ed. *The Emerging Physics of Consciousness.* Berlin: Springer-Verlag, 193–253.

———. (2006b) "The entwined mysteries of anesthesia and consciousness." *Anesthesiology* (2006) 105: 400–12 (http://www.quantumconsciousness.org/documents/informationprocessing_hameroff_000.pdf).

———. (2007) "The brain is both neural computer and quantum computer." *Cognitive Science* 31:1035–45 (http://www.quantumconsciousness.org/documents/CogScipub.pdf).

———. (2008) "That's life: The geometry of pi electron resonance clouds," D. Abbott, P. Davies, and A. Pati, eds. *Quantum Aspects of Life.* Singapore: World Scientific, 403–34.

———. (2009) "The conscious pilot: Dendritic synchrony moves through the brain to mediate consciousness." *Journal of Biological Physics* 10.1007/s10867–009–9148-x.

Hameroff, S. R., J. E. Dayhoff, R. Lahoz-Beltra, A. Samsonovich, and S. Rasmussen. (1992) "Conformational automata in the cytoskeleton: models for molecular computation." *IEEE Computer* (October Special Issue on Molecular Computing): 30–39.

Hameroff, S. R., and R. Penrose. (1995) "Orchestrated reduction of quantum coherence in brain microtubules: A model for consciousness." *Neural Network World* 5 (5): 793–804.

———. (1996a) "Conscious Events as Orchestrated Space-Time Selections." *Journal of Consciousness Studies* 3(1): 36–53.

———. (1996b) "Orchestrated reduction of quantum coherence in brain microtubules: A model for consciousness," in *Toward a Science of Consciousness—The First Tucson Discussions and Debates.* S. R. Hameroff, A. Kasaniak, and A. C. Scott, eds., Cambridge: MIT Press (http://www.quantumconsciousness.org/penrose-hameroff/orchOR.html).

Hameroff, S., and R. C. Watt. (1982) "Information processing in microtubules." *Journal of Theoretical Biology* 98: 549–61.

Hodgson, D. (2007) "Making our own luck." *Ratio* 20: 278–92.

Jibu, M., S. Hagan, S.R. Hameroff, K. H. Pribram, and K. Yasue. (1994) "Quantum optical coherence in cytoskeletal microtubules: implications for brain function." *BioSystems* 32: 195–209.

Jibu, M., K. Yasue, and S. Hagan. (1995) "Water laser as cellular 'vision,'" submitted.

Koch, C., and F. C. R. Crick. (2001) "The zombie within." *Nature* 411: 893.

Krolhzy, F., A. Frenkel, and B. Lukacs. 1986) "On the possible role of gravity on the reduction of the wave function," in R. Penrose and C.J. Isham, eds. *Quantum Concepts in Space and Time.* Oxford: Oxford University Press.

Kurzweil, R. (2005) *The Singularity is Near: When Humans Transcend Biology.* New York: Viking Press.

Libet, B., E. W. Wright, Jr., B. Feinstein, and D. K. Pearl. (1979) "Subjective referral of the timing for a conscious sensory experience." *Brain* 102: 193–224.

Louria, D., and S. Hameroff. (1996) "Computer simulation of anesthetic binding in protein hydrophobic pockets," in S. R. Hameroff, A. Kaszniak, and A. C. Scott, eds. *Toward a Science of Consciousness—The First Tucson Discussions and Debates.* Cambridge: MIT Press.

Marshall, I. N. (1989) "Consciousness and Bose-Einstein condensates." *New Ideas in Psychology* 7: 73–83.

Marshall, W., C. Simon, R. Penrose, and D Bouwmeester. (2003) "Towards quantum superpositions of a mirror." *Physical Review Letters* 91:13.

Margulis, L. (1975) *Origin of Eukaryotic Cells.* New Haven: Yale University Press.

Nagel, Thomas. (1974) "What is it like to be a bat?" in The *Mind's I: Fantasies and Reflections on Self and Soul.* D.R. Hofstadter and D.C. Dennett, eds. New York: Basic Books, 1981, 391–403 (originally published in *The Philosophical Review,* October, 1974).

Pearle, P. (1989) "Combining stochastic dynamical state vector reduction with spontaneous localization." *Physical Review* D 13: 857–68.

Pearle, P., and E. Squires. (1994) "Bound-state excitation, nucleon decay experiments and models of wave-function collapse." *Physical Review Letters* 73(1): 1–5.

Penrose, R. (1987) "Newton, quantum theory and reality," in *300 Years of Gravity.* S.W. Hawking and W. Israel, eds. Cambridge: Cambridge University Press, 17–49.

———. (1989) *The Emperor's New Mind.* Oxford: Oxford University Press.

———. (1993) "Gravity and quantum mechanics," in R.J. Gleiser, C.N. Kozameh, and O.M. Moreschi, eds., *General Relativity and Gravitation: Proceedings of the Thirteenth International Conference on General Relativity and Gravitation* (held at Cordoba, Argentina 28 June–4 July 1992) Part 1: Plenary Lectures. Bristol, UK: Institute of Physics Publications.

———. (1994) *Shadows of the Mind.* Oxford: Oxford University Press.

———. (1996) "On gravity's role in quantum state reduction." *General Relativity and Gravitation* 28(5): 581–600.

Penrose, R., and S. R. Hameroff. (1995) "What gaps? Reply to Grush and Churchland." *Journal of Consciousness Studies* 2(2): 99–112. (http://www.quantumconsciousness.org/penrose-hameroff/ whatgaps.html.

Pokorny, J. (2004) "Excitations of vibrations in microtubules in living cells." *Bioelectrochemistry* 63: 321–26.

Rasmussen, S., H. Karampurwala, R. Vaidyanath, K.S. Jensen, and S. Hameroff. (1990) "Computational connectionism within neurons: A model of cytoskeletal automata subserving neural networks." *Physica* D 42: 428–49 (http://www.quantumconsciousness.org/documents/Computational_ Rasmussen_000. pdf).

Reimers, J. R., L. K. McKemmish, R. H. McKenzie, A. E. Mark, and N. S. Hush. (2009) "Weak, strong, and coherent regimes of Frohlich condensation and their applications to terahertz medicine and quantum consciousness." *Proceedings of the National Academy of Sciences USA,* doi/10.1073/pnas. 0806273106 (see http://www.quantumconsciousness.org/PNAS.htm).

Rensch, B. (1960) *Evolution Above the Species Level.* New York: Columbia University Press.

Russell, B. (1954) *The Analysis of Matter.* New York: Dover Publications (originally published 1927).

Ryle, G, (1949) *The Concept of Mind.* Chicago: University of Chicago Press.

Samsonovich, A., A. C. Scott, and S. Hameroff. (1992) "Acousto-conformational transitions in cytoskeletal microtubules: Implications for intracellular information processing." Nanobiology 1:457–68 (http://www.quantumconsciousness.org/documents/acousto_samsonovich_000.pdf).

Schrödinger, E. (1935) "Die gegenwarten situation in der quantenmechanik." *Naturwissenschaften* 23: 807–12, 823–8, 844–9 (translation by J. T. Trimmer, in *Proceedings of the American Philosophical Society* 124 (1980): 323–38; reprinted in *Quantum Theory and Measurement,* J.A. Wheeler and W.H. Zurek, eds. Princeton: Princeton University Press, 1983).

Shimony, A. (1993) *Search for a Naturalistic World View—Volume II: Natural Science and Metaphysics.* Cambridge: Cambridge University Press.

Spinoza, B. (1677) *Ethica in Opera quotque reperta sunt* (3rd edition). J. van Vloten and J.P.N. Land, eds. Netherlands: Den Haag.

Stubenberg, L. (1996) "The place of qualia in the world of science," in *Toward a Science of Consciousness—The First Tucson Discussions and Debates*. S.R. Hameroff, A. Kaszniak, and A.C. Scott, eds., Cambridge: MIT Press.

Tart, C. T. (1995) Personal communication and information gathered from "Buddha-1 newsnet."

Tegmark, M. (2000) "The importance of quantum decoherence in brain processes." *Physical Review* E 61: 4194–206.

Tollaksen, J. (1996) "New insights from quantum theory on time, consciousness, and reality," in S. R. Hameroff, A. Kaszniak, and A. C. Scott, eds. *Toward a Science of Consciousness—The First Tucson Discussions and Debates*. Cambridge: MIT Press.

von Rospatt, A. (1995) *The Buddhist Doctrine of Momentariness: A survey of the origins and early phase of this doctrine up to Vasubandhu*. Stuttgart: Franz Steiner Verlag.

Vos, M.H., J. Rappaport, J. Ch. Lambry, J. Breton, and J.L. Martin. (1986) "Visualization of coherent nuclear motion in a membrane protein by femtosecond laser spectroscopy." *Nature* 363: 320–25.

Walleczek, J. (1995) "Magnetokinetic effects on radical pairs: a possible paradigm for understanding sub-kT magnetic field interactions with biological systems," in *Biological Effects of Environmental Electromagnetic Fields* (Advances in Chemistry, no. 250), M. Blank, ed. Washington, DC: American Chemical Society Books.

Wheeler, J.A. (1957) "Assessment of Everett's 'relative state' formulation of quantum theory." *Reviews of Modern Physics* 29: 463–65.

———. (1990) "Information, physics, quantum: The search for links," in W. Zurek, ed. *Complexity, Entropy, and the Physics of Information*. Reading, MA: Addison-Wesley.

Whitehead, A. N. (1925) *Science and the Modern World*. New York: Macmillan.

———. (1929) *Process and Reality*. New York: Macmillan.

———. (1933) *Adventures of Ideas*. New York: Simon and Schuster.

Yu, W., and P. W. Baas. (1994) "Changes in microtubule number and length during axon differentiation." *Journal of Neuroscience* 14(5): 2818–29.

36

Quantum Interactive Dualism:
The Libet and Einstein-Podolsky-Rosen Causal Anomalies[1]

Henry P. Stapp

1. Introduction

We all feel that certain of our conscious thoughts can *cause* our voluntary bodily actions to occur. Our lives, our institutions, and our moral codes are largely based on that intuition. The whole notion of "cause" probably originates in that deep-seated feeling.

The strongest argument against this basic intuition—that our thoughts *cause* our voluntary bodily actions—stems from an experiment performed by Benjamin Libet.[2] In this experiment, a subject is instructed to voluntarily perform, during a certain time interval, a simple physical action, such as raising a finger. Libet found that a measurable precursor of the physical action, known as the "readiness potential," occurs in the brain about one-third of a second prior to the occurrence of the psychologically described act of willing that action to occur.

This empirical result appears to show, on the face of it, that the conscious act of *willing* must be a *consequence* of this associated brain activity, not the *cause* of it; for, according to the normal idea of cause, nothing can cause a prior happening to occur.

This example is just one instance of a general feature of mind–brain phenomena—namely, the fact that conscious experiences always seem to occur after a lot of preparatory work has already been done by the brain. This feature accords with the classical physics concept of the causal closure of the physical, and it leads plausibly to the conclusion that the felt causal efficacy of our conscious thoughts is an illusion.

One of the most intensely studied aspects of quantum mechanics is the occurrence of correlations, in which a "voluntary" choice made at one time appears to affect events that occurred earlier than this choice, or simultaneously with it, yet far away. These correlations were the basis of a famous paper published in 1935 by Albert Einstein and two younger colleagues, Boris Podolsky and Nathan Rosen. The existence of certain puzzles associated with these correlations is called the Einstein-Podolsky-Rosen (EPR) paradox. These correlations are correctly predicted by quantum mechanics, but they cannot be comprehended within the conception of the physical world postulated by classical mechanics.

In both the Libet and EPR cases, the existence of these apparent causal anomalies suggests that what seems to us to be "voluntary" free choices are actually mechanically determined by the physically described aspects of nature, in keeping with the precepts of classical physics. However, in their search for a rationally coherent understanding of various twentieth-century data, the founders of quantum theory were driven to a theory that consistently treats our voluntary choices as "free choices." They are "free" in the sense that they are not determined by any currently known laws, even though, according to the laws of quantum mechanics, they have specified physical consequences. This essay describes how orthodox quantum mechanics reconciles this idea of physically effective voluntary "free choices" with the Libet and EPR data

2. From Classical Mechanics to Orthodox Quantum Mechanics

During the seventeenth century, Isaac Newton created the foundations for what developed during the eighteenth and nineteenth centuries into what is now called classical physics, or classical mechanics. Classical mechanics conceives the physical world to be composed of classically conceived particles and classically conceived fields. Classically conceived particles are like miniature planets that move though space under the influence of fields of force generated by the other particles. This entire physical structure develops in time in a way fixed by mechanical laws that *entail the causal closure of the physical:* The whole physically described structure is determined for all time by these mechanical laws—which refer only to these physically described elements themselves—together with initial conditions on these physically described parts.

Around the beginning of the twentieth century, it was discovered that this classical-mechanical conception of the physical world was incompatible with the behaviors of large (visible) systems whose activities depended sensitively upon the behaviors of their atomic constituents. The classical conception of physical reality was therefore abandoned by physicists, at the fundamental level, and was replaced by a vastly different conceptual arrangement.

The logical basis of this conceptual change is a curious mathematical change. To pass from a classically conceived physical system to its quantum generalization, the numbers that described the classically conceived physical properties are replaced by mathematical actions, called "operators."

A principal difference between numbers and mathematical actions/operators is that the order in which one multiplies numbers does not matter—two times three is equal to three times two—but the order in which one applies actions does matter: for two actions, A and B, the action of A followed by the action of B—which, for historical reasons, physicists represent as BA—is not equal, in general, to AB.

The paradigmatic example is this: An important number in classical physics is the number x that represents how far some object has been displaced, in some direction, from an initial point $x = 0$. An equally important number is the number p that represents the momentum $p = mv$ of the object, where m is the mass of the object, and v is its velocity in the direction associated with x. In classical physics, x and p are *numbers,* and hence $xp - px = 0$, but in the quantum counterpart of the classical system $\mathbf{xp} - \mathbf{px} = i\hbar$, where $\hbar$ is a number discovered and measured by Max Planck in 1900, and i is a number that multiplied by itself gives minus one.

This difference between classical mechanics and quantum mechanics might seem to be a mere mathematical technicality, one having no deep conceptual import. Indeed, on the scale of human activities, the smallness of the effective difference between numbers and the corresponding mathematical actions might naturally lead one to expect that the *conceptual* changes needed to cope with this mathematical change would be unimportant at the level of human beings and their actions. But this is apparently not the case. The founders of quantum theory—in order to secure a rationally coherent and consistent way of dealing, in a scientifically satisfactory manner, with the technical problems introduced by the replacement of numbers by actions—were forced to formulate their theory in terms of *actions,* and in particular the actions of human investigators. Specifically, their theory is formulated in terms of predictions about the observable responses to actions that are chosen *by human agents,* with the intent to probe certain properties of systems described in the mathematical language of quantum mechanics. But this means that the basic physical theory deals *no longer with intrinsic properties of physically described systems,* but fundamentally, with the interplay between observed and observing systems. And these observing systems are, paradigmatically, conscious human participants. Here, the word "conscious" highlights the fact that the theory involves, basically, not solely the physical language of the quantum mathematics, but, equally importantly, the concepts and language that we human beings use to communicate to our colleagues "what we have done and we have learned." Moreover, in a fundamental way, the theory also involves the so-called "free choices on the part of the experimenter," which are experienced by experimenters as conscious choices.

Any physical theory, to be relevant to our lives, must link certain mathematical features of the theory to the streams of consciousness of human beings. Quantum theory is built squarely upon the recognition of this fact. To see how this works, consider the mathematical action $\mathbf{x}$ discussed above. As already mentioned, this mathematical action $\mathbf{x}$ replaces the number x that in classical mechanics specifies where (along a straight line) the (center of an) object is located. The postulated correspondence between the quantum mathematics and experienced perceptions ties the *mathematical action $\mathbf{x}$ to the empirical probing action that would yield, as its perceived outcome, the number x that would specify the location of the object being probed, insofar as that object has a well-defined location.* Similarly, the mathematical action $\mathbf{p}$ is tied to a physical probing action that would yield, as its perceived outcome, the number p that specifies the momentum of the observed object, insofar as that momentum is well defined.

Not every possible mathematical action has a perceptual counterpart. But the basic interpretive assumption in orthodox contemporary physics is that every possible probing action with a perceivable outcome has, in the quantum mathematics, an action counterpart: an associated operator. *Thus, an intrinsic mind–matter connection is built directly into the fabric of our basic physical theory.*

This profound difference between contemporary physical theory and the classical physical theories of the eighteenth and nineteenth centuries would appear, prima facie, to be relevant to issues pertaining to the relationship between mind and matter. The earlier theories are approximations to the newer theory, and these approximations systematically exorcize, in a rationally coherent but physically inaccurate way, dynamical connections between mind and matter that the newer theory incorporates.

The connection between mind and matter occurring in the original, *pragmatic* formulation of quantum mechanics—which is known as the Copenhagen interpretation—was converted to a connection between mind and brain by an elaboration upon the Copenhagen interpretation developed by the renowned logician and mathematician John von Neumann. This developed form was named "the orthodox interpretation" by von Neumann's close colleague Eugene Wigner, and it is the starting point of most, if not all, investigations into the nature of the reality that lies behind the pragmatically successful rules of quantum mechanics.

In spite of this seemingly relevant twentieth-century development in physics, contemporary neuroscience and philosophy of mind continue to base their quests to understand consciousness on an essentially nineteenth-century conceptualization of the human brain, ignoring the facts that the older conception of reality has been known to be false for almost a century, and that, in stark contrast to the nineteenth-century conceptualization, contemporary orthodox physics has specified dynamical connections between brains and minds built intrinsically into it.

Planck's constant is a very tiny number on the scale of human activities. Consequently, the replacement of a classical system by its quantum counterpart turns out to be unimportant for predictions pertaining to the observable properties of physical systems whose behaviors are insensitive to the behaviors of their atomic-sized constituents. But the behaviors of brains are understood in terms the behaviors of the ions flowing into and out of neurons. So it is not clear, *a priori*, that the behavior of a conscious brain will, in every case, be essentially nondependent upon how its atomic-sized constituents behave. Indeed, quantum calculations pertaining to the release of neurotransmitter molecules into the synaptic clefts separating communicating neurons show that quantum effects are important in principle.[3] According to the principles of contemporary physics, the behavior of a living brain must *in principle* be treated as a quantum mechanical system, with classical concepts applied only when justified by special circumstances.

No computations have ever shown that a conscious human brain can be validly treated in the classical approximation. On the other hand, the three-centuries-old effort to understand the connection between mind and brain within the conceptual framework of materialist classical physics has led to profound conceptual difficulties. These difficulties have provided fertile ground for philosophical disputes that have enlivened the fields of philosophy of mind and neurophilosophy without producing much consensus. But one point of near unanimity is the conclusion that materialism is surely the adequate and appropriate theoretical foundation for the scientific study of consciousness: that the injection by twentieth-century physics of the effects of conscious choices made by observer-participants into the basic dynamics of physical systems can safely be ignored. Still, a rationally coherent conceptualization that has *specified* mind–brain dynamical connections—which arise from the basic precepts of empirically valid physics—could conceivably provide a more adequate foundation for the scientific study of the behavior of real mind-brain systems than a nineteenth-century approximation that is inadequate in principle for systems whose behaviors depend significantly upon the dynamics of their atomic constituents, an approximation that systematically exorcises the quantum-physics-mandated dynamical effects of conscious choices made by conscious agents.

Over the past few years, I have been engaged in an effort to introduce into the scientific studies of consciousness certain basic results pertaining to the dynamics of the mind–brain system that are entailed by orthodox contemporary physics. Numerous applications have been made in the domains of psychology, psychiatry, and neuroscience.[4] I shall give here first a brief summary of some of the key elements of this quantum approach, and then use the theory to give a unified treatment of the Libet and the Einstein-Podolsky-Rosen data.

Classical physics is nominally about the *internal properties* of physical systems, but it is known to be fundamentally false. It has been replaced by quantum physics, which is about the *interplay* between *observed systems* (described in terms of mathematical quantities attached to space-time points, i.e., *res extensa*), and *observing systems* (described in terms of elements of streams of consciousness, i.e., *res cogitans*).

Although the various *effects* of a probing action made by a probing system upon a probed system are specified by quantum theory, the *cause* of the probing action is not specified by the theory. There is, therefore, a causal gap! The quantum-theoretic laws determine neither when a probing action will occur, nor which aspects of the observed system will be probed. Niels Bohr emphasizes this key feature of quantum mechanics when he says:

> The freedom of experimentation, presupposed in classical physics, is of course retained and corresponds to the free choice of experimental arrangement for which the mathematical structure of the quantum mechanical formalism offers the appropriate latitude.[5]
>
> To my mind there is no other alternative than to admit in this field of experience, we are dealing with individual phenomena and that our possibilities of handling the measuring instruments allow us only to make a choice between the different complementary types of phenomena that we want to study.[6]

In practical applications, in both classical and quantum mechanics, physicists treat the human experimenter as an agent who sets up experiments on the basis of his reasons. In neither classical nor quantum theory does anyone actually use the dynamical equations to determine what a real experimenter will actually do. The brain is too complex and too inaccessible to non-disturbing observations at the needed level of accuracy to permit this. In classical physics, there is the *presumption* that the physical laws determine *in principle* what an experimenter will do. But this presumption goes far beyond what has been scientifically tested and confirmed. In the more accurate contemporary orthodox quantum theory, the conclusion is just the opposite: In principle, the known laws definitely *do not determine how the experimenter will act,* nor even place statistical conditions on these choices. To fill this lacuna, the founders of quantum mechanics brought into the theory certain inputs from conscious human beings, namely their choices of their own actions. This introduction of physically efficacious conscious choices into the physical theory in a fundamental way was the most radical of the breaks with precedent introduced by the founders of quantum theory, and it is the one most vigorously opposed by physicists seeking a closer-to-tradition alternative to the Copenhagen and orthodox (von Neumann) approaches. However, none of the proposed alternatives appears to be satisfactory as yet, even to its supporters.[7]

Specifically, quantum theory brings into the causal description, in addition to the (sometimes violated) deterministic continuous evolution in accordance with the quantum generalization of the deterministic classical process of evolution, also *choices of two kinds,* both of which are implemented, or represented, by abrupt "quantum jumps" in the continuous deterministic evolution. One of the two kinds of choices determines the familiar collapse of the wave function (or reduction of the wave packet). It is called by Dirac a "choice on the part of nature," and it is a choice—from among the several alternative possible outcomes of a probing action performed upon an observed/probed system—of *one particular outcome.* These "choices on the part of nature" are "random": They are asserted by the theory to conform to certain statistical conditions. These choices on the part of nature are precisely where the randomness enters (irreducibly) into contemporary physics.

But, according to the orthodox precepts, this statistically governed "choice on the part of nature" *must be preceded* by another choice: a choice of which (probing) experiment is to be performed, and when it will be performed. No known laws constrain *this choice of the probing action,* and it is consistently treated in orthodox quantum theory as "a free choice on the part of the experimenter." This "choice on the part of the experimenter" fixes the form of the physically/mathematically described probing action. The representation *within the quantum mathematics* of this probing action is called by von Neumann "Process 1."

The *logical* need for this choice, which is not specified by any known law, persists, even when the quantum-mathematically described part of the universe—which in the original Copenhagen interpretation does *not* include either the body or the brain of the observer, nor even his or her measuring devices—is expanded (by von Neumann) to include the entire physical universe, *including* the bodies and brains of the observers. The essential point is that the inclusion of the body and brain of the human agent/participant into the quantum-mechanically described universe *still leaves undetermined* the choices made by that human person.

This logically needed choice is relegated, in von Neumann's words, to the experimenter's "abstract ego." But no matter what words are used, the fact remains that the inclusion of the body and brain of the observer into the physically described quantum world leaves undetermined the *logically needed choice* of which physical Process 1 probing action actually occurs. No known law, statistical or otherwise, specifies which probing action—Process 1 action—actually occurs.

The choosing process, whatever it is, that specifies this choice of the actually occurring Process 1 is called Process 4. Process 2, so named by von Neumann, is the continuous deterministic evolution via the Schroedinger equation, whereas Process 3 is the choice on the part of nature of which outcome/feedback from the probing action actually occurs. Process 2 reigns only during the intervals *between* the various abrupt Process 1 and Process 3 quantum jumps.

This need for the occurrence of physically efficacious Process 4 choices that are not determined by any known law, statistical or otherwise, constitutes a prima facie breakdown, within orthodox quantum mechanics, of the doctrine of the causal closure of the physical. As it is taught to physicists in their university courses, quantum theory is presented as a set of rules that allow scientists to form expectations about the feedbacks they will receive by performing any *one* of many possible probing actions, among which they are free to choose. This practical format is the basis of the conceptual structure of quantum theory.

To prepare the way for the analysis to follow, I need to spell out in a bit more detail the structure compactly summarized above.

The conversion of the classically conceived universe to its quantum generalization—obtained by replacing numbers by actions—is called "quantization." It converts the classical deterministic equation of motion into its quantum counterpart, von Neumann's Process 2. Like its classical counterpart, this quantum law of evolution is deterministic: Left alone, it would determine the quantum state of the universe for all times from its primordial form. The relativistic (quantum field theoretic) form of this law is moreover *local:* the changes in the quantum state associated with any region are determined by the properties associated with very nearby regions, and no influence propagates faster than the speed of light.

By itself, this Process 2 evolution is dynamically insufficient. Given some initial conditions, it produces at a later time, not the mathematical counterpart of *one single perceptual probing action,* but rather the counterparts of a *continuous smear* of alternative possible probing actions. Orthodox quantum theory resolves this difficulty by supplementing the Process 2 evolution by certain abrupt changes, which von Neumann calls "Process 1 interventions." Each such mathematical intervention is tied by the quantum laws to a particular perceivable probing action performed upon the observed system by an observing system external to it.

Neither the property of the observed system that is probed by this intervention, nor the time when this probing action occurs, is fixed by the mechanical Process 2. These two features are considered to be fixed by the observing system. This assignment of responsibility, or of causal origin, accords with the fact that in actual scientific practice, it is the human experimenter that selects, by conscious choice, which particular probing action will be performed upon the system he or she is observing, and when that probing action will be performed. Of course, an agent's conscious choices are not independent of what is going on in his brain, but orthodox contemporary physics does not determine how the psychic and physical components of reality combine to *cause* the Process 1 events to be what they turn out to be.

The *effect* of the Process 1 intervention upon the observed system *is* specified by the quantum laws. This intervention selects from the smear of possible probing actions some particular one. The effect of this singled-out probing action upon the mathematically described state of the observed system is this: It separates this state into a set of disjoint (i.e., non-overlapping) components in a way such that (1) the statistical weights assigned by the theory to these individual components adds to unity; and (2) each component corresponds to a *phenomenologically distinct* outcome of that probing action.

After this Process 1 separation has been made, nature picks out and saves *one* of the possible psychophysical outcomes of the chosen probing action, eradicating the rest. *Nature's* selection of outcomes—called Process 3 in my terminology—is asserted by the theory to respect the statistical weights assigned to the alternative possible outcomes. The quantum mathematical structure becomes tied in this way to phenomenology, and the theory generates practical rules that allow statistical predictions pertaining to experiences to be deduced from the postulated mathematical structure.

This injection of human volitional choices into the physical dynamics *at a basic level* is completely contrary to the precepts of classical physics. But this change accommodates the fact that we human beings do in fact inject our conscious intentions into the physically described world whenever we act intentionally. Accepting quantum mechanics opens the door to the possibility of a more detailed, *and more useful,* putative understanding of this effect of conscious intent than classical mechanics can provide.

3. The Libet Causal Anomalies

In the Libet experiment, the initial intentional act is to choose willfully to perform at some future time, say within the next minute, the act of raising a finger. We often make such resolves to act in some specified way at some future time, and these commitments are often met with great precision. However, in the Libet case, the resolve is rather imprecise as regards the exact time of the specified action. It is doubtful that any person, informed even by a multitude of probing devices about the state of the subject's brain at the beginning of the specified interval, could predict with good accuracy just when the choice to move the finger will occur. And even if every neurophysiological-level feature of the brain were given at the outset, it is still questionable whether, even in a world that obeyed the deterministic laws of classical physics, this macroscopic data would fix the time at which the conscious choice occurs. There is just too much latitude for initially small-scale variations to develop over the course of time into significant macroscopic effects. Even within deterministic classical physics, the best one could do with actual macroscopic data would be to make a statistical model based on that data and the known general properties of the brain.

In the case of the dynamics of a warm, wet, living human brain interacting with its environment, almost all quantum interference effects *connecting appreciably different locations* will (almost certainly) be washed out, and the quantum model will become similar to a classical *statistical* model that features a collection of parallel classically conceived worlds, each with some statistical weight. However, in the classical case one can imagine that exactly one of the statistically weighted, alternative, classically conceived possibilities is the "real" one, and that the statistical smearing represents a mere lack of knowledge as to which of the weighted possibilities represents the "actual real world."

This "lack of knowledge" interpretation cannot be carried over to quantum theory. However, to a good approximation, the various weighted classically conceived worlds of classical statistical theory can be understood to represent *simultaneously existing potentialities,* some subset of which will eventually be selected by some Process 1 probing event. This Process 1 action will be followed by a Process 3 choice (on the part of nature) that specifies which of the alternative possible outcomes of the chosen probing action actually occurs. *All potentialities that do not lead to the outcome that actually occurs are eradicated by these collapse or reduction events, leaving only those which lead to the psychophysical event that actually does occur.*

In the Libet experiment, the mind–brain "set," fixed by the initial conscious intention to raise the finger within the next minute, should cause the quantum mechanically described brain to generate classically describable potentialities corresponding to the various alternative times at which the specified conscious act could occur. Thus, the following scenario is compatible with quantum mechanics and is suggested by it:

The initial intent (to raise the finger within the next minute) will lead to the production of a collection of parallel potentialities, each corresponding to a possible time at which the readiness potential can start its build-up. Shortly after some of the classically described potentialities have developed to the point of specifying a certain possible perceivable probing action, the question will pop into the stream of consciousness: "Shall I perform this action?" If the answer is, "No," as it is likely to be right at the beginning, then the potentialities *leading up to the performance of that action at that time* will be eradicated. A short time later, a similar Process 1 question will be posed. The outcome is again likely to be "no," and the batch of potentialities leading to the "yes" option will again be erased. Eventually, in accordance with the statistical rules, a "yes" outcome will be selected by nature, and the set of potentialities leading to the "no" outcome will be wiped out. Only the (essentially classically described) potentialities *leading to* this "yes" outcome will remain.

The "yes" event is a psychophysical event that is felt or experienced as the feeling or knowledge, "I shall now raise my finger," and it is represented in the physically described world as the actualization, at that moment, of the neurological activity that constitutes the template for the action of raising the finger. (This template is a neural/brain activity that, if held in place for a sufficiently long interval, will tend to cause the finger to rise.) All brain activities—which have the ontological character of potentialities—that are incompatible with this intent are *eliminated by this event* from the quantum mechanical state of the brain. Hence, they are eliminated from the statistical mixture of classically described states that approximately represents this quantum state.

Now suppose there is in place some measuring device that can, in the approximately correct classical description of what is (possibly) going on, detect the time at which the readiness potential starts its build-up. This time of the inception of the build-up is long (one-third of a second) before the psychophysical event that will, *only later,* actualize this particular classically described world. Now suppose, furthermore, that the classically described measuring device activates a classically described timer that records the time of the beginning of the build-up of the readiness potential. This classically described *record* of the time of the start of the build-up of the readiness potential will continue to exist alongside the increasing readiness potential.

When some person, at some later time, after the occurrence of the psychophysical event that determines which of the classically described worlds survives—and hence also determines the time at which the build-up of the readiness potential began—reads the timer, he will find out that the start of the build-up of the readiness potential occurred *before the occurrence of the psychophysical event that selected the classical world that specifies the time when that build-up began.*

The key point here is that the *record* of the time of the start of the build-up of the readiness potential is *a causal off-shoot* of this build-up, and this record will be actualized along with the actualization of the potentiality *represented (to a good approximation) by* the classically described process that the actualization event selects. Thus the recorded time of the beginning of the build up of the readiness potential will be earlier than the time of the event that actually determined (according to this quantum ontology) the time of the beginning of this build up: the recorded time of the beginning of the build up will be fixed by an event that occurs only later.

Such seeming causal anomalies have been a prime point of attack on orthodox quantum theory, and they continue to fascinate physicists even today, under the names "quantum nonlocality," or "Bell's theorem," or "EPR paradox." Although this quantum ontological way of understanding the quantum correlation tends to upset people accustomed to thinking about the world in classical mechanical terms, no logical inconsistency or conflict with empirical data has ever been established. One can be quite confident in accepting that all of the known empirical evidence is compatible with this non-classical but logically consistent "quantum ontological" conception of how the world works.

On the other hand, one can certainly adhere, alternatively, to the *pragmatic* point of view, which holds that, even though this quantum ontology accords with all of the empirically verified relationships between human experiences, and seems to provide a coherent putative "understanding" of what is going on, such success by no means implies that this understanding is veridical. For one can express the empirical predictions in compact ways that avoid any commitment concerning what is "really happening." Thus many—and probably most—quantum physicists hold that, as scientists, the pragmatic option is all they need to commit to. On the other hand, for those who seek something more than merely "a set of rules that work" the quantum ontological model is a viable (i.e., not yet disproven) and logically coherent conception of the way that Nature actually works. The same cannot be said of local deterministic materialism.

Human agents play a very special role in this quantum ontology. This feature is a holdover from the pragmatic stance of the original Copenhagen formulation of the theory, which was concerned principally with establishing a rationally coherent basis for practical applications. However, von Neumann's analysis shows that there is no empirical evidence that *every* occurring collapse event is associated with an event in a human stream of consciousness. It is certainly more plausible, from a scientific perspective, to assume that there are similar events associated with other biological organisms, and there is no empirical evidence that confutes that position. Indeed, von Neumann's analysis reveals, more generally, that collapse events that act *macroscopically* on physical systems that are interacting strongly with their environments would be virtually impossible to detect. There is presently no evidence that rules out the possibility that enormous numbers of macroscopic collapse events are occurring all the time in large systems that are strongly connected to their environments. Hence the special role originally assigned to human beings is no part of the general quantum ontological model being described here.

The main cause of reservations about the actual truth of this quantum ontology is that it entails faster-than-light transfer of information. These faster-than-light issues are essentially those that arise in the much-discussed EPR paradox.

4. The Einstein-Podolsky-Rosen Causal Anomalies

In 1935, Albert Einstein, Boris Podolsky, and Nathan Rosen published what is perhaps the most discussed scientific paper of the twentieth century. Entitled "Can Quantum Mechanical Description of Physical Reality Be Considered Complete?" the paper argues that Copenhagen quantum theory does not give a complete description of physical reality. The argument depends on a specific way of identifying what is meant by "physical reality." This identification depends on an assumption about the absence of influences that act backward in time or faster than the speed of light. Niels Bohr wrote a rebuttal that essentially admitted that the strong notion of no faster-than-light influence used in classical physics does indeed fail in quantum theory, but claimed that an adequate replacement holds within the epistemological framework of quantum mechanics.[8]

The Einstein-Podolsky-Rosen argument is based on an examination of the predictions of quantum theory pertaining to certain correlations between *macroscopic* observable events that occur at essentially the same time in laboratories that lie far apart.

A simple classical example of a correlation between events occurring at essentially the same time in far-apart laboratories is this. Suppose one has a pair of balls, one red, the other green. Suppose one loads them into two rifles, and fires them in opposite directions into two far-apart laboratories, in which the balls will be caught and examined at essentially the same time. The colors found in the two regions will obviously be correlated: If red is found in one lab then green will be found in the other, and vice-versa. There is nothing strange or peculiar about a correlation of this kind.

The simplest quantum example is similar, and is, again, not in itself a problem. We can set up a certain experimental arrangement of the macroscopic preparing and measuring devices that will produce a situation analogous to the one with the two colored balls. Quantum mechanics predicts and empirical evidence confirms that, under these *macroscopically specified* experimental conditions, if a red light flashes on the detector in one laboratory, then a green light will flash at essentially the same time on the detector in the other laboratory, and vice-versa.

Einstein and his colleagues (henceforth EPR) considered a slightly more complex situation, in which there are two alternative possible settings of the measuring device in the first lab and two alternative possible settings of the device in the second lab. If the *first* setting is chosen in both labs, then, as before, green in either lab entails red in the other, and vice-versa. Moreover, if the second setting is chosen in both labs then, as before, green in either lab entails red in the other, and vice-versa.

A basic feature of quantum theory is this: the theory is mathematically incompatible with the idea that there exists *both* a property, P1, that fixes which outcome will occur if the measurement in, say, the second lab specified by the *first* possible setting of the device in that (second) lab is performed, and also, *simultaneously*, a property, P2, that fixes which outcome will occur if the measurement in the second lab specified by the *second* possible setting of the device in that (second) lab is performed. Quantum theory regards two such properties, P1 and P2, as *complementary* properties that cannot both exist simultaneously.

Quantum Interactive Dualism

EPR devised an argument that seemed to show that these two properties P1 and P2 *do* exist simultaneously. Their argument produced consternation in Copenhagen. Bohr's close colleague, Leon Rosenfeld, described the situation as follows:

> This onslaught came down upon us like a bolt from the blue. Its effect on Bohr was remarkable. We were then in the midst of groping attempts at exploring . . . [another problem]. . . A new worry could not come at a less propitious time. Yet as soon as Bohr had heard my report of Einstein's argument, everything else was abandoned: we had to clear up such a misunderstanding at once. We should reply by taking up the same example and showing the right way to speak about it. Bohr immediately started dictating to me the outline of such a reply. Very soon, however, he became hesitant: "No, this will not do, we must try all over again . . . we must make it quite clear. . . ." So it went on for a while with growing wonder at the unexpected subtlety of the argument. . . . Eventually he broke off with the familiar remark that he "must sleep on it." The next morning he at once took up the dictation again, . . . the real work now began in earnest: day after day, week after week, the whole argument was patiently scrutinized.[9]

What is the argument that set off this huge commotion, which reverberates even to this day? Einstein and his colleagues introduced the following "criterion of physical reality":

> If, without in any way disturbing a system, we can predict with certainty (i.e., with probability equal to unity) a physical property of a system, then there exists an element of physical reality corresponding to that property.

This criterion seems completely reasonable, and completely in line with the Copenhagen philosophy, which is built upon the idea of predictions of properties of systems as revealed by the observed outcomes of experiments performed upon those systems.

In the experimental situation just mentioned, the setting of each device can be chosen and fixed *just before* the outcome at that device appears. The distance between the two labs can then be made so large that there is no time (according to the claim of the theory of relativity that *nothing* can travel faster than the speed of light) for a choice of setting in either lab to have any effect at all on the faraway outcome, red or green.

However, the experimenter in the first lab *can predict with certainty* the property P1 that is measured by using the first setting in the (faraway) second lab. He can do this simply by choosing the first setting in his own (first) lab and observing the outcome, red or green, and then inferring that P1 must be, respectively, green or red. The impossibility assumed (by EPR) of any faster-than-light or backward-in-time influence entails that this *action and act of observation in the first lab* cannot disturb in any way this property P1 measurable in the second lab. Thus, according to the EPR criterion, there is an element of physical reality P1 corresponding to the property that is measured in the second lab when one uses the *first* setting there.

By choosing the *second* setting in the first lab, one finds that a property P2 corresponding to the *second setting* in the second lab is, likewise, an element of physical reality. But—for inescapable mathematical reasons—quantum theory cannot accommodate the simultaneous existence of these two elements of physical reality, P1 and P2. Hence, as a description of physical reality, quantum theory must, according to EPR, be incomplete.

EPR finish off their argument with the following crucial remark:

One could object to this conclusion on the grounds that our criterion of reality is not sufficiently restrictive. Indeed, one would not arrive at our conclusion if one insisted that two or more physical quantities can be regarded as simultaneous elements of reality only when they can be simultaneously measured or predicted. On this point of view, since either one or the other, but not both simultaneously of the quantities P and Q can be predicted they are not simultaneously real. This makes the reality of P and Q depend upon the process of measurement carried out on the first system, which does not disturb the second system in any way. No reasonable definition of reality could be expected to permit this.[10] [Note: EPR's P and Q are essentially equivalent to our P1 and P2.]

Bohr accepts that the orthodox principles of quantum theory demand that P and Q cannot, within that theory, both be assigned well-defined values. How does he reconcile this fact with the EPR argument that both are elements or physical reality?

The essence of Bohr's reply is the following passage:

From our point of view we now see that the wording of the above-mentioned criterion of physical reality proposed by Einstein, Podolsky, and Rosen contains an ambiguity as regards the meaning of the expression "without in any way disturbing the system." Of course there is in a case like that just considered no question of a mechanical disturbance of the system under investigation during the last critical stage of the measuring procedure. But even at this stage there is essentially the question of *an influence on the very conditions which describe the possible types of predictions regarding the future behaviour of the system.* [Bohr's italics.] Since these conditions constitute an inherent element of any phenomenon to which the term "physical reality" can be properly attached, we see that the argumentation of the above-named authors does not justify their conclusion that quantum-mechanical description is essentially incomplete.[11]

If Bohr's argument strikes you as obscure, you are not alone. Many philosophers and physicists have judged Bohr's reply to be insufficient, and have concluded that Einstein won the debate. Bohr himself says, in his contribution to the Einstein volume, "Reading these passages, I am deeply aware of the inefficiency of expression which must have made it very difficult to appreciate the trend of the argumentation. . . ."[12]

That is an accurate statement. Yet his later arguments do not seem to help.

One feature of Bohr's answer does come across clearly: His reply rejects, *at some level,* Einstein's idea of "without in any way disturbing the system;" Bohr rejects, at some level, Einstein's assumption that the freely chosen measurement process performed in the nearby lab *does not disturb in any way* the system in the faraway lab, even though any such disturbance would have to act essentially instantaneously. That is, in order to rationally counter the Einstein argument, Bohr found himself forced to reject Einstein's principle that *all* causal actions act only forward in time, and no faster than the speed of light. If that principle fails, the EPR argument collapses.

Bohr's point, in essence, is that once the experimenter in the first lab chooses to do one of the two possible measurements in his lab—for example, the one specified by the first (respectively second) setting in his own lab—he loses the capacity to make any prediction about the outcome of a measurement in the other lab associated with the second (respectively first) setting in that faraway lab. Thus the experimenter's choice of what to do here has changed *what he*

Quantum Interactive Dualism

can know about events in the faraway region. In an essentially epistemological theory in which the basic reality is "our knowledge," a *reality* associated with the faraway lab can therefore be said to depend upon one's choice, made here, about what one will freely choose to do here. But then the EPR claim that no reality "there" can depend upon what one can freely choose to do "here" fails: The EPR argument goes down the drain.

Of course, an epistemologically based conception of reality goes against Einstein's more traditional idea of reality. But this issue of the need for basic physical theory to deal with non-epistemologically based realities is the core issue in the Bohr-Einstein dispute. Hence, Einstein cannot simply assert, without in some way begging the central question, that "reality" must be defined non-epistemologically.

Bohr's argumentation is basically philosophical and about what we can know. It dodges the ontological issues usually associated with the phrase "physical reality," which is normally *contrasted* to what we know or can know. But the von Neumann-based quantum ontology, described above, explains the workings of this "action at a distance" in "ontological" terms. This ontology incorporates Heisenberg's idea of *potentia* as an objective tendency for a physically describable event to occur in association with an increment in human knowledge. This ontology that is based not on *substances,* but rather on psychophysical *events* and mathematically described "objective tendencies" for such events to occur. These tendencies are nonsubstantive because they can change abruptly whenever a new psychophysical event occurs, perhaps far away. It is, basically, the acceptance of such "tendencies as objective realities" that differentiates this Heisenberg-type quantum ontology from substance-based ontologies.

The quantum ontological explanation of the EPR-type correlations is similar to the explanation of the Libet back action. In the EPR case, the actualization in one region of some particular probing action and its outcome actualizes also the particular causal chain that leads up to that outcome, *along with its causal off-shoots,* and it eliminates the *potentialities* that would have produced the possible outcomes that were not actualized. But then a conscious choice of probing action made at one time and place can have ontological consequences in faraway regions. These *faraway consequences* are *effects* of causal off-shoots of possible processes that are actualized by events in the *nearby region* that depend on choices freely made in the nearby region.

These words are more than verbal hand waving. They are descriptions in ordinary words of exactly what the von Neumann mathematical representation of the evolving state is doing. Insofar as one accepts the idea that the reality is represented by the mathematics, and that our words and concepts should conform to what the mathematics is saying, this quantum ontology follows. It is an accurate description of what the quantum mathematics is saying.

This ontology accords with the orthodox quantum principle that the properties P1 and P2 do not exist simultaneously, and that the existence or nonexistence of such a property in one region can depend upon what a faraway experimenter does in a region that is space-like separated from the first. That is, this ontological conceptualization is in accord with the orthodox quantum principles, and, in agreement with Bohr's answer to EPR, it rejects the strong version of the principle of no faster-than-light effect *of any kind*. As already mentioned, Bohr's rejection was essentially epistemological, and the quantum ontology translates this into a nonclassical, nonsubstantive ontological conceptualization that does bring into the dynamics the effects of our "free" choices of how we will act.

One essential point needs to be emphasized. Von Neumann's formulation of quantum theory, which provides the mathematical foundation for this ontology, was first published in 1932, and it is nonrelativistic. A state of the universe is given for each "instant of time."

However, this formalism was generalized by Tomonaga and by Schwinger around the middle of the twentieth century to *relativistic quantum field theory,* with the quantum states now defined not on fixed-time surfaces but on space-like surfaces.[13] (Every point of a fixed-time surface lies at the same time, whereas points on a space-like surface can lie at different times, but every point of a space-like surface is separated by a space-like interval from every other point on that surface.)

In this relativistic generalization, a Process 1 event, freely chosen and acting on a local (nearby) portion of a space-like surface, followed by some local (nearby) Process 3 outcome, can "instantly" affect the part of the state associated with a distant (faraway) portion of that space-like surface. And this "faraway" effect can depend upon which Process 1 event was locally chosen. Thus Einstein's demand that such choices of probing actions can have no faster-than-light influence of any kind is violated, in accord with Bohr's denial of the validity of that condition. However, the relativistic formulation does satisfy the basic requirement of the theory of relativity that no "signal" can be transmitted faster than light. (A "signal" is a message such that the decipherable content received is influenced by the sender.) Thus in the robust practical sense of communicating what one knows (here) to distant receivers, there are no faster-than-light actions, even though the Tomonaga-Schwinger quantum ontology does explicitly exhibit faster-than-light transfers to faraway regions of information that is influenced by nearby free choices. The reason that this explicit faster-than-light transfer of information cannot carry a message intended by the local experimenter is that the faraway effects of the nearby choice depends *jointly* upon the experimenter's choice of the local experiment and nature's choice of the local outcome in such a way that if the faraway receiver knows nothing about nature's local (nearby) choice, then he cannot acquire from his observations any information about the experimenter's local (nearby) choice. This result is a direct consequence of the quantum rules.

The relativistic (Tomonaga-Schwinger) von Neumann ontology satisfies the demands of the theory of relativity, yet explicitly exhibits the sort of faster-than light effects alluded to in Bohr's answer to EPR. This rationally coherent conception of nature resolves the mysteries of the seeming causal anomalies by setting forth a new "quantum-theoretic" way of understanding nature, an understanding based not on substances but on psychophysical events and objective tendencies for such events to occur.

The fact that this particular orthodox ontology involves faster-than-light effects does not imply that *no* rationally coherent theory can agree with the quantum predictions unless it allows transfer of information about a free choice made in one region to a space-like separated region. But that strong result can be proved.

Certain theories entail the validity of certain statements of the form:

If experiment 1 is performed and the outcome is A, then if, instead, experiment 2 had been performed the outcome would necessarily be B.

For example, according to classical physics, if we shoot a charged particle into a region with, say, uniform magnetic field H and it follows a semi-circle of radius R, then if we had chosen magnetic field 2H, with every other relevant condition unchanged, then the particle would have followed a semi-circle of radius R/2.

To establish the unavoidable need in any adequate theory of nature for some sort of faster-than-light transfer, we may consider an experiment of the kind first investigated by Lucien Hardy. As in the EPR case, there are two space-time regions, situated so that nothing can get from either region to the other one without traveling either faster than light or backward in time. In each region, either one or the other of two alternative possible probing actions can be chosen and performed. And for each performed experiment, one or the other of two alternative possible outcomes of that experiment will appear in the region in which that measurement is performed.

Let one of the two regions be called R and the other be called L, and let the space-time region R lie *later* than the space-time region L (in some specified coordinate frame). The first needed assumption is this:

> The choices of which of two possible experiments will be performed in regions R and L *can be treated* as independent free variables.

This does not mean that, in the total scheme of things, each of these two choices is undetermined until it actually occurs, only that the choice of which experiment to perform can be fixed in so many alternative possible ways by systems so disconnected, prior to the probing action, from the system being probed, that the choice of which probing action is performed *can be treated* as a free variable in the context of the analysis of this experiment. This free choice assumption is endorsed by Bohr, and is used by EPR.

The second assumption is this:

> No matter which experiment is chosen and performed in the earlier region L, whatever outcome *appears and is recorded there* is independent of which probing action will be chosen and performed later in the faraway region R.

These two assumptions—along with the assumed validity of four simple predictions of quantum theory for a Hardy-type experiment—allow one to *prove* some interesting properties of the following statement, which I have named SR, because it is a statement that refers to possible happenings in region R.

> SR: If the first of the two alternative possible probing actions in region R gives the first of the two possible outcomes, then the second of the two alternative possible probing actions in region R, if it had been performed instead of the first one, would necessarily have given the first possible outcome of that second probing action.

This statement does not involve two coexisting incompatible properties: The two incompatible properties in R exist only under incompatible conditions in R. Statement SR is *logically entailed by the two assumptions described above and the validity of four predictions of quantum theory* to be true or false, according to whether the experimenter in region L chooses to perform in L one or the other of the two alternative possible actions available to him.[14]

The conditions that logically determine whether this statement SR is true or false are conditions on outcomes appearing in region R under the alternative possible conditions that can be freely chosen in that region R. But this statement is required by the laws of quantum mechanics to be true or false according to which choice is freely made by the experimenter in region L, which is space-like separated from region R. This demand cannot be met by a theory that allows no information about the free choice made in L to get to region R.

A rationally coherent understanding of natural phenomena that allows our choices of which experiments we perform to be treated as free variables is logically possible, but any such theory that strictly enforces the principle of no faster-than-light or backward-in-time transfer of information appears to be excluded by this argument, which thereby removes an important barrier to the acceptance of the quantum ontology described above.

5. Application to Libet

Numerous applications of this quantum ontology to the understanding of phenomena in psychology, psychiatry, and neuroscience, related to the connection of mind to brain have been described in Schwartz *et al.*[15] The central idea is to begin to fill the lacuna in the causal structure associated with Process 4—the process of choosing *which* Process 1 will occur, and *when* it will occur—by distinguishing two kinds of Process 4 choices: passive choices and active choices. The passive choices are entailed by brain activity alone: For these passive choices, the Process 1 action occurs when an associated threshold in brain activity is reached. The expression of this physically described threshold remains to be specified.[16] Once this initial psychophysical event occurs, and the follow-up Process 3 outcome has produced a "yes" response, there can be a felt evaluation. The key assumption is that if this felt evaluation is sufficiently positive, then there may be an *active* effort to attend to this idea, which, if sufficiently strong, will produce an almost immediate repeat of the original psychophysical event associated with Process 1. If the repetitions are sufficiently rapid, then a well-known quantum effect, the quantum Zeno effect, will cause a long string of essentially identical Process 1–Process 3 pairs to occur. This rapid sequence of events will, by virtue of the known quantum rules, tend to hold in place the associated template for action, which will tend to cause the intended action to occur. Thus, conscious intentions motivated by felt valuations become injected into the brain dynamics in a way that tends to cause consciously intended actions to occur.[17]

This conception of what is going on is in close accord with William James's assertions:

> I have spoken as if our attention were wholly determined by neural conditions. I believe that the array of things we can attend to is so determined. No object can catch our attention except by the neural machinery. But the amount of the attention which an object receives after it has caught our attention is another question. It often takes effort to keep mind upon it. We feel that we can make more or less of the effort as we choose. If this feeling be not deceptive, if our effort be a spiritual force, and an indeterminate one, then of course it contributes coequally with the cerebral conditions to the result. Though it *introduce* no new idea, it will deepen and prolong the stay in consciousness of innumerable ideas which else would fade more quickly away. The delay thus gained might not be more than a second in duration—but that second may be critical; for in the rising and falling considerations in the mind, where two associated systems of them are nearly in equilibrium it is often a matter of but a second more or less of attention at the outset, whether one system shall gain force to occupy the field and develop itself and exclude the other, or be excluded itself by the other. When developed it may make us act, and that act may seal our doom. When we come to the chapter on the Will we shall see that the whole drama of the voluntary life hinges on the attention, slightly more or slightly less, which rival motor ideas may receive. . . . Consent to the idea's undivided presence, this is effort's sole achievement.[18]

This understanding is in line also with James's later assertion: "[Y]our acquaintance with reality grows literally by buds or drops of perception. Intellectually and on reflection you can divide them into components, but as immediately given they come totally or not at all."[19]

Turning to the Libet situation, we see that there is an important difference between it and the EPR situation. In the Libet case, the initial action that initiates the agent's later action—namely, the agent's commitment to raise the finger some time during the next minute—occurs *before* the development of the causal off-shoot, and it generates the chain of events associated with both the creation of the causal off-shoots (namely the creation of the *records* of the beginnings of the various parallel build-ups of the readiness potential) and also the subsequent conscious probing actions, one of which will eventually lead to the actualization of *one* of these records. In the Libet case, this causal linkage breaks the control of the *active* conscious choice (to raise the finger now) upon the causal off-shoot (the record); these *active* conscious choices act *only* to hold the template for action in place long enough to cause the finger to rise, or by failing to so act, to effectively *veto* that physical action. Thus the *active* conscious choices do *not* influence the causal off-shoots in the efficacious way that they do in the EPR case. They act only either to consent to the process of raising the finger, caused by the initial commitment to do so and nature's subsequent "yes," or to veto this physical action by refusing to initiate the repetitions needed to produce the action.[20] However, in the generation of correlations between two phenomena occurring in different regions, the key role of an actualization of a potentiality having a causal off-shoot is the same in both the Libet and EPR cases, as is the explanation of the capacity of a person's conscious choices, unconstrained by any yet-known laws, to influence his physical actions.

6. Conclusion

The quantum mechanical understanding of the mind–brain dynamical system explained and defended by Schwartz *et al.*,[21] and further elaborated in two essays of mine,[22] both accommodates and ventures an explanation of the ability of our conscious intentions to influence our physical behavior. This theory also covers the Libet data in a natural way. It reconciles Libet's empirical findings with the capacity of our conscious intentions to influence our actions, without these intentions being themselves determined by the physically described aspects of the theory. This separation is achieved by exploiting a causal gap in the mathematically expressed laws of quantum mechanics. This gap is filled in actual scientific practice by invoking the conscious intentions of the human participants. This practical and intuitively felt role of conscious intentions is elevated, within the proposed quantum ontology, to the status of an ontological reality coherently and consistently integrated into quantum laws.[23]

Notes

1. Adapted, with permission, from *Erkenntnis* Vol. 65, No. 1, 2006 (July), 117–42.
2. Benjamin Libet 1985; 2003.
3. Stapp 2004a.
4. Stapp 2004a, 2005, 2006a-d; Schwartz, Stapp, and Beauregard 2005.
5. Bohr 1958: 73.
6. Ibid., 51.
7. See Appendix A of Schwartz *et al.*, and the references cited therein, most particularly Stapp (2002), but also Stapp (2006a-d).
8. Bohr 1935.
9. Rosenfeld 1967.
10. Einstein, Podolsky, and Rosen 1935.
11. Bohr 1935.
12. Einstein 1951: 234.
13. Tomonaga 1946; Schwinger 1951.
14. Stapp 2004b.
15. Schwartz, Stapp, and Beauregard 2005.
16. Cf. Stapp 1999.
17. See Stapp (2004a), Chapter 12, for the mathematical details.
18. James 1892.
19. James 1911.
20. Schwartz *et al.* (2005), for example.
21. Ibid.
22. Stapp 2005; 2006.
23. I thank Wolfram Hinzen, for helpful suggestions pertaining to the presentation of this material, and Jeff Barrett and Peter Molenaar for their comments on this article.

References

Bohr, N. (1935) "Can quantum mechanical description of physical be considered complete?" *Physical Review* 48: 696–702.

———. (1958) *Atomic Physics and Human Knowledge.* New York: Wiley.

———. (1963) *Essays 1958–1962 on Atomic Physics and Human Knowledge.* New York: Wiley.

Einstein, A., Podolsky, B. & Rosen, N. (1935) "Can quantum mechanical description of physical reality be considered complete?" *Physical Review* 47: 777–80.

———. (1951) "Remarks to the essays appearing in this collected volume," in P.A. Schilpp, ed., *Albert Einstein: Philosopher-Scientist.* New York: Tudor.

James, W. (1892) "Psychology: the briefer course," in *William James, Writings 1879–1899.* New York: Library of America.

———. (1911) "Some Problems in Philosophy," in *William James, Writings 1902–1910.* New York: Library of America.

Libet, B. (1985) "Unconscious cerebral initiative and the role of conscious will in voluntary action." *Behavioral and Brain Sciences* 8: 529–66.

———. (2003) "Cerebral physiology of conscious experience: Experimental Studies," in N. Osaka, ed., *Neural Basis of Consciousness [Advances in consciousness research series, volume 49].* New York: John Benjamins.

Rosenfeld, L. (1967) "Niels Bohr in the thirties: consolidation and extension of the conception of complementarity," in S. Rozental, ed., *Niels Bohr: His life and work as seen by his friends and colleagues.* Amsterdam: North-Holland, 114–36.

Schwartz, J., Stapp, H. & Beauregard, M. (2005) "Quantum physics in neuroscience and psychology: a neurophysical model of mind/brain interaction," *Philosophical Transactions of the Royal Society* B 360 (1458): 1308–27 (http://wwwphysics.lbl.gov/~stapp/PTRS.pdf).

Schwinger, J. (1951) "Theory of quantized fields I," *Physical Review* 82: 914–27.

Stapp, H. P. (1999) "Attention, Intention, and Will in Quantum Physics." *Journal of Consciousness Studies* 6(8–9): 143–64.

———. (2002) "The basis problem in many-worlds theories." *Canadian Journal of Physics* 80: 1043–52.

———. (2004a) *Mind, Matter, and Quantum Mechanics* (Second Edition). Berlin: Springer.

———. (2004b) "A Bell-type theorem without hidden variables." *American Journal of Physics* 72: 30–33.

———. (2005) "Quantum Interactive Dualism: An Alternative to Materialism," *Journal of Consciousness Studies* 12(11): 43–58.

———. (2006a) "Quantum Approaches to Consciousness," in M. Moskovitch and P. Zelago, eds., *Cambridge Handbook of Consciousness.* Cambridge: Cambridge University Press (http://wwwphysics.lbl.gov/~stapp/ stappfiles.html).

———. (2006b) "Quantum Mechanical Theories of Consciousness," in M. Velmans and S. Schneider, eds., *Blackwell Companion to Consciousness.* Oxford: Blackwell Publishers (http://www-physics.lbl.gov/~stapp/stappfiles.html).

———. (2006c) "The Quest for Consciousness: A Quantum Neurobiological Approach," see http://www-physics.lbl.gov/~stapp/Quest.pdf.

———. (2006d) *Mindful Universe: Quantum Mechanics and the Participating Observer.* Berlin: Springer (see http://www-physics.lbl.gov/~stapp/MU.pdf).

Tomonaga, S. (1946) "On a relativistically invariant formulation of the quantum theory of wave fields." *Progress of Theoretical Physics* 1: 27–42.

Von Neumann, J. (1955/1932) *Mathematical Foundations of Quantum Mechanics*. Princeton: Princeton University Press (translated by Robert T. Beyer from the 1932 German original, *Mathematische Grundlagen der Quantenmechanik*. Berlin: J. Springer).

37

THE PHYSICAL SCIENCES, NEUROSCIENCE, AND DUALISM

JAMES P. MORELAND

Throughout history, most people have been substance and property dualists. Thus, Jaegwon Kim's concession seems right: "We commonly think that we, as persons, have a mental and bodily dimension.... Something like this dualism of personhood, I believe, is common lore shared across most cultures and religious traditions."[1]

Today it is widely held that, while broadly logically possible, dualism is no longer plausible in light of the advances of modern science, especially neuroscience. This attitude is especially prominent outside Christian circles. Thus, John Searle says that it is an obvious fact of physics that "the world consists entirely of physical particles in fields of force...."[2] He goes on to say that much of the justification for the various forms of physicalism that dominate philosophy of mind is the assumption that

> they represent the only scientifically acceptable alternatives to the anti-scientism that went with traditional dualism, the belief in the immortality of the soul, spiritualism, and so on. Acceptance of the current views is motivated not so much by an independent conviction of their truth as by a terror of what are apparently the only alternatives. That is, the choice we are tacitly presented with is between a "scientific" approach, as represented by one or another of the current versions of "materialism" and an "antiscientific" approach, as represented by Cartesianism or some other traditional religious conception of the mind.[3]

This attitude is not limited to non-Christian thinkers. Christian philosopher Nancey Murphy claims that physicalism is not primarily a philosophical thesis, but the hard core of a scientific research program for which there is ample evidence. This evidence consists in the fact that "biology, neuroscience, and cognitive science have provided accounts of the dependence on physical processes of *specific* faculties once attributed to the soul."[4] Dualism cannot be *proven* false—a dualist can always appeal to correlations or functional relations between soul and brain/body—but advances in science make it a view with little justification. According to Murphy, "science has provided a massive amount of evidence suggesting that we need not postulate the existence of an entity such as a soul or mind in order to explain life and consciousness."[5]

I disagree. My purpose in what follows is not to argue directly for some form of dualism. I have done that elsewhere.[6] Rather, I shall argue that *once we get clear on the central first and second order issues in philosophy of mind, it becomes evident that stating and resolving those issues is basically a (theological and) philosophical matter for which discoveries in the hard sciences are largely irrelevant.* Put differently, *these philosophical issues are, with rare exceptions, autonomous from (and authoritative with respect to) the so-called deliverances of the hard sciences.*

In what follows, I shall 1) clarify certain preliminary notions; 2) defend my central thesis by focusing on two paradigm cases—one regarding property and one regarding substance dualism—that are representative of the actual dialectic in the literature in philosophy of mind; and 3) respond to two defeaters of my thesis.

1. Clarification of Important Preliminaries Relevant to the Autonomy Thesis

Two preliminaries need clarification: identification of the central issues in philosophy of mind and the nature of the Autonomy and Authority Theses. The central topics tend to revolve around three interrelated families of issues constituted by the following kinds of representative questions:[7]

(1) *Ontological Questions:* To what is a mental or physical property or event identical? To what is the owner of mental properties/events identical? What is a human person? How are mental properties related to mental events? (e.g., Do the latter exemplify or realize the former?) Are there essences and, if so, what is the essence of a mental event or of a human person?

(2) *Epistemological Questions:* How do we acquire knowledge about other minds and about our own minds? Is there a proper epistemic order to first-person knowledge of one's own mind and third-person knowledge of other minds? How reliable is first-person introspection and what is its nature (e.g., a non-doxastic seeming or a disposition to believe)? If reliable, should first-person introspection be limited to providing knowledge about mental states or should it be extended to include knowledge about one's own ego?

(3) *Semantic Questions:* What is a meaning? What is a linguistic entity and how is it related to a meaning? Is thought reducible to or a necessary condition for language use? How do the terms in our commonsense psychological vocabulary get their meaning?

The main second-order topics in philosophy of mind are in a final category:

(4) *Methodological Questions:* How should one proceed in analyzing and resolving the first-order issues that constitute the philosophy of mind? What is the proper order between philosophy and science? Should we adopt some form of philosophical naturalism, set aside so-called first philosophy, and engage topics in philosophy of mind within a framework of our empirically best-attested theories relevant to those topics? What is the role of thought experiments in philosophy of mind and how does the "first-person point of view" factor into generating the materials for formulating those thought experiments?

In order to clarify the Autonomy and Authority Theses, I can do no better than cite advocate George Bealer's statement of them:

I wish to recommend two theses. [1] *The autonomy of philosophy:* Among the central questions of philosophy that can be answered by one standard theoretical means or another, most can in principle be answered by philosophical investigation and argument without relying substantively on the sciences. [2] *The authority of philosophy:* Insofar as science and philosophy purport to answer the same central philosophical questions, in most cases the support that science could in principle provide for those answers is not as strong as that which philosophy could in principle provide for its answers. So, should there be conflicts, the authority of philosophy in most cases can be greater in principle.[8]

Of the two, the Autonomy Thesis is less controversial and, in my view, clearly correct, at least in certain areas outside philosophy of mind. Debates about universals, the status of the identity of indiscernibles, and so forth are carried out with virtually no regard whatever for the latest findings in science. Most of the first- and second-order topics in philosophy of mind are similarly autonomous, or so I shall shortly argue.

The Principle of Authority is more controversial but not for the reason that may first come to mind. At first glance, ambivalence towards the principle may arise from the idea that science is a superior guide to joint areas of exploration. I disagree. The controversial nature of the Authority Principle derives from the fact that, in those cases where philosophical considerations carry more weight than scientific ones, it is usually open to someone to adopt an antirealist depiction of the relevant scientific view, operationalize the relevant terms that constitute it, and avoid epistemic conflict by resorting to an autonomy depiction of the philosophical and scientific aspects of the disputed area.

2. Two Paradigm Case Studies on Behalf of the Autonomy Thesis

Once we get before us the four families of questions listed above, it becomes evident that scientific discoveries play virtually no role at all in formulating or resolving those issues. To support this claim further, I have selected, almost at random, a paradigm case debate about the nature of mental properties/events in philosophy-of-mind literature to serve as an illustration of the Autonomy Thesis: Paul Churchland's treatment of two different approaches to closely related semantic and epistemic issues.[9]

According to Churchland, a popular physicalist approach to these issues is the network theory of meaning for the terms in our psychological vocabulary. Here, one looks not for an ontological analysis of meaning itself, but rather for a theory about how psychological terms get meaning. On this view, the best way to proceed is to start with a third-person perspective and focus on publicly accessible language to see how terms in folk psychology get their usage. These terms primarily function as theoretical terms used to explain/predict other people's behavior. Moreover, says Churchland, as theoretical terms, they get their meaning by their relations to laws, principles, and other terms in the entire theory in which they are embedded.

For Churchland, the epistemic approach most suited to this semantic theory starts with third person questions about knowledge of other minds and assimilates first-person to third-person knowledge. We are justified in applying a mental term to another creature just in case this provides the best explanation for and prediction of the creature's behavior. Churchland claims that one's justification need owe nothing at all to one's examination of one's own case and, thus, one could justifiably apply a mental term such as "pain" to a creature and, thus, know its meaning, even if one had never had the relevant experience.

Churchland characterizes self-consciousness as the ability to use a linguistic network to judge that one's various mental states satisfy the interlocking network of folk psychology. Thus, self-consciousness is largely something that is learned. Moreover, for Churchland, all perception is theory-laden, including self- "perception," and self-consciousness is essentially linguistic behavior of a certain sort.

Space considerations prevent me from presenting Churchland's largely accurate depiction of a dualist approach to these questions, but it involves a commitment to such things as irreducible self-presenting properties, first-person introspection and ostensive definition, epistemic movement from the first- to the third-person, non-doxastic mental states as temporally and epistemically prior to concepts and judgments, and meanings that are not essentially linguistic.

Who is right and what factors are relevant to this question? The answer is complicated and the dialogue involves thought experiments that, in my view, derive their force from first-person introspection, debates about private languages, analyses of the relationship between thought and language, and so on. What is less complicated is that factual information in the hard sciences is virtually irrelevant to these issues. Almost no book in philosophy of mind where these issues are discussed contains any detailed scientific information that plays a role in the discussion. Curiously, while Churchland himself is a physicalist and an advocate of naturalism as a second-order methodological thesis, and while he does include scientific information in *Matter and Consciousness,* that scientific information comes in the second half of the book and plays absolutely no role whatever in presenting the core philosophical issues and arguments in the first half of the book. Thus, his actual practice underscores the Autonomy Thesis.

The Autonomy Thesis is also justified by the actual nature of the debates about the ontological status of the entity that contains consciousness. Space considerations forbid a presentation of the dialectic between substance dualists and their physicalist opponents, but this is not necessary here. I believe that by listing the five main arguments for substance dualism, it will become virtually self-evident to an honest reader that the nature of the task of stating, defending, or criticizing these arguments underscores the Autonomy Thesis.

Currently, there are three main forms of substance dualism being debated. First, there is Cartesian dualism, according to which the mind is a substance with the ultimate capacities for consciousness and is connected to its body by way of an external causal relation.[10] Second, there is Thomistic substance dualism, one important version of which takes the soul to be broader than the mind in containing not merely the capacities for consciousness, but also those which ground biological life and functioning. On this view, the (human) soul diffuses, informs, unifies, animates, and makes human the body. The body is not a physical substance, but rather, an ensouled physical structure such that if it loses the soul, it is no longer a human body in a strict, philosophical sense.[11] According to the third form, a substantial immaterial self emerges from the functioning of the brain and nervous system, but once it emerges, it exercises its own causal powers and continues to be sustained by God after death.[12]

At least five arguments have been offered in the recent literature for some form of substance dualism:

Argument 1: In acts of introspection, one is aware of 1) one's self as an unextended center of consciousness; 2) various capacities of thought, sensation, belief, desire, and volition which one exercises and which are essential, internal aspects of the kind of thing one is; and 3) one's sensations as being such that there is no possible world in which they could exist and not be one's own. The best explanation for this fact is to take mental states to be modes of the self and mental properties to be kind-defining properties.

Argument 1 is actually two arguments that draw their force from what substance dualists claim people know about themselves from attending to themselves and their conscious states. Put more formally and in the first person, these two variants of an argument from introspection look like this:

Variant One:

(1) I am an unextended center of consciousness (justified by introspection).
(2) No physical object is an unextended center of consciousness.
(3) Therefore, I am not a physical object.
(4) Either I am a physical object or an immaterial substance.
(5) Therefore, I am an immaterial substance.

Variant Two:

(1) My sensations (and other states of consciousness) are either externally or internally related to me.
(2) If I am a physical object, then my sensations are externally related to me such that there is a possible world in which those sensations exist and are not so related to me.
(3) There is no possible world in which my sensations exist without being mine (justified by introspection).
(4) Therefore, I am not a physical object and my sensations are internally related to me.
(5) If a sensation is internally related to me, then it is a mode of my self.
(6) If an entity x is a mode of some entity y, then x is an inseparable entity dependent for its existence on y such that (a) x is modally distinct from and internally related to y and (b) x provides information about the nature of the thing y of which it is a mode.
(7) Therefore, I am a thing whose nature is to have sensations (and other states of consciousness).

Stewart Goetz and Geoffrey Madell have advanced versions of Argument 1.[13]

Argument 2: Personal identity at and through time is primitive and absolute. Moreover, counterexamples exist that show that the various bodily or psychological (e.g., memory) conditions proffered for personal identity are neither necessary nor sufficient. Put linguistically, talk about persons is not analyzable into talk about their bodies or connected mental lives. Furthermore, the primitive unity of consciousness cannot be accounted for if the self is a bodily or physical mereological compound. These facts are not innocuous but, rather, have important metaphysical implications. Substance dualism, according to which the soul is taken as a substance with an essence constituted by the potential for thought, belief, desire, sensation, and volition, is the best explanation of these facts. Different versions of this argument have been advanced by Richard Swinburne and William Hasker.[14] Some non-reductive physicalists who advocate a material composition view of human persons have offered responses to some of these points.[15]

Argument 3: The indexicality of thought provides evidence for the truth of substance dualism. A complete, third-person physical description of the world will fail to capture the fact expressed by "I am J. P. Moreland." No amount of information non-indexically expressed captures the content conveyed by this assertion. The first-person indexical "I" is irreducible and ineliminable, and this feature of "I" is not innocuous, but rather, is explained by claiming that "I" refers to a nonphysical entity—the substantial self with at least the power of self-awareness. Moreover, if mental predicates are added to the third-person descriptive language, this still fails to capture the state of affairs expressed by statements like "I am thinking that P." Finally, the system of indexical reference (e.g., "I," "here," "there," "this," "that") must have a unifying center that underlies it.[16] This unifying center is the same entity referred to by "I" in expressions like "I am thinking that P," namely, the conscious substantial subject taken as a self-conscious, self-referring particular.[17] More formally:

(1) Statements using the first-person indexical "I" express facts about persons that cannot be expressed in statements without the first-person indexical.
(2) If I am a physical object, then all the facts about me can be expressed in statements without the first-person indexical.
(3) Therefore, I am not a physical object.
(4) I am either a physical object or an immaterial substance.
(5) Therefore, I am an immaterial substance.

Geoffrey Madell and H. D. Lewis have advocated this type of argument.[18]

Argument 4: Some have argued for substance dualism on the grounds that libertarian freedom is true, and either a necessary condition for libertarian freedom is substance dualism or the latter is the best explanation for the former. The argument may be put this way (using only the form in which substance dualism is a necessary condition for libertarian freedom):

(1) Human beings exercise libertarian agency.
(2) No material object (one which is such that all of its properties, parts, and capacities are at least and only physical) can exercise libertarian agency.
(3) Therefore, human beings are not material objects.[19]
(4) Human beings are either material objects or immaterial substances.
(5) Therefore, they are immaterial substances.

Substance dualist John Foster has employed this sort of argument.[20]

Argument 5: A modal argument for substance dualism has been advanced by Keith Yandell and by Charles Taliaferro, and while it comes in many forms, it may be fairly stated as follows:[21]

(1) If x is identical to y, then whatever is true of x is true of y and vice-versa.
(2) I can strongly conceive of myself as existing disembodied or, indeed, without any physical particular existing.
(3) If I can strongly conceive of some state of affairs S that S possibly obtains, then I have good grounds for believing of S that S is possible.

(4) Therefore, I have good grounds for believing of myself that it is possible for me to exist and be disembodied.

(5) If some entity x is such that it is possible for x to exist without y, then (i) x is not identical to y and (ii) y is not essential to x.

(6) My physical body is not such that it is possible for it to exist disembodied or without any physical particular existing.

(7) Therefore, I have good grounds for believing of myself that I am not identical to a physical particular, including my physical body, and that no physical particular, including my physical body, is essential to me.

A parallel argument can be developed to show that possessing the ultimate capacities for consciousness is essential to one's self. Issues in the physical sciences have virtually nothing at all to do with stating and resolving the issues that constitute the core of the debate about substance dualism.

In a way, the property or substance dualist is in a dialectical disadvantage because he takes his view to be obvious in light of first-person introspection. Thus, many dualist arguments, e.g., the Knowledge Argument for property dualism or the Simple Argument for substance dualism, involve thought experiments that point to our direct knowledge of mental entities, and the dualist invites others to attend to what he believes is a matter of commonsense knowledge.[22] The dualist will likely agree with Searle's remark that if one is unwilling to admit that one is conscious, one needs therapy, not an argument.[23]

In a similar manner, an advocate of the Autonomy Thesis is in a dialectical disadvantage. He takes the thesis to be fairly obvious and invites others to attend to the actual dialogical issues as they pepper the pages of literature in philosophy of mind, believing that one will simply be able to see that those issues are largely philosophical and not scientific. This is precisely what I have tried to do in this section. Rather than elaborate further on this dialectical situation, I shall instead turn to a consideration of two prominent counterarguments offered by Nancey Murphy to the Autonomy Thesis.

3. Response to Two Counterarguments

First, Murphy claims that while substance dualism cannot be proven false, nevertheless "biology, neuroscience, and cognitive science have provided accounts of the dependence on physical processes of *specific* faculties once attributed to the soul."[24] For Murphy, "science has provided a massive amount of evidence suggesting that we need not postulate the existence of an entity such as a soul or mind in order to explain life and consciousness."[25] Thus, since advances in science have provided detailed accounts of mental/physical dependencies that make postulation of the soul otiose, the Autonomy Thesis is false, at least in this case.

I offer three responses. First, many substance dualists do not believe in a substantial ego primarily because it is a theoretical postulate with superior explanatory power. Rather, they take the ego to be something of which people are directly aware. The point is not that dualists are right about this. Given this dualist approach, the point is that advances in our knowledge of mental/physical dependencies are simply beside the point. And the further debate about which approach is the fundamental one for defending substance dualism is not something for which advances in scientific knowledge are relevant.

Second, in those cases where substance dualism *is* postulated as the best explanation for a range of purported facts, typically those facts are distinctively philosophical and not the scientific ones Murphy mentions. Arguments from the unity of consciousness, the possibility of disembodied survival or body switches, the best view of an agent to support agent causation, and the metaphysical implications from the use of the indexical "I" (see above) are typical of arguments offered by substance dualists, and the facts Murphy mentions are not particularly relevant for assessing these arguments. Those and related scientific facts (e.g., split-brain phenomena) may provide difficulties for certain versions of substance dualism, but they are not decisive—dualists have provided reasonable responses to them—and, in any case, they are less important than the philosophical issues mentioned above.

Finally, the discovery of "the dependence on physical processes of *specific* faculties once attributed to the soul" does not provide sufficient grounds for attributing those faculties to the brain rather than to the soul. (After all, are dualists supposed to think that mental/physical correlations or causal relations are vague and unwieldy and not specific and regular?) To see this, it is important to get clear on the use of "faculty" as the term has been historically used in discussions of substances in general, and the soul in particular.[26] Roughly, a faculty of some particular substance is a natural grouping of resembling capacities or potentialities possessed by that thing. For example, the various capacities to hear sounds would constitute a person's auditory faculty. Moreover, a capacity gets its identity and proper metaphysical categorization from the type of property it actualizes. The nature of a capacity-to-exemplify-F is properly characterized by F itself. Thus, the capacity to reflect light is properly considered a physical, optical capacity. For property dualists, the capacities for various mental states are mental and not physical capacities. Thus, the faculties that are constituted by those capacities are mental and not physical faculties.

Now, arguably, a particular is the kind of thing it is in virtue of the actual and potential properties/faculties essential and intrinsic to it. Thus, a description of the faculties of a thing provides accurate information about the kind of particular that has those faculties. For example, a description of the (irreducible) dispositions of gold provides us with information about the sort of thing gold is.

A description of a particular's capacities/faculties is a more accurate source of information about its nature than is an analysis of the causal/functional conditions relevant for the particular to act in various ways. The latter can either be clues to the intrinsic nature of that particular or else information about some other entity that the particular relates to in exhibiting a particular causal action.

For example, if Smith needs to use a magnet to pick up certain unreachable iron filings, information about the precise nature of the magnet and its role in Smith's action does not tell us much about the nature of Smith (except that he is dependent in his functional abilities on other things, e.g., the magnet). We surely would not conclude that the actual and potential properties of a magnet are clues to Smith's inner nature.

Similarly, functional dependence on causal relations to the brain are of much less value in telling us what kind of thing a human person is than is a careful description of the kind-defining mental capacities—i.e., faculties—human persons as such possess. In this case, various forms of physicalism and substance dualism are empirically equivalent theses, and in fact there is no non-question-begging theoretical virtue (e.g., simplicity) that can settle the debate if it is limited to being a scientific debate. But it should not be so limited and, indeed, paradigm case substance dualists such as F. R. Tennant approached the subject of the nature of the self and its relationship to faculties from a distinctively first-person introspective point of view. The choice to side with Murphy over against Tennant cannot be made on the basis of detailed scientific correlations. Rather, it must be made on the basis of factors such as one's evaluation of the strength of first-person awareness of the self and its conscious life.[27]

Murphy's second argument is that we should take physicalism not merely as a philosophical thesis, but primarily as the hard core of a scientific research program. If we look at physicalism, not as a philosophical thesis, but as a scientific theory, then there is ample scientific evidence for it.[28]

If one follows Murphy's advice, then the Autonomy Thesis will have to be set aside. But for at least two reasons, I think Murphy's recommendation is ill-advised. For one thing, it is unclear how physicalism in any of its forms is actually used as the "hard core of a scientific research program" in a way relevant to debates in philosophy of mind. To see this, it will be helpful to get before us some important points made by Alvin Plantinga and Bas C. van Fraassen.

Plantinga contrasts Duhemian and Augustinian science, derived, respectively, from the ideas of Pierre Duhem and St. Augustine.[29] According to Duhem, religious and, more importantly, metaphysical doctrines have often entered into physical theory. Many scientists have sought explanations of the phenomena (the appearances) in terms of underlying material causes. A proffered characterization of those causes often employs divisive metaphysical commitments, as when Aristotelians, Cartesians, and atomists gave disparate accounts of the phenomenon of magnetism.

If the aim of physical theory is to explain phenomena in terms of the ultimate nature of their causes, says Duhem, then physical science becomes subordinate to metaphysics and is no longer an autonomous science. Thus, estimates of the worth of a physical theory will depend upon the metaphysics one adopts. When practitioners of an area of physical science embrace different metaphysical schemes, progress is impeded because there is a compromise in the cooperation needed for progress. Successful science, if it is to be common to all, should not employ religious or metaphysical commitments only acceptable to some, including theism or physicalist naturalism. For Duhem, it is not the absence of metaphysics as such that serves the prudential interests of science, but of metaphysical views that divide us.

Augustinian science stands in contrast to Duhemian science. An Augustinian approach to science eschews methodological naturalism and employs religious or metaphysical commitments specific to a group of practitioners, commitments not widely shared throughout the scientific community. Augustinian science sanctions the use of scientific data to justify a religious or metaphysical proposition specific to a group of practitioners.

According to Plantinga, Duhemian science will not "employ assumptions like those, for example, that seem to underlie much cognitive science. For example, it could not properly assume that mind-body dualism is false, or that human beings are material objects; these are metaphysical assumptions that divide us."[30] More generally, the fact that there is a distinction between Duhemian and Augustinian science and that the former can be practiced at all seems to justify the Autonomy Thesis by showing that the progress of and data derived in accordance with Duhemian science are usually not of fundamental importance for resolving the deeper metaphysical issues that divide practitioners into different Augustinian camps.

Aspects of van Fraassen's philosophy of science lead to a similar conclusion. Van Fraassen argues that the theoretical postulates of a scientific theory typically go beyond the observational evidence, and that several different metaphysical characterizations are empirically equivalent.[31]

For van Fraassen, the primary goal of a scientific theory is to be empirically adequate, and acceptance of the unobservable metaphysical postulates of a theory is merely a pragmatic stance taken by advocates of a research program to continue searching for greater empirical adequacy.

This is what is actually going on when scientists employ physicalism as the hard core of a scientific research program. They are simply proffering either physically detectable operational definitions of mental states or are straightforwardly searching for physical correlates/causal relations for those mental states. There is not a single discovery in neuroscience (or cognitive science) that requires or even provides adequate justification for abandoning property or substance dualism, since the main issues in neuroscience and philosophy of mind conform to the Autonomy Thesis. The actual success of, say, neuroscience is strictly due to its Duhemian nature. This is why, in the last few decades, three Nobel Prize winners in neuroscience or related fields were a substance dualist (John C. Eccles), an emergent-property dualist (Roger Sperry), and a strict physicalist (Francis Crick). What divided them were philosophical differences, not differences about scientific facts.

In fact, in their contribution to this volume, Crick and Christof Koch acknowledge that one of the main attitudes among neuroscientists is that the nature of consciousness is "a philosophical problem, and so best left to philosophers."[32] This posture comports perfectly with Duhemian science. Elsewhere, they claim that "scientists should concentrate on questions that can be experimentally resolved and leave metaphysical speculations to 'late-night conversations over beer.'"[33] Methodologically, they set aside philosophical questions about the nature of consciousness and study the neural correlates and causal/functional roles of conscious states. If this is all it means to say that physicalism is "the hard core of a scientific research program," a dualist will heartily agree. In any case, such a Duhemian appropriation of physicalism underscores and does not provide a counterargument to the Autonomy Thesis.

The mistaken notion that progress in neuroscience requires an Augustinian commitment to physicalism as an essential component of that progress derives not from the actual physical facts of neuroscience or the actual way neuroscience is practiced, as evidenced by the Duhemian approach of Crick and Koch, but from the sociological fact that many contemporary neuroscientists just happen to be physicalists, and many people, including some philosophers, seem overly impressed with the cultural authority of science.

Second, when scientists study the causal correlates/functional relations between conscious states and the brain, they must rely on first-person reports about those states. To see this, consider the binding problem delineated by John Searle:

> I need to say something about what neurobiologists call "the binding problem." We know that the visual system has cells and indeed regions that are especially responsive to particular features of objects such as color, shape, movement, lines, angles, etc. But when we see an object we have a unified experience of a single object. How does the brain bind all of these different stimuli into a single, unified experience of an object? The problem extends across the different modes of perception. All of my experiences at present are part of one big unified conscious experience (Kant, with his usual gift for catchy phrases, called this "the transcendental unity of apperception").[34]

Scientists are seeking to find a region of the brain that "unifies" all the different stimuli that activate various parts of the brain. But exactly why would anyone think that such unification should be sought? Certainly not from an empirical investigation of the brain itself. Rather, we know from first-person introspection—in my view, of our own substantial selves and our conscious states—that all of our experiences are unified into one field of consciousness and, in fact, are possessed by one unified I, and it is on the basis of this knowledge that the scientific research program is justified and motivated. Moreover, William Hasker has argued that the phenomena that underlie this research are best explained by (emergent) substance dualism.[35] Whether Hasker is right or not is itself a philosophical matter that illustrates the Autonomy Thesis.

Given that (1) substance/property dualism are widely acknowledged to be the commonsense positions based on first-person introspection, and (2) the task of arguing for or against dualism, so grounded, is a philosophical one, and (3) neuroscientific research must rely on first-person introspective reports, the Autonomy Thesis seems to capture adequately the role of pre-philosophical intuitions and distinctively philosophical issues in neuroscience. The debate between dualists and physicalists is not about scientific facts. It is about things such as the status of first-person introspection as a source of justification for commonsense beliefs about the self and consciousness, the status of philosophical knowledge, and the proper philosophical interpretation of the role of physicalism in scientific research.

The truth of the Autonomy Thesis is what philosophers should have expected all along, and it constitutes philosophical self-understanding throughout the history of philosophy up to and including the present. In his 1886 lectures on the limitations of scientific materialism, John Tyndall claimed that "The chasm between the two classes of phenomena" is of such a nature that we might establish empirical association between them, but it

> . . . would still remain intellectually impassable. Let the consciousness of love, for example, be associated with a right-handed spiral motion of the molecules in the brain, and the consciousness of hate with a left-handed spiral motion. We should then know when we love that the motion is in one direction, and when we hate that the motion is in the other; but the "WHY" would remain as unanswerable as before.[36]

Little has changed since Tyndall made this remark. Specifically, no advance in knowledge of the specificity of detail regarding the correlations between mental and physical states provides any evidence against dualism or, more importantly, against the Autonomy Thesis. When philosophers discuss topics in philosophy of mind, they do not employ specific information in the hard sciences, because that information is not relevant to their issues. In evaluating functionalism, it does not matter if one claims that a functional state is realized by brain-state alpha or by a more detailed description of the relevant brain state.

Scientific data play virtually no role at all in philosophy-of-mind literature. In fact, it is rare for a philosophical text in philosophy of mind to include any scientific information. The same cannot be said, however, of scientific discussions of topics in these areas. For example, after claiming to set aside philosophical issues in order to focus on the more important empirical issues, Crick and Koch's discussion of consciousness and neuroscience is literally teeming with philosophical claims about philosophical topics with which they, qua scientists, are inadequately equipped to deal. For example, they claim that "Philosophers, in their carefree way, have invented a creature they call a 'zombie,' who is supposed to act just as normal people do but to be completely *un*conscious. This seems to us to be an untenable scientific idea."[37]

Relatedly, in considering whether two people in a similar brain state would experience the same quale, they say that "One is therefore tempted to use the philosopher's favorite tool, the thought experiment. Unfortunately, this enterprise is fraught with hazards, since it inevitably makes assumptions about how brains behave, and most of these assumptions have so little experimental support that conclusions based on them are valueless."[38]

Crick and Koch have a poor grasp on the role of thought experiments in philosophical argumentation. (Does the Knowledge Argument advocate make assumptions about how brains work in the actual world?) Moreover, when compared to philosophical treatments of topics in philosophy of mind, the discussion by Crick and Koch illustrates an asymmetry between neuroscience and philosophy of mind and, therefore, the Autonomy Thesis: Scientists cannot adequately discuss the central topics in philosophy of mind without making substantive philosophical claims, but philosophers need not discuss scientific data to treat adequately these same philosophical issues.

If I am right about all this, then if someone is going to be a mind/body physicalist, he or she cannot appeal to science to justify that commitment. It may well be that in first-person introspection, one discovers oneself to be constituted by animality, or there may be overriding philosophical and theological arguments for physicalism, though I suspect that even these arguments are not the real reason for physicalism's popularity among many academics. Explaining why I have these suspicions must be left for another occasion, but I suspect this candid admission by Thomas Nagel is on the mark: Speaking of his own fear of religion, Nagel says, "It isn't just that I don't believe in God and, naturally, hope that I'm right in my belief. It's that I hope there is no God! I don't want there to be a God; I don't want the universe to be like that. My guess is that this cosmic authority problem is not a rare condition and that it is responsible for much of the scientism and reductionism of our time."[39] Obviously, both the philosophical/theological issues mentioned above and the task of assessing Nagel's admission cry out for further dialogue. But one thing seems clear. Whenever and wherever that dialogue takes place, it will be a nice illustration of the Autonomy Thesis.

Notes

1. Jaegwon Kim, "Lonely Souls: Causality and Substance Dualism," in *Soul, Body and Survival* ed. by Kevin Corcoran (Ithaca, NY: Cornell University Press, 2001), 30.
2. John Searle, *The Rediscovery of the Mind* (Cambridge, MA: MIT Press, 1992), xii.
3. Ibid., 3–4.
4. Nancey Murphy, "Human Nature: Historical, Scientific, and Religious Issues," in Warren S. Brown, Nancey Murphy and H. Newton Malony, *Whatever Happened to the Soul?* (Minneapolis: Fortress Press, 1998), 17. Cf. 13, 27, 139–43.
5. Ibid., 18.
6. See J. P. Moreland and Scott Rae, *Body and Soul* (Downers Grove, IL: 2000).
7. Paul Churchland orders the first half of his book *Matter and Consciousness* (Cambridge, MA: MIT Press, rev. ed., 1988) around these families of issues.
8. George Bealer, "On the Possibility of Philosophical Knowledge," in *Philosophical Perspectives 10: Metaphysics, 1996,* ed. by James E. Tomberlin (Cambridge, MA: Blackwell, 1996), 1.
9. Paul Churchland, *Matter and Consciousness,* chapters three and four.
10. See Richard Swinburne, *The Evolution of the Soul* (Oxford: Clarendon, rev. ed., 1997).
11. See Moreland and Rae, *Body and Soul.*
12. See William Hasker, *The Emergent Self* (Ithaca, NY: Cornell University Press, 1999).
13. Stewart Goetz, "Modal Dualism," delivered at the Midwestern Meeting of the Society of Christian Philosophers, March 9, 1996; Geoffrey Madell, *The Identity of the Self* (Edinburgh: Edinburgh University Press, 1981; cf., Madell, *Mind and Materialism* (Edinburgh: Edinburgh University Press, 1988), 103–25. Madell claims that argument one (and two and three below) support either a substantial, immaterial self or a view of the self in which it is taken to be an immaterial property of *being a self.* Madell opts for the latter. For a critique of Madell on this point, see J. P. Moreland, "Madell's Rejection of a Substantial, Immaterial Self," *Philosophia Christi* 1 (1999): 111–14. Space forbids a defense of these arguments, but it may be useful to clarify certain notions central to them, e.g., "being internally related to me," "an external relation." To begin with, let us take as primitive the notion of a constituent/whole relation. A constituent/whole relation obtains between two entities just in case one entity is in the other as a constituent. So understood, there are two main types of constituent/whole relations: the standard separable part/whole relation of mereology and the accidental or essential predication relation. When a whole has a part or an accidental or essential property, the part or property is a constituent in the whole. In the sense used here, when one entity is a constituent of a whole, it is internally related to that whole. By contrast, "an external relation" in this context is one which relates one entity to another without the former becoming a constituent of the latter. Thus, "to the left of" is an external relation in this sense. Next, it is important to clarify the notion of a mode. Here is a sufficient condition of some entity's being a mode of another entity: If, for some substance S and property P, S exemplifies P, then the state of affairs—S's exemplifying P (call it A)—is a mode of S. As such, the mode is dependent on S for its existence and is internally related to S. There is no possible world where A exists and S does not. Moreover, if at some time t, S exemplifies P, then at all times prior to t, S had the (first or higher order) potentiality to exemplify P. And part of what makes S the kind of substance it is, is its potentialities. Now the substance dualist argues that sensations (and other mental states) are modes of the substantial self. As such, they are constituted by kind-defining mental properties and, thus, their possessors are members of mental kinds.
14. Richard Swinburne, *The Evolution of the Soul*, 145–73; William Hasker, *The Emergent Self,* 122–46.
15. Cf. Peter van Inwagen, *Material Beings* (Ithaca, NY: Cornell University Press, 1990), especially chapters 2 and 9; "Dualism and Materialism: Athens and Jerusalem?" *Faith and Philosophy* 12 (October 1995): 475–488;

Lynn Rudder Baker, "Need a Christian Be a Mind/Body Dualist?" *Faith and Philosophy* 12 (October 1995): 489–504; Trenton Merricks, "A New Objection to A Priori Arguments for Dualism," *American Philosophical Quarterly* 31 (January 1994): 81–85; "The Resurrection of the Body and the Life Everlasting," in *Reason for the Hope Within,* ed. by Michael J. Murray (Grand Rapids: Eerdmans, 1999), 261–86); Kevin J. Corcoran, "Persons and Bodies: The Metaphysics of Human Persons," Ph.D. Dissertation, Purdue University, 1997.

16. I omit temporal indexicals like "now" and "then" because in my view, there are two primitive indexicals that cannot be reduced to or eliminated in favor of the other: "I" and "now." "Now" expresses an irreducible fact about temporal reality—presentness—and it implies an A series view of time. The fact that "I" and "now" are both primitive may have something to do with the fact that finite, conscious beings are intrinsically temporal entities.

17. See Madell, *Mind and Materialism,* 103–25.

18. See Madell, *The Identity of the Self;* H. D. Lewis, *The Elusive Self* (Philadelphia: Westminster Press, 1982).

19. If human beings exercise libertarian agency, then i) they have the power to initiate change as a first mover and ii) they have the power to refrain from exercising their power to initiate change and iii) they act for the sake of reasons as irreducible, teleological ends for the sake of which they act. It has been argued that these features of libertarian agency are not physical properties and powers and, thus, libertarian agents are not physical objects. See John Foster, *The Immaterial Self* (London: Routledge, 1991), 266–80; Grant Gillett, "Actions, Causes, and Mental Ascriptions," in *Objections to Physicalism,* ed. by Howard Robinson (Oxford: Clarendon, 1993), 81–100; J. P. Moreland, "Naturalism and Libertarian Agency," *Philosophy and Theology* 10, 1997): 351–81. But cf. Timothy O'Conner, "Agent Causation," in *Agents, Causes, & Events* (New York: Oxford, 1995), 178–80.

20. Foster, *The Immaterial Self,* 266–80. Cf. Tim O'Connor, *Persons & Causes* (New York: Oxford University Press, 2000); J. P. Moreland, "Timothy O'Connor and the Harmony Thesis: A Critique," *Metaphysica* 3, No. 2 (2002): 5–40.

21. Cf. Keith Yandell, "A Defense of Dualism," *Faith and Philosophy* 12 (1995): 548–66; Charles Taliaferro, "Animals, Brains, and Spirits," *Faith and Philosophy* 12 (1995): 567–81.

22. For a recent discussion of the Knowledge Argument, see J. P. Moreland, "The Knowledge Argument Revisited," *International Philosophical Quarterly* 43 (June 2003): 219–28. For an exposition and defense of the Simple Argument, see Stewart Goetz, "Modal Dualism: A Critique," in *Soul, Body & Survival,* ed. by Kevin Corcoran (Ithaca, NY: Cornell University Press, 2001), 89–104.

23. Searle, *The Rediscovery of the Mind,* 8–9.

24. Nancey Murphy, "Human Nature: Historical, Scientific, and Religious Issues," in Warren S. Brown, Nancey Murphy and H. Newton Malony, *Whatever Happened to the Soul?* (Minneapolis: Fortress Press, 1998), 17. Cf. 13, 27, 139–43.

25. Ibid., 18.

26. For example, see F. R. Tennant, *Philosophical Theology I: The Soul and Its Faculties* (Cambridge: Cambridge University Press, 1956), 1–138, especially pages 33–43.

27. The Autonomy Thesis and the epistemic authority of first-person introspective knowledge relative to scientific claims is powerfully woven into Edmund Husserl's practice of bracketing the world and proffering phenomenological descriptions of various intentional objects as experienced and of the intrinsic features of the various mental acts directed upon those objects. For a detailed description of a paradigm case of Husserl in this regard, see J. P. Moreland, "Naturalism, Nominalism, and Husserlian Moments," *The Modern Schoolman* 79 (January/March 2002): 199–216.

28. Nancey Murphy, "Nonreductive Physicalism: Philosophical Issues," in *Whatever Happened to the Soul,* 127–48.

29. Alvin Plantinga, "Methodological Naturalism," in *Facts of Faith and Science Vol. 1: Historiography and Modes of Interaction,* ed. by Jitse M. van der Meer (Lanham, Maryland: University Press of America, 1996), 177–221.

30. Ibid., 209–10.

31. Bas C. van Fraassen, *The Scientific Image* (Oxford: Oxford University Press, 1980); "To Save the Phenomena," in *Scientific Realism,* ed. by Jarrett Leplin (Berkeley: University of California Press, 1984), 250–59.

32. Francis Crick and Christof Koch, "Consciousness and Neuroscience," reprinted from *Cerebral Cortex* 8 (1998): 97–107.

33. Cf. John Horgan, "Can Science Explain Consciousness?" *Scientific American* (July 1994): 91.

34. John Searle, "The Mystery of Consciousness: Part I," *The New York Review of Books,* November 1995, 60–66. The quote is from page 64.

35. See William Hasker, *The Emergent Self* (Ithaca, NY: Cornell University Press, 1999), 122–46, 171–203.

36. John Tyndall, "Scientific Materialism," in his *Fragments of Science Vol. II.*

37. Francis Crick, Christof Koch, Ibid., 98.

38. Ibid., 104.

39. Thomas Nagel, *The Last Word* (New York: Oxford University Press, 1997), 130–31.

Part VII

Science, Ethics, and Religion

Introduction

This seventh and final section focuses on the prospects for naturalistic ethics and the relationship between science and religion. Evolutionary naturalism is examined as a foundation for ethics in contrasting essays, and the tenability of religious belief in light of modern science is then evaluated in two more essays, again to contrary conclusions.

In "Evolution and Ethics," **Michael Ruse** defends an evolutionary approach to ethics, by which he means that Darwinism provides an adequate explanation of the origin and structure of ethics and obviates the contention that morality needs a foundation. Ruse argues that the key breakthrough in this enterprise was the rise of sociobiology in the 1970s, with various models of kin selection and reciprocal altruism and the like, showing how Darwinian advantage could be gained by helping others. He goes on to consider whether a naturalistic thread in the history of ethics—encompassing such figures as Aristotle, Hobbes, and Hume—provides a means whereby Darwinism can give a proper grounding for metaethical justifications as well. This is *not* the case, he concludes; rather, substantive morality is an illusion: ethics completely lacks an objective foundation and Darwinian evolutionary theory leads to a moral skepticism, a kind of moral nonrealism. Nonetheless, in the final analysis, Ruse makes this concession to theism: if you stay with naturalism, then there is no foundation, and in this sense substantive ethics is an illusion; but if you are, for example, a Christian, then you should subscribe to some kind of natural law position, seeing morality as doing things not so much because God has ordered them, but because this is human nature as created by God, who wants us to be true to what we are. The question of whether a Darwinian ethicist of the kind Ruse represents could be a Christian is thereby answered in the affirmative: it is possible, though he does not himself adopt this position. What he does suggest is that natural law theory is the only defensible Christian position on ethics, and it cries out for a Darwinian foundation.

Dallas Willard sees things in a considerably different light, and argues strongly for a rejection of evolutionary ethics in his essay "Naturalism's Incapacity to Capture the Good Will." Willard begins with a discussion of how naturalism and other ethical aspects of human life are to be understood, since he thinks these topics are bound up in much present confusion. He then argues that intentions are the fundamental locus of moral value and that, as Kant advocated, the only unconditional good in the domain of moral values is a good will.

Part VII

Furthermore, he contends that the good will is something for which no consistent form of naturalism can give an account, or even comprehend.

The final two essays in the volume take a hard look at supernaturalism in the light of modern biology and cosmology. In "Naturalism, Science, and Religion," **Michael Tooley** argues that neither science nor rational argumentation lends support to religion, most especially theism; rather, he advocates a robust naturalism that he maintains should *not* be equated with physicalism or materialism. **William Lane Craig** takes an opposing view in "Theism Defended," arguing that traditional natural theology, supplemented by insights from modern science and mathematics, provides ample defense of the theistic worldview, and that naturalism is deeply problematic and implausible.

38

EVOLUTION AND ETHICS

MICHAEL RUSE

Ethics is an illusion put in place by natural selection to make us good cooperators.
—Michael Ruse and Edward O. Wilson, 1985

I want here to defend an evolutionary approach to ethics, by which I mean the nature and foundation of morality.[1] I shall not spend time critiquing other positions, but at the end of my discussion I shall say something about the relationship between my thinking and that of the Christian.

1. NORMATIVE ETHICS

Let us distinguish two levels in moral discourse: *normative* or *substantive ethics*, which deals with what one ought to do ("Love your neighbor as yourself"), and *metaethics*, which deals with why one ought to do what one ought to do ("God wants you to love your neighbor as yourself"). If one links evolution and normative ethics, then one is engaged in a naturalistic enterprise—one is trying to show how and why people feel about moral statements. One is not judging the moral statements as such. Of course, within the system, one can discuss consistency and consequences—was the invasion of Iraq a morally wise action?—but, ultimately, one is in the business of description and scientific explanation.

There has been much work done in the past twenty years trying to show how Darwinism does explain (in the sense of showing the origins of) normative ethics.[2] The key breakthrough was the rise of *sociobiology* in the 1970s, with the various models of kin selection and reciprocal altruism and the like, showing how Darwinian advantage could be gained by helping others.[3] It was all a kind of enlightened self-interest on the part of the genes. "You scratch my back and I'll scratch yours." Naturally, thinkers differ on the precise details of the appropriate scientific approach. The bottom line is that all are attempting to explain normative ethics as the result of evolutionary processes, and by this is meant that natural selection of some kind is the chief causal force.

I have myself for at least two decades been arguing for such a naturalistic, evolution-based approach to normative ethics.[4] As a philosopher, I recognize that my contribution to the discussion at this level is necessarily limited. A naturalistic approach means just that—one puts oneself in the hands of the scientists. These would include primatologists, students of comparative cultures, game theorists, evolutionary psychologists, economists (perhaps), and others. All I will say here is that I find the results thus far very encouraging, although I am sure my critics would say that they would hardly expect me to find otherwise. But, you may properly ask, what is the content of a normative ethics for an evolutionist like me? One of the reasons why, in the past, evolutionary ethics tended to have a very bad reputation is that it seemed to have such vile implications. Given that natural selection depends on a bloody struggle for existence, most obviously, the normative ethics of an evolutionary ethicist is going to translate into some kind of laissez-faire socioeconomics—every person for themselves, and widows and children to the wall. So-called Social Darwinism. Listen to Herbert Spencer:

> We must call those spurious philanthropists, who, to prevent present misery, would entail greater misery upon future generations. All defenders of a Poor Law must, however, be classed among such. That rigorous necessity which, when allowed to act on them, becomes so sharp a spur to the lazy and so strong a bridle to the random, these pauper's friends would repeal, because of the wailing it here and there produces. Blind to the fact that under the natural order of things, society is constantly excreting its unhealthy, imbecile, slow, vacillating, faithless members, these unthinking, though well-meaning, men advocate an interference which not only stops the purifying process but even increases the vitiation—absolutely encourages the multiplication of the reckless and incompetent by offering them an unfailing provision, and *discourages* the multiplication of the competent and provident by heightening the prospective difficulty of maintaining a family.[5]

Well, leaving aside historical issues—see my book *The Evolution-Creation Struggle*[6] for a discussion where I try to put Spencer into more context—let me say flatly that I do not see that today's understanding of the workings of natural selection implies anything like this at all. The whole point of today's approach is reciprocation—in this life, often you get much further by cooperating than by fighting, and this is especially true of a social animal like us humans. You are out hunting and see an antelope. Better to combine forces and hunt as a pair or a group. Better to have sentiments of giving, rather than unenlightened selfishness. Of course, this means sharing the spoils, but as the old saying goes, a bird in the hand is worth two in the bush. Does this not all lay us open to cheating? Certainly it does, but again humans are good at detecting cheating and punishing transgressors. The person who cheats gets spotted and isolated. Another old saying: honesty is the best policy.

Are there any maxims of the evolutionary ethicist, like: "Love your neighbor as yourself"? My sense is that the evolutionary ethicist can and will appropriate all of the usual general maxims! Love your neighbor. Be kind and tolerant. Give 'til it hurts. Respect people even if you do not much like them. And so forth. You would expect this. A maxim like "Love your neighbor as yourself" is something generally accepted, even by nonbelievers like me—it is the kind of thing that humans agree on—and so it is precisely something like this that I expect natural selection to have promoted. And I think—and here I am referring to the empirical work of the biologists—it is precisely this kind of reciprocation maxim that selection produces.

Note that the maxim does not ask you to be a sucker. You expect a certain dignity and reciprocation. It means that you expect your neighbor to do good by you as you would to your neighbor, and as you would if you were in the neighbor's position—and conversely. Also, do note that I do not expect people always to obey the maxim. That is the point about ethics. We have a sense of obligation. This pushes us towards helping others. We are also selfish—too obvious a consequence of selection to need discussion. Life is a balance between the two. There are big advantages to being social animals, so we have features that help us to live in such a way, including normative morality. We also have features to look after our own self-interest. Sometimes, therefore, we do what we feel we should do. Sometimes we do not.

For myself, and I speak now as the father of five children, I think that fairness is a very important aspect of normative morality. We all know that children resent unfairness more than anything. But so do adults. I suspect we would all happily pay another penny on the income tax if we knew that all of the fat cats were likewise to pay their share of the tax. This, of course, is precisely the position of one of the leading ethicists of the last half century, namely John Rawls in his 1971 book *A Theory of Justice*. Rawls argues for a position that he calls "justice as fairness." He argues that, in order to be just, one ought to be fair. For Rawls, being fair does not necessarily mean giving everybody absolutely equal shares of everything. Rather, he invites us to put ourselves behind what he calls a "veil of ignorance," not knowing what position we might find ourselves occupying in society: whether we will be male or female, rich or poor, black or white, healthy or sick, or any of these things.

Then Rawls asks what position self-interest dictates as the best kind of society to find oneself in, and Rawls's answer is that it is a society where in some sense everybody does as well as one might possibly expect, given our various talents. It may well be that we will be born female and rich and powerful and healthy and so forth. If we knew we were going to find ourselves in that position, then we would want maximally to reward people in that position. But we may be male and poor and helpless, in which case we would lose out. So there is a kind of initial presumption of equality. Yet this is overthrown as soon as one recognizes that something like the availability of good medical care is going to be of benefit to everybody. And if, in fact, the only way that you can get the most talented people to become doctors is by paying them more than twice what you pay university professors, then "justice as fairness" dictates the propriety of this kind of inequality. So what Rawls ends up with is a society with inequalities, but in some sense a society where the inequalities benefit each and every individual in the group.

I would suggest that this is very much the kind of society that one expects evolutionary biology to have produced. That is to say, a group of people who think that one ought to be just, meaning that one ought to be fair. A group that will recognize that there will be inequalities, but will also recognize that these inequalities will in some sense be of benefit to all. I am not suggesting that every actual society has turned out like this, but that is not really the point. One recognizes there are going to be all sorts of ways in which biology will fail to match what the genes might dictate as best. There may be inequalities brought about by particular circumstance or fortune or whatever. But the point is how we think that a society ought to be, even if it is not necessarily always that way. That is to say, how we think morally that a society ought to be, even though it does not always work out that way.

Let me finish this discussion by noting that Rawls is promoting what is generally known as a "social contract" theory of morality. It is as if a group of people had got together and decided that they would mutually benefit from cooperation. As Rawls points out, the weakness of social contract theories—from that of Plato's *Republic* on—is that in actual fact it is highly unlikely that such a group meeting ever did occur. The biologist offers an alternative, much more plausible theory, namely that it was natural selection working on the genes that did the job. We are in a social contract theory because of biology. No actual planning was needed.

2. Metaethics

Let us turn now to foundations. Even those philosophers (like Rawls) who think that evolution can help with normative ethics tend to pull back here. It is one thing to turn normative ethics over to the empiricists. It is quite another to think that the results of empirical science can truly answer questions that are so fundamentally philosophical—so dear to the hearts of those of us who stand in the tradition of Plato and Aquinas and Kant. Let me rush in where angels fear to tread. There is another philosophical tradition to ethics—that of Aristotle, Hobbes, and Hume—where the natural world is considered relevant, all the way down (or up). I believe we do now have enough material to make some judgments and decisions at the metaethical level.

Biology—let us now agree, for the sake of argument, natural selection—has played some significant role in making us moral beings. Morality is an adaptation like hands and teeth and penises and vaginas. Obviously biology does not play the only role, and we must certainly allow culture some significant part also. How significant we can leave more or less open, between two false extremes—that everything is basically cultural (the blank-slate hypothesis) and that everything is basically biological (the genetic determinism hypothesis). The point is that morality has come through human evolution and it is adaptive. But what about justification? What kind of metaethical justification can one give for the love commandment or a Rawlsian justice-as-fairness? I would argue that ultimately there is no justification that can be given! That is to say, I argue that at some level one is driven to a kind of moral skepticism: a skepticism, please note, about foundations rather than about substantive dictates. What I am saying therefore is that, properly understood, the Darwinian approach to ethics leads one to a kind of moral nonrealism.[7]

In this respect, the Darwinian metaethics I am putting forward in this paper differs very dramatically from the approach taken traditionally by Darwinian metaethicists—especially those who used to fly under the banner of Social Darwinism. There, the foundational appeal was to the very fact of evolution. People like Herbert Spencer and his modern-day representatives, like Edward O. Wilson, argue that one ought to do certain things because by so doing one is promoting the welfare of evolution itself. Specifically, one is promoting human beings as the apotheosis of the evolutionary process—a move condemned by philosophers as a gross instance of the naturalistic fallacy, or as a flagrant violation of Hume's Law (that which denies that one can move legitimately from the way that things are to the way that things ought to be). My kind of evolutionary metaethics agrees with the philosopher that the naturalistic fallacy is a fallacy and so also is the violation of Hume's Law. My kind of evolutionary metaethics also agrees that Social Darwinism is guilty as charged. But my kind of evolutionary metaethics takes this failure as a springboard of strength to its own position. My Darwinian metaethics avoids the fallacy, not so much by denying that the fallacy is a fallacy, but by doing an end run around it, as it were. There is no fallacious appeal to evolution as foundations because there are no foundations to appeal to!

3. Objectification

To be blunt, my Darwinian metaethics says that substantive morality is a kind of illusion, put in place by our genes, in order to make us good social cooperators.[8] I would add that the reason why the illusion is such a successful adaptation is that not only do we believe in substantive morality, but we also believe that substantive morality does have an objective foundation. An important part of the phenomenological experience of substantive ethics is not just that we feel that we that ought to do the right and proper thing, but that we feel that we ought to do the right and proper thing because it truly is the right and proper thing. As John Mackie (1979) argued before me, an important part of the moral experience is that we objectify our substantive ethics.[9] There are in fact no foundations, but we believe that there are, in some sense, foundations.

There is a good biological reason why we do this. If, with the emotivists, we thought that morality was just simply a question of emotions without any sanction or justification behind them, then pretty quickly morality would collapse into futility. I might dislike you stealing my money, but ultimately why should you not do so? It is just a question of feelings. But in actual fact, the reason why I dislike you stealing my money is not simply because I do not like to see my money go, but because I think that you have done wrong. You really and truly have done wrong in some objective sense. This gives me and others the authority to criticize you. Substantive morality stays in place as an effective illusion because we think that it is no illusion but the real thing. Thus, I am arguing that the epistemological foundation of evolutionary ethics is a kind of moral nonrealism, but that it is an important part of evolutionary ethics that we think it is a kind of moral realism.

4. Spiritualism

In a way, what has been given thus far is just a statement rather than a proof. What justification can I offer for my claim that evolution points towards ethical skepticism (about foundations)? Why should one not say that there truly is a moral reality underlying morality at the substantive level, and that our biology has led us to it? After all, we would surely want to say that we are aware of the speeding train bearing down on us because of our biology, but this in no sense denies the reality of the speeding train.[10] Why should we not say, in a like fashion, that we are aware of right and wrong because ultimately there is an objective right and wrong lying behind moral intuitions?

However, things are rather different in the moral case from the speeding-train case. A more insightful analogy can be drawn from spiritualism. In the First World War, when so many young men were killed, the bereaved—the parents, the wives, the sweethearts, on both sides of the trenches—often went to spiritualists, hoping to get back in touch with the departed dead. And indeed they would get back in touch. They would hear the messages come through the Ouija boards, or whatever, assuring them of the happiness of the now deceased. Hence, the people who went to spiritualists would go away comforted. Now, how do we explain this sort of thing? Cases of fraud aside, we would say that people were not listening to the late departed, but rather were hearing voices created by their own imaginations, voices that were in some sense helping them to compensate for their loss. What we have here is some kind of individual illusion brought about by powerful social circumstances. No one would think that the late Private Higgins was really speaking to his mum and dad. Indeed, there are notorious cases where people were reported killed and then found not to be dead. How embarrassing it would be to have heard the late departed assure you of his well-being, and then to find out that the late departed was in fact lying injured in a military field hospital.

In the spiritualism case, once we have got the causal explanation as to why people hear as they do, we recognize that there is no further call for ultimate foundations. I would argue that the biological case is very similar. That there are strong biological reasons for cooperation; naturally, we are going to be selfish people, but as cooperators we need some way to break through this selfishness, and so our biology has given us morality in order to help us do it. I stress that this is not to say that we are always going to be moral people: in fact, we are an ambivalent mixture of good and bad, as the Christian well knows.[11] It is to say that we do have genuine moral sentiments that we think are objective, and that these were put in place by biology. Once we recognize this, we see the sentiments as illusory—although, because we objectify, it is very difficult to recognize this fact. That is why I am fairly confident that my having told you of this fact will not now mean that you will go off and rape and pillage, because you now know that there is no objective morality. The truth does not always set you free.

5. Progress

But still you might protest that what I have said does not mean that there is no objective morality behind all of this: either an objective morality of a Platonic ilk that actually exists out there, or an objective morality of the Kantian form that is a kind of necessary condition for rational beings getting along. Here, however, the Darwinian can come back with a further argument, namely one based on doubts about biological progress. There is no natural climb upwards from the blob up to the human, from the monad to the man, as people used to say in the nineteenth century. Rather, evolution is a directionless process, going nowhere rather slowly.[12] What this means in this particular context is that there is really no reason why humans might not have evolved in a very different sort of way, without the kind of moral sentiments that we have. From the Darwinian perspective, there is no ontological compulsion about moral thinking.

It is true that, as Kant stressed, it may possibly be that social animals may necessarily have to have certain formal rules of behavior.[13] But it is not necessarily the case that these formal rules of behavior have to incorporate what we would understand as commonsense (substantive) morality. In particular, we might well have evolved as beings with what I like to call the "John Foster Dulles system of morality," so named after Eisenhower's secretary of state during the Cold War in the 1950s. Dulles hated the Russians, and he knew that the Russians hated him. He felt he had a moral obligation to hate the Russians, because if he did not, everything would come tumbling down. But because there was this mutual dislike, of a real obligation-based kind, there was in fact a level of cooperation and harmony. The world did not break down into war and destruction. As a Darwinian, it is plausible to suggest that humans might have evolved with the John Foster Dulles kind of morality, where the highest ethical calling would not be love your neighbor, but to hate your neighbor. But remember that your neighbor hates you, and so you had better not harm him or her, because they are going to come straight back at you and do the same.

Now, at the very least, this means that we have the possibility not only of our own (substantive) morality, but of an alternative, very different kind of morality: a morality that may have the same formal structure, but which certainly has a different content. The question now is, if there is an objective foundation to substantive morality, which of the two is right? At a minimum, we are left with the possibility that we humans now might be behaving in the way that we do, but that in fact, what is objective morality is something quite different from what we believe. We believe what we do because of our biology, and we believe that because of our biology, our substantive morality is objectively justified. But the true objective morality is something other from what we have.

Obviously, this is a sheer contradiction to what most people mean by objective morality. What most people mean by objective morality incorporates the fact that it is going to be self-revealing to human beings. Not necessarily to all human beings but—like Descartes' clear and distinct ideas—certainly self-revealing to all decent human beings who work hard at it. So, given Darwinism, we have a refutation of the existence of such a morality. Darwinian evolutionary biology is nonprogressive, pointing away from the possibility of our knowing objective morality. We might be completely deceived, and since objective morality could never allow this, it cannot exist. For this reason, I argue strongly that Darwinian evolutionary theory leads one to a moral skepticism, a kind of moral nonrealism.

6. Christianity

I seem to have drenched the Christian altar pretty thoroughly! At this late stage, is it possible still to light it? Can I light it without the aid of Jehovah? Since I have long argued that a Darwinian can be a Christian, and since I have just presented my account of Darwinian ethics, I had better have a crack at the job.[14] The normative part is easy, because I am not subscribing to traditional Social Darwinism. I am promoting a normative ethics that has prescriptions like "Love your neighbor as yourself" right at its heart. I am onside with the Christian in this respect. I am onside also in arguing that the prescriptions present themselves to us as genuine prescriptions, and not just as rules of self-interest. I am also very much in line with the Christian in my claim that we do not always obey the call of morality, and that humans are often an unhappy mix of wanting to do good but not doing it.

It is the metaethics that seems troublesome. How can someone who says, almost proudly, that ethics is an illusion of the genes mesh with Christianity, a religion that puts obedience to God's word and will right at the heart? In fact, it is not as difficult as it seems, so long as you remember that I am offering a naturalistic account of ethics, and Christianity is a supernatural religion. I am saying that if you ask, "Take God out of the equation and can you still get ethics?" my answer is, "Yes, you can, if you are talking about normative ethics, but when you enter the metaethical realm you find that it is all biology and psychology, with no further meaning. The thought that there is something more—the thought needed to make normative ethics function—is an illusion."

But what if you now add God to the equation? If you are a Christian, I take it that you believe that, even though we humans and our ethics occurred by natural means, in some deeper sense we are all part of God's plan. He wanted us to appear and He wanted us to be as we are. We are made in God's image. How God could have made us appear is a matter of some debate (among those who take a Darwinian stance). I have argued against progress in selection-fueled evolution, but others (scientists) would disagree. Richard Dawkins thinks that progress occurs through competition between lines, leading to a kind of arms-race effect.[15] Simon Conway-Morris thinks that there are niches pre-existing, and intelligence just appears naturally at the end of the process.[16] Robert J. Russell argues that God puts direction in at the quantum level.[17] (Richard Dawkins is an atheist, whereas Simon Conway Morris and Robert J. Russell are Christians. But I do not see why the Christian should not avail him or herself of Dawkins's science.)

Personally, I would not go this route at all. I prefer a more Augustinian solution that sees God as outside time and where, for him, the thought of creation, the act of creation, and the product of creation are as one.[18] God just knew that the process of natural selection could and would lead to humankind. End of argument. As with any Augustinian argument there is here an odor of determinism, but that I will leave to the theologians. The point I would make is that I think that the appearance of humans, with the features we have, is compatible with a fairly conservative Christian theology. (In other words, I am not adopting something like Whiteheadian process theology, which strikes me as deeply anti-Christian.)

But now think about ethics—substantive ethics—against this God-given background. We do what we should do not simply because God has ordered it—we all know that there are notorious problems with simple-minded divine command theory—but because this is our nature as formed by God. We should not be cruel to small children because for humans such behavior is unnatural. We should love our neighbor as ourselves because we are social animals and God has designed us for this very situation. We are not male orangutans who swing through the jungle, only meeting up with females for a near-rape experience. Marriage is good for humans because we are male and female, we need love and support, and children are a good thing and they too need love and support. (For once, let us leave on one side issues like homosexuality and abortion.)

In other words, my position is that, if you stay with naturalism, then there is no foundation, and in this sense substantive ethics is an illusion. If you are a Christian, then you should subscribe to some kind of natural law position, seeing morality not so much as a matter of doing things because God has ordered them, but because this is human nature as created by God. He wants us to be true to what we are. (Natural law theory is the creation of Aquinas, out of Aristotle, hardly a surprise, given what a naturalistic philosopher Aristotle truly was.)

I am not a Christian. That is not the point. The question is whether a Darwinian ethicist of the kind I have been articulating and defending could be a Christian. My conclusion is that it is possible. More strongly, I think that the only really defensible Christian position on ethics—natural law theory—cries out for a Darwinian backing.

7. Conclusion

This, then, is my thinking about ethics. If you point out that, far from being very original, my whole position starts to sound very much like that of David Hume,[19] who likewise thought that morality was a matter of psychology, rather than reflection of non-natural objective properties, I shall take this as a compliment, not a criticism. I regard my position as that of David Hume—brought up to date via the science of Charles Darwin.[20] What better tradition could one have than that?

Notes

1. As I acknowledge in this paper, over the years I have written much on evolution and ethics. This account is indebted to several discussions, including one of my Gifford Lectures given in Glasgow in 2001 and one of the lectures I gave in 2004 as a guest of the Dai Woo Foundation in Seoul, Korea.
2. Sober and Wilson (1997); Wright (1994); Gibbard (1990); and Skyrms (1998); to take some examples.
3. Ruse 1985.
4. Ruse 1986; 1995; 2001a.
5. Spencer 1851, 323–324.
6. Ruse 2005.
7. Ruse 1986.
8. Ruse and Wilson 1985; 1986.
9. John Mackie 1979.
10. Nozick 1981.
11. Ruse 2001b.
12. Ruse 1993; McShea 1991.
13. Kant 1959.
14. Ruse 2001b.
15. Dawkins 1986.
16. Conway Morris 2003.
17. Russell *et al.* 1988.
18. McMullin 1986; 1996.
19. Hume 1978.
20. Darwin 1859.

References

Conway Morris, S. (2003) *Life's Solution: Inevitable Humans in a Lonely Universe.* Cambridge: Cambridge University Press.

Darwin, C. (1859) *On the Origin of Species.* London: John Murray.

Dawkins, R. (1986) *The Blind Watchmaker.* New York, NY: Norton.

Gibbard, A. (1990) *Wise Choices, Apt Feelings: A Theory of Normative Judgment.* Cambridge, MA: Harvard University Press.

Hume, D. (1978) *A Treatise of Human Nature.* Oxford: Oxford University Press.

Kant, I. (1959) *Foundations of the Metaphysics of Morals.* Indianapolis: Bobbs-Merrill.

Mackie, J. (1979) *Hume's Moral Theory.* London: Routledge and Kegan Paul.

McMullin, E. (1986) "Introduction: Evolution and Creation," in *Evolution and Creation,* E. McMullin, ed. Notre Dame: University of Notre Dame Press, 1–58.

———. (1996) "Evolutionary Contingency and Cosmic Purpose," in *Finding God in All Things.* M. Himes, and S. Pope, eds., New York: Crossroad, 140–61.

McShea, D W. (1991) "Complexity and Evolution: What Everybody Knows." *Biology and Philosophy* 6 (3): 303–25.

Nozick, R. (1981) *Philosophical Explanations.* Cambridge, MA: Harvard University Press.

Rawls, J. (1971) *A Theory of Justice.* Cambridge, MA: Harvard University Press.

Ruse, M. (1985) *Sociobiology: Sense or Nonsense?* Second ed. Dordrecht: Reidel.

———. (1986) *Taking Darwin Seriously: A Naturalistic Approach to Philosophy.* Oxford: Blackwell.

———. (1993) "Evolution and Progress." *Trends in Ecology and Evolution* 8 (2): 55–59.

———. (1995) *Evolutionary Naturalism: Selected Essays.* London: Routledge.

———. (2001a) "Altruism: A Darwinian Naturalist's Perspective," in *Altruism.* J. Schloss, ed., New York: Oxford University Press.

———. (2001b) *Can a Darwinian Be a Christian?* Cambridge: Cambridge University Press.

———. (2005) *The Evolution-Creation Struggle.* Cambridge, Mass.: Harvard University Press.

Ruse, M., and E. O. Wilson, (1985) "The Evolution of Morality." *New Scientist* 1478: 108–28.

———. (1986) "Moral Philosophy as Applied Science." *Philosophy* 61: 173–92.

Russell, R. J., W. R. Stoeger, and G. V. Coyne, eds. (1988.) *Physics, Philosophy, and Theology: A Common Quest for Understanding.* Vatican City: Vatican Observatory.

Skyrms, B. (1998) *Evolution of the Social Contract.* Cambridge: Cambridge University Press.

Sober, E., and D. S. Wilson. (1997) *Unto Others: The Evolution of Altruism.* Cambridge, MA: Harvard University Press.

Spencer, H. (1851) *Social Statics; Or the Conditions Essential to Human Happiness Specified and the First of them Developed.* London: J. Chapman.

Wright, R. (1994) *The Moral Animal: Evolutionary Psychology and Everyday Life.* New York: Pantheon.

39

Naturalism's Incapacity to Capture the Good Will[1]

Dallas Willard

One area that has been a problem for the naturalistic outlook has been the ethical. Judgments about who is a good or bad person, what is the right or wrong act, and what ought or ought not to be done have proven resistant, to say the least, to translation into or replacement by judgments about material or physical reality. Moral judgments frankly seem, on almost any reading, to be about something other than that reality. Conversely, one can say that naturalism (in the modern sense of the term) has presented a problem for morality, and has seemed to many to undermine any prospect of a moral basis for individual or collective human life.

But it is very difficult, I find, clearly to join the issue or issues involved here. My initial clarifications, in the effort to get at those issues, will be rather lengthy, so let me state my position at the outset in order that we can be clear about where we are going.

1. Four Understandings of Naturalism

There are four different understandings of "naturalism" that need to be kept in mind in a discussion of "naturalism and ethics." The first is the one that identifies the ethical course of life as one that is "according to nature." In this sense, Socrates, Plato, Aristotle, Epicurus, Epictetus, Aquinas, Butler, and Kant—and I—adhere to naturalism in ethics. Naturalism in this sense makes human nature the point of reference for the understanding of ethical (moral) reality, but it does not necessarily or usually restrict human nature to the categories of the sense-perceptible, quantifiable, or causal. This sense of "naturalism" can be set aside for present purposes, because "nature" in this more inclusive sense does not pose the issues for ethics (the reductive issues, we may say) that are of interest to us, and it may even stand in opposition to "naturalism" as most commonly understood today.

The second understanding of naturalism is defined in opposition to the "intuitionist" ethical theorists of the first half of the twentieth century: Moore, Prichard, Ross, and Ewing. Naturalism, here, is the view that there are no irreducibly moral properties. This is close to the "naturalism" at issue in this volume, but retains a certain distance nonetheless. This is because there might be no irreducible moral properties, and the properties to which moral properties reduce might still not be parts of "nature" as many would understand the term.

The third understanding of naturalism is closely related to the second. It attempts to take science as its point of reference and holds that ethical distinctions must, somehow, fall within or be accounted for from within the sciences and the domains of reality they deal with. There is a great deal of semantical and logical "wiggle" in this position. For example, one may or may not insist on the primacy of physics in such a way that naturalism becomes, in fact, "physicsism"—a contemporary form of scientism. And there are several ways of doing or not doing this. I shall argue that this third understanding of naturalism cannot be stated in such a way that it is of any use in philosophical or theoretical discussions.

The fourth understanding of naturalism amounts to straightforward materialism or physicalism. Sometimes the attempt is made to deduce materialism or physicalism from physicsism. If I am right, that really can't be done. So naturalism in this sense must certify itself rationally by engaging in *a priori* argument or "first philosophy." But it is hard to see any way to make its thesis consistent with that mode of self-justification. Its justification would have to come from the sciences to be self-consistent. But it really can't. In any case, I shall argue that the basic moral distinctions cannot be drawn from within the resources of a straightforward physicalism or materialism. That will be my main point.

If I am right about all of this, one might conclude that the discussion of "naturalism and ethics" was over and we could go home. But much will also depend on what we take the basic moral distinctions to be. And here, I believe, the disagreements are currently so deep that it is impossible to canvass the field in any really illuminating manner—at least I could not begin to do it in the space (and possibly intelligence) available here. So I will take the course of simply explaining and historically locating what I take "the basic moral distinctions" to be—they will have to do with will and intention, on my view—and then explaining why I think straightforward physicalism is incapable of coming to terms with these distinctions from within its resources.

2. Why "Scientific" Naturalism is Useless to Philosophy

First, let me explain why I think the third understanding of naturalism cannot be stated in such a way that it is of any use in philosophical or theoretical discussions.

Problems with Invoking "Science"

Methodological monism is an enduring aspect of generic naturalism, and modern naturalism is often specified simply in terms of an exclusive application of "scientific method" in all inquiries. But how can that method support claims about the nature of reality as a whole? For example, one might state that the only realities are atoms (quarks, strings, etc.) and derivatives thereof. But how would one support this claim? It certainly cannot be derived from any specific science (physics, chemistry, on up to, say, anthropology) or from any conjunction of specific sciences. And it is not to be derived through any application of experimental techniques within any science.

The naturalist must then have recourse to that popular but philosophically suspect abstraction, "science" itself, which says even less than the individual sciences about the nature of reality as a whole, because it says nothing at all. It isn't the kind of thing that can say anything, though many individuals—usually, I think, not themselves scientists, and certainly not scientists expressing truths within the competence of their profession—present themselves as speaking for science, and thus as being "scientific" in some extended but, hopefully, still authoritative sense.

John Searle seems to be in this position. He speaks of "our scientific view of the world," which, according to him, every informed person with her wits about her now believes to be true. He speaks of a view of the world which includes "all of our generally accepted theories about what sort of place the universe is and how it works."[2] "It includes," he continues, "theories ranging from quantum mechanics and relativity theory to the plate tectonic theory of geology and the DNA theory of hereditary transmission," etc. We might imagine a very long conjunctive sentence—containing the specific theories he has in mind as conjuncts—that would, supposedly, express the "worldview" in question.

But this will hardly do what he wants. One thing that will not show up in such a conjunctive sentence is any claim about reality as a whole, "the universe," or about knowledge in general. Such specific scientific theories as those just mentioned—and no matter how many of them we may list—cannot provide an ontology. They never even attempt to determine what it is to exist or what existence is, and cannot, by the nature of their content, provide an exhaustive list of what ultimate sorts of things there are. Their existential claims are always restricted to specific types of entities as indicated in their basic concepts. This we might have known at least since Aristotle.

I emphasize the point that to suppose that a given scientific theory or conjunction of such theories provides an ontology constitutes a logical mistake, a misreading of what the theories say and imply. Those theories, and the bodies of knowledge wherein they are situated, actually say nothing whatsoever about the universe or about how it—the whole "thing"—works. This is a merely semantical point about the meaning or logical content of the claims or sentences that make up the sciences. It is to be established or refuted by examining, precisely, those claims and sentences. It turns out that they do not even mention the universe, the totality of all that exists, nor do they say anything about the boundaries of knowledge in general. Such matters simply do not fall within the purview of their methods or findings. They all tacitly specify a delimited "universe of discourse" by the basic concepts they employ.

In support of this claim we ask: Could anyone possibly find the place in some comprehensive and duly accredited scientific text or treatment, or some technical paper, where it is demonstrated or even necessarily assumed by the science concerned that all that exists consists of particles or fields or strings—or of language, culture and "meanings"—or whatever the proper subject matter of the science is? Would anyone be able to mention the name of the physicist who established this as an "obvious fact of physics"?[3] Exactly where in the "atomic theory of matter" is the claim about what "the universe consists entirely of" to be found?

"After all," Searle rhetorically asks, "do we not know from the discoveries of science that there is really nothing in the universe but physical particles and fields of forces acting on physical particles?" The answer, contrary to his assumption, is surely, "No, we do not." Again, could he possibly just point out when, where, how, and by whom this "discovery of science" was made. Has it actually been made?

Also, before the philosopher can use "the discoveries of science" he must determine what "science" says. But this is to reify science, to treat it as an entity that issues "results." Science, as already indicated, says nothing at all. Particular scientists do. Unfortunately, they also make unscientific statements. How can we tell when an individual scientist is making scientific statements ex cathedra, as it were, and "science" is therefore speaking, and when he is not? And can a "scientific" statement be false or perhaps illicitly derived and still be scientific?

If a scientific statement can be false or based on logical errors, then a scientific statement may be less than knowledge. How, then, could it be required that we accept such statements as a basis or framework for philosophical work? History shows that statements accepted as "scientific" have been both false and based on logical errors. Is the advocate of naturalism, then, one who works under an authority that may be and has been wrong? He himself would rarely, if ever, have the competence to do the scientific work and therefore must be taking the statements of "science" on authority. But authority is in fact one of the things we would expect naturalism to stand against. Historically it has done so, and that has been one of its virtues. How can it avoid resting on blind authority, however, if what Searle and others say is true? And is a philosopher's statement about science, a scientific theory, or a scientist, to be automatically regarded as itself scientific? What can its status be? Is it a "naturalistic" statement?

The words "science" and "scientific" frankly do not mean very much in many contexts where they are used. The old problem of "demarcation," discussed so intensely some decades ago, has not really gone away. A good rule might be to never use those words in premises intended to support a conclusion.

The Dilemma of Naturalism

Naturalism staggers back and forth between physicalism (materialism) as a general ontology or first philosophy, and outright physicsism or scientism (which need not take the form of physicsism)—often, though not always, trying to derive physicsism from scientism and then physicalism from physicsism. This continues up to the present.

In a recent review, Patricia Kitcher chides Stephen Stich for "philosophical Puritanism" when he takes naturalism to hold that the only real entities are physical.[4] Such a position apparently has now led Stich to give up naturalism "in favor of an open-ended pluralism." Pluralism, as he takes it, is a position that counts as legitimate all properties "invoked in successful scientific theories." But for Kitcher, it seems, such "pluralism," tied to "successful science," is just the naturalism we want. She points out how "the obvious authorities" on naturalistic epistemology (Quine, Goldman) counsel us to "make free use of empirical psychology" and to "reunite epistemology with psychology."[5] Forget physicalism, her point seems to be. A loose scientism is enough to secure naturalism for us. Indeed, many of the "generous" naturalists of the mid-twentieth century, gathered around Dewey and Sidney Hook, identified naturalism precisely with acceptance of science and only science as the arbiter of truth and reality, and they seemed, at least, to accept whatever came out the end of the pipe of "scientific inquiry" as knowledge and reality.

But if the points made above about science, even "successful science," and about psychology in particular, are true, Kitcher's advice—similar to the advice of a Dewey or Hook—simply cannot be followed. It is vacuous in practice, for there is no way of identifying and accessing the "successful science" which is proposed as defining naturalism. At most, you get "science now," which is really only "some scientist(s) now." And certainly no science (including psychology) that was not naturalistic in some strongly physicalistic or at least empiricist sense would be accepted as "successful" by those inclined to naturalism. Then we are back in the circle: naturalism in terms of science—but, of course, naturalistic science.

For these reasons I take it that the appeal to science cannot serve to specify naturalism. There are, then, good reasons to be a "Puritan" if you want to advocate naturalism. Naturalism has to be an honest metaphysics; and that metaphysics has to be "unqualified physicalism," as mentioned above. But then a thinker who would be naturalist would feel pressure to resort to some specific *a priori* analyses to render his ontological specification of naturalism plausible. Short of that, one simply can find no reason why naturalistic monism with respect to reality, knowledge, or method should be true: no reason why there should not be radically different kinds of realities with correspondingly radically different kinds of knowledge and inquiry. Why, *a priori,* should one suppose the sciences could be "unified"? And why should we think that the identifiable sciences together could exhaust knowledge and reality? It is simply a hope that some people have shared. But hopes often lack reason. The lack of reason in the case at hand is, I think, what made A. E. Murphy conclude at the mid-century, in his review of a very important collection of essays at the time, *Naturalism and the Human Spirit,* "that the naturalists, who have so much that is good to offer, still lack and need a philosophy."[6]

In addition to the difficulty of coming up with the required *a priori* analyses, however, to turn to such inquiry as might produce them would (as I have already indicated) be to break with the epistemological monism essential to naturalism and introduce something like a "first philosophy." This would be discontinuous with the empirical methods of the sciences. In showing its justification through *a priori* analysis, naturalism would simply give up the game.

In specifying what naturalism is, therefore, one seems to be faced with an inescapable dilemma. Either one must turn to *a priori* (nonempirical and extra-scientific) analyses to establish its monism (which will refute naturalism's basic claim about knowledge and inquiry), or its claim will have to rest upon a vacuous appeal to what "science" says.

That might seem to end the discussion about naturalism as a philosophical alternative. But there may be a way to keep it going. One could retreat to a mere methodological naturalism and say that scientific method—identified somehow—is our only hope as human beings. Whether or not we can adequately specify naturalism or know it to be true, one might say, the "scientific method" must be exclusively followed for the sake of human well-being. Naturalism would then be a humane proposal, not a philosophical claim—and that would, in fact, do justice to a great deal of its history. (Not infrequently one picks up the suggestion that one ethically ought to be a naturalist, that the "right kind of people" are naturalists. And likewise for the other side, that the right kind of people are anti-naturalists.) The proposal would be to assume in our inquiries that only the physical (or the empirical) exists and to see if inquiry based upon that assumption is not more successful in promoting human ends (and hence is more morally praiseworthy) than any other type of inquiry. We would not need to insist that non-naturalistic explanations of some or all events are impossible, just that they are not possibilities that need to be seriously considered in scientific work. (As Steven Weinberg has decided not to worry about fairies.)

But even if we regard naturalism as merely a humane proposal, we must still raise the issue of whether straightforward physicalism (the only version of naturalism that makes sense) can deal with ethical phenomena or provide an adequate interpretation of the moral life and moral principles.

3. Naturalism and Ethics

So now we turn to the other side of the naturalism/ethics contrast. What are the distinctions—with the corresponding properties and relations—that naturalism (as physicalism) would have to account for if it were to encompass the field of ethics successfully? This question requires me to take a position on issues that run very deep in the philosophical understanding of the moral life and of the evaluations of various kinds that accompany it and direct it. To prepare the way for that I need to make some observations about where we stand today in our usual approach to ethical theory.

I think it is not easy to understand why we undertake ethical theory as we now usually do: why we start with the questions we do and confine the discussion in the manner now characteristic in ethical thought. Historical process, it seems to me, has much more to do with all this than does insight into moral phenomena—including moral judgments or, if you like, moral utterances—with their distinctive features.

Since Hume, at least, discussions in ethical theory have been driven by epistemology: by views concerning what can and cannot be known, and how knowledge must work in particular areas. Not, of course, that reflections on ethical matters before Hume had nothing essentially to do with epistemological considerations. The two areas have always been intertwined. But, in Hume and afterwards, epistemological considerations became the dominant (though often silent) ones in determining what the moral judgment is and what it is about. Hedonistic utilitarianism (Bentham, Mill), for example, or emotivism could never have risen to prominence as a theory of the moral life had it not been for the assumption that value must be something empirical, something feelable. And that is, of course, precisely Hume's assumption, which took a lengthy period of time after Hume to develop into an assumption that could be relied upon in public discussions of what ought to be done and what is right or wrong.

The emergence of noncognitivism in ethical theory was, I believe, quite inevitable, given the ascendancy of empiricism to dominance in the theory of knowledge and the domination of ethical reflections by the theory of knowledge. Naturalism is the current reformulation of classical empiricism. One might easily suspect that if empiricism is the correct analysis of knowledge, there will certainly be no moral knowledge, because the substance of the moral life is not empirical. It is not something that is feeling or sensation or is of what can be felt or sensed.

The fateful alternative that G. E. Moore so innocently posed was bound to be resolved in the direction he thought unthinkable: "If it is not the case that 'good' denotes something simple and indefinable, only two alternatives are possible: either it is a complex, a given whole, about the correct analysis of which there may be disagreement; or else it means nothing at all, and there is no such subject as Ethics."[7] "Okay," the response was, "so there is no such subject." And with that, we have emotivism and noncognitivism.

Just a little more recent history of the field, as I perceive it: To be blunt, we have not really begun to recover from noncognitivism in the field of ethical theory, for all its frenzied activity. The growth industry of applied, professional, and social-issues "ethics" conceals the fact that matters of principle raised by noncognitivism have remained unbudged. The effort to dig out by means of the analysis of language really did not succeed; frankly, the multitude of claims made about ethical language are hardly more "empirical" or "naturalistic" than claims made about ethical phenomena themselves, and hardly more a matter of general agreement. It is not easy to find a good reason for not being an "emotivist" with reference to semantics and logic, if you have already accepted the grounds that drove people to emotivism in ethical theory.

Then (in the mid-sixties and following) the demands of social, legal, and professional life in this country forced the issue of the justification of various and scattered moral claims having to do with justice (primarily civil rights: discrimination, the draft, etc.). These claims, it was thought, had to have a cognitive right or wrong, "ought" or "ought not" to them. Had to. But the urgency of those demands did not resolve the underlying issue—namely, the issue of the nature of moral phenomena and knowledge thereof. Those issues are still hanging fire today, and that fact also makes the problem of understanding the connection between ethics (or the moral life) and naturalism difficult to state in any satisfactory way. There is very little agreement as to what it is we are trying to relate naturalism to when we try to relate it to the moral life and, by extension, to ethics.

A striking illustration of the obscurity about what the central moral phenomena are is provided, it seems to me, by the career of John Rawls. Possibly no work in the field of ethics received more attention during the last quarter century than did his *A Theory of Justice* (1971). This fact is inseparable from the just-noted emergence of the demands of social, legal, and professional issues in the sixties. Rawls, and others such as Robert Nozick, received the hearing they did because of the pressing need to determine what, morally, was right or wrong, obligatory or not, with reference to certain social and professional issues.

But by the mid-eighties or thereabouts, Rawls's view was "that his theory of justice is best understood as a political rather than as a moral doctrine—and as such is committed to no metaphysical theses. . . . Rawls draws a new distinction which is basic to his discussions: a distinction 'between a political conception of justice and a comprehensive religious, philosophical, or moral doctrine.' What he seeks to provide is a public philosophy which does not incorporate or amount to any comprehensive doctrine."[8]

One cannot but wonder what kind of mistake or shift this represents. How could one make a mistake like this? Apparently Rawls (and certainly multitudes of others) thought he was advancing a moral theory of justice, and then discovered that he was not, that the theory was not a moral theory of justice after all.

How could one be mistaken about something like that? I believe this is explained by how moral reflection was taken up after the noncognitivism eruption and the, to me at least, pretty obvious inability of the analysis of language, and of moral language in particular, to do anything to counter noncognitivism and provide a cognitive basis for ethical claims about social and "applied" issues. "Rights" talk was exempted from noncognitivive status by the force of social events, and philosophers stepped into the "rights" arena as if it were the arena of moral phenomena and ethical analysis. Rationality (understood in one way or another, but always "formally") then became the ultimate point of reference for the certification of specific normative claims, insofar as any certification was thought possible. The good person or society was to be the (in some sense) rational person or society, and the bad person (the Nazi is the set case) or the wrong action (discrimination, for example) must be irrational (in some formal sense). And if they are not irrational they are morally okay, at least. And rationality here was not a deep matter of some kind, such as one might find in Plato, Aristotle, or Kant. It was to be a matter of behavior, for most thinkers, and one in which it will, supposedly, make sense to discuss the question of whether and under what conditions a computer or computer program or robot could be rational or irrational.[9]

Moral phenomena are, accordingly, pushed into the public arena of the behavioral and the social. Needless to say, this is a definite advantage for anyone who wants to naturalize ethics—though by no means is that the end of the story. One can see here, I believe, the continued subservience of ethical theory to the "empiricist imperative," as we might call it. Richard Brandt in his *Ethical Theory* (1959) says: "The essential thesis of naturalism is the proposal that ethical statements can, after all, be confirmed, ethical questions answered, by observation and inductive reasoning of the very sort that we use to confirm statements in the empirical sciences. . . . [T]he meaning of ethical statements is such that we can verify them just like the statements of psychology or chemistry."[10] That is, they refer, in the end, to sense-perceptible or at least "feelable" facts (such as desire or pleasure or pain or social behavior). The appeal to rationality as the ultimate point of reference in moral judgment might with some justification be seen as the most recent effort to "save" moral phenomena from empiricism, currently called naturalism. By it, moral phenomena are completely externalized.

Well, but one might also say that this is only "saving" moral phenomena by abandoning them altogether and substituting something else. (Like "liberating" villages by destroying them.) A well-known statement by Elizabeth Anscombe from 1958 was "that it is not profitable for us at present to do moral philosophy; that it should be laid aside at any rate until we have an adequate philosophy of psychology, in which we are conspicuously lacking." She went on to say that "the differences between the well-known English writers on moral philosophy from Sidgwick to the present day are of little importance," and that we should try to stop using "right," "wrong," "ought," etc. in a moral sense, because they are derivative "from an earlier conception of ethics which no longer generally survives, and are only harmful without it."[11] I don't think anything that has happened since she published her paper in 1958 would have changed her view of the difficulties into which ethical theory has fallen.

These claims by Anscombe should be placed alongside the "Disquieting Suggestion" of chapter 1 of Alasdair MacIntyre's *After Virtue,* that "we have—very largely, if not entirely—lost our comprehension, both theoretical and practical, of morality."[12] Both what Anscombe has to say about "giving up" attempts to theorize about morality and the very language of morality until something (an "adequate philosophy of psychology"!) is developed, and what MacIntyre has to say about the loss of a functional language of morality, and therewith the power of moral concepts and principles to govern life (until "community" is somehow restored)—both of these points would make considerable sense if one simply assumed that our attempts to theorize and live the moral life had shifted away from its actual center or basic subject matter, objectively considered. And this might be understood as having happened because of the adoption of a theory of knowledge according to which the genuine organizational center of moral reality and moral phenomena is unknowable. That would leave us a choice between adopting explicit noncognitivism (which seems to me to be impossible in practice) or trying to deal with the moral life and ethical theory in terms of aspects of it ("rights," "justice," "professional ethics," "applied ethics," whatever) that are completely peripheral and therefore incapable of providing practical or theoretical unity to the moral life.

This, I think, is what has actually happened. Whether or not naturalism is compatible with ethics in this external and fragmentized sense is an issue that very likely cannot be clearly stated or resolved, and, in any case, it might not much matter. If we consider ethics and the moral life from another perspective, however, and one that is not constantly worrying about meeting the demands of the empiricist imperative, a clear issue can be joined with naturalism, and clearly resolved, and it can be seen that this is an issue that matters a great deal to the understanding and conduct of the moral life.

4. On Good and Evil in the Moral Realm

All of this is to suggest how the primary moral phenomena could have been lost from view and replaced by the externalities of actions and social structures and processes. What then should we say about the basic good and evil in the moral realm? "The external performance," Hume says, "has no merit. We must look within to find the moral quality. . . ."[13] I agree. For him, the moral distinctions fall between what he calls "qualities of mind." These are his virtues and vices. Not actions but the sources of action in the human system are the fundamental subjects of moral appraisal. Moral appraisal is not basically about what people do, but about what they would do, could do. What they actually do is, from the moral point of view, of interest primarily because it is revelatory of what they would or would not do, could or could not bring themselves to do, and therefore of their moral identity. (Of course, actions have interests and values other than moral ones.)

Hume never arrived at a unitary conceptualization of virtue (or vice) precisely because he tried to confine his investigation to an empirical survey. All he could come up with was: "Personal Merit [virtue] consists altogether in the possession of mental qualities, useful or agreeable to the person himself or to others."[14] But Kant was not so restricted, and he identifies the central moral phenomenon as the good will. This, he famously says, is the only thing good without qualification, good regardless of whatever else may be true. Again, I believe he was entirely correct about this. The good will is the primary moral phenomenon. Kant's efforts to characterize the good will in merely formal terms may have been less than spectacularly successful; but that is not the only way he characterizes it, and he insists in his doctrine of virtue that the good will has two *a priori* (nonempirical) ends: one's own moral perfection and the happiness of others. These are the material ends of the good will for Kant, imposing obligations in their own right.

I mention Hume and Kant not to enter into exposition of them, but simply to locate a broad tradition of ethical theorizing that locates moral value not in action but in the sources of action, and not in the formal features of moral experience, but in the material aims of action and dispositions organized around them. This is a tradition that reached a sort of maturity in the work of late nineteenth-century thinkers such as Sidgwick, Bradley, and especially T. H. Green, and I want to identify with that tradition. For the following one hundred years after these thinkers, this tradition has been paralyzed, if not killed off, by the effects of Moore and his followers and critics. It was a tradition that focused upon the will and the role of the will in the organization of the "ideal self." The "ideal self" was, of course, the good person, which everyone finds themselves obliged to be.

5. The Good Person: A Matter of the Heart

The morally good person, I would say, is a person who is intent upon advancing the various goods of human life with which they are effectively in contact, in a manner that respects their relative degrees of importance and the extent to which the actions of the person in question can actually promote the existence and maintenance of those goods.

The person who is morally bad or evil is one who is intent upon the destruction of the various goods of human life with which they are effectively in contact, or who is indifferent to the existence and maintenance of those goods.

Being morally good or evil clearly will be a matter of degree, and there surely will be few, if any, actual human beings who exist at the extreme ends of the scale. (An interesting but largely pointless question might be how humanity distributes on the scale: a nice bell curve or . . . what?)

Here, I submit, is the fundamental moral distinction: the one that is of primary human interest, and from which all the others, moving toward the periphery of the moral life and ethical theory, can be clarified. For example: the moral value of acts (positive and negative); the nature of moral obligation and responsibility; virtues and vices; the nature and limitations of rights, punishment, rewards, justice, and related issues; the morality of laws and institutions;

and what is to be made of moral progress and moral education. A coherent theory of these matters can, I suggest, be developed only if we start from the distinction between the good and bad will or person—which, admittedly, almost no one is currently prepared to discuss. That is one of the outcomes of ethical theorizing through the twentieth century.

I believe that this is the fundamental moral distinction, because I believe that it is the one that ordinary human beings constantly employ in the ordinary contexts of life, both with reference to themselves (a touchstone for moral theory, in my opinion) and with reference to others (where it is employed with much less clarity and assurance). And I also believe that this is the fundamental moral distinction because it seems to me the one most consistently present at the heart of the tradition of moral thought that runs from Socrates to Sidgwick—all of the twists and turns of that tradition notwithstanding.

Just consider the role of "the good" in Plato, Aristotle, and Augustine, for example, stripped, if possible, of all the intellectual campaigns and skirmishes surrounding it. Consider Aquinas's statement that "this is the first precept of law, that good is to be done and promoted, and evil is to be avoided. All other precepts of the natural law are based upon this; so that all the things which the practical reason naturally apprehends as man's good belong to the precepts of the natural law under the form of things to be done or avoided."[15] Or consider how Sidgwick arrives at his "maxim of Benevolence"—"that each one is morally bound to regard the good of any other individual as much as his own, except in so far as he judges it to be less, when impartially viewed, or less certainly knowable or attainable by him."[16] A few further clarifications must be made before turning to my final argument:

1. I have spoken of the goods of human life in the plural, and have spoken of goods with which we are in effective contact, i.e., can do something about. The good will is manifested in its active caring for particular goods that we can do something about, not in dreaming of "the greatest happiness of the greatest number" or even of my own "happiness" or of "duty for duty's sake." Generally speaking, thinking in high-level abstractions will always defeat moral will. As Bradley and others before him clearly saw, "my station and its duties" is nearly, but not quite, the whole moral scene, and can never be simply bypassed on the way to "larger" things. One of the major miscues of ethical theory since the sixties has been, in my opinion, its almost total absorption in social and political issues. Of course, these issues also concern vital human goods. But moral theory simply will not coherently and comprehensively come together from their point of view. They do not essentially involve the center of moral reality, the will.

2. Among human goods—things that are good for human beings and enable them to flourish—are human beings and certain relationships to them, and, especially, good human beings. That is, human beings that fit the above description. One's own wellbeing is a human good, to one's self and to others, as is what Kant called the moral "perfection" of oneself. Of course non-toxic water and food, a clean and safe environment, opportunities to learn and to work, stable family and community relations, and so forth, all fall on the list of particular human goods. (Most of the stuff for sale in our society probably does not.)

There is no necessity of having a complete list of human goods or a tight definition of what something must be like to be on the list. Marginal issues, "lifeboat" cases, and the finer points of conceptual distinction are interesting exercises and have a point for philosophical training; but it is not empirically confirmable, to say the least, that the chances of having a good will or being a good person improve with philosophical training in ethical theory, as that has been recently understood. It is sufficient to become a good or bad person that one have a good general understanding of human goods and how they are effected by action. And that is also sufficient for the understanding of the good will and the goodness of the individual. We do not have to know what the person would do in a lifeboat situation to know whether or not they have good will, though what they do in such situations may throw light on who they are, or on how good (or bad) they are. The appropriate response to actions in extreme situations may not be a moral judgment at all, but one of pity or admiration, of the tragic sense of life, or amazement at what humans are capable of, etc.

3. The will to advance the goods of human life with which one comes into contact is inseparable from the will to find out how to do it and do it appropriately. If one truly wills the end, one wills the means, and coming to understand the goods that we effect, and their conditions and interconnections, is inseparable from the objectives of the good person and the good will. Thus, knowledge, understanding, and rationality are themselves human goods, to be appropriately pursued for their own sakes, but also because they are absolutely necessary for moral self-realization. Formal rationality is fundamental to the good will, but is not sufficient to it. It must be acknowledged that one of the moral strong points of naturalism is its concern about advancing the goods of human life and about combating the forces of ignorance and superstition that work against those goods. One cannot understand naturalism as a historical reality, or as a present fact, if one does not take this point into consideration.

4. Thus, the morally good (or evil) will or person will necessarily incorporate the following elements at least:
    a. Consciousness, the various intentional states that make up the mental life.
    b. Knowledge of the various goods of human life and of their conditions and interconnections. This will include much knowledge of fact, but also logical relations, as well as the capacity to comprehend them to form hypothetical judgments and to reach conclusions on the basis of premises.
    c. The capacity to form and sustain long-range, even lifelong intentions. One is not a morally good person by accident or drift, but by a choice settled into character: a choice to live as a person who is intent upon advancing the various goods of human life with which they are effectively in contact, etc. The corresponding is true of a morally evil person. Intention—settled intention, or disposition—is the fundamental locus of moral value, deeper than will as a mere faculty (which does not by itself yield moral value) or as an act of will or choice (which is momentary, as character is not). It is this type of intention, worked into the substance of one's life, that is moral identity. And it is the moral identity of persons that naturalism would have to account for if it were successfully to accommodate the moral life and ethical theory.

6. The Argument—Finally

Can the moral identity of the good (or evil) person be captured within the categories of naturalism as physicalism? I believe it clearly cannot. The argument against it is an old and simple one.

Suppose that we have an acceptable list of physical properties and relations. We might take them from physical theory, as the properties and relations corresponding to the concepts of current physics: location, mass, momentum, and so forth. (Who knows what the future, or ultimate physics, will look like?) Or, moved by the above doubts about what philosophy can soundly derive from the sciences, we could turn to the "primary qualities" of modern philosophy, and, for that matter, add on the "secondary" ones as well: color, odor, etc. I don't think we need, for present purposes, to be very scrupulous about this list, either. Let us agree that whatever goes on such a list will count as physical properties, and that narrow naturalism is the proposal to confine our inquiries and conclusions to whatever shows up on the list and to combinations thereof.

The argument, then, is simply that no such physical property or combination thereof constitutes the basic components of the good will or person, such as intentionality, knowledge, choice, or the settled intentions that make up moral identity and character. At the simplest level, none of those properties or their combinations constitutes a representation of anything, or qualifies their bearer as being of or about anything. The properties of those properties (and of combinations thereof) are not the same as the properties of representations (ideas, thoughts, propositions, beliefs, statements), much less of intentions, decisions, and the permanent inclinations that make up character. If this is correct, and if the narrower naturalism admits only these "physical" properties, then there are no good or bad wills or persons in the world of the narrower naturalism.

Of course, if there are no representations, there is no knowledge or choice, and if there is no knowledge or choice, there are no settled intentions with reference to anything, much less the goods of human life. The logical relations required in thought, knowledge, and choice also will not show up in the world of naturalism. The ontological structure of the good will therefore cannot be present in the world of narrower naturalism—nor, for that matter, in the world of the actual sciences as now commonly understood.

Note that my claim is that such physical properties never constitute the good (or bad) will and its subcomponents. I say nothing here about the latter not emerging from the physical properties of, say, the human brain. This is not because I think they may so emerge, although some form of interaction between them and the brain, body, and social world, for example, surely does take place. Rather, it is because I can only regard talk of the emergence of irreducibly mental properties from the brain or the central nervous system as mere property dualism *cum* apologies. I accept that emergence can be employed as a valid and useful concept in numerous domains, e.g., chemistry, sociology, and the arts. But its valid employment requires some degree of insight into why this emerges from that. Such insight is lacking, in my opinion, in the case of the brain and experiences generally, and certainly with respect to the substructures of the morally good (or bad) will.

Finally, naturalism as a worldview lives today on promises. "We are going to show how all personal phenomena, including the moral, emerges from the chemistry (brain, DNA) of the human body." And, of course, the actual sciences (specific investigative practices) have made many wonderful discoveries and inventions. But after three hundred years or so of promises to "explain everything," the grand promises become a little tiresome, and the strain begins to show. And anyway, nothing in actual practice by scientists going about their work depends upon the grand promises—which can and do force sensible people to say things that have nothing to do with sense or science.[17]

7. Parting Salvo

Thomas Nagel has remarked: "Something peculiar happens when we view action from an objective or external standpoint. Some of its most important features seem to vanish under the objective gaze. Actions seem no longer assignable to individual agents as sources, but become instead components of the flux of events in the world of which the agent is a part. . . . The essential source of the problem is a view of persons and their actions as part of the order of nature. . . . That conception, if pressed, leads to the feeling that we are not agents at all, that we are helpless and not responsible for what we do. Against this judgment, the inner view of the agent rebels. The question is whether it can stand up to the debilitating effects of a naturalistic view."[18]

Really, all I have said in my basic argument is that a close adherence to science as that would be commonly understood, or to naturalism as a "first philosophy" (physicalism/materialism), has the effect that the primary structures and properties of the moral domain—those involved in the good (or bad) will—are lost sight of, and hence cannot function in the coherent organization of either the understanding (ethical theory) or the practice of the moral life. In Nagel's fine phrase, "Some of its most important features seem to vanish under the objective gaze."

They have vanished at present, and that has led to the current situation (deplored by Anscombe and MacIntyre) where there is no moral knowledge that is publicly accessible in our culture, i.e., that could be taught to individuals by our public institutions as a basis for their development into morally admirable human beings who can be counted on to do the "right thing" when it matters. This is what I call "The Disappearance of Moral Knowledge." That disappearance is now a fact in North American society.

The challenge to naturalism is, therefore, not just to come up with a convincing theory of the moral life (an analysis of moral concepts, utterances, and so forth). If what I have said is true, naturalism will not be able to do that. But suppose it could. Its work would not be done, but would hardly have begun. It would still have to create a moral culture by which people could live. It would still have the task of providing a body of moral understanding by which ordinary as well as extraordinary human beings could direct their own lives. Naturalism has always promised to do this through its leading spokespersons, and continues to do so today—through individuals such as Professor Larry Arnhart (the author of *Darwinian Natural Right* and other books and articles). Theirs is a tremendously ambitious undertaking. But they have much to do. Let them attempt it.

Naturalism has managed to occupy the intellectual high ground, and in the minds of many the moral high ground, in contemporary society—especially within the academy. It has put the Inquisition as well as the Moral Majority in its place. It is now the authority.[19]

So, let it lead, if it can. Not-being-superstitious-any-more will hardly serve as an adequate positive basis of moral understanding and moral development. Having been saved from the Moral Majority, how will we be saved from the immoral minority—or is that the majority? From Spinoza to Voltaire to Condorçet to Büchner and Häckel, to Dewey and Hook, and into the present, the promise of naturalism has been one of genuine moral enlightenment. But we cannot any longer live on promises. If naturalism is to be taken seriously in the capacities it wishes to be taken seriously, the promissory notes have come due. Naturalism must now turn them into cash. The need now is to stand and deliver. Let concrete and abstract, individual and social moral understanding and guidance, come forth from the views of Darwin, Dawkins, Dennett, Ruse, Searle, Wilson, and Arnhart. Let them tell us, corporately and individually, how to become persons of good will, reliably guided by moral obligation to do what is right and honorable.

This is especially relevant because the ultimate human issue underlying naturalism as a movement, and one seldom out of sight in discussions involving it, is: Who shall determine policy? If I am anywhere close to right in my main argument, the naturalists do not stand a chance at developing a body of knowledge to serve in concrete moral guidance. Certainly "survival" or "natural selection" will never suffice, though it has a point to make. But actuality would prove possibility. If they do what they have promised (or anything close) we will know it can be done. Since so much is at stake for humanity and we have good reason to believe it cannot be done, we must search for guidance elsewhere.

Notes

1. This essay is based on a paper read at "The Nature of Nature" conference. An earlier version of it appeared in *Philosophia Christi* 4 (1), 2002: 9-28. The author is grateful for permission to use the material here.
2. John R. Searle, *The Rediscovery of the Mind* (Cambridge, MA: MIT Press, 1992), 85.
3. *The Rediscovery of the Mind*, xii.
4. Patricia Kitcher, "Review of Stephen Stich's *Deconstructing the Mind*," *Journal of Philosophy* 95 (December 1998): 641–44 (see pages 641–42).
5. Kitcher 1998: 642.
6. A.E. Murphy, "Review of Y.H. Krikorian, ed. *Naturalism and the Human Spirit*," *The Journal of Philosophy* 42 (1945): 400–17 (see page 417).
7. G.E. Moore *Principia Ethica* (Cambridge: Cambridge University Press, 1903), § 15.
8. From Chandran Kukathas and Philip Pettit, eds. *Rawls: A Theory of Justice and It's Critics* (Stanford University Press, 1990), 134–35; referring to Rawls's "Justice as Fairness: Political not Metaphysical" and "The Priority of Right and Ideas of the Good," in *Philosophy and Public Affairs,* vols. 14 and 17.
9. Peter Unger is one of very few contemporary thinkers to question the idea of a tight connection between morality and rationality [see Peter Unger, *Living High and Letting Die* (Oxford: Oxford University Press, 1996), 21–22].
10. Richard Brandt, *Ethical Theory: The Problems of Normative and Critical Ethics* (Englewood Cliffs, NJ: Prentice-Hall, 1959), 152.
11. See Elizabeth Anscombe, "Modern Moral Philosophy," *Philosophy* 33 (January 1958): opening discussion.
12. Alasdair MacIntyre, *After Virtue,* second edition (Notre Dame: University of Notre Dame Press, 1984), 2.
13. David Hume, *A Treatise of Human Nature*, second edition, P.H. Nidditch, ed. (Oxford: Clarendon Press, 1978; originally published 1739-40), 477–78.
14. David Hume, opening sentence to Section IX, "Conclusion," of *An Enquiry Concerning the Principles of Morals* (first published [posthumously] in 1777).
15. St. Thomas Aquinas, *Treatise on Law*, Question XCIV, Second Article.
16. Henry Sidgwick, *Methods of Ethics*, Book III, Chap. XIII, 7th edition (New York: Dover Publishers,1966 [originally published 1907), 382. Sidgwick was, of course, very careful to incorporate his intuitions of justice and prudence into this crowning maxim.
17. At "The Nature of Nature" conference, where this paper was originally presented, a justifiably well regarded worker in the field of cosmology was heard to say: "It all begins in a state of absolute nothing, which makes a quantum transition to something very small, and then 'inflation' sets in. . . ." The proper response here is: what "which"? The antecedent of the relative pronoun was stated to be non-existent.
18. Thomas Nagel, *The View from Nowhere* (Oxford: Oxford University Press. 1986), 110.
19. If you want to see how true this is, just consider that the leading question of "The Nature of Nature" conference was posed as follows: "Is the universe self-contained or does it require something beyond itself to explain its existence and internal function?" If, now, you were to sit in on courses at the host institution for that conference—or a good many other institutions of higher learning having an association with religion—you would find that the courses are all taught on the assumption (possibly excepting religion, but that is not necessarily so) that the universe is self-contained and does *not* require something beyond itself to explain its existence and internal function. The course content in most schools with a religious association is *exactly* the same as in schools with none. This is, commonly, a point of pride among faculty at schools with a religious association. That is exactly what I mean when I say that naturalism has now won and is the authority.

40

NATURALISM, SCIENCE, AND RELIGION

MICHAEL TOOLEY

1. INTRODUCTION

In this essay, I shall begin by considering alternative definitions of "naturalism," and by asking how the term is best understood in the present context. In answering this question, I shall distinguish between anti-naturalism on the one hand, and supernaturalism on the other.

Next, I shall discuss the relation between science and supernaturalism, and I shall argue, first, that a commitment to scientific method does not in itself presuppose a rejection of supernaturalism, and secondly, that scientific investigation and theorizing could in principle show either that supernaturalism is true, or that it is false.

I shall then go on to survey a number of pro-supernaturalist arguments and discussions, both traditional and more recent. Given the number of such arguments and discussions, my critical commentary will necessarily be rather brief, though I shall comment in more detail upon certain recent arguments. The basic conclusion for which I shall argue is that there is no good reason for thinking that supernaturalism is true.

2. NATURALISM, ANTI-NATURALISM, AND SUPERNATURALISM

How should *naturalism* be understood? Here are two common definitions:

(1) Nothing exists outside of the spatiotemporal universe (David Armstrong).

(2) Our spatiotemporal universe is causally closed.

These definitions differ in important ways. Naturalism, understood in the first way, excludes Platonic realms, including such things as numbers, uninstantiated properties, and intentional entities. By contrast, if naturalism is understood in the second way, it is compatible with such Platonic realms, provided that the entities in question do not affect the spatiotemporal world.

Which definition is best depends upon the issues one is considering. Since my focus is going to be on the question of whether it is reasonable to view the spatiotemporal world as causally closed, I shall adopt the second of the above definitions.

Next, *supernaturalism*: this I shall take to be the view that there exists at least one nonembodied mind that stands in some causal relation to the spatiotemporal universe. The question I want to address, accordingly, is whether naturalism or supernaturalism is the more reasonable view, given our present evidence, and given relevant philosophical arguments.

3. Naturalism versus Physicalism/Materialism

A preliminary issue that calls for comment concerns the distinction between *naturalism*, on the one hand, and *physicalism* (or *materialism*) on the other. Here the point is simply that it is an error to equate naturalism with physicalism (or materialism), where the latter position is defined in terms of the following thesis:

(3) Everything that exists is purely physical.

The error of equating naturalism—in either of the senses defined above—with materialism (or physicalism) is quite widespread, and it is crucial to avoid it, since there are objections that, if sound, refute physicalism, but not naturalism. One crucial issue, for example, is whether the sensuous qualities that are present in one's conscious experiences—such as greenness, or the smell of lilac, or the taste of vegemite, for example—can be reduced to the properties and relations that enter into the fundamental theories in physics. If not, then reductive physicalism at least is false, and perhaps physicalism, in general, is false—depending upon whether it is plausible to hold that phenomenal qualities are emergent physical properties.

Another crucial issue concerns the existence of objective values. If physicalism is true, then such values can exist only if they are reducible to physical states of affairs. By contrast, naturalism, in either of the above senses, is perfectly compatible with the existence of objective values that are not reducible to physical states of affairs.

4. Science, Naturalism, and Supernaturalism

A number of people think that there is some close relationship either between science and naturalism or, even, between science and physicalism. Here are six theses that are rather common:

(1) Science rests upon the assumption that naturalism is true.

(2) Science rests upon the assumption that physicalism is true.

(3) While science is neutral on the question of whether naturalism is true, the goal of science is to discover natural causes of events.

(4) While science is neutral on the question of whether physicalism is true, the goal of science is to discover physical causes of events.

(5) While the use of the scientific method might show that naturalism is false, and supernaturalism true, it could never show that supernaturalism is false.

(6) Questions concerning the existence and nature of supernaturalistic entities lie outside the scope of scientific method.

None of these six theses seems to me acceptable. My reason for thinking this is a matter of three claims. First, any hypothesis that has empirical or experiential implications, either deductive or probabilistic, is open to both scientific confirmation and disconfirmation via the method of hypothesis. Secondly, many supernaturalistic hypotheses do have empirical or experiential implications. So supernaturalistic hypotheses can, at least in many cases, be confirmed or disconfirmed. Finally, if a hypothesis is open to scientific confirmation or disconfirmation, and if it deals with basic laws, or with the ultimate constituents of reality, then I see no reason to view it as falling outside the scope of scientific investigation.

If some supernaturalistic hypotheses can be confirmed—and I believe that there is excellent reason for holding that this is logically possible—then scientific inquiry could lead to a rejection of naturalism. But it is equally true that science can establish not only that various, *specific* supernaturalistic hypotheses are false, but also that the *general* claim that supernaturalistic entities exist is without any evidential support, and so is unlikely to be true.

5. Traditional Pro-Supernaturalist Arguments

A large number of arguments have been offered in support of supernaturalism, usually in the form of arguments in support of the existence of some sort of god. In this section, I'll mention some of the more important traditional pro-supernaturalist arguments, and comment on some of them *very* briefly.

First, there are a number of traditional proofs of the existence of God. These include both purely philosophical arguments and arguments that rest upon fairly detailed empirical or experiential claims:

(1) Ontological arguments

(2) Cosmological arguments

(3) Arguments from order to design

(4) Moral arguments

(5) Arguments from religious experience

(6) Arguments from miracles

Ontological arguments come in different forms, but all of them fall prey, I would argue, to a variant of an objection advanced by Gaunilo, who argued that Anselm's ontological argument for a perfect being could be paralleled by an argument for a perfect island. The way that I would strengthen Gaunilo's objection is to consider, not a perfect island, but certain entities that have incompatible properties—such as a perfectly immovable object, and a perfectly irresistible force.

Cosmological arguments generally involve two weaknesses. First, they usually involve an unsound argument against the possibility of an infinite causal or explanatory regress. Secondly, such arguments usually provide no sound reason for concluding that the object whose existence is being argued for lies outside of the spatiotemporal world.

Arguments from order to design are best left to later, when I shall comment on a modern version advanced by Michael Behe.

Moral arguments come in a number of versions. One very common type involves the idea that God is needed to provide a foundation for objective moral values. Such versions, however, typically run afoul of Plato's *Euthyphro* argument. Some present-day philosophers—such as Robert Adams—have tried to escape this objection, for example, by identifying what is morally right with what is commanded by a *benevolent* deity. But this sort of ad hoc move does not really work. For one thing, if there are moral truths, then at least some of them must be *basic,* and any basic moral truths must be necessary. There are good reasons for holding, however, that it is logically and metaphysically possible that God does not exist, and from these two propositions it follows that the existence of objective values does not entail the existence of God.

Religious experiences are many and varied, and arguments that attempt to provide evidence for the existence of a deity, or for the truth of a particular religion, by appealing to religious experience, can raise quite different issues, depending upon the type of experience involved. One of the major divides here is between arguments that appeal to mystical experiences and arguments that appeal to non-mystical experiences—such as visions, various sorts of emotional experiences, and so on. Mysticism is a rather complicated issue, and I shall not comment on it here. In the case of arguments that do not appeal to mystical experiences, however, a feature that is both very striking, and epistemologically very important, is that the content of such experiences is typically closely tied to the religious beliefs of the person having the experiences: Hindus, for example, tend not to have visions of the Virgin Mary. This fact provides a very good reason for concluding that such experiences, rather than providing information about the supernatural realm, are purely subjective.

In the case of the appeal to miracles, the question that one needs to begin with is whether it is reasonable to believe that any candidate miracles have actually occurred—where a candidate miracle is an event that, first of all, appears to be very improbable, given only natural causes, and, secondly, whose nature is such that its occurrence could contribute to some divine purpose.

Here, too, I shall have to confine myself to very cursory remarks. I do believe, however, that a careful survey of the relevant evidence supports the conclusion that there are no candidate miracles that we have good reason to believe have occurred. Among the relevant points here are, first of all, a number made by David Hume in his discussion, dealing specifically with the very low quality of the evidence for candidate miracles that is typically offered:

(1) The reports in question are usually second-hand ones, by people far removed from the time of the purported event.

(2) The witnesses are often unreliable.

(3) Typically, the witnesses have a strong desire to believe that the purported event occurred, since, for example, it would provide support for religious beliefs that they accept.

(4) In a large number of cases of purported candidate miracles, it has been established that deception has been involved.

(5) Reports of miracles are frequent in "ignorant and barbarous nations," and then decline dramatically in times of more careful observation and reporting.

In addition, more recent studies provide strong support for Hume's claim that it is not reasonable to believe that candidate miracles actually occur:

(1) Andrew D. White, in a chapter entitled "The Growth of Legends of Healing" in his book *A History of the Warfare of Science with Theology within Christendom* (1896), took a very close look at the miracles attributed to St. Francis Xavier by reading Xavier's own letters, those of his contemporaries, and then later biographies of Xavier. White—who was himself a Christian, and not an unbeliever—showed, first, how events that are described by Xavier, and that are not in any way miraculous, gradually get transformed in biographies of Xavier, first becoming miraculous, and then ever more impressively so, as the biography is further removed from the time of Xavier, and, secondly, how, though neither Xavier nor his contemporaries ever claimed that Xavier had resurrected anyone from the dead, later biographers do advance such claims, and offer more and more evidence for such claims the further removed the biography is from Xavier's time.

(2) A committee formed by the Archbishop of Canterbury carried out a careful study of faith healing, and in their 1920 report they indicated that they had found no evidence to support the claim that miraculous healings take place.

(3) The British Medical Association reached the same conclusion thirty years later

(4) A British psychiatrist, Louis Rose, devoted almost twenty years of his life to attempting to find candidate miracles, and he describes that search in detail in his 1971 book *Faith Healing*. Rose also began his study thinking that there would be no difficulty in finding such events. But the result was that he found none at all.

(5) An American doctor, William A. Nolen, author of *Healing: A Doctor in Search of a Miracle*, carried out a very detailed and thorough investigation in the early 1970s of some well-known people who had claimed to perform faith healings—including the very famous Kathryn Kuhlman. Nolen, like Rose before him, was unable to find even a single miraculous cure.

(6) The magician James Randi carefully investigated a number of ministers who claimed to be able to perform faith healings, and the results of his investigations are described in his 1989 book *The Faith Healers*. What Randi discovered was not miracles, but deliberate fraud.

The conclusion, in short, is that there is no reason even for thinking that there are any candidate miracles. As a consequence, the argument from miracles does not even make it to the starting blocks.

6. More Recent Pro-Supernaturalist Arguments

In present discussions of the issue of naturalism versus supernaturalism, one encounters both traditional arguments and some novel arguments. Of the latter, I think that the following are the most important:

(1) Arguments that appeal to consciousness and/or nonphysical properties.

(2) The fine-tuning of physical constants argument.

(3) Anti-evolution arguments.

My discussion of the first two arguments will be rather brief. I shall then comment at much greater length on attempts to support supernaturalism by arguing against evolution.

The Appeal to Consciousness and/or Emergent Properties

Many philosophers would reject arguments that attempt to establish the existence of God by appealing to consciousness, or to emergent properties involved in mental states, by arguing that a reductionist, physicalist account can be given of the mind and of mental states. I myself am not convinced that that is so, since I believe that experiences involve sensuous qualities that cannot be reduced to the properties and relations that enter into theories in physics. But even if one assumes that there are such emergent properties—be they physical or nonphysical—it does not seem to me that this provides any reason for thinking it is likely that a supernatural deity exists. For how does the argument go? The thought often seems to be, first, that if there are such emergent properties, then there must be laws of nature linking such emergent properties to physicalistic states of the brain, and, secondly, that such laws are much more likely to exist if there is a deity than if no deity exists. My response to this is to ask how likely it is that there is such a deity, and here the point is, first, that such a deity, being nonphysical, must also involve conscious states with relevant properties, and secondly, that it is not easy to see how one can argue that the probability of such a being's existing is greater than the probability that there are laws linking physicalistic states to states of consciousness, or to sensuous qualities: if the latter are for some reason improbable, mustn't that reason also be a reason why the existence of a nonembodied being possessing consciousness is just as improbable?

The Fine-Tuning Argument

The basic thrust of the fine-tuning argument is that the probability of there being laws of nature that would make possible the development of a world containing living things is extraordinarily low, and thus that it is much more likely that there is a supernatural being who wanted to create a world containing living things, and so who chose laws of nature that would make this possible.

This argument strikes me as more interesting than most arguments for theism, and it deserves more extended discussion than I can give it here. In addition, it strikes me as an argument where physicists may well have important things to contribute. (Here I am thinking, for example, of some of Steven Weinberg's discussion in his article "A Designer Universe?") Very briefly, however, I do want to mention four objections to the fine-tuning argument as it presently stands.

The first is that one cannot determine how likely it is, for example, that there is a universe whose laws are such that the universe will give rise to living things by considering only *possible modifications of the physical constants involved in the actual laws;* one needs to consider *all possible laws*—all but an infinitesimal proportion of which will be radically different from the actual laws. Until this is done—and no proponent of the fine-tuning argument has done this—it has not been shown that it is extremely unlikely that a universe without a creator would contain laws that, for example, make possible the existence of living things.

The second objection turns upon the fact that any deity that exists has powers. Powers, however, presuppose laws, and, in the case of the deity's own powers, laws that the deity in question cannot have created. Thus, in the case of an omnipotent deity, there will be a law that entails that, for some unknown property A, it is a law that whenever anything that has property A wills that some proposition p be true, p is true. But then one can ask *how likely* it is that reality will contain whatever laws must exist if it is to be possible for there to be a being that is, say, omnipotent and omniscient. How does the probability of reality's having such laws compare with the probability of its containing laws that make it possible for inanimate matter to give rise to living things?

This is not an easy question to answer. One line of thought, however, is this. For any property A, and any number k from 0 to 1, the following expresses a logically possible law: If anything that has property A wills that some proposition p be true, then the probability that p is true is equal to k.

Unless one can somehow argue that the probability of the existence of such a law has a non-infinitesimal value in the special case where $k = 1$, the probability that property A is the categorical ground of the dispositional property of omnipotence will be infinitesimal, and the same will be true with regard to any finite set of properties.

The fine-tuning argument need not, of course, involve the hypothesis that there is an omnipotent and omniscient designer. My basic point here, however, is that it is crucial to compare the probability that there are laws that make life possible with the probability that there are laws of the sort needed for the existence of an extremely powerful designer who can bring life-supporting laws into existence, and if one is to have a successful fine-tuning argument, one needs to show that the latter probability is greater than the former. No one, so far, has done this.

The third objection is this. If L is any logically possible set of laws, there is a logically possible deity, GL, who is such that, if that deity existed, and had the power to actualize any set of laws, the set of laws that would be actualized would be L. But once this relation is noticed, the following problem is immediately evident. Let S be the set of all sets of life-supporting laws, and T be the set of all sets of not-life-supporting laws, and let U be the set of all logically possible deities that would actualize sets of laws belonging to S, and V be the set of all logically possible deities that would, instead, actualize sets of laws belonging to set T. Given the relationship between possible sets of laws and possible deities, if the probability is low that the set of actual laws is a member of set S, how can it fail to be the case that the probability that there is a deity belonging to set U is precisely as low? But if it is precisely as low, then the fine-tuning argument cannot provide any grounds for thinking that it is likely that there is a deity.

The fourth and final objection is this. Fine-tuning arguments have been based on a consideration of the extent to which one can vary a *single constant* and still have laws that would be compatible with the existence of carbon-based life. As the first objection above made clear, this does not provide a good basis for a conclusion about the probability of life in an undesigned universe, since the laws of nature could be radically different in form from those in our universe. But if one is going to focus upon laws that differ from actual laws only with regard to physical constants, the least one can do, as Victor Stenger points out in his book *Has Science Found God?* is to consider a random variation of *all of the constants at once*.

Stenger carried out a preliminary investigation of this matter. In particular, he focused upon the question of the likelihood that a universe with laws of the relevant sort would contain long-lifetime stars, since that is one crucial property that a universe must have if life is to be able to evolve. Stenger then considered simultaneous random variations, over ten orders of magnitude, of the following four constants: the strength of the electromagnetic force, the mass of the proton, the mass of the electron, and the strong interaction strength. What he found was that the distribution of stellar lifetimes for one hundred simulated universes was such that most of the universes would "allow time for stellar evolution and heavy element nucleosynthesis."[1]

The upshot is that advocates of fine-tuning arguments need to consider what the probability of life is when all of the fundamental physical constants of the laws of our world are allowed, simultaneously, to vary randomly. Stenger's initial results suggest that when that is done, it may well be that fine-tuning arguments collapse.

7. ANTI-EVOLUTION ARGUMENTS

Critics of evolution come in three main varieties:

(1) Young-Earth creationists.

(2) Old-Earth creationists who claim that species (or "kinds") could not, or did not, evolve;

(3) Creationists who claim that evolution cannot account for biochemical systems that exhibit "irreducible complexity."

These three varieties advance incompatible claims, but it is noteworthy that they tend to direct their criticisms against the common enemy of evolution, rather than against each other. If the basic goal here were to arrive at truth, one would expect that criticisms would be directed at *all* erroneous views in an unbiased way. But this is rarely the case in such areas, and what it rather suggests is that there are certain beliefs—such as the belief that human life has a cosmic meaning—that drive the arguments, and which would be abandoned only with the greatest reluctance.

Young-Earth creationists include such people as Henry Morris and Duane Gish. The main thing to be said about the young-Earth creationists is that they have only two options. Either they must reject large areas of science—including physics, astronomy, and geology—or they must, as Gosse did, hold that God is a massive deceiver who created a world that, to all appearances, is very old indeed, by deliberately creating a world with fossils and fossil sequences, canyons that involve massive erosion, light that appears to have traveled from very distant galaxies, elements in rocks that appear to belong to radioactive decay series, etc. Neither option seems satisfactory.

Old-Earth creationists, on the other hand, by rejecting "macroevolution," are forced to postulate numerous acts of creation, occurring at different times and places. As a result, they are unable to explain many, many facts that the theory of evolution can explain, such as why the fossil record is as it is, or why species are distributed as they are.

8. Michael Behe and Darwin's Black Box

In his book *Darwin's Black Box*, Michael Behe claims that there are complex biochemical systems for which there is no naturalistic explanation, and which it is reasonable to view as having been designed. What sort of complex system does Behe have in mind here?

Behe's answer begins with a quote from Charles Darwin: "If it could be demonstrated that any complex organ existed which could not possibly have been formed by numerous, successive, slight modifications, my theory would absolutely break down."[2] Behe then asks, "What type of biological system could not be formed by 'numerous, successive, slight modifications'?" The answer that he immediately offers is this: "Well, for starters, a system that is irreducibly complex. By *irreducibly complex* I mean a single system composed of several well-matched, interacting parts that contribute to the basic function, wherein the removal of any one of the parts causes the system to effectively cease functioning."[3]

But why should one think that irreducibly complex systems could not have arisen naturalistically, without design, by a series of slight modifications? In this section, I shall both examine the logic of Behe's argument, and consider some scientific responses to it.

Behe's Argument

What is Behe's basic argument? Behe states it only very briefly, on pages 39–40, but it appears to be essentially as follows:

(1) It is *impossible* for an irreducibly complex system to have arisen by a *direct* route: "An irreducibly complex system cannot be produced directly (that is, by continuously improving the initial function, which continues to work by the same mechanism) by slight, successive modifications of a precursor system, because any precursor to an irreducibly complex system that is missing a part is by definition nonfunctional."[4]

(2) It is *possible* for an irreducibly complex system to have arisen by an *indirect,* circuitous route, but the more complex the system, the less likely this is: "As the complexity of an interacting system increases, though, the likelihood of such an indirect route drops precipitously."[5]

(3) Moreover, the likelihood that *all* such systems have arisen naturalistically—as Darwinism requires—is lowered even more by the large number of such systems: "And as the number of unexplained, irreducibly complex biological systems increases, our confidence that Darwin's criterion of failure has been met skyrockets toward the maximum that science allows."[6]

(4) The conclusion, accordingly, is that it is *extremely unlikely* that all such irreducibly complex systems could have arisen by an indirect route.

(5) It is also extremely unlikely that such irreducibly complex systems could have arisen by sheer chance. But even if they did, such an occurrence would not be in accordance with the gradual development via small modifications that Darwin envisaged.

A PURELY LOGICAL RESPONSE ON BEHE'S ARGUMENT

The most satisfactory response to Behe's argument involves showing in a reasonably detailed way that specific, irreducibly complex systems can come about by evolution. I shall turn to that sort of response in the next subsection. Here, however, I want to argue that Behe's argument is not even initially plausible.

To see why this is so, consider three biological systems—{B*}, {A, B*}, and {A, B}—which are as follows. First, {A, B*} has arisen from {B*}, via a minor modification; similarly, {A, B} has arisen from {A, B*}, also via a minor modification. Secondly, B* plays the same role in the {A, B*}-system that B plays in the {A, B}-system. Thirdly, B is not capable of doing what A does, nor is A capable of doing what B does, so the {A, B}-system is irreducibly complex. Finally, B*, in contrast to B, is also capable of doing, albeit less efficiently, what A does, and this is one of the things that B* does in the initial {B*}-system.

Given that B* can play the causal role that A plays in the {A, B*}-system, the latter is not an irreducibly complex system. But given that in the {A, B}-system, B cannot play the role that A does, the {A, B}-system is irreducibly complex. Therefore it would seem that we have a case where an irreducibly complex system arises directly from a system that is not irreducibly complex. If this is right then, Behe's claim that it is impossible for an irreducibly complex system to have arisen by a direct route does not appear to be a logical truth.

Behe might contend, however, that the transition involved in moving from the {B}-system to the {A, B}-system via the {A, B*}-system is not a direct transition in his sense, since in his characterization of what is involved in what he refers to as a "direct" route, Behe mentions two features: (a) a direct route linking a later system to an earlier one is one where the same function is performed by all of the systems along the route; (b) a direct route linking a later system to an earlier one is one where the same function is performed by the same (type of?) mechanism in all of the systems along the route. Given this characterization of a direct route, a route could be indirect either because it involved a change in function, or, alternatively, because, while it involved no change in function, it did involve a change in the type of mechanism.

Let us concentrate on the change of mechanism alternative. Although Behe does not really say what counts as a different mechanism, I presume that he would view the transition from the original system {B*} to the system {A, B*} as indirect, on the grounds that a different mechanism must be involved, given that the original, {B*}-system involves only a single element that is causally active, whereas the {A, B*}-system involves two elements, each of which is causally active. But this rejoinder would not save Behe's argument. In the first place, even if the route from the {B}-system to the {A, B}-system via the {A, B*}-system is not a direct transition in Behe's sense, it is a route in which an irreducibly complex system arises in a gradual way by a (very short) series of minor modifications. Secondly, the transition from the {A, B*}-system to the {A, B}-system will be direct, since we can assume that B not only performs the same function as B*, but does it in the same way. The fact that B* possesses a power that B does not—namely, to play the causal role that A plays—does not make the transition from the {A, B*}-system to the {A, B}-system an indirect one, since B* does not perform that extra function within the {A, B*}-system: it does no more than what B does within the {A, B*}-system.

But if a route such as

$$\{B^*\} \S \{A, B^*\} \S \{A, B\}$$

is a possible route to an irreducibly complex system, and one that involves only minor modifications, what reason can Behe offer in support of his claim that "as the complexity of an interacting system increases, . . . the likelihood of such an indirect route drops precipitously"? There does not seem to be any reason why either the transition from {B*} to {A, B*} or the transition from {A, B*} to {A, B} should be especially improbable. Indeed, in appropriate circumstances the laws of nature might render such transitions extremely likely. But if this can be so in the case of a move from a one-element system to a two-element system, what grounds can Behe have for claiming that the likelihood "drops precipitously" when one considers a more complex system?

From a logical point of view, then, the basic point here is that an irreducibly complex system—*no matter how complicated it is*—may be *only one step removed from a system that it is not irreducibly complex*. Thus, suppose that the structure $\{A_1, A_2, \ldots A_i \ldots A_n\}$ has a certain function, and that it is not an irreducibly complex structure, since A_2, say, could be eliminated without the resulting structure ceasing to have the function in question. Then $\{A_1, A_2, \ldots A_i \ldots A_n\}$ could give rise to a different structure $\{A_1, A_2, \ldots A_i* \ldots A_n\}$, where A_i has been replaced by A_i^*, and where $\{A_1, A_2, \ldots A_i^* \ldots A_n\}$ is irreducibly complex because while, in the original structure, what was done by A_2 could also be done, albeit less efficiently, by A_i, A_i^* cannot do what A_2 does at all

Responses by Biologists to Behe's Argument

There have been a number of responses by biologists to Behe's argument. One of the most extended is by Kenneth Miller in his book *Finding Darwin's God*. There Miller makes a number of points. One is that Behe's complaint that no evolutionary explanations are available for biochemical systems that exhibit irreducible complexity is *premature in the extreme* in some cases, since the details of some of these systems—such as the bacterial flagellum, are not yet fully understood.[7]

Secondly, Miller next goes on to cite a number of relevant studies that have been published since Behe's book appeared:

(1) "In January 1998, in the lead article in *American Scientist*, a journal of general science for technical audiences, Anthony M. Dean described in great detail the structural and biochemical changes that took place millions of years ago to produce two enzymes, two biochemical machines, essential to the subsequent evolution of life."[8]

(2) "In 1997, John M. Logsdon and Ford Doolittle reviewed in detail how [certain] . . . mechanisms could have produced, in strictly Darwinian fashion, the remarkable 'anti freeze' proteins of Antarctic fish."[9]

(3) "Now for the evolution of a complex multipart biochemical structure. In 1998, Siegfried Musser and Sunney Chan described the evolution of the cytochrome c oxidase protein pump, a complex, multipart molecular machine that plays a key role in energy transformation in the cell."[10]

(4) "In 1996, Enrique Melédez-Hevia and his colleagues published . . . a paper in the *Journal of Molecular Evolution*. As every high school biology student learns, the Krebs cycle is an extremely complex series of interlocking reactions that release chemical energy from food. If there was any doubt that this paper represented exactly the kind of study that Behe demanded, the authors removed it in their first two sentences: The evolutionary origin of the Krebs citric acid cycle has been for a long time a model case in the understanding of the origin and evolution of metabolic pathways. How can the emergence of such a complex pathway be explained?"[11]

After describing the work in question, Miller concludes by saying: "The Krebs cycle is a complex biochemical pathway that requires the interlocking, coordinated presence of at least nine enzymes and three cofactors. And a Darwinian explanation for its origin has now been crafted."[12]

Finally, Miller examines one of the irreducibly complex biochemical systems on which Behe specifically focuses in his book—blood clotting. In his book, Behe concluded his discussion of blood clotting by focusing upon the work of Russell Doolittle—a scientist who has devoted his life to studying blood-clotting mechanisms in different animals. In doing this, Behe chose to comment upon an article by Doolittle with the title "The Evolution of Vertebrate Blood Coagulation: A Case of Yin and Yang"—and, as one might expect, Behe had a good deal of fun with the terms "yin" and "yang."

The impression one gets from reading Behe's discussion of Russell Doolittle's work is that it contributes nothing to showing that the irreducibly complex blood clotting system could not have evolved. In addition, it is very hard to understand, from Behe's discussion, what Doolittle's account is. But if one turns instead to Kenneth Miller's discussion, a *very* different picture emerges.[13] Among other things, Miller focuses upon a different article by Doolittle, one that he coauthored with D. F. Feng, and titled "Reconstructing the Evolution of Vertebrate Blood Coagulation from a Consideration of the Amino Acid Sequences of Clotting Proteins"—an article that Behe doesn't even mention in his book, but which Miller says contains a much more formal presentation of the relevant process than is found in the article on which Behe focuses.[14]

A final point is this. Biologists have given evolutionary explanations of irreducibly complex systems at the macroscopic level—for example, the evolution of the bone structure involved in the human ear, and the evolution of echolocation in the case of bats. Behe dismisses such explanations on the grounds that such systems *presuppose* the existence of irreducible biochemical systems. But this response involves a clear mistake about the overall logic of the discussion. The point is that a given biological system, S, may involve irreducible complexity at a number of different places and/or levels. Suppose, for example, that system S involves parts $\{A_1, A_2, \ldots A_n\}$ and is irreducibly complex, in that if any of the A_i were to be removed from S, the resulting system, S*, could not function in the way that S does. Suppose, further, that A_1 is itself an irreducibly complex system. What if one can show how S could have evolved from a system T that involves an irreducibly complex part—namely, A_1—but no other irreducible complexity? Behe treats this as irrelevant, since one has not completely eliminated *all* of the irreducible complexity involved in the original system S. But it is *not* irrelevant: one has shown how *one instance of irreducible complexity* could have arisen by evolution. So Behe's contention that cases where one shows that the amount of irreducible complexity can be reduced are irrelevant is simply unsound. Such cases are relevant, and the accumulation of such cases provides excellent inductive reason for thinking that the origin of other systems that exhibit irreducible complexity can also be explained in evolutionary terms.

To sum up: In the first place, Behe's contention that biochemical systems that involve irreducible complexity could not have arisen via a series of small modifications involves a *philosophical* argument that is simply unsound, and, given the unsoundness of that argument the upshot is that Behe has done *nothing* to show that irreducible systems cannot arise via evolution. Secondly, given the complexity of such systems, it is not surprising that the construction of explanations is a challenging task, and thus there is nothing striking about the existence of irreducibly complex systems for which no explanation is as yet available. Thirdly, work that has been done over the past few years, and which is described by Miller, seems to make it eminently reasonable to believe that such explanations will be forthcoming. Finally, Behe's dismissal as irrelevant cases where one can show how a system that exhibits irreducible complexity could have developed in an evolutionary fashion from a system that involves *less* irreducible complexity is simply unjustified.

9. The Argument from Imperfection and the Argument from Vestigial Organs and Pseudogenes

The Argument from Imperfection

The basic claim advanced in the Argument from Imperfection is that the existence of design faults *makes it less likely* that life was created by a very intelligent and very powerful person. Behe mentions this argument,[15] but fails to address serious formulations of the argument—which are probabilistic in nature—preferring to focus instead on obviously weak syllogistic formulations.[16]

How might a probabilistic version of the Argument from Imperfection be formulated? Here is one way of doing so. Let us use the following abbreviations:

D = The eye was designed by a *very* intelligent and *very* powerful being.

E = The eye developed via evolution.

F = The eye has design faults.

Prob(X/Y) = Probability of X given Y.

By the definition of conditional probabilities:

$$\text{Prob}(D/F) \times \text{Prob}(F) = \text{Prob}(D\&F) = \text{Prob}(F/D) \times \text{Prob}(D)$$

Therefore:

$$\frac{\text{Prob}(D/F)}{\text{Prob}(D)} = \frac{\text{Prob}(F/D)}{\text{Prob}(F)}$$

Assume, for simplicity, that

D or E

(This leaves out the following possibilities: (a) the eye arose by chance; (b) the eye arose by natural law in a nonevolutionary way; (c) the eye was designed by a being that was either not very intelligent, or not very powerful. These possibilities could be taken into account in a more complicated formulation of the argument.)

Since D and E are mutually exclusive, we therefore have:

(1) $\text{Prob}(D) = 1 - \text{Prob}(E)$

Since we are also assuming, for simplicity, that D or E, we also have:

(2) $\text{Prob}(F) = \text{Prob}(F/E) \times \text{Prob}(E) + \text{Prob}(F/D) \times \text{Prob}(D)$

Substitution of (1) in (2) then gives us

$\text{Prob}(F) = \text{Prob}(F/E) \times \text{Prob}(E) + \text{Prob}(F/D) \times \text{Prob}(D)$
$= \text{Prob}(F/E) \times \text{Prob}(E) + \text{Prob}(F/D) \times [1 - \text{Prob}(E)]$
$= [\text{Prob}(F/E) - \text{Prob}(F/D)] \times \text{Prob}(E) + \text{Prob}(F/D)$

But it is surely true both that

$\text{Prob}(E) > 0$

and that

$\text{Prob}(F/E) > \text{Prob}(F/D)$

and from this it follows from

$\text{Prob}(F) = [\text{Prob}(F/E) - \text{Prob}(F/D)] \times \text{Prob}(E) + \text{Prob}(F/D)$

that

$\text{Prob}(F) > \text{Prob}(F/D)$

This, together with

$$\frac{\text{Prob}(D/F)}{\text{Prob}(D)} = \frac{\text{Prob}(F/D)}{\text{Prob}(F)}$$

then entails that

Prob(D/F) < Prob(D)

So F *lowers* the probability of D.

The Argument from Vestigial Organs and Pseudogenes

The basic claim advanced in the Argument from Vestigial Organs and Pseudogenes is that the existence of vestigial organs and pseudogenes *makes it less likely* that life was created by a very intelligent and very powerful person.

Let us use the following abbreviations:

D = Living organisms were designed by a *very* intelligent and *very* powerful being.

E = Living organisms developed via evolution.

V = There are vestigial organs and pseudogenes.

The probabilistic argument just set out in the case of imperfections can be paralleled exactly in the case of vestigial organs and pseudogenes, thereby generating the conclusion that Prob(D/V) < Prob(D). So V *lowers* the probability of D.

Here, as in the case of the Argument from Imperfection, Behe briefly mentions objections to design that appeal to vestigial organs and pseudogenes,[17] but he fails to grapple with serious, probabilistic formulations of this crucial type of objection to anti-evolutionary design hypotheses.

10. William Dembski and the Design Inference

With the publication in 1998 of his book *The Design Inference: Eliminating Chance Through Small Probabilities,* William Dembski quickly gained a reputation in creationist circles as a sophisticated and powerful critic of a naturalistic and evolutionary view of the world. Space does not permit me to discuss, in an extended fashion, even that initial book, let alone Dembski's many later publications in defense of the design hypothesis, including a number of articles and two further books—*Intelligent Design: The Bridge between Science and Theology* (1999) and *No Free Lunch: Why Specified Complexity Cannot Be Purchased without Intelligence* (2002). But readers who are interested in an extended critical discussion of Dembski's work as a whole, including how his arguments have altered over time, can turn to the excellent, very lengthy, and very detailed treatment found in Chapter 1—"A Consistent Inconsistency: How William Dembski Infers Intelligent Design"—of Mark Perakh's book *Unintelligent Design*.

I shall confine myself, then, to Dembski's first book, *The Design Inference,* and to commenting, very briefly on some of the most important flaws. First of all, then, there is Dembski's central idea—the idea of an "explanatory filter." Dembski's contention here is the way to establish the presence of design is by ruling out alternative possibilities. In particular, according to the explanatory filter approach, the way to determine whether an event should be explained by postulating a designer is by first considering whether the event could be due to a regularity or whether it could be due to chance. If neither of these competing possibilities is acceptable, one can conclude that design was present.

This whole approach to the detection of design is unsatisfactory, and for a number of reasons. In the first place, and contrary to what Dembski supposes, one cannot really investigate the question of whether an event is due to some unknown regularity or other: one needs to consider *specific* regularities. There is no way to rule out an explanation of an event in terms of laws unless one surveys the laws that might serve to provide such an explanation.

Secondly, assume that one is able, in a given case, to rule out both a regularity explanation and a chance explanation. Then, according to the explanatory-filter approach, a design hypothesis wins by default. The problem here is that the probability of the design hypothesis does not enter into the investigation at any point: the design hypothesis can win by default regardless of how *intrinsically improbable* that hypothesis is.

That this is unacceptable can be seen by considering a variant on one of Dembski's cases (the Nicholas Caputo election fraud case). Suppose that a man—John—has to select people for certain positions, and that he is supposed to do this in a politically unbiased way. It turns out, however, that all of a very large number of the people whom John has chosen belong to the far left of the political spectrum. It might very well be reasonable, at that point, to think that design is present, and let us assume that that is the result that Dembski's explanatory-filter approach generates. But now suppose that John's selection method is very carefully examined, and that it is determined that his method of selection does in fact involve a type of process that is completely random. This latter fact does not, given Dembski's explanatory-filter approach, undermine the design conclusion that was arrived at, since the observations in question do not supply us with any explanation in terms of regularities, nor do they reduce the improbability of the events in question. But on a sound approach to the detection of design, those observations are, of course, crucial. Before they are made, the design hypothesis is quite reasonable because all that is required is human design in the form of cheating by John, and that is not especially improbable. But once it is established that John is using a process that involves no intervention by him that is aimed at procuring the outcome in question, nor any intervention by any other human, acceptance of a design hypothesis involves acceptance of the idea that there is some intelligent *nonhuman* agent who is controlling the apparently random process and who wants people on the far left to be selected. But now we have a design hypothesis that is much less likely than the sort of design hypothesis that was initially possible.

A third central weakness in Dembski's approach becomes apparent if one considers the case of lotteries. Suppose that Mary wins a lottery in which the winner has been chosen by some process that involves quantum indeterminism. Then, applying Dembski's explanatory-filter approach, explanation via regularities can be ruled out. So we move on to consider the possibility of chance. The lottery, however, can involve as many tickets as one likes, so Mary's chance of winning can be as small as one wants. Accordingly, it would seem that chance can also, on Dembski's approach, be rejected. Are we then justified in concluding that an intelligent being with the power to intervene in quantum events brought it about that Mary won?

Dembski's way of attempting to avoid this problem is by introducing the idea of an event's being *specified*. Dembski's explanation of his idea of specification leaves a good deal to be desired with regard to clarity, but the basic thought can be grasped by comparing the case where a person is dealt a hand containing thirteen specific cards with the case where this happens and when the person also states in advance what those cards will be. Each event is equally unlikely. But the second event provides good grounds for concluding that trickery is involved, whereas the second does not. As a result, Dembski is led to advance the following thesis: "Specified events of low probability do not occur by chance."[18]

But suppose that Mary, on buying the lottery ticket, declares, "This ticket will win the lottery." Now the event of Mary's winning the lottery is not only an event of very low probability, it is also a specified event. According to Dembski, then, we can conclude that Mary's winning the lottery did not occur by chance. Hence, given that the winning ticket was arrived at by a process involving quantum indeterminism, we are justified in concluding that an intelligent being with the power to intervene in quantum events brought it about that Mary won.

The source of the problem here, once again, is that the probability of the relevant design hypothesis never gets considered by Dembski's explanatory-filter approach, and it is precisely this which is crucial. For the relevant design hypothesis in the lottery case is that there is an intelligent being with the power to intervene in quantum events who wants it to be the case that *Mary* wins, and as there are as many precisely parallel design hypotheses concerning all the other people who have bought lottery tickets as there are tickets in the lottery, the initial probability of the hypothesis that there is a certain powerful intelligent being who wants *Mary* to win will be extraordinarily low, and the fact that Mary has won, together with the fact that she declared in advanced that she would, certainly do not make it reasonable to accept the design hypothesis in question.

Finally, what is the correct approach to the question of whether a design hypothesis should be accepted? The answer is that it involves assigning prior probabilities to *all* of the hypotheses, including the design hypothesis, and then using *standard probability theory*—including Bayes's Theorem—to work out the posterior probabilities of the different hypotheses. When this is done, one avoids the unacceptable conclusions in the modified Caputo case mentioned above, and in lottery cases where someone specifies an outcome in advance.

Why, then, doesn't Dembski adopt this obvious and natural approach? Why doesn't he use Bayes's Theorem, and the apparatus of standard probability theory? This is a crucial question, and a detailed response to it should occupy a central place in Dembski's book. But what one finds, instead, is that Dembski's answer to this question occupies exactly seven lines, and is as follows:

> Bayes's theorem is therefore of particular interest if one wants to understand the degree to which evidence confirms one hypothesis over another. . . . It follows that Bayes's theorem has little relevance to the design inference. Indeed, confirming hypotheses is precisely what the design inference does not do. The design inference is in the business of eliminating hypotheses, not confirming them.

This argument is very weak. One cannot be *justified* in eliminating, say, the hypothesis that some event E is due either to regularity or to chance, and thus of embracing, in virtue of the explanatory-filter approach, the hypothesis that E is due to design, unless one can show that the design hypothesis has a *higher* probability than the "regularity or chance" alternative. So probability theory in general—and Bayes's theorem in particular—is crucial, and Dembski's failure to employ probability theory, including Bayes's theorem, means that his whole approach to the detection of design is unsound.

11. Naturalism, Supernaturalism, God, and Evil

In the preceding, I have focused on the question of whether there are good grounds for rejecting naturalism in favor of supernaturalism. It is important to note, however, that virtually all contemporary philosophers who argue against naturalism are interested in offering reasons for accepting, not just supernaturalism in general, but a particular form of supernaturalism—namely, classical monotheism, understood as the view that there is a person who is the creator of everything else that exists, and who is eternal, omniscient, omnipotent, and perfectly good. Given that this is so, it is important to ask what the prospects would be, *if* one could mount a successful case against naturalism, of moving from the proposition that supernaturalism is true to the further conclusion that monotheism is true.

For convenience, let us speak of an argument for the existence of God as indirect if it goes by way of an intermediate conclusion that naturalism is false, and, if this is not so, as direct. Direct arguments for the existence of God were once the norm. This does not seem to be so today. Present-day theists, on the whole, rather than either offering direct arguments for the existence of God, or defending the view that belief in the existence of God is noninferentially known, or noninferentially justified, tend to offer arguments against naturalism. Such philosophers rarely go on to address, however, the crucial question of how one is to move from a refutation of naturalism to the conclusion that monotheism is true.

The logical gulf between supernaturalism in general and monotheism in particular is enormous. Perhaps some form of polytheism is true, and there are many deities with limited powers. Perhaps there is only a single deity, whose powers or knowledge is limited. Perhaps deism is true, and there is a creator who is omnipotent and omniscient, who views the fate of his creation and its inhabitants as a matter of little concern. Perhaps there is an omnipotent and omniscient deity who is perfectly evil.

Given the great range of possibilities that might obtain if supernaturalism were true, why is it that the move from supernaturalism in general to monotheism in particular is the focus of so little philosophical discussion? Perhaps the visibility, in Western society, of the three great monotheistic religions—Judaism, Christianity, and Islam—is part of the answer. But one has only to consider Eastern religions, or a multitude of past religions, to see that a successful case against naturalism leaves one very far indeed from a justification of theistic belief.

What, then, are the prospects of bridging this gulf? I think they are very slim indeed. First, as I have just emphasized, supernaturalism includes an enormous number of alternatives. Secondly, even if polytheistic and limited-deity alternatives can somehow be set aside, one is still left with an infinite range of possibilities with regard to a creator with unlimited attributes. Thirdly, within that infinite range, there are deities whose *a priori* probability does not appear to be significantly less than that of the God of monotheism—such as an omnipotent and omniscient and morally indifferent creator, or an omnipotent and omniscient and perfectly evil creator. How, in the face of such alternatives, does the theist propose to argue that theism is likely to be true? Finally, there is the argument from evil. Many present-day theists, of course, believe that there are satisfactory ways of responding to the argument from evil—for example, by appealing to great goods of which we humans have no knowledge. I would argue, however, that no satisfactory response is, in fact, available. Among other things, I believe that one can show, for example, using basic principles of inductive logic, that if there are N states of affairs, each of which, judged by the rightmaking and wrongmaking properties that we are aware of, it would be morally wrong to allow, then the probability that none of them is such that it is morally wrong to allow that state of affairs, given the totality of rightmaking and wrongmaking properties, both known and unknown, must be less than $1/(N + 1)$. But a proof of this must be left for another time and place.

12. Summing Up: Naturalism versus Supernaturalism

Though it has not been possible to do more than briefly survey some of the most important arguments in support of supernaturalism, I have tried to indicate why those arguments for supernaturalism are unsound. In addition, and as I emphasized in the final section, even if a successful attack upon naturalism could, contrary to what I have argued, be mounted, there would still be an enormous chasm between that conclusion and the further contention that *monotheism* is true. Nor do the prospects of bridging that gulf seem at all bright, especially if, as I would argue, there is a sound version of the evidential argument from evil. The upshot, accordingly, is that, even if one could somehow establish that there was a realm of non-Platonic entities beyond the natural world, neither God—understood as an omnipotent, omniscient, and morally perfect being—nor even a close approximation to such a deity could be among the supernatural beings inhabiting such a realm.

Notes

1. Stenger 2003: 156–57.
2. Darwin 1872: 154.
3. Behe 1996: 39.
4. Ibid.
5. Behe, 40.
6. Ibid.
7. Miller 1999: 147.
8. Miller, 148.
9. Miller, 149.
10. Ibid.
11. Miller, 150.
12. Miller, 151–52.
13. Miller, 152–58.
14. Miller, 305 (footnote 38).
15. Behe, 222–25.
16. Behe, 224.
17. Behe, 225–27.
18. Dembski 1998: 48.

References

Behe, Michael J. (1996) *Darwin's Black Box: The Biochemical Challenge to Evolution*. New York: The Free Press.

———. (1997). "Michael Behe's Response to Boston Review Critics," http://www.discovery.org/scripts/viewDB/index.php?command=view&id=47.

———. (2000). "A Response to Critics of Darwin's Black Box," http://www.iscid.org/papers/Behe_ReplyToCritics_121201.pdf.

Coyne, Jerry (1997). "More Crank Science," *Boston Review*, February/March http://www.bostonreview.net/BR22.1/coyne.html.

Darwin, Charles. (1872 [1988]) *Origin of Species*, 6th edition. New York: New York University Press.

Dembski, William A. (1998) *The Design Inference: Eliminating Chance Through Small Probabilities*. Cambridge: Cambridge University Press.

———. (1999) *Intelligent Design: The Bridge between Science and Theology*. Downers Grove, IL: InterVarsity Press.

———. (2002) *No Free Lunch; Why Specified Complexity Cannot Be Purchased without Intelligence*. Lanham, MD: Rowman and Littlefield.

Miller, Kenneth R. (1999) *Finding Darwin's God*. New York: HarperCollins Publishers.

Nolen, William A. (1974) *Healing: A Doctor in Search of a Miracle*. New York: Random House Inc.

Pennock, Robert T. (1991) *Tower of Babel: The Evidence Against the New Creationism*. Cambridge, MA: MIT Press.

Perakh, Mark. (2004) *Unintelligent Design*. Amherst, New York: Prometheus Books.

Randi, James. (1989) *The Faith Healers*. Buffalo: Prometheus Books.

Rose, Louis (1971) *Faith Healing*. Baltimore: Penguin Books.

Stenger, Victor J. (2003) *Has Science Found God?* Amherst: Prometheus Books.

Weinberg, Steven. (1999) "A Designer Universe?" in *The New York Review of Books*, Vol. 46, No. 16, October 21, 1999.

White, Andrew D. (1896 [1993]) *A History of the Warfare of Science with Theology in Christendom*. Buffalo: Prometheus Books.

41

Theism Defended

William Lane Craig

1. Introduction

During the last half century professional philosophers have witnessed a veritable revolution in Anglo-American philosophy that has transformed the face of their discipline. In order to appreciate the sea change that has occurred, one has only to cast a glance backward at the broad contours of philosophy as practiced at our most prominent universities during the 1950s and '60s. In a recent retrospective, the eminent Princetonian Paul Benacerraf recalls what it was like doing philosophy at Princeton during that time. The overwhelmingly dominant mode of thinking was scientific naturalism. Physical science was taken to be the final, and really only, arbiter of truth. Metaphysics had been vanquished, expelled from philosophy like an unclean leper. "The philosophy of science," says Benacerraf, "was the queen of all the branches" of philosophy since "it had the tools . . . to address all the problems."[1] Any problem that could not be addressed by science was simply dismissed as a pseudo-problem. If a question did not have a scientific answer, then it was not a real question—just a pseudo-question masquerading as a real question. Indeed, part of the task of philosophy was to clean up the discipline from the mess that earlier generations had made of it by endlessly struggling with such pseudo-questions. There was thus a certain self-conscious, crusading zeal with which philosophers carried out their task. The reformers, says, Benacerraf, "trumpeted the militant affirmation of the new faith . . . in which the fumbling confusions of our forerunners were to be replaced by the emerging science of philosophy. This new enlightenment would put the old metaphysical views and attitudes to rest and replace them with the new mode of doing philosophy."

The book *Language, Truth, and Logic* by the British philosopher A. J. Ayer served as a sort of manifesto for this movement. As Benacerraf puts it, it was "not a great book," but it was "a wonderful exponent of the spirit of the time." The principal weapon employed by Ayer in his campaign against metaphysics was the vaunted Verification Principle of Meaning. According to that principle, which went through a number of revisions, a sentence, in order to be meaningful, had to be capable in principle of being empirically verified. Since metaphysical statements were beyond the reach of empirical science, they could not be verified and were therefore dismissed as meaningless combinations of words.

Ayer was quite explicit about the theological implications of this verificationism.[2] Since God is a metaphysical object, Ayer says, the possibility of religious knowledge is "ruled out by our treatment of metaphysics." Thus, there can be no knowledge of God. Now, someone might say that we can offer evidence of God's existence. But Ayer will have none of it. If, by the word "God," you mean a transcendent being, says Ayer, then the word "God" is a metaphysical term, and so "it cannot be even probable that a god exists." He explains, "To say that 'God exists' is to make a metaphysical utterance which cannot be either true or false. And by the same criterion, no sentence which purports to describe the nature of a transcendent god can possess any literal significance." Theism was, as they say, not even false—it was nonsense.

Philosophers soon began to realize, however, that the Verification Principle was a double-edged sword. For it would force us to dismiss as meaningless not only theological statements but also a great many scientific statements as well, so that the principle cuts off at the knees science itself, in whose service this weapon had been wielded. As it turns out, physics is filled with metaphysical statements that cannot be empirically verified. As philosopher of science Bas van Fraassen nicely puts it: "Do the concepts of the Trinity [and] the soul . . . baffle you? They pale beside the unimaginable otherness of closed space-times, event-horizons, EPR correlations, and bootstrap models."[3] If the ship of scientific naturalism was not to be scuttled, verificationism had to be cut loose. But there was a price to be paid for abandoning the Verification Principle. Since verificationism had been the principal means of barring the door to metaphysics, the abandonment of verificationism meant that there was no longer anyone at the door to prevent this unwelcome visitor from making a reappearance.

Worse, and even more fundamentally, it was also realized that the Verification Principle is self-refuting. One need only ask oneself whether the sentence "A meaningful sentence must be capable in principle of being empirically verified" is *itself* capable of being empirically verified. Obviously not; no amount of empirical evidence would serve to verify its truth. The Verification Principle is therefore by its own lights a meaningless combination of words, which need hardly detain us, or at best an arbitrary definition, which we are at liberty to reject. Therefore, the Verification Principle and the theory of meaning that it supported have been almost universally abandoned by philosophers of all stripes.

The collapse of verificationism during the second half of the twentieth century was undoubtedly the most important philosophical event of the century. Its demise brought about a resurgence of metaphysics, along with other traditional problems of philosophy that had been hitherto suppressed. Accompanying this resurgence has come something new and altogether unanticipated: a renaissance in Christian philosophy.

The face of Anglo-American philosophy has been transformed as a result. Theism is on the rise; atheism is on the decline. Atheism, though perhaps still the dominant philosophical viewpoint at the American university, is a philosophy in retreat. In a recent article in the secularist journal *Philo,* Quentin Smith laments what he calls "the desecularization of academia that evolved in philosophy departments since the late 1960s." He complains,

> Naturalists passively watched as realist versions of theism . . . began to sweep through the philosophical community, until today perhaps one-quarter or one-third of philosophy professors are theists, with most being orthodox Christians. . . . [I]n philosophy, it became, almost overnight, "academically respectable" to argue for theism, making philosophy a favored field of entry for the most intelligent and talented theists entering academia today.[4]

Smith concludes, "God is not 'dead' in academia; he returned to life in the late 1960s and is now alive and well in his last academic stronghold, philosophy departments."[5]

I think that Smith is probably exaggerating when he estimates that one-quarter to one-third of American philosophers are theists; but what his estimates do reveal is the *perceived impact* of theists upon this field. The principal error that Smith makes is calling philosophy departments God's "last stronghold" at the university. On the contrary, philosophy departments are a foothold, affording access to yet further reaches. Since philosophy is foundational to every discipline of the university, revolutions in philosophy are apt eventually to have reverberations in other fields. For example, one of the consequences of the resurgence of metaphysics and the turn toward theism in philosophy is the intelligent design movement in the sciences, as young philosophers of science have begun to question the implicit assumption of naturalism that served to exclude appeal to intelligent causes from the pool of live explanatory options.

The renaissance of Christian philosophy over the last half century has been accompanied by a re-appreciation of the traditional arguments for the existence of God. An argument is a set of statements that serve as premises leading to a conclusion. What makes for a good argument? Three crucial elements are involved. First, a good argument must have true premises, and the conclusion must follow from the premises. In the case of a deductive argument the conclusion will follow from the premises by logical necessity; in the case of an inductive argument, the conclusion will be rendered more probable than not by the premises. Second, a good argument must not be question-begging; that is to say, the reasons one believes the premises to be true must be independent of the argument's conclusion. Otherwise one would simply be reasoning in a circle, since the only reason one believes at least one of the premises is that one already believes the conclusion. Finally, third, the premises of a good argument must be more plausible than their denials. For an argument to be a good one, it is not required that we have one hundred percent certainty of the truth of its premises. Some of the premises in a good argument may strike us as only slightly more plausible than their denials; other premises may seem to us highly plausible. But so long as a statement is more plausible than its negation, then we should believe it rather than its negation, and so it may serve as a premise in a good argument.

Thus, a good argument for God's existence need not make it *certain* that God exists. Certainty is what most people have in mind when they deny that there are any proofs of God's existence. If we equate "proof" with one hundred percent certainty, then we may agree and yet insist that there are still good arguments to think that God exists. The question is not whether the denial of a particular premise in the argument is possible (or even plausible); the question is whether the denial is as plausible, or more plausible, than the premise. If it is not, then we should believe the premise.

Now plausibility is, to a great extent, a person-dependent notion. Some people may find a premise plausible and others not. Accordingly, some people will agree that a particular argument is a good one, while others will say that it is a bad argument. Given our diverse backgrounds and biases, we should expect such disagreements. In cases of disagreement, we simply have to dig deeper and explore what reasons we each have for thinking a premise to be true or false. When we do so, we shall be asking what warrant there is for the argument's premises. Rational persons should accept a premise as more plausible than not if it has greater warrant for them than the premise's negation.

2. Five Arguments on Behalf of Theism

In this paper, I shall examine five of the traditional arguments for God's existence that have enjoyed a revival of interest in our day. Space does not permit me to go into responses to various critiques of theism, which I have sought to address in another place.[6]

Contingency Argument. A simple statement of this argument might run as follows:

1. Anything that exists has an explanation of its existence (either in the necessity of its own nature or in an external cause).

2. If the universe has an explanation of its existence, that explanation is God.

3. The universe exists.

4. Therefore the explanation of the existence of the universe is God.

Premise (1) is a modest version of the Principle of Sufficient Reason. It circumvents the typical objections to strong versions of that principle (e.g., that there must be an explanation of every fact or truth). For (1) requires merely that any existing *thing* have an explanation of its existence. This premise is compatible with there being brute *facts* about the world. What it precludes is that there could exist things that just exist inexplicably. According to (1) there are two kinds of things: necessary beings, which exist of their own nature and so have no external cause of their existence, and contingent beings, whose existence is accounted for by causal factors outside themselves. Numbers might be prime candidates for the first sort of being, while familiar physical objects fall under the second kind of being.

This premise seems quite plausible, at least more so than its contradictory. One thinks of Richard Taylor's illustration of finding a translucent ball while walking in the woods.[7] One would find the claim that the ball just exists inexplicably quite bizarre; and increasing the size of the ball, even until it becomes coextensive with the cosmos, would do nothing to obviate the need for an explanation of its existence.

Premise (2) is, in effect, the contrapositive of the typical atheist retort that, on the atheistic worldview, the universe simply exists as a brute contingent thing. Moreover, (2) seems quite plausible in its own right. For if the universe, by definition, includes all of physical reality, then the cause of the universe must (at least causally prior to the universe's existence) transcend space and time and therefore cannot be temporal or material. But there are only two kinds of things that could fall under such a description: either an abstract object or else a mind. But abstract objects do not stand in causal relations. Whether we are talking about mathematical objects, universals, propositions, or any other of a host of abstract objects that many philosophers deem to exist, it belongs conceptually to such objects that they are causally effete and so cannot effect anything. By contrast, minds are the causes with which we are the most intimately acquainted of all causes, as we experience the causal efficacy of our own willings. Therefore, it follows that the explanation of the existence of the universe is an external, transcendent, personal cause—which is one meaning of "God."

Finally, (3) states the obvious, that there is a universe. It follows that God exists.

It is open to the atheist to retort that while the universe has an explanation of its existence, that explanation lies not in an external ground but in the necessity of its own nature; in other words, (2) is false. This is, however, an extremely bold suggestion that atheists have not been eager to embrace. We have, one can safely say, a strong intuition of the universe's contingency. A possible world in which no concrete objects exist certainly seems conceivable. We generally trust our modal intuitions on other familiar matters; if we are to do otherwise with respect to the universe's contingency, then the atheist needs to provide some reason for such skepticism other than his desire to avoid theism. Moreover, as we shall see below, we have good reason to think that the universe does not exist by a necessity of its own nature.

The premises of the argument thus all seem more plausible than their negations. It therefore follows logically that the explanation why the universe exists is to be found in God.

Cosmological Argument. A simple version of this argument might go as follows:

1. Whatever begins to exist has a cause.

2. The universe began to exist.

3. Therefore, the universe has a cause.

Conceptual analysis of what it means to be a cause of the universe then helps to establish some of the theologically significant properties of this being.

Premise (1) seems obviously true—at the least, more so than its negation. It is rooted in the metaphysical intuition that something cannot come into being from nothing. To suggest that things could just pop into being uncaused out of nothing is to quit doing serious metaphysics and to revert to magic. Moreover, if things could really come into being uncaused, out of nothing, then it becomes inexplicable why just anything and everything do not come into existence uncaused from nothing. Furthermore, the conviction that an origin of the universe requires a causal explanation seems quite reasonable, for on the atheistic view, if the universe began at the Big Bang, there was not even the *potentiality* of the universe's existence prior to the Big Bang, since nothing is prior to the Big Bang. But then how could the universe become actual if there was not even the potentiality of its existence? The theist may plausibly claim that it makes much more sense to say that the potentiality of the universe lay in the power of God to create it. Finally, the first premise is constantly confirmed in our experience. Atheists who are scientific naturalists thus have the strongest of motivations to accept it.

Premise (2), the more controversial premise, may be supported by both deductive, philosophical arguments and inductive, scientific arguments. In my paper "Naturalism and the Origin of the Universe" (included in this volume), I lay out some of the scientific evidence in support of (2), so here I shall focus on the philosophical arguments.

The first argument we shall consider is the argument based on *the impossibility of the existence of an actual infinite*. We may formulate the argument as follows:

2.11. An actual infinite cannot exist.
2.12. An infinite temporal regress of events is an actual infinite.
2.13. Therefore, an infinite temporal regress of events cannot exist.

Since the universe is not distinct from the temporal series of past events, it follows that the universe must have begun to exist. Let us examine more closely each of the two premises.

2.11. *An actual infinite cannot exist.* By an actual infinite, one means any collection having at a time t a number of definite and discrete members that is greater than any natural number 0, 1, 2, 3, ... An infinite set in standard Zermelo-Fraenkel axiomatic set theory is defined as any set R that has a proper subset that is equivalent to R. A proper subset is a subset that does not exhaust all the members of the original set, which is to say, at least one member of the original set is not also a member of the subset. Two sets are said to be equivalent if the members of one set can be related to the members of the other set in a one-to-one correspondence, that is, such that a single member of the one set corresponds to a single member of the other set, and vice-versa. I shall call this convention the Principle of Correspondence. Equivalent sets are regarded as having the same number of members. Thus, an infinite set is one in which the whole set has the same number of members as a proper part. The notion of an actual infinite is to be contrasted with a potential infinite, which is any collection having at any time t a number of definite and discrete members that is equal to some natural number but which over time increases endlessly toward infinity as a limit.

Premise (2.11) asserts, then, that an actual infinite cannot exist. It is usually alleged that this sort of argument has been invalidated by Georg Cantor's work on the actual infinite and by subsequent developments in set theory. But the ontological finitist can legitimately respond that this allegation not only begs the question against intuitionistic denials of the mathematical existence of the actual infinite, but, more seriously, it begs the question against non-Platonist views of the ontology of mathematical objects. These are distinct questions, and most non-Platonists would not go to the intuitionistic extreme of denying mathematical legitimacy to the actual infinite; they would simply insist that acceptance of the mathematical existence of certain entities does not imply an ontological commitment to the metaphysical reality of such objects. Cantor's system and set theory may be taken to be simply a universe of discourse, a mathematical system based on certain adopted axioms and conventions. On antirealist views of mathematical objects such as Balaguer's fictionalism or Chihara's constructivism, there are no mathematical objects at all, let alone an infinite number of them.[8] One may consistently hold that while the actual infinite is a fruitful and consistent concept within the postulated universe of discourse, it cannot be transposed into the spatiotemporal world, for this would involve counterintuitive absurdities.

Perhaps the best way to support this premise is by way of thought experiments—like the famous Hilbert's Hotel, a brainchild of the great German mathematician David Hilbert—which illustrate the various absurdities that would result if an actual infinite were to be instantiated in the real world.[9] The absurdity in this case is not merely practical and physical; it seems ontologically absurd that a hotel exist that is completely full and yet can accommodate untold infinities of new guests just by moving people around. Worse, if such a hotel could exist in reality, then it would lead to situations that would not be warranted by transfinite arithmetic, and finally to logical contradictions. For, while in transfinite arithmetic the inverse operations of subtraction and division of infinite quantities from or by infinite quantities are not permitted, these could occur with a real hotel occupied by real people. In such cases one winds up with logically impossible situations, such as subtracting identical quantities from identical quantities and finding non-identical differences.

Howard Sobel observes that such situations bring into conflict two "seemingly innocuous" principles, namely,

(i) There are not more things in a multitude M than there are in a multitude M' if there is a 1–1 correspondence of their members; and

(ii) There are more things in M than there are in M' if M' is a proper submultitude of M.

We cannot have both of these principles along with

(iii) An actually infinite multitude exists.

For Sobel the choice is clear: "The choice we have taken from Cantor is to hold on to (i) while restricting the proper submultiplicity condition to finite multiplicities. In this way we can 'have' comparable infinite multitudes."[10]

But the choice taken from Cantor, of which Sobel speaks, is a choice on the part of the mathematical community to reject intuitionism in favor of infinite set theory. Intuitionism would too radically truncate mathematics to be acceptable to most mathematicians. But, as already indicated, that choice does not validate metaphysical conclusions. The metaphysician wants to know why, in order to resolve the inconsistency among (i)–(iii), it is (ii) that should be jettisoned (or restricted). Why not instead reject (i), which is a mere set-theoretical convention, or restrict it to finite multiplicities? More to the point, why not reject (iii) instead of the apparently innocuous (i) or (ii)? It certainly lacks the innocuousness of those principles, and giving it up would enable us to affirm both (i) and (ii). We can "have" comparable infinite multiplicities in mathematics without admitting them into our ontology. In view of the absurdities that ontologically real actual infinite multitudes would engender, perhaps we are better off without them.

Sobel thus needs some *argument* for the falsity of (ii). It is insufficient merely to point out that if (i) and (iii) are true, then (ii) is false, for that is merely to reiterate the incompatibility of the three propositions. The opponent of (iii) may argue that if (iii) were true, then (i) would be true with respect to such a multitude, as Sobel believes; and that if (i) and (iii) were true, then the various counterintuitive situations would result; therefore if (iii) were true, the various counterintuitive situations would result. But because these situations are really impossible, it follows that (iii) is not possibly true. In order to refute this reasoning, one must do more than point out that if (i) and (iii) are true, then (ii) is false, for that is merely to reiterate that if an actual infinite were to exist and the Principle of Correspondence were valid with respect to it, then the relevant situations would result, which is not in dispute.

2.12. *An infinite temporal regress of events is an actual infinite.* This second premise seems obviously true. If the universe never began to exist, then prior to the present event, there have existed an actually infinite number of previous events.

2.13. *Therefore an infinite temporal regress of events cannot exist.* If (2.11) and (2.12) are true, then (2.13) follows logically. The series of past events must be finite and have a beginning, Q.E.D.

The second philosophical argument we shall consider on behalf of premise (2) is the argument based on *the impossibility of the formation of an actual infinite by successive addition.* We may formulate the argument as follows:

2.21. The series of events in time is a collection formed by successive addition.
2.22. A collection formed by successive addition cannot be actually infinite.
2.23. Therefore, the series of events in time cannot be actually infinite.

Since the series of events in time is not distinct from the universe, it follows that the universe began to exist. Let us take a closer look at each of the two premises.

2.21. *The series of events in time is a collection formed by successive addition.* This may seem rather obvious. The past did not spring into being whole and entire but was formed sequentially, one event occurring after another. As obvious as (2.21) might seem, however, it does presuppose the truth of a tensed theory of time.[11] Since advocates of a tenseless theory of time deny the reality of temporal becoming, they deny that the past series of events was formed by successive addition. There are, however, powerful reasons to think that the tenseless theory of time is mistaken and that temporal becoming is real.[12] We may take it, then, that time is dynamic and therefore that the past has been formed sequentially, one event elapsing after another.

2.22. *A collection formed by successive addition cannot be actually infinite.* Sometimes this problem is described as the impossibility of traversing the infinite. In order for us to have "arrived" at today, temporal existence has, so to speak, traversed an infinite number of prior events. But before the present event could arrive, the event immediately prior to it would have to arrive; and before that event could arrive, the event immediately prior to it would have to arrive; and so on *ad infinitum*. No event could ever arrive, since before it could elapse there will always be one more event that will had to have happened first. Thus, if the series of past events were beginningless, the present event could not have arrived, which is absurd.

Again, illustrations can help to drive home the point. Consider Tristram Shandy, who, in the novel by Sterne, writes his autobiography so slowly that it takes him a whole year to record the events of a single day. Though he write forever, Tristram Shandy would only get farther and farther behind, so that instead of finishing his autobiography, he would progressively approach a state in which he would be *infinitely* far behind. But suppose instead that Tristram Shandy has been writing from eternity past at the rate of one day per year. Should he not now be infinitely far behind? For if he has lived for an infinite number of years, Tristram Shandy has recorded an equally infinite number of past days. Given the thoroughness of his autobiography, these days are all consecutive days. At any point in the past or present, therefore, Tristram Shandy has recorded a beginningless, infinite series of consecutive days. But now the question inevitably arises: *Which* days are these? Where in the temporal series of events are the days recorded by Tristram Shandy at any given point? The answer can only be that *they are days infinitely distant from the present.*[13] For there is no day on which Tristram Shandy is writing that is finitely distant from the last recorded day. If Tristram Shandy has been writing for one year's time, then the most recent day he could have recorded is one year ago. But if he has been writing two years, the most recent day he could have recorded is the day immediately after a day at least two years ago. This is because it takes a year to record a day, so that to record two days he must have two years. In other words, the longer he has written the further behind he has fallen. In fact, the recession into the past of the most recent recordable day can be plotted according to the formula (present date $-n$ years of writing) $+ n - 1$ days. But what happens if Tristram Shandy has, *ex hypothesi,* been writing for an infinite number of years? The most recent day of his autobiography recedes to infinity, that is to say, to a day infinitely distant from the present. Nowhere in the past at a finite distance from the present can we find a recorded day, for by now Tristram Shandy is infinitely far behind. The beginningless, infinite series of days that he has recorded are days that lie at an infinite temporal distance from the present. But there is no way to traverse the temporal interval from an infinitely distant event to the present, or, more technically, for an event that was once present to recede to an infinite temporal distance. Since the task of writing one's autobiography at the rate of one year per day seems obviously coherent, what follows from the Tristram Shandy story is that an infinite series of past events is absurd.

It is frequently objected that this sort of argument illicitly presupposes an infinitely distant starting point in the past and then pronounces it impossible to travel from that point to today, whereas in fact from any given point in the past, there is only a finite distance to the present, which is easily traversed.[14] But proponents of the argument have not in fact assumed that there was an infinitely distant starting point in the past. To traverse a distance is to cross every proper part of it. As such, traversal does not entail that the distance traversed has a beginning or ending point, or a first or last part. The fact that there is *no beginning* at all, not even an infinitely distant one, seems only to make the problem worse, not better. To say that the infinite past could have been formed by successive addition is like saying that someone has just succeeded in writing down all the negative numbers, ending at -1. And, proponents of the argument may ask, how is the claim that, from any given moment in the past, there is only a finite distance to the present even relevant to the issue? For the question is how the *whole* series can be formed, not a finite portion of it. To think that because every *finite* segment of the series can be formed by successive addition the whole *infinite* series can as well is to commit the fallacy of composition.

It seems, then, that to try to form an actually infinite collection of things by successive addition is impossible. The only way an actual infinite could come to exist in the real world would be by being created all at once, simply in a moment. It would be a hopeless undertaking to try to form it by adding one member after another.

2.23. *Therefore, the series of events in time cannot be actually infinite.* Given the truth of the premises, the conclusion logically follows. If the universe did not begin to exist a finite time ago, then the present moment would never arrive. But obviously it has arrived. Therefore, the universe is finite in the past, and the temporal series of past events is not beginningless.

We thus have good philosophical (as well as scientific) grounds for affirming the second premise of the cosmological argument. It is noteworthy that since a being that exists by a necessity of its own nature must exist either timelessly or sempiternally (otherwise its coming into being or ceasing to be would make it evident that its existence is not necessary), it follows that the universe cannot be metaphysically necessary, which fact confirms the claim of the contingency argument above that the universe does not exist by a necessity of its own nature.

It follows logically that the universe has a cause. Conceptual analysis of what properties must be possessed by such an ultra-mundane cause enables us to recover a striking number of the traditional divine attributes, revealing that if the universe has a cause, then an uncaused, personal creator of the universe exists, who *sans* the universe is beginningless, changeless, immaterial, timeless, spaceless, and enormously powerful.[15]

Teleological Argument. We may formulate a design argument as follows:

1. The fine-tuning of the universe is due either to physical necessity, chance, or design.

2. It is not due to physical necessity or chance.

3. Therefore, it is due to design.

What is meant by "fine-tuning"? The physical laws of nature, when given mathematical expression, contain various constants, such as the gravitational constant, whose values are independent of the laws themselves; moreover, there are certain arbitrary quantities that are simply put in as boundary conditions on which the laws of nature operate, for example, the initial low entropy condition of the universe. By "fine-tuning" one means that the actual values assumed by the constants and quantities in question are such that small deviations from those values would render the universe life-prohibiting or, alternatively, that the range of life-permitting values is exquisitely narrow in comparison with the range of assumable values.

Laypeople might think that if the constants and quantities had assumed different values, then other forms of life might well have evolved. But so long as the laws of nature are held constant, this claim is quite mistaken. By "life" scientists mean that property of organisms to take in food, extract energy from it, grow, adapt to their environment, and reproduce. The point is that in order for the universe to permit life so defined, whatever form organisms might take, the constants and quantities have to be incomprehensibly fine-tuned. In the absence of fine-tuning, not even matter or chemistry would exist, not to speak of planets where life might evolve.

It might be objected that in universes governed by different laws of nature, such deleterious consequences might not result from varying the values of the constants and quantities. The teleologist need not deny the possibility, for such universes are irrelevant to his argument. All he needs to show is that among possible universes governed by the same laws (but having different values of the constants and quantities) as the actual universe, life-permitting universes are extraordinarily improbable. John Leslie gives the illustration of a fly, resting on a large, blank area of the wall.[16] A shot is fired, and the bullet strikes the fly. Now even if the rest of the wall outside the blank area is covered with flies, such that a randomly fired bullet would probably hit one, nevertheless it remains highly improbable that a single, randomly fired bullet would strike the solitary fly within the large, blank area. In the same way, we need only concern ourselves with the universes in our nomological neighborhood in determining the probability of the universe's being finely tuned.

Now premise (1) states the three alternatives in the pool of live options for explaining cosmic fine-tuning. The question is which is the best explanation?

On the face of it the alternative of physical necessity seems extraordinarily implausible. If the primordial matter and antimatter had been differently proportioned, if the universe had expanded just a little more slowly, if the entropy of the universe were marginally greater, any of these adjustments and more would have prevented a life-permitting universe, yet all seem perfectly possible physically. The person who maintains that the universe must be life-permitting is taking a radical line that requires strong proof. But as yet there is none; this alternative is put forward as a bare possibility.

Sometimes physicists do speak of a yet to be discovered Theory of Everything (T.O.E.), but such nomenclature is, like so many of the colorful names given to scientific theories, quite misleading. A T.O.E. actually has the limited goal of providing a unified theory of the four fundamental forces of nature, but it will not even attempt to explain literally everything. Stephen Hawking, in addressing the question whether string theory or M-Theory, the most promising candidate for a T.O.E. to date, predicts the distinctive features of our universe, observes, "Most physicists would rather believe string theory uniquely predicts the universe, than [accept] the alternatives. These are that the initial state of the universe is prescribed by an outside agency, code named God. Or that there are many universes, and our universe is picked out by the anthropic principle."[17] If that is the case, then Hawking has bad news: "M-theory cannot predict the parameters of the standard model. Obviously, the values of the parameters we measure must be compatible with the development of life.... But within the anthropically allowed range, the parameters can have any values. So much for string theory predicting the fine structure constant." Hawking concludes,

> [E]ven when we understand the ultimate theory, it won't tell us much about how the universe began. It cannot predict the dimensions of spacetime, the gauge group, or other parameters of low energy effective theory.... It won't determine how this energy is divided between conventional matter, and a cosmological constant, or quintessence. The theory will... predict a nearly scale free spectrum of fluctuations. But it won't determine the amplitude. So to come back to the question.... Does string theory predict the state of the universe? The answer is that it does not. It allows a vast landscape of possible universes, in which we occupy an anthropically permitted location.

Furthermore, it seems likely that any attempt to significantly reduce fine-tuning will itself turn out to involve fine-tuning. This has certainly been the pattern in the past. In light of the specificity and number of instances of fine-tuning,[18] it is unlikely to disappear with the further advance of physical theory.

What, then, of the alternative of chance? Teleologists seek to eliminate this hypothesis either by appealing to the specified complexity of cosmic fine-tuning (a statistical approach to design inference) or by arguing that the fine-tuning is significantly more probable on design (theism) than on the chance hypothesis (atheism) (a Bayesian approach).[19] Common to both approaches is the claim that the universe's being life-permitting is highly improbable.

In order to save the hypothesis of chance, defenders of that alternative have increasingly recurred to the Many Worlds Hypothesis, according to which a World Ensemble of concrete universes exists, thereby multiplying one's probabilistic resources. In order to guarantee that by chance alone a universe like ours will appear somewhere in the Ensemble, an actually infinite number of such universes is usually postulated. But that is not enough; one must also stipulate that these worlds are randomly ordered with respect to the values of their constants and quantities, lest they be of insufficient variety to include a life-permitting universe. Finally, appeal is made to the anthropic principle, according to which we should not be surprised at the extraordinary fine-tuning of the universe, since observers can exist only in universes in the World Ensemble that exhibit such fine-tuning.

Is the Many Worlds Hypothesis as good an explanation as the Design Hypothesis? It seems doubtful. In the first place, as a metaphysical hypothesis, the Many Worlds Hypothesis is arguably inferior to the Design Hypothesis because the latter is simpler. According to Ockham's Razor, we should not multiply causes beyond what is necessary to explain the effect. But it is simpler to postulate one Cosmic Designer to explain our universe than to postulate the infinitely bloated and contrived ontology of the Many Worlds Hypothesis. Only if the Many Worlds theorist could show that there exists a single, comparably simple mechanism for generating a World Ensemble of randomly varied universes would he be able to elude this difficulty.

Second, there is no known way of generating a World Ensemble. No one has been able to explain how or why such a collection of varied universes should exist. Some proposals, like Lee Smolin's cosmic evolutionary scenario, actually served to weed out life-permitting universes,[20] while others, like Andre Linde's chaotic inflationary scenario,[21] turned out to require fine-tuning themselves.

Third, there is no evidence for the existence of a World Ensemble apart from the fine-tuning itself. But the fine-tuning is equally evidence for a Cosmic Designer. Indeed, the hypothesis of a Cosmic Designer is again the better explanation because we have independent evidence of the existence of such a being in the other theistic arguments.

Fourth, the Many Worlds Hypothesis faces a severe challenge from evolutionary biology.[22] According to the prevailing theory of biological evolution, intelligent life like ourselves, if it evolves at all, will do so as late in the lifetime of its star as possible. Given the complexity of the human organism, it is overwhelmingly more probable that human beings will evolve late in the lifetime of the Sun rather than early.[23] Hence, if our universe is but one member of a World Ensemble, then it is overwhelmingly more probable that we should be observing a very old sun rather than a relatively young one. In fact, adopting the Many Worlds Hypothesis to explain away fine-tuning results in a strange sort of illusionism: it is far more probable that all our astronomical, geological, and biological estimates of age are wrong, that we really do exist very late in the lifetime of the Sun, and that the Sun and the Earth's appearance of youth is a massive illusion. Thus, the Many Worlds Hypothesis fails as a plausible explanation of cosmic fine-tuning.

It therefore seems that the fine-tuning of the universe is plausibly due neither to physical necessity nor to chance. It is therefore due to design. The implication of this hypothesis is that there exists a Cosmic Designer who fine-tuned the initial conditions of the universe for intelligent life. Is this alternative implausible?

Detractors of design sometimes object that the Designer himself remains unexplained. It is said that an intelligent Mind also exhibits complex order, so that if the universe needs an explanation, so does its Designer. If the Designer does not need an explanation, why think that the universe does?

This popular objection is based on a misconception of the nature of explanation.[24] In order for an explanation to be the best, one need not have an explanation of the explanation (indeed, such a requirement would generate an infinite regress, so that everything becomes inexplicable). If astronauts should find traces of intelligent life on some other planet, we need not be able to explain the origin of such extraterrestrials in order to recognize that they are the best explanation of the artifacts. In the same way, the design hypothesis's being the best explanation of the fine-tuning does not depend on our being able to explain the Designer.

Moreover, the complexity of a mind is not really analogous to the complexity of the universe. A mind's *ideas* may be complex, but a mind itself is a remarkably simple thing, being an immaterial entity not composed of parts. Moreover, a mind, in order to be a mind, must have certain properties like intelligence, consciousness, and volition. These are not contingent properties that it might lack, but are essential to its nature. So it is difficult to see any analogy between the contingently complex universe and a mind. Postulating an uncreated Mind behind the cosmos thus seems the best of the alternative explanations of the fine-tuning of the universe for intelligent life.

Moral Argument. Theists have presented a wide variety of moral justifications for belief in a deity. One such argument may be formulated as follows:

1. If God does not exist, objective moral values and duties do not exist.

2. Objective moral values and duties do exist.

3. Therefore, God exists.

Consider premise (1). To speak of objective moral values and duties is to say that moral distinctions between what is good/bad or right/wrong hold independently of whether any human being holds to such distinctions. Many theists and atheists alike agree that if God does not exist, then moral values and duties are not objective in this sense.

For if God does not exist, human beings are just animals, relatively advanced primates; and animals are not moral agents. Richard Taylor invites us to imagine a group of people living in a state of nature without any customs or laws. Suppose one person kills another and takes his goods. Taylor comments,

> Such actions, though injurious to their victims, are no more unjust or immoral than they would be if done by one animal to another. A hawk that seizes a fish from the sea *kills* it, but does not *murder* it; and another hawk that seizes the fish from the talons of the first *takes* it, but does not *steal* it—for none of these things is forbidden. And exactly the same considerations apply to the people we are imagining.[25]

If God does not exist, then it is difficult to see any reason to think that human beings are special or that their morality is objectively valid. It follows that our perceived moral obligations are no more binding upon us than other sociocultural conventions and accoutrements of civilization. As a result of sociobiological pressures, there has evolved among *homo sapiens* a sort of "herd morality" that functions well in the perpetuation of our species in the struggle for survival. But there does not seem to be anything about *homo sapiens* that makes this morality objectively binding. If the film of evolutionary history were rewound and shot anew, very different creatures with a very different set of values might well have evolved. By what right do we regard our morality as objective rather than theirs? The humanist philosopher Paul Kurtz sees the dilemma: "The central question about moral and ethical principles concerns this ontological foundation. If they are neither derived from God nor anchored in some transcendent ground, are they purely ephemeral?"[26]

Some philosophers, equally averse to transcendently existing moral values as to theism, try to evade the dilemma by simply affirming the existence of objective moral principles or supervenient moral properties in the context of a naturalistic worldview. But the advocates of such theories are typically at a loss to justify their starting point. If there is no God, then it is hard to see any ground for thinking that the herd morality evolved by *homo sapiens* is objectively true or that moral goodness supervenes on certain natural states of such creatures.

If our approach to meta-ethical theory is to be serious metaphysics rather than just a "shopping list" approach, whereby one simply helps oneself to the supervenient moral properties or principles needed to do the job, then some sort of explanation is required for why moral properties supervene on certain natural states or why such principles are true.[27] It is insufficient for the naturalist to point out that we do, in fact, apprehend the goodness of some feature of human existence, for that only goes to establish the objectivity of moral values and duties, which just is premise (2) of the moral argument.

We therefore need to ask whether moral values and duties can be plausibly anchored in some transcendent, non-theistic ground. Let us call this view "atheistic moral realism." Atheistic moral realists affirm that objective moral values and duties do exist and are not dependent upon evolution or human opinion, but they insist that they are not grounded in God. Indeed, moral values have no further foundation. They just exist.

It is difficult, however, even to comprehend this view. What does it mean to say, for example, that the moral value *justice* just exists? It is hard to know what to make of this. It is clear what is meant when it is said that a person is just; but it is bewildering when it is said that in the absence of any people, *justice* itself exists.

Second, the nature of moral obligation seems incompatible with atheistic moral realism. Suppose that values like *mercy, justice, forbearance,* and the like just exist. How does that result in any moral obligations for me? Why would I have a moral duty, say, to be merciful? Who or what lays such an obligation on me? On this view moral vices such as *greed, hatred,* and *selfishness* also presumably exist as abstract objects, too. Why am I obligated to align my life with one set of these abstractly existing objects rather than any other? In contrast with the atheist, the theist can make sense of moral obligation because God's commands can be viewed as constitutive of our moral duties.

Thirdly, it is fantastically improbable that just the sort of creatures who would correspond to the abstractly existing realm of moral values would emerge from the blind evolutionary process. This seems to be an utterly incredible coincidence when one thinks about it. It is almost as though the moral realm *knew* that we were coming. It is far more plausible to regard both the natural realm and the moral realm as under the hegemony of a divine creator and lawgiver than to think that these two entirely independent orders of reality just happened to mesh.

As for premise (2), I take it that, in moral experience, we do apprehend a realm of objective moral values, just as in sensory experience we apprehend a realm of objectively existing physical objects. For any argument against the objectivity of the moral realm, a parallel argument can be constructed against the objective existence of the external world of physical objects. Just as it is impossible for us to get outside our sensory perceptions to test their veridicality, so there is no way to test independently the veridicality of our moral perceptions. But in the absence of some defeater, we rationally trust our perceptions, whether sensory or moral.

It follows, then, that God exists. Moral values can be understood as grounded in God's moral nature and our moral duties in his moral commands, for example, "You shall love your neighbor as yourself." Our moral obligations are thus constituted by the commands of a just and loving God. For any action A and moral agent S, we can explicate the notions of moral requirement, permission, and forbiddenness of A for S:

A is required of S *iff* a just and loving God commands S to do A.

A is permitted for S *iff* a just and loving God does not command S not to do A.

A is forbidden to S *iff* a just and loving God commands S not to do A.

Since our moral duties are grounded in the divine commands, they are neither independent of God nor are they arbitrary, since God's commands are determined by his just and loving nature.

The question might be pressed as to why God's nature should be taken to be definitive of goodness. But unless we are nihilists, we have to recognize some ultimate standard of value, and God seems to be the least arbitrary stopping point. Moreover, God's nature is singularly appropriate to serve as such a standard. For, by definition, God is a being *worthy of worship*. And only a being that is the locus and source of all value is worthy of worship.

Ontological Argument. The common thread in ontological arguments is that they attempt to deduce the existence of God from the very concept of God, together with certain other necessary truths. Proponents of the argument claim that once we understand what God is—the greatest conceivable being or the most perfect being or the most real being—then we shall see that such a being must in fact exist. The argument tends to sharply polarize philosophers. Many would agree with Schopenhauer's dismissal of the argument as "a charming joke," but a number of recent prominent philosophers such as Norman Malcolm, Charles Hartshorne, and Alvin Plantinga not only take the argument seriously but consider it to be sound.

In his version of the argument,[28] Plantinga appropriates Leibniz's insight that the argument assumes that the concept of God is possible, that is to say, that it is possible that a being falling under that concept exists or—employing the semantics of possible worlds—that there is a possible world in which God exists. Plantinga conceives of God as a being who is "maximally excellent" in every possible world, where maximal excellence entails such excellent-making properties as omniscience, omnipotence, and moral perfection. Such a being would have what Plantinga calls "maximal greatness." Now maximal greatness, Plantinga avers, is possibly instantiated, that is to say, there is a possible world in which a maximally great being exists. But then this being must exist in a maximally excellent way in every possible world, including the actual world. Therefore, God exists.

We can formulate Plantinga's version of the ontological argument as follows:

1. It is possible that a maximally great being exists.

2. If it is possible that a maximally great being exists, then a maximally great being exists in some possible world.

3. If a maximally great being exists in some possible world, then it exists in every possible world.

4. If a maximally great being exists in every possible world, then it exists in the actual world.

5. If a maximally great being exists in the actual world, then a maximally great being exists.

The principal issue to be settled with respect to Plantinga's ontological argument is what warrant exists for thinking the key premise (1) to be true. It is crucial in this regard to keep in mind the difference between metaphysical and merely epistemic possibility. One is tempted to say, "It's possible that God exists, and it's possible that he doesn't exist!" But this is true only with respect to epistemic possibility; if God is conceived as a maximally great being, then his existence is either metaphysically necessary or metaphysically impossible, regardless of our epistemic uncertainty. Thus, the epistemic entertainability of the key premise (or its denial) does not guarantee its metaphysical possibility.

It might be said that the idea of a maximally great being is intuitively a coherent notion and, hence, possibly instantiated. In this respect, the idea of God differs from supposedly parallel notions traditionally put forward by the argument's detractors, like the idea of a maximally great island or of a necessarily existent lion. The properties that go to make up maximal excellence have intrinsic maxima, whereas the excellent-making properties of things like islands do not. There could always be more palm trees or native dancing girls! Moreover, it is far from clear that there even are objective excellent-making properties of things like islands, for the excellence of islands seems to be relative to one's interests—does one prefer a desert island or an island boasting the finest resort hotels? The idea of something like a necessarily existent lion also seems incoherent. For, as a necessary being, such a beast would have to exist in every possible world that we can conceive. But any animal that could exist in a possible world in which the universe is comprised wholly of a singularity of infinite density just is not a lion. By contrast, a maximally excellent being could transcend such physical limitations and so be conceived as necessarily existent.

Sobel has objected that the intuitive possibility of a maximally great being ought to carry no weight in our assessment of the metaphysical possibility of the crucial premise "Possibly, a maximally great being exists."[29] For, he points out, a concept need not be prima facie incoherent, as in the case of a married bachelor, in order to be impossible. Thus, we cannot know *a priori* whether maximal greatness is possibly instantiated. But the illustration Sobel uses to undermine our confidence in our modal intuitions concerning the possibility of a maximally great being—namely, a "dragoon," a thing that, if it is possible, is a dragon in whichever world is the actual world—is not at all analogous to the concept of a maximally great being. On the contrary, while the concept of a necessary being seems intuitively possible, the concept of a dragoon, a contingent thing that mysteriously manages to track whichever world is actual so as to exist in the actual world while not existing in all possible worlds, seems intuitively impossible. Sobel's dragoon illustration therefore does little to undermine the force of our modal intuitions concerning the possibility of maximal greatness.

Perhaps the greatest challenge to the appeal to modal intuition to warrant the premise that maximal greatness is possibly exemplified is that it seems intuitively coherent in the same way to conceive of a quasi-maximally great being, say, one that is in every other respect maximally excellent save that it does not know truths about future contingents. Why is the key premise of the ontological argument more plausibly true than a parallel premise concerning quasi-maximal greatness? Are we not equally warranted in thinking that a quasi-maximally great being exists? Perhaps not; for maximal greatness is logically incompatible with quasi-maximal greatness. Since a maximally great being is omnipotent, no concrete object can exist independently of its creative power. As an omnipotent being, a maximally great being must have the power to freely refrain from creating anything at all, so that worlds must exist in which nothing other than the maximally great being exists. But that entails that if maximal greatness is possibly exemplified, then quasi-maximal greatness is not. A quasi-excellent being may exist in many worlds (worlds in which the maximally great being has chosen to create it), but such a being would lack necessary existence and thus not be quasi-maximally great. Hence, if maximal greatness is possibly exemplified, quasi-maximal greatness is impossible. Thus, our intuition that a maximally great being is possible is not undermined by the claim that a quasi-maximally great being is also intuitively possible, for we see that the latter intuition depends on the assumption that maximal greatness is not possibly exemplified, which begs the question.

Still, modal skeptics will insist that we have no way of knowing *a priori* whether maximal greatness or quasi-maximal greatness is possibly exemplified. It cannot be both, but we have no idea if either is possible. Our intuitions about modality are unreliable guides. Can anything more be said in defense of the ontological argument's key premise? Plantinga provides a clue when he says that if we "carefully ponder" the key premise and the alleged objections to it, if we "consider its connections with other propositions we accept or reject" and we still find it compelling, then we are within our rational rights in accepting it.[30] Such a procedure is a far cry from the sort of *a priori* speculations decried by the modal skeptic. Even if we cannot determine *a priori* whether maximal greatness is possibly exemplified, we may come to believe on the basis of *a posteriori* considerations that it is possible that a maximally great being exist. For example, other theistic arguments like the contingency argument and the moral argument may lead us to think that it is plausible that a maximally great being exists. The one argument leads to a metaphysically necessary being that is the ground of existence for any conceivable finite reality and the other to a locus of moral value that must be as metaphysically necessary as the moral values it grounds. Considerations of simplicity might also come into play here. For example, it is simpler to posit one metaphysically necessary, infinite, morally perfect being than to think that distinct necessary beings exist instantiating these various excellent-making properties. Similarly, with respect to quasi-maximally great beings, Richard Swinburne's contention seems plausible: it is simpler (or perhaps less ad hoc) to posit either zero or infinity as the measure of a degreed property than to posit some inexplicably finite measure. Thus, it would be more plausible to think that maximal greatness is possibly instantiated than quasi-maximal greatness. On the basis of considerations like these, we might well consider ourselves to be warranted in believing that it is possible that a maximally great being exists.

The question that arises at this point is whether the ontological argument has not then become question-begging. For it might seem that the reason one thinks that it is possible that a maximally great being exists is that one has good reasons to think that a maximally great being does exist. But this misgiving may arise as a result of thinking of the project of natural theology in too linear a fashion. The theistic arguments need not be taken to be like links in a chain, in which one link follows another so that the chain is only as strong as its weakest link. Rather, they are like links in a coat of chain-mail, in which all the links reinforce one another so that the strength of the whole exceeds that of any single link. The ontological argument might play its part in a cumulative case for theism, in which a multitude of factors simultaneously conspire to lead one to the global conclusion that God exists.

3. Conclusion

Traditional arguments for God's existence such as the above, not to mention creative new arguments, are alive and well on the contemporary scene in Anglo-American philosophy. Together with the failure of antitheistic arguments,[31] they help to explain the current renaissance of interest in theism.

Notes

1. Paul Benacerraf, "What Mathematical Truth Could Not Be—I," in *Benacerraf and His Critics,* ed. Adam Morton and Stephen P. Stich (Oxford: Blackwell: 1996), 18.
2. A. J. Ayer, *Language, Truth, and Logic* (New York: Dover Publications, 1952), Chapter VI: "Critique of Ethics and Theology."
3. Bas van Fraassen, "Empriricism in the Philosophy of Science," in *Images of Science,* P. Churchland and C. Hooker, eds. (Chicago: University of Chicago Press, 1985), 258.
4. Quentin Smith, "The Metaphilosophy of Naturalism" *Philo* 4/2 (2001): 3–4. A sign of the times: *Philo* itself, unable to succeed as a secular organ, has now become a journal for general philosophy of religion.
5. Ibid., 4.
6. See J. P. Moreland and William Lane Craig, *Philosophical Foundations for a Christian Worldview* (Downer's Grove, Ill.: Intervarsity, 2003), chaps. 7, 25–27; see also William Lane Craig, ed., *Philosophy of Religion: A Reader and Guide* (Edinburgh: Edinburgh University Press, 2002; New Brunswick, NJ: Rutgers University Press, 2002).
7. Richard Taylor, *Metaphysics,* 4th ed., Foundations of Philosophy (Englewood Cliffs, NJ: Prentice-Hall, 1991), 100–1.
8. Mark Balaguer, *Platonism and Anti-Platonism in Mathematics* (New York: Oxford University Press, 1998), part II; idem, "A theory of Mathematical Correctness and Mathematical Truth," *Pacific Philosophical Quarterly* 82 (2001): 87–114; *Stanford Encyclopedia of Philosophy, s.v.* "Platonism in Metaphysics," by Mark Balaguer (Summer 2004), ed. Edward N. Zalta, http://plato.stanford.edu/archives/sum2004/entries/platonism/; Charles S. Chihara, *Constructibility and Mathematical Existence* (Oxford: Clarendon Press, 1990); idem, *A Structural Account of Mathematics* (Oxford: Clarendon Press, 2004).
9. The story of Hilbert's Hotel is related in George Gamow, *One, Two, Three, Infinity* (London: Macmillan, 1946), 17.
10. Jordan Howard Sobel, *Logic and Theism: Arguments for and against Beliefs in God* (Cambridge: Cambridge University Press, 2004), 186–87.
11. Such a theory, often called the A-Theory of time, affirms that presentness is an objective, nonrelational feature of things or events and that temporal becoming is similarly absolute and objective. The so-called B-Theory of time denies these affirmations of the A-Theory, affirming that tenses are ascribed only relationally to things or events and that temporal becoming is purely phenomenal.
12. See William Lane Craig, *The Tensed Theory of Time: A Critical Examination,* Synthese Library 293 (Dordrecht: Kluwer Academic Publishers, 2000); idem, *The Tenseless Theory of Time: A Critical Examination,* Synthese Library 294 (Dordrecht: Kluwer Academic Publishers, 2000).
13. See Robin Small, "Tristram Shandy's Last Page," *British Journal for the Philosophy of Science* 37 (1986): 214–5.
14. John L. Mackie, *The Miracle of Theism* (Oxford: Clarendon Press, 1982), 93; Sobel, *Logic and Theism,* 182. *N.B.* that the objectors just assume that the ordinal type of the infinite series of past events is ω^*, the order type of the negative numbers . . . , -3, -2, -1, whereas the Tristram Shandy Paradox would compel us to ascribe to the series of past events the order type $\omega^* + \omega^*$, the order type of the series . . . , -3, -2, -1, . . . , -3, -2, -1. In handling the objection I shall follow the critics in their assumption.
15. See the argument in my "Naturalism and Cosmology," in *Analytic Philosophy without Naturalism,* ed. A. Corradini, S. Galvan, and J. Lowe (London: Routledge, 2005).
16. John Leslie, *Universes* (London: Routledge, 1989), 17.
17. S. W. Hawking, "Cosmology from the Top Down," paper presented at the Davis Cosmic Inflation Meeting, University of California, Davis, May 29, 2003.

18. Ernan McMullin observes, "It seems safe to say that any later theory, no matter how different it may be, will turn up approximately the same . . . numbers. And the numerous constraints that have to be imposed on these numbers . . . seem both too specific and too numerous to evaporate entirely" (Ernan McMullin, "Anthropic Explanation in Cosmology," paper presented at the conference "God and Physical Cosmology," University of Notre Dame, January 30–February 1, 2003).

19. The statistical approach is championed by William A. Dembski, *The Design Inference: Eliminating Chance through Small Probabilities,* Cambridge Studies in Probability, Induction, and Decision Theory (Cambridge: Cambridge University Press, 1998), the Bayesian approach by Robin Collins, *The Well-Tempered Universe* (forthcoming).

20. See T. Rothman and G. F. R. Ellis, "Smolin's Natural Selection Hypothesis," *Quarterly Journal of the Royal Astronomical Society* 34 (1993): 201–12.

21. See John Earman and Jesus Mosterin, "A Critical Look at Inflationary Cosmology," *Philosophy of Science* 66 (1999): 1–49; Collins, *Well-Tempered Universe.*

22. I owe this point to Robin Collins.

23. In fact, Barrow and Tipler list ten steps in the evolution of *homo sapiens*, each of which is so improbable that before it would occur the Sun would have ceased to be a main sequence star and incinerated the Earth (John Barrow and Frank Tipler, *The Anthropic Cosmological Principle* [Oxford: Clarendon Press, 1986], 561–5)!

24. See Peter Lipton, *Inference to the Best Explanation* (London: Routledge, 1991).

25. Richard Taylor, *Ethics, Faith, and Reason* (Englewood Cliffs, NJ: Prentice-Hall, 1985), 14.

26. Paul Kurtz, *Forbidden Fruit* (Buffalo, N.Y.: Prometheus Books, 1988), 65.

27. Some philosophers seem to suppose that moral truths, being necessarily true, cannot have an explanation of their truth. The crucial presupposition that necessary truths cannot stand in relations of explanatory priority to one another is not merely not evidently true, but seems plainly false. For example, on a non-fictionalist account *2+3=5* is necessarily true because the Peano axioms for standard arithmetic are necessarily true. Or again, *No event precedes itself* is necessarily true because *Temporal becoming is an essential and objective feature of time* is necessarily true. It would be utterly implausible to suggest that the relation of explanatory priority obtaining between the relevant propositions is symmetrical.

28. Alvin Plantinga, *The Nature of Necessity,* Clarendon Library of Logic and Philosophy (Oxford: Clarendon Press, 1974), chap. X.

29. Sobel, *Logic and Theism,* 92–4.

30. Plantinga, *Nature of Necessity,* 221.

31. See note 6.

Acknowledgments

The book you hold in your hands would never have seen the light of day were it not for the generous financial support of a variety of institutions, and the encouragement of many colleagues and friends. The initial grant that provided the research leave at Baylor University necessary to get this book project off the ground came from Grace Valley Christian Center in Davis, California. The editors extend their grateful thanks to its pastor, P. G. Mathew, and to Richard Spencer and Gerrit Buddingh', for understanding the significance of this work, and for their clarity of vision and commitment to it. The editors are supremely grateful as well for the unflagging support of Discovery Institute's Center for Science and Culture in helping us see this work through to completion, and especially for the personal encouragement and advice of Stephen Meyer, John West, and Bruce Chapman.

We thank the financial sponsors of the original academic conference at Baylor University, held in April 2000, from which this book draws its name and most of its contributors. In particular, we offer our heartfelt thanks to Baylor University's Institute for Faith and Learning and its director at the time, Michael Beaty, as well as to Baylor's former president, Robert Sloan, and former provost, Donald Schmeltekopf. We are also very grateful to the Discovery Institute, the Templeton Foundation, and to *Touchstone Magazine* and its editor James Kushiner, for their financial support and promotion of the conference.

Our gratitude is due to each of the contributors to this volume, especially for essays written specifically for this book, but also for permission to reprint here recent work that is representative of their views and relevant to the theme under consideration. Additional information regarding reprinted essays is available on the sources and permissions page.

The editors also wish to thank Janet Oberembt and Denyse O'Leary, respectively, for extensive help in formatting the original manuscript, and for supplementary and comprehensive editorial and proofreading assistance with the electronic galley proofs. Special thanks are due to Denyse for wrestling to the ground the mammoth task of preparing the name and subject indices for the volume. We also want to thank the editors at ISI Books, particularly Jed Donahue and Jennifer Fox, for their persistence in seeing this project through to completion.

Last, but certainly not least, we thank our wives, Mari-Anne Gordon and Jana Dembski, for keeping the home fires burning, and for their constant encouragement and loving support.

Contributors

Douglas D. Axe, Director of Biologic Institute in Redmond, Washington, where he oversees fundamental biological and computational research bearing on intelligent design. With a doctorate from Caltech in biochemical engineering, he spent a decade at the University of Cambridge working on the evolution of protein folds before assuming his present position, and is the author of a variety of papers in molecular biology journals of seminal significance for intelligent design research.

Michael Behe, Professor of Biological Sciences at Lehigh University and Senior Fellow with Discovery Institute's Center for Science and Culture. In addition to a solid record of peer-reviewed publications in the biological literature, his books *Darwin's Black Box* and *The Edge of Evolution* have broken new ground in the field of intelligent design. His notion of irreducible complexity continues to be intensely debated.

David Berlinski, widely acclaimed author and Senior Fellow with Discovery Institute's Center for Science and Culture. With a doctorate in philosophy from Princeton University, he is the author of many books and essays and has become a trenchant critic of scientific pretentions, directing his skepticism toward both scientific materialism and intelligent design. His book *The Devil's Delusion* addresses the neo-atheists.

Leda Cosmides, Professor of Psychology at the University of California, Santa Barbara. With an undergraduate degree in biology and a doctorate in cognitive psychology (both from Harvard), she, along with her anthropologist husband John Tooby, have been seminal figures and prolific researchers in the development of evolutionary psychology, which attempts to understand human psychology in terms of evolution.

Contributors

William Lane Craig, Research Professor of Philosophy at Talbot School of Theology in La Mirada, California. The author and editor of over thirty books, including *The Kalam Cosmological Argument*; *Theism, Atheism, and Big Bang Cosmology*; *God, Time, and Eternity*; and *The Blackwell Companion to Natural Theology*, he has also published widely in the professional journals of philosophy and theology.

Francis Crick, the late J. W. Kieckhefer Distinguished Research Professor at the Salk Institute for Biological Studies in La Jolla, California. Crick shared the Nobel Prize for his work uncovering the structure of DNA. Later in his career, he turned to research in neuroscience and the physical basis of consciousness.

Christian de Duve, Professor Emeritus at the Medical Faculty of the University of Louvain, Belgium and Andrew W. Mellon Professor Emeritus at the Rockefeller University in New York. He shared the Nobel Prize for discoveries on the structural and functional organization of cells. His books *Vital Dust* and *Singularities: Landmarks on the Pathways of Life* defend a materialist vision of life's origin.

William A. Dembski, Research Professor in Philosophy at Southwestern Seminary in Ft. Worth and Senior Fellow with Discovery Institute's Center for Science and Culture. With doctorates in both mathematics and the philosophy of science, he is a prolific author who published the first academic monograph on intelligent design (*The Design Inference*, Cambridge) and founded the first intelligent-design research center at a major university (Baylor's Michael Polanyi Center).

Steve Fuller, Professor of Sociology at the University of Warwick, England, with a doctorate in the history and philosophy of science from the University of Pittsburgh. A prolific author, he has become a key defender of intelligent design on grounds of academic freedom, arguing that academic freedom must allow vigorous dissent from orthodoxy and consensus.

Alvin Goldman, Board of Governors Professor of Philosophy and Cognitive Science at Rutgers University. Trained in philosophy at Columbia and Princeton, his work has focused on naturalistic accounts of human action and knowledge. A cross-disciplinary scholar, he has explored connections between cognitive science and analytic epistemology.

Guillermo Gonzales, Associate Professor of Physics at Grove City College and Senior Fellow with Discovery Institute's Center for Science and Culture. His work on the galactic habitable zone was featured on the cover of *Scientific American*. His book *The Privileged Planet* (co-authored with Jay Richards) examined new lines of evidence for design in astronomy and cosmology.

Contributors

Bruce L. Gordon, Associate Professor of Science and Mathematics at The King's College in New York and past Research Director for Discovery Institute's Center for Science and Culture. Formerly on the faculty of Baylor University, he has written on the interpretation of modern physics, the history and philosophy of science, metaphysics, and the philosophy of religion.

Alan Guth, Victor F. Weisskopf Professor of Physics at the Massachusetts Institute of Technology. The inventor of a new cosmological model called "inflation," he applies theoretical particle physics to understanding the early universe. A prolific researcher in theoretical physics and author of the acclaimed popular science book *The Inflationary Universe: The Quest for a New Theory of Cosmic Origins*, much of his current work focuses on the density fluctuations arising from inflation, and studying the possibility of inflation in brane-world models.

Stuart Hameroff, Professor Emeritus of Anesthesiology and Psychology and Director of the Center for Consciousness Studies, the University of Arizona, Tucson. In attempting to uncover a quantum-theoretic basis for consciousness, his research has focused on the molecular mechanisms of anesthetic gas molecules and information processing in cytoskeletal microtubules.

Philip Kitcher, John Dewey Professor of Philosophy and James R. Barker Professor of Contemporary Civilization at Columbia University. A philosopher of science and mathematics, he is especially interested in how ethics and politics constrain scientific research, how altruism and morality evolve, and how science and religion come into apparent conflict. He is the author of *Abusing Science: The Case Against Creationism*; *The Nature of Mathematical Knowledge*; *Vaulting Ambition: Sociobiology and the Quest for Human Nature*; *Science, Truth and Democracy*; and recently, *Living with Darwin: Evolution, Design, and the Future of Faith*.

Christof Koch, Lois and Victor Troendle Professor of Cognitive and Behavioral Biology at Caltech. Straddling the fields of biology, physics, and engineering, he attempts to understand the neurophysiological basis of consciousness. Besides tracking the neuronal correlates of consciousness, he also investigates the biophysical mechanisms underlying computation (information processing) at the level of synapses, channels and membranes. In addition to many articles in scientific journals and a number of books, he is most recently the author of *The Quest for Consciousness: A Neurobiological Approach*.

Robert C. Koons, Professor of Philosophy at the University of Texas at Austin. Specializing in formal semantics, decision theory, and causation, he has authored a variety of philosophical articles and two important books: *Paradoxes of Belief and Strategic Rationality* (Cambridge) and *Realism Regained: An Exact Theory of Causation, Teleology, and the Mind* (Oxford). He organized an early seminal conference on the topic of this volume: *Naturalism, Theism, and the Scientific Enterprise* (1997).

Contributors

Robert J. Marks II, Distinguished Professor in the Department of Electrical and Computer Engineering at Baylor University and Director of the Evolutionary Informatics Lab (www.evoinfo.org), he is the author of numerous technical articles and a number of textbooks. A pioneer in the field of computational intelligence, he has broken new ground for intelligent design through his work on conservation of information as applied to evolutionary processes.

Ernan McMullin, John Cardinal O'Hara Professor of Philosophy Emeritus at the University of Notre Dame. An ordained Roman Catholic priest, he is a wide-ranging and prolific historian and philosopher of science. His numerous books include *Evolution and Creation*, *The Inference That Makes Science*, *The Church and Galileo*, and (co-edited with James Cushing) *Philosophical Consequences of Quantum Theory: Reflections on Bell's Theorem*.

Stephen C. Meyer, Director of Discovery Institute's Center for Science and Culture in Seattle, Washington. A Cambridge-University trained philosopher of science, he is the author of peer-reviewed publications in technical, scientific, and philosophical journals. His widely acclaimed *Signature in the Cell: DNA and the Evidence for Intelligent Design* (HarperOne) marks a signal advance for ID theory.

James P. Moreland, Distinguished Professor of Philosophy at Biola University and Director of the Eidos Christian Center. He publishes widely in the philosophical literature. Among his many books are *Christianity and the Nature of Science*, *Philosophical Foundations of a Christian Worldview* (co-authored with William Lane Craig), *Naturalism: A Critical Analysis* (co-edited with William Lane Craig), *Body and Soul: Human Nature and the Crisis in Ethics* (with Scott Rae), and *Consciousness and the Existence of God*.

Nancey Murphy, Professor of Christian Philosophy at Fuller Theological Seminary. With doctorates in theology and philosophy, she has concentrated on expanding theological inquiry to incorporate insights from philosophy and science, as evident in her book *Theology in the Age of Scientific Reasoning*, and her co-edited works *Downward Causation and the Neurobiology of Free Will*; *Physics and Cosmology: Scientific Perspectives on the Problem of Natural Evil*; and *Evolution and Emergence: Systems, Organisms, Persons*.

Ronald Numbers, Hilldale Professor of the History of Science and Medicine at the University of Wisconsin, Madison. An expert on how Darwinism has impacted American thought, he has written extensively on the continuing challenges to Darwinism within American culture, notably, in his books *Darwinism Comes to America* and *The Creationists: From Scientific Creationism to Intelligent Design*.

Contributors

Roger Penrose, Rouse Ball Professor Emeritus of Mathematics at the Mathematical Institute, University of Oxford. A preeminent mathematical physicist and recipient of numerous awards (such as the Wolf Prize, shared with Stephen Hawking), he has also written books on the broader significance of science, notably, *The Emperor's New Mind: Concerning Computers, Minds, and the Laws of Physics*, and most recently, a masterful survey of modern physics titled *The Road to Reality: A Complete Guide to the Laws of the Universe*.

Alvin C. Plantinga, John A. O'Brien Professor of Philosophy Emeritus at the University of Notre Dame. For many years on faculty at Calvin College, he specializes in epistemology, metaphysics, and philosophy of religion and is the author of a variety of influential books, including *The Nature of Necessity*, and a trilogy on epistemology comprised of the volumes *Warrant: The Current Debate*, *Warrant and Proper Function*, and *Warranted Christian Belief*, all from Oxford University Press. The renaissance of a deeply influential, high-caliber Christian philosophy within the mainstream academy is credited, above all, to Plantinga.

Mark Ptashne, formerly the Herschel Smith Professor of Molecular Biology at Harvard and now for many years the Ludwig Chair of Molecular Biology at Memorial Sloan-Kettering Cancer Center, New York. His work on gene regulation has received numerous awards, including the Lasker Prize for Basic Research. He is the author of numerous scientific articles and two books, *A Genetic Switch* and *Genes and Signals* (co-authored with Alexander Gann).

Fazale Rana, Executive Vice President of Research and Apologetics at *Reasons to Believe*. With a doctorate in chemistry from Ohio University, his research has focused on the naturalistic barriers facing life's origin and subsequent development. His work has appeared in *Origins of Life and Evolution of Biospheres*, *Biochemistry*, *Applied Spectroscopy*, *FEBS Letters*, *Journal of Microbiological Methods*, and *Journal of Chemical Education*.

Michael Ruse, Lucyle T. Werkmeister Professor of Philosophy at Florida State University. A philosopher of biology, his testimony in *McLean v. Arkansas* (1981) was critical in overturning an equal-time law requiring the teaching of creationism. A prolific author, his books include *Darwin and Design: Does Evolution have a Purpose?*; *Can a Darwinian be a Christian?*; *Defining Darwin: Essays on the History and Philosophy of Evolutionary Biology*; and the edited collection *Philosophy of Biology*.

John Searle, Slusser Professor of Philosophy at the University of California, Berkeley. Educated at Oxford, where he was a Rhodes Scholar, he is a preeminent philosopher of mind and language, focusing on speech acts, consciousness, intentionality, and the construction of social reality. A prolific author, his most recent book is *Making the Social World: The Structure of Human Civilization*.

Contributors

Michael Shermer, Founding Publisher of *Skeptic* magazine, Executive Director of the Skeptics Society, monthly columnist for *Scientific American*, host of the Skeptics Distinguished Science Lecture Series at Caltech, and Adjunct Professor of Economics at Claremont Graduate University. His many books include *The Mind of the Market* and *Why Darwin Matters*.

Henry Stapp, research scientist with the Theoretical Physics Group at the Lawrence Berkeley National Laboratory, University of California, Berkeley. A post-doctoral researcher with Wolfgang Pauli, he has focused on the foundations of quantum mechanics, especially in connection with the problem of consciousness. In addition to many scientific articles and essays, he is the author of the books *Mind, Matter, and Quantum Mechanics* and *Mindful Universe: Quantum Mechanics and the Participating Observer*.

Mark Steiner, Professor of Philosophy at the Hebrew University of Jerusalem. With a doctorate in philosophy from Princeton, he writes on the application of mathematics to the natural sciences and is especially interested in the epistemological implications of the match between formal methods and physical reality. He is the author of *The Applicability of Mathematics as a Philosophical Problem* (Harvard).

William Talbott, Professor of Philosophy at the University of Washington, Seattle. A student of Robert Nozick with a doctorate from Harvard, he specializes in epistemology, ethics, rational choice theory, social and political philosophy, the philosophy of human rights, and the philosophy of law. He is the author of *Which Rights Should Be Universal?* (Oxford).

John Tooby, Professor of Anthropology at the University of California, Santa Barbara. With his wife Leda Cosmides, he is a prolific researcher, bringing together cognitive science, cultural anthropology, evolutionary biology, paleoanthropology, cognitive neuroscience, and hunter-gatherer studies to develop the new field of evolutionary psychology.

Michael Tooley, Distinguished College Professor of Philosophy at the University of Colorado, Boulder. Known for his work on metaphysics and applied ethics, he is the author of several books. He edited a comprehensive five-volume collection (from Routledge) titled *Analytical Metaphysics*, and is the author of *Causation: A Realist Approach*, as well as *Time, Tense, and Causation* (both from Oxford).

Howard Van Till, Professor Emeritus of Astronomy and Physics at Calvin College. His books in the 1980s (notably, *Science Held Hostage* and *The Fourth Day*) played a powerful role in moving evangelical higher education to accept theistic evolution over against creationism. He emphasizes a form of "theological naturalism" that assimilates divine governance to natural processes in terms of a "robust formational economy principle."

Contributors

Steven Weinberg, Jack S. Josey-Welch Foundation Chair in Science and Regental Professor at the University of Texas at Austin. For his work on unifying the fundamental forces of nature, he shared the Nobel Prize. An extraordinarily prolific and influential theoretical physicist, he is the author of countless research articles and numerous textbooks. He believes a primary cultural role for science should be the attenuation of religious belief, and frequently emphasizes this point in his popular writings.

Dallas Willard, Professor in the School of Philosophy at the University of Southern California in Los Angeles. His philosophical publications focus on epistemology, the philosophy of mind and logic, and the philosophy of Edmund Husserl. He has also written extensively on Christian spirituality, including *The Spirit of the Disciplines* and *Renovation of the Heart*.

Michael Williams, Krieger-Eisenhower Professor and Chair of the Department of Philosophy at Johns Hopkins University. A student of Richard Rorty with a doctorate from Princeton, he is a prominent epistemologist whose work has focused on responding to philosophical skepticism. He is the author of *Groundless Beliefs: An Essay on the Possibility of Epistemology*; *Unnatural Doubts: Epistemological Realism and the Basis of Scepticism*; and *Problems of Knowledge: A Critical Introduction to Epistemology*.

Nicholas Wolterstorff, Noah Porter Professor Emeritus of Philosophical Theology at Yale University. A graduate of Calvin College—where he also taught for thirty years—with a doctorate from Harvard, his interests in philosophy have been wide-ranging, focusing on universals, philosophy of art, the epistemology of John Locke and Thomas Reid, philosophy of religion, and political philosophy. His books include *On Universals: An Essay in Ontology*; *Thomas Reid and the Story of Epistemology*; *John Locke and the Ethics of Belief*; and *Justice: Rights and Wrongs*.

SOURCES AND PERMISSIONS

The editors wish to thank the authors and the following publishers for permission to reprint previously published materials:

Ronald L. Numbers, "Science without God: Natural Laws and Christian Beliefs," from *When Science and Christianity Meet*, edited by David C. Lindberg and Ronald L. Numbers, 265–85. Copyright © 2003 by the University of Chicago. Reprinted by permission of University of Chicago Press.

Robert C. Koons, "The Incompatibility of Naturalism and Scientific Realism," from *Naturalism: A Critical Analysis*, edited by William Lane Craig and J. P. Moreland, 43–69. Copyright © 2000 by Routledge. Reprinted by permission of Taylor & Francis Books, UK.

"Introduction" and "Truth and Realism" reprinted by permission of the publisher from EPISTEMOLOGY AND COGNITION by Alvin I. Goldman, 1–5, 7, 142–61, Cambridge, MA: Harvard University Press, Copyright © 1986 by the President and Fellows of Harvard College.

David Berlinski, "On the Origins of Life" (February 2006). Reprinted from COMMENTARY, Vol. 121, No. 2, February 2006, 22–33, by permission. Copyright © 2006 by Commentary, Inc.

Christian de Duve, "Mysteries of Life: Is There 'Something Else'?" *Perspectives in Biology and Medicine*, 1–15. © 2002 by Johns Hopkins University Press–JOURNALS. Reprinted by permission of JOHNS HOPKINS UNIVERSITY PRESS in accordance with Copyright Clearance Center Confirmation Number 2320539.

Mark Ptashne, "Regulated recruitment and cooperativity in the design of biological regulatory systems," *Philosophical Transactions of the Royal Society A: Mathematical, Physical and Engineering Sciences*, Vol. 361, 1223–234. Copyright © 2003 by The Royal Society of London. Reprinted by permission of The Royal Society.

Figures 1.6a (19), 1.7 (21), and 1.8 (23) from Mark Ptashne and Alexander Gann, *Genes and Signals* (Cold Spring Harbor: Cold Spring Harbor Laboratory Press, Copyright © 2002), are reproduced as Figures 1, 2, and 3 (respectively) in Chapter 17 with the permission of Cold Spring Harbor Laboratory Press.

Sources and Permissions

Figure 7 in Chapter 17 is adapted from Mark Ptashne and Alexander Gann, "Imposing specificity by localization: mechanism and evolvability," *Current Biology*, Vol. 8, No. 22 (November 1998), 812–22. Copyright © 1998 by Elsevier Limited. Used by permission of Elsevier in accordance with Copyright Clearance Center License Number 2327840017854.

Figure 9 in Chapter 17 is reproduced from Brenton R. Graveley, Klemens J. Hertel, and Tom Maniatis, "SR proteins are 'locators' of the RNA splicing machinery," *Current Biology*, Vol. 9, R6–R7. Copyright © 1999 by Elsevier Limited. Reproduced by permission from Elsevier in accordance with Copyright Clearance Center License Number 2377820824573.

Fazale Rana, *The Cell's Design* (Grand Rapids: Baker Books, a division of Baker Publishing Group, 2008). Copyright © 2008 by Baker Publishing Group. Material from pages 203–224 modified with the author's approval and used by permission of Baker Publishing Group.

Alan H. Guth, "Eternal Inflation and Its Implications," *Journal of Physics A: Mathematical and Theoretical*, Vol. 40, 2007, 6811–826. Copyright © 2007. Reprinted by permission of Institute of Physics Publishing Ltd., http://www.iop.org/journals/jphysa.

Steven Weinberg, "Living in the Multiverse," from Bernard Carr, ed. *Universe or Multiverse?* Cambridge: Cambridge University Press, 2007, 29–42. Copyright © 2007 Cambridge University Press. Reprinted with the permission of Cambridge University Press.

Philip Kitcher, "Mathematical Naturalism," in "History and Philosophy of Modern Mathematics", *Minnesota Studies in the Philosophy of Science, Vol. XI*, edited by William Aspray and Philip Kitcher, 293–325. Copyright © 1988 by the University of Minnesota. Reproduced by permission of The University of Minnesota Press.

OXFORD HANDBOOK OF PHILOSOPHY OF MATHEMATICS AND LOGIC, edited by Stewart Shapiro. Chapter 20, "Mathematics—Application and Applicability," by Mark Steiner. Oxford: Oxford University Press, 2005, 625–50. Copright © 2005 by Oxford University Press. Reproduced by permission of Oxford University Press, Inc.

John Tooby and Leda Cosmides, "Toward Mapping the Evolved Functional Organization of Mind and Brain," in Michael Gazzaniga, editor, *The New Cognitive Neurosciences, Second Edition*. Cambridge: MIT Press, 1999. Copyright © 1999 Massachusetts Institute of Technology. Reprinted by permission of The MIT Press.

David Berlinski, "On the Origins of Mind" (November 2004). Reprinted from COMMENTARY, Vol. 118, No. 4, November 2004, 26–35, by permission. Copyright © 2004 by Commentary, Inc.

John R. Searle, "Consciousness," *Annual Review of Neuroscience*, Vol. 23, 2000, 557–78. Reprinted with permission, from *Annual Review of Neuroscience*, Volume 23, © 2000 by Annual Reviews, www.annualreviews.org.

Francis Crick and Christof Koch, "Consciousness and Neuroscience," from *Cerebral Cortex*, Vol. 8, 1998, 97–107. Copyright © 1998. Reprinted by permission of Oxford University Press in accordance with Copyright Clearance Center License Number 2300990044932.

Figure 1 in Chapter 33 is adapted from *Neuron*, Vol. 30, No. 2, May 2001, Joaquín M. Fuster, "The Prefrontal Cortex—An Update: Time Is of the Essence," Figure 10, page 329. Copyright © 2001 by Elsevier Limited. Used by permission of Elsevier in accordance with Copyright Clearance Center License Number 2327941299982.

Figure 2 in Chapter 33 is taken from FIG. 4B, page 3411 of D.L. Sheinberg and N.K. Logothetis, "The role of temporal cortical areas in perceptual organization," *Proceedings of the National Academy of Sciences U.S.A.*, Vol. 94, 3408–413, April 1997. Copyright © 1997, National Academy of Sciences, U.S.A. Used with permission.

Sources and Permissions

Figure 3 of Chapter 33 is reproduced from Sheng He, Patrick Cavanagh, and James Intriligator, "Attentional resolution and the locus of visual awareness," *Nature,* Vol. 383, 343–37. Copyright © 1996 by Nature Publishing Group. Reproduced by permission of Nature Publishing Group in accordance with Copyright Clearance Center License Number 2346600602779.

The major part of Chapter 34 was adapted by the author, Nancey Murphy, from her essay "Supervenience and the Downward Efficacy of the Mental: A Nonreductive Physicalist Account of Human Action," in Robert J. Russell *et al.*, eds. *Neuroscience and the Person: Scientific Perspectives on Divine Action* (Vatican City State: Vatican Observatory Press, 1999), 147–64. Copyright © 1999. The author and editors thank the Vatican Observatory for permission to use this material here.

Stuart Hameroff and Roger Penrose, "Conscious Events as Orchestrated Space-Time Selections," *Journal of Consciousness Studies* 1996, Vol. 3, No. 1, 36–53. Copyright © 1996. Adapted and reprinted with permission from Imprint Academic.

Figure 3 in Chapter 35 is a reproduction of Figure 7.4, entitled "A Microtubule," from page 359 of *Shadows of the Mind,* by Roger Penrose. Copyright © 1994 by Oxford University Press. Used by permission of Oxford University Press, Inc.

Figure 5 in Chapter 35 is reproduced from *Physica D: Nonlinear Phenomena,* Vol. 42, June 1990, Steen Rasmussen, Hasnain Karampurwala, Rajesh Vaidyanath, Klaus S. Jensen, and Stuart Hameroff, "Computation connectionism with neurons: A model of cytoskeletal automata subserving neural networks," 428–49. Copyright © 1990. Used with permission from Elsevier in accordance with Copyright Clearance Center License Number 2331001467945.

Figure 7 in Chapter 35 is reproduced from B. Libet, E. W. Wright Jr., B. Feinstein, and D.K. Pearl, "Subjective referral of the timing for a conscious sensory experience: A functional role for the somatosensory specific projection system in man," *Brain* 1979, Vol. 102 (March), 193–224. Copyright © 1979. Used by permission of Oxford University Press in accordance with Copyright Clearance Center License Number 2330340739767.

Henry P. Stapp, "Quantum Interactive Dualism II: The Libet and Einstein-Podolsky-Rosen Causal Anomalies," *Erkenntnis,* Vol. 65, No. 1, 2006, 117–42. Copyright © 2006. Reprinted with the permission of Springer in accordance with Copyright Clearance Center License Number 2300980825495.

Chapter 39 was adapted by the author, Dallas Willard, from his paper read at *The Nature of Nature* conference at Baylor University (April 12–15, 2000) and subsequently published as "Naturalism's Incapacity to Capture the Good Will," in *Philosophia Christi,* Vol. 4, No. 1, 2002, 9–28. Copyright © 2002. The author and editors are grateful to *Philosophia Christi* for permission to use the material here.

NAMES INDEX

Adami, Christoph, 369, 370, 372, 373
Adams, F., 622
Adams, Robert, 883
Adelard of Bath, 63
Aguirre, A., 500
Akaike, Hirotugu, 223–25
Albert of Saxony, 15
Albertus Magnus, 19, 717, 718
Albrecht-Buehler, Guenter, 805
al-Dïn al-Färisï, Kamäl, 20
al-Haytham, Ibn, 20–21
Alifanov, Oleg, 723
Allen, A. P., 604
Allman, John, 753
Alston, William P., 147, 326
Altman, Sidney, 282
Anscombe, Elizabeth, 872, 877
Anselm, 882
Aquinas, Thomas, xiii, 19, 713, 858, 865, 874
Arbib, Michael, 782, 783, 785
Aristotle, xiii, 362, 364, 506, 679, 781, 853, 858, 865, 867, 871, 874
Armani-Hamed, N., 551, 552
Armitage, P., 607
Armstrong, David, 137, 880
Arnhart, Larry, 877
Arrhenius, Gustaf, 286
Ashok, S. K., 548
Aspect, Alain, 181
Astbury, William, 295
Atkins, Peter, 355, 542, 543
Atobe, K., 608

Augustine, 19, 843, 874
Averroes, xiii
Avery, Oswald, 296
Axe, Douglas, 273, 306–7
Ayala, Francisco, 101, 365, 776
Ayer, A. J., 902

Bacon, Francis, 12–13, 64
Bacon, Roger, 21
Baird, Robert, xxi
Baker, David, 201
Balaguer, Mark, 906
Baltimore, David, 363, 448
Banks, Thomas, 570
Barberini, Matteo, 16–17
Barricelli, Nils, 368, 387
Barrow, John D., 330, 508, 516–17
Bartels, A., 736, 739
Bateson, Gregory, xiv
Bealer, George, 836–37
Begelman, M. C., 618
Behe, Michael J., 74, 273, 349–50, 351, 352–53, 883, 888–92, 894
Bell, John, 183
ben Gerson, Levi, 21
Benacerraf, Paul, 901
Benitez, N., 613
Bentham, Jeremy, 870
Bergmann, Max, 295
Bergson, Henri, 352
Berkeley, George, 203, 653, 667, 668
Berlinski, David, 273, 369, 370, 692

Index

Blakeslee, Sandra, 369
Bohr, Niels, 807, 819, 824, 825, 826–27, 828, 829
Boltzmann, Ludwig, 521
Bonaparte, Napoleon, 68
Bond, G., 614
BonJour, Laurence, 173
Boorstin, Daniel, 14
Borde, Arvind, 514–15, 520, 562, 574
Borel, Emile, 308
Born, Max, 793
Bousso, R., 486, 548, 559, 575, 578, 579
Boyd, Richard, 241
Boyle, Robert, 22, 64
Bozza, V., 577
Bracht, John, 370
Bradley, D. C., 753–54, 873, 874
Bragg, Lawrence, 296
Brandom, Robert, 249, 252
Brandt, Richard, 871
Brecht, Bertolt, 16
Brickner, Richard M., 757
Brillouin, Leon, 371
Brothers, Leslie, 786
Bruno, Giordano, 14, 15, 17
Büchner, Ludwig, 877
Buffon, Georges-Louis Leclerc de, 67–68
Bulterow, Alexander, 284
Burckhardt, Jacob, 13
Buridan, Jean, 15, 63
Burley, Walter, 15
Buss, David, 714
Buss, Leo, 446
Butler, Joseph, 865

Cairns-Smith, A. G., 304
Caldera, K., 606
Calura, F., 621
Calvin, John, 8, 9, 771
Calvin, Melvin, 448
Campbell, Donald, 228, 235, 772, 778, 779, 780–81, 782
Campbell, George Douglas, 72
Cantor, Georg, 654, 906, 907
Capellaro, E., 621
Caputo, Nicholas, 895, 896
Carroll, Sean, 439–40
Cauchy, Augustin-Louis, 644, 646, 653, 661
Cech, Thomas, 282
Chalmers, David, 759, 790, 806
Chalmers, Thomas, 68

Chamberlain, Thomas, 324
Chan, Sunney, 891
Changeux, Jean-Pierre, 780
Chargaff, Erwin, 296–97, 303
Charron, Pierre, 11
Chauncy, Charles, 67
Chesterton, G. K., 360–61
Chihara, Charles S., 906
Christina (Sweden), 11
Churchland, Patricia, 139–40, 216
Churchland, Paul, 216, 781, 837, 838
Cicero, 21
Clarke, J. N., 616
Clarke, Samuel, 9
Clerk Maxwell, James, 717
Clifton, Rob, 199
Cohen, I. Bernard, 16
Cohen, Jack, 447, 449
Cole, M. N., 313
Coleman, S., 553
Collins, Francis, 365, 393
Columbus, Christopher, 14, 15
Condillac, Étienne Bonnot de, 558
Condorçet, Marquis de, 13, 877
Conway Morris, Simon, xv, xvi, 323, 447–48, 462, 479, 861
Copernicus, Nicolaus, 14, 15, 19, 216, 648
Copleston, Frederick, 506
Correia, A. C. M., 609
Cosmides, Leda, 691–92, 714, 724
Cotterill, R., 727
Coyne, Jerry, 361, 391
Craig, William Lane, 485, 559–60, 854
Crick, Francis, 277–79, 280–82, 296, 300, 303, 304, 448, 692, 693, 727, 735, 736, 737–38, 844, 845–46
Crosby, J. L., 368
Cudworth, Ralph, 9

d'Alembert, Jean, 646
Damasio, A., 727
Dar, A., 615
Darwin, Charles, xiii, 71–73, 101–2, 107, 115, 139, 140, 251, 276, 294, 325, 327, 332n6, 350, 352, 353, 361, 429, 431, 432, 440, 445, 460, 462, 554, 704, 862, 877, 888
Davidson, Donald, 772–73
Davies, P. C. W., 303, 352, 508, 523, 527, 533n107
Davis, Edward B., 64

Index

Dawkins, Richard, xvi, 24, 75, 101, 137, 302, 312, 365–67, 371–372, 373, 389–90, 392–93, 720, 861, 877
De Rujula, A., 615
De Vito, Scott, 224
De Witt, Bryce, 564, 565
de Duve, Christian, 273, 311, 313, 314, 315, 322
de Maricourt, Petrus Peregrinus, 21
de Molina, Luis, 28
de Morgan, Augustus, 426, 428*n35*
de Muris, Johannes, 21
de Muynck, Willem, 196–98
Dean, Anthony M., 890
Dedekind, Richard, 643, 654, 659
de'Luzzi, Mondino, 16
Dembski, William A., xiv, xix–xxiv, 32, 99, 273, 301, 307–8, 310, 330, 335*n61*, 336*n62,63*, 339*n128*, 339–40*n141*, 351, 370, 414, 894–97
Democritus, 21, 225, 682
Dennett, Daniel C., 137, 216, 447, 448, 541, 543, 714, 877
Denton, Michael, 353, 354, 355
Descartes, René, 9, 11, 12, 22, 64, 67, 253, 649, 667, 713, 734–35, 771, 861
Devlin, Keith, 385
Dewey, John, xi, 868, 877
Dicke, R. H., 489
Dimopoulos, S., 551, 552
Dine, Michael, 550, 580
Dirac, Paul, 217, 793
Dirichlet, Peter, 659
Distler, J., 552
Dobzhansky, Theodosius, xvi, 311, 695
Doolittle, Ford, 890
Doolittle, Russell, 891
d'Oresme, Nicole, 15, 22
Douglas, M., 548, *557n28*, 580
Douglas, M. R., 550
Douglas, Mary, 718
Douglass, William, 65, 66
Drake, Frank, 444
Draper, J. W., 13
Dretske, Fred, 151*n17*, 218, 360
Duhem, Pierre, 843
Dulles, John Foster, 860
Dummett, Michael, 230–31, 232, 234, 236, 681
Dunn, James, 771
duVair, Guillaume, 11
Dyson, F., 319
Dyson, L., 582

Earman, J., 569
Eccles, John C., 726, 758, 844
Eddington, Arthur, 509
Edelman, Gerald M., 727, 740, 741, 750, 780, 783
Edwards, Harold, 660
Edwards, Jonathan, 203
Eigen, Manfred, 318, 319, 350, 363, 448
Einstein, Albert, 23, 179, 216, 217, 350, 507, 547, 565, 670, 695, 793, 824, 825–26, 828
Eisner, Thomas, 448
Eliot, T. S., 3
Enç, Berent, 774
Engels, Friedrich, xiii–xv
Engle, S., 756
English, Thomas, 382
Epictetus, 865
Epicurus, 21, 225, 865
Erasmus, 10
Ertem, G., 285–86
Eschenmoser, Albert, 285
Euler, Leonhard, 644, 646
Eusebius, 724
Ewing, D. G., 865

Fahr, H., 618
Feigenbaum, M., 660, 661
Feng, D. F., 891
Fermat, Pierre de, 644
Ferris, J. P., 285–86
Fersht, Alan, 306
Feynman, Richard, 23, 680
Field, Hartry, 667, 679–80, 681
Fine, Arthur, 180, 186, 188
Fisher, Ronald, 722
Fitzgerald, G. F., 547
Fleming, Gordon, 194, 201
Fodor, Jerry, 219
Fontana, Walter, 446
Forrest, Barbara, 95, 102
Forster, Malcolm, 223–25
Foster, Lewis, 840
Foster, M. B., 8
Fourier, Joseph, 661
Franck, S., 604, 606
Franklin, Benjamin, 66
Franklin, Rosalind, 296
Freeman, Charles, 14
Freeman, W., 727
Frege, Gottlob, 169, 644, 678, 679, 680–81

Index

French, Steven, 676
Friedman, Alexander, 507
Fröhlich, Herbert, 807
Fuster, J. M., 751, 752

Galilei, Galileo, xiii, 14, 16–17, 19, 63–64, 84, 251, 682, 734–35
Gallie, W. B., 326
Galois, Évariste, 649, 675
Galton, Francis, 23
Gamow, George, 510
Gärdenfors, Peter, 714
Gasperini, Maurizio, 518, 519, 574, 576
Gassendi, Pierre, 21–22
Gates, Bill, 302
Gaunilo, 882
Gauss, Carl Friedrich, 217, 644, 649, 670
Gazzaniga, M. S., 727, 729
Gehrels, N., 614
Gehring, Walter, 353
Gell-Mann, Murray, 676–77
Gibbon, Edward, 15
Giddings, S. B., 548
Gilbert, Walter, 282
Gilbert, William, 21
Gish, Duane, 887
Gödel, Kurt, 718
Goetz, Stewart, 839
Goldman, Alvin, 136, 868
Gonzalez, Guillermo, 486, 603, 613, 617
Goodale, M., 748, 751
Goodman, Nelson, 237, 238–39
Goodwin, Brian, 388
Gordon, Bruce L., xix–xxiv, 135–36, 486, 602
Gosse, Philip H., 887
Goswami, A., 790
Gould, Stephen Jay, xvi, 27, 108, 137, 323, 356, 446, 448, 449, 460
Grant, B. R., 721
Grant, P. R., 721
Gratton, S., 500
Gray, Asa, 72, 327
Green, Michael, 573
Green, T. H., 873
Greenfield, Susan, 727
Griffin, David Ray, 544, 545
Grimes, Thomas, 774
Gromov, Mikhael, 715
Gross, David, 554
Gutenberg, Johannes, 18

Guth, Alan, 485, 486, 514, 520, 562, 574
Guyot, Arnold, 70

Haag, Rudolf, 199
Hadamard, Jacques, 715
Haeckel, Ernst, 294, 432, 877
Hagan, S., 805
Häggström, Olle, 390
Haldane, J. B. S., 276–77, 722
Halvorson, Hans, 199, 201
Hameroff, Stuart, 693
Hammer, F., 621
Harak, G. Simon, 785
Hardy, G. H., 658
Hardy, Lucien, 828
Hare, R. M., 772
Hart, M. H., 610
Hartle, James, 515, 516
Hartshorne, Charles, 915
Hasker, William, 839, 845
Hawking, Stephen, 510, 511, 515, 516–17, 525, 526, 563, 566–67, 569, 584, 910–11
He, S., 756
Hegerfeldt, G. C., 198–99
Heisenberg, Werner, 676, 793, 827
Herschel, John, 70
Heytesbury, Walter, 15
Hilbert, David, 562, 906
Hirstein, W., 748
Ho, Yu-Chi, 382
Hobbes, Thomas, 853, 858
Hobson, J., 727
Hodge, Charles, 70
Hogan, C., 551
Hook, Sidney, 868, 877
Hooker, Joseph, 276
Hoyle, Fred, 304, 508, 509, 552, 559, 563, 624
Hubbard, J. L., 715
Hubble, Edwin, 507
Hume, David, xiii, 158, 163, 164, 172–74, 204, 330, 506, 527, 853, 858, 862, 870, 872–73, 884
Humphreys, Paul, 192–93
Hunter, Michael, 64
Huxley, Thomas H., 62, 71–72, 73, 432
Hwang, J., 577

Ida, S., 608, 611, 617
Igel, Christian, 381
Ingle, D., 748
Isham, Christopher, 512, 513, 565

Index

Jacob, François, 279
James, William, 249, 830
Jamieson, Kathleen Hall, 713–14
Jauch, J. M., 196
Jibu, M., 805
John I (England), 19
Johnson, Phillip E., 74
Johnson, Victor, 721–22
Jones, John E., 95–96, 102, 113–14
Joyce, Gerald F., 286, 287, 322

Kachru, S., 548, 552, 575
Kaku, Michio, 568
Kallosh, S., 548, 575, 576
Kaloper, N., 578, 579
Kamminga, Harmke, 293
Kant, Immanuel, 247, 256–69, 651, 668, 671, 740, 844, 853, 858, 860, 865, 871, 873, 874
Kasting, J. K., 605, 606, 610
Kauffman, Stuart, xiv, 314, 319–21, 352, 388–89, 446–47
Kautsky, K., xiv
Kay, Lily, 303
Kelley, Shana O., 290
Kelly, Kevin, 447
Kendrew, John, 295
Kennedy, John F., 326
Kenyon, Dean, 107, 313, 314
Kepler, Johannes, 9, 15–16, 19, 84, 554, 670
Kettlewell, Bernard, 431
Khoury, J., 576
Kieseppä, I. A., 224
Kim, H., 577
Kim, Jaegwon, 191, 192–93, 773, 835
Kimura, Motoo, 722–23
Kirkwood, Daniel, 69
Kitcher, Patricia, 868
Kitcher, Philip, 113, 641
Kleban, M., 582
Klein, Richard, 453
Knie, K., 613
Koch, Christof, 692, 693, 736, 737–38, 844, 845–46
Kochen, S., 183
Kofman, L., 576
Koonin, Eugene, 582–83
Koons, Robert, 136
Kronecker, Leopold, 654, 660
Kronz, Fred, 190, 192–93, 194
Kuchner, M. J., 608

Kuhn, Thomas, xi
Kukla, Andre, 224
Kummer, Ernst, 660
Kuppers, Bernd-Olaf, 293, 312, 314
Kurtz, Paul, 913
Kushiner, James, xxiii$n$3

Lacroix, Sylvestre, 646
Laertius, Diogenes, 21
Lagrange, Joseph Louis, 643–44, 646, 649, 656
Lamarck, Jean-Baptiste, 71
La Mettrie, Julien Offray de, 67
Landsberg, P. T., 312
Langton, Stephen, 19
Laplace, Pierre-Simon de, 68–69, 717
Laskar, J., 608
Laudan, Larry, xii–xiii
Laughlin, F., 622
Laughlin, G., 611
Lavoisier, Antoine, 350
Le Conte, John, 69
Le Conte, Joseph, 73
Leakey, Richard, 453, 454
Léger, A., 608
Leibniz, Gottfried Wilhelm, 9, 506, 646, 649, 652, 660, 716, 758, 790
Lemaître, Georges, 507
Lenardic, A., 610
Lenin, V. I., xiv
Lenski, Richard, 273, 369, 436, 437, 479–81
Lepine, J. R. D., 619
Leslie, John, 910
Leuba, James, 24
Leucippus, 21
Leven, P. A., 296
Levin, Benjamin, 719
Levins, R., 235
Lewis, C. S., 4
Lewis, David, 187, 204, 205, 773
Lewis, H. D., 840
Lewontin, Richard, 101, 584, 723
l'Huilier, Simon, 646
Libet, Benjamin, 81–86, 693, 727, 755, 803, 821–24
Lin, D. N. C., 611, 617
Lindberg, David, 14, 21
Linde, André D., 486, 496–97, 513–14, 519, 520, 524, 548, 555, 559, 561, 575, 576, 578, 579, 911
Linde, D., 496–97

Index

Lineweaver, C. H., 613, 614, 617
Linnaeus, Carolus, xv
Lipsius, Justus, 10–11
Lissauer, J. J., 610
Liu, David, 282
Livio, M., 554
Llinas, R., 739, 740, 741
Lloyd, Seth, 308
Loar, Brian, 235
Locke, John, xii, 683
Loewer, Barry, 191
Logothetis, N., 753
Logsdon, John M., 890
Lorentz, Hendrik, A., 547
Lovelock, James, 352
Lucretius, 21, 225
Lunine, J. I., 607
Lyell, Charles, 70–71, 325, 327

MacKay, Donald, 781, 784, 785, 786
Mackie, John, 859
Maclaurin, Colin, 646
Macready, William, 382
Madell, Geoffrey, 839, 840
Malament, David, 198–99
Malcolm, Norman, 915
Maldecena, J., 575
Maloney, A., 548
Marcus, Gary, 714
Margulis, Lynn, 352, 387–88
Marks, Robert J. III, xiv, 99, 273
Marochnik, L. S., 620
Martel, H., 549, 553
Marx, Karl, xiv, 364, 365
Matese, J. J., 613
Mather, Cotton, 65–66
Mathew, P. G., xxi
Matteucci, F., 621
Matthaei, J. Heinrich, 279
Maxwell, James Clerk, 717
Maynard Smith, John, 141, 319, 363, 369
Mayr, Ernst, 430, 431, 445, 446, 702
McAllister, L., 575
McDonough, Thomas R., 444, 456n6
McDowell, John, 262–67
McGinn, Colin, 782
McIntyre, Alasdair, 872
McLaughlin, Brian P., 191
McMullin, Ernan, 2
McRae, Murdo William, 447, 449

Medawar, Peter, 371
Meester, Ronald, 390–92
Melédez-Hevia, Enrique, 891
Melott, A. L., 615
Mendel, Gregor, xv
Mendelson, Elliot, 672
Mendez, A., 604–5
Meselson, Matthew, 448
Meyer, Stephen, xv, 2, 99, 102–3, 273
Meyering, Theo, 773
Mezhlumian, A., 496–97
Mill, John Stuart, 204, 386, 388, 646, 647, 668,
    678, 679, 681, 870
Miller, Kenneth, 113, 370–71, 387, 388,
    395n20, 890, 891, 892
Miller, Stanley, 277, 283, 284, 349
Millikan, Ruth Garrett, 218, 220–21
Milner, D., 748, 751
Mishkin, M., 751
Mivart, St. George, xvi
Monod, Jacques, 279, 297, 300, 303, 355
Moore, G. E., 865, 870
Mora, P. T., 309, 319
More, Henry, 9
Moreland, James P., 693–94
Morgenbesser, Sidney, 668
Morowitz, Harold, 290–91, 304, 312
Morris, Henry, 887
Morris, Simon Conway, xv, xvi, 323, 447–48,
    462, 479, 861
Mosterin, J., 569
Mulder, Gerardus Johannes, 350
Murphy, A. E., 869
Murphy, Nancey, 692, 835, 841, 842, 843
Musser, Siegfried, 891

Nagel, Thomas, 728, 846, 877
Ne'eman, Yuval, 676, 678
Neurath, Otto, 229
Newsome, W. T., 755
Newton, Isaac, xii, xiii, 8, 9, 64–65, 67, 115,
    216, 251, 554, 641, 644, 646, 647, 649, 652,
    660, 668–69, 671, 715, 717, 718, 794, 816
Newton-Smith, W. H., 241
Nichol, John Pringle, 68–69
Nicolas of Cusa, 15
Nicolis, G., 313
Niemann, Carl, 295
Nirenberg, Marshall, 279
Noether, Emmy, 659, 675

Index

Nolen, William A., 884
Noller, Harry, 282
Novikov, I. D., 524
Nozick, Robert, 871
Numbers, Ronald, 1–2

Oakley, Francis, 8
O'Connor, Timothy, 191
O'Neill, C., 610
Ooguri, H., 553
Oparin, Aleksandr I., 105, 276–77, 293, 294, 310–11, 349
Oresme, Nicole, 63
Orgel, Leslie, 281, 286, 287, 303, 322

Page, Don, 569
Pagels, Heinz, 368
Paley, William, 349
Papineau, David, 218, 220–21
Paré, Ambroise, 65
Parfait, Derek, 506
Pascal, Blaise, 64
Pasteur, Louis, 350
Paul of Taranto, 21
Pauling, Linus, 287, 448
Pavlov, A. A., 618–19
Pawson, T., 410
Peano, Guiseppe, 673
Peebles, P. J. E., 489
Peirce, C. S., 232, 323–24, 326, 671
Penfield, Wilder, 755
Pennock, Robert, 95, 109, 110, 369
Penrose, Roger, 526–27, 510, 569–70, 571–72, 693, 793, 795, 806, 807
Pepyne, David, 382
Perakh, Mark, 894
Perrett, R. S., 751
Philoponus, John, 20
Picasso, Pablo, 460
Pinker, Steve, 376, 714, 721
Planck, Max, 816
Plantinga, Alvin, 4, 7–8, 135, 154–64, 240, 915, 916
Plato, xiii, 667, 668, 679, 857, 858, 865, 871, 874
Plotkin, Henry, 714
Podolsky, Boris, 824, 825–26
Polanyi, John, xx
Polanyi, Michael, xxi, xxiii*n*2, 316–17
Polchinski, Joseph, 486, 548, 559, 575, 579

Pollard, D., 610
Preskill, James, 491, 525
Prichard, H. A., 865
Prigogine, Ilya, 304, 313, 317
Prince, Thomas, 66
Prothero, Donald, 447–48
Prusiner, Stanley, 350
Ptashne, Mark, 273
Ptolemy, 20, 718–19, 720
Putnam, Hilary, 230, 231, 233–34, 235, 239–40, 241, 242
Pythagoras, 682

Quastler, Henry, 318–19, 322, 329
Quine, Willard V., xi, 226, 228, 248, 868

Ramachandran, V. S., 748
Ramsey, Frank, 204, 249
Rana, Fazale, 273
Randi, James, 884
Raup, David, 448
Raven, J. A., 612
Rawls, John, 857, 858, 871
Ray, Thomas, 369, 370, 372, 373
Rees, Martin J., 355, 552, 554–55, 618
Reichenbach, Hans, 225
Reid, Thomas, 138, 147, 259, 269
Rensch, B., 790
Rey-Benayas, S. J., 604
Rice, William North, 72
Riemann, Bernhard, 217, 670
Roberts, Jon H., 72–73
Robertson, Douglas, 360
Robertson, M. P., 284
Rode, B. M., 283
Rolston, Holmes, 364
Rose, Louis, 884
Rosen, Nathan, 824, 825–26
Rosenfeld, Leon, 825
Ross, W. D., 865
Roulle, H. M., 276
Rousseau, Jean-Jacques, 13
Rowe, Mary, 186
Ruetsche, Laura, 201
Ruse, Michael, 104, 105, 106, 853, 855, 877
Russell, Bertrand, 137, 258, 260, 506, 645, 682, 790
Russell, Robert J., 861
Rutherford, Ernest, 254
Ryle, Gilbert, 11

Index

Sagan, Carl, 14, 74, 82, 444, 445, 446, 448, 613
Sanger, Fred, 295, 297
Sarkar, Sahotra, 300, 303
Sauer, Robert, 305–6, 307, 368
Scalo, J., 615
Schaffer, Cullen, 382
Scherer, K., 618
Schimmel, Paul, 283, 290
Schklovsky, J. S., 613
Schlegeli, Richard, 521
Schneider, S. M., 604
Schneider, Thomas, 370, 371, 372, 373
Schönborn, Christoph, 554
Schopf, J. William, 478–79
Schrödinger, Erwin, 279, 303, 350, 793
Schwartz, Jeffrey, 830, 831
Schwarz, John, 572, 573
Schwinger, J., 828
Scott, Eugenie, 107
Scriven, Michael, 325, 326, 327
Searle, John, 693, 771–72, 835, 841, 844, 867, 868, 877
Segal, Irving, 201
Sellars, Wilfrid, 249, 250–51, 252
Semmelweis, Ignaz, 735
Seneca, Lucius Annaeus, 63
Shannon, Claude, 298–99
Shapiro, P., 549, 553
Shapiro, Robert, 284, 319
Shaviv, N. J., 620
Shermer, Michael, 112, 113, 273
Sherrington, Arthur, 726
Shimony, Abner, 183–84
Shoemaker, Sydney, 191
Shostak, Seth, 444
Sidgwick, Henry, 872, 873, 874
Silk, Joseph, 524
Silverstein, E., 548
Simon, H. A., 714
Simpson, George Gaylord, 445
Skinner, B. F., 722
Slipher, Vesto, 507
Sloan, Robert, xxi, xxiii
Smith, D. S., 612
Smith, John Maynard, 141, 319, 363, 369
Smith, Quentin, 902–3
Smolin, Lee, 355
Sobel, Howard, 906, 907, 916
Sober, Elliott, 223–25, 325, 326
Socrates, 865, 874

Sparacio, Micah, 370
Specker, E., 183
Spencer, Herbert, 856, 858
Spencer, Richard, xxi–xxii
Sperry, Roger, 726, 844
Spiegelman, Sol, 388
Spinoza, Baruch, 667, 790
Spitzer, Robert, 581
Sporns, O., 740
Stalnaker, Robert, 316, 360
Stapp, Henry, 693
Stark, Rodney, 14, 18, 37
Starr, Kenneth, xxi
Steiner, Mark, 641–42
Steinhardt, Paul, 491, 500, 519, 520, 574, 576, 577
Steinman, G., 313
Stenger, Victor, 887
Sternberg, Shlomo, 676
Stewart, Ian, 447, 449
Stich, Stephen, 216
Strominger, A., 548
Stubenberg, L., 790
Suga, Horoaki, 288–89
Susskind, Leonard, 486, 548, 559, 575, 579, 580, 582, 585
Swinburne, Richard, 839, 917
Szathmary, Eörs, 363
Szilard, Leo, 386

Talbott, William, 135
Taliaferro, Charles, 840
Tarski, Alfred, 231–32, 248, 718
Tattersall, Ian, 453–54
Taylor, Charles, 6, 7, 10–12, 35–36
Taylor, Richard, 4, 904, 913
Tegmark, Max, 552, 554, 562, 568
Teilhard de Chardin, Pierre, xvi, 352
Teller, Paul, 190, 194, 774–75
Tennant, F. R., 842, 843
Theodoric of Freiburg, 21
Thom, René, 661, 715
Thomas, B. C., 615
Thomas, Keith, 67
Tiehen, Justin, 190, 192–93, 194
Tinsley, Beatrice, 522
Tipler, Frank J., 330, 508
Tomonaga, S., 828
Tononi, Giulio, 740, 741
Tooby, John, 691–92, 714

Index

Tooley, Michael, 854
Touma, J., 609
Toussaint, Marc, 381
Traub, J. F., 660, 661
Trivedi, S. P., 548, 575
Tryon, Edward, 512
Turing, Alan, 716
Turner, Frank M., 73
Turner, M., 578
Turney, Peter, 225
Turok, N. G., 500, 574, 576, 577
Tyndall, John, 73, 845

Ungerleider, L. G., 751
Urban VIII, 16–17
Urey, Harold, 277, 283

Vafa, C., 553
Valencia, D., 610
Valtonen, M. J., 617
Van de Sande, Bret, 308
Van Till, Howard, 485
van der Waerden, B. L., 659
van Fraassen, Bas, 180, 186, 187, 195, 196, 843–44, 902
van Inwagen, Peter, 137
Vanchurin, V., 492
Varadarajan, U., 552
Varela, Francisco, 352
Veneziano, Gabriele, 518, 519, 574, 576, 577, 591n116
Veras, D., 607
Verlinde, E., 553
Vesalius, Andreas, 14, 16
Vilenkin, Alexander, 491, 492, 514–15, 520, 549, 561, 562, 563, 564, 574, 582, 592n137
Voltaire, 13, 877
Von Boh, W., 620
von Neumann, John, 311–12, 793, 817, 819, 820, 823, 827

Wald, George, 304
Wallace, Alfred Russel, 445, 554
Wallner, C., 613
Walton, J. C., 319
Ward, W. R., 608
Warfield, Benjamin B., 72
Watson, James, 277–78, 296, 297, 303
Watson, John, 722
Watson, Richard, 432, 433, 440

Weaver, Charles, xxi
Weaver, Richard, 5
Weaver, Warren, xv, 300
Weber, Max, 251
Weierstrass, Karl, 643, 644
Weinberg, E., 578
Weinberg, Steven, 217–18, 224, 443, 485–86, 558, 869, 886
Weiskrantz, L., 727, 737, 751
West, A. A., 612
West, B. H., 715
Weyl, Hermann, 676
Wheeler, John C., 507, 615, 790
White, Andrew Dickson, 13–14, 15, 24, 884
Whitehead, Alfred North, 23, 562, 645, 790, 808
Wickramasinghe, C., 304
Wiener, Norbert, xiv
Wightman, Arthur, 199
Wigner, Eugene, 184, 312, 671–72, 677, 681, 817
Wilczek, Frank, 568
Wilkins, Maurice, 278, 296
Willard, Dallas, 853–54
William of Ockham, 7, 15, 19
Williams, D. M., 608, 610
Williams, Michael, 136
Wills, Christopher, 453
Wilson, David Sloan 139
Wilson, Edward O., 448, 855, 858, 877
Wilson, John, xx
Wimsatt, William, 235
Winitzki, S., 492
Winthrop, John IV, 66
Wisdom, J., 609
Wise, Steven, 751
Witten, Edward, 573
Woese, Carl, 281, 288, 289
Wöhler, Friedrich, 276, 277, 291
Wolpert, David, 382
Wolstencroft, R. D., 612
Wolterstoff, Nicholas, 136
Wong, Hong Yu, 191
Wright, Frank Lloyd, 460
Wright, George Frederick, 72
Wright, Robert, 451, 452
Wright, Sewall, 722

Xavier, Francis, 884

Index

Yandell, Keith, 840
Yeghikyan, A., 618
Yockey, Hubert, 288, 304, 317

Zeki, S., 736, 739
Zeldovich, Ya B., 524
Zermelo, Ernst, 644–45
Zheng, J. Q., 617
Zubrin, Robert, 444

Subject Index

abduction *(Meyer)*, 100–101, 323–24
abiogenesis, 273. *See also* origin of life (OOL)
academic freedom, xxiv
Akaike criterion, 223–25
Almagest, 719
American Civil Liberties Union (ACLU), 96, 113
amino acids, 333*n26*
    chance production *(Meyer)*, 304–10
    construction *(Axe)*, 412–13
    DNA and acids *(Berlinski)*, 278–80
    left- vs. right-handed, 305
anthropic principle, anthropic reasoning
    anthropic multiverse *(Weinberg)*, 486, 548–50, 554
    criticism of anthropic explanations *(Gordon)*, 578–83, 591*n116*
    Denton's approach *(de Duve)*, 354, 355
    eternal inflation and string landscape *(Guth)*, 495
    Pre-Big Bang scenario *(Craig)*, 519
anti-essentialism, 9
anti-materialism, 180–81. *See also* quantum theory (QT)
antibiotic resistance *(Behe)*, 437–38
antirealism, *(Goldman)*, 230–37. *See also* realism
appearance of design, 101–2. *See also* design
apriorism *(Williams)*, 248–49
Aristotelianism, 21–22
astrobiology, 603. *See also* Search for Extraterrestrial Intelligence (SETI)

atheism
    18th century *(McMullin)*, 82
    vs. naturalism *(Plantinga)*, 137
    See also naturalism
atomism *(Gordon)*, 21–22
Augustine's doctrine of creation *(McMullin)*, 87
Averroism, xiii
AVIDA, 369, 372–73
Axe, Douglas
    functional proteins, 306–7
    protein fold challenge to Darwinism, 274–75

"baby universes," 525–26
Baylor University, xx
Behe, Michael, 103
    Darwin's theory outlined, 429–31
    multigenerational evolution fails, 275
    Origin of Species demonstrates no speciation, 429
    other biologists' criticism *(Tooley)*, 890–92
    philosophical issues *(Tooley)*, 888–90
    random mutation's efficacy unproven, 431
    random mutations mostly deleterious, 433–35
    randomness plays "limited" evolution role, 438, 440
belief
    naturalism and evidence *(Plantinga)*, 147–48
    facts vs. propositions *(Wolterstorff)*, 266
    false belief *(Goldman)*, 242–43
Bell's theorem *(Gordon)*, 180, 203
    Bell Inequality derived, 184–89
    counterfactual excluded middle, 186–89

Index

Berlinski, David
  EP as "pseudoscience," 692
  EP's "palpably unreliable" stories, 722–23
  origin of life, 273
Big Bang cosmology, 486
  Big Bang Model, 488, 507–9, 905
  confirmed repeatedly, 520
  disturbing to some *(Craig)*, 509–10, 512, 515, 529n23
  Steady State vs. Big Bang, 509–10
  why disliked, 559–60 *(Gordon)*
  See also universe, origin of universe
Big Crunch *(Craig)*, 510–12. *See also* universe, end of universe
biochemistry. *See* life; molecular biology; origin of life (OOL)
biological information *(Meyer)*
  complexity and specificity, 300–302
  information problem *(Dembski/Marks)*, 363–67
  LCI applied *(Dembski/Marks)*, 386–93
  nature, 294–304
  origin, 273, 293 304–22
  See also information; origin of life (OOL)
biology
  not teleology (Coyne), 361
  regulated systems *(Ptashne)*, 400–403
  See also biological information; computational biology; molecular biology
black holes
  at end of universe, 522
  white holes, 526–27
Blind Watchmaker, The, 366
blindsight, 737, 751
Boltzmann
  "Boltzmann brain" *(Gordon)*, 582
  hypothesis *(Craig)*, 519
brain
  brain adaptations *(Cosmides/Tooby)*, 700–707
  evolution explains mind *(Cosmides/Tooby)*, 696–97, 699–707
  evolves for fitness, not truth (Churchland), 139–40
  See also consciousness; mind
"brain in a vat" *(Talbott)*, 160–61, 163
branes, 519
Brief History of Time, A, 563
"bubble universes," 561

capitalism, 18–20
Carnegie Commission, 24
Cartesianism, 11
cassette mutagenesis, 305–6, 307
Catholic Church
  Galileo affair *(Gordon)*, 16–17
  Giordano Bruno affair *(Gordon)*, 14, 15, 17
  qualified methodological naturalism *(McMullin)*, 87
causation
  agent causation only true cause *(Gordon)*, 203
  causal adequacy *(Meyer)*, 99–100, 324–25
  common cause *(Gordon)*, 183
  local hidden variables *(Gordon)*, 189–90
  material causation, 9
  See also causation, intelligent causation
causation, intelligent causation
  biological convergence *(Rana)*, 481
  intelligent cause of universe *(Gordon)*, 559
  repeated occurrences *(Rana)*, 461
  transcendent cause *(Gordon)*, 585, 586n4
  See also causation, mental causation
causation, mental causation, 692
  bottom up causation of mind *(Murphy)*, 771–72
  bottom up causation of mind *(Searle)*, 733–74
  top down causation of mind *(Murphy)*, 778–80
  See also causation, quantum theory (QT)
causation, quantum theory (QT)
  physical causality's demise *(Gordon)*, 189–90
  QT causation *(Gordon)*, 562–63
  QT causation *(Stapp)*, 819–21
  See also causation; cosmology; origin of life (OOL)
cell
  "coded chemistry," 281
  early views, 432
  minimal cell, 286, 293
  sequence of information *(Meyer)*, 300–302
  See also life
Center for Science and Culture (DI), xx
  Nature of Nature conference, xxii
central dogma (Crick), 280–281
chance origin of life *(Meyer)*, 304–10. *See also* origin of life (OOL)
chemical bonds of life, 287
chemical evolution
  chemistry to life proposals *(Berlinski)*, 282–90
  observability *(Meyer)*, 107–8
  See also origin of life (OOL)

Index

chimpanzees, 30
Christianity
  Christian Darwinism *(Ruse)*, 861–62
  Christian philosophy *(Craig)*, 902–3
  cultural advances *(Gordon)*, 18–22
  freedom and Christianity *(Gordon)*, 34n12
  See also Christianity, science and Christianity
Christianity, science and Christianity
  coexistence with naturalism *(Numbers)*, 75
  myths detailed *(Gordon)*, 14–17
  Ratzsch's view *(McMullin)*, 94n19
  response to materialist science, 36n35
  science progress *(Gordon)*, 12–23
  Stark's view, 37n62
cladistics, 447
Climbing Mount Improbable, 389, 392
cognition
  cognitive design *(Plantinga)*, 170–71
  cognitive design *(Talbott)*, 157–59
  See also cognition, cognitive faculties *(Talbott)*
cognition, cognitive faculties *(Talbott)*
  global reliability, 156–59
  local reliability, 154–56
  separable cognitive faculties, 154
  separable vs. non-separable, 155
  See also cognition, cognitive neuroscience *(Cosmides/Tooby)*
cognition, cognitive neuroscience *(Cosmides/Tooby)*
  brain as evolved organ, 696–97
  cognition explained, 695
  pivotal science, 707
  See also cognition, cognitive science *(Goldman)*
cognition, cognitive science *(Goldman)*
  how term used, 229
  purpose, 228
  See also brain; cognition; consciousness; mind
coincidence *(Koons)*, 221–22
common descent, 273–74
  DNA's independent origins *(Rana)*, 275
  "incontrovertibly established" *(de Duve)*, 347–48
  pattern, not law, 120
  science and commonsense *(Fuller)*, xi
  See also convergence *(Rana)*
complex systems *(Ptashne)*, 274
complexity *(Meyer)*, 300–302. *See also* specified complexity
computational biology
  biology vs. computer *(Dembski/Marks)*, 367–70

  Darwinism tested *(Dembski/Marks)*, 365–67
  input equals output *(Dembski/Marks)*, 371
  limitations *(Behe)*, 432, 440
  prebiotic natural selection *(Meyer)*, 312–13
  simulations *(Dembski/Marks)*, 370–72
  See also Darwinism; evolution
concepts *(Wolterstorff)*, 256–69
  realism and concepts, 256–58
consciousness, 691–94
  animal consciousness *(Hameroff/Penrose)*, 805–6
  animal consciousness *(Koch/Crick)*, 747
  arising from physical circumstances, 4
  characteristics *(Searle)*, 729–33
  definition *(Searle)*, 727–28
  future experiments *(Koch/Crick)*, 760–62
  hard science irrelevant, 693–94
  history of research *(Hameroff/Penrose)*, 790, 804
  microtubule role *(Hameroff/Penrose)*, 797–800
  neural correlates *(Searle)*, 735–36
  not evidence for God *(Tooley)*, 885
  objective reduction model *(Hameroff/Penrose)*, 693
  philosophy and consciousness *(Koch/Crick)*, 758
  purpose *(Koch/Crick)*, 748
  QM and consciousness *(Hameroff/Penrose)*, 790–92
  QT and consciousness, 693
  research *(Searle)*, 726–27, 735–36
  space-time effect *(Hameroff/Penrose)*, 794–97
  transcendence and meaning, 4
  See also consciousness, models of consciousness
consciousness, models of consciousness
  building block model *(Searle)*, 736–39
  dualist model *(Moreland)*, 835–45
  neuroscience approach needed *(Koch/Crick)*, 746–47
  Orch OR *(Hameroff/Penrose)*, 800–808
  unified-field theory *(Searle)*, 738–41
  See also consciousness, problems
consciousness, problems
  binding problem *(Searle)*, 730–31
  meaning and consciousness *(Koch/Crick)*, 760
  qualia *(Hameroff/Penrose)*, 790–92
  qualia *(Koch/Crick)*, 758–60
  qualia *(Murphy)*, 780–82
  self-consciousness *(Koch/Crick)*, 747

Index

consciousness, problems *(continued)*
  See also consciousness, quantum interactive dualism (QID) *(Stapp)*
consciousness, quantum interactive dualism (QID) *(Stapp)*, 815–16
  brain as quantum system, 818–19
  causation, 819–21
  EPR causal anomalies, 824–30
  Libet causal anomalies, 821–24, 830–31
  quantum collapse, 823–24
  See also consciousness, visual consciousness *(Koch/Crick)*
consciousness, visual consciousness *(Koch/Crick)*, 692, 747–48
  bistable percepts, 753–54, 760–62
  electrical brain stimulation, 755–57
  frontal lobe hypothesis, 757–58
  nature and function, 748–52
  neural correlates (NCC) *(Searle)*, 737–38
  on-line system model, 751–52
  V1 hypothesis, 755–57
  See also consciousness; mind
Conservation of Information Theorems *(Dembski/Marks)*, 373–81
  fitness-theoretic version, 378–81
  function-theoretic version, 374–76
  measure-theoretic version, 376–78
conservation of information, 105, 274. *See also* Law of Conservation of Information (LCI)
constructivism, 9
contingency
  contingency argument for God *(Craig)*, 904–5
  contingency hypothesis *(Shermer)*, 450–55
  contingency of life *(Shermer)*, 445–48
  contingency vs. convergence *(Rana)*, 461–62
  evolution and contingency, xvi
  "gospel of contingency" *(de Duve)*, 356–57
  radical contingency *(Shermer)*, 275
convergence *(Rana)*, 447
  convergence vs. contingency, 461–62
  evolutionary convergences, 275
  macroevolution shows design, 275
  molecular convergence, 460–81
Cosmides, Leda
  cognitive science research faulted, 706–7
  evolution explains mind, 691
  valued activities are side effects, 699–700
cosmological constant, 522–23
cosmology
  "baby universes," 525–26
  branes, 519
  "bubble universes," 561
  cosmic Darwinism, 535–36
  cosmic imaginary, 6, 7
  inflaton field *(Gordon)*, 561
  initial singularity *(Craig)*, 526–27
  Many Worlds Hypothesis, 521
  post-Enlightenment cosmologies *(Numbers)*, 68–74
  quantum fluctuations' role *(Guth)*, 502n6
  quantum tunneling, 524–25
  Second Law of Thermodynamics, *(Craig)*, 522–27
  singularities *(Gordon)*, 586–87n10
  universe expands forever *(Craig)*, 511
  white holes, 526–27
  wormholes, 522, 553
  See also cosmology, fine-tuning
cosmology, fine-tuning
  design and fine-tuning *(Gordon)*, 30
  fine-tuning doubted *(Tooley)*, 882–83, 885–87
  fine-tuning, explanations *(McMullin)*, 84–85
  See also cosmology puzzles
cosmology puzzles
  CBR anisotropy, 490
  cosmological argument for God *(Craig)*, 905–9
  cosmological constant *(Craig)*, 522–23
  flatness, of universe *(Guth)*, 488–90
  homogeneity, of universe *(Guth)*, 488
  horizon problem *(Gordon)*, 569
  isotropy of universe *(Guth)*, 488
  See also Big Bang cosmology; habitable zones; inflationary cosmology; universe
Cosmos (TV documentary), 82
Cours d'Analyse, 646
Craig, William Lane
  atheism in retreat, 902–3
  naturalism implausible, 854
creationism *(Van Till)*, 537–38
Crick, Francis
  consciousness problem ripe for attack, 761–62
  Crick on protein synthesis *(Berlinski)*, 278–80
  philosophers' "very poor record," 758
Critique of Pure Reason, 267
Crucible of Creation, 447
cryptography, 336n65
cybernetics, xiv

Index

Dark Ages, 12–23
dark energy, 511
Darwin, Charles
  causation theories, 325
  Darwin's circle's social goals *(Numbers)*, 73–74
  Darwin's Doubt, 139, 140, 144–45
  his early beliefs, 23
  information without intelligence *(Dembski/Marks)*, 361
  irreducible complexity, 29
  origin of life views, 276, 294, 332n6
  use of abduction *(Meyer)*, 327
  See also Darwinian evolution; naturalism
Darwin's Black Box, 74, 103, 349–50, 352
Darwin's Dangerous Idea, 447
Darwinian evolution
  computer simulations *(Dembski/Marks)*, 365–67
  origin of life *(Berlinski)*, 286–87
  science criteria, observability *(Meyer)*, 107–8
  tree of life, xv
  See also Darwinism; naturalism
Darwinian Natural Right, 877
Darwinism, 26
  functional protein folds problem *(Axe)*, 420–25
  impact *(Gordon)*, 24–26
  mathematics problems *(Gordon)*, 28
  morality as illusion *(Ruse)*, 858–59
  nominalism and Darwinism *(Gordon)*, 26
  prebiotic NS *(Meyer)*, 310–13
  random mutation *(Behe)*, 431–41
  rescues natural law theory *(Ruse)*, 862
  secularization role, 1
  teleology and Darwinism, 274, 429
  vast information required, 338–39n126
  See also Darwinian evolution; naturalism
Dawkins, Richard
  appearance of design, 24–25
  computational Darwinism *(Dembski/Marks)*, 365–67
  computer simulation of prebiotic NS *(Meyer)*, 312–13
  evolution has no goal, 367
  LCI and Dawkins, 392–93
  WEASEL, 371–72
de Broglie-Bohm theory, 189, 192, 203–4
de Duve, Christian
  Behe's work, 349–52, 354

Dembski's work, 351
Denton's work, 354, 355
  humanity part of cosmic pattern, 357
  methodological naturalism necessary, 273
  origin of life, 274
  "something else," 354
  supernatural, 346
Dembski, William
  complexity-theoretic specification, 300–302
  continuous evolution scenario, 27
  flaws in argument *(Tooley)*, 894–97
  junk DNA prediction, 112–13
  origin of life, 32
  probabilistic resources, 307–10
  See also Dembski, William, theories
Dembski, William, theories
  design inferences *(Meyer)*, 330
  explanatory filter, 99
  "no free lunch," xiv, 105, 378, 381, 382–83
  universal probability bound, 339n128
  See also design; intelligent design (ID)
Dennett, Daniel, 543
deoxyribose nucleic acid (DNA). *See* DNA
Descartes, René
  evil demon, 159
  nominalism, 22
Design Inference, The, 99, 103, 894
design
  appearance of design, 101–2
  arguments against design, 887–94
  Aristotle's view, 362
  challenges historical contingency *(Rana)*, 478–81
  common design, convergence, 27–28
  design arguments in science, 115–16
  design detection, 115
  design vs. randomness, 24–25
  distinct form of inference *(Meyer)*, 320–28
  Design Hypothesis *(Craig)*, 911–12
  design hypothesis *(Meyer)*, 322–23
  design inference, 28
  nature of design *(Dembski/Marks)*, 361–94
  See also intelligent design (ID)
determinism *(Gordon)*
  de Broglie-Bohm theory, 189, 203–4
  local determinism and quantum theory (QT), 188
  pilot-wave theory, 189

Index

directed panspermia (Crick), 281–82
Discovery Institute, xx
divine command theory, 7–8
DNA
  "ancient repetitive elements," 29–30
  complexity and specificity *(Meyer)*, 296–98, 300–302
  development and DNA, 29
  informational specified complexity, xv, 273
  "98% junk" thesis, 29–30
  structure explained, 277–78
  "pseudogenes," 29
  self-organization theories *(Meyer)*, 313–18
  See also DNA, replication
DNA, replication *(Rana)*, 476, 477
  evolved twice, 478
  origin of replication, 474–75
  signature style, 475–78
Dover trial
  Discovery Institute's view, 119*n42*
  ID as not science, 118*n33*
  methodological naturalism *(Meyer)*, 113–14
  See also education; naturalism
Dreams of a Final Theory, 217
dualism, 758
  dualism explained *(Moreland)*, 838
  early Christians not dualists *(Murphy)*, 771
  historically widespread *(Moreland)*, 835
  interactive dualism, 693
  materialism vs. dualism *(Searle)*, 734
  quantum interactive dualism (QID) *(Stapp)*, 815–31
  unneeded *(Murphy)*, 770–771
  See also dualism, substance dualism *(Moreland)*
dualism, substance dualism *(Moreland)*, 847*n13*
  supporting arguments, 538–43
  three main forms, 538
  See also mind-body problem

E. coli (Behe), 275, 436–38
EcoCyc, 413
education, on nature
  creationism in schools, 85
  Darwinism in schools, 85
  Darwinism's shortcomings taught, 91
  intelligent design in schools, 85, 89–90, 103
Einstein, Albert, 217, 695
electron, 547
  nonlocalizability, 181
  spin-singlet state, 184–89
electroweak theory, 217
elegance, in physics *(Weinberg)*, 217–18
eliminativism, 216
embryogenesis, 29
emergence *(Gordon)*, 4
  classical mechanical limit (CM), 190–91
  dynamic emergence, 190–95
  relational holism, 194
  supervenience, 191–92
empiricism, 9
  empirical predictions, 179
  philosophy and empiricism *(Goldman)*, 140–41
End of Christianity, 27
Enlightenment *(Numbers)*, 67
entanglement, of quantum systems, 180. *See also* quantum theory (QT)
entropy, xv–xvi
  end of universe *(Craig)*, 522–27
  universal entropy *(Gordon)*, 581–83
epiphenomenalism *(Plantinga)*, 140–42
epistemology *(Goldman)*
  epistemic justification, 229
  evaluative vs. descriptive, 228–29
  Humean view not dominant *(Williams)*, 248
  individual vs. social, 228
  naturalism-based, 228–29
  true belief *(Goldman)*, 229
  See also belief
Essay (Locke), 683
Essays on the Intellectual Powers of Man, 147
Ethical Theory, 871
ethics
  good and evil *(Willard)*, 872–73
  metaethics *(Ruse)*, 858
  moral argument for God *(Craig)*, 912–14
  naturalism's dilemma *(Willard)*, 868–69
  naturalist ethics *(Willard)*, 869–78
  noncognitivism *(Willard)*, 870–71
  normative ethics *(Ruse)*, 855–57
  objective morality *(Craig)*, 913–14
  objective morality *(Ruse)*, 858–59, 860–61
  progress in ethics *(Ruse)*, 860–61
  substantive ethics *(Ruse)*, 859–60
Euthyphro, 883
ev, 370–71, 372–73

Index

evolution
  goal-directed evolution, xvi
  irreducible complexity, 29
  large bacterial populations *(Behe)*, 275
  neutral evolution, 722–23
  protein sampling problem *(Axe)*, 412–15
  See also evolution, proposed mechanisms *(Ptashne)*
evolution, proposed mechanisms *(Ptashne)*
  cooperativity, 400–406
  genetic regulation, 405–6
  hypothetical steps, 405
  lambda switch, 406–8
  proteolysis, 408
  regulated recruitment, 401–3, 408–10
  RNA splicing, 408
  verified predictions, 402–4
  See also evolution, views
evolution, views
  complex systems highly evolvable *(Ptashne)*, 274
  conflict with naturalism *(Plantinga)*, 138–48
  evolution, contingency, xvi
  enough time for Darwinism *(de Duve)*, 350
  explanation *(Ptashne)*, 400
  "life must evolve" view *(Shermer)*, 444–45
  "life needn't evolve" view *(Shermer)*, 445–48
  non-naturalist theories *(de Duve)*, 352–53
  randomness ineffective *(Behe)*, 439
  transitions require design *(Gordon)*, 29
  See also Darwinism; naturalism
Evolution-Creation Struggle, 856
Evolutionary Creation (Lamoureux), 27
evolutionary fitness landscapes, *441*
evolutionary psychology
  explains brain function *(Cosmides/Tooby)*, 700–707
  changes mind/brain research *(Cosmides/Tooby)*, 705–7
  explains mind, 695
  helps neuroscience *(Cosmides/Tooby)*, 697
  materialist's choice *(Berlinski)*, 721–22
evolutionary theories
  compatible with naturalism *(Talbott)*, 163–64
  events explained re prior events *(Meyer)*, 105–6
  naturalism opposes it *(Plantinga)*, 137–48
  naturalism supports it *(Talbott)*, 152–64
exobiology. *See* Search for Extraterrestrial Intelligence (SETI)
exoplanets, 607. *See also* habitable zones
Expelled: No Intelligence Allowed, xxivn7
experience *(Wolterstorff)*, 262–66
experiment, 20–21
extraterrestrial intelligence
  global tuning *(Gonzalez)*, 625
  skepticism *(Shermer)*, 443, 444–45
  See also habitable zones

Faith Healers, The, 884
Faith Healing, 884
fallibilism *(Williams)*, 250
Fifth Miracle, The, 352
fine-tuning
  fine-tuning dismissed *(de Duve)*, 353
  fine-tuning doubted *(Tooley)*, 882–83
  global tuning *(Gonzalez)*, 623–25
  global vs. local tuning *(Gonzalez)*, 602–3
  Many Worlds Hypothesis (MWH) *(Craig)*, 911–12
  multiverse vs. fine-tuning, 485
  possible causes outlined *(McMullin)*, 84–85
  string theory and fine-tuning *(Gordon)*, 576–83
  Theory of Everything (TOE) *(Craig)*, 910–12
  See also cosmology
First Three Minutes, The, 443
folk science. *See* science, popular culture and science
formalism, 217
"free lunch" *(Steiner)*, 681. *See also* "no free lunch"

Galileo affair, 16–17
Gasperini-Veneziano model, 486
Genes and Signals, 400, 401, 403
Genesis, Book of
  naturalism and Genesis, 84
  nebular hypothesis interpretation, 70
genetic code *(Berlinski)*, 287–90
geocentrism, 15–16
"ghost in the machine," 11
Giordano Bruno affair, 14, 15, 17
God
  arguments for God *(Craig)*, 903–17
  divine will and nature *(Gordon)*, 7–10
  evil and God *(Tooley)*, 897–98
  ghost in machine concept, 11
  laws of nature as evidence, 68–69
  middle knowledge concept, 28

Index

God *(continued)*
   nature and God *(Van Till)*, 535–45
   nominalism and God, 7–8
   See also God, belief in God
God, belief in God
   Freudian wish-fulfillment *(Talbott)*, 154
   polls, 75
   proofs not convincing *(Tooley)*, 882–87
   scientists' belief in God, 24, 75, 80n38
   underwriting true beliefs *(Plantinga)*, 138–39
   wish-fulfillment *(Talbott)*, 154
   See also Christianity; religion
God Delusion, The, 392
Gordon, Bruce
   Hartle-Hawking's "no boundary," 563, 566–69
   multiverse explains everything, nothing, 582–83
   quantum theory inconsistent with naturalism, 135–36, 181
   Tegmark's multiverses, 562–63
   Vilenkin's cosmology, 563–65
Gould, Stephen Jay, 108
   contingency, 446–48
   de Duve's view, 356
   "loving concordat," 83–84
   miracles excluded, 83
   NOMA problems *(McMullin)*, 84–86
   non-overlapping magisteria (NOMA), 83–84
Grace Valley Christian Center, xxi–xxii
gravity
   Newton's mathematical description, 9, 22–23
   no proposed mechanism, 23
Grundlagen, 679
Guth, Alan
   "bubble universes," 485
   on probability of ETs, 497, 498

habitable zones *(Gonzalez)*
   circumstellar habitable zone (CHZ), 603, 605–13
   cosmic habitable ages (CHA), 603, 620–24
   exoplanets, 607
   galactic habitable zone (GHZ), 603, 613–20
   global tuning, *623–25*
   Milky Way, 616–20
   needs of life, 603–5
   Solar system, 617–20
   Sun and Earth, 611–13

Hameroff, Stuart
   evolution and platonic information, 808
   objective reduction of consciousness, 693
Has Science Found God?, 887
Hawking, Stephen
   "baby universes," 525–26
   "no boundary" proposal, 567–68
Healing: A Doctor in Search of a Miracle, 884
heliocentrism, 15–16
hidden variables *(Gordon)*
   hidden variable theories, 184–89
   quantum theory (QT), 180–81
   stochastic hidden variables, 182, 184
   See also quantum theory (QT)
historical progressivism. *See* convergent realism
historical sciences *(Meyer)*
   causal explanations not laws, 101
   ID as historical science, 98
   natural law and historical sciences, 105
History of the Conflict between Religion and Science (1874), 13
History of the Warfare of Science with Theology in Christendom (1896), 13–14, 884
   White's motivations, 14
homology, 29
horizontal gene transfer (HGT) *(de Duve)*, 348
Hubble expansion, 487–88
human immunodeficiency virus (HIV) *(Behe)*, 437–38
humanism *(Gordon)*, 26
humans
   human nature, MN view *(McMullin)*, 85
   "quirk of evolution" *(Shermer)*, 455
   "special" status a temptation *(Shermer)*, 443
   unique *(Fuller)*, xiii

ICAM, 369
"Illusion of Defeat" *(Talbott)*, 153, 155
improbability, 299. *See also* probability
induction, 386–87
inference to best explanation *(Meyer)*, 324
   argument from ignorance, 328–31
infinity, 905–9
inflationary cosmology, 485, 486
   arguments for *(Guth)*, 487–90
   beginning required *(Gordon)*, 576
   BGV theory *(Gordon)*, 561–63
   "Boltzmann brain" *(Gordon)*, 582
   Boltzmann hypothesis *(Craig)*, 519

Index

"bubble universes," 570–72
explanatory *(Guth)*, 501
explanatory failure *(Gordon)*, 569–70
false vacuum, *491–92*, 497
matter vs. mind *(Gordon)*, 583–85
nonanthropic explanation *(Guth)*, 495
requires fine-tuning *(Gordon)*, 570–72
special assumptions *(Gordon)*, 570–72
why proposed *(Gordon)*, 560–61
youngness paradox *(Guth)*, 496–98
See also inflationary cosmology, inflaton field inflationary cosmology, inflaton field
inflationary cosmology, inflaton field
  inflation field *(Gordon)*, 561, 570–72, 588–89
  inflaton field *(Guth)*, 490, 491, *492*
  See also inflationary cosmology, models and scenarios
inflationary cosmology, models and scenarios
  Chaotic Inflationary Model, *513–14*
  chaotic inflation theory *(Weinberg)*, 553, 561–63
  eternal chaotic inflation (Guth), 492–94
  eternal inflation *(Guth)*, 490–501
  eternal inflation probability *(Guth)*, 495–96
  eternal new inflation *(Guth)*, 491–92, 494
  past-eternal inflation *(Gordon)*, 561–63
  past eternal inflation *(Guth)*, 498–500
  pre-Big Bang inflation *(Gordon)*, 574–75, 591*n116*
  successful model *(Guth)*, 487
  See also Big Bang cosmology; inflationary cosmology; string cosmology; inflationary string landscape; quantum cosmology
inflationary string landscape, 486. *See also* string landscape
information
  biology's problem *(Dembski/Marks)*, 363–64
  creation of information, 360–63
  Darwin's solution *(Dembski/Marks)*, 364–67
  intelligence and information, 328
  metaphor in biology, 303–4
  molecular biology vs. information theory (IT) *(Meyer)*, 293–94
  origin of information, 293, 363
  origin of life *(Meyer)*, 293–94
  See also information, characteristics
information, characteristics, 390
  identifies by excluding, 360
  information vs. probability, 337*n103*

nature shuffles, doesn't create *(Dembski/Marks)*, 363
design produces information *(Dembski/Marks)*, 361–62
information displacement *(Meyer)*, 318–23
nature of information *(Dembski/Marks)*, 361–94
presupposes possibilities, 360
redundancy *(Meyer)*, 316–17
specified vs. semantic, 302
See also information theory (IT)
information theory (IT)
  cybernetics, xiv
  molecular biology impact *(Meyer)*, 298–99
  origin of information, 293
  Shannon information, 298–99
  See also information; intelligence
Inherit the Wind, 95–96
Institute for Faith and Learning (Baylor), xx, xxi
intelligence
  contingent evolution of intelligence *(Shermer)*, 447–48
  evolution timespan *(Shermer)*, 445–46
  fully natural thesis *(Dembski/Marks)*, 362–63
  information and intelligence *(Meyer)*, 328
  intelligent cause of universe *(Gordon)*, 559, 585, 586*n4*
  reverses entropy *(Dembski/Marks)*, 386
  sufficient cause *(Gordon)*, 30
  undergirding rational science *(Gordon)*, 5–6
  See also intelligent causation; knowledge
intelligent causation
  biological convergence *(Rana)*, 481
  origin of life *(Gordon)*, 31–32
  origin of universe *(Gordon)*, 568–69
  See also causation; design; intelligent design (ID)
Intelligent Design, 894
intelligent design (ID)
  1980s challenge to naturalism, 74–75
  as heuristic, 30
  current historical sciences and ID, 100–101
  intelligent design theory, 9
  "science" label *(McMullin)*, 88, 89
  See also intelligent design (ID), characteristics
intelligent design (ID), characteristics
  abduction use, 327–28
  early response to ID *(Numbers)*, 74–75
  explanation of events, not laws *(Meyer)*, 105
  not supernaturalist *(Dembski/Marks)*, 398*n77*

Index

intelligent design (ID), characteristics *(continued)*
  uses conservation of information law, 105
  uses methods of historical sciences *(Meyer)*, 100–101
  weds engineering, natural sciences, 394
  See also intelligent design (ID), opponents
intelligent design (ID), opponents
  criteria applied with bias *(Meyer)*, 106–7
  ID cites no mechanism, 124–*125n100, 101*
  ID not testable, 117*n*7, 118*n*26
  intelligent agent unobservable, 108
  intelligent agent unpredictable, 108
  not science by definition, 95
  not science, reasons listed *(Meyer)*, 106
  not testable, 108–13
  peer review issues, 102–4
  philosophical naturalism and ID, 30
  rooted in natural theology *(Numbers)*, 74
  See also intelligent design (ID), popular detraction
intelligent design (ID), popular detraction
  brings back supernatural *(de Duve)*, 346
  episodic creationism *(Van Till)*, 537–38
  God as designer *(de Duve)*, 354
  ends search for free lunch, 394
  media reports *(Meyer)*, 95
  methodological naturalism and ID, 30
  natural history and design, xv
  natural law and ID *(Meyer)*, 104–6
  not testable *(McMullin)*, 89–90
  omnipotent deity posited *(Meyer)*, 108, 110
  secret ID motivations *(Van Till)*, 540
  stifles research *(de Duve)*, 354
  See also intelligent design (ID), science
intelligent design (ID), science, 331
  addresses specific questions *(Meyer)*, 101–102
  best explanation for DNA *(Meyer)*, 100
  best explanation for fine-tuning, origins *(Gordon)*, 486
  causal adequacy *(Meyer)*, 99–100
  definitions of science *(Meyer)*, 96–98
  empirical evidence *(Meyer)*, 98
  engineering as ID science, 361
  events explained re prior events, 105–6
  junk DNA and ID *(Meyer)*, 112–13
  key publications, 120*n*46
  methods explained *(Meyer)*, 101
  observability *(Meyer)*, 107–8
  peer review *(Meyer)*, 102–4
  predictability *(Meyer)*, 100, 111–12

  research program, 2
  See also intelligent design (ID), testability *(Meyer)*
intelligent design (ID), testability *(Meyer)*, 99–100, 108–13
  comparative explanatory power, 99–100
  "failed test" claim, 121*n*68
  how refutable, 109–10
  how tested, 109–10
  See also design; intelligence; intelligent design (ID)
intentionality *(Koons)*, 215, 216, 218. *See also* consciousness
intrinsic properties, 182
Introduction to Mathematical Logic, 672
intuition, 259–61. *See also* consciousness
irrationalism, 5. *See also* rationalism
irreducible complexity, 98
  doubted *(Tooley)*, 890–92
  homeogenes disprove *(de Duve)*, 353
  mechanical analogies poor images *(de Duve)*, 349
  neo-Darwinism and complexity *(Gordon)*, 29
Islam, xiii

Jones, John E., 95–96
junk DNA
  ID and junk DNA *(Meyer)*, 112–13
  ID predictions, 123*n*95

kalām argument, 562
Kant, Immanuel *(Wolterstoff)*, 256–69
  achievability of knowledge, 136
  human nature, 267–269
  intuitions, 258–61, 270*n*4
  legacy, 256–58
  misinterpreted, 262–66
  objects vs. acts of acquaintance, 266
  perception vs. experience, 258–61
  pragmatists' revision, 257–58
  representations, 260–61
Kauffman, Stuart
  on Law of Conservation of Information (LCI), 388–89
  origin of life scenario, 319–21
kenosis, 540, 543–44
Kitzmiller v. Dover. *See* Dover trial
knowledge
  achievable *(Wolterstorff)*, 136
  knowing God *(McMullin)*, 94*n*18

Index

naturalism vs. knowledge *(Koons)*, 215
naturalist's view, 218
true belief, 218–20, 248
"wishful thinking" *(Williams)*, 250
Koch, Christof
consciousness problem ripe for attack, 761–62
philosophers' "very poor record," 758

Lamarckism, 71
Language, Truth, and Logic, 901
language, 248
Laplace, Pierre-Simon de
atheism, 68
his cosmology, 68–69
nebular hypothesis, 68–70
Law of Conservation of Generalization Performance, 386
Law of Conservation of Information (LCI), 363, 381–86
applied to biology *(Dembski/Marks)*, 386–93
Church-Turing Thesis analogue, 383
critiques, 390–93
Dawkins' mechanism and LCI, 389–90
entropy and LCI, 385–86
experimental verification proposal, 393–94
explained, 274
mathematically demonstrable *(Dembski/Marks)*, 274
Maxwell's Demon paradox, 386
not Kolmogorov complexity, 383–384
regress, 384–85
searches and LCI, 384
stated, 381
support from conservation theorems, 382
Meester's view, 390–93
law of nature *(Meyer)*, 120
laws of nature *(Gordon)*, 203–4
Lenski, Richard
computer sims vs. biology, 369
E. coli experiment *(Rana)*, 479–81
E. coli, long term evolution, 275
E. coli studies *(Behe)*, 436–38
life
coded chemistry *(Berlinski)*, 287–90
comes only from life, 281
history *(de Duve)*, 347–48
needs of life *(Gonzalez)*, 603–5
See also life forms; origin of life (OOL)

life forms
codes in life forms *(Berlinski)*, 279–80
complexity, specificity *(Meyer)*, 293–94
Life's Solution, 447, 479
local hidden variables, 189–90
locality (LOC) *(Gordon)*, 184, 185–89. *See also* nonlocality
Long Term Evolution Experiment (LTEE), 275, 479–81. *See also* Lenski, Richard
Low Probability Thesis *(Talbott)*, 152, 163, 164
LUCA (last universal common ancestor), 474

M theory. *See* string theory
macroscopic emergence (QT) theory, 190–95
malaria studies *(Behe)*, 433–35, 437–38
malicious demon, 175–76
thought experiment *(Talbott)*, 159, 163
man. *See* humans
Man's Place in the Universe, 445
Many Worlds Hypothesis (MWH) *(Craig)*, 521, 911–12
Mars, vs. Earth, 610–11
materialism
deficient worldview due to QT *(Gordon)*, 205
dematerialized materialism, xiii–xv
dualism vs. materialism, *(Searle)*, 734
Engels on materialism, xxx–xiv
pointless *(Steiner)*, 682
quantum theory (QT) *(Gordon)*, 189–90
reductive vs. non-reductive, 145–47
See also materialism, requirements for materiality
materialism, requirements for materiality *(Gordon)*
common cause (CC), 183
intrinsic properties, 182
material individuality criterion (CMI), 182
precise value principle (PVP), 182, 183
spin, 183
state-dependent properties, 182
See also naturalism
mathematics
constructivism *(Craig)*, 906
evolution of math *(Kitcher)*, 663n11–13
five components of practice *(Kitcher)*, 647
historical product *(Kitcher)*, 646–50, 658–61
how progress measured *(Kitcher)*, 654–58
logical applications *(Steiner)*, 678–80
mathematical description, 179
philosophy of mathematics *(Kitcher)*, 643–46
physics and mathematics *(Steiner)*, 682

mathematics *(continued)*
  reducing epistemology to science *(Kitcher)*, 649–50
  science and mathematics, 20
  See also mathematics, application *(Steiner)*
mathematics, application *(Steiner)*
  application to physics, 667–68
  canonical nonempirical, 672–78
  empirical applications, 668–72
  empirical canonical, 668–70
  empirical noncanonical, 670–72
  See also mathematics, approaches
mathematics, approaches
  fictionalism *(Craig)*, 906
  group theory *(Steiner)*, 675–78, 684–85n19–22
  mathematical naturalism *(Kitcher)*, 641, 643–61, 657–58
  Platonism vs. nominalism *(Steiner)*, 678–83
  Platonist mathematics *(Kitcher)*, 655–57
  Pythagoreanism *(Steiner)*, 642, 682–83, 686–87n43
  See also mathematics, apriorism *(Kitcher)*
mathematics, apriorism *(Kitcher)*, 662n8
  explained, 644–46
  rejected, 650
  See also mathematics, conflicts
mathematics, conflicts
  Darwinism's problems, 28
  epistemology vs. practicality *(Kitcher)*, 661
  Newton vs. Leibniz *(Kitcher)*, 647–48, 649, 652–54
  philosophical problems *(Kitcher)*, 641–42
  pure vs. applied *(Steiner)*, 678–79
  true vs. useful *(Steiner)*, 679–80
  See also mathematics
matter
  matter vs. mind question *(Gordon)*, 583–85
  nature of matter *(Dembski/Marks)*, 361–94
  See also materialism; naturalism
Maxwell's Demon paradox, 386, 398n63
meaning *(Williams)*, 255 n1
measure-theoretic theorem, 274
mechanical philosophy
  nominalism's influence *(Gordon)*, 21–22
  origin *(Numbers)*, 64
mechanics, classical vs. QM *(Stapp)*, 816–21. *See also* quantum mechanics (QM)
mechanism, 9

medieval science *(Gordon)*, 12–23
  Bacon's view, 12–13
  continuity, 38–39n69
Meditations, 667
mental causation, 692. *See also* causation, mental causation
mental properties, 4. *See also* mind
MESA, 369
metaphilosophical naturalism *(Williams)*, 248, 252
metaphysical naturalism *(Gordon)*, 26. *See also* philosophical naturalism (PN)
metaphysics
  realistic naturalists *(Williams)*, 248
  return from exile *(Craig)*, 901–2
  See also philosophy
methodological naturalism (MN)
  Dover Trial *(Meyer)*, 113–14
  evolution in schools *(McMullin)*, 85
  explained *(McMullin)*, 83
  ID vs. MN, 330, 331
  origin of term, 76n3
  supernatural intervention vs. MN, 541
  support for philosophical naturalism (PN) *(Willard)*, 869
  theists' role, 1–2
  See also methodological naturalism (MN), types
methodological naturalism (MN), types *(McMullin)*
  qualified, Version One, 86–88
  qualified, Version Two, 88–91
  strong methodological naturalism, 83–86
  See also methodological naturalism (MN), views
methodological naturalism (MN), views
  hindrance as ground rule *(Meyer)*, 115–16
  inadequate constraint *(Gordon)*, 558–59
  incoherent *(McMullin)*, 82–83
  legitimate role *(Meyer)*, 125n103
  Lyell's assumption, 42n105
  necessary *(de Duve)*, 274
  Newton's rejection, 23
  unexamined convention *(Meyer)*, 113–14
  See also methodological naturalism (MN); philosophical naturalism (PN)
methodology *(Goldman)*, 234
Meyer, Stephen
  digital information in cell, 98
  DNA and informational specified complexity, 273

Index

Smithsonian journal controversy, 102–3, 119*n*43, 44
Michael Polanyi Center (MPC), xx, xxiii*n*2
  shutdown, xxi
Middle Ages, 12–23
  cultural advances, 18–22
Miller, Ken
  defends *ev*, 370–71
  design as undetectable, 395*n*20
Mind and World (McDowell), 262
Mind of God, The, 352
mind, characteristics
  "inclined not compelled" (Aquinas), 713
  mind vs. matter *(Gordon)*, 583–85
  naturalism and evidence *(Plantinga)*, 147–48
  origins of mind *(Berlinski)*, 714–19
  origins of mind *(Cosmides/Tooby)*, 699–707
  quantum theory and mind, 202–3
mind, philosophy of mind *(Moreland)*
  authority thesis, 836–37
  autonomy thesis, 836–37, 848*n*27
  autonomy thesis studied, 837–41
  central questions, 836
  hard sciences irrelevant, 836, 845–46
mind, similes *(Berlinski)*
  biological artifact, 721–24
  body organ, 718–21
  computer, 714–18
  See also mind, views of mind
mind, views of mind
  Darwinism explains mind *(Murphy)*, 778–80, 783
  folk psychology *(Moreland)*, 837–38
  materialism and beliefs *(Plantinga)*, 138–48, 150*n* 12–15, 151*n*17
  naturalism undermines knowing *(Plantinga)*, 137–48
  top down causation *(Murphy)*, 778–80, 782–85
  See also consciousness; mind-body relationship
mind-body relationship
  MN view, 85
  models, Descartes, 11
  problem, 730–31
  problem's history *(Searle)*, 734–35
  See also consciousness; mind
minimal cell, 286, 293. *See also* cell
miracles, 545
modern science. *See* science, modern science

Moderne Algebra, 659
molecular biology
  central dogma (Crick), 280–281
  complexity, specificity of life, 294–95
  information theory (IT) *(Meyer)*, 293–94, 298–99
  origin of life *(Berlinski)*, 282–83
  Shannon information, 298–99
molecular clock, xv
molecular convergence *(Rana)*
  bacteria vs. archaea/eukaryotes, 474
  examples and types, 462–74
Molinism, 28
monism, 4
monotheism *(Tooley)*, 897–98
morality. *See* ethics
multiverse
  fine-tuning vs. multiverse, 485
  our universe naturally selected, 84
  types *(Weinberg)*, 549–55
  See also inflationary cosmology; naturalism; quantum cosmology; string cosmology; string landscape
Murphy, Nancey, 692–93
  Christians should be physicalists, 772–76
  Darwinism explains mind, 778–80, 783
  soul, mind not needed, 836, 841–43
  substantial mind unneeded, 776–77
  views summarized, 786

National Academy of Sciences, 276
natural law
  description vs. explanation *(Meyer)*, 105
  historical sciences and natural law *(Meyer)*, 105
  ID and natural law *(Meyer)*, 104–6
  necessitarianism untenable, *(Gordon)*, 181
natural philosophy *(Numbers)*, 64–74
natural selection (NS)
  creates all functional complexity *(Cosmides/Tooby)*, 697
  pre-biotic NS *(Meyer)*, 310–13
  produces mind *(Cosmides/Tooby)*, 696–98
  promotes love of neighbor *(Ruse)*, 856–57
  selects for behavior, not belief *(Plantinga)*, 139
  universe as selected, 84
  See also Darwinism
Natural Theology, 429
natural theology, xiii. *See also* theology
Naturalism and the Human Spirit, 869

Index

naturalism, characteristics
- atheism vs. naturalism *(Plantinga)*, 137
- coexistence with Christianity *(Numbers)*, 75
- controls science practice *(Numbers)*, 62
- creed *(Plantinga)*, 137, 138
- defined *(McMullin)*, 82
- defined *(Tooley)*, 880–82
- distaste for metaphysics *(Gordon)*, 180
- holds moral high ground *(Willard)*, 877–78
- intelligence as fully natural *(Dembski/Marks)*, 362–63
- key common doctrines *(Williams)*, 253
- mental properties and naturalism, 4
- naturalistic ethics *(Willard)*, 869–72
- provides worldview, 137
- rational explanation and naturalism *(Gordon)*, 3
- rule in academy, 35–36n27–28
- supernatural vs. natural *(Numbers)*, 62
- supernaturalism vs. naturalism *(Tooley)*, 898
- use of term, xi
- will and morality *(Willard)*, 873–78
- *See also* naturalism, evolutionary theory

naturalism, evolutionary theory
- compatible with evolutionary theory *(Talbott)*, 163–64
- conflicts with evolution *(Plantinga)*, 138–48
- evolutionary theory opposes it *(Plantinga)*, 137–48
- evolutionary theory supports it *(Talbott)*, 152–64
- *See also* naturalism, history

naturalism, history, 6–7
- ancient world, 63
- early America *(Numbers)*, 65–67
- early modern world *(Numbers)*, 64–65
- 18th century, new meanings *(McMullin)*, 82
- medieval world *(Numbers)*, 63
- 1980s ID challenge *(Numbers)*, 74
- social sciences embracing naturalism *(Numbers)*, 73–74
- supreme in late 20th century, 74
- *See also* naturalism, popular culture

naturalism, popular culture
- attitude in science, 180
- "natural," usage *(Williams)*, 251–53
- "Scientific Worldview," 137
- *See also* naturalism, types

naturalism, types
- atheistic naturalism *(Van Till)*, 539
- biological naturalism *(Searle)*, 692
- evolutionary naturalism, 163
- four understandings *(Willard)*, 865–66
- mathematical naturalism *(Kitcher)*, 641, 643–61
- naturalistic theism *(Van Till)*, 485
- ontological naturalism *(Gordon)*, 179
- realism vs. pragmatism *(Williams)*, 247, 252–53
- realistic naturalism, explained *(Williams)*, 247–51
- reliabilism, 248
- religiously agnostic naturalism *(Van Till)*, 539–40
- remains promising approach *(de Duve)*, 355–57
- RFEP as naturalism *(Van Till)*, 538–39
- scientific naturalism *(Van Till)*, 540
- *See also* naturalism, views

naturalism, views
- affirm to last plausible model *(de Duve)*, 351–52
- Christianity and naturalism *(Numbers)*, 64–67
- contradicts scientific realism *(Koons)*, 226
- deficient worldview due to QT *(Gordon)*, 205
- experiments underestimated *(Fuller)*, xiii
- "highly malleable" *(Gordon)*, 4
- inadequacy *(Fuller)*, xiii–xiv
- Kant rejected today, 247
- mathematics problems, 28
- naturalism undercuts realism *(Koons)*, 136
- naturalists should embrace realism *(Goldman)*, 136
- nonstarter as physical theory *(Gordon)*, 201
- poses ethical dilemma *(Willard)*, 868–69
- postulate, not belief *(de Duve)*, 346–57
- purposeless universe *(Gordon)*, 40n99, 100
- science and naturalism *(Fuller)*, xii
- science's base *(de Duve)*, 346–57
- seen as God's laws *(Numbers)*, 72–73
- "self-defeating" *(Gordon)*, 4–5
- smallpox and naturalism *(Numbers)*, 65–66
- successful for science *(de Duve)*, 347
- "view from nowhere" *(Nagel)*, xii
- "weasel word" *(Gordon)*, 206n2
- *See also* intelligent design (ID); methodological naturalism (MN); naturalism, characteristics; philosophical naturalism (PN); pragmatism

Index

Nature of Mathematical Knowledge, The, 658, 665n34, 35, 666n38
Nature of Nature conference
  balance, xx
  Dembski's role, xx
  Gordon's role, xx
  history, xx
  key theme, xxiii–xxivn3
  repercussions, xxi
nature
  capital "N" *(McMullin)*, 82
  disenchantment of nature *(Weber)*, 251–53
  intelligence as factor *(Fuller)*, xiv–xv
  nature of nature *(Dembski/Marks)*, 361–94
  not causally closed *(Koons)*, 216
  shuffles information *(Dembski/Marks)*, 363
  See also causation; information
Nature's Destiny, 353
Nature of Mathematical Knowledge, The, 658
Neanderthals *(Shermer)*, 452–54
nebular hypothesis, 68–70
necessitarianism, 203
necessity
  SETI and necessity of life *(Shermer)*, 443–45
  theory choice *(Koons)*, 221–22
neo-Darwinism
  evidence against, 28–30
  theistic evolution's acceptance, 27
  See also Darwinism
neuroscience. *See* cognitive science; evolutionary psychology
Neutral Theory of Molecular Evolution, 722
new atheism *(Gordon)*, 33–34n9
Newton, Isaac
  on gravity, 22–23
  solar system as device *(Numbers)*, 64–65
98% human-chimp DNA similarity, 30
98% junk gene thesis, 29–30
"no free lunch" (NFL), xiv, 105
  NFL theorems, 378, 381
  theorems' history, 382–83
  See also Conservation of Information Theorems
nominalism
  Darwinism and nominalism *(Gordon)*, 26
  defined, 7
  doctrine of creation *(McMullin)*, 87–88
  effect on science *(Gordon)*, 8
  mechanical philosophy and nominalism, 22
  voluntarism, 35n21
  See also realism

non-reductive
  non-reductive materialism *(Plantinga)*, 145–47
  non-reductive naturalism, 4
  non-reductive physicalism *(Murphy)*, 692
  See also physicalism, non-reductive *(Murphy)*
nonlocality *(Gordon)*
  dynamic emergence, 190–95
  Humean supervenience, 204–5
  nonlocal correlations, 180
  prism models as alternative, 186
  quantum systems, 180
  relational holism, 190–95
  status of loopholes, 186
  See also nonlocalizability
nonlocalizability *(Gordon)*
  algebraic QFT, 198–201
  explanation, 181
  quantum systems, 180
  See also nonlocality *(Gordon)*; quantum mechanics (QM)
nucleic acids, 278–79. *See also* amino acids; DNA; proteins; RNA

old earth creationists *(Tooley)*, 888
On the Origin of Species, 99, 101–2, 431, 461
One Long Argument, 430–31
ontological naturalism (ON), 82, 83
  representational naturalism vs. ON *(Koons)*, 216, 220
  challenge of quantum theory *(Gordon)*, 179
  defined *(Koons)*, 216
  explained *(Gordon)*, 179–80
  origin of life *(McMullin)*, 85–86
  physical explanation paramount *(Gordon)*, 179–80
  scientific realism vs. ON *(Koons)*, 215
  See also methodological naturalism (MN); ontology
ontology
  Fock space and ontology *(Gordon)*, 195–98
  ontological argument for God *(Craig)*, 915–17
  quantum theory (QT) challenge *(Gordon)*, 195–96
"order for free," xiv. *See also* "no free lunch"
Origin of Order, The, 446
origin of life (OOL)
  methodological naturalism as hindrance *(Meyer)*, 116

Index

origin of life (OOL) *(continued)*
    origin of biological information *(Meyer)*, 293
    progress claimed *(de Duve)*, 349
    protein synthesis *(Berlinski)*, 278–80
    right vs. left-handed sugars, 285
    See also origin of life (OOL), Darwinian origin
origin of life (OOL), Darwinian origin
    "all-out Darwinian perspective" *(Woese)*, 289–90
    Darwin's views, 276
    Darwinian evolution *(Berlinski)*, 286–87
    prebiotic natural selection *(Meyer)*, 312–13
    See also origin of life (OOL), history of field *(Berlinski)*
origin of life (OOL), history of field *(Berlinski)*
    1773 urea discovery, 276
    DNA, double helix discovery, 277–78
    Haldane-Oparin hypothesis, 276–77
    Miller-Urey experiment, 277
    See also origin of life (OOL), intelligent design (ID)
origin of life (OOL), intelligent design (ID)
    impossible if design-free *(Gordon)*, 28–30, 29
    information displacement problem *(Meyer)*, 318–23
    requires intelligent cause *(Gordon)*, 31–32
    See also origin of life (OOL), Oparin's hypothesis
origin of life (OOL), Oparin's hypothesis
    coacervates, 311
    modern doubts about hypothesis, 311–12
    prebiotic natural selection, 310–13
    See also origin of life (OOL), theories
origin of life (OOL), theories
    chance origin *(Meyer)*, 304–10
    "chemical minestrone," 319–21
    comets' role *(de Duve)*, 349
    computational evolution research *(Dembski/Marks)*, 368–70
    early OOL, 348–52
    highly probable, 27–28
    inflationary multiverse invoked, 582–83
    lagoon OOL, 284
    metabolism first, 291
    Morowitz's view, 290–91
    peptides first *(de Duve)* 350–52
    physics and chemistry explain, 349
    prebiotic soup, 277
    Quastler's scenario, 318–19
    RNA world, 321–22
    self-organization theories, *(Meyer)*, 313–18
    simple to complex *(Meyer)*, 294–95
    Suga's work, 288–90
    swift OOL, 348–49
    See also minimal cell; origin of life (OOL); RNA world
Origins of Order, The, 319
origin of mind *(Berlinski)*, 724. *See also* consciousness

particles *(Gordon)*
    indexed particles, 196–98
    do not really exist, 194–95
    spin-singlet states, 184–89
    See also quantum theory (QT)
peer review, 102–4
Penrose, Roger
    entropic fine-tuning of Big Bang, 526–27, 571–72
    evolution and platonic information, 808
    objective reduction of consciousness, 693
phenomenalism, 202–3
philosophical naturalism (PN)
    ID and PN, 30
    Darwinism's contribution *(Gordon)*, 26
    defined *(Koons)*, 216
    dominant 1950s on *(Craig)*, 901–5
    "irrational" *(Gordon)*, 5
    mathematical naturalism *(Kitcher)*, 641, 643–61
    natural sciences and PN, 215, 225
    quantum field theory and PN *(Gordon)*, 201
    rules academy *(Gordon)*, 3, 40–41n100
    scientism's role, 215
    undermining rationality *(Gordon)*, 1
    verificationism *(Craig)*, 901–2
    See also methodological naturalism (MN)
philosophy
    Christian philosophy, 902–3
    empiricism and philosophy *(Goldman)*, 140–41
    metaphilosophy *(Williams)*, 247
    natural science vs. philosophy *(Williams)*, 247
    ontological argument for God, 915–17
    scientific naturalism useless *(Willard)*, 866–69
    See also consciousness; mind
physicalism
    deficient worldview *(Gordon)*, 205
    unneeded by neuroscience *(Moreland)*, 844, 846
    See also physicalism, non-reductive *(Murphy)*

Index

physicalism, non-reductive *(Murphy)*, 692
   defined, 771–72
   top down mind causation, 778–80, 782–85
physics
   "different physics," 486, 562
   Theory of Everything (TOE) *(Craig)*, 910–12
   universal knowledge and physics, xi
   "physicsism" *(Willard)*, 868
   See also quantum theory (QT)
pilot-wave theory, 189
Plantinga, Alvin
   alien thought experiment, 140–41, 144–45
   argues by analogy *(Talbott)*, 153
   Ernan McMullin's view, 93–94n14, 15
   evolution vs. naturalism, 135
   Talbott's EAAN, 166–72
   reliance on Hume *(Talbott)*, 163, 176–77
Platonism, 216
   God's will and Platonism, 8
Polanyi Center. *See* Michael Polanyi Center (MPC)
polynucleotides, 388–89
positron, 217
pragmatism *(Williams)*, 5
   anti-supernatural, 253–54
   correspondence theory, 250
   denial of absolutes, 33n6, 7, 8
   explained, 248–50
   key characteristics, 253–54
   more naturalistic than realism, 253–54
   objections, to pragmatism, 252–53
   pragmatic naturalism, 247
   realism vs. pragmatism, 248–51, 252–53
   simplicity and pragmatism *(Koons)*, 225
   See also realism
prebiotic atmosphere, 283–85
   prebiotic natural selection *(Meyer)*, 310–13
   prebiotic soup, 277
   See also origin of life (OOL)
Precise Value Principle (PVP), 182, 183
prediction, empirical prediction, 179
Principia Mathematica, 22
Principles of Geology, 70–71, 325
probability
   distributions vs. outcomes, 383
   freak events, 310
   information vs. probability, 299, 337n103
   probabilistic resources *(Meyer)*, 335n61, 336n62–64
   reliability vs. probability, *(Koons)*, 220
   universal limit, 336n65

Proceedings of the Biological Society of Washington, 102–3
Prolegomena, 258, 267
protein folds
   challenge to Darwinism *(Axe)*, 274–75
   sampling problem *(Axe)*, 412–15
   See also proteins
proteins
   amino acids, 412–13
   complexity and specificity *(Meyer)*, 295
   Darwinian origin "implausible" *(Axe)*, 420
   described, 412
   domains vs. chains *(Axe)*, 424–25
   protein synthesis *(Berlinski)*, 278–80
   regulation, 400–403
   sampling problem magnified *(Axe)*, 425–26
   self-organization theories, *(Meyer)*, 313–18
   sequence possibilities, *(Axe)*, 413–14
   sequence's importance *(Axe)*, 422–23
   size, coupling importance *(Axe)*, 415–20
   size, function vs. size *(Axe)*, 415–20
   stepwise evolution *(de Duve)*, 350–52
pseudogenes, 29
   evidence against ID *(Tooley)*, 894
Ptolemy, 718–19, 720
punctuated equilibria, 108

quantum cosmology, 486
   criticisms of *(Gordon)*, 563–69
   Hartle-Hawking model, 563, 565–69
   imaginary time, 515–16, 565–68
   "new" or "different physics" before the Planck time, 500, 563
   superspace *(Gordon)*, 563–66
   Vacuum Fluctuation Models, 512–13
   Vilenkin's approach, 563–65
   See also Big Bang cosmology, inflationary cosmology, quantum gravity, string cosmology
quantum field theory (QFT) *(Gordon)*
   algebraic QFT, Newton-Wigner localization, 198, 200–201
   algebraic QFT, Reeh-Schlieder theorem, 198, 200–201
   indexed particle, 196–98
   particle interpretations impossible, 198–201
   See also quantum mechanics (QM)
quantum gravity
   Hartle-Hawking model *(Craig)*, 515–16

Index

quantum gravity *(continued)*
  imaginary time, 515–16
  presupposed by quantum cosmology *(Gordon)*, 563–69
  Quantum Gravity Models (QGM), 514–17
  See also Big Bang cosmology; inflationary cosmology; string theory
quantum mechanics (QM)
  chemical bonds of life, 287
  classical mechanics vs. QM *(Stapp)*, 816–21
  consciousness and QM *(Hameroff/Penrose)*, 790–92
  Humean supervenience *(Gordon)*, 204–5
  many worlds *(Gordon)*, 587
  See also quantum theory (QT)
quantum theory (QT), 9
  achievements, 179
  action at a distance *(Stapp)*, 827
  classical mechanical limit, 190–91
  consciousness and QT, 693
  correct explanation anti-materialist, 180–81
  correlations, no physical explanation *(Gordon)*, 183–84, 195
  determinism, de Broglie-Bohm theory *(Gordon)*, 189, 203–4
  determinism, pilot-wave theory, 189
  empirical predictions, 179
  explanatory demand *(Gordon)*, 204–5
  Humean supervenience, 204–5
  incompatible with materialism *(Gordon)*, 181
  law and QT, 203–4
  local determinism and QT, 188
  loopholes, 186
  mind and QT, 202–3
  nature of theory *(Gordon)*, 179
  necessitarian law and QT *(Gordon)*, 203–4
  nonlocality, 181
  objective properties lacking, 203
  offers mathematical description, 179
  ontological deflation as myth *(Gordon)*, 195–96
  ontological naturalism challenged *(Gordon)*, 179
  prism models, 186
  quantum coincidence *(Gordon)*, 204–5
  spin of particles, 183
  stochastic vs. deterministic models, 188
  structural realism and QT *(Gordon)*, 202–3
  von Neumann interpretation *(Stapp)*, 819–21, 827–30
  See also causation; consciousness; emergence theory; materialism; quantum cosmology; supervenience

Rana, Fazale
  contingency vs. convergence, 461–62
  design vs. contingency, 478–81
  evolutionary convergences, 275
random
  explained *(Plantinga)*, 149n7
  evolution theory's use, 24–25
  random mutation, efficacy unproven *(Behe)*, 431, 433–35
Rare Earth hypothesis, 604
rationality, 691
  alternative to Hume *(Talbott)*, 176–77
  arising from physical circumstances, 4
  Humean rationality *(Talbott)*, 172–74
  nonreductive rationalists, 4
  types of rationality *(Kitcher)*, 650–54
  See also consciousness; mind
Rawls, John
  justice as fairness *(Ruse)*, 857
  political justice *(Willard)*, 871
realism
  bivalence component *(Goldman)*, 230–31
  concepts and realism *(Wolterstorff)*, 256–58
  disentangling meanings *(Goldman)*, 230
  evidence, truth, and realism *(Goldman)*, 236–37
  fittingness *(Goldman)*, 237–40
  metaphysical realism *(Goldman)*, 240
  nonepistemic approach to truth *(Goldman)*, 232
  pragmatism vs. realism *(Williams)*, 248–51, 252–53
  scientific realism *(Goldman)*, 240–44
  See also antirealism; realistic naturalism; scientific realism
realistic naturalism (RN)
  explained *(Williams)*, 247–51
  physicalism favored *(Williams)*, 248
  reliable discriminative responsiveness *(Williams)*, 250–51
  scientific realism and RN, 248
reality *(Wolterstorff)*, 265–66
reason
  top down causation *(Murphy)*, 782–85
  See also consciousness; mind; rationality

Index

reductionism, 4
  beliefs and reductionism *(Plantinga)*, 145–47
  reductive naturalism, 4
  See also materialism
Reformation
  expanding the sacred (Taylor), 35n23, 24
  Reformed vs. Thomist theology *(McMullin)*, 87–88
relativity *(Stapp)*, 827–30
reliability
  central to representational naturalism *(Koons)*, 218–20
  conditional reliability *(Koons)*, 219–20
  explained 220 *(Koons)*, 220–21
  non-separable cognitive faculties *(Talbott)*, 158
  outside information sources *(Talbott)*, 153–54
  science's reliability *(Koons)*, 215
  theory choice *(Koons)*, 221–22
  See also reliability, coherence
reliability, coherence
  coherence *(Plantinga)*, 166–72
  coherence *(Talbott)*, 173–74
  T-coherence *(Plantinga)*, 166–72
  See also reliability, global reliability *(Talbott)*
reliability, global reliability *(Talbott)*, 156–62
  "brain in a vat" thought experiment *(Talbott)*, 160–61
  evaluations *(Talbott)*, 159–61
  global unreliability *(Talbott)*, 159
  "malicious demon" thought experiment *(Talbott)*, 159, 175–76
  See also reliability, local reliability *(Talbott)*
reliability, local reliability *(Talbott)*, 154, 155
  non-separable cognitive faculties *(Talbott)*, 154, 158–59
  See also reliability, thought experiments
reliability, thought experiments
  "augmented XX"*(Talbott on Plantinga)*, 157–61, 174–75
  "augmented ZZ"*(Talbott on Plantinga)*, 175–76
  "brain in a vat" *(Talbott)*, 160–61
  "malicious demon" *(Talbott)*, 159, 175–76
  Muller-Lyre illusion *(Plantinga on Talbott)*, 168–70
  "tardy revelation" *(Talbott)*, 155–56
  "XX example" *(Plantinga on Talbott)*, 166–72
  "XX example" *(Talbott on Plantinga)*, 156–57, 161, 162–63

"ZZ example" *(Plantinga on Talbott)*, 171
"ZZ example" *(Talbott on Plantinga)*, 161–62
See also cognition; reliability
religion
  adaptive false beliefs *(Plantinga)*, 142–143
  magic and religion (Taylor), 10
  natural origin claims, 73–74
  warfare myth, 12–23
  See also Christianity
representational naturalism (RN) *(Koons)*
  defined, 216
  explained, 218
  ontological naturalism vs. RN, 216, 220
  See also naturalism
Republic, 667
retrodiction, 101. *See also* prediction
RNA
  explained *(Berlinski)*, 279–80. *See also* RNA world
RNA world
  Berlinski on RNA world, 281–82
  growing doubt, 290–91
  Meyer on RNA world, 321–22
  ribozyme discovery, 282
  self-organizing RNA, *(Meyer)*, 313–18
  self-replicating RNA *(Berlinski)*, 285–90
  See also DNA
Royal Society, 8
rRNA, 280, 282. *See also* RNA
Ruse, Michael
  Christian can be Darwinist, 861–62
  substantive morality as illusion, 853

schools. *See* education
science, characteristics
  aspirations *(Fuller)*, xii
  Duhemian vs. Augustinian *(Moreland)*, 843–44
  philosophy vs. natural science *(Williams)*, 247
  popular culture perception, 14–17
  presuppositions, xix
  problem-solving skills *(Fuller)*, xi
  proscriptive generalization, 330
  purposes *(Fuller)*, xii
  transcendent purpose *(Gordon)*, 5–6, 27–30
  universality, xii, 89
  See also science, defined
science, defined
  ID and definitions *(Meyer)*, 104
  definition vs. methodology *(Meyer)*, 98

Index

science, defined *(continued)*
   no single definition *(Meyer)*, 97–98
   methods as criterion *(Meyer)*, 97
   non-science vs. science *(McMullin)*, 88
   "other ways of knowing"*(McMullin)*, 90–91
   See also science, history of science
science, history of science
   discoveries about science, 547–48
   Enlightenment science *(Numbers)*, 67
   Judeo-Christian foundations, 6–8
   medieval science *(Gordon)*, 12–23, 20–22
   science myths, 9–10
   scientific creationism *(Numbers)*, 74
   See also science, models of inquiry
science, models of inquiry, 292n2
   abduction, 323–24
   evaluation, 116
   induction, xii–xiii, 386–87
   methodological naturalism *(Meyer)*, 331
   See also science, modern science
science, modern science
   Aristotelianism rejected *(Williams)*, 251
   beginnings *(Numbers)*, 67–74
   debt to medieval science *(Gordon)*, 23
   rejects appeals to God *(Numbers)*, 62
   See also science, naturalism and science
science, naturalism and science
   "naturalism" as Huxley's term, 62
   naturalism as science's base *(de Duve)*, 346–57
   naturalism useless to philosophy *(Willard)*, 866–69
   philosophical naturalism's setback *(Koons)*, 225
   theology and science *(Fuller)*, xiii
   warfare myth, 12–23
   See also science, popular culture and science
science, popular culture and science
   consensus science, 119n45
   influences popular ideas, xi
   invoking "science" *(Willard)*, 866–68
   "junk" DNA, 29–30
   label "science" *(McMullin)*, 90–91
   98% human-chimp DNA similarity, 30
   See also science, characteristics
scientific realism (SR)
   convergent realism approach *(Goldman)*, 241–43
   defined *(Koons)*, 216
   described *(Goldman)*, 241–44
   justification *(Goldman)*, 243–44
   metaphysical realism *(Goldman)*, 243–44

   Millikan's view *(Koons)*, 221–22
   naturalism vs. *(Koons)*, 215
   ontological naturalism vs. *(Koons)*, 215
   Papineau's view *(Koons)*, 221–22
   representational naturalism vs. *(Koons)*, 216, 220
   scientific antirealism *(Koons)*, 225–26
   semantic thesis *(Goldman)*, 241
   See also antirealism; realism; realistic naturalism
scientism
   defined *(Gordon)*, 3–4
   philosophical naturalism's role *(Koons)*, 215
Scopes Monkey Trial, 95–96
Search for Extraterrestrial Intelligence (SETI), 115
   global tuning *(Gonzalez)*, 625
   necessity of life *(Shermer)*, 443–45
   skepticism *(Shermer)*, 445–48
Searle, John
   consciousness as holistic, 740
   consciousness as key biology problem, 741–42
   physicalism's necessity, 835
   "scientific view" *(Willard)*, 867–68
   See also consciousness
Second Law of Thermodynamics
   conservation of information *(Dembski/Marks)*, 385–86
   explained *(Craig)*, 522–27
secularism
   belief in God polls, 75
   Darwinism and secularism, 26
   distortions *(Gordon)*, 34n11
   history, 6–7
   Taylor's view, 35n23, 35–36n27
   See also atheism; naturalism
self, 10–12. *See also* consciousness; mind
self-organization theories *(Meyer)*
   DNA, proteins, RNA, 313–18
   information displacement problem, 319–23
   new laws proposed, 318
   order and complexity conflated, 317
   origin of life, 313–18
semantics
   consciousness and semantics, 4
   semantic epiphenomenalism *(Plantinga)*, 141–42
sequence, 422
   nucleotide sequence odds *(Berlinski)*, 286–87

Index

sequences *(Meyer)*, 334n39, 335n57
  See also proteins
SETI. *See* Search for Extraterrestrial Intelligence (SETI)
Shadows of the Mind, 791
Shermer, Michael
  civilizations' lifespans, 456–58n6, 7
  contingency hypothesis, 449, 450–55
  contingency vs. necessity, 443–55
  contingent-necessity model, 448–51
  humanity "quirk of evolution," 455
  Neanderthals' inferiority, 452–54
  radical contingency thesis, 275
sickle cell anemia *(Behe)*, 433–35
Signature in the Cell, 99, 113, 114
simplicity *(Koons)*
  Akaike vs. Forster-Sober, 223–25
  overfitting, 223–25
  pervasiveness, 216–18
  supernatural cause, 220–21
slavery, 37n
Social Darwinism *(Ruse)*, 856
sociobiology *(Ruse)*, 855
special relativity, 547
specificity *(Meyer)*, 300–302
specified complexity (SC)
  activity of intelligent agents *(Meyer)*, 323
  flaws in argument *(Tooley)*, 894–97
  origin of life and SC *(Meyer)*, 297–331
spin, of particles, 183
  spin-singlet state, 184–89
Standard Big Bang model/Standard Model. *See* Big Bang cosmology
Stapp, Henry
  brain not classical mechanics, 818
  consciousness as events, not substance, 693
  quantum theory (QT) empirically verified, 823
  See also brain; mind
stasis, 29
state-dependent properties, 182
Steady State Model, 509–10
Steinhardt-Turok model, 486
stochastic hidden variables, 182, 184
string cosmology
  cyclic ekpyrotic model *(Gordon)*, 562, 573–74
  cyclic ekpyrotic model *(Guth)*, 500
  cyclic ekpyrotic scenario *(Craig)*, 519–20
  fine-tuning issues *(Gordon)*, 576–83

pre-Big Bang inflationary scenario *(Gordon)*, 574–75
pre-Big Bang scenario *(Craig)*, 518–19
string landscape *(Gordon)*, 575–76, 592n135
string landscape *(Guth)*, 494–96. *See also* string landscape
Veneziano-Gasperini model *(Craig)*, 518–19
See also inflationary cosmology; quantum cosmology; string landscape; string theory
string landscape
  conjoined with eternal inflation *(Guth)*, 494–96
  explained and criticized *(Gordon)*, 575–76, 579–83, 592n135
  explained and defended *(Weinberg)*, 548, 550–55
  naturalistic motivations evaluated *(Gordon)*, 583–85
  successful prediction required *(Weinberg)*, 548
string theory, 485–86
  central concepts explained *(Gordon)*, 572–73
  has many solutions (vacua), 486, 548, 573
  M-theory *(Gordon)*, 573–74, 579–81
strong methodological naturalism (SMN) *(McMullin)*, 83–86
structural realism, and QT *(Gordon)*, 202–3
Summer for the Gods, 96
SUPERFAMILY database, 424
supernaturalism
  defined *(Tooley)*, 881
  naturalism vs. *(Tooley)*, 898
  proofs not convincing *(Tooley)*, 882–87
  RFEP and supernatural *(Van Till)*, 539–41
supervenience, 4, 692
  defined, explained *(Murphy)*, 772–76
  emergence and supervenience "unenlightening" *(Gordon)*, 191–92
  Humean supervenience *(Gordon)*, 204–5
  non-reductive supervenience *(Murphy)*, 776–78
  "sleight of hand" *(Gordon)*, 4
  types *(Murphy)* 775–76
  See also Gordon, Bruce; Murphy, Nancey
symbiogenesis *(Dembski/Marks)*, 387–88
symbiosis, vs. competition *(Wright)*, 451–52
symmetry
  in physics *(Weinberg)*, 217
  symmetry principles, 547
Symplectic Techniques in Physics, 667
System of Logic, A, 678

Index

Talbott, William
- evolution and naturalism, 135
- Plantinga and Humean rationality, 193
- propositions naturalist accepts, 152–53

targeted search, 367, 368. *See also* computational evolution

teleology
- biology not teleology *(Coyne)*, 361
- Darwinian evolution teleological *(Dembski/Marks)*, 274, 363, 390–93
- Denton's approach *(de Duve)*, 354, 355
- modern science rejects teleology *(Williams)*, 251
- OOS vs. teleology, 429–31
- representational naturalism's view *(Koons)*, 218–20
- teleological argument for God *(Craig)*, 909–12
- teleology is measurable *(Dembski/Marks)*, 274
- *See also* design

Templeton Foundation, xx

testability, ID and testability, 108–13
- intelligent design not testable *(McMullin)*, 89–90
- unobservable *(Meyer)*, 109–10

Tetrabiblos, 719

textbooks, 40*n100*

thalassemia *(Behe)*, 433–35

Theaetatus, 679

theism
- kenotic theism *(Van Till)*, 540, 543–44
- naturalistic theism *(Van Till)*, 540, 544–45
- reason's foundation *(Gordon)*, 33*n9*
- supernatural theism *(Van Till)*, 540–41
- theism without supernatural *(Van Till)*, 543–44
- theistic science *(McMullin)*, 88
- *See also* theistic evolution

theistic evolution
- accommodates neo-Darwinism *(Gordon)*, 27
- claims God created via Darwinism *(Dembski/Marks)*, 365–67

theology
- process theology *(Van Till)*, 544–45
- qualified methodological naturalism and theology *(McMullin)*, 87–88
- science and theology *(Fuller)*, xiii

Theory of Everything (TOE) *(Craig)*, 910–11

Theory of Justice, A, 871

thermodynamics *(Craig)*, 522–27

Thomism *(McMullin)*, 87

Tierra, 369, 372–73

Tooby, John
- cognitive science research faulted, 706–7
- evolution explains mind, 691
- valued activities are side effects, 699–700

Tooley, Michael
- irreducible complexity doubted, 888–92
- religion unsupported by science, reason, 854

Touchstone Magazine, xx, xxiii*n3*

transcendence
- defenders of neo-Darwinism, 28–30
- science and transcendence, 5–6

transitions in evolution. *See* evolution

tree of life, xv
- "nonsense," *(Gordon)*, 29

tRNA, 279–80, 297. *See also* RNA

truth
- alethic design and truth *(Talbott)*, 163–64
- brain evolves for fitness, not truth (D. S. Wilson), 139
- correspondence and truth, 237–40
- deflationary theories *(Williams)*, 249–50
- epistemic approach, weaknesses *(Goldman)*, 232–34
- false beliefs *(Goldman)*, 242–43
- intelligence, and truth *(Goldman)*, 230
- justification and truth *(Goldman)*, 230, 234
- nonepistemic view, explained *(Goldman)*, 235
- pragmatism and truth *(Williams)*, 250
- Putnam's classifications, 231–32
- realist view *(Goldman)*, 235
- simplicity and truth *(Koons)*, 220–21
- triangulation and truth *(Goldman)*, 235–36
- verification and truth *(Goldman)*, 230, 235

Turing machine, 716–17

Unintelligent Design, 894

universal probability bound (UPB), 339–40*n141*

universe, characteristics
- composition, 511
- entropy, 571–72
- flatness, 560–61, 569–70
- horizon, 560–61, 569
- listed *(Guth)*, 488–90
- no magnetic monopoles, 569
- space as expanding, 528–29*n15*
- *See also* universe, end of universe

universe, end of universe *(Craig)*
- Big Crunch, 510–12
- end of universe *(Davies)*, 523

Index

end of universe theory, 522–27
expands forever, 511
oscillating models, 510–12
See also universe, origin of universe
universe, origin of universe
  design and origin *(Craig)*, 533n107
  inflationary cosmology model *(Guth)*, 487–501
  intelligent cause *(Gordon)*, 568–69
  naturalist proposals failed *(Gordon)*, 558–59
  quantum fluctuations' role *(Guth)*, 502n6
  See also universe, views of universe
universe, views of universe
  historicization *(Craig)*, 507
  nature of universe, xix
  oscillating models *(Craig)*, 510–12
  self-creation *(Van Till)*, 535–45
  why exists *(Craig)*, 506–7, 527, 529n24
  See also cosmology; habitable zones; universe, characteristics

Van Till, Howard
  change of views, 544–45, 546n10
  fully-gifted Creation, 543–44
  gaps in Darwin's story, 537–38
  naturalistic theism, 485–86, 544–45
  RFEP, 536–39
  RFEP misunderstood, 541–43
verification *(Goldman)*, 230–31, 235. *See also* truth
vestigial organs *(Tooley)*, 894
Visual Brain in Action, 751
voluntarism, 35n21

warfare myths, 12–23
Warrant and Proper Function, 152
WEASEL, 371–73, 392–93, 397n49
Weinberg, Steven
  anthropic reasoning, 548–50
  discoveries about science, 485–86
  multiverse, 558
  multiverse uncertain, 554–55
  simplicity in physics, 217–18
  string theory, 548
  turning point of physics, 548
Whig history, 33n1
Wigner, Eugene
  Bell Inequality and Wigner, 184–89
  group theory pioneer, 677
  mathematics "unreasonably effective," 671–72, 681
will, human will, 11. *See also* God, God's will
Willard, Dallas
  naturalism lives on promises, 876
  naturalist ethics merely "science now," 868
  naturalism useless to philosophy, 866–69
wish-fulfillment *(Talbott)*, 154
witchcraft *(Stark)*, 35n24
Wonderful Life, 446–447, 448, 449, 461
worldview *(McMullin)*, 94n20
wormholes, 522, 553

young earth creationists *(Tooley)*, 887–88